$$\cos\theta = \frac{1 - \tan^2\frac{\theta}{2}}{1 + \tan^2\frac{\theta}{2}}$$

$$\sin\theta = \frac{2\tan\frac{\theta}{2}}{1 + \tan^2\frac{\theta}{2}}$$

Calculus of One & Several Variables

Ka

In the editorial series of

I. M. SINGER

Massachusetts Institute of Technology

ROBERT T. SEELEY
Brandeis University

Calculus of One & Several Variables

SCOTT, FORESMAN AND COMPANY
GLENVIEW, ILLINOIS LONDON

Library of Congress Catalog Card Number 79-189443
ISBN: 0-673-07779-9
AMS 1970 Subject Classification 98A20

Printed in the United States of America.
Regional offices of Scott, Foresman and Company are located in Dallas, Oakland, N. J., Palo Alto, and Tucker, Ga.

Preface

This large volume combines the second edition of *Calculus of One Variable* with *Calculus of Several Variables*, and thus covers the standard material for three or four semesters. The organization of the book is explained at the end of the Introduction, but a few points deserve emphasis here.

All the elementary functions are introduced and differentiated in Chapter II; this is convenient for the rest of the text, for instructors, and for readers who might well be encountering these functions elsewhere at the same time.

Differential equations appear early and often. The equations governing constant acceleration, exponential growth, and periodic motion are discussed and solved in Chapter III, and others are solved as the necessary techniques become available.

Definite integrals are introduced first as areas, then as Riemann sums, and finally are evaluated by the Fundamental Theorem of Calculus. It is possible, perhaps, to short-cut Riemann sums, but it then becomes more difficult to explain the applications of integration, and the Fundamental Theorem.

Chapters 9–11 develop the ideas of approximations, sequences, and series. If the appendix material on limits and continuity is to be covered in any detail, it should follow approximations and sequences; without this background, the definitions and proofs are, inevitably, hard to understand.

Part II, Several Variables, works toward an intuitive geometric understanding of vectors, gradients, line integrals, and surface integrals by stressing analytic calculations with a geometric or physical interpretation. At the end we suggest the unified mathematical framework embracing all these various concepts, laying the groundwork for the general Stokes' theorem given in more sophisticated approaches.

The level of maturity rises gradually in the text as a whole, in individual chapters, and even within some chapter sections; every instructor should know how far to go with his particular class.

The significance of many theorems is easily hidden even from the best students. I make a special point to show why we need theorems on maximum values, convergence of Riemann sums, and convergence of infinite series. These need not be proved, but they must be understood. Those who do follow the proofs will find the Least Upper Bound Axiom presented with similar care.

This volume owes much to the reviews and suggestions by readers and users. I thank all these collaborators anonymously; the entire list is too long, and a partial list has no logical stopping place. I am grateful to the publishers for their help in collecting this advice, and their generous attitude toward last-minute changes. Special thanks are due to the editor for his hard work in arranging page breaks and figures, and in rooting out errors and cloudy passages. In spite of all this, doubtless some errors remain. The author and publishers will be grateful to readers who bring these to our attention, or who have other comments to make.

Robert T. Seeley

Contents

Chapter II *Computation of Derivatives*

Chapter III *Applications of Derivatives*

Chapter IV *Theory of Maxima*

Chapter V *Introduction to Integrals*

Chapter IX *Approximations*

Chapter X *Infinite Sequences*

Chapter XI *Infinite Series*

Part Two. Calculus of Several Variables

Appendix II *How to Prove the Basic Propositions of Calculus*

Answers to Selected Problems

Table of Natural Logarithms

Index

Part One

Calculus of One Variable

Introduction

The road to understanding in calculus is a long one. Before setting out, we would like to sketch how calculus developed, and introduce briefly some of the men who contributed to it. With this background, we outline the plan of the book, and suggest several ways to use it.

A THUMBNAIL SKETCH OF THE HISTORY OF CALCULUS

Calculus rests on mathematical developments that go back as far as four thousand years to the Babylonians and the Egyptians, but we won't start there. The immediate contributions came at the end of the Renaissance.

Preliminary steps

First came the development of algebra. In the 1500's, the Italians achieved spectacular results in the solution of the equation $ax^4 + bx^3 + cx^2 + dx + e = 0$, and in the process they advanced the use of negative and complex numbers. In 1585, Simon Stevin of Bruges published *La Disme,* the first proposal of a systematic use of decimal expansions. And there were improvements in notation by many, including René Descartes, abandoning clumsy verbal communication in favor of symbols very much like ours today. (This step is more important than it seems. A good notation not only saves space on the paper; it saves space in the memory and effort in the mind, both of which are at a premium.)

Second came the development of analytic geometry, due in large part to Descartes' *La Géometrie* (1637), which made it obvious that the new efficiency in algebra could be well employed in geometry. Pierre Fermat, a contemporary of Descartes, also made important contributions in this direction, finding equations of straight lines and conic sections.

Third, there were various methods for determining tangents, areas, and volumes that led naturally to the ideas of differentiation and integration, the two main ideas of calculus. The best known of these methods are Fermat's way of drawing tangents (discovered about 1629) and B. Cavalieri's determination of areas and volumes, published in 1635. And Isaac Barrow showed an important connection between these two ideas, tangents and area, that was a direct forerunner of the fundamental theorem relating differentiation and integration.

Fourth, there were results on infinite series and products, primarily due to John Wallis, which appeared in 1655.

Finally, in physics there was a concerted study of motion, primarily the motion of falling bodies and of the planets. The most famous men in this work were Galileo, the astronomical observer Tycho Brahe, and Johann Kepler, who analyzed Tycho's observations and deduced three simple but profound empirical laws governing planetary motion.

The invention of calculus

Isaac Newton, a student of Barrow, gathered the mathematical developments together into one general theory, calculus, and applied it to solve the physical problems of the motion of falling bodies and of the planets. He showed that Kepler's laws imply the "inverse square" law of gravitation: Each planet is attracted to the sun by a force proportional to m/r^2, where m is the mass of the planet and r is the distance from planet to sun.

Newton did not shout *Eureka!* and run into the streets to announce his discovery. Perhaps his caution was due to an error in the commonly accepted distance to the moon, which was not corrected until 1679: because of the error, the force of gravity at the surface of the earth (as found from falling bodies) did not agree well enough with the force that Newton derived to explain the motion of the moon. Whatever the cause of the delay, by the time his invention of calculus was finally made known, much of the same theory had been found independently by Gottfried Leibniz, his contemporary. Thus Newton and Leibniz are both considered the inventors of calculus.

Neither Newton nor Leibniz succeeded in making the logic of their methods understood. Their reasoning was so mysterious that George Berkeley, an Irish bishop, published in 1734 the famous pamphlet *The Analyst* in which he defended his own faith by pointing out that Newton and his followers treated objects no more substantial than "ghosts of departed quantities," and that the foundations of religion were every bit as secure as those of Newton's analysis.

Development and application

In spite of the logical difficulties, both Newton and Leibniz had strong evidence that their methods contained some essential truth. Newton could explain the motion of the planets. And Leibniz had expressed his discoveries in a notation so apt that, although nobody understood exactly why, it led automatically to results that were seen to be correct.

From the end of the seventeenth century to the beginning of the nineteenth, calculus developed in the notation and outlook of Leibniz, but continued to find its inspiration and application in the project of explaining the physical world by mathematics, so successfully begun by Newton. The greatest mathematicians of this period were Leonhard Euler (1707–1783) and Joseph Louis Lagrange (1736–1813). Euler wrote the first widely read texts on calculus and others equally popular on algebra and trigonometry. He made advances in all fields of mathematics, and in dynamics, in the study of "least action" and energy, in the three-body problem of astronomy (the effect of mutual attractions between the earth, moon, and sun), in hydraulics, and in optics. Lagrange pursued these same questions, achieving greater unity and generality. His greatest work is the monumental *Mécanique analytique* (1788), which brought the science of mechanics close to its present form.

The great wealth of mathematical results, consistent with itself and with physical observations, proved beyond a doubt that calculus had abstracted certain essential features of the universe in which we live. But the logical foundations were still poorly understood, and even Euler was occasionally led by his formal manipulations to results that can hardly be considered correct. (One of these aberrations serves as a bad example in Chapter XI below. Unfortunately, most of Euler's voluminous and outstanding work is beyond the scope of this book, so we are not able to balance the bad impression created by this one example.)

The first great mathematician of the nineteenth century was Carl Friedrich Gauss (1777–1855), who made important contributions in the theory of the integers, use of infinite series, theory of surfaces, complex numbers, difficult numerical computations, astronomy, electricity and magnetism, surveying, and development of the telegraph.

Securing the foundations

A further contribution of Gauss was to the underlying logic of calculus, overcoming the valid objections to the work of the founders. This development continued with Augustin Cauchy's book *Cours d'analyse* (1821) and culminated in the work of Karl Weierstrass (1815–1897) and Richard Dedekind (1831–1916). Dedekind's contribution was a penetrating analysis of the nature of the real numbers; Weierstrass pointed out subtle logical oversights in the work of his predecessors, and in his own work he adopted the standards of rigor and logic that still apply today.

The nature of the logical problems in calculus, and the means of overcoming them, are illustrated by the contrast between classical Euclidean geometry (which achieves geometric results by exclusively geometric means) and analytic geometry (which employs algebraic methods to achieve geometric results). The link between algebra and geometry is a coordinate system, an idea thousands of years old. Two axes are drawn, as in Fig. 1, and to each point P is assigned the ordered pair of numbers (x,y), x being the signed distance from P to one axis, and y the signed distance from P to the other axis. This process can be reversed: given x and y, you can measure off the corresponding distances from the axes and thus find P. Hence every statement about points can be translated into a corresponding statement about ordered pairs of numbers, and vice versa.

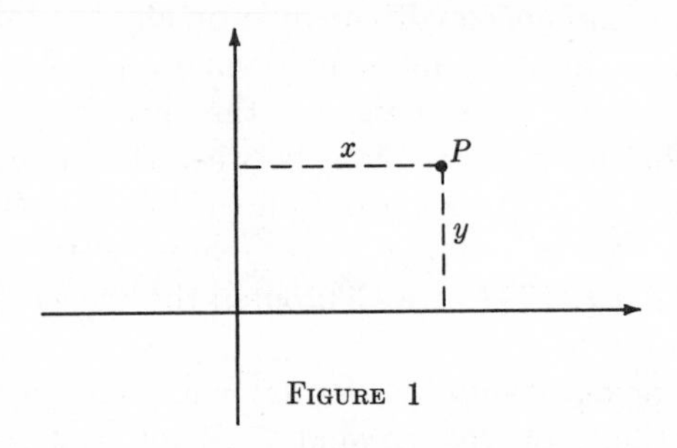

FIGURE 1

The correspondence is only valuable when algebra is well enough understood so that we can solve the algebraic translation of the geometric problem. By the time of Descartes, algebra was up to this challenge, and when he proposed as a general method the translation of problems from geometry to algebra, the idea was taken up so enthusiastically that there were actually complaints about the "clatter of the coordinate mill."

But for over a century there were no complaints about the underlying logic of analytic geometry. It was well based on the geometry of Euclid, and this was considered above reproach. However, in the course of soul-searching over the foundations of calculus, weaknesses were found even in Euclid.

One of the weak points shows up in Euclid's construction of an equilateral triangle on a given base. Take the given base AB as in Fig. 2. The problem is to show that there is a point C such that the lengths AC and BC equal the length of the given segment AB. Euclid solves this by drawing about each endpoint A and B a circle of radius $r = AB$, as in Fig. 3. Let C be a point of intersection of the two circles. Then BC and AC are radii of the two circles, so $BC = r = AB$ and $AC = r = AB$; hence ABC is an equilateral triangle.

The picture is clear, but the proof is not complete! Euclid's axioms and postulates *do not guarantee* that the two circles will intersect, so this argument does not provide the desired point C.

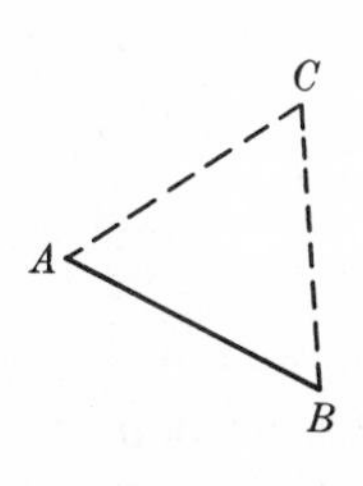

FIGURE 2

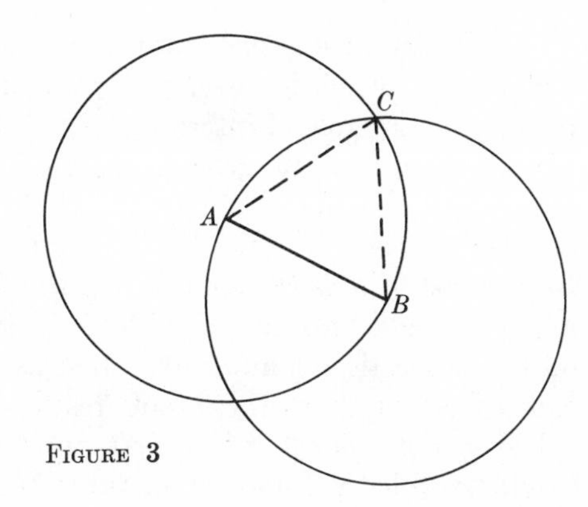

FIGURE 3

But it is easy to prove by analytic geometry that there *is* a point C forming an equilateral triangle ABC. Let $A = (x_1, y_1)$ and $B = (x_2, y_2)$ be the given points, and let $C = (x,y)$ be any other point. Then, by the distance formula of analytic geometry, we have for the three distances AB, AC, BC,

$$\begin{aligned} AB &= \sqrt{(x_1 - x_2)^2 + (y_1 - y_2)^2}\,, \\ AC &= \sqrt{(x_1 - x)^2 + (y_1 - y)^2}\,, \\ BC &= \sqrt{(x_2 - x)^2 + (y_2 - y)^2}\,. \end{aligned} \tag{1}$$

If we set

$$\begin{aligned} x &= \frac{x_1 + x_2}{2} + \sqrt{3}\,\frac{y_1 - y_2}{2}\,, \\ y &= \frac{y_1 + y_2}{2} + \sqrt{3}\,\frac{x_2 - x_1}{2}\,, \end{aligned} \tag{2}$$

it is a routine matter to check that the three distances in (1) are indeed the same, so we have found the desired point C.

This simple example illustrates the most basic problem in calculus and geometry, the *existence* problem. Arguing geometrically, we inferred the existence of the point C by looking at the picture, which is not really legitimate. But the problem is easily solved in analytic geometry, for we deduce the existence of C from a certain property of numbers: For every positive number a, there exists a positive square root b such that $b^2 = a$. We use this property in writing the distance formulas (1) and in using $\sqrt{3}$ in the formulas (2) for x and y.

In calculus, besides this existence question, there is the problem of giving precise definitions of the subtle processes of differentiation and integration. These problems, too, can be solved by appeal to the properties of numbers. Thus we are led to the currently accepted view of the logical structure of calculus, which is briefly this:

We start with a system of numbers, the real numbers (rationals and irrationals), which is developed without any appeal to geometry except, perhaps, in the role of interpreter.

Then, in analytic geometry, we do not think of the plane as existing first, with axes and coordinates imposed on it as a means of proving difficult results. Rather, we take this position: *The plane is nothing more or less than the set of ordered pairs of real numbers.* Distance and straight lines are defined algebraically, and all the proofs are based directly or indirectly on the properties of numbers. Just as the use of algebra in geometry is called analytic geometry, proofs resting on the properties of numbers are called *analytic proofs.* The first proof we gave above, based on Fig. 3, would be called a "proof by picture," an "intuitive proof," or a "heuristic proof"; and the second proof, based on the formulas (2), is an analytic one.

When we come to calculus, the sometimes mysterious arguments of the founding fathers—arguments which could be understood only with a highly developed intuition, and accepted only with a great deal of insight and faith—are replaced by analytic proofs based on the properties of the real numbers, proofs that can be checked by any diligent student.

We need this firm logical basis of calculus not just for reasons of philosophy or esthetics. There are important questions, particularly in connection with infinite series, that cannot be satisfactorily answered by arguments based on pictures and geometric intuition. Then, practical considerations as well as logic require a frame of reference in which it is really possible to decide whether a result is true or false.

THE PLAN OF THIS BOOK

We follow, roughly, the historical development outlined above, introducing the main ideas intuitively at first, and gradually exposing the main lines of the theory as we develop computational techniques and applications. Finally, when we come to infinite sequences and series, we take the plunge, and deduce everything from the properties of the real numbers. Two appendices discuss these properties in detail, and fill in the proofs that have been by-passed in the earlier chapters. To make it clear as we go just what is being taken temporarily on faith, we use two different labels:

Proposition for a fact whose proof is deferred, and

Theorem for a fact to be proved on the spot.

The table of contents shows the basic outline, but a few comments are appropriate.

Chapter III is the heart of the book. Starting with simple, important examples of increasing and decreasing functions, it goes on to solve the differential equations for proportional growth, for motion with constant acceleration, and for simple harmonic oscillation (of springs or electrical circuits). The goal is to introduce significant applications of calculus early, by concentrating on simple special cases well suited to a clean mathematical presentation. These applications require, of course, the ability to differentiate the elementary functions. This is the topic of

Chapter II, techniques of differentiation. The explanations are sometimes algebraic, sometimes geometric. It is important that none of the elementary functions is postponed, so we have on hand a wide range of examples for theorems like the chain rule, and for applications arising anywhere after this early chapter.

Chapter I is a first introduction to limits, derivatives of polynomials, and simple applications such as maxima, and velocity and acceleration. For those who have already seen such material, a reading should suffice.

Chapter 0 summarizes "precalculus" material. Problems are provided at the end of the chapter to test which parts need more than a rapid reading.

Turning now to the other chapters:

Chapter V introduces the definite integral and the Fundamental Theorem of Calculus. All but the simplest proofs are postponed to an Appendix. A variety of applications provide grist for the mill of

Chapter VI, on techniques of integration.

Chapter VII concerns two-dimensional vectors and polar coordinates, climaxing in the relation between Kepler's laws of planetary motion and Newton's laws of mechanics.

Chapter VIII concerns complex numbers, and their use in solving constant-coefficient linear differential equations.

Chapter IX on approximations, though important in its own right, is placed here as preparation for the deeper waters of

Chapter X, which proves the basic theorems about infinite sequences. How many of the proofs to study in detail is, of course, a matter of taste. The Bolzano-Weierstrass theorem is included for those few who will ultimately be doing all the theory, though it is not required before the end of Appendix II.

Chapter XI presents infinite series, stressing power series as a means of computation, and of solving differential equations.

Appendix I gives background about summation, induction, inequalities, and the Least Upper Bound Axiom. It can be covered ad lib, whenever it appears necessary.

Appendix II gives the nitty-gritty proofs that are by-passed in the text.

Chapter 0

Prerequisites

This chapter contains the few basic facts required to begin calculus, and establishes our notation. It is not expected to teach you much that is new; in fact, if you can do the review problems on pages 50–51, you can safely skip the whole chapter.

0.1 THE REAL NUMBERS AS POINTS ON A LINE

The long and fruitful collaboration between algebra and geometry is based on one simple idea: numbers can be represented as the points on a line.

Take a horizontal line and a unit of distance. Fix some point on the line, and call it the origin [Fig. 1]. Given any real number a, plot it $|a|$ units from the origin, to the right if a is positive and to the left if a is negative, as in the figure.

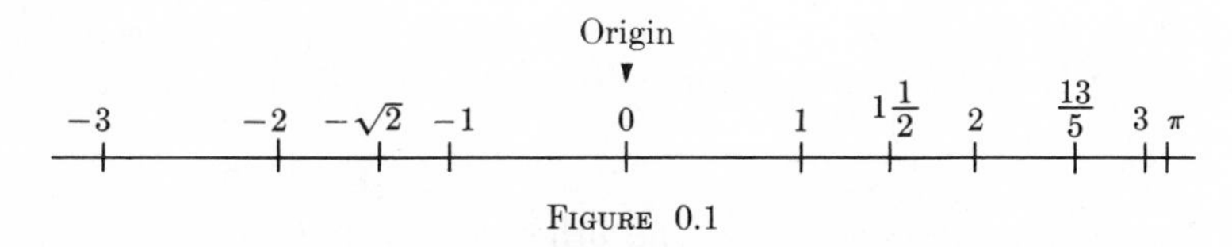

FIGURE 0.1

The symbol $|a|$ used here denotes the *absolute value* of a, defined by

ABSOLUTE VALUE

$$|a| = \begin{cases} a & \text{if } a \text{ is positive} \\ 0 & \text{if } a = 0 \\ -a & \text{if } a \text{ is negative.} \end{cases}$$

Thus $|1| = 1$, $|-1| = 1$, $|-3/2| = 3/2$. It is easy to check that $|-a| = a$; $|ab| = |a| \cdot |b|$; and $|a/b| = |a|/|b|$ if $b \neq 0$. [See Problems 9 and 10 below.]

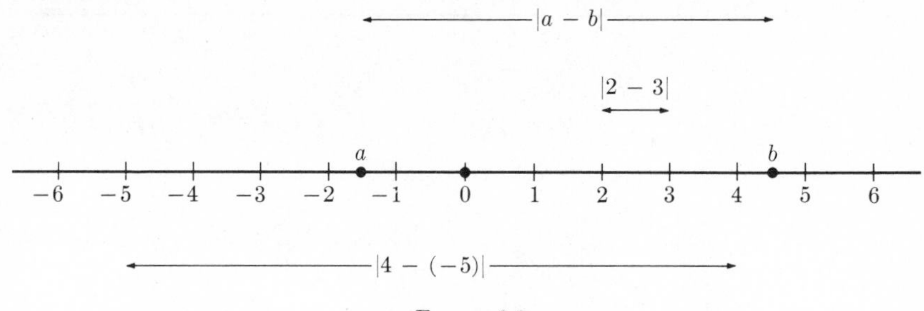

FIGURE 0.2

The geometric interpretation of absolute value is implicit in the prescription above for plotting points on the line, namely: $|a|$ is the distance between a and 0. More generally, $|a - b|$ is the *distance between a and b* [Fig. 2]. For example,

$|a - b|$ IS DISTANCE BETWEEN a AND b

$$|2 - 3| = |-1| = 1 = \text{the distance from 2 to 3}$$

$$|4 - (-5)| = |9| = 9 = \text{the distance from 4 to } -5.$$

The representation as points on a line reveals clearly the *ordering* of the real numbers: *when a is less than b* (written $a < b$), *then a appears to the left of b.*

If $a < b$, the set of numbers lying between a and b is called the *open interval* from a to b, denoted (a,b) [Fig. 3]. Precisely, (a,b) is the set of all numbers x such that $a < x < b$. (The notation $a < x < b$ means that both $a < x$ and $x < b$.) The numbers a and b are called the *endpoints* of the interval (a,b). Notice that the open interval (a,b) does not include its endpoints.

ENDPOINTS

If we add to the open interval (a,b) its endpoints, we obtain the *closed interval* $[a,b]$, the set of all numbers x such that $a \leq x \leq b$ [Fig. 4]. (The notation $a \leq x$ is read "a is less than or equal to x," and means that either $a < x$ or $a = x$. Thus $2 \leq 3$, and $2 \leq 2$.) Notice that in sketching an interval, a square bracket or a dot indicates an endpoint that is included, while a parenthesis or the absence of a dot indicates an endpoint that is excluded.

SET NOTATION

The definition of intervals (and other sets) can be given briefly and concisely in a standard *set notation.* The phrase "the set of all numbers x such that $1 < x < 2$" is written

$$\{x: 1 < x < 2\}.$$

Inside the braces we give first the symbol x that is to stand for a general number, and then the conditions on x that characterize the numbers in the set. Thus, for example, the set of all numbers with absolute value less than 1 is written $\{x: |x| < 1\}$. For another example, $\{x: x > 0\}$ denotes the set of all positive numbers.

In this notation, open and closed intervals can be defined as follows:

$$(a,b) = \{x\colon a < x < b\}$$
$$[a,b] = \{x\colon a \le x \le b\}.$$

Similarly, we define the half-open intervals

INTERVALS: OPEN, CLOSED, INFINITE

$$(a,b] = \{x\colon a < x \le b\}$$
$$[a,b) = \{x\colon a \le x < b\} \qquad \text{[Fig. 5]}$$

and the infinite closed intervals

$$[a, +\infty) = \{x\colon a \le x\}$$
$$(-\infty,b] = \{x\colon x \le b\}.$$

We leave it to you to define the infinite open intervals $(a, +\infty)$ and $(-\infty,b)$.

Notations such as $[a, +\infty)$ and $(-\infty,b]$ are suggestive, but don't let them suggest too much. We are not introducing numbers $+\infty$ and $-\infty$, and we will not discuss sums such as "$a + \infty$" or "$\infty + -\infty$". Note that $(-\infty,b]$ is called a *closed* interval because it contains the only endpoint in question, namely b.

The use of x to stand for a general number is common, but not compulsory. The set $\{x\colon x \le b\}$ could just as well be written $\{t\colon t \le b\}$ or $\{a\colon a \le b\}$ or $\{\#\colon \# \le b\}$ or even, in an emergency, $\{?\colon ? \le b\}$. Each of these describes exactly the same set, the closed infinite interval $(-\infty,b]$.

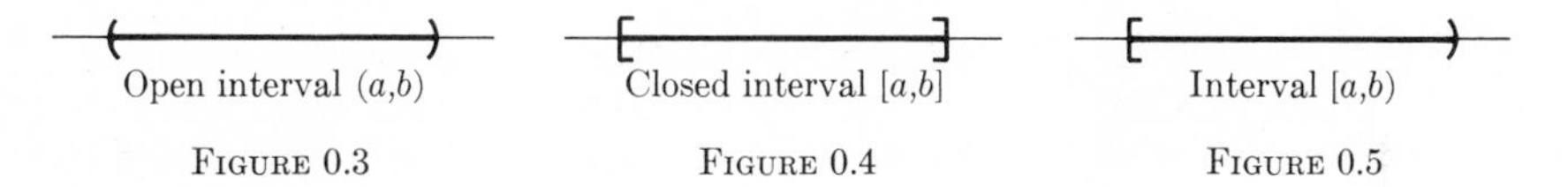

Open interval (a,b) — FIGURE 0.3

Closed interval $[a,b]$ — FIGURE 0.4

Interval $[a,b)$ — FIGURE 0.5

We conclude this section by introducing a simple *sign convention.* It is generally understood that "a profit of -5 dollars" means "a loss of 5 dollars," or that "-2 steps forward" means "2 steps backward." In the same vein, the prescription for plotting numbers on the line can be given simply as follows: Each number a is plotted a distance a to the right of the origin. If a is positive, the meaning is clear; if a is negative, we understand "a to the right" as "$|a|$ to the left." More generally, $b + a$ is plotted a distance a to the right of b, and $b - a$ is plotted a distance a to the left of b.

Example 1. The set $\{t\colon -1 < t \le 5\}$ is the interval $(-1,5]$, sketched in Fig. 6.

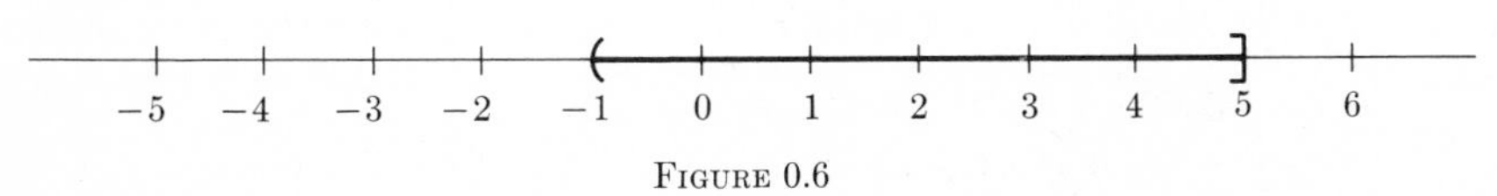

FIGURE 0.6

Example 2. The set $\{x: 1 < x < 0\}$ contains no points at all, since no number which is >1 can also be <0. This is an empty interval.

IMPORTANT EXAMPLE

Example 3. The set $\{y: |y - 2| < 1\}$ consists of all numbers y that are within distance 1 of the number 2. As Fig. 7 shows, this is the interval $(1,3)$.

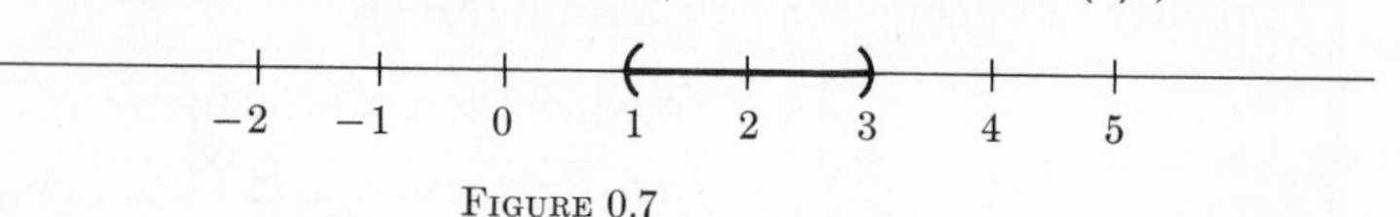

FIGURE 0.7

Example 4. Describe the set $\{x: x(x - 1) > 0\}$. The key to this example is the *rule of signs:* a product ab is positive if and only if a and b are both positive or both negative. The solution is illustrated in Fig. 8. The first line shows where x is positive and where negative; the second shows where $x - 1$ is positive and where negative; the third shows the same for the product $x(x - 1)$.

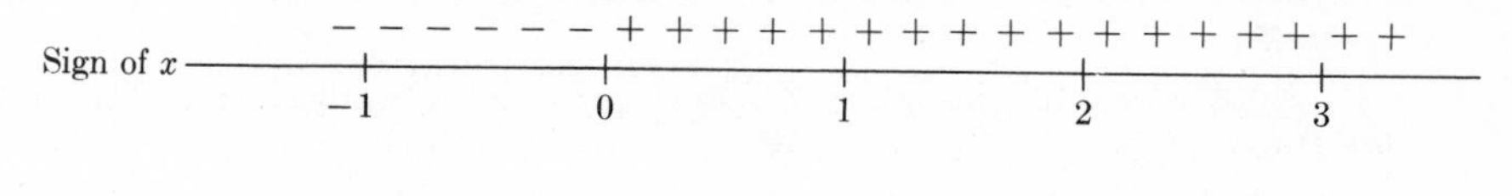

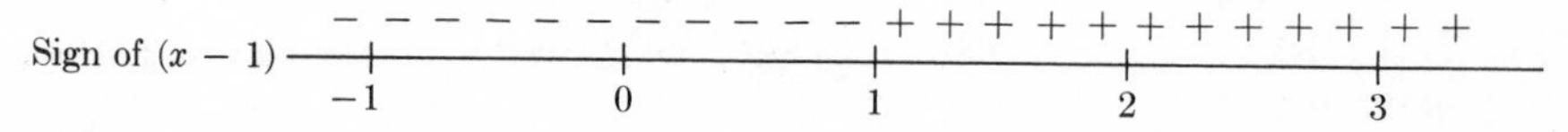

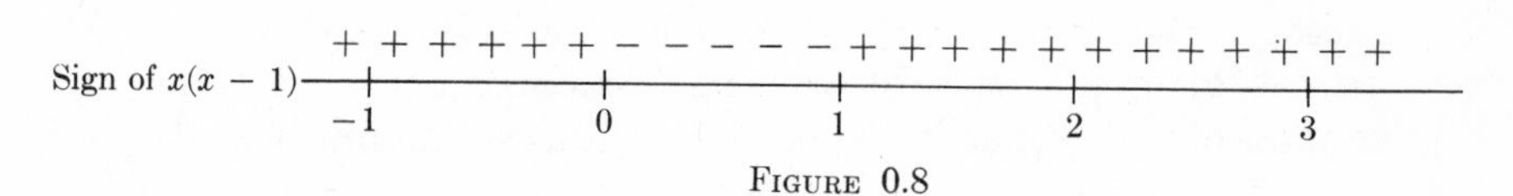

FIGURE 0.8

Looking at the first two lines, you can easily see where both factors have the same sign; the corresponding points are marked with plus signs on the third line. Similarly, where the first two lines show opposite signs, the third line shows minus signs. The third line shows that the set $\{x: x(x - 1) > 0\}$ consists of the *two* intervals $(-\infty,0)$ and $(1,+\infty)$. The set is sketched in Fig. 9. [*Question*: Why is it not the two intervals $(-\infty,0]$ and $[1,\infty)$?]

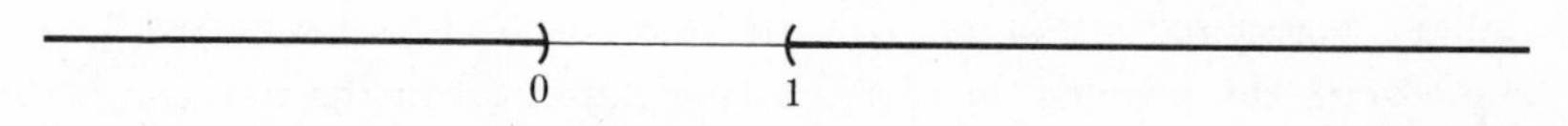

FIGURE 0.9 Sketch of $\{x: x(x - 1) > 0\}$

Example 5. Sketch the set $\{s: s < 1 \text{ or } s \geq 2\}$. A number s is in this set if it is <1, or if it is ≥ 2. The set thus consists of the two intervals $(-\infty,1)$ and $[2,+\infty)$. [See Fig. 10.]

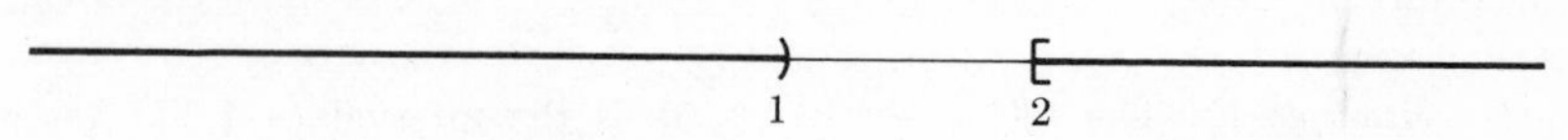

FIGURE 0.10 Sketch of $\{s: s < 1 \text{ or } s \geq 2\}$

Example 6. Sketch the set $\{s: -1 \leq s < 1 \text{ or } s > 0\}$. *Solution.* In Fig. 11, we show on the top line the numbers s satisfying $-1 \leq s < 1$; and on the second line, the numbers s satisfying $s > 0$. Then the third line shows all the numbers satisfying either or both of these two conditions, i.e. the set $\{s: -1 \leq s < 1 \text{ or } s > 0\}$. This turns out to be the interval $[-1,\infty)$.

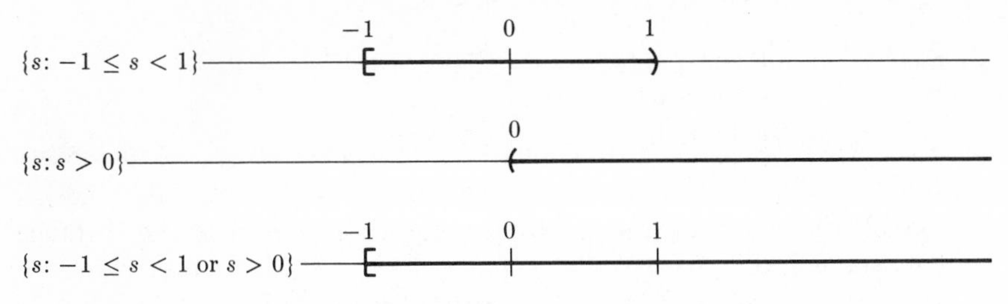

FIGURE 0.11 Sketch of $\{s: -1 \leq s < 1 \text{ or } s > 0\}$

Remark. In Example 6, we took "$-1 \leq s < 1$ or $s > 0$" to mean "either $-1 \leq s < 1$ or $s > 0$ *or both*." In mathematics, "or" is always used in this sense: "A or B" means "A or B or both."

Example 7. Sketch the set $\{z: |z| \leq 1 \text{ and } z > 0\}$. In Fig. 12, the first line shows where $|z| \leq 1$, the second line shows where $z > 0$, and the third line shows where *both* conditions hold. This is the set $\{z: |z| \leq 1 \text{ and } z > 0\}$.

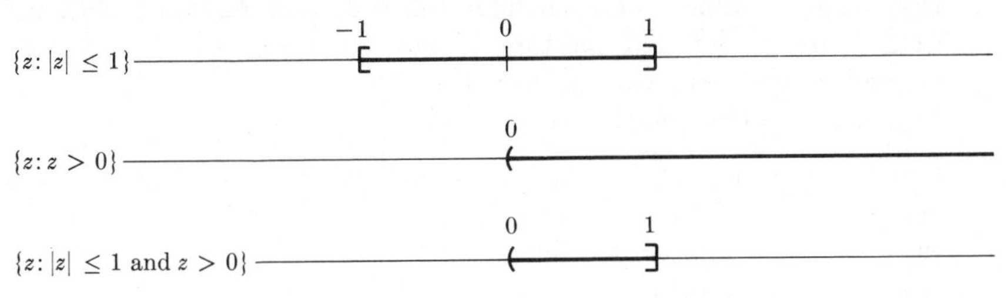

FIGURE 0.12 Sketch of $\{z: |z| \leq 1 \text{ and } z > 0\}$

PROBLEMS A dot before a problem indicates that it is answered (at least partly) in the back of the book.

• **1.** Plot the following numbers, and give their absolute values
(a) -2 (b) $-3/-5$ (c) $8/3$
(d) $-(-2/3)$ (e) $-\sqrt{2}$ (f) $-a$ for $a = -1/8$

• **2.** For the given numbers a and b, plot a, b, and $a - b$. Check that $|a - b|$ is the distance *between a and b*.
(a) $a = 2, b = 3$ (c) $a = 2, b = -3$
(b) $a = -2, b = 3$ (d) $a = -2, b = -3$

3. Do Problem 2 for the following pairs.
(a) $a = -3, b = -1$ (c) $a = 3/2, b = 1$
(b) $a = -1/2, b = 1/4$ (d) $a = 2, b = -1\frac{1}{2}$

4. Sketch the following sets. [Notice that there are only four distinct sets in the whole list of ten.]
•(a) $\{x: x > 1 \text{ and } x < 2\}$ (f) $[1,2]$
•(b) $\{x: x = 0 \text{ or } x = 1\}$ (g) $\{x: x(x-1) = 0\}$
(c) $(1,2)$ (h) $\{z: z^2 = z\}$
•(d) $\{y: y > 1 \text{ or } y < 2\}$ (i) $\{t: t = 0 \text{ or } t = 1\}$
(e) $\{t: |t - \frac{3}{2}| < \frac{1}{2}\}$ •(j) $\{x: 1 \leq x \leq 2\}$

5. Sketch the following sets.

• (a) $\{x: x(x+1) > 0\}$
(b) $\{x: x(x+1) < 0\}$
• (c) $\{x: x(x+1) = 0\}$

6. Sketch the following sets. (Remember that $|a - b|$ is the distance between a and b.)

(a) $\{x: |x| < 1\}$
(b) $\{x: |x-2| < 1\}$
(c) $\{\theta: |\theta - \pi| < 3\}$
(d) $\{s: |s+2| < 1\}$
• (e) $\{x: |x-a| < b\}$
(f) $\{x: 0 < |x-1| < 2\}$
• (g) $\{t: 1 < |t-3| < 2\}$
• (h) $\{x: 0 < |x-a| < b\}$
• (i) $\{z: |z-1| < |z|\}$
(j) $\{z: |z-1| \leq |z|\}$
• (k) $\{x: |x-2| + |x-3| = 1\}$
• (l) $\{x: |x-2| + |x-3| = 2\}$
(m) $\{x: |x-2| + |x-3| = 1/2\}$

7. How many numbers does each of the following intervals contain? (This is not a profound question, it just requires a careful reading of the definitions of the various intervals.)

(a) $[a,a]$ (b) $[a,a)$ (c) $(a,a]$ (d) (a,a)

8. Plot the integers -2, -1, 0, 1, 2 on a horizontal line. Then plot the points that are in the given positions.

• (a) distance -2 to the right of $1/2$
• (b) distance $1/2$ to the right of -2
(c) distance -2 to the left of 0
(d) distance $-1/3$ to the right of $-2/3$
• (e) distance $-1/2$ to the left of $-1/2$

9. Show that $|-a| = |a|$ for every number a. [Hint: Treat separately the three cases: a positive, $a = 0$, and a negative.]

10. Show that $|a \cdot b| = |a| \cdot |b|$. [Hint: Remember how multiplications are usually performed. To compute $a \cdot b$, you form the product of the absolute values and then set

$$a \cdot b = |a| \cdot |b| \quad \text{if } a \text{ and } b \text{ have the same sign,}$$

$$a \cdot b = -|a| \cdot |b| \quad \text{if } a \text{ and } b \text{ have opposite signs.}$$

Check the desired formula in each of these two cases.]

11. Show that if $b \neq 0$, then $\left|\frac{a}{b}\right| = \frac{|a|}{|b|}$. [Hint: By Problem 10,

$$\left|b \cdot \frac{a}{b}\right| = |b| \cdot \left|\frac{a}{b}\right|.]$$

12. Show that if $a \neq 0$, then a^2 is positive. [Hint: Use the hint in Problem 10.]

0.2 THE SYMBOLS $\Rightarrow$ AND $\Leftrightarrow$

Statements of the form "if ... then ..." are so common in mathematics that a special abbreviation has been introduced, the arrow $\Rightarrow$. As a simple example, the statement

$$\text{if } a < b \text{ and } b < c, \quad \text{then} \quad a < c$$

is written

$$a < b \text{ and } b < c \quad \Rightarrow \quad a < c. \tag{1}$$

You can read the symbol $\Rightarrow$ as "implies" or "implies that"; thus (1) is read "$a < b$ and $b < c$ implies $a < c$." The arrow delivers its message much quicker than the words do, showing at a glance that the hypothesis "$a < b$ and $b < c$" leads to the conclusion "$a < c$." Notice that the one-way arrow $\Rightarrow$ is appropriate in (1), since $a < c$ does *not* imply $a < b$ and $b < c$; in fact, from $a < c$ you cannot deduce anything at all about b.

The corresponding abbreviation for "if and only if" is a double arrow $\Leftrightarrow$. For example,

$$a < b \quad \text{if and only if} \quad b > a$$

is written

$$a < b \quad \Leftrightarrow \quad b > a.$$

The double arrow shows clearly that the statement on the left implies the statement on the right, *and conversely* the statement on the right implies the one on the left. In other words, the two statements are equivalent; they are two different ways to say exactly the same thing.

The symbols $\Rightarrow$ and $\Leftrightarrow$ replace a variety of words; an explicit list will keep both the symbols and the words straight:

$A \Rightarrow B$ means A implies B; if A then B;
B follows from A; A only if B.

$A \Leftrightarrow B$ means $A \Rightarrow B$ and conversely $B \Rightarrow A$;
A is equivalent to B; A if and only if B.

Another common notation for "if and only if" is the odd-looking contraction "iff".

Some final examples:

Triangle ABC is equilateral $\Leftrightarrow$ triangle ABC is equiangular.

Two lines in the plane intersect in a single point iff they are not parallel.

Line A is parallel to B and B is parallel to C $\Rightarrow$ A is parallel to C.

You don't get it yet $\Rightarrow$ you should go back and try again.

You get it already $\Rightarrow$ you should go on to the next section.

0.3 POINTS IN THE PLANE

We extend the representation of numbers as points on a line, and represent *ordered pairs* of numbers as *points in a plane.*

An ordered pair of numbers is two numbers with one of them designated as the first and the other as the second. Suppose that x is the first number and y is the second; then the ordered pair is denoted (x,y).

The order is important; for example, (2,3) is not the same *ordered* pair as (3,2). In general, (x,y) is different from (y,x), unless it happens that $x = y$.

The same symbol (2,3) can now (unfortunately) denote two different things: an ordered pair or the open interval $\{x: 2 < x < 3\}$. Fortunately, the context of discussion usually makes it clear whether we mean a pair or an interval. And when we want to remove all shadow of doubt, we say "the interval (a,b)" or "the pair (a,b)."

The connection between ordered pairs and points in the plane is suggested by the house numbering of many cities.* In the plane, draw a horizontal line and a vertical line, intersecting in a point O called the origin [Fig. 13]. On each line, plot the real numbers as in §0.1; on the vertical axis plot positive numbers above O, and negative numbers below O. The unit of distance is to be the same on both axes.

Now, given any ordered pair of real numbers (x,y), we obtain a corresponding point P as in Fig. 13. The first number x, measured along the horizontal axis, gives the signed distance to the right of the vertical axis; the second number y, measured along the vertical axis, gives the signed distance above the horizontal axis. By "signed distance" we mean that the sign convention applies; for example, the point $(-2,1)$ is 2 units to the *left* of the vertical axis, and 1 unit above the horizontal axis.

The use of x for the first number and y for the second goes all the way back to Descartes, the founder of analytic geometry. However, we can just as well use other letters, and label points as (s,t), (a,b), and so on.

When using (x,y), the horizontal axis is called the *x axis,* and the vertical axis is called the *y axis.*

Traditional analytic geometry uses the correspondence between points and pairs of numbers to solve geometric problems by algebraic methods. However, the algebraic methods have so many applications other than geometry that this "traditional" point of view has given way; *we now view the ordered pairs of numbers as the essential mathematical objects,* and the correspondence illustrated in Fig. 13 serves mainly to portray algebraic statements in an intuitively appealing geometric form, by means of graphs

---

* Chicago, for one. There are two "base-line" streets, Madison Street running east-west, and State Street running north-south. There are 800 street numbers per mile. An address such as "2245 North Kedzie (3200 West)" is 2245/800 miles north of Madison Street, and 3200/800 miles west of State Street.

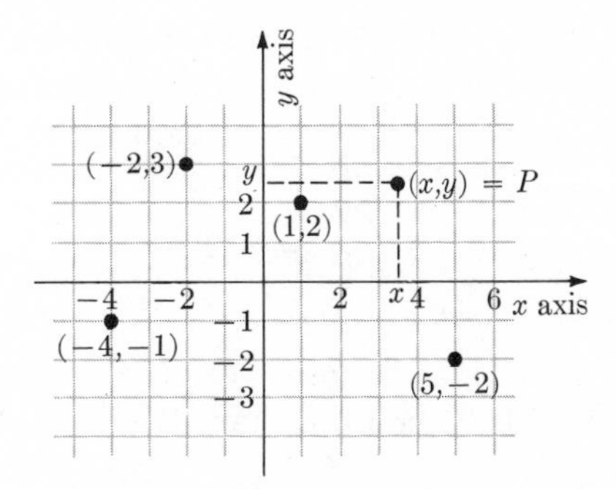

FIGURE 0.13

of all kinds. For us, *the plane is the set of all ordered pairs of real numbers.* Thus we call an ordered pair (x,y) a *point*; the numbers x and y are called the first and second coordinates of the point.

All our definitions and proofs rest ultimately on relations between these coordinates. Thus, for example, we define distance as follows:

DISTANCE IN THE PLANE

Definition 1. The *distance* between two points $P_1 = (x_1, y_1)$ and $P_2 = (x_2, y_2)$ is denoted P_1P_2, and is defined by the formula

$$P_1P_2 = \sqrt{(x_1 - x_2)^2 + (y_1 - y_2)^2}\,.$$

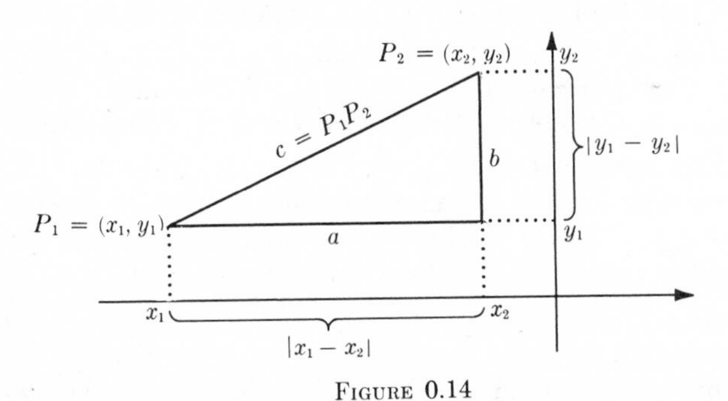

FIGURE 0.14

Figure 14 explains this definition. The segment from P_1 to P_2 is the hypotenuse of a right triangle, and the sides of that triangle have the lengths $a = |x_2 - x_1|$ and $b = |y_1 - y_2|$. According to classical plane geometry, the length c of the hypotenuse satisfies the relation

$$c^2 = a^2 + b^2 = |x_2 - x_1|^2 + |y_2 - y_1|^2 = (x_2 - x_1)^2 + (y_2 - y_1)^2.$$

Thus it is consistent with classical plane geometry to define the distance from P_1 to P_2 to be

$$P_1P_2 = c = \sqrt{(x_2 - x_1)^2 + (y_2 - y_1)^2}\,.$$

We come now to the elementary but fundamental concept of *straight lines*. The algebraic definition may not be obvious at first, so we begin with a few simple examples. You should be able to check the following statements by a moment's thought, and a glance at Fig. 15.

(a) The x axis is the set $\{(x,y) : y = 0\}$.

(b) The y axis is the set $\{(x,y) : x = 0\}$.

(c) Any vertical line has the form $\{(x,y) : x = c\}$, where the constant c is the signed distance from the y axis to the line.

(d) Any horizontal line has the form $\{(x,y) : y = c\}$, where the constant c is the signed distance from the x axis to the line.

MAIN DIAGONAL

(e) The line bisecting the first and third quadrants is called the *main diagonal*. Algebraically, it is the set of points whose first and second components are equal, i.e. the set $\{(x,y) : y = x\}$.

(f) Assuming the laws of similar triangles, you can show that the line through the origin (0,0) and the point (1,2) [Fig. 16] is characterized by the equation $y/x = 2/1$, or $y = 2x$. In other words, this line is the set $\{(x,y) : y = 2x\}$.

(g) To raise the line given in (f) by three units, we increase all the y values by 3, that is, we replace $y = 2x$ by $y = 2x + 3$. The resulting line is the set $\{(x,y) : y = 2x + 3\}$ [Fig. 16].

The general pattern emerges from these examples, and we can formulate the algebraic definition of straight lines as follows:

STRAIGHT LINE

Definition 2. A *vertical line* is any set of the form $\{(x,y) : x = c\}$, where c is a given constant.

A *nonvertical line* is any set of the form $\{(x,y) : y = mx + b\}$, where m and b are constants. This set is called the *graph* of the equation

$$y = mx + b. \tag{1}$$

SLOPE

By *straight line* we mean either a vertical line or a nonvertical line. The number m in equation (1) is called the *slope* of the line.

The examples below will show that Definition 2 agrees with our intuitive understanding of straight lines, and also with Euclid's concept.

Example 1. Sketch the line $L = \{(x,y) : y = -2x + 5\}$. *Solution.* To find particular points on the line, assign values to one coordinate, say x, and solve the equation $y = -2x + 5$ for the other coordinate. Thus, for example, $x = 0$ gives $y = 5$, so (0,5) is on L; and $y = 0$ gives $x = \frac{5}{2}$, so $(\frac{5}{2},0)$ is on L. Plot these two points, and draw L through them with a straightedge, obtaining the sketch in Fig. 17. If you plot any other point (x,y) whose coordinates x and y satisfy $y = -2x + 5$, it will lie on this same line. [Try some.]

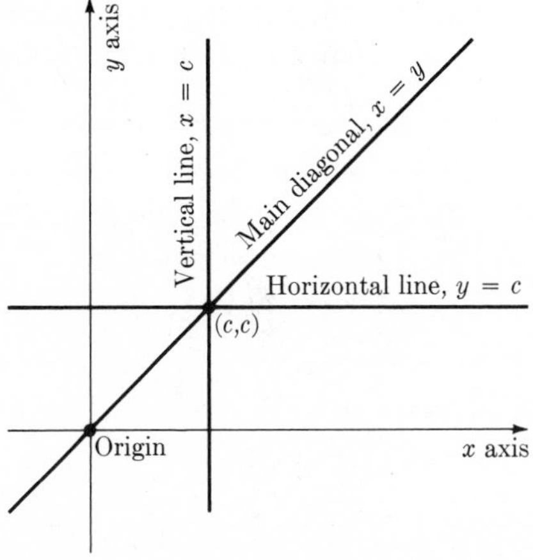

FIGURE 0.15

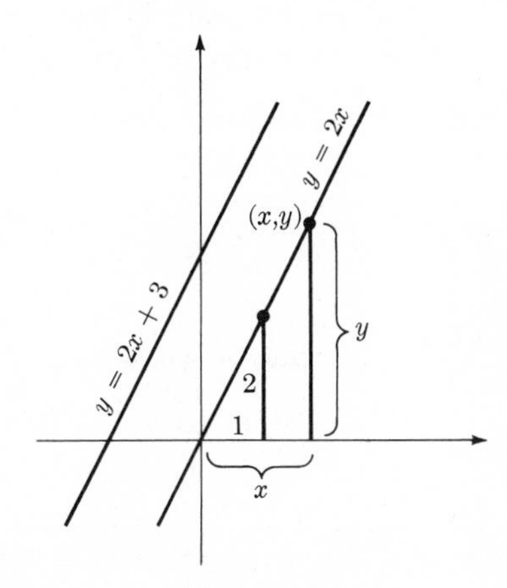

FIGURE 0.16

Example 2. Find a line passing through the points $P_1 = (2,3)$ and $P_2 = (3,2)$.
Solution. The points P_1, P_2 and the line are sketched in Fig. 18. The two given x values ($x = 2$ for P_1, $x = 3$ for P_2) are not the same; therefore the line L we seek must be nonvertical, so it has an equation of the form (1) above, $y = mx + b$. The point $P_1 = (2,3)$ lies on L if and only if equation (1) is satisfied with $x = 2$ and $y = 3$, i.e.

$$3 = m \cdot 2 + b. \tag{2}$$

Similarly, $P_2 = (3,2)$ lies on L if and only if

$$2 = m \cdot 3 + b. \tag{3}$$

Subtracting equation (2) from equation (3) gives $-1 = m$, and substituting this back in (2) gives $b = 5$. Thus, if P_1 and P_2 lie on any line, it must have the equation $y = mx + b$ with $m = -1$ and $b = 5$, i.e.

$$y = -x + 5.$$

Conversely, it is easy to check that P_1 and P_2 do indeed lie on this line, so the problem is solved. The slope of our line is $m = -1$.

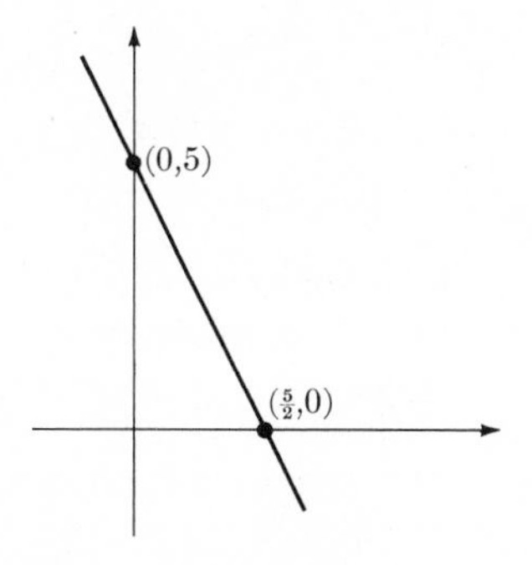

FIGURE 0.17

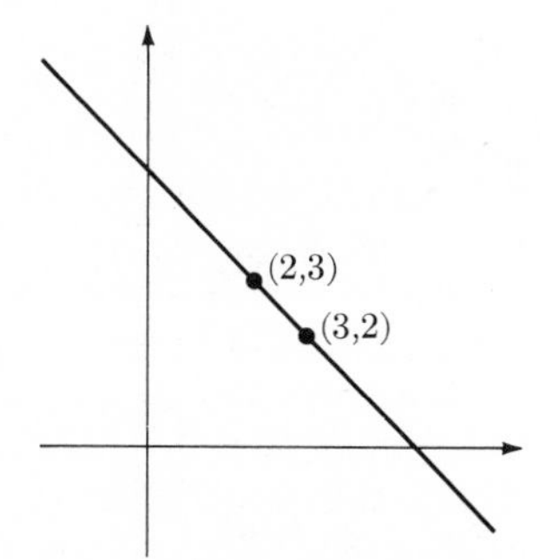

FIGURE 0.18

The solution of Example 2 can be carried out with *any* pair of distinct points $P_1 = (x_1, y_1)$ and $P_2 = (x_2, y_2)$ [see Problem 11], thereby proving:

STRAIGHT LINE THROUGH TWO GIVEN POINTS

Theorem 1. *Through any two distinct points P_1 and P_2, there passes one and only one straight line L. When $x_1 = x_2$, the line L is the set*

$$\{(x,y) : x = x_1\}.$$

When $x_1 \neq x_2$, L is the graph of

$$y = \frac{y_2 - y_1}{x_2 - x_1} x + \frac{y_1 x_2 - y_2 x_1}{x_2 - x_1}. \tag{4}$$

Equation (4) is equivalent to

$$y - y_1 = \frac{y_2 - y_1}{x_2 - x_1}(x - x_1) \tag{5}$$

since (5) is transformed to (4) by transposing y_1 and putting the result on the common denominator $x_2 - x_1$. Each of these equivalent forms is useful; equation (5) is easier to remember, but (4) is in the standard form

$y = mx + b$, with $m = \dfrac{y_2 - y_1}{x_2 - x_1}$. It thus follows from (4) that:

SLOPE OF LINE THROUGH TWO POINTS

The line through (x_1, y_1) and (x_2, y_2) has slope

$$m = \frac{y_2 - y_1}{x_2 - x_1} = \frac{\text{difference in “}y\text{ values”}}{\text{difference in “}x\text{ values”}}. \tag{6}$$

Substituting (6) back in (5), it follows that:

the line through a given point $P_1 = (x_1, y_1)$ with slope m has the equation

LINE WITH GIVEN SLOPE THROUGH GIVEN POINT

$$y - y_1 = m(x - x_1). \tag{7}$$

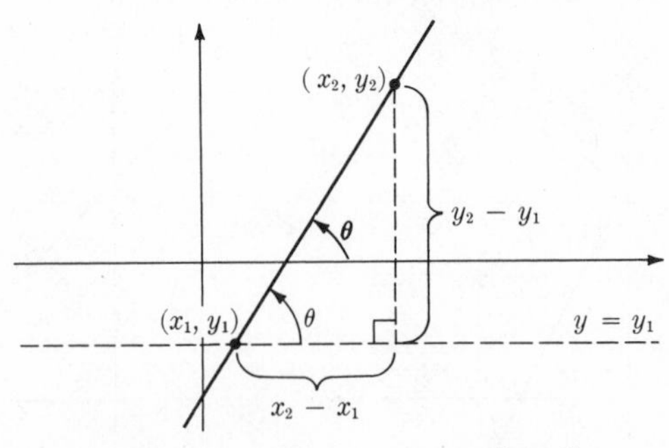

FIGURE 0.19 $\tan\theta = \dfrac{y_2 - y_1}{x_2 - x_1}$

From Fig. 19, you can see that the slope satisfies

$$m = \frac{y_2 - y_1}{x_2 - x_1} = \tan \theta, \tag{8}$$

where θ is the angle at which the line L cuts the horizontal line

$$\{(x,y) : y = y_1\}.*$$

The word slope suggests a hillside, and in fact this is not a bad way to think about it. Imagine the graph of $y = mx + b$ as a hillside, and imagine a point P moving from left to right along the graph (hillside), starting at (x_1, y_1) and ending at (x_2, y_2). Then the slope m is the *increase in y per unit increase in x*. If x increases by 1 unit, then y increases by m units, as you can see from (6); setting $x_2 - x_1 = 1$ gives $y_2 - y_1 = m$. If m is positive, then the increase in y is positive, and the line slopes up from left to right [Fig. 20]. Moreover, the larger m is, the larger is the increase in y, and the steeper is the line [Fig. 22].

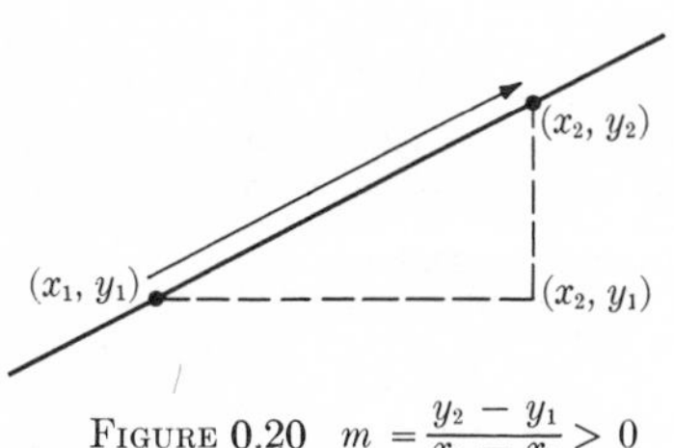

FIGURE 0.20 $m = \dfrac{y_2 - y_1}{x_2 - x_1} > 0$

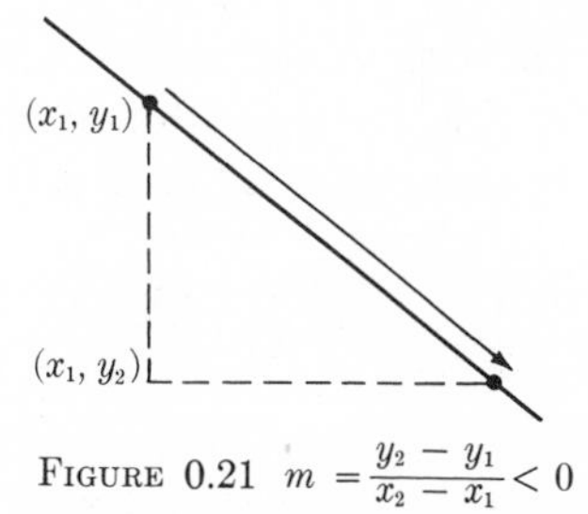

FIGURE 0.21 $m = \dfrac{y_2 - y_1}{x_2 - x_1} < 0$

In just the same way you can see that *when $m < 0$, the line slopes down from left to right*, as in Fig. 21. And the larger $|m|$ is, the steeper is the line [Fig. 22]. When $m = 0$, the line is called *horizontal*.

---

* Sometimes it is convenient to use *different* scales on the two axes, for example in plotting population vs. time. Then the slope is still defined by $\dfrac{y_2 - x_1}{x_2 - x_1}$, but this no longer is equal to $\tan \theta$. Different scales also invalidate the distance formula [Def. 1] and the slope condition for perpendicular lines [Def. 4 below].

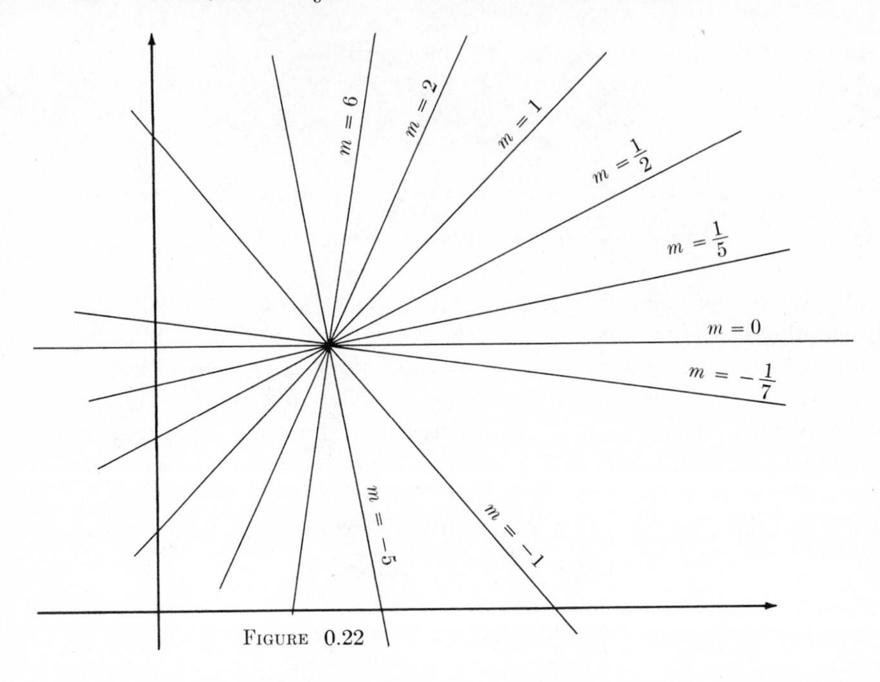

FIGURE 0.22

Formula (8) and Fig. 23, giving the slope as $\tan\theta$, make it easy to remember equation (7) for the line through P_1 with slope m. If $P = (x,y)$ is any point on the line, distinct from P_1, then we have

$$m = \tan\theta = \frac{y - y_1}{x - x_1},$$

and this leads immediately to (7). Similarly, to remember equation (5) for the line through two given points P_1 and P_2, note [Fig. 24] that m can be computed either from the points $P = (x,y)$ and $P_1 = (x_1, y_1)$ or from the points $P_1 = (x_1, y_1)$ and $P_2 = (x_2, y_2)$; thus

$$m = \tan\theta = \frac{y - y_1}{x - x_1} = \frac{y_2 - y_1}{x_2 - x_1},$$

and this leads immediately to (5).

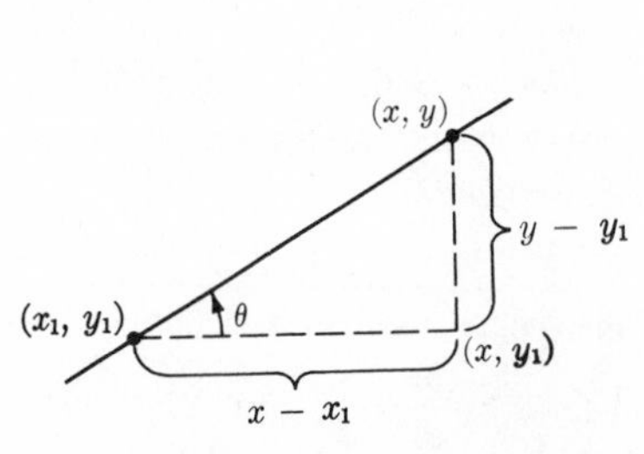

FIGURE 0.23 $m = \tan\theta = \dfrac{y - y_1}{x - x_1}$

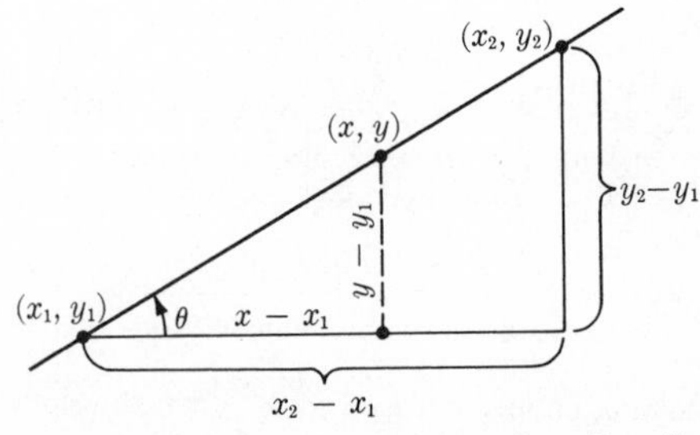

FIGURE 0.24 $m = \tan\theta = \dfrac{y - y_1}{x - x_1} = \dfrac{y_2 - y_1}{x_2 - x_1}$

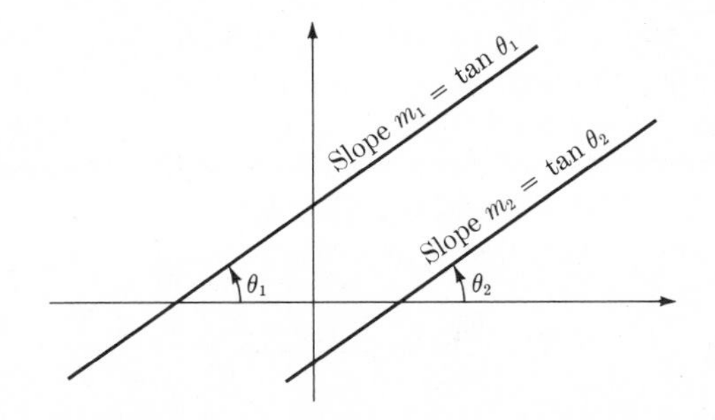

FIGURE 0.25 Lines are parallel when $\theta_1 = \theta_2$, or $m_1 = m_2$

Definition 3. Two lines are *parallel* iff

PARALLEL LINES

(i) both are vertical, or

(ii) both are nonvertical and have the same slope.

With this definition, it is easy to prove the fundamental facts about parallel lines: *If lines L_1 and L_2 are not parallel, then they intersect in a single point. If L_1 and L_2 are parallel, then either they have no points in common or they coincide.* [See Example 5.]

Perpendicular lines can also be defined algebraically:

Definition 4. Two lines are *perpendicular* iff

PERPENDICULAR LINES

(i) one is vertical and the other has slope 0, or

(ii) both are nonvertical, and their slopes m_1 and m_2 satisfy

$$m_1 m_2 = -1.$$

The reason for the condition $m_1 m_2 = -1$ is not immediately obvious, but the trigonometric identity*

$$\tan\left(\theta_1 + \frac{\pi}{2}\right) = -\frac{1}{\tan\theta_1}$$

explains it. For, a line L_1 cutting the x axis at an angle θ_1 is perpendicular to a line L_2 cutting the axis at an angle $\theta_2 = \theta_1 + \pi/2$; and then the slopes m_1 and m_2 of the two lines satisfy

$$m_2 = \tan\theta_2 = \tan\left(\theta_1 + \frac{\pi}{2}\right) = -\frac{1}{\tan\theta_1} = -\frac{1}{m_1},$$

so $m_1 m_2 = -1$.

---

* We use radian measure for angles; the radian measure of a right angle is $\pi/2$. For a review of trigonometry, see the Appendix to §2.3.

Example 3. Find the line through $(-3,-2)$ parallel to the x axis.
Solution. Parallel lines are lines of equal slope; and the slope of the x axis is 0, because its equation is $y = 0 = 0\cdot x + 0$. Hence we want the line through $(-3,-2)$ with slope 0. This is given by equation (7) with $x_0 = -3$, $y_0 = -2$, and $m = 0$. The resulting equation is

$$y - (-2) = 0\cdot(x - (-3)), \qquad \text{or} \qquad y = -2.$$

[This result could be read directly from Fig. 26.]

Example 4. Find the intersection of the two lines with equations $y = 2x + 5$ and $y = -3x - 4$. *Solution.* If the point (x,y) is on both lines, then it satisfies both equations

$$y = 2x + 5 \tag{9}$$

and

$$y = -3x - 4. \tag{10}$$

Equating the two different expressions for y and solving the result for x gives $x = -9/5$; plugging back in (9) gives $y = -18/5 + 5 = 7/5$. Thus $(-9/5, 7/5)$ is the intersection. [As a check, see that these coordinates satisfy (10) as well as (9).]

The technique in Example 4 can be used to prove a general result, as follows:

INTERSECTION OF TWO LINES

Example 5. Show that two nonvertical lines with different slopes intersect in exactly one point. *Solution.* The equations of the lines are

$$y = m_1x + b_1 \tag{11}$$

$$y = m_2x + b_2\,, \tag{12}$$

where $m_1 \neq m_2$. Any point (x,y) is on both lines iff it satisfies both these equations. Suppose that (x,y) lies on both lines. Subtracting (12) from (11), we obtain

$$0 = (m_1 - m_2)x + b_1 - b_2\,.$$

Since $m_1 - m_2 \neq 0$, this implies that $x = \dfrac{b_2 - b_1}{m_1 - m_2}$, and hence from (11),

$$y = m_1\frac{b_2 - b_1}{m_1 - m_2} + b_1 = \frac{m_1b_2 - m_2b_1}{m_1 - m_2}\,.$$

Thus if the given lines intersect at all, they intersect in the point

$$P = \left(\frac{b_2 - b_1}{m_1 - m_2}, \frac{m_1b_2 - m_2b_1}{m_1 - m_2}\right).$$

This point P actually does lie on both lines (as you can easily check), so the problem is solved.

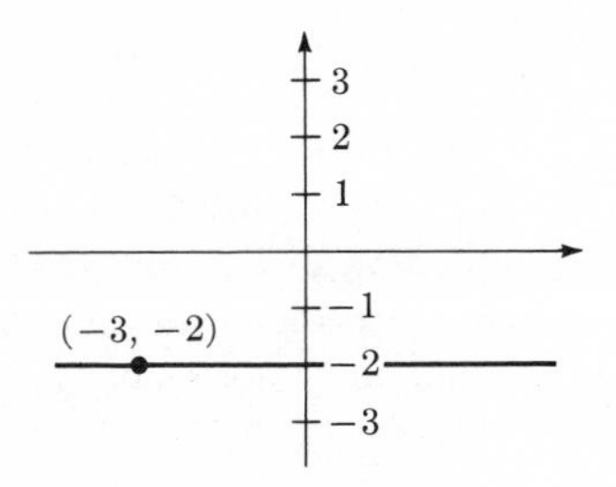

FIGURE 0.26

Example 6. Let $P_1 = (2,3)$ and $P_2 = (3,2)$. Show that the set $\{P: PP_1 = PP_2\}$ is a line perpendicular to the line through P_1 and P_2 [Fig. 27]. [*Recall:* $\{P: PP_1 = PP_2\}$ denotes the set of all points P such that $PP_1 = PP_2$, i.e. the distance from P to P_1 equals the distance from P to P_2.] *Solution.* Let $P = (x,y)$. Then the following equations are all equivalent:

$PP_1 = PP_2$

$\Leftrightarrow$

distance from (x,y) to $(2,3)$ = distance from (x,y) to $(3,2)$

$\Leftrightarrow$

$$\sqrt{(x-2)^2 + (y-3)^2} = \sqrt{(x-3)^2 + (y-2)^2} \qquad \text{[Definition 1]}$$

$\Leftrightarrow$

$$(x-2)^2 + (y-3)^2 = (x-3)^2 + (y-2)^2$$

[for positive numbers a and b, $a = b \Leftrightarrow a^2 = b^2$]

$\Leftrightarrow$

$$-4x - 6y = -6x - 4y \qquad \text{[simple algebra]}$$

$\Leftrightarrow$

$$y = x.$$

Hence the given set $\{P: PP_1 = PP_2\}$ is the line $L = \{(x,y): y = x\}$, which we recognize as the main diagonal [Fig. 27].

It remains to check that this line L is perpendicular to the line L' through P_1 and P_2. But in Example 2, we found the equation $y = -x + 5$ for the line through P_1 and P_2, so L' has slope $m' = -1$. The main diagonal L has slope $m = 1$, so $mm' = -1$, and the lines are indeed perpendicular.

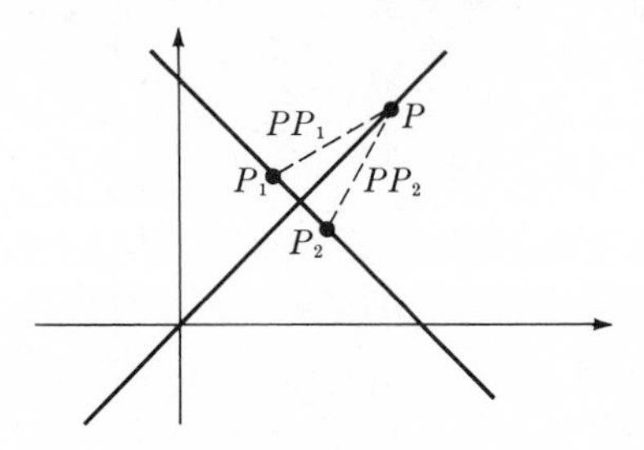

FIGURE 0.27

PERPENDICULAR BISECTOR

Remark 1. The set $\{P:PP_1 = PP_2\}$ is called the *perpendicular bisector* of the line segment from P_1 to P_2 [Fig. 28]. You probably recall from Euclidean geometry that this set is perpendicular to the line through P_1 and P_2 at a point midway between P_1 and P_2. Problems 14 and 15 below ask you to prove this in the present context, i.e. on the basis of Definitions 1–4 above.

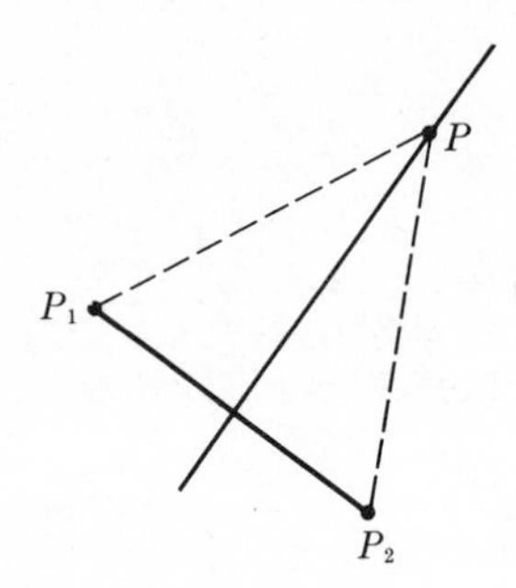

FIGURE 0.28 The perpendicular bisector of the segment from P_1 to P_2 is $\{P:PP_1 = PP_2\}$

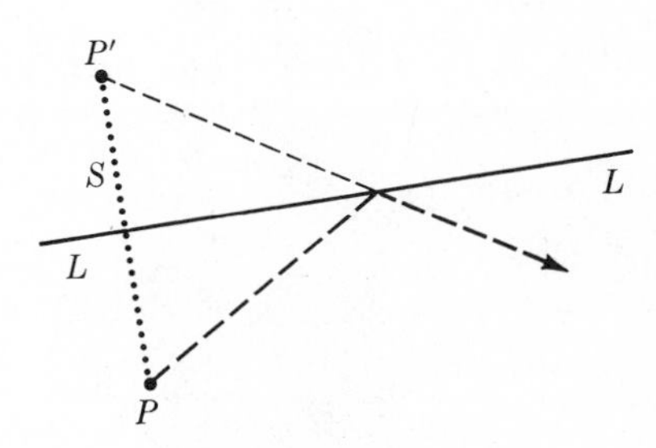

FIGURE 0.29

REFLECTION IN A LINE

Remark 2. Given a line L and a point P not on L, a point P' is called the *reflection of P in L* if L is the perpendicular bisector of the segment from P to P' [Fig. 29]. Example 6 shows that (3,2) is the reflection of (2,3) in the main diagonal. You can show in general that *(b,a) is the reflection of (a,b) in the main diagonal* [Fig. 30; Problem 12].

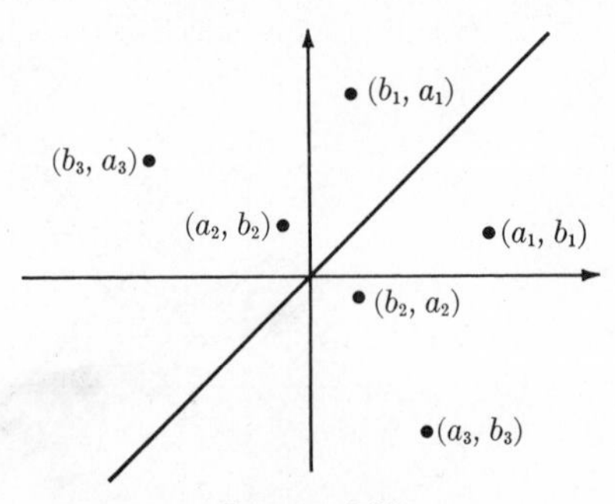

FIGURE 0.30

PROBLEMS

1. Plot the points (1,1), (−1,−1), (1,−1), (−1,1), (0,3), (−3,0), (1/2,−1/2), (1/3,−1/3).

2. Let $P_0 = (0,0)$, $P_1 = (2,-1)$, $P_2 = (3,1)$, $P_3 = (1,-3)$.
(a) Compute the distances P_0P_2, P_1P_2, P_2P_3.
(b) Plot the points carefully, measure the distances, and compare your results with the results obtained in part (a). [Hint: In making the comparison, you need not extract the square roots in (a). Square the measured distances, and compare these with the squares of the computed distances.]

3. Sketch the following lines, and find the slope of each. [In some examples, we have abandoned the traditional x and y, just for practice.]
(a) $\{(x,y) : y = -x\}$
• (b) $\{(x,y) : x + y = 3\}$ [Hint: You can solve for y; $x + y = 3 \Leftrightarrow y = -1 \cdot x + 3$.]
(c) $\{(s,t) : s + t = 0\}$
• (d) $\{(x,y) : x + y = -3\}$
(e) $\{(x,y) : x - y = 4\}$
• (f) $\{(x,y) : 2x + 3y = 1\}$
(g) $\{(x,y) : -4x - 6y = 1\}$
(h) $\{(a,b) : 3a - 2b = 0\}$

• **4.** You are given two of the lines in Problem 3; using Definitions 3 and 4, state whether the two lines are parallel, perpendicular, or neither. See if the results are consistent with the sketches in Problem 3.
(a) 3(a) and 3(c)
(b) 3(a) and 3(e)
(c) 3(a) and 3(f)
(d) 3(f) and 3(g)
(e) 3(f) and 3(h)
(f) 3(a) and 3(b)

5. For each of the following pairs of points, find (i) the slope of the line through the points and (ii) the equation of the line through the points. Sketch the points and the line carefully (preferably on graph paper), and see whether the slope you computed seems correct. If the line is vertical, do not assign it any slope.
• (a) (0,0) and (0,1)
• (b) (1,1) and (2,1)
• (c) (1/2,1/3) and (−2,0)
• (d) (−5,1) and (11,−1)
(e) (3,−1) and (5,0)
(f) (3,−5) and (−1,5)
(g) (4,2) and (8,−3)
(h) (−6,3) and (7,−6)
(i) (−6,3) and (8,−3)
(j) (12,8) and (6,0)
(k) (5,1) and (12,0)
(l) (7,2) and (2,7)

6. Find the line parallel to the given line and passing through the given point. Sketch the given line, the given point, and the new line.
• (a) Parallel to the graph of $y = 2x + 3$, through (0,0)
• (b) Parallel to the graph of $y = 4x$, through (1,2)
(c) Parallel to the graph of $y = \frac{1}{3}x + 2$, through $(-\frac{1}{3},2)$
(d) Parallel to the graph of $y = 111x - \pi$, through (1,−1)

7. Repeat Problem 6 for the following lines and points. [Hint: To find the slope, rewrite the given equation in the form $y = mx + b$.]

• (a) Parallel to the graph of $x + 3y = 10$, through $(0,10)$

• (b) Parallel to the graph of $30x - 3y = 1$, through $(10,0)$

(c) Parallel to the graph of $2y + 3x = 4$, through $(1,-3)$

(d) Parallel to the graph of $x = 4y + 2$, through $(3,8)$

• **8.** Figure 31 shows the graphs of four straight lines. From the graphs, compute the slope of each line with reasonable accuracy.

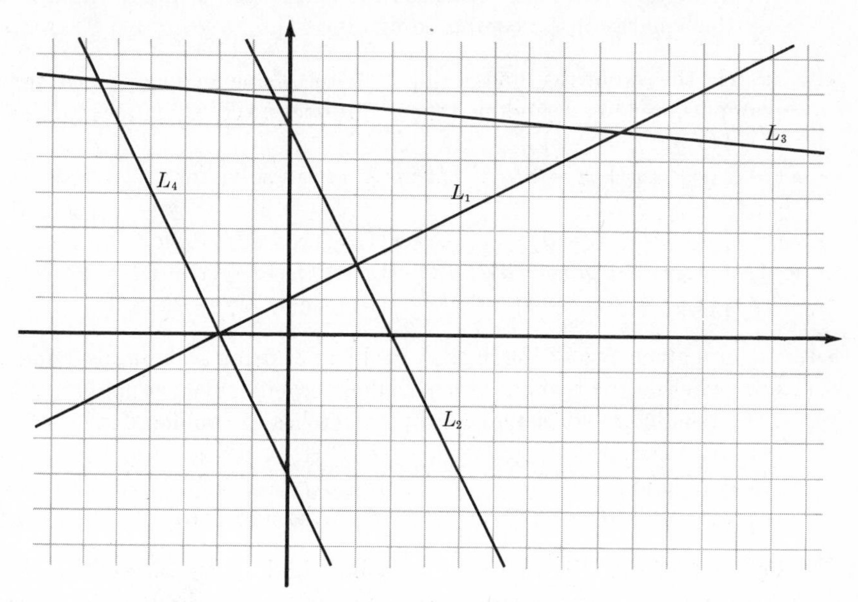

FIGURE 0.31

• **9.** For the following points P_1 and P_2, deduce from $PP_1 = PP_2$ the equation of the perpendicular bisector. Check that the relation $m_1m_2 = -1$ holds between the slope m_1 of the bisector and the slope m_2 of the line through P_1 and P_2.

(a) $P_1 = (1,1)$, $P_2 = (-1,-1)$

(b) $P_1 = (0,3)$, $P_2 = (5,-1)$

(c) $P_1 = (a,b)$, $P_2 = (a,-b)$

•**10.** Write the equation of the line through the origin and perpendicular to the graph of $4y - 3x = 11$. [Hint: To find the slope of the given line, rewrite the given equation in the form $y = mx + b$.]

11. This problem proves Theorem 1: *Through two distinct points* $P_1 = (x_1, y_1)$ *and* $P_2 = (x_2, y_2)$ *there passes one and only one line* L. [See Example 2 above.]

(a) Prove that if $x_1 \neq x_2$, and P_1 and P_2 are both on the graph of $y = mx + b$, then

$$m = \frac{y_2 - y_1}{x_2 - x_1} \quad \text{and} \quad b = \frac{y_1x_2 - y_2x_1}{x_2 - x_1}.$$

(b) Prove conversely that P_1 and P_2 lie on the graph of

$$y = \frac{y_2 - y_1}{x_2 - x_1}x + \frac{y_1x_2 - y_2x_1}{x_2 - x_1}.$$

(c) Prove that if $x_1 \neq x_2$, then P_1 and P_2 lie on no vertical line $\{(x,y): x = c\}$.

(d) Prove that if $x_1 = x_2$ but $y_1 \neq y_2$, then P_1 and P_2 lie on the graph of $x = x_1$, and lie on *no* nonvertical line.

12. Prove that if $a \neq b$, then the main diagonal is the perpendicular bisector of the segment from (a,b) to (b,a). [Thus (b,a) is the reflection of (a,b) in the main diagonal. Recall that the main diagonal is the line $\{(x,y): y = x\}$.]

13. Prove that if $P_1 \neq P_2$, then the set $\{P: PP_1 = PP_2\}$ is a straight line. [See Example 6 and Remark 1.]

14. (a) Prove that the line in Problem 13 is perpendicular to the line through P_1 and P_2.

(b) Prove that the two lines in part (a) intersect in the point

MIDPOINT BETWEEN (x_1, y_1) AND (x_2, y_2)

$$M = \left(\frac{x_1 + x_2}{2}, \frac{y_1 + y_2}{2}\right).$$

(c) Prove that the two lines in part (a) are perpendicular.

(d) Prove that $P_1M = MP_2$. [Thus the perpendicular bisector in Problem 13 intersects the line through P_1 and P_2 in a point M equidistant from P_1 and P_2.]

15. (a) Prove that $(a,-b)$ is the reflection of (a,b) in the horizontal axis. [See Remark 2 above.]

(b) Prove that $(-b,-a)$ is the reflection of (a,b) in an appropriate line.

16. (a) Show that $\{(x,y): x^2 + y^2 = r^2\}$ is a circle of radius r centered at the origin. [Hint: If P is in the given set, and O is the origin, show that $OP = r$; show also the converse.]

(b) Show that $\{(x,y): (x - x_0)^2 + (y - y_0)^2 = r^2\}$ is a circle of radius r centered at $P_0 = (x_0, y_0)$.

17. (a) Sketch the set $\{(x,y): |x| + |y| = 1\}$. [It's a square.]
(b) Sketch $\{(x,y): |x| + |y| \leq 1\}$.

18. [*The Pythagorean Theorem*] Prove that $(P_1P_2)^2 + (P_1P_3)^2 = (P_2P_3)^2$ if and only if the line through P_1 and P_2 is perpendicular to the line through P_1 and P_3. [*Remark*: This important fact was known empirically, centuries before Pythagoras and Euclid. We used this empirical knowledge in defining distance [Definition 1 above]. Our definition makes the Pythagorean Theorem automatically true in the geometric system we have set up, at least in the special case where one line is vertical and the other horizontal; now you are asked to prove it in general.]

19. This problem concerns Euclid's construction of an equilateral triangle [see page 4].
(a) Given two points $P_1 = (x_1, y_1)$ and $P_2 = (x_2, y_2)$, prove that the point $P = (x,y)$ given by

$$x = \frac{1}{2}(x_1 + x_2) - \frac{\sqrt{3}}{2}(y_2 - y_1),$$

$$y = \frac{1}{2}(y_1 + y_2) + \frac{\sqrt{3}}{2}(x_2 - x_1)$$

satisfies $PP_1 = PP_2 = P_1P_2$. In other words, the triangle with vertices P, P_1, and P_2 is equilateral.
(b) Draw a sketch showing that there should be *two* different points, P and P', such that $PP_1 = PP_2 = P_1P_2$ and $P'P_1 = P'P_2 = P_1P_2$.
(c) Find the coordinates of a point P', different from the point P in part (a), such that the triangle with vertices P', P_1, and P_2 is equilateral.

Remark 3. You may object that the coordinates of the point P given in Problem 19(a) were "pulled out of a hat." These coordinates can be found by solving the simultaneous equations $PP_1 = PP_2 = P_1P_2$; or they can be found by trigonometry, as follows.

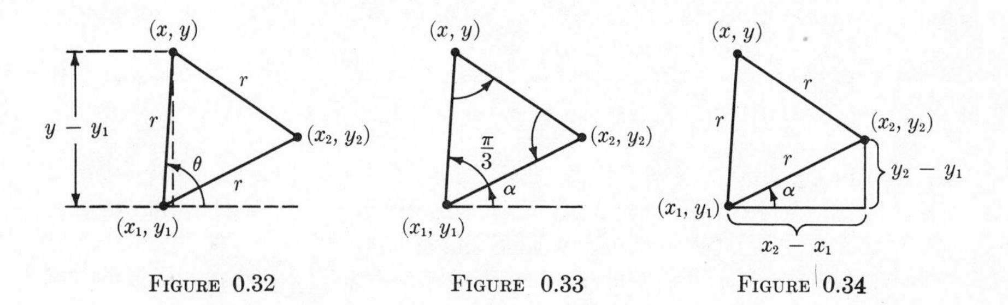

FIGURE 0.32 FIGURE 0.33 FIGURE 0.34

Let $r = P_1P_2$. Then from Fig. 32 you can read that $x = x_1 + r\cos\theta$ and $y = y_1 + r\sin\theta$; Figure 33 shows that $\theta = \alpha + \pi/3$; and Fig. 34 gives $\cos\alpha = (x_2 - x_1)/r$ and $\sin\alpha = (y_2 - y_1)/r$. Then from the relations

$$\cos(\alpha + \beta) = \cos\alpha\cos\beta - \sin\alpha\sin\beta,$$

$$\sin(\alpha + \beta) = \sin\alpha\cos\beta + \cos\alpha\sin\beta,$$

$$\cos\frac{\pi}{3} = \frac{1}{2}, \qquad \text{and} \qquad \sin\frac{\pi}{3} = \frac{\sqrt{3}}{2},$$

you can deduce immediately that

$$x = x_1 + r\cos\theta = x_1 + r\cos\alpha\cos\frac{\pi}{3} - r\sin\alpha\sin\frac{\pi}{3}$$

$$= x_1 + \frac{1}{2}(x_2 - x_1) - \frac{\sqrt{3}}{2}(y_2 - y_1) = \frac{1}{2}(x_1 + x_2) - \frac{\sqrt{3}}{2}(y_2 - y_1),$$

$$y = y_1 + r\sin\theta = y_1 + r\sin\alpha\cos\frac{\pi}{3} + r\cos\alpha\sin\frac{\pi}{3}$$

$$= y_1 + \frac{1}{2}(y_2 - y_1) + \frac{\sqrt{3}}{2}(x_2 - x_1) = \frac{1}{2}(y_1 + y_2) + \frac{\sqrt{3}}{2}(x_2 - x_1),$$

as given in Problem 19.

Remark 4. We have before us in Problem 19 an example of the following standard operating procedure in analytic geometry and calculus:

1. A problem is stated in physical, geometric, or some other nonnumerical terms. [In Problem 19: Given a line segment, construct an equilateral triangle with the segment as one side.]

2. The problem is translated into purely numerical terms. [In this example: Given (x_1, y_1) and (x_2, y_2), find (x,y) such that $(x - x_1)^2 + (y - y_1)^2 = (x - x_2)^2 + (y - y_2)^2 = (x_1 - x_2)^2 + (y_1 - y_2)^2$.]

3. A solution to the numerical problem is suggested by a picture. [In this example, the solution is suggested by Figs. 32–34, as explained in Remark 3.]

4. An analytic proof is given to show that the solution suggested by the picture actually solves the problem. [In this example, you supplied the analytic proof in Problem 19 above.]

Step 4 serves two purposes. It provides a check of any computations made in step 3. More important, it assures us that we have not been led astray by the picture on which step 3 was based.

Step 3 is usually the one where the greatest imagination is required. Unless the problem is of a standard type that you have learned, you have to *think*: What do I know that could be used here? The solution will not always be suggested by a picture; sometimes it will be a formula that you remember, or an analogous problem.

0.4 FUNCTIONS AND GRAPHS

A prominent milestone in the scientific Renaissance was Galileo's discovery of the law of falling bodies: An object falling freely from rest for t seconds falls $s(t) = 16t^2$ feet. (Of course, Galileo used other units of time and distance, but with any units the law still has the form $s(t) = ct^2$, where c is a constant.) Thus to each time t there is associated a specific number $s(t) = 16t^2$, giving the distance the object has fallen. Such an association is a *function*. As this example may suggest, functions are essential in every quantitative science.

The role of calculus, both in applications and in pure mathematics, is that of a powerful means of analyzing functions. Obviously, then, the elementary facts about functions outlined here are essential background for the study of calculus.

We begin by defining our terms. Let D be a given set, for example:

the set of numbers in an interval (a,b);

a set of points in the plane;

the set of all rectangles;

the set of all candidates for Boston City Council in 1968.

REAL-VALUED FUNCTION; DOMAIN

A *real-valued function with domain D* assigns to each member of D a unique real number. For example, if D is the set of all candidates in an election, the "vote function" assigns to each candidate the number of votes he receives. If D is the set of all real numbers, there is a "squaring function" which assigns to each number x its square, x^2. If D is the set of all positive numbers, there is a "square root function" assigning to each positive number x its *positive* square root $\sqrt{x}$. (The square root function cannot assign "$\pm\sqrt{x}$", because this would assign two *different* numbers to x, and a function can't do that.)

Functions are generally denoted by letters, like f, g, φ, etc.

Let f be a function, and let x be a member of its domain D. The number assigned to x is denoted $f(x)$ [read "f of x"], and is called the *value* of the function at x.

The set of all values assigned by f is called the *range* of f:

$$\text{range of } f = \{y : y = f(x) \text{ for some } x \text{ in the domain of } f\}.$$

RANGE

For example, if f is the squaring function, $f(x) = x^2$, then the range of f is the interval $[0, +\infty)$. This may be obvious, but in any case it is easy to prove, as follows:

y is in the ~~domain~~ range of f $\Leftrightarrow$ $y = f(x)$ for some real x [definition of range]

$\Leftrightarrow$ $y = x^2$ for some real x [since $f(x) = x^2$]

$\Leftrightarrow$ $y \geq 0$ [$y \geq 0$ iff y is a square]

$\Leftrightarrow$ y is in the interval $[0, +\infty)$.

Real-valued functions arise naturally in geometry, and in any field of study where things can be measured.

Example 1. D is the set of all rectangles, and f is the area function; $f(R) =$ area of R, for every rectangle R. The range of f is the interval $(0,+\infty)$, since every positive number is the area of some rectangle.

Example 2. At time $t = 0$, an object is dropped from rest, and falls freely for two seconds before landing. Then at any time t, $0 \leq t \leq 2$, the object has fallen a certain distance $d(t)$ feet. We thus have a real-valued function d defined on the domain $D = [0,2]$. According to Galileo's law, $d(t) = 16t^2$, so in two seconds the object drops $16 \cdot 4 = 64$ feet. Thus the range of d is the interval $[0,64]$; the distance ranges from 0 to 64 as t varies from 0 to 2.

Example 3. Let $P(t)$ denote the number of people alive at time t, where t is measured in years A.D. [Thus $P(1984)$ denotes the population on New Year's Eve 1983, and $P(-46)$ denotes the population about the time when Julius Caesar introduced his calendar.] This function P has domain $D = (-\infty,+\infty)$; the range of P is as yet unknown. Surely $P(t) = 0$ when t is sufficiently large negative, and probably $P(t) = 0$ when t is sufficiently large positive. (But let us hope that $P(t) > 0$ for $0 \leq t \leq 2100$, at least.)

Remark 1. Examples 2 and 3 can be challenged: Since it is impossible to determine $d(t)$ and $P(t)$ exactly, how can we claim that d and P are well-defined functions?

Similar problems arise in all but the most sophisticated applications of mathematics to physics, biology, and so on. Nevertheless, each of these fields accepts the idealized but very useful assumption that there are functions such as those described in these examples.

Graph of a function

GRAPH DEFINED Every function f has a *graph*, defined as $\{(x,y): y = f(x)\}$, the set of all points (x,y) in the plane such that x is in the domain of f and $y = f(x)$. A sketch of the graph generally gives a useful picture of the function. The domain of f can be visualized by projecting the graph onto the x axis [Fig. 35]; projecting the graph onto the y axis gives the range.

Given the graph, you can find the value $f(x)$ by starting at the point x in the domain, moving vertically to the graph of f, then moving horizontally to the y axis [Fig. 35].

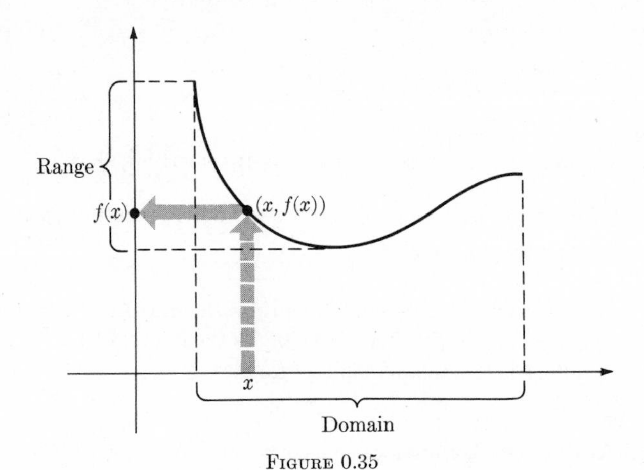

FIGURE 0.35

Example 4. The domain D is all real numbers, and the function f assigns to any number x the number $2x - 3$; thus $f(x) = 2x - 3$. We define this function *and its domain* briefly by writing

$$f(x) = 2x - 3, \qquad -\infty < x < \infty.$$

The graph of f is the set $\{(x,y): y = f(x)\}$ or $\{(x,y): y = 2x - 3\}$, which is a straight line of slope 2 [Fig. 36]. From the graph, you can see that the range of f is the whole real line; every horizontal line intersects the graph, thus the projection of the graph on the y axis includes every point on the y axis.

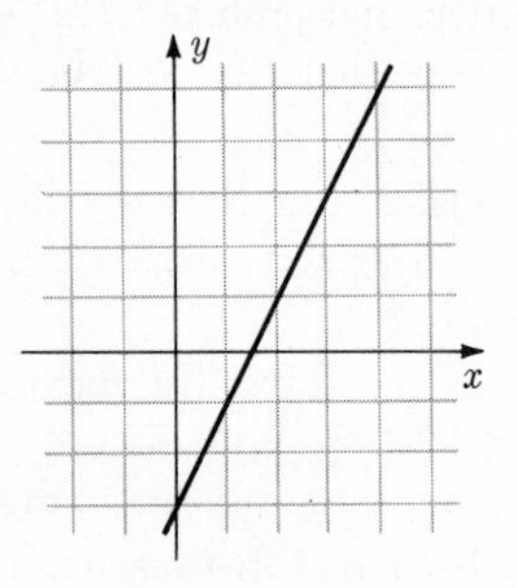

FIGURE 0.36

Example 5. $f(x) = \frac{1}{2}x^2 + 1, \quad -\infty < x < \infty$. This function is called a *polynomial.* Generally, a polynomial is any function of the form

POLYNOMIAL FUNCTION

$$f(x) = a_0 + a_1x + \cdots + a_nx^n = \sum_0^n a_jx^j,$$

where the a_j are given constants.† If $a_n \neq 0$, the polynomial has degree n; thus $\frac{1}{2}x^2 + 1$ is a polynomial of degree 2. The graph of this polynomial is a parabola [Fig. 37], as we will show in §1.4. From Fig. 37 you can see that the range of f is the interval $[1,\infty)$. We sketched this graph by making the following table of function values, plotting the corresponding points $(x,f(x))$ on the graph, and drawing a smooth curve through these points.

x	$f(x) = \frac{1}{2}x^2 + 1$	$(x,f(x))$
0	1	$(0,1)$
1	$1\frac{1}{2}$	$(1,1\frac{1}{2})$
-1	$1\frac{1}{2}$	$(-1,1\frac{1}{2})$
2	3	$(2,3)$
-2	3	$(-2,3)$
3	$5\frac{1}{2}$	$(3,5\frac{1}{2})$
-3	$5\frac{1}{2}$	$(-3,5\frac{1}{2})$

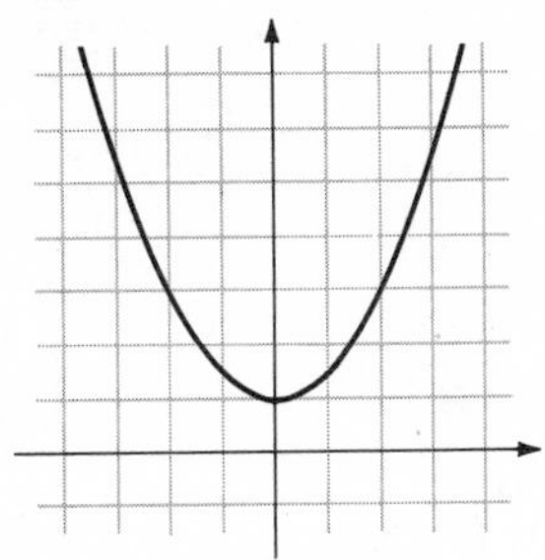

FIGURE 0.37

Example 6. On the domain $D = \{x : x \neq 0\}$, define a function f by $f(x) = 1/x$. The function and its domain are given briefly by

$$f(x) = \frac{1}{x}, \qquad x \neq 0,$$

To graph f, we construct the following table, plot the corresponding points, and sketch the graph in Fig. 38. [The graph is actually a hyperbola with the x and y axes as asymptotes; see Problem 14.]

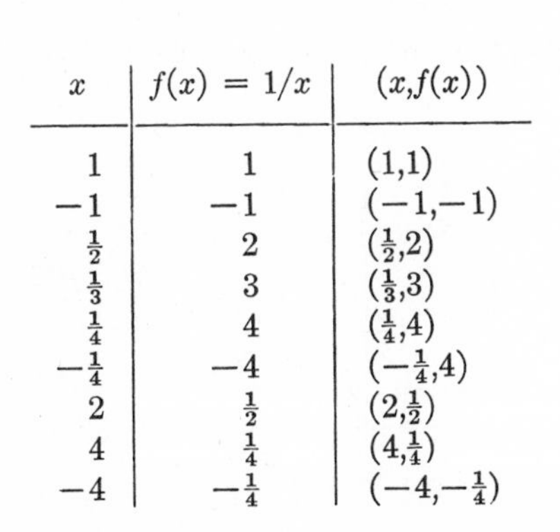

x	$f(x) = 1/x$	$(x,f(x))$
1	1	$(1,1)$
-1	-1	$(-1,-1)$
$\frac{1}{2}$	2	$(\frac{1}{2},2)$
$\frac{1}{3}$	3	$(\frac{1}{3},3)$
$\frac{1}{4}$	4	$(\frac{1}{4},4)$
$-\frac{1}{4}$	-4	$(-\frac{1}{4},4)$
2	$\frac{1}{2}$	$(2,\frac{1}{2})$
4	$\frac{1}{4}$	$(4,\frac{1}{4})$
-4	$-\frac{1}{4}$	$(-4,-\frac{1}{4})$

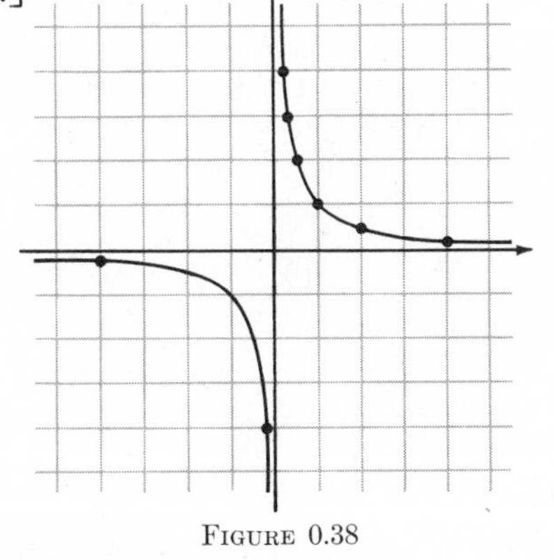

FIGURE 0.38

† The "summation notation" $\sum_0^n$ is explained in the Appendix, §AI.1.

This method of sketching a graph by plotting a few points is rather crude, and not always reliable. For instance, with $f(x) = 1/x$, we might have plotted the points (1,1), (2,1/2), (3,1/3), (−1,−1), (−2,−1/2), (−3,−1/3), and then drawn something like Fig. 39. Obviously, that figure grossly misrepresents the behavior of f near $x = 0$, which is shown correctly in Fig. 38. One use of calculus is to improve on the crude method of plotting points at random.

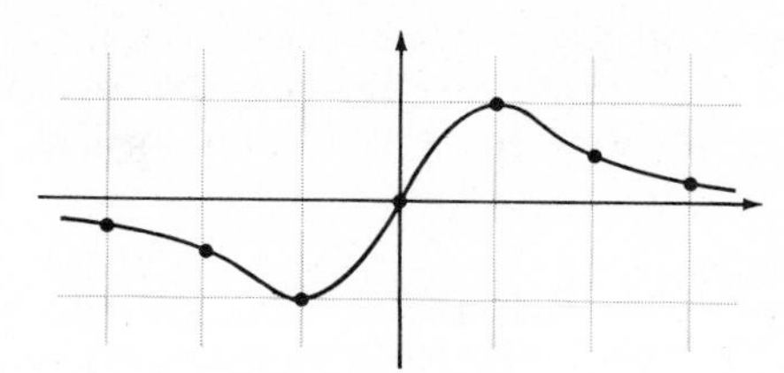

FIGURE 0.39 Wrong graph of $f(x) = 1/x$, based on insufficient information

Example 7. The function

RATIONAL FUNCTION

$$f(x) = \frac{x^2 + x + 1}{x + 1}, \qquad x \neq -1$$

is called a *rational function,* since it is the ratio of two polynomials. Notice that the domain D is the largest set on which the ratio is defined; when $x = -1$, the denominator is zero.

The sketch of the graph [Fig. 40] shows that the range of f consists of two intervals. [Chapter IV shows how these intervals can be determined exactly.]

SQUARE ROOT FUNCTION

Example 8. The *square root function,* denoted $\sqrt{\ }$, is defined on the domain $D = [0,+\infty)$, as follows: $\sqrt{\ }(x)$ is that number y such that $y \geq 0$ and $y^2 = x$. The graph [Fig. 41] is the upper half of the parabola $\{(x,y) : x = y^2\}$. [We have written $\sqrt{\ }(x)$ to conform to the general function notation $f(x)$. Actually, for this particular function it is customary to write $\sqrt{x}$ instead of $\sqrt{\ }(x)$.]

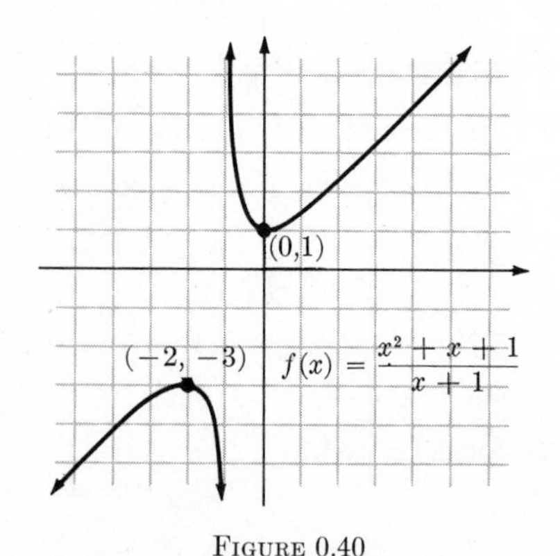

FIGURE 0.40

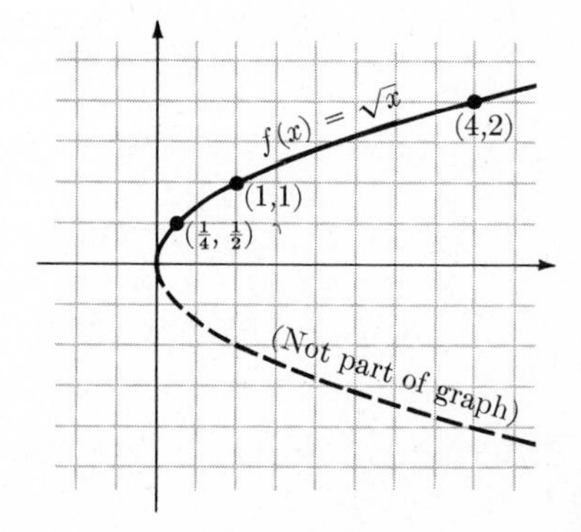

FIGURE 0.41

We have not defined $\sqrt{x}$ if $x < 0$, for if $x < 0$ then there *is* no real number y such that $y^2 = x$, and hence x has no real square root. Thus the domain of $\sqrt{\ }$ is the interval $[0,\infty)$. The range is also $[0,\infty)$; for, every number $y \geq 0$ can be written as $y = \sqrt{x}$ by setting $x = y^2$.

Example 9. $g(x) = |x|, \quad -\infty < x < \infty$. This is the ***absolute value function:*** its graph consists of two half-lines joined at the origin [Fig. 42]. The range of g is $[0,\infty)$.

Example 10. $\psi(t) = \sqrt{t^2}, \quad -\infty < t < \infty$. Since $\sqrt{\ }$ means ***nonnegative*** square root, you can easily check that $\psi(t) = t$ if $t \geq 0$, and $\psi(t) = -t$ if $t \leq 0$; hence $\psi(t) = |t|$.

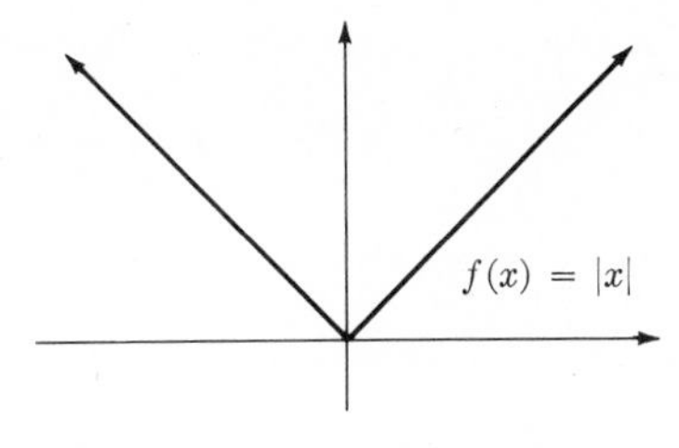

FIGURE 0.42

The formula defining ψ in Example 10 looks different from the formula defining g in Example 9; however, the two functions are equal, $\psi = g$. For they have the same domain D, and they assign the same numbers to each point of D, i.e. $\psi(x) = g(x)$ for every x.

WHICH PLANE SETS ARE GRAPHS?

It is clear from Fig. 43 that *a set F in the plane is the graph of a function f if and only if each vertical line intersects F in at most one point.* For example, the set F in Fig. 43(a) is the graph of the function f defined as follows: $f(x_0) = y_0$, where (x_0, y_0) is the intersection of F with the vertical line through $(x_0, 0)$. In Fig. 43(b), on the other hand, the set F does not define $f(x_0)$; put differently, no matter how we pick $f(x_0)$, we cannot recover the whole set F as the graph of f, since either P or P' will be missing.

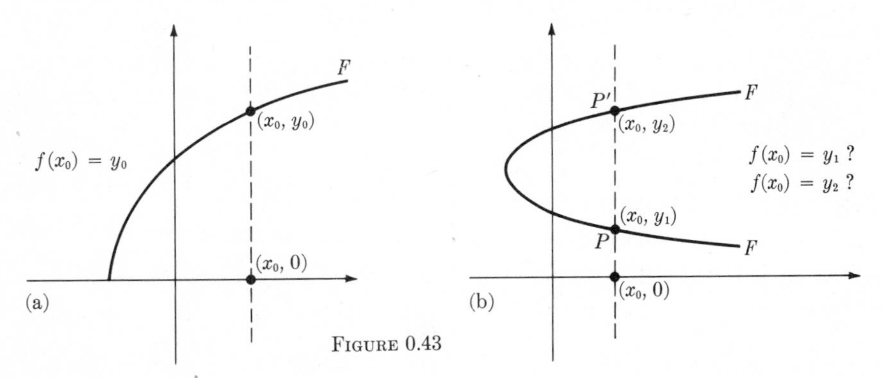

FIGURE 0.43

Reflection in the x axis

If you reflect the graph of a function f in the x axis, the new set obtained is again the graph of a function, call it g [Fig. 44]. Let $P = (x,f(x))$ be any point in the graph of f. Its reflection in the axis is the point $P' = (x,-f(x))$; since this point is in the graph of g, we have $(x,-f(x)) = (x,g(x))$, hence

$$g(x) = -f(x).$$

This function g is called the *negative* of f, denoted $-f$.

Inverse functions

Consider the two functions in Fig. 45, the square

$$s(x) = x^2, \qquad -\infty < x < \infty$$

and the cube

$$f(x) = x^3, \qquad -\infty < x < \infty.$$

The second function, f, has the following three properties:

PROPERTIES OF 1-1 FUNCTION

(i) Each horizontal line intersects the graph of f at most once.

(ii) For each number y in the range of f, the equation $y = f(x) = x^3$ has exactly one solution. For example, taking $y = -8$, we have $-8 = x^3 \Leftrightarrow x = -2 = (-8)^{1/3}$; and generally $y = x^3 \Leftrightarrow x = y^{1/3}$.

(iii) If you reflect the graph of f in the main diagonal, the new set obtained is the graph of a function [Fig. 46]. In fact, it is the graph of the *cube root function*

$$g(x) = x^{1/3}, \qquad -\infty < x < \infty.$$

The square function s, on the other hand, has none of these properties:

Many horizontal lines intersect the graph *twice* [Fig. 45].

The equation $y = s(x) = x^2$ has *two* solutions (for $y > 0$), namely $x = +\sqrt{y}$ and $x = -\sqrt{y}$.

If you reflect the graph of s in the main diagonal, the new set obtained is *not* the graph of any function [Fig. 46], since many vertical lines intersect the graph twice.

A little thought should persuade you that conditions (i), (ii), and (iii) are all equivalent. Any function f satisfying these conditions is called *one-to-one*, because of condition (ii); each number y in the range of f comes from just one number x in the domain of f, that is, the equation

$$y = f(x) \tag{1}$$

has just one solution x.

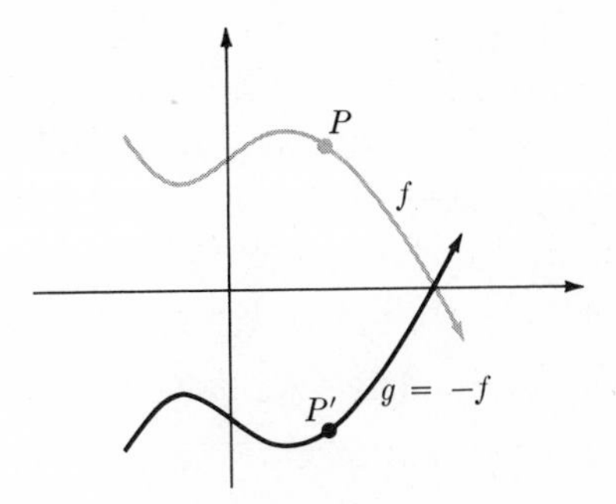

FIGURE 0.44

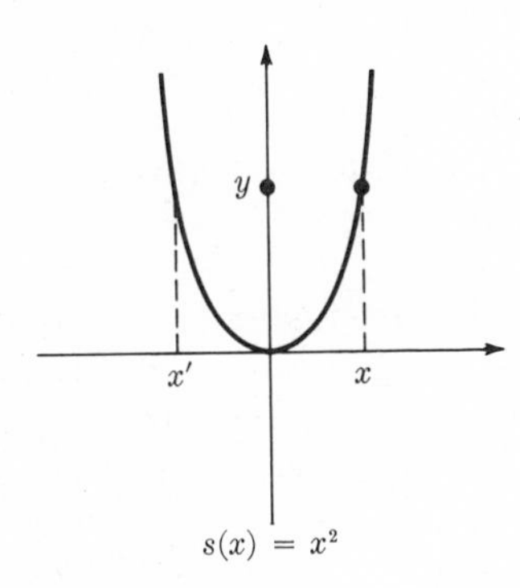

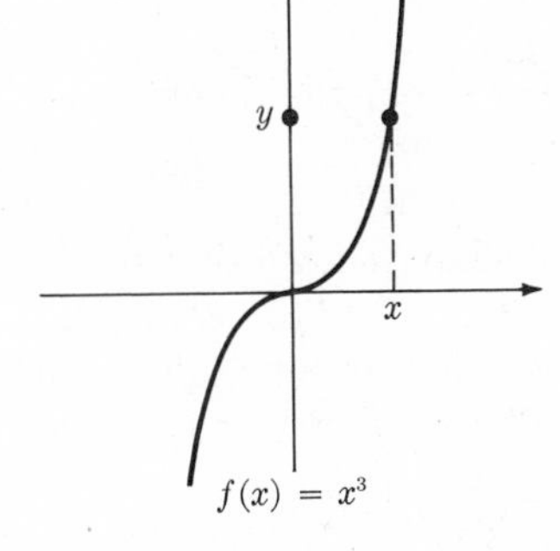

FIGURE 0.45

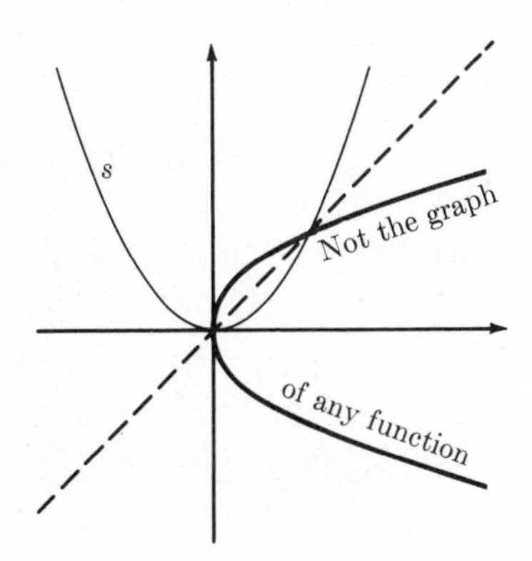

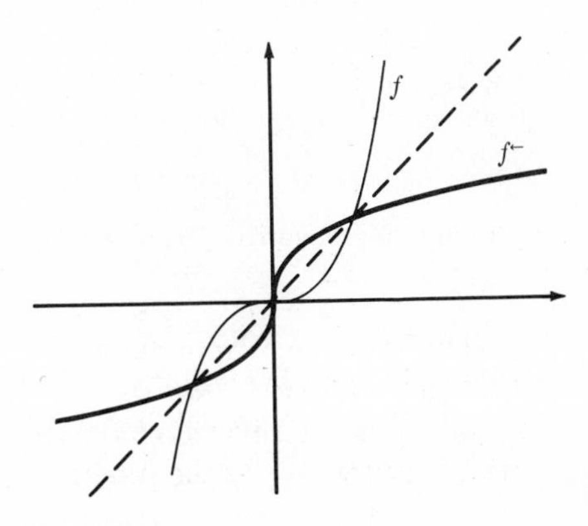

FIGURE 0.46

GRAPH OF INVERSE FUNCTION

Every function f which is one-to-one has an *inverse*, denoted $f^{\leftarrow}$ or f^{-1}. The graph of $f^{\leftarrow}$ is obtained by reflecting the graph of f in the main diagonal. This is legitimate, in view of condition (iii). The domain of $f^{\leftarrow}$ is the range of f; for each y in the domain of $f^{\leftarrow}$, $f^{\leftarrow}(y)$ is defined to be the unique solution of the equation (1) for x. That is, when f is one-to-one, then

RELATION BETWEEN f AND $f^{\leftarrow}$

$$y = f(x) \quad \Leftrightarrow \quad x = f^{\leftarrow}(y). \tag{2}$$

For the cube function f given above, we have

$$y = f(x) \quad \Leftrightarrow \quad y = x^3 \quad \Leftrightarrow \quad x = y^{1/3}. \tag{3}$$

Comparing (3) with (2), we see that the inverse of the cube function f is the cube root function,

$$f^{\leftarrow}(y) = y^{1/3}, \qquad -\infty < y < \infty.$$

Another familiar example is the logarithm; $\log_{10} y$ is traditionally defined as "that number x such that $10^x = y$," in other words,

$$y = 10^x \quad \Leftrightarrow \quad x = \log_{10} y.$$

It follows that the function $\log_{10}$ is the inverse of the function

$$f(x) = 10^x, \qquad -\infty < x < \infty.$$

We have shown that the cube root is an inverse function, and the same is true for square roots, fourth roots, and so on. But some care is required with even roots, since the even power functions are not one-to-one on the whole line. The next example covers the important case of square roots.

Example 11. The square function $s(x) = x^2$, $-\infty < x < \infty$, is not one-to-one, and thus does not have an inverse [Fig. 46]. However, if we restrict the domain of x^2 to the interval $[0, +\infty)$, we can obtain the positive square root as an inverse function. Consider the function

$$f(x) = x^2, \qquad 0 \le x < \infty.$$

Then f *is* one-to-one, and reflecting the graph *does* yield the graph of a function [Fig. 47]. This inverse function is the positive square root. A formal proof goes as follows: $\sqrt{y}$ is defined as "that number x which is ≥ 0, and whose square is y"; in symbols, for $y \ge 0$

$$y = x^2 \text{ and } x \ge 0 \quad \Leftrightarrow \quad x = \sqrt{y}.$$

Comparing this with (2), we see that the inverse to f is given by $f^{\leftarrow}(y) = \sqrt{y}$, $y \ge 0$.

Thus the square root function is the inverse of the square function *with domain restricted to* $[0, +\infty)$. Similarly, the $2n$th root function $g(y) = \sqrt[2n]{y}$ is the inverse of the $2n$th "power function" with its domain restricted to $[0, +\infty)$, the function

$$f(x) = x^{2n}, \qquad x \ge 0.$$

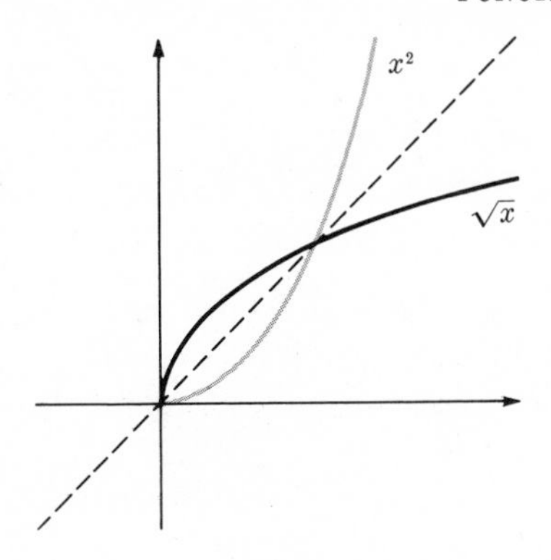

FIGURE 0.47

When a function is tabulated, its inverse can be obtained simply by "reading the table backward," as in the following example.

INVERSE FUNCTION FROM A TABLE

Example 12. Here is a simplified table of the function $f(x) = 10^x$:

x	-1.0	-0.9	-0.8	-0.7	-0.6	-0.5	-0.4	-0.3	-0.2	-0.1	0
10^x	0.1	0.126	0.158	0.2	0.251	0.316	0.398	0.501	0.631	0.794	1

The inverse function $\log_{10}$ is defined by "$\log_{10} y$ is that number x such that $10^x = y$"; thus $\log_{10} 1 = 0$, $\log_{10} .501 = -0.3$, etc. A table for the inverse function is not really necessary, but it could be obtained simply by setting $y = 10^x$ and $x = \log_{10} y$ in the table above, and placing the y row first:

y	0.1	0.126	0.158	0.2	0.251	0.316	0.398	0.501	0.631	0.794	1
$\log_{10} y$	-1.0	-0.9	-0.8	-0.7	-0.6	-0.5	-0.4	-0.3	-0.2	-0.1	0

In Fig. 48 the graph of $f(x) = 10^x$ was constructed from the first table, and the graph of $f^{\leftarrow}(y) = \log_{10} y$ from the second table. Notice that each graph is the reflection of the other in the main diagonal.

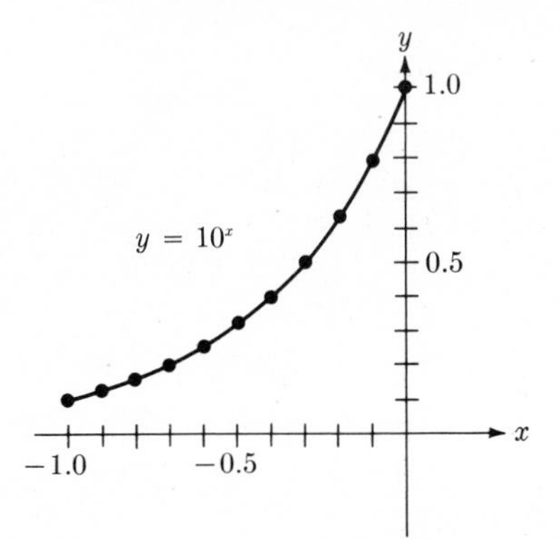

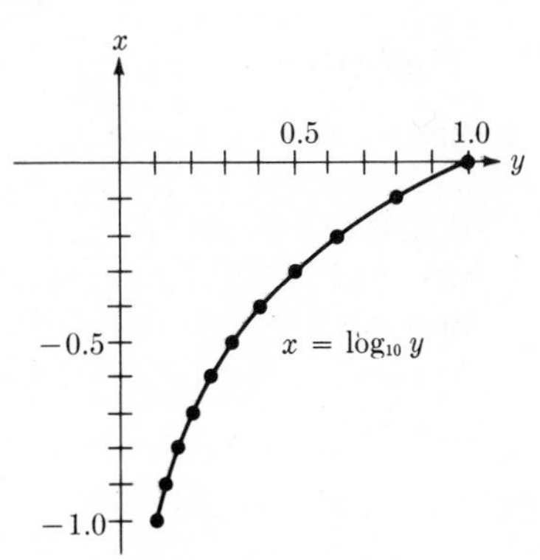

FIGURE 0.48

Remark. We have not really defined what a function *is*, but merely said what it does; it assigns a unique number to each member of its domain. No formal definition is given here because, as far as I can see, none of the currently acceptable definitions (i) tells clearly and unequivocally what a function is, and yet (ii) remains consistent with current mathematical usage and terminology. Forgetting about (i), we could say "a function is an assignment" or "a transformation" or "a correspondence" or "a mapping" or "a rule," but these are either vague, or conjure up irrelevant visions.

The problems related with (i) would be solved by the following definition:

FUNCTION: VERY FORMAL DEFINITION

A *function* is a set f of ordered pairs (a,b), with the following property: if (a,b) and (a',b') are in the set f, and $a = a'$, then $b = b'$. The domain D of f is the set of all first elements a,

$$D = \{a\colon (a,b) \text{ is in } f, \text{ for some } b\}.$$

The *range* of f is the set of all second elements b,

$$\text{range of } f = \{b\colon (a,b) \text{ is in } f, \text{ for some } a\}.$$

If a is in the domain of f, then $f(a)$ denotes the unique element b such that (a,b) is in f.

But notice that what this defines is usually called the *graph*, not the function, so point (ii) above is violated. Moreover, functions generally are given directly in the form of a transformation or rule (such as $f(x) = x^2$), not as a set of ordered pairs (such as $f = \{(x,y) : y = x^2\}$).

Fortunately, these problems are not crucial; with the descriptions and examples in the text, you should be able to recognize a function when you see one, and that is what really matters.

PROBLEMS

Problems 9–11 introduce *odd* and *even* functions.

1. Sketch the graphs of the following functions, and find the range of each function. [If the domain is not given explicitly, take the largest domain to which the given formula can be applied; for example, in part (a) take the domain to be $[-1, \infty)$, since $\sqrt{x+1}$ is defined only for $x \geq -1$.]

(a) $a(x) = \sqrt{x+1}$

(b) $b(x) = x^3 - 1$

(c) $c(s) = \dfrac{1}{s+2}$

(d) $d(t) = \dfrac{t}{|t|}$

(e) $e(t) = |t - 1|$

(f) $f(t) = \dfrac{1}{|t - 1|}$

(g) $g(x) = 3, \quad -\infty < x < \infty$

(h) $h(\theta) = 5\theta - 2$

2. Let a, b, c, etc., denote the functions in Problem 1. Find the following numbers. [*Example*: Find $a(b(1))$. *Solution*: $b(1) = 1^3 - 1 = 0$, so $a(b(1)) = a(0) = \sqrt{0 + 1} = 1$.] If the required number is not defined, say so.

(a) $b(a(1))$

(b) $b(a(3))$

(c) $f(e(0))$

(d) $f(c(1))$

(e) $f(a(0))$

(f) $f(a(x))$

(g) $e(f(-a(x)))$

(h) $h(h(\theta))$

3. For each function a, b, c, etc., in Problem 1, sketch the graph of $-a$, $-b$, $-c$, etc.

4. (a) For which of the functions in Problem 1 does there exist an inverse function?

(b) Sketch the graph of each inverse function, and find its domain and range. [Sketch the original graph, the main diagonal, and then the reflection in the main diagonal.]

(c) Find a formula for each inverse function. [*Example*: Find the inverse of $m(s) = 1/s^3$. *Solution*: From (2) above, $t = m(s) \Leftrightarrow s = m^{\leftarrow}(t)$; that is, to find the inverse $m^{\leftarrow}(t)$, solve $t = m(s)$ for s. We find $t = m(s) \Leftrightarrow t = 1/s^3 \Leftrightarrow s^3 = 1/t \Leftrightarrow s = \sqrt[3]{1/t}$, so $m^{\leftarrow}(t) = \sqrt[3]{1/t}$.]

(d) Check in each case that $f(f^{\leftarrow}(y)) = y$ and $f^{\leftarrow}(f(x)) = x$.

5. (a) Let $f(x) = a_1x + a_0$. Prove that the graph of f is a straight line, and find its slope.

(b) Supposing that $a_1 \neq 0$, find the inverse function $f^{\leftarrow}$.

(c) Show that the slope of the inverse function is $1/a_1$.

6. The simple table below gives an important function in statistics, the so-called "error function," denoted erf. Thus, for example, $\mathrm{erf}(0.5) = .52$ and $\mathrm{erf}(1.5) = .97$.

(a) For the inverse function $\mathrm{erf}^{\leftarrow}$, find $\mathrm{erf}^{\leftarrow}(0.2)$ and $\mathrm{erf}^{\leftarrow}(0.8)$, to one decimal.

(b) Construct a table for $\mathrm{erf}^{\leftarrow}$.

(c) Sketch graphs of erf and $\mathrm{erf}^{\leftarrow}$ for the ranges and domains covered in the table.

Table of erf(t)

t	.0	.1	.2	.3	.4	.5	.6	.7	.8	.9
0	0.0	0.11	.22	.33	.43	.52	.60	.68	.74	.80
1	.84	.88	.91	.93	.95	.97	.98	.98	.99	.99

range - answers
- what y is equal to

7. The relation between degrees Fahrenheit and degrees centigrade is given by a function F, such that if c is the temperature in degrees centigrade, then $F(c)$ is the temperature in degrees Fahrenheit.
   (a) Find the function F, given that: the graph of F is a straight line; $F(0) = 32$ (temperature of melting ice); and $F(100) = 212$ (temperature of boiling water).
   (b) The function F "converts" centigrade into Fahrenheit. Find the inverse function $F^{\leftarrow}$, which "converts" Fahrenheit into centigrade.
   (c) Graph F and $F^{\leftarrow}$ together; note the intersection at $(-40, -40)$.

8. A function f is called "self-inverse" if (i) its domain and range are equal, and (ii) $f(f(x)) = x$. Which of the following functions are self-inverse? What symmetry does the graph of a self-inverse function have?

(a) $f(x) = x, \quad -\infty < x < \infty$

(b) $f(x) = -x, \quad -\infty < x < \infty$

(c) $f(x) = -x, \quad 0 \leq x < \infty$

(d) $f(x) = \dfrac{1}{x}, \quad x \neq 0$

(e) $f(x) = \dfrac{1}{x^2}, \quad x \neq 0$

(f) $f(x) = -\dfrac{1}{x}, \quad x \neq 0$

(g) $f(x) = \dfrac{1}{x^3}, \quad x \neq 0$

(h) $f(x) = 0, \quad -\infty < x < \infty$

(i) $$f(x) = \begin{cases} -\sqrt{1 - x^2}, & 0 < x \leq 1 \\ \sqrt{1 - x^2}, & -1 \leq x < 0 \end{cases}$$

(j) $$f(x) = \begin{cases} x + 1, & -1 \leq x < 0 \\ 0, & x = 0 \\ x - 1, & 0 < x \leq 1 \end{cases}$$

•9. ODD, EVEN FUNCTIONS Suppose that the domain D of f is symmetric about the origin, that is, x is in $D \Leftrightarrow -x$ is in D. Then f is called *even* if $f(-x) = f(x)$ for all x in D, and *odd* if $f(-x) = -f(x)$ for all x in D. Which of the following functions are even, which are odd, and which are neither?

(a) $a(x) = x^2 + 1$

(b) $b(x) = x^2 + x + 1$

(c) $c(x) = x^3 + x$

(d) $d(x) = x^{11} + 3x^3 - x$

(e) $e(x) = x^{100} - 5x^{50} + 1$

(f) $f(x) = x^{100} - x^{99}$

(g) $g(x) = |x|$

(h) $h(x) = |x + 1|$

(i) $i(x) = \sqrt[3]{x^2}$

(j) $j(x) = \sqrt[3]{x}$

10. A set S in the plane is called [see Fig. 49]:

(i) symmetric with respect to the y axis if

$$(x,y) \text{ is in } S \iff (-x,y) \text{ is in } S,$$

(ii) symmetric with respect to the x axis if

$$(x,y) \text{ is in } S \iff (x,-y) \text{ is in } S,$$

(iii) symmetric with respect to the origin if

$$(x,y) \text{ is in } S \iff (-x,-y) \text{ is in } S,$$

(iv) symmetric with respect to the main diagonal if

$$(x,y) \text{ is in } S \iff (y,x) \text{ is in } S.$$

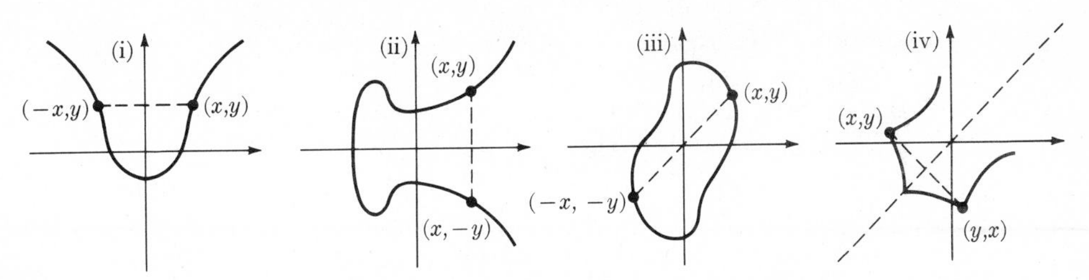

FIGURE 0.49

• (a) Which of these symmetries does the graph of an even function have? [A glance at the graphs of one or two special cases will easily suggest the answer.]

• (b) Which symmetry does the graph of an odd function have?

(c) Which symmetry does the graph of a self-inverse function have? [See Problem 8.]

(d) Which functions have graphs symmetric with respect to the x axis?

11. Figure 50 shows the part of the graph of a function f lying to the right of the y axis.

(a) Complete the graph if f is an even function [see Problem 10(a)].

(b) Complete the graph if f is an odd function [see Problem 10(b)].

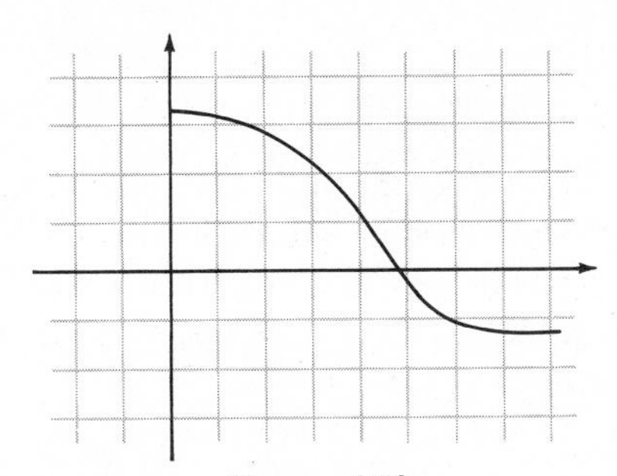

FIGURE 0.50

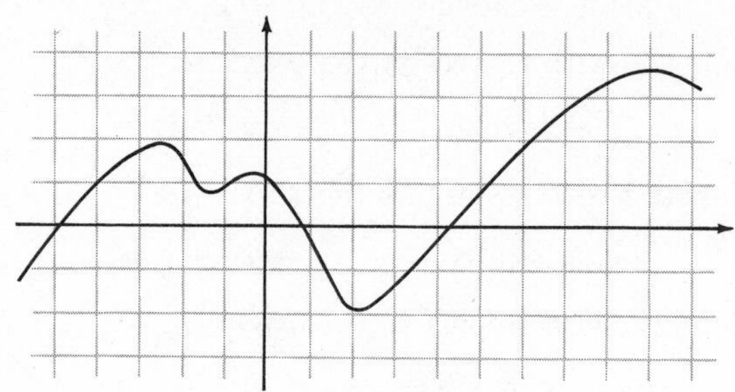

FIGURE 0.51

12. Figure 51 gives the graph of a function f.

(a) Sketch the graph of $-f$.

(b) Sketch the graph of the function g defined by $g(x) = |f(x)|$. [This function is denoted $|f|$.]

(c) Sketch the graph of the function $h(x) = f(-x)$.

(d) Sketch the graph of the function $i(x) = -f(-x)$.

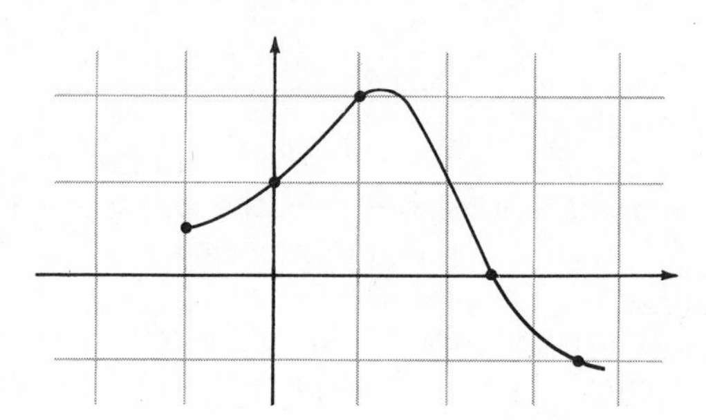

FIGURE 0.52

13. Figure 52 gives the graph of a function f and plots five points on the graph. Sketch the graphs of the following functions, and plot the corresponding points.

(a) $f_1(x) = 2f(x)$

(b) $f_2(x) = \frac{1}{3}f(x)$

(c) $f_3(x) = f(x) + 1$

(d) $f_4(x) = f\left(\frac{x}{2}\right), \quad -2 \leq x \leq 7$

(e) $f_5(x) = f(x-1), \quad 0 \leq x \leq 4\frac{1}{2}$

(f) $f_6(x) = -f(x)$

14. (a) Let F_1 and F_2 be points in the plane, and let c be a positive number. The *hyperbola* with foci F_1 and F_2 and axis of length c is defined to be the set $\{P: |PF_1 - PF_2| = c\}$ [see Fig. 53]. Prove that the hyperbola with foci $F_1 = (a,a)$ and $F_2 = (-a,-a)$ and axis of length $2a$ is the graph of $f(x) = a^2/2x$.

(b) Find the foci of the hyperbola in Example 6.

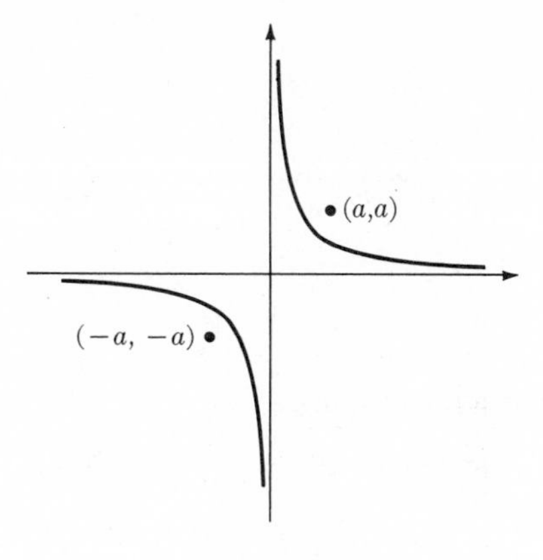

FIGURE 0.53

15. Explain as carefully as possible the following:

(a) Each horizontal line intersects the graph of f in at most one point iff for each number y_0, the equation $y_0 = f(x)$ has at most one solution x.

(b) A set in the plane is the graph of a function iff each vertical line intersects the set in at most one point.

(c) A function f is one-to-one iff the reflection of the graph of f in the main diagonal is again the graph of a function. [Use (a) and (b).]

(d) A point (x,y) is on the graph of $f \Leftrightarrow (y,x)$ is on the graph of $f^{\leftarrow}$. [See the relation (2) above.]

(e) The graph of $f^{\leftarrow}$ is the reflection of the graph of f in the main diagonal. [What is the relation between the points (a,b) and (b,a)?]

SUMMARY

Here is a brief list of the essential notation and ideas in Chapter 0. If you understand this, and can do the review problems below, you can skip the chapter.

Section 1

Intervals are denoted as follows:

$$[a,b] = \{x: a \le x \le b\} \qquad \text{(closed interval)}$$

$$(a,b) = \{x: a < x < b\} \qquad \text{(open interval)}$$

$$[a,b) = \{x: a \le x < b\} \qquad \text{(neither open nor closed)}$$

$$[a,\infty) = \{x: x \ge a\} \qquad \text{(infinite interval)}$$

The *absolute value* $|a - b|$ means, geometrically, the distance from a to b. The following formulas are valid:

$$|ab| = |a| \cdot |b|, \qquad \text{and} \qquad |a/b| = |a|/|b| \quad (\text{if } b \ne 0).$$

Section 2

$A \Rightarrow B$ means A implies B; if A then B; B follows from A; A only if B.

$A \Leftrightarrow B$ means $A \Rightarrow B$ and conversely $B \Rightarrow A$; A is equivalent to B; A if and only if B.

Section 3

The plane is simply the set of all ordered pairs of real numbers.

For each real number c, the set $\{(x,y): x = c\}$ is a *vertical line.* For each pair of real numbers m and b, the set $\{(x,y): y = mx + b\}$ is a *nonvertical line of slope* m. This line is called the *graph of the equation* $y = mx + b$.

The line through two points (x_1, y_1) and (x_2, y_2), where $x_1 \ne x_2$, is the graph of the equation

$$y - y_1 = \frac{y_2 - y_1}{x_2 - x_1}(x - x_1).$$

The slope of this line is

$$m = \frac{y_2 - y_1}{x_2 - x_1} = \frac{\text{difference in } y \text{ values}}{\text{difference in } x \text{ values}}.$$

The line through (x_1, y_1) with slope m is the graph of $y - y_1 = m(x - x_1)$.

Two nonvertical lines with slopes m_1 and m_2 are

$$\textit{parallel} \quad \text{if } m_1 = m_2$$

$$\textit{perpendicular} \text{ if } m_1 m_2 = -1.$$

The graph of $y = x$ is called the *main diagonal.* If $P = (a,b)$ is any point, then $P' = (b,a)$ is the reflection of P in the main diagonal.

Section 4

The *graph* of a function f is the set $\{(x,y) : y = f(x)\}$. The *domain* of f is the set of all x for which $f(x)$ is defined, and the *range* of f is the set of all values $f(x)$ taken on by f.

Two functions f and g are equal iff f and g have the same domain D, and $f(x) = g(x)$ for every x in D.

A function f has an inverse g iff every horizontal line intersects the graph of f in at most one point. The fundamental relation between f and its inverse g is

$$y = f(x) \quad \Leftrightarrow \quad x = g(y).$$

The graph of g is obtained by reflecting the graph of f in the main diagonal. The inverse function is generally denoted $f^{\leftarrow}$, or f^{-1}.

REVIEW PROBLEMS Answers on page A-94

Section 1

1. Sketch each of the following sets of real numbers.
(a) $\{x: |x - 2| < 1\}$
(b) $(1,3)$
(c) $[-3,-1)$
(d) $\{t: |t + 2| \leq 1\}$
(e) $\{z: 0 < |z - 1| < 1/2\}$
(f) $\{x: x^2 - x < 0\}$
(g) $\{x: (x - 1)(x - 2)(x - 3) \geq 0\}$

2. With the real numbers plotted in the usual way on a horizontal line, what number is:
(a) 1 to the right of -3
(b) -1 to the right of -3
(c) -3 to the left of -1
(d) 10 to the left of -10

Section 2

3. Which of the following implications are true for all x and y?
(a) $y = 3x + 4 \quad \Leftrightarrow \quad 2y = 6x + 8$
(b) $y = x^2 \quad \Rightarrow \quad y \geq 0$
(c) $y = x^2 \quad \Rightarrow \quad x \geq 0$
(d) $x^2 - x < 0 \quad \Leftrightarrow \quad 0 < x < 1$
(e) $y = x^2$ iff $x = \pm\sqrt{y}$

Section 3

4. Sketch the graph of the given equation. (The graph is the set of all points (x,y) satisfying the equation.)
(a) $y = x$
(b) $y = 0$
(c) $y = 3$
(d) $y = 2x - 1$
(e) $y = \frac{1}{2}x + 1$
(f) $y = -x - 2$

5. Find the equation of the line L satisfying the given condition. Find the slope of each line.
(a) L passes through $(1,2)$ and $(5,3)$
(b) L passes through $(1,1)$, parallel to the line in Problem 4(e)
(c) L passes through $(1,1)$, perpendicular to the line in Problem 4(a)

6. Find all points of intersection of the graph of $y = 2x + 1$ with:
(a) the x axis
(b) the y axis
(c) the main diagonal
(d) the graph of $y = 2x - 1$

7. Let $P_1 = (1,0)$ and $P_2 = (5,-1)$. Find the reflections P_1' and P_2' of P_1 and P_2 in the given line, and sketch them.
(a) in the main diagonal
(b) in the y axis
(c) in the x axis

Section 4

8. Sketch the graph of the given function, and indicate its domain D and range R.

(a) $f(x) = 1 - 2x^2$
(b) $f(x) = 2x + 3$
(c) $f(x) = \sqrt{x + 1}$
(d) $f(x) = |x + 1|$

9. Which of the following functions f has an inverse $f^{\leftarrow}$? Sketch each such function, together with its inverse. Give a formula for the inverse function.

(a) $f(x) = x^2, \quad x \geq 0$
(b) $f(x) = x^2, \quad -\infty < x < \infty$
(c) $f(x) = x^3, \quad -\infty < x < \infty$
(d) $f(x) = 2/x, \quad x \neq 0$
(e) $f(x) = \sqrt{x}, \quad x \geq 0$
(f) $f(x) = 3x - 1$

An Introduction to Derivatives

The origin of differential calculus can be traced back at least to 1629, when Pierre Fermat found an interesting way to construct the tangents to a parabola. Fermat's method contains implicitly the idea of derivative of a function, and is therefore a natural place to begin the study of this fundamental concept.

Although Fermat was concerned with parabolas, we need not restrict ourselves to that case. Let f be any function, and let $P_0 = (x_0, f(x_0))$ be any point on the graph of f. We want to find the equation of the line tangent to the graph at P_0 [Fig. 1].

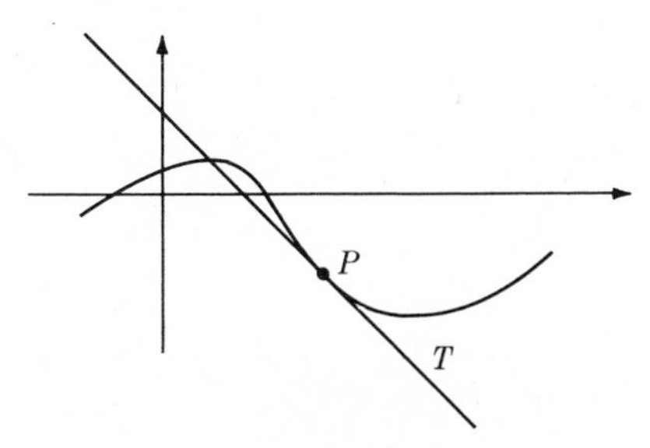

FIGURE 1.1 T is tangent to the graph of f at $P = (x_0, f(x_0))$

First, the line must pass through the given point P_0. Since any non-vertical line through (x_0, y_0) has an equation of the form $y - y_0 = m(x - x_0)$, the equation of the desired tangent line through $(x_0, f(x_0))$ takes the form

$$y - f(x_0) = m(x - x_0),$$

where m is the slope of the line. This number m is what must be determined, for the other constants in the equation (namely x_0 and $f(x_0)$) are already given.

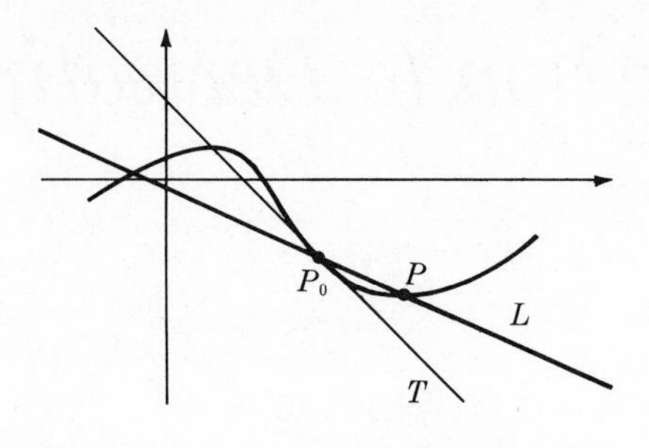

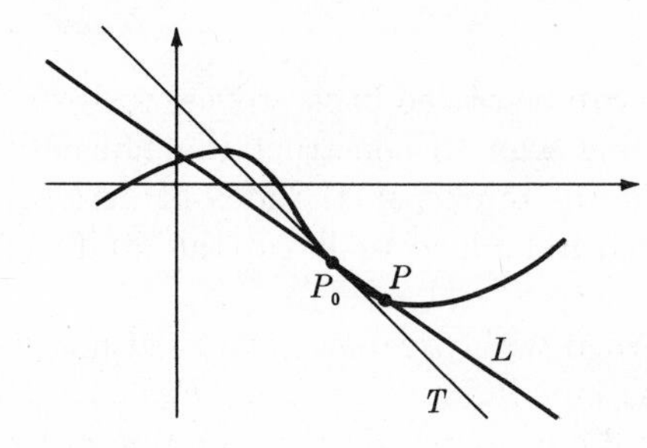

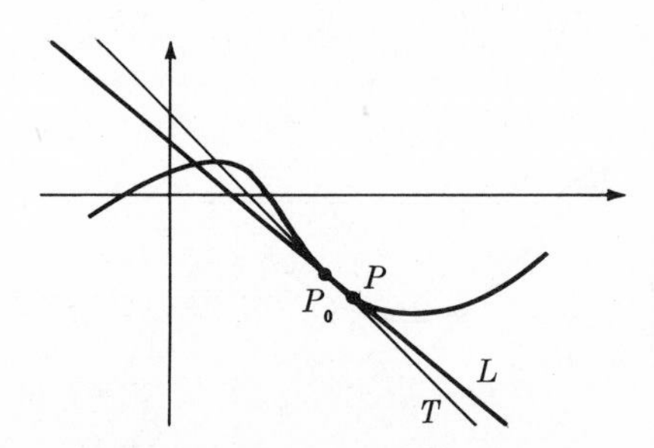

FIGURE 1.2

Fermat's idea for the determination of m is illustrated in Fig. 2. Take the line L through the given point $P_0 = (x_0, f(x_0))$ and some other point $P = (x, f(x))$ on the graph. Let m_L denote the slope of this line,

$$m_L = \frac{\text{difference of } y \text{ values}}{\text{difference of } x \text{ values}} = \frac{f(x) - f(x_0)}{x - x_0}, \qquad x \neq x_0 .$$

As the point P approaches P_0, the line L approaches the tangent line, and therefore the slope m_L should approach the desired slope m of the tangent line.

To see how this works in a simple case, take $f(x) = x^2$, and $x_0 = \frac{1}{2}$. The corresponding point on the graph is $P_0 = (\frac{1}{2}, f(\frac{1}{2})) = (\frac{1}{2}, \frac{1}{4})$. Let $P = (x, f(x))$ be any other point on the graph. The slope m_L of the line L through P_0 and P [Fig. 3] is given by

$$m_L = \frac{f(x) - f(\frac{1}{2})}{x - \frac{1}{2}} = \frac{x^2 - \frac{1}{4}}{x - \frac{1}{2}}, \qquad x \neq \tfrac{1}{2}. \tag{1}$$

As the point $P = (x, x^2)$ approaches $P_0 = (\frac{1}{2}, \frac{1}{4})$, clearly x approaches $\frac{1}{2}$ (abbreviated "$x \to \frac{1}{2}$"). The real problem, then, is to find out what happens to the slope m_L in (1), as $x \to \frac{1}{2}$. Our first impulse is to set $x = \frac{1}{2}$ in formula (1) for m_L; but unfortunately $x = \frac{1}{2}$ is precisely the point where m_L is not defined, so we *cannot* set $x = \frac{1}{2}$, as it stands. But note that

$$m_L = \frac{x^2 - \frac{1}{4}}{x - \frac{1}{2}} = \frac{(x - \frac{1}{2})(x + \frac{1}{2})}{x - \frac{1}{2}} = x + \tfrac{1}{2}, \qquad x \neq \tfrac{1}{2}.$$

With this new expression for m_L, the desired slope is easily found. As $x \to \frac{1}{2}$, then $m_L = x + \frac{1}{2} \to \frac{1}{2} + \frac{1}{2} = 1$; thus *the tangent line at* $(\frac{1}{2}, \frac{1}{4})$ *has slope* $m = 1$. Putting this value of m in the equation $y - f(x_0) = m(x - x_0)$ and recalling that $x_0 = \frac{1}{2}$, $f(x_0) = \frac{1}{4}$, we find the equation of the tangent line to be $y - \frac{1}{4} = x - \frac{1}{2}$, graphed in Fig. 4.

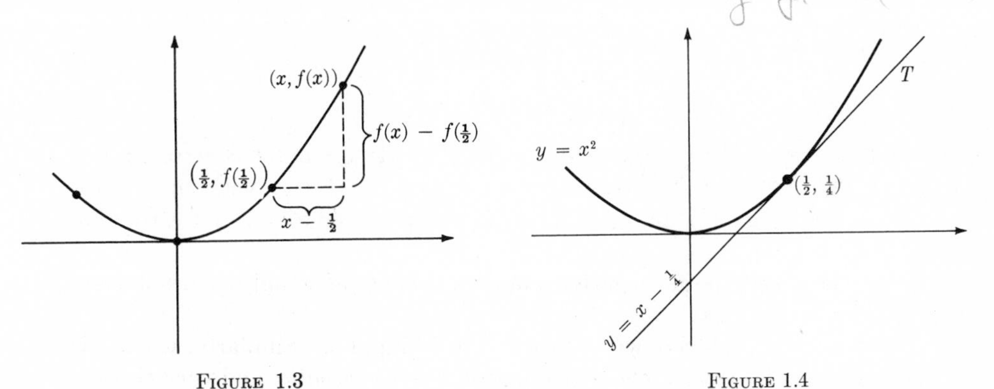

FIGURE 1.3

FIGURE 1.4

Every application of Fermat's method leads to the same problem that arose in this example; in considering the quotient

$$m_L = \frac{f(x) - f(x_0)}{x - x_0} \qquad \text{as} \quad x \to x_0,$$

we cannot simply set $x = x_0$, since the denominator is zero when $x = x_0$. This kind of problem is handled by the theory of *limits*, which is accordingly the first topic in this chapter.

1.1 LIMITS

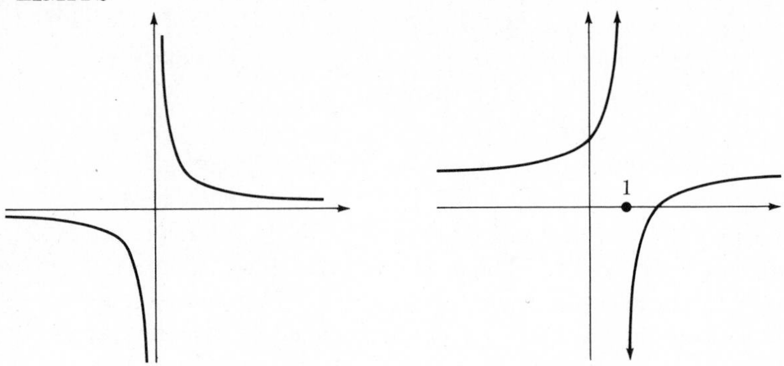

FIGURE 1.5 Type $-\infty$, $+\infty$

FIGURE 1.6 $g(x) = \dfrac{x-2}{x-1}$, type $+\infty$, $-\infty$

The function $f(x) = 1/x$, $x \neq 0$, in Fig. 5 is not defined at $x = 0$ since the denominator in the quotient $1/x$ is zero there. Near this exceptional point $x = 0$, the graph of f displays its most prominent feature; it "blows up." That is, as x approaches 0, the values $f(x) = 1/x$ grow large without bound, and the graph runs off the top of the page. We describe this situation by saying that the graph of f has a *vertical asymptote* at $x = 0$.

The function

$$g(x) = \frac{x-2}{x-1}, \qquad x \neq 1,$$

has a vertical asymptote at $x = 1$, where the denominator is zero. When x is slightly larger than 1, the denominator $x - 1$ is a very small positive number, and the numerator $x - 2$ is nearly equal to -1, so the fraction $\dfrac{x-2}{x-1}$ is a very large negative number. Thus, as x approaches 1 from the right, $g(x)$ tends down "toward $-\infty$" [Fig. 6]. Similarly, when x is slightly smaller than 1, the denominator $x - 1$ is *negative* and very close to zero, while the numerator is nearly equal to -1, so the fraction $\dfrac{x-2}{x-1}$ is a large *positive* number; thus, as x approaches 1 from the left, $g(x)$ tends up "toward $+\infty$" [Fig. 6].

To distinguish the behavior in Fig. 5 from that in Fig. 6, we say that the graph of f has a vertical asymptote at 0 "of type $-\infty$, $+\infty$", that is, tending to $-\infty$ from the left and to $+\infty$ from the right. On the other hand, the graph of g in Fig. 6 has a vertical asymptote at 1 of type $+\infty$, $-\infty$. The function $1/x^2$ has a vertical asymptote at 0 of type $+\infty$, $+\infty$ [Fig. 7], and $-1/x^2$ has an asymptote of type $-\infty$, $-\infty$ [Fig. 8].

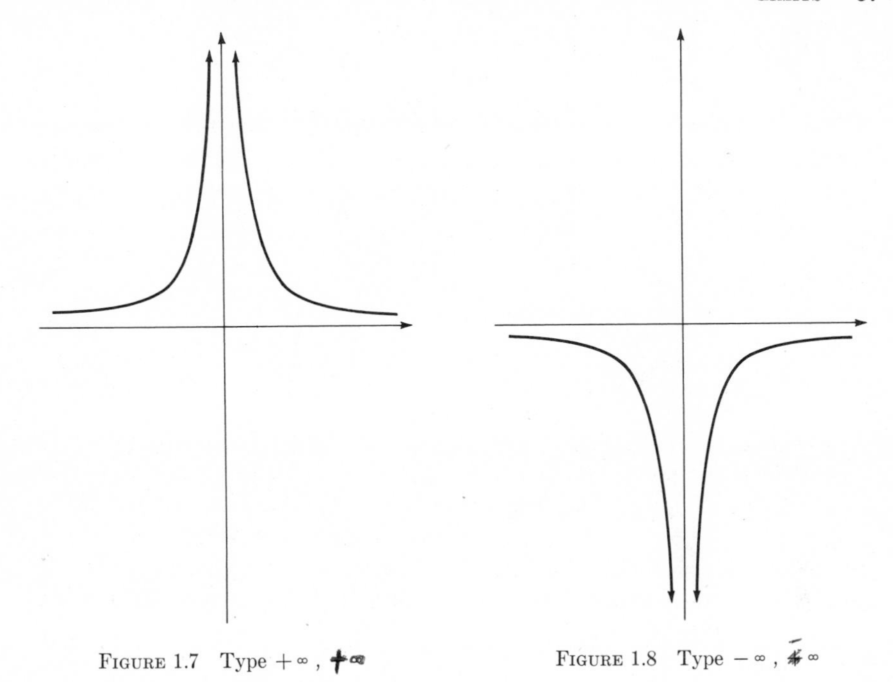

FIGURE 1.7 Type $+\infty$, $+\infty$ FIGURE 1.8 Type $-\infty$, $-\infty$

At first glance, the function

$$h(x) = \frac{x^2 - 1}{x - 1}, \qquad x \neq 1 \tag{1}$$

appears to have a vertical asymptote at $x = 1$, since the denominator is zero there. However, at $x = 1$ the numerator is also zero, so there may be no asymptote after all; multiplying a huge number $\dfrac{1}{x-1}$ by a very small number $x^2 - 1$ does not *necessarily* yield a huge product, or a very small one either. To see what really happens, notice that $x^2 - 1 = (x - 1)(x + 1)$; hence cancelling $x - 1$ from the numerator and denominator in (1) gives

$$h(x) = x + 1, \qquad x \neq 1. \tag{2}$$

Thus the graph of h is a straight line with the point (1,2) deleted [Fig. 9]. As x tends toward 1 from either the right or the left, our function $h(x) = x + 1$ tends toward the value 2. We describe this by saying

as x tends to 1, the limit of $h(x)$ is 2,

which is abbreviated

$$\lim_{x \to 1} h(x) = 2.$$

(1,2) is not on the graph

FIGURE 1.9 $h(x) = \dfrac{x^2 - 1}{x - 1}, x \neq 1$

Remark. It may seem strange to give the function h in the form (1) instead of the simpler form (2). However, this is precisely what we find in computing the slope of the tangent line to the graph of a polynomial, a rational function in which a zero in the denominator is cancelled by a corresponding zero in the numerator. In those cases, we will have to decide whether the fraction has a finite limit (as in Fig. 9), or not (as in Figs. 5–8, or Fig. 10 below).

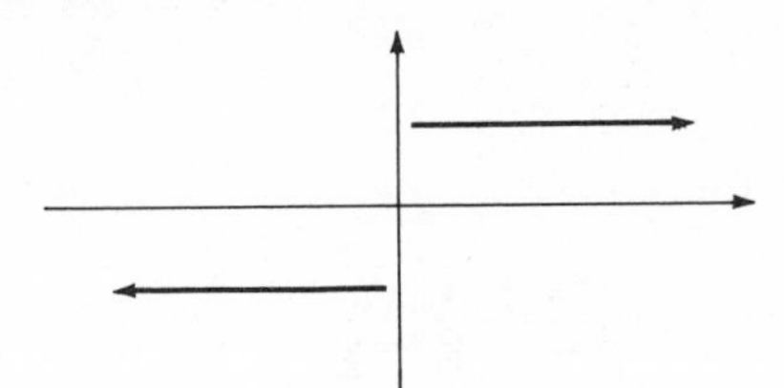

FIGURE 1.10 $s(x) = \dfrac{x}{|x|}, x \neq 0$

The function in Fig. 10,

$$s(x) = \frac{x}{|x|}, \qquad x \neq 0,$$

shows a slightly different behavior than the previous examples at its exceptional point $x = 0$. As x tends to 0 from the right, $s(x)$ tends to $+1$, but as x tends to 0 from the left, $s(x)$ tends to -1. The notation for this situation is

$$\lim_{x \to 0+} s(x) = 1, \qquad \lim_{x \to 0-} s(x) = -1.$$

The $+$ appearing after 0 indicates that x is approaching 0 from the right, and the $-$ indicates approach from the left. The symbol

$$\lim_{x \to 0+}$$

is read "the limit as x approaches zero from the right."

***Limits at* ∞**

Another important feature of a graph is its behavior for very large positive x or very large negative x. In discussing this, we abbreviate "x grows large positive" by $x \to +\infty$, and "x grows large negative" by $x \to -\infty$. Figure 11 shows various possibilities:

As $x \to +\infty$, then

$$x^n \to +\infty \text{ for } n = 1, 2, 3, \ldots$$

(x^n grows large positive, as x does)

$$x^0 \to 1$$

$$x^n \to 0 \text{ for } n = -1, -2, -3, \ldots.$$

As $x \to -\infty$, then

$$x^n \to +\infty \text{ for } n = 2, 4, 6, \ldots$$

(x^n grows large *positive* as x grows large *negative*)

$$x^n \to -\infty \text{ for } n = 1, 3, 5, \ldots$$

(x^n grows large negative, as x does)

$$x^0 \to 1$$

$$x^n \to 0 \text{ for } n = -1, -2, -3, \ldots.$$

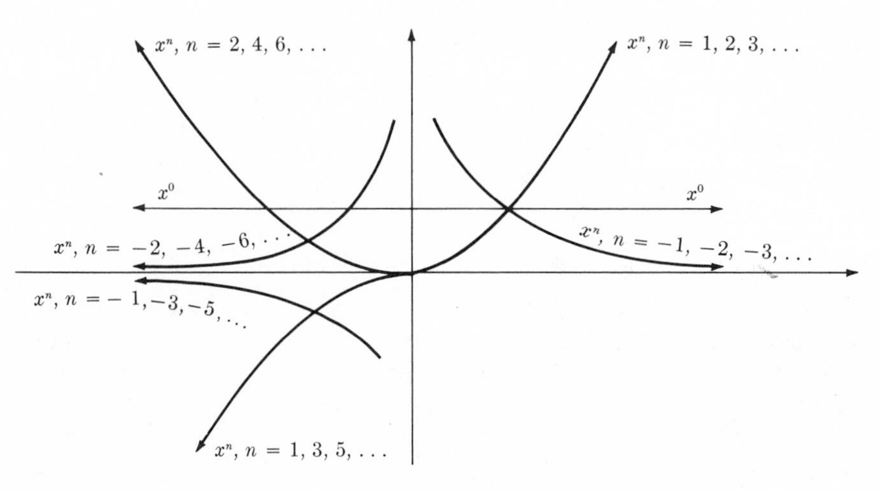

FIGURE 1.11 Powers of x as $x \to \pm\infty$

From these observations we can deduce the behavior of any *rational function*

$$f(x) = \frac{a_n x^n + a_{n-1}x^{n-1} + \cdots + a_0}{b_m x^m + b_{m-1}x^{m-1} + \cdots + b_0},$$

as in some of the examples below.

Example 1. Describe the function

$$f(x) = \frac{x^3 - 3x + 2}{x^2 - 4} \tag{3}$$

near the points where f is not defined. *Solution.* The denominator factors, $x^2 - 4 = (x - 2)(x + 2)$. Hence the denominator is zero, and f is not defined, for $x = \pm 2$. We first investigate $x = 2$.

When x is slightly larger than 2, the denominator $x^2 - 4$ is positive and very small, while the numerator $x^3 - 3x + 2$ is nearly equal to $2^3 - 6 + 2 = 4$, so the fraction $f(x)$ is positive and very large; thus $f(x) \to +\infty$ as $x \to 2+$ (i.e. as $x \to 2$ from the right). As $x \to 2-$ (i.e. as $x \to 2$ from the left), the denominator is negative and very small, while the numerator is nearly equal to 4, so $f(x) \to -\infty$ as $x \to 2-$. Thus at $x = 2$, f has a vertical asymptote of type $-\infty, +\infty$.

Next, we investigate f near the point $x = -2$. The denominator is zero there, but so is the numerator, $(-2)^3 - 3(-2) + 2 = 0$. Hence there may be a finite limit as $x \to -2$. To check, divide numerator and denominator by $x + 2$, (which is the factor that makes the denominator vanish at $x = -2$):

$$\begin{array}{rl}
 & x^2 - 2x + 1 \\
x + 2 \,\big) & x^3 + 0x^2 - 3x + 2 \\
 & x^3 + 2x^2 \\
\hline
 & -2x^2 - 3x \\
 & -2x^2 - 4x \\
\hline
 & x + 2 \\
 & x + 2 \\
\hline
 & 0
\end{array}
\qquad
\begin{array}{rl}
 & x - 2 \\
x + 2 \,\big) & x^2 + 0x - 4 \\
 & x^2 + 2x \\
\hline
 & -2x - 4 \\
 & -2x - 4 \\
\hline
 & 0
\end{array}$$

Thus dividing numerator and denominator by $x + 2$ yields

$$f(x) = \frac{x^3 - 3x + 2}{x^2 - 4} = \frac{x^2 - 2x + 1}{x - 2}, \qquad x \neq \pm 2. \tag{4}$$

Now, when x is near -2, the numerator on the right in (4) is near 9 and the denominator is near -4, so $f(x)$ is near $-9/4$, and

$$\lim_{x \to -2} f(x) = -\frac{9}{4}.$$

Example 2. Study the function f in (3) for large values of x. *Solution.* Factor the highest degree term x^3 out of the numerator, and the highest degree term x^2 out of the denominator, obtaining

$$f(x) = x\,\frac{1 - 3x^{-2} + 2x^{-3}}{1 - 4x^{-2}}, \qquad x \neq 0, 2, -2. \tag{5}$$

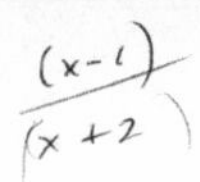

When x is very large, the terms x^{-2} and x^{-3} are very small, so the fraction $\frac{1 - 3x^{-2} + 2x^{-3}}{1 - 4x^{-2}}$ is nearly equal to 1. Hence from (5), when x is very large, $f(x)$ is nearly the same as x. We conclude that $f(x) \to +\infty$ as $x \to +\infty$, and $f(x) \to -\infty$ as $x \to -\infty$.

The graph in Fig. 12 is based on the results in Examples 1 and 2, together with the fact that $f(x) = 0$ only for $x = 1$. $\left(\text{It also helps to notice that } f(x) = x + \frac{1}{x-2},\ x^2 \neq 4.\right)$

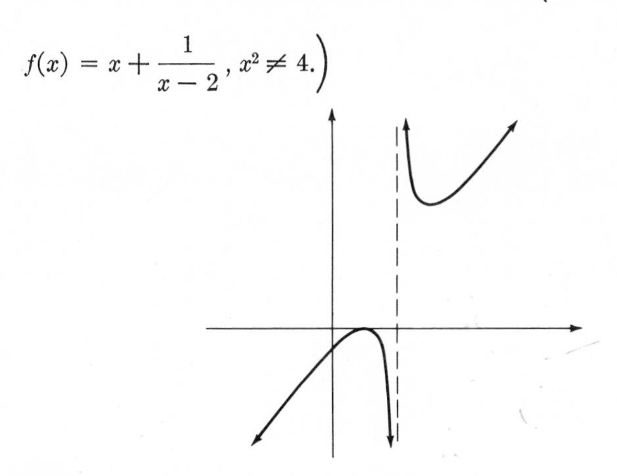

FIGURE 1.12

Example 3. Let $f(x) = x^2$, and find $\lim_{x \to 2} \frac{f(x) - f(2)}{x - 2}$. [Recall that $\frac{f(x) - f(2)}{x - 2}$ is the slope m_L of the line through the points $(2, f(2))$ and $(x, f(x))$, and the limit of m_L as $x \to 2$ gives the slope of the tangent line by Fermat's method, as outlined in the introduction to this chapter. *Solution.*

$$\lim_{x \to 2} \frac{f(x) - f(2)}{x - 2} = \lim_{x \to 2} \frac{x^2 - 2^2}{x - 2} = \lim_{x \to 2} \frac{(x-2)(x+2)}{x - 2} = \lim_{x \to 2} (x + 2) = 4.$$

Example 4. Let $f(x) = x^2$, and let x_0 be any fixed number. Find

$$\lim_{x \to x_0} \frac{f(x) - f(x_0)}{x - x_0}.$$

[This limit gives the slope of the tangent line to the graph of f at the point $(x_0, f(x_0))$.] *Solution.*

$$\lim_{x \to x_0} \frac{f(x) - f(x_0)}{x - x_0} = \lim_{x \to x_0} \frac{x^2 - x_0^2}{x - x_0} = \lim_{x \to x_0} \frac{(x - x_0)(x + x_0)}{x - x_0}$$

$$= \lim_{x \to x_0} (x + x_0) = 2x_0.$$

Example 5. Let $f(x) = 3x^2 + 2x + 1$, and find $\lim\limits_{x \to x_0} \dfrac{f(x) - f(x_0)}{x - x_0}$.

Solution.

$$\frac{f(x) - f(x_0)}{x - x_0} = \frac{3x^2 + 2x + 1 - 3x_0{}^2 - 2x_0 - 1}{x - x_0}$$

$$= 3\,\frac{x^2 - x_0{}^2}{x - x_0} + 2\,\frac{x - x_0}{x - x_0} = 3(x + x_0) + 2,$$

so

$$\lim_{x \to x_0} \frac{f(x) - f(x_0)}{x - x_0} = \lim_{x \to x_0} [3(x + x_0) + 2] = 6x_0 + 2.$$

Example 6. Investigate $g(x) = \dfrac{2x^2 - 3}{3x^2 + 5}$. *Solution.* The denominator is never zero, so we have only the behavior at $\pm\infty$ to consider. Factor out the highest term $2x^2$ from the numerator, and the highest term $3x^2$ from the denominator, obtaining

$$g(x) = \left(\frac{2}{3}\right) \cdot \frac{1 - (\frac{3}{2})x^{-2}}{1 + (\frac{5}{2})x^{-2}}.$$

When x is very large, the fraction $\dfrac{1 - (\frac{3}{2})x^{-2}}{1 + (\frac{5}{3})x^{-2}}$ is slightly less than, but nearly equal to 1, so

$$g(x) \to \frac{2}{3} \text{ as } x \to +\infty, \qquad \text{and} \qquad g(x) \to \frac{2}{3} \text{ as } x \to -\infty. \qquad \text{[Fig. 13]}$$

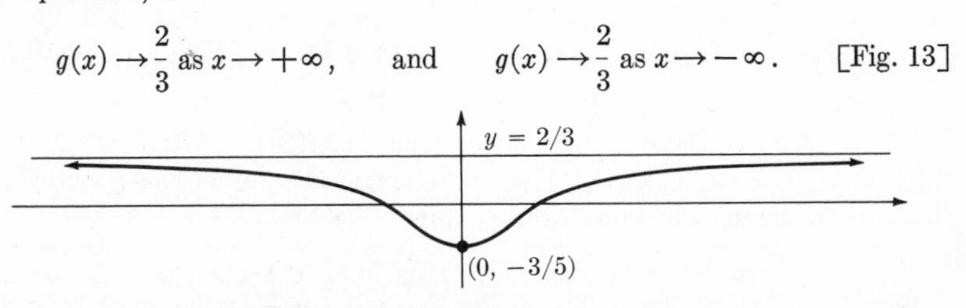

FIGURE 1.13

Example 7. $h(x) = \dfrac{x - 1}{1 + 3x^2}$. Again, there are no points where h is not defined, so we consider the behavior at $\pm\infty$. Factor the highest term x out of the numerator, and the highest term $3x^2$ out of the denominator, obtaining

$$h(x) = \frac{1}{3x}\,\frac{1 - x^{-1}}{1 + \frac{1}{3}x^{-2}}.$$

The fraction $\dfrac{1 - x^{-1}}{1 + \frac{1}{3}x^{-2}}$ is nearly equal to 1 for large values of x, so h behaves like $1/3x$. Thus $h(x) \to 0$ as $x \to \pm\infty$. [Fig. 14]

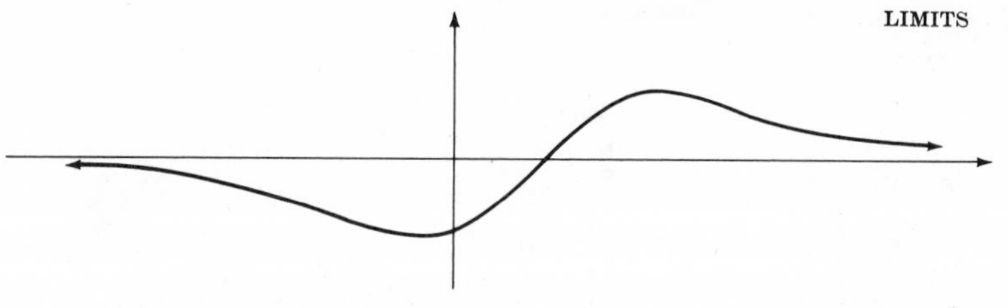

FIGURE 1.14

From the last two of these examples it is easy to see, in general, how a rational function

$$f(x) = \frac{a_n x^n + \cdots + a_0}{b_m x^m + \cdots + b_0}, \qquad a_n \neq 0, \quad b_m \neq 0, \tag{6}$$

will behave as $x \to \pm\infty$. Factor out the highest degree term from the numerator, and the highest degree term from the denominator, and thus rewrite f in the form

$$f(x) = \frac{a_n}{b_m} x^{n-m} \frac{1 + \left(\frac{a_{n-1}}{a_n x}\right) + \cdots + \left(\frac{a_0}{a_n x^n}\right)}{1 + \left(\frac{b_{m-1}}{b_m x}\right) + \cdots + \left(\frac{b_0}{b_m x^m}\right)}.$$

When this is done, you can see that:

(i) If $n < m$, then $f(x) \to 0$ as $x \to \pm\infty$ [Fig. 14].

(ii) If $n = m$, then $f(x) \to a_n/b_m$ as $x \to \pm\infty$ [Fig. 13].

(iii) If $n > m$, then $f(x) \to \pm\infty$ as $x \to \pm\infty$ [Fig. 12].

In case (ii), we say that "f has the limit a_n/b_n at ∞," and write

$$\lim_{x \to +\infty} f(x) = \frac{a_n}{b_n}, \qquad \lim_{x \to -\infty} f(x) = \frac{a_n}{b_n}.$$

Remark. This informal "pictorial" introduction to limits will give way to a more formal approach as we go along. Rigorous analytic definitions and proofs of the basic theorems on limits are given in Appendix II.

PROBLEMS A dot before a problem indicates that it is answered (at least partly) in the back of the book.

1. In the following problems, (i) determine all points where f is not defined, (ii) at each point a found in (i), determine whether f has a vertical asymptote, or a finite limit $\lim_{x \to a} f$; or finite but unequal left and right-hand limits, $\lim_{x \to a+} f$ and $\lim_{x \to a-} f$. In the case of an asymptote, determine its type.

• (a) $\dfrac{x^2-1}{(x-1)^2}$ • (e) $\dfrac{3x^5+2x^2-5}{1+2x}$ • (i) $\dfrac{x+1/x}{1+x^2}$

• (b) $\dfrac{(x-1)^2}{x^2-1}$ (f) $\dfrac{3x^3-2x^2-x-14}{2(x-2)(x-3)}$ • (j) $2\dfrac{x-1}{|x-1|}$

• (c) $\dfrac{x^3-27}{x^2-9}$ (g) $\dfrac{3x^3-2x^2-x-10}{2(x-2)(x-3)}$ (k) $\dfrac{x}{|x-1|}$

• (d) $\dfrac{2x^3+1}{x^2}$ (h) $\dfrac{3x^4+2x^2-\frac{7}{4}}{2x^2-1}$ (l) $\dfrac{(x-1)^2}{|x-1|}$

• (m) $f(x)=\begin{cases} x^2+3x+2, & x<6 \\ x-5, & x>6 \end{cases}$

• **2.** For each of the functions in Problem 1, analyze the behavior as $x \to +\infty$ and as $x \to -\infty$.

3. For each of the functions in Problem 1, sketch a graph based on the results in Problems 1 and 2. If there are any easily discovered points at which $f(x) = 0$, show this additional information on the graph.

4. For each of the following functions, find $\lim\limits_{x\to 3} \dfrac{f(x)-f(3)}{x-3}$.

• (a) $f(x) = 2x^2$ • (f) $f(x) = 1/x, \quad x \neq 0$

• (b) $f(x) = 3x^2$ (g) $f(x) = \dfrac{1}{x+1}, \quad x \neq -1$

• (c) $f(x) = \frac{1}{2}x^2$

• (d) $f(x) = mx$ (m a constant) (h) $f(x) = x^3$

• (e) $f(x) = 2x^2+3x+1$ (i) $f(x) = 2x^3+x^2+3x+1$

• **5.** For the functions in Problem 4, find $\lim\limits_{x\to x_0} \dfrac{f(x)-f(x_0)}{x-x_0}$.

6. This problem illustrates the dangers in an intuitive approach to limits. Consider $\lim\limits_{x\to+\infty} (\sqrt{x^2+x}-x)$.

(a) Before making any computations, guess whether the above limit exists, and if so, what it is.

(b) Compute $f(x) = \sqrt{x^2+x}-x$ to two decimals, for $x = 1, 2, 3, 4, 5$ (and possibly beyond).

(c) Show that $f(x) = (1+\sqrt{1+1/x})^{-1}$, $x > 0$, and deduce from this what the limit $\lim\limits_{x\to+\infty} f(x)$ actually is.

1.2 DERIVATIVES

Recall Fermat's method of finding the slope of the tangent line at a point $P_0 = (x_0, f(x_0))$ on the graph of f. Compute the quotient

$$m_L = \frac{f(x) - f(x_0)}{x - x_0}, \tag{1}$$

giving the slope of the line L between P_0 and another nearby point $P = (x, f(x))$ on the graph. As x approaches x_0, this quotient approaches the desired slope m. In other words, the slope is given by the *limit* of the quotient (1),

$$m = \lim_{x \to x_0} \frac{f(x) - f(x_0)}{x - x_0}. \tag{2}$$

This limit has found many applications aside from the construction of tangents. To give only a sampling, in physics, limits of the form (2) occur as velocity, acceleration, density, and electrical current; in economics, as "marginal profit," "marginal cost," and "marginal revenue"; in chemistry as rate of reaction; in probability, as "probability density" [see §1.7]. So, naturally, a neutral non-geometric terminology is required.

The quotient (1) is called a *difference quotient* for the function f, since it is the quotient of two differences; the denominator is the difference $x - x_0$, and the numerator is the corresponding difference in f, namely $f(x) - f(x_0)$. **The limit of this quotient as $x \to x_0$ is called the *derivative* of f at x_0, denoted $f'(x_0)$:**

☆☆ DEFINITION OF DERIVATIVE

$$\mathbf{f'(x_0) = \lim_{x \to x_0} \frac{f(x) - f(x_0)}{x - x_0}}. \tag{3}$$

For example, if $f(x) = x^2$, then $f'(x_0) = 2x_0$ [§1.1, Example 4].

The process of computing the derivative $f'(x_0)$ is called *differentiation*, because of the difference quotient on the right-hand side of (3).

The notation $f'(x_0)$ suggests that differentiation produces a new function f', and indeed it does: the function f' assigns to the number x_0 the limit in (3). The *domain* of f is the set of numbers x_0 for which this limit exists.

Example 1. Take the function $f(x) = 1/x$, $x \neq 0$. To compute the derivative $f'(x_0)$, form the difference quotient

$$\frac{f(x) - f(x_0)}{x - x_0} = \frac{1/x - 1/x_0}{x - x_0}, \qquad x \neq x_0.$$

It is not immediately clear what happens to this expression as $x \to x_0$, but the situation is clarified by some algebraic manipulation. Begin by putting the numerator $1/x - 1/x_0$ on a common denominator,

$$\frac{1}{x} - \frac{1}{x_0} = \frac{x_0 - x}{xx_0}.$$

Then we find

$$\frac{f(x) - f(x_0)}{x - x_0} = \frac{x_0 - x}{xx_0(x - x_0)} = \frac{-1}{xx_0};$$

hence $f'(x_0) = \lim_{x \to x_0} (-1/xx_0) = -1/x_0^2$.

We can check this computation of f' by looking at the graph of f [Fig. 15]. Since x_0^2 is always ≥ 0, the derivative $f'(x_0) = -1/x_0^2$ is always negative. This corresponds to the fact that the line tangent to the graph slopes down to the right at each point; for the derivative is the slope of the tangent line. Moreover, when x_0 is close to zero, $-1/x_0^2$ is a *very large* negative number, and the tangent line is nearly vertical; when x_0 is far from zero, $-1/x_0^2$ is nearly 0, and the tangent line is nearly horizontal.

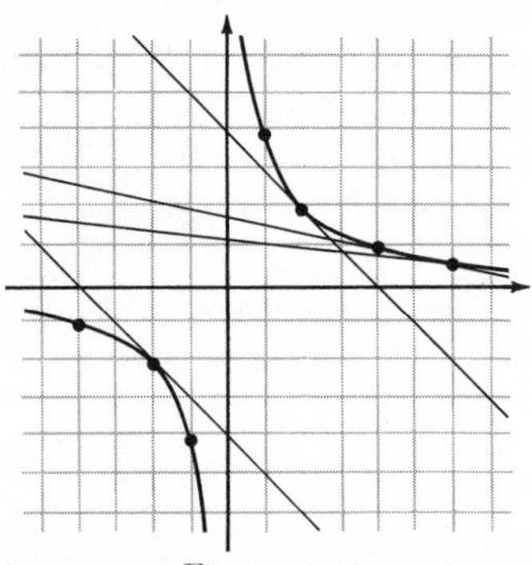

FIGURE 1.15

When you become more familiar with derivatives, the roles of computation and graphing are reversed; instead of using the graph to check that the derivative has been found correctly, you first compute the derivative and then use it to help construct the graph, as in §1.4 below.

Example 2. Compute the derivative of the general quadratic $f(x) = ax^2 + bx + c$, where a, b, and c are constants.
Solution.

$$\frac{f(x) - f(x_0)}{x - x_0} = \frac{ax^2 + bx + c - ax_0^2 - bx_0 - c}{x - x_0} = a\frac{x^2 - x_0^2}{x - x_0} + b\frac{x - x_0}{x - x_0}$$

$$= a(x + x_0) + b, \qquad x \neq x_0.$$

Thus

$$f'(x_0) = \lim_{x \to x_0} [a(x + x_0) + b] = a(x_0 + x_0) + b = 2ax_0 + b.$$

This limit exists for every x_0, so that the domain of f' is the whole real line.

The formula for f' should be memorized:

$$\text{If } f(x) = ax^2 + bx + c,$$
$$\text{then } f'(x_0) = 2ax_0 + b. \tag{4}$$

A reasonable shorthand for this is

$$(ax^2 + bx + c)' = 2ax + b.$$

Example 3. Let $f(x) = x^2 + x$. This fits formula (4) if we set $a = 1$, $b = 1$, and $c = 0$. Thus $f'(x_0) = 2ax_0 + b = 2x_0 + 1$. The table to the left of Fig. 16 gives the slope $f'(x_0)$ of the tangent line at various points $(x_0, f(x_0))$ on the graph, and the tangent lines at these points are shown in the figure. You can check the slopes of these lines and thus see that the graph in Fig. 16 is consistent with the general formula (4) by which we computed $f'(x_0)$.

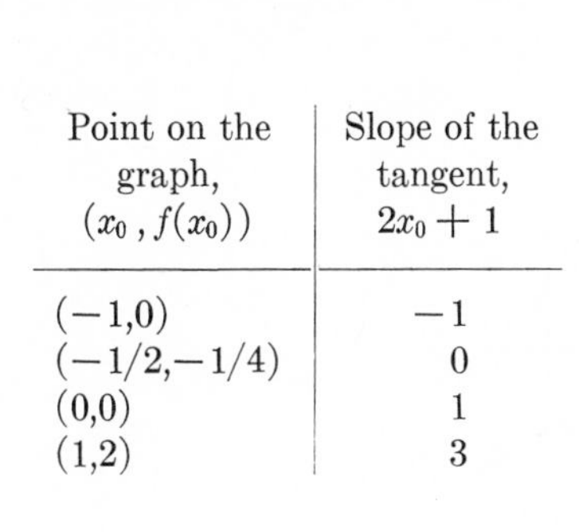

Point on the graph, $(x_0, f(x_0))$	Slope of the tangent, $2x_0 + 1$
$(-1,0)$	-1
$(-1/2,-1/4)$	0
$(0,0)$	1
$(1,2)$	3

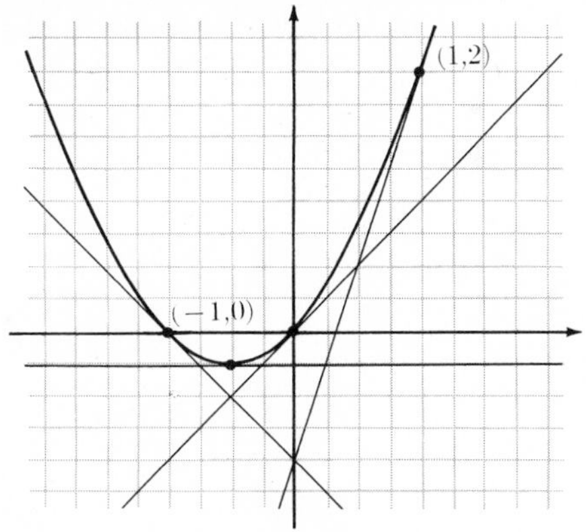

FIGURE 1.16

Example 4. Let $f(x) = x^3$ [Fig. 17]. The general difference quotient is

$$\frac{f(x) - f(x_0)}{x - x_0} = \frac{x^3 - x_0^3}{x - x_0}, \qquad x \neq x_0 .$$

To take the limit as $x \to x_0$, divide $x - x_0$ into the numerator. Since $x^3 - x_0^3 = (x - x_0)(x^2 + xx_0 + x_0^2)$, we get

$$\frac{f(x) - f(x_0)}{x - x_0} = \frac{(x - x_0)(x^2 + xx_0 + x_0^2)}{x - x_0}$$
$$= x^2 + xx_0 + x_0^2.$$

Hence the derivative is

$$f'(x_0) = \lim_{x \to x_0} \frac{f(x) - f(x_0)}{x - x_0}$$
$$= x_0^2 + x_0x_0 + x_0^2 = 3x_0^2. \tag{5}$$

FIGURE 1.17

Figure 17 shows the tangent line at (1,1), so our formula should apply with $x_0 = 1$. You can check from the figure that the tangent line actually does have the slope $f'(x_0) = 3x_0^2 = 3$, in agreement with (5).

Question: What is a tangent line? We have shown how to compute tangent lines by taking derivatives, but haven't really explained what a tangent line is. An elementary explanation is not easy, but we shall try.

Figure 18 shows some lines that you will probably agree are tangent to the given graphs at the points shown.

And Fig. 19 shows other lines that pass through some of those same points, but are *not* tangent.

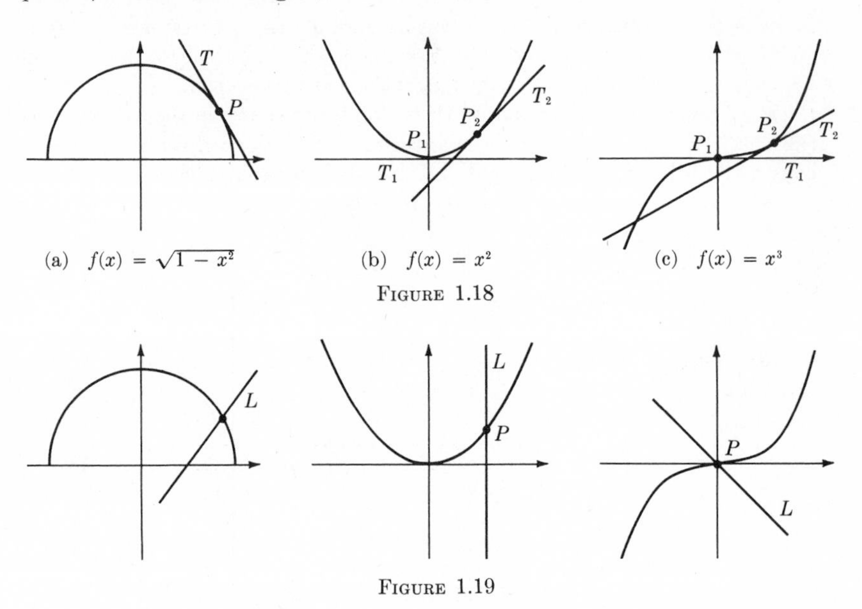

(a) $f(x) = \sqrt{1 - x^2}$ (b) $f(x) = x^2$ (c) $f(x) = x^3$

FIGURE 1.18

FIGURE 1.19

What are the special properties of the lines in Fig. 18 that make them tangent? From geometry you may remember some criteria for tangents, of which we recall two:

(a) The tangent line at P is the line that intersects the curve only at P.

(b) The tangent line intersects the curve at P, but does not cross it.

These criteria have the virtue of simplicity, but unfortunately they apply only in special cases. For instance, in Fig. 19 all the lines marked L intersect the graph only at P, so they appear to meet criterion (a); but these lines are *not* tangent. And the line T_2 in Fig. 18(c) does not meet criterion (a) but it *is* tangent. Thus criterion (a) is not a good one. And criterion (b) is not much better; for in Fig. 18(c), the line T_1 is tangent at P_1, but crosses the curve at P_1, thus violating criterion (b).

The failure of such simple criteria forces us to consider a more complicated idea. Suppose we plot the graph of f near the point P to a larger and larger scale. [This is illustrated in Fig. 20 for the function $f(x) = x^3$ and the point $P = (1/2, 1/8)$.] **As the scale is increased, the graph looks more and more like part of a certain straight line; and this is precisely the *tangent line.***

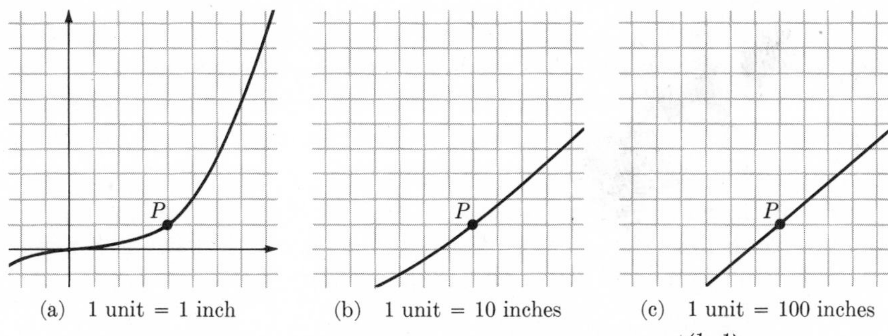

FIGURE 1.20 Graphs of $f(x) = x^3$ near the point $P = \left(\frac{1}{2}, \frac{1}{8}\right)$

Fermat's method for computing the slope of the tangent is directly related to this idea. If the graph of f is a very good approximation to the tangent line at $(x_0, f(x_0))$, *then when x is near x_0*, the slope of the line through $(x_0, f(x_0))$ and $(x, f(x))$ is very nearly equal to the slope of the tangent. The nearer x comes to x_0, the better is the approximation, and thus taking the limit as $x \to x_0$, we find the slope of the tangent line itself.

Warning: *There are functions f that do not have a derivative at some points where they are defined.* A simple example is the absolute value function $f(x) = |x|$. To find $f'(0)$ in this case, we would form the difference quotient $\dfrac{f(x) - f(0)}{x - 0} = \dfrac{|x|}{x}$. This is $+1$ if $x > 0$ and -1 if $x < 0$, and hence *does not tend to any single number as x tends to zero.* In this case, we say that $f'(0)$ does not exist; or that f is not differentiable at 0; or that the function f' is not defined at 0. In fact, the domain of f' is the set $\{x : x \neq 0\}$, and

$$f'(x) = \begin{cases} 1 & \text{if } x > 0 \\ -1 & \text{if } x < 0. \end{cases}$$

That $f'(0)$ does not exist corresponds to the fact that the graph has no tangent line, in the sense discussed above. Figure 21 shows the function $(x) = |x|$ sketched to larger and larger scales; it is clear that as the scale increases, the graph does not become anything like a straight line, and so there is no tangent.

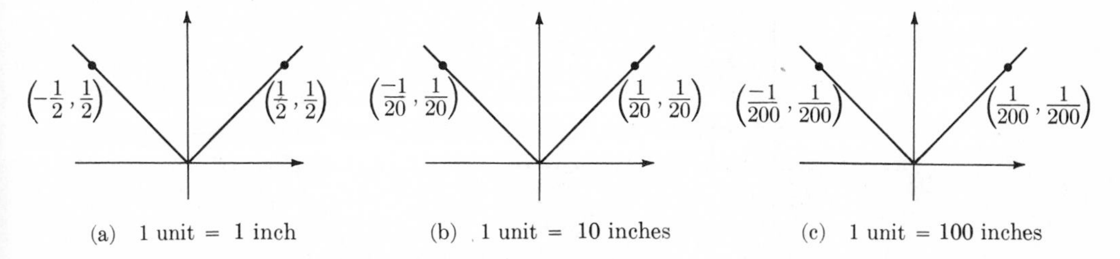

FIGURE 1.21 Graphs of $f(x) = |x|$ near $P = (0, 0)$

We conclude with a less peculiar example, the square root function $s(x) = x^{1/2}$. We find $s'(1)$ by forming the difference quotient

$$\frac{s(x) - s(1)}{x - 1} = \frac{x^{1/2} - 1}{x - 1}. \tag{6}$$

Recall the formula $(a - b)(a + b) = a^2 - b^2$. This suggests how to obtain a factor in the numerator of (6) to cancel the $x - 1$ in the denominator: multiply top and bottom by $x^{1/2} + 1$, obtaining

$$\frac{s(x) - s(1)}{x - 1} = \frac{(x^{1/2} - 1)(x^{1/2} + 1)}{(x - 1)(x^{1/2} + 1)}$$

$$= \frac{x - 1}{(x - 1)(x^{1/2} + 1)} = \frac{1}{x^{1/2} + 1}.$$

Thus $\lim\limits_{x \to 1} \dfrac{s(x) - s(1)}{x - 1} = \dfrac{1}{1 + 1} = \dfrac{1}{2}$, so $s'(1) = \dfrac{1}{2}$. In the same way, we find for any $x_0 > 0$ that

$$s'(x_0) = \frac{1}{2x_0^{1/2}} = \frac{1}{2} x_0^{-1/2},$$

since

$$\lim_{x \to x_0} \frac{s(x) - s(x_0)}{x - x_0} = \lim_{x \to x_0} \frac{x^{1/2} - x_0^{1/2}}{x - x_0} \cdot \frac{x^{1/2} + x_0^{1/2}}{x^{1/2} + x_0^{1/2}}$$

$$= \lim_{x \to x_0} \frac{1}{x^{1/2} + x_0^{1/2}} = \frac{1}{2x_0^{1/2}} = \frac{1}{2} x_0^{-1/2}.$$

Note that $s'(x_0) = \frac{1}{2}x_0^{-1/2}$ is positive and grows steadily larger as $x_0 \to 0$. Again, this is reflected in the behavior of the graph of s [Fig. 22]; the tangent is always rising to the right and is nearly vertical at points near zero.

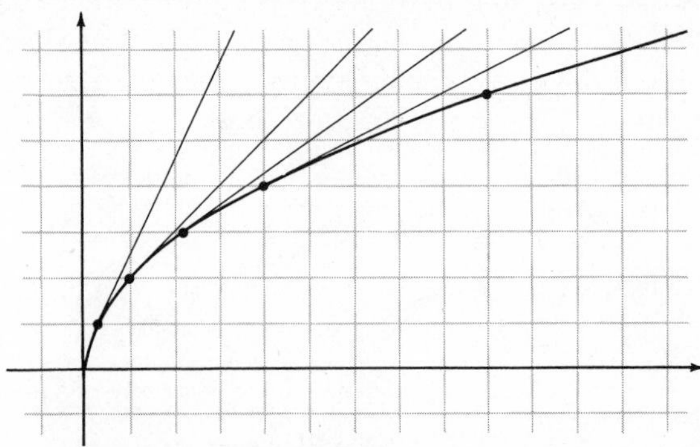

FIGURE 1.22 Tangents to the graph of $f(x) = \sqrt{x}$

PROBLEMS

1. For the given function f, find $f'(2)$, and $f'(x_0)$ in general.

•(a) $f(x) = x^2 + 1$

•(b) $f(x) = \dfrac{1}{x + 1}$

•(c) $f(x) = x^{-1/2}$

(d) $f(x) = x^{1/3}$ [Hint: $(a - b)(a^2 + ab + b^2) = a^3 - b^3$]

•(e) $f(x) = \dfrac{1}{x^2 + 1}$

(f) $f(x) = \dfrac{x}{x + 1}$

(g) $f(x) = \frac{1}{3}x^3$

(h) $f(x) = \frac{1}{4}x^4$

(i) $f(x) = \dfrac{1}{2x + 1}$

2. In Problem 1(d), show that $f'(0)$ does not exist. Sketch the graph of $x^{1/3}$ to see why this is so.

3. In Problems 1(b), 1(c), and 1(e), sketch the graph of f. Check its behavior in connection with the formula for f'.

4. The table below gives values of $\sin x$ (in radian measure) for $x = 0$, 0.2, 0.4, 0.6, 0.8, 1.0.

(a) Sketch the graph of $m(x,0) = \dfrac{\sin x - \sin 0}{x - 0}$ by plotting carefully the points on the graph of m corresponding to $x = \pm 0.2, \pm 0.4, \ldots$. [Recall that $\sin(-x) = -\sin x$; thus $\sin(-0.2) = -\sin(0.2) = -0.199$, etc.]

•(b) Does the graph suggest what $m(0,0)$, i.e. what the derivative $\sin'(0)$, should be?

x	0.0	0.2	0.4	0.6	0.8	1.0
$\sin x$	0.00	0.199	0.39	0.56	0.72	0.84
$\cos x$	1.00	0.98	0.92	0.83	0.70	0.54

• **5.** Repeat Problem 4 with sin replaced by cos. (Recall that $\cos(-x) = \cos x$.)

6. Compute the derivative log′ (1) graphically, as in Problems 4 and 5, from the following table of the function $\log x$ (the "natural logarithm").

x	0.4	0.6	0.8	1.0	1.2	1.4	1.6
$\log x$	−0.92	−0.52	−0.22	0.00	0.18	0.34	0.47

7. Compute $f'(x_0)$ in the following cases.
(a) $f(x) = ax^3 + bx^2 + cx + d$
• (b) $f(x) = a_4x^4 + a_3x^3 + a_2x^2 + a_1x + a_0$
(c) $f(x) = a_nx^n + a_{n-1}x^{n-1} + \cdots + a_1x + a_0$.
[Hint: $x^k - x_0{}^k = (x - x_0)(x^{k-1} + x_0x^{k-2} + x_0{}^2x^{k-3} + \cdots + x_0{}^{k-1})$]

8. Compute $f'(x_0)$, $x_0 \neq 0$, in the following cases.
(a) $f(x) = x^{-2}$
• (b) $f(x) = x^{-3}$
(c) $f(x) = x^{-n}$

9. (a) Plot $f(x) = x^3 + x$ near the point (0,0) to each of the following scales:

1 inch = 1 unit on the axis;
5 inches = 1 unit on the axis;
10 inches = 1 unit on the axis.

(The scales need not be exactly these, but the graphs should be drawn by plotting three or four points carefully.)

(b) From your graphs, guess the slope of the straight line which approximates the graph near this point (0,0).

10. (a) Some of the graphs in Fig. 23 have tangent lines which intersect the graph in more than one point. Give sketches showing this.
(b) Some of the graphs in Fig. 23 have tangent lines which cross the graph at the point of tangency. Give sketches showing this. (Points where the tangent line cross the curve are called *inflection points.*)

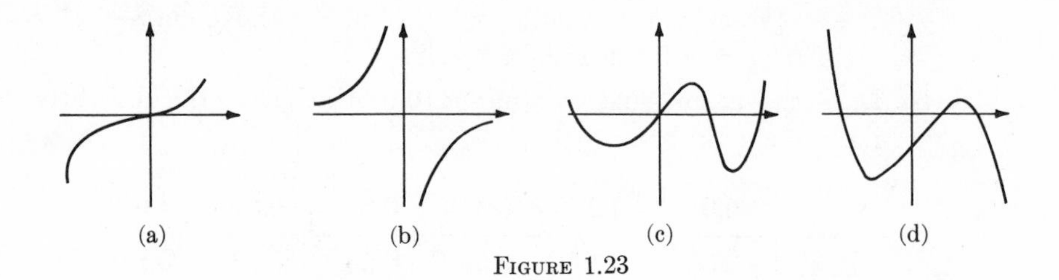

FIGURE 1.23

11. For some of the graphs in Fig. 23, there are lines that intersect the graph exactly once, but which are *not* tangent to the graph. Give sketches showing this.

• **12.** There are two lines passing through the point $(0, -\frac{1}{4})$ that are tangent to the parabola $\{(x,y) : y = x^2\}$. Find their equations.

13. The parabola $\{(x,y) : y = x^2\}$ determines three sets of points R_0, R_1, R_2 in the plane. R_0 consists of those points through which there passes *no* line tangent to the parabola; R_1 consists of those points through which exactly one such line passes; and R_2 consists of those points through which exactly two such lines pass. Find R_0, R_1, R_2 by inspection. Can you verify this analytically?

1.3 REFLECTION IN A PARABOLIC MIRROR

When light is reflected from a plane mirror, the angle from the mirror to the incoming ray equals the angle from the reflected ray to the mirror [Fig. 24]. When the mirror is *curved* [Fig. 25], the tangent line determines how the ray is reflected. Near the point of reflection P, the mirror looks very much like the tangent line T, and the light is reflected so that the angles α and β between the rays and this line are equal.

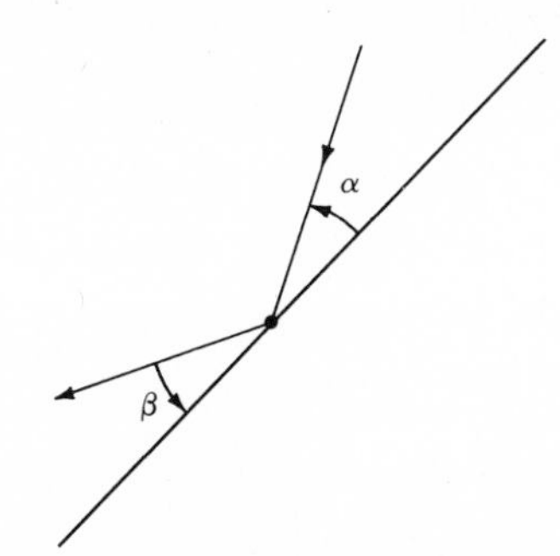

FIGURE 1.24

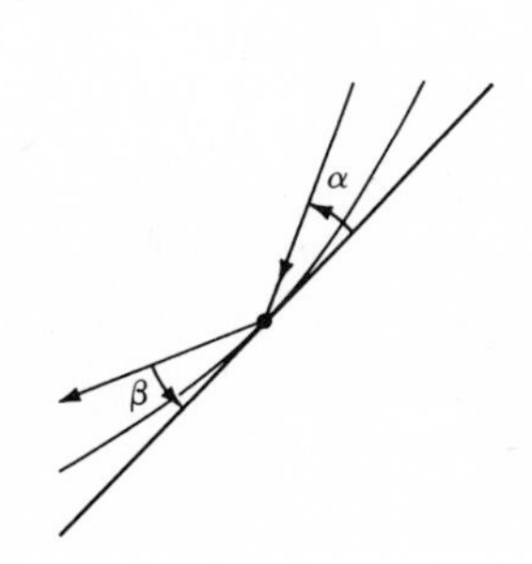

FIGURE 1.25 $\alpha = \beta$

We use this principle to study the reflective properties of the mirror sketched in Fig. 27. Take the graph of $f(x) = ax^2$ [Fig. 26] and revolve it about the vertical axis, generating the surface in Fig. 27. The graph is a parabola (as we shall prove below), and the surface is called a "paraboloid of revolution." Suppose that the inside of this surface is silvered to form a mirror. Imagine a ray of light coming into the mirror parallel to its axis (the vertical axis in Fig. 27). In what direction does the reflected ray go?

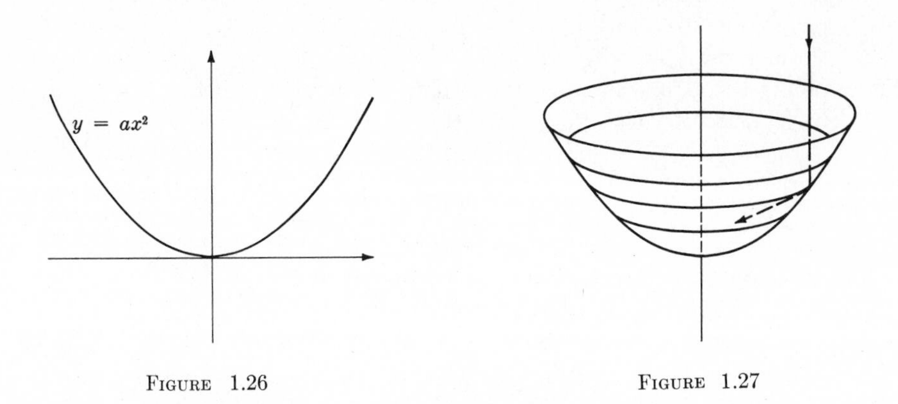

FIGURE 1.26

FIGURE 1.27

To answer this, consider once more the graph of $f(x) = ax^2$, which is a cross section of the mirror. Suppose the incoming ray lies in the plane of this cross section. Then the reflected ray remains in this plane, and we can pose the question like this: At what point $(0,p)$ does the reflected ray cross the vertical axis?

The situation is shown in Fig. 28. We have to determine the number p such that the angle β equals the angle α. We do this by computing $\tan \alpha$ and $\tan \beta$, and equating the results.

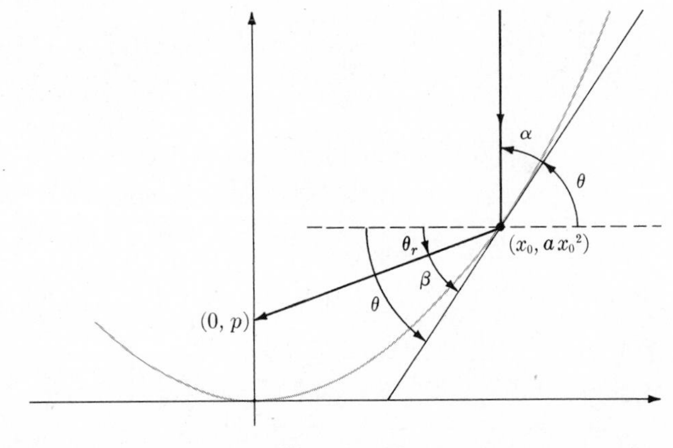

FIGURE 1.28

First, draw the horizontal line $\{(x,y) : y = ax_0^2\}$, and let θ be the angle from this to the tangent line, as in Fig. 28. Then $\tan\theta = 2ax_0$, for $\tan\theta$ is the slope of the tangent line (§0.3), the slope of the tangent line is the derivative, and the derivative of ax^2 is $2ax$. Now it is clear from Fig. 28 that $\alpha + \theta$ is a right angle, so we find

$$\tan\alpha = \cot\theta = \frac{1}{\tan\theta} = \frac{1}{2ax_0}. \tag{1}$$

[See §2.3, Appendix, for this particular bit of trigonometry.]

Next, compute $\tan\beta$. Let θ_r be the angle from the graph of the horizontal line $y = ax_0^2$ to the reflected ray of light, as in Fig. 28. Then $\beta = \theta - \theta_r$, as the figure shows, so

$$\tan\beta = \tan(\theta - \theta_r) = \frac{\tan\theta - \tan\theta_r}{1 + \tan\theta\tan\theta_r}. \tag{2}$$

[This formula for the tangent of the difference of two angles is derived in the problems of the appendix following §2.3.] We already know that

$$\tan\theta = 2ax_0, \tag{3}$$

so we only need to find $\tan\theta_r$. This is easy; $\tan\theta_r$ is the slope of the line through $(0,p)$ and (x_0, ax_0^2) [see Fig. 28 again], so

$$\tan\theta_r = \frac{ax_0^2 - p}{x_0 - 0}. \tag{4}$$

Putting (3) and (4) into (2), we find

$$\tan\beta = \frac{2ax_0 - \dfrac{ax_0^2 - p}{x_0}}{1 + 2ax_0\dfrac{ax_0^2 - p}{x_0}} = \frac{ax_0^2 + p}{x_0 + 2a^2x_0^3 - 2ax_0p}. \tag{5}$$

Finally, since $\tan\alpha = \tan\beta$, we find p by equating the expressions (1) and (5):

$$\frac{1}{2ax_0} = \frac{ax_0^2 + p}{x_0 + 2a^2x_0^3 - 2ax_0p}. \tag{6}$$

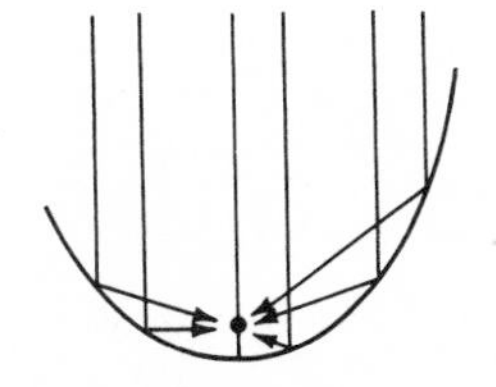

FIGURE 1.29

You can easily solve (6) to find that

$$p = 1/4a.$$

It is significant that this result $p = 1/4a$ does not depend on x_0; it is the *same* for *every* point of reflection $P = (x_0, ax_0^2)$. [See Fig. 29.] Thus, for instance, if the mirror is aimed at the sun, then all the rays are reflected to the same point, and a great deal of heat can be produced there. This point $(0, 1/4a)$ is called the *focus* of the graph. (The word "focus" means *hearth.*)

Geometric characterization of the parabola

The focus $F = (0, 1/4a)$ of the graph of $f(x) = ax^2$, which we found by studying the reflecting properties, is equally important in the geometric characterization of the graph.

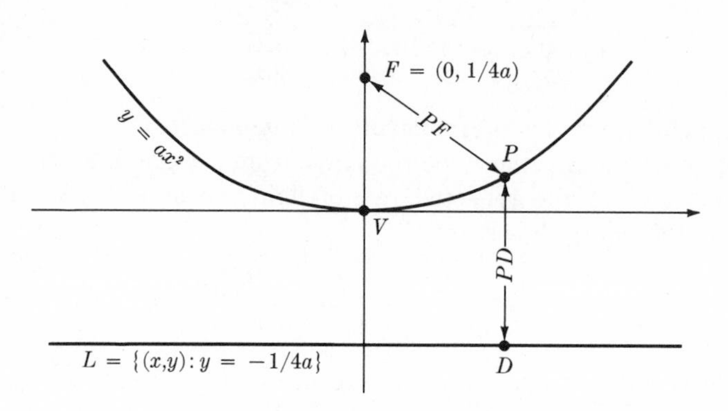

FIGURE 1.30

Let L [Fig. 30] be the horizontal line $\{(x,y): y = -1/4a\}$; this is called the *directrix* of the graph of $f(x) = ax^2$. Notice that the distance from L to the vertex V equals the distance from V to the focus F. Let $P = (x,y)$ be any point in the plane, and let $D = (x, -1/4a)$ be the foot of the perpendicular from P to the directrix L. We will prove that *the graph of f consists of all points P which are equidistant from the focus F and the directrix L*, i.e. all points P such that $PF = PD$. What we have to prove is that

$$PF = PD \iff P \text{ is on the graph of } f(x) = ax^2. \qquad (7)$$

This can be done straightforwardly:

$$PF = PD \iff (PF)^2 = (PD)^2 \qquad [\text{since } PF \geq 0 \text{ and } PD \geq 0]$$

$$\iff (x-0)^2 + \left(y - \frac{1}{4a}\right)^2 = (x-x)^2 + \left(y - \left(-\frac{1}{4a}\right)\right)^2$$

$$[\text{since } F = (0, 1/4a) \text{ and } D = (x, -1/4a)]$$

$$\iff x^2 + y^2 - \frac{1}{2a}y + \frac{1}{16a^2} = y^2 + \frac{1}{2a}y + \frac{1}{16a^2}$$

$$\iff x^2 = \frac{1}{a}y$$

$$\iff y = ax^2$$

$$\iff (x,y) \text{ is on the graph of } f(x) = ax^2.$$

The condition $PF = PD$ is the geometric defining property of the parabola, so we have now proved that *the graph of* $f(x) = ax^2$ *is a parabola with focus* $(0, 1/4a)$, as claimed above. More generally, you can prove that the graph of *any* quadratic $f(x) = ax^2 + bx + c$ is a parabola [§1.4, Problem 5].

Example 1. Find the focus and directrix of the graph of $f(x) = \frac{1}{2}x^2$.
Solution. Since $f(x) = ax^2$ has focus $F = (0, 1/4a)$, and the given equation $f(x) = \frac{1}{2}x^2$ has $a = \frac{1}{2}$, the focus is $F = (0,\frac{1}{2})$. The directrix is just as far below the vertex as the focus is above; since the vertex is the origin in this case, the directrix is the line $\{(x,y) : y = -\frac{1}{2}\}$ [Fig. 31].

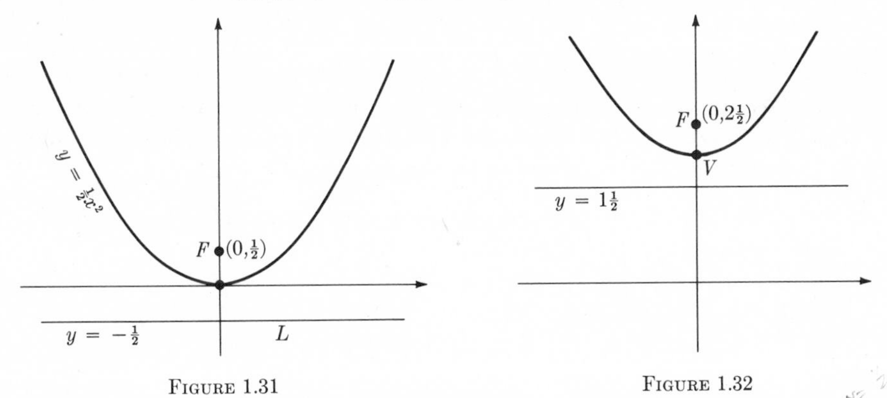

FIGURE 1.31

FIGURE 1.32

Example 2. Find the focus and directrix of the graph of $f(x) = \frac{1}{2}x^2 + 2$.
Solution. This is the same graph as in Example 1, but moved up two units; the vertex is at (0,2) instead of (0,0). Thus the focus and directrix are also moved up two units; they are $F = (0,2\frac{1}{2})$ and $L = \{(x,y) : y = 1\frac{1}{2})\}$ [Fig. 32].

PROBLEMS

1. Solve equation (6) for p.

2. Sketch each of the following parabolas, together with its focus and directrix.

• (a) $f(x) = \frac{1}{4}x^2$	(d) $f(x) = 8x^2$
(b) $f(x) = x^2$	• (e) $f(x) = \frac{1}{4}x^2 + 1$
(c) $f(x) = 4x^2$	(f) $f(x) = 4x^2 - 2$

• **3.** A ray of light enters a parabola along the line $x = x_0$ (as in Fig. 28), is reflected at $P = (x_0, ax_0^2)$, continues through the focus F, and is reflected from the other side of the parabola. Along what line does the reflected ray go?

4. Suppose that a ray of light, parallel to the axis of a parabola, is reflected from the *outside*, as in Fig. 33. Where does the reflected ray go?

FIGURE 1.33

5. The graph of $f(x) = a^2/2x$ is a hyperbola with foci (a,a) and $(-a,-a)$. [See Problem 14(a), §0.4.] Show that if light starts at the first focus (a,a) and is reflected from the hyperbola, then the reflected ray lies along a line emanating from the second focus $(-a,-a)$.

1.4 THE DERIVATIVE AS AN AID TO GRAPHING

The graphs and tangent lines we have sketched illustrate a simple but useful fact:

> **Where the derivative f' is positive, both the tangent line and the graph of f slope up to the right; where the derivative is negative, they slope down to the right; and where the derivative is zero, the tangent is horizontal.**

We now apply this to sketch graphs of quadratic and cubic polynomials (polynomials of degree 2 and 3).

Quadratic polynomials

Let

$$f(x) = ax^2 + bx + c, \qquad a \neq 0. \tag{1}$$

From formula (6), §1.2, we have

$$f'(x) = 2ax + b. \tag{2}$$

This derivative is zero for just one value of x, the value $x_0 = -b/2a$. The corresponding point $V = (-b/2a, f(-b/2a))$ is the most important point on the graph, and should be plotted first because [Problem 5 below] *the graph is a parabola with vertex V.* The horizontal line through V is tangent to the graph (since $f' = 0$ at that point) and the vertical line through V is the axis of the parabola, which is a line of symmetry for the graph. By plotting V and one or two other points and using the symmetry, you can quickly construct a good sketch of the graph.

Example 1. Sketch the graph of $f(x) = -2x^2 + 4x + 3$. *Solution.* By formula (2), $f'(x) = 2\cdot(-2)x + 4 = -4x + 4$. Thus $f'(x) = 0$ when $x = 1$, and the vertex of the parabola is $V = (1,f(1)) = (1,5)$, shown in Fig. 34. We sketch the tangent and the line of symmetry through V, plot two other points P_1 and P_2 corresponding to $x = 0$ and $x = -1$, and thus obtain the graph in Fig. 34. Notice that the derivative $f'(x) = -4x + 4$ grows more and more negative as x increases from 1, corresponding to the fact that the graph slopes down more and more steeply to the right of the line of symmetry. (If the coefficient of x^2 had been positive, instead of negative, the graph would open upward, as in Fig. 35.)

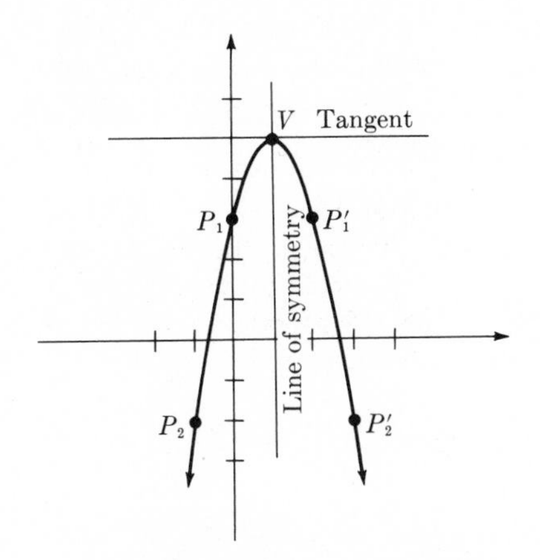

FIGURE 1.34 $f(x) = -2x^2 + 4x + 3$

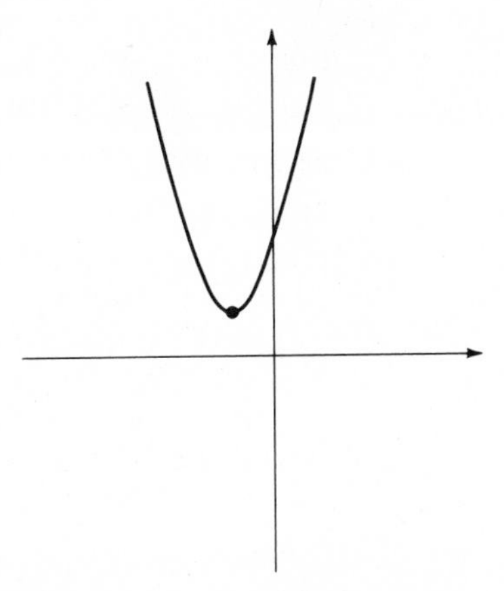

FIGURE 1.35 $f(x) = 2x^2 + 4x + 3$

Cubic polynomials

We begin by taking the derivative of a general cubic polynomial $f(x) = ax^3 + bx^2 + cx + d$. [Cf. Problem 7, p. 72.] The difference quotient is

$$\frac{f(x) - f(x_0)}{x - x_0} = \frac{ax^3 + bx^2 + cx + d - (ax_0^3 + bx_0^2 + cx_0 + d)}{x - x_0}$$

$$= \frac{a(x^3 - x_0^3) + b(x^2 - x_0^2) + c(x - x_0)}{x - x_0}$$

$$= a(x^2 + xx_0 + x_0^2) + b(x + x_0) + c.$$

[Recall the factorizations $x^2 - x_0^2 = (x - x_0)(x + x_0)$ and $x^3 - x_0^3 = (x - x_0)(x^2 + xx_0 + x_0^2)$.] Thus

$$\lim_{x \to x_0} \frac{f(x) - f(x_0)}{x - x_0} = a(x_0^2 + x_0x_0 + x_0^2) + b(x_0 + x_0) + c$$

$$= 3ax_0^2 + 2bx_0 + c.$$

In other words,

$$\begin{aligned} \text{if} \quad f(x) &= ax^3 + bx^2 + cx + d, \\ \text{then} \quad f'(x) &= 3ax^2 + 2bx + c. \end{aligned} \tag{3}$$

Our program for graphing a cubic polynomial is now the following: Take the derivative, using formula (3). Find the places, if any, where the derivative is zero, i.e. where the tangent line is horizontal. Evaluate f at these places, and plot the corresponding points on the graph. Investigate the sign of the derivative f' at the points where $f' \neq 0$ and, finally, sketch the graph.

Example 2. Let $f(x) = 2x^3 + 3x^2 - 2$. Then according to formula (3), we have $f'(x) = 3 \cdot 2x^2 + 2 \cdot 3x = 6x(x+1)$. Hence $f'(x) = 0$ when $x = 0$ or $x = -1$, and the points on the graph where the tangent is horizontal are $(0, f(0)) = (0, -2)$, and $(-1, f(-1)) = (-1, -1)$. These are plotted in Fig. 37. To see how the graph looks in the intervals between these points, we investigate the sign of the derivative. As Fig. 36 shows, $f'(x) = 6x(x+1)$ is negative between $x = -1$ and $x = 0$, and is positive outside this interval. So the graph of f rises up to the point $(-1, -1)$, descends to $(0, -2)$, and then rises again. For good measure, we have plotted a point to the left of $(-1, -1)$ and another to the right of $(0, -2)$. The graph is sketched in Fig. 37.

Sign of x: − − − − − − − − + + + + +
−3 −2 −1 0 1 2 3

Sign of $x + 1$: − − − − + + + + + + + +
−3 −2 −1 0 1 2 3

Sign of $6x(x + 1)$: + + + + + − − − + + + + +
−3 −2 −1 0 1 2 3

FIGURE 1.36

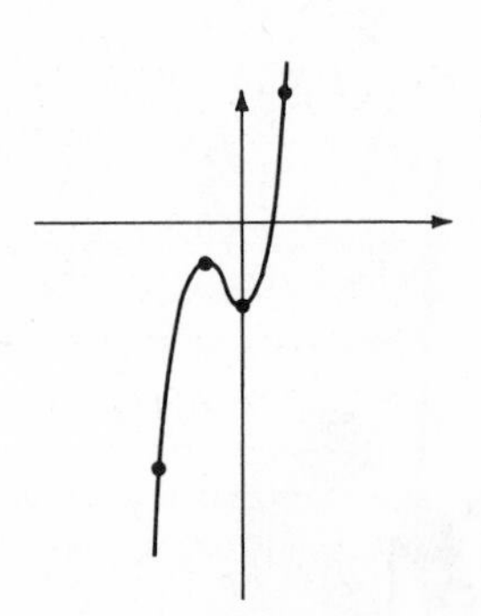

FIGURE 1.37 Graph of $2x^3 + 3x^2 - 2$

The graph in Fig. 37 crosses the x axis exactly once, at some point between $x = 0$ and $x = 1$. So $2x^3 + 3x^2 - 2 = 0$ for just one real number x, and this number lies between 0 and 1. This fact, though obvious from the graph, was not at all obvious from the original algebraic expression $2x^3 + 3x^2 - 2$.

Example 3. Let $f(x) = x^3 + x$. Then $f'(x) = 3x^2 + 1$. This expression is always ≥ 1, so the graph is always rising to the right. The slope of the tangent line at $(0,0)$ is $3 \cdot 0^2 + 1 = 1$, and the tangent line at every other point has a slope greater than 1. [See Fig. 38.]

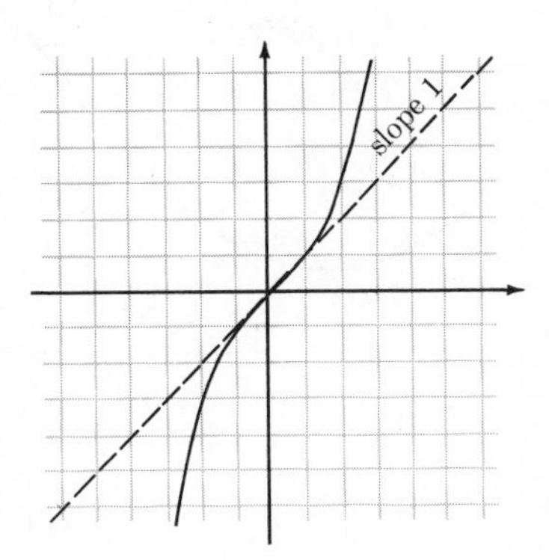

FIGURE 1.38 Graph of $x^3 + x$

PROBLEMS

1. For the following quadratic polynomials f, find f', plot the vertex and two other points, and sketch the graph, as in Example 1 above.

•(a) $f(x) = -x^2 + 3x + 2$
•(b) $f(x) = 5x^2 - 2x + 1$
(c) $f(x) = 2x^2 + 3x + 4$
(d) $f(x) = x^2 + x + 1$
(e) $f(x) = -x^2 - 2x + 2$
(f) $f(x) = 10x^2 - 5x$

2. For the following cubic polynomials f, find f', plot the points where there is a horizontal tangent, decide where the graph is rising and where descending, and sketch the graph, as in Examples 2 and 3 above.

•(a) $f(x) = -x^3 + x + 2$
•(b) $f(x) = x^3 - 3x - 1$
(c) $f(x) = x^3 + 2x + 1$
(d) $f(x) = x^3 - 3x^2 + 3x + 2$
(e) $f(x) = x^3 + 3x^2 + 3x + 1$
(f) $f(x) = 4 - x^3 - 2x$

3. Let $f(x) = ax^2 + bx + c$, and suppose $a > 0$.

•(a) Show that $f'(x) > 0 \iff x > -b/2a$, and $f'(x) < 0 \iff x < -b/2a$.
(b) Interpret the result (a) graphically.

4. Let $f(x) = ax^2 + bx + c$, and suppose $a < 0$.

•(a) Determine where $f' > 0$ and where $f' < 0$.
(b) Interpret the result (a) graphically.

5. We claimed above that the graph of $f(x) = ax^2 + bx + c$ has vertex $V = (-b/2a, f(-b/2a))$.

(a) Show that $V = \left(-\dfrac{b}{2a}, \dfrac{4ac - b^2}{4a}\right)$.

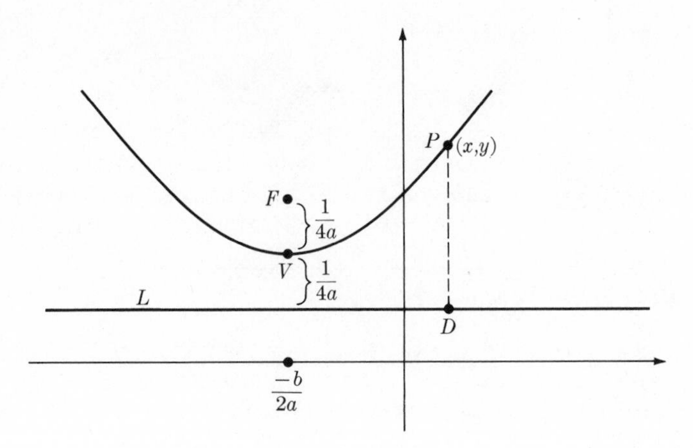

FIGURE 1.39

- (b) Find the coordinates of the point F which is $1/4a$ units directly above V. [Fig. 39]
- (c) Find the equation of the horizontal line L which is $1/4a$ units below V. [Fig. 39]
- (d) Let $P = (x,y)$, and find the coordinates of the point D at the foot of the perpendicular from P to L.

(e) Prove that P is on the graph of $f(x) = ax^2 + bx + c$ iff $PF = PD$. [Thus the graph is a parabola with focus F and directrix L.]

6. If you have done Problem 7, §1.2, correctly, you will know that the tangent to the graph of $f(x) = x^4 + 2x^3 + 2$ at an arbitrary point $(x_0, f(x_0))$ has slope $4x_0^3 + 6x_0^2$.
   (a) Find the points at which the tangent is horizontal.
   (b) Find where the graph is rising and where it is descending.
   (c) Sketch the graph.

7. On the graph of every cubic there is a *center* $C = (x_0, y_0)$, such that for every point $P_1 = (x_1, y_1)$ on the curve, there is a corresponding point $P_1' = (2x_0 - x_1, 2y_0 - y_1)$; see Fig. 40.
   (a) Show that (x_0, y_0) is the midpoint of the segment between (x_1, y_1) and $(2x_0 - x_1, 2y_0 - y_1)$.

   Find the center of the following cubics.
   - (b) x^3
   - (c) $x^3 + x$
   - (d) $x^3 + x + 1$
   - (e) $x^3 + x^2$
   - (f) $ax^3 + bx^2 + cx + d$

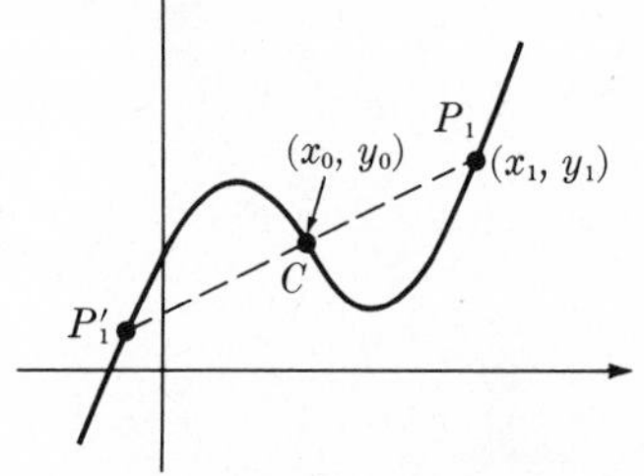

FIGURE 1.40

1.5 MAXIMUM PROBLEMS

Sketching graphs is not an idle academic pastime; it answers a surprising number of concrete questions. The simplest of these asks "What is the most efficient way to . . . ?" For example, what is the most efficient way to make a box out of a square piece of cardboard? To be very specific:

Example 1. Given a piece of cardboard $12'' \times 12''$ [Fig. 41], what size should the corner notches be so that the resulting open box has maximum volume? Small notches make a very flat box with small volume, and large notches make a tall thin box, again with small volume. The most efficient shape must occur somewhere between these extremes, but where?

The answer is found by carefully writing out the question in algebraic form, and then sketching an appropriate graph. Let x be the edge-length of the small flap at each corner [Fig. 41]. The volume of the resulting box will be

$$\text{length} \times \text{width} \times \text{height} = (12 - 2x)(12 - 2x)x.$$

Denote this volume by $V(x)$,

$$V(x) = (12 - 2x)(12 - 2x)x = 4x^3 - 48x^2 + 144x.$$

We want to make $V(x)$ as large as possible by choosing the appropriate value of x. Notice that although the above expression for $V(x)$ makes sense for every number x, the nature of the problem puts certain restrictions on x, namely

$x \geq 0$ [since x is the edge of a square]

$x \leq 6$ [since if $x > 6$, two squares from adjacent corners overlap; see Fig. 41].

So the problem is reduced to this: Find the maximum value of the function V defined by

$$V(x) = 4x^3 - 48x^2 + 144x, \qquad 0 \leq x \leq 6. \tag{1}$$

The maximum is easily found by sketching the graph of V on the interval [0, 6]. The derivative is

$$V'(x) = 12x^2 - 96x + 144 = 12(x^2 - 8x + 12) = 12(x - 2)(x - 6),$$

and therefore

$$V'(x) = 0 \qquad \text{when } x = 2 \text{ or } x = 6.$$

It is easy to check that

$$V'(x) > 0, \qquad x < 2$$

$$V'(x) < 0, \qquad 2 < x < 6$$

$$V'(x) > 0, \qquad x > 6.$$

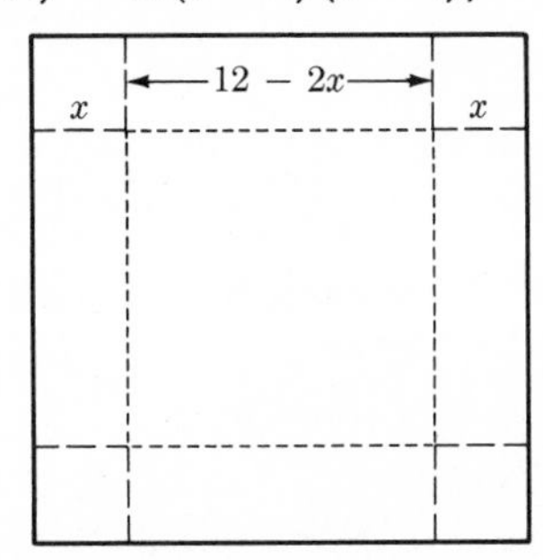

FIGURE 1.41 Cut corners out, and fold along dotted lines

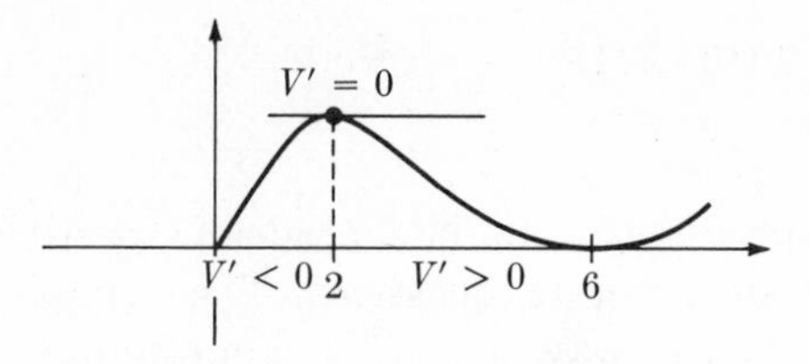

FIGURE 1.42

Therefore, the function (1) has a graph as in Fig. 42. The graph shows that the maximum value of V occurs at $x = 2$. Therefore, you should cut out squares of edge-length 2; the resulting box has height $x = 2$, and sides of length $12 - 2x = 8$, so the resulting volume is

$$V(2) = 8 \cdot 8 \cdot 2 = 128.$$

Example 2. Of all rectangles having a given perimeter P, find the one with maximum area. Your intuition [supported by Fig. 43] may tell you it must be a square, but can you prove it? We begin, as before, with a sketch; Fig. 43

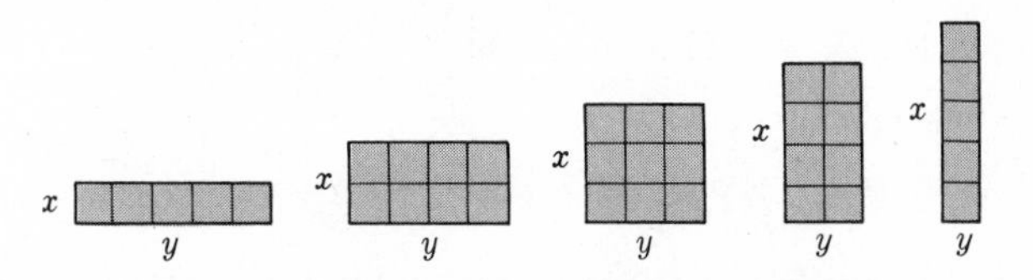

FIGURE 1.43 Five rectangles with the same perimeter, $2x + 2y = 12$

will do. The quantity to be maximized is the area

$$A = xy$$

which appears to depend on *two* unknown quantities x and y. However, we are also told that the perimeter must be some given number P, so

$$2x + 2y = P.$$

This equation allows us to express y in terms of x,

$$y = \tfrac{1}{2}P - x, \tag{2}$$

and therefore to express the area $A = xy$ in terms of x only,

$$A(x) = xy = x \cdot (\tfrac{1}{2}P - x) = \tfrac{1}{2}Px - x^2.$$

This is the function to be maximized. The nature of the problem puts restrictions on x, namely

$x \geq 0$ [since x is the edge of a rectangle]

$\tfrac{1}{2}P - x \geq 0$, or $x \leq \tfrac{1}{2}P$ [since $\tfrac{1}{2}P - x = y$ is the edge of a rectangle]

so we maximize the function

$$A(x) = \tfrac{1}{2}Px - x^2, \qquad 0 \leq x \leq \tfrac{1}{2}P.$$

The derivative is

$$A'(x) = \tfrac{1}{2}P - 2x,$$

and therefore

$$A'(x) = 0 \quad \text{when } x = \tfrac{1}{4}P.$$

Since $A(x)$ is a quadratic with the coefficient of x^2 negative, it has a *maximum* where $A'(x) = 0$, at $x = \frac{1}{4}P$. This is one edge of the rectangle; the other edge is given by (2):

$$y = \tfrac{1}{2}P - x = \tfrac{1}{2}P - \tfrac{1}{4}P = \tfrac{1}{4}P = x.$$

It follows that maximum area requires that $y = x$, and the rectangle in question is indeed a square, as expected.

These two examples illustrate how to solve problems of this type. First, draw a sketch, labeling all the relevant known and unknown quantities. Write out the quantity to be maximized, and any further information that is given, in terms of the labels in the sketch. Then express the quantity to be maximized as a function of just one variable. [We did this in Example 2 when we rewrote $A = xy$ as $A(x) = x \cdot (\frac{1}{2}P - x)$. In Example 1 the volume was already a function of just one variable, x.] Next, find the appropriate domain for the function, as determined by the nature of the problem. Finally, find the maximum of this function. In the problems below, the function is a polynomial and the maximum is found by taking the derivative, as in the previous sections.

Example 3. Compute the distance from the point (0,1) to the line $\{(x,y) : y = 2x - 1\}$. *Solution.* This is a minimum problem; the distance from a point P_0 to a line is defined to be the minimum of the distances from P_0 to all the points on the line. The distance D from (0,1) to any point (x,y) in the plane is

$$D = \sqrt{(x-0)^2 + (y-1)^2}\,.$$

This depends on both x and y; but we have not yet used the fact that the point (x,y) must be on the given line. This happens iff $y = 2x - 1$, so we are interested in the function

$$D(x) = \sqrt{(x-0)^2 + (2x-1-1)^2} = \sqrt{5x^2 - 8x + 4}\,.$$

This function is hard to differentiate by the limited techniques we have developed so far, but the problem can be simplified by an observation: Since all values of D are *positive*, the minimum value of D is the square root of the minimum value of D^2. So we consider the function

$$f(x) = (D(x))^2 = 5x^2 - 8x + 4, \qquad -\infty < x < \infty.$$

There are no restrictions on x, since every x corresponds to a point on the line. We have a quadratic opening upward, so there is a minimum where the derivative is zero, i.e. where

$$0 = f'(x) = 10x - 8, \qquad \text{or } x = \frac{4}{5}\,.$$

The minimum of f is therefore

$$f\left(\frac{4}{5}\right) = 5\left(\frac{4}{5}\right)^2 - 8\left(\frac{4}{5}\right) + 4 = \frac{16 - 32 + 20}{5} = \frac{4}{5}\,.$$

The minimum of D is $\sqrt{4/5}$, and this is the distance from the point (0,1) to the line $\{(x,y) : y = 2x - 1\}$.

PROBLEMS

More difficult problems of this type appear in §4.1.

1. A topless box is to be made out of a square piece of cardboard, as in the example above. What dimensions give maximum volume if the dimensions of the square are
   • (a) 10×10 ? (b) $s \times s$?

2. A box with a top is to be cut out of a rectangular piece of cardboard [Fig. 44] just as in Example 1. How deep should the box be for maximum volume, if the rectangle is
   • (a) 2 feet $\times$ 3 feet?
   (b) $a \times b$? [Do not expect a neat answer; check it by comparing with part (a).]

3 • (a) Find the point P_0 on the parabola $\{(x,y): y = x^2\}$ which is closest to the point $P_1 = (3,0)$. [Compare Example 3 above.]
   (b) Show that the line through P_0 and P_1 is perpendicular to the parabola at P_0. ["Perpendicular to the parabola at P_0" means perpendicular to its tangent line at P_0. The condition for perpendicular lines is given in §0.3.]

• 4. A track of total length 400 meters consists of two equal semicircles and two equal straightaways. What are the dimensions of the track enclosing the largest area?

• 5. The track in Problem 4 encloses three areas, a rectangle and two semicircles. What are the dimensions of the track enclosing the *rectangle* of largest area?

• 6. The postoffice in New Chaos does not accept a package if the sum of the length and the perimeter of the cross section exceeds 60 inches.
   (a) Find the rectangular box of largest volume that meets this restriction. [Note that the cross section must be a square if the volume is to be maximized.]
   (b) Find the cylindrical box of largest volume that meets this restriction.

7. Let L be the graph of $y = mx + b$, and $P_0 = (x_0, y_0)$ be any point in the plane.
   (a) Show that the point P_1 on L nearest to P_0 is

$$P_1 = \left(\frac{x_0 + my_0 - mb}{1 + m^2}, \frac{mx_0 + m^2y_0 + b}{1 + m^2}\right).$$

   (b) Show that the distance from P_0 to the line L is

$$\frac{|y_0 - mx_0 - b|}{\sqrt{1 + m^2}}.$$ [Keep cool in the calculations.]

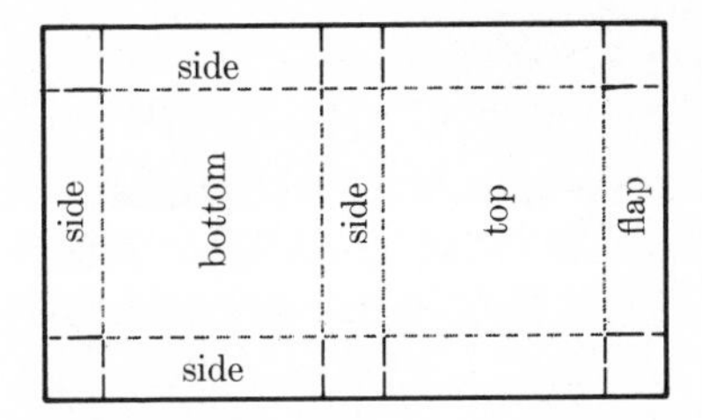

FIGURE 1.44 Cut along dashed lines, then fold along dotted lines

1.6 NEWTON'S METHOD FOR SQUARE ROOTS (optional)

As a final example of the use of tangent lines, we present an efficient method of extracting roots of polynomials, discovered by Isaac Newton. This section presents the method for the simplest case, the extraction of square roots; the appendix to the section shows how efficient the method is.

To take a specific example, suppose we want to compute $\sqrt{3}$. This amounts to solving the equation $x^2 = 3$ or, equivalently, to finding the point where the graph of x^2 cuts the line $\{(x,y) : y = 3\}$. [See Fig. 45.] Since $1^2 < 3 < 2^2$, we know that the graph cuts the line $y = 3$ somewhere between $x = 1$ and $x = 2$, probably closer to $x = 2$. [See Fig. 45 again.]

To apply Newton's method, start with a reasonable approximation to the root $\sqrt{3}$; call it x_0. For instance, you could take $x_0 = 2$. Then draw the tangent line to the graph at the point (x_0, x_0^2) and determine the next approximation x_1 from the intersection of this tangent line with the line $\{(x,y) : y = 3\}$ [Fig. 46].

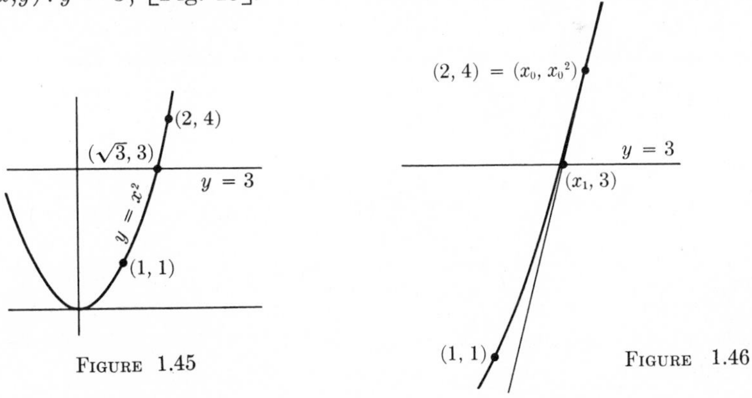

FIGURE 1.45

FIGURE 1.46

This number x_1 is not quite the same as the actual root. But it is a closer approximation than x_0, and it can be computed easily, while the actual root cannot. Once you have found x_1, you can repeat the process to find a still better approximation x_2, then x_3, and so on, obtaining a whole sequence of approximations $x_1, x_2, x_3, \ldots, x_n, \ldots$. Before proceeding to the general formula, it would be useful to do the following example.

Example 1. [Do it yourself; solution below.]
(a) Find the equation of the line tangent to the graph of $f(x) = x^2$ at the point (2,4).
(b) Find the intersection of this tangent line with the line $\{(x,y): y = 3\}$, and thus find the number x_1 in Fig. 46.
(c) Compute x_1^2. [The result suggests that x_1 is much closer to $\sqrt{3}$ than x_0 is, since $x_0^2 = 4$.]

Just as you found x_1 from x_0 in part (b), we find the formula expressing x_{n+1} in terms of x_n. The equation of the tangent line at (x_n, x_n^2) is

$$y - x_n^2 = 2x_n(x - x_n). \tag{1}$$

The intersection of the graph of (1) with the line $\{(x,y): y = 3\}$ is found by solving $3 - x_n^2 = 2x_n(x - x_n)$ for x; the result is $x = (3 + x_n^2)/2x_n$. This gives the next approximation to $\sqrt{3}$,

$$x_{n+1} = \frac{3 + x_n^2}{2x_n} = \frac{3}{2x_n} + \frac{x_n}{2}. \tag{2}$$

For example, taking $x_1 = \frac{7}{4} = 1.75$ [as found in part (b) of Example 1], we obtain

$$x_2 = \frac{3 + x_1^2}{2x_1} = \frac{97}{56} = 1.7321\ldots,$$

which is much more accurate than x_1.

For any number $c > 0$, you can find good approximations to $\sqrt{c}$ in just the same way. Beginning with a first approximation x_0, compute $x_1, x_2, x_3, \ldots$ in succession until you have achieved the desired accuracy. [The Appendix below shows how to determine the accuracy.] Just as in the special case $c = 3$, where we found the equation (2), we obtain the general formula

APPROXIMATIONS TO $\sqrt{c}$

$$x_{n+1} = \frac{c + x_n^2}{2x_n} = \frac{c}{2x_n} + \frac{x_n}{2}, \tag{3}$$

which gives successively better approximations to $\sqrt{c}$.

Remark. Formula (3) is actually much older than Newton; apparently it was known even by the ancient Babylonians. Newton's contribution is the geometric idea in Fig. 46, namely, to replace the curve by its tangent line. This same idea applies equally well to finding cube roots, fourth roots, etc., and more generally to finding the roots of any polynomial; see Problems 5–7.

Solution to Example 1

(a) $f'(x) = 2x$, $f'(2) = 4$, so the equation of the tangent line is $y - 4 = 4(x - 2)$, or $y = 4x - 4$. (b) Set $3 = 4x - 4$, and find $x = \frac{7}{4}$.
(c) $x_1^2 = \frac{49}{16} = 3\frac{1}{16}$. [This is much closer to 3 than $2^2 = 4$ is.]

PROBLEMS

1. Derive formula (3) above.

• 2. Compute $\sqrt{10}$ to two decimals by Newton's method, i.e. find the digits a and b in $\sqrt{10} = 3.ab \pm 0.01$. Check your answer by computing $(3.ab)^2$, and then either $(3.ab + 0.01)^2$ [if it turns out that $(3.ab)^2 < 10$] or $(3.ab - 0.01)^2$ [if it turns out that $(3.ab)^2 > 10$].

3. Use Newton's method to give approximations to the following square roots, accurate to two decimals.
(a) $\sqrt{2}$ •(b) $\sqrt{5}$ •(c) $\sqrt{8.5}$ •(d) $\sqrt{3.14}$

4. By appropriate sketches, show the following properties of the approximations $x_0, x_1, \ldots$ to $\sqrt{c}$, determined by Newton's method as above.
(a) If $x_0 = \sqrt{c}$, then $x_n = \sqrt{c}$ for each succeeding approximation x_n.
(b) If $x_0 > 0$, but $x_0 \neq \sqrt{c}$, then x_1 is too large, $x_1 > \sqrt{c}$. Also, $x_2 > \sqrt{c}$, $x_3 > \sqrt{c}$, ...
(c) $x_1 > x_2$, $x_2 > x_3$, $x_3 > x_4$, $\ldots$, $x_n > x_{n+1}$.

• 5. Deduce Newton's method for cube roots, and obtain an approximation of $3^{1/3}$ that you think is accurate to two decimal places.

6. Let f be any function, and let x_n be an approximation to the solution of $f(x) = c$. Show that Newton's method [replacing the graph of f by its tangent line at $(x_0, f(x_0))$] gives the next approximation as

$$x_{n+1} = x_n + \frac{c - f(x_n)}{f'(x_n)}.$$

[Assume that $f'(x)$ exists for all x.]

• 7. Use Newton's method (Problem 6) to approximate the solution of $2x^3 + 3x^2 - 2 = 0$, which lies between $x = 0$ and $x = 1$. [See §1.4.] Start with $x_0 = 1$, and compute x_1 and x_2.

APPENDIX: THE ACCURACY OF NEWTON'S METHOD FOR SQUARE ROOTS*

The previous section §1.6 neglected an important aspect of the calculation of square roots: It gave no efficient way of determining *how accurate* the approximations are. (Problem 2 above suggested an inefficient way.) We now supply this part of the discussion.

---

* Practice with inequalities; preparation for rigorous work with infinite sequences, and with limits in general. The basic rules for inequalities are outlined in §AI.3.

We suppose x_n has been found, and try to see how accurate x_{n+1} will be. Recall from §1.6 that

$$x_{n+1} = \frac{x_n^2 + c}{2x_n}. \tag{1}$$

Hence

$$x_{n+1} - \sqrt{c} = \frac{x_n^2 + c}{2x_n} - \sqrt{c} = \frac{x_n^2 - 2\sqrt{c}x_n + c}{2x_n} = \frac{(x_n - \sqrt{c})^2}{2x_n}. \tag{2}$$

This relates the error $x_{n+1} - \sqrt{c}$ to the error of the previous approximation, $x_n - \sqrt{c}$. Suppose, for simplicity, that $1/2x_n \leq 1$; then from (2), the error $x_{n+1} - \sqrt{c}$ is less than the square of the previous error $x_n - \sqrt{c}$. For instance if x_n is accurate within 10^{-m}, then x_{n+1} is accurate within 10^{-2m}; i.e. it is accurate to twice as many decimal places.

Example 1. Since $(1.7)^2 < 3 < (1.8)^2$, the number $x_1 = 1.7$ approximates $\sqrt{3}$ within 10^{-1}, that is, $|x_1 - \sqrt{3}| < 10^{-1}$. Hence

$$x_2 = \frac{x_1^2 + 3}{2x_1} = \frac{5.89}{3.4} = 1.7323\ldots$$

approximates $\sqrt{3}$ within an error which is given in (2) as

$$\frac{(x_1 - \sqrt{3})^2}{2x_1} < \frac{10^{-2}}{3.4} < 3 \times 10^{-3}.$$

Thus $x_2 = 1.732\ldots$ gives $\sqrt{3}$ with an error of at most 3 in the third decimal place.

Our next method estimates the error even more effectively. To estimate $x_{n+1} - \sqrt{c}$ by (2), we need to know beforehand an estimate of the error, $x_n - \sqrt{c}$, in the previous approximation. By a little algebra, we can rework the formula to obtain an estimate for $x_{n+1} - \sqrt{c}$ that does *not* require any knowledge of $x_n - \sqrt{c}$.

Starting with formula (2), we have

$$x_{n+1} - \sqrt{c} = \frac{(x_n - \sqrt{c})^2}{2x_n}$$

$$= \frac{(x_n - \sqrt{c})^2(x_n + \sqrt{c})^2}{2x_n(x_n + \sqrt{c})^2} = \frac{(x_n^2 - c)^2}{2x_n(x_n + \sqrt{c})^2}. \tag{3}$$

This is the formula used to estimate the error $x_{n+1} - \sqrt{c}$. You may object that the appearance of $\sqrt{c}$ on the right side of (3) makes it useless since $\sqrt{c}$ is not known exactly. The following example shows how this objection is overcome.

Example 2. Take $x_1 = 1.7$ as an approximation to $\sqrt{3}$, and estimate the accuracy of the next approximation $x_2 = 1.7323\ldots$ found in Example 1.
Solution

$$x_2 - \sqrt{c} = \frac{(x_1^2 - c)^2}{2x_1(x_1 + \sqrt{c})^2}$$

$$= \frac{(2.89 - 3)^2}{3.4(1.7 + \sqrt{3})^2} \qquad [\text{since } x_1 = 1.7 \text{ and } c = 3]$$

$$< \frac{(.11)^2}{3.4(1.7 + 1.7)^2} \qquad [\text{since } 1.7 < \sqrt{3}, \text{ and a smaller denominator makes a larger fraction}]$$

$$= \frac{1}{3.4}\left(\frac{.11}{3.4}\right)^2 < \frac{1}{3.4}(.033)^2 < 3.3 \times 10^{-4}.$$

[The last line was obtained by computing the first few digits in the decimal expansions of .11/3.4 and $(.033)^2$.] Thus $x_2 = 1.7323\ldots$ gives $\sqrt{3}$ with an error of at most 3.3 in the fourth decimal place.

Now it is clear why formula (3) is useful, even though the *unknown* $\sqrt{c}$ appears on the right-hand side. We can replace the right-hand side by a slightly larger *known* quantity, by replacing $\sqrt{c}$ with any slightly smaller known quantity.

Example 3. Take 1.732 as an approximation to $\sqrt{3}$. [We obtain this either from a table or from the previous example.] We use (1) to obtain a new approximation, and (3) to show how good it is. [The calculations can be done with paper and pencil, but a desk calculator would be handy.]

$$x_0 = 1.732, \qquad x_0^2 = 2.999824,$$

$$x_1 = \frac{x_0^2 + c}{2x_0} = \frac{5.999824}{3.464}. \tag{4}$$

Now we use (3) to estimate the error in the approximation (4):

$$3 - x_0^2 = 0.000176 < 2 \cdot 10^{-4}, \tag{5}$$

$$x_1 - \sqrt{3} = \frac{(x_0^2 - 3)^2}{2x_0(x_0 + \sqrt{3})^2} \tag{6}$$

$$< \frac{4 \cdot 10^{-8}}{2 \cdot \frac{3}{2}(\frac{3}{2} + \frac{3}{2})^2} = \frac{4}{27} \cdot 10^{-8} < \frac{1}{6} \cdot 10^{-8}. \tag{7}$$

Here, for convenience, we have replaced the numerator $(x_0^2 - 3)^2$ in (6) by a larger number, $4 \cdot 10^{-8}$ (see (5)), and have replaced the terms x_0 and $\sqrt{3}$ in the denominator by a smaller number, $\frac{3}{2}$. The effect of this is to *increase* the fraction [see §AI.3]; hence we obtain the inequality (7).

We have just shown that the number x_1 approximates $\sqrt{3}$ with an error $< \frac{1}{6}\cdot 10^{-8}$, and hence we can obtain from it the decimal expansion of $\sqrt{3}$ to eight places. In fact, we find by long division that

$$x_1 = 1.73205081 \pm 2\cdot 10^{-9}. \tag{8}$$

Combining the errors in approximating $\sqrt{3}$ by x_1 and x_1 by the decimal expansion (8), we find that the total error is $< \frac{1}{6}\cdot 10^{-8} + 2\cdot 10^{-9} < \frac{2}{5}\cdot 10^{-8}$, and hence

$$\sqrt{3} = 1.73205081 \pm \tfrac{2}{5}\cdot 10^{-8}. \tag{9}$$

In Example 3, the simple formula (1) led from a three-decimal expansion to an eight-decimal expansion of $\sqrt{3}$. By applying the formula once more, we would obtain an expansion valid to at least sixteen decimals, as noted in connection with formula (2). Thus the successive approximations in Newton's method improve by leaps and bounds.

PROBLEMS

[Do *at most* two of the first four problems!]

1. (a) Use (2) and the inequality $(1.41)^2 < 2 < (1.42)^2$ to compute $\sqrt{2}$ to four decimals. [See Example 1.]

(b) Use (2) and the inequality $(1.414)^2 < 2 < (1.415)^2$ to compute $\sqrt{2}$ to six decimals.

(c) Use (3) with $x_0 = 1.41$, and show that x_1 is accurate within 1.5×10^{-5}.

2. Compute the following numbers to four decimals. [Compare Problem 3, §1.6; now you should *guarantee* the accuracy.]

(a) $\sqrt{10}$ (b) $\sqrt{5}$ • (c) $\sqrt{8.5}$ (d) $\sqrt{3.1416}$

3. Compute $\sqrt{10}$ to eight decimals. Prove that your answer is this accurate.

4. Find a rational number p/q (where p and q are integers) such that $0 < p/q - \sqrt{2} < 10^{-10}$.

5. Newton's method of extracting cube roots leads to a sequence of approximations for $c^{1/3}$ determined by $x_{n+1} = (2x_n^3 + c)/3x_n^2$. [See Problem 5, §1.6.]

(a) Show that

$$x_{n+1} - c^{1/3} = \frac{2x_n + c^{1/3}}{3x_n^2}(x_n - c^{1/3})^2.$$

[Compare formula (2) above.]

(b) Show that

$$x_{n+1} - c^{1/3} = \frac{2x_n + c^{1/3}}{3x_n^2(x_n^2 + x_n c^{1/3} + c^{2/3})^2} (x_n^3 - c)^2.$$

[Compare formula (3) above.]

(c) Conclude from part (a) that if $x_n \geq c^{1/3}$ and $x_n \geq 1$, then x_{n+1} is accurate to twice as many decimals as x_n.

(d) Conclude from part (b) that if $x_n \geq c^{1/3}$, then $x_{n+1} - c^{1/3} \leq (x_n^3 - c)^2/9c^{5/3}$.

6. Compute $3^{1/3}$ to four decimals, using Problem 5(c) or 5(d) to prove the accuracy.

7. A common procedure with iterative techniques (such as Newton's method) is to compute $x_1, x_2, \ldots, x_n$, until the differences between successive approximations are negligible. A careful analysis shows that you can generally stop *even before* this happens.

(a) Show that in Newton's method for square roots,

$$x_{n+1} - \sqrt{c} = (x_{n+1} - x_n)^2 \frac{2x_n}{(x_n + \sqrt{c})^2}.$$

[In particular, when $x_n > \sqrt{c}$ (as it generally is),

$$x_{n+1} - \sqrt{c} < \frac{x_n}{2c} (x_{n+1} - x_n)^2.$$

Roughly, when x_{n+1} agrees with x_n to k decimals, then x_{n+1} is accurate to $2k$ decimals.]

(b) Take $x_1 = 2\frac{1}{4}$ as a first approximation to $\sqrt{5}$, compute x_2, and show that $x_2 - \sqrt{5} < 10^{-4}$.

(c) Show that the next approximation x_3 is accurate to at least twelve decimal places. [Use part (a).]

1.7 VELOCITY AND OTHER APPLICATIONS

The generality and significance of differentiation can be seen only in the light of a wide variety of applications beyond the simple one we have stressed so far, the construction of tangents. To be sure, the tangent line provides most of our geometric understanding of derivatives, and suggests how to use them. But no one picture tells the whole story; the concepts we are about to discuss (velocity, acceleration, density, etc.) suggest the many applications of calculus, and should increase your intuitive understanding of derivatives.

The first two (velocity and acceleration) played an important role in Newton's simultaneous development of calculus and the theory of motion; we use them frequently to illustrate mathematical results. Some of the other topics can be postponed, if you wish, until they arise in problems and examples later on in the text.

Velocity

The main concept in the study of motion is *velocity*. Imagine an object moving along a line, and imagine the real numbers spread out along this line (as in §0.1) so that each position on the line is described by a real number. Two examples:

"Object"	Position numbers
Car on a road	Distance from one end of the road (as shown by mileposts)
Falling weight	Height above ground

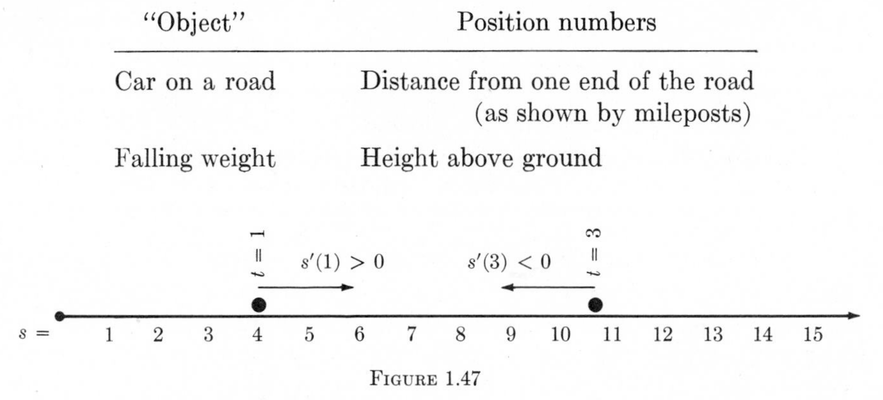

FIGURE 1.47

To be specific, imagine the line as horizontal [Fig. 47], and position measured as distance from the left end. At each time t, the position of the object is described by a number $s(t)$; notice that s will increase with motion to the right, and decrease with motion to the left. The velocity tells *how fast* the position is changing, and in which direction; velocity is positive for motion to the right, and negative for motion to the left [Fig. 47].

The precise mathematical definition of velocity is simple; *the velocity* $v(t)$ *at time* t *is the derivative of the position function,* $v(t) = s'(t)$. The reason for this definition can be seen by considering the *average velocity* over intervals of time. At time t_1 the object is at position $s(t_1)$, and at time t_2 it is at $s(t_2)$; the change of position is $s(t_2) - s(t_1)$; and the *average rate* of change of position per unit time (average velocity) is the change divided by the time elapsed,

$$\text{average velocity from time } t_1 \text{ to } t_2 \quad = \quad \frac{s(t_2) - s(t_1)}{t_2 - t_1}.$$

For example, if a car is at milepost 58 at 1 o'clock and at milepost 90 at 1:30, we have $s(1) = 58$, $s(1\frac{1}{2}) = 90$, and the average velocity over this interval is

$$\frac{s(1\frac{1}{2}) - s(1)}{1\frac{1}{2} - 1} = \frac{32 \text{ miles}}{\frac{1}{2} \text{ hour}} = 64 \text{ miles per hour.}$$

To find the *instantaneous velocity* at 1 o'clock, we would compute the average velocity over shorter and shorter intervals:

$$\frac{s(1\frac{1}{60}) - s(1)}{1\frac{1}{60} - 1} \qquad \text{(average for one minute)}$$

$$\frac{s(1\frac{1}{3600}) - s(1)}{1\frac{1}{3600} - 1} \qquad \text{(average for one second)}$$

etc.

The velocity *at* 1:00 o'clock is the *limit* of these averages; denoting the velocity at 1:00 o'clock by $v(1)$, we thus have

$$v(1) = \lim_{t \to 1} \frac{s(t) - s(1)}{t - 1}.$$

But this limit is precisely the definition of the derivative $s'(1)$, so $v(1) = s'(1)$.

Similar reasoning applies in the general case, and justifies our definition of velocity: For any position function s, *the velocity is the derivative of* s; for any time t_0,

$$v(t_0) = s'(t_0).$$

Example 1. A billiard ball is dropped beside a scale, and its position is recorded on film by a light flashing every 1/30 second, with the results shown in Fig. 48. Suppose we start measuring time when the ball is in the position marked $t = 0$. Then, as accurately as we can read the figure, we have $s(6/30) = 22$, $s(7/30) = 30$, $s(8/30) = 39$, and so on. Hence the average velocity from $t = 6/30$ to $t = 8/30$ is

$$\frac{s(8/30) - s(6/30)}{8/30 - 6/30} = \frac{17}{1/15} = 255,$$

and the average velocity from $t = 1/3$ to $t = 1/2$ is

$$\frac{s(1/2) - s(1/3)}{1/2 - 1/3} = \frac{131 - 60}{1/6} = 426.$$

We cannot compute the velocity itself at any particular time from a picture such as Fig. 48, but we obtain an approximation by taking the *average* velocity over short intervals. For instance, we can approximate the velocity $v(7/30) = s'(7/30)$ by taking the average velocities

$$\frac{s(8/30) - s(7/30)}{1/30} = 270$$

and

$$\frac{s(6/30) - s(7/30)}{0 - 1/30} = \frac{-8}{-1/30} = 240.$$

Then we expect that the velocity at time $t = 7/30$ lies somewhere between these two numbers, say $v(7/30)$ is about 255.

This is a little crude. To obtain better approximations, we would require measurements over shorter intervals of time; and as the intervals grew shorter and shorter, these approximations would tend to the velocity $v(7/30)$.

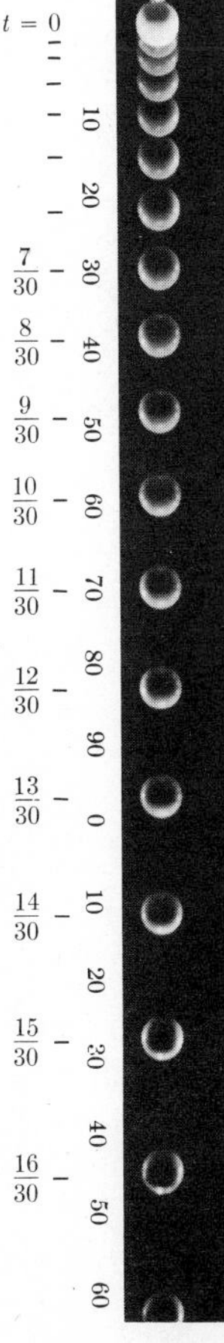

FIGURE 1.48*

From PSSC PHYSICS, D.C. Heath and Company, Boston, 1965

In applications, we think of the derivative as the "rate of change." The difference $s(t) - s(t_0)$ is a change of position. Dividing this by the time $t - t_0$ taken to reach the new position gives the *average rate of change* in s over the interval from t_0 to t. Taking the limit of this rate as the interval from t to t_0 decreases to zero, we find the rate of change at the time t_0.

Acceleration

When you push down the accelerator of a car, its velocity increases (if all goes well); and if the velocity increases very fast, the car is said to have good *acceleration*. In physics, acceleration is defined to be the *rate of change of the velocity*; i.e. if the velocity at time t is $v(t)$, then the acceleration at time t is $v'(t)$.

In the case of a freely falling object, we shall find that the velocity is a polynomial of the first degree, $v(t) = a + bt$. In this case, the acceleration is

$$v'(t_0) = \lim_{t \to t_0} \frac{v(t) - v(t_0)}{t - t_0} = \frac{a + bt - a - bt_0}{t - t_0} = \lim_{t \to t_0} b = b.$$

Acceleration enters into Newton's famous laws of motion. If a force F moves an object having a fixed mass m, then the acceleration v' is related to the force by $F = mv'$.

Example 2. Figure 49 shows a velocity function v. Sketch the graph of the corresponding acceleration function $a = v'$. *Solution.* From $t = 0$ to $t = 1$ the v curve is horizontal, so the slope v' is zero. From $t = 1$ to $t = 2$ the slope grows increasingly negative; at $t = 2$ we have sketched a tangent line, and calculated the derivative $v'(2)$ as the $\dfrac{\text{change in } v}{\text{change in } t}$ along this tangent line, which is approximately -14. As t increases from 2 to 3, the slope returns to zero. A graph with these features in sketched in Fig. 50.

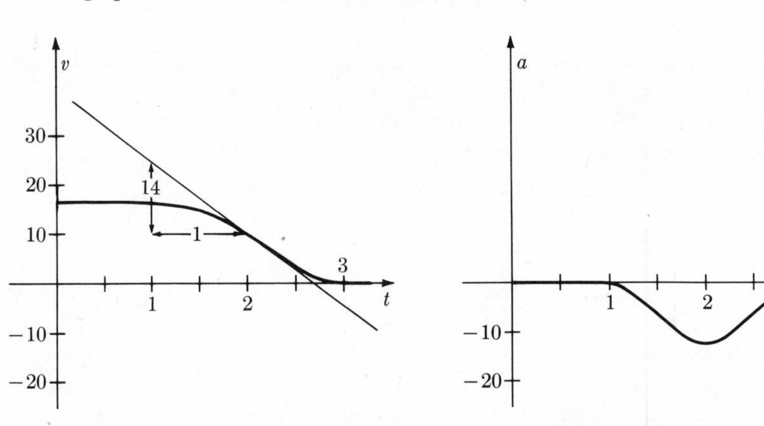

FIGURE 1.49 Graph of v

FIGURE 1.50 Graph of v'

Example 3. Figure 51 shows $v(t) = s'(t)$. Given $s(0) = -10$, sketch the graph of s for $0 \leq t \leq 4$. *Solution.* For $0 \leq t \leq 1$, $s' = 20$, so for these values s rises with a constant slope 20, as in Fig. 52. For $1 < t < 2$, the slope decreases to 0, as in the figure. Finally, for $2 < t < 3$, the velocity becomes *negative*; the slope of the graph of s is negative and s decreases. The object turns around at $t = 3$ and heads back toward its starting position.

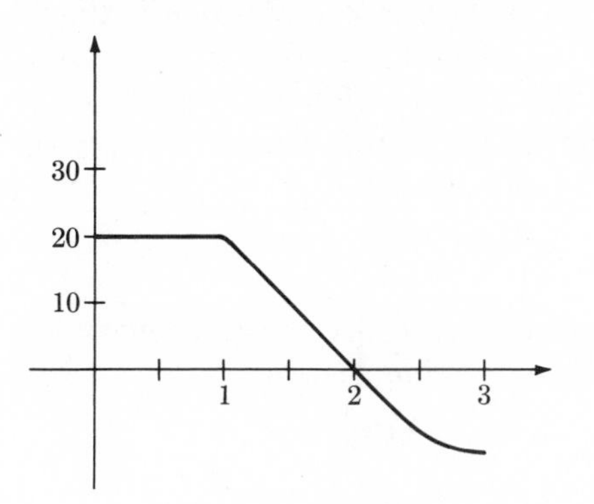

FIGURE 1.51 Graph of s'

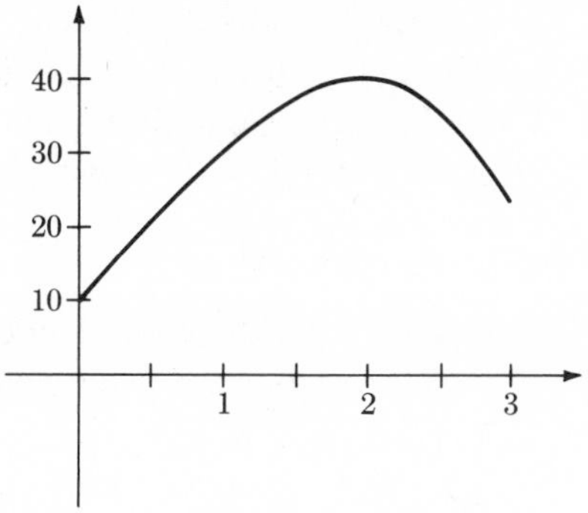

FIGURE 1.52 Graph of s

Density

Suppose that a straight rod lies along the x axis with its left end at the origin. Suppose the rod is not homogeneous, but that some parts are heavier per unit length than others. To find the linear density (i.e. the mass per unit length) of the rod at a point $x_0 > 0$, you could cut out a small piece of the rod (say the piece from x_0 to x), weigh it, and divide the mass by the length of the piece $x - x_0$. The smaller the piece, the more accurately it gives the density at the point x_0.

This process is directly related to differentiation. Let $M(x)$ be the mass of that part of the rod between 0 and x. Then $M(x) - M(x_0)$ is the mass of that part between x_0 and x [Fig. 53], and $\dfrac{M(x) - M(x_0)}{x - x_0}$ approximates the density at x_0. The approximation gets better when x is closer to x_0, and thus the density at x_0 is precisely the derivative

$$M'(x_0) = \lim_{x \to x_0} \frac{M(x) - M(x_0)}{x - x_0}.$$

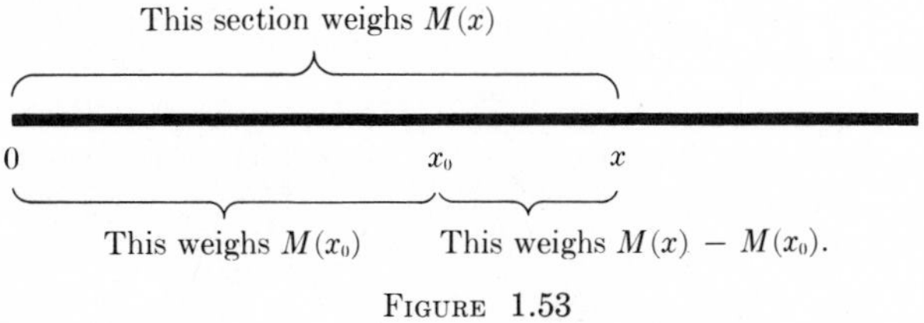

This section weighs $M(x)$

0 x_0 x

This weighs $M(x_0)$ This weighs $M(x) - M(x_0)$.

FIGURE 1.53

Current

Suppose water is flowing through a pipe. Denote by $Q(t)$ the quantity of water that has passed a certain point P from time 0 to time t. The rate of flow, or *current,* is then $Q'(t)$. To measure $Q'(t_0)$ approximately, we might collect all the water that passes P from time t_0 to time t, and divide by the time $t - t_0$ it takes to pass. The amount of water in question is simply $Q(t) - Q(t_0)$. When we divide by $t - t_0$ and take t closer and closer to t_0, the quotient tends to $Q'(t_0)$, which is thus the rate of flow at time t_0.

Exactly the same discussion applies to electric current if we replace "pipe" by "wire," and "water" by "electric charge."

Population growth

A function giving the number of objects in some collection over some interval of time is called a *population function.* Examples are:

$P(t)$ = the number of people on the earth at time t
$R(t)$ = ... rabbits on a farm ...
$D(t)$ = ... dollars in a savings account ...
$B(t)$ = ... bacteria in a culture ...
$E(t)$ = ... electrons in an electrical capacitor ...
$A(t)$ = ... radioactive atoms in a sample of radium ...

The rate of change of population functions is often given as an increase or decrease by a certain *percentage per unit time.* For example, $P(t)$, the number of people on the earth at time t, is currently increasing at the rate of 2 percent per year; the amount $A(t)$ of radioactive radium in a given sample decreases by 35 percent per millenium; when no deposits or withdrawals are made, $D(t)$ increases by a fixed percent per year, called the *true yearly interest rate.*

These rates are given as percentage increases rather than absolute increases because, at least in the short run, percentage rates are more nearly constant than absolute rates. In fact, in the case of the radium sample, the basic law of radioactive decay states that the percentage increase in $A(t)$ per unit time *really is* a constant. According to this law of decay, the percentage *decrease* in time h,

$$\frac{A(t_0) - A(t_0 + h)}{A(t_0)},$$

depends only on h, but not on t_0 or $A(t_0)$; thus the ratio is a function of h only,

$$\frac{A(t_0) - A(t_0 + h)}{A(t_0)} = f(h) \tag{1}$$

or, setting $t_0 + h = t$, $\quad \dfrac{A(t_0) - A(t)}{A(t_0)} = f(t - t_0).$

This leads to a relation between $A(t)$ and its derivative, as follows:

$$\frac{A(t) - A(t_0)}{t - t_0} = \frac{A(t_0) - A(t)}{t_0 - t} = -\frac{f(t - t_0)}{t - t_0} A(t_0).$$

Letting t approach t_0, we get

$$A'(t_0) = \lim_{t \to t_0} \frac{A(t) - A(t_0)}{t - t_0} = -\lim_{t \to t_0} \frac{f(t - t_0)}{t - t_0} A(t_0)$$

$$= -kA(t_0), \tag{2}$$

where k denotes the number $\lim_{t \to t_0} \frac{f(t - t_0)}{t - t_0}$ or, since $t - t_0 = h$, $\lim_{h \to 0} (f(h)/h)$.

The relation we have just obtained,

$$A'(t) = -kA(t), \qquad k \text{ constant}, \tag{3}$$

is better than (1) since it describes the process of radioactive decay by a single constant k, rather than by a function $f(h)$.

Remark 1. We will show in Chapter III how to reverse this derivation, and determine the function $f(h)$ in (1) from the constant k in (3).

Remark 2. There is a serious objection to taking the limit as $t \to t_0$ in (2). In a sufficiently small time interval, the change in population is likely to be either 1 or 0, so that $\frac{A(t) - A(t_0)}{t - t_0}$ is either a very large number or zero, and does not tend to any number as the interval is further shortened. Another way to pose the same problem is to imagine the graph of A, as in Fig. 54. The tangent lines to this graph are all either horizontal or vertical—not very revealing in any case. To take the derivative in such a situation requires a simplifying assumption whereby the actual graph of A is replaced by a smoothed-out version, represented by the light line in Fig. 54.

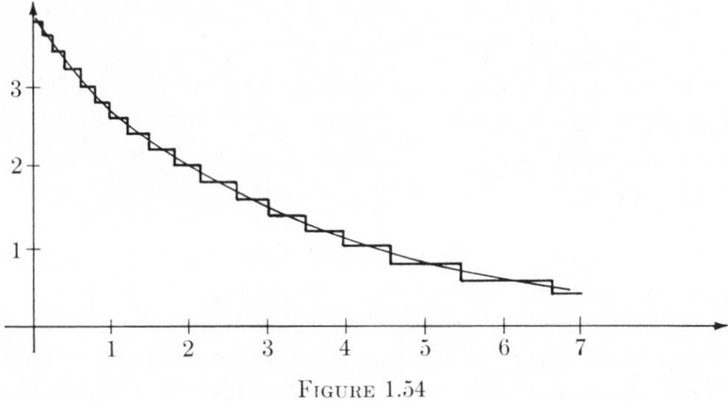

FIGURE 1.54

With relevant simplifying assumptions, the same sort of law applies to the growth of populations such as people or bacteria, for in a given time interval a certain number of births per thousand population and a certain number of deaths occur. Hence, in the short run, the growth of a population P is proportional to its size. We arrive as before at the law

$$P'(t) = kP(t),$$

where k may be either positive or negative, according as the birth rate or the death rate is higher. This law enters into the forecasts of population growth, but, as Malthus pointed out, it is not valid in the long run. For as the size of the population changes, its relation to the environment changes, sometimes in a quite disastrous way. When the population becomes very large in relation to its environment, the rates of birth and death are affected, and the "law" $P'(t) = kP(t)$ is "broken."

Marginal Revenue

Consider a sales operation in which the quantities to be measured are the *number of items sold*, the *cost of producing them*, and the *revenue* (the amount of money collected for that number of items). Let the revenue from the sale of x items be $R(x)$. (It is difficult to determine the function R, for the number of items sold is somehow related to the price per item—but let us suppose the function R is given.) Then the derivative R' is called the *marginal revenue*. If the number of items sold is quite large, we might think of R' as the increase in revenue from the sale of one more item. In the same way, if we suppose the cost $C(x)$ of producing x items is given, then C' is called the *marginal cost*.

As long as R' is greater than C', the profit can be increased by producing (and selling) more items; for $R' > C'$ means simply that a slight increase in the number of items produced and sold causes a larger increase in revenue than in costs. If $R' < C'$, fewer items should be produced. When $R' = C'$, we can hope that the profit $R - C$ has been maximized.

The objection to taking derivatives that was raised in the discussion of population growth applies here even more strongly. The rebuttal is the same: The procedure requires a simplifying assumption that is reasonable if a large number of items is sold.

Actually, the same problem arises with current, for electrical current consists of a (large) number of electrons, and a current of water consists of a (large) number of water molecules. The problem is less obvious in these cases because the number of items is extremely large.

Temperature gradient

Imagine a metal rod of uniform composition and cross section, insulated along its length, and arranged so that the temperature can be measured at various points along the rod [Fig. 55]. Suppose the rod lies along the

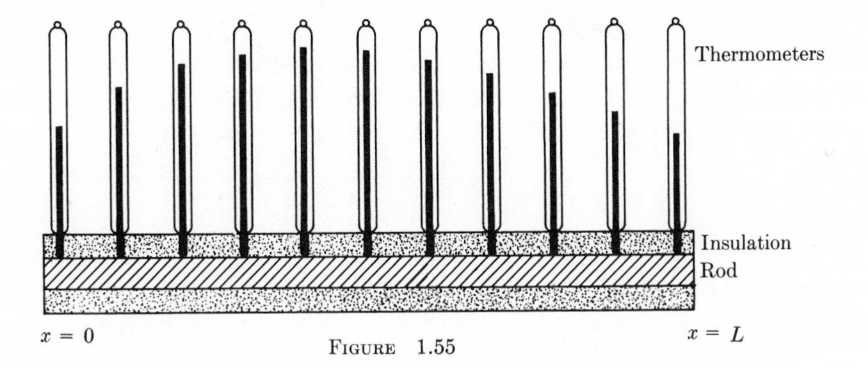

FIGURE 1.55

interval $[0,L]$ on the x axis, and that the temperature at x is $T(x)$. Then the derivative $T'(x)$ is called the *temperature gradient.* An experiment described in elementary physics texts shows that if the temperature gradient T' is *constant*, then the amount of heat in calories per second passing through the rod from left to right is

$$KS\frac{T(0) - T(L)}{L},$$

where S is the cross-sectional area of the rod and K is the thermal conductivity.

If T' is not constant, then a complete picture of the heat flow in the rod requires more information than the flow in and out at the ends. In this case, we ask at what rate and in what direction does heat flow through an arbitrary section of the rod, say at $x = x_0$. You could imagine how to measure this by considering a very short piece of the rod from x_0 to x. The gradient in this short piece does not vary much, and the section should behave more or less as though the gradient were constant. Then through the short section from x_0 to x, the rate of heat flow in calories per second is approximately

$$KS\frac{T(x_0) - T(x)}{x - x_0} = -KS\frac{T(x) - T(x_0)}{x - x_0}.$$

The supposition of a constant gradient becomes more realistic as the section of rod becomes shorter; taking the limit as $x \to x_0$, we find the rate of flow through the section of the rod at $x = x_0$ to be

$$-KST'(x_0).$$

This is a basic law of heat conduction. The minus sign reflects the fact that heat flows from the hotter to the colder part. In Fig. 56 we have graphed T and indicated that heat flows to the left when $T' > 0$ and to the right when $T' < 0$; that is, heat flows "downhill."

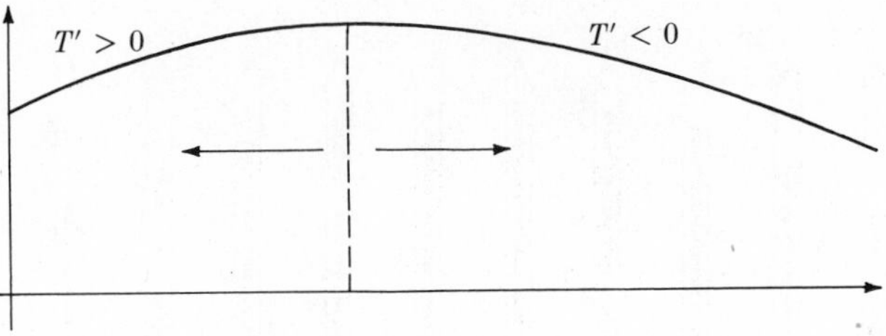

$T' > 0$ $T' < 0$

FIGURE 1.56

PROBLEMS

Velocity and acceleration

• **1.** Suppose that $s(t) = 16t^2 + 5t + 1$.
(a) Find the velocity $v(t) = s'(t)$. [Use a formula, or take the limit of the difference quotient.]
(b) From your answer in part (a), find the acceleration $a(t) = v'(t)$.

2. The table below gives the position $s(t)$ of a falling billiard ball at various times t. [See Fig. 48, p. 95.] Find the velocity $s'(2/5)$ as follows.

(a) Plot $\dfrac{s(t) - s(2/5)}{t - 2/5}$ for $t = 9/30, 10/30, 11/30, 13/30, 14/30, 15/30$.

(b) Sketch a curve through these points.

• (c) From the curve in (b), estimate the limit of the difference quotient

$$\frac{s(t) - s(2/5)}{t - 2/5} \text{ as } t \to 2/5.$$

t	9/30	10/30	11/30	12/30	13/30	14/30	15/30
$s(t)$	49	60.0	72	85	99	115	131.0

• **3.** The following table gives the position of an object moving in a straight line. Find $s'(3)$, approximately, by the method of Problem 2.

t	0	1	2	3	4	5	6	7	8
$s(t)$	10.0	9.9	9.2	8.3	7.0	5.4	3.6	1.7	0

• **4.** Figure 57 gives graphs of three position functions $s(t)$ and three velocity functions $v(t)$, but the velocity in each line does not necessarily match the position function in the same line. For each distance function in the first column, select the matching velocity function in the second column.

5. In Fig. 57, for each velocity function in the second column, select the matching acceleration function $a = v'$ in the third column.

6. Figure 58 gives four examples of velocity functions. Try to construct the corresponding position function s such that $s' = v$, by drawing a graph which has slope $v(t)$ at each point t. Begin the motion at the point $s(0)$ as follows:

Fig. 58(a): $s(0) = 0$ Fig. 58(c): $s(0) = -1$
Fig. 58(b): $s(0) = 1$ Fig. 58(d): $s(0) = 0$

7. Repeat Problem 6, with the following choices of $s(0)$:

Fig. 58(a): $s(0) = -1$ Fig. 58(c): $s(0) = -2$
Fig. 58(b): $s(0) = 2$ Fig. 58(d): $s(0) = 1$

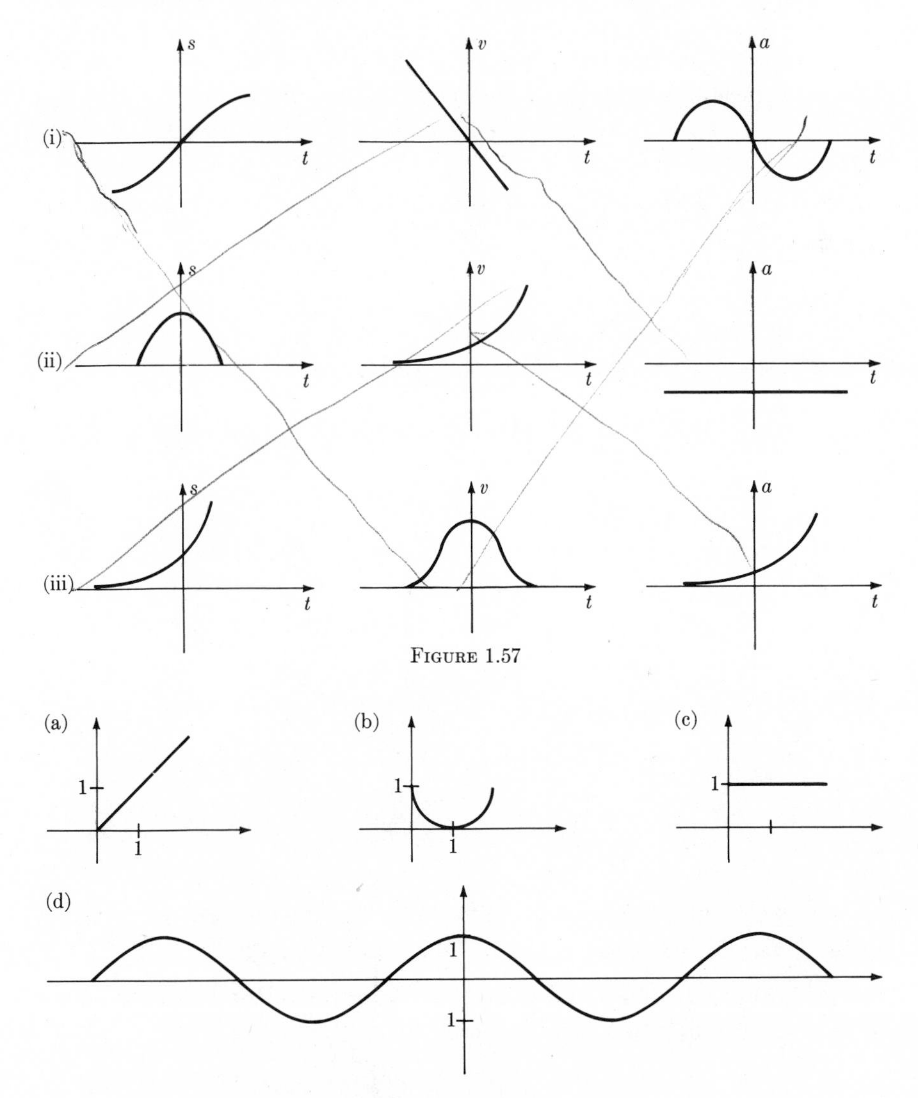

FIGURE 1.57

FIGURE 1.58 Graphs of $v = s'$

Density

• **8.** (a) A rod has mass 2 grams per centimeter, i.e. x centimeters of the rod weighs $2x$ grams. Using the definition of density given above, prove that the density of this rod is constant. [Hint: $M(x) = 2x$.]

(b) Suppose that a rod has mass D grams per centimeter, i.e. x centimeters weighs Dx grams, where D is a constant. Find the density of the rod, as in part (a).

9. A rod lies between $x = 0$ and $x = 1$, and the section in the interval $[0,x]$ has mass $M(x) = 5x - 2x^2$.

(a) Find the density at point x.

• (b) Which end of the rod is denser, $x = 0$ or $x = 1$?

Current

10. Figure 59 gives the graph of the electrical charge $Q(t)$ in a capacitor at time t.

(a) By sketching tangents to the graph, estimate the current $Q'(t)$ for $t = 0, 1, 2, 3$.

(b) Sketch a graph of the current.

• (c) Does the current $Q'(t)$ appear to be proportional to the charge $Q(t)$?

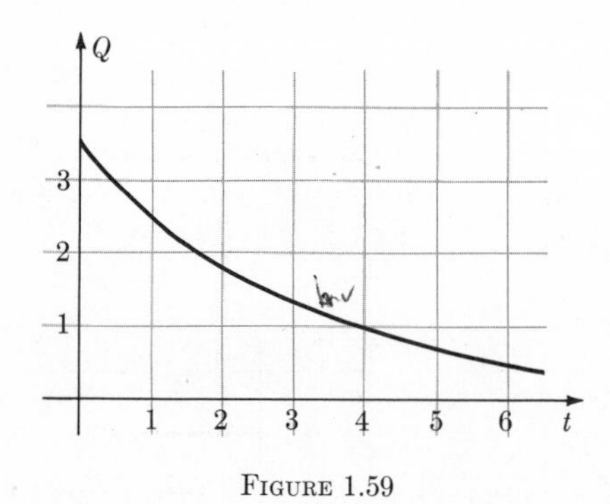

FIGURE 1.59

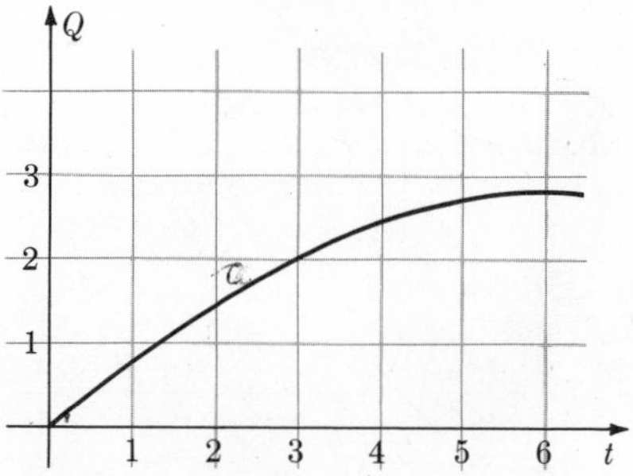

FIGURE 1.60

• **11.** Repeat Problem 10 for the graph in Fig. 60. [Note: Figure 59 is for a *discharging* capacitor, and Fig. 60 for a *charging* one.]

12. Each of the graphs in Figs. 59–62 satisfies *one* of the following equations. Which graph satisfies which equation? [The letters A and B stand for *positive* constants.]

(a) $Q' = AQ$

(b) $Q' = -AQ$

(c) $Q' + AQ = B$

(d) $Q' - AQ = B$.

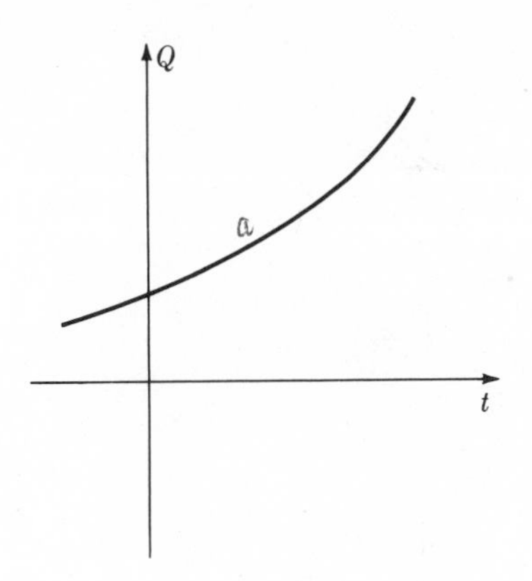

FIGURE 1.61

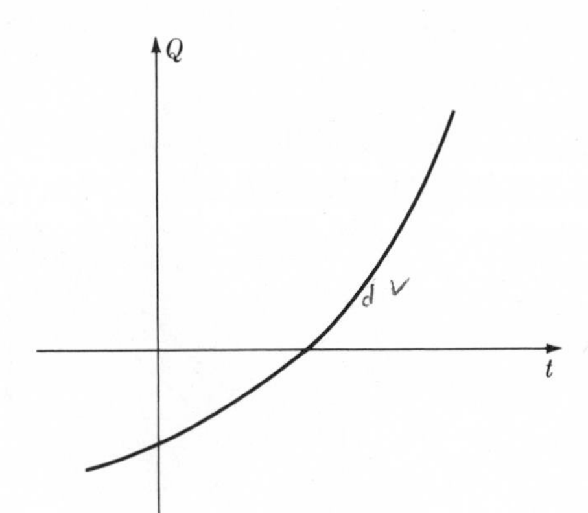

FIGURE 1.62

Population growth

13. The light line in Fig. 54, page 99, gives the graph of a function $A(t)$ obeying the law of radioactive decay, $A' = -kA$. Obtain three determinations of the constant k by measuring the slope A' and the height A at three points on the graph and solving $A' = -kA$ for k. Are these determinations in good agreement with each other?

14. The following table gives the number of bacteria $B(t)$ at time t, growing according to the law $B' = kB$.

(a) Estimate $B'(t)$ for several values of t by forming difference quotients (using the values in the table), and thus determine the constant k from the equation $B'(t) = kB(t)$.

(b) Sketch an accurate graph of B, determine graphically the slope $B'(t)$ for several values of t, and thus determine once more the constant k.

[You will know you are right if the various determinations of k are in reasonable agreement.]

Time	0	0.1	0.2	1.0	1.1	1.2	4.9	5.0	5.1
Bacteria	1000	1023	1047	1259	1288	1318	3000	3162	3236

Marginal revenue

15. Figure 63 gives the graph of the cost $C(x)$ of producing x items and the graph of the revenue $R(x)$ from selling x items. Copy the graphs, and indicate (a) when the operation is profitable and (b) when the profit is a maximum. Look for the maximum profit at first directly, then by using the condition $R' = C'$.

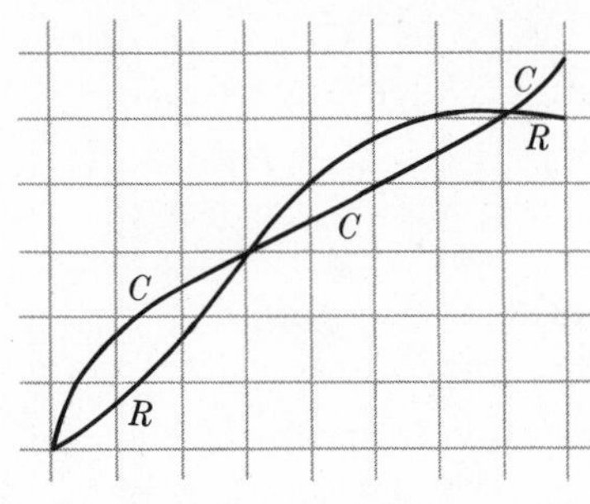

FIGURE 1.63

16. (a) From Fig. 63, construct a graph of the profit $P(x) = R(x) - C(x)$.
(b) On the graph in part (a), indicate the maximum profit and minimum profit. [Minimum profit = maximum loss.]
(c) Sketch a graph of the "marginal profit" $P'(x)$.

Temperature gradient

17. An insulated rod lies along the x axis from $x = 0$ to $x = L$; the temperature at x is $T(x) = x(L - x)$. Is heat flowing to the right or to the left at the point $x = L/4$? At $x = 3L/4$?

1.8 LEIBNIZ NOTATION

Leibniz, in developing his version of the calculus (about 1675), denoted derivatives by the symbol df/dx, rather than $f'(x)$. His notation comes from the difference quotient in the definition of derivative,

$$\lim_{x \to x_0} \frac{f(x) - f(x_0)}{x - x_0}.$$

The denominator here is the difference in two values of x; the numerator is the corresponding difference in the values of f; the limit of this quotient is denoted $\frac{df}{dx}(x_0)$, in Leibniz notation. For example,

$$\text{if } f(x) = 2x^2 + x, \quad \text{then } \frac{df}{dx}(x_0) = 4x_0 + 1. \tag{1}$$

There are many variations on this, chosen for convenience. For example, the derivative given in (1) can also be written

$$\frac{df}{dx} = 4x + 1 \qquad \text{or} \qquad \frac{d(2x^2 + x)}{dx} = 4x + 1.$$

Yet another version comes from the equation $y = f(x)$ for points (x,y) on the graph of f; given the formula for the graph,

$$y = 2x^2 + x, \qquad \text{we write} \qquad \frac{dy}{dx} = 4x + 1$$

for the derivative. These are all acceptable ways of saying that the derivative of the function defined by $f(x) = 2x^2 + x$ is given by $f'(x) = 4x + 1$.

Similarly,

$$\frac{d(3x^3 - 2x^2)}{dx} = 9x^2 - 4x; \qquad \frac{d(5t^4 - 4t)}{dt} = 20t^3 - 4;$$

$$\text{if} \quad z = 4x^3 - 2x^2, \quad \text{then} \quad \frac{dz}{dx} = 12x^2 - 4x.$$

Leibniz notation is particularly appropriate in applications. For example, let $s(t)$ denote position at time t; then the velocity at time t (the derivative of the position function) is denoted ds/dt, in Leibniz notation. This has the advantage of displaying the proper units; with s in meters and t in seconds, the velocity ds/dt is in meters/sec, as the notation suggests. Similarly, we have:

$$v(t) = \text{VELOCITY at time } t \text{ seconds} \left(\text{in } \frac{\text{meters}}{\text{sec}}\right),$$

$$\frac{dv}{dt} = \text{ACCELERATION} \left(\text{in } \frac{\text{meters}}{(\text{sec})^2}\right)$$

$$M(x) = \text{MASS of first } x \text{ meters of rod (in grams)},$$

$$\frac{dM}{dx} = \text{DENSITY} \left(\text{in } \frac{\text{grams}}{\text{meter}}\right)$$

$$R(x) = \text{REVENUE from selling } x \text{ items (in dollars)},$$

$$\frac{dR}{dx} = \text{MARGINAL REVENUE} \left(\text{in } \frac{\text{dollars}}{\text{item}}\right)$$

$$Q(t) = \text{CHARGE at time } t \text{ seconds (in coulombs)},$$

$$\frac{dQ}{dt} = \text{CURRENT} \left(\text{in } \frac{\text{coulombs}}{\text{sec}}, \text{ or amperes}\right).$$

Leibniz notation has mathematical advantages as well; these show up mainly in the systematic computing of derivatives [§2.4 below], but one minor advantage is apparent now; it is easier to write

$$\frac{d(z^3 + 5z^2 + 1)}{dz} = 3z^2 + 10z$$

than to write

$$\text{if } f(z) = z^3 + 5z^2 + 1, \quad \text{then } f'(z) = 3z^2 + 10z.$$

PROBLEMS

1. Compute the following.

•(a) $y = 3x^2 + 1, \quad dy/dx = ?$
•(b) $z = 5y^2 + 2y - 1, \quad dz/dy = ?$
•(c) $s = 2t, \quad ds/dt = ?$
•(d) $t = \frac{1}{2}s, \quad dt/ds = ?$
(e) $x = 3y^2 - 5y, \quad dx/dy = ?$

2. The area of a circle of radius r is given by $A = \pi r^2$, and the circumference by $C = 2\pi r$. Show that $dA/dr = C$. Can you explain this geometrically, by considering difference quotients?

3. The volume of a ball of radius r is given by $V = \frac{4}{3}\pi r^3$, and the surface area of the ball by $S = 4\pi r^2$. Show that $dV/dr = S$. Can you explain this geometrically by considering difference quotients?

4. The volume of a cone of angle α and height h [Fig. 64] is $V = \frac{1}{3}\pi h^3(\tan \alpha)^2$, and the area of the base is $A = \pi(h \tan \alpha)^2$. Show that $dV/dh = A$. Can you explain this geometrically by considering difference quotients?

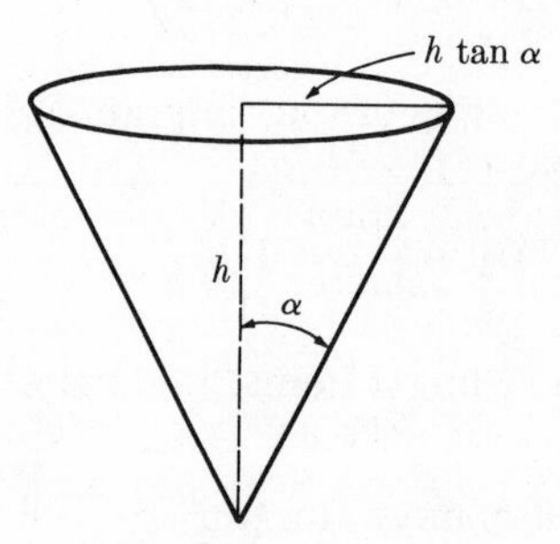

FIGURE 1.64

Computation of Derivatives

Chapter I sketched just a few of the many applications of differentiation. Before taking up more applications, you should learn how to compute derivatives efficiently.

This chapter shows how to differentiate the so-called *elementary functions,* which are built up by performing certain operations on four basic functions: the *constant* function 1, the *identity* function [the function f such that $f(x) = x$ for every x], the trigonometric *sine* function and the *logarithm* (defined below). The operations are: multiplying by a constant; adding, multiplying, or dividing two functions; taking inverse functions; and composing two functions. We have already seen examples of these operations, although we have not studied them all explicitly.

The rules for differentiation reflect precisely how the elementary functions are built up. We find the derivatives of the four basic functions; and, supposing the derivatives of certain functions are known, we show how to compute the derivatives of any function built up from these by performing the operations mentioned above.

The general formulas for derivatives are based primarily on the following facts about limits:

Proposition A. *Suppose that the limits* $\lim_{x\to x_0} F(x)$ *and* $\lim_{x\to x_0} G(x)$ *exist; and let c be any number. Then*

$$\lim_{x\to x_0} [cF(x)] = c \lim_{x\to x_0} F(x) \tag{1}$$

$$\lim_{x\to x_0} [F(x) + G(x)] = \lim_{x\to x_0} F(x) + \lim_{x\to x_0} G(x) \tag{2}$$

$$\lim_{x\to x_0} [F(x)G(x)] = \Big[\lim_{x\to x_0} F(x)\Big]\Big[\lim_{x\to x_0} G(x)\Big] \tag{3}$$

$$\lim_{x\to x_0} \frac{F(x)}{G(x)} = \frac{\lim_{x\to x_0} F(x)}{\lim_{x\to x_0} G(x)} \qquad \text{if } \lim_{x\to x_0} G(x) \neq 0. \tag{4}$$

The following simple examples illustrate these formulas:

$$\lim_{x\to x_0} 2x^2 = 2\left[\lim_{x\to x_0} x^2\right] = 2\cdot x_0^2 \qquad \text{[Formula (1)]}$$

$$\lim_{x\to x_0} (2x^2 + 3x) = \left[\lim_{x\to x_0} 2x^2\right] + \left[\lim_{x\to x_0} 3x\right] = 2x_0^2 + 3x_0 \qquad \text{[Formula (2)]}$$

$$\lim_{x\to 0} [(2x^2 + 1)(x + 2)] = \left[\lim_{x\to 0} (2x^2 + 1)\right]\cdot\left[\lim_{x\to 0} (x + 2)\right] = 1\cdot 2$$

$$\lim_{x\to 2}\left[\frac{2x^2 + 1}{x + 2}\right] = \frac{\lim_{x\to 2} (2x^2 + 1)}{\lim_{x\to 2} (x + 2)} = \frac{9}{4}$$

Formulas (2) and (3) extend easily to sums and products of three or more functions. Thus, for example,

$$\begin{aligned}\lim_{x\to x_0} \{[F(x)G(x)]H(x)\} &= \left\{\lim_{x\to x_0} [F(x)G(x)]\right\}\left\{\lim_{x\to x_0} H(x)\right\}\\ &= \left[\lim_{x\to x_0} F(x)\right]\left[\lim_{x\to x_0} G(x)\right]\left[\lim_{x\to x_0} H(x)\right].\end{aligned}$$

It is not hard to see why Proposition A is correct. Let A denote $\lim_{x\to x_0} F(x)$ and B denote $\lim_{x\to x_0} G(x)$. The statement $\lim_{x\to x_0} F(x) = A$ means essentially that when x is very close to x_0, then $F(x)$ is very close to A; similarly, $\lim_{x\to x_0} G(x) = B$ means that when x is very close to x_0, then $G(x)$ is very close to B. Thus, when x is very close to x_0, we must have

$cF(x)$ is very close to cA,

$F(x) + G(x)$ is very close to $A + B$,

$F(x)G(x)$ is very close to AB,

$F(x)/G(x)$ is very close to A/B $\quad (B \neq 0)$.

This is essentially what formulas (1)–(4) mean.

The actual proofs of (1)–(4) are precise "quantitative" versions of these slightly vague "qualitative" arguments. [See §AII.1.]

Formulas (1) and (2) bear a formal similarity to certain properties of straight lines. Consider the function $L(t) = mt$, whose graph is a straight line through the origin. Then L satisfies the equations

$$L(ct) = c\cdot L(t) \qquad [\text{since } m\cdot ct = c\cdot mt] \qquad (1')$$

$$L(t + s) = L(t) + L(s) \qquad [\text{since } m\cdot(t + s) = mt + ms]. \qquad (2')$$

Now, the properties (1) and (2) are formally the same as (1′) and (2′); simply replace $F(x)$ by t, $G(x)$ by s, and $\lim\limits_{x\to x_0}$ by L. Since (1′) and (2′) are the properties of a function whose graph is a straight line, the analogous relations (1) and (2) are called properties of *linearity*. On account of (1) and (2), taking limits is called a *linear operation*.

Linearity is so common and useful that it is occasionally invoked without any justification. It is tempting to think that

$$\frac{1}{x+y} \stackrel{?}{=} \frac{1}{x} + \frac{1}{y},$$

$$\cos(x+y) \stackrel{?}{=} \cos x + \cos y,$$

$$\log(a+b) \stackrel{?}{=} \log a + \log b,$$

and so on, although on reflection we realize that the correct formulas are somewhat different.

2.1 DERIVATIVES OF SUMS AND PRODUCTS

The constant function $u(x) = 1$ and the identity function $f(x) = x$ have been differentiated in Chapter I; the derivatives are $g'(x) = 0$, and $f'(x) = 1$. The formulas in this section allow you to deduce from these two particular results the derivative of any polynomial.

Multiplication by a constant

Let c be any number and f be any function, and define a new function cf by the rule $(cf)(x) = c \cdot f(x)$. For example, if $f(x) = x^2$, then the function $3f$ is defined by $(3f)(x) = 3x^2$.

Theorem 1. *If f has a derivative at some point x_0, then so does cf, and $(cf)'(x_0) = c \cdot f'(x_0)$. Briefly,*

$$(cf)' = c \cdot f'. \tag{1}$$

Example 1. Compute the derivative of the function $g(x) = 2x$. *Solution.* Take $c = 2$ and $f(x) = x$; then $g = cf$. Hence, by (1), $g' = (cf)' = c \cdot f' = 2$, since $c = 2$ and $f' = 1$.

Example 2. If $g(x) = ax$ for any constant a, then g is obtained from the function $f(x) = x$ by multiplying by the constant a; hence $g' = (af)' = a \cdot f' = a \cdot 1 = a$. This agrees with the fact that the graph of g is a line of slope a.

Formula (1) is easy to prove. To find $(cf)'(x_0)$, form the difference quotient

$$\frac{(cf)(x) - (cf)(x_0)}{x - x_0} = \frac{c \cdot f(x) - c \cdot f(x_0)}{x - x_0} = c \cdot \frac{f(x) - f(x_0)}{x - x_0}.$$

By hypothesis,

$$\lim_{x \to x_0} \frac{f(x) - f(x_0)}{x - x_0} = f'(x_0).$$

Now apply Proposition A(1) with $F(x) = \dfrac{f(x) - f(x_0)}{x - x_0}$; the result is

$$\lim_{x \to x_0} c \frac{f(x) - f(x_0)}{x - x_0} = c \lim_{x \to x_0} \frac{f(x) - f(x_0)}{x - x_0} = cf'(x_0). \quad Q.E.D.$$

Rule for sums

Let f and g be any functions, and define a new function $f + g$ by the rule $(f + g)(x) = f(x) + g(x)$. For example, if $f(x) = 3x^2$ and $g(x) = 2x$, then $(f + g)(x) = 3x^2 + 2x$.

Theorem 2. *If f and g have derivatives at some point x_0, then so does $f + g$, and $(f + g)'(x_0) = f'(x_0) + g'(x_0)$. Briefly, the derivative of a sum is the sum of the derivatives:*

$$(f + g)' = f' + g'. \tag{2}$$

Example 3. If $f(x) = ax^2$ and $g(x) = bx + c$, then $f + g$ is the quadratic $(f + g)(x) = ax^2 + bx + c$. Since $f'(x) = 2ax$ and $g'(x) = b$, $f + g$ has the derivative $(f + g)'(x) = f'(x) + g'(x) = 2ax + b$. [This agrees with the formula deduced above, in §1.2.]

Formula (2) is easy to prove. The difference quotient for $f + g$ is

$$\frac{(f + g)(x) - (f + g)(x_0)}{x - x_0} = \frac{[f(x) + g(x)] - [f(x_0) - g(x_0)]}{x - x_0}$$

$$= \frac{f(x) - f(x_0)}{x - x_0} + \frac{g(x) - g(x_0)}{x - x_0}.$$

Now apply Proposition A(2) with $F(x) = \dfrac{f(x) - f(x_0)}{x - x_0}$ and $G(x) =$

$\dfrac{g(x) - g(x_0)}{x - x_0}$. The result is

$$\lim_{x \to x_0} \left[\frac{f(x) - f(x_0)}{x - x_0} + \frac{g(x) - g(x_0)}{x - x_0} \right] = \lim_{x \to x_0} \frac{f(x) - f(x_0)}{x - x_0} + \lim_{x \to x_0} \frac{g(x) - g(x_0)}{x - x_0}$$

$$= f'(x_0) + g'(x_0). \qquad Q.E.D.$$

In view of (1) and (2), differentiation is called a *linear operation* [see page 111].

Product rule

Noting how closely the formulas for derivatives have followed those for limits, one could easily guess a rule for the derivative of a product: $(fg)' = f'g'$. Let's check this eminently reasonable guess, taking $f(x) = x$, $g(x) = x$. Then the product fg is defined by

$$(fg)(x) = f(x) \cdot g(x) = x \cdot x = x^2, \qquad \text{so} \qquad (fg)'(x) = 2x.$$

On the other hand, in the present case, $f'(x) = 1$ and $g'(x) = 1$, so $f'g' = 1$. Thus $f'g'$ is not the same as $(fg)'$, and the "obvious" formula is false!

Is there, then, a correct formula for the derivative of a product? To find out, we go back to the definition of derivative as the limit of a difference quotient. The difference quotient for fg is

$$\frac{(fg)(x) - (fg)(x_0)}{x - x_0} = \frac{f(x)g(x) - f(x_0)g(x_0)}{x - x_0}$$

[since $(fg)(x) = f(x)g(x)$ by definition], but it is not obvious how to relate this to the difference quotients for f and g. The interpretation of products as areas gives a clue. The large rectangle in Fig. 1 represents

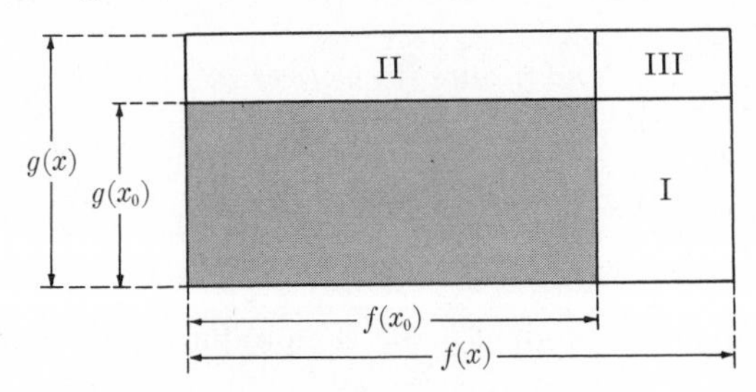

FIGURE 2.1

$f(x)g(x)$, and the shaded rectangle represents $f(x_0)g(x_0)$. The difference is the sum of rectangles I, II, and III, whose areas add up to

$$[f(x) - f(x_0)]\cdot g(x_0) + f(x_0)\cdot[g(x) - g(x_0)] + [f(x) - f(x_0)]\cdot[g(x) - g(x_0)].$$

Thus we can write the difference quotient for fg in terms of the difference quotients for f and g as follows:

$$\frac{(fg)(x) - (fg)(x_0)}{x - x_0} = \frac{f(x) - f(x_0)}{x - x_0}\, g(x_0) + f(x_0)\, \frac{g(x) - g(x_0)}{x - x_0} + (x - x_0)\, \frac{f(x) - f(x_0)}{x - x_0}\, \frac{g(x) - g(x_0)}{x - x_0}\, .$$

By Proposition A,

$$\begin{aligned}\lim_{x \to x_0} \frac{(fg)(x) - (fg)(x_0)}{x - x_0} &= \lim_{x \to x_0} \left[\frac{f(x) - f(x_0)}{x - x_0}\, g(x_0)\right] \\ &\quad + \lim_{x \to x_0} \left[f(x_0)\, \frac{g(x) - g(x_0)}{x - x_0}\right] \\ &\quad + \lim_{x \to x_0} \left[(x - x_0)\, \frac{f(x) - f(x_0)}{x - x_0}\, \frac{g(x) - g(x_0)}{x - x_0}\right] \\ &= \left[\lim_{x \to x_0} \frac{f(x) - f(x_0)}{x - x_0}\right] g(x_0) + f(x_0) \lim_{x \to x_0} \frac{g(x) - g(x_0)}{x - x_0} \\ &\quad + \left[\lim_{x \to x_0} (x - x_0)\right]\left[\lim_{x \to x_0} \frac{f(x) - f(x_0)}{x - x_0}\right]\left[\lim_{x \to x_0} \frac{g(x) - g(x_0)}{x - x_0}\right] \\ &= f'(x_0)g(x_0) + f(x_0)g'(x_0) + 0\cdot f'(x_0)\cdot g'(x_0).\end{aligned}$$

We have proved:

Theorem 3. *If f and g have derivatives at a point x_0, then fg has a derivative also, and*

$$(fg)'(x_0) = f'(x_0)\cdot g(x_0) + f(x_0)\cdot g'(x_0).$$

The derivative of a product is *not* the product of the derivatives; it is a sum, and in each term *one* of the original functions has been differentiated:

PRODUCT RULE

$$(fg)' = f'g + fg'. \tag{3}$$

Example 4. Let $f(x) = x$, $g(x) = x$. Then $(fg)(x) = x \cdot x = x^2$. By (3), $(fg)'(x) = f'(x) \cdot g(x) + f(x) \cdot g'(x) = 1 \cdot x + x \cdot 1 = 2x$.

Example 5. The function $h(x) = x^3$ can be written as the product of the two functions $f(x) = x$ and $g(x) = x^2$. Using the product rule (3) and Example 4, we find $h'(x) = f'(x)g(x) + f(x)g'(x) = 1 \cdot x^2 + x \cdot 2x = 3x^2$.

Example 6. [Do it yourself; solution below.]
(a) Write x^4 as the product $x^2 \cdot x^2$, and compute its derivative, using (3).
(b) Write x^4 as the product $x \cdot x^3$, and compute its derivative, using (3).

Example 7. Differentiate $(x^2 + 1)(x^2 + 2x + 3)$. *Solution.* Set $f(x) = x^2 + 1$, $g(x) = x^2 + 2x + 3$. Then we want the derivative $(fg)'$, which is

$$\begin{aligned}(fg)'(x) &= f'(x)g(x) + f(x)g'(x) \\ &= (2x)(x^2 + 2x + 3) + (x^2 + 1)(2x + 2).\end{aligned}$$

Sums and products of more than two functions

The rule for sums applies when more than two functions are involved. For example, a sum of three functions can be written as a sum of two functions, one of which is itself a sum:

$$f + g + h = (f + g) + h.$$

Applying the rule for sums *twice*, we have

$$[(f + g) + h]' = (f + g)' + h' = f' + g' + h'.$$

This can be extended to cover the sum of any number of functions; **the derivative of a sum is the sum of the derivatives:**

$$(f_1 + f_2 + \cdots + f_n)' = f_1' + f_2' + \cdots + f_n'.$$

[See Problem 9 below.]

There is a corresponding extension of the product rule to more than two factors. For three factors, we have $(fgh)' = f'gh + fg'h + fgh'$, and in general the derivative of the product of n functions is a sum of n terms, in each of which one of the n functions has been differentiated:

$$(f_1 f_2 \cdots f_n)' = (f_1' f_2 \cdots f_n) + (f_1 f_2' \cdots f_n) + \cdots + (f_1 \cdots f_{n-1} f_n').$$

[See Problem 10 below.]

Derivative of x^n

Examples 4–6 suggest that for any integer $n \geq 0$,

$$f(x) = x^n \quad \Rightarrow \quad f'(x) = nx^{n-1}; \tag{4}$$

to differentiate x^n, multiply by the exponent n, then decrease the exponent by 1.

Formula (4) is indeed correct, and the way to prove it is by induction. [See Appendix §I.2 if you are not familiar with mathematical induction.] The proof consists of two steps:

(i) Formula (4) is true when $n = 1$. For in this case, $f(x) = x^1 = x$, so $f'(x) = 1$, and thus indeed $f'(x) = nx^{n-1}$ [since $n = 1$].

(ii) If formula (4) is true for $n = m$, then it is true for $n = m + 1$. To prove this, we suppose that x^m has the derivative mx^{m-1} and try to find the derivative of x^{m+1}. This is done by the product rule (3):

$$x^{m+1} = x \cdot x^m = f \cdot g,$$

where $f(x) = x$ and $g(x) = x^m$, so the derivative is

$$f'g + fg' = 1 \cdot x^m + x \cdot mx^{m-1} = (m + 1)x^m.$$

Thus if (4) is true for $n = m$, it is also true for $n = m + 1$. This completes the proof by induction.

Combining formulas (1) and (4), we find that the derivative of cx^n is cnx^{n-1}. This, together with the rule for sums, gives the derivative of any polynomial:

$$f(x) = a_n x^n + a_{n-1}x^{n-1} + \cdots + a_1 x + a_0$$

$$\Rightarrow \quad f'(x) = na_n x^{n-1} + (n - 1)a_{n-1}x^{n-2} + \cdots + a_1 \,. \tag{5}$$

Thus, multiply each term by its exponent, and decrease all the exponents by 1.

Example 8. The derivative of $4x^3 - \frac{1}{3}x^2 + 3$ is

$$3 \cdot 4x^2 - 2 \cdot \tfrac{1}{3}x + 0 = 12x^2 - \tfrac{2}{3}x.$$

The derivative of $11x^{15} + \frac{1}{3}x^2 + x + 1$ is

$$15 \cdot 11x^{14} + 2 \cdot \tfrac{1}{3}x + 1 + 0 = 165x^{14} + \tfrac{2}{3}x + 1.$$

Example 9. To take the derivative of $(x^2 + 1)(x^2 - 1)$, let $f(x) = x^2 + 1$ and $g(x) = x^2 - 1$. Then

$$(fg)' = f'g + fg' = (2x + 0) \cdot (x^2 - 1) + (x^2 + 1) \cdot (2x - 0) = 4x^3.$$

You can check this by multiplying out the original expression: the product is $x^4 - 1$, which has the derivative $4x^3$.

Example 10. [Do it yourself; solution below.] Find the derivative of $f(x) = \frac{1}{4}x^4 + x^3$ and of $g(x) = x^5 - 1$.

Example 11. [Do it yourself; solution below.] Find the derivative of $(1 + x)^2$ first by multiplying out, and again by applying the product rule to $(1 + x)(1 + x)$. Check that the results agree.

Rule for powers

If f is any function and n is any positive integer, then f^n is defined by

$$f^n(x) = [f(x)]^n.$$

This is a familiar convention in trigonometry, where $\sin^2 x$ means $(\sin x)^2$.
For positive integer powers we have:

Theorem 4. ***The derivative of*** f^n ***is*** $nf^{n-1} \cdot f'$.

DERIVATIVE OF A POWER

More precisely, if $f'(x_0)$ exists, then f^n has a derivative at x_0 given by $(f^n)'(x_0) = nf^{n-1}(x_0)f'(x_0)$. There are three or four ways to prove this; Problem 11 below asks you to find one of them.

Example 12. Find the derivative of $(1 + x^2)^n$. *Solution.* Theorem 4 fits this, with $f(x) = 1 + x^2$:

$$(f^n)' = \underbrace{n}\ \underbrace{f^{n-1}}\ \underbrace{f'}$$

so

$$((1 + x^2)^n)' = n\ (1 + x^2)^{n-1}(2x).$$

Example 13. Differentiate $(x^2 + x + 1)^3$. *Solution.* You could multiply out, and differentiate the result as a polynomial. However, it is much easier like this:

$$(f^n)' = \underbrace{n}\ \underbrace{f^{n-1}}\ \underbrace{f'}$$

so

$$\frac{d((x^2 + x + 1)^3)}{dx} = 3(x^2 + x + 1)^2(2x + 1).$$

Example 14. Differentiate $[(x^2 - 1)^3 + x]^3$. *Solution.* Apply Theorem 4 twice:

$$\frac{d([(x^2 - 1)^3 + x]^3)}{dx} = 3[(x^2 - 1)^3 + x]^2 \cdot \frac{d[(x^2 - 1)^3 + x]}{dx}$$

$$= 3[(x^2 - 1)^3 + x]^2 \cdot [3(x^2 - 1)^2 2x + 1].$$

In the first line we applied Theorem 4 with $f(x) = (x^2 - 1)^3 + x$ and $n = 2$; in the second line, we applied it with $f(x) = x^2 - 1$ and $n = 3$.

Example 15. Find dy/dx, if $y = (5x^3 + 3)(2x^2 + 1)^2$.

Solution. $$\frac{dy}{dx} = \frac{d(5x^3+3)}{dx}(2x^2+1)^2 + (5x^3+3)\frac{d(2x^2+1)^2}{dx}$$
$$= 15x^2(2x^2+1)^2 + (5x^3+3)2(2x^2+1)^1(4x).$$

Solutions

Example 6 (a) $$\frac{d(x^2 \cdot x^2)}{dx} = (2x \cdot x^2) + (x^2 \cdot 2x) = 4x^3$$

(b) $$\frac{d(x \cdot x^3)}{dx} = 1 \cdot x^3 + x \cdot 3x^2 = 4x^3$$

Example 10 $$\frac{d(\frac{1}{4}x^4 + x^3)}{dx} = \tfrac{1}{4} \cdot 4x^3 + 3x^2 = x^3 + 3x^2$$

$$\frac{d(x^5 - 1)}{dx} = 5x^4 - 0 = 5x^4$$

Example 11 $$\frac{d[(1+x)^2]}{dx} = (1 + 2x + x^2)' = 2 + 2x = 2(1+x)$$

$$\frac{d[(1+x)(1+x)]}{dx} = (0+1)(1+x) + (1+x)(0+1) = 2(1+x).$$

PROBLEMS

Problem 5 gives a preview of the next section. Problems 6 and 7 concern "double roots" of polynomials.

1. Find the derivatives of the following polynomials.

• (a) $5x^3 - 3x^5 + 1$
• (b) $4x^2 - 8x$
(c) $\frac{x^5}{5} - \frac{x^4}{4} + \frac{x^3}{3} - \frac{x^2}{2} + x - \pi$
• (d) $8t^4 - 4t^2 - 8$
(e) $96t - \frac{1}{3}t^3$
(f) $256 + 96t - 16t^2$
• (g) $\frac{1}{5}(x^3 - 9x^2) - \frac{2}{3}(x+1)$

2. Find the derivatives of the given functions in each of two ways: (i) apply the product rule directly and (ii) multiply out and then differentiate, using (5). Check that the results agree.

(a) $(x-1)(x+1)$
(b) $(x-1)(x^2+1)$
(c) $(x^2+x)(2x^2+3x+1)$
(d) $(x+2)^2(x-\pi)$
(e) $(x-1)(x-2)(x-3)$
(f) $(x-1)(x+2)(x-3)(x+4)$

3. If $f(x) = \frac{1}{3}x^3 - \frac{1}{2}x^2 - 2x$, find the following sets.

• (a) $\{x : f'(x) = 0\}$
(b) $\{t : f'(t) = -3\}$
(c) $\{z : f'(z) = 10\}$
• (d) $\{s : f'(s) > 0\}$

4. Find dy/dx.

(a) $y = (x^2 + 1)^3$
(b) $y = (5x^4 + 3x^2 + 1)^2$
(c) $y = (x + 1)^n$
(d) $y = [(x^2 + 1)^2 + 2]^3$
(e) $y = [(x + 1)^2 + 1]^2$
(f) $y = ([(x^3 + 1)^2 + 2]^3 + 3)^4$

5. Suppose that f is a quotient with numerator N and denominator D, that is, $f(x) = N(x)/D(x)$. Suppose further that N, D, and f are all differentiable at the point x_0.

(a) From $N = fD$, deduce a formula for $N'(x_0)$.

(b) Supposing $D(x_0) \neq 0$, deduce that

$$f'(x_0) = \frac{D(x_0)N'(x_0) - N(x_0)D'(x_0)}{D(x_0)^2}.$$

(c) Compute f' from part (b), when $f(x) = \dfrac{x^2 - 1}{x - 1}$, $x \neq 1$.

(d) Compute f' directly (noting that $f(x) = x + 1$, $x \neq 1$), and show that the result is consistent with part (c).

6. Let $P(x) = (x - r)(x - s)$.

(a) Show that if $r \neq s$, then $P(r) = P(s) = 0$, but $P'(r) \neq 0$ and $P'(s) \neq 0$.

(b) Show that if $r = s$, then $P(r) = 0$ and $P'(r) = 0$. [The numbers r and s are called the *roots* of the polynomial P; they are the solutions of $P(x) = 0$. If $r = s$, then r is called a *double root*. The problem shows that r is a double root iff $P(r) = 0$ *and* $P'(r) = 0$. Thus, at a double root the graph of P is *tangent* to the x axis.]

7. Consider the polynomial

$$P(x) = (x - r_1)(x - r_2) \cdots (x - r_m),$$

with $r_1, \ldots, r_m$ some real numbers, called the roots of P.

(a) Show that if no two of the roots are equal, then $P(r_j) = 0$, but $P'(r_j) \neq 0$ for each j.

(b) Show that if $r_j = r_k$ and $k \neq j$, then $P'(r_j) = 0$ and $(x - r_j)$ is a factor of both P and P'.

8. Sketch the graph of x^3 on the interval $[-1,1]$, and then sketch the graph of $2x^3$ to the same scale. Sketch tangent lines to each graph at corresponding points (say at $x = \frac{1}{2}$ and $x = -\frac{1}{2}$). If your sketches are accurate, the tangent to $2x^3$ will have twice the slope of the tangent to x^3, in accord with formula (1).

9. Prove by induction that $\left(\sum_1^n f_j\right)' = \sum_1^n f_j'$. [See §AI.2 on induction.]

10. (a) Prove $(fgh)' = f'gh + fg'h + fgh'$.

(b) Prove the corresponding rule for the product of n factors.

(c) Obtain the derivative of f^n by applying the result in (b).

11. This problem proves that $(f^n)' = nf^{n-1}\cdot f'$, by induction.
(a) Check the formula for $n = 1$. [In this formula, we take $[f(x)]^0 = 1$, even when $f(x) = 0$.]
(b) Check it for $n = 2$, by writing $f^2 = f\cdot f$ and applying the product rule for $(fg)'$ with $g = f$.
(c) Check it for $n = 3$, by writing $f^3 = f\cdot f^2$ and applying the product rule with $g = f^2$.
(d) Supposing that $(f^m)' = mf^{m-1}\cdot f'$, prove that

$$(f^{m+1})' = (m+1)f^m\cdot f'.$$

[This, together with part (a), completes a proof by induction that $(f^n)' = nf^{n-1}\cdot f'$ for *every* integer $n = 1, 2, \ldots$. See §AI.2.]

12. A point (x,y) lies on the straight line $\{(x,y): x + y = 1\}$. Under this restriction, find the maximum of:
(a) xy (c) x^3y (e) x^2y^3
(b) x^2y (d) x^3y^2 (f) x^my^n

2.2 THE DERIVATIVE OF A QUOTIENT

Problem 5(b) above gives the formula for the derivative of a quotient with numerator N and denominator D,

QUOTIENT RULE

$$\left(\frac{N}{D}\right)' = \frac{DN' - ND'}{D^2}. \tag{1}$$

The development of this formula in Problem 5 is not quite complete; we *assumed that* $(N/D)'$ *exists,* and then deduced what it must be. In the quotient rule, we assume only that N and D have derivatives at x_0, and that $D(x_0) \neq 0$; then we conclude that N/D *has* a derivative at x_0, and this derivative is given by (1).

Theorem 5. *If* $N'(x_0)$ *and* $D'(x_0)$ *exist, and* $D(x_0) \neq 0$, *then*

$$\left(\frac{N}{D}\right)'(x_0) = \frac{D(x_0)N'(x_0) - N(x_0)D'(x_0)}{D(x_0)^2}. \tag{2}$$

Proof. The difference quotient for N/D is

$$\frac{[N(x)/D(x)] - [N(x_0)/D(x_0)]}{x - x_0} = \frac{D(x_0)N(x) - N(x_0)D(x)}{(x - x_0)D(x)D(x_0)}.$$

The numerator here presents a difficulty like that in the product rule, and the resolution is similar. Rewrite the numerator as

$$D(x_0)N(x) - N(x_0)D(x) = D(x_0)[N(x) - N(x_0)] - N(x_0)[D(x) - D(x_0)],$$

so the difference quotient becomes

$$\frac{\dfrac{N(x)}{D(x)} - \dfrac{N(x_0)}{D(x_0)}}{x - x_0} = \frac{D(x_0)\dfrac{N(x) - N(x_0)}{x - x_0} - N(x_0)\dfrac{D(x) - D(x_0)}{x - x_0}}{D(x)D(x_0)}. \quad (3)$$

It seems clear that as $x \to x_0$, the right-hand side of (3) will have as its limit the right-hand side of (2). However, to prove this on the basis of Proposition A we need the following simple result:

Lemma 1. *If $D'(x_0)$ exists, then* $\lim_{x \to x_0} D(x) = D(x_0)$.

IMPORTANT FACT

Proof. We can write

$$D(x) = D(x_0) + [D(x) - D(x_0)]$$

$$= D(x_0) + (x - x_0)\frac{D(x) - D(x_0)}{x - x_0}, \quad \text{if } x \neq x_0 .$$

Hence by Proposition A

$$\lim_{x \to x_0} D(x) = \lim_{x \to x_0} D(x_0) + \lim_{x \to x_0} (x - x_0) \lim_{x \to x_0} \frac{D(x) - D(x_0)}{x - x_0}$$

$$= D(x_0) + 0 \cdot D'(x_0) = D(x_0). \qquad Q.E.D.$$

Returning to (3), we thus find that the denominator on the right has the limit

$$\lim_{x \to x_0} [D(x)D(x_0)] = D(x_0) \lim_{x \to x_0} D(x) = [D(x_0)]^2.$$

Since $D(x_0)^2 \neq 0$ by hypothesis, we can apply the rule for the limit of a quotient, $\lim (F/G) = (\lim F)/(\lim G)$, and thus obtain from (3)

$$\lim_{x \to x_0} \frac{\dfrac{N(x)}{D(x)} - \dfrac{N(x_0)}{D(x_0)}}{x - x_0} = \frac{D(x_0) \lim_{x \to x_0} \dfrac{N(x) - N(x_0)}{x - x_0} - (Nx_0) \lim_{x \to x_0} \dfrac{D(x) - D(x_0)}{x - x_0}}{\lim_{x \to x_0} [D(x)D(x_0)]}$$

$$= \frac{D(x_0)N'(x_0) - N(x_0)D'(x_0)}{D(x_0)^2}. \qquad Q.E.D.$$

Any function of the form N/D, where N and D are *polynomials*, is called a *rational function*, because it is the ratio of two polynomials. The rules for sums and products yield an easy formula for the derivative of any polynomial; combining this with the quotient rule, we can differentiate any rational function.

Example 1. Compute the derivative of $f(x) = \dfrac{x}{x+1}$.

Solution. We have $N(x) = x$ and $D(x) = x + 1$, so

$$f' = (N/D)' = \frac{DN' - ND'}{D^2} = \frac{(x+1)\cdot 1 - x(1+0)}{(x+1)^2} = \frac{1}{(x+1)^2}.$$

Example 2. If $f(x) = (x^3 + 10x^2 + 1)/(x^2 + 1)$, then

$$f'(x) = \frac{DN' - ND'}{D^2} = \frac{(x^2+1)(3x^2+20x) - (x^3+10x^2+1)(2x)}{(x^2+1)^2}.$$

Recall the formula

$$(f^n)' = nf^{n-1}f' \tag{4}$$

for the derivative of the nth power of a function f. With the quotient rule we can extend this formula to the case where n is a *negative* integer, as follows:

(i) Write $f^n = \dfrac{1}{f^{-n}}$.

(ii) By the quotient rule,

$$\left(\frac{1}{f^{-n}}\right)' = \frac{0\cdot f^{-n} - 1\cdot (f^{-n})'}{(f^{-n})^2}. \tag{5}$$

(iii) Since n is a negative integer, $-n$ is a positive integer; and since (4) is true for positive integers, $(f^{-n})' = -nf^{-n-1}f'$.

(iv) Substitute this expression for $(f^{-n})'$ in the right-hand side of (5), obtaining

$$(f^n)' = \left(\frac{1}{f^{-n}}\right)' = \frac{(-1)(-nf^{-n-1})f'}{f^{-2n}} = nf^{n-1}f'.$$

Thus (4) remains true when n is a negative integer. (Of course, it is *not* true at points where $f = 0$, for there, $f^{-1}, f^{-2}, \ldots$ are not defined.)

Example 3. If $f(x) = x$, then $x^n = f^n(x)$ has the derivative $nf^{n-1}(x)f'(x) = nx^{n-1}$. For instance, the derivative of x^{-1} is $-x^{-2}$, that of x^{-2} is $-2x^{-3}$, that of x^{-3} is $-3x^{-4}$, and so on.

Example 4.

If $g(x) = (x^4 + 2x + 1)^{-3}$, then $g'(x) = (-3)(x^4 + 2x + 1)^{-4}(4x^3 + 2)$,
For $g(x) = f(x)^n$, so $g'(x) = n \cdot f(x)^{n-1} \cdot f'(x)$.

Example 5. Let $h(x) = (2x^2 + 3)(x^2 + x + 1)^{-4}$. This has the form of a product $h = fg$; thus

$$h' = [\ f\ ]\cdot[\ g'\ ] + [\ f'\ ]\cdot[\ g\ ]$$

$$= [2x^2 + 3]\,[(-4)(x^2 + x + 1)^{-5}(2x + 1)] + [4x + 0]\cdot[(x^2 + x + 1)^{-4}].$$

Remark 1. The most common difficulty with the quotient rule is to remember which term is plus and which is minus. You can keep this straight by memorizing (1) verbally, like a nonsense rhyme:

"N over D prime equals DN' minus ND' over D^2."

A second way to keep the signs straight is to use (4) and the product rule:

$$\left(\frac{N}{D}\right)' = (ND^{-1})' = N'D^{-1} + N(D^{-1})' \qquad \text{[product rule]}$$

$$= N'D^{-1} + N(-1)D^{-2}D' \qquad [(f^{-1})' = (-1)f^{-2}f']$$

$$= N'/D - ND'/D^2 = \frac{DN' - ND'}{D^2}.$$

From this point of view, the minus before the D' comes from differentiating D^{-1}, which "brings down" the exponent -1.

Example 6. Differentiate $f(x) = \dfrac{x+1}{x-1}$ by the product rule.

Solution. $f(x) = (x + 1)(x - 1)^{-1}$,

$$f'(x) = 1\cdot(x - 1)^{-1} + (x + 1)\cdot(-1)(x - 1)^{-2} = \frac{-2}{(x - 1)^2}.$$

Remark 2. Lemma 1 above deduced the relation

$$\lim_{x\to x_0} D(x) = D(x_0) \tag{6}$$

merely as a step in evaluating the limit of (3). Relation (6) turns out to be crucial in proving many fundamental results of calculus. Any function D which satisfies (6) is said to be *continuous at* x_0. Intuitively, D is continuous at x_0 if the graph of D runs in a continuous (but not necessarily straight or smooth) curve through the point $(x_0, D(x_0))$; see Fig. 2.

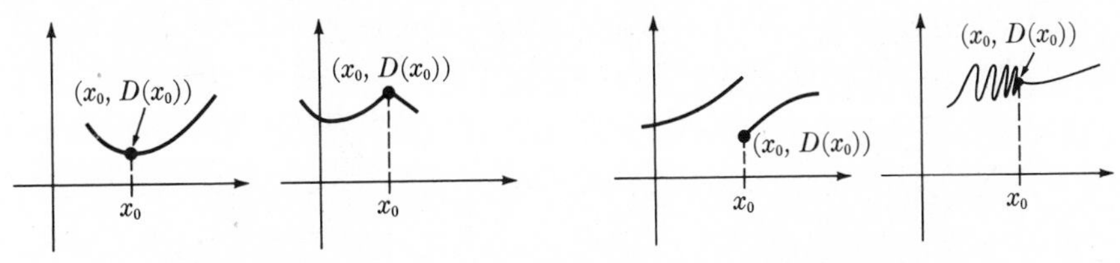

Continuous at x_0 Discontinuous at x_0

FIGURE 2.2

PROBLEMS

[Problems 7 and 8 prepare for the differentiation of the trigonometric functions.]

• **1.** Compute the derivatives of the following rational functions.

(a) $f(x) = \dfrac{1}{x+1}$

(b) $f(x) = \dfrac{x}{x+1}$

(c) $f(t) = \dfrac{t^2}{t^2+1}$

(d) $f(z) = \dfrac{z^3-1}{z-1}$

(e) $f(x) = \dfrac{x^2+2x+3}{x^4+x^2+1}$

(f) $f(t) = \dfrac{at+b}{ct+d}$ (for $t \neq -d/c$)

(g) $f(y) = y^{-1} + y^{-2} + y^{-3}$ (for $y \neq 0$)

(h) $f(s) = \dfrac{s}{s-1} - \dfrac{1}{s-1}$ (for $s \neq 1$)

(i) $f(a) = \dfrac{at+b}{(a+1)^2}$ (for $a \neq -1$; here t and b are some fixed numbers)

2. Compute dy/dx:
- •(a) $y = (x^2+1)^3$
- •(b) $y = 2(x^2+1)^{-3}$
- •(c) $y = 3(10x^5+5x^4-2)^{-4}$
- (d) $y = 4(8x^5+7x^{-2})^{-1}$
- (e) $y = 5(7x^{-5}+8x^2)^{-4}$
- (f) $y = (x^3+x^2+1)^{-1000}$

3. Compute the derivatives of the following.
- •(a) $f(x) = (x^2-1)(x^2+1)^{-3}$
- •(b) $f(x) = (3x^2+2x)^3(2x^2+2x+2)^{-5}$
- •(c) $\varphi(u) = (u^2-3)^2(u^2+3)^{-1}(u+1)^{-2}$
- (d) $h(w) = (w+1)/(w-1) + (w+2)(w+3)^2(w+4)^{-3}$
- (e) $F(t) = \dfrac{(t+1)(t+2)}{t+3}$
- (f) $G(s) = \dfrac{(s^2+1)^3}{(s-1)^4}(s^2+2)$

4. Recall that $\{(x,y): y = f(x)\}$ is the graph of f, and $y = f(x)$ is called the *equation of the graph*. Find the slope at an arbitrary point $(x_0, f(x_0))$ on the graphs with the following equations.
- •(a) $y = (x^2+1)^4$
- •(b) $y = (x+1)^2(x^2+1)^{-4}$
- •(c) $y = [(x-1)/(x+1)]^2$
- •(d) $y = (x^3-x)^{-3}$
- (e) $y = (x^{-1}-x)^2$
- (f) $y = (x^{-2}-x)^{-2}$
- (g) $y = x^2(x-1)^{-1}$

★5. (a) Graph the function $f(x) = x/(1 + x^2)$ by locating the points where $f = 0$ and where $f' = 0$, and by locating the intervals where $f' > 0$ and where $f' < 0$.
(b) Do the same for $f(x) = x^2/(1 + x)$. In this case, locate also the points where f is not defined. Investigate the behavior of f and f' just to the right and just to the left of these points.

6. The formula $1 + x + x^2 + \cdots + x^n = \dfrac{1 - x^{n+1}}{1 - x}$ can be proved by induction, or by multiplying each side by $1 - x$. Assuming this result, differentiate each side of the formula to obtain an expression for $1 + 2x + 3x^2 + \cdots + nx^{n-1}$.

7. The definition of derivative can be given in the form

$$f'(x_0) = \lim_{h \to 0} \frac{f(x_0 + h) - f(x_0)}{h} \tag{7}$$

instead of the form we have used up to now,

$$f'(x_0) = \lim_{x \to x_0} \frac{f(x) - f(x_0)}{x - x_0}.$$

To see the connection, simply put $h = x - x_0$; then $x_0 + h = x$, and $h \to 0$ means the same as $x \to x_0$.

Actually, in the form (7), it is no longer necessary to put a subscript on the x; without any change of meaning, we can write

$$f'(x) = \lim_{h \to 0} \frac{f(x + h) - f(x)}{h}. \tag{8}$$

For example, if $f(x) = ax^2 + bx + c$, then computing $f'(x)$ by means of (8) gives

$$f'(x) = \lim_{h \to 0} \left[\frac{a(x + h)^2 + b(x + h) + c - ax^2 - bx - c}{h}\right]$$

$$= \lim_{h \to 0} \frac{2axh + ah^2 + bh}{h} = \lim_{h \to 0} [2ax + ah + b] = 2ax + b,$$

in agreement with our familiar formula.

Use (8) to compute the derivatives of the following functions.
(a) $f(x) = x^3$
(b) $g(x) = 1/x$
(c) $c(\theta) = 2\theta^2$
(d) $s(\theta) = 3/\theta^2$

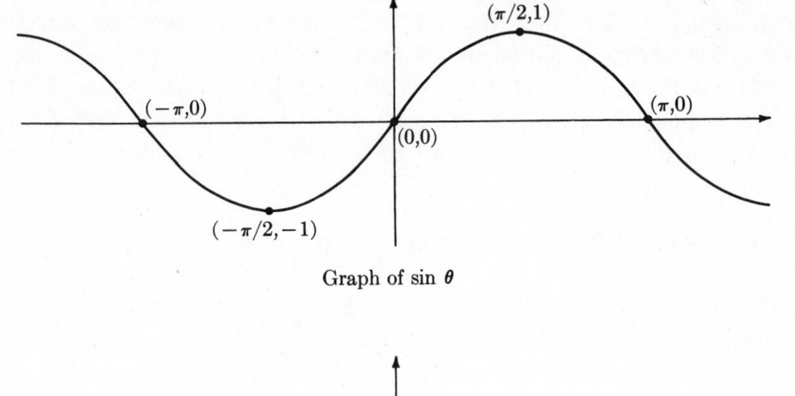

Graph of sin θ

(0,1)
(−π/2,0)
(π/2,0)
(−π,−1)
(π,−1)

Graph of cos θ

FIGURE 2.3

8. Figure 3 gives the graphs of two functions, the *trigonometric functions* sin (θ) and cos (θ), with θ in radians. By looking carefully at the graphs, you can see that *most of the formulas listed below must be false.* For each false formula, give the evidence showing why it is false.

(a) $\sin'(\theta) = 1 - \theta^2$
(b) $\sin'(\theta) = \cos(\theta)$
(c) $\sin'(\theta) = \sin(\theta)$
(d) $\sin'(\theta) = 2\cos(\theta)$
(e) $\sin'(\theta) = -\cos(\theta)$
(f) $\cos'(\theta) = -\theta$
(g) $\cos'(\theta) = \sin(\theta)$
(h) $\cos'(\theta) = \cos(\theta)$
(i) $\cos'(\theta) = 2\cos(\theta)$
(j) $\cos'(\theta) = -\sin(\theta)$

2.3 DERIVATIVES OF THE TRIGONOMETRIC FUNCTIONS

The familiar trigonometric functions (sine, cosine, tangent, etc.) are far more important than their simple geometric origins might suggest. They arise naturally in periodic phenomena (such as vibrations and crystal structure), and are pressed into service in almost every branch of mathematical analysis. So, it will not surprise you to hear that it is essential to know their derivatives.

In elementary trigonometry it is convenient to work with degrees because the number 360 has so many factors. In calculus, however, *radian measure* is universally employed, for reasons that will soon be clear. Thus we define $\sin\theta$ and $\cos\theta$ as suggested in Fig. 4. Draw the unit circle, the circle about the origin with radius 1. Given a real number θ, go θ units of arc length counterclockwise around the unit circle, starting at the point (1,0). Let (x,y) be the point obtained in this way [Fig. 4(a)]; then

$$x = \cos\theta \qquad \text{and} \qquad y = \sin\theta.$$

If θ is negative, use the sign convention; go $|\theta|$ units of arc length *clockwise* around the unit circle [Fig. 4(c)], arriving at a point (x,y); then again $x = \cos\theta$, $y = \sin\theta$.*

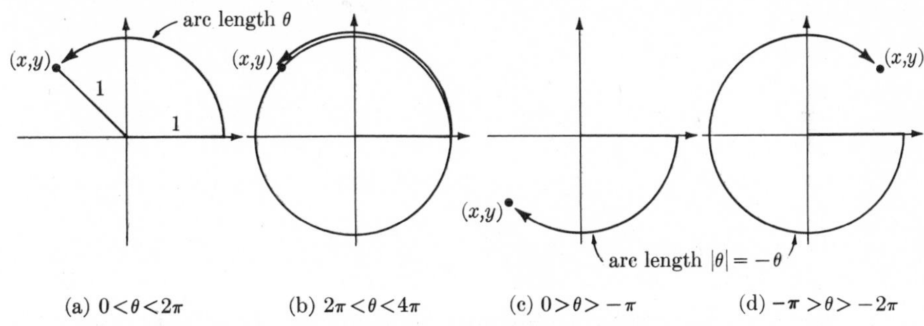

FIGURE 2.4 In each case, $x = \cos\theta$ and $y = \sin\theta$. In part (b), the path goes all the way around the circle once, and then continues on to the point (x,y).

Our prescription gives two functions, sin and cos, each defined for all real numbers θ. [Figure 3 shows the graphs.] The appendix to this section reviews trigonometry from this "unit circle" point of view, in case it is unfamiliar or forgotten. The most important items are the identity $\cos^2\theta + \sin^2\theta = 1$, and the addition formulas

ADDITION FORMULAS

$$\sin(\theta + \varphi) = \sin\theta\cos\varphi + \cos\theta\sin\varphi$$

$$\cos(\theta + \varphi) = \cos\theta\cos\varphi - \sin\theta\sin\varphi.$$

Our object here is to compute the derivatives of the trigonometric functions; we begin by finding $\sin'(0)$. Since $\sin(0) = 0$, the definition of derivative gives

$$\sin'(0) = \lim_{\theta\to 0}\frac{\sin\theta - \sin 0}{\theta - 0} = \lim_{\theta\to 0}\frac{\sin\theta}{\theta}.$$

---

* Nobody seems to know why "clockwise" is taken as negative and "counterclockwise" as positive; it's just a mildly confusing historical accident.

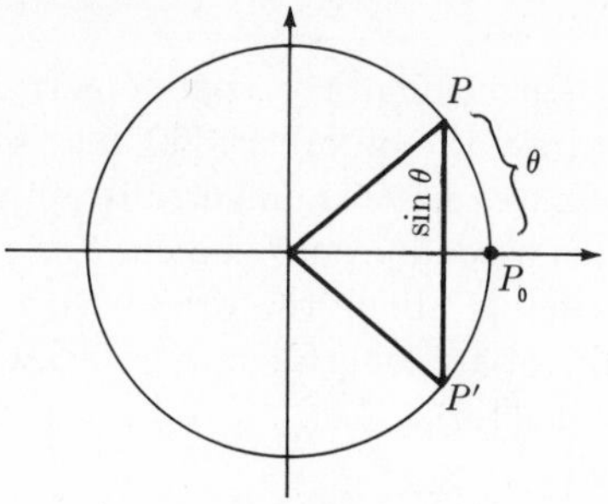

As Fig. 5 shows,

$$\frac{\sin\theta}{\theta} = \frac{2\sin\theta}{2\theta} = \frac{\text{chord } PP'}{\text{arc } PP'}.$$

FIGURE 2.5 Chord $PP' = 2\sin\theta$, arc $PP' = 2\theta$

A basic geometric principle asserts that as points P and P' on a given circle tend to a common point P_0, then the ratio of chord length to arc length tends to 1:

$$\frac{\text{chord } PP'}{\text{arc } PP'} \to 1 \quad \text{as } P \text{ and } P' \text{ tend to a common point } P_0.$$

Now, as $\theta \to 0$, clearly [Fig. 5] the two points $P = (\cos\theta, \sin\theta)$ and $P' = (\cos\theta, -\sin\theta)$ tend to the common point $P_0 = (1,0)$. Therefore,

$$\sin'(0) = \lim_{\theta\to 0} \frac{\sin\theta}{\theta} = \lim \frac{\text{chord } PP'}{\text{arc } PP'} = 1. \tag{1}$$

The derivative $\cos'(0)$ can now be computed by a little algebraic juggling:

$$\cos'(0) = \lim_{\theta\to 0} \frac{\cos\theta - \cos(0)}{\theta - 0} = \lim_{\theta\to 0} \frac{\cos\theta - 1}{\theta}$$

[because $\cos(0) = 1$]

$$= \lim_{\theta\to 0} \frac{\cos^2\theta - 1}{\theta(\cos\theta + 1)} = \lim_{\theta\to 0} \frac{-\sin^2\theta}{\theta(\cos\theta + 1)}$$

[because $\cos^2\theta + \sin^2\theta = 1$]

$$= -\left[\lim_{\theta\to 0} \sin\theta\right] \cdot \left[\lim_{\theta\to 0} \frac{\sin\theta}{\theta}\right] \cdot \left[\lim_{\theta\to 0} \frac{1}{\cos\theta + 1}\right]$$

[limit of a product]

so

$$\cos'(0) = 0 \times 1 \times \tfrac{1}{2} = 0. \tag{2}$$

In arriving at this last line we used formula (1) above, as well as the relations

$$\lim_{\theta\to 0} \sin\theta = 0, \qquad \lim_{\theta\to 0} \cos\theta = 1, \tag{3}$$

which are evident from the graphs, and also from Fig. 4.

Now we can compute $\sin'(\theta)$, for every real θ, by using the addition formula

$$\sin(\theta + \varphi) = \sin\theta\cos\varphi + \cos\theta\sin\varphi.$$

We find (using the notation of Problem 7 above)

$$\begin{aligned}\sin'(\theta) &= \lim_{h\to 0}\frac{\sin(\theta+h) - \sin\theta}{h} = \lim_{h\to 0}\frac{\sin\theta\cos h + \cos\theta\sin h - \sin\theta}{h}\\ &= \lim_{h\to 0}\sin\theta\,\frac{\cos h - 1}{h} + \lim_{h\to 0}\cos\theta\,\frac{\sin h}{h}\\ &= (\sin\theta)\cdot 0 + (\cos\theta)\cdot 1 \qquad [\text{by (2) and (1)}]\end{aligned}$$

so

$$\mathbf{\sin'(\theta) = \cos\theta\,.} \tag{4}$$

The derivative of cos can be deduced in exactly the same way as that of sin [Problem 3 below]; the result is

$$\mathbf{\cos'(\theta) = -\sin\theta\,.} \tag{5}$$

This agrees with the graph [Fig. 3]; at $\theta = 0$ the cos has zero slope; as θ goes from 0 to $\pi/2$ the slope *decreases* to $-1 = -\sin(\pi/2)$; as θ goes from $\pi/2$ to π, the slope returns to $0 = -\sin(\pi)$; and so on.

The derivative of tan follows from (4), (5), and the quotient rule:

$$\tan' = \left(\frac{\sin}{\cos}\right)' = \frac{\cos\cdot\sin' - \sin\cdot\cos'}{\cos^2} = \frac{\cos^2 + \sin^2}{\cos^2} = \frac{1}{\cos^2} = \sec^2.$$

Once again, the graph [Fig. 6] agrees with the formula; the graph of tan always slopes up, having a minimum slope of 1 at $\theta = 0$, and growing infinitely steep at $\theta = \pm\pi/2$, where $\sec^2\theta$ grows infinitely large.

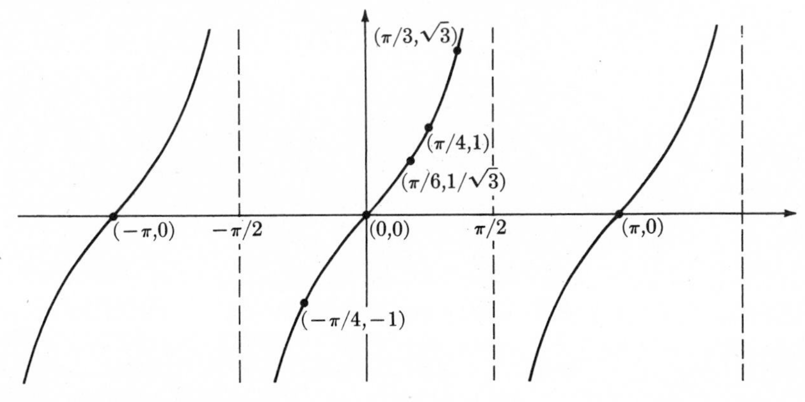

FIGURE 2.6 Graph of tan θ

We leave it to you [Problem 2 below] to compute the rest of the entries in the following table. Notice that the derivatives of all the "co" functions have minus signs.

DERIVATIVES OF TRIG FUNCTIONS

f	f'	$f(\theta)$	$f'(\theta)$
sin	cos	$\sin\theta$	$\cos\theta$
cos	$-\sin$	$\cos\theta$	$-\sin\theta$
tan	$\sec^2$	$\tan\theta$	$\sec^2\theta$
cot	$-\csc^2$	$\cot\theta$	$-\csc^2\theta$
sec	sec tan	$\sec\theta$	$\sec\theta\tan\theta$
csc	$-\csc\cot$	$\csc\theta$	$-\csc\theta\cot\theta$

Remark 1. The simplicity of formulas (4) and (5) explains why radian measure is right for calculus; any other angular measure would introduce an inconvenient numerical factor. For example, if $s(\alpha)$ denotes the sine of the angle of α degrees [so $s(90) = 1$, $s(30) = 1/2$, etc.] then $s'(\alpha) = (\pi/180)\cdot c(\alpha)$, where $c(\alpha)$ is the cosine of the angle of α degrees. [See Problem 8.] By using radian measure, we avoid this inconvenient factor of $\pi/180$.

Remark 2. The arguments here and in the appendix below are all based on pictures—in fact, our *definition* of the trigonometric functions is based on a picture, Fig. 4. Rigorous analytic definitions and proofs are not so elementary, and will be postponed to Chapter XI. The essential points which are explained geometrically here and in the appendix below [and will be proved analytically in Chapter XI] are summarized on the facing page.

PROBLEMS [continued on page 132]

1. Find the derivative of each of the following functions f.

•(a) $f(x) = \sin x \cos x$

•(b) $f(x) = x \sin x$

•(c) $f(t) = \dfrac{\sin t}{t}$

(d) $f(\theta) = \sin^2\theta - \cos^2\theta$

(e) $f(\varphi) = \dfrac{1}{\sin^2\varphi}$

(f) $f(x) = x + \sin x$

2. Compute the derivatives of cot, sec, and csc. [Recall cot = cos/sin, sec = 1/cos, csc = 1/sin.]

3. Prove that $\cos' = -\sin$. [Imitate the proof of (4), using

$$\cos(\theta + \varphi) = \cos\theta\cos\varphi - \sin\theta\sin\varphi.]$$

4. Compute the derivative of $f(x) = \cos^2 x + \sin^2 x$, using formulas (4) and (5), but *not* the identity $\cos^2\theta + \sin^2\theta = 1$. Show that the result is identically 0.

Proposition B. *There are two functions* sin *and* cos, *defined for all real numbers, with the following properties.*

(i) *Symmetries*

$$\sin(-\theta) = -\sin\theta, \qquad \cos(-\theta) = \cos\theta$$

(ii) *Addition formulas*

$$\sin(\theta + \varphi) = \sin\theta\cos\varphi + \cos\theta\sin\varphi$$

$$\cos(\theta + \varphi) = \cos\theta\cos\varphi - \sin\theta\sin\varphi$$

(iii) *Short table of values*

θ	0	$\pi/6$	$\pi/4$	$\pi/3$	$\pi/2$
$\sin\theta$	0	$1/2$	$1/\sqrt{2}$	$\sqrt{3}/2$	1
$\cos\theta$	1	$\sqrt{3}/2$	$1/\sqrt{2}$	$1/2$	0

(iv) *Periodicity*

$$\sin(\theta + 2\pi) = \sin\theta, \qquad \cos(\theta + 2\pi) = \cos\theta$$

(v) *The circle property*

Given any point (x,y) with $x^2 + y^2 = r^2 \neq 0$, there is one and only one number θ such that

$$0 \leq \theta < 2\pi \quad \text{and} \quad x/r = \cos\theta, \quad y/r = \sin\theta.$$

(vi) *Derivatives*

$$\sin' = \cos, \qquad \cos' = -\sin$$

The circle property (v) implies that

$$\cos^2\theta + \sin^2\theta = \frac{x^2}{r^2} + \frac{y^2}{r^2} = \frac{x^2 + y^2}{r^2} = 1,$$

for $0 \leq \theta < 2\pi$. By the periodicity (iv), $\cos^2\theta + \sin^2\theta = 1$ for *all* $\theta \geq 0$, and in view of the symmetry (i), it holds for *all real* θ. This basic formula, together with (i) and (ii), yields all the well-known trigonometric identities.

5. Sketch the graphs of $f(x) = \sqrt{2}\sin x$ and $g(x) = \sqrt{2}\cos x$ on the same set of axes. Show that they always intersect at right angles.

6. (a) Sketch the graphs of $\sin x$ and $\sin(2x)$ to the same scale.
(b) Notice that the graph of $\sin(2x)$ at any point x_0 looks "just like" the graph of $\sin x$ at the point $2x_0$, but twice as steep. Explain this.
(c) From (b), deduce that the derivative of $\sin(2x)$ at $x = x_0$ should be $2\cos(2x_0)$.
(d) Show, by taking difference quotients, that the derivative of $\sin(2x)$ at $x = x_0$ really is $2\cos(2x_0)$.

7. Repeat Problem 6, replacing 2 everywhere by 1/2.

8. Let $s(\alpha) = \sin\left(\frac{\pi}{180}\alpha\right)$, and $c(\alpha) = \cos\left(\frac{\pi}{180}\alpha\right)$.

(a) Explain why $s(\alpha)$ and $c(\alpha)$ are respectively the sine and the cosine of an angle of α degrees.

(b) Show that $s'(\alpha) = \frac{\pi}{180}c(\alpha)$ and $c'(\alpha) = -\frac{\pi}{180}s(\alpha)$. [Hint:

$$\frac{s(\alpha+h)-s(\alpha)}{h} = \frac{\pi}{180}\,\frac{\sin\left(\frac{\pi}{180}\alpha+\frac{\pi}{180}h\right)-\sin\left(\frac{\pi}{180}\alpha\right)}{\frac{\pi}{180}h}.]$$

9. Let $f(x) = x + \sin x$.
(a) Find the points where $f' = 0$.
(b) Show that at all other points $f' > 0$.
(c) Sketch the graph.

APPENDIX: THE TRIGONOMETRIC FUNCTIONS

The sine and cosine are introduced in geometry as quantities associated with angles. In calculus, on the other hand, it is more useful to think of them as functions assigning numbers to numbers. We define the functions as they are used in calculus, and later relate them with angles.

The definitions are based on the arc lengths of circular segments [Fig. 4, page 127].

Definition 1. Let θ be any real number. Starting at the point $(1,0)$, go a distance θ counterclockwise around the unit circle, arriving at a point (x,y) as shown in Fig. 4. Then $\sin\theta = y$, $\cos\theta = x$. Further, if $x \neq 0$, then $\tan\theta = y/x$ and $\sec\theta = 1/x$; if $y \neq 0$, then $\cot\theta = x/y$ and $\csc\theta = 1/y$.

A sign convention applies here: if θ is negative, then "a distance θ counterclockwise" means "a distance $|\theta|$ clockwise." [See the circles (c) and (d) in Fig. 4.]

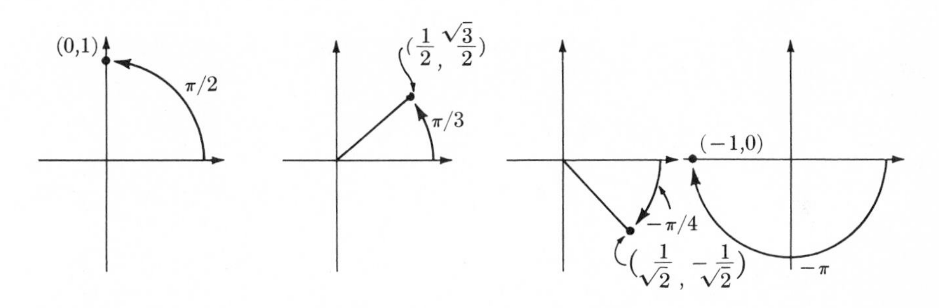

FIGURE 2.7

Figure 7 shows some particular values of $\sin\theta$ and $\cos\theta$, found in the following way. Start with the fact that a circle of radius 1 has circumference of length 2π, and hence a quarter of a circle has length $\frac{1}{4}(2\pi) = \pi/2$. Then notice that the point $(0,1)$ is one-fourth of the way around the circle, starting at $(1,0)$ [Fig. 7(a)]. Thus starting at $(1,0)$ and going a distance $\pi/2$ counterclockwise around the unit circle, you arrive at $(0,1)$, which means that

$$\cos\frac{\pi}{2} = 0, \qquad \sin\frac{\pi}{2} = 1. \tag{1}$$

Further, the point $(1/2, \sqrt{3}/2)$ is one-sixth of the way around the circle [Problem 2 below]. Since one-sixth of a circle has length $\frac{1}{6}(2\pi) = \pi/3$, it follows that $\sin(\pi/3) = \sqrt{3}/2$, as in Fig. 7(b). The other values given in Fig. 7 are found in the same way. The graphs of sin and cos [Fig. 3] can be sketched by plotting these and other similarly deduced values.

Although we have based our definition on the unit circle, it is possible (and sometimes useful) to use a circle about the origin of arbitrary radius $r > 0$ [Fig. 8]. Given θ, begin at the point $(r,0)$ and go counterclockwise through an arc length $r\theta$ along the circle of radius r, arriving at a point (x,y); then

$$\frac{x}{r} = \cos\theta, \qquad \frac{y}{r} = \sin\theta,$$

$$\frac{x}{y} = \tan\theta, \qquad \frac{y}{x} = \cot\theta,$$

$$\frac{r}{x} = \sec\theta, \qquad \frac{r}{y} = \csc\theta.$$

By the laws of similar triangles and the formula for circular arc lengths, these definitions give the same values of $\cos\theta$, $\sin\theta$, etc., for every value of r.

Connection with angles

The trigonometric functions as here defined are related to the trigonometric functions of angles through *radian measure*. Given an angle $\angle P_1P_2P_3$ [Fig. 9], extend both sides indefinitely, draw a circle of radius 1 with center at P_2, and let P_1' and P_3' be the points of intersection of this circle with the extended sides of the angle [Fig. 9]. Then the length θ of the shortest circular arc from P_1' to P_3' is the radian measure of the angle $\angle P_1P_2P_3$.

Now it is easy to see the connection between the functions defined in Definition 1, and the trigonometric functions of angles as they are given in plane geometry. Take a point $P = (x,y)$ on the unit circle, with $x \geq 0$ and $y \geq 0$, and draw the right triangle shown in Fig. 10. Then the arc length θ indicated in the figure is the radian measure of the angle $\angle POP_0$, and using the definition of sine as "opposite side over hypotenuse," we find

$$\text{sine } (\angle POP_0) = \frac{y}{1} = y = \sin\theta,$$

and similarly

$$\text{cosine } (\angle POP_0) = \frac{x}{1} = x = \cos\theta.$$

In other words, the sine and cosine of a given angle are respectively equal to $\sin\theta$ and $\cos\theta$, where θ is the *radian measure* of the given angle.

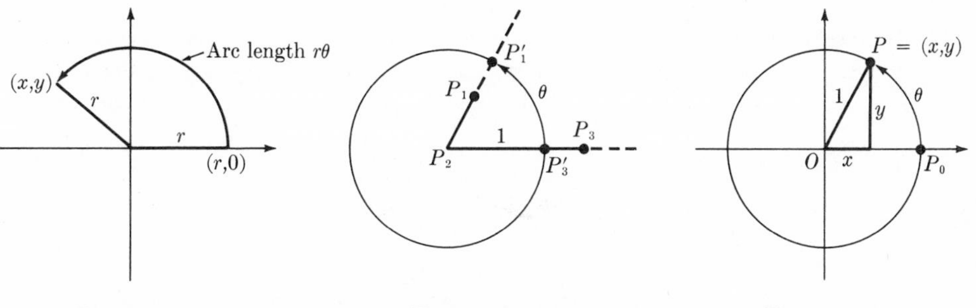

FIGURE 2.8 FIGURE 2.9 FIGURE 2.10

Trigonometric identities

By definition, $(\cos\theta, \sin\theta)$ is a point on the unit circle $\{(x,y):x^2 + y^2 = 1\}$, so we obtain immediately the identity

$$\cos^2\theta + \sin^2\theta = 1. \tag{2}$$

Figure 11 illustrates two other simple identities. Part (a) shows a point (x,y) on the unit circle and an arc length θ, related to (x,y) as in Definition 1; part (b) shows the corresponding situation when θ is replaced by $-\theta$. You can see that replacing θ by $-\theta$ does not change x, but it changes y to $-y$. Since $x = \cos\theta$ and $y = \sin\theta$, we find that

$$\begin{aligned} \cos(-\theta) &= \cos\theta, \\ \sin(-\theta) &= -\sin\theta. \end{aligned} \tag{3}$$

Thus cos is an *even* function, and sin an *odd* function [see Problem 9, §0.4].

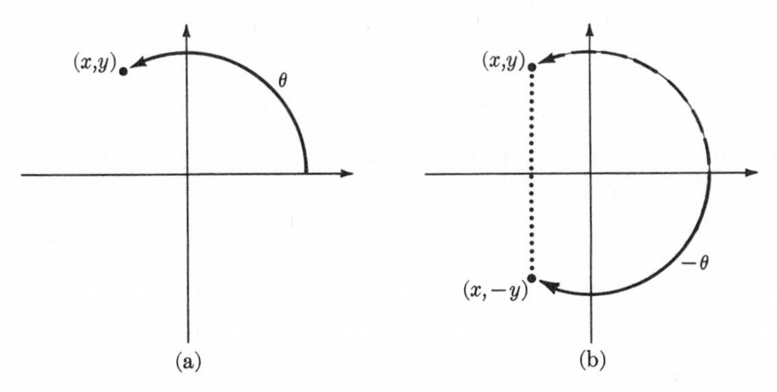

FIGURE 2.11

The *addition formulas* for cos and sin express $\cos(\alpha \pm \beta)$ and $\sin(\alpha \pm \beta)$ in terms of $\cos \alpha$, $\cos \beta$, $\sin \alpha$, and $\sin \beta$. These formulas can be derived as follows. In Fig. 12(a), we have $A = (\cos \alpha, \sin \alpha)$ and $B = (\cos \beta, \sin \beta)$. In Fig. 12(b), each point has been moved a distance α clockwise along the circle, so that A is moved to $A' = (1,0)$, and B is moved to $B' = (\cos(\beta - \alpha), \sin(\beta - \alpha))$. Now, the distance AB between A and B is the same as the distance $A'B'$ between A' and B', so

$$(AB)^2 = (A'B')^2.$$

Using the formula for the distance between two points [§0.3], this equation becomes

$$(\cos \beta - \cos \alpha)^2 + (\sin \beta - \sin \alpha)^2$$
$$= (\cos(\beta - \alpha) - 1)^2 + (\sin(\beta - \alpha) - 0)^2,$$

and then, multiplying out each side,

$$(\cos \beta)^2 - 2 \cos \beta \cos \alpha + (\cos \alpha)^2 + (\sin \beta)^2 - 2 \sin \beta \sin \alpha + (\sin \alpha)^2$$
$$= (\cos(\beta - \alpha))^2 - 2 \cos(\beta - \alpha) + 1 + (\sin(\beta - \alpha))^2.$$

Since $(\cos \theta)^2 + (\sin \theta)^2 = 1$, this last equation reduces to

$$\cos(\beta - \alpha) = \cos \beta \cos \alpha + \sin \beta \sin \alpha, \tag{4}$$

which is the *addition formula for the cosine.*

We obtain an important special case of the addition formula by setting $\beta = \pi/2$ and $\alpha = \theta$, and using equation (1). The result is

$$\cos\left(\frac{\pi}{2} - \theta\right) = \left(\cos \frac{\pi}{2}\right)(\cos \theta) + \left(\sin \frac{\pi}{2}\right)(\sin \theta) = \sin \theta. \tag{5}$$

Similarly, setting $\beta = \pi/2$ and $\alpha = (\pi/2 - \theta)$, we find

$$\cos \theta = \sin\left(\frac{\pi}{2} - \theta\right). \tag{6}$$

The number $(\pi/2) - \theta$ is called the *complement* of θ; the sum of θ and its complement equals $\pi/2$, a "right angle." Equation (5) says that $\sin \theta$ is the cos of the complement of θ, and (6) says that $\cos \theta$ is the sin of the complement of θ.

The remaining formulas for sums and differences all follow from (3) and (4). For example, setting $\alpha = -\theta$ in (4) and invoking (3) yields

$$\cos(\beta + \theta) = \cos(\beta - (-\theta)) = \cos \beta \cos(-\theta) + \sin \beta \sin(-\theta)$$
$$= \cos \beta \cos \theta - \sin \beta \sin \theta. \tag{7}$$

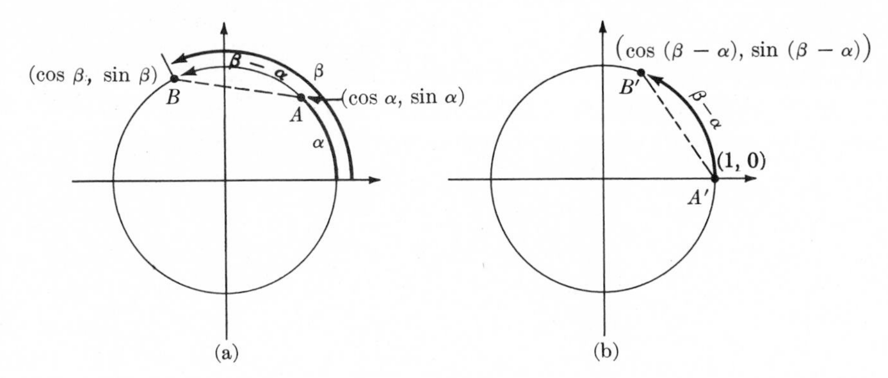

FIGURE 2.12

Problems 8 and 9 below ask you to derive the remaining two formulas:

$$\sin(\beta + \theta) = \sin\beta\cos\theta + \cos\beta\sin\theta \tag{8}$$

and

$$\sin(\beta - \theta) = \sin\beta\cos\theta - \cos\beta\sin\theta. \tag{9}$$

Now we can easily derive identities for the other trigonometric functions, tan, sec, and so on. In Fig. 4, we have $\cos\theta = x$, $\sin\theta = y$, and $\tan\theta = y/x$, so

$$\tan\theta = \frac{\sin\theta}{\cos\theta}, \quad \text{when } \cos\theta \neq 0.$$

In the same way, we find

$$\cot\theta = \frac{\cos\theta}{\sin\theta}, \qquad \sec\theta = \frac{1}{\cos\theta}, \qquad \csc\theta = \frac{1}{\sin\theta},$$

each formula being valid when the denominator is not zero. Next, dividing the formula $\cos^2\theta + \sin^2\theta = 1$ by $\cos^2\theta$, we obtain

$$1 + \tan^2\theta = \sec^2\theta.$$

For the tangent of $\beta - \alpha$ we have the formula

$$\tan(\beta - \alpha) = \frac{\sin(\beta - \alpha)}{\cos(\beta - \alpha)} = \frac{\sin\beta\cos\alpha - \cos\beta\sin\alpha}{\cos\beta\cos\alpha + \sin\beta\sin\alpha}.$$

Dividing both the numerator and denominator of the second fraction by $\cos\beta\cos\alpha$ yields an expression involving only tangents,

$$\tan(\beta - \alpha) = \frac{\tan\beta - \tan\alpha}{1 + \tan\beta\tan\alpha}. \tag{10}$$

These are the identities used in the text; the problems at the end of this section give other identities that you can easily derive.

Graphs of the trigonometric functions

The graph of sin can be sketched easily by plotting the points computed in the text and in Problems 1–4 below. The work is simplified by using the fact that $\sin(-\theta) = -\sin\theta$; hence the graph for $\theta < 0$ has the same shape as the graph for $\theta > 0$, reflected in the origin. It also helps to use the identity

$$\sin(\theta + 2\pi) = \sin\theta, \tag{11}$$

which follows from (8), since $\cos 2\pi = 1$ and $\sin 2\pi = 0$. Because of (11), sin is called *periodic*, with period 2π. This formula shows that the part of the graph on the interval $0 \leq \theta \leq 2\pi$ is repeated on each of the intervals $2\pi \leq \theta \leq 4\pi$, $4\pi \leq \theta \leq 6\pi$, ..., $-2\pi \leq \theta \leq 0$, $-4\pi \leq \theta \leq -2\pi$, etc. All this shows up in Fig. 13(a).

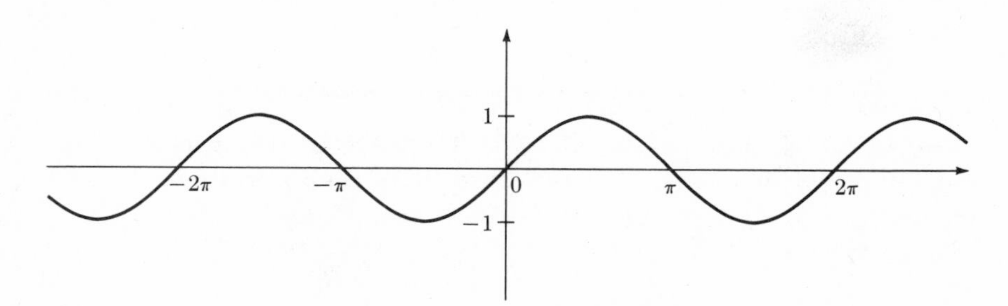

FIGURE 2.13(a) $\sin\theta$

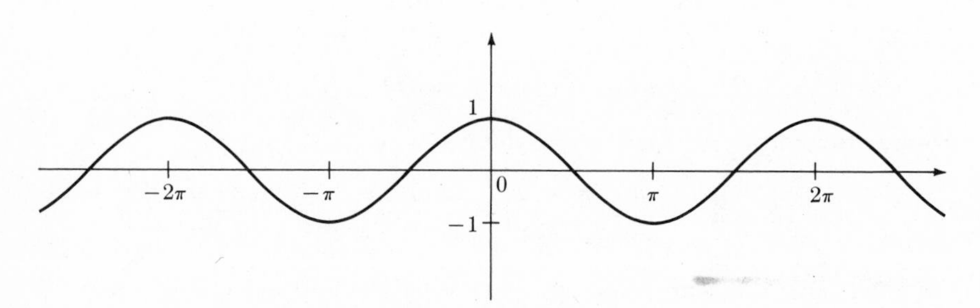

FIGURE 2.13(b) $\cos\theta$

Formula (11) reflects the fact that the circumference of the unit circle has arc length 2π [Fig. 14]. Suppose two ants, Ant_1 and Ant_2, start at the point (1,0) on the unit circle, and Ant_1 goes arc length θ, arriving at a point (x_1, y_1), while Ant_2 goes arc length $\theta + 2\pi$, arriving at a point (x_2, y_2). By going the extra length 2π, Ant_2 has taken exactly one extra turn around the circle, so he must arrive at the same point. Thus $(x_1, y_1) = (x_2, y_2)$, and it follows that

$$\sin(\theta) = y_1 = y_2 = \sin(\theta + 2\pi).$$

The same argument shows, of course, that $\cos(\theta) = x_1 = x_2 = \cos(\theta + 2\pi)$. Thus cos, like sin, has period 2π, and the graph of cos consists of an endless repetition of its shape on the interval $[0,2\pi]$.

Notice that cos has a *different* symmetry from sin; the graph of cos is symmetric about the y axis, not about the origin [Fig. 13(b)].

The function

$$\tan\theta = \frac{\sin\theta}{\cos\theta} \tag{12}$$

is not defined when $\cos\theta = 0$, i.e. when $\theta = \pi/2, 3\pi/2, 5\pi/2, \ldots, -\pi/2, -3\pi/2, -5\pi/2, \ldots$. Thus, we look for vertical asymptotes at these points. Consider a number θ slightly less than $\pi/2$; from Fig. 13, $\sin\theta$ is nearly 1 and $\cos\theta$ is a very small positive number, so $\tan\theta = \sin\theta/\cos\theta$ is a huge positive number. Similarly, when θ is slightly larger than $\pi/2$, then $\tan\theta$ is a huge negative number. Hence at $\theta = \pi/2$ the graph of tan has a vertical asymptote of type $+\infty$, $-\infty$, as in Fig. 6 above. There is a vertical asymptote of the same type at each of the other points where $\tan\theta$ is not defined.

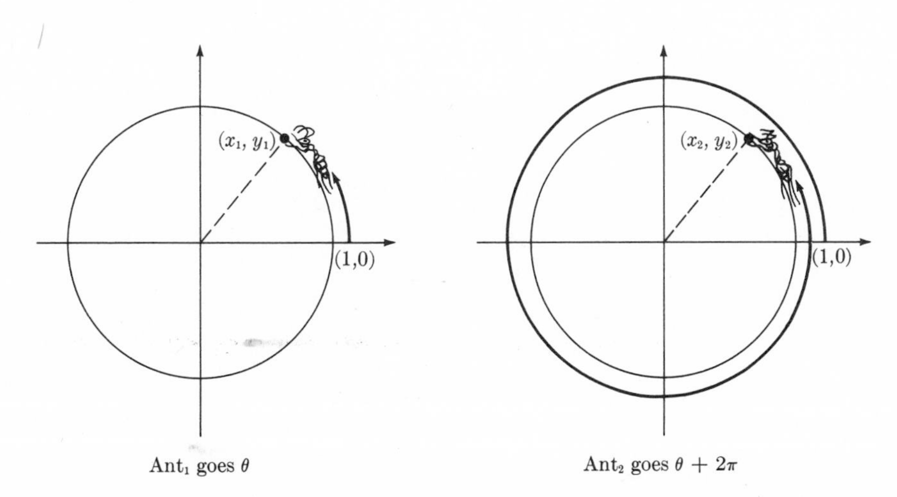

FIGURE 2.14 Two Ants

PROBLEMS

Problems 4, 6, 7, and 11–14 give formulas used elsewhere in the text.

1. From a sketch like Fig. 7, evaluate $\sin \pi$, $\cos \pi$, $\sin (3\pi/2)$, $\sin (-\pi)$, $\cos (-\pi)$, $\sin (-\pi/2)$, $\cos (-\pi/2)$.

2. Show that the three line segments inside the semicircle in Fig. 15(a) are all equal [and hence divide the semicircle into three equal arcs of length $\pi/3$].

3. (a) Show that $\cos \pi/6 = \sqrt{3}/2$ and $\sin \pi/6 = 1/2$. [Hint: Show that the six line segments inside the semicircle in Fig. 15(b) are all equal.]
(b) Show that $\cos \pi/4 = 1/\sqrt{2}$ and $\sin \pi/4 = 1/\sqrt{2}$.
(c) Find $\cos (3\pi/4)$ and $\sin (3\pi/4)$.

4. Fill out the following table. Plot the corresponding points on the unit circle.

θ	$\sin \theta$	$\cos \theta$	$\tan \theta$
$\pi/6$	$1/2$	$\sqrt{3}/2$	$1/\sqrt{3}$
$\pi/4$	$1/\sqrt{2}$		
$\pi/3$			
$\pi/2$			
$2\pi/3$			
$3\pi/4$			
$5\pi/6$			
π			
$3\pi/2$			
2π			
$-\pi/6$			
$-\pi/4$			
$-\pi/3$			
$-\pi 2$			
$-\pi$			

5. Draw the sketch corresponding to Fig. 11 in the following cases:
(a) $0 > \theta > -\pi/2$
(b) $\pi < \theta < 3\pi/2$
(c) $-2\pi > \theta > -2\pi - \pi/2$

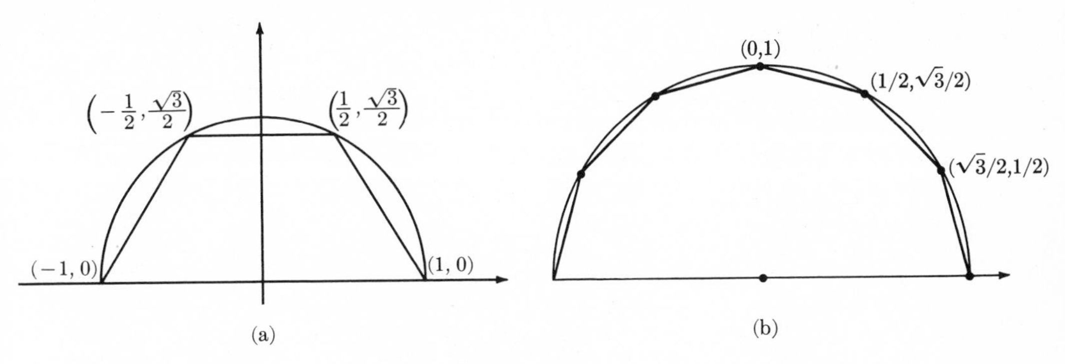

FIGURE 2.15

6. Sketch the graphs of the following functions by finding all zeros and asymptotes.
   (a) $\cos 3x$
   (b) $\cos(x/3)$
   (c) $\tan 2x$
   (d) $\tan \pi x$
   (e) $\cot \pi x$
   (f) $\sec 2\pi x$
   (g) $\cos(x - \pi/4)$
   (h) $\cos(3x - \pi/4)$
   (i) $\tan(2x - \pi/2)$
   (j) $\cot(x - \pi)$

7. Let A and B be constants, and let $r = \sqrt{A^2 + B^2}$.
   (a) Show that $(A/r, B/r)$ is a point on the unit circle.
   (b) Let θ be an angle such that $\cos\theta = A/r$ and $\sin\theta = B/r$. Show that $A\cos t + B\sin t = r\cos(t - \theta)$.

8. Prove formula (8) by setting $\alpha = \pi/2 - \theta$ in formula (4).

9. Prove formula (9).

10. Prove the following formulas.
   (a) $\sin(\theta \pm \pi/2) = \pm\cos\theta$
   (b) $\cos(\theta \pm \pi/2) = \mp\sin\theta$

   (that is, $\cos(\theta + \pi/2) = -\sin\theta$ and $\cos(\theta - \pi/2) = +\sin\theta$)

   (c) $\sin(\theta + \pi) = -\sin\theta$
   (d) $\cos(\theta + \pi) = -\cos\theta$
   (e) $\sin(\theta + 2\pi) = \sin\theta$
   (f) $\cos(\theta + 2\pi) = \cos\theta$

11. Prove the following, by using $\tan\theta = \sin\theta/\cos\theta$, together with results in the text and in previous problems.
   (a) $\tan(-\theta) = -\tan\theta$
   (b) $\tan(\theta \pm \pi/2) = -1/\tan\theta$
   (c) $\tan(\theta \pm \pi) = \tan\theta$
   (d) $\tan(\alpha + \beta) = \dfrac{\tan\alpha + \tan\beta}{1 - \tan\alpha\tan\beta}$

12. Prove that:
   (a) $(\csc\theta)^2 = 1 + (\cot\theta)^2$
   (b) $(\sec\theta)^2 = 1 + (\tan\theta)^2$

13. Prove that:

(a) $2 \cos \alpha \cos \beta = \cos (\alpha - \beta) + \cos (\alpha + \beta)$
(b) $2 \sin \alpha \sin \beta = \cos (\alpha - \beta) - \cos (\alpha + \beta)$
(c) $2 \cos \alpha \sin \beta = \sin (\alpha + \beta) - \sin (\alpha - \beta)$
(d) $2(\cos \theta)^2 = 1 + \cos (2\theta)$
(e) $2(\sin \theta)^2 = 1 - \cos (2\theta)$

14. Prove that $\cos (\theta/2) = \sqrt{\dfrac{1 + \cos \theta}{2}}, \ -\pi \leq \theta \leq \pi.$

[Hint: Since $-\pi/2 \leq \theta/2 \leq \pi/2$, you have $\cos (\theta/2) \geq 0$, so the given equation is equivalent to the one obtained by squaring both sides. After squaring, use the addition formula on $\cos (\theta/2 + \theta/2)$.]

15. Prove that $\sin (\theta/2) = \sqrt{\dfrac{1 - \cos \theta}{2}}, \ 0 \leq \theta \leq 2\pi.$

16. Each of the following function definitions can be applied for certain values of t. Find these values of t, and sketch the graph of the function. Show the vertical asymptotes and the points where the function equals zero.

(a) $a(t) = \cot t$
(b) $b(t) = \sec t$
(c) $c(t) = \csc t$
(d) $d(t) = \sin (1/t)$
(e) $e(t) = \tan (1/t)$
(f) $f(t) = \dfrac{\cos t}{t}$

2.4 COMPOSITE FUNCTIONS AND THE CHAIN RULE

By far the most important of the general rules for computing derivatives is the *chain rule*. It applies to a general way of combining functions called *composition*.

Consider the function

$$f(x) = \sin (\cos x).$$

In order to compute the value of f at some point, you have to do two operations in succession: first find the cosine of x, then find the sine of the result.

The function in this example is called the *composition* of the sine with the cosine, and is denoted sin∘cos. You can think of the little circle between the two functions as standing for "of." Thus $\sin \circ \cos(x)$ is the sine of the cosine of x. Notice that the function to be computed *first* (the cos) appears on the *right*.

In general, the composition $g \circ h$ of two functions g and h is defined by

$$g \circ h(x) = g(h(x)).$$

Recall that the *domain* of a function is the set of all numbers for which the function is defined. The domain of $g \circ h$ is taken to be the set

$$\{x: x \text{ is in the domain of } h, \text{ and } h(x) \text{ is in the domain of } g\}.$$

These are the numbers x for which $g \circ h(x)$ is defined.

Example 1. Let $f(x) = x^2$, $g(x) = \sin x$, and $h(x) = \sqrt{x}$. Then $g \circ f(x) = \sin(x^2)$; $f \circ g(x) = (\sin x)^2 = \sin^2 x$; $h \circ f(x) = \sqrt{x^2} = |x|$; $f \circ h(x) = x$, defined only for $x \geq 0$; $h \circ (f + g)(x) = \sqrt{x^2 + \sin x}$, defined for all x with $x^2 + \sin x \geq 0$; $g \circ (h \circ f)(x) = \sin(|x|)$; $(g \circ h) \circ f(x) = \sin(\sqrt{x^2}) = \sin(|x|)$.

Example 2. [Do it yourself; solution below.] Let $f(x) = x + 1$, $g(x) = \cos x$, and $h(x) = x^3$. Find (a) $f \circ g(x)$ (b) $g \circ h(1)$ (c) $h \circ g(1)$ (d) $h \circ f(t)$.

Notice from Examples 1 and 2 that *the order in which the functions are written makes a difference:* $f \circ g$ is not the same as $g \circ f$, since $f \circ g(x) = f(g(x))$, but $g \circ f(x) = g(f(x))$. In the above examples, you can see that these are not the same. [For one thing, $\sin(x^2)$ is negative when x^2 is between π and 2π, while $(\sin x)^2$ is never negative.]

On the other hand, when composing three functions f, g, and h, the *way in which the terms are grouped* makes *no* difference: $f \circ (g \circ h)(x)$ and $(f \circ g) \circ h(x)$ are both equal to $f[g(h(x))]$.

As a final caution, note that the composition $f \circ g$ is not at all related to the product $f \cdot g$, although the notations look similar.

The Chain Rule

To see what the general formula for the derivative of a composite function should be, consider some examples.

Example 3. Differentiate $\cos(2x)$. *Solution.* The limit of the difference quotient is

$$\lim_{x \to x_0} \frac{\cos(2x) - \cos(2x_0)}{x - x_0} = \lim_{x \to x_0} 2 \cdot \frac{\cos(2x) - \cos(2x_0)}{2x - 2x_0}$$

$$= 2 \lim_{x \to x_0} \frac{\cos(2x) - \cos(2x_0)}{2x - 2x_0}. \tag{1}$$

Now, as $x \to x_0$, then $2x \to 2x_0$; and as $2x \to 2x_0$, the quotient

$$\frac{\cos(2x) - \cos(2x_0)}{2x - 2x_0} \tag{2}$$

has the limit $\cos'(2x_0) = -\sin(2x_0)$, by definition of the derivative. Hence from (1),

$$\lim_{x \to x_0} \frac{\cos(2x) - \cos(2x_0)}{x - x_0} = 2 \lim_{x \to x_0} \frac{\cos(2x) - \cos(2x_0)}{2x - 2x_0} = 2 \cdot [-\sin(2x_0)].$$

The derivative of $\cos(2x)$ is thus $2(-\sin(2x))$.

Notice that the trick in (1) was to rewrite the difference quotient for $\cos 2x$ as the difference quotient for the cosine itself [as in (2)]; the limit of this rewritten form can be evaluated, since the derivative of the cosine is known.

Example 4. Differentiate $\sin(x^2)$. *Solution.* We rewrite the difference quotient for $\sin(x^2)$ in terms of a difference quotient for the sine:

$$\frac{\sin(x^2) - \sin(x_0^2)}{x - x_0} = \frac{\sin(x^2) - \sin(x_0^2)}{x^2 - x_0^2} \cdot \frac{x^2 - x_0^2}{x - x_0}.$$

Since $\sin' = \cos$, we have

$$\lim_{x^2 \to x_0^2} \frac{\sin(x^2) - \sin(x_0^2)}{x^2 - x_0^2} = \cos(x_0^2); \tag{3}$$

for the quotient on the left in (3) is precisely the difference quotient for the derivative of the sine function at the point x_0^2, with the "nearby point" being called x^2. Further, as $x \to x_0$, then $x^2 \to x_0^2$, so

$$\lim_{x \to x_0} \frac{\sin(x^2) - \sin(x_0^2)}{x^2 - x_0^2} = \lim_{x^2 \to x_0^2} \frac{\sin(x^2) - \sin(x_0^2)}{x^2 - x_0^2} = \cos(x_0^2).$$

Hence we can compute the derivative of $\sin(x^2)$ as follows:

$$\begin{aligned}\lim_{x \to x_0} \frac{\sin(x^2) - \sin(x_0^2)}{x - x_0} &= \lim_{x \to x_0} \frac{\sin(x^2) - \sin(x_0^2)}{x^2 - x_0^2} \cdot \frac{x^2 - x_0^2}{x - x_0} \\ &= \lim_{x \to x_0} \frac{\sin(x^2) - \sin(x_0^2)}{x^2 - x_0^2} \cdot \lim_{x \to x_0} \frac{x^2 - x_0^2}{x - x_0} \\ &= \cos(x_0^2) \cdot 2x_0 .\end{aligned}$$

The general rule for the derivative of a composite function $f \circ g$ follows the same pattern as the two preceding examples:

$$\begin{aligned}\lim_{x \to x_0} \frac{f \circ g(x) - f \circ g(x_0)}{x - x_0} &= \lim_{x \to x_0} \frac{f(g(x)) - f(g(x_0))}{g(x) - g(x_0)} \cdot \frac{g(x) - g(x_0)}{x - x_0} \\ &= \lim_{x \to x_0} \frac{f(g(x)) - f(g(x_0))}{g(x) - g(x_0)} \lim_{x \to x_0} \frac{g(x) - g(x_0)}{x - x_0} \\ &= f'(g(x_0))g'(x_0).\end{aligned}$$

This is the *chain rule.* Stated in full, it says:

Theorem 6. *If g is differentiable at x_0, and f is differentiable at $g(x_0)$, then $f \circ g$ is differentiable at x_0, and the derivative is*

$$(f \circ g)'(x_0) = f'(g(x_0))g'(x_0). \tag{4}$$

The rule for the derivative of the nth power of a function [derived in §2.1]

$$(g^n)' = ng^{n-1}g' \tag{5}$$

is an example of the chain rule. Setting $f(u) = u^n$, we have

$$g(x)^n = f(g(x)) = f\circ g(x).$$

We find $f'(u) = nu^{n-1}$, so $f'(g(x)) = ng(x)^{n-1}$; hence by the chain rule

$$(g^n)'(x) = (f\circ g)'(x) = f'(g(x))g'(x) = ng(x)^{n-1}g'(x),$$

which is the same as (5).

Example 5. Find the derivative of $\sin(5x)$. *Solution.* The given function is computed in two steps: given x, first compute $5x$, then $\sin(5x)$. Thus

$$\sin(5x) = f(g(x)),$$

where

$$f(u) = \sin u, \qquad g(x) = 5x.$$

We find

$$f'(u) = \cos u, \qquad f'(g(x)) = \cos(5x), \qquad g'(x) = 5.$$

Hence, by the chain rule, the derivative of $\sin(5x)$ is

$$f'(g(x))g'(x) = [\cos(5x)]5 = 5\cos 5x.$$

Example 6. Find the derivative of $\sin(2x^2 + 1)$. *Solution.* The given function is a composite; first compute $2x^2 + 1$, then $\sin(2x^2 + 1)$. Thus

$$\sin(2x^2 + 1) = f(g(x))$$

where

$$f(u) = \sin u, \qquad g(x) = 2x^2 + 1.$$

We find

$$f'(u) = \cos u, \qquad f'(g(x)) = \cos(2x^2 + 1), \qquad g'(x) = 4x.$$

Hence, by the chain rule, the derivative of $\sin(2x^2 + 1)$ is

$$f'(g(x))g'(x) = [\cos(2x^2 + 1)][4x] = 4x\cos(2x^2 + 1).$$

Example 7. Find the derivative of $\sec(1 + \sin^2 x)$. *Solution.* We have

$$\sec(1 + \sin^2 x) = f(g(x)),$$

where

$$f(u) = \sec u, \qquad g(x) = 1 + \sin^2 x.$$

We find

$$f'(u) = \sec u \tan u, \qquad f'(g(x)) = \sec(1 + \sin^2 x)\tan(1 + \sin^2 x),$$

$$g'(x) = [2\sin x]\cos x.$$

Hence, by the chain rule, the derivative of $\sec(1 + \sin^2 x)$ is

$$f'(g(x))g'(x) = [\sec(1 + \sin^2 x)\tan(1 + \sin^2 x)][2\sin x\cos x].$$

Leibniz notation [§1.8] makes it impossible to forget the chain rule. As an example, consider the function $F(x) = \cos(x^2)$. This is a composite function; first, compute x^2, then $\cos(x^2)$. Let us assign a letter to the result of the first computation, say

$$u = x^2, \tag{6}$$

and another letter to the result of the second computation,

$$y = \cos u = \cos(x^2) = F(x). \tag{7}$$

By the chain rule,

$$F'(x) = -\sin(x^2)\cdot 2x. \tag{8}$$

To put this in Leibniz notation, note that

$$\frac{dy}{dx} = F'(x) \qquad [\text{since } y = F(x)]$$

$$\frac{du}{dx} = \frac{dx^2}{dx} = 2x \qquad [\text{since } u = x^2]$$

$$\frac{dy}{du} = \frac{d\cos u}{du} = -\sin u = -\sin(x^2) \qquad [\text{since } y = \cos u \text{ and } u = x^2].$$

Therefore, (8) takes the form

CHAIN RULE IN LEIBNIZ NOTATION

$$\frac{dy}{dx} = \frac{dy}{du}\frac{du}{dx}. \tag{9}$$

In exactly the same way, the general chain rule formula (4) reduces to (9) if we set $y = f(u)$ and $u = g(x)$.

Two comments about formula (9) are in order. *First comment*: The formula looks obvious; simply cancel the two factors of du on the right-hand side, and you obtain the left-hand side. *Second comment*: The notation is misleading on a conceptual level, because in fact dy/du and du/dx are not really fractions; they are *limits* of fractions. There is no number "du" which can be cancelled according to the rules of ordinary algebra to justify (9). So we have to accept the anomalous situation: the formula looks obvious, and it *is* true; but the proof is more complicated than (9) suggests.

Example 8. Compute the derivative of $F(x) = \sin(x^3 + 1)$, using Leibniz notation. *Solution.* F can be computed in two steps; first

$$u = x^3 + 1 \tag{10}$$

then

$$y = \sin(u) = \sin(x^3 + 1). \tag{11}$$

We have $du/dx = 3x^2$ from (10) and $dy/du = \cos u$ from (11), so

$$\frac{dy}{dx} = \frac{dy}{du}\frac{du}{dx} = (\cos u)(3x^3) = 3x^2 \cos(x^3 + 1).$$

When more than two functions are involved, the origin of the term "chain rule" becomes more obvious; each function contributes a link to the chain.

Example 9. Differentiate $F(x) = \tan(\sin(x^2))$. *Solution.* The function F is computed in three steps:

$$\begin{aligned} u &= x^2 \\ v &= \sin u = \sin(x^2) \\ y &= \tan v = \tan(\sin(x^2)). \end{aligned}$$

The derivative is then

$$\frac{dy}{dx} = \frac{dy}{dv}\frac{dv}{dx}; \quad \text{but} \quad \frac{dv}{dx} = \frac{dv}{du}\frac{du}{dx}, \quad \text{so} \quad \frac{dy}{dx} = \frac{dy}{dv}\frac{dv}{du}\frac{du}{dx} = (\sec^2 v)(\cos u)(2x)$$

$$= \sec^2(\sin(x^2)) \cdot \cos(x^2) \cdot 2x.$$

[We computed du/dx, etc., by looking back at the definitions of u, v, and y.]

Example 10. Differentiate $F(x) = \tan(\cos(\sin(x^{10} + 1)))$.
Solution. The function is computed in four steps:

$$\begin{aligned} u &= x^{10} + 1, & du/dx &= 10x^9 \\ v &= \sin u = \sin(x^{10} + 1), & dv/du &= \cos u \\ w &= \cos v = \cos(\sin(x^{10} + 1)), & dw/dv &= -\sin v \\ y &= \tan w = \tan(\cos(\sin(x^{10} + 1))) = F(x) & dy/dw &= \sec^2 w. \end{aligned}$$

Therefore

$$\frac{dy}{dx} = \frac{dy}{dw}\frac{dw}{dv}\frac{dv}{du}\frac{du}{dx} = (\sec^2 w)(-\sin v)(\cos u)(10x^9).$$

The final step, which may be omitted, is to substitute for u, v, and w their expressions in terms of x.

In all these examples we have introduced symbols for every step in the computation of the composite function; either letters like f and g for the functions involved (Examples 5–7), or letters like u, v, w for the results of each successive computation (Examples 8–10). With practice this is no longer necessary. You can generally see the various steps in the composition, and simply write down the product of the derivatives of all these steps. Thus, for example, we write the derivative of $F(x) = \cos(2x^2 + x)$ directly as follows:

$$F(x) = \cos\,(2x^2 + x)$$

$$F'(x) = [-\sin\,(2x^2 + x)][4x + 1].$$

This method is most efficient for the composition of several functions, in which case it is known* as "peeling the onion":

$$\frac{df\big(g(h(x))\big)}{dx} = f'(g(h(x)))g'(h(x))h'(x).$$

Example 11. The given function appears on one line, and below it appears the derivative as computed by the chain rule:

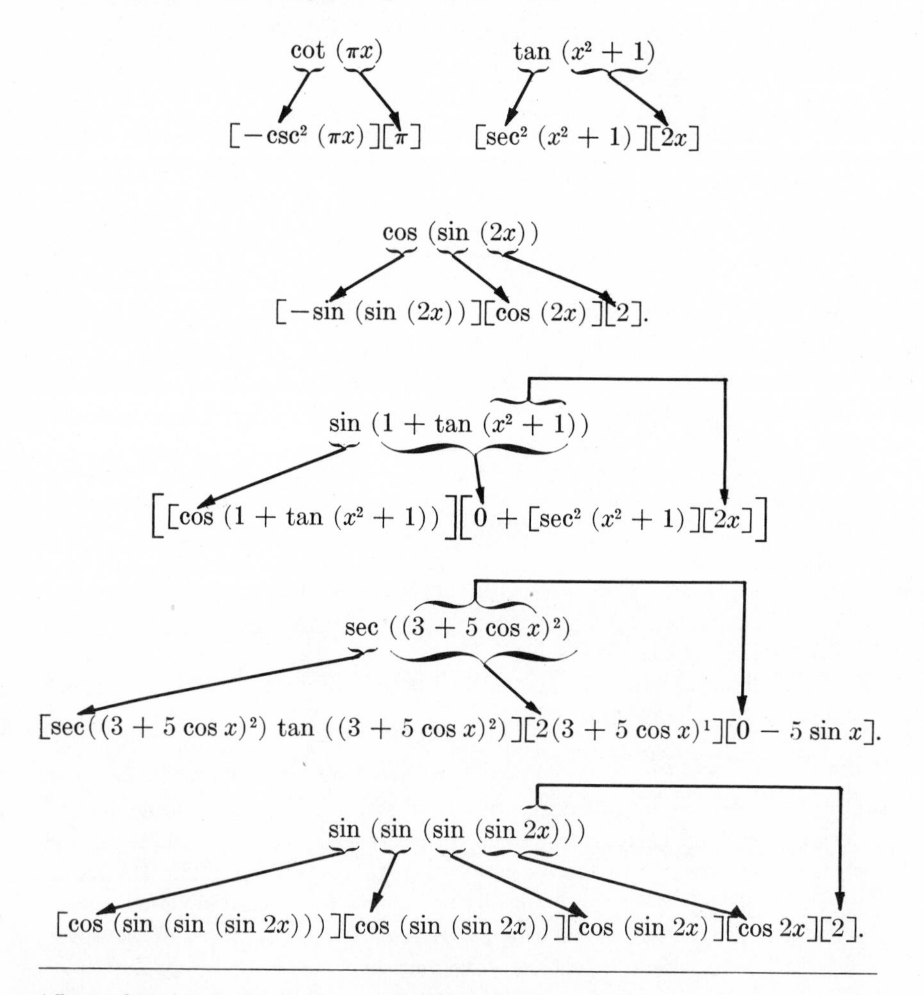

* I owe this name to Herb Gross, a colorful American pedagog.

Remark. Our informal derivation of the chain rule (4) raises two theoretical questions. First of all, how can we write $\dfrac{f(g(x)) - f(g(x_0))}{g(x) - g(x_0)}$ when there is no guarantee that $g(x) \neq g(x_0)$? Second, what is really going on when, for example, from the limit

$$\lim_{x^2 \to x_0^2} \frac{\sin(x^2) - \sin(x_0^2)}{x^2 - x_0^2} = \cos(x_0^2)$$

we draw the conclusion that

$$\lim_{x \to x_0} \frac{\sin(x^2) - \sin(x_0^2)}{x^2 - x_0^2} = \cos(x_0^2)?$$

This change from $\lim\limits_{x^2 \to x_0^2}$ to $\lim\limits_{x \to x_0}$ is justified by the following general fact about "composition of limits":

Proposition C. *Suppose that* $\lim\limits_{x \to x_0} g(x) = y_0$, *and* $\lim\limits_{y \to y_0} D(y) = D(y_0)$.

Then

$$\lim_{x \to x_0} D(g(x)) = D(y_0).$$

[Recall that a function D satisfying $\lim\limits_{y \to y_0} D(y) = D(y_0)$ is called *continuous* at y_0.] It is easy to see why Proposition C is reasonable; when x is very close to x_0, then $g(x)$ is very close to y_0, so $D(g(x))$ is very close to $D(y_0)$. The actual proof of the proposition [§AII.1] is a precise "quantitative" version of this informal "qualitative" argument.

Returning to the first question, concerning division by $g(x) - g(x_0)$, we need a reasonable replacement for $\dfrac{f(g(x)) - f(g(x_0))}{g(x) - g(x_0)}$ when $g(x) = g(x_0)$. Put more simply, what should $\dfrac{f(y) - f(y_0)}{y - y_0}$ be replaced by when $y = y_0$? The answer is obvious; when $y = y_0$, then $\dfrac{f(y) - f(y_0)}{y - y_0}$ should be replaced by

$$\lim_{y \to y_0} \frac{f(y) - f(y_0)}{y - y_0} = f'(y_0). \tag{12}$$

With this background, we give the *complete proof of Theorem 6 (the chain rule)*:

Let $y_0 = g(x_0)$, and define

$$D(y) = \frac{f(y) - f(y_0)}{y - y_0}, \qquad \text{if } y \neq y_0, \tag{13}$$
$$D(y_0) = f'(y_0).$$

Then, by (12), $\lim_{y \to y_0} D(y) = D(y_0)$. Moreover, it is easy to check that

$$\frac{f(g(x)) - f(g(x_0))}{x - x_0} = D(g(x)) \frac{g(x) - g(x_0)}{x - x_0}, \quad x \neq x_0; \tag{14}$$

when $g(x) \neq g(x_0)$, this follows from (13) with $y = g(x)$ and $y_0 = g(x_0)$; and in the remaining case where $g(x) = g(x_0)$, both sides of (14) reduce to 0, so (14) remains valid. Since $g'(x_0)$ is assumed to exist, it follows that

$$\lim_{x \to x_0} g(x) = g(x_0) = y_0 \qquad [\text{Lemma 1, §2.2}].$$

Hence by Proposition C above,

$$\lim_{x \to x_0} D(g(x)) = D(y_0) = f'(y_0). \tag{15}$$

Hence, from (14),

$$\begin{aligned}\lim_{x \to x_0} \frac{f(g(x)) - f(g(x_0))}{x - x_0} &= \lim_{x \to x_0} D(g(x)) \lim_{x \to x_0} \frac{g(x) - g(x_0)}{x - x_0} \\ &= f'(y_0) \cdot g'(x_0) \qquad [\text{from (15)}] \\ &= f'(g(x_0)) \cdot g'(x_0). \qquad \textit{Q.E.D.}\end{aligned}$$

Solution to Example 2

(a) $f \circ g(x) = \cos x + 1$ (b) $g \circ h(1) = \cos(1)$
(c) $h \circ g(1) = (\cos(1))^3$ (d) $h \circ f(t) = \cos(t + 1)$

PROBLEMS

1. Let $f(x) = x^2$, $g(x) = \sin x$, $h(x) = \sqrt{x}$, and $k(x) = x + 1$. Find the following.

- • (a) $g \circ h(x)$
- • (b) $g \circ h(t)$
- • (c) $g \circ h(1)$
- • (d) $h \circ g(y)$
- (e) $h \circ g(1)$
- (f) $h \circ (f \circ g)(x)$
- (g) $(h \circ f) \circ g(x)$
- (h) $g \circ (h \circ f)(t)$
- • (i) $g \circ k(t)$
- (j) $k \circ g(z)$
- • (k) $h \circ (k \circ f)(s)$
- (l) $h \circ (f \circ k)(t)$

2. Write each of the following functions in the form $f \circ g$, where you can compute the derivatives of f and g. Then find the derivative of the given function.

(a) $\sin(2x)$
(b) $\cos^2 x$
(c) $\cos(2 \tan x)$
(d) $\tan(1 + x^2)$
(e) $\sin^2 x + 2$
(f) $\sec(\cos x)$
(g) $\sin(2 + x^2)$
(h) $(x^2 + 1)^3$

3. Compute the derivatives of the following.

(a) $\cos(2x)$
(b) $\cos(x^2)$
(c) $\cos(3x + 1)$
(d) $\cos\left(\dfrac{3x+1}{4x-2}\right)$
(e) $\sin(2x)$
(f) $\sin(x^2 + 5x)$
(g) $\sin\left(\dfrac{x^2+5x}{x^2+1}\right)$
(h) $\tan(x^3 + 3x)$

4. Write down the derivatives of the following functions either in Leibniz notation [Examples 8–10] or directly [as in Example 11].

(a) $(\cos 2x)^{-2}$
(b) $\cos(\sin x)$
(c) $\cos(\sin 5x)$
(d) $\dfrac{x-2}{\sin 2x}$
(e) $\sin((2 \cos x + 5)^2)$
(f) $\sin(\tan x)$
(g) $\sec(\sin^2 2x + 1)$
(h) $\sin(\cos(\tan 5x))$

5. Let k be a constant. Show that the derivative of $f(kx)$ is $kf'(kx)$, assuming that $f'(kx)$ exists. [See Examples 2 and 4.]

6. Suppose that f and g are functions such that $f \circ g(x) = x$ for all x, and $f'(x)$ and $g'(x)$ exist for all x. Show that $f'(g(x)) = 1/g'(x)$.

7. [*Reflection in an ellipse*] The graph of

$$f(x) = \frac{b}{a}\sqrt{a^2 - x^2}, \qquad |x| \le a,\ a > b,$$

is the upper half of an ellipse of major axis $2a$ and minor axis $2b$ [Fig. 16]. The point $F_1 = (c,0)$, where $c = \sqrt{a^2 - b^2}$, is a *focus* of the ellipse. Show that light leaving F_1 and reflected from the graph crosses the x axis again at the second focus $F_2 = (-c,0)$. [In §1.3 the analogous problem is solved for a parabola.]

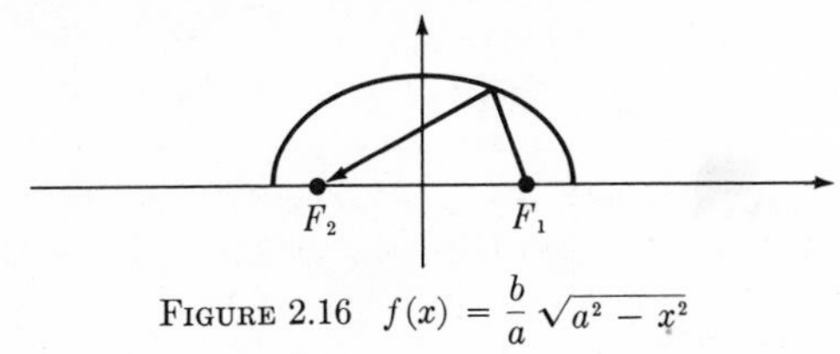

FIGURE 2.16 $f(x) = \dfrac{b}{a}\sqrt{a^2 - x^2}$

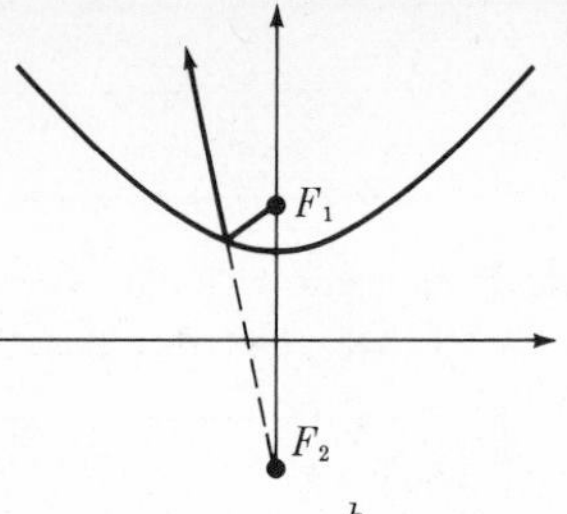

8. [*Reflection in a hyperbola*] The graph of

$$y = \frac{b}{a}\sqrt{x^2 + a^2}$$

FIGURE 2.17 $y = \frac{b}{a}\sqrt{x^2 + a^2}$

is the upper half of a hyperbola [Fig. 17]. The point $F_1 = (0,c)$, where $c = \sqrt{a^2 + b^2}$, is a *focus* of the hyperbola. Show that light leaving F_1 and reflected from the graph goes to infinity along a ray whose backward extension crosses the x axis at the second focus $F_2 = (0,-c)$. [See §1.3 for an analogous problem.]

9. Let $f(u) = 3u + 2$, $g(x) = (x - 2)/3$.
(a) Compute $(f \circ g)'$ by the chain rule.
(b) Compute $f \circ g(x)$, and differentiate this to check the result in (a).

2.5 DERIVATIVES OF INVERSE FUNCTIONS

Every one-to-one function g has an inverse f [§0.4]. When the derivative of g is known, the derivative of the inverse, f, is easily computed. Before discussing the method in general, consider some special cases.

The cube root function

$$f(x) = x^{1/3}, \qquad -\infty < x < \infty$$

satisfies the equation

$$f(x)^3 = (x^{1/3})^3 = x, \qquad -\infty < x < \infty, \tag{1}$$

reflecting the fact that f is the inverse of $g(y) = y^3$, $-\infty < y < \infty$. Assuming that f has a derivative, we can compute it by differentiating each side of formula (1). On the right-hand side is the simple function $h(x) = x$, $-\infty < x < \infty$, whose derivative is $h'(x) = 1$. On the left-hand side is the composite function $f(x)^3$, whose derivative is $3f(x)^2f'(x)$. Since, according to (1), the two functions are equal, their derivatives must also be equal: $3f(x)^2f'(x) = 1$, $-\infty < x < \infty$. Therefore $f'(x) = 1/[3f(x)^2]$, $-\infty < x < \infty$. Since $f(x) = x^{1/3}$, we obtain at last

$$f'(x) = \frac{1}{3(x^{1/3})^2} = \frac{1}{3}x^{-2/3}, \qquad -\infty < x < \infty.$$

In Leibniz notation, $dx^{1/3}/dx = \frac{1}{3}x^{-2/3} = \frac{1}{3}x^{1/3-1}$. Notice that this looks just like the formula for the derivative of x^n for integers $n = 1, 2, 3, \ldots$, $dx^n/dx = nx^{n-1}$.

The nth root function

$$f(x) = x^{1/n}, \qquad 0 < x < \infty$$

satisfies the equation

$$f(x)^n = x, \qquad 0 < x < \infty,$$

for f is defined as the inverse of $g(y) = y^n$, $0 < y < \infty$. Assuming that f has a derivative, we can differentiate each side of the equation and obtain

$$nf(x)^{n-1}f'(x) = 1, \qquad 0 < x < \infty;$$

hence

$$f'(x) = \frac{1}{nf(x)^{n-1}} = \frac{1}{nx^{(n-1)/n}} = \frac{1}{n}x^{(1/n)-1}, \qquad 0 < x < \infty. \tag{2}$$

When n is odd, exactly the same argument works for $-\infty < x < \infty$, not just for $0 < x < \infty$.

Formula (2) for the derivative of $x^{1/n}$ leads directly to an analogous formula for $x^{m/n}$; in Leibniz notation,

$$\frac{dx^{m/n}}{dx} = \frac{m}{n}x^{(m/n)-1} \qquad \text{[Problem 2 below].}$$

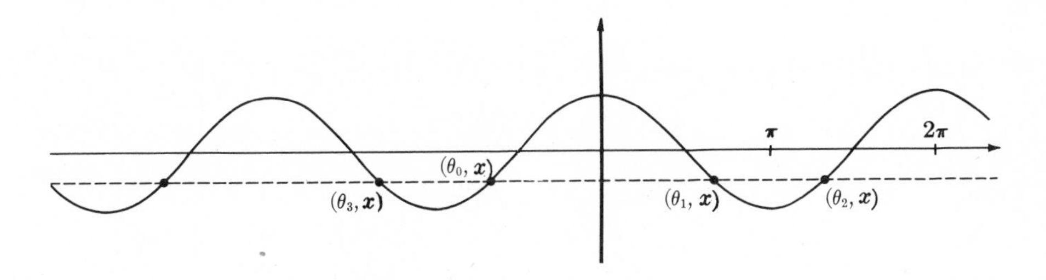

FIGURE 2.18 $x = \cos\theta_0 = \cos\theta_1 = \cos\theta_2 = \cos\theta_3$; $\theta_1 = \arccos x$

The arccos

Given a number x between -1 and 1, the equation

$$x = \cos\theta$$

has infinitely many solutions [Fig. 18]. However, there is a standard way of picking one solution, by restricting our attention to numbers θ between 0 and π; for each x in the interval $[-1,1]$, there is *exactly one* θ such that $x = \theta$ *and* $0 \le \theta \le \pi$. This number θ is called *arccos* x:

$$\theta = \arccos x \qquad \Leftrightarrow \qquad x = \cos\theta \quad \text{and} \quad 0 \le \theta \le \pi. \tag{3}$$

Making the restriction $0 \le \theta \le \pi$ is sometimes called "choosing principal values" for the *arccos*.

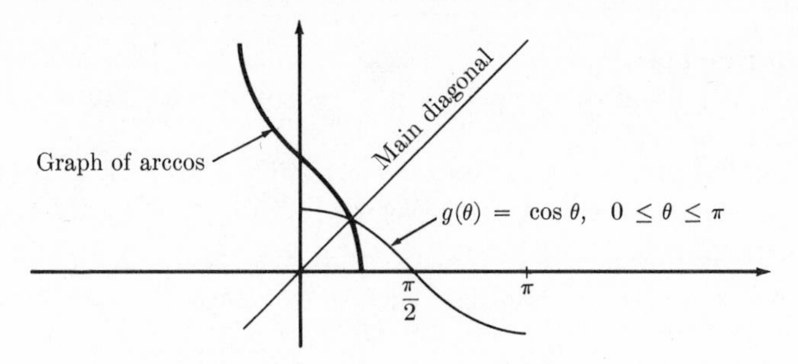

FIGURE 2.19

Our prescription for the arccos shows that it is the inverse of the cos *with the domain appropriately restricted* [Fig. 19], i.e. the inverse of the function

$$g(\theta) = \cos\theta, \qquad 0 \leq \theta \leq \pi.$$

Thus, assuming that the function

$$f(x) = \arccos x, \qquad -1 \leq x \leq 1$$

has a derivative, we can compute it as in the previous examples. From (3) it follows that

$$\cos(\arccos x) = x$$

since in (3) both are equal to $\cos\theta$. Differentiating each side, using the chain rule on the left, gives

$$-\sin(\arccos x)\cdot\arccos'(x) = 1;$$

hence

$$\arccos'(x) = -1/\sin(\arccos x). \tag{4}$$

In principle, the problem of differentiating the arccos is solved by (4), but $-\sin(\arccos x)$ can be simplified. Setting $\arccos x = \theta$, $0 \leq \theta \leq \pi$, as in (3), we find $x = \cos\theta$, and

$$\sin(\arccos x) = \sin\theta = \pm\sqrt{1-\cos^2\theta} = \pm\sqrt{1-x^2}.$$

Since $0 \leq \theta \leq \pi$, it follows that $\sin\theta \geq 0$, so we must choose the + sign. We thus obtain from (4)

$$\arccos'(x) = -1/\sqrt{1-x^2}. \tag{5}$$

These computations should be checked against the graph of arccos. To obtain this graph, we begin with the graph of

$$g(\theta) = \cos\theta, \qquad 0 \leq \theta \leq \pi,$$

draw in the main diagonal, and sketch the reflection of the graph of g in this diagonal [Fig. 19]. Notice that the slope is always <0, so we have chosen the proper sign in arriving at (5). Moreover, as x tends to $+1$ or -1, the slope becomes infinite, and this, too, agrees with (5).

Note: Many books use $\cos^{-1}$ instead of arccos to denote the inverse cos function. We avoid this because $\cos^{-1} x$ is ambiguous; it might mean either $1/\cos x$ or $\arccos x$.

The name "arccos" comes from the unit circle definition of $\cos \theta$, in which θ is the length of an arc on the circle. In other words, $x = \cos \theta$ means geometrically that x is the cos of a certain arc, θ; conversely, θ is the "arc whose cos is x". Analogously, the other inverse trigonometric functions will be denoted arcsin, arctan, and so on.

The Inverse Function Theorem

The examples above suggest that derivatives of inverse functions can be rather easily computed; but in each case we had to *assume* at the outset that the derivative exists. This assumption is justified by the following proposition, which says that in all reasonable cases the inverse function does indeed have a derivative.

☆☆ INVERSE FUNCTION THEOREM

Proposition D. *Suppose that the domain of g is an open interval I, and either*

(i) $g'(y) > 0$ *for all points y in I, or*

(ii) $g'(y) < 0$ *for all points y in I.*

Then g is one-to-one (so it has an inverse), and the inverse function has a derivative at each point in its domain.

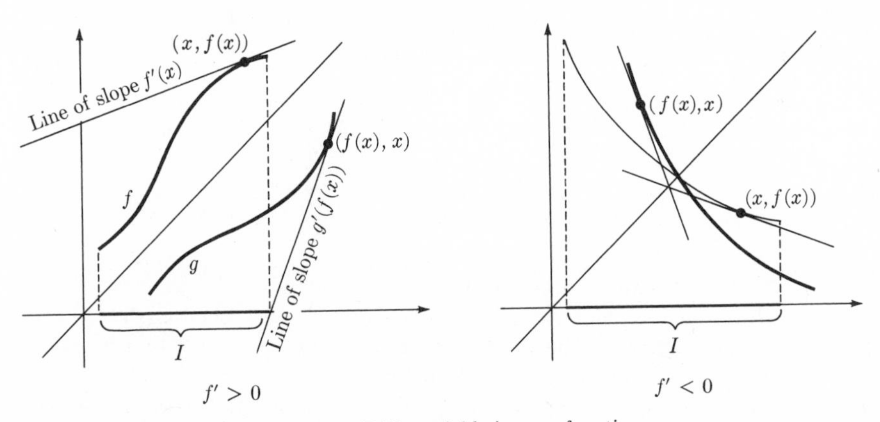

FIGURE 2.20 Differentiable inverse functions

Figure 20 illustrates the proposition. When $g'(y) > 0$ for all y in I, then the graph of g is always rising to the right and has a nonhorizontal tangent line at every point, whose slope is given by g'. The reflected graph therefore has a nonvertical tangent line at every point, and the slope of this tangent line gives f'. [The actual proof, in §AII.6, is considerably more involved than these two sentences suggest!]

Granted the proposition, the derivative of an inverse function can be computed as in the examples above. Suppose that f is the inverse of g; thus

$$y = f(x) \quad \Leftrightarrow \quad x = g(y). \tag{6}$$

It follows that

$$g(f(x)) = x \quad \text{for all } x \text{ in the domain of } f. \tag{7}$$

To prove (7), suppose that x is given, denote $f(x)$ by y, and apply (6), obtaining $x = g(y) = g(f(x))$. Finally, to compute f', differentiate (7), obtaining

$$g'(f(x))f'(x) = 1,$$

$$f'(x) = 1/g'(f(x)).$$

The final step is to simplify $g'(f(x))$, if possible; the details depend on the particular functions involved.

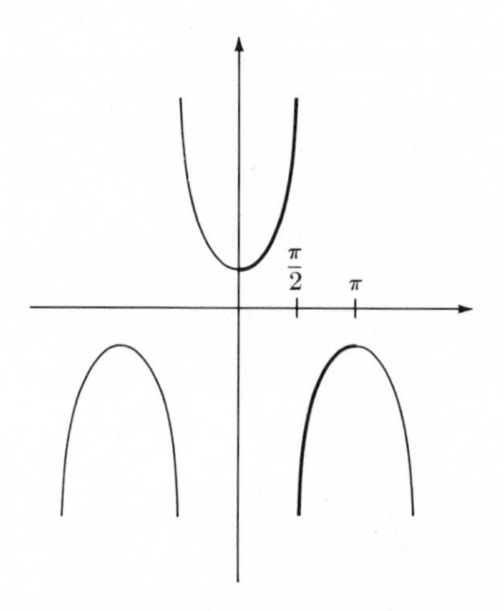

FIGURE 2.21

Example 1. Compute arcsec$'(x)$. *Solution.* First of all, we must decide what function g we should apply Proposition D to; it will be the sec, on some appropriate interval. The graph of sec [Fig. 21] consists of various pieces to which the proposition applies, for example, one for $0 < \theta < \pi/2$, and another for $\pi/2 < \theta < \pi$. [We leave out $\theta = 0$ and $\theta = \pi$, since the proposition requires an open interval.] We will take the first piece and thus define

$$g(\theta) = \sec\theta, \qquad 0 < \theta < \pi/2.$$

Since $g'(\theta) = \sec\theta\tan\theta > 0$ for $0 < \theta < \pi/2$, the inverse function (which we denote by $f(x) = \text{arcsec}\, x$) has a derivative. We compute it by differentiating (7), which in the present case is

$$\sec(\text{arcsec}\, x) = x.$$

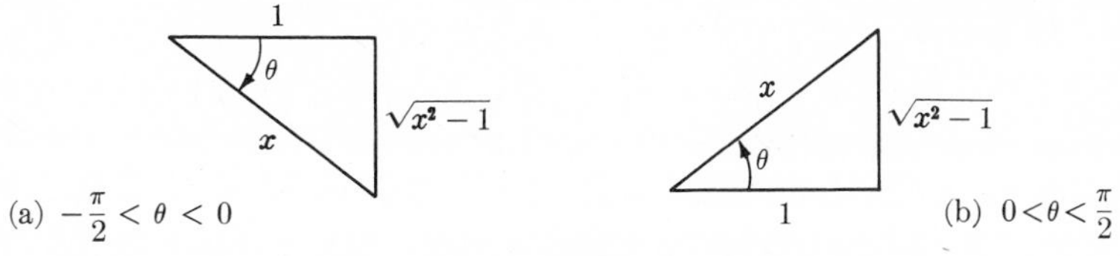

FIGURE 2.22 $x = \sec\theta$

Differentiating,

$$\sec(\operatorname{arcsec} x)\tan(\operatorname{arcsec} x)\operatorname{arcsec}'(x) = 1,$$

$$\operatorname{arcsec}'(x) = \frac{1}{\sec(\operatorname{arcsec} x)\tan(\operatorname{arcsec} x)}.$$

We already know that $\sec(\operatorname{arcsec} x) = x$. The *tan* term simplifies too, as suggested by Fig. 22. Set $\theta = \operatorname{arcsec} x$, so $\sec\theta = x$, and $\tan\theta = \pm\sqrt{\sec^2\theta - 1} = \pm\sqrt{x^2-1}$. Since $0 < \theta < \pi/2$, we have $\tan\theta > 0$, and we must choose the $+$ sign. Thus

$$\operatorname{arcsec}'(x) = \frac{1}{x\sqrt{x^2-1}}, \qquad 1 < x < \infty. \tag{8}$$

Taking $\sec\theta$ on the interval $\pi/2 < \theta < \pi$, we would obtain

$$\operatorname{arcsec}'(x) = \frac{1}{-x\sqrt{x^2-1}}, \qquad -\infty < x < -1.$$

Combining these two, we obtain a function [Fig. 23] whose domain consists of the two intervals $(-\infty, -1)$ and $(1, \infty)$; notice that the term $1/\sqrt{x^2-1}$ in (8) is defined only for x in these intervals, and not for $|x| \le 1$. On each interval, the derivative can be given by

$$\operatorname{arcsec}'(x) = \frac{1}{|x|\sqrt{x^2-1}}. \tag{9}$$

The conventional arcsec function is the one in Fig. 23, with the two endpoints $x = -1$ and $x = 1$ *included* in the domain; however, the derivative at these points does not exist, since the tangent line is vertical.

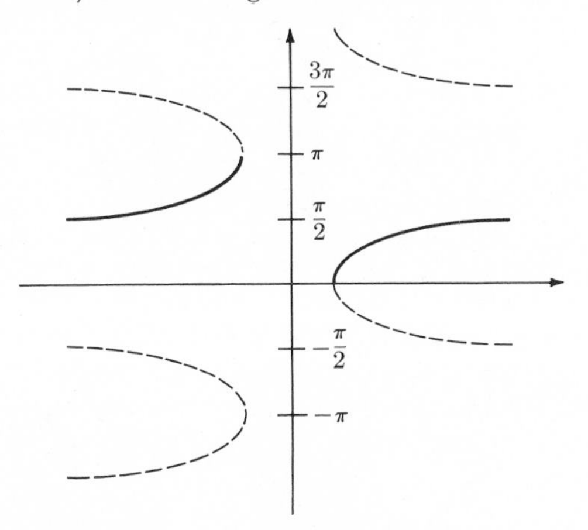

FIGURE 2.23 arcsec

PROBLEMS

1. For each of the following functions, sketch the graph of the inverse, derive the given derivative formula by direct computation, and check it by comparing it to the graph.

(a) $g(\theta) = \sin\theta,\ -\pi/2 \le \theta \le \pi/2;$

$$\arcsin'(x) = \frac{1}{\sqrt{1 - x^2}},\ |x| < 1$$

(b) $g(\theta) = \tan\theta,\ -\pi/2 < \theta < \pi/2;$

$$\arctan'(x) = \frac{1}{1 + x^2},\ -\infty < x < \infty$$

(c) $g(\theta) = \cot\theta,\ 0 < \theta < \pi;$

$$\operatorname{arccot}'(x) = \frac{-1}{1 + x^2},\ -\infty < x < \infty$$

(d) $g(\theta) = \csc\theta,\ 0 < |\theta| < \pi/2;$

$$\operatorname{arccsc}'(x) = \frac{-1}{|x|\sqrt{x^2 - 1}},\ |x| > 1$$

(e) $g(y) = 1/y,\ y \ne 0;\quad d(1/x)/dx = -1/x^2,\ x \ne 0$

(f) $g(y) = y^4,\ y > 0;\quad d(x^{1/4})/dx = \frac{1}{4}x^{-3/4}$

2. If $f(x) = x^{m/n}$, $x > 0$, with m and n integers, show that $f'(x) = (m/n)x^{(m/n)-1}$, $x > 0$. [Hint: $x^{m/n} = (x^{1/n})^m$ is a composite function.]

3. (a) Show that for any function f,

$$(f^{m/n})'(x) = \frac{m}{n} f^{(m/n)-1}(x) f'(x). \qquad \text{[See Problem 2.]}$$

(b) Using part (a), show that

$$\frac{d(x^2 + 1)^{2/3}}{dx} = \frac{4x}{3}(x^2 + 1)^{-1/3}.$$

4. (a) Show that

$$(\arctan f)'(x) = \frac{f'(x)}{1 + f^2(x)}.$$

(b) Find $\dfrac{d \arctan(2x^3 - 1)}{dx}$, using part (a).

(c) Find $\dfrac{d \arctan\left(\dfrac{x - a}{b}\right)}{dx}$, where a and b are constants.

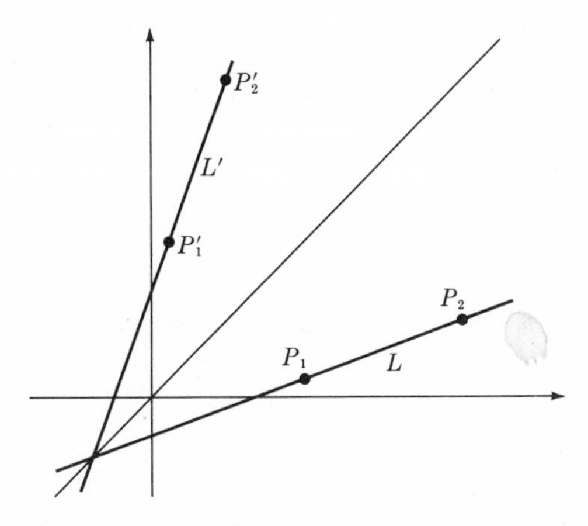

FIGURE 2.24

5. (a) Let L be a nonvertical line, and L' be its reflection in the main diagonal [Fig. 24]. Show that $m' = 1/m$, where m and m' are the slopes of L and L' respectively. [Hint: If $P_1 = (x_1, y_1)$ and $P_2 = (x_2, y_2)$ are on L, then $P_1' = (y_1, x_1)$ and $P_2' = (y_2, x_2)$ are on L'.]
   (b) Show geometrically that if f and g are inverse functions, then $f'(x) = 1/g'(f(x))$. [See Fig. 20.]

6. The following problems constitute "finger exercises" in differentiation techniques. In each case, find the derivative of the given function. Do not try to simplify your answers, unless you can see some particular simple steps to take.
   • (a) $(\cos\theta)(\arcsin\theta) + \theta$
   • (b) $\arctan(2x + 1)$
   • (c) $\sqrt{x^2 + 1}$
   • (d) $[\arcsin(1 - x^2)]^{-1/2}$
   • (e) $\cos(2\arcsin(3t))$
   (f) $\arccos(4\sin(5t))$
   (g) $(\cos x)^{-1} + \arccos x$
   (h) $\dfrac{\cos x}{\arctan x}$
   (i) $\sec[(\arctan x)(x^2 + 1)^{-1/3} + x^{-5/3}]$
   (j) $\sqrt{(x - 1)^2 + 1} + \sqrt{(x - 2)^2 + 3}$

2.6 THE NATURAL LOGARITHM

Early in the seventeenth century, John Napier conceived the idea of logarithms as an aid in trigonometric computations. The basic idea was to convert products and quotients into sums and differences, which are easier to calculate. The success of the project was a great boon to men like Johann Kepler, whose analysis of astronomical observations required laborious computations.

The logarithm function used in calculus today is not quite the same as the one tabulated by Napier, nor is it the same as the "common logarithm," or "log to the base 10", but it does have the same basic property of reducing multiplication to addition. The function we use is denoted *log*, and its basic property is that

$$\log(ab) = \log a + \log b.$$

There is a simple geometric quantity having this property, and the following definition of the function log is based upon it. [The connection of this definition with Napier's logarithms is outlined in §6.8, Problem 8. The connection of our log with $\log_{10}$ is given below in §2.7.]

log a DEFINED

If a is any number ≥ 1, then $\log a$ *is the shaded area in Fig. 25*(i). *If $0 < a < 1$, then* $\log a$ *is the negative of the shaded area in Fig. 25*(ii). The function so defined is called the *natural logarithm*, or *logarithm to the base e*, and is sometimes denoted *ln* instead of *log*. We always use natural logarithms in calculus, just as we always consider the trigonometric functions in radian measure. Thus, in this book, *log always means the function defined here, and not the "common logarithm"* $\log_{10}$.

Logarithms turn out to be more than a mere computational device; they are the key to the equation of proportional population growth, $A' = kA$ [Problem 5 below]. Moreover, the methods used here to derive the fundamental properties of log foreshadow the theory of *integration*, taken up more systematically in Chapter V.

Fundamental formula, $\log ab = \log a + \log b$

In Fig. 26(i), the sum of the two shaded areas is $\log ab$, by definition. Since the left-hand area is $\log a$, the right-hand area must be $\log ab - \log a$. By carefully comparing this area to the one in Fig. 26(ii), we will show that $\log ab - \log a = \log b$, and this is equivalent to the fundamental formula.

To compare the two areas in question, approximate each of them by n rectangles of equal base. Figure 27 does this with $n = 6$. We will show that each of the rectangles in Fig. 27(i) has the same area as the corresponding rectangle in Fig. 27(ii).

Consider Fig. 27(i). The interval $[1,b]$ has been divided into n equal subintervals of length $(b - 1)/n$ by the points

$$1, 1 + \frac{b-1}{n}, 1 + 2\frac{b-1}{n}, \ldots, 1 + k\frac{b-1}{n}, \ldots b.$$

The corresponding rectangles have their upper right-hand corners on the graph of $1/x$; thus the heights of the 1st, 2d, ..., kth, ..., nth rectangles are respectively

$$\frac{1}{1 + \frac{b-1}{n}}, \frac{1}{1 + 2\frac{b-1}{n}}, \ldots, \frac{1}{1 + k\frac{b-1}{n}}, \ldots, \frac{1}{b}.$$

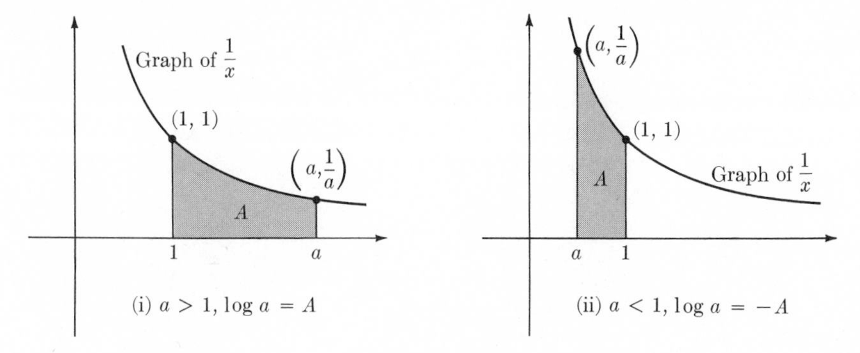

FIGURE 2.25

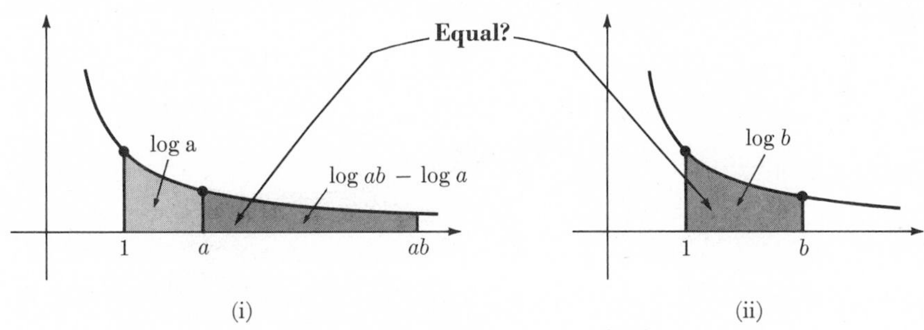

FIGURE 2.26 $\log ab - \log a = \log b$

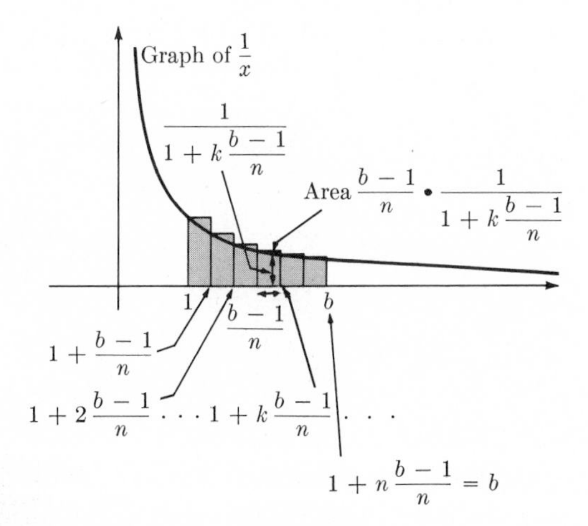

FIGURE 2.27(i)

Multiplying each height by the base $\dfrac{b-1}{n}$, we find that the sum of the areas of all these rectangles is

$$\frac{b-1}{n}\cdot\frac{1}{1+\dfrac{b-1}{n}}+\frac{b-1}{n}\cdot\frac{1}{1+2\dfrac{b-1}{n}}+\cdots$$

$$+\frac{b-1}{n}\cdot\frac{1}{1+k\dfrac{b-1}{n}}+\cdots+\frac{b-1}{n}\cdot\frac{1}{b}. \quad (1)$$

Now consider Fig. 27 (ii). The interval $[a, ab]$ has been divided into n equal subintervals of length $\dfrac{ab-a}{n}$ by the points

$$a,\ a+\frac{ab-a}{n},\ \ldots,\ a+k\frac{ab-a}{n},\ \ldots,\ ab.$$

The corresponding rectangles have heights

$$\frac{1}{a+\dfrac{ab-a}{n}},\quad \frac{1}{a+2\dfrac{ab-a}{n}},\ \ldots,\quad \frac{1}{a+k\dfrac{ab-a}{n}},\ \ldots,\quad \frac{1}{ab}.$$

The base of each rectangle is $\dfrac{ab-a}{n}$; multiplying bases by heights, and *cancelling the factor a in each term,* we find the sum of the areas of all the rectangles in Fig. 27 (ii) to be

$$\frac{b-1}{n}\cdot\frac{1}{1+\dfrac{b-1}{n}}+\frac{b-1}{n}\cdot\frac{1}{1+2\dfrac{b-1}{n}}+\cdots$$

$$+\frac{b-1}{n}\cdot\frac{1}{1+k\dfrac{b-1}{n}}+\cdots+\frac{b-1}{n}\cdot\frac{1}{b}. \quad (2)$$

Comparing this with the sum (1) for Fig. 27 (i), we find they are exactly the same!

Now, if we let n increase [Fig. 28], the sum (1) of the rectangles on the left tends to $\log b$, and the sum (2) of the rectangles on the right tends to the heavily shaded area in Fig. 26 (ii) which is $\log ab - \log a$. Since the sums are equal for each n, the areas are equal; thus $\log ab - \log a = \log b$, or

$$\log ab = \log a + \log b.$$

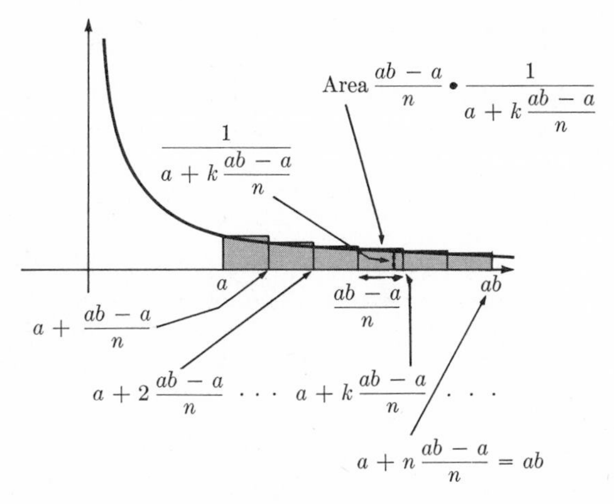

FIGURE 2.27(ii)

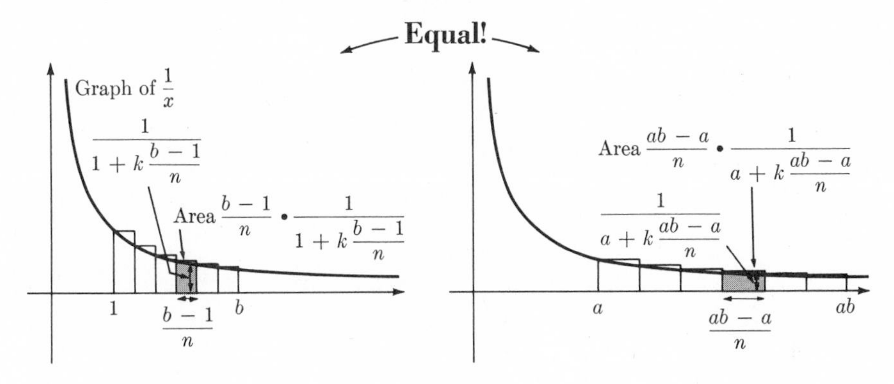

FIGURE 2.27

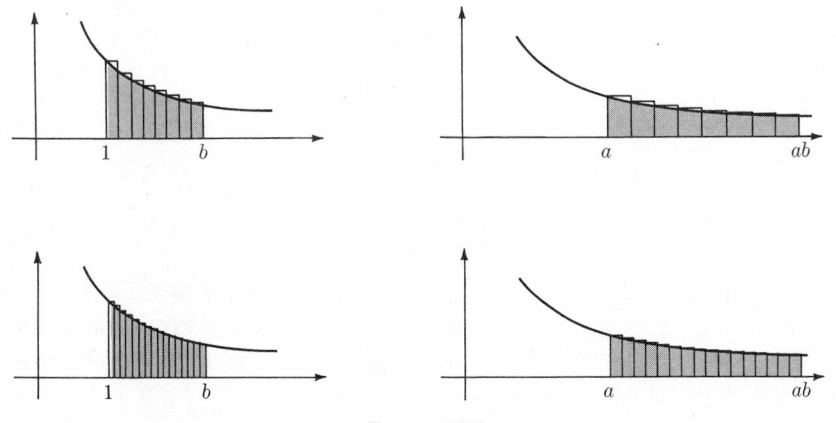

FIGURE 2.28

Our derivation covers the case where $a > 1$ and $b > 1$. The remaining cases, such as $a < 1$ and $b < 1$, are left to you.

The derivation shows why the graph $\{(x,y): y = 1/x\}$ has this particular property. The rectangles on the right in Fig. 27 are obtained from those on the left by multiplying the *bases* by a and the *heights* by $1/a$, so the areas are unchanged.

Remark 1. The calculation of the areas in Fig. 27, using sums of many thin rectangles, is our first encounter with one of the fundamental ideas of calculus. For more examples, see §5.1 on Riemann sums and integrals.

Monotonic property

If $1 \leq a < b$, then $\log a$ is a smaller area than $\log b$. Similar arguments for the cases $a < 1 \leq b$ and $a < b < 1$ show that in *all* cases

$$a < b \quad \Rightarrow \quad \log a < \log b,$$

as Fig. 32 below shows. We can also derive the converse:

$$\log a < \log b \quad \Rightarrow \quad a < b.$$

For if a were greater than or equal to b, then we would have $\log a \geq \log b$ instead of $\log a < \log b$. Similarly,

$$\log a = \log b \quad \Rightarrow \quad a = b;$$

i.e. two numbers with the same log are equal.

Derivative of the log function

The geometric definition of log leads immediately to its derivative. In Fig. 29, the sum of the two shaded areas is $\log x$, while the lightly shaded area is $\log x_0$. The heavily shaded area, the difference of these, is $\log x - \log x_0$. This heavily shaded area contains a small rectangle of base

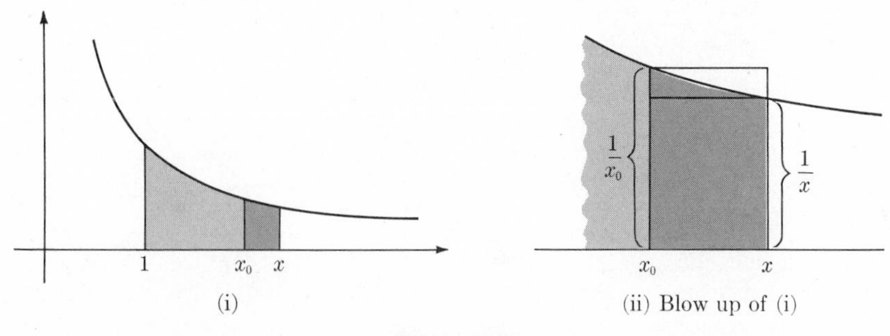

FIGURE 2.29

$x - x_0$ and height $1/x$, and it lies within a larger rectangle of base $x - x_0$ and height $1/x_0$ [Fig. 29(ii)]. Hence

$$\frac{1}{x} \cdot (x - x_0) < \log x - \log x_0 < \frac{1}{x_0} \cdot (x - x_0).$$

Since we have taken $x - x_0 > 0$, the inequality above survives a division by $x - x_0$, and hence

$$\frac{1}{x} < \frac{\log x - \log x_0}{x - x_0} < \frac{1}{x_0}.$$

If we had taken $x < x_0$, we would have

$$\frac{1}{x} > \frac{\log x - \log x_0}{x - x_0} > \frac{1}{x_0}.$$

In either case, the difference quotient

$$\frac{\log x - \log x_0}{x - x_0}$$

lies between $1/x$ and $1/x_0$; since $\lim_{x \to x_0} 1/x = 1/x_0$, we conclude from Fig. 30 that

$$\lim_{x \to x_0} = \frac{\log x - \log x_0}{x - x_0} = \frac{1}{x_0}.$$

$\log'(x) = \dfrac{1}{x}$

Hence *the derivative of* log *is given by* $\log'(x) = 1/x$.

It is this simple formula for the derivative that makes our definition of log the right one for calculus. You will see in §2.7, Problem 5, that $\log_{10}$ has a more complicated derivative.

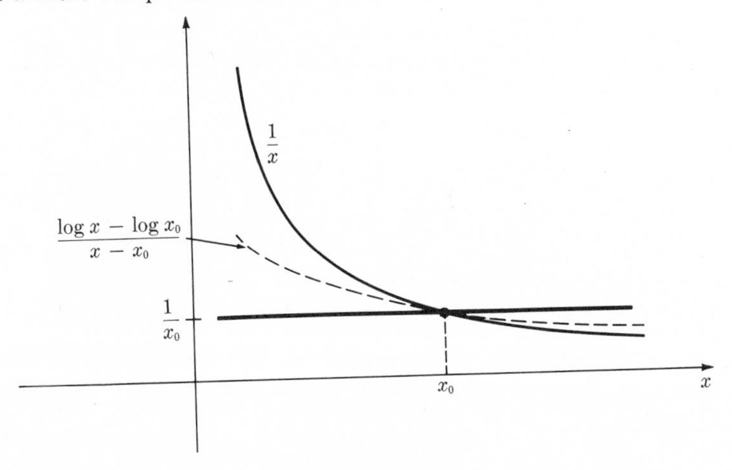

FIGURE 2.30

Graph of the log function

From the definition, we have $\log 1 = 0$; $\log x > 0$ if $x > 1$; and $\log x < 0$ if $x < 1$. To get a more complete picture, we need more values of the function. We begin by finding log 2, which is the area under the graph of $1/x$ between $x = 1$ and $x = 2$. The area can be approximated by three trapezoids, as in Fig. 31. The area of each trapezoid is the base times the average of the two heights, so log 2 is approximately

$$\frac{1}{3}\left(\frac{1+\frac{3}{4}}{2}\right)+\frac{1}{3}\left(\frac{\frac{3}{4}+\frac{3}{5}}{2}\right)+\frac{1}{3}\left(\frac{\frac{3}{5}+\frac{1}{2}}{2}\right)=0.7.$$

In Problems 4–6 below you will show that log 2 differs from 0.7 by less than 0.02, so 0.7 is close enough to log 2 for the purposes of graphing.

This approximation to log 2 allows us to approximate other values by the fundamental formula $\log ab = \log a + \log b$. For example, $\log 4 = \log (2 \cdot 2) = \log 2 + \log 2$; and, in general, $\log (2^n) = n \log 2$ for any integer n. Hence there are arbitrarily large logarithms (with n a large positive integer and 2^n a *very* large number) and arbitrarily large *negative* logarithms (with n a large negative integer and 2^n nearly zero). By plotting a number of these points, we arrive at the graph in Fig. 32. Notice that the shape of the graph is consistent with the derivative $\log' (x) = 1/x$. As $x \to 0$, the slope becomes very steep, and as x grows large, the slope tends to zero.

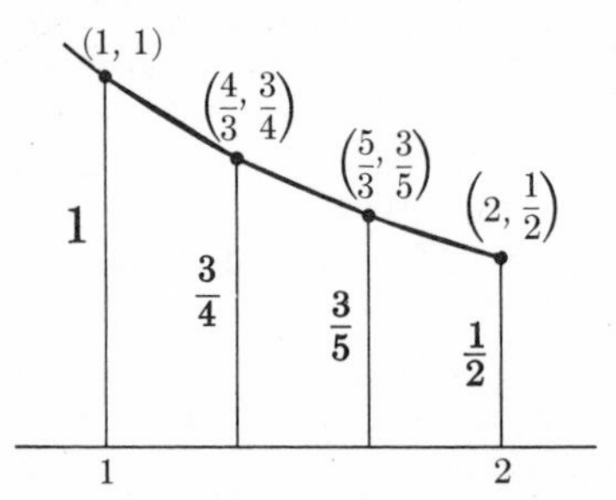

FIGURE 2.31 Approximation of log 2

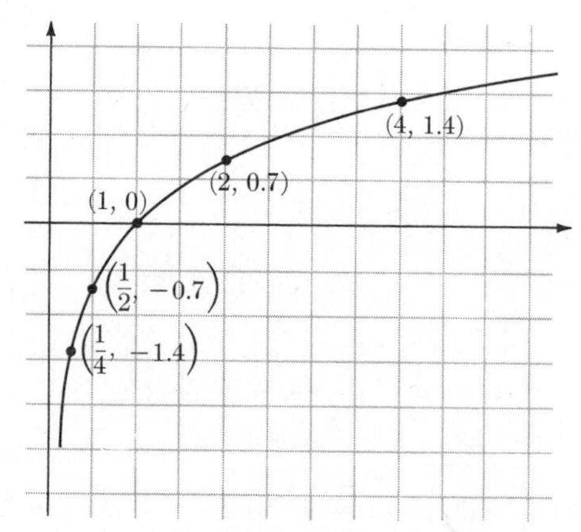

FIGURE 2.32 Graph of log (values rounded off to one decimal place)

Since the graph is unbounded above and below, it is clear from Fig. 32 that the graph crosses each horizontal line $\{(x,y): y = y_0\}$ in exactly one point. Thus *given any real number* y_0, *there is exactly one number* $x_0 > 0$ *such that* $\log x_0 = y_0$.

Summary. The essential content of this section is summed up in

Proposition E. *There is a function* log, *defined for* $x > 0$, *with the following properties*:

(i) *Multiplicative property*

$$\log ab = \log a + \log b$$

(ii) *The derivative is*

$$\log'(x) = 1/x$$

(iii) *Monotonic property*

$$a < b \Leftrightarrow \log a < \log b, \qquad \text{and} \qquad a = b \Leftrightarrow \log a = \log b$$

(iv) *Range property*

Given any real number y_0, there is exactly one positive number x_0 such that $\log x_0 = y_0$.

Besides these analytic facts, it is helpful to remember the shape of the graph [Fig. 32].

Notice that the construction of log given here, and the derivation of properties (i)–(iv), was based largely on pictures. The theoretical justification comes in Chapter V [§5.3, Problem 6].

Remark 2. The proof that

$$\lim_{x \to x_0} \frac{\log x - \log x_0}{x - x_0} = \frac{1}{x_0},$$

based on Fig. 30, illustrates a general method of proving limit relations, which we call:

The Poe Principle.* *Suppose that* x_0 *lies in an open interval* I, *and the following conditions hold:*

POE PRINCIPLE FOR LIMITS

$f(x) \leq g(x) \leq h(x)$ *for all points* x *in* I, *except perhaps* $x = x_0$;

$\lim_{x \to x_0} f(x) = L$ *and* $\lim_{x \to x_0} h(x) = L$.

Then $\lim_{x \to x_0} g(x) = L$.

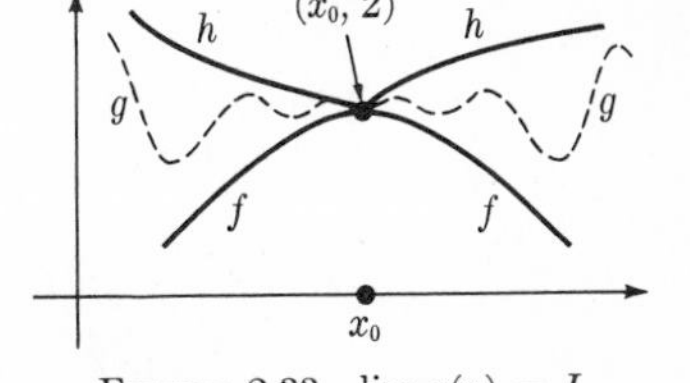

FIGURE 2.33 $\lim_{x \to x_0} g(x) = L$

* I owe this name to D. van Dantzig, a versatile Dutch mathematician.

The result is intuitively clear from Fig. 33; as $x \to x_0$, the graphs of f and h close in on the graph of g (as the walls of the pit close in on the victim in Edgar Allen Poe's "The Pit and the Pendulum"), so $g(x)$ is forced to the value L as $x \to x_0$.

For a proof of the Poe Principle see §AII.2.

PROBLEMS

Problems 1–4 concern derivative formulas; 5 solves the equation $A' = kA$; 6–8 compute logarithms carefully; 9 and 10 relate to the proof that $\log'(x) = 1/x$, and 11 applies the Poe Principle.

1. Find the derivatives of the following functions.

- • (a) $\log(2x)$
- • (b) $\log(ax)$ $(ax > 0, a \text{ constant})$
- • (c) $\log(x^2)$
- • (d) $(\log x)^2$
- • (e) $\log(\cos x)$
- • (f) $\log(\cos(2x+1))$
- • (g) $\log\log x^2 = \log(\log(x^2))$
- (h) $x \arctan x - \frac{1}{2}\log(1+x^2)$
- (i) $\log[\log(\log(\log \sin x))]$

2. Show that if f is differentiable at x_0 and $f(x_0) > 0$, then

$$(\log \circ f)'(x_0) = \frac{f'(x_0)}{f(x_0)}.$$

3. (a) Obtain the product rule for $(fg)'$ by differentiating the formula $\log(fg) = \log f + \log g$.

(b) Do the same for the product of n functions.

(c) Obtain the quotient rule by taking logs.

4. (a) Let $f(x) = \log(-x)$, $-\infty < x < 0$. Show that $f'(x) = 1/x$.

(b) If $f(x) = \log|x|$ for $x \neq 0$, show that $f'(x) = 1/x$.

(c) Graph the function f in part (b), and check that the derivative formula is consistent with the graph. [Does it have the proper sign for $x < 0$?]

(d) If u is a function differentiable at x_0, and $u(x_0) \neq 0$, show that

$$\frac{d \log|u(x)|}{dx} = \frac{u'(x)}{u(x)}.$$

5. Log is closely related to the equation $A'(t) = kA(t)$ for proportional population growth [§1.7]. Assume $A(t) > 0$ for all times t, so the equation can be written $A'/A = k$. This quotient looks like the u'/u in Problem 4(d), and suggests that you prove:

(a) If $A' = kA$ and $A > 0$, then $(\log A)' = k$.

(b) If $\log A(t) = kt$, and A' exists, then $A'(t) = kA(t)$.

(c) If $\log A(t) = kt + c$, for any constant c, and if A' exists, then $A' = kA$.

6. Explain geometrically why $0.7 > \log 2$. [See Fig. 31.]

7. We are going to obtain a lower bound for log 2, to go with the upper bound 0.7 obtained in Problem 6.

(a) Let $0 < a < b$. Find the equation of the tangent line to the graph of $1/x$ at $\frac{1}{2}(a+b)$, the midpoint of the segment from a to b.

• (b) Observe geometrically that the graph of $1/x$ lies *above* its tangent line at each point. Conclude that the area under the graph of $1/x$ between a and b is *greater* than the area of the trapezoid bounded by the line in part (a), and show that the area of this trapezoid is $2\,\dfrac{b-a}{a+b}$.

(c) Apply part (b) to the three intervals [1,4/3], [4/3,5/3], [5/3,2], and show that

$$\log 2 > \frac{2}{3}\left[\frac{1}{7/3} + \frac{1}{9/3} + \frac{1}{11/3}\right] = \frac{478}{693} = 0.689\ldots.$$

[Hence $0.689 < \log 2 < 0.7$, and 0.7 approximates log 2 with an error <0.011.]

8. This problem requires a calculating machine.

(a) Approximate log 2 by the sum of the areas of ten trapezoids.

(b) Use Problem 7(b) to show how accurate your result is.

• **9.** Let $A(t)$ be the shaded area in Fig. 34. Find $A'(t)$.

10. Let $A(t)$ be the shaded area in Fig. 35, for $0 < t < 2$. Find $A'(t)$.

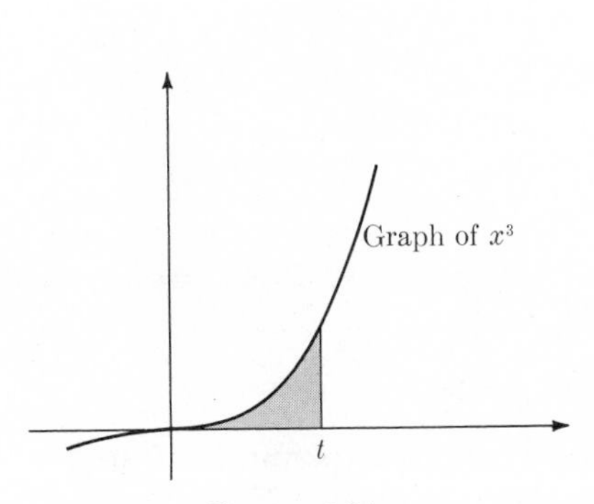

FIGURE 2.34

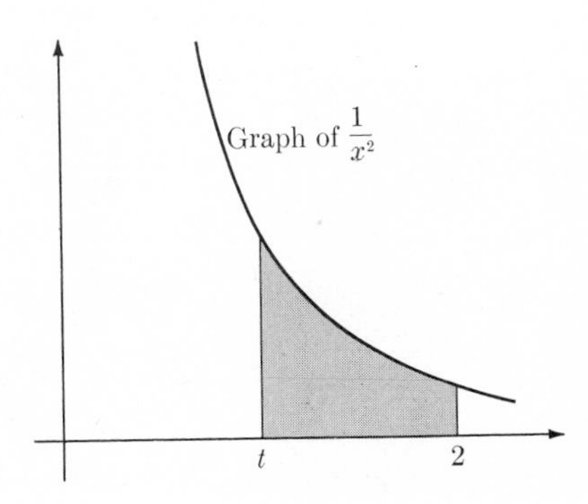

FIGURE 2.35

11. The relation $\lim_{\theta \to 0} \frac{\sin \theta}{\theta} = 1$ can be deduced from the Poe Principle, using the fact that in a circle of radius r, a sector of arc length ℓ has area $\frac{1}{2}r\ell$ [Fig. 36].

(a) Comparing the shaded sector in Fig. 37 to an inner and an outer triangle, show that $\frac{1}{2} \sin \theta \cos \theta < \frac{1}{2}\theta < \frac{1}{2} \tan \theta$, $0 < \theta < \pi/2$.

(b) Deduce that [Fig. 38]

$$\cos \theta < \frac{\sin \theta}{\theta} < \sec \theta, \qquad -\pi/2 < \theta < \pi/2, \quad \theta \neq 0.$$

[Hint: Use part (a) for $0 < \theta < \pi/2$, and note that $\cos \theta$, $(\sin \theta)/\theta$, and $\sec \theta$ are all even functions.]

(c) Conclude that $\lim_{\theta \to 0} \frac{\sin \theta}{\theta} = 1$.

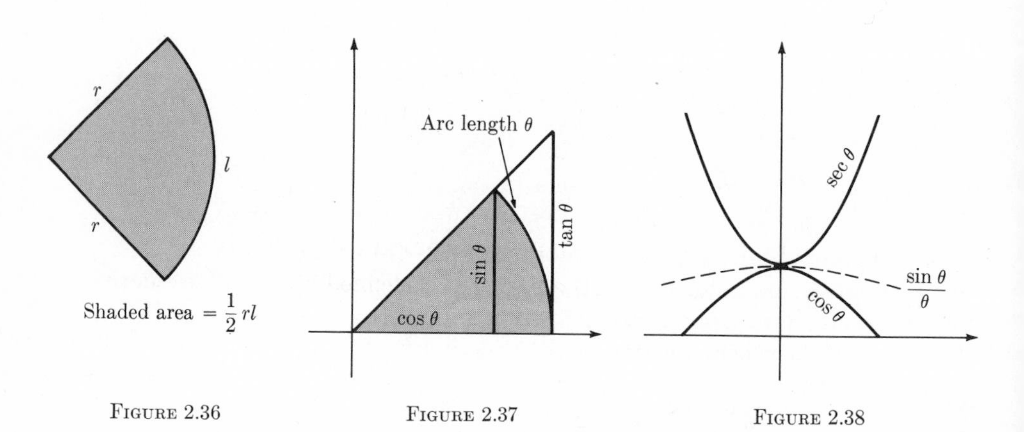

FIGURE 2.36 FIGURE 2.37 FIGURE 2.38

2.7 LOGS AND EXPONENTIALS

The last (but surely not least) of the elementary functions we investigate are the exponential functions, of the form

$$f(x) = c^x, \qquad -\infty < x < \infty,$$

where c is a positive constant. The central role of these functions in the study of growth and decay is demonstrated in the next chapter [§3.3]; here we lay the groundwork for these applications by establishing the basic relations between exponentials and logarithms, and computing their derivatives.

In any such development, certain algebraic formulas must be established:

$$c^x c^y = c^{x+y} \tag{1}$$

$$c^{-x} = 1/c^x \tag{2}$$

$$\log_c (xy) = \log_c x + \log_c y \tag{3}$$

$$\log_c (a^b) = b \log_c a \tag{4}$$

$$a^b = c^{b \log_c(a)} \tag{5}$$

In addition, to do calculus, we need the derivatives of the *exponential function to the base c*

$$\frac{d(c^x)}{dx} = ?$$

the *log to the base c*

$$\frac{d \log_c x}{dx} = ?$$

and the *power function with exponent b*

$$\frac{dx^b}{dx} = ?$$

Traditionally, the algebraic formulas (2)–(5) are all derived from (1). We shall begin, instead, with the natural log defined above, whose derivative we already know; some of formulas (1)–(5) will be used as definitions, and the others, together with the derivative formulas, will be deduced from them.

Since logs and exponentials are inverse to each other, we introduce the inverse of the natural log. Notice that log is defined on an open interval $(0,+\infty)$ and $\log'(x) = 1/x > 0$ throughout that interval, so [Proposition D] log *has* a differentiable inverse, which we call *exp*. This function actually has the form $\exp(x) = e^x$ for a certain number e, called the *base of the natural logarithms*. To prove this requires a few definitions and some algebraic juggling.

Definition 1. The function exp is the inverse to the function log. EXPONENTIAL FUNCTION

By the fundamental relations between inverse functions [§2.5, §0.4],

$$x = \log y \Leftrightarrow y = \exp(x), \tag{6}$$

and therefore

$$\exp(\log y) = y, \qquad \log(\exp(x)) = x. \tag{7}$$

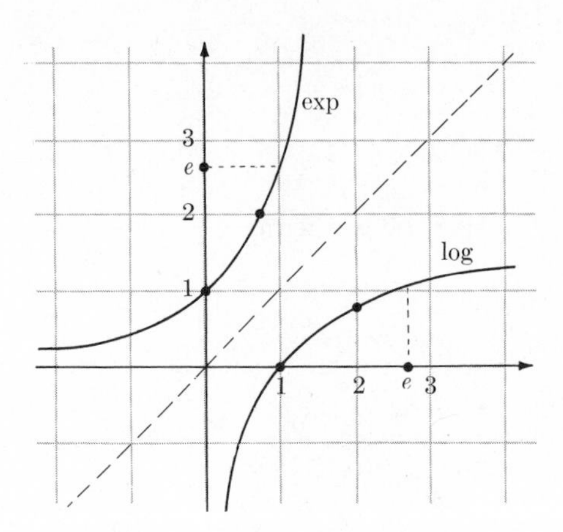

FIGURE 2.39 Graph of log and exp

The graph of exp [Fig. 39] is obtained by reflecting the graph of log in the main diagonal. As the figure shows, $\exp(0) = 1$ [since $\log(1) = 0$], and $\exp(x) > 0$ for every x [since $\log y$ is defined only for $y > 0$].

The general method of finding derivatives of inverse functions shows that *exp is its own derivative*:

$$\exp' = \exp. \tag{8}$$

You should be able to show this yourself; if you need help, look up the answer to Problem 2 below.

Fundamental formulas for the function exp

From $\log(ab) = \log a + \log b$ follows a corresponding formula for exp. In fact, using (7),

$$\log[(\exp x)(\exp y)] = \log(\exp x) + \log(\exp y) = x + y.$$

Applying exp to both sides of this, and using (7) again, we obtain

$$(\exp x)(\exp y) = \exp(x + y). \tag{9}$$

Taking $y = -x$ in (9) gives

$$\exp(x)\exp(-x) = \exp(x - x) = \exp(0).$$

Since $\exp(0) = 1$, we find $\exp(x)\exp(-x) = 1$, and dividing by $\exp(x)$ gives

$$\exp(-x) = \frac{1}{\exp(x)}. \tag{10}$$

Formulas (9) and (10) look just like (1) and (2), suggesting that *exp* really is connected to the familiar concept of exponents. We make this connection explicit in the following *definition of* a^b:

Definition 2. If $a > 0$, and b is real, then the number a^b is defined by

$$a^b = \exp(b \log a). \tag{11}$$

a^b DEFINED

If we take the log of each side of (11), the left becomes $\log(a^b)$, but the right simplifies to $b \log a$, by (7). Thus

$$\log(a^b) = b \log a. \tag{12}$$

Further, we obtain the familiar law of exponents:

$$\begin{aligned} a^b a^c &= \exp(b \log a) \exp(c \log a) && [\text{by } (11)] \\ &= \exp((b + c) \log a) && [\text{by } (9)] \\ &= a^{b+c}. && [\text{by } (11)] \end{aligned}$$

Definition 2 simplifies if the base a is chosen so that $\log a = 1$. This occurs for exactly one number, which is called e:

e BASE OF NATURAL LOGARITHMS

Definition 3. The number e is defined by the equation $\log e = 1$, where log is the natural logarithm.

Setting $y = e$ in (7) entails $\log y = 1$, so $\exp(1) = e$. More generally, setting $a = e$ and $b = x$ in (11) yields

$$e^x = \exp(x) \tag{13}$$

and thus exp is the *exponential function to the base e*.

Moreover, the natural logarithm *log* is the *logarithm to the base e*, denoted $\log_e$. To show this, we use the standard definition of $\log_a$:

Definition 4. For any constant $a > 0$, $a \neq 1$, we define $\log_a x$ by the equation

$$a^{\log_a x} = x.$$

For example, $2^{\log_2 8} = 8$ [by Definition 4], and $2^3 = 8$ [well-known fact], so $\log_2 8 = 3$. Similarly, $e^{\log_e x} = x$ [Definition 4] and $e^{\log x} = \exp(\log x) = x$ [by (13) and (11)], so $\log x = \log_e x$, as we claimed.

You can show [Problem 3] that Definition 4 implies the formula

$$\log_a x = \frac{\log x}{\log a},$$

so $\log_a x$ can be easily found when the natural logs are known. Conversely, natural logs can be computed from logs to any other base by the formula

$$\log x = \frac{\log_a x}{\log_a e}.$$

And of course, $\log_a$ has the basic property of logarithms [Problem 4]

$$\log_a (xy) = \log_a x + \log_a y.$$

Finally, knowing the derivatives of the "natural" log and exp, we can deduce the derivatives of the general exponential, log, and power functions, as suggested in the problems below:

$$\frac{dc^x}{dx} = c^x \log c, \qquad c > 0$$

$$\frac{d \log_a x}{dx} = \frac{1}{x \log a} = \frac{\log_a e}{x}, \qquad a > 0, a \neq 1$$

$$\frac{dx^b}{dx} = bx^{b-1}.$$

This last formula should look familiar; we proved it in §2.1 for the case where b is a positive integer, in §2.2 for negative integers, and in §2.4 for rational numbers. The first two formulas should *not* be memorized; remember instead the more basic formulas summarized on the facing page, and deduce the others from the basics, when needed.

Example 1. If $g(x) = e^{x^2}$, then $g'(x) = 2xe^{x^2}$.
If $g(x) = e^{2x}$, then $g'(x) = 2e^{2x}$.

Example 2. $\dfrac{d3^x}{dx} = \dfrac{de^{x \log 3}}{dx} = (\log 3)e^{x \log 3} = 3^x \log 3.$

Example 3. If $f(x) = \log_2 x$, then $f'(x) = ?$
First solution. By the definitions,

$$x = 2^{\log_2 x} = e^{(\log 2)(\log_2 x)} = e^{(\log 2)f(x)}.$$

Differentiating each side of this equation gives $1 = e^{(\log 2)f(x)}(\log 2)f'(x)$, so

$$f'(x) = \frac{1}{\log 2e^{(\log 2)f(x)}} = \frac{1}{x \log 2}.$$

Second solution. Taking logs in $2^{\log_2 x} = x$ gives $(\log_2 x) \log 2 = \log x$, or

$$f(x) = \log_2 x = \frac{\log x}{\log 2}.$$

Since $\log' (x) = 1/x$, we get $f'(x) = \dfrac{1}{x \log 2}.$

Summary

Definitions: Log a is plus or minus a certain area, as in Fig. 25.

Exp is the inverse of the log function.

$e = \exp(1)$ and $e^x = \exp(x)$.

$a^b = e^{b \log a} = \exp(b \log a)$ for $a > 0$.

$a^{\log_a x} = x$ [defining $\log_a x$].

Basic formulas: $\log(ab) = \log a + \log b$.

$a^{b+c} = a^b a^c$.

$\log(a^b) = b \log a$.

$(a^b)^c = a^{bc}$.

Derivatives: $\log'(x) = 1/x$.

$\exp'(x) = \exp(x)$.

The derivative of $e^{f(x)}$ is $f'(x)e^{f(x)}$.

[This formula for the derivative of $e^{f(x)}$ follows from the chain rule. Set $g(x) = e^{f(x)} = \exp(f(x))$. Then, differentiating,

$$g'(x) = [\exp(f(x))][f'(x)] = f'(x)e^{f(x)}.]$$

Example 4. $\dfrac{dx^\pi}{dx} = ?$ *Solution.* We expect the answer $\pi x^{\pi-1}$, but up to now this general form has been proved only for *rational* exponents, $x^{m/n}$. In this case, we can use Definition 2:

$$\frac{dx^\pi}{dx} = \frac{de^{\pi \log x}}{dx} = e^{\pi \log x}\frac{d\pi \log x}{dx} \qquad \text{[chain rule]}$$

$$= x^\pi \cdot \pi \cdot \frac{1}{x} = \pi x^{\pi-1}.$$

Example 5. Let $f(x) = x^x$, $x > 0$. Find $f'(x)$. [Note how this differs from the previous examples; here *both* the base *and* the exponent vary. Nevertheless, proceeding with due caution, we can find the derivative.]

First method. $f(x) = x^x = e^{x \log x}$, so by the chain rule

$$f'(x) = \left(1 \cdot \log x + x \cdot \frac{1}{x}\right) e^{x \log x} = (\log x + 1)x^x.$$

Second method. $\log f(x) = x \log x$. Differentiate (using the chain rule on the left and the product rule on the right) to find

$$\frac{1}{f(x)} f'(x) = x \cdot \frac{1}{x} + 1 \cdot \log x = 1 + \log x,$$

and hence $f'(x) = (1 + \log x)x^x$.

Remark 1. The number e can be computed, approximately, from the defining relation $\log e = 1$. From §2.6, $\log 2 < 0.7$; by definition, $\log e = 1$; and Problem 6 below shows that $\log 3 > 1$. Thus

$$\log 2 < \log e < \log 3,$$

and from the monotonic property of log and exp it follows that $2 < e < 3$ [see Fig. 39]. This approach *can* produce accurate approximations, but it is not very efficient [see Problem 6(d)]. Chapter XI on infinite series gives a much more efficient way.

Remark 2. The definition of general exponents a^b by formula (11) raises an important question. We already have a definition of a^b when b is a *rational* number m/n, namely

$$a^{m/n} = \sqrt[n]{\underbrace{a \cdot a \cdots a}_{m \text{ factors}}}\,. \tag{14}$$

The question is, when b is a rational number, does (11) agree with this old definition? We will show that it does, by accepting (11) as the basic definition and deriving (14) as a consequence. First, recall that Definition 1 implies $a^b a^c = a^{b+c}$, and also $a^1 = \exp(1 \log a) = \exp(\log a) = a$. Therefore

$$a \cdot a = a^1 \cdot a^1 = a^{1+1} = a^2$$
$$(a \cdot a) \cdot a = a^2 \cdot a^1 = a^{2+1} = a^3$$
$$(a \cdot a \cdot a) \cdot a = a^3 \cdot a^1 = a^{3+1} = a^4$$

and generally

$$\underbrace{a \cdot a \cdots a}_{m \text{ factors}} = a^m. \tag{15}$$

Thus Definition 2 agrees with the standard meaning of a^m for integer exponents m.

To go further, we need the formula

$$(a^b)^c = a^{bc}, \tag{16}$$

which follows easily from Definition 1 [see Problem 9]. In view of this, we can show that $a^{1/n}$ [as defined by (11)] *is actually the* nth *root of* a; replace a by $a^{1/n}$ in (15), obtaining

$$\underbrace{a^{1/n} \cdots a^{1/n}}_{n \text{ factors}} = (a^{1/n})^n$$

$$= a^1 = a \qquad [\text{by (16)}].$$

Finally, by (16) again,

$$a^{m/n} = (a^{1/n})^m,$$

which shows that $a^{m/n}$ [as defined by (11)] is the mth power of the nth root of a, as it should be.

PROBLEMS

1. Find the derivatives of the following functions.

• (a) e^{3x}
(b) e^{x^3}
• (c) e^{x^2+5x}
(d) $e^{\sin x}$
• (e) $\sin(e^x)$
(f) 2^x
• (g) a^x
(h) $\exp(\cos(2x+1))$
• (i) $\log[\cos(\exp(t))]$
(j) $\exp(e^x)$
• (k) $x \arcsin x + (a^2 - x^2)^\pi$
(l) $x \arctan x - \frac{1}{2}\log(1+x^2)$
• (m) $e^{ax}\cos bx$
(n) $e^{ax}\sin bx$
(o) $\log(e^x + e^{2x})$
• (p) $\exp(e^{\exp x})$

• **2.** Prove that $\exp' = \exp$, using the definition of exp as the inverse to log.

3. Suppose that $a > 0$ and $a \neq 1$.
(a) Prove: $a^y = x \iff y = \log x/\log a$, hence $\log_a x = \log x/\log a$. [Use (12).]
(b) Prove that $\log x = \log_a x/\log_a e$.

4. Prove that $\log_a(xy) = \log_a x + \log_a y$.

5. (a) Find $f'(x)$ if $f(x) = \log_a x$.
(b) Find $g'(x)$ if $g(x) = x^b$. [Use (11).]

6. (a) Show that $\log 2 < 0.7 < 1$.
(b) Show that $\log 3 > \frac{1}{5} + \frac{1}{6} + \frac{1}{7} + \frac{1}{8} + \frac{1}{9} + \frac{1}{10} + \frac{1}{11} + \frac{1}{12} > 1$. [Hint: Compare log 3 to the sums of the areas of eight rectangles, each of base $\frac{1}{4}$, lying under the graph of $1/x$ between $x = 1$ and $x = 3$.]
(c) Conclude that $2 < e < 3$.
(d) Show that $2.7 < e < 2.75$. [The computations are too extensive without computer; simply suggest how they should be made.]

• **7.** Let $a > 0$, $a \neq 1$. Prove that $f(x) = a^x$ and $g(y) = \log_a y$ are inverse functions to each other.

8. We have defined e by the equation $\log e = 1$. This problem gives another formula for e, namely $e = \lim_{h\to 0}(1+h)^{1/h}$.

IMPORTANT LIMIT

(a) Prove that $\lim_{h\to 0}\frac{1}{h}\log(1+h) = 1$.

$$\left[\text{Hint: } \log'(1) = \lim_{h\to 0}\frac{\log(1+h) - \log 1}{h}.\right]$$

(b) Conclude that $\lim_{h\to 0}(1+h)^{1/h} = e$.

[Apply exp to the relation in (a), and use Proposition C, §2.4, and Lemma 1, §2.2.]

(c) Prove similarly that $e^x = \lim_{h\to 0}(1+xh)^{1/h}$.

9. Use (11) to prove that $(a^b)^c = a^{bc}$, $a > 0$.

10. Problem 5, §2.6, gave solutions of the equation $A' = kA$ in the form $\log A(t) = kt + c$.

(a) Assuming that $\log A(t) = kt + c$, prove that $A(t) = Ce^{kt}$, for a constant $C > 0$, and that $A'(t) = kA(t)$.

(b) Find a function A such that $A'(t) = 5A(t)$ for all t, and $A(0) = 2$.

[*Warning:* We have now found solutions of the equation $A' = kA$; but are these the only solutions? This important question is answered in the next chapter.]

2.8 HYPERBOLIC FUNCTIONS (optional)

The trigonometric functions are sometimes called the *circular* functions because of their close connection with the circle $\{(x,y): x^2 + y^2 = 1\}$. [See Fig. 40.] Certain combinations of exponentials are related in the same way to the *hyperbola* $\{(x,y): x^2 - y^2 = 1\}$ [See Fig. 41.] These functions are called *hyperbolic sine, hyperbolic cosine,* etc., and denoted sinh (pronounced "sinch"), cosh, tanh, etc. The definitions are

$$\sinh \theta = \frac{e^{\theta} - e^{-\theta}}{2}, \qquad \cosh \theta = \frac{e^{\theta} + e^{-\theta}}{2}, \qquad \tanh \theta = \frac{\sinh \theta}{\cosh \theta},$$

$$\coth \theta = \frac{1}{\tanh \theta}, \qquad \operatorname{sech} \theta = \frac{1}{\cosh \theta}, \qquad \operatorname{csch} \theta = \frac{1}{\sinh \theta}.$$

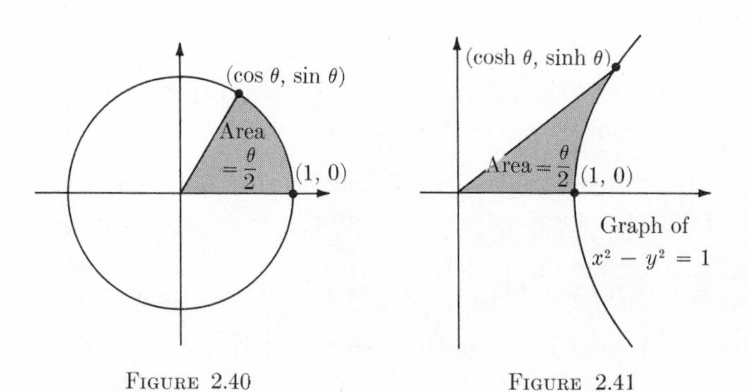

FIGURE 2.40 FIGURE 2.41

The first three are graphed in Fig. 42. For $|\theta| \leq 1$, the graph of $\cosh \theta$ looks much like the parabola $f(\theta) = 1 + \theta^2/2$, but for large values of θ the sides of the $\cosh \theta$ graph rise much more steeply than the sides of the parabola.

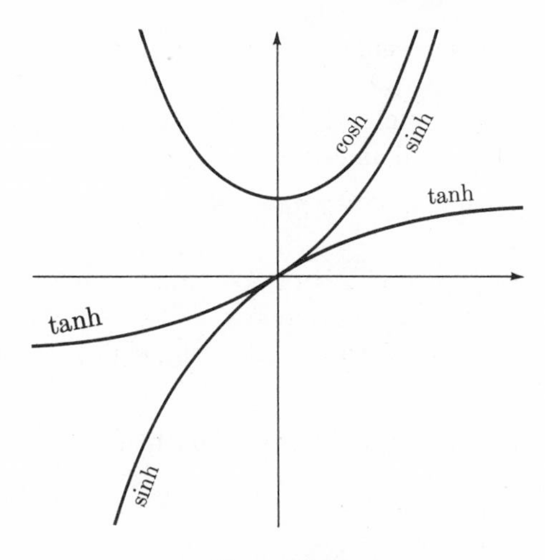

FIGURE 2.42

A flexible chain or rope supported at two points [Fig. 43] will hang in the shape of the graph of $\frac{1}{a}\cosh(ax)$, where a is a constant depending on the chain and the units of distance [see §7.5, Problem 10]. Hence the graph is called a *catenary* ("catena" being the Latin for "chain").

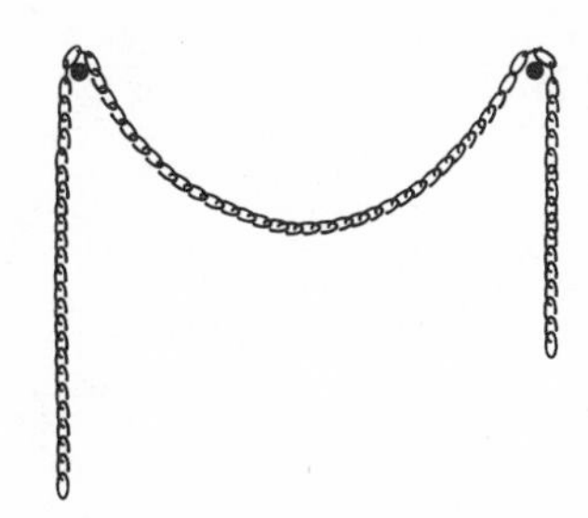

FIGURE 2.43 Catenary

The problems below list the elementary properties of the hyperbolic functions; the verification is left to you as a review of this chapter. Unfortunately, we must wait until §6.6, Problems 13 and 14, to verify that the shaded area in Fig. 41 is really $\theta/2$.

PROBLEMS

1. Prove the following identities.

(a) $\cosh^2\theta - \sinh^2\theta = 1$

(b) $1 - \tanh^2\theta = \operatorname{sech}^2\theta$

(c) $\coth^2\theta - 1 = \operatorname{csch}^2\theta$

2. Prove the following symmetry formulas.
(a) $\cosh(-\theta) = \cosh\theta$
(b) $\sinh(-\theta) = -\sinh\theta$
(c) $\tanh(-\theta) = -\tanh\theta$.

3. Prove the following addition formulas.
(a) $\cosh(\theta + \varphi) = \cosh\theta\cosh\varphi + \sinh\theta\sinh\varphi$
(b) $\cosh(\theta - \varphi) = \cosh\theta\cosh\varphi - \sinh\theta\sinh\varphi$
[Use Problem 2 and part (a).]
(c) $\sin(\theta \pm \varphi) = \sinh\theta\cosh\varphi \pm \cosh\theta\sinh\varphi$
(d) $\tanh(\theta \pm \varphi) = ?$

4. Prove the following "double angle" formulas.
(a) $\sinh 2\theta = 2\sinh\theta\cosh\theta$
(b) $\cosh 2\theta = 1 + 2\sinh^2\theta$

5. Prove that $(\cosh\theta + \sinh\theta)^n = \cosh(n\theta) + \sinh(n\theta)$.

6. Prove the following derivative formulas.
(a) $\sinh' = \cosh$
(b) $\cosh' = \sinh$
(c) $\tanh' = \text{sech}^2$
(d) $\coth' = ?$
(e) $\text{sech}' = ?$
(f) $\text{csch}' = ?$

7. Show the following. [These facts, together with the symmetries in Problem 2, are the basis of the graphs in Fig. 42.]

(a) $\cosh\theta \geq 1$ $\left[\text{Hint: } x + \dfrac{1}{x} \geq 2, \text{ for } x > 0.\right]$

(b) $\cosh\theta \geq \frac{1}{2}e^{|\theta|}$

(c) For $\theta > 0$, $\dfrac{e^\theta - 1}{2} < \sinh\theta < \cosh\theta$

(d) $-1 < \tanh\theta < 1$

8. Sketch graphs of the following inverse hyperbolic functions, restricting domains where necessary.
(a) arctanh
(b) arcsinh
(c) arccosh
(d) arcsech

9. Compute the derivatives of the inverse functions in Problem 8.

10. (a) Solve $\dfrac{e^\theta - e^{-\theta}}{e^\theta + e^{-\theta}} = y$, first for e^θ, and then for θ.

(b) Show that

$$\tanh\theta = y \quad \Leftrightarrow \quad \theta = \frac{1}{2}\log\frac{1+y}{1-y}, \qquad |y| < 1.$$

[This gives the formula for the inverse hyperbolic tangent, denoted arctanh or, sometimes, $\tanh^{-1}$.]

(c) Show that $\tanh \theta$ assumes every value between -1 and $+1$ for exactly one value of θ. [This fact shows up in the graph, Fig. 42.]

11. Obtain formulas for the following inverse functions, as in Problem 10.
(a) arcsinh (b) arccosh (c) arcsech

12. Check the derivatives found in Problem 9, by differentiating the formulas from Problems 10 and 11.

2.9 SUMMARY OF DERIVATIVE FORMULAS

The formulas derived in this chapter are summarized in the following three tables. *Table A* should definitely be known from memory. The results in *Table B* should be available to you without outside reference, but you may prefer to remember how to derive them rather than what they are. The various placements of sign are difficult to remember precisely, and it is often safer to carry out the derivation. *Table C* is primarily for reference. In each formula, u stands for an arbitrary differentiable function; this is included to remind you of the chain rule.

Example 1. $\dfrac{d\cos(3x^2+1)}{dx} = ?$ *Solution.* Table A gives the derivative of $\cos u$ as $-(\sin u)u'$. To make this fit our present case, set $u(x) = 3x^2 + 1$, which gives $u'(x) = 6x$. We thus get

$$\frac{d\cos(3x^2+1)}{dx} = -(\sin u)u' = -(\sin(3x^2+1))\cdot 6x = -6x\sin(3x^2+1).$$

Example 2. $\dfrac{d\arctan x}{dx} = ?$ *Solution.* Table A gives the derivative of $\arctan u$ as $u'/(1+u^2)$. To make this fit our present case, set $u(x) = x$, which gives $u'(x) = 1$. We thus get

$$\frac{d\arctan x}{dx} = \frac{u'}{1+u^2} = \frac{1}{1+x^2}.$$

Table A

f	f'
u^n	$nu^{n-1}u'$ (n an integer; $u \neq 0$ if $n < 0$)
u^a	$au^{a-1}u'$ (a any constant; $u > 0$)
cu	cu' (c any constant)
uv	$u'v + uv'$
uvw	$u'vw + uv'w + uvw'$
$\sum_1^n u_j$	$\sum_1^n u_j'$
$\dfrac{u}{v}$	$\dfrac{u'v - uv'}{v^2}$
$u \circ v$	$(u' \circ v)v'$
$\sin u$	$(\cos u)u'$
$\cos u$	$-(\sin u)u'$
$\tan u$	$(\sec^2 u)u'$
$\cot u$	$-(\csc^2 u)u'$
$\sec u$	$(\sec u)(\tan u)u'$
$\csc u$	$-(\csc u)(\cot u)u'$
$\arctan u$	$\dfrac{u'}{1 + u^2}$
$\log \lvert u \rvert$	u'/u ($u \neq 0$) [See Problem 4, §2.6]
$e^u = \exp(u)$	$e^u u'$

TABLE B

f	f'	Principal value and domain restrictions
$u^{\leftarrow}$ [u inverse]	$\dfrac{1}{u' \circ u^{\leftarrow}}$	
a^u	$(a^u)(\log a)u'$	
$\log_a \lvert u \rvert$	$\dfrac{u'}{u \log a}$	
$\arcsin u$	$\dfrac{u'}{\sqrt{1 - u^2}}$	$-\pi/2 < \arcsin u < \pi/2$; $\lvert u \rvert < 1$
$\operatorname{arcsec} u$	$\dfrac{u'}{\lvert u \rvert \sqrt{u^2 - 1}}$	$0 < \operatorname{arcsec} u < \pi$; $\lvert u \rvert > 1$

TABLE C

$\arccos u$	$\dfrac{-u'}{\sqrt{1 - u^2}}$	$0 < \arccos u < \pi$; $\lvert u \rvert < 1$
$\operatorname{arccot} u$	$\dfrac{-u'}{1 + u^2}$	$0 < \operatorname{arccot} u < \pi$
$\operatorname{arccsc} u$	$\dfrac{-u'}{\lvert u \rvert \sqrt{u^2 - 1}}$	$-\pi/2 < \operatorname{arccsc} u < \pi/2$; $\lvert u \rvert > 1$
$\sinh u$	$\cosh u \cdot u'$	
$\cosh u$	$\sinh u \cdot u'$	
$\tanh u$	$\operatorname{sech}^2 u \cdot u'$	
$\coth u$	$-\operatorname{csch}^2 u \cdot u'$	
$\operatorname{sech} u$	$-\operatorname{sech} u \tanh u \cdot u'$	
$\operatorname{csch} u$	$-\operatorname{csch} u \coth u \cdot u'$	
$\operatorname{arcsinh} u$	$\dfrac{u'}{\sqrt{1 + u^2}}$	
$\operatorname{arccosh} u$	$\dfrac{u'}{\sqrt{u^2 - 1}}$	$\operatorname{arccosh} u > 0$; $u > 1$
$\operatorname{arctanh} u$	$\dfrac{u'}{1 - u^2}$	$\lvert u \rvert < 1$
$\operatorname{arcsech} u$	$\dfrac{-u'}{u\sqrt{1 - u^2}}$	$0 < u < 1$; $\operatorname{arcsech} u > 0$

PROBLEMS

Compute the derivatives of the following functions. [Problems 36 and 37 are background for Chapter VI.]

- • **1.** $e^{\cos x}$
- **2.** $\cos (e^x)$
- • **3.** $1 + \cos^2 x$
- **4.** $2 \cos 2x$
- • **5.** $\log \left(\dfrac{1+x}{1-x}\right)$
- **6.** $\log (x + \sqrt{x^2 + 1})$
- • **7.** $\log (x + \sqrt{x^2 - 1})$
- **8.** $\log (x^{-1} + \sqrt{x^{-2} - 1})$
- • **9.** $\log (x^{-1} + \sqrt{x^{-2} + 1})$
- **10.** $\arctan (a^2 + (x - b)^2)$
- • **11.** $x(ax^2 + bx + c)^{-2}$
- **12.** $(ax^2 + bx + c)^{-k}$
- • **13.** e^{kx} (k a constant)
- **14.** e^{kx^2}
- • **15.** $(e^{kx})^2$
- **16.** $\cos^2 (e^x) + \sin^2 (e^x)$
- **17.** $\cosh^2 (e^x) + \sinh^2 (e^x)$
- • **18.** $\exp (e^x)$
- **19.** $\exp (\exp' (e^t))$
- • **20.** $\exp (1 + \log x)$
- • **21.** $\exp (\log^2 x)$
- • **22.** $\exp [(\log x)(\cos x)]$
- • **23.** $\log (x(x - 1))$
- **24.** $\log (e^x + e^{2x})$
- • **25.** $\log (\cos x + \sin x + \tan x)$
- • **26.** $\log (\tan \theta)$
- • **27.** $\cos (\log x + \log x^2)$
- **28.** $\cos (\arcsin x)$
- • **29.** $\sqrt{1 - x^2}$
- • **30.** $\sin (\arccos x)$
- **31.** $\sin (\arcsin x + \arcsin 2x)$
- **32.** $\tan (2 \arctan x)$
- **33.** $\sec (\log (x^2 + 1))$
- **34.** $\sec (1 + \sqrt{1 + \sqrt{x}})$
- • **35.** $\arctan (\csc \theta)$

36. Compute the derivatives of the following functions. Each can be reduced to a surprisingly simple form.

- • (a) $\log (\sec \theta + \tan \theta)$
- (b) $\log \dfrac{x + a}{x + b}$
- • (c) $e^{bx}(b \cos ax + a \sin ax)$
- (d) $x^2(2 \log x - 1)$
- • (e) $(x^2 - 2x + 2)e^x$
- (f) $[(ax)^3 - 3(ax)^2 + 6ax - 6]e^{ax}$

37. You are given a function F with undetermined constants $A, B, C, \ldots$. Differentiate the function, and select the constants so that F' satisfies the given equation.

(a) $F(x) = Ae^{2x}, \quad F'(x) = e^{2x}$

(b) $F(x) = A \cos kx, \quad F'(x) = -\sin kx \quad$ [Give A in terms of k.]

(c) $F(x) = Ae^{3x^2}, \quad F'(x) = xe^{3x^2}$

(d) $F(x) = A \log |x + 1| + B \log |1 - x|, \qquad F'(x) = \dfrac{1}{x^2 - 1}$

(e) $F(x) = A \log |x + 1| + B \log |1 - x|, \qquad F'(x) = \dfrac{x + 2}{x^2 - 1}$

(f) $F(x) = A \log |x - 2| + B(x - 2)^{-1}, \qquad F'(x) = \dfrac{x + 1}{(x - 2)^2}$

(g) $F(x) = A \arctan\left(\dfrac{x - 1}{2}\right) + B \log (x^2 - 2x + 5),$

$$F'(x) = \frac{x + 1}{x^2 - 2x + 5}$$

2.10 IMPLICIT DIFFERENTIATION AND RELATED RATES

In many simple applications of calculus we are given a relation involving one or more functions, and by differentiation we obtain a relation between their derivatives.

For example, let x and y stand for the coordinates of a point on the circle of radius r about the origin; then

$$x^2 + y^2 = r^2. \tag{1}$$

Suppose we think of y as related to x by some differentiable function, $y = f(x)$, such that (1) is valid for every x. Then differentiating (1) gives

$$2x + 2y \frac{dy}{dx} = 0;$$

hence

$$\frac{dy}{dx} = -\frac{x}{y}. \tag{2}$$

This means that for *any* function relating y to x so that (1) holds, formula (2) for the derivative is also valid. In this case, there are two obvious functions, one given by

$$y = \sqrt{r^2 - x^2} = (r^2 - x^2)^{1/2}, \tag{3}$$

and the other by

$$y = -\sqrt{r^2 - x^2} = -(r^2 - x^2)^{1/2}. \tag{4}$$

We can check the general result (2) by differentiating (3) or (4). In the case of (3), we get

$$\frac{dy}{dx} = \frac{1}{2}(r^2 - x^2)^{-1/2}(-2x) = \frac{-x}{\sqrt{r^2 - x^2}} = -\frac{x}{y},$$

since $y = \sqrt{r^2 - x^2}$. In the case of (4), we get

$$\frac{dy}{dx} = -\frac{1}{2}(r^2 - x^2)^{-1/2}(-2x) = \frac{x}{\sqrt{r^2 - x^2}},$$

which again has the form $-x/y$, since now $y = -\sqrt{r^2 - x^2}$ as in (4). In either case, formula (2) checks (as of course it must).

We can change the example slightly by supposing that (1) is valid, but that *both* x and y are related to another quantity t by differentiable functions. The derivatives of these functions are denoted dx/dt and dy/dt. Differentiating the given relation

$$x^2 + y^2 = r^2$$

with this understanding, we get

$$2x\frac{dx}{dt} + 2y\frac{dy}{dt} = 0. \tag{5}$$

This equation must hold *no matter what* functions relate x and y to t, as long as (1) remains valid for every t. We can check it by trying particular relations, for example

$$x = r\cos t, \qquad y = r\sin t.$$

This is consistent with (1), for $x^2 + y^2 = r^2\cos^2 t + y^2\sin^2 t = r^2$. And (5) checks, since $dx/dt = -r\sin t$ and $dy/dt = r\cos t$, so

$$2x\,dx/dt + 2y\,dy/dt = 2r\cos t(-r\sin t) + 2r\sin t(r\cos t) = 0.$$

Equations (2) and (5) were obtained by what is called *implicit differentiation.* Given a relation such as (1) between certain functions, you can differentiate both sides to obtain a new relation between these functions and their derivatives (assuming, of course, that the derivatives exist!) This is the basis of one of the elementary applications of calculus, the solution of problems in "related rates."

Example 1. A point $P = (x,y)$ moves on the circle (1). At a certain time t_0, P is at the point $(3r/5, 4r/5)$, and $dx/dt = 4$. Find dy/dt at that time t_0.
Solution. Since (1) holds at every time t, the relation (3), obtained by differentiating (1), must also hold at every time t. If we substitute in (5) the values $x = 3r/5$, $y = 4r/5$, $dx/dt = 4$, which hold at the given time t_0, we obtain

$$2\frac{3r}{5}\cdot 4 + 2\frac{4r}{5}\cdot\frac{dy}{dt} = 0.$$

Solving this equation for dy/dt, we find $dy/dt = -3$.

Example 2. From the shore, a man M observes a rotating beacon light L. The light rotates once a minute, and the spot of light crosses a wall directly behind the man at a rate of 10 yards per second. The wall is perpendicular to the line from the man to the beacon. How far is the beacon from the wall?

A problem like this requires a picture first of all. We draw Fig. 44 and assign letters to whatever seems appropriate. [This is the nonmathematical part of the problem. It's an interesting part, but we don't consider it our job to explain in detail how to do it: you are strictly on your own here.] Once this is done, we can write out the given information:

$$\frac{d\theta}{dt} = \frac{2\pi}{60}$$ [since the light rotates through an angle of 2π radians in sixty seconds],

$$\frac{dx}{dt} = 10 \text{ when } \theta = 0$$ [since dx/dt is the rate at which the spot of light moves along the wall, and this is given as 10 yards per second when $\theta = 0$].

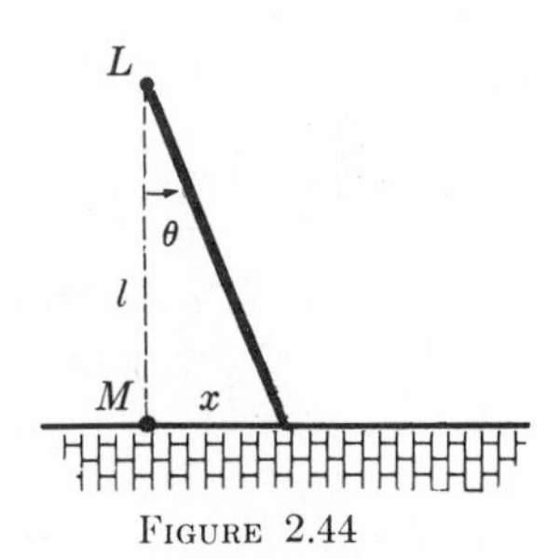

FIGURE 2.44

The relation between x, l, and θ can be seen from the figure:

$$x = l \tan\theta. \tag{6}$$

To deduce from this the general relation between x, l, θ, and the derivatives of these quantities with respect to time, differentiate (6) by the chain rule:

$$\frac{dx}{dt} = l \sec^2\theta\,\frac{d\theta}{dt}. \tag{7}$$

[Notice that l is a constant in this problem; if it were variable, the right-hand side of (7) would have an additional term $\frac{dl}{dt}\tan\theta$, by the product rule.] Finally, substitute into (7) the particular values given above for $\theta = 0$, and solve for l:

$$10 = l \cdot 1 \cdot \frac{2\pi}{60}, \qquad \text{so} \qquad l = \frac{300}{\pi} \text{ yards.}$$

PROBLEMS

• **1.** A point moves along the circle $\{(x,y)\colon x^2 + y^2 = 1\}$. At some particular time t_0, it is known that $x = 1/\sqrt{2}$, $y = -1/\sqrt{2}$, and $dx/dt = 3$. Find dy/dt at time t_0.

In Problems 2–4, some quantities are supposed to satisfy a given relation. Then it is further supposed that there are functions relating each of these quantities to the time t. Find the relation between the various quantities and their rates of change. For example, if $a^2 + b^2 + c^2 = 1$, then $2a\, da/dt + 2b\, db/dt + 2c\, dc/dt = 0$.

• **2.** $e^{xy} = \sin x$ (x and y functions of t)

• **3.** $x^2/a^2 + y^2/b^2 = c^2$ (a, b, and c constants; x and y functions of t)

4. $x = r\cos\theta$, $y = r\sin\theta$ (x, y, r, and θ all functions of t)

5. Let $x = r\cos\theta$, $y = r\sin\theta$ (x, y, and r all functions of θ). Express $dx/d\theta$ and $dy/d\theta$ in terms of r, θ, and $dr/d\theta$.

• **6.** Suppose $x^2 + 3xy + y^2 = 1$, and y is a function of x. Find dy/dx.

• **7.** A planet moves along the path $x^2/a^2 + y^2/b^2 = 1$.
(a) Find the relation between x, y, dx/dt, and dy/dt.
(b) At some time t_0, $x = a/2$, $y = b\sqrt{3}/2$, and $dy/dt = 3^{10}$. Find dx/dt at that time.

• **8.** A ship S is traveling near the coast at an unknown speed and direction. An observer in a lighthouse L takes measurements of the distance r from himself to the ship, and the angle θ from the ship to the coast. At a particular time t_0, he finds that $r = 6$, $dr/dt = 3$, $\theta = \pi/3$, and $d\theta/dt = -3$. Find dy/dt, the rate of increase of the distance from ship to shore, at time t_0. [See Fig. 45.]

9. Do Example 2 if the angle between the wall and the line ML to the beacon is $\pi/3$ instead of $\pi/2$. [In Fig. 44, the angle is $\pi/2$.]

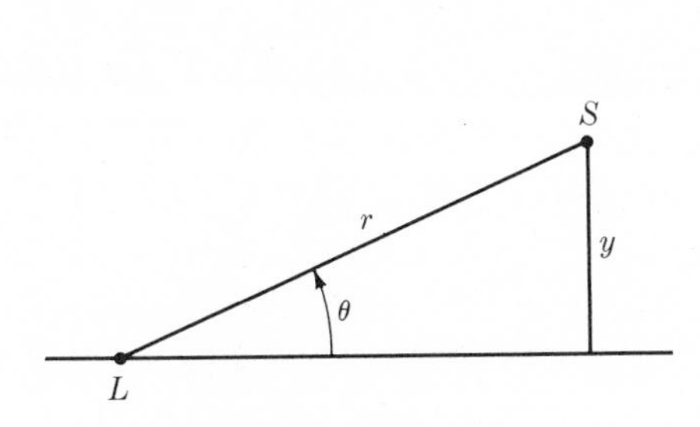

FIGURE 2.45

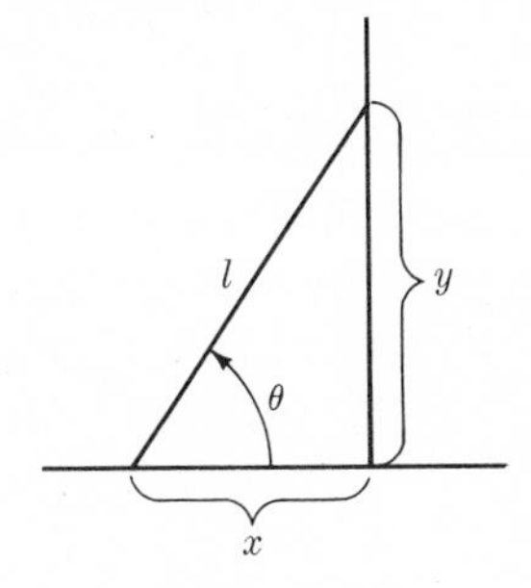

FIGURE 2.46

10. A ladder of length l leans against a wall at an angle θ, as in Fig. 46.
(a) Express x and y as functions of θ, and find $dy/d\theta$.
(b) Show that $1 = -l \sin \theta \, d\theta/dx$.
• (c) Use the results in (a) and (b) and the chain rule to find dy/dx. [Your answer will express dy/dx as a function of θ.]
• (d) Express dy/dx as a function of x.

• **11.** A ladder 13 feet long leans against a high, vertical wall. At time t_0, the lower end, five feet from the wall, is slipping away from the wall at 2 feet per second.
(a) How fast is the top of the ladder slipping down at time t_0?
(b) A man is on the ladder, 8 feet above the ground, at time t_0. How fast is he approaching the ground?

12. A ladder 13 feet long leans against a wall 7 feet high, projecting over the top. As in Problem 11, the bottom of the ladder is slipping away from the wall. If the ladder slides down the top edge of the wall at 2 feet per second, how fast is the bottom slipping at the moment when 3 feet of ladder project over the top of the wall?

• **13.** A container is shaped like the cone in Fig. 47. Water flows in at the constant rate of 2 cm³/min. How fast is the water level rising when the container is filled to height h cm? [Your answer, of course, will depend on h, and on the angle α between the axis and the side of the cone. The volume of a cone is $1/3 \times$ height $\times$ area of base.]

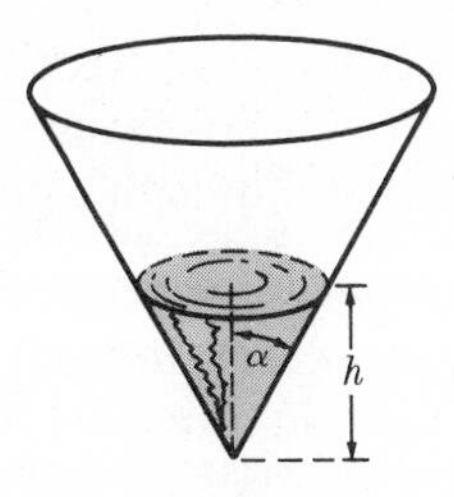

FIGURE 2.47

14. A sawmill dumps sawdust on a conical pile at the rate of 100 ft³/day. The side of the pile has slope 1, i.e. it makes an angle of $\pi/4$ with the ground. At what rate does the height of the pile increase when it is h ft. high?

• **15.** In a freezing rain, ice is forming on a telephone line, and the thickness of ice increases at the rate of $\frac{1}{8}$ in./hr. Consider a 1-foot long section of wire on which ice has formed to produce a cylinder of radius r_0 inches. At what rate is the volume of ice increasing on this 1-foot section of wire?

16. A tractor lifts a load of hay by pulling a rope over a pulley, as in Fig. 48. Given the following data, find the rate at which the load of hay is rising:

Length of rope, tractor hitch to pulley: 15 yards
Height of pulley above tractor hitch: 9 yards
Speed of tractor: 30 yards per minute.

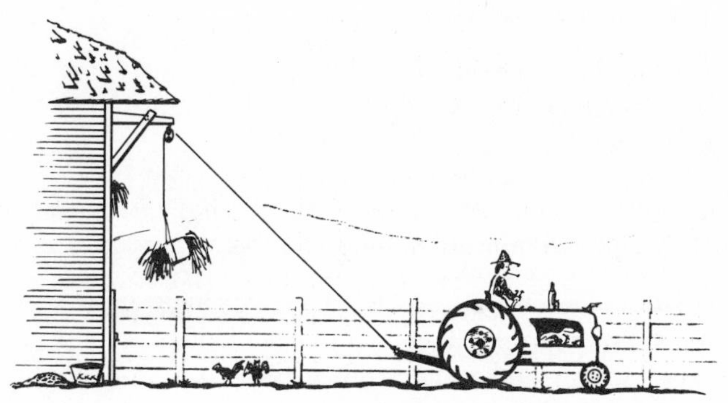

FIGURE 2.48

17. Two trains are on parallel tracks a quarter of a mile apart, one going 40 miles per hour, and the other 60 miles per hour in the same direction. Passenger A in the slow train is watching passenger B in the fast train. What is the rate of change of the distance between them

• (a) when they are directly opposite each other?
• (b) when A is an eighth of a mile in front of B?
(c) when B is an eighth of a mile in front of A?

18. Two cars are on roads that intersect at right angles, each car headed for the intersection. At what rate is the distance between them increasing if car A is one mile from the intersection and going 60 miles per hour, while car B is one-half mile from the intersection and going 80 miles an hour?

19. Do Problem 18 if the roads intersect at an angle of $\pi/3$.

20. Ship A is sailing out along the x axis; at time t hours it is at the point $(0,20t)$, $20t$ miles from the origin. A sailor sights ship B at the time $t = 1/2$ and makes these observations:

The distance r from A to B is 3 miles

The angle θ between the x axis and the line from A to B is $\pi/3$ radians

The distance r is increasing at -30 miles per hour

The angle θ is increasing at $-\pi$ radians per hour.

(a) Express the coordinates (x,y) of ship B in terms of r, θ, and t.

(b) At $t = 1/2$, find these coordinates (x,y) and find dx/dt and dy/dt.

(c) Assuming that dx/dt and dy/dt are constant, find x and y as functions of t.

2.11 SOME GEOMETRIC EXAMPLES (optional)

The circle of radius r has area $A = \pi r^2$, hence $dA/dr = 2\pi r$. But the circumference C of the circle is precisely $2\pi r$; thus

$$\frac{dA}{dr} = C. \tag{1}$$

This relation can actually be seen geometrically by taking difference quotients [Fig. 49], *without* knowing beforehand the formulas for A and C. Let $A(r)$ denote the area of the circle of radius r, and $C(r)$ its circumference. When h is a small positive number, then $A(r+h) - A(r)$ is the area of a narrow annulus. Intuitively, it is not hard to see that the area of such an annulus is larger than

$$h \times (\text{inner circumference}) = h \cdot C(r)$$

and is smaller than

$$h \times (\text{outer circumference}) = h \cdot C(r+h).$$

Thus

$$C(r) \le \frac{A(r+h) - A(r)}{h} \le C(r+h),$$

and by the Poe Principle, [§2.6]

$$\frac{dA}{dr} = \lim_{h\to 0} \frac{A(r+h) - A(r)}{h} = C(r).$$

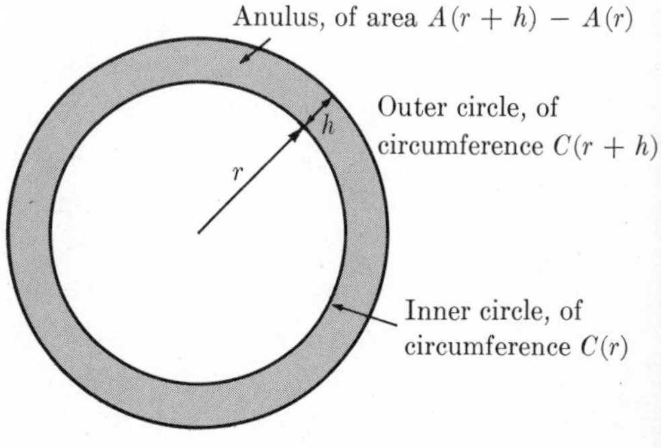

Anulus, of area $A(r+h) - A(r)$

Outer circle, of circumference $C(r+h)$

h

r

Inner circle, of circumference $C(r)$

FIGURE 2.49

Thus we recover (1) by a direct geometric argument.

This section is devoted to further examples where derivatives are computed by appeal to geometric intuition.

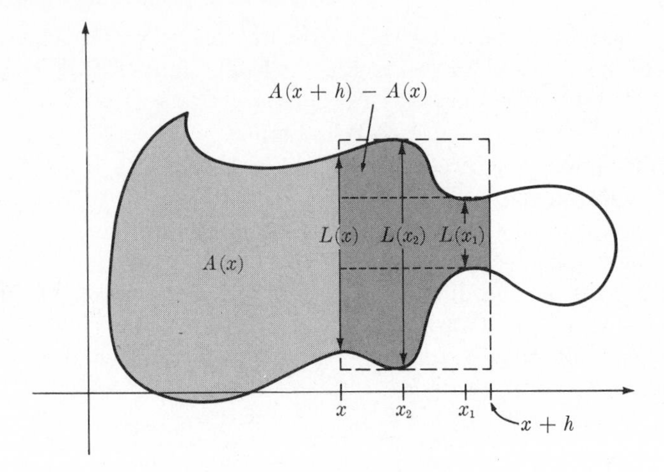

FIGURE 2.50

Areas

Consider a region R in the plane. Let $L(x)$ denote the length of the vertical cross section of R through the point $(x,0)$, and let $A(x)$ denote the area of that part of R lying to the left of this cross section [Fig. 50]. We compute dA/dx as follows. Given a small positive number h, let $L(x_1)$ denote the smallest cross section with $x \leq x_1 \leq x + h$, and $L(x_2)$ denote the largest cross section with $x \leq x_2 \leq x + h$. Then it is intuitively clear that

$$h \cdot L(x_1) \leq A(x + h) - A(x) \leq h \cdot L(x_2);$$

for, the region of area $A(x + h) - A(x)$ *contains* a rectangle of base h and height $L(x_1)$, and *is contained in* a rectangle of base h and height $L(x_2)$. Therefore, dividing by h,

$$L(x_1) \leq \frac{A(x + h) - A(x)}{h} \leq L(x_2).$$

Now, as $h \to 0$, we have $x_1 \to x$ and $x_2 \to x$; hence $L(x_1) \to L(x)$ and $L(x_2) \to L(x)$. Thus by the Poe Principle

$$\lim_{h \to 0} \frac{A(x + h) - A(x)}{h} = L(x),$$

or

$$\frac{dA}{dx} = L. \tag{2}$$

In words, the derivative of the "area to the left of x" is the length of the cross section at x.

The formula for the derivative of the logarithm is a special case of (2). Let R be the region bounded on the left by the line $\{(x,y) : x = 1\}$, bounded below by the x axis, and bounded above by the graph of $f(x) = 1/x$

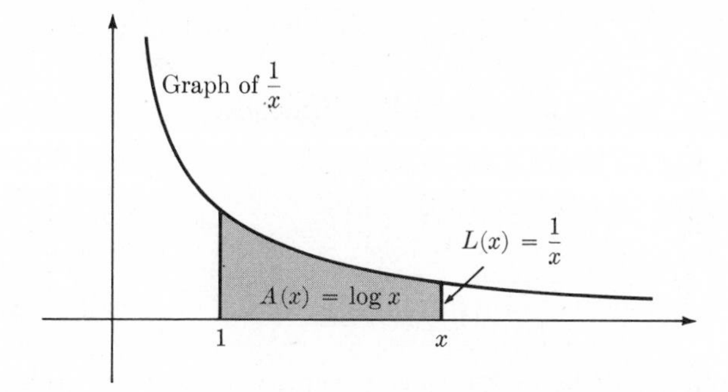

FIGURE 2.51

[Fig. 51]. Then for $x \geq 1$, the area $A(x)$ of that part of R lying to the left of x is precisely $\log x$, and the length of the cross section at x is $1/x$. Thus (2) implies that

$$\frac{d \log x}{dx} = \frac{1}{x}.$$

Volume of revolution

A volume symmetric about an axis (such as the shapes produced on a lathe or a potter's wheel) is called a *volume of revolution*. Suppose that the axis of symmetry is the x axis, as in Fig. 52. Let $A(x)$ denote the area of the perpendicular cross section at x, and let $V(x)$ denote the partial volume lying to the left of this cross section. Given a small positive number h, let $A(x_1)$ denote the smallest cross-sectional area with $x \leq x_1 \leq x + h$, and $A(x_2)$ denote the largest cross-sectional area with $x \leq x_2 \leq x + h$. Then $V(x + h) - V(x)$ is the volume of a region lying inside a cylinder of width h and cross section $A(x_2)$, and containing a cylinder of height h and cross section $A(x_1)$; thus

$$hA(x_1) \leq V(x + h) - V(x) \leq hA(x_2).$$

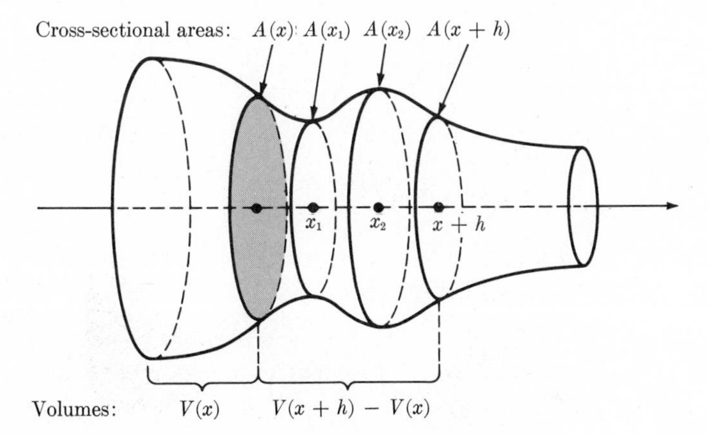

FIGURE 2.52

As $h \to 0$, then $x_1 \to x$ and $x_2 \to x$, hence $\lim_{h\to 0} A(x_1) = A(x)$ and $\lim_{h\to 0} A(x_2) = A(x)$. Thus by the Poe Principle,

$$\lim_{h\to 0} \frac{V(x+h) - V(x)}{h} = A(x), \qquad \text{or} \qquad \frac{dV}{dx} = A.$$

In words, the derivative of the "volume to the left of x" is the cross-sectional area at x.

The same result holds for more general volumes; see Problem 3.

Arc length

Let f be a differentiable function defined on an interval $[a,b]$, and let $s(x)$ denote the arc length of that part of the graph lying over the interval $[a,x]$. Our method of computing ds/dx is based not on the Poe Principle, but on the intuitive geometric principle that

$$\lim_{\text{arc}\to 0} \left(\frac{\text{arc length}}{\text{chord length}}\right) = 1. \tag{3}$$

The arc length between $P = (x,f(x))$ and $P' = (x+h, f(x+h))$ is simply $s(x+h) - s(x)$, and the length of the chord between P and P' is [Fig. 53]

$$l = \sqrt{h^2 + [f(x+h) - f(x)]^2}. \tag{4}$$

We can therefore take the derivative of the arc length as follows:

$$\lim_{h\to 0} \frac{s(x+h) - s(x)}{h} = \lim_{h\to 0} \frac{\text{arc length}}{h}$$

$$= \lim_{h\to 0} \frac{\text{arc length}}{\text{chord length}} \lim_{h\to 0} \frac{\text{chord length}}{h}$$

$$= 1 \cdot \lim_{h\to 0} \sqrt{1 + \left(\frac{f(x+h) - f(x)}{h}\right)^2}$$

[by (3) and (4)]

$$= \sqrt{1 + \lim_{h\to 0} \left(\frac{f(x+h) - f(x)}{h}\right)^2}$$

[by Lemma 1, §2.4]

$$= \sqrt{1 + f'(x)^2}.$$

Thus:

$$\frac{ds}{dx} = \sqrt{1 + \left(\frac{df}{dx}\right)^2}.$$

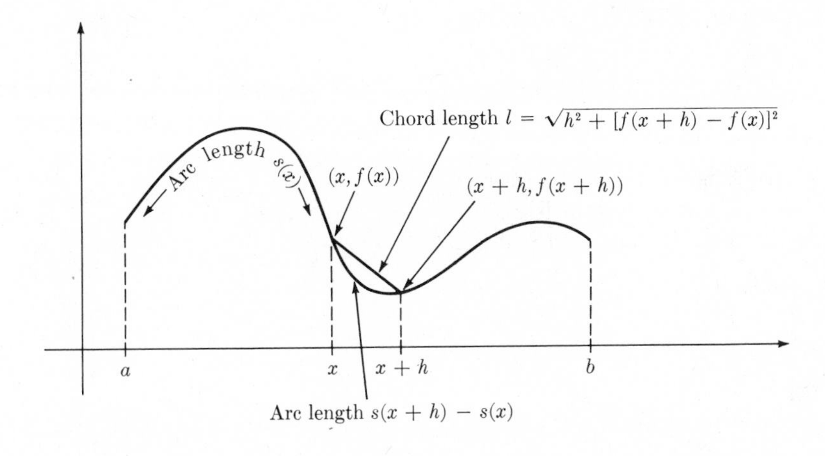

FIGURE 2.53

PROBLEMS

1. Find dA/dx, the derivative of the area of R to the left of x, when R is the given region, and x varies in the given range. Make sketches.
   (a) The triangle with vertices (0,0), (10,0), (10,10m), where m is constant; $0 < x < 10$.
   (b) The region bounded above by the graph of e^x, on the left by the y axis, and below by the x axis; take $0 \leq x$.
   (c) The disk $\{(x,y): x^2 + y^2 \leq 1\}$; $-1 < x < 1$.
   •(d) The ellipse $\left\{(x,y): \dfrac{x^2}{a^2} + \dfrac{y^2}{b^2} = 1\right\}$; $-a < x < a$.

2. (a) Check the answer to Problem 1(a) by computing $A(x)$ explicitly, and taking its derivative.
   (b) Check the answer to Problem 1(c) by showing that $A(x) = x\sqrt{1 - x^2} + \pi - \arccos x$, and differentiating this function. [Hint: To find the area formula, notice that for $x > 0$, $A(x)$ is the area of a large piece of pie plus a triangle, Fig. 54.]

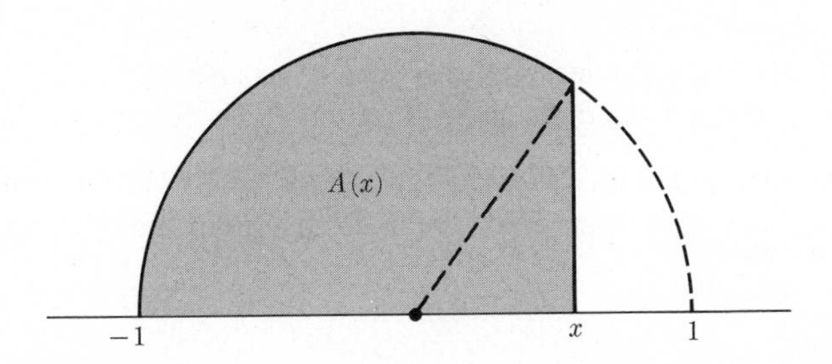

FIGURE 2.54

3. Compute dV/dx, the derivative of the "volume of R lying to the left of x," in the following cases:
   • (a) R is a cube of edge e, the x axis is perpendicular to one face at $x = 0$ and to another face at $x = e$. Compute the derivative for $0 < x < e$.
   (b) R is a solid ball of radius r; $-r < x < r$.
   (c) R is the cone in Fig. 55.
   • (d) R is the pyramid in Fig. 56.

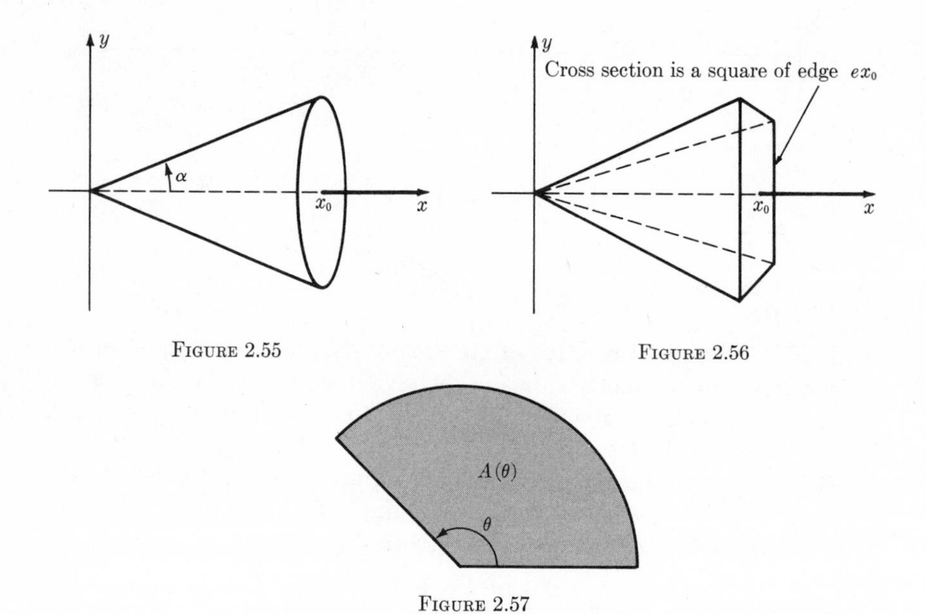

FIGURE 2.55

FIGURE 2.56

FIGURE 2.57

4. Let $A(\theta)$ denote the area of a section of a circle of radius r with angle θ [Fig. 57]. *Without* assuming any formula for the area of a sector, prove that $dA/d\theta = r\theta$. [Hint: Compare $A(\theta + h) - A(\theta)$ to the areas of an inscribed and a circumscribed triangle, and apply the Poe Principle.]

5. Let V and S denote respectively the volume and surface area of a sphere of radius r.
   (a) Using formulas for V and S, show that $dV/dr = S$.
   (b) Explain the result in part (a) geometrically.

6. Suppose that A and B are plane regions, and A is contained in B. Is the perimeter of A necessarily less than the perimeter of B? Explain.

Chapter III

Applications of Derivatives

The applications in this chapter, beginning with elementary graphs and ending with an analysis of oscillating springs and electric circuits, are all based on one simple idea; the sign of f' determines whether f is increasing or decreasing, i.e. whether the graph of f is rising to the right or falling to the right.

This fundamental idea deserves a precise formulation.

Definition 1. Let the domain of f include an interval I. Then: **INCREASING FUNCTIONS**

f is *increasing* on I iff for all x_1 and x_2 in I,

$$x_1 < x_2 \quad \Rightarrow \quad f(x_1) \leq f(x_2). \tag{1}$$

f is *strictly increasing* on I iff for all x_1 and x_2 in I,

$$x_1 < x_2 \quad \Rightarrow \quad f(x_1) < f(x_2). \tag{2}$$

Figure 1 illustrates *increasing* and *strictly increasing*.

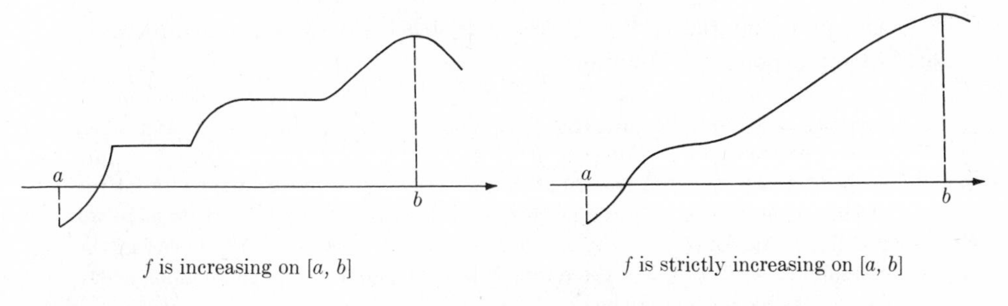

f is increasing on $[a, b]$ f is strictly increasing on $[a, b]$

FIGURE 3.1

The corresponding definition of *decreasing* replaces "$f(x_1) \leq f(x_2)$" in (1) by "$f(x_1) \geq f(x_2)$", and *strictly decreasing* replaces "$f(x_1) < f(x_2)$" by "$f(x_1) > f(x_2)$" [Fig. 2].

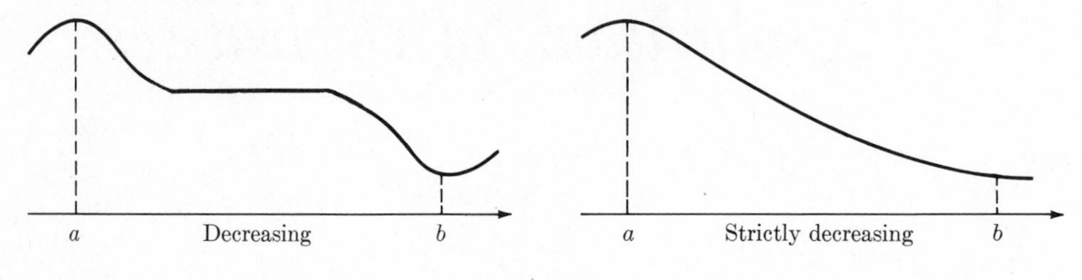

FIGURE 3.2

The basic relation between the sign of f' and the intervals on which f is increasing or decreasing is:

BASIS OF CHAPTER III

Proposition F. *Suppose that $f'(x)$ exists for every x in an interval I. Then:*

$$f' \geq 0 \text{ throughout } I \Rightarrow f \text{ is increasing on } I$$
$$f' > 0 \text{ throughout } I \Rightarrow f \text{ is strictly increasing on } I$$
$$f' \leq 0 \text{ throughout } I \Rightarrow f \text{ is decreasing on } I$$
$$f' < 0 \text{ throughout } I \Rightarrow f \text{ is strictly decreasing on } I.$$

This is easy to believe, but hard to prove. [The proof is in §4.2.] In this chapter we take the proposition for granted and explore its consequences.

3.1 INCREASING AND DECREASING FUNCTIONS

Our first example applying Proposition F leads to an important property of the exponential function.

Example 1. Find the intervals where $f(x) = x^2e^{-x}$ is increasing, and those where it is decreasing. *Solution.* $f'(x) = 2xe^{-x} - x^2e^{-x} = x(2 - x)e^{-x}$. Since $e^{-x} > 0$ for all x, f' has the same sign as $x(2 - x)$. We find the sign of the product from the signs of the factors, as in Fig. 3. Hence $f'(x)$ is *negative* for $x < 0$, *positive* for $0 < x < 2$, and *negative* for $x > 2$; and $f'(x)$ vanishes at the endpoints $x = 0$, $x = 2$ of these intervals. This determines the regions where f is increasing and decreasing.

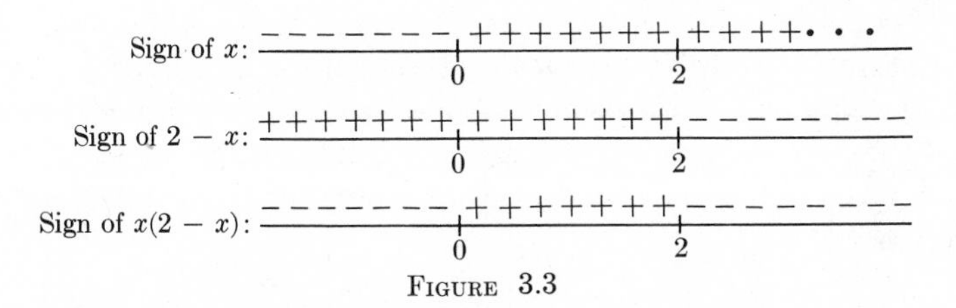

FIGURE 3.3

Knowing the regions of increase and decrease gives some idea of the graph. In Example 1, the graph might resemble any of those in Fig. 4(a), but not the one in Fig. 4(b).

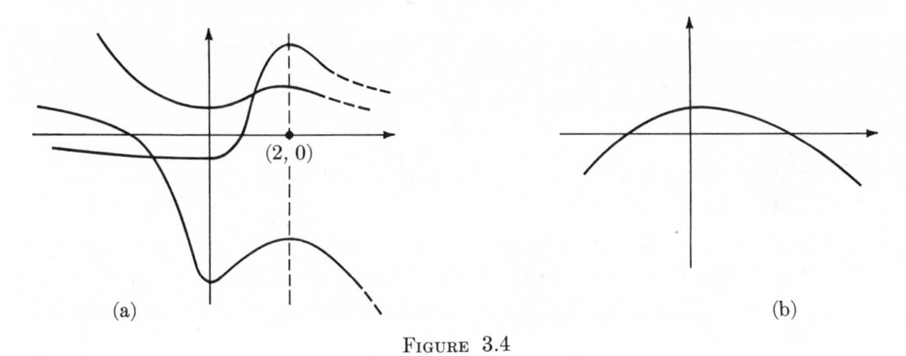

FIGURE 3.4

We get a still better idea by finding *the values of f at the points where f' changes sign.* To illustrate this, we continue the analysis of x^2e^{-x} begun in Example 1.

Example 1 (cont.). Let $f(x) = x^2e^{-x}$, as above. Since f' changes sign at $x = 0$ and $x = 2$, we compute $f(0) = 0$ and $f(2) = (2/e)^2$. [The value of $(2/e)^2$ is approximately $\frac{1}{2}$.]

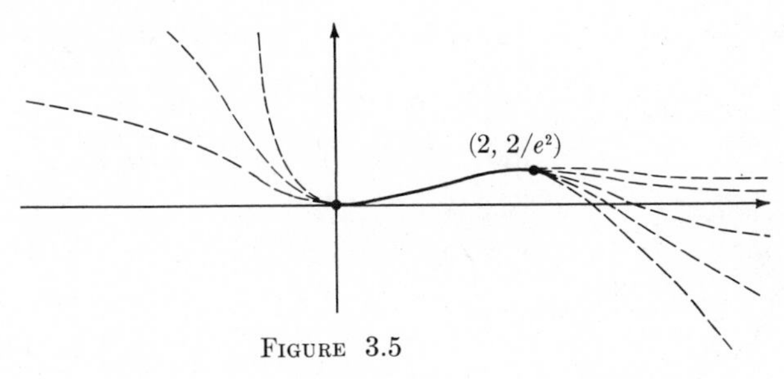

FIGURE 3.5

The graph now could be like any in Fig. 5, but not like any in 4(a). To determine its behavior to the left of $x = 0$, we note that when $x < 0$, then $e^{-x} > 1$, so $x^2e^{-x} > x^2$. The graph rises to the left with increasing steepness, always above the graph of x^2.

For the behavior of the graph to the right of $x = 2$, we can check *where f is positive and where negative.* (It is a good general rule to determine this in graphing any function.) We see that $f(x) = x^2e^{-x}$ is never negative (since $x^2 \geq 0$ and $e^{-x} > 0$), and $f(x)$ vanishes only when $x = 0$. With all this information, we obtain the good sketch of the graph of x^2e^{-x} shown in Fig. 6.

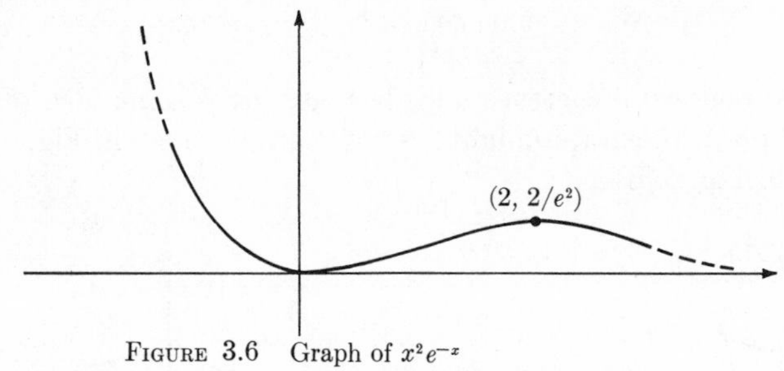

FIGURE 3.6 Graph of x^2e^{-x}

Notice from the graph that when $x > 0$, then $x^2e^{-x} \leq (2/e)^2$. It follows that $e^x \geq (e/2)^2x^2$ when $x > 0$, and we see that *as x becomes large and positive, e^x grows faster than x^2.* In the problems, you will show that e^x grows faster than *any* power of x. This comparison between e^x and x^n follows from either of the following two inequalities, to be proved in Problem 16:

e^x GROWS FASTER THAN x^n

$$e^x \geq \left(\frac{e}{n}\right)^n x^n \quad \text{if } x \geq 0, \tag{1}$$

$$e^x > x^n \qquad \text{if } x > \left(\frac{n+1}{e}\right)^{n+1}. \tag{2}$$

Hence anything that grows exponentially grows very fast indeed.

The graph of x^2e^{-x} in Fig. 6 suggests that the positive x axis is an asymptote, i.e. that $\lim_{x\to+\infty} x^2e^{-x} = 0$. This is true, and in fact we have more generally

$$\lim_{x\to+\infty} x^me^{-x} = 0 \quad \text{for any real } m. \tag{3}$$

The limit (3) is proved by applying (1) with $n = m + 1$, as follows. From (1), we get

$$x^{m+1}e^{-x} \leq \left(\frac{m+1}{e}\right)^{m+1};$$

hence

$$0 < x^me^{-x} < \frac{1}{x}\left(\frac{m+1}{e}\right)^{m+1}, \qquad x > 0. \tag{4}$$

Since $\lim_{x \to +\infty} (1/x) = 0$, the limit (3) follows from (4), by the Poe Principle [§2.6].

The limit (3) is yet another way to say mathematically that e^x grows faster than x^m as $x \to +\infty$; since $\lim_{x \to +\infty} (x^m/e^x) = 0$, the denominator e^x must grow faster than the numerator x^m.

Snell's law

A less abstract application of Proposition F is Snell's law of refraction, concerning "the quickest route from A to B." The straight line from A to B is always the *shortest* route, but not always the *quickest.*

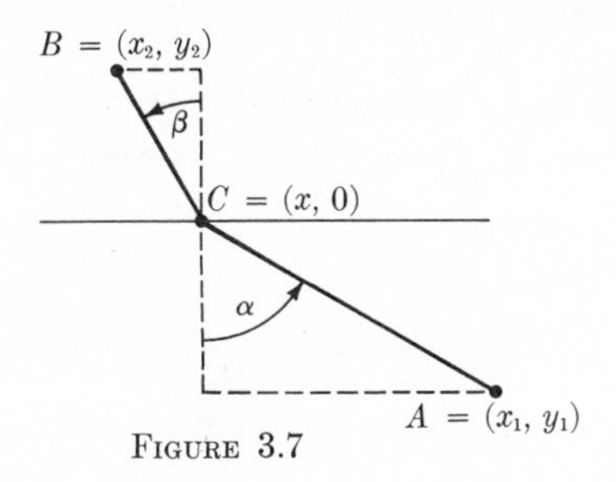

FIGURE 3.7

For example, in Fig. 7, suppose that the x axis is the edge of a pool, A is a lifeguard, and B is a swimmer in distress. Because running is faster than swimming, it is faster *not* to take the straight line from A to B, but to take a broken-line path (as suggested in the figure) that calls for less swimming and more running. Given the speeds of running and swimming, Snell's law tells which broken-line path is the quickest. The same law applies when a ray of light goes from A to B, if air occupies the region below the x axis and glass occupies the region above; for the speed of light in air (call it s_1) is greater than the speed of light in glass (call it s_2).

We will suppose the quickest path follows a straight line from A to some point C on the x axis, and continues on another straight line from C to B. Our problem is to determine C. Since C lies on the axis, it has the coordinates $(x,0)$ for some x. We compute the distances AC and CB, remember that the time spent is (distance)/(speed), and thus find that the time T spent in going from A to C to B is

$$T(x) = \frac{AC}{s_1} + \frac{CB}{s_2} = \frac{[(x - x_1)^2 + y_1^2]^{1/2}}{s_1} + \frac{[(x - x_2)^2 + y_2^2]^{1/2}}{s_2}.$$

We compute the derivative,

$$T'(x) = \frac{x - x_1}{s_1[(x_1 - x)^2 + y_1^2]^{1/2}} + \frac{x - x_2}{s_2[(x_2 - x)^2 + y_2^2]^{1/2}}$$

$$= -\frac{\sin \alpha}{s_1} + \frac{\sin \beta}{s_2} \qquad [\text{Fig. 7}].$$

The derivative is zero precisely when

SNELL'S LAW

$$\frac{\sin \alpha}{s_1} = \frac{\sin \beta}{s_2}. \tag{5}$$

Moving C to the right increases β and decreases α [see Fig. 7], so $T'(x) = -\sin \alpha/s_1 + \sin \beta/s_2$ is an increasing function. Hence if x_0 is the point where $T'(x_0) = 0$, then $T'(x)$ is negative to the left of x_0, positive to the right of x_0. It follows that T is a decreasing function on the interval $(-\infty, x_0]$, and an increasing function on $[x_0, +\infty)$, so x_0 gives the path requiring minimum time. [See Fig. 8.]

We referred to Fig. 7 to show that T' is increasing; this can also be done using Proposition F, as in Problem 21 below.

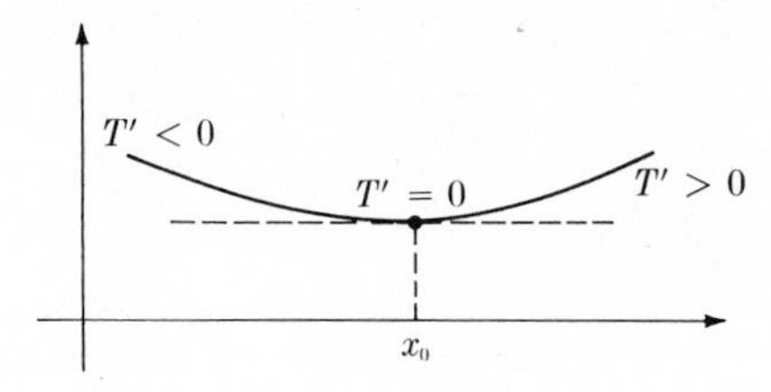

FIGURE 3.8

Snell's law explains the interesting phenomenon of refraction of light. According to *Fermat's principle*, a ray of light travels from A to B by the quickest route. Moreover, light travels more slowly in a dense medium such as glass or water than it does in a thin medium such as air. Hence, on crossing a boundary between glass and air, or between water and air, rays of light are "broken," or "refracted," in the way described by formula (5). In the case of glass and air, refraction is exploited to make lenses. [See Fig. 9.] In the case of water and air, refraction causes the familiar effect shown in Fig. 10.

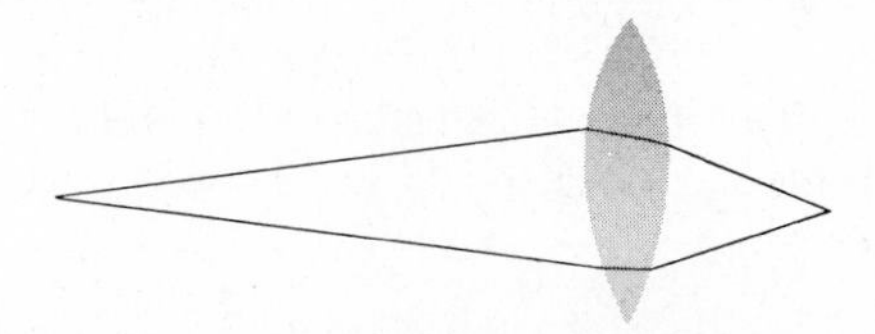

FIGURE 3.9 Bending of light by a lens

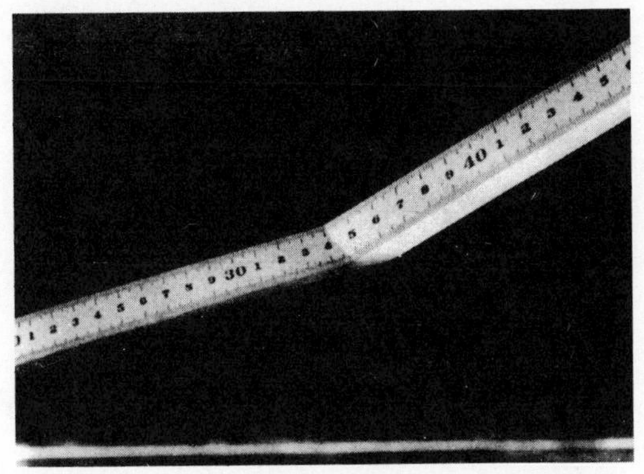

FIGURE 3.10*

PROBLEMS

1. Sketch the following functions, using Proposition F, together with the information in §1.7 on asymptotes.

• (a) $f(x) = \dfrac{x^2 - 2x + 1}{x - 2}, \quad x \neq 2$ [Examples 1 and 2, §1.1]

• (b) $g(x) = \dfrac{2x^2 - 3}{3x^2 + 5}$ [Example 3, §1.1]

(c) $h(x) = \dfrac{x - 1}{1 + 3x^2}$ [Example 4, §1.1]

In Problems 2–15, find the intervals of increase and decrease of f, the values of f at the endpoints of these intervals, and (where practical) the intervals in which f is positive or negative. Sketch a graph of f based on this information.

2. $f(x) = x^3 + 1$

• **3.** $f(t) = 3t^4 - 5t^3 + 3t^2 + 10$

4. $f(z) = z^3 - 3z^2 + 3z + 1$

• **5.** $f(x) = x + \sin x$

6. $f(y) = y - \sin y$

• **7.** $f(x) = (x + 1)(x + 2)(x + 3)$

8. $f(x) = \sin x + \cos x$

• **9.** $f(x) = x + 1/x$

10. $f(x) = \sqrt{x} \log x \quad (x > 0)$

From PSSC PHYSICS, D.C. Heath and Company, Boston, 1965

• **11.** $f(x) = x^{1/3} \log x \qquad (x > 0)$

12. $f(x) = x^{1/n} \log x \qquad (x > 0;\ n \text{ an integer} > 0)$

• **13.** $f(x) = x^3 e^{-x}$

14. $f(x) = x^n e^{-x}$
[The general shape of the graph depends on whether n is odd or even.]

15. $f(x) = xe^{-x^2}$

16. Let n be any positive integer and x be any positive number.
(a) Show that $e^x \geq [xe^{n+1}/(n+1)^{n+1}]x^n$. [Hint: Graph $x^{n+1}e^{-x}$ for $x > 0$.]
(b) Deduce from (a) that when $x > (n+1)^{n+1}/e^{n+1}$, then $e^x > x^n$ and $e^{-x} < x^{-n}$.

17. Let a be any positive number, and set $f(x) = x^a \log x$ for $x > 0$.

AS $x \to 0$, $\log x \to -\infty$ SLOWER THAN $-x^{-a}$ FOR EVERY $a > 0$

(a) Show that f is decreasing on the interval $(0, e^{-1/a}]$ and increasing on the interval $[e^{-1/a}, 1]$, and hence that for $0 < x < 1$ we have $0 > x^a \log x \geq -1/ae$.
(b) From part (a), deduce that $0 < |\log x| < (1/ae)x^{-a}$ for every $a > 0$, when $0 < x < 1$.
(c) For any $a > 0$, $\lim\limits_{x \to 0+} x^a \log x = 0$. [Hint: $x^a \log x = x^{a/2}(x^{a/2} \log x)$, and $\lim\limits_{x \to 0+} x^{a/2} = 0$.]

18. (a) Show that for any number $a > 0$, $x^{-a} \log x$ is bounded for $x \geq 1$.
(b) Find $\lim\limits_{x \to +\infty} x^{-a} \log x$, $a > 0$.

19. (a) Show that $x - \sin x \geq 0$ for $x \geq 0$. [Hint: show that the function $x - \sin x$ vanishes for $x = 0$ and is increasing for $x \geq 0$.]

(b) Show that $\cos x + \dfrac{x^2}{2} - 1 \geq 0$ for $x \geq 0$.

(c) Show that $\sin x + \dfrac{x^3}{3 \cdot 2} - x \geq 0$ for $x \geq 0$.

(d) Show that $-\cos x + \dfrac{x^4}{4 \cdot 3 \cdot 2} - \dfrac{x^2}{2} + 1 \geq 0$ for $x \geq 0$.

(e) Show that $-\sin x + \dfrac{x^5}{5 \cdot 4 \cdot 3 \cdot 2} - \dfrac{x^3}{3 \cdot 2} + x \geq 0$ for $x \geq 0$.

20. (a) Combine parts (c) and (e) of Problem 19 to show that for $x > 0$,

APPROXIMATING $\sin x$ BY POLYNOMIALS

$$x - \frac{x^3}{6} \leq \sin x \leq x - \frac{x^3}{6} + \frac{x^5}{120}.$$

(b) Show that $\sin(1) = (1 - \frac{1}{6} + \frac{1}{240}) \pm \frac{1}{240}$, that is, $\sin(1) = \frac{201}{240}$ with a maximum error of $\frac{1}{240}$.

(c) Show that $\sin(\frac{1}{2}) = \frac{23}{48} + \frac{1}{7680}$ with a maximum error of $\frac{1}{7680}$.

21. Let $f(x) = \dfrac{x - x_1}{s_1[(x_1 - x)^2 + y_1^2]^{1/2}} + \dfrac{x - x_2}{s_2[(x_2 - x)^2 + y_2^2]^{1/2}}$, where s_1 and s_2 are positive. [This is the function $T'(x)$ in the discussion of Snell's law.] Show that f is strictly increasing on the whole line.

22. Two points (x_1, y_1) and (x_2, y_2) are above the x axis. Show that the shortest path from one point to the other via the x axis is the one such that $\alpha = \beta$ in Fig. 11. [Using Fermat's principle, we deduce from this that light is reflected in such a way that the angle of incidence α equals the angle of reflection β.]

REFLECTION IN A PLANE MIRROR

FIGURE 3.11

23. A traveling merchant came to Minsk with his scale out of balance; one side was shorter than the other [Fig. 12]. However, he compensated for this by weighing part of every order on each side. If a customer wanted, say, two pounds of sugar, the merchant gave him one "skimpy" pound [Fig. 13(a)] and one "generous" pound [Fig. 13(b)]. Was this unfair to anyone? If so, to whom? [Hint: By the law of the lever, the weight P_s of the "skimpy" pound is given by $l \cdot 1 = xl \cdot P_s$.]

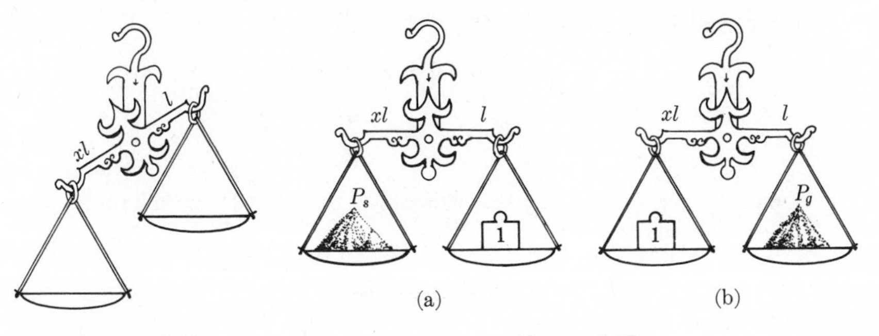

FIGURE 3.12

FIGURE 3.13

3.2 PARALLEL GRAPHS

From the very definition of derivative, every constant function has derivative zero. The converse is also true, and is one of the most useful facts in the theory of calculus.

f IS CONSTANT IFF $f' = 0$

Theorem 1. *Suppose that f is defined on an interval I, and that $f'(x) = 0$ for every x in I. Then f is constant on I; there is a constant C such that $f(x) = C$ for every x in I.*

Proof. Take any point x_1 in I, and set $C = f(x_1)$. Let x be any other point in I. Suppose, first, that $x > x_1$. Since $f' = 0$, it follows that $f' \geq 0$, so f is increasing on I; hence

$$f(x) \geq f(x_1). \tag{1}$$

But $f' \leq 0$ also, so f is decreasing; hence

$$f(x) \leq f(x_1). \tag{2}$$

The inequalities (1) and (2) together imply that $f(x) = f(x_1) = C$. Thus $f(x) = C$ for every $x > x_1$. A nearly identical argument for $x < x_1$ completes the proof [Problem 12].

Example 1. Let $f(x) = \arctan x + \operatorname{arccot} x$, defined for all real x. Then

$$f'(x) = \frac{1}{1+x^2} + \frac{-1}{1+x^2} = 0,$$

so f is a constant C,

$$\arctan x + \operatorname{arccot} x = C. \tag{3}$$

Once it is known that a function is constant, a natural question is: what is the exact value of the constant? This is usually answered by substituting any convenient value for x. In this case, we take $x = 0$; since (3) holds for all x, we have in particular, on substituting $x = 0$,

$$\arctan 0 + \operatorname{arccot} 0 = C,$$

$$0 \quad + \quad \frac{\pi}{2} \quad = C;$$

hence $C = \pi/2$. Putting this value of C back in (3), we obtain an identity for arctan and arccot,

$$\arctan x + \operatorname{arccot} x = \frac{\pi}{2}.$$

Theorem 1 has the following important generalization.

Theorem 2. *Suppose that f and g are differentiable on an interval I, and that $f' = g'$ throughout I. Then there is a constant C such that $f(x) = g(x) + C$ for every x in I.*

"LAW OF PARALLEL GRAPHS"

Proof. The hypotheses guarantee that $(f - g)' = f' - g' = 0$ throughout I. By Theorem 1, $f - g$ is a constant C; hence $f = g + C$. *Q.E.D.*

Geometrically, $f = g + C$ says that the graph of f is obtained by simply raising the graph of g through a distance C [Fig. 14], while $f' = g'$ says that the tangents at corresponding points are parallel [Fig. 15]. In either case, we are inclined to call the graphs of f and g parallel. Theorem 2 shows that the second of these notions of "parallel" implies the first, so we call this theorem the "law of parallel graphs."

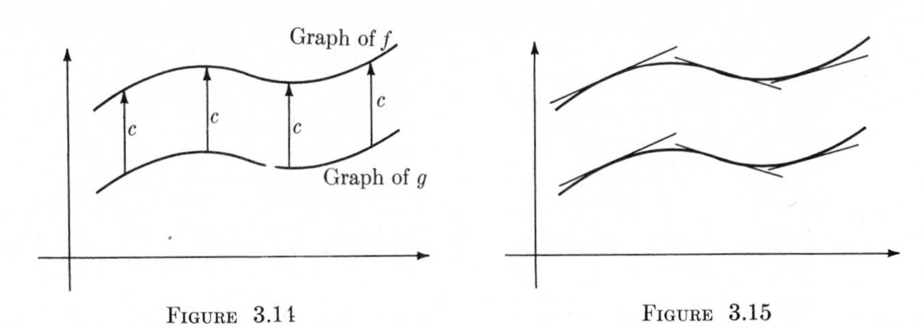

FIGURE 3.14

FIGURE 3.15

LAW OF PARALLEL GRAPHS REPHRASED

The significance of Theorem 2 in applications is this: *Up to an additive constant C, any function f on an interval I is determined by its derivative f'.*

Example 2. Given that $f'(x) = 2$ for all x, what can be said about f? *Solution.* You can say that $f(x) = 2x + C$ for some constant C. For, the function $g(x) = 2x$ satisfies $g' = 2 = f'$; hence, by Theorem 2, $f(x) = g(x) + C = 2x + C$.

More generally, for any constant a,

$$f'(x) = a \text{ for all } x \Rightarrow f(x) = ax + C.$$

For, the function $g(x) = ax$ satisfies $g' = a = f'$, so $f(x) = g(x) + C = ax + C$.

Example 3. A car accelerates at the constant rate of $1/10$ miles per minute per minute, for time $t \geq 0$. If at $t = 0$ it is traveling at $1/2$ mile per minute, find its velocity at all times $t \geq 0$. *Solution.* Let $v(t)$ be the velocity at time t. Since the acceleration is the rate of change of velocity, we can sum up the given information in the following two equations:

$$v'(t) = \frac{1}{10}, \qquad t \geq 0, \tag{4}$$

$$v(0) = \frac{1}{2}. \tag{5}$$

From (4), it follows that

$$v(t) = \frac{1}{10}t + C, \qquad t \geq 0. \tag{6}$$

[For, $g(t) = \frac{1}{10}t$ has the derivative $g'(t) = \frac{1}{10} = v'(t)$, so $v(t) = g(t) + C = \frac{1}{10}t + C$.] The constant C can now be determined by (5):

$$\frac{1}{2} = v(0) \qquad \text{[by (5)]}$$

$$= \frac{1}{10}\cdot 0 + C = C. \qquad \text{[by (6)]}$$

Hence $C = \frac{1}{2}$; and from (6), $v(t) = \frac{1}{10}t + \frac{1}{2}$, $t \geq 0$.

Remark. The equation

$$f'(x) = a, \qquad -\infty < x < \infty. \tag{7}$$

is called a *differential equation for f,* i.e. an equation involving derivatives of f, which is to hold at all points in some given interval I. The equation is *solved* when we have found all functions f satisfying the equation in the given interval. Thus the differential equation (7) is solved in Example 2; we showed that f satisfies (7) iff f has the form $f(x) = ax + C$, where C is a constant.

In Example 3, $v'(t) = 1/10$ is a differential equation, and $v(0) = 1/2$ is called an *initial condition,* i.e. a condition bearing not on the whole time interval in question, but only at the very beginning. The differential equation (4) has infinitely many solutions, all differing by constants; the initial condition (5) picks out a particular one.

Many "laws of nature" take the form of differential equations, as in Example 3. One of the main objects of calculus is to study these equations, and to solve them, if possible.

Some equations can be solved "by inspection," such as the following:

SOME DIFFERENTIAL EQUATIONS SOLVED

$$f'(x) = a, \text{ all } x, \quad \Leftrightarrow \quad f(x) = ax + C$$

$$f'(x) = ax + b, \text{ all } x, \quad \Leftrightarrow \quad f(x) = \tfrac{1}{2}ax^2 + bx + C$$

$$f'(x) = ax^2 + bx + c, \text{ all } x, \quad \Leftrightarrow \quad f(x) = \tfrac{1}{3}ax^3 + \tfrac{1}{2}bx^2 + cx + C.$$

In each case, the result is proved simply by differentiating the proposed expression for f, observing that the result agrees with the given expression for f', and appealing to Theorem 2.

Our final example applies Theorem 2 in a different sort of problem.

Example 4. The equation $\log ab = \log a + \log b$ can be deduced from the formula $\log'(x) = 1/x$. Take the two functions

$$f(x) = \log x, \qquad g(x) = \log ax.$$

The derivatives are

$$f'(x) = \frac{1}{x}$$

and, by the chain rule,

$$g'(x) = \frac{1}{ax} \cdot a = \frac{1}{x}.$$

Since $f' = g'$, we find that $g = f + C$; that is,

$$\log ax = \log x + C. \tag{8}$$

In this case, we evaluate the constant C by setting $x = 1$, which gives

$$\log a = \log 1 + C = C.$$

Putting this back in (8) yields

$$\log ax = \log x + \log a.$$

Thus the fundamental algebraic property of the logarithm and the derivative formula $\log'(x) = 1/x$ are closely related; the second implies the first.

PROBLEMS

1. Show that the derivatives of $f(x) = 1/(x + 1)$ and $g(x) = -x/(x + 1)$ are equal. By what constant do these functions differ?

• 2. A particle is accelerated at a constant rate 32 ft/sec^2, and at time $t = 0$ its velocity is 0. Find $v(t)$.

3. A particle is accelerated at a constant rate a, and at time $t = 0$ its velocity is v_0. Show that $v(t) = v_0 + at$. [Compare Example 3.]

4. A particle has velocity $s'(t) = v_0 + at$, where v_0 and a are constants. At time $t = 0$ its position is $s_0 = s(0)$. Show that its position at any time t is given by $s(t) = (a/2)t^2 + v_0t + s_0$.

5. Determine the functions f defined by the following conditions. [The differential equation is supposed to hold for all real values of x, t, or z.]

(a) $f'(x) = x, \quad f(0) = 1$
(b) $f'(x) = 2x + 5, \quad f(0) = 0$
(c) $f'(x) = 4x^3 - 3x^2 + 2x - 1, \quad f(0) = 10$
(d) $f'(t) = 8t + 1, \quad f(0) = 0$
(e) $f'(z) = z^2 + 3z - 1, \quad f(1) = 5$

6. Suppose that $f'(x) = ax^3 + bx^2 + cx + d$, where a, b, c, and d are constants. What can you say about f? Prove your answer.

7. Solve the differential equation

$$f'(x) = a_n x^n + a_{n-1}x^{n-1} + \cdots + a_1 x + a_0 .$$

8. Assume that $\sin' = \cos$, $\cos' = -\sin$, and $\cos 0 = 1$, $\sin 0 = 0$, but assume no other trigonometric identities.

(a) Prove that $\sin^2 x + \cos^2 x$ is constant.
(b) Prove that $\sin^2 x + \cos^2 x = 1$.

9. Let $f(x) = \log x$ when $x > 0$, and $f(x) = (\log |x|) + 1$ when $x < 0$. Let $g(x) = \log |x|$ when $x \neq 0$.

(a) Show that $f'(x) = g'(x)$ for $x \neq 0$, but there is *no* constant C such that $f(x) = g(x) + C$ for all $x \neq 0$. [Sketch the graphs of f and g.]
(b) Show that the result in part (a) does not contradict Theorem 2.

10. Each of the following functions f is determined, up to an additive constant, by specifying its derivative. Consult Table A, §2.9, to find f.

(a) $f'(x) = -x^{-2}, \quad x > 0$
(b) $f'(x) = \dfrac{1}{x}, \quad x > 0$
(c) $f'(x) = \dfrac{1}{1 + x^2}$
(d) $f'(x) = \dfrac{2}{1 + x^2}$

11. Find the functions satisfying the following conditions by determining the constant C occurring in the solutions to Problem 10.

(a) $f'(x) = -x^{-2}, \quad x > 0; \quad f(1) = 0$
(b) $f'(x) = \dfrac{1}{x}, \quad x \neq 0; \quad f(1) = f'(1)$
(c) $f'(x) = \dfrac{1}{1 + x^2}, \quad \lim\limits_{x \to +\infty} f(x) = 0$
(d) $f'(x) = \dfrac{2}{1 + x^2}, \quad f(0) = -10$

12. Complete the proof of Theorem 1; show that if $f' = 0$ on I, then $f(x) = f(x_1)$ for $x < x_1$.

13. Let $A(\theta)$ be the area of a sector of a circle with angle θ (in radians) and radius r [Fig. 16]. Do not assume any formula for $A(\theta)$; the formula will be deduced in this problem.

(a) Show that for $\theta_0 < \theta$,

$$\tfrac{1}{2}r^2 \cos(\theta - \theta_0) \sin(\theta - \theta_0) < A(\theta) - A(\theta_0) < \tfrac{1}{2}r^2 \tan(\theta - \theta_0),$$

by comparing $A(\theta) - A(\theta_0)$ to appropriate triangles.

(b) Show that

$$\tfrac{1}{2}r^2 \cos(\theta - \theta_0) \frac{\sin(\theta - \theta_0)}{\theta - \theta_0} < \frac{A(\theta) - A(\theta_0)}{\theta - \theta_0} < \tfrac{1}{2}r^2 \frac{\tan(\theta - \theta_0)}{\theta - \theta_0},$$

whether $\theta_0 < \theta$ or $\theta < \theta_0$.

(c) Deduce that $A'(\theta_0) = \frac{1}{2}r^2$. [Hint: As $\alpha \to 0$, then $\dfrac{\sin \alpha}{\alpha} \to \sin'(0) = \cos(0) = 1$.]

(d) Notice that $F(\theta) = \frac{1}{2}r^2\theta$ has the derivative $F'(\theta) = \frac{1}{2}r^2$. Deduce that $A(\theta) = \frac{1}{2}r^2\theta + C$ for some constant C.

(e) Find the constant C.

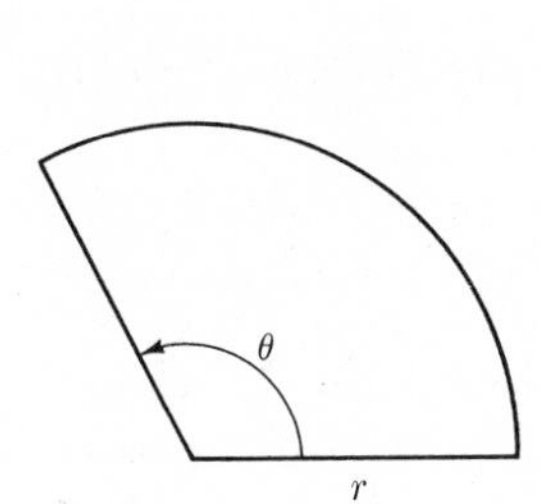

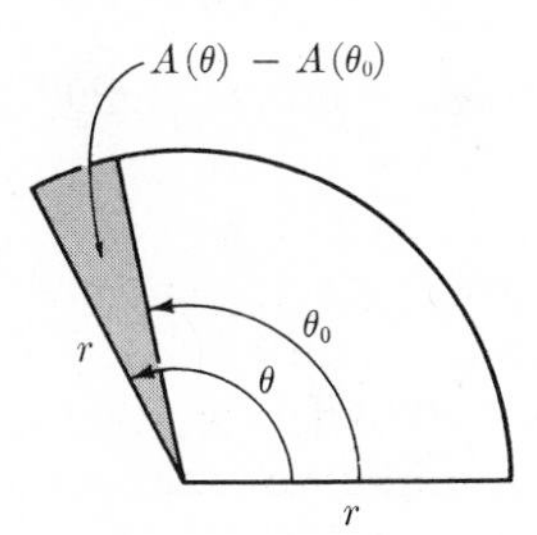

FIGURE 3.16

3.3 EXPONENTIAL GROWTH AND DECAY

Many processes involving growth or decay (for example, decay of radioactive substances, growth of bacterial colonies, and discharge of an electric capacitor through a resistor) are described by the differential equation

PROPORTIONAL GROWTH, DECAY

$$A'(t) = kA(t). \tag{1}$$

In this equation, k is a constant, and $A(t)$ denotes the amount of some substance (e.g. radium, bacteria, or electrical charge) at time t. Equation (1) says that the rate of change of A is proportional to the amount of A.

Suppose that $A > 0$. If the constant k is positive, then $A' = kA > 0$, so A is *increasing* (as in the case of bacteria); if $k < 0$, then $A' = kA < 0$, so A is *decreasing* (as in the case of radioactive decay, or electrical discharge). The larger $|k|$, the greater the rate of growth or decay. The number k is called the *growth* constant.

Assuming still that $A > 0$, we can solve equation (1) and determine A as a function of t. First, divide (1) by $A(t)$, obtaining

$$\frac{A'(t)}{A(t)} = k. \tag{2}$$

Next, recall [from Table A, §2.9] that $(\log A)' = A'/A$. Combining this with (2), we get

$$(\log A)'(t) = A'(t)/A(t) = k.$$

But the function f defined by $f(t) = kt$ also solves $f'(t) = k$. Since the two functions $\log A$ and f have the same derivative, they must differ by a constant c,

$$(\log A)(t) = f(t) + c = kt + c.$$

Taking exponentials, we get

$$A(t) = e^{kt+c} = e^c e^{kt} = Ce^{kt}, \tag{3}$$

where $C = e^c$ is again a constant. Finally, we can determine the constant C by setting $t = 0$ in formula (3); $A(0) = Ce^0 = C$, so $C = A(0)$. Thus

SOLUTION OF (1)

$$A(t) = A(0)e^{kt}, \tag{4}$$

where $A(0)$ is the amount present at time 0.

From the solution (4), we can see again that for $k > 0$, A is growing, and for $k < 0$, A is decaying; and the larger $|k|$ is, the faster is the growth or decay. But these are merely qualitative facts that we knew when we started; equation (4) is much more precise and quantitative, and therefore more useful.

Example 1. Suppose that a radioactive element decays according to the law $A' = -3A$. Find $A(t)$ for all t. *Solution.* Rather than appeal to a memorized formula, we follow the line of argument above. We deduce from $A' = -3A$ that

$$\frac{A'(t)}{A(t)} = -3$$

$$(\log A)'(t) = -3$$

$$\log A(t) = -3t + c$$

$$A(t) = e^{-3t+c} = Ce^{-3t}$$

$$A(0) = C \qquad \text{[by setting } t = 0\text{]}$$

$$A(t) = A(0)e^{-3t}.$$

$\log \frac{1}{2} = \log 1 - \log 2 \quad = \ -\log 2$

Example 2. The *half-life* of a radioactive element is the time in which it is reduced to half its original amount. Thus if $A(t)$ is the amount at time t, then the half-life L is determined by the equation $A(L) = \frac{1}{2}A(0)$. Find the half-life of the element in Example 1. *Solution.* From Example 1, $A(t) = A(0)e^{-3t}$. Thus solving the equation $A(L) = \frac{1}{2}A(0)$, we get

$$\begin{aligned} A(L) = \tfrac{1}{2}A(0) \quad &\Leftrightarrow \quad A(0)e^{-3L} = \tfrac{1}{2}A(0) \\ &\Leftrightarrow \quad e^{-3L} = \tfrac{1}{2} \\ &\Leftrightarrow \quad -3L = \log \tfrac{1}{2} = -\log 2 \\ &\Leftrightarrow \quad L = \tfrac{1}{3} \log 2. \end{aligned}$$

So the half-life is $\frac{1}{3} \log 2$.

Example 3. A certain radioactive variety of iodine (called Iodine ${}_{53}\mathrm{I}^{128}$) has a half-life of 25 minutes. What is the decay constant if time is measured in minutes? *Solution.* If the decay constant is k, then the amount at time t is $A(t) = A(0)e^{kt}$. Since the half-life is 25, we have $\frac{1}{2}A(0) = A(25) = A(0)e^{25k}$, so

$$\frac{1}{2} = e^{25k} \quad \text{and} \quad \log \frac{1}{2} = 25k, \qquad k = \frac{1}{25} \log \frac{1}{2} = \frac{-\log 2}{25}.$$

Remark. In solving equation (1), we assumed that $A > 0$. There is a more sophisticated solution that does not require this assumption.

Theorem 3. *Suppose that f is differentiable on an interval I, and for some constant k,*

$$f'(x) = kf(x) \quad \text{for all } x \text{ in } I. \tag{6}$$

Then there is a constant C such that

$$f(x) = Ce^{kx} \quad \text{for all } x \text{ in } I. \tag{7}$$

GENERAL SOLUTION OF $f' = kf$

Proof. The equation (7) which we are to prove is equivalent to

$$f(x)e^{-kx} = C \quad \text{on } I. \tag{8}$$

By Theorem 1, (8) can be proved simply by showing that the derivative of $f(x)e^{-kx}$ is 0. By the product rule and (6), the derivative of $f(x)e^{-kx}$ is

$$f'(x)e^{-kx} + f(x)[-ke^{-kx}] = kf(x)e^{-kx} - kf(x)e^{-kx} = 0. \qquad Q.E.D.$$

Notice that Theorem 3 allows the possibility that $C > 0$ or $C \leq 0$, as of course it must; $f(x) = -e^{kx}$ solves (6), and this f has the form Ce^{kx} with $C < 0$.

PROBLEMS

1. Suppose that $A' = kA$, with $k < 0$, and that L is the half-life. Show that $L = (\log 2)/|k|$.

2. Suppose that $A' = kA$ and that L is the half-life. Show that for *any* time t, $A(t + L) = \frac{1}{2}A(t)$.

3. The following table gives the half-life of various radioactive elements. What is the growth constant k, if time is measured in years? What percentage of each element remains after decaying for one year? [See the LOG TABLES, immediately preceding the Index.]

• (a) Uranium 238; 4.4×10^9 years
(b) Radium; 1590 years
(c) Radon; 3.82 days
(d) Lead; ∞ years

• **4.** A colony of bacteria grows at a rate proportional to the amount of the substance. At the end of ten minutes it has increased by 3 percent.
(a) Determine the constant k in $A' = kA$. [Hint: You are given that $A(10) = (1.03)A(0)$; use this with equation (4).]
(b) How long does it take for the colony to double?

5. Suppose that a population grows according to the law $A' = kA$ and that in one year it increases by 10 percent.
(a) In how many years does it double? [See Problem 4.]
(b) In how many years does it grow ten times as large?

6. Suppose a population grows according to the law $A' = k\sqrt{A}$.
(a) Find an expression for A analogous to formula (3) in the text. [Hint: Notice that $A'/(2\sqrt{A}) = k/2$, and consult Table A, §2.9.]
• (b) Suppose that $A(0) = 3$, $k = 2$. Find $A(t)$.

7. Suppose a population grows according to the law $A' = 10A^2$.
(a) Show that $A(t) = 1/(C - 10t)$ for some constant C.
(b) Suppose $A(0) = 10^3$. Solve for C.
(c) Sketch a graph of A. Notice that there is a *real* population explosion in this case. Exactly when is the explosion at its peak?

Problems 8–12 concern the differential equation

$$f' + af = b, \qquad a \text{ and } b \text{ constants, } a \neq 0, \tag{9}$$

which is slightly more general than (6).

8. What choice of a and b reduces (9) to (6)?

9. (a) Suppose that f solves (9), and let $g = f - b/a$. Prove that $g' = -ag$.
(b) Prove that if f solves (9) on an interval I, then $f(x) = b/a + Ce^{-ax}$ for some constant C. [Hint: From part (a), $f = g + b/a$.]

10. Let $f(x) = A + Be^{-ax}$, where A, B, and a are constants.

(a) If $a > 0$, find $\lim_{x\to+\infty} f(x)$.

(b) If $a < 0$, find $\lim_{x\to-\infty} f(x)$.

(c) Sketch a graph of f, assuming that $A > 0$, $B > 0$, and $a > 0$. Show the y intercept [the point where the graph crosses the y axis], and the behavior as $x \to +\infty$.

(d) Repeat part (c) if $A > 0$, $B < 0$, $a > 0$.

(e) Repeat part (c) if $A > 0$, $B > 0$, $a < 0$; this time show the behavior as $x \to -\infty$.

(f) Repeat part (e) if $A > 0$, $B < 0$, $a < 0$.

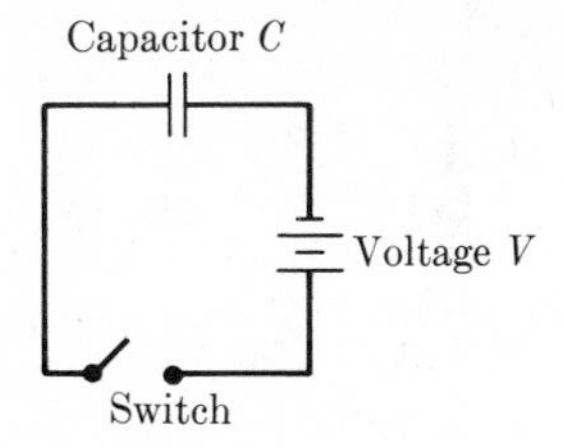

FIGURE 3.17

11. A capacitor carrying no charge for time $t < 0$ is suddenly attached at time $t = 0$ to a constant voltage V, by closing a switch [Fig. 17]. The charge $Q(t)$ in the capacitor then satisfies the equation

$$RQ' + \frac{Q}{C} = V,$$

where $C > 0$ is a constant, called the *capacitance* of the capacitor, and $R > 0$ is the resistance of the circuit.

(a) Show that for $t > 0$, $Q(t) = VC + ce^{-t/RC}$ for some constant c. [Notice that we have replaced the constant C in Problem 9(b) by c, to distinguish it from the traditional C for "capacitance" occurring in this problem.]

(b) Evaluate the constant c in part (a), using the given fact that $Q(0) = 0$.

(c) Show that $\lim_{t\to+\infty} Q(t) = VC$.

(d) At what time t is the capacitor 99% charged, i.e. when is $Q(t) = (.99)VC$? [Notice that the greater the resistance R, the longer it takes to be 99% charged.]

MORTGAGES

12. [*Mortgages*] Although mortgages are paid monthly, not continuously, there are so many payments that it is realistic to study mortgages *as though* they were paid continuously. Let $P(t)$ be the principal (the amount of money still owed at time t). Payments are made at a constant rate of c dollars/month. Part of this payment is interest on P, which is due at the rate $r \cdot P$ dollars/month, where r, the monthly interest rate, is a constant. (If the annual interest rate is 6%, then r is about $.06/12 = .005$.) The rest of the monthly payment reduces the principal, so the principal is *reduced* by $c - rP$ dollars/month. Thus we arrive at the *mortgage differential equation*

$$P' = -(c - rP) = rP - c,$$

$$r = \text{monthly interest rate,}$$

$$c = \text{monthly payment.}$$

(a) Solve the mortgage differential equation, using Problem 9(b).
(b) Show that $P(t) = c/r + [P(0) - c/r]e^{rt}$.
(c) Sketch a graph of P, assuming $P(0) < c/r$.
(d) Show that the mortgage is paid off at time

$$t = \frac{1}{r}\log\frac{c}{c - rP(0)}.$$

(e) Assuming $r = .005$, what is the monthly payment on a mortgage of \$20,000, to be paid in 240 months? [See the LOG TABLES.]

3.4 SECOND DERIVATIVES

When f is a function, its derivative f' is likewise a function. Thus we can consider the derivative of f', which is called the *second derivative* of f, denoted f'' (read "f double prime" or "f second").

Example 1. Let $g(t) = e^{-t^2}$; find $g''(t)$. [Solution below.]

The first derivative f' shows up in the graph of f as the slope; the second derivative f'' shows up as the *rate of change* of the slope. When f'' is positive, the slope is increasing; and when f'' is negative, the slope is decreasing [Fig. 18]. The function f is called *convex* when the slope is increasing and *concave* when the slope is decreasing. From Fig. 18 you can see that this corresponds to the usual meaning of "convex" and "concave" if the graph is viewed from below; this is why the labels are placed *below* the graphs.

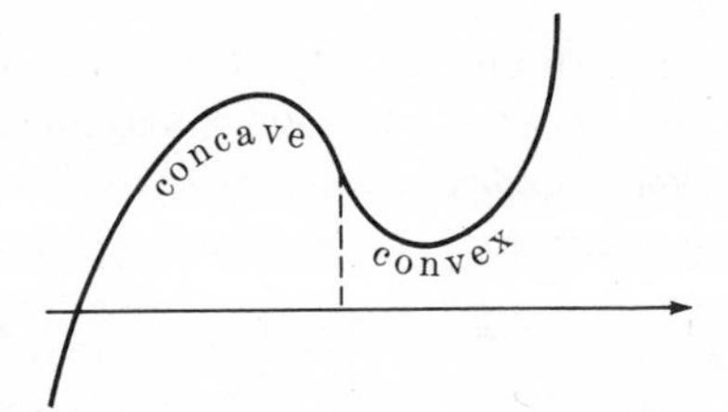

FIGURE 3.18

Example 2. If $f(x) = \sin x$, then $f'(x) = \cos x$, and $f''(x) = -\sin x$. Thus the slope is *decreasing* when $\sin x > 0$ and *increasing* when $\sin x < 0$. This can be seen in Fig. 19.

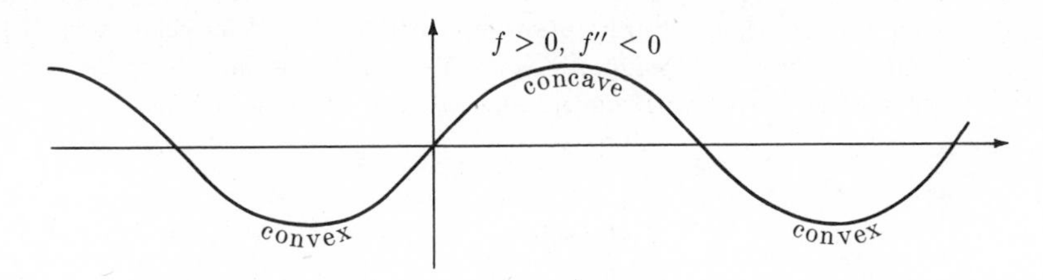

FIGURE 3.19

Example 3. Sketch the graph of the function $g(t) = e^{-t^2}$ above, showing the regions where g is convex and where it is concave. [Solution below; this is the famous "bell-shaped curve" in statistics.]

We don't have to stop at the second derivative. The third derivative of f, i.e. the derivative of f'', is denoted f''' (read "f triple prime"). But more than three primes are unwieldy, and we use $f^{(n)}$ as an alternate notation for the nth derivative; $f = f^{(0)}$, $f' = f^{(1)}$, $f'''' = f^{(4)}$, and so on. Notice the distinction between $f^{(n)}$ [the nth derivative] and f^n [the nth power]. For example, if $f(x) = x^3$, then $f^{(2)}(x) = 6x$, whereas $f^2(x) = f(x)^2 = x^6$.

Acceleration and Newton's law

Let s be a position function; $s(t)$ gives the position at time t of some object moving along a straight line [§1.7]. Then the *velocity* v is the derivative of s,

$$v = s',$$

and the *acceleration* a is the derivative of the velocity,

$$a = v' \qquad [\S 1.8].$$

ACCELERATION
$\boldsymbol{a = s''}$

Thus *the acceleration* $a = (s')' = s''$ *is the second derivative of the position function.* For instance, if $s(t) = t^2$, then the velocity is $s'(t) = 2t$ and the acceleration is $s''(t) = 2$.

Suppose that the mass of our moving object is m, and the force acting on it is F. Then F, m, and the position function s are related by the formula known as *Newton's second law of motion:* $F = (ms')'$. In simple examples the mass m is constant, and Newton's second law becomes

☆☆☆ NEWTON'S LAW

$$F = ms'' = ma, \tag{1}$$

force equals mass times acceleration. When the force F is given, we try to solve (1) for the position function s. This equation is called a *differential equation of second order*—a differential equation because it involves derivatives of the unknown function s, and of *second order* because it involves the second derivative s'', but no derivatives s''', $s^{(4)}, \ldots$ of higher order.

Example 4. An object of constant mass $m = 2$ is moved by a constant force $F = 10$. At time $t = 0$ its position is $s(0) = 3$, and its velocity is $s'(0) = 4$. Find $s(t)$ for all t. *Solution.* Newton's law (1) becomes $10 = 2s''$, or $s'' = 5$. Since s'' is the derivative of s', we can write this equation as

$$(s')'(t) = 5.$$

By the method in §3.2, the solution of this equation for the function s' is

$$s'(t) = 5t + A, \qquad A \text{ a constant.} \tag{2}$$

Solving (2) for s, we find

$$s(t) = \tfrac{1}{2}5t^2 + At + B, \qquad B \text{ a constant.} \tag{3}$$

The constants A and B are determined from the given *initial values* $s(0) = 3$ and $s'(0) = 4$. Setting $t = 0$ in (2) gives

$$4 = s'(0) = 0 + A; \qquad \text{hence} \quad A = 4.$$

Similarly, setting $t = 0$ in (3) gives

$$3 = s(0) = 0 + B; \qquad \text{hence} \quad B = 3.$$

Putting these values of A and B back in (3) gives, finally,

$$s(t) = \tfrac{1}{2}(5t^2) + 4t + 3.$$

Proceeding exactly as in Example 4, you can easily show that:

SOLUTIONS OF $s'' = c$

(i) Any solution of $s'' = c$ (where c is a given constant) has the form

$$s(t) = \tfrac{1}{2}ct^2 + At + B, \tag{4}$$

where A and B are arbitrary constants [compare equation (3)]; conversely, every function of the form (4) satisfies the equation $s'' = c$. Briefly,

$$s'' = c \iff s(t) = \tfrac{1}{2}ct^2 + At + B \quad \text{for some constants } A \text{ and } B.$$

Further,

(ii) any given initial conditions $s(0) = s_0$ and $s'(0) = v_0$ can be satisfied by the appropriate choice of A and B in (4). In fact, choosing $A = v_0$ and $B = s_0$ makes $s(0) = s_0$ and $s'(0) = v_0$.

In physical terms, this means that the equation of motion $s'' = c$, together with the initial position and velocity $s_0 = s(0)$ and $v_0 = s'(0)$, provide exactly enough information to determine the motion s completely. Less information would not be enough, and more would be superfluous.

This statement is still true when the equation $s'' = c$ is replaced by much more general second order equations; other examples are given in the problems below, and in the next section.

Solution to Examples 1 and 3.

$g(t) = e^{-t^2}$, $g'(t) = -2te^{-t^2}$,
$g''(t) = -2e^{-t^2} - 2t(-2t)e^{-t^2} = 2(2t^2 - 1)e^{-t^2}$.
$g'(t) > 0$ when $t < 0$, $g'(t) < 0$ when $t > 0$;
$g''(t) > 0$ when $|t| > 1/\sqrt{2}$, $g''(t) < 0$ when $|t| < 1/\sqrt{2}$,
$g(0) = 1$. The function is even. [See Fig. 20.]

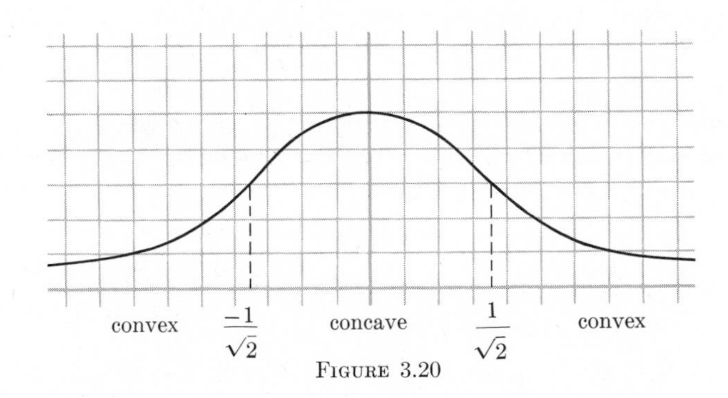

FIGURE 3.20

PROBLEMS

Problems 4 and 5 prepare for the next section; Problems 6–13 prepare for the study of Taylor polynomials, §9.1. Problems 14 and 15 concern roots of polynomials. Problem 16 gives a theorem about maxima and minima, and 17 applies this to compare the *arithmetic* and *geometric means*.

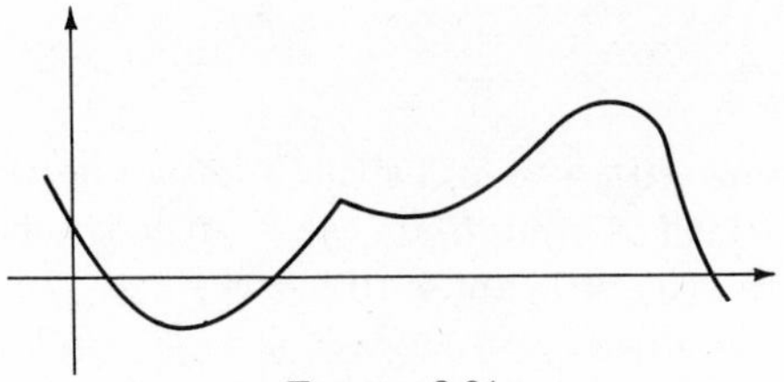

FIGURE 3.21

1. Indicate in Fig. 21 where $f'' > 0$ and where $f'' < 0$.

2. For each of the following functions, find where $f' > 0$, where $f' < 0$, where $f'' > 0$, and where $f'' < 0$. Plot the points on the graph where $f' = 0$ or $f'' = 0$. Sketch the graph.

(a) $f(x) = 3x^2 + 1$
(b) $f(x) = 1 - 3x^2$
(c) $f(x) = e^x$
(d) $f(x) = e^{-x}$
•(e) $f(x) = x^3 + 3x^2 + 2x + 1$
•(f) $f(x) = x^4 - 4x^3 + 6x^2 + 1$
•(g) $f(x) = (x^2 - 1)/x$
•(h) $f(x) = x^{1/3}(x^2 - 9)$
•(i) $f(0) = 0$ and $f(x) = x \log |x|$ if $x \neq 0$
•(j) $f(0) = 0$ and $f(x) = x^3 \log |x|$ if $x \neq 0$
(k) $f(x) = \dfrac{x - 3}{x^2 - 16}$
•(l) $f(x) = \dfrac{x^2 - 9}{x^2 - 1}$
(m) $f(x) = \sin 2x$
•(n) $f(x) = \tan (x/2)$
(o) $f(x) = \cot (x/2)$

3. You are given the derivative f' of a function f. Find where f is increasing, where decreasing, where convex, and where concave. Sketch a graph displaying these features.

(a) $f'(x) = x^2 - 4$
(b) $f'(x) = x^3 + 2x$
(c) $f'(x) = \dfrac{1}{1 - x}$, $x \neq 1$
(d) $f'(x) = \dfrac{x}{1 - x}$, $x \neq 1$
•(e) $f'(x) = e^{-x^2/2}$
•(f) $f'(x) = e^{-x^3}$

•4. (a) Let $f(t) = e^{\omega t}$, where ω is some constant. Find f', f'', f''', and generally $f^{(n)}$.
(b) Choose ω so that f in part (a) satisfies the differential equation $f'' = cf$, where c is a positive constant.

5. (a) Let $f(t) = \sin \omega t$, where ω is some constant. Find f', f'', f''', and generally $f^{(n)}$.
(b) Choose ω so that f satisfies the differential equation $f'' + cf = 0$, where c is a positive constant.

6. (a) Let f be a polynomial of degree ≤ 1, $f(x) = a_1x + a_0$. Prove that $f'' = 0$.
(b) Prove, conversely, that if f is any function such that $f''(x) = 0$ for all x, then f is a polynomial of degree ≤ 1. Precisely, $f(x) = a_1x + a_0$, where $a_1 = f'(0)$ and $a_0 = f(0)$. [Hint: First find f', then f.]

7. (a) Let f be a polynomial of degree ≤ 2, $f(x) = a_2x^2 + a_1x + a_0$. Prove that $f''' = 0$.
(b) Prove, conversely, that if f is any function such that $f'''(x) = 0$ for all x, then f is a polynomial of degree ≤ 2. In fact, $f(x) = f(0) + xf'(0) + \frac{1}{2}x^2f''(0)$. [Hint: First find f'', then f', then f. Compare Problems 5–7, §3.2.]

8. Suppose that $f^{(n)}(x) = 0$ for all x. What form does f take? Prove your answer.

9. In each of the following parts, determine the function f.
(a) $f' = 0,\quad f(0) = 1$
(b) $f'' = 0,\quad f(0) = \frac{1}{2},\quad f'(0) = \frac{1}{3}$
• (c) $f''' = 0,\quad f(0) = 2,\quad f'(0) = 3,\quad f''(0) = 4$
(d) $f^{(4)} = 5,\quad f(0) = 2,\quad f'(0) = 1,\quad f''(0) = -3,\quad f'''(0) = -1$
(e) $f^{(n)} = 0,\quad f(0) = c_0,\quad f'(0) = c_1, \ldots,\quad f^{(n-1)}(0) = c_{n-1}$

10. [*Modification of Problem 9*] Suppose that $f(1) = 1$, $f'(1) = 3$, $f''(1) = 6$, and $f'''(x) = 0$ for all x.
(a) Prove that $f''(x) = 6$ for all x.
(b) Prove that $f'(x) = 6x - 3$.
(c) Prove that $f(x) = 3x^2 - 3x + 1$.
[Compare Problems 5–7, §3.2.]

11. [*Generalization of Problem 7*] Suppose that c is a constant, and $f(c) = a_0$, $f'(c) = a_1$, $f''(c) = a_2$, and $f'''(x) = 0$ for all x.
(a) Prove that $f''(x) = a_2$.
(b) Prove that $f'(x) = a_2(x - c) + a_1$.
(c) Prove that $f(x) = \frac{1}{2}a_2(x - c)^2 + a_1(x - c) + a_0$.

$$\left[\text{Hint: Notice that } \frac{d(x-c)^2}{dx} = 2(x - c).\right]$$

12. [*Generalization of Problem 11*] Suppose that c is a constant, and $f(c) = a_0$, $f'(c) = a_1$, $f''(c) = a_2$, $f'''(c) = a_3$, and $f^{(4)}(x) = 0$ for all x. Find an expression for f analogous to the one in Problem 11(c).

EVERY POLYNOMIAL $P(x)$ CAN BE EXPRESSED IN POWERS OF $x - c$

13. [*Generalization of Problems 8 and 12*] Suppose that c is a constant, and $f(c) = a_0$, $f'(c) = a_1, \ldots, f^{(n)}(c) = a_n$, while $f^{(n+1)}(x) = 0$ for all x.

(a) Find an expression for f analogous to the one in Problem 11(c).

(b) Show that

$$f(x) = f(c) + (x-c)f'(c) + \tfrac{1}{2}(x-c)^2 f''(c) + \cdots + \frac{1}{n!}(x-c)^n f^{(n)}(c),$$

where $n!$ denotes the product of the first n integers, $n! = 1 \cdot 2 \cdot 3 \cdots n$.

(c) Show that if P is any polynomial of degree $\leq n$, then $P^{(n+1)} = 0$, hence the formula in part (b) is valid with $f = P$.

14. [*Roots of a polynomial*] Let $P(x) = \sum_0^n a_j x^j$ be a polynomial. A number r is called a *root* of P if P is divisible by $x - r$, i.e. if $P(x) = (x - r)Q(x)$ for some polynomial Q.

(a) Show that if r is a root, then $P(r) = 0$.

(b) Show, conversely, that if $P(r) = 0$, then r is a root, i.e. $\dfrac{P(x)}{x-r}$ is a polynomial. [Hint: By Problem 13,

$$P(x) = P(r) + (x-r)P'(r) + \cdots + \frac{1}{n!}(x-r)^n P^{(n)}(r).$$

If $P(r) = 0$, then....]

15. [*Multiple roots of a polynomial*] Let $P(x) = \sum_0^n a_j x^j$ be a polynomial. A number r is called a *double root* of P if P is divisible by $(x - r)^2$, a *triple root* if P is divisible by $(x - r)^3$, and an *m-fold root* if P is divisible by $(x - r)^m$, i.e., if $P(x) = (x - r)^m Q(x)$ for some polynomial Q.†

(a) Prove that if r is an m-fold root, then $0 = P(r) = P'(r) = \cdots = P^{(m-1)}(r)$. [If the general case is too hard, begin with $m = 2$ and $m = 3$.]

(b) Prove, conversely, that if $0 = P(r) = P'(r) = \cdots = P^{(m-1)}(r)$, then r is an m-fold root. [See Problem 14(b).]

---

† This is the "loose" sense of m-fold root; the "strict" sense requires that P be divisible by $(x - r)^m$ and *not* by $(x - r)^{m+1}$.

16. (a) Prove the following THEOREM. *Suppose that f is defined on an interval I, and $f'(c) = 0$ for some point c in I, and $f'' > 0$ throughout I. Then $f(c) < f(x)$ for every x in I, except $x = c$.* [Hint: Prove that $f' > 0$ to the right of c, and $f' < 0$ to the left of c.]

(b) State and prove the corresponding result when $f'' > 0$ is replaced by $f'' \leq 0$.

(c) Prove Snell's law [§3.1] by applying part (a).

17. Let $a_1, \ldots, a_n$ be n positive numbers. The *arithmetic mean*, or *average*, of these is $A = \dfrac{a_1 + \cdots + a_n}{n}$, and the *geometric mean* is $G = (a_1 \cdots a_n)^{1/n}$. When all the numbers $a_1, \ldots, a_n$ equal the same number a, then $A = a$ and $G = a$. The question is: When the a_j are *not* all equal, which is larger, the arithmetic mean A or the geometric mean G? This can be answered by gradually changing the a_j into their geometric mean G, and applying Problem 16. Define

ARITHMETIC MEAN EXCEEDS GEOMETRIC MEAN

$$f_1(t) = a_1{}^t G^{1-t}, \ldots, f_j(t) = a_j{}^t G^{1-t}, \ldots, f_n(t) = a_n{}^t G^{1-t}.$$

Notice that as t varies from 0 to 1, then $f_j(t)$ varies from $f_j(0) = G$ (the geometric mean) to $f_j(1) = a_j$ (the jth given positive number).

(a) Prove that $[f_1(t) \cdots f_n(t)]^{1/n} = G$ for every t; the geometric mean of $f_1(t), \ldots, f_n(t)$ is the same for every t.

(b) Let $F(t) = \dfrac{1}{n}[f_1(t) + \cdots + f_n(t)]$ be the average of the $f_j(t)$.

Prove that

$$F'(t) = \frac{1}{n} G^{1-t} \left(a_1{}^t \log \frac{a_1}{G} + \cdots + a_n{}^t \log \frac{a_n}{G} \right).$$

(c) Prove that $F'(0) = 0$, $F(0) = G$, and $F(1) = A$.

(d) Prove that

$$F''(t) = \frac{1}{n} G^{1-t} \left[a_1{}^t \left(\log \frac{a_1}{G} \right)^2 + \cdots + a_n{}^t \left(\log \frac{a_n}{G} \right)^2 \right],$$

and that $F''(t) > 0$ for all t, *unless* $a_1 = a_2 = \cdots = a_n = G$.

(e) Conclude that $F(0) = G$ is an absolute minimum, i.e. that $G = F(0) \leq F(t)$ for all t; and if not all a_j are equal, then $F(0) < F(t)$ for all $t \neq 0$. [See Problem 16.]

(f) Conclude that $G < A$, i.e.

$$(a_1 \cdots a_n)^{1/n} < \frac{1}{n}(a_1 + \cdots + a_n),$$

unless all the a_j are equal.

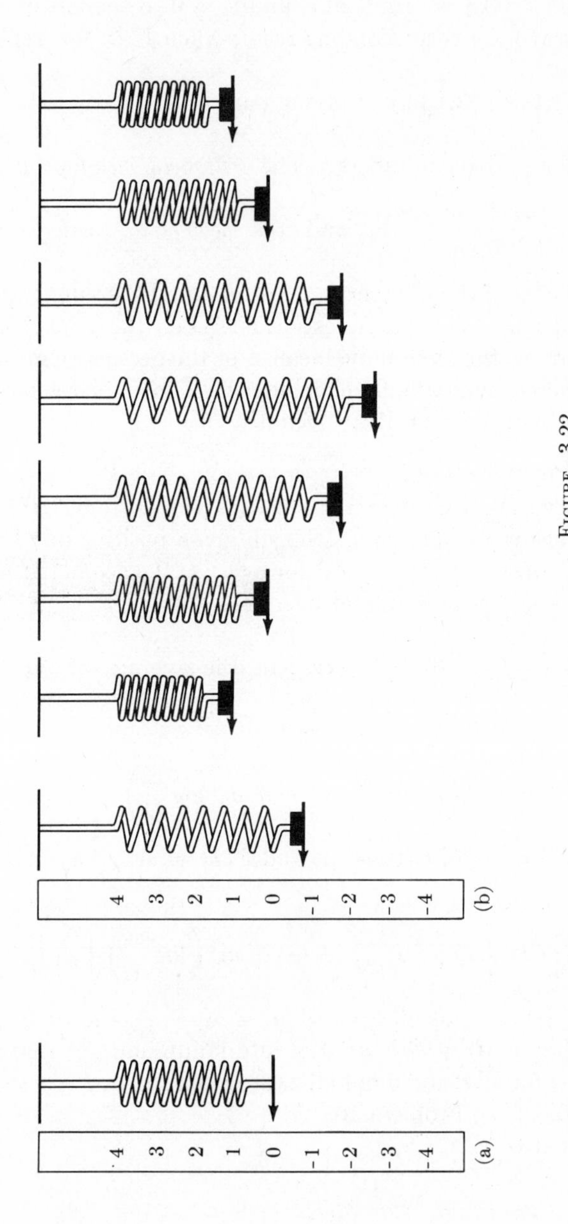

FIGURE 3.22

3.5 PERIODIC MOTION (Optional)

The motion of a mass hung on a spring [Fig. 22] and the current in the simple electrical circuit of Fig. 23 are both described by an equation of the form

$$(1) \qquad f'' + cf = 0, \qquad c > 0.$$

The law of motion for the mass on a spring can be deduced from Newton's second law and *Hooke's law:* If a spring is stretched from its natural length by an amount S, then the spring pulls back with a force kS, where k is a constant describing the stiffness of the spring. (Robert Hooke was an English experimenter, contemporary with Newton.) Figure 22(a) shows the spring with no mass attached, together with a scale to measure the stretching of the spring; if $s(t)$ denotes the position of the indicator arrow at time t, then on our scale $s(t) < 0$ when the spring is stretched downward. There are two forces acting on the mass m:

$$-mg \quad \text{[gravity]}$$
$$-ks \quad \text{[the pull of the spring].}$$

[Check the signs here: The force of gravity is taken negative, since it pulls downward. The pull by the spring is $-ks$, since when the spring is stretched we have $s < 0$, so $-ks > 0$, which corresponds to the fact that the stretched spring pulls upward.] By Newton's law, the sum of these forces is ms'':

$$(2) \qquad -mg - ks = ms''.$$

This equation can be reduced to (1) by setting

$$f = mg + ks.$$

Since k and mg are constants, the derivatives of f and $f' = ks'$, $f'' = ks''$; hence from (2)

$$-f = -mg - ks = \frac{m}{k} f''$$

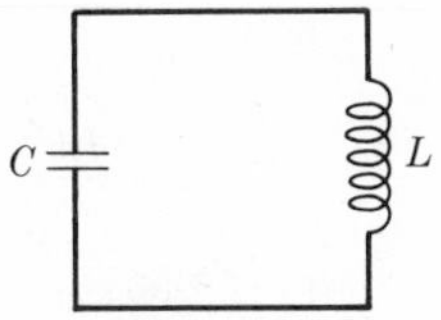

which reduces to (1) with $c = k/m$.

C L

FIGURE 3.23

The equation for the electrical circuit in Fig. 23 is

$$(3) \qquad LCI'' + I = 0,$$

where $I(t)$ is the current at time t, L is the *inductance* of the circuit, and C the *capacitance;* the resistance of the circuit is assumed to be negligible. We will not attempt to derive (3).

Equation (1) looks similar to the equation $f'' = a$ which we solved in the previous section; but it is actually much more difficult. The notation will be simplified by introducing a new constant ω, defined by $\omega = \sqrt{c}$ (recall that $c > 0$); then equation (1) becomes

$$(4) \qquad f'' + \omega^2 f = 0, \qquad \omega \neq 0.$$

We analyze this equation in four simple steps. The first two amount to a solution "by inspection"; using our experience with derivatives, we guess at possible solutions, and check that they work.

I. *For any constant A, the function* $f(t) = A \cos \omega t$, *solves* (4). *So does* $f(t) = A \sin \omega t$. PROOF. Differentiate $A \cos \omega t$ twice, and substitute the result in (4). Similarly for $A \sin \omega t$.

II. *For any constants A and B, the function*

$$f(t) = A \cos \omega t + B \sin \omega t \tag{5}$$

solves (4). PROOF. Differentiate (5) twice, and substitute in (4).

It may appear that (5) completes the solution of equation (4), but actually it does not. We have proved "by inspection" that

$$f(t) = A \cos \omega t + B \sin \omega t \quad \Rightarrow \quad f'' + \omega^2 f = 0.$$

It remains to prove the converse,

$$f'' + \omega^2 f = 0 \quad \Rightarrow \quad f(t) = A \cos \omega t + B \sin \omega t \tag{6}$$

for some constants A and B. The example of the oscillating spring shows the significance of this. We know that the motion is governed by the equation $f'' + \omega^2 f = 0$, but it is not immediately obvious what this equation tells us about f. Once (6) is proved, however, the equation implies that f is a combination of *sin* and *cos* functions, and this tells a great deal.

The next two steps lead to an ingenious proof of the implication (6).

III. *The difference between any two solutions of* (4) *is again a solution of* (4). PROOF. Let f and g be any solutions of (4), i.e. $f'' = -\omega^2 f$ and $g'' = -\omega^2 g$. Then

$$(f - g)' = f' - g';$$

hence

$$\begin{aligned}(f - g)'' &= (f' - g')' = f'' - g'' \\ &= -\omega^2 f + \omega^2 g = -\omega^2 (f - g).\end{aligned}$$

It follows that $(f - g)'' + \omega^2 (f - g) = 0$, as was to be proved.

The next step is crucial, and a little surprising.

IV. *If f is any function satisfying* (4), *then f satisfies an identity of the form*

$$(\omega f)^2 + (f')^2 = \text{constant}. \tag{7}$$

The proof is simple: The function on the left-hand side of (7) has the derivative

$$2(\omega f)\omega f' + 2(f')f'' = 2f'(\omega^2 f + f'') = 0,$$

[by (4)] and this proves (7).

The question is, where did formula (7) come from? It was suggested by the familiar identity

$$\sin^2 \omega t + \cos^2 \omega t = 1; \tag{8}$$

for, $f(t) = A \sin \omega t$ is a solution of (4), and with this particular f, equations (7) and (8) say the same thing.

Now at last we can give the complete solution of equation (4).

Theorem 4. *Suppose that $f'' + \omega^2 f = 0$ on the whole line. Then*

GENERAL SOLUTION OF $f'' + \omega^2 f = 0$

$$f(t) = A \cos \omega t + B \sin \omega t, \tag{9}$$

where $A = f(0)$ and $B = \dfrac{1}{\omega} f'(0)$.

Proof. Suppose that f is a function satisfying $f'' + \omega^2 f = 0$. Let $g(t) = A \cos \omega t + B \sin \omega t$, with $A = f(0)$ and $B = \omega^{-1} f'(0)$. We want to prove that $f = g$, in other words, that $f - g$ is identically zero. By step II, the function g solves (4). Then by step III, $f - g$ also solves (4). Then by step IV,

$$\omega^2 (f - g)^2 + (f' - g')^2 = \text{constant}. \tag{10}$$

By our choice of $A = f(0)$ and $B = \omega^{-1} f'(0)$, we have

$$g(0) = A \cos 0 + B \sin 0 = A = f(0)$$
$$g'(0) = -\omega A \sin 0 + \omega B \cos 0 = \omega B = f'(0),$$

so $f(0) - g(0) = 0$ and $f'(0) - g'(0) = 0$. Hence substituting $t = 0$ in (10) shows that the constant in (10) is zero, and we obtain

$$\omega^2 (f - g)^2 + (f' - g')^2 = 0.$$

Since a sum of squares can be zero only if each term is zero, it follows that $\omega(f - g)$ and $f' - g'$ are identically zero. Since $\omega \neq 0$, we find at last that $f - g = 0$. *Q.E.D.*

To visualize the graph of the solution (9), we rewrite (9) in the form

GENERAL SOLUTION REWRITTEN

$$f(t) = \sqrt{A^2 + B^2} \sin (\omega t - \varphi), \tag{11}$$

where φ is chosen so that

$$\sin \varphi = -\frac{A}{\sqrt{A^2 + B^2}}, \qquad \cos \varphi = \frac{B}{\sqrt{A^2 + B^2}}.$$

To see that (11) is the same as (9), apply the addition formulas for sin:

$$\sqrt{A^2+B^2}\sin(\omega t-\varphi) = \sqrt{A^2+B^2}\,(\sin\omega t\cos\varphi - \cos\omega t\sin\varphi)$$

$$= \sqrt{A^2+B^2}\left((\sin\omega t)\frac{B}{\sqrt{A^2+B^2}} - (\cos\omega t)\frac{-A}{\sqrt{A^2+B^2}}\right)$$

$$= A\cos\omega t + B\sin\omega t.$$

The graph of (11) has the shape of a sine curve [Fig. 24] which has been

AMPLITUDE (i) amplified by a factor $\sqrt{A^2+B^2}$ [since the maximum is $\sqrt{A^2+B^2}$, not 1];

PERIOD (ii) compressed toward the y axis by a factor ω [since f has period $2\pi/\omega$, not 2π,

$$f\left(t+\frac{2\pi}{\omega}\right) = \sqrt{A^2+B^2}\sin\left(\omega\left(t+\frac{2\pi}{\omega}\right)-\omega\varphi\right) = f(t)];$$

"PHASE SHIFT" (iii) shifted to the right by φ/ω [since $f(t) = 0$ when $t = \varphi/\omega$, not when $t = 0$].

According to Theorem 4, *any solution of* (4) *has this shape.*

Notice that the differential equation (4) determines the period $2\pi/\omega$, but not the *amplitude* $\sqrt{A^2+B^2}$ or the "phase shift" φ/ω; these two depend on the constants A and B, which are generally to be determined by the initial conditions $f(0)$ and $f'(0)$.

Example 1. A mass m hangs on a spring with stiffness constant k, as in Fig. 22. Find the period of the resulting oscillation. *Solution.* The spring equation is $f'' + \frac{k}{m}f = 0$, which is the same as (4) with $\omega = \sqrt{k/m}$. Hence f has period $2\pi/\omega = 2\pi\sqrt{m/k}$. Thus increasing the mass m will increase the period and make the oscillation slower; increasing the stiffness k will decrease the period and make the oscillation faster. These qualitative results would be obvious from a few simple tests with springs; but the precise *quantitative* result, that the period varies as $\sqrt{m/k}$, would require many experiments to be clearly established.

PROBLEMS

1. Find all solutions of $f'' + \omega^2 f = 0$ satisfying the following conditions.
   - •(a) $f(0) = 1,\ f'(0) = 0$
   - (b) $f(0) = 0,\ f'(0) = 1$
   - •(c) $f(\pi/\omega) = 2,\ f'(\pi/\omega) = 0$
   - (d) $f(\pi/\omega) = f'(\pi/\omega) = -1$
   - •(e) $f(0) = f(2\pi/\omega) = 0$
   - (f) $f(0) = f(\pi/\omega) = 0$
   - •(g) $f(0) = f(2\pi/\omega) = 1$
   - (h) $f(0) = f(\pi/\omega) = 1$

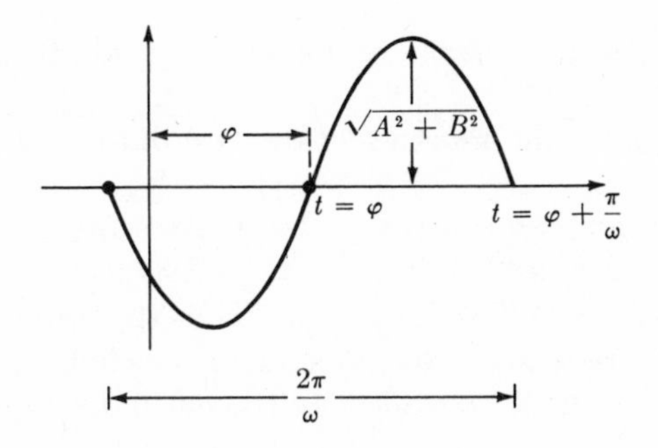

FIGURE 3.24

• **2.** Show that for any numbers t_0, A_0, and B_0, there is a solution of $f'' + \omega^2 f = 0$ satisfying $f(t_0) = A_0, f'(t_0) = B_0$, and only one solution.

3. Prove that the constant in the identity (6) is $\omega^2(A^2 + B^2)$, where $A^2 + B^2$ is the square of the amplitude of f. [Use (9) or (11).]

4. Let the current I in a circuit satisfy $LCI'' + I = 0$. Show that I oscillates with frequency $2\pi/\omega = 2\pi\sqrt{LC}$.

5. The addition formula $\cos(\theta + \varphi) = \cos\theta \cos\varphi - \sin\theta \sin\varphi$ can be deduced from Theorem 4, as follows.

(a) Let θ be fixed; show that the function $f(t) = \cos(\theta + t)$ satisfies $f'' + f = 0$.

(b) Evaluate $f(0)$ and $f'(0)$, and apply Theorem 4.

6. Let $\theta(t)$ denote the angle of a swinging pendulum at time t [Fig. 25]. The equation of motion is

$$l\theta'' = -g \sin\theta, \tag{12}$$

PENDULUM EQUATION

where l is the length of the pendulum and g is the gravity constant. [This equation is derived in §7.10. However, you can see from Fig. 25 that it is reasonable; $l\theta''$ is the acceleration to the right *along the circular arc,* and $mg \sin\theta$ is the component of the gravitational force acting to the left along this arc, so Newton's law suggests $ml\theta'' = -mg \sin\theta$.]

(continued)

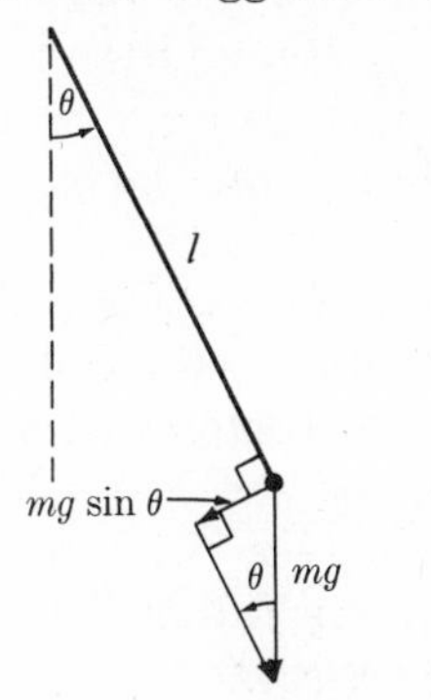

FIGURE 3.25

When the deflections of the pendulum are small, then θ is small; hence $\sin\theta$ is virtually the same as θ $\left(\text{because } \lim_{\theta\to 0} \frac{\sin\theta}{\theta} = 1\right)$. Hence, for small deflections, we can reasonably replace the actual equation (12) by the simpler equation $l\theta'' = -g\theta$. *Show that the solutions θ of this equation have period $T = 2\pi\sqrt{l/g}$.* Thus the period is increased (and the oscillation slowed) by lengthening the pendulum, or by transporting it to a place where g is smaller (to the moon, for example).

FIGURE 3.26

7. A fragile object of mass m is to be cushioned by a spring of stiffness k, to protect it against shocks when dropped [Fig. 26]. Suppose that the object lands on the spring at time $t = 0$ with velocity v_0. Arrange the coordinates so that the function f describing the position of the object at time $t \geq 0$ satisfies $f(0) = 0$, $f'(0) = v_0$. The equation of motion for $t \geq 0$ is then $f'' + \frac{k}{m} f = 0$, and the solutions for $t \geq 0$ are given by (9), with $\omega = \sqrt{k/m}$.
   (a) Given $f(0) = 0$, $f'(0) = v_0$, find $f(t)$.
   (b) Show that the maximum compression of the spring [the maximum value of $|f(t)|$ for $t \geq 0$] is $v_0\sqrt{m/k}$.
   (c) Given $f(0) = 0$, $f'(0) = v_0$, show that the maximum acceleration of the object [the maximum value of $|f''(t)|$ for $t \geq 0$] is $v_0\sqrt{k/m}$.
   •(d) Suppose that the maximum velocity of the fall is expected to be M (thus $|v_0| \leq M$), and the maximum allowable acceleration of the object is A (thus we must keep $|f''(t)| \leq A$ for $t \geq 0$). What is the stiffest spring (largest value of k) that will keep $|f''(t)| \leq A$ whenever $|v_0| \leq M$?
   •(e) An actual spring has limited "compressibility" L; if it is compressed by the amount L, the coils lie directly on top of each other, the spring can be compressed no further, and the cushioning effect is lost. For the stiffness k found in part (d), find the maximum compression [see part (b)], and thus determine the minimum compressibility L that the spring must have if it is not to lose its cushioning effect.

GENERAL SOLUTION OF $f'' = \omega^2 f$

8. Theorem 4 solves the equation $f'' + cf$ for $c > 0$, and §3.4 solves it for $c = 0$. This problem solves it for $c < 0$.

(a) Show that for any constants A and B, the function $f(t) = Ae^{\omega t} + Be^{-\omega t}$ solves $f'' = \omega^2 f$. [The remaining parts (b)–(e) prove the converse.]

(b) Suppose that $f'' = \omega^2 f$, and let $g = f' + \omega f$. Prove that $g' = \omega g$.

(c) Continuing part (b), prove that $g(t) = Ce^{\omega t}$ for some constant C, hence $f' + \omega f = Ce^{\omega t}$.

(d) Continuing part (c), prove that

$$\frac{d(e^{\omega t}f)}{dt} = d\left(\frac{C}{2\omega}e^{2\omega t}\right)\Big/ dt, \qquad \text{hence} \qquad e^{\omega t}f = \frac{C}{2\omega}e^{2\omega t} + B$$

for some constant B.

(e) Conclude, finally, that $f = Ae^{\omega t} + Be^{-\omega t}$ for constants A and B.

9. (a) Prove that the hyperbolic cosine function $f(t) = \cosh t = \frac{1}{2}(e^t + e^{-t})$ satisfies

$$f'' - f = 0, \qquad f(0) = 1, \qquad f'(0) = 0.$$

[Notice that the trigonometric cosine function satisfies the analogous conditions $f'' + f = 0, f(0) = 1, f'(0) = 1.$]

(b) Prove that $f(t) = \sinh t = \frac{1}{2}(e^t - e^{-t})$ satisfies

$$f'' - f = 0, \qquad f(0) = 0, \qquad f'(0) = 1.$$

[What are the analogous conditions for $\sin t$?]

Theory of Maxima

This brief chapter concerns two central theorems, the *Maximum Value Theorem* and the *Mean Value Theorem,* which form the theoretical basis for the applications given in Chapters I and III.

4.1 THE MAXIMUM VALUE THEOREM

We begin by defining terms. Suppose that a function f is defined for every x in some set S. The maximum value of f on S, denoted $\max_S f$, is the largest value assumed by f on the set S. Precisely:

Definition 1. $M = \max_S f$ if

MAXIMUM VALUE

(i) $f(x) \leq M$ for every x in S, and

(ii) $f(c) = M$ for some point c in S.

A point c where f assumes the maximum value is called a *maximum point.* Figure 1 shows the maximum value M and a maximum point c for a function f on an interval $[a,b]$. The maximum may occur at an endpoint, as in Fig. 2.

Analogously, c is a minimum point for f on the set S if c is in S, and $f(c) \leq f(x)$ for every x in S. And then $f(c) = m$ is the *minimum value* of f on S, denoted $\min_S f$.

For example, $\max_{[-1,1]} x^2 = 1$, and $\min_{[-1,1]} x^2 = 0$ [Fig. 3].

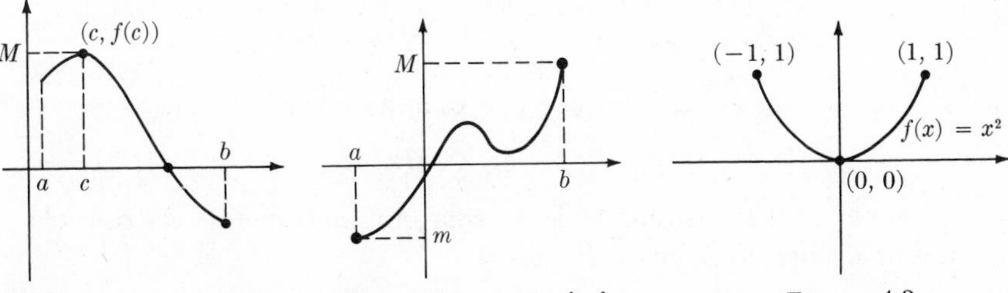

FIGURE 4.1 $M = \max_{[a,b]} f$ FIGURE 4.2 $m = \min_{[a,b]} f$ FIGURE 4.3

In locating maxima (or minima), we generally rely on derivatives; the maximum is likely to come where the derivative is zero, and the slope is zero [Fig. 1]. Proposition G puts this the other way around; at an interior point (not an endpoint), if the derivative is *not* zero you can have neither a maximum nor a minimum [Fig. 4].

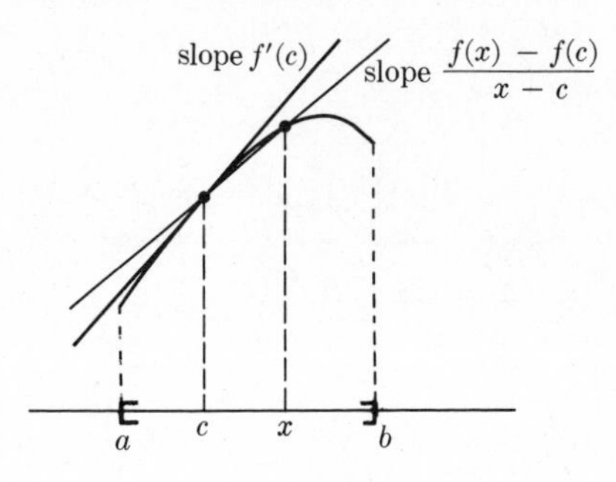

FIGURE 4.4

Proposition G. *Suppose that $f'(c) > 0$. Then $f(x) < f(c)$ if x is just to the left of c, and $f(x) > f(c)$ if x is just to the right of c. Precisely, there is a number $\delta > 0$ such that*

$$c - \delta < x < c \quad \Rightarrow \quad f(x) < f(c)$$

and

$$c < x < c + \delta \quad \Rightarrow \quad f(c) < f(x).$$

This is easy to see from Fig. 4. The proof [§AII.2] can be summarized as follows. We are given

$$\lim_{x \to c} \frac{f(x) - f(c)}{x - c} = f'(c) > 0;$$

hence when x is near c, we must have

$$\frac{f(x) - f(c)}{x - c} > 0. \tag{1}$$

When x is to the left of c, then $x - c < 0$, so multiplying (1) by $x - c$ reverses the inequality [§AI.3], yielding

$$f(x) - f(c) < 0, \qquad \text{or} \qquad f(x) < f(c).$$

When x is to the right of c, then $x - c > 0$, and (1) yields

$$f(x) - f(c) > 0, \qquad \text{or} \qquad f(x) > f(c).$$

The effect of Proposition G is to rule out most points as possible minimum or maximum points. Precisely:

Theorem 1. *Suppose that $a < c < b$, and $f'(c) \neq 0$. Then c is neither a minimum nor a maximum point for f on the interval $[a,b]$.*

Proof. If $f'(c) \neq 0$, then either $f'(c) > 0$ or $f'(c) < 0$. Suppose that $f'(c) > 0$ [Fig. 4]. Then by Proposition G there are points just to the left of c where $f(x) < f(c)$; since $a < c$, some of these points lie in the interval $[a,b]$, hence $f(c)$ is not a minimum for f on $[a,b]$. Similarly, to the right of c there are points in the interval $[a,b]$ where $f(c) < f(x)$, so $f(c)$ is not a maximum for f on $[a,b]$. This completes the proof in the case where $f'(c) > 0$.

If $f'(c) < 0$, then $(-f)'(c) > 0$, so c is neither a minimum nor a maximum for $-f$; hence c is neither a maximum nor a minimum for f. *Q.E.D.*

Theorem 1 leaves only three candidates for maximum and minimum points of a function f defined on a closed interval $[a,b]$; the endpoints, the points c where $f'(c) = 0$, and the points x where $f'(x)$ does not exist.

Example 1. Take the function $f(x) = \dfrac{3}{x^2 - 2x + 2}$ on the interval $[0,2]$.

Here, f' exists everywhere, and $f'(x) = -3\dfrac{2x-2}{(x^2 - 2x + 2)^2} = 0$ if and only if $x = 1$. Thus the candidates for maximum and minimum points are $x = 1$, and the endpoints $x = 0$, $x = 2$. Since $f(0) = \frac{3}{2}$, $f(2) = \frac{3}{2}$, and $f(1) = 3$, we conclude that 0 and 2 are minimum points and $\min_{[0,2]} f = \frac{3}{2}$, while 1 is a maximum point and $\max_{[0,2]} f = 3$.

Example 2. Define f on the interval $[-4,4]$ by setting $f(x) = x + 1/x$ if $0 < |x| < 4$ and setting $f(0) = 0$ [Fig. 5]. Here, $f'(0)$ does not exist; but when $x \neq 0$, $f'(x) = 1 - 1/x^2$, and this is zero when $x = \pm 1$. Thus the candidates for maximum and minimum points are $x = \pm 1$, $x = \pm 4$ (the endpoints), and $x = 0$ (where f' does not exist). Since $f(-4) = -4\frac{1}{4}$, $f(-1) = -2$, $f(0) = 0$, $f(1) = 2$, and $f(4) = 4\frac{1}{4}$, we conclude that $x = -4$ is a minimum point and $\min_{[-4,4]} f = -4\frac{1}{4}$, while $x = 4$ is a maximum point and $\max_{[-4,4]} f = 4\frac{1}{4}$.

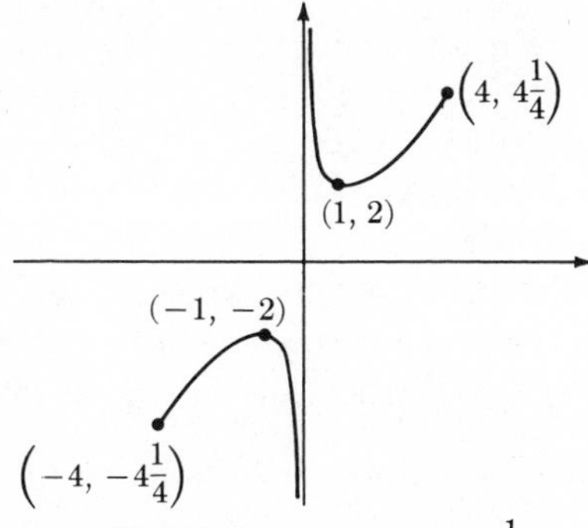

FIGURE 4.5 $f(x) = x + \dfrac{1}{x}$

Something is wrong in Example 2! Figure 5 shows that $4\frac{1}{4}$ is *not* a maximum value of f, for $f(x)$ is much larger than $4\frac{1}{4}$ when x is near zero. And $-4\frac{1}{4}$ is not a minimum value either. What is wrong?

The trouble lies not in Theorem 1, but in the way we applied it. Theorem 1 rules out certain points as maximum points. But *it does not follow* that any of the remaining points gives a maximum, for there may be *no maximum point whatsoever*, and the function may have no maximum value. This is exactly the situation in Fig. 5.

What we can conclude from Theorem 1 is merely this: The maximum and minimum points, *if there are any*, are found among the endpoints, the points where $f' = 0$, and the points where f' fails to exist.

The Maximum Value Theorem

Our situation now is not very satisfactory. Theorem 1 only helps to find a maximum point *if* we know beforehand that there *is* one; but how can we tell whether there is a maximum or not?

To see what conditions might guarantee the existence of a maximum, we begin by looking for examples in which a maximum *fails* to exist; then we will know what our conditions should rule out.

Look at the graphs in Figures 6, 7, and 5. In each case, there is *no point* c at which $f(c)$ is greater than or equal to all other values of f.

In Fig. 6(a), the reason is that the interval is *unbounded.* In 6(b), the interval is *not closed;* the maximum and minimum in this case should occur at the endpoints, but these are left out of the open interval (0,1). The moral of these two sketches is clear: A decent function may fail to attain a maximum on an interval I unless that interval is *bounded* and *closed.*

Figure 7 illustrates another sort of difficulty; each function shown there is defined on a closed bounded interval $[a,b]$, and yet does not attain a maximum. In 7(a), the function builds up toward a maximum at 0, but, instead of attaining it, drops down to -1. In 7(b), the function assumes values closer and closer to 1 as $x \to 0$ from the right, but it does not attain at any point a *maximum* value, a value $\geq$ all the others. Similarly, in Fig. 5, the function builds up as $x \to 0$ from the right, but does not attain any maximum.

In each case there is a point where a maximum might be expected, namely at $x = 0$; but it is not attained because *the function is not continuous* at 0, i.e. $\lim_{x\to 0} f(x) \neq f(0)$ [Figs. 5 and 7(a)], or $\lim_{x\to 0+} f(x) \neq f(0)$ [Fig. 7(b)].

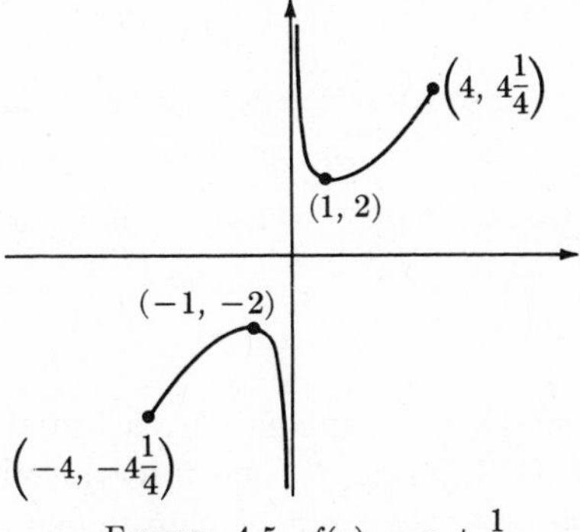

FIGURE 4.5 $f(x) = x + \frac{1}{x}$

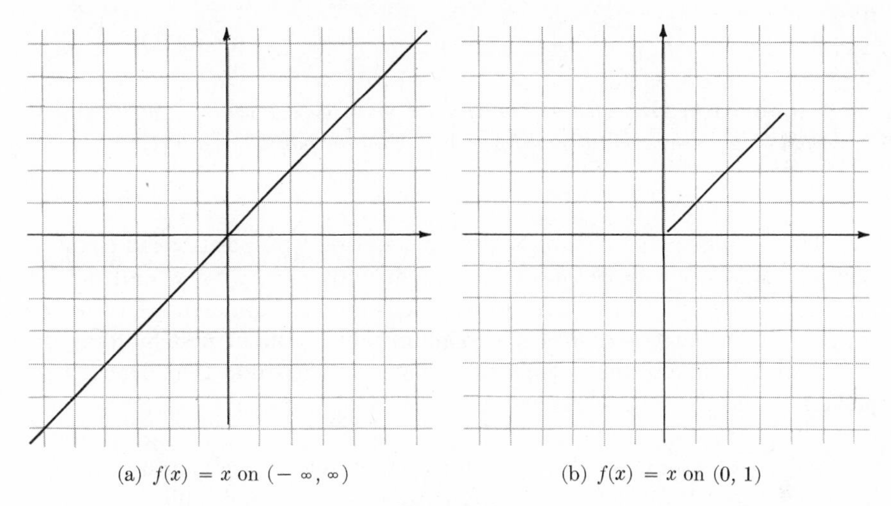

(a) $f(x) = x$ on $(-\infty, \infty)$

(b) $f(x) = x$ on $(0, 1)$

FIGURE 4.6

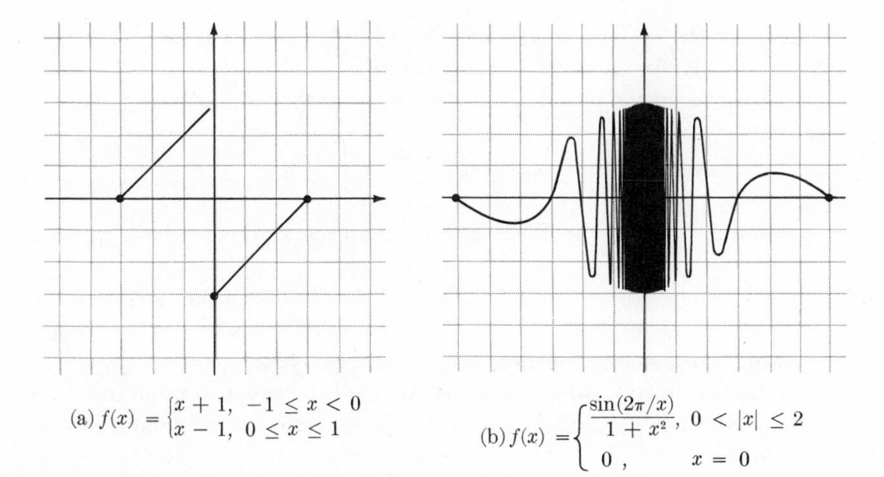

(a) $f(x) = \begin{cases} x + 1, & -1 \le x < 0 \\ x - 1, & 0 \le x \le 1 \end{cases}$

(b) $f(x) = \begin{cases} \dfrac{\sin(2\pi/x)}{1 + x^2}, & 0 < |x| \le 2 \\ 0\ , & x = 0 \end{cases}$

FIGURE 4.7

"CONTINUOUS ON $[a,b]$"

Definition 2. A function f is *continuous* on $[a,b]$ iff

$$\lim_{x \to x_0} f(x) = f(x_0) \quad \text{for } a < x_0 < b,$$

$$\lim_{x \to a+} f(x) = f(a), \qquad \text{and} \qquad \lim_{x \to b-} f(x) = f(b).$$

We have discovered two reasons why a function f might not have a maximum value on an interval I: The interval might not be closed and bounded, or the function might not be continuous. The *Maximum Value Theorem* says that these are the *only* reasons why a maximum and minimum might not occur.

☆☆☆ MAXIMUM VALUE THEOREM

Proposition H. *If f is continuous on a closed finite interval $[a,b]$, then f has a maximum point and a minimum point in $[a,b]$.*

This is one of the most difficult and important propositions in elementary calculus. We have shown that a maximum may very well *not* be attained if the hypotheses are not fulfilled; but we cannot hope to prove that the hypotheses *guarantee* a maximum point without first building up the analytic foundations. Thus the proof of Proposition H is deferred to §AII.4.

We now return to the problem of locating maxima and minima. Theorem 1 and Proposition H combine to give us a simple rule:

HOW TO LOCATE MAXIMA, MINIMA

Theorem 2. *If f is continuous on a finite closed interval $[a,b]$, then there* is *a maximum point and a minimum point, and these occur at the endpoints, or at the points where $f' = 0$, or at the points where f' fails to exist.*

Theorem 2 justifies the discussion in Example 1 above, where we considered $f(x) = \dfrac{3}{x^2 - 2x + 2}$ on $[0,2]$; for, this function has a derivative everywhere, and hence is continuous everywhere [§2.2, Lemma 1].

The discussion in Example 2 is *not* justified, however, for the function f considered there is *not* continuous at 0. That's why we reached a ridiculous conclusion.

Theorem 2 is the basic theoretical tool in maximum problems over *closed finite* intervals. For other types of interval there is no such complete method, but the following two theorems cover most cases:

Theorem 3. *Suppose that f is defined on an interval I, and $f'(c) = 0$.*

(a) *If $f'' \geq 0$ throughout I [thus f is* convex, *Fig.* 8(a)], *then $f(c) = \min_I f$.*

(b) *If $f'' \leq 0$ throughout I [f is* concave, *Fig.* 8(b)], *then $f(c) = \max_I f$.*

ANOTHER WAY TO LOCATE MAXIMA, MINIMA

This theorem is "obvious" from the figures, and easy to prove—see the problems of §3.4. Another closely related theorem is almost equally obvious, but considerably harder to prove:

Theorem 4. *Suppose that f is differentiable throughout an interval I, and $f'(c) = 0$ for* exactly one *point c in I. Then*

$$f''(c) > 0 \quad \Rightarrow \quad f(c) < f(x) \quad \text{for all } x \neq c \text{ in } I,$$
$$f''(c) < 0 \quad \Rightarrow \quad f(c) > f(x) \quad \text{for all } x \neq c \text{ in } I.$$

YET ANOTHER WAY

This is proved in §4.2 below. Figure 8 shows why $f''(c) > 0$ corresponds to a minimum, and $f''(c) < 0$ to a maximum. Figure 9 shows why the theorem does *not* apply if there is more than one point where $f'(c) = 0$.

When $f(x) > f(c)$ [not merely $f(x) \geq f(c)$] for all $x \neq c$, then c is called a *strict* minimum point. Thus Theorem 4 gives strict minimum and strict maximum points.

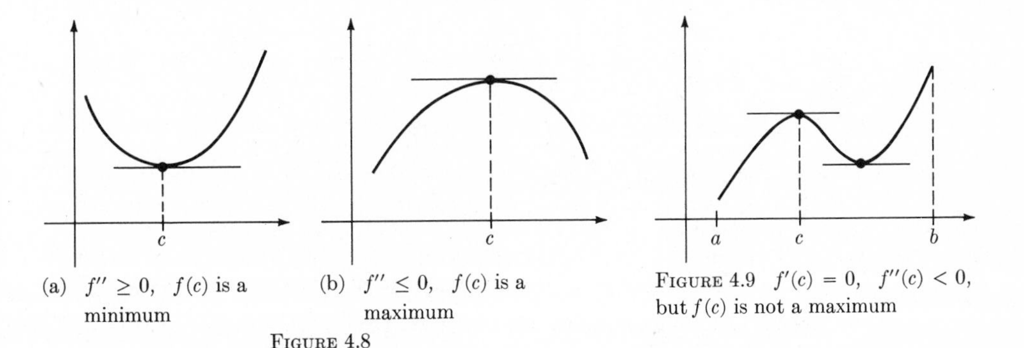

(a) $f'' \geq 0$, $f(c)$ is a minimum

(b) $f'' \leq 0$, $f(c)$ is a maximum

FIGURE 4.8

FIGURE 4.9 $f'(c) = 0$, $f''(c) < 0$, but $f(c)$ is not a maximum

Example 3. Find the maximum or minimum of $f(x) = x + 1/x$ on the interval $(0,\infty)$. *Solution.* Since the interval $(0,\infty)$ is not closed and finite, Theorem 2 does not apply; hence we try either Theorem 3 or 4. We find

$$f'(x) = 1 - \frac{1}{x^2} = 0 \quad \text{when } x = \pm 1$$

$$f''(x) = \frac{2}{x^3}. \tag{2}$$

From (2), $f'(c) = 0$ for exactly one point in the interval $(0,\infty)$, the point $c = 1$. Since $f''(c) = f''(1) = 2$, this point gives a *minimum* of f over the interval $(0,\infty)$, by Theorem 4. [Theorem 3 also applies.]

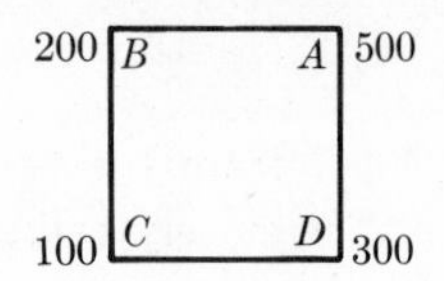

FIGURE 4.10

Example 4. Supplies (of food, books, tires, coal, policemen, strikers, demonstrators, whatever) are distributed at four depots, A, B, C, D, situated at the four corners of a square [Fig. 10]. They can be moved vertically or horizontally, but not diagonally. If there are 500 at A, 200 at B, 100 at C, 300 at D, how can they be equally redistributed with the least amount of moving? *Solution.* When they have been equally redistributed, the number at each corner will be one-fourth of the total, that is to say

$$\frac{500 + 200 + 100 + 300}{4} = 275.$$

Let

$$(\text{number moved from } A \text{ to } B) = x.$$

[If $x < 0$, this means that $|x|$ are moved in the other direction, from B to A.] Then there are $200 + x$ at B; in order to end up with only 275 at B, the remainder must be moved on to C. Thus

$$(\text{number moved from } B \text{ to } C) = (200 + x) - 275 = x - 75.$$

Then there are $100 + (x - 75) = 25 + x$ at C, so

$$(\text{number moved from } C \text{ to } D) = (25 + x) - 275 = x - 250.$$

[Presumably $x - 250$ will be negative, which means that supplies are actually moved from D to C, not from C to D.] Finally, at D, there are now $300 + (x - 250) = x + 50$, so

$$(\text{number moved from } D \text{ to } A) = (x + 50) - 275 = x - 225.$$

As a check, the number finally remaining at A is

$$500 - (\text{number moved from } A \text{ to } B) + (\text{number moved from } D \text{ to } A)$$

$$= 500 - x + (x - 225) = 275.$$

The total number of supplies moved is

$$T(x) = |x| + |x - 75| + |x - 250| + |x - 225|. \tag{3}$$

Since the number x moved from A to B must clearly be ≤ 500 and $\geq (-200)$ [Fig. 10], we consider T on the interval $[-200,500]$. The graph of T can be found by "adding" the graphs of the four absolute value functions in (3), as in Fig. 11. Apparently there is some latitude in the solution; moving any number between 75 and 225 from A to B yields the minimum total moving: 400 items moved.

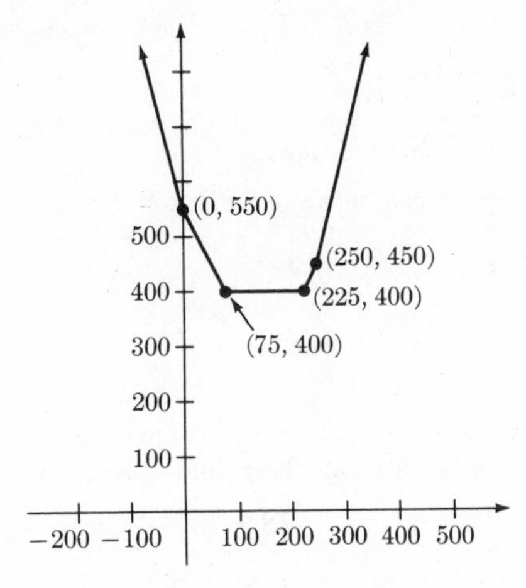

FIGURE 4.11

In this example, the minimum occurs at a zero of the derivative; $T'(x) = 0$ for all x in the open interval $(75,225)$. However, if there had been five supply depots, not four, and transfers were allowed only between adjacent depots [Problem 6 below], the relevant graph would look like Fig. 12, and the minimum would occur only at a point where the derivative fails to exist. Thus Theorem 2 is not just being pedantic when it includes such points as possible maximum or minimum points.

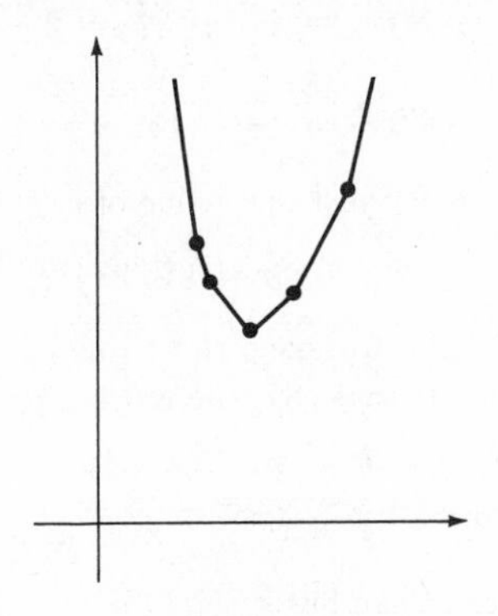

FIGURE 4.12

Many applied maximum and minimum problems [like those in §1.5] can be solved by appeal to one of Theorems 2, 3, or 4.

Example 5. A tin can in the shape of a circular cylinder is to contain a volume V. What are the dimensions of such a can having minimum surface area (and therefore requiring the least material to make)? *Solution.* Let r be the radius of the can, and h its height [Fig. 13]. The volume of the cylinder is (area of base) $\times$ (height) $= \pi r^2 h$, so

$$\pi r^2 h = V, \qquad V \text{ a given constant.} \tag{4}$$

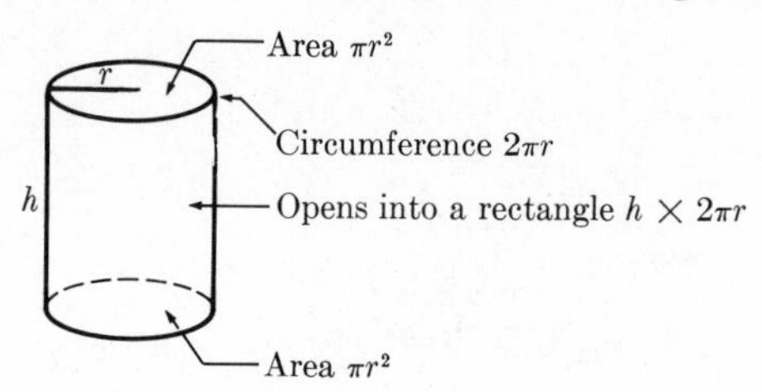

FIGURE 4.13

The can is made of two disks of radius r and a rectangle of dimensions $2\pi r \times h$, curved to make the walls of the cylinder; so the surface area is

$$S = 2\pi r^2 + 2\pi rh.$$

We use (4) to eliminate one of the variables; in this case it is simplest to solve for h,

$$h = V/\pi r^2. \tag{5}$$

Hence the function to be minimized is

$$S(r) = 2\pi r^2 + 2\pi r(V/\pi r^2) = 2\pi r^2 + 2V/r.$$

The allowable values of r are $0 < r < \infty$. Since $(0, \infty)$ is not a closed finite interval, Theorem 2 does not apply, and we must appeal to Theorem 3 or 4. The derivatives of S are

$$S'(r) = 4\pi r - \frac{2V}{r^2} = 2\,\frac{2\pi r^3 - V}{r^2}, \tag{6}$$

$$S''(r) = 4\pi + 4V/r^3. \tag{7}$$

From (6), $S'(r) = 0$ for just one value of r,

$$r = (V/2\pi)^{1/3}, \tag{8}$$

and from (7), $S''(r) > 0$; hence the value of r in (8) gives a minimum, by Theorem 3. From (5) and (8), the most efficient can has the proportions

$$\frac{h}{r} = \frac{V}{\pi r^3} = 2\,\frac{V}{\pi(V/2\pi)} = 2. \tag{9}$$

Thus the most efficient can has a height equal to its diameter. [Actually, not many cans are made this way; ecologists, take notice.]

Example 6. A wholesaler has contracted to supply a retailer with 10 bottles (of something or other) per day. If the wholesaler buys from the factory x bottles at a time, he pays $27 + 5x$ dollars for them. The storage of the undelivered portion of the batch of x bottles, averaged over the whole period of storage, costs x cents a day (in rent, insurance, etc.). He can store at most 300 bottles. What size order minimizes the cost to the wholesaler? *Solution.* We will minimize the total cost per day (call it T) considered as a function of x, the number of bottles bought from the factory at one time. A batch of x bottles costs $27 + 5x$ dollars and lasts $\frac{1}{10}x$ days, so

$$\text{purchase cost per day} = \frac{27 + 5x}{\frac{1}{10}x} = \frac{270}{x} + 50.$$

Since the storage cost is x cents a day, or $x/100$ dollars a day, the total cost per day (purchase plus storage) is

$$T(x) = x/100 + 50 + 270/x.$$

Since he can store at most 300 bottles, we consider this function T on the interval $0 \leq x \leq 300$. We find

$$T'(x) = 1/100 - 270x^{-2}, \qquad T''(x) = 540x^{-3}.$$

Since $T'(c) = 0$ only for $c = \sqrt{27{,}000}$, and $T''(c) > 0$, it follows that $c = \sqrt{27{,}000}$ gives a minimum for the total cost T. Thus the best amount to buy from the factory is 164 bottles, since 164 is the nearest integer to $\sqrt{27{,}000}$.

PROBLEMS

1. Each of the following functions is continuous at every point of the given interval $[a,b]$. Apply Theorem 2 to find the maximum and minimum.

• (a) $f(x) = \dfrac{x}{1 + x^2}$ on $[-2,2]$

(b) $f(x) = \dfrac{x}{1 + 2x^2}$ on $[-1,1]$

(c) $f(x) = x^2e^x$ on $[-2,2]$

(d) $f(x) = e^x - e^{-x}$ on $[-2,2]$

• (e) $f(x) = \dfrac{e^x - e^{-x}}{e^x + e^{-x}}$ on $[-2,2]$

(f) $f(x) = \log(1 + x^2)$ on $[-3,3]$

• (g) $f(x) = |x|$ on $[1,3]$

(h) $f(x) = (x - 1)^{2/3}$ on $[0,2]$

2. For the following functions f and intervals I, determine whether Theorem 3 yields (i) a minimum over I, (ii) a maximum over I, or (iii) no result. Answer the same questions about Theorem 4.

• (a) $f(x) = \dfrac{x}{1+x^2}$ on $[0,+\infty)$

• (b) $f(x) = \dfrac{x}{1+x^2}$ on $(-\infty,+\infty)$

(c) $f(x) = xe^x$ on $(-\infty,+\infty)$

• (d) $f(x) = x^3$ on $[-1,1]$

(e) $f(x) = x^4$ on $[-1,1]$

3. For each of the following functions and intervals, find at least one hypothesis of Theorem 2 which is violated. Further, in each case decide whether f has a maximum and/or a minimum on the given interval I.

• (a) $f(x) = \sin x$ on $(-\infty,+\infty)$
(b) $f(x) = 1/x^2$ on $(0,1]$
• (c) $f(x) = \arctan x$ on $(-\infty,\infty)$
(d) f is the function shown in Fig. 14.

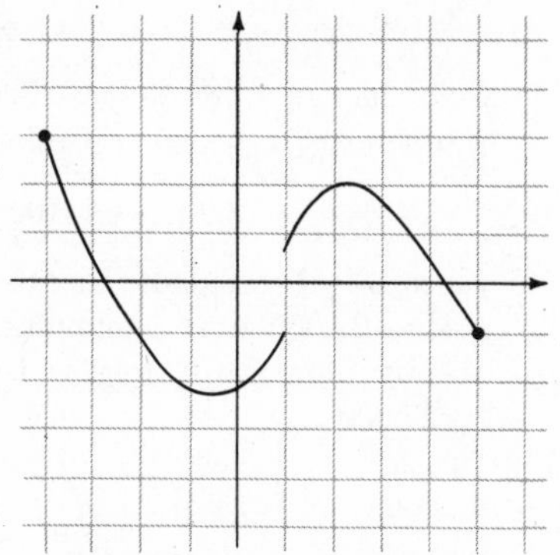

FIGURE 4.14

4. Outline a proof of the following variation of Proposition G: *If $f'(c) < 0$, then there is an interval to the left of c on which $f(x) > f(c)$, and an interval to the right of c on which $f(x) < f(c)$.* Draw an appropriate sketch.

5. Theorem 4 suggests the following conjecture: "Suppose that f' exists throughout an interval I, and f' is zero at c_1 and c_2, but at no other points; and $f''(c_1) < 0$, $f''(c_2) > 0$. Then $f(c_1)$ is a maximum and $f(c_2)$ is a minimum." Is this conjecture true or false?

6. Five depots A, B, C, D, E arranged in a circle around the edge of a lake have the following supplies of item Z: 50 at A, 30 at B, 0 at C, 100 at D, 10 at E. Items may be moved from any depot to either adjacent depot. Moving the smallest possible number of items, how should they be equally redistributed?

• 7. Three depots A, B, C, at distances a, b, c from each other [Fig. 15], have the following supplies: 100 at A, 30 at B, 20 at C. The cost of moving an item is proportional to the distance moved. What is the cheapest plan for equal redistribution?

8. (a) The side of the can in Example 5 is to be made of cardboard costing a cents per square inch (cpi²), and the top and bottom are aluminum costing b cpi². What dimensions give minimum cost of materials? Show that for this minimum, the side costs twice as much as the top and bottom together.
   (b) Repeat part (a) for a can whose top and bottom are squares, like a milk carton.

• 9. The top and bottom of the can in Example 5 have to be cut out of a rectangular piece of material, and the scrap [shaded area in Fig. 16] is thrown away. How should r and h be chosen so that the least material is used?

10. Suppose that the material for the can in Example 5 costs a cents per unit area, and the scrap [shaded area in Fig. 16] can be sold for b cents per unit area, $b < a$. How should r and h be chosen to minimize the net cost of materials?

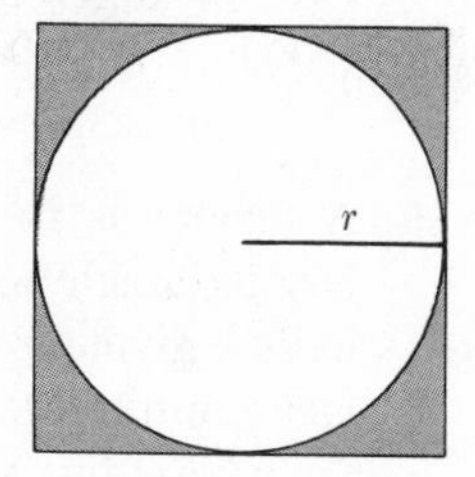

FIGURE 4.16

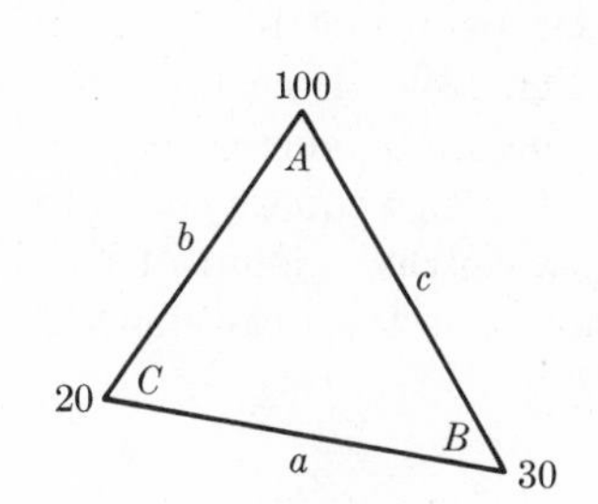

FIGURE 4.15

• 11. (a) Find the point P_0 on the hyperbola $H = \{(x,y): xy = 1\}$ which is closest to the origin.
   (b) Find the point P_0 on H which is closest to the point $(0,1)$.

12. A wire of length L is cut into two pieces, one of which is bent into a circle, and the other into a square.
   (a) How should the wire be cut to make the sum of the two enclosed areas a maximum?
   (b) How should the wire be cut to make the sum of the enclosed areas a minimum?

• 13. A conical drinking cup has radius r and height h. The volume is $\frac{1}{3}\pi r^2 h$, and the surface area is $\pi r\sqrt{h^2 + r^2}$.
   (a) For a *given volume*, find the ratio r/h giving least surface area.
   (b) For a *given surface area*, find the ratio r/h giving greatest volume.

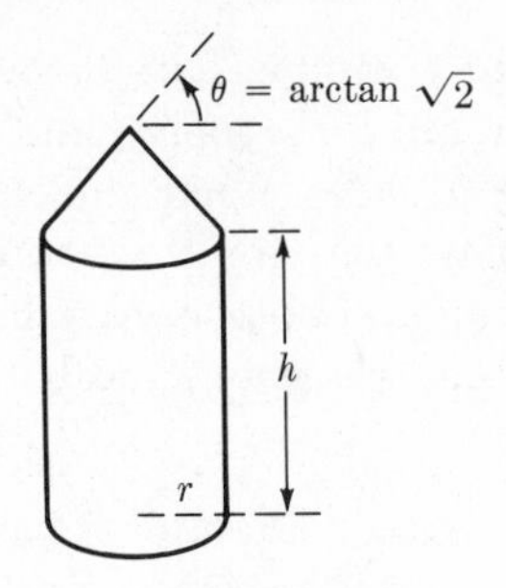

FIGURE 4.17

14. A wheat storage program calls for the construction of many circular silos with conical tops [Fig. 17]. Let h be the height and r be the radius of the cylindrical part, and assume that the conical top has slope $\sqrt{2}$ (slopes at an angle of $\arctan\sqrt{2}$). The surface area of the *cone* is then $\sqrt{3}\pi r^2$, and its volume is $\sqrt{2}\pi r^3/3$.

(a) For a given volume V, what ratio r/h gives the least surface area?

(b) For a given surface area S, what ratio r/h gives the greatest volume?

• **15.** A farmer is to make two equal rectangular pens side by side, each having a side in common with the other [Fig. 18]. If each pen is to have a given area A, what is the minimum amount of fencing required? [The minimum is neither two squares ($a = b$) nor a square cut in half ($a = 2b$); but there is a simple relation between the amount of "horizontal" fencing, (which is $4a$), and the amount of "vertical" fencing.]

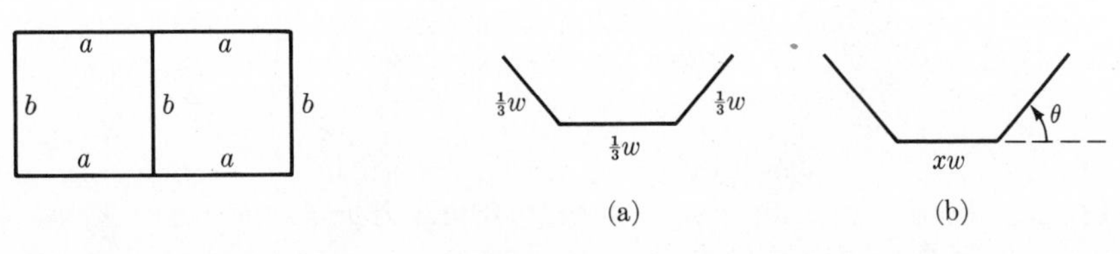

FIGURE 4.18 FIGURE 4.19

16. A trough is to be made with a flat bottom and two equal sloping sides by bending a long piece of sheet metal of width w.

(a) If the sides and bottom each have width $w/3$ [Fig. 19(a)], what angle of sides gives the largest cross-sectional area?

(b) If the angle between side and bottom is a given angle θ, $0 \leq \theta \leq \pi/2$ [Fig. 19(b)] what width xw should the bottom have?

• **17.** What is the largest rectangle that can be inscribed in a semicircle of given radius r, with two corners on the diameter and two corners on the semicircle? [Compare your answer to the largest rectangle inscribed in a given complete circle.]

18. Given the two points $F_1 = (1,0)$ and $F_2 = (-1,0)$, and the circle $C = \{(x,y): x^2 + y^2 = 4\}$.
(a) Find the points P on C such that $PF_1 + PF_2$ is a minimum.
(b) Find the points P on C such that $PF_1 + PF_2$ is a maximum.
[If you know much about ellipses, this problem has an obvious geometric solution.]

• **19.** Given the graph G of a differentiable function f, and a point P_0 not on G, suppose that the distance from P_0 to G achieves a minimum at a point P_1 on G, not an endpoint of G. Show that the line through P_0 and P_1 is perpendicular to G at P_1.

20. A large desk top of length L is slid on edge around a rectangular hallway corner, from a hall of width a to another of width b [Fig. 20]. What is the minimum width b for which this maneuver is possible?

• **21.** A cable has a core of radius r, and an outer casing of radius R. What proportions give the "speediest" cable, given that the speed of signaling varies as $(r/R)^2 \log(R/r)$. [Set $x = r/R$; then the speed is $-kx^2 \log x$.]

22. The stiffness of a rectangular beam of width w and depth d varies as wd^3, according to a standard formula from mechanical engineering. [Note that stiffness (resistance to bending) is different from strength (resistance to breaking); compare Problem 8, §1.5.] Find the dimensions of the stiffest beam that can be cut from a round log of radius r, and show that it has the proportions $d/w = \sqrt{3}$.

• **23.** The brightness at a distance r from a point source of light of intensity I is I/r^2. Suppose that on the x axis there is a light of intensity A at the origin, and a light of intensity B at the point $x = 1$. For what ratio A/B does the darkest spot occur at $x = 1/3$?

24. An island 20 miles from a relatively straight coast is to arrange permanent car ferry service to a city 50 miles down the coast [Fig. 21].
(a) If the ferry goes 15 mph and cars average 45 mph, where should the mainland ferry terminal be located to make the trip as quick as possible?
(b) If the ferry goes F mph and cars average C mph, for what values of F/C should the terminal be located right in the mainland city to make the trip as quick as possible?

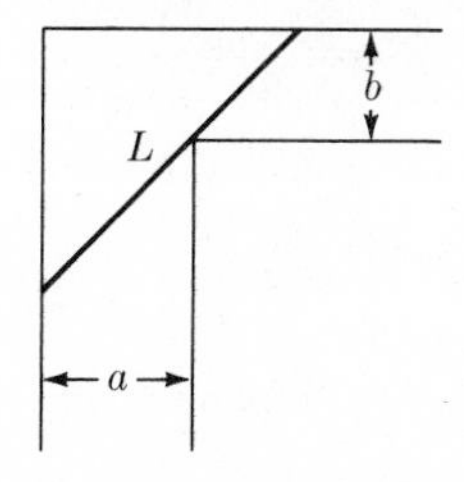

FIGURE 4.20

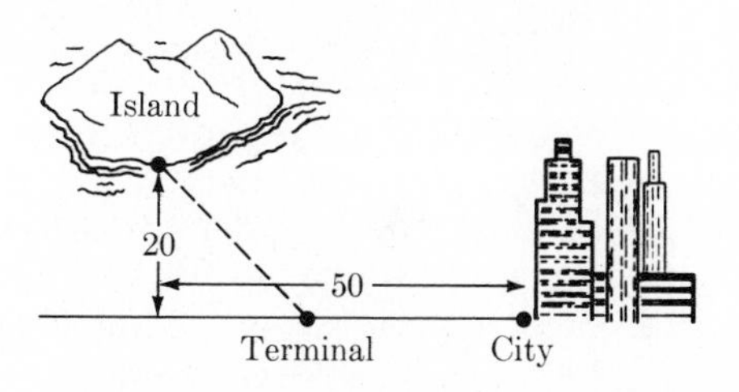

FIGURE 4.21

25. (a) In Example 6 above suppose that the factory revises its price, charging $30 + 4x$ dollars for x bottles; and the wholesaler has less storage available, a maximum of 170 bottles can be stored. What is now the cheapest batch?

(b) In Example 6 above, suppose the factory price is $a + bx$ dollars for x bottles, the storage costs average cx cents per day for a batch of x bottles, and at most N bottles can be stored. Show that the size of the cheapest batch depends on a, c, and N, but not on b.

26. A farmer has 100 pigs, each weighing 300 pounds. It costs 50¢ a day to keep one pig. The pigs gain weight at 5 pounds a day. They sell today for 50¢ a pound, but the price is falling by 1¢ a day. How many days should the farmer wait to sell his pigs in order to maximize his profit?

27. The total cost of producing and marketing x framuses is $1000 + 10x^{1/2} + 10^{-4}x^{3/2}$ dollars, and to sell x framuses they must be priced at $20x^{-1/2}$ dollars per framus.

(a) Show that the profit made by producing and selling x framuses priced in this way is $P(x) = 10x^{1/2} - 10^{-4}x^{3/2} - 1000$.

(b) What amount of production maximizes the profit?

28. Suppose it costs $A + Bx$ to produce x units of an item, and they can all be sold at a price $C - Dx$ per unit, where A, B, C, D are positive constants.

(a) What value of x maximizes profit, and what is the corresponding price?

(b) If a tax T is added to the production cost, what is the new profit-maximizing price? How much of the tax is "passed on" to the buyer?

(c) What is the situation if the tax T is added to the selling price?

29. In a study of the economics of dike building in Holland [Report of the Delta Commission, 1960], it was estimated that the construction cost of raising the dikes to height x was $A + Bx$, and the expected long run losses due to floods occurring when the sea rises above height x was $Ce^{-\alpha x}$. [Here, A, B, C, and α are empirically determined constants.] What height minimizes the total cost, $A + Bx + Ce^{-\alpha x}$?

4.2 THE MEAN VALUE THEOREM

The Maximum Value Theorem has obvious applications in maximum problems. Less obvious, but more important, is its role in developing the logical structure of calculus. This section gives a chain of theorems leading from the Maximum Value Theorem to the proofs of Proposition F (that $f' \geq 0 \Rightarrow f$ is increasing) and of Theorem 3 (the second-derivative criterion for maxima and minima).

The first link in the chain is *Rolle's Theorem* [Fig. 22]:

ROLLE'S THEOREM

Theorem 5. *Suppose that*

(1) *f is continuous on the* closed *finite interval* $[a,b]$,

(2) *$f'(x)$ exists for every x in the* open *interval* (a,b), *and*

(3) $f(a) = f(b)$.

Then there is a point c such that

(4) $a < c < b$ *and*

(5) $f'(c) = 0$.

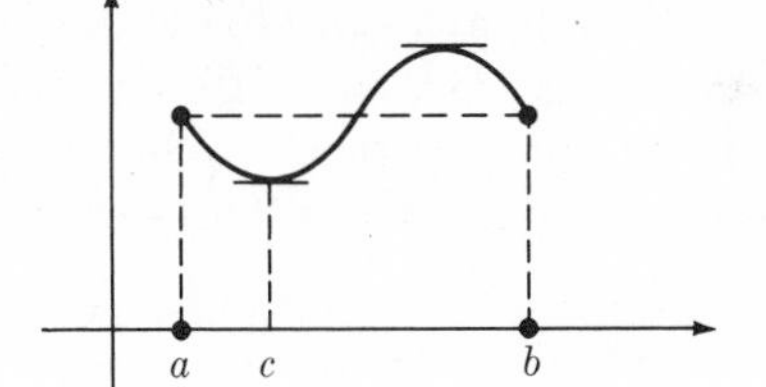

FIGURE 4.22 Rolle's Theorem

This conclusion is "obvious" from Fig. 22; you can see that the graph will always be flat on the top, or the bottom, or both. In the rigorous proof, we invoke the Maximum Value Theorem to guarantee that the graph really has a "top" and a "bottom."

Proof of Theorem 5. By hypothesis (1) and the Maximum Value Theorem, f has a minimum point c_1 and a maximum point c_2 in $[a,b]$ such that

$$f(c_1) \leq f(x) \leq f(c_2) \quad \text{for } a \leq x \leq b. \tag{6}$$

Consider three cases.

Case (i): c_1 is not an endpoint. Then set $c = c_1$. Conclusion (4) is valid, since c_1 is not an endpoint. Hence, by (2), $f'(c_1)$ exists. Since c_1 is a minimum point, we have $f'(c_1) = 0$ [Theorem 1, §4.1], and (5) is proved.

Case (ii): c_2 is not an endpoint. Then set $c = c_2$, and complete the argument exactly as in case (i).

Case (iii): c_1 and c_2 are both endpoints. Then, by hypothesis (3), $f(c_1) = f(c_2)$. Hence from (6), f is *constant* on $[a,b]$, so $f'(c) = 0$ for *every* c in (a,b). *Q.E.D.*

Figure 23 illustrates the next link in the chain, the *Mean Value Theorem:*

☆☆☆ MEAN VALUE THEOREM

Theorem 6. *Suppose that*

(7) *f is continuous on the closed finite interval $[a,b]$, and*

(8) *$f'(x)$ exists for every x in the open interval (a,b).*

Then there is a point c such that

(9) $a < c < b$ *and*

$$(10) \quad f'(c) = \frac{f(b) - f(a)}{b - a}.$$

The quotient in (10) is the slope of the line L through the ends of the graph [Fig. 23], and (10) says that some point on the graph has a tangent which is parallel to L. When $f(a) = f(b)$, then (10) says that $f'(c) = 0$; thus the Mean Value Theorem includes Rolle's Theorem as a special case.

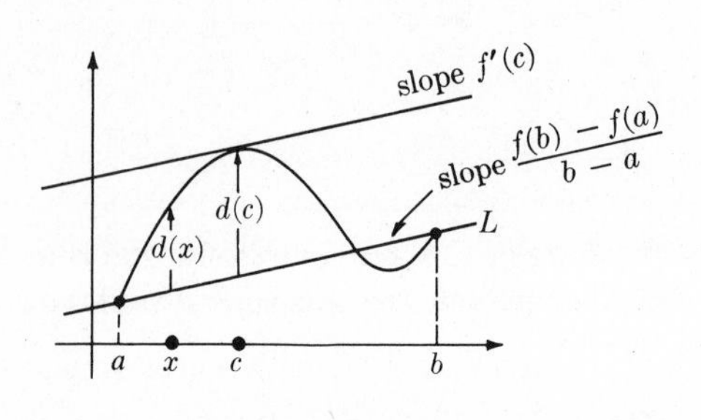

FIGURE 4.23

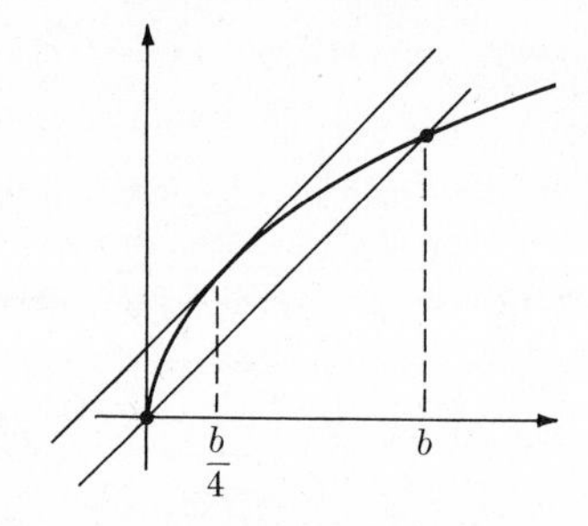

FIGURE 4.24

Example 1. Take $f(x) = \sqrt{x}$ on the interval $[0,b]$, where $b > 0$. Then $f'(x)$ exists for $0 < x < b$, and $\lim_{x \to b-} f(x) = f(b)$ and $\lim_{x \to 0-} f(x) = f(0) = 0$ [Fig. 24], so the Mean Value Theorem applies. We can find exactly which point c satisfies (9) and (10), as follows:

$$\frac{f(b) - f(a)}{b - a} = \frac{\sqrt{b} - 0}{b} = \frac{1}{\sqrt{b}},$$

$$f'(x) = \frac{1}{2} x^{-1/2} = \frac{1}{2\sqrt{x}}.$$

Hence $f'(x) = 1/\sqrt{b}$ iff $x = b/4$. Taking $c = b/4$, both (9) and (10) are satisfied.

To prove the Mean Value Theorem, we apply Rolle's Theorem to the difference between f and the function g whose graph is the line L [Fig. 23],

$$g(x) = f(a) + \frac{f(b) - f(a)}{b - a}(x - a). \tag{11}$$

Proof of Theorem 6. Let $d(x) = f(x) - g(x)$ [Fig. 23], where g is defined as in (11). We will show that d satisfies the hypotheses of Rolle's Theorem. Hence there is a point c satisfying (9), and

$$d'(c) = 0. \tag{12}$$

But $d = f - g$, so

$$d'(c) = (f - g)'(c) = f'(c) - g'(c)$$

$$= f'(c) - \frac{f(b) - f(a)}{b - a}, \tag{13}$$

as you can check by computing g' from (11). But (12) and (13) imply (10), which is what we had to prove.

It remains to show that the function d satisfies the hypotheses of Rolle's Theorem.

(i) $d(a) = d(b)$. From (11), $g(a) = f(a)$ and $g(b) = f(b)$ [as Fig. 23 shows; L is the graph of g]; hence

$$d(a) = f(a) - g(a) = 0 \quad \text{and} \quad d(b) = f(b) - g(b) = 0.$$

(ii) *d is differentiable on (a,b).* From (11), $g'(x)$ exists for *every* x, hence $d'(x) = f'(x) - g'(x)$ exists wherever $f'(x)$ does; in particular, $d'(x)$ exists for x in (a,b).

(iii) *d is continuous on $[a,b]$.* From (11), $\lim_{x \to c} g(x) = g(c)$ for *every* c, and $\lim_{x \to c} f(x) = f(c)$ for $a < c < b$, by hypothesis. Hence, since

$$d(x) = f(x) - g(x),$$

$$\lim_{x \to c} d(x) = \lim_{x \to c} f(x) - \lim_{x \to c} g(x) = f(c) - g(c) = d(c).$$

This proves that d is continuous at every point c in the open interval (a,b). Similarly, $\lim_{x \to a+} d(x) = d(a)$ and $\lim_{x \to b+} d(x) = d(b)$, so d is continuous on the whole interval $[a,b]$.

Thus all hypotheses of Rolle's Theorem are satisfied. *Q.E.D.*

A typical application of the Mean Value Theorem is:

f IS STRICTLY INCREASING IF $f' > 0$

Theorem 7. *Suppose that f satisfies the hypotheses (7) and (8) in the Mean Value Theorem, and, in addition, f' is strictly positive on (a,b),*

(14) $$a < x < b \Rightarrow f'(x) > 0.$$

Then f is strictly increasing on $[a,b]$. Similarly, if $f'(x) < 0$ for $a < x < b$, then f is strictly decreasing on $[a,b]$.

Proof. Let $\bar{a}$ and $\bar{b}$ be any two points in $[a,b]$, with $\bar{a} < \bar{b}$; we must prove that $f(\bar{a}) < f(\bar{b})$. Since the hypotheses of the Mean Value Theorem apply on $[a,b]$, they also apply on any smaller interval $[\bar{a},\bar{b}]$. Hence there is a point $\bar{c}$ such that

$$\frac{f(\bar{b}) - f(\bar{a})}{\bar{b} - \bar{a}} = f'(\bar{c}) \quad \text{and} \quad \bar{a} < \bar{c} < \bar{b}. \tag{15}$$

Since $a \leq \bar{a} < \bar{c} < \bar{b} \leq b$, the hypothesis (14) shows that $f'(\bar{c}) > 0$. Since $\bar{b} - \bar{a} > 0$, it follows from (15) that $f(\bar{b}) - f(\bar{a}) > 0$. *Q.E.D.*

The proof when $f'(c) < 0$ on (a,b) is virtually the same, and is left as Problem 4.

Theorem 7 is quite similar to Proposition F, on which the entire Chapter III is based. The proof of Proposition F itself is left as Problem 5.

Finally, we prove Theorem 4, which was stated in §4.1:

Theorem 4. *Suppose that f is differentiable throughout an interval I, and $f'(a) = 0$ for exactly one point a in I. Then*

(16) $f''(a) > 0 \Rightarrow f(a) < f(x)$ *for all $x \neq a$ in I,*

(17) $f''(a) < 0 \Rightarrow f(a) > f(x)$ *for all $x \neq a$ in I.*

The first step in the proof (Lemma 1 below) is illustrated in Fig. 25.

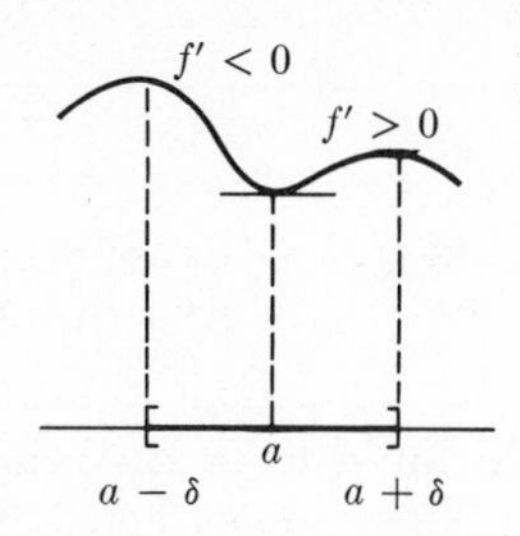

FIGURE 4.25 Lemma 1

Lemma 1. *If $f'(a) = 0$ and $f''(a) > 0$, then there is an open interval $I' = (a - \delta, a + \delta)$ containing a, such that*

(18) $$f(a) < f(x) \quad \text{for every } x \neq a \text{ in } I'.$$

Proof. Apply Proposition G, §4.1, to f' rather than to f. Since $(f')'(a) > 0$, there is a number $\delta > 0$ such that

$$a - \delta < x < a \quad \Rightarrow \quad f'(x) < f'(a), \tag{19}$$

$$a < x < a + \delta \quad \Rightarrow \quad f'(a) < f'(x). \tag{20}$$

But $f'(a) = 0$, so (19) implies that $f' < 0$ in the interval $(a - \delta, a)$, and (20) implies that $f' > 0$ in the interval $(a, a + \delta)$. Hence [Theorem 7] f is strictly decreasing on $[a - \delta, a]$ and strictly increasing on $[a, a + \delta]$. It follows that

$$a - \delta \leq x < a \quad \Rightarrow \quad f(x) > f(a),$$

$$a < x \leq a + \delta \quad \Rightarrow \quad f(a) < f(x),$$

which is the same as (18). *Q.E.D.*

When an open interval I' exists as in (18), then a is called a *strict local minimum point* for f. *Strict,* because $f(a) < f(x)$ [not merely $f(a) \leq f(x)$] for $x \neq a$ in I'; and *local,* because $f(a)$ need not be a minimum with respect to the entire domain of f, but only with respect to points near a, i.e. points in the small interval I'.

Returning to the proof of Theorem 4, we establish (16) by contradiction. Suppose that

$$f''(a) > 0 \tag{21}$$

and at the same time

$$\text{there is a point } b \neq a \text{ such that } f(b) \leq f(a). \tag{22}$$

We must show that this leads to an impossible conclusion.

Case (i): The number b in (22) is $>a$. Then f is differentiable on $[a,b]$, hence continuous on $[a,b]$; hence, by the Maximum Value Theorem, f has a maximum point c in $[a,b]$,

$$f(c) = \max_{[a,b]} f.$$

By Lemma 1, $f(a) < \max_{[a,b]} f = f(c)$, so $c \neq a$. By (22), $f(b) \leq f(a) < f(c)$, so $c \neq b$. Hence by Theorem 1, §4.1, $f'(c) = 0$. But this is impossible, since it contradicts the hypothesis in Theorem 4 that $f'(x) = 0$ *only at the one point* $x = a$.

Case (ii): $b < a$. Now f has a maximum on $[b,a]$, and the proof goes as in case (i).

The proof of (17) [concerning a strict local maximum] is left as Problem 6.

PROBLEMS

1. In each of the following cases, decide whether the Mean Value Theorem applies. If it does, find a number c in (a,b) such that $f'(c) = \dfrac{f(b) - f(a)}{b - a}$. Sketch a graph showing the tangent at $(c,f(c))$ and the line through the ends of the graph on $[a,b]$.

- •(a) $f(x) = 1/x;\ \ a = 1,\ \ b = 2$
- (b) $f(x) = 1/x;\ \ a = -1,\ \ b = 2$
- •(c) $f(x) = x^3;\ \ a = 0,\ \ b = 1$
- (d) $f(x) = x^3;\ \ a = -1,\ \ b = 0$
- •(e) $g(\theta) = \sin\theta;\ \ a = 0,\ \ b = \pi/2$
- (f) $h(\theta) = \tan\theta;\ \ a = \pi/4,\ \ b = 3\pi/4$
- •(g) $f(x) = \sqrt{1 - x^2};\ \ a = -1,\ \ b = 0$
- (h) $f(t) = t^2(t - 1);\ \ a = 0,\ \ b = 1$
- •(i) $f(x) = x^{2/3} = (x^2)^{1/3};\ \ a = -1,\ \ b = 1$
- (j) $f(x) = 1$ for $x \geq 0$ and $f(x) = 0$ for $x < 0;\ \ a = -1,\ \ b = 1$

2. (a) Let $f(x) = x^2$. In this case, show that for any interval $[a,b]$ the point c given in the Mean Value Theorem is actually the midpoint, $c = \frac{1}{2}(a + b)$.

(b) Repeat part (a) for any polynomial of second degree, $f(x) = c_2x^2 + c_1x + c_0$.

(c) Find a function f for which the "Mean Value" point c is *not* the midpoint of $[a,b]$.

3. Locate the strict local maximum and minimum points of the following functions f on the following given intervals I, using Lemma 1 for the minima, and the analogous condition for a strict local maximum: $f'(a) = 0$ and $f''(a) < 0$. In which cases is the strict local maximum actually a maximum over the whole interval I?

- •(a) $f(x) = 1 - x^2$ on $[-1,1]$
- (b) $f(x) = x - x^3$ on $[-1,1]$
- •(c) $f(x) = x - x^3$ on $[-2,2]$
- (d) $f(x) = x^4 - 2x^2 + 2$
- •(e) $f(x) = \sin x$ on $(-4\pi,4\pi)$
- •(f) $f(t) = \cos(1/t)$ on $(0,1)$
- (g) $f(t) = (1 + t^2)^{-1}\cos(1/t)$ [Find the local maxima and minima approximately, from the graph, not by solving $f'(a) = 0$.]
- (h) $f(t) = t\cos(1/t)$ [See the remark in part (g).]

4. In Theorem 7 above, prove that if $f' < 0$ on (a,b), then f is strictly decreasing on $[a,b]$.

5. Prove Proposition F, page 198.

6. In Theorem 4, prove (17). [Hint: You can either imitate the proof of (16), or apply (16) to $-f$.]

7. Use the Mean Value Theorem directly to prove that if f is continuous on $[a,c]$, and $f' = 0$ throughout (a,c), then f is constant on $[a,c]$. [Hint: Prove that $f(b) - f(a) = 0$ for $a \leq b \leq c$.]

8. A (not necessarily differentiable) function f defined on an interval I is called *convex* on I if

CONVEX FUNCTIONS

$$\frac{f(x_2) - f(x_1)}{x_2 - x_1} \leq \frac{f(x_3) - f(x_2)}{x_3 - x_2}$$

whenever $x_1 < x_2 < x_3$ are three points in I [Fig. 26].

(a) Prove that if f' exists on I, and is increasing, then f is convex. [This reconciles the above definition of "convex" with the earlier one, §3.4, which applied only to differentiable functions.]

(b) Prove that if $f'' \geq 0$ throughout I, then f is convex on I.

(c) Prove that for $x_1 < x_2 < x_3$, the following two conditions are equivalent:

$$\frac{y_2 - y_1}{x_2 - x_1} \leq \frac{y_3 - y_2}{x_3 - x_2} \quad \Leftrightarrow \quad y_2 \leq y_1 + \frac{y_3 - y_1}{x_3 - x_1}(x_2 - x_1).$$

[See Figs. 26 and 27. This gives an alternate geometric definition of convexity; between any two points x_1 and x_3, the graph of f lies below the straight line through $P_1 = (x_1, f(x_1))$ and $P_3 = (x_3, f(x_3))$.]

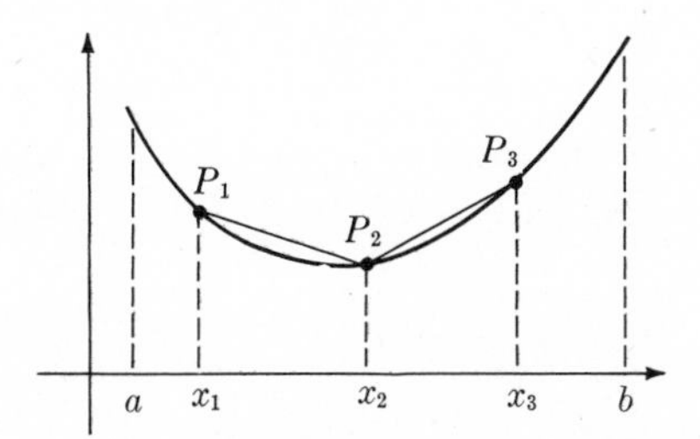

FIGURE 4.26

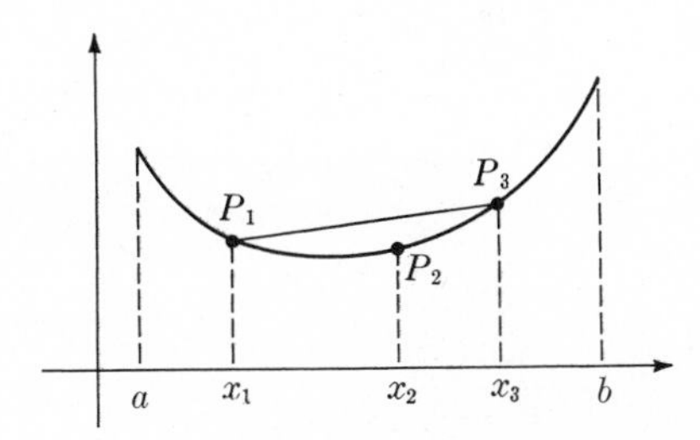

FIGURE 4.27

Introduction to Integrals

The two main concepts in calculus are developed from geometric ideas related to curves. The derivative comes from the construction of tangents to a curve; the subject of this chapter, the *integral,* comes from the computation of the *area* of a curved region. The integral, like the derivative, finds many applications beyond its geometric origins.

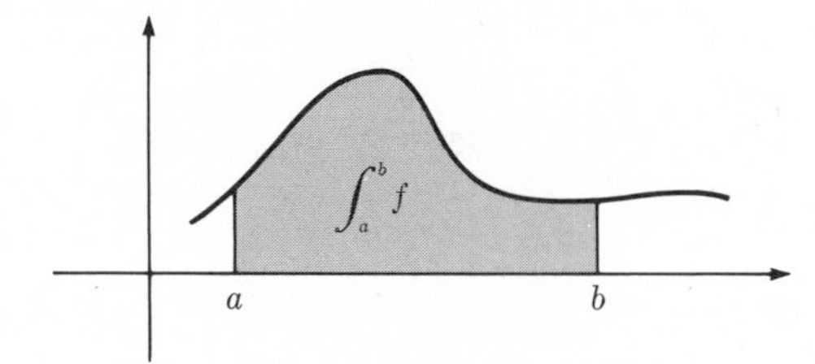

FIGURE 5.1

5.1 THE DEFINITE INTEGRAL

Let f be a positive, continuous function on $[a,b]$. The *integral of f from a to b* is, intuitively, the *area* shown in Fig. 1, lying below the graph of f, above the x axis, and between the vertical lines $\{(x,y): x = a\}$ and $\{(x,y): x = b\}$. This integral is denoted

$$\int_a^b f.$$

An alternate notation, due to Leibniz, is

$$\int_a^b f(x)\, dx.$$

The function f is called the *integrand* (the thing being integrated), and $[a,b]$ is the *interval of integration.* The number a is the *lower limit of integration,* and b is the *upper limit of integration.*

From Fig. 2 we find:

$$(1) \quad \int_a^b c = c \cdot (b - a) \qquad \text{[where } c \text{ denotes a constant function]}$$

$$(2) \quad \int_0^b cx\,dx = \frac{1}{2}cb^2 \qquad \text{[area of a right triangle]}$$

$$(3) \quad \int_a^b cx\,dx = (b - a)\frac{cb + ca}{2} = \frac{1}{2}c(b^2 - a^2) \qquad \text{[area of a trapezoid]}$$

$$(4) \quad \int_1^b \frac{1}{x}\,dx = \log b \qquad \text{[definition of the natural logarithm]}$$

When f is *negative,* a sign convention applies; $\int_a^b f$ is *minus* the area between the graph of f and the x axis. One justification of this choice of sign is that formulas (1)–(3) remain valid when $c < 0$. For example, if $f(x) = -2$, the area in question is $2 \cdot (b - a)$ [Fig. 3], but it lies below the x axis, so the integral is $-2 \cdot (b - a)$, and we get

$$\int_a^b (-2)\,dx = -2 \cdot (b - a),$$

which agrees with (1) when $c = -2$.

INTEGRAL AS DIFFERENCE OF AREAS

When f takes on both positive and negative values [Fig. 4], then $\int_a^b f$ is the sum of the areas above the axis, *minus* the areas *below* the axis. Thus [Fig. 5]

$$\int_{-1}^2 x\,dx = 2 - \tfrac{1}{2} = 1\tfrac{1}{2}, \qquad \int_{-1}^1 x\,dx = \tfrac{1}{2} - \tfrac{1}{2} = 0.$$

Similarly, in view of the symmetry of the sin function, we get [Fig. 5]

$$\int_{-\pi}^{\pi} \sin x\,dx = 0.$$

Just as the equations

$$f(x) = cx, \quad -\infty < x < \infty \qquad \text{and} \qquad f(t) = ct, \quad -\infty < t < \infty$$

define exactly the same function f, the symbols

$$\int_a^b f(x)\,dx \qquad \text{and} \qquad \int_a^b f(t)\,dt$$

denote exactly the same integral. In fact, we can use any other letter

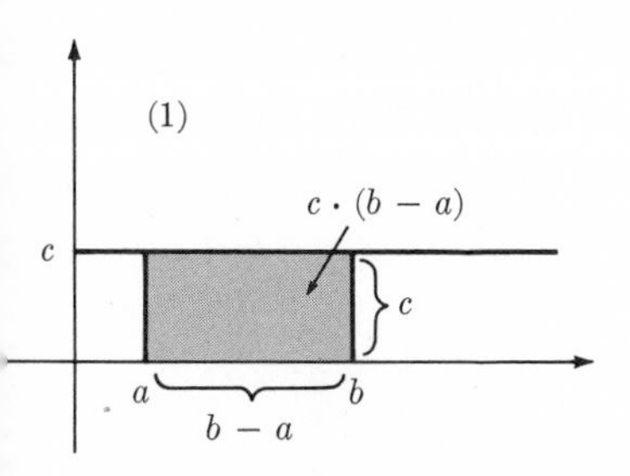

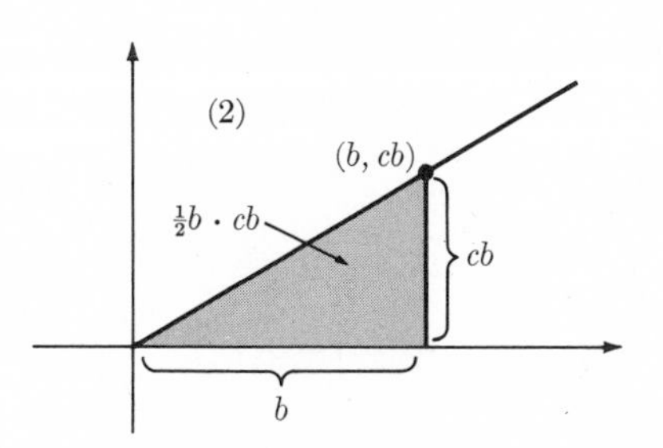

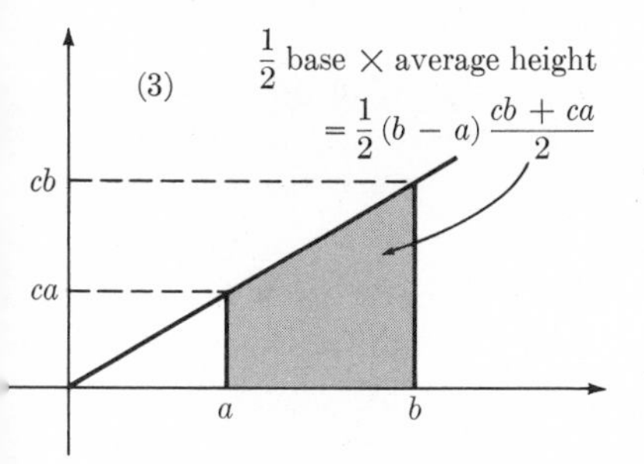

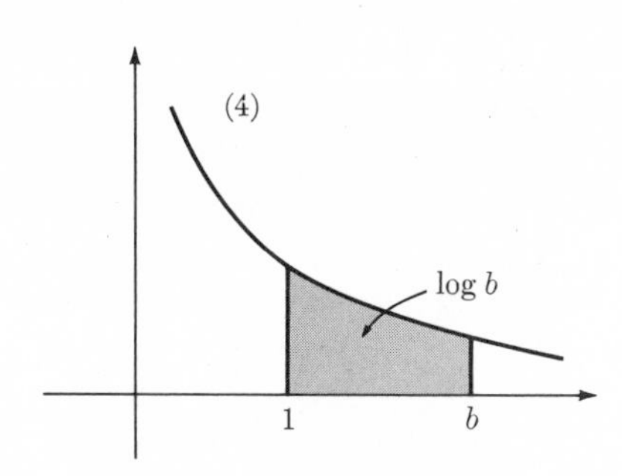

FIGURE 5.2

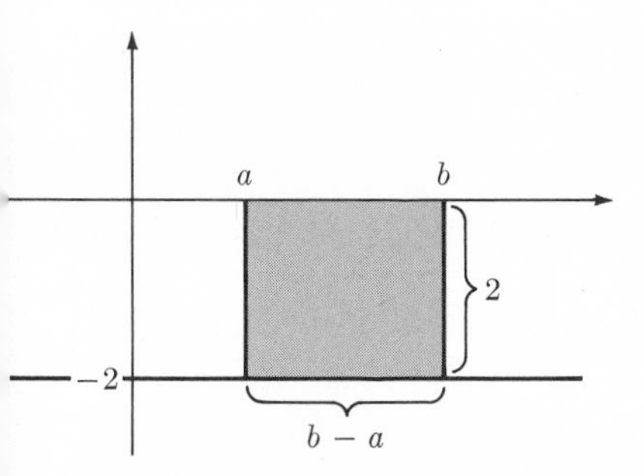

FIGURE 5.3

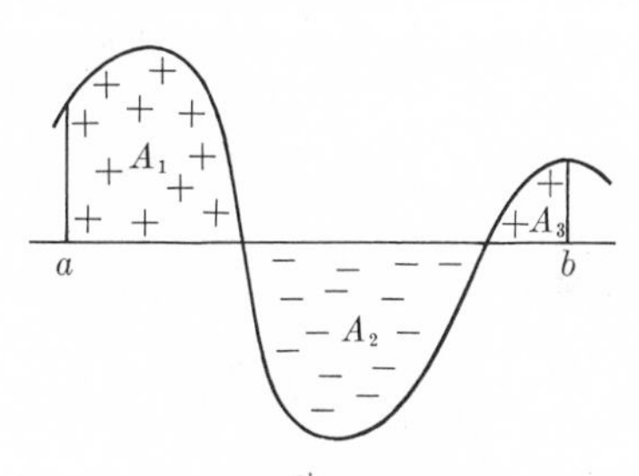

FIGURE 5.4 $\int_a^b f = A_1 - A_2 + A_3$

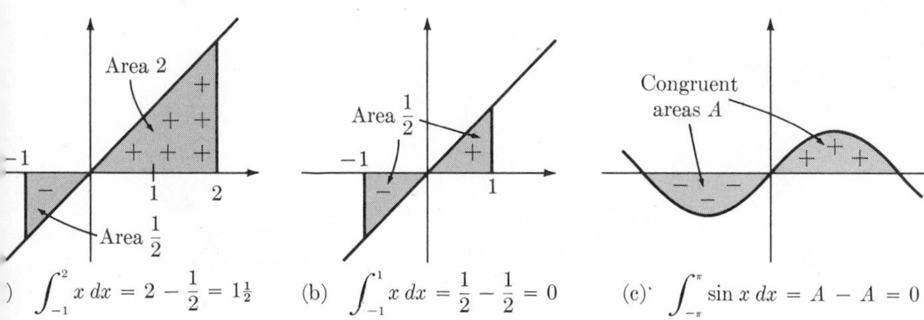

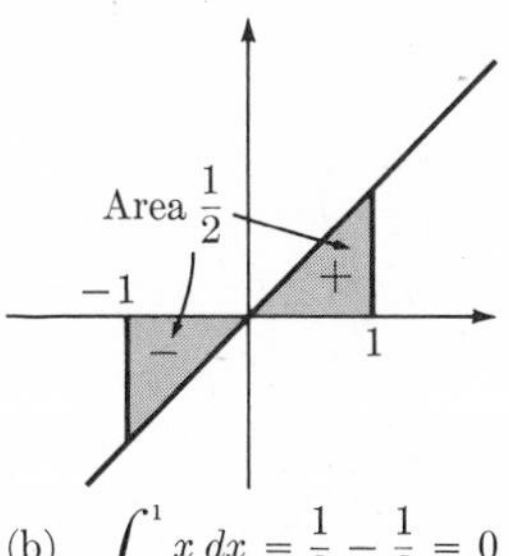

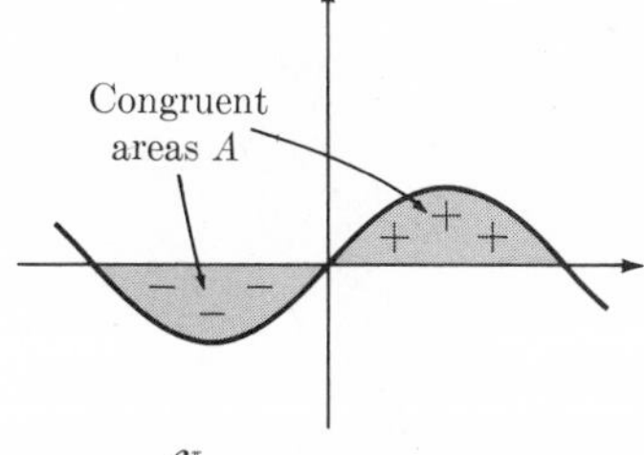

(a) $\int_{-1}^{2} x\,dx = 2 - \frac{1}{2} = 1\frac{1}{2}$

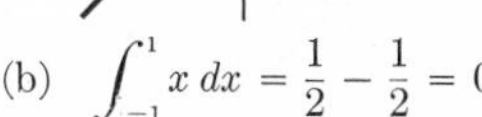

(b) $\int_{-1}^{1} x\,dx = \frac{1}{2} - \frac{1}{2} = 0$

(c) $\int_{-\pi}^{\pi} \sin x\,dx = A - A = 0$

FIGURE 5.5

instead of x or t, so long as it is not being simultaneously used for something else. Thus formulas (1)–(4) can be rewritten

$$\int_a^b c\,dt = c\cdot(b-a) \qquad \int_a^b c\alpha\,d\alpha = \tfrac{1}{2}c(b^2-a^2)$$

$$\int_0^b cs\,ds = \tfrac{1}{2}cb^2 \qquad \int_1^b \frac{1}{Z}\,dZ = \log b.$$

But a collection of symbols such as $\int_a^b f(b)\,db$ is considered ungrammatical since b is already being used for an endpoint of the interval of integration.

The elementary area formulas (for triangles, circles, etc.) are rather few, so we need other ways to evaluate integrals. The most basic is the method of *Riemann sums*, which we illustrate by evaluating $\int_0^b x^2\,dx$. This is the area of the shaded region in Fig. 6. Imagine this area sliced into n strips of equal width b/n, as in the figure. Now, *replace each strip by a rectangle*, as in Fig. 7. The jth rectangle has base b/n, and since its upper right-hand corner lies on the graph of the function $f(x) = x^2$ [Fig. 7], the rectangle has height $f(jb/n) = (jb/n)^2$. The area of this rectangle is

$$\text{base} \times \text{height} = \frac{b}{n} f\left(\frac{jb}{n}\right) = \frac{b}{n}\left(\frac{jb}{n}\right)^2 = \frac{b^3}{n^3}j^2.$$

Hence the sum of the areas of these rectangles, call it S_n, is

$$S_n = \frac{b}{n} f\left(\frac{b}{n}\right) + \frac{b}{n} f\left(\frac{2b}{n}\right) + \cdots + \frac{b}{n} f(b)$$

$$= \frac{b^3}{n^3}1^2 + \frac{b^3}{n^3}2^2 + \cdots + \frac{b^3}{n^3}j^2 + \cdots + \frac{b^3}{n^3}n^2 = \frac{b^3}{n^3}\sum_{j=1}^n j^2. \tag{5}$$

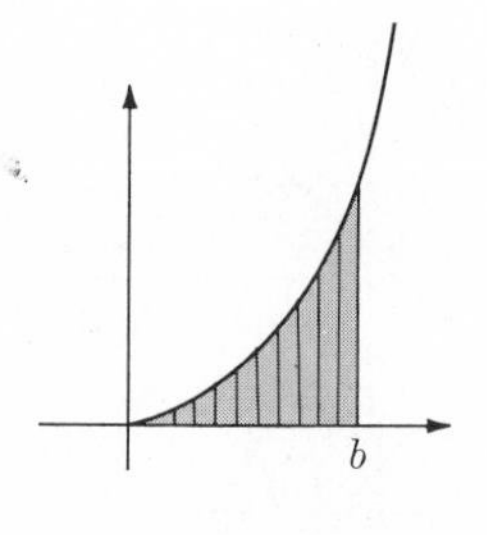

FIGURE 5.6

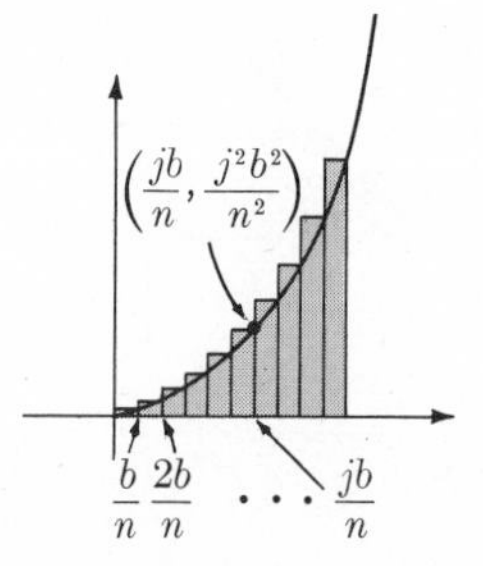

FIGURE 5.7

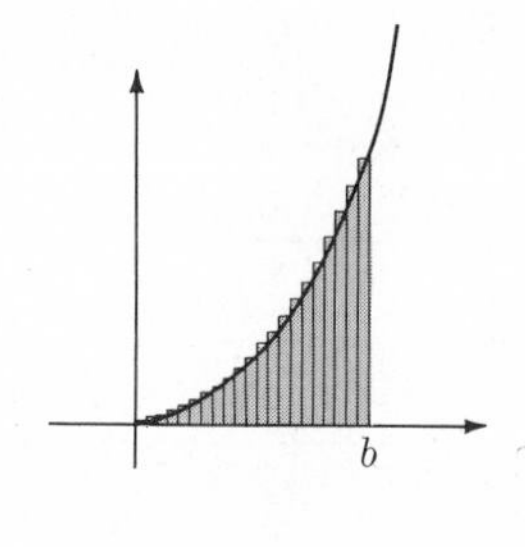

FIGURE 5.8

The notation $\sum_{j=1}^{n} j^2$ is shorthand for $1^2 + 2^2 + \cdots + n^2$ [§AI.1]. There happens to be a convenient formula for this sum [§AI.2], namely

$$\sum_{j=1}^{n} j^2 = \frac{n^3}{3} + \frac{n^2}{2} + \frac{n}{6}. \tag{6}$$

Therefore, from (5)

$$S_n = \frac{b^3}{n^3}\left(\frac{n^3}{3} + \frac{n^2}{2} + \frac{n}{6}\right) = \frac{b^3}{3} + \frac{b^3}{2n} + \frac{b^3}{6n^2}.$$

As n grows large, the terms $b^3/2n$ and $b^3/6n^2$ both tend to zero, and hence

$$S_n \to b^3/3.$$

At the same time, as Fig. 8 suggests, S_n tends to the area under the graph,

$$S_n \to \int_0^b x^2\, dx.$$

Therefore

$$\int_0^b x^2\, dx = b^3/3. \tag{7}$$

The expression (5), which is a sum of areas of rectangles approximating the area under the graph of f, is called a *Riemann sum.* Riemann sums can be formed for any function f defined on a finite interval $[a,b]$. Divide the area under the graph of f [Fig. 9(i)] into strips by choosing a number of points x_0; $x_1, \ldots, x_n$ ranging from a to b [Fig. 9(ii)]. In each interval $[x_{j-1}, x_j]$ choose a point ξ_j, and form a rectangle of base $x_j - x_{j-1}$ and height $f(\xi_j)$ [Fig. 9(iii)]. Denote the base by Δx_j,

$$x_j - x_{j-1} = \Delta x_j.$$

The area of our rectangle is then $f(\xi_j)\,\Delta x_j$; and the sum of all these areas,

RIEMANN SUM

$$S_n = \sum_{j=1}^{n} f(\xi_j)\,\Delta x_j \tag{8}$$

is an approximation to the area under the graph, i.e. to $\int_a^b f$.

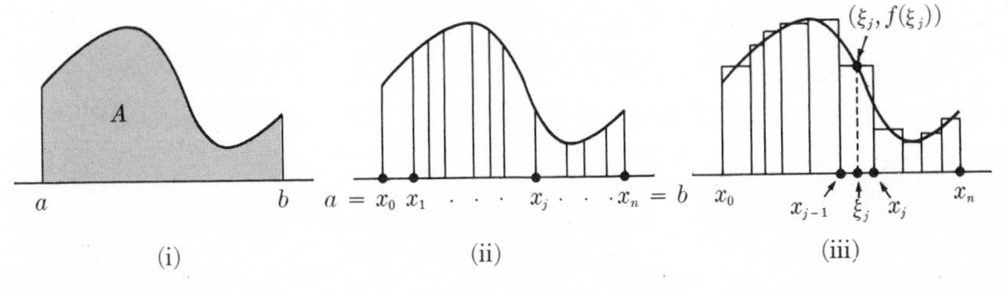

FIGURE 5.9

The collection of points $x_0, x_1, \ldots, x_n$ is called a *partition* of the interval $[a,b]$; we denote such a partition by writing

PARTITION

$$a = x_0 < x_1 < x_2 < \cdots < x_n = b.$$

The points ξ_j are called *points of evaluation*. The sum

$$S_n = \sum_{j=1}^{n} f(\xi_j)\,\Delta x_j$$

is called a *Riemann sum* for f.

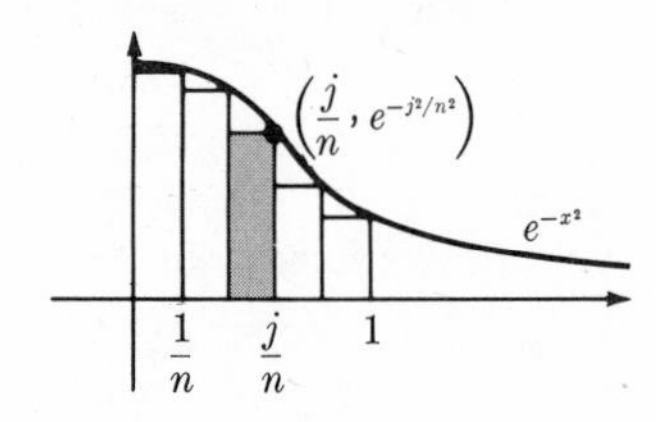

FIGURE 5.10

To deduce the integral formula $\int_0^b x^2\,dx = \frac{1}{3}b^3$, we needed the convenient formula (6) to simplify the Riemann sum (5). Formulas like (6) are available only for very simple functions f; but even without such a formula, the Riemann sums can still be used to give reasonable approximations to the integral. We illustrate this with an integral arising in statistics, $\int_0^b e^{-x^2}\,dx$. To be specific, take $b = 1$. Form the partition [Fig. 10]

$$0 = x_0 < \frac{1}{n} < \frac{2}{n} < \cdots < \frac{j}{n} < \cdots < \frac{n}{n} = 1.$$

In the jth subinterval $\left[\frac{j-1}{n}, \frac{j}{n}\right]$ choose the point of evaluation ξ_j to be the right endpoint, $\xi_j = j/n$. Each subinterval has length

$$\Delta x_j = x_j - x_{j-1} = \frac{j}{n} - \frac{j-1}{n} = \frac{1}{n},$$

and $f(\xi_j) = e^{-(j/n)^2} = e^{-j^2/n^2}$, so the Riemann sum is

$$S_n = \sum_{j=1}^{n} f(\xi_j)\,\Delta x_j = \sum_{j=1}^{n} (e^{-j^2/n^2})\,\frac{1}{n}. \tag{9}$$

With $n = 5$, the sum is

$$S_5 = \frac{1}{5}\sum_{j=1}^{5} e^{-j^2/25} = \frac{1}{5}\left[e^{-1/25} + e^{-4/25} + e^{-9/25} + e^{-16/25} + e^{-1}\right]$$

$$= 0.68\ldots. \tag{10}$$

From Fig. 10, this expression appears to be a reasonable approximation to the integral $\int_0^1 e^{-x^2}\,dx$, but is clearly *smaller* than the actual integral, since each approximating rectangle lies inside the corresponding area under the graph. For this reason, the sums (9) and (10) are called *lower Riemann sums* for f on the interval $[a,b]$.

LOWER, UPPER RIEMANN SUMS

By choosing the points ξ_j differently, we could form an *upper* Riemann sum, which is *larger* than the actual integral. In general, $\sum_{j=1}^{n} f(\xi_j)\,\Delta x_j$ is an upper Riemann sum if $f(\xi_j)$ is the *maximum* of f over the jth interval $[x_{j-1}, x_j]$. In the present example, the maximum occurs at the left endpoint [Fig. 10], so we form an upper sum by taking $\xi_j = x_{j-1} = (j-1)/n$. Thus, an upper Riemann sum for $\int_0^1 e^{-x^2}\,dx$ is

$$\sum_{j=1}^{n} \left(e^{-(j-1)^2/n^2}\right) \frac{1}{n}.$$

When $n = 5$, this upper sum is

$$\frac{1}{5}\sum_{j=1}^{5} e^{-(j-1)^2/25} = \frac{1}{5}[1 + e^{-1/25} + e^{-4/25} + e^{-9/25} + e^{-16/25}]$$

$$= 0.80\ldots. \qquad (11)$$

The difference between (10) and (11) is less than 0.12, so each of these sums approximates $\int_0^1 e^{-x^2}\,dx$ within 0.12. Their average, 0.74, approximates the integral within 0.06.

Notice that the upper and lower sums (10) and (11) agree in all terms except the first and the last; their difference is thus $\frac{1}{5}(1 - e^{-1})$. In general, with a subdivision into n equal parts, the difference would be $\frac{1}{n}(1 - e^{-1})$. Therefore, taking n very large, we can approximate the integral very accurately; to approximate it within 10^{-3}, we take $\frac{1}{n} < 10^{-3}$, or $n > 10^3$.

Of course, 10^3 is a lot of terms, and you must be wondering if there aren't more efficient ways to compute integrals. There are indeed; the next section gives a method that computes many integrals exactly, and §9.3 gives efficient ways to approximate those integrals that cannot be computed exactly.

Remark. In this section we have been using, informally, the idea of *limit of a sequence.* The approximating areas S_n in formulas (5), (8), and (9) form an *infinite sequence of numbers*

$$S_1, S_2, S_3, S_4, \ldots, S_n, \ldots.$$

The statement "$S_n \to S$ as $n \to \infty$" means, roughly, that S_n is a very good approximation to S when n is very large. S is called the *limit* of the sequence $S_1, S_2, \ldots$, and we write

LIMIT OF A SEQUENCE

$$S_n \to S, \qquad \text{or} \qquad \lim_{n\to\infty} S_n = S.$$

Chapter X studies infinite sequences and their limits, and shows that sequence limits behave very much like function limits. In particular, if A_n and B_n are sequences having limits $\lim\limits_{n\to\infty} A_n$ and $\lim\limits_{n\to\infty} B_n$, then:

$$\lim_{n\to\infty} (cA_n) = c \lim_{n\to\infty} A_n .$$

$$\lim_{n\to\infty} (A_n + B_n) = \lim_{n\to\infty} A_n + \lim_{n\to\infty} B_n .$$

If $A_n \leq B_n$, then $\lim\limits_{n\to\infty} A_n \leq \lim\limits_{n\to\infty} B_n$.

If $A_n \leq C_n \leq B_n$, and $\lim\limits_{n\to\infty} A_n = \lim\limits_{n\to\infty} B_n = A$, then $\lim\limits_{n\to\infty} C_n = A$.

[Poe Principle]

PROBLEMS

1. Verify the following graphically.

(a) $\int_2^4 5\,dx = 10$

(b) $\int_3^5 -4\,dx = -8$

(c) $\int_{-1}^1 2x\,dx = 0$

(d) $\int_{-2}^1 t\,dt = -1\frac{1}{2}$

(e) $\int_{-r}^r \sqrt{r^2 - x^2}\,dx = \frac{1}{2}\pi r^2$

(f) $\int_0^{2\pi} \cos\theta\,d\theta = 0$

Summary. Let f be defined on an interval $[a,b]$. Then, intuitively, the *integral* $\int_a^b f$ is the area above the x axis and below the graph of f, *minus* the area below the axis but above the graph of f. It can be approximated by *Riemann sums*

$$S_n = \sum_{j=1}^{n} f(\xi_j)\,\Delta x_j\,,$$

where $a = x_0 < x_1 < \cdots < x_n = b$ is a partition of the interval $[a,b]$ into n subintervals $[x_{j-1}, x_j]$, and each point of evaluation ξ_j is chosen in the subinterval $[x_{j-1}, x_j]$. The Riemann sum S_n is called an *upper sum* if $f(\xi_j) = \max_{[x_{j-1},x_j]} f$, and is called a *lower sum* if $f(\xi_j) = \min_{[x_{j-1},x_j]} f$. Every upper sum is greater than or equal to $\int_a^b f$, and every lower sum is less than or equal to $\int_a^b f$. By computing both upper and lower sums, you can see how accurately these sums approximate the integral [as we did for $\int_0^1 e^{-x^2}\,dx$ above]. With luck, you can find the limit of the sums S_n, and thus determine the integral exactly [as we did for $\int_0^b x^2\,dx$].

2. This problem evaluates $\int_0^{\pi/2} \sin^2\theta\, d\theta$ and $\int_0^{\pi/2} \cos^2\theta\, d\theta$.

(a) Show that $\int_0^{\pi/2} \sin^2\theta\, d\theta = \int_0^{\pi/2} \cos^2\theta\, d\theta$. [Show that the two areas in question are congruent by reflecting in the line $\theta = \pi/4$.]

(b) Show that $\int_0^{\pi/2} (1 - \sin^2\theta)\, d\theta = \frac{\pi}{2} - \int_0^{\pi/2} \sin^2\theta\, d\theta$. [See Fig. 11, and reflect in the line $y = 1/2$.]

(c) Prove that $\int_0^{\pi/2} \sin^2\theta\, d\theta = \int_0^{\pi/2} \cos^2\theta\, d\theta = \frac{\pi}{4}$. Use parts (a) and (b).

(d) Evaluate $\int_0^{\pi} \sin^2\theta\, d\theta$ and $\int_0^{2\pi} \cos^2\theta\, d\theta$.

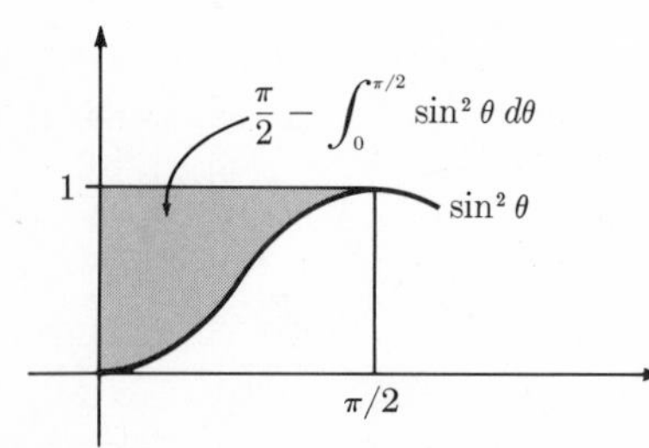

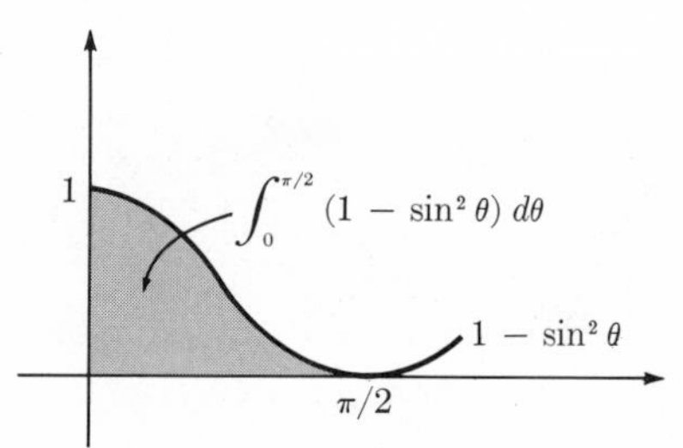

FIGURE 5.11

3. You are given a function and a partition of an interval $[a,b]$. Sketch the function, form the upper and lower sums, and find the difference between the upper and lower sums.

(a) $f(x) = x$, $a = 0 < \frac{1}{5} < \frac{2}{5} < \frac{3}{5} < \frac{4}{5} < \frac{5}{5} = b$

(b) $f(x) = x$, $a = 0 < \frac{b}{n} < \frac{2b}{n} < \cdots < \frac{nb}{n} = b$

• (c) $f(x) = x^{-2}$,

$$a = 1 < 1 + \frac{3}{n} < 1 + \frac{2\cdot 3}{n} < 1 + \frac{3\cdot 3}{n} < \cdots < 1 + \frac{n\cdot 3}{n} = b$$

(d) $f(x) = x^2$, $a = 0 < \frac{1}{4} < \frac{2}{4} < \frac{3}{4} < \frac{4}{4} = b$

(e) $f(x) = \frac{1}{x^2}$, $a < a + \frac{b-a}{n} < a + 2\frac{b-a}{n} < a + 3\frac{b-a}{n}$

$$< \cdots < a + n\frac{b-a}{n} = b, \text{ where } a > 0.$$

4. This problem uses Riemann sums to show that $\int_0^b cx\,dx = \frac{1}{2}cb^2$. Take $f(x) = cx$, and partition the interval $[0,b]$ into n equal subintervals.
(a) Form the upper sum, and show that as n grows large, this sum tends to $\dfrac{cb^2}{2}$.

(b) Form the lower sum, and show that as n grows large, this sum, like the upper sum, tends to $\dfrac{cb^2}{2}$. $\left[\text{Hint: } \sum_{j=1}^{n} j = \dfrac{n(n+1)}{2}.\right]$

5. Repeat Problem 4 with $f(x) = cx^2$, and thus show that $\int_0^b cx^2\,dx = \dfrac{cb^3}{3}$.

$\left[\text{Hint: } \sum_{j=1}^{n} j^2 = \dfrac{n^3}{3} + \dfrac{n^2}{2} + \dfrac{n}{6}.\right]$

6. Repeat Problem 4 with $f(x) = cx^3$, and thus show that $\int_0^b cx^3\,dx = \dfrac{cb^4}{4}$.

$\left[\text{Hint: } \sum_{j=1}^{n} j^3 = \dfrac{n^4}{4} + \dfrac{n^3}{2} + \dfrac{n^2}{4}.\right]$

7. Repeat Problem 4 for $f(x) = cx^4$. $\left[\text{Hint: } \sum_{j=1}^{n} j^4 = \dfrac{n^5}{5} + c_4n^4 + c_3n^3 + c_2n^2 + c_1n + c_0\right.$, where c_0, c_1, c_2, c_3, c_4 are some constants.$\Big]$

8. Repeat Problem 4 for $f(x) = e^x$.

$\left[\text{Hint 1: } \sum_{0}^{n-1} r^k = \dfrac{r^n - 1}{r - 1} \text{ and } \sum_{1}^{n} r^k = \dfrac{r(r^n - 1)}{r - 1}, \text{ if } r \neq 1\right]$

$\left[\text{Hint 2: } \dfrac{h}{\exp(h) - \exp(0)} \to \dfrac{1}{\exp'(0)} = 1 \text{ as } h \to 0; \text{ set } h = b/n.\right]$

9. Repeat Problem 4 for $f(x) = e^{cx}$, where c is any constant.

10. Show that $\int_0^2 e^{-x^2}\,dx$ can be approximated within 1/10 by using a partition of [0,1] into 10 equal subintervals. [Average the upper and lower sums.]

11. Let f be a decreasing function on $[a,b]$.

(a) Show that for any partition, $\sum_1^n f(x_j)\,\Delta x_j$ is a lower sum.

(b) Show that $\sum_1^n f(x_{j-1})\,\Delta x_j$ is an upper sum.

(c) Take a partition into n equal subintervals, so $x_j - x_{j-1} = (b-a)/n$. Show that the difference between the upper and lower sums is

$$[f(a) - f(b)]\frac{b-a}{n}.$$

[Note that this tends to 0 as n grows large.]

• **12.** Use Problem 11, with $f(x) = \sqrt{1-x^2}$ and $[a,b] = [0,1]$, to obtain approximations to $\pi/4$ that are accurate to $\pm(1/10)$. [It is not required to evaluate the square roots that arise.]

5.2 A PROBLEM OF EXISTENCE

Riemann sums can be formed for any function f defined on an interval $[a,b]$, and there appears to be no reason why the integral $\int_a^b f$ should not also be defined for every f. But "every f" includes some strange possibilities, such as the following "Dirichlet function," named for its inventor, P. G. L. Dirichlet:

$$f(x) = \begin{cases} 1 & \text{if } x \text{ is rational} \\ 0 & \text{if } x \text{ is irrational} \end{cases} \qquad 0 \le x \le 1. \tag{1}$$

It is impractical to sketch the graph of this function; every interval, no matter how small, contains both rational numbers, where f is 1, and irrational numbers, where f is 0. Thus the graph consists of parts of two lines [Fig. 12], one with a break at each irrational number, and the other with a break at each rational number.

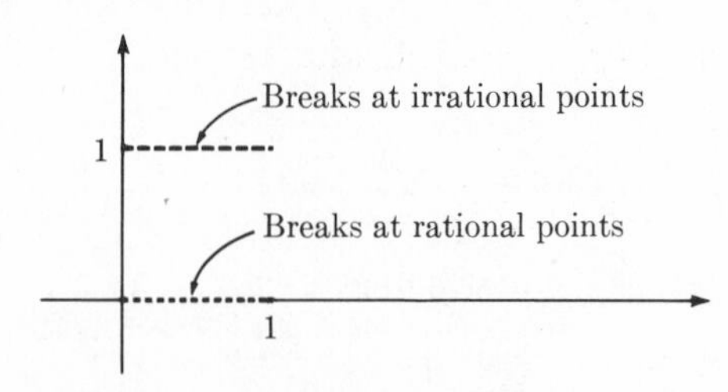

FIGURE 5.12

It is also difficult to see what the "area under the graph of f" should be; we are taking the area of a square with infinitely many lines missing, one for each irrational number. Perhaps the Riemann sums will help, as they did with the area under the parabola. Partition $[0,1]$ into n equal subintervals, by the points

$$0 = x_0 < \frac{1}{n} < \frac{2}{n} < \cdots < 1.$$

Take $\xi_j = j/n$. By the definition (1), $f(\xi_j) = 1$, and this is clearly the maximum of f in the interval $[x_{j-1}, x_j]$. We thus obtain the *upper sum*

$$\sum_{j=1}^{n} f(\xi_j)\,\Delta x_j = \sum_{j=1}^{n} 1 \cdot \frac{1}{n} = \underbrace{\frac{1}{n} + \frac{1}{n} + \cdots + \frac{1}{n}}_{n \text{ terms}} = 1.$$

To form a *lower sum*, take ξ_j irrational, say $\xi_j = j/n - 1/\sqrt{2}n$. By the definition (1), $f(\xi_j) = 0$, and this is clearly the minimum of f in the interval $[x_{j-1}, x_j]$. The resulting lower sum is

$$\sum_{j=1}^{n} f(\xi_j)\,\Delta x_j = \sum_{j=1}^{n} 0 \cdot \frac{1}{n} = 0.$$

Thus no matter *how* small the subintervals are, the upper sum is 1, and the lower sum is 0. As the length of the subintervals tends to 0, the upper sums "tend to" 1, and the lower sums "tend to" 0. The limit thus depends on how the points of evaluation ξ_j are chosen. In this case, *the Riemann sums give no satisfactory value for* $\int_0^1 f$.

This schizophrenic example undermines two tacit assumptions of the previous section:

(i) that "the area under the graph of f" is always a meaningful concept, and

(ii) that the upper and lower Riemann sums always tend to a definite number as the lengths of the subintervals decrease to 0.

The following proposition assures us that the second assumption, though invalid for the "Dirichlet" example, *is* valid for *continuous* functions f.

Recall that f is *continuous* on $[a,b]$ if $\lim_{x \to x_0} f(x) = f(x_0)$ for $a < x_0 < b$, $\lim_{x \to a+} f(x) = f(a)$, and $\lim_{x \to b-} f(x) = f(b)$.

INTEGRAL AS LIMIT OF RIEMANN SUMS

Proposition I. *Let f be continuous on $[a,b]$. Let*

$$a = x_0 < x_1 < \cdots < x_n = b$$

be the partition of $[a,b]$ into n equal subintervals. In each subinterval choose any *point ξ_j, and form the Riemann sum*

$$S_n = \sum_{j=1}^{n} f(\xi_j)\,\Delta x_j\,.$$

Then S_n tends to a definite limit S, which is independent of the choice of the points ξ_j.

This proposition clears up the difficulty in defining the "area under the graph of f." When f is continuous and $f \geq 0$, we simply *define* the area under the graph to be the limit S in Proposition I. More generally, *when f is continuous* (but not necessarily positive) *we define the integral of f to be the limit of the Riemann sums:*

$$\int_a^b f = \lim \sum_1^n f(\xi_j)\,\Delta x_j\,.$$

For example, the function $f(x) = e^{-x^2}$ is continuous everywhere; in particular, it is continuous on the interval $[0,b]$. Therefore, the integral $\int_0^b e^{-x^2}\,dx$ exists, and in fact

$$\int_0^b e^{-x^2}\,dx = \lim_{n\to\infty} \sum_{j=1}^{n} f(\xi_j)\,\Delta x_j = \lim_{n\to\infty} \sum_{j=1}^{n} f\left(j\frac{b}{n}\right)\frac{b}{n} = \lim_{n\to\infty} \frac{b}{n} \sum_{j=1}^{n} e^{-j^2b^2/n^2}.$$

Here we have taken the partition

$$0 = x_0 < \frac{b}{n} < 2\frac{b}{n} < \cdots < j\frac{b}{n} < \cdots < n\frac{b}{n} = b,$$

and points of evaluation $\xi_j = j\dfrac{b}{n}$ = right-hand endpoint of $[x_{j-1}, x_j]$.

PROBLEMS

In Problems 1–9, identify each limit as the definite integral of an appropriate continuous function over an interval.

1. $\displaystyle \lim_{n\to\infty} \frac{1}{n} \sum_{j=1}^{n} \frac{1}{1 + j/n}$

[Hint: $x_j = \xi_j = 1 + j/n$; find $a = x_0$, $b = x_n$, and f.]

• **2.** $\lim_{n\to\infty} \frac{b-1}{n} \sum_{j=1}^{n} \frac{1}{1 + j\frac{b-1}{n}}$

3. $\lim_{n\to\infty} \frac{1}{n} \sum_{j=1}^{n} e^{j/n}$

• **4.** $\lim_{n\to\infty} \frac{b}{n} \sum_{n=1}^{n} e^{-jb/n}$

5. $\lim_{n\to\infty} \frac{b-a}{n} \sum_{j=1}^{n} e^{-j(b-a)/n}$

• **6.** $\lim_{n\to\infty} \frac{1}{n^5} \sum_{j=1}^{n} j^4$

7. $\lim_{n\to\infty} \sum_{j=1}^{n} \frac{1}{n+j}$ $\left[= \frac{1}{n} \sum_{j=1}^{n} \frac{1}{1+j/n}\right]$

• **8.** $\lim_{n\to\infty} \sum_{j=1}^{n} \frac{n}{(n+j)^2}$ $\left[= \frac{1}{n} \sum_{j=1}^{n} \frac{1}{(1+j/n)^2}\right]$

9. $\lim_{n\to\infty} \sum_{j=1}^{n} \frac{n^2}{(n+j)^3}$

10. This problem is hard, but not impossible. Define a function f on the interval [0,1] by

$$f(x) = \begin{cases} \frac{1}{q} & \text{if } x \text{ is rational and } x = \frac{p}{q} \text{ in lowest terms} \\ 0 & \text{if } x \text{ is irrational.} \end{cases}$$

(a) Sketch the graph of f, plotting all points corresponding to rational numbers with denominators of 6 or less.
(b) Show that every lower Riemann sum for f on [0,1] is zero.
(c) Show that there are arbitrarily small upper sums for f.
(d) Conclude that the area under the graph is 0.

5.3 THE FUNDAMENTAL THEOREM OF CALCULUS

The evaluation of integrals by taking limits of Riemann sums is a tricky matter; the simple integral $\int_0^b x^2\,dx = \frac{1}{3}b^3$ requires the formula $\sum_{j=1}^{n} j^2 = \frac{n^3}{3} + \frac{n^2}{2} + \frac{n}{6}$, and most examples are even more difficult.

Fortunately, the Riemann sums can be bypassed in many cases. To illustrate the method, we begin with an example.

Example 1. Evaluate $\int_2^3 x^5\,dx$. *Solution.* Instead of attacking this single problem directly, as in §5.1, we consider a whole family of problems: Compute $\int_2^t x^5\,dx$ for *every* $t \geq 2$. Let $A(t)$ denote this integral [intuitively, the area under the graph of $f(x) = x^5$ on the interval $[2,t]$, Fig. 13]

$$A(t) = \int_2^t x^5\,dx.$$

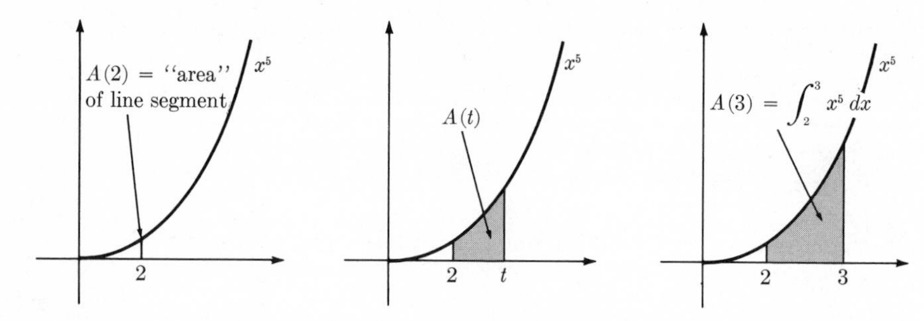

FIGURE 5.13 (Not drawn to scale, for practical reasons)

It is clear from the figure that $A(2) = 0$ [since $A(2)$ is the "area" of a line segment] and that as t moves to the right, the area increases; thus A increases with t. Although we have no formula for $A(t)$, *we can find a formula for the rate of increase of A*, from the definition of the derivative

$$A'(t) = \lim_{h\to 0} \frac{A(t+h) - A(t)}{h}.$$

Suppose that $h > 0$. Then $A(t+h) - A(t)$ is the area of the shaded region in Fig. 14. Comparing this to the areas of the two rectangles in the figure, we have

$$ht^5 < A(t+h) - A(t) < h(t+h)^5,$$

and therefore [since $h > 0$]

$$t^5 < \frac{A(t+h) - A(t)}{h} < (t+h)^5.$$

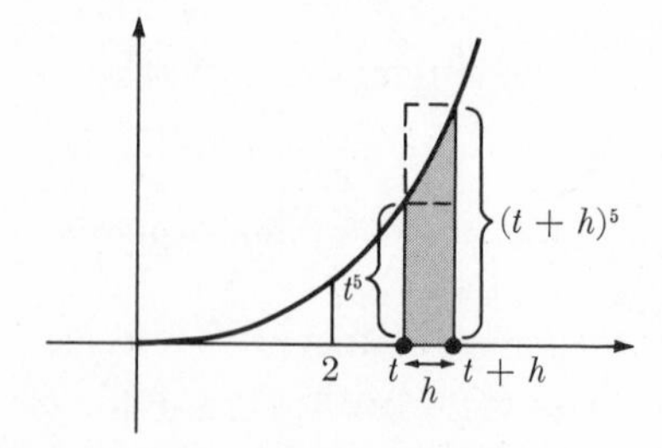

FIGURE 5.14

Taking the limit as $h \to 0$, the far left-hand side and far right-hand side of this inequality each has the limit t^5. Therefore, by the Poe Principle, the middle term has the same limit,

$$A'(t) = \lim_{h \to 0} \frac{A(t+h) - A(t)}{h} = t^5.$$

It follows that

$$A(t) = \tfrac{1}{6}t^6 + C \tag{1}$$

for some constant C. [Recall the line of argument in §3.2; since $\frac{1}{6}t^6$ and $A(t)$ both have the same derivative t^5, they must differ by a constant C.] But $A(2) = 0$, as noted above, so setting $t = 2$ in formula (1) yields

$$0 = A(2) = \tfrac{1}{6} \cdot 2^6 + C.$$

Hence $C = -\frac{1}{6} \cdot 2^6$, and

$$A(t) = \tfrac{1}{6}t^6 - \tfrac{1}{6} \cdot 2^6.$$

In particular,

$$A(3) = \tfrac{1}{6} \cdot 3^6 - \tfrac{1}{6} \cdot 2^6 = 665/6,$$

and we have computed the given integral $\int_2^3 x^5\,dx = A(3) = 665/6$.

The crucial step in Example 1 was to show that the function

$$A(t) = \int_2^t x^5\,dx$$

has the derivative $A'(t) = t^5$. This is a special case of a general formula for functions defined by integrals, called the *Fundamental Theorem of Calculus:*

Proposition J. *Let f be continuous on $[a,b]$. Define a function A by*

$$A(t) = \int_a^t f = \int_a^t f(x)\,dx, \qquad a \le t \le b.$$

☆☆☆ FUNDAMENTAL THEOREM OF CALCULUS

Then A is continuous, $A(a) = 0$, and

$$A'(t) = f(t), \qquad a < t < b.$$

It is not hard to explain why the proposition should be true. Take $h > 0$. Then [Fig. 15]

$$A(t+h) - A(t) = \int_t^{t+h} f$$

(area under the graph on $[t, t+h]$).

Let $m = f(x_1)$ be the minimum of f on $[t, t+h]$, and $M = f(x_2)$ be the maximum. Then [Fig. 16]

$$hf(x_1) \leq A(t+h) - A(t) \leq hf(x_2).$$

Dividing by $h > 0$, we get

$$f(x_1) \leq \frac{A(t+h) - A(t)}{h} \leq f(x_2). \tag{2}$$

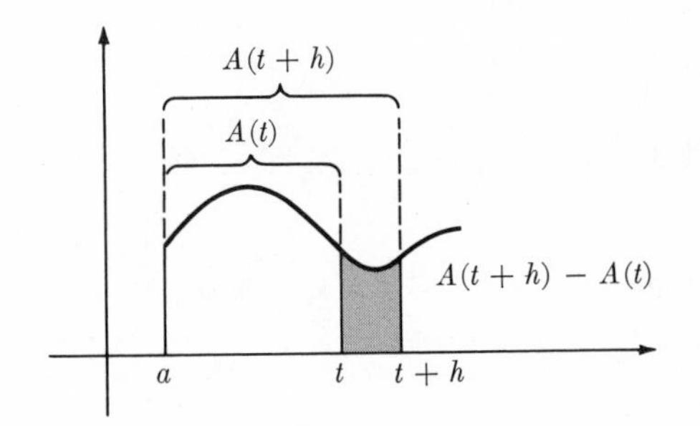

FIGURE 5.15

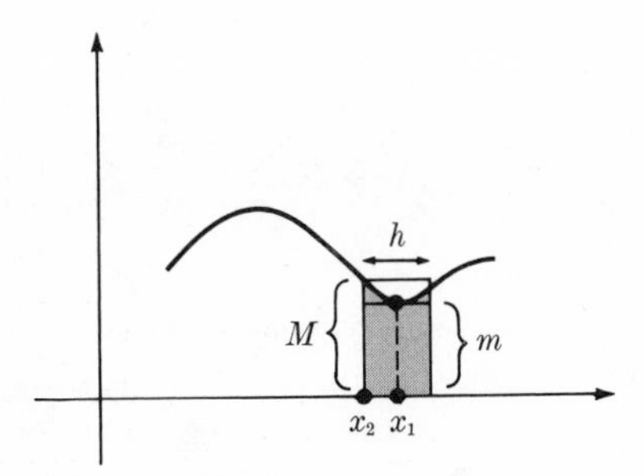

FIGURE 5.16 $mh \leq$ shaded area $\leq Mh$

Now, x_1 and x_2 are both in the interval $[t, t+h]$, so in the limit as $h \to 0$ we have $x_1 \to t$ and $x_2 \to t$. Since f is continuous, it follows that

$$\lim_{h\to 0} f(x_1) = f(t) \qquad \text{and} \qquad \lim_{h\to 0} f(x_2) = f(t).$$

Therefore, applying the Poe Principle to (2), we find

$$A'(t) = \lim_{h\to 0} \frac{A(t+h) - A(t)}{h} = f(t).$$

In our picture we took $h > 0$ and $f \geq 0$, but you can check that the inequality (2) remains valid when $h < 0$ or $f \leq 0$, so the same argument works in all cases.

The rigorous proof of Proposition J [§AII.8] consists in justifying analytically the various steps where we referred to pictures in the argument above.

The logarithm gives a familiar special case of the Fundamental Theorem [see Problem 6 below]. We have

$$A(t) = \log t = \int_1^t \frac{1}{x}\,dx.$$

Therefore, by the Fundamental Theorem, $A'(t) = \log'(t) = 1/t$.

The Fundamental Theorem has many uses; the one that concerns us now is the evaluation of integrals. This is based on the so-called *Second Fundamental Theorem of Calculus:*

Theorem 1. *Let f be continuous on $[a,b]$. Suppose we have a function F that solves the equation $F' = f$. Then*

$$\int_a^b f = F(b) - F(a).$$

SECOND FUNDAMENTAL THEOREM

Proof. We imitate the second half of Example 1. Let $A(t) = \int_a^t f$, $a \le t \le b$. Then $A(0) = 0$, and $A(b) = \int_a^b f$ is the integral we are trying to evaluate. By the (first) Fundamental Theorem, $A' = f$. By assumption, $F' = f = A'$, so A and F must differ by a constant,

$$A(t) = F(t) + C, \qquad a \le t \le b.$$

Setting $t = a$, we get $0 = A(a) = F(a) + C$, so $C = -F(a)$, and

$$A(t) = F(t) - F(a).$$

Finally, setting $t = b$, we get

$$\int_a^b f = A(b) = F(b) - F(a). \qquad Q.E.D.$$

Example 2. Evaluate $\int_a^b x^2\,dx$. *Solution.* We are integrating the function $f(x) = x^2$. To apply Theorem 1, we need a solution F of the equation $F'(x) = f(x) = x^2$. Clearly, $F(x) = \frac{1}{3}x^3$ solves this equation. Therefore,

$$\int_a^b x^2\,dx = F(b) - F(a) = \tfrac{1}{3}b^3 - \tfrac{1}{3}a^3.$$

Notice that this agrees (as of course it must) with the formula laboriously derived in §5.1, using Riemann sums:

$$\int_0^b x^2\,dx = \tfrac{1}{3}b^3.$$

Note: The notation

$$F\Big|_a^b, \quad \text{or} \quad \Big[F\Big]_a^b,$$

is used to abbreviate the expression $F(b) - F(a)$ arising in Theorem 1.

Example 3. Evaluate $\int_{-1}^{3} e^{-x}\,dx$. *Solution.* The function $F(x) = -e^{-x}$ solves $F'(x) = e^{-x}$. Therefore,

$$\int_{-1}^{3} e^{-x}\,dx = F(3) - F(-1) = F\Big|_{-1}^{3} = -e^{-3} + e^{1}.$$

Example 4. Evaluate $\int_{-r}^{r} \sqrt{r^2 - x^2}\,dx$. *Solution.* Believe it or not, the function

$$F(x) = \frac{1}{2}x\sqrt{r^2 - x^2} + \frac{r^2}{2}\arcsin\frac{x}{r}$$

solves $F'(x) = \sqrt{r^2 - x^2}$. [Check this; differentiate the given expression for f.] Therefore,

$$\int_{-r}^{r} \sqrt{r^2 - x^2}\,dx = F\Big|_{-r}^{r} = \frac{r^2}{2}\arcsin(1) - \frac{r^2}{2}\arcsin(-1)$$

$$= \frac{r^2}{2}\left(\frac{\pi}{2} - \left(-\frac{\pi}{2}\right)\right) = \frac{1}{2}\pi r^2.$$

[Notice that the integral in question is the area of a half-circle of radius r.]

The function F produced in Example 4 must appear far-fetched indeed; it works, but where did it come from? We leave this mystery to be cleared up in the next chapter, which gives standard techniques for solving equations of the form $F' = f$.

Example 5. $\int_{0}^{2} e^{-x^2}\,dx$. *Solution* (?) Perhaps the previous example made you confident that we could always solve $F' = f$, no matter what f is given; if so, that confidence is about to be shattered. Try as you will, you *cannot* find an elementary function (concocted out of polynomials, trigonometric, exponential, and log functions, as in Chapter II) that solves the equation $F'(x) = e^{-x^2}$. At the moment, we can only approximate the integral by Riemann sums. Chapters IX and XI give more efficient approximations.

Theorem 1 can be used to prove some elementary but important facts about integrals of continuous functions. The first property is called *additivity with respect to intervals.* Figure 17 makes Theorem 2 obvious:

Theorem 2. *Let f be continuous on $[a,b]$, and let $a < c < b$. Then*

ADDITIVITY OVER INTERVALS

$$\int_a^b f = \int_a^c f + \int_c^b f.$$

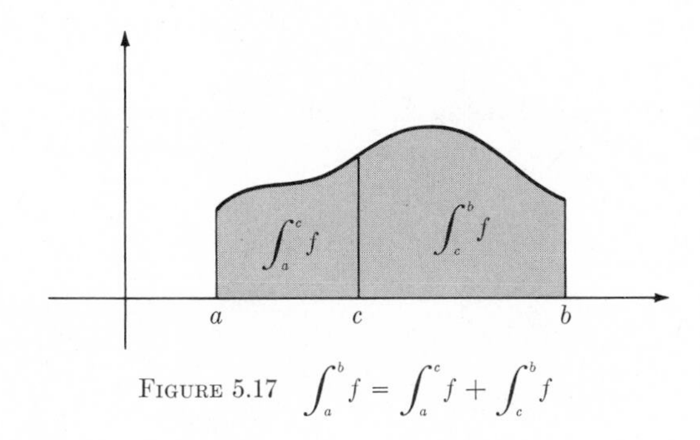

FIGURE 5.17 $\int_a^b f = \int_a^c f + \int_c^b f$

Proof. By the Fundamental Theorem [Proposition J] the function $F(t) = \int_a^t f$ solves $F' = f$. By Theorem 1,

$$\int_a^c f = F(c) - F(a) \qquad \text{and} \qquad \int_c^b f = F(b) - F(c).$$

Adding these integrals, the $F(c)$ terms cancel, yielding

$$\int_a^c f + \int_c^b f = F(b) - F(a) = \int_a^b f. \qquad Q.E.D.$$

The next theorem is not so obvious from a picture, but its form recalls the familiar linearity property of limits and derivatives:

Theorem 3. *Let f and g be continuous on $[a,b]$, and let c_1 and c_2 be any constants. Then*

INTEGRATION IS LINEAR

$$\int_a^b (c_1 f + c_2 g) = \left(c_1 \int_a^b f\right) + \left(c_2 \int_a^b g\right).$$

Proof. Let $F' = f$ and $G' = g$. Then

$$(c_1F + c_2G)' = c_1F' + c_2G' = c_1 f + c_2 g,$$

so we can use $c_1F + c_2G$ to evaluate $\int_a^b (c_1 f + c_2 g)$ by Theorem 1:

$$\int_a^b (c_1 f + c_2 g) = \Big[c_1 F + c_2 G\Big]_a^b = c_1 F(b) + c_2 G(b) - c_1 F(a) - c_2 G(a)$$

$$= c_1[F(b) - F(a)] + c_2[G(b) - G(a)]$$

$$= c_1 \int_a^b f + c_2 \int_a^b g. \qquad Q.E.D.$$

[This can also be proved using Riemann sums, and that approach gives some added insight into the theorem; see Problem 4 below.]

Example 6. $\int_2^3 (10x^5 + 4x^2)\,dx = ?$ *Solution.*

$$\int_2^3 (10x^5 + 4x^2)\,dx = 10 \int_2^3 x^5 + 4 \int_2^3 x^2\,dx \qquad \text{[Theorem 3]}$$

$$= 10\Big[\tfrac{1}{6}x^6\Big]_2^3 + 4\Big[\tfrac{1}{3}x^3\Big]_2^3 \qquad \text{[Examples 1 and 2]}$$

$$= 10[\tfrac{1}{6}3^6 - \tfrac{1}{6}2^6] + 4[\tfrac{1}{3}3^3 - \tfrac{1}{3}2^3].$$

As a final consequence of the Fundamental Theorem, we prove the following fact [which is "obvious" from Fig. 18]:

BASIC INEQUALITY FOR INTEGRALS

Theorem 4. *If f and g are continuous on $[a,b]$, then*

$$f \le g \quad \Rightarrow \quad \int_a^b f \le \int_a^b g.$$

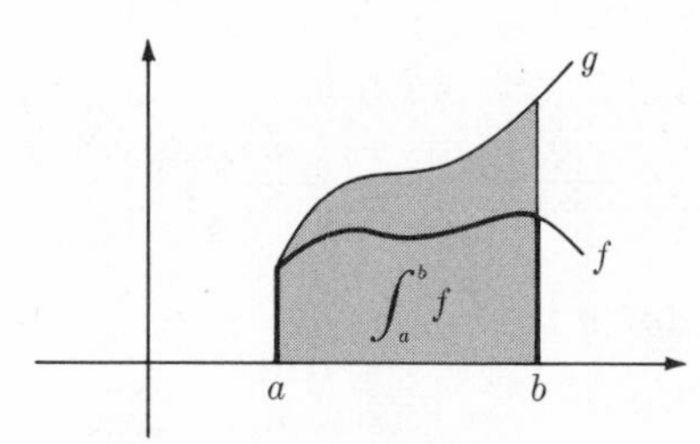

FIGURE 5.18 $f \le g \Rightarrow \int_a^b f \le \int_a^b g$

Proof. We have to show that $\int_a^b g - \int_a^b f \ge 0$. By Theorem 3,

$$\int_a^b g - \int_a^b f = \int_a^b (g - f).$$

Let $A(t) = \int_a^t (g - f)$. Then $A'(t) = g(t) - f(t) \geq 0$ [by hypothesis], so A is an *increasing* function, and $A(b) \geq A(a)$. Therefore,

$$\int_a^b g - \int_a^b f = \int_a^b (g - f) = A(b) - A(a) \geq 0. \qquad Q.E.D.$$

Remark. The function

INDEFINITE INTEGRAL

$$A(t) = \int_a^t f, \qquad a \leq t \leq b$$

is called an *indefinite integral,* because the upper limit of integration, t, is not a "definite number," but varies between a and b. By contrast, when we think of the upper limit as fixed, as in $\int_a^b f$, we call this a *definite integral.* The definite integral defines a number, and the indefinite integral defines a function.

With this terminology, the Fundamental Theorem reads as follows: Given a continuous function f, form the indefinite integral $A(t) = \int_a^t f$; then $A' = f$. Thus *indefinite integration is the reverse of differentiation.*

PROBLEMS

Problem 5 introduces a convenient sign convention: $\int_a^b f = -\int_b^a f$.

Problem 6 proves Proposition E, §2.6, concerning log.

1. Use Theorem 1 to evaluate the following integrals.

(a) $\int_0^1 2x\,dx$

(b) $\int_0^1 cx\,dx$ [c a constant]

(c) $\int_2^3 100x^{99}\,dx$

•(d) $\int_3^5 x^{99}\,dx$

(e) $\int_2^3 5\,dx$

(f) $\int_0^h y^2\,dy$

(g) $\int_{-r}^r 2r\,dx$ [r is a constant]

• (h) $\int_{-1}^{0} e^{x}\,dx$ (i) $\int_{-1}^{0} e^{-x}\,dx$ (j) $\int_{0}^{100} e^{-3x}\,dx$

(k) $\int_{a}^{b} e^{cx}\,dx = \dfrac{e^{cb} - e^{ca}}{c}$ [c is a constant]

(l) $\int_{0}^{\pi/2} \cos\theta\,d\theta = 1$ (m) $\int_{0}^{\pi/2} \sin\theta\,d\theta = 1$

(n) $\int_{a}^{b} \cos\omega t\,dt = \dfrac{\sin\omega b - \sin\omega a}{\omega}$ [ω is a constant, $\neq 0$]

(o) $\int_{a}^{b} \sin\omega t\,dt = ?$ (p) $\int_{2}^{3} r^{-2}\,dr$

2. Use Theorems 1 and 2 (and in some cases Problem 1) to evaluate the following integrals.

• (a) $\int_{0}^{1} (1 + x + x^2)\,dx$ (d) $\int_{a}^{b} \frac{1}{2}(e^{x} + e^{-x})\,dx$

(b) $\int_{2}^{3} (y^2 - 5)\,dy$ • (e) $\int_{0}^{\pi} (1 + \cos\theta)\,d\theta$

• (c) $\int_{0}^{r} (r^2 - x^2)\,dx$ (f) $\int_{0}^{\pi/4} (1 + \cos 2\theta)\,d\theta$

3. Let $0 < a < b$. Use Theorem 4 and Problem 1(k) to show that

$$\frac{e^{-ab} - e^{-b^2}}{b} = \int_{a}^{b} e^{-bx}\,dx \leq \int_{a}^{b} e^{-x^2}\,dx \leq \int_{a}^{b} e^{-ax}\,dx = \frac{e^{-a^2} - e^{-ab}}{a}.$$

[These inequalities can be used for large a and b to estimate the integral $\int_{a}^{b} e^{-x^2}\,dx$ arising in statistics.]

4. Prove that $\int_{a}^{b} (f + g) = \left(\int_{a}^{b} f\right) + \left(\int_{a}^{b} g\right)$ by using Riemann sums and the definition $\int_{a}^{b} (f + g) = \lim\limits_{n\to\infty} \sum\limits_{j=1}^{n} [f(\xi_j) + g(\xi_j)]\Delta x_j$. Figure 19 illustrates this method of proof.

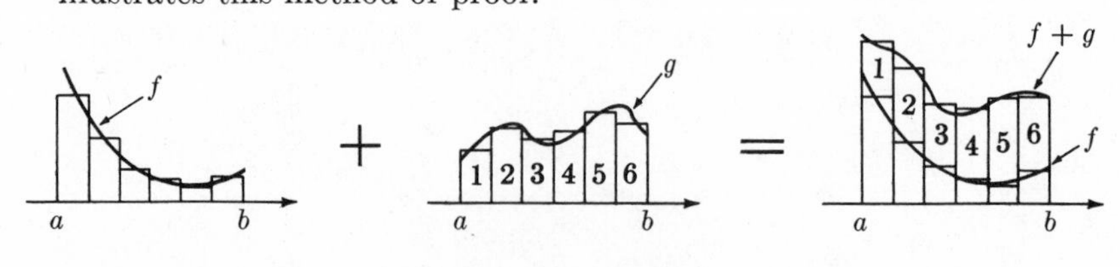

FIGURE 5.19 $\int_a^b f + \int_a^b g = \int_a^b (f + g)$

Note that sums can be rearranged as follows:

$$\sum_{j=1}^{n} (c_j + d_j) = \sum_{j=1}^{n} c_j + \sum_{j=1}^{n} d_j .$$

• **5.** It is convenient to define integrals such as $\int_1^{1/2} \frac{1}{x}\,dx$, where the "lower limit" is actually greater than the "upper limit," by introducing an appropriate *sign convention*: $\int_1^{1/2} \frac{1}{x}\,dx$ is taken to mean $-\int_{1/2}^{1} \frac{1}{x}\,dx$. Generally we define

REVERSING LIMITS OF INTEGRATION CHANGES THE SIGN

$$\int_a^b f = -\int_b^a f, \qquad \text{when } a > b \text{ and } f \text{ is continuous on } [b,a].$$

Which of the following remain true with this extended definition of integral? In each case, assume that f and g are continuous on any intervals in question.

(a) $\int_1^b \frac{1}{x}\,dx = \log b$, for $b < 1$

(b) $\int_a^b f = F(b) - F(a)$

(c) $\int_a^b f + \int_b^c f = \int_a^c f$

(d) $\int_a^b (c_1 f + c_2 g) = c_1 \int_a^b f + c_2 \int_a^b g$

(e) $\int_a^b f \le \int_a^b g$ if $f \le g$

(f) If $a < c < b$, and $A(t) = \int_c^t f$, then $A' = f$.

6. [*Definition of log*] The informal discussion of the natural logarithm finds its formal justification in the Fundamental Theorem of Calculus. Note that the function $f(x) = 1/x$, $x \neq 0$, has a derivative at every point, and is therefore continuous [Lemma 1, §2.2]. We therefore can (and do) define the natural logarithm by

log *t*: FORMAL DEFINITION

$$\log t = \int_1^t \frac{dx}{x}, \qquad 0 < t < \infty.$$

When $0 < t \le 1$, this involves the sign convention in Problem 5. The following points outline a proof of Proposition E, §2.6.

(a) Prove that $\log'(t) = 1/t$. [See Problem 5(f) for the case $0 < t \le 1$.]

(b) Prove that $\log at = \log a + \log t$. [Hint: Compare the derivatives of $\log at$ and $\log t$, and recall §3.2.]

(c) Prove that log is strictly increasing.

(d) Prove that log is one-to-one, and has an inverse defined on an interval. [See §2.5.]

(e) Prove that $\log(2^n) = n \log 2$, for $n = 0, \pm 1, \pm 2, \ldots$. [A strict proof requires mathematical induction, §AI.2.]

(f) Prove that the range of log is the whole interval $(-\infty, +\infty)$. [By part (d), the range is an interval; by part (e), the interval contains arbitrarily large numbers, both positive and negative.

5.4 SOME APPLICATIONS OF INTEGRALS

Integrals arose out of the study of area, but, like derivatives, they turn out to have a great many applications. Here we show how they enter in the calculation of position, area, volume, arc length, mass, probability, torque and center of gravity, and work.

This is a long list, perhaps too long to digest all at once; you may wish to skip some sections now, and return to them as they are required later in various problems. But you should follow the derivation of three or four of these integrals now, to get the general idea: The quantity in question is approximated by a sum, which turns out to be a Riemann sum for an integral.

Distance

Consider a simple "navigational" problem: A particle moves along a straight line; its velocity $v(t)$ is given for all times t, $a \le t \le b$; but the position $s(t)$ is *not* given. The problem is to deduce the *change* in position $s(b) - s(a)$. There are two ways to approach this.

First method: Divide the time interval $[a,b]$ into small subintervals $[t_{j-1}, t_j]$. In each short interval the velocity does not change very much, so the change in position from time t_{j-1} to time t_j is nearly

$$v(\tau_j)(t_j - t_{j-1}) = v(\tau_j)\,\Delta t_j\,,$$

where τ_j is any point in the subinterval. Then the total change of position is nearly

$$\sum_1^n v(\tau_j)\,\Delta t_j\,. \tag{1}$$

As the partitions are made finer, this sum tends to the actual change in position and, at the same time, to the integral

$$\int_a^b v(t)\,dt;$$

for, the sum (1) is a Riemann sum for this integral. We conclude that when a particle moves with a variable velocity v, the *net change in position from* $t = a$ *to* $t = b$ *is given by the integral* $\int_a^b v(t)\,dt$.

CHANGE IN POSITION IS THE VELOCITY INTEGRAL

Second method: We know that $s'(t) = v(t)$. Therefore, by the (second) Fundamental Theorem of Calculus, we obtain

$$s(b) - s(a) = \int_a^b s' = \int_a^b v,$$

the same result as before.

It is important to notice that when $v < 0$, the particle moves to the *left*, and the position function *decreases*. The integral $\int_a^b v$ gives the *net change in position*. By contrast, the *total distance traveled* is $\int_a^b |v|$.

Example 1. If the velocity is steadily increasing according to the formula $v(t) = ct$, where c is a constant, then the change in position from $t = 0$ to $t = T$ is

$$\int_0^T ct\,dt = \tfrac{1}{2}ct^2\Big|_0^T = \tfrac{1}{2}cT^2.$$

[See §5.3 for the evaluation of the integral.]

Area between two curves

Suppose that $f \geq g$ on an interval $[a,b]$, and consider the area between the graphs of f and g on $[a,b]$. [See Fig. 20.] Divide the area into strips by a partition $a = x_0 < \cdots < x_n = b$. Then the area of each strip is approximately $[f(\xi_j) - g(\xi_j)]\Delta x_j$, so the area itself is approximated by

$$\sum [f(\xi_j) - g(\xi_j)]\Delta x_j\,. \tag{2}$$

As the subintervals of the partition are made smaller, these approximations tend to the actual area, and the sums (2) tend to $\int_a^b (f - g)$, since they are

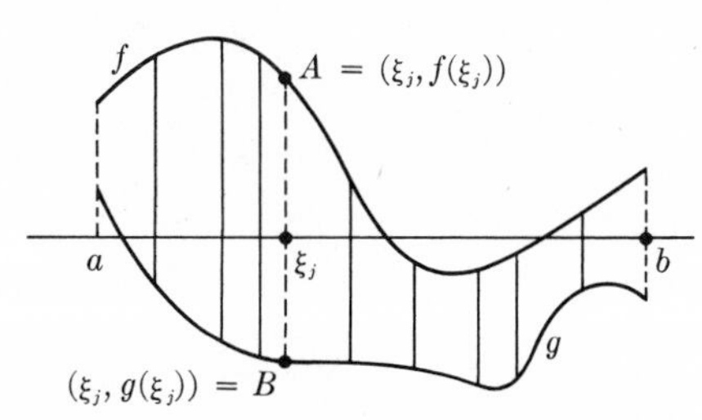

FIGURE 5.20 The length of the line AB is $f(\xi_j) - g(\xi_j)$

precisely the Riemann sums for the integral of $f - g$. Hence this integral gives the area between the graphs and the lines $x = a$ and $x = b$.

Example 2. A disk of radius r is bounded by the graphs of

$$f(x) = \sqrt{r^2 - x^2}, \qquad |x| \leq r$$

and

$$g(x) = -\sqrt{r^2 - x^2}, \qquad |x| \leq r.$$

Hence its area is

$$\int_{-r}^{r} (f - g) = \int_{-r}^{r} (\sqrt{r^2 - x^2} + \sqrt{r^2 - x^2})\, dx$$

$$= \int_{-r}^{r} 2\sqrt{r^2 - x^2}\, dx = \Big[x\sqrt{r^2 - x^2} + r^2 \arcsin (x/r)\Big]_{-r}^{r}$$

$$= r^2 \arcsin (1) - r^2 \arcsin (-1) = \pi r^2.$$

[See Example 4, §5.3, for the evaluation of the integral.]

Arc length

Suppose that f is a function with a continuous derivative f'. We can find the length of its graph on the interval $[a,b]$ as follows. Partition $[a,b]$ into small subintervals $[x_{j-1}, x_j]$, and add up the lengths of the straight line segments from $(x_{j-1}, f(x_{j-1}))$ to $(x_j, f(x_j))$ as in Fig. 21. As the intervals $[x_{j-1}, x_j]$ are made smaller, these approximations tend to the length of the curve. Now, the length of the segment from $(x_{j-1}, f(x_{j-1}))$ to $(x_j, f(x_j))$ is $\sqrt{(x_j - x_{j-1})^2 + (f(x_j) - f(x_{j-1}))^2}$, so the sum

$$\sum_{j=1}^{n} \sqrt{(x_j - x_{j-1})^2 + (f(x_j) - f(x_{j-1}))^2} \tag{3}$$

tends to the length of the curve as the intervals are made smaller. But from the Mean Value Theorem [§4.2]

$$f(x_j) - f(x_{j-1}) = f'(\xi_j)(x_j - x_{j-1}) = f'(\xi_j)\Delta x_j,$$

where ξ_j is some number in $[x_{j-1}, x_j]$. Hence we can write the approximating sum (3) as $\sum_{1}^{n} \sqrt{1 + f'(\xi_j)^2}\Delta x_j$. This is precisely a Riemann sum for the integral

$$\int_a^b \sqrt{1 + f'(x)^2}\, dx,$$

so this integral gives the arc length.

FIGURE 5.21 A curve approximated by sums of straight line segments

This formula can also be obtained without appealing to the Mean Value Theorem; §7.7 suggests one way, and yet another can be based on §2.11, as follows. Let $s(t)$ denote the length of the graph on the interval $[a,t]$. Then [§2.11] $s'(t) = \sqrt{1 + f'(t)^2}$. It follows from the (second) Fundamental Theorem of Calculus that the length of the graph on $[a,b]$ is given by

ARC LENGTH INTEGRAL

$$s(b) - s(a) = \int_a^b s'(t)\,dt = \int_a^b \sqrt{1 + f'(t)^2}\,dt.$$

Example 3. The graph of $f(x) = \frac{1}{2}(e^x + e^{-x})$ is a catenary, the shape assumed by a chain (or a telephone line) hanging freely between two supports. The length of that part of the curve from $x = 0$ to $x = b$ is

$$\int_0^b \sqrt{1 + f'(x)^2}\,dx = \int_0^b \sqrt{1 + \tfrac{1}{4}(e^x - e^{-x})^2}\,dx$$

$$= \int_0^b \tfrac{1}{2}\sqrt{2 + e^{2x} + e^{-2x}}\,dx$$

$$= \int_0^b \tfrac{1}{2}(e^x + e^{-x})\,dx$$

$$= \tfrac{1}{2}\int_0^b e^x\,dx + \tfrac{1}{2}\int_0^b e^{-x}\,dx$$

[Theorem 2, §5.3]

$$= \frac{1}{2}\,e^x\Big|_0^b + \frac{1}{2}\Big[-e^{-x}\Big]_0^b = \frac{1}{2}\,(e^b - e^{-b}).$$

Volume of a solid of revolution

The volume of the sphere was first computed by Archimedes. This was a great accomplishment in 250 B.C., but now it can be done quite easily by a general method, which applies to solids of revolution.

A solid of revolution is formed by revolving a plane surface about an axis. Spheres, cones, footballs, and tires are solids of revolution; so are most shapes produced on a lathe or a potter's wheel.

Suppose that the surface to be revolved is the area under the graph of a function $f \geq 0$ on an interval $[a,b]$, and that the x axis is the axis of revolution [Fig. 22]. Slice the volume (like a salami) into small sections by planes perpendicular to the x axis [Fig. 22]. If the planes pass through the points $(x_j, 0)$, then each small section can be approximated by a disk of thickness Δx_j and radius $f(\xi_j)$, with ξ_j in the interval $[x_{j-1}, x_j]$. The volume of such a disk is

$$\text{area of base} \times \text{thickness} = \pi r^2 h = \pi f(\xi_j)^2 \Delta x_j ,$$

so the entire volume of the solid of revolution is approximated by the sum

$$\sum_1^n \pi f(\xi_j)^2 \Delta x_j . \tag{4}$$

This sum is the volume of a row of disks that approximates the actual volume better and better as the thickness of the disks decreases to zero [Fig. 23], and we conclude that the sum (4) tends to the volume of the solid. But the sum (4) is a Riemann sum for the integral

VOLUME OF SOLID OF REVOLUTION ABOUT x AXIS

$$\int_a^b \pi f(x)^2 \, dx, \tag{5}$$

so this integral gives the volume of the solid.

Remark 1. Notice that each term $\pi f(\xi_j)^2 \Delta x_j$ in the Riemann sum for the volume is generated by revolving about the x axis one of the rectangles [with base Δx_j and height $f(\xi_j)$] in the Riemann sum $\sum f(\xi_j) \Delta x_j$ for the area that was revolved.

Remark 2. Problem 10 below gives an alternate formula for volumes of revolution.

Example 4. Let $f(x) = \sqrt{r^2 - x^2}$, $-r \leq x \leq r$. The area under the graph of f is the upper half of a disk of radius r, and rotating this about the x axis forms a ball of radius r, the solid studied by Archimedes. The volume of this ball is, by (5),

$$\begin{aligned}
\int_{-r}^{r} \pi f(x)^2 \, dx &= \int_{-r}^{r} \pi (r^2 - x^2) \, dx \\
&= \pi \int_{-r}^{r} r^2 \, dx - \pi \int_{-r}^{r} x^2 \, dx && \text{[Theorem 3]} \\
&= \pi \Big[r^2 x \Big]_{-r}^{r} - \pi \Big[\tfrac{1}{3} x^3 \Big]_{-r}^{r} && \text{[Theorem 1]} \\
&= \pi[r^2 r - r^2(-r)] - \pi[\tfrac{1}{3} r^3 - \tfrac{1}{3}(-r)^3] \\
&= \tfrac{4}{3} \pi r^3,
\end{aligned}$$

and this is Archimedes' formula for the volume of the ball of radius r.

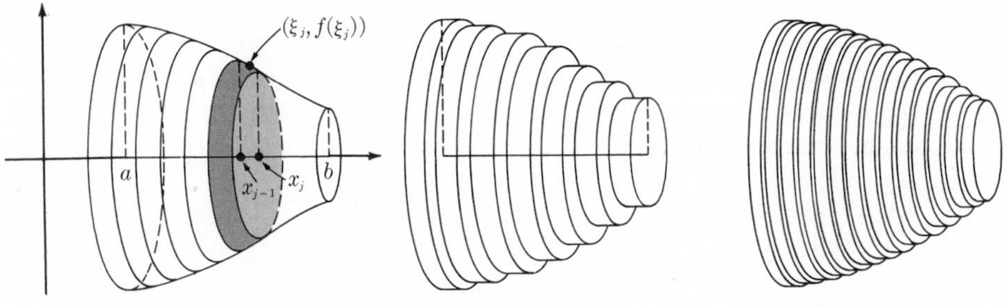

FIGURE 5.22

FIGURE 5.23

Area of a surface of revolution

A surface is generated by revolving about the x axis the graph of a function f on an interval $[a,b]$, as in Fig. 24(i). To obtain an approximation to the area of the surface, consider the surface generated by revolving about the x axis one of the polygonal lines used to approximate the length of the graph [Fig. 24(ii)]. This surface is a frustum of a cone. Now, the lateral area of a frustum of a cone with radii r_1 and r_2 and height h is $\pi(r_1 + r_2) \times \sqrt{h^2 + (r_1 - r_2)^2}$. [See Problems 13 and 14 below.] Thus the approximating surface has the area

$$\sum \pi[f(x_j) + f(x_{j-1})]\sqrt{1 + f'(\xi_j)^2}\Delta x_j . \tag{6}$$

Unfortunately, this is not exactly a Riemann sum for any function, since (6) does not have the form $g(\xi_j)\Delta x_j$ for any function g. But when the interval $[x_{j-1}, x_j]$ is small, then $f(x_j) + f(x_{j-1})$ differs very little from $2f(\xi_j)$, and with this change the sum (6) becomes

$$\sum 2\pi f(\xi_j)\sqrt{1 + f'(\xi_j)^2}\Delta x_j , \tag{7}$$

which is a Riemann sum for the integral

$$\int_a^b 2\pi f(x)\sqrt{1 + f'(x)^2}\, dx. \tag{8}$$

AREA OF SURFACE OF REVOLUTION ABOUT x AXIS

We conclude that the surface area is given by the integral (8).

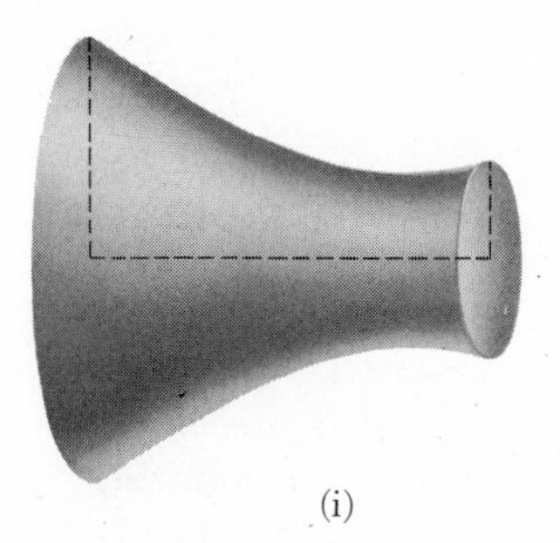

(i)

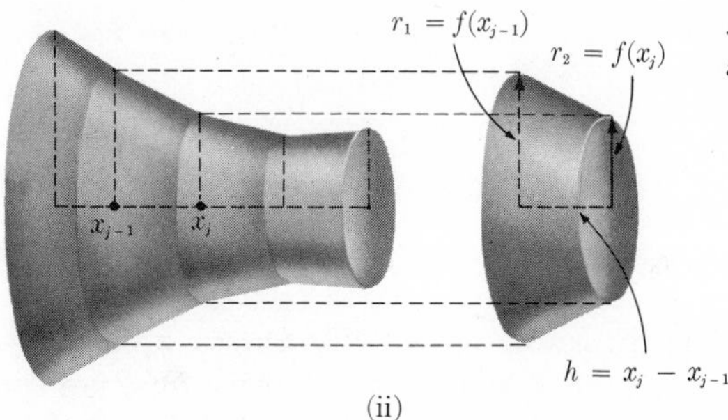

(ii)

FIGURE 5.24

You may justifiably feel that this derivation is too loose to be completely convincing. It would take a long general discussion of surface area to provide an argument that the sums of conical frustra tend to the actual surface area, so we will say no more about that aspect. The rest of the argument, particularly the transition from (6) to (7), is investigated further in Problem 15 below.

The integral (8) leads easily to the formula for the surface area of a sphere [Problem 12].

There is a formula similar to (8) for the area of a surface revolved about the y axis, namely

AREA OF SURFACE OF REVOLUTION ABOUT y AXIS

$$2\pi \int_a^b x\sqrt{1 + f'(x)^2}\, dx.$$

[See Problem 16 below.]

Mass

Suppose that a rod lying between $x = 0$ and $x = L$ has a varying density δ; near any point x, the density of the rod is $\delta(x)$ grams per centimeter. Then [Fig. 25] the mass of rod lying between two nearby points x_{j-1} and x_j is nearly $\delta(\xi_j)\Delta x_j$, where ξ_j is any point in the interval $[x_{j-1}, x_j]$, and the mass of the *whole* rod is nearly

$$\sum_1^n \delta(\xi_j)\Delta x_j . \tag{9}$$

As the lengths $\Delta x_j \to 0$, these approximations tend to the actual mass, and the sums (9) tend to $\int_0^L \delta(x)\, dx$. Thus the total mass is $\int_0^L \delta(x)\, dx$.

Similarly, the mass of rod between any two points a and b is $\int_a^b \delta(x)\, dx$.

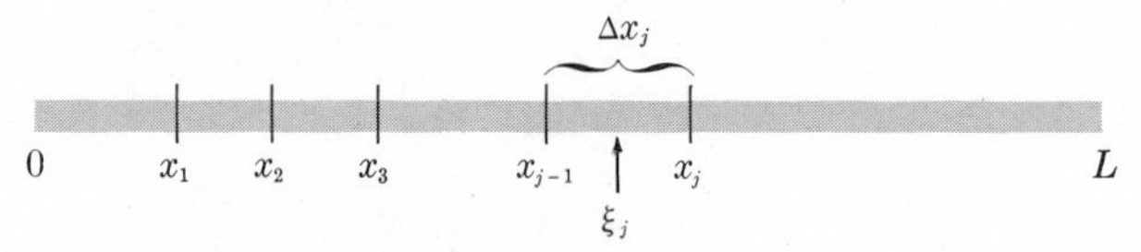

FIGURE 5.25 The mass of the rod between x_{j-1} and x_j is approximately $\delta(\xi_j)\Delta x_j$

Probability

The theory of probability uses a fundamental concept called *probability density*, analogous to the density function considered above in computing mass. As an example, the probability that a television tube will fail before it is b months old is given by an integral

$$\int_0^b \theta e^{-\theta t}\, dt,$$

where θ is a constant. More generally, the probability of failure after a months but before b months is $\int_a^b \theta e^{-\theta t}\, dt$. The function

$$f(t) = \theta e^{-\theta t}, \qquad t > 0$$

is called the *probability density function* for failure of the tube.

Another formula of the same general type is found for errors in measurement. Suppose many repeated measurements are made of the same quantity, such as the mass of the electron, or the distance from Marathon to Athens. Given two numbers a and b, with $a < b$, the ratio

$$\frac{\text{number of measurements giving values between } a \text{ and } b}{\text{total number of measurements made}}$$

[intuitively, the probability that a measurement yields a value between a and b] is given rather accurately by the integral

"NORMAL" PROBABILITY INTEGRAL

$$\int_a^b \frac{1}{\sqrt{2\pi}\,\sigma}\, e^{-\frac{1}{2}[(x-\mu)/\sigma]^2}\, dx,$$

where μ is the average of all the measurements, and the constant σ describes how widely the measurements vary above and below the average.

$$f_{\mu,\sigma} = \frac{1}{\sqrt{2\pi}\,\sigma}\, e^{-\frac{1}{2}[(x-\mu)/\sigma]^2}$$

is called the *normal* probability density function with *mean* μ and *deviation* σ. Figure 26 shows these "normal curves" for large and small values of σ. The curve has a maximum at $x = \mu$ (the mean), and falls off symmetrically on both sides. When the deviation σ is small, then the area under the curve is concentrated near the mean μ; that is, most of the measurements are near μ.

Quantitative results about the normal distribution are postponed to Chapters VI and XI, where they can be obtained more easily than here.

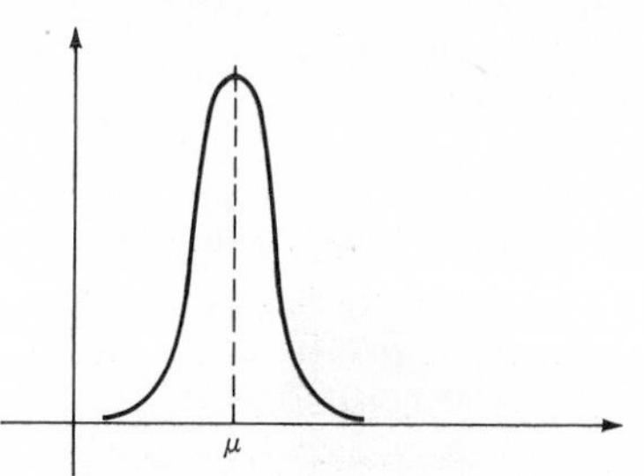

Small deviation σ, density concentrated near the mean μ

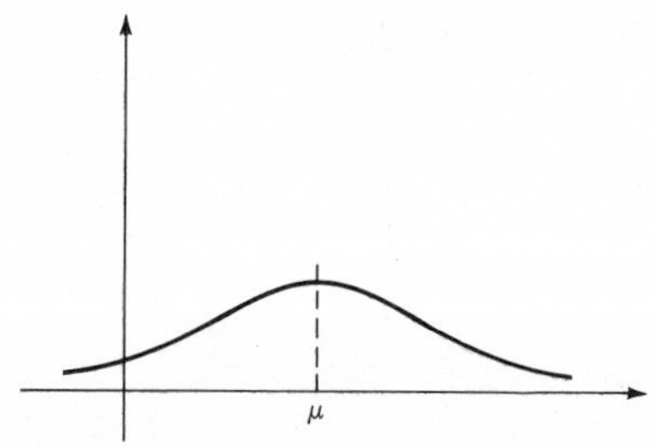

Large deviation σ, density spread out around the mean μ

FIGURE 5.26 Normal densities $f_{\mu,\sigma}$

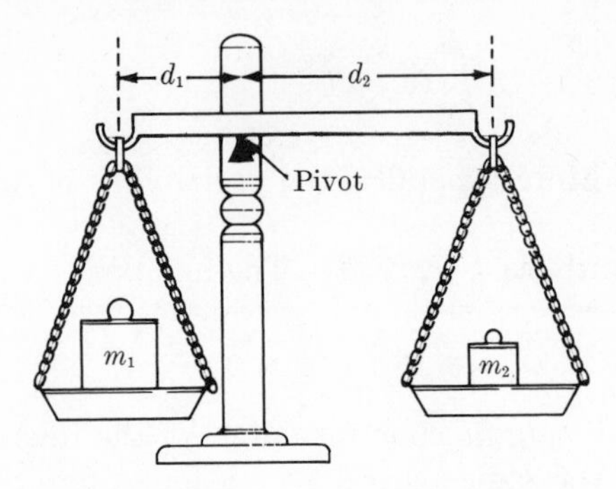

FIGURE 5.27 $m_1d_1 = m_2d_2$ for equilibrium

Torque and center of gravity

In an old-fashioned scale [Fig. 27] the arm is not necessarily in equilibrium when the masses m_1 and m_2 on both sides are the same. What must be the same on both sides is the *product* of the mass m by its distance d from the pivot point,

$$m_1d_1 = m_2d_2 . \tag{10}$$

Hence the notion of *torque* (turning force) is introduced,

$$\text{torque} = \pm(\text{force} \times \text{distance from pivot}).$$

The + or − sign is determined by taking clockwise torques as positive, and counterclockwise torques as negative.

The choice of signs becomes automatic if we assign coordinates to the points on the balance arm, with zero at the pivot, positive numbers to the right, and negative to the left. Then [Fig. 28] *a mass m at position x exerts a torque γmx,* where γm is the force of gravity on the mass m; the number γ is a gravity constant. When $x > 0$, then m is to the right of the pivot and exerts a clockwise (i.e. positive) torque γmx; when $x < 0$, then m is to the left of the pivot and exerts a counterclockwise (i.e. negative) torque γmx.

With this convention, consider again the two masses, m_1 at position $x_1 = -d_1$ and m_2 at position $x_2 = d_2$. The torques are γm_1x_1 and γm_2x_2; *the scale is in equilibrium when the sum of these torques is* 0,

$$\gamma m_1x_1 + \gamma m_2x_2 = 0.$$

Since $x_1 = -d_1$ and $x_2 = d_2$, you can see that this is the same as condition (10).

More generally, if there are n masses $m_1, \ldots, m_n$ at positions $x_1, \ldots, x_n$, the total torque is

$$\gamma(m_1x_1 + \cdots + m_nx_n),$$

and the scale is balanced when this sum is zero.

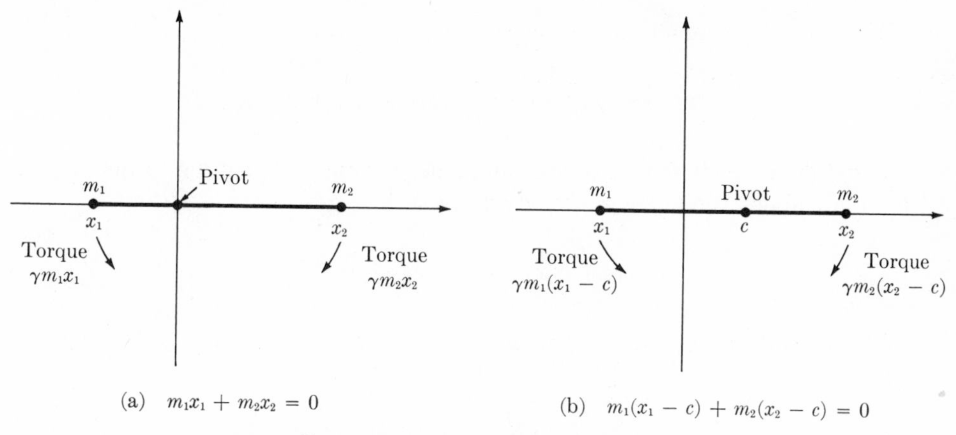

(a) $m_1x_1 + m_2x_2 = 0$ (b) $m_1(x_1 - c) + m_2(x_2 - c) = 0$

FIGURE 5.28 Equilibrium conditions

Suppose now that the pivot point is moved from $x = 0$ to $x = c$ [Fig. 28(b)]. Now, a mass m at position x exerts a (clockwise) torque of $\gamma m(x - c)$; and n masses $m_1, \ldots, m_n$ at positions $x_1, \ldots, x_n$ are in balance when

$$\gamma(m_1(x_1 - c) + \cdots + m_n(x_n - c)) = 0.$$

Finally, instead of a simple balance arm, consider a horizontal plate of uniform thickness and density, pivoted along the line $x = c$ [Fig. 29]. What is the condition for balance in this case?

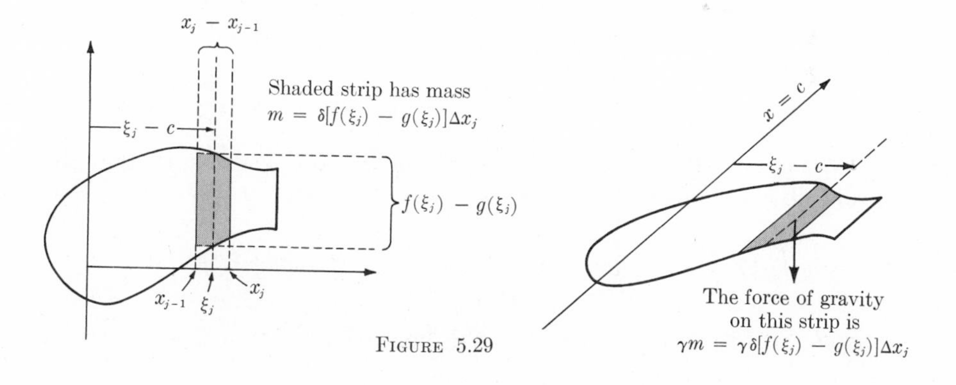

FIGURE 5.29

Let the plate be bounded by the graphs of f and g on $[a,b]$, where $f \geq g$. Divide the plate into strips parallel to the line $x = c$. Each strip has area approximately $[f(\xi_j) - g(\xi_j)]\Delta x_j$; and mass approximately $\delta \cdot [f(\xi_j) - g(\xi_j)]\Delta x_j$ (where δ is the mass per unit area of the plate). It thus exerts a torque of approximately

$$\gamma\delta[f(\xi_j) - g(\xi_j)]\Delta x_j(\xi_j - c).$$

The total torque of all the strips is thus approximately

$$\gamma\delta \sum_{1}^{n} (\xi_j - c)[f(\xi_j) - g(\xi_j)]\Delta x_j ,$$

and as $\Delta x_j \to 0$, this becomes the actual torque. Hence the torque of the plate about the line $x = c$ is given by the integral

$$\gamma\delta \int_a^b (x - c)[f(x) - g(x)]\, dx.$$

The condition for equilibrium is

$$\int_a^b (x - c)[f(x) - g(x)]\, dx = 0,$$

or

$$\int_a^b \Big(x[f(x) - g(x)] - c[f(x) - g(x)]\Big)\, dx = 0.$$

By Theorem 2, this is the same as

$$\int_a^b x[f(x) - g(x)]\, dx - c\int_a^b [f(x) - g(x)]\, dx = 0.$$

Solving this equation for c, we find that the plate balances when $c = \bar{x}$, where

$$\bar{x} = \frac{\displaystyle\int_a^b x[f(x) - g(x)]\, dx}{\displaystyle\int_a^b [f(x) - g(x)]\, dx}.$$

A similar argument (outlined in the problems below) shows that the torque about the line $y = c$ is given by

$$\gamma\delta \int_a^b \left(\frac{f + g}{2} - c\right)(f - g) = \gamma\delta\left[\int_a^b \frac{f^2 - g^2}{2} - c\int_a^b (f - g)\right].$$

Therefore, the plate is balanced on this line when $c = \bar{y}$, where

$$\bar{y} = \frac{\displaystyle\frac{1}{2}\int_a^b (f^2 - g^2)}{\displaystyle\int_a^b (f - g)}.$$

CENTER OF GRAVITY The point $(\bar{x},\bar{y})$ is called the *center of gravity* of the plate. It can be shown that the plate is balanced along *any* line through this point.

Work

When a constant force F acting in a straight line moves an object along that line through a distance l, then by definition the *work* done by the force is Fl. If now a *variable* force F acting along the x axis moves a particle from a to b, the work done is by definition

$$\int_a^b F(x)\,dx = \text{work done by force } F(x) \text{ from } x = a \text{ to } x = b.$$

WORK DONE BY *F*

The Riemann sums show where this integral comes from. We can approximate the work done by F by dividing the interval $[a,b]$ into n small intervals $[x_{j-1}, x_j]$. In each such interval, the force does not vary too much, and the work done is nearly $F(\xi_j)\,\Delta x_j$, since $F(\xi_j)$ is the force and Δx_j is the length of the interval. Thus over the whole interval the work done is nearly $\sum_1^n F(\xi_j)\Delta x_j$. As the length of the small intervals is reduced to zero, the approximations tend to the work done, which is thus $\int_a^b F(x)\,dx$.

An important example is the work done by a spring with spring constant k. When the spring is at position x, the force it exerts is $F(x) = -kx$ [see §3.5]. Hence the work done by the spring in moving from rest position at $x = 0$ to any other position $x = x_0$ is

$$\int_0^{x_0} F(x)\,dx = \int_0^{x_0} (-k)x\,dx = -\tfrac{1}{2}kx_0^2$$

by §5.1. Notice that the work done by the spring is always negative. This is reasonable; the spring *resists* displacement from rest in either direction, so its contribution to the work of displacement is negative.

Concluding Remark. The discussions in this section were brief and sketchy in order to present rapidly a variety of interpretations of the integral. None of these discussions can be considered a mathematical proof. Indeed, without prior definitions of arc length, surface area, volume, and so on, we are not in a position to prove that any of these geometric quantities equals a certain integral. [§AII.9 gives such a definition for arc length, and then proves rigorously the integral formula for it.]

We may, however, consider the arguments as justifying the interpretation of $\int\sqrt{1 + (f')^2}$ as the length of a curve, and $\int \pi f^2$ as a volume of revolution, and so on. Aside from the arguments, there is a practical justification for the interpretations; the numbers given by these integrals agree with actual measurements of lengths, volumes, and so on.

PROBLEMS

The problems are grouped by topics. Most of the integrals have already been evaluated in the problems in §5.3.

Position

1. Given v, find the change in position for the given time interval.
(a) $v(t) = -2t$ for $1 \leq t \leq 3$
• (b) $v(t) = e^{3t}$ for $0 \leq t \leq 10$
(c) $v(t) = \sin \omega t$ for $0 \leq t \leq \pi/\omega$ [ω is a constant]
• (d) $v(t) = \sin \omega t$ for $0 \leq t \leq 2\pi/\omega$.

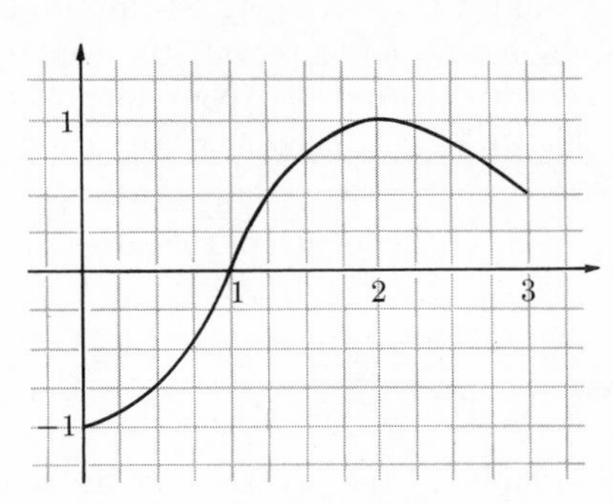

FIGURE 5.30

2. Figure 30 gives a graph of the velocity v of a particle moving in a straight line.
(a) Estimate the net change in position from $t = 0$ to $t = 3$, i.e.

estimate $\int_0^3 v(t)\, dt$. [Estimate areas by counting squares.]

(b) Assuming $s(0) = 1$, sketch a graph of the position of the particle as a function of time.

3. (a) For the particle in Problem 2, estimate the *total distance traveled* from $t = 0$ to $t = 3$, rather than the change in position.
(b) Sketch a graph of the distance $D(t)$ traveled from time 0 to time t.

4. Secret agent 00X is blindfolded and abducted on a train by six sinister sexpots. Two and a half minutes out of St. Andrews station he is thrown off the train, along with a short-fused time-bomb. He has a two-way radio concealed on his person (as the sinister six suspected); but due to the dense fog, the police will not find him in time unless he can locate himself to the nearest quarter mile. How can he do this? *Solution.* With his perfect pitch and his knowledge of trains, 00X can tell from the sound of the wheels how fast the train is going; with his keen sense of timing and fantastic memory, he records the following information:

Time (seconds)	0	10	20	30	40	50	60	70	80	90	100	110	120	130	140	150	(thrown out of train)
Speed (mph)	0	15	30	40	30	15	5	−5	10	20	30	40	40	30	20	15	

Noting that the time intervals above are $10/3600 = 1/360$ of an hour, estimate his position by forming upper and lower sums. [He is more than $25/36 = .69\ldots$ miles, and less than $9/8 = 1.125$ miles out of the station. The rescuers fan out from a point 0.9 miles from the station; 00X is saved, but two of the rescue party are blown up.]

Area

5. A triangle has vertices (a,y_1), (b,y_2), and (b,y_3), where $a < b$ and $y_2 < y_3$ [Fig. 31].
(a) Find equations for the functions f and g whose graphs are the straight lines indicated in Fig. 31.
• (b) Apply the formula for area between two graphs to show that the triangle has area $\frac{1}{2}(b - a)(y_3 - y_2)$.

6. The graphs of $f(x) = \dfrac{b}{a}\sqrt{a^2 - x^2}$ and $g(x) = -\dfrac{b}{a}\sqrt{a^2 - x^2}$ on the interval $[-a,a]$ together form an ellipse. Show that the area bounded by this ellipse is πab. [Hint: Recognize $\int_{-a}^{a} \sqrt{a^2 - x^2}\,dx$ as the area of the upper half of a circle.]

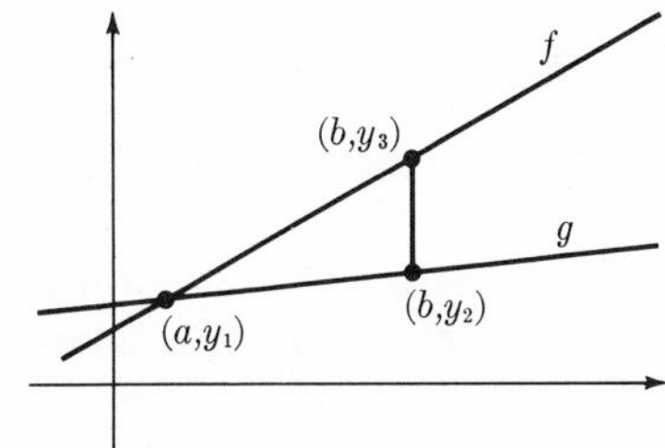

FIGURE 5.31

Volume of revolution

• 7. A football-shaped solid called an *ellipsoid* is obtained by revolving the set

$$\left\{(x,y): 0 \le y \le \frac{b}{a}\sqrt{a^2 - x^2},\ -a \le x \le a\right\}$$

about the x axis.
(a) Write an integral for the volume of the solid.
(b) Write an integral for the volume of that part to the right of $x = 0$.
(c) Evaluate the integrals.

8. Using formula (5), show that the volume of the right circular cone in Fig. 32 is $\frac{1}{3}\pi ab^2$.

9. A region lies between two planes perpendicular to the x axis, one at $x = a$ and one at $x = b$ [Fig. 33]. Let $A(x_0)$ denote the area of the cross section at $x = x_0$. Justify the integral $\int_a^b A(x)\,dx$ for the volume of the region. [No formal proof is requested; an appropriate sketch and a few remarks will do.]

10. The most general cone [Fig. 34] is generated by all the lines from a given point V (the vertex) to a given plane region B (the base). This problem outlines a proof that the volume of such a cone is $\frac{1}{3}Ah$, where A is the area of the base B, and h is the height of the cone [i.e. h = distance from V to the plane through B].

(a) Imagine a vertical y axis with the origin at V, and the base B lying in the plane $y = h$. Show that the area of the cross section at y_0 is $(y_0/h)^2 A$ where A is the area of the given base B. [Use the geometric principle that the area of similar figures varies as the square of their linear dimensions. If this *general* principle is unfamiliar, you may assume that the base is a triangle.]

(b) Show that the volume of the cone is $\int_0^h (y/h)^2 A\,dy$. [See Problem 9.]

(c) Evaluate the integral in part (b).

11. [*Volume by "cylindrical shells" method*]

(a) Show that the area of a circular anulus of radii r_1 and r_2 is $\pi(r_2^2 - r_1^2) = \pi(r_2 - r_1)(r_2 + r_1)$. [Do *not* use integrals for this! Use the formula for the area of a circle.]

(b) Show that a cylindrical shell of radii r_1 and r_2 and height h has volume $\pi h(r_1 + r_2)(r_2 - r_1)$. [See Fig. 35(i).]

(c) Let A be the set $\{(x,y): a \leq x \leq b,\ g(x) \leq y \leq f(x)\}$, where $a \geq 0$ and $g \leq f$ on the interval $[a,b]$. A solid of revolution is generated by revolving A about the y axis. Explain why the volume of the solid is $\int_a^b 2\pi x[f(x) - g(x)]\,dx$. [Hint: The "back half" of the solid is shown in Fig. 35(ii). The darkly shaded area is very nearly a cylindrical shell.]

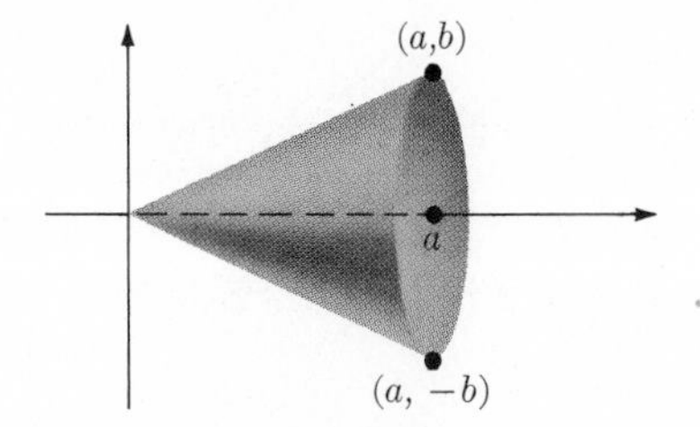

FIGURE 5.32

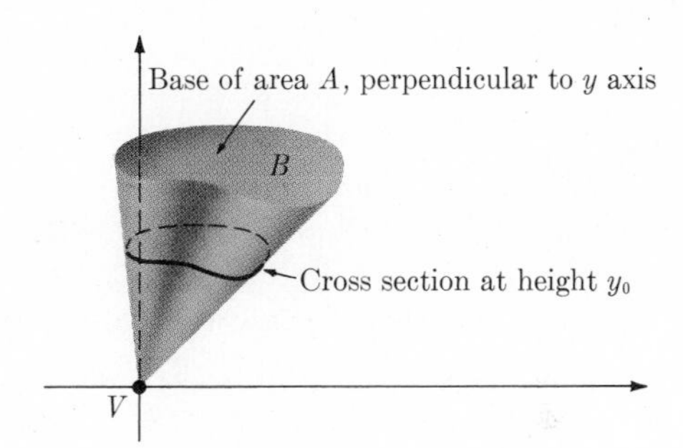

FIGURE 5.34

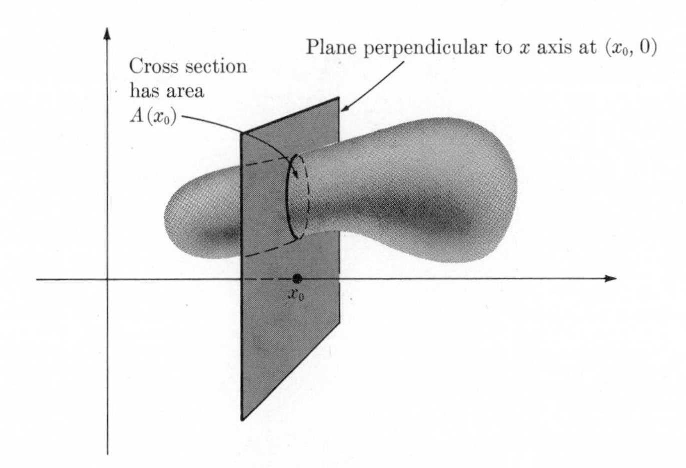

FIGURE 5.33

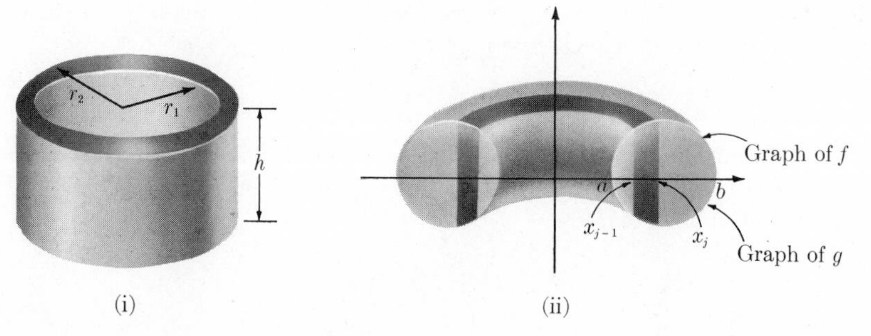

FIGURE 5.35

Surface area

12. You can obtain a sphere of radius r by revolving about the x axis the graph of $f(x) = \sqrt{r^2 - x^2}$ on $[-r,r]$.

(a) Show that the area of this sphere is $\int_{-r}^{r} 2\pi r\, dx$. [Use formula (8).]

(b) Evaluate $\int_{-r}^{r} 2\pi r\, dx$.

13. Show that the area of a cone of slant height l and radius r is equal to πrl, by each of the following methods.

(a) Cut the cone along the dotted line in Fig. 36(i), and roll it out flat. It then forms a certain fraction of a circle of radius l, whose area you can easily compute.

(b) Imagine the cone to be constructed of n triangles of altitude l and base $2\pi r/n$, as in Fig. 36(ii). [This assumption becomes more and more accurate as n increases. Explain why the base of each triangle is taken to be $2\pi r/n$.]

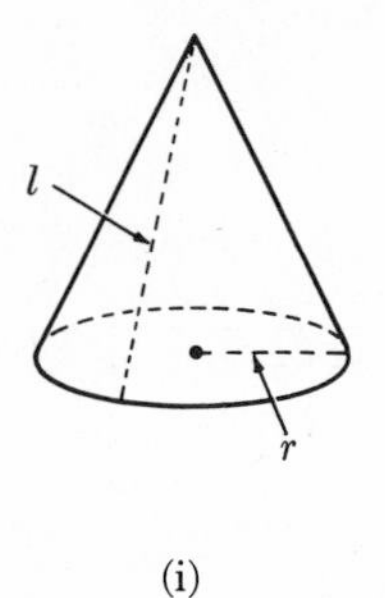

(i)

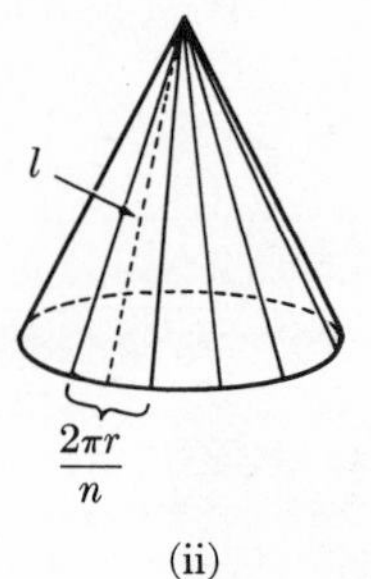

(ii)

FIGURE 5.36

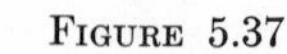

FIGURE 5.37

14. Derive from the result in Problem 13 a formula for the area of a frustum of a cone with radii r_1 and r_2 and height h [Fig. 37].

15. This problem concerns the formula for the area of a surface of revolution. You may find it difficult, but try it anyway.

(a) Suppose $|f'(x)| \leq M$ on $[a,b]$. Show from the Mean Value Theorem that $|f(x_1) - f(x_2)| \leq M\,|x_1 - x_2|$ if x_1 and x_2 are in $[a,b]$.

(b) Suppose $x_{j-1} \leq \xi_j \leq x_j$. Show that $|f(x_j) + f(x_{j-1}) - 2f(\xi_j)| \leq 2M\,|x_j - x_{j-1}|$; i.e. that $f(x_j) + f(x_{j-1})$ cannot differ from $2f(\xi_j)$ by more than $2M(x_j - x_{j-1})$.

(c) Show that if all intervals in the partition $a = x_0 < \cdots < x_n = b$ have length $\leq l$, then each term of the sum (6) in the text differs from the corresponding term of the sum (7) by no more than $2\pi Ml\sqrt{1 + M^2}\,\Delta x_j$.

(d) Show that the difference between (6) and (7) is $\leq 2\pi Ml\sqrt{1 + M^2}(b - a)$ [which is negligible when l is made small; as $l \to 0$, both (6) and (7) tend to the same limit.]

16. The graph of f on $[a,b]$ for $a > 0$ is revolved about the y axis. Show that the surface area should be

$$2\pi \int_a^b x\sqrt{1 + f'(x)^2}\,dx.$$

Mass and Density

17. A straight canal, of constant cross sectional area A square meters and of length L meters, leads from a freshwater lake to the salty sea. By measuring the salinity at various points in the canal, it is found that the salt concentration at distance x from the lake is cx grams per cubic meter.

(a) Show that the number of grams of salt in the canal is $\int_0^L cxA\,dx$.

(b) Show that this integral equals $\frac{1}{2}cAL^2$.

(c) Rework the problem if the concentration is $10e^{-4x}$, instead of cx.

Probability

18. (a) Evaluate the probability of picture tube failure in the first b months, supposing this is given by the integral $\int_0^b \theta e^{-\theta t}\,dt$, where θ is a constant.

(b) Show that as $b \to +\infty$, the probability of failure $\to 1$, and as $b \to 0$, the probability of failure $\to 0$.

19. Suppose the probability of picture tube failure in the first b months is $\int_0^b \theta e^{-\theta t}\,dt$. Are the tubes more durable when θ is large, or when θ is small?

Torque

20. A horizontal rectangle of density δ with sides h and w is pivoted about a line parallel to side w, at distance l from the center of the rectangle [Fig. 38]. Show that the torque is $\gamma\delta hwl$.

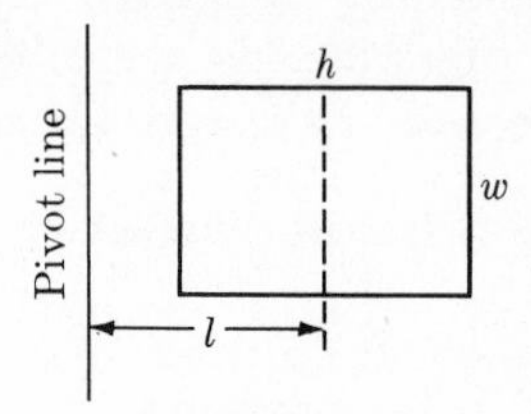

FIGURE 5.38

21. A horizontal plate is bounded by the graphs of f and g on the interval $[a,b]$, and $f \geq g$ on the interval. Show that the net torque exerted by the plate about the line $y = c$ is

$$\gamma\rho \int_a^b \left[\left(\frac{f+g}{2} - c\right)(f - g)\right],$$

where ρ is the density of the plate (assumed constant), and where γ is a constant of gravity. [Use Problem 20.]

Work

22 •(a) An object is moved from $x = 1$ to $x = 100$ by a force F that varies inversely as the square of the distance from the origin, $F(x) = cx^{-2}$ for a constant c. Find the work done.

(b) Repeat part (a) for an object moved from $x = -1$ to $x = -100$. [The answer should be the same!]

(c) What if it is moved from $x = 1$ to $x = \infty$?

23. A mass m moves on a spring; at time t its position is $s(t)$. The total force exerted on the mass (by the spring and by gravity) is $-ks$, where k is the spring constant. The equation of motion is then $ms'' + ks = 0$ [§3.5].

CONSERVATION OF ENERGY

• (a) Show that the work done by these forces is $-\frac{1}{2}ks^2 +$ constant.

(b) The *kinetic energy* is defined to be

$$KE = \tfrac{1}{2}mv^2 = \tfrac{1}{2}m(s')^2.$$

The *potential energy* is defined to be *minus* the work done,

$$PE = \tfrac{1}{2}ks^2.$$

Using the equation of motion $ms'' + ks = 0$, prove that $PE + KE$ is constant. [This is a special case of the so-called *Law of Conservation of Energy*. Notice that proving this law $PE + KE =$ constant was the crucial Step IV in analyzing the equation $ms'' + ks = 0$, §3.5.]

5.5 UNBOUNDED INTERVALS AND DISCONTINUOUS FUNCTIONS

We have discussed $\int_a^b f$, where $[a,b]$ is a finite interval and f is continuous on $[a,b]$. Other cases arise, and are handled by appropriate limiting processes.

Unbounded intervals

The work done against gravity in moving an object of mass m from the surface of the earth (4000 miles from the center) to a height h above the surface is given by the integral

$$\int_{4000}^{4000+h} \frac{cm}{r^2}\,dr = -\frac{cm}{r}\bigg|_{4000}^{4000+h} = cm\left[\frac{1}{4000} - \frac{1}{4000+h}\right]$$

where c is a constant. Although in actual practice it is impossible to raise the object "infinitely far," it is still reasonable to ask how much work it would take if it *were* to be raised infinitely far. The answer is given by the limit

$$\lim_{h\to+\infty} \int_{4000}^{4000+h} \frac{cm}{r^2}\,dr = \lim_{h\to+\infty} cm\left[\frac{1}{4000} - \frac{1}{4000+h}\right] = \frac{cm}{4000}\,.$$

This limit is denoted, naturally, by $\int_{4000}^{\infty} \frac{cm}{r^2}\,dr$.

IMPROPER INTEGRAL—INFINITE INTERVAL

In general, let f be continuous on $[a,+\infty)$. Then the expression $\int_a^\infty f$ is called an *improper integral.* If $\lim_{b\to\infty} \int_a^b f$ exists, it is said that $\int_a^\infty f$ is *convergent*, and $\int_a^\infty f = \lim_{b\to\infty} \int_a^b f$. If $\lim_{b\to\infty} \int_a^b f$ does *not* exist, then $\int_a^\infty f$ is called *divergent.*

Example 1. Investigate the improper integral $\int_0^\infty e^{-x}\,dx$ [which arises in "probability of breakdown" studies]. *Solution.* Since $f(x) = e^{-x}$ is the derivative of $F(x) = -e^{-x}$, we have

$$\int_0^b e^{-x}\,dx = -e^{-x}\bigg|_0^b = 1 - e^{-b},$$

and therefore

$$\int_0^{\infty} e^{-x}\,dx = \lim_{b\to+\infty}\int_0^b e^{-x}\,dx = \lim_{b\to+\infty}(1 - e^{-b}) = 1.$$

The geometric interpretation of this result is surprising: The region under the graph of $f(x) = e^{-x}$ for $x \geq 0$ is *unbounded* but has *finite area.*

Example 2. Investigate the improper integral $\int_1^{\infty} e^{-x^2}\,dx$ [which arises in "distribution of error" studies]. *Solution.* As noted above, we cannot evaluate this integral by appeal to the formula

$$\int_a^b f = F(b) - F(a), \qquad \text{where } F' = f.$$

The reason is simple: The only function F available is given by $F(t) = \int_a^t e^{-x^2}\,dx$, and this doesn't help in evaluating the integral. Thus it is impractical to obtain an explicit formula for $\int_1^b e^{-x^2}\,dx$.

However, it *is* possible to decide whether or not the improper integral is convergent. Notice that for $x \geq 1$, we have $x^2 \geq x$, and therefore $e^{-x^2} \leq e^{-x}$. Thus the area under the graph of e^{-x^2} on the infinite interval $[1,\infty)$ should be *less* than the corresponding area for e^{-x}, which is given by

$$\lim_{b\to\infty}\int_1^b e^{-x}\,dx = \lim_{b\to\infty} -e^{-x}\Big|_1^b = \lim_{b\to\infty}(e^{-1} - e^{-b}) = e^{-1}.$$

It is reasonable to conclude that the area under the graph of e^{-x^2} on the infinite interval $[1,\infty)$ is less than e^{-1}. This is true, but the proof depends on the following proposition, which we have not yet proved.

BOUNDED INCREASING FUNCTION HAS A LIMIT AS $x \to +\infty$

Proposition K. *Let F be an increasing function on $[a,\infty)$, and suppose there is a constant M such that $F(x) \leq M$ for $a \leq x < +\infty$. Then F has a limit as $x \to +\infty$, and $\lim_{x\to+\infty} F(x) \leq M$.*

[Sketch, Fig. 39; Proof, §10.3, Problem 6.]

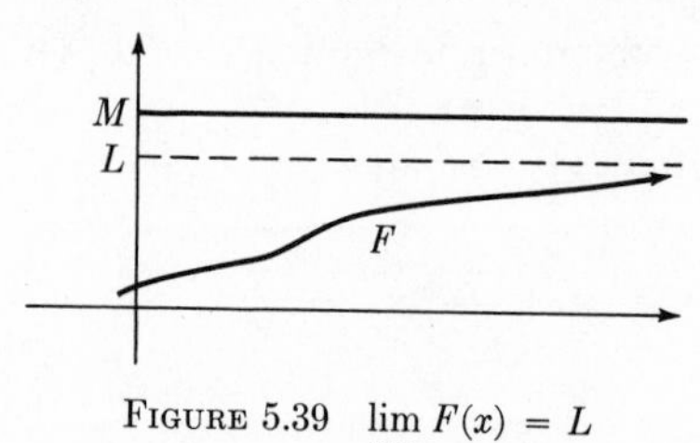

FIGURE 5.39 $\lim_{x\to\infty} F(x) = L$

In the present case, we have

$$F(b) = \int_1^b e^{-x^2}\,dx \le \int_1^b e^{-x}\,dx = e^{-1} - e^{-b} \le e^{-1}.$$

Furthermore, as b increases, then $F(b)$ also increases, since the integrand e^{-x^2} is positive. Therefore, we can apply Proposition K with $M = e^{-1}$, and conclude that

$$\lim_{b\to\infty} \int_1^b e^{-x^2}\,dx \le e^{-1}.$$

Thus the improper integral $\int_1^\infty e^{-x^2}\,dx$ is convergent, and the area under the graph of e^{-x^2} on $[1, \infty)$ is finite.

Unbounded functions

Two electrons at distance r repel each other with a force c/r^2, where c is a constant. If they are originally 1 cm apart $[r = 1]$, how much work is required to bring them together $[r = 0]$? The work in question is given by the integral

$$\int_0^1 \frac{c}{r^2}\,dr.$$

This is called an *improper integral* because the function $f(r) = c/r^2$ is *not* continuous on the interval of integration $[0,1]$; it is unbounded at the left end [Fig. 40]. Such an improper integral is interpreted as the limit

$$\lim_{h\to0+} \int_h^1 \frac{c}{r^2}\,dr = \lim_{h\to0+} \frac{-c}{r}\bigg|_h^1 = \lim_{h\to0+} \left[\frac{c}{h} - \frac{c}{1}\right] = +\infty.$$

This improper integral is *divergent*. The work required in bringing the electrons together is infinite, which means that it can't be done.

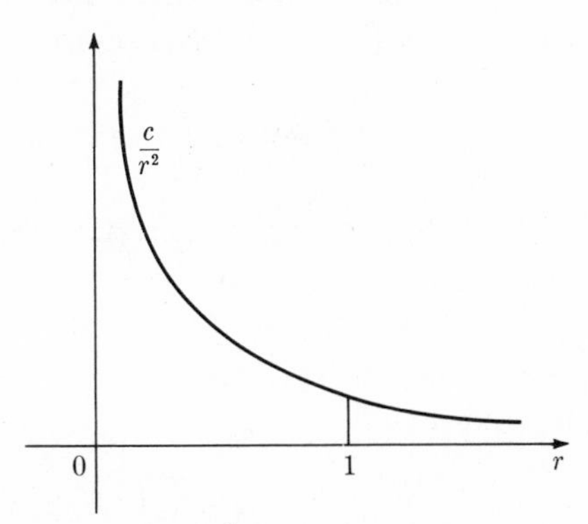

FIGURE 5.40 $\int_0^1 \frac{c}{r^2}\,dr$ is an improper integral

IMPROPER INTEGRAL—DISCONTINUOUS FUNCTION

In general, if f is not continuous on $[a,b]$, but *is* continuous on the smaller intervals $[a+h,b]$ for every $h > 0$, then $\int_a^b f$ is called an *improper integral* [Fig. 41].

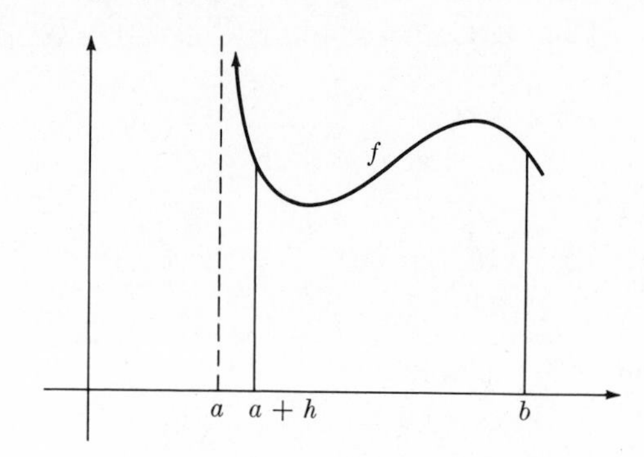

FIGURE 5.41 $\int_a^b f = \lim_{h\to0+} \int_{a+h}^b f$

If the limit $\lim_{h\to0+} \int_{a+h}^b f$ exists, then the improper integral is called *convergent*, and we write

$$\int_a^b f = \lim_{h\to0+} \int_{a+h}^b f.$$

[Sometimes the notation $\int_{a+}^b f$ is used to denote this kind of improper integral.] If the limit fails to exist, then $\int_a^b f$ is called *divergent*.

Similar remarks obviously apply when f is continuous on $[a, b-h]$ for every $h > 0$, but not on $[a,b]$. In this case,

$$\int_a^b f = \lim_{h\to0+} \int_a^{b-h} f,$$

if the limit exists. If the limit does not exist, the integral is called divergent.

Example 3. The region between the x axis and the graph of $f(x) = x^{-1/2}$ on $(0,1]$ is unbounded. Does it have finite area? *Solution.* We take the area to be the improper integral

$$\int_{0+}^1 x^{-1/2}\,dx = \lim_{h\to0+} \int_h^1 x^{-1/2}\,dx.$$

Notice that $F(x) = 2x^{1/2}$ solves $F'(x) = x^{-1/2}$, so

$$\lim_{h\to 0+} \int_h^1 x^{-1/2}\,dx = \lim_{h\to 0+} 2x^{1/2}\Big|_h^1 = \lim_{h\to 0+} (2 - 2\sqrt{h}) = 2.$$

The area in question is finite.

There may also be discontinuities of f *within* the interval of integration. If, for example, f is continuous on $[a,b]$ except for a discontinuity at the point c, then $\int_a^b f$ is broken into two parts $\int_a^c f + \int_c^b f$, and each of these is treated as above. For example,

$$\int_{-1}^{1} x^{-1/3}\,dx = \int_{-1}^{0} x^{-1/3}\,dx + \int_{0}^{1} x^{-1/3}\,dx$$

$$= \lim_{h\to 0+} \int_{-1}^{-h} x^{-1/3}\,dx + \lim_{h\to 0+} \int_{h}^{1} x^{-1/3}\,dx$$

$$= \lim_{h\to 0+} \tfrac{3}{2}x^{2/3}\Big|_{-1}^{-h} + \lim_{h\to 0+} \tfrac{3}{2}x^{2/3}\Big|_{h}^{1} = 0.$$

Piecewise continuous functions

Suppose that current of constant strength is flowing in an electrical circuit, but a switch reverses the direction every second. The graph of the current then looks like Fig. 42. Such a function is called *piecewise continuous;* its graph consists of pieces of continuous functions on closed intervals, with jumps at the ends of the intervals. This particular example can be integrated by sight over any interval, simply by computing the area of a number of rectangles; the result is the same as that given by the limiting process outlined above. In particular, the indefinite integral is easily found to be the function F in Fig. 43. Notice that $F' = f$, *except at the points where f is discontinuous;* at these points, F' does not exist. [Recall, in the Fundamental Theorem of Calculus, the important hypothesis that f be continuous.]

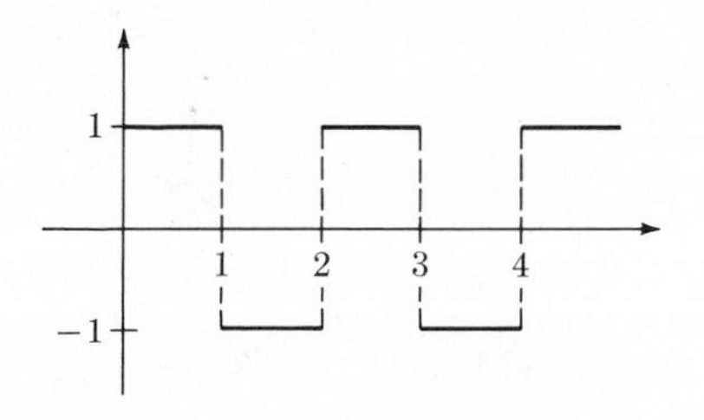

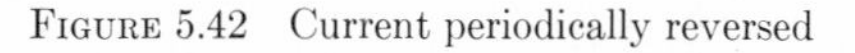

FIGURE 5.42 Current periodically reversed

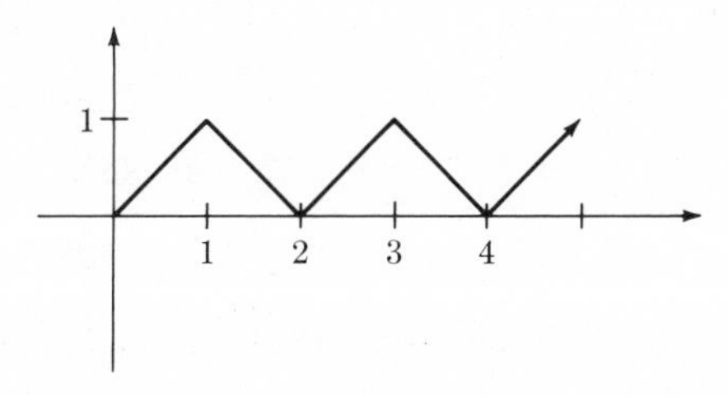

FIGURE 5.43 Indefinite integral of Fig. 42

PROBLEMS

1. *Gabriel's horn* is obtained by revolving the graph of $f(x) = 1/x$ on the interval $[1, \infty)$ about the x axis. Show that the volume of the horn, $\int_1^\infty \pi f^2$, is finite, but the surface area $\int_1^\infty 2\pi f\sqrt{1 + (f')^2}$ is infinite.

   $\left[\text{Hint: Notice that } \int_1^b f\sqrt{1 + (f')^2} \geq \int_1^b f, \text{ since } f > 0.\right]$

2. Show that $\int_a^\infty e^{-x^2}\,dx \leq \int_a^\infty e^{-x}\,dx = e^{-a}$.

3. Show that $\int_0^\infty \theta e^{-\theta t}\,dt = 1$. [What is the derivative of $e^{-\theta t}$?]

4. (a) Evaluate $\int_1^b \frac{dt}{t}$, $b > 0$.

   (b) Evaluate $\int_1^b \frac{dt}{t^2}$, $b > 0$.

   (c) Show that $\lim_{b \to +\infty} \int_1^b \frac{dt}{t}$ does not exist, but $\lim_{b \to +\infty} \int_1^b \frac{dt}{t^2} = 1$.

   (d) Can you explain, by interpreting $\int_1^b \frac{dt}{t}$ and $\int_1^b \frac{dt}{t^2}$ as areas, why the second integral might have a limit as $b \to +\infty$, while the first does not?

   (e) For what values of the constant a does $\int_1^\infty \frac{dt}{t^a}$ exist?

5. For what values of the constant a does $\int_0^1 \frac{dt}{t^a}$ exist? [Compare Problem 4.]

6. Find $\int_0^\infty \frac{dt}{1 + t^2}$. [This is the area of an unbounded region.]

7. Show that Fig. 43 gives the indefinite integral $F(t) = \int_0^t f$ for the "step function" f in Fig. 42.

8. Graph the indefinite integral $F(t) = \int_0^t f$ for each of the piecewise continuous functions f in Fig. 44.

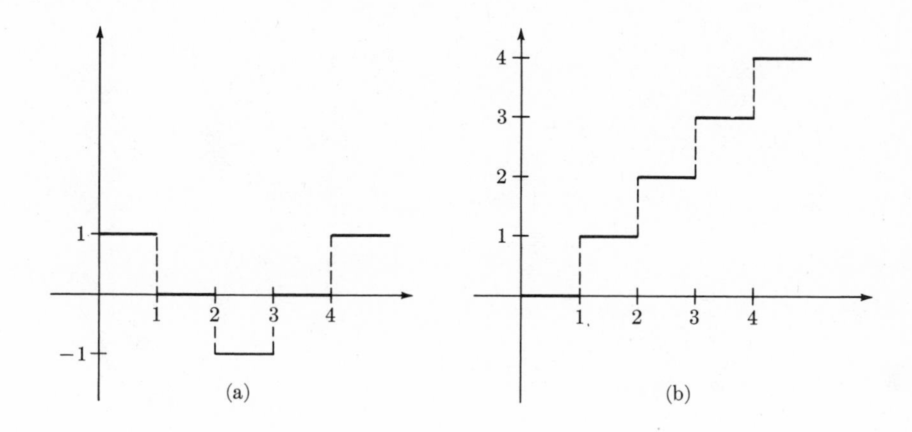

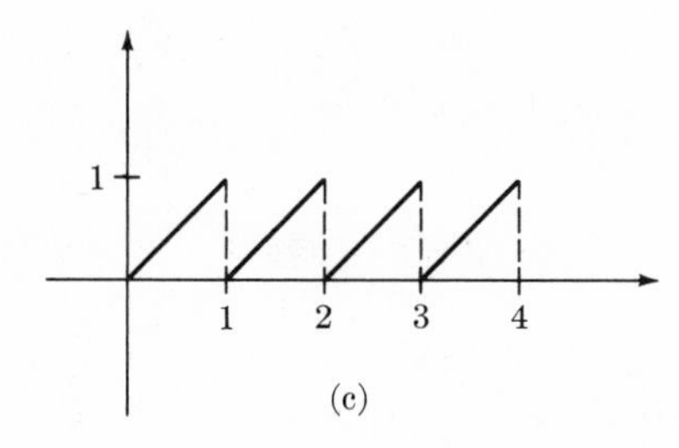

FIGURE 5.44

Chapter VI

Techniques of Integration

Many different questions arising in the various applications of mathematics can be reduced to the following general problem: Given a function f, find another function F whose derivative is the given function f. In other words, given f, "solve" the equation $F' = f$ by finding F. For example:

Given the velocity v, find a position function s such that $s' = v$ [§2.2].

Given a continuous function f, evaluate the definite integral $\int_a^b f$ by the Second Fundamental Theorem of Calculus,

$$\int_a^b f = F(b) - F(a), \qquad \text{where } F' = f \qquad [\S5.3].$$

This last, rather general example includes the numerous applications to areas, volumes, surface areas, probability, and so on, outlined in §5.4.

The relation between f and the function F solving $F' = f$ can be viewed in various ways, each giving rise to a different terminology:

(i) Since f is the *derivative* of F, it is reasonable to call F a *primitive* of f.

(ii) Finding a function F such that $F' = f$ is the reverse of differentiation; the process can therefore be called *antidifferentiation*, and F can be called an *antiderivative* of f.

(iii) When f is continuous, then the *indefinite integral* $F(x) = \int_a^x f$ satisfies $F' = f$. Hence it is reasonable to give the name *indefinite integration* to the process of finding an F such that $F' = f$, and to call any solution of this equation an *indefinite integral* of f. This is the terminology we will adopt.

NOTATION FOR INDEFINITE INTEGRALS

An indefinite integral F of f is denoted by $F = \int f$ or [in Leibniz notation] $F(x) = \int f(x)\,dx$.

As with Leibniz notation for derivatives and definite integrals, the x appearing here is a "dummy variable," and can be replaced throughout by any other letter that will not lead to confusion. For example, the indefinite integral can be denoted by

$$F(t) = \int f(t)\,dt, \qquad F(z) = \int f(z)\,dz, \qquad \text{etc.};$$

but certainly *not* by $F(f) = \int f(f)\,df$, since this requires the single letter f to play two different roles.

When F is an indefinite integral of f, so is $F + c$ for any constant c, since $(F + c)' = F' + 0 = F' = f$. Conversely, by the law of parallel graphs, any two indefinite integrals *defined on the same interval* differ by a constant. Hence indefinite integrals are usually written in the form $F + c$; c is called the "constant of integration." Examples:

$$\int 1\,dx = x + c,$$

$$\int x\,dx = \frac{1}{2}x^2 + c,$$

$$\int t\,dt = \frac{1}{2}t^2 + c,$$

$$\int \cos u\,du = \sin u + c,$$

$$\int \frac{1}{x}\,dx = \log|x| + c.$$

These formulas can be checked immediately by differentiating the right-hand side.

Clearly, the first prerequisite for computing indefinite integrals is a good mastery of differentiation. In fact, this chapter simply rephrases results for derivatives as results for indefinite integrals, and shows how to work with them from this new point of view.

CAUTION

Remark 1. The last example above, $\int(1/x)\,dx = \log|x| + c$, requires a word of caution. The law of parallel graphs applies only to functions defined on an *interval*. Now, $\log|x|$ is not defined on an interval, but rather on the two separate intervals $(-\infty, 0)$ and $(0, \infty)$. In each of these intervals we can write $\int(1/x)\,dx = \log|x| + c$; but the constant c may be *different* in each interval. For instance, if we set $F(x) = (\log x) + 1$ for $x > 0$ and $F(x) = (\log |x|) - 1$ for $x < 0$, then $F'(x) = 1/x$ wherever $F'(x)$ and $1/x$ are defined; but $F(x)$ is not equal to $\log |x| + c$ for any single constant c [Fig. 1].

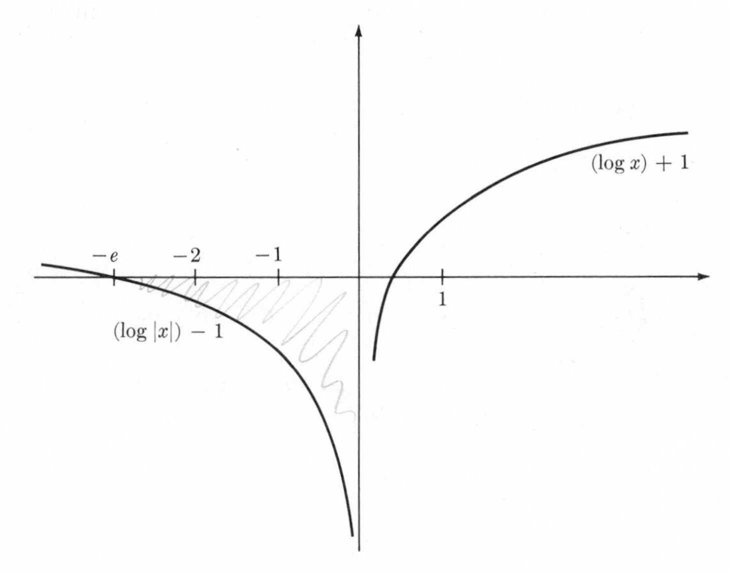

FIGURE 6.1 $F'(x) = 1/x, \quad x \neq 0$

Remark 2. As noted above, the First Fundamental Theorem of Calculus gives an indefinite integral of any continuous function f, in the form $F(x) = \int_a^x f$. But we want something more. Suppose f is written explicitly as an elementary function (i.e. a function built up out of the polynomials, the trigonometric functions, and the log and exponential functions by the processes of arithmetic, composition, and the taking of inverse functions). Then we would like to write the indefinite integral $\int f$ likewise in such an explicit form. This is precisely what we did in the case of differentiation in Chapter II.

For integration, we do not give such a complete set of rules—and for a very good reason. No such complete set of rules can possibly exist! It can happen that f is an elementary function, but $\int f$ is not; for example, e^{-x^2} is an elementary function, but $\int e^{-x^2}$ is not. (The proof of this is too long to take up here.)

Indefinite integration is basically a game, like chess, bridge, or poker. It has techniques and stratagems, but lacks a complete program guaranteeing success in every case.

6.1 LINEAR COMBINATIONS

If a and b are any constants, and F and G are differentiable functions, then the function $aF + bG$ is differentiated by the simple formula

$$(aF + bG)' = aF' + bG'. \tag{1}$$

This yields a corresponding formula for indefinite integrals

$$\int (af + bg) = a\int f + b\int g, \tag{2}$$

INDEFINITE INTEGRATION IS LINEAR

or in Leibniz notation

$$\int \Big(af(x) + bg(x)\Big)\,dx = a\int f(x)\,dx + b\int g(x)\,dx. \tag{3}$$

Thus knowing indefinite integrals for f and g, you can write down immediately an indefinite integral for $af + bg$.

The proof of (2) is easy:

$$\left(a\int f + b\int g\right)'$$

$$= a\left(\int f\right)' + b\left(\int g\right)' \qquad \text{[by (1)]}$$

$$= af + bg. \qquad \text{[by definition; } \textstyle\int f \text{ has the derivative } f, \text{ and } \int g \text{ has the derivative } g\text{]}$$

Since $a\int f + b\int g$ has the derivative $af + bg$, it is an indefinite integral of $af + bg$, and this proves (2).

Example 1. $\int (2x + 3)\,dx = ?$ *Solution.*

$$\int (2x + 3)\,dx = 2\int x\,dx + 3\int dx \qquad \text{[by (3)]}$$

$$= 2\left(\frac{x^2}{2} + c_1\right) + 3(x + c_2) \qquad \left[\text{since } \frac{d(x^2/2)}{dx} = x \text{ and } \frac{dx}{dx} = 1\right]$$

$$= x^2 + 3x + 2c_1 + 3c_2$$

$$= x^2 + 3x + c.$$

Notice the role of the "constants of integration" c_1, c_2, and c. The constants in $\int x\,dx$ and $\int 1\,dx$ need not be the same, so we distinguish them by labeling the first c_1, and the second c_2. Then in the last step, $2c_1 + 3c_2$ is an arbitrary constant, and we simplify the final expression by relabeling it, setting $2c_1 + 3c_2 = c$.

Example 2. $\int(1 + 2\cos\theta)\,d\theta = ?$ *Solution.*

$$\int (1 + 2\cos\theta)\,d\theta = \int 1\,d\theta + 2\int \cos\theta\,d\theta \qquad \text{[by (3)]}$$

$$= (\theta + c_1) + 2(\sin\theta + c_2)$$

$$[\text{since } d\theta/d\theta = 1 \text{ and } (d\sin\theta)/d\theta = \cos\theta]$$

$$= \theta + 2\sin\theta + c,$$

where $c = c_1 + 2c_2$.

Formulas (2) and (3) extend to any number of factors, by induction; if $a_1, \ldots, a_n$ are any constants, then

$$\int (a_1 f_1 + a_2 f_2 + \cdots + a_n f_n) = a_1 \int f_1 + a_2 \int f_2 + \cdots + a_n \int f_n .$$

INTEGRAL OF A LINEAR COMBINATION

[See Problem 5 below.]

The expression $a_1 f_1 + a_2 f_2 + \cdots + a_n f_n$ is called a *linear combination* of the functions $f_1, f_2, \ldots, f_n$. Thus the formula above says that the integral of a linear combination of $f_1, \ldots, f_n$ is that same linear combination of the integrals $\int f_1, \ldots, \int f_n$.

Remark. Recall the standard notation

$$\Big[F(x)\Big]_a^b = F(b) - F(a),$$

which is convenient in making calculations based on the Second Fundamental Theorem of Calculus. Thus, for example,

$$\int_1^2 x\,dx = \frac{x^2}{2}\bigg|_1^2 = \frac{2^2}{2} - \frac{1^2}{2} = 1\tfrac{1}{2} .$$

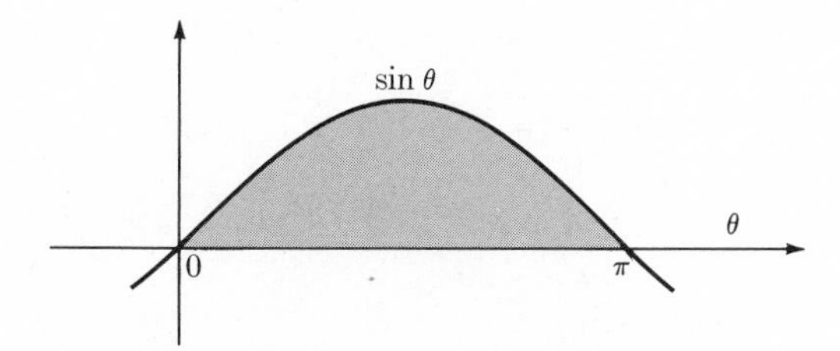

FIGURE 6.2 Area under one "arch" of the sin curve

Example 3. Compute the area under one "arch" of the sin curve. *Solution.* As in Fig. 2, one "arch" of the graph of $\sin\theta$ runs from $\theta = 0$ to $\theta = \pi$. The area under this arch is $\int_0^\pi \sin\theta\,d\theta$, which we evaluate as follows:

$$\int \sin\theta\, d\theta = -\cos\theta + c, \qquad \left[\text{since } \frac{d(-\cos\theta)}{d\theta} = \sin\theta\right];$$

hence

$$\int_0^\pi \sin\theta\, d\theta = \Big[-\cos\theta + c\Big]_0^\pi = (-\cos\pi + c) - (-\cos 0 + c)$$

$$= -(-1) + c - (-1 + c)$$

$$= 2.$$

NO CONSTANT OF INTEGRATION FOR DEFINITE INTEGRALS

Notice that the constant c in $F(\theta) + c$ drops out when the definite integral $\int_a^b f$ is evaluated:

$$(F(b) + c) - (F(a) + c) = F(b) - F(a).$$

For this reason, it is not necessary to include the constant of integration when evaluating definite integrals; simply write

$$\int_0^\pi \sin\theta\, d\theta = \Big[-\cos\theta\Big]_0^\pi = -\cos\pi - (-\cos 0) = 1 + 1 = 2.$$

PROBLEMS

1. Show that $\int x^a\, dx = \dfrac{x^{a+1}}{a+1} + c$ if $a \neq -1$.

• 2. Find the following integrals by consulting the tables in §2.9. *Memorize the results.*

BASIC INTEGRALS TO KNOW

(a) $\int u^a\, du \quad [a \neq -1]$

(b) $\int \dfrac{du}{u} \quad \left(\text{This means } \int \dfrac{1}{u}\, du.\right)$

(c) $\int e^u\, du$

(d) $\int \sin u\, du \quad$ [What is $(\cos)'$? $(-\cos)'$?]

(e) $\int \cos u\, du$

(f) $\int \sec^2 u\, du$

(g) $\int \csc^2 u\, du$

(h) $\int \sec u \tan u\, du$

(i) $\int \csc u \cot u\, du$

(j) $\int \dfrac{du}{1 + u^2}$

3. Evaluate the following indefinite and definite integrals.

• (a) $\int (2x^2 + 3)\, dx; \quad \int_{-1}^{2} (2x^2 + 3)\, dx$

(b) $\int (1 + 5x + 10x^2)\, dx; \quad \int_{0}^{10} (1 + 5x + 10x^2)\, dx$

• (c) $\int (-1 + 2x)^2\, dx; \quad \int_{2}^{-3} (-1 + 2x)^2\, dx$

• (d) $\int (1 + x + x^2 + x^3 + x^4)\, dx; \quad \int_{0}^{1} (1 + x + x^2 + x^3 + x^4)\, dx$

(e) $\int x(x^5 + x^{100})\, dx; \quad \int_{0}^{2} x(x^5 + x^{100})\, dx$

(f) $\int (A + Bx + Cx^2)\, dx; \quad \int_{-h}^{h} (A + Bx + Cx^2)\, dx$

4. Evaluate the following integrals, using the rule for sums and a table of derivatives, or the simple table of integrals constructed in Problem 2.

• (a) $\int_{0}^{1} \frac{\pi\, dt}{1 + t^2}$

(b) $\int_{a}^{b} (3 \sin \theta + 2 \cos \theta)\, d\theta$

• (c) $\int (\sec^2 \theta + \csc^2 \theta)\, d\theta$

(d) $\int \sec t(\sec t + \tan t)\, dt$

• (e) $\int u(1 + u^2)\, du$

(f) $\int \frac{2\, dx}{\sqrt{1 - x^2}}$

5. Show by induction [§AI.2] that $\int \left(\sum_{1}^{n} a_j f_j\right) = \sum_{1}^{n} a_j \int f_j$.

6. If P is a polynomial, so is $\int P$. Write the formula for $\int \sum_{0}^{n} a_j t^j\, dt$.

7. (a) Find the derivative of $\sin b\theta$, where b is any constant.

(b) Find $\int \cos(b\theta)\, d\theta$. [It's not quite $\sin b\theta$, but . . . .]

(c) Find $\int \sum_{1}^{n} a_j \cos(j\theta)\, d\theta$, where $a_1, \ldots, a_n$ are constants.

• (d) Find $\int \sum_{0}^{n} a_j \cos(j\theta)\, d\theta$, where $a_0, \ldots, a_n$ are constants.

VOLUMES OF REVOLUTION

8. If the graph of f on $[a,b]$ is revolved about the x axis, the volume enclosed is $\int_a^b \pi f(x)^2\,dx$ [§5.4]. Compute this volume in the following cases. Sketch each volume.

• (a) $f(x) = \sqrt{r^2 - x^2}$ on $[-r,r]$ (ball of radius r)

(b) $f(x) = cx$ on $[0,b]$ (cone)

(c) $f(x) = x^2 + 1$ on $[-b,b]$ (section of a paraboloid)

• (d) $f(x) = 1/x$ on $[a,b]$, $a > 0$ (section of a hyperboloid)

(e) $f(x) = \sqrt{cx}$ on $[a,b]$, $0 \leq a < b$ (section of a parabolic saucer.)

9. A spherical storage tank of radius r contains liquid to a depth h. Find the volume of the liquid. [Hint: If you "freeze" the liquid and roll the tank on its side, you get a volume very similar to the one in Problem 8(a).]

CENTER OF GRAVITY

10. Let A be the area bounded below by the x axis, on the right by the vertical line $\{(x,y) : x = 2\}$, and above by the segment of the parabola $\{(x,y) : y = x^2\}$ from (0,0) to (2,4).

(a) Compute the area of A.

• (b) Compute the "net torque" of A with respect to the y axis [given by $\int x l(x)\,dx$, where $l(x)$ is the length of the vertical cross section of A through the point $(x,0)$; see §5.4.]

(c) Compute the "net torque" of A with respect to the x axis [given by the integral in Problem 21, §5.4.]

• (d) Find the center of gravity of A [see §5.4.]

11. Find the center of gravity of the triangle bounded below by the x axis, above by the graph of $y = mx$ ($m > 0$, $0 \leq x \leq b$), and on the right by the vertical line $\{(x,y) : x = b\}$. [See §5.4.]

12. Find the center of gravity of the area $\{(x,y) : 0 \leq y \leq 1 - x^2\}$. [See §5.4.]

WORK

• **13.** A space ship raises a 2-pound box of rocks from the surface of the moon up to a winch which is 10 feet above the surface, by pulling up a rope that weighs 1/10 pound per foot. How much energy does this operation require? [Hint: Energy required = work done (assuming negligible loss through friction, etc.). When the box is x feet below the winch, the force required to raise box and rope is $\gamma(2 + x/10)$, where γ is the gravity constant for the surface of the moon; $\gamma = 5$ (approximately) if the force is measured in poundals. See §5.4 for the integral for work.]

6.2 SUBSTITUTION

The integral $\int \cos(3x + 1)\, dx$ appears related, somehow, to the known integral $\int \cos u\, du = \sin u + c$. To see the relation, we set $u = 3x + 1$, which converts $\cos(3x + 1)$ into $\cos u$; but what is the relation between dx and du? The Leibniz notation for derivatives suggests the following:

$$u = 3x + 1 \qquad \frac{du}{dx} = 3 \qquad dx = \frac{1}{3}\, du,$$

hence

$$\int \cos(3x + 1)\, dx = \int \cos u \frac{1}{3}\, du = \frac{1}{3} \sin u + c = \frac{1}{3} \sin(3x + 1) + c.$$

Although this juggling with dx and du is not justified by ordinary algebra [since dx and du do not stand for numbers], the final result is correct, as you can see by differentiating $\frac{1}{3} \sin (3x + 1)$.

The juggling is actually justified by the chain rule. A composite function $F(g(x))$ has the derivative $F'(g(x))g'(x)$, so

$$\int F'(g(x))g'(x)\, dx = F(g(x)) + c. \tag{1}$$

On the other hand, if we let f stand for the derivative F', we have

$$\int f(u)\, du = F(u) + c. \tag{2}$$

Thus (1) and (2) combine to give

$$\int f(g(x))g'(x)\, dx = \int f(u)\, du, \qquad \text{where } u = g(x). \tag{3}$$

SUBSTITUTION FORMULA

[In the example above, $\int \cos (3x + 1)3\, dx = \int \cos u\, du = \sin u + c$, where $u = 3x + 1$.] Comparing the equal integrals in (3), you can see the essential point to remember in making substitutions:

Setting $u = g(x)$ entails $du = g'(x)\, dx$.

Fortunately, the Leibniz notation makes this "obvious"; if $u = g(x)$, then $du/dx = g'(x)$, "hence", multiplying through by dx, we find $du = g'(x)\, dx$.

Example 1. Evaluate $\int \cos(x^2)\cdot 2x\,dx$. *Solution.* Set $u = x^2$. Then $du/dx = 2x$, so $du = (2x)\,dx$, and the integral is evaluated as follows:

$$\begin{array}{ccccc}
\int \underbrace{\cos(x^2)}\cdot\underbrace{2x\,dx} & & = & & \sin(x^2) + c \\
u = x^2 \Big\downarrow & \Big\downarrow du = 2x\,dx & & & \Big\uparrow u = x^2 \\
\int \cos(u) & du & = & & \sin\ u + c.
\end{array}$$

In this "flow diagram," start at the upper left-hand corner, follow the arrows down [indicating the substitution of u for x^2], evaluate the resulting integral $\int \cos u\,du$, and then follow the arrow up on the right-hand side [indicating the substitution of x^2 back for u]. The result is the equality in the top line, $\int \cos(x^2)\cdot 2x\,dx = \sin(x^2) + c$.

Example 2. $\int x^2 e^{1+x^3}\,dx = ?$ *Solution.* Let $u = 1 + x^3$; then $du = 3x^2\,dx$, so $x^2\,dx = \frac{1}{3}\,du$, and we get

$$\begin{array}{ccccc}
\int \underbrace{e^{1+x^3}}\ \underbrace{x^2\ dx} & & & = & \frac{1}{3}e^{1+x^3} + c \\
u = 1 + x^3 \Big\downarrow & \Big\downarrow du = 3x^2\,dx & & & \Big\uparrow u = 1 + x^3 \\
\int e^u & \frac{1}{3}du & = \frac{1}{3}\int e^u\,du & = & \frac{1}{3}e^u + c.
\end{array}$$

[Check by differentiating $\frac{1}{3}e^{1+x^3} + c$.]

Example 3. $\int (3x + 1)^{10}\,dx = ?$ *Solution.* You could multiply out $(3x + 1)^{10}$ and integrate the resulting polynomial. However, it is much easier to set $u = 3x + 1$, $du = 3\,dx$; hence

$$\begin{array}{ccccc}
\int \underbrace{(3x + 1)^{10}}\ \underbrace{dx} & & & = & \frac{1}{33}(3x + 1)^{11} + c. \\
u = 3x + 1 \Big\downarrow & \Big\downarrow du = 3\,dx & & & \Big\uparrow u = 3x + 1 \\
\int u^{10} & \frac{1}{3}du = & \frac{1}{3}\int u^{10}\,du & = & \frac{1}{3}\cdot\frac{1}{11}u^{11} + c.
\end{array}$$

[Check by differentiating.]

Example 4. $\int e^{-x^2}\,dx = ?$ *Solution.* Set $u = -x^2$; then $du = -2x\,dx$, and we get

$$\int \underbrace{e^{-x^2}}_{u=-x^2}\ \underbrace{dx}_{du=-2x\,dx} \longrightarrow \int (e^u)\left(\frac{-1}{2x}\,du\right) = ?$$

This motley expression is useless, as it stands. The factor $-1/2x$ is *not* a constant, hence it cannot be "factored" out of the integral. The logical next step is to convert x into u; since $u = -x^2$, set $x = \sqrt{-u}$ (and worry later about the choice of $+\sqrt{-u}$ or $-\sqrt{-u}$). Alas, the resulting integral $\frac{1}{2}\int e^u(-u)^{-1/2}\,du$ is no improvement over the given integral $\int e^{-x^2}\,dx$; the substitution has failed. In fact, as we remarked before, this integral *cannot* be expressed in terms of elementary functions, so no substitution will evaluate it in those terms.

It is difficult, in fact impossible, to give a general rule telling in every case what substitution to make, i.e. how to choose the function u so that the integral is simplified best. Certain obvious composite functions suggest u; for instance, $\cos(x^2)$ suggests $u(x) = x^2$. And you should generally look for some part of the integrand that is the derivative of some other part, as in Example 1, where $2x$ is the derivative of x^2. More generally, look for some part of the integrand that is the derivative of some other part *up to a constant factor*, for constant factors can always be adjusted, as in Examples 2 and 3.

Some more examples:

Example 5. $\int \sin x \cos x\,dx = ?$ *Solution.* Set $u = \sin x$,

$$\begin{array}{ccc} \int \underbrace{\sin x}_{u=\sin x}\ \underbrace{\cos x\,dx}_{du=\cos x\,dx} & = & \frac{1}{2}\sin^2 x + c \\ \downarrow & & \uparrow\ u=\sin x \\ \int u\ du & = & \frac{1}{2}u^2 + c. \end{array}$$

Example 6.

$$\int \frac{e^{\sqrt{x}}}{\sqrt{x}}\,dx \ldots\ldots\ldots\ldots = 2e^{\sqrt{x}} + c$$

$$u = x^{1/2}\ \Big\downarrow \qquad \Big\downarrow \begin{aligned} du &= \tfrac{1}{2}x^{-1/2}\,dx \\ dx &= 2u\,du \end{aligned} \qquad \Big\uparrow\ u = \sqrt{x}$$

$$\int \frac{e^u}{u}\,2u\,du = 2\int e^u\,du = 2e^u + c.$$

Example 7. $\displaystyle\int_1^2 e^{3x}\,dx = ?$ *Solution.* First, evaluate the indefinite integral by the substitution $u = 3x$,

$$\int \underset{\sim}{e^{3x}}\ \ \underset{\sim}{dx} \ldots\ldots\ldots\ldots = \tfrac{1}{3}e^{3x} + c$$

$$u = 3x\ \Big\downarrow \qquad \Big\downarrow\ du = 3\,dx \qquad \Big\uparrow\ u = 3x$$

$$\int (e^u)\left(\tfrac{1}{3}du\right) = \tfrac{1}{3}\int e^u\,du = \tfrac{1}{3}e^u + c.$$

Hence

$$\int_1^2 e^{3x}\,dx = \tfrac{1}{3}e^{3x}\Big|_1^2 = \tfrac{1}{3}(e^6 - e^3) = \tfrac{1}{3}e^3(e^3 - 1).$$

Example 8.

$$\int \frac{x+1}{x^2+2x}\,dx \quad = \quad \frac{1}{2}\log|x^2 + 2x| + c$$

$$u = x^2 + 2x\ \Big\downarrow\ du = (2x+2)\,dx \qquad \Big\uparrow\ u = x^2 + 2x$$

$$\int \frac{1}{u}\frac{1}{2}\,du \quad = \quad \frac{1}{2}\log|u| + c.$$

Notice that the integrand $(x+1)/(x^2+2x)$ is not defined on the whole line, since $x^2 + 2x = 0$ when $x = 0$ or $x = -2$. In each of the three open intervals $(-\infty,-2)$, $(-2,0)$, $(0,+\infty)$, we can write

$$\int \frac{x+1}{x^2+2x}\,dx = \frac{1}{2}\log|x^2+2x| + c,$$

but the constant c may be different in each of these intervals.

Example 9. $\displaystyle\int_1^6 \frac{x+1}{x^2+2x}\,dx = ?$ *Solution.* The interval of integration [1,6] lies in the interval $(0,+\infty)$, and Example 8 shows that on $(0,+\infty)$ we have

$$\int \frac{x+1}{x^2+2x}\,dx = \frac{1}{2}\log|x^2+2x| + c.$$

Hence

$$\int_1^6 \frac{x+1}{x^2+2x}\,dx = \frac{1}{2}\log|x^2+2x|\Big|_1^6$$

$$= \frac{1}{2}\log(48) - \frac{1}{2}\log(3)$$

$$= 2\log 2 = \text{(approx.) } 1.4.$$

Example 10. $\displaystyle\int_{-1}^1 \frac{x+1}{x^2+2x}\,dx = ?$ *Solution.* Proceeding as in Example 9, we obtain

$$\int_{-1}^1 \frac{x+1}{x^2+2x}\,dx = \frac{1}{2}\log|x^2+2x|\Big|_{-1}^1 = \frac{1}{2}\log 3 - \frac{1}{2}\log 1 = \frac{1}{2}\log 3.$$

In this case, however, these calculations are not justified, and the "equation"

$$\int_{-1}^1 \frac{x+1}{x^2+2x}\,dx = \frac{1}{2}\log 3$$

that we have obtained does not really mean anything. For, the function $f(x) = (x+1)/(x^2+2x)$ has a vertical asymptote at $x = 0$; thus f is not bounded on the interval of integration $[-1,1]$, and therefore $\int_{-1}^1 f(x)\,dx$ must be evaluated as an improper integral [§5.5]. Doing this, you will find that

$$\int_{-1}^0 \frac{x+1}{x^2+2x}\,dx \quad \text{and} \quad \int_0^1 \frac{x+1}{x^2+2x}\,dx$$

do not exist as improper integrals. Thus the given integral has no meaning as it stands. [However, if it arose in some natural context, that context might suggest how to make sense of the improper integral.]

PROBLEMS

Each of the first six problems concerns a particular pattern that is easily integrated by substitution, and should (with practice) be recognized at sight. Problem 15 discusses the important normal distribution in statistics.

1 • (a) $\int \cos 2x\, dx$ (d) $\int e^{-2t}\, dt$

(b) $\int \sin 3x\, dx$ • (e) $\int \sec^2 10\theta\, d\theta$

• (c) $\int e^{5x}\, dx$ (f) $\int F'(kx)\, dx$

---

2 • (a) $\int \cos(3 + \theta)\, d\theta$ [Set $u = 3 + \theta$.]

(b) $\int \sin(2 + 5x)\, dx$ • (g) $\int \sqrt{1 + 3x}\, dx$

• (c) $\int \frac{dx}{3 + 7x}$ (h) $\int e^{at+b}\, dt, \quad a \neq 0$

(d) $\int \frac{dx}{(3x - 4)^2}$ (i) $\int (ax + b)^n\, dx, \quad a \neq 0$

• (e) $\int \frac{dt}{\sqrt{a + bt}}, \quad b \neq 0$ • (j) $\int \sin(a\theta + b)\, d\theta$

(f) $\int e^{5t+1}\, dx$ (k) $\int \cos(a\theta + b)\, d\theta$

---

3 • (a) $\int \theta \cos \theta^2\, d\theta$ (d) $\int t^n \exp(at^{n+1})\, dt$

(b) $\int x^2 \sin(2x^3)\, dx$ • (e) $\int xe^{-x^2}\, dx$

• (c) $\int x^3 e^{3x^4}\, dx$ (f) $\int t^n \cos(at^{n+1})\, dt$

---

4 • (a) $\int \theta \sin(\theta^2 + 1)\, d\theta$ • (e) $\int \frac{x + 1}{x^2 + 2x}\, dx$

(b) $\int \theta^2 \cos(2\theta^3 + 3)\, d\theta$ (f) $\int x\sqrt{a^2 + x^2}\, dx$

• (c) $\int (x^2 + 1)e^{x^3+3x}\, dx$ • (g) $\int x(a^2 - x^2)^n\, dx, \quad n \neq -1$

(d) $\int \sqrt{x^2 + 2x}(x + 1)\, dx$ (h) $\int x \sin(ax^2 + b)\, dx$

5 • (a) $\int (\sin\theta)^2 \cos\theta \, d\theta \quad [u = \sin\theta]$

(b) $\int \sin\theta \cos\theta \, d\theta$

• (c) $\int \sin^n \theta \cos\theta \, d\theta, \quad n \neq -1$

(d) $\int \sin^n ax \cos ax \, dx, \quad n \neq -1$

• (e) $\int \cos^n ax \sin ax \, dx, \quad n \neq -1$

(f) $\int \sin ax \, e^{\cos ax} \, dx$

• (g) $\int \cos^3 \theta \, d\theta \quad [\cos^2\theta = 1 - \sin^2\theta]$

(h) $\int \sin^3 2\theta \, d\theta \quad [\sin^2 2\theta = 1 - \cos^2 2\theta]$

---

6 • (a) $\int \frac{t\,dt}{a^2 + t^2}$

(b) $\int \frac{z^2\,dz}{a + bz^3}$

(c) $\int \frac{dx}{x \log x} \quad \left[= \int \frac{1/x}{\log x}\,dx\right]$

• (d) $\int \tan\theta \, d\theta \quad \left[= \int \frac{\sin\theta}{\cos\theta}\,d\theta\right]$

(e) $\int \frac{\sin\theta + \cos\theta}{\sin\theta - \cos\theta}\,d\theta$

(f) $\int \frac{\sin\theta - \cos\theta}{\sin\theta + \cos\theta}\,d\theta$

• (g) $\int \sec\theta \, d\theta \quad \left[= \int \frac{\sec\theta(\sec\theta + \tan\theta)}{\sec\theta + \tan\theta}\,d\theta\right]$

---

7. (a) $\int \frac{dt}{a^2 + t^2} \quad [\text{Set } t = au]$

• (b) $\int \frac{dt}{a^2 + (t - b)^2} \quad [t - b = au]$

(c) $\int \frac{dt}{t^2 - 2t + 2} \quad [t^2 - 2t + 2 = 1 + (t - 1)^2]$

(d) $\int \frac{dx}{3x^2 - 6x + 5}$ $[3x^2 - 6x + 5 = 2 + 3(x-1)^2 = 2(1 + \frac{3}{2}(x-1)^2)]$

• (e) $\int \frac{dx}{Ax^2 + Bx + C}$, where $B^2 < 4AC$ [Complete the square, writing $Ax^2 + Bx + C = A(r + (x-p)^2)$; see the appendix to this section, page 327.]

8. [*Some special trigonometric integrals*]

(a) $\int 2\cos^2\theta\, d\theta$ $[1 + \cos 2\theta = (\cos^2\theta + \sin^2\theta) + (\cos^2\theta - \sin^2\theta) = 2\cos^2\theta]$

• (b) $\int \cos^2 3\theta\, d\theta$

(c) $\int \sin^2\theta\, d\theta$ $[1 - \cos 2\theta = \cdots]$

• (d) $\int (2\sin^2\theta + 3\cos^2 4\theta)\, d\theta$

• (e) $\int \cos^4\theta\, d\theta$ $\left[= \int (\cos^2\theta)^2\, d\theta\right]$

(f) $\int \sin^4\theta\, d\theta$

9 • (a) $\int x\sqrt{3x+5}\, dx$, $[u = 3x + 5]$

(b) $\int x\sqrt{Ax+B}\, dx$

• (c) $\int t(t-1)^{1/5}\, dt$

(d) $\int x(2x-t)^{2/3}\, dx$

• (e) $\int x^2(2x+3)^\pi\, dx$

10. Each of the following integrals is easy to evaluate for some particular value of n. [For example, $\int x^n e^{x^3}\, dx$ is easy to evaluate by substituting $u = x^3$, if $n = 2$; $\int x^2 e^{x^3}\, dx = \int e^u \cdot \frac{1}{3}\, du = \frac{1}{3}e^u + c = \frac{1}{3}e^{x^3} + c$.] Find an "easy" value of n, and integrate.

• (a) $\int x^n \cos(x^2)\, dx$

(b) $\int x^2 \cos(x^n)\, dx$

• (c) $\int x^n \log x\, dx$

(d) $\int x^n \sin(x^4)\, dx$

• (e) $\int \sqrt{x}\sin(x^n)\, dx$

(f) $\int x^n e^{\sqrt{x}}\, dx$

11. Some of the calculations below have errors. Decide which are wrong, and correct them if possible.

(a) $\int \cos 3x\, dx = \int u\, du = \sin u + c = \sin 3x + c, \quad [u = \cos 3x]$

(b) $\int \frac{dx}{x^2 - 1} = \int \left(\frac{1}{x-1} - \frac{1}{x+1}\right) dx = \log|x-1| - \log|x+1| + c$

(c) $\int \frac{dx}{4x^2 + 1} = \int \frac{2\, du}{u^2 + 1} = 2 \arctan u = 2 \arctan 2x, \quad [u = 2x]$

(d) $\int \frac{dx}{4x^2 + 1} = \int \frac{du}{u} = \log u = \log(4x^2 + 1), \quad [u = 4x^2 + 1]$

12. Evaluate the following definite integrals; some of the corresponding indefinite integrals are given [with hints] in the problems above.

• (a) $\int_0^b \frac{dt}{a^2 + t^2}$

(b) $\int_0^{\pi/2} \sin\theta \cos\theta\, d\theta$

• (c) $\int_0^{\pi/2} \sin\theta \cos^2\theta\, d\theta$

(d) $\int_0^b \frac{dt}{\sqrt{a + t}}, \quad (a > 0, b > -a)$

• (e) $\int_0^b \frac{dx}{x^2 + x + 1}$

13. Evaluate the following improper integrals, provided they exist.

IMPROPER INTEGRALS

(a) $\int_0^\infty \frac{dt}{a^2 + t^2}$

• (b) $\int_{-\infty}^\infty \frac{dt}{a^2 + t^2}$

• (c) $\int_a^{a+1} \frac{dt}{\sqrt{t - a}}$

(d) $\int_0^\infty xe^{-x^2}\, dx$

• (e) $\int_{-\infty}^\infty xe^{-x^2}\, dx$ [Explain the answer geometrically, in terms of areas.]

14. Find $\lim_{a \to +\infty} \int_0^b \frac{dt}{a^2 + t^2}$. Does $\lim_{a \to 0} \int_0^b \frac{dt}{a^2 + t^2}$ exist? Try to explain these results by showing what happens to the graph of $1/(a^2 + t^2)$ as $a \to +\infty$ or as $a \to 0$.

15. This problem concerns the normal probability density function of mean μ and standard error σ,

"NORMAL" PROBABILITY CURVE

$$f_{\mu,\sigma}(x) = \frac{1}{\sqrt{2\pi}\,\sigma} \exp\left(-\frac{(x-\mu)^2}{2\sigma^2}\right)$$

[§5.4]. For measurements subject to chance variations according to this particular law, the probability of a measurement between any two given values a and b is $\int_a^b f_{\mu,\sigma}(x)\,dx$. This integral cannot be evaluated as an elementary function. It can be evaluated, however, by referring to tables of values for the "error function" $\operatorname{erf}(x) = \frac{2}{\sqrt{\pi}} \int_0^x e^{-t^2}\,dt$. You can prove that erf is an odd function; further, $\lim_{x\to+\infty} \operatorname{erf}(x) = 1$, but you will have to accept this on our authority until §9.2, Problem 16.

(a) Show that $\int_a^b f_{\mu,\sigma}(x)\,dx = \frac{1}{2}\operatorname{erf}\left(\frac{b-\mu}{\sigma\sqrt{2}}\right) - \frac{1}{2}\operatorname{erf}\left(\frac{a-\mu}{\sigma\sqrt{2}}\right)$. [Hint 1: Set $u = \frac{x-\mu}{\sigma\sqrt{2}}$. Hint 2: $\frac{2}{\sqrt{\pi}} \int e^{-u^2}\,du = \operatorname{erf}(u) + c$. Why?]

(b) Show that $\lim_{b\to+\infty} \int_{-b}^{b} f_{\mu,\sigma}(x)\,dx = 1$.

Remarks. Part (a) shows that probabilities based on any of the above density functions $f_{\mu,\sigma}$ can all be computed from the tables for just one function, erf. [Statisticians often use tables for $\int_0^x f_{0,1}$ rather than $\operatorname{erf}(x)$, but the method is essentially the same.] Part (b) implies the following reassuring fact: The probability that a measurement lies between $-\infty$ and $+\infty$ is 1. In other words, 100% of all measurements lie between $-\infty$ and $+\infty$, as we would surely expect. The constant $1/(\sqrt{2\pi}\,\sigma)$ is in the definition of $f_{\mu,\sigma}$ for precisely this reason; any other constant would make the probability of a measurement between $-\infty$ and $+\infty$ either less than 1 or greater than 1.

• **16.** Find the length of the curve $\{(x,y): (y+1)^2 = 4x^3\}$ from $(0,-1)$ to $(1,1)$. [See §5.4.]

ARC LENGTH

17. The force of gravity on an object x miles above the Earth is $Am/(x+R)^2$, where m is the mass of the object, R is the radius of the Earth, and A is a constant. Find the work done by gravity when an object falls to Earth from 6 miles high. [See §5.4.]

WORK

• **18.** The portion of the parabola $y = cx^2$ between $x = 0$ and $x = b$ [where $0 < b$] is revolved about the y axis. Find the area of the resulting surface, given by $2\pi \int x\sqrt{1 + f'(x)^2}\, dx$. **SURFACE AREA**

19. The curve in Problem 16 is revolved about the y axis. Find the area of the resulting surface, given by $2\pi \int x\sqrt{1 + f'(x)^2}\, dx$.

20. Find the area of the surface generated by revolving about the x axis the segment of the curve $\{(x,y) : y = x^3\}$ between $x = 0$ and $x = 1$.

APPENDIX: COMPLETING THE SQUARE

It is a simple but useful fact that *every quadratic* $ax^2 + bx + c$ *can be written in the form* $a[(x - p)^2 + r]$. For example,

$$x^2 + 2x + 2 = (x + 1)^2 + 1,$$

$$2x^2 - 3x - 2 = 2[(x - \tfrac{3}{4})^2 - \tfrac{25}{16}].$$

HOW TO COMPLETE THE SQUARE

It is easy to obtain these decompositions, simply by multiplying out $(x - p)^2$ and comparing terms. For example, to rewrite $3x^2 - 5x + 1$ in this form, we set

$$3x^2 - 5x + 3 = a[(x - p)^2 + r] = a(x^2 - 2px + p^2 + r)$$

$$= ax^2 - 2apx + ap^2 + ar.$$

Comparing the x^2 terms [$3x^2$ on the left, and ax^2 on the right] yields $a = 3$; comparing the x terms shows that $-5 = -2ap = -6p$, so $p = \frac{5}{6}$; finally, comparing constant terms, we have $3 = a(p^2 + r) = 3 \cdot \frac{25}{36} + 3r$, so $r = 1 - \frac{25}{36} = \frac{11}{36}$. Thus

$$3x^2 - 5x + 3 = 3[(x - \tfrac{5}{6})^2 + \tfrac{11}{36}].$$

Another easy method is to factor out the coefficient of x^2, and note that after this step, p must be $-\frac{1}{2}$ times the coefficient of x; finally, r is added to adjust the constant terms. For example, to complete the square for $3x^2 - 5x + 3$ we would first write

$$3x^2 - 5x + 3 = 3(x^2 - \tfrac{5}{3}x + 1).$$

Clearly, $x^2 - \frac{5}{3}x$ agrees with the first two terms of $(x - \frac{5}{6})^2$, so we continue as follows:

$$3x^2 - 5x + 3 = 3(x^2 - \tfrac{5}{3}x + 1) = 3[(x - \tfrac{5}{6})^2 + 1 - (\tfrac{5}{6})^2];$$

the term $1 - (\frac{5}{6})^2 = \frac{11}{36}$ must be added to change the constant term $(\frac{5}{6})^2$ coming from $(x - \frac{5}{6})^2$ into the given constant term, 1.

This process of rewriting quadratics is called *completing the square.* It is useful for integration [Problem 7(e) above]. It also yields the familiar quadratic formula, as follows:

$$ax^2 + bx + c = 0 \iff x^2 + \frac{b}{a}x + \frac{c}{a} = 0$$

$$\iff \left(x + \frac{b}{2a}\right)^2 + \frac{c}{a} - \frac{b^2}{4a^2} = 0$$

$$\iff \left(x + \frac{b}{2a}\right)^2 = \frac{b^2 - 4ac}{4a^2}$$

$$\iff x + \frac{b}{2a} = \pm\frac{\sqrt{b^2 - 4ac}}{2a} \qquad [\text{if } b^2 - 4ac \geq 0]$$

QUADRATIC FORMULA

$$\iff x = \frac{-b \pm \sqrt{b^2 - 4ac}}{2a}.$$

Completing the square can also be used to find the center of any circle given by an equation of the form

CENTER OF A CIRCLE

$$x^2 + y^2 + Ax + By = C.$$

This is equivalent to

$$\left(x + \frac{A}{2}\right)^2 + \left(y + \frac{B}{2}\right)^2 = C + \frac{A^2}{4} + \frac{B^2}{4},$$

which says that the distance from (x,y) to $(-A/2, -B/2)$ is $\sqrt{C + \frac{A^2}{4} + \frac{B^2}{4}}$ [if the number in the radical is ≥ 0]. Thus (x,y) lies on a circle of radius $r = \sqrt{C + \frac{A^2}{4} + \frac{B^2}{4}}$ with center $(-A/2, -B/2)$.

PROBLEMS

Each of the following equations defines a *circle,* a *point* (a "circle of radius 0"), or an *empty set.* Determine which, and find the center and radius of each circle.

• **1.** $x^2 + y^2 - 2x - 4y = 0$

• **2.** $x^2 + y^2 + 1 = 0$

• **3.** $x^2 + y^2 + 2x + 4y + 5 = 0$

4. $x^2 + y^2 + x + y + 6 = 0$

5. $2x^2 + 2y^2 + 4x = 0$

6.3 INTEGRATION BY PARTS

The product formula for derivatives, $(FG)' = F'G + FG'$, says that FG is an indefinite integral of $F'G + FG'$:

$$FG + c = \int (FG)' \qquad \text{[definition of indefinite integral]}$$

$$= \int (F'G + FG') = \int F'G + \int FG'.$$

Transposing terms, we obtain the formula for *integration by parts,*

$$\int FG' = FG - \int F'G. \tag{1}$$

INTEGRATION BY PARTS FORMULA

[No "constant of integration" is required in (1), since the indefinite integrals in (1) already imply a constant of integration on each side.]

The formula is usually written in Leibniz notation, as follows:

$$\int \boldsymbol{u}\,\boldsymbol{dv} = \boldsymbol{uv} - \int \boldsymbol{v}\,\boldsymbol{du}, \tag{2}$$

IN LEIBNIZ NOTATION

where u, du, v, and dv are treated just as in substitution; setting $u = F(x)$, $du = F'(x)dx$, $v = G(x)$, $dv = G'(x)dx$ shows that (2) is the same as (1).

Integration by parts converts the problem of evaluating $\int u\,dv$ into the similar problem of evaluating $\int v\,du$; we naturally hope that the new integral will be easier to evaluate than the given one.

Example 1. $\int \log x\,dx = ?$ *Solution.* Set $u = \log x$ and $dv = dx$. Then $du = \log'(x)\,dx = \frac{1}{x}\,dx$, and we can take $v = x$, since that entails $dv = dx$. Hence, by (2),

$$\int \;\; u \;\; dv = \;\; u \cdot v \; - \int v\,du$$

$$\uparrow \; \uparrow \qquad \downarrow \; \downarrow \qquad \downarrow \; \downarrow$$

$$\longrightarrow \quad \int \underbrace{\log x}\,\underbrace{dx} = [\log x][x] - \int x\,\frac{1}{x}\,dx$$

$$= x\log x - \int 1\cdot dx$$

$$= x\log x - x + c.$$

Remark. Instead of $v = x$, we could take $v = x + C$, since that entails $dv = dx$. But this is not useful, since the constant C cancels out:

$$\int u \quad dv = \qquad u \cdot v \qquad - \int v \, du$$

$$\int \log x \, dx = [\log x][x + C] - \int (x + C) \frac{1}{x} \, dx$$

$$= x \log x + C \log x - \int 1 \cdot dx - C \int \frac{1}{x} \, dx$$

$$= x \log x + C \log x - [x + c_1] - C[\log x + c_2]$$

$$= x \log x - x - c_1 - Cc_2 = x \log x - x + c_3.$$

Example 2. $\int x \sin x \, dx = ?$ *Solution.* Set $x = u$ and $\sin x \, dx = dv$. Then $dx = du$, and we can take $v = -\cos x$, since that entails $dv = \sin x \, dx$. Hence, by (2),

$$\int u \quad dv \quad = u \cdot v \quad - \int \quad v \quad du$$

$$\uparrow \quad \uparrow \quad \downarrow \quad \downarrow \qquad \downarrow \quad \downarrow$$

$$\int x \, \overbrace{\sin x \, dx} = x[-\cos x] - \int [-\cos x] \, dx$$

$$= -x \cos x + \int \cos x \, dx$$

$$= -x \cos x + \sin x + c.$$

From these two examples, you can see that integration by parts really converts the problem $\int u \, dv$ into *two* problems: Having chosen $dv = G'(x) \, dx$, first integrate G' to obtain $v = G(x)$; then apply (2), and complete the job by integrating $\int v \, du$. In Example 1, having set $dv = dx = 1 \cdot dx$, you integrate the constant 1 to get $v = x$. In Example 2, having set $dv = \sin x \, dx$, you integrate $\sin x$ to get $v = -\cos x$. [As remarked above, we do not write $v = -\cos x + C$; *any* v such that $dv = -\cos x \, dx$ will do, so there is no reason to include a constant of integration.]

The motives behind the choice of u and dv are clear. First, for $dv = G'(x)dx$ you must choose something that can be integrated to find $v = G(x)$; and for u, generally choose something that simplifies when it is differentiated.

Example 3. $\int x \log x \, dx = ?$ *Solution.* There are three conceivable choices of u and dv:

(a) $u = x$ and $dv = \log x \, dx$; then $du = dx$, $v = x \log x - x$ (from Example 1), and the given integral $\int u \, dv$ is "reduced" to the evaluation of $\int v \, du = \int (x \log x - x) \, dx$.

(b) $u = \log x$ and $dv = x \, dx$; then $du = (1/x) dx$, $v = x^2/2$, and the new integral that we have to evaluate is $\int v \, du = \int \frac{1}{2} x \, dx$.

(c) $u = x \log x$ and $dv = dx$; then $du = (1 + \log x) \, dx$, $v = x$, and the new integral is $\int v \, du = \int (x + x \log x) \, dx$.

Of these, the second choice (b) simplifies the integral best. Continuing (b), we have

$$\int x \log x \, dx = \frac{1}{2} x^2 \log x - \int \frac{1}{2} x \, dx = \frac{1}{2} x^2 \log x - \frac{1}{4} x^2 + c.$$

Example 4. For $\int x e^x \, dx$, make three choices of u and dv. If possible, find du and v in each case. Complete the integration using one of the three choices. [Solution below.]

Sometimes one integration by parts does not completely solve the problem, and we have to repeat the process.

Example 5. $\int x^2 e^x \, dx = ?$ *Solution.* Since x^2 simplifies to $2x$ when differentiated, and e^x is easy to integrate, we set $u = x^2$, $dv = e^x \, dx$:

$$\begin{array}{ccccccc} \int u & dv & = u & v & - \int v & du & \\ \uparrow & \uparrow & \downarrow & \downarrow & \downarrow & \downarrow & \\ \int x^2 & \underbrace{e^x \, dx} & = x^2 & e^x & - \int e^x & 2x \, dx & = x^2 e^x - 2 \int x e^x \, dx. \end{array} \tag{3}$$

To complete the evaluation, integrate $\int x e^x dx$ by parts:

$$\begin{array}{ccccccc} \int u & dv & = u & v & - \int u & du & \\ \uparrow & \uparrow & \downarrow & \downarrow & \downarrow & \downarrow & \\ \int x & \underbrace{e^x \, dx} & = x & e^x & - \int e^x & dx & = x e^x - e^x + c. \end{array}$$

Hence from (3), $\int x^2 e^x \, dx = x^2 e^x - 2[x e^x - e^x + c]$.

From Example 5, it is clear that for any integer $n > 0$, the integral $\int x^n e^{ax} \, dx$ can be evaluated by integrating by parts n times. The essence of the calculation is given conveniently in the following *reduction formula*:

REDUCTION FORMULA

$$\int x^n e^{ax}\,dx = \frac{1}{a}x^n e^{ax} - \frac{n}{a}\int x^{n-1}e^{ax}\,dx. \tag{4}$$

This formula is deduced directly from the basic integration by parts formula (2), with $u = x^n$ and $dv = e^{ax}\,dx$. It "reduces" the problem of integrating $\int x^n e^{ax}\,dx$ to that of integrating $\int x^{n-1}e^{ax}\,dx$, which is one step simpler.

Example 6. $\displaystyle\int x^3e^{2x}\,dx = \frac{1}{2}x^3e^{2x} - \frac{3}{2}\int x^2e^{2x}\,dx$

[by (4), with $n = 3$, $a = 2$]

$$= \frac{1}{2}x^3e^{2x} - \frac{3}{2}\left(\frac{1}{2}x^2e^{2x} - \frac{2}{2}\int xe^{2x}\,dx\right)$$

[by (4), with $n = 2$]

$$= \frac{1}{2}x^3e^{2x} - \frac{3}{4}x^2e^{2x} + \frac{3}{2}\left(\frac{1}{2}xe^{2x} - \frac{1}{2}\int e^{2x}\,dx\right)$$

[by (4), with $n = 1$]

$$= \frac{1}{2}x^3e^{2x} - \frac{3}{4}x^2e^{2x} + \frac{3}{4}xe^{2x} - \frac{3}{8}e^{2x} + c.$$

In the next example, integration by parts seems to lead to a vicious circle; but a closer look leads us out of it.

Example 7.

$$\int (\sin x)^n\,dx = \int (\sin x)^{n-1}\sin x\,dx \qquad [\text{choose } u = (\sin x)^{n-1},\ dv = \sin x\,dx]$$

$$= -(\sin x)^{n-1}\cos x + (n-1)\int (\sin x)^{n-2}(\cos x)^2\,dx$$

$$= -(\sin x)^{n-1}\cos x + (n-1)\int (\sin x)^{n-2}(1 - \sin^2 x)\,dx$$

$$= -(\sin x)^{n-1}\cos x + (n-1)\int (\sin x)^{n-2}\,dx - (n-1)\int (\sin x)^n\,dx.$$

We seem to have gotten nowhere; but actually we can view this as *an equation to be solved for* $\int(\sin x)^n\,dx$. Transposing $-(n-1)\int(\sin x)^n\,dx$ from the right to the left, we find

$$(1 + n - 1)\int (\sin x)^n\,dx = -(\sin x)^{n-1}\cos x + (n-1)\int (\sin x)^{n-2}\,dx,$$

or

$$\int (\sin x)^n\,dx = -\frac{1}{n}(\sin x)^{n-1}\cos x + \left(1-\frac{1}{n}\right)\int (\sin x)^{n-2}\,dx. \quad (5)$$

ANOTHER REDUCTION FORMULA

Repeated applications of this formula reduce the integral of any positive integer power of $\sin x$ to either $\int dx$ (when n is even) or $\int \sin x\,dx$ (when n is odd). In either case, the problem of evaluating $\int (\sin x)^n\,dx$ is solved by the reduction formula (5).

In Example 7 we obtained an equation that could be solved algebraically for the original integral. The same method, combined with repeated integration by parts, applies to certain products of sines, cosines, and exponentials.

Example 8. $\int e^{bx}\cos ax\,dx = ?$ *Solution.* Let $u = \cos ax$, and $dv = e^{bx}dx$, so

$$du = -a\sin ax\,dx \qquad \text{and} \qquad v = \frac{1}{b}e^{bx}.$$

Thus

$$\int e^{bx}\cos ax\,dx = \frac{1}{b}e^{bx}\cos ax + \frac{a}{b}\int e^{bx}\sin ax\,dx. \quad (6)$$

We cannot yet solve for $\int e^{bx}\cos ax$, but integrating by parts once more will suffice. In $\int e^{bx}\sin ax\,dx$, set $u = \sin ax$, $dv = e^{bx}dx$; then $du = a\cos ax\,dx$, $v = \frac{1}{b}e^{bx}$, and therefore

$$\int e^{bx}\sin ax\,dx = \frac{1}{b}e^{bx}\sin ax - \frac{a}{b}\int e^{bx}\cos ax\,dx. \quad (7)$$

Setting (7) in (6) yields

$$\int e^{bx}\cos ax\,dx = \frac{1}{b}e^{bx}\cos ax + \frac{a}{b^2}e^{bx}\sin ax - \frac{a^2}{b^2}\int e^{bx}\cos ax\,dx,$$

$$\frac{a^2+b^2}{b^2}\int e^{bx}\cos ax\,dx = \frac{1}{b}e^{bx}\cos ax + \frac{a}{b^2}e^{bx}\sin ax + c, \quad (8)$$

$$\int e^{bx}\cos ax\,dx = \frac{be^{bx}\cos ax + ae^{bx}\sin ax}{a^2+b^2} + c_1.$$

Note that a constant of integration c appears explicitly in line (8); in the previous line it was implicit in the indefinite integral $\int e^{bx}\cos ax\,dx$.

Naturally, various methods can be combined.

Example 9. $\int e^{\sqrt{x}}\,dx = ?$ *Solution.* Begin with a substitution: $w = \sqrt{x}$, $dw = \frac{1}{2}x^{-1/2}\,dx$, or $dx = 2\sqrt{x}\,dw = 2w\,dw$. Thus

$$\int e^{\sqrt{x}}\,dx = \int e^w 2w\,dw = 2\left[we^w - \int e^w\,dw\right] \qquad [\text{by (4) with } n = 1]$$

$$= 2we^w - 2e^w + c = 2(\sqrt{x}-1)e^{\sqrt{x}} + c.$$

Solution to Example 4.
(a) $u = xe^x$, $dv = dx$; $du = (xe^x + e^x)\,dx$, $v = x$.
(b) $u = x$, $dv = e^x\,dx$; $du = dx$, $v = e^x$.
(c) $u = e^x$, $dv = x\,dx$; $du = e^x\,dx$, $v = x^2/2$.
Continuing (b), $\int xe^x\,dx = xe^x - \int e^x\,dx = xe^x - e^x + c$.
Check: $(xe^x - e^x)' = e^x + xe^x - e^x = xe^x$.

PROBLEMS

The reduction formula in Problem 13 is used in the next section.

1. Evaluate the following integrals, using the reduction formulas (4) and (5).

• (a) $\int t^2e^{-t}\,dt$

• (c) $\int \sin^2 x\,dx$

(b) $\int_0^b t^2e^{-t}\,dt$

(d) $\int \sin^4 x\,dx$

• (e) $\int \sin^3 ax\,dx$ [Hint: Substitute $u = ax$; a is a constant.]

2. Evaluate the following [where a and b are not 0].

• (a) $\int x \cos ax\,dx$

• (d) $\int x^2 \cos ax\,dx$

(b) $\int \theta \sin b\theta\,d\theta$

(e) $\int t^2 \sin^2 t\,dt$

• (c) $\int \theta \sec^2 \theta\,d\theta$

(f) $\int \theta \tan^2 \theta\,d\theta$, [$\tan^2 = \sec^2 - 1$]

3. Evaluate the following.

• (a) $\int \arctan x\,dx$ $\left[\text{In} \int \frac{x}{1+x^2}\,dx \quad \text{set } w = 1 + x^2.\right]$

• (b) $\int \arcsin x\,dx$

(c) $\int \arccos x\,dx$

• (d) $\int \operatorname{arcsec} x\,dx$ $\left[\text{In} \int \frac{dx}{\sqrt{x^2-1}}\right.$, set $x = \sec\theta$, $dx = \sec\theta\tan\theta\,d\theta$.

For $\left.\int \sec\theta\,d\theta, \text{ see problems in §6.2.}\right]$

• (e) $\int x \arctan x \, dx$

(f) $\int x^2 \log x \, dx$

(g) $\int (\log x)^2 \, dx$

• (h) $\int x(\log x)^2 \, dx$

4. Evaluate:

• (a) $\int x^3 e^{-x^2} \, dx$ • (b) $\int x^3 \arctan(x^2) \, dx$ (c) $\int x^3 \cos(x^2) \, dx$

5. Evaluate $\int \cos(\log x) \, dx$ in two different ways:
(a) by a double integration by parts
(b) by the substitution $w = \log x$

6. (a) Evaluate $\int x \sin ax \, dx$.
(b) Evaluate $\int x^2 \sin ax \, dx$.
• (c) Obtain reduction formulas for $\int x^n \sin ax \, dx$ and $\int x^n \cos ax \, dx$.

7. Evaluate $\int (\cos ax)(\cos bx) \, dx$ by each of the following methods, and reconcile the results.
(a) Note that $2(\cos ax)(\cos bx) = \cos(ax + bx) + \cos(ax - bx)$.
(b) Integrate by parts as in Example 8.

8. Evaluate $\int (\cos ax)(\sin bx) \, dx$ in two different ways:
(a) by using an appropriate trigonometric formula, and
(b) by integrating by parts. [See Problem 7.]

9. (a) Evaluate $\int x^2 (\log x)^2 \, dx$.
• (b) Obtain a reduction formula relating $\int x^m (\log x)^n \, dx$ to

$$\int x^m (\log x)^{n-1} \, dx.$$

REDUCTION FORMULAS

10 •(a) Obtain a reduction formula relating $\int (\cos ax)^m \, dx$ to

$$\int (\cos ax)^{m-2} \, dx. \qquad \text{[See Example 7.]}$$

(b) Use the formula obtained in (a) to evaluate $\int (\cos 2x)^2 \, dx$.
(c) Evaluate $\int (\cos 3\theta)^3 \, d\theta$.

11. (a) Obtain the reduction formula

$$\int (\sec x)^n \, dx = \frac{(\sec x)^{n-2} \tan x}{n-1} + \frac{n-2}{n-1} \int (\sec x)^{n-1} \, dx.$$

(b) Evaluate $\int (\sec x)^4 \, dx$.
(c) Evaluate $\int (\sec 2x)^4 \, dx$. [Set $u = 2x$.]

12 • (a) Obtain a reduction formula for $\int x^n e^{-ax^2}\,dx$.

(b) Show that when n is odd, the reduction formula provides a complete evaluation; and if n is even, it reduces the integral to $\int e^{-ax^2}\,dx$.

(c) If $\operatorname{erf}(x) = \dfrac{2}{\sqrt{\pi}} \displaystyle\int_0^x e^{-t^2}\,dt$, show that

$$\int e^{-ax^2}\,dx = \frac{1}{2}\sqrt{\frac{\pi}{a}}\operatorname{erf}(\sqrt{a}\,x) + c.$$

[Set $t^2 = ax^2$.]

13. Obtain the reduction formula

$$(2n-2)\int \frac{du}{(u^2+1)^n} = \frac{u}{(u^2+1)^{n-1}} + (2n-3)\int \frac{du}{(u^2+1)^{n-1}}.$$

IMPROPER INTEGRALS

[Hint: Integrate $\displaystyle\int \frac{du}{(u^2+1)^{n-1}}$ by parts, differentiating $\dfrac{1}{(u^2+1)^{n-1}}$ and integrating du; and notice that $\dfrac{u^2}{(u^2+1)^n} = \dfrac{(u^2+1)}{(u^2+1)^n} - \dfrac{1}{(u^2+1)^n}$.]

14. Evaluate the following improper integrals [§5.5].

• (a) $\displaystyle\int_0^\infty xe^{-x}\,dx$

• (b) $\displaystyle\int_0^\infty x^2e^{-x}\,dx$

• (c) $\displaystyle\int_0^\infty x^3e^{-x}\,dx$

(d) $\displaystyle\int_0^\infty x^ne^{-x}\,dx$

15. Evaluate the following improper integrals [§5.5].

(a) $\displaystyle\int_0^1 \log x\,dx$

(b) $\displaystyle\int_0^1 x\log x\,dx$

(c) $\displaystyle\int_0^1 x^n \log x\,dx$ [See Problem 9.]

16. Given that $\displaystyle\int_0^\infty e^{-x^2}\,dx = \sqrt{\pi}/2$, evaluate:

• (a) $\displaystyle\int_0^\infty x^2e^{-x^2}\,dx$

(b) $\displaystyle\int_0^\infty x^3e^{-x^2}\,dx$

(c) $\displaystyle\int_0^\infty xe^{-x^2}\,dx$

6.4 RATIONAL FUNCTIONS

The techniques presented so far (integration of linear combinations, substitution, integration by parts) are general; you should consider them as possibilities when confronted with almost any integrand. This section gives special techniques applying to a particular type of integrand, a rational function P/Q (where P and Q are polynomials).

In principal, the integral $\int P/Q$ can always be evaluated in a straightforward way; it is particularly easy when Q has degree 1, but grows progressively more laborious for denominators Q of degree 2, 3, 4, and so on. We will work up patiently from the simplest case.

Denominators of degree 1

If Q has degree 1, then $Q(x) = ax + b$, $a \neq 0$. By long division, any fraction $P(x)/Q(x)$ can be written in the form

$$\frac{P}{Q} = D + \frac{R}{Q}, \tag{1}$$

where D is a polynomial and R is a constant. Hence

$$\int \frac{P(x)}{Q(x)}\,dx = \int D(x)\,dx + \int \frac{R\,dx}{ax+b}.$$

Any polynomial D can be integrated by inspection, and since R is a constant, we get

$$\int \frac{R}{ax+b}\,dx = \frac{R}{a}\log|ax+b| + c, \qquad [\text{substitute } u = ax+b]$$

so the integral is completely evaluated.

Example 1. $\displaystyle\int \frac{x^2+1}{2x+3}\,dx = ?$ *Solution.* Divide $Q(x) = 2x + 3$ into $P(x) = x^2 + 1$, continuing until a constant remainder R is obtained:

$$\begin{array}{lrll}
 & & \tfrac{1}{2}x - \tfrac{3}{4} & \longrightarrow D(x) \\
Q(x) \longrightarrow & 2x+3 \,\big) & x^2 + 0x + 1 & \leftarrow P(x) \\
 & & x^2 + \tfrac{3}{2}x & \\
 & & -\tfrac{3}{2}x + 1 & \\
 & & -\tfrac{3}{2}x - \tfrac{9}{4} & \\
 & & \tfrac{13}{4} & \longrightarrow R
\end{array}$$

LONG DIVISION: EXAMPLE

This division shows that

$$\frac{x^2+1}{2x+3} = \frac{1}{2}x - \frac{3}{4} + \frac{13/4}{2x+3} \tag{2}$$

$\left[\text{that is, } \frac{P(x)}{Q(x)} = D(x) + \frac{R}{Q(x)}, \text{ as required in formula (1).}\right]$ From (2), it follows that

$$\begin{aligned}
\int \frac{x^2+1}{2x+3}\,dx &= \int \left(\frac{1}{2}x - \frac{3}{4} + \frac{13/4}{2x+3}\right) dx \\
&= \int \left(\frac{1}{2}x - \frac{3}{4}\right) dx + \frac{13}{4}\int \frac{dx}{2x+3} \\
&= \frac{1}{4}x^2 - \frac{3}{4}x + \frac{13}{4}\int \frac{\frac{1}{2}\,du}{u} \qquad [u = 2x+3] \\
&= \frac{1}{4}x^2 - \frac{3}{4}x + \frac{13}{8}\log|u| + c \\
&= \frac{1}{4}x^2 - \frac{3}{4}x + \frac{13}{8}\log|2x+3| + c.
\end{aligned}$$

Quadratic denominators

Suppose now that the denominator Q is quadratic, $Q(x) = ax^2 + bx + c$, where $a \neq 0$. The first step is to reduce P/Q, by long division, to the form

$$\frac{P}{Q} = D + \frac{R}{Q},$$

where D is a polynomial, and now *R is a polynomial of degree <2*, $R(x) = \alpha x + \beta$. The polynomial D can be integrated at sight; the treatment of the "remainder term" R/Q depends on the number of distinct real roots of the denominator Q. We consider the three cases: two distinct real roots, one such root, or no real roots.

Case (i) **Two distinct roots**

When Q has two distinct real roots r_1 and r_2, then Q is *reducible*, i.e. it can be factored,

$$Q(x) = ax^2 + bx + c = a(x - r_1)(x - r_2),$$

into a product of polynomials of lower degree, with real-valued constants a, r_1, and r_2. In this case, it is always possible to write R/Q as the sum of two "partial fractions" $A/(x - r_1)$ and $B/(x - r_2)$,

$$\frac{R(x)}{Q(x)} = \frac{\alpha x + \beta}{a(x - r_1)(x - r_2)} = \frac{A}{x - r_1} + \frac{B}{x - r_2} \tag{3}$$

PARTIAL FRACTIONS FOR A REDUCIBLE QUADRATIC DENOMINATOR

where A and B are constants; the resulting terms have known integrals as logarithms.

Example 2. $\int \frac{dx}{x^2 - 1} = ?$ *Solution.* It is easy to check that

$$\frac{1}{x^2 - 1} = \frac{1}{(x - 1)(x + 1)} = \frac{1/2}{x - 1} - \frac{1/2}{x + 1}.$$

The integral is therefore

$$\int \frac{dx}{x^2 - 1} = \int \left(\frac{1/2}{x - 1} - \frac{1/2}{x + 1}\right) dx = \frac{1}{2}\int \frac{dx}{x - 1} - \frac{1}{2}\int \frac{dx}{x + 1}$$

$$= \frac{1}{2}\log|x - 1| - \frac{1}{2}\log|x + 1| + c$$

$$= \frac{1}{2}\log\left|\frac{x - 1}{x + 1}\right| + c.$$

Example 3. $\int \frac{2x\,dx}{x^2 - 4} = ?$ *Solution.* In this case, formula (3) becomes

$$\frac{2x}{x^2 - 4} = \frac{2x}{(x - 2)(x + 2)} = \frac{1}{x - 2} + \frac{1}{x + 2},$$

so

$$\int \frac{2x\,dx}{x^2 - 4} = \int \frac{dx}{x - 2} + \int \frac{dx}{x + 2} = \log|x - 2| + \log|x + 2| + c$$

$$= \log|x^2 - 4| + c.$$

In the preceding two examples we obtained the partial fractions expression (3) "by inspection." Sometimes this is easy to see, but you don't *have* to rely on a lucky guess; the constants A and B in (3) can be determined in a perfectly straightforward way, as in the next example.

Example 4. Write $\frac{x - 1}{x^2 - 4}$ in the form (3). *Solution.* The denominator $Q(x) = x^2 - 4 = (x - 2)(x + 2)$ has two distinct roots $r_1 = 2$ and $r_2 = -2$, so we write, as in (3),

$$\frac{x - 1}{x^2 - 4} = \frac{A}{x - 2} + \frac{B}{x + 2}; \tag{4}$$

the problem now is to determine the constants A and B so that this equation is valid. Putting the right-hand side on a common denominator gives

$$\frac{x-1}{x^2-4} = \frac{A(x+2)+B(x-2)}{x^2-4} = \frac{(A+B)x+2(A-B)}{x^2-4},$$

where we have collected like powers of x in the numerator. This equality will be valid if the numerator on the far left equals the one on the far right. You can see that they will be equal if

$$1 = A + B \qquad \text{(this makes the two } x \text{ terms equal)}$$

$$-1 = 2(A-B) \qquad \text{(this makes the two constant terms equal)}.$$

These simultaneous equations for A and B have the unique solution $A = 1/4$, $B = 3/4$. Putting these values back in (4) gives

$$\frac{x-1}{x^2-4} = \frac{1/4}{x-2} + \frac{3/4}{x+2}.$$

Therefore, $\dfrac{x-1}{x^2-4}$ can be integrated as follows:

$$\int \frac{x-1}{x^2-4}\,dx = \int \left(\frac{1/4}{x-2} + \frac{3/4}{x+2}\right) dx = \frac{1}{4}\int \frac{dx}{x-2} + \frac{3}{4}\int \frac{dx}{x+2}$$

$$= \frac{1}{4}\log|x-2| + \frac{3}{4}\log|x+2| + c.$$

Example 5. $\displaystyle\int \frac{x^2+x+1}{x^2-1}\,dx = ?$ *Solution.* By long division,

$$\begin{array}{r} 1 \\ x^2-1\overline{)x^2+x+1} \\ \underline{x^2\qquad-1} \\ x+2 \end{array} \qquad \text{so} \qquad \frac{x^2+x+1}{x^2-1} = 1 + \frac{x+2}{x^2-1}.$$

We proceed to the integration of $(x+2)/(x^2-1)$. The denominator x^2-1 factors into $(x+1)(x-1)$, hence has two distinct real roots $r_1 = -1$ and $r_2 = 1$. We are thus in Case (i) [distinct roots], and can set

$$\frac{x+2}{x^2-1} = \frac{A}{x+1} + \frac{B}{x-1}. \tag{5}$$

To determine the constants A and B, put the right-hand side of (5) on a common denominator, obtaining

$$\frac{x+2}{(x+1)(x-1)} = \frac{A(x-1)+B(x+1)}{(x+1)(x-1)} = \frac{(A+B)x+(B-A)}{(x+1)(x-1)}. \tag{6}$$

Comparing the numerator on the far left with the one on the far right, you can see that they will be the same if

$$1 = A + B \qquad \text{(this makes the } x \text{ terms equal)}$$

$$2 = B - A \qquad \text{(this makes the constant terms equal).}$$

These simultaneous equations for A and B have the unique solution

$$A = -\frac{1}{2}, \qquad B = \frac{3}{2}.$$

Putting these values of A and B in (5) yields

$$\frac{x+2}{(x-1)(x+1)} = \frac{-1/2}{x+1} + \frac{3/2}{x-1}.$$

Combining this with the long division made at the very beginning, we finally obtain

$$\int \frac{x^2+x+1}{x^2-1}\,dx = \int \left(1 + \frac{x+2}{x^2-1}\right) dx \qquad \text{[from long division]}$$

$$= \int \left(1 + \frac{-1/2}{x+1} + \frac{3/2}{x-1}\right) dx$$

[from the decomposition (5)]

$$= x - \frac{1}{2}\log|x+1| + \frac{3}{2}\log|x-1| + c.$$

DETERMINING THE CONSTANTS

The constants A and B in the partial fractions decomposition (3) above,

$$\frac{R(x)}{Q(x)} = \frac{A}{x - r_1} + \frac{B}{x - r_2}, \tag{3}$$

can always be determined as we did in these examples. Put the right-hand side of (3) on the denominator $Q(x)$ [see e.g. (6) above]. The resulting numerator $N(x) = (A + B)x - r_1B - r_2A$ contains the constants A and B, and *there is one and only one way to choose A and B such that $N(x)$ is identical with the given numerator $R(x)$*. [See Problem 10 for the proof of this.]

Case (ii) ***Double root***

When a real quadratic Q has just one real root, this must be a double root, i.e. $x - r$ occurs twice as a factor of Q,

$$Q(x) = a(x - r)^2.$$

In this case, *it is always possible to write R/Q as the sum of two partial fractions, as follows:*

PARTIAL FRACTIONS FOR A DOUBLE ROOT QUADRATIC

$$\frac{R(x)}{Q(x)} = \frac{\alpha x + \beta}{a(x-r)^2} = \frac{A}{x-r} + \frac{B}{(x-r)^2}, \tag{7}$$

where A and B are to be determined as in Case (i). [See Problem 11 below.]

Example 6. $\int \frac{x\,dx}{(x+2)^2} = ?$ *Solution.* The numerator has degree <2, so the integrand is already in the form R/Q, where degree$(R) <$ degree(Q), and long division is not required. To evaluate $\int R/Q$, apply (7). We obtain

$$\frac{x}{(x+2)^2} = \frac{A}{x+2} + \frac{B}{(x+2)^2} = \frac{A(x+2)+B}{(x+2)^2} = \frac{Ax + (2A+B)}{(x+2)^2}. \tag{8}$$

Comparing numerators on the far left and far right, we get

$$1 = A \qquad \text{(comparing terms in } x\text{)}$$

$$0 = 2A + B \qquad \text{(comparing constant terms)}.$$

Hence $A = 1$, $B = -2$. Putting these values in (8) yields

$$\int \frac{x\,dx}{(x+2)^2} = \int \left(\frac{1}{x+2} + \frac{-2}{(x+2)^2}\right) dx$$

$$= \log|x+2| + 2(x+2)^{-1} + c.$$

Notice in (8), as in (5), that there are *two* constants A and B to be determined, and there are two equations to determine them; one from the term in x, and one from the constant term. Thus it is essential that the numerator R has degree $<2 =$ degree(Q); if the degree were higher, we would have more equations to satisfy (coming from terms in x^2, x^3, etc.), and this would prove impossible.

Case (iii) ***Q is irreducible***

When Q has no real roots, it is called *irreducible;* for it cannot be factored into real polynomials of lower degree. Instead of factoring, complete the square [appendix to §6.2, page 327]:

$$Q(x) = ax^2 + bx + c = a[(x-p)^2 + q^2].$$

The substitution $u = \frac{x-p}{q}$ (or $x = qu + p$) now converts the denominator to the form $aq^2(u^2+1)$, and the integral is evaluated as follows:

$$\int \frac{R(x)}{Q(x)}\,dx = \int \frac{R(qu+p)}{aq^2(u^2+1)}\,d(qu+p) = A\int \frac{u\,du}{u^2+1} + B\int \frac{du}{u^2+1}$$

PARTIAL FRACTIONS FOR AN IRREDUCIBLE QUADRATIC

[A and B are constants]

$$= \frac{1}{2}A\log(u^2+1) + B\arctan u + c.$$

[Do *not* try to memorize this formula; in any given example, you will arrive at it automatically by carrying out the substitution.]

Since a quadratic must have 2, 1, or 0 distinct real roots, the three cases above complete the discussion of quadratic denominators.

Example 7. $\int \frac{x\,dx}{2x^2+4x+3} = ?$ *Solution.* We seek the roots of the denominator by the quadratic formula: $x = \frac{-4 \pm \sqrt{16-24}}{4}$. This entails the square root of a negative number, so there are *no* real roots, and we are in the irreducible case (iii). Complete the square:

$$2x^2+4x+3 = 2(x^2+2x+3/2) = 2[(x+1)^2+1/2].$$

To put the denominator in the form u^2+1, set $u = \frac{x+1}{\sqrt{1/2}} = \sqrt{2}(x+1)$, which entails

$$du = \sqrt{2}\,dx, \qquad 2\left[(x+1)^2+\frac{1}{2}\right] = 2\left(\frac{u^2}{2}+\frac{1}{2}\right) = u^2+1, \qquad x = \frac{1}{\sqrt{2}}u - 1.$$

Hence

$$\int \frac{x\,dx}{2x^2+4x+3} = \int \frac{\frac{1}{\sqrt{2}}u-1}{u^2+1}\,\frac{1}{\sqrt{2}}\,du = \frac{1}{2}\int \frac{u}{u^2+1}\,du - \frac{1}{\sqrt{2}}\int \frac{du}{u^2+1}$$

$$= \frac{1}{4}\log(u^2+1) - \frac{1}{\sqrt{2}}\arctan u + c$$

$$= \frac{1}{4}\log(2x^2+4x+3) - \frac{1}{\sqrt{2}}\arctan(\sqrt{2}\,x+\sqrt{2}) + c.$$

You might now do some of Problems 1, 2, and 9–11 before going on to:

The general case

INTEGRATION BY PARTIAL FRACTIONS: GENERAL METHOD

Any rational function can be integrated, in principle, by combining and extending the methods given for the special cases above. Evaluate P/Q by the following steps I–V.

I By long division, write

$$P/Q = D + R/Q,$$

where degree$(R) <$ degree(Q). [Of course, if degree$(P) <$ degree(Q), this first step is unnecessary.]

II *Factor* the denominator Q. This is sometimes difficult, but in principle it is always possible to write Q as a *product of first degree polynomials and irreducible quadratic polynomials,*

$$Q(x) = a(x - r_1)^{m_1}(x - r_2)^{m_2}\cdots(x^2 + b_1x + c_1)^{n_1}(x^2 + b_2x + c_2)^{n_2}\cdots. \tag{9}$$

[This can be proved by using complex numbers; see §8.5.] Collect all like factors, so that in (9) we have $r_1 \neq r_2$, etc.

III *Write* R/Q as a sum of *partial fractions*, as follows:

Let $x - r$ be a first degree factor of Q, and let m be the power to which it occurs in the factorization (9). Then the sum of partial fractions for R/Q includes the following terms:

TERMS IN INTEGRAND FROM $\dfrac{1}{(x-r)^m}$

$$\frac{A_1}{x - r} + \frac{A_2}{(x - r)^2} + \cdots + \frac{A_m}{(x - r)^m}. \tag{10}$$

[Thus in Example 4 above, the factor $x - 2$ occurred to the first power, $m = 1$, so we included the partial fraction $\dfrac{A}{x - 2}$; similarly, $x + 2$ occurred to the first power, so we included a corresponding fraction $\dfrac{B}{x + 2}$. In Example 6, $x + 2$ occurred to the second power, $m = 2$, so we included the fractions $\dfrac{A}{x + 2}$ and $\dfrac{B}{(x + 2)^2}$.]

Next, let $x^2 + bx + c$ be an irreducible quadratic factor, and let n be the power to which it occurs in the factorization (9). Then the sum of partial fractions for R/Q includes

TERMS IN INTEGRAND FROM $\dfrac{1}{(x^2+bx+c)^n}$

$$\frac{B_1x + C_1}{x^2 + bx + c} + \frac{B_2x + C_2}{(x^2 + bx + c)^2} + \cdots + \frac{B_nx + C_n}{(x^2 + bx + c)^n}. \tag{11}$$

[In Example 7 above, the integrand $\dfrac{x}{2x^2 + 4x + 3} = \dfrac{x/2}{x^2 + 2x + 3/2}$ was already in the form suggested in (11). Hence there was nothing to be gained by a partial fractions decomposition.]

IV Combine the partial fractions introduced in III on the denominator Q; this is possible since all the denominators in (10) and (11) are factors of Q. Let N denote the numerator of the combined partial fractions. Then *there is one and only one choice of the constants* $A_1, A_2, \ldots, B_1, B_2, \ldots, C_1, C_2, \ldots$ *such that the numerator* N *is precisely equal to the given numerator* R. [This claim (which we shall not prove) assures us that every fraction R/Q, where degree(R) < degree(Q), can be decomposed into partial fractions. It is not really necessary to know this, however; in any given case you can check whether or not your particular decomposition works, by actually putting it on a common denominator.] The coefficients $A_1, \ldots,$ etc. are determined by comparing like powers of x in the expressions for R and N [see Examples 4–6 above], or by substituting particular values of x [see Example 8 below], or by a combination of these methods.

V Integrate the terms introduced in (10) and (11). Those in (10) reduce to the form $\int u^k\,du$ with the substitution $u = x - r$. For the terms in (11), complete the square, as in Case (iii) above, writing $x^2 + bx + c = (x - p)^2 + q^2$. Then set $u = \dfrac{x - p}{q}$, obtaining terms of the form

$$\int \frac{u\,du}{(u^2 + 1)^k} \quad \text{and} \quad \int \frac{du}{(u^2 + 1)^k}.$$

The first form yields to the substitution $w = u^2 + 1$, and the second can be integrated by the reduction formula

REDUCTION FORMULA FOR $\displaystyle\int \frac{du}{(u^2 + 1)^n}$

$$(2n - 2)\int \frac{du}{(u^2 + 1)^n} = \frac{u}{(u^2 + 1)^{n-1}} + (2n - 3)\int \frac{du}{(u^2 + 1)^{n-1}}. \tag{12}$$

[See Problem 13, §6.3. The integrals arising from (10) and (11) can also be found in most tables of integrals.]

Example 8. The crux of the method is in step III, where the appropriate partial fractions are set up. Here are some factored denominators, with the corresponding decompositions:

(a) $Q(x) = (x - 1)(x - 2)(x - 3)$ [three terms of type (10)]

$$\frac{R(x)}{Q(x)} = \frac{A}{x + 1} + \frac{B}{x + 2} + \frac{C}{x + 3}, \qquad \text{degree } (R) < 3 = \text{degree } (Q)$$

(b) $Q(x) = (x-1)(x^2+x+1)^2$

[one term of type (10) and one of type (11)]

$$\frac{R(x)}{Q(x)} = \frac{A}{x-1} + \frac{Bx+C}{x^2+x+1} + \frac{Dx+E}{(x^2+x+1)^2},$$

degree $(R) < 5 =$ degree (Q)

(c) $Q(x) = (x+2)^3(x^2+1)^2$

$$\frac{R(x)}{Q(x)} = \frac{A}{x+2} + \frac{B}{(x+2)^2} + \frac{C}{(x+2)^3} + \frac{Dx+E}{x^2+1} + \frac{Fx+G}{(x^2+1)^2}.$$

Notice that *the number of undetermined constants* corresponding to a given factor *equals the degree* of that factor, and the total number of undetermined constants $A, B, \ldots$ equals the degree of the given denominator Q.

Example 9. $\displaystyle\int \frac{x\,dx}{(x-1)(x-2)(x-3)} = ?$ *Solution.* The numerator has lower degree than the denominator, and the denominator is factored, so we are ready to assign partial fractions. As in Example 8(a), set

$$\frac{x}{(x-1)(x-2)(x-3)} = \frac{A}{x-1} + \frac{B}{x-2} + \frac{C}{x-3}.$$

Put the right-hand side on the given denominator Q, obtaining

$$\frac{x}{(x-1)(x-2)(x-3)}$$

$$= \frac{A(x-2)(x-3) + B(x-1)(x-3) + C(x-1)(x-2)}{(x-1)(x-2)(x-3)}.$$

There is one and only one way to choose the constants such that the numerators are identical, i.e. such that

$$x = A(x-2)(x-3) + B(x-1)(x-3) + C(x-1)(x-2), \quad (13)$$

for every real number x. In particular, (13) must hold with $x = 1$, which gives

$$1 = A(-1)(-2) + B\cdot 0 + C\cdot 0; \qquad \text{hence } A = 1/2.$$

The identity (13) must also hold with $x = 2$, so

$$2 = A\cdot 0 + B\cdot 1\cdot(-1) + C\cdot 0; \qquad \text{hence } B = -2.$$

Similarly, setting $x = 3$ yields

$$3 = A\cdot 0 + B\cdot 0 + C\cdot 2\cdot 1; \qquad \text{hence } C = 3/2.$$

Thus

$$\int \frac{x}{(x-1)(x-2)(x-3)}\,dx$$

$$= \int \left(\frac{1/2}{x-1} + \frac{-2}{x-2} + \frac{3/2}{x-3}\right) dx$$

$$= \frac{1}{2}\log|x-1| - 2\log|x-2| + \frac{3}{2}\log|x-3| + c.$$

Remark 1. In Example 9, we did not bother to collect like powers of x on the right-hand side of (13), but determined the coefficients A, B, and C by using particular values of x that simplified the equations. This is legitimate, since it is known beforehand that A, B, and C can be chosen, in one and only one way, to make the two sides of (13) identical.

Remark 2. Factoring the denominator is not always easy; the next example gives some useful hints.

HINTS FOR FACTORING

Example 10. Factor $Q(x) = x^4 - x^3 - x + 1$. *Solution.* Begin by dividing out all first degree factors. Since $Q(1) = 0$, $(x - 1)$ must be a factor. Dividing $(x - 1)$ into $Q(x)$ yields

$$Q(x) = (x-1)(x^3-1). \tag{14}$$

The cubic $x^3 - 1$ is zero when $x = 1$, so it must have the factor $x - 1$; indeed, $x^3 - 1 = (x-1)(x^2+x+1)$, so from (14),

$$Q(x) = (x-1)(x-1)(x^2+x+1) = (x-1)^2(x^2+x+1). \tag{15}$$

Finally, the roots of the quadratic $x^2 + x + 1$ can be found by the quadratic formula; in this case the formula does not produce any *real* roots, so $x^2 + x + 1$ is irreducible. Therefore, (15) factors Q into a product of the type required for partial fractions [step II above].

Example 11. $\int \frac{dx}{x^3-1} = ?$ *Solution.* Long division is not required, so we factor,

$$x^3 - 1 = (x-1)(x^2+x+1).$$

The second factor is irreducible, so we have one term of the form (10) and one of the form (11), and we express R/Q as follows:

$$\frac{R(x)}{Q(x)} = \frac{1}{x^3-1} = \frac{A}{x-1} + \frac{Bx+C}{x^2+x+1}$$

$$= \frac{A(x^2+x+1) + (Bx+C)(x-1)}{x^3-1}. \tag{16}$$

Equate numerators,

$$1 = A(x^2 + x + 1) + (Bx + C)(x - 1).$$

Set $x = 1$ (why is this convenient?) and find

$$1 = 3A, \qquad A = 1/3.$$

Set $x = 0$ and find

$$1 = A - C, \qquad C = A - 1 = -2/3.$$

Set $x = -1$ and find

$$1 = A + (-B + C)(-2); \qquad 2B = 1 - A + 2C = -2/3; \qquad B = -1/3.$$

Put these values of A, B, and C in (16), and integrate:

$$\int \frac{dx}{x^3 - 1} = \int \left(\frac{1/3}{x - 1} + \frac{-x/3 - 2/3}{x^2 + x + 1}\right) dx$$

$$= \frac{1}{3} \log |x - 1| - \frac{1}{3} \int \frac{x + 2}{x^2 + x + 1} dx. \tag{17}$$

In the last integral, complete the square and reduce the denominator to the form $u^2 + 1$:

$$x^2 + x + 1 = \left(x + \frac{1}{2}\right)^2 + \frac{3}{4} = \left(x + \frac{1}{2}\right)^2 + \left(\frac{\sqrt{3}}{2}\right)^2 = (x - p)^2 + q^2;$$

$$u = \frac{x - p}{q} = \frac{x + 1/2}{\sqrt{3}/2} = \frac{2x + 1}{\sqrt{3}}, \qquad x = \frac{\sqrt{3}u - 1}{2}, \qquad dx = \frac{\sqrt{3}}{2} du;$$

$$x^2 + x + 1 = \left(x + \frac{1}{2}\right)^2 + \left(\frac{\sqrt{3}}{2}\right)^2 = \frac{3}{4}(u^2 + 1), \qquad x + 2 = \frac{\sqrt{3}}{2} u + \frac{3}{2}.$$

Hence

$$\int \frac{x + 2}{x^2 + x + 1} dx = \int \frac{u\, du}{u^2 + 1} + \sqrt{3} \int \frac{du}{u^2 + 1} = \frac{1}{2} \log (u^2 + 1) + \sqrt{3} \arctan u + c$$

$$= \frac{1}{2} \log \left(\frac{4}{3}(x^2 + x + 1)\right) + \sqrt{3} \arctan \left(\frac{2x + 1}{\sqrt{3}}\right) + c.$$

Substituting this for the last integral in (17) completes the integration:

$$\int \frac{dx}{x^3 - 1} = \frac{1}{3} \log |x - 1| - \frac{1}{6} \log \left(\frac{4}{3}(x^2 + x + 1)\right) - \frac{1}{\sqrt{3}} \arctan \left(\frac{2x + 1}{\sqrt{3}}\right) + c.$$

Remark 3. By a slight change in the general procedures I–V, it is possible to replace the tedious task of *integrating* the partial fractions by the somewhat simpler task of *differentiating*. Suppose you want to integrate P/Q. Carry out steps I (writing $P/Q = D + R/Q$) and II (factoring Q) as before; but in III, do *not* write an expression for R/Q; instead, *write directly an expression for the indefinite integral* $\int R/Q$ involving undetermined constants $A, B, C, \ldots$. (The appropriate expression is given in (18) and (19) below.) Call this expression F; you then differentiate F, and choose the constants so that $F' = R/Q$, as desired.

INTEGRATION BY PARTIAL FRACTIONS: ALTERNATE METHOD

This method is often better than the standard one outlined above; since no substitution or integration by parts is required, there are fewer opportunities to make trivial computational errors. On the other hand, the terms in (18) and (19) below are harder to remember than those in (10) and (11).

The terms in $F = \int R/Q$ are determined as follows: If Q contains a factor $(x - r)^m$, then the indefinite integral F includes

TERMS IN INTEGRAL FROM $\dfrac{1}{(x-r)^m}$

$$A_1 \log|x - r| + \frac{A_2}{x - r} + \cdots + \frac{A_m}{(x - r)^{m-1}}. \tag{18}$$

If Q contains a factor $[(x - p)^2 + q^2]^n$, then F includes

TERMS FROM $\dfrac{1}{[(x-p)^2+q^2]^n}$

$$B_1 \log((x - p)^2 + q^2) + C_1 \arctan\left(\frac{x - p}{q}\right) + \frac{B_2 x + C_2}{(x - p)^2 + q^2} + \cdots + \frac{B_n x + C_n}{[(x - p)^2 + q^2]^{n-2}}. \tag{19}$$

The terms in (18) and (19) are precisely those that arise from integrating (10) and (11), with the constants relabeled for convenience.

Example 12. Rework Example 11 by this method. *Solution.* The factorization

$$x^3 - 1 = (x - 1)(x^2 + x + 1) = (x - 1)\left[\left(x + \frac{1}{2}\right)^2 + \left(\frac{\sqrt{3}}{2}\right)^2\right]$$

shows that $1/(x^3 - 1)$ has an indefinite integral F of the form

$$F(x) = A \log|x - 1| + B \log\left(\left(x + \frac{1}{2}\right)^2 + \left(\frac{\sqrt{3}}{2}\right)^2\right) + C \arctan\left(\frac{x + 1/2}{\sqrt{3}/2}\right). \tag{20}$$

The A term comes from the factor $x - 1$ [see (18)], and the B and C terms from the irreducible quadratic factor $x^2 + x + 1$ [see (19)]. We now must choose A, B, and C so that $F'(x) = 1/(x^3 - 1)$. Differentiating (20), we find

$$F'(x) = \frac{A}{x-1} + \frac{B(2x+1)}{x^2+x+1} + \frac{2}{\sqrt{3}} \frac{C}{\left(\frac{x+1/2}{\sqrt{3}/2}\right)^2 + 1}$$

$$= \frac{A}{x-1} + \frac{B(2x+1)}{x^2+x+1} + \frac{\sqrt{3}}{2} \cdot \frac{C}{x^2+x+1}$$

$$= \frac{A(x^2+x+1) + B(2x+1)(x-1) + \frac{\sqrt{3}}{2} C(x-1)}{x^3-1}.$$

To make $F'(x) = 1/(x^3-1)$, we equate numerators:

$$A(x^2+x+1) + B(2x+1)(x-1) + \frac{\sqrt{3}}{2} C(x-1) = 1.$$

Setting $x = 1$ yields $3A = 1$, so $A = 1/3$.

Setting $x = -\frac{1}{2}$ yields $A \cdot \frac{3}{4} - C \cdot \frac{3\sqrt{3}}{4} = 1$, so $C = -1/\sqrt{3}$.

Setting $x = 0$ yields $A - B - \frac{\sqrt{3}}{2} C = 1$, so $B = -1/6$.

Putting these back in (20), we find for $\int \frac{dx}{x^3-1}$ the expression

$$F(x) = \frac{1}{3} \log |x-1| - \frac{1}{6} \log (x^2+x+1) - \frac{1}{\sqrt{3}} \arctan \left(\frac{2x+1}{\sqrt{3}}\right) + c.$$

This is equivalent to the solution found in Example 11.

Remark 4. The integration of rational functions follows one of the standard mathematical "plot outlines": Once the *general form* of a solution is known, it is usually easy to determine specific solutions in specific cases. A much simpler example of this is found in completing the square: Once you know that $x^2 + bx + c$ can be written in the form $(x-p)^2 + r$, it is easy to determine the constants p and r for any given quadratic $x^2 + bx + c$; simply multiply out $(x-p)^2 + r$ and compare it with the given quadratic polynomial. Similarly, once you know that R/Q can be written as a sum of fractions of the types in (10) and (11), it is fairly easy to determine the constants appearing there for any given numerator R.

The method outlined in Remark 3 applies the same general idea, but a little closer to the end result; once you know that the *integral* $\int R/Q$ can be written as a sum of functions of the types in (18) and (19), it is fairly easy to determine the constants appearing there, and the integration problem is solved.

Concluding Remark. Do not be so impressed by these fancy methods that you overlook simpler ones, when they apply. For example, to integrate

$$\int \frac{3x^2 - 6x + 2}{x^3 - 3x^2 + 2x + 1}\, dx = \log |x^3 - 3x^2 + 2x + 1| + c,$$

it is much better to use the substitution $u = x^3 - 3x^2 + 2x + 1$ than to factor the denominator and go through the method of partial fractions.

PROBLEMS

[Note: Problems 4 and 5 are based on the "standard" method of partial fractions, whereas Problem 6 is based on the method in Remark 3 above.]

1. Write the following quotients P/Q in the form $D + R/Q$, where D and R are polynomials, and degree$(R) <$ degree(Q).

• (a) $\dfrac{x^2 + x}{x + 2}$

• (b) $\dfrac{x^2}{(x-1)(x-2)}$

(c) $\dfrac{x^3}{x^3 - x^2 - x + 1}$

(d) $\dfrac{x^4}{(x+1)^2}$

(e) $\dfrac{x^5}{(x^2+1)^2}$

• (f) $\dfrac{x}{(x^2+9)^2}$

2. Integrate the following rational functions with denominators of degree ≤ 2.

• (a) $\displaystyle\int \frac{dx}{(x+3)(x-2)}$

(b) $\displaystyle\int \frac{x^2 + x}{x + 2}\, dx$

• (c) $\displaystyle\int \frac{dx}{x^2 + 2x + 4}$

(d) $\displaystyle\int \frac{x\, dx}{x^2 + 9}$

• (e) $\displaystyle\int \frac{x^4\, dx}{(x+1)^2}$

(f) $\displaystyle\int \frac{x^2\, dx}{(x-1)(x-2)}$

• (g) $\displaystyle\int \frac{dx}{x^2 + 4x + 4}$

(h) $\displaystyle\int \frac{2x - 3}{(x+1)(x-7)}\, dx$

3. Compute the following. [a, b, n, p, and q denote constants.]

• (a) $\displaystyle\int \frac{dx}{(ax+b)^n},\quad n > 1$

(b) $\displaystyle\int \frac{dx}{(x-p)^2 + q^2}$

• (c) $\displaystyle\int \frac{dx}{((x-p)^2 + q^2)^2}$ [See (12) above]

(d) $\displaystyle\int \frac{dx}{((x-p)^2+q^2)^3}$

• (e) $\displaystyle\int \frac{(x-p)\,dx}{(x-p)^2+q^2}$ (f) $\displaystyle\int \frac{(x-p)\,dx}{[(x-p)^2+q^2]^n}, \quad n > 1$

4. For the given polynomial Q, determine the general partial fractions decomposition of R/Q for degree$(R) <$ degree(Q), as in Example 8. [The number of constants in your answer should equal the degree of Q.]

• (a) $Q(x) = (x^2-1)^2$ (d) $Q(x) = (x^2+16)^3$
(b) $Q(x) = x^3 - x^2 - x + 1$ • (e) $Q(x) = (x^2+2x+2)^2$
• (c) $Q(x) = (x^2+1)^2$ (f) $Q(x) = x^4 - a^4, \quad a \neq 0$

5. Determine the constants in the partial fractions decompositions of the following functions.

• (a) $\displaystyle\frac{1}{(x^2-1)^2}$ (d) $\displaystyle\frac{x^5}{(x^2+16)^3}$

(b) $\displaystyle\frac{x^2+x-1}{x^3-x^2-x+1}$ • (e) $\displaystyle\frac{x^3}{(x^2+2x+2)^2}$

• (c) $\displaystyle\frac{1}{(x^2+16)^3}$ (f) $\displaystyle\frac{1}{x^4-a^4}, \quad a \neq 0$

• **6.** [*Alternate to Problems* 4 *and* 5] For the polynomials Q in Problem 4, determine the general form of $\int R/Q$ for degree$(R) <$ degree(Q). [See Remarks 3 and 4. Your answer should have n undetermined constants, where $n =$ degree (Q).]

• **7.** Integrate the functions in Problem 5. [You should do *either* Problem 5 or Problem 6 before this.]

8. Integrate the following. Check your own answers by differentiating the result.

• (a) $\displaystyle\int \frac{x^3\,dx}{x^3-x^2-x+1}$

(b) $\displaystyle\int \frac{x\,dx}{(x^2+1)^2}$ [Look for an easy way.]

• (c) $\displaystyle\int \frac{\sin\theta\,d\theta}{\cos^2\theta + 3\cos\theta + 2}$

(d) $\displaystyle\int \frac{e^{2t}\,dt}{e^{2t}+e^t-2}$

• (e) $\displaystyle\int \frac{2u^2 + 3u + 1}{(u+3)(u+2)(u-1)}\,du$ (h) $\displaystyle\int \frac{z^5 - z^4 - 3z + 5}{z^4 - 2z^3 + 2z^2 - 2z + 1}\,dz$

(f) $\displaystyle\int \frac{x^3 - 3x + 2}{(x+1)(x-1)^3}\,dx$ (i) $\displaystyle\int \frac{dx}{x^4 + 5x^2 + 4}$

(g) $\displaystyle\int \frac{2t^3 + 3t^2 + t - 1}{(t+1)(t^2 + 2t + 2)^2}\,dt$ (j) $\displaystyle\int \frac{4x^3 + 10x}{x^4 + 5x^2 + 4}\,dx$

(k) $\displaystyle\int \frac{3x\,dx}{x^5 + 2x^4 - 10x^3 - 20x^2 + 9x + 18}$

(l) $\displaystyle\int \frac{t^3 - t^2 + 2t + 3}{(t^2 + 2t + 2)^2}\,dt$

(m) $\displaystyle\int \frac{2x^7 - 6}{(x^2+1)^4}\,dx$ (n) $\displaystyle\int \frac{t^4 + 1}{(t^2+4)^3}\,dt$

9. (a) Show that for $x > |h|$, $h \neq 0$,

$$\int \frac{dx}{(x-h)(x+h)} = -\frac{\log(x+h) - \log(x-h)}{2h} + c.$$

(b) Show that $\displaystyle\lim_{h\to 0} \frac{\log(x+h) - \log(x-h)}{2h} = \frac{1}{x}$. [Hint: Rewrite the numerator as $[\log(x+h) - \log x] + [\log x - \log(x-h)]$, and notice you have two difference quotients for the logarithm.]

(c) Conclude that $\displaystyle\lim_{h\to 0} \int \frac{dx}{(x-h)(x+h)} = \int \left[\lim_{h\to 0} \frac{1}{(x-h)(x+h)}\right] dx.$

10. Let $r_1 \neq r_2$. Show that for any given first degree polynomial $\alpha x + \beta$, there are unique constants A and B such that

$$\frac{\alpha x + \beta}{(x - r_1)(x - r_2)} = \frac{A}{x - r_1} + \frac{B}{x - r_2}.$$

[Hint: Put the right-hand side on a common denominator, and determine A and B by comparing numerators.] Is this still true if $r_1 = r_2$?

11. Let r be given. Show that for any given first degree polynomial $\alpha x + \beta$, there are unique constants A and B such that

$$\frac{\alpha x + \beta}{(x - r)^2} = \frac{A}{x - r} + \frac{B}{(x - r)^2}.$$

12. This problem shows that if two polynomials agree in any open interval, then they have precisely the same coefficients. Let $R(x) = r_n x^n + r_{n-1}x^{n-1} + \cdots + r_1 x + r_0$ and $S(x) = s_n x^n + \cdots + s_0$ be two polynomials of degree $\leq n$, and let a be any real number.

(a) Show that if $R^{(n)}(a) = S^{(n)}(a)$, then the coefficients of x^n agree, i.e. $r_n = s_n$.

(b) Show that if $R^{(n)}(a) = S^{(n)}(a)$ *and* $R^{(n-1)}(a) = S^{(n-1)}(a)$, then $r_n = s_n$ and $r_{n-1} = s_{n-1}$.

(c) Show that if $R^{(j)}(a) = S^{(j)}(a)$ for $j = 0, 1, 2, \ldots, n$, then R and S have precisely the same coefficients, i.e. $r_k = s_k$ for $k = 0, \ldots, n$.

(d) Show that if $R(x) = S(x)$ for all x in some open interval containing a, then R and S have the same coefficients.

13. Let Q be a nonzero polynomial, and suppose that R and N are polynomials such that $R(x)/Q(x) = N(x)/Q(x)$, wherever $Q(x) \neq 0$. Prove that R and N have exactly the same coefficients. [See Problem 12. This result is not actually used in the method of partial fractions, but it answers a natural question: Do two equal rational functions with the same denominator necessarily have the same numerator?]

6.5 SPECIAL TRIGONOMETRIC INTEGRALS

We already have a number of special techniques for evaluating trigonometric integrals. Here we collect them, together with some new ones.

Odd powers of sine and cosine

$\int (\sin a\theta)^j (\cos a\theta)^k \, d\theta$ can be evaluated easily if either j or k is odd. For example, if j is odd, then $j - 1$ is even, and we can write the integral as

$$\int (\sin a\theta)^{j-1} (\cos a\theta)^k \sin a\theta \, d\theta = \int (\sin^2 a\theta)^{(j-1)/2} (\cos a\theta)^k \sin a\theta \, d\theta$$

$$= \int (1 - \cos^2 a\theta)^{(j-1)/2} (\cos a\theta)^k \sin a\theta \, d\theta$$

$$= -\frac{1}{a} \int (1 - u^2)^{(j-1)/2} u^k \, du. \qquad [u = \cos a\theta]$$

Thus to integrate $\int (\sin a\theta)^j (\cos a\theta)^k \, d\theta$ where j is odd, we split off one factor $\sin a\theta$, so that the remaining factor $(\sin a\theta)^{j-1}$ is an *even* power. This is converted to $\cos a\theta$ by the formula $\sin^2 a\theta = 1 - \cos^2 a\theta$, and we can reduce the problem to the integration of a polynomial by the substitution $u = \cos a\theta$.

If j were even but k were odd, we would split off a factor $\cos a\theta$, convert $(\cos a\theta)^{k-1}$ by using the formula $\cos^2 a\theta = 1 - \sin^2 a\theta$, and then reduce to a polynomial by setting $u = \sin a\theta$.

Example 1. $\int \sin a\theta \cos^k a\theta \, d\theta = ?$ *Solution.* The sin appears to an odd power, so set $u = \cos a\theta$, $du = -a \sin a\theta \, d\theta$; hence

$$\int \sin a\theta \cos^k a\theta \, d\theta = \int u^k \left(-\frac{1}{a}\right) du = -\frac{u^{k+1}}{a(k+1)} + c \qquad [k \neq -1]$$

$$= -\frac{\cos^{k+1} a\theta}{a(k+1)} + c.$$

Example 2. $\int \cos^3 2\theta \, d\theta = ?$ *Solution.* Since cos appears to an odd power, we split off one factor of cos and convert the rest to sin:

$$\int \cos^3 2\theta \, d\theta = \int (\cos^2 2\theta)(\cos 2\theta \, d\theta) = \int (1 - \sin^2 2\theta) \cos 2\theta \, d\theta$$

$$= \int (1 - u^2) \frac{1}{2} \, du \qquad [u = \sin 2\theta, \; du = 2 \cos 2\theta \, d\theta]$$

$$= \frac{1}{2} u - \frac{1}{6} u^3 + c = \frac{1}{2} \sin 2\theta - \frac{1}{6} \sin^3 2\theta + c.$$

Even powers of sine and cosine

The integral $\int (\sin a\theta)^j (\cos a\theta)^k \, d\theta$ cannot be handled as above if both j and k are even. In this case, we use one of the identities $\cos^2 a\theta = 1 - \sin^2 a\theta$ or $\sin^2 a\theta = 1 - \cos^2 a\theta$ to obtain an integral involving only even powers of sine or cosine. Then it is possible to reduce the powers by one of the reduction formulas derived by integration by parts (Example 7, page 332):

$$\int (\sin ax)^n \, dx = -\frac{(\sin ax)^{n-1} \cos ax}{na} + \frac{n-1}{n} \int (\sin ax)^{n-2} \, dx,$$

$$\int (\cos ax)^n \, dx = \frac{(\cos ax)^{n-1} \sin ax}{na} + \frac{n-1}{n} \int (\cos ax)^{n-2} \, dx. \tag{1}$$

Another method uses the formulas

$$2 \sin^2 x = 1 - \cos 2x,$$

$$2 \cos^2 x = \cos 2x + 1. \tag{2}$$

(These follow immediately from $\cos 2x = \cos^2 x - \sin^2 x$ and $1 = \cos^2 x + \sin^2 x$.) Any even power of sine or cosine can be reduced by successive applications of these formulas.

Example 3. $\cos^4 x = (\cos^2 x)^2 = \left(\dfrac{\cos 2x + 1}{2}\right)^2$

$$= \frac{1}{4}\cos^2 2x + \frac{1}{2}\cos 2x + \frac{1}{4}$$

$$= \frac{1}{8}(\cos 4x + 1) + \frac{1}{2}\cos 2x + \frac{1}{4}.$$

So $\int \cos^4 x\, dx = (\sin 4x)/32 + x/8 + (\sin 2x)/4 + x/4 + c.$

This second method, based on (2), does not require you to memorize the complicated and rarely used reduction formulas (1); however, it is a little more efficient to use the reduction formulas when they are available in a handy reference.

Integer powers of the tangent

Positive integer powers of $\tan\theta$ can be evaluated by the reduction formula

$$\int (\tan\theta)^n\, d\theta = \int (\tan\theta)^{n-2}(\sec^2\theta - 1)\, d\theta$$

$$= \int (\tan\theta)^{n-2}\sec^2\theta\, d\theta - \int (\tan\theta)^{n-2}\, d\theta$$

$$= \frac{(\tan\theta)^{n-1}}{n-1} - \int (\tan\theta)^{n-2}\, d\theta,$$

together with

$$\int \tan\theta\, d\theta = \int \frac{\sin\theta}{\cos\theta}\, d\theta = \int \frac{-du}{u} = -\log|u| + c$$

$$= -\log|\cos\theta| + c. \qquad [u = \cos\theta]$$

Negative integer powers of $\tan\theta$ are *positive* integer powers of $\cot\theta$, and these are evaluated by

$$\int (\cot\theta)^n\, d\theta = -\frac{(\cot\theta)^{n-1}}{n-1} - \int (\cot\theta)^{n-2}\, d\theta,$$

together with

$$\int \cot\theta\, d\theta = \log|\sin\theta| + c.$$

[For the derivation of these formulas, see Problem 17 below.]

Example 4. $\int \tan^3 2\theta \, d\theta = \int \tan 2\theta(\sec^2 2\theta - 1)\, d\theta$

$$= \int \tan 2\theta \sec^2 2\theta \, d\theta - \int \frac{\sin 2\theta}{\cos 2\theta}\, d\theta$$

$$= \int u \cdot \frac{1}{2}\, du - \int \frac{-\frac{1}{2}\, dv}{v} \qquad [u = \tan 2\theta,\ v = \cos 2\theta]$$

$$= \frac{u^2}{4} + \frac{1}{2}\log|v| + c = \frac{\tan^2 2\theta}{4} + \frac{1}{2}\log|\cos 2\theta| + c.$$

Integer powers of secant

Even powers of the secant can be easily reduced to rational functions:

$$\int (\sec\theta)^{2n}\, d\theta = \int (\sec\theta)^{2n-2}\sec^2\theta\, d\theta$$

$$= \int (1 + \tan^2\theta)^{n-1}\sec^2\theta\, d\theta$$

$$= \int (1 + u^2)^{n-1}\, du. \qquad [u = \tan\theta]$$

The integration is completed by expanding $(1 + u^2)^{n-1}$.

Odd powers of the secant are essentially odd powers of the cosine; they can be integrated as follows:

$$\int (\sec\theta)^{2n-1}\, d\theta = \int \frac{\cos\theta\, d\theta}{(\cos\theta)^{2n}}$$

$$= \int \frac{\cos\theta\, d\theta}{(1 - \sin^2\theta)^n}$$

$$= \int \frac{du}{(1 - u^2)^n}\,.$$

Another method is to obtain the reduction formula [Problem 16 below]

$$(n - 1)\int (\sec\theta)^n\, d\theta = (\sec\theta)^{n-2}\tan\theta + (n - 2)\int (\sec\theta)^{n-2}\, d\theta.$$

When n is odd, this reduces the integral ultimately to $\int \sec\theta\, d\theta$, which can be integrated as in the following example.

Example 5. $\int \sec\theta\, d\theta = ?$

First solution. Treat $\sec\theta$ as $(\cos\theta)^{-1}$:

$$\int (\cos\theta)^{-1}\, d\theta = \int \frac{\cos\theta\, d\theta}{\cos^2\theta} = \int \frac{\cos\theta\, d\theta}{1-\sin^2\theta}$$

$$= \int \frac{du}{1-u^2} \qquad [u = \sin\theta]$$

$$= -\frac{1}{2}\log|1-u| + \frac{1}{2}\log|1+u| + c \qquad [\text{partial fractions}]$$

$$= \frac{1}{2}\log\left(\frac{1+\sin\theta}{1-\cos\theta}\right) + c.$$

Second solution [a trick that handles only this particular integral]:

$$\int \sec\theta\, d\theta = \int \sec\theta\, \frac{\sec\theta + \tan\theta}{\sec\theta+\tan\theta}\, d\theta$$

$$= \int \frac{\sec^2\theta + \sec\theta\tan\theta}{\tan\theta+\sec\theta}\, d\theta$$

$$= \int \frac{du}{u} \qquad [u = \sec\theta + \tan\theta]$$

$$= \log|u| + c = \log|\sec\theta+\tan\theta| + c.$$

It is easy to show that $(\sec\theta+\tan\theta)^2 = \dfrac{1+\sin\theta}{1-\sin\theta}$, so the two solutions agree.

Sines and cosines of different arguments

Integrals of the form $\int \cos a\theta \cos b\theta\, d\theta$, $\int \cos a\theta \sin b\theta\, d\theta$, or $\int \sin a\theta \sin b\theta\, d\theta$ can be evaluated by two integrations by parts as in Example 8, §6.3. But it is easier to use the identities

$$2(\cos a\theta)(\cos b\theta) = \cos(a+b)\theta + \cos(a-b)\theta,$$

$$2(\cos a\theta)(\sin b\theta) = \sin(a+b)\theta - \sin(a-b)\theta,$$

$$2(\sin a\theta)(\sin b\theta) = \cos(a-b)\theta - \cos(a+b)\theta.$$

["Argument" is an old word for the point where a function is evaluated. The "different arguments" referred to here are $a\theta$ and $b\theta$.]

Example 6. $\displaystyle\int_0^{\pi/2} (\cos 2\theta)(\cos 3\theta)\, d\theta = ?$ [Solution below.]

The substitution $u = \tan(x/2)$

Every trigonometric function can be expressed in terms of $\sin x$ and $\cos x$, and somebody once noticed that the substitution $u = \tan(x/2)$ expresses each of these as a rational function of u, while $dx = 2\,du/(1+u^2)$ is likewise rational in u. Hence this substitution is a potentially powerful method of evaluating those trigonometric integrals not covered by the simple methods listed above. The relevant formulas are easy to derive:

$$u = \tan\left(\frac{x}{2}\right), \qquad x = 2 \arctan u, \qquad dx = \frac{2\,du}{1+u^2}; \tag{3}$$

$$\sin x = 2 \sin\left(\frac{x}{2}\right) \cos\left(\frac{x}{2}\right) = \frac{2 \tan(x/2)}{\sec^2(x/2)} = \frac{2u}{1+u^2}; \tag{4}$$

$$\cos x = 2\cos^2\left(\frac{x}{2}\right) - 1 = \frac{2}{\sec^2(x/2)} - 1 = \frac{1-u^2}{1+u^2}. \tag{5}$$

Example 7. $\displaystyle\int \frac{dx}{\sin x + \cos x} = ?$ *Solution.* We have from (4) and (5)

$$\sin x + \cos x = \frac{(2u + 1 - u^2)}{(1+u^2)},$$

so the integral is $\int (-2\,du)/(u^2 - 2u - 1)$. This is a rational function, and we integrate it by factoring the denominator:

$$u^2 - 2u - 1 = (u - r_1)(u - r_2),$$

where the roots are $r_1 = 1 + \sqrt{2}$ and $r_2 = 1 - \sqrt{2}$. So

$$\int \frac{-2\,du}{u^2 - 2u - 1} = A \log|u - r_1| + B \log|u - r_2| + c,$$

where A and B are easily determined by differentiation: $A = -1/\sqrt{2}$ and $B = 1/\sqrt{2}$. Hence

$$\int \frac{dx}{\sin x + \cos x}$$

$$= -\frac{1}{\sqrt{2}} \log|\tan(x/2) - 1 - \sqrt{2}| + \frac{1}{\sqrt{2}} \log|\tan(x/2) - 1 + \sqrt{2}| + c.$$

Solution to Example 6. $\displaystyle\int_0^{\pi/2} \cos 2\theta \cos 3\theta \, d\theta = \frac{1}{2} \int_0^{\pi/2} (\cos 5\theta + \cos\theta)\, d\theta$

$$= \left[\frac{1}{5} \sin 5\theta + \sin\theta\right]_0^{\pi/2} = \frac{6}{5}.$$

PROBLEMS

• **1.** $\int \sin\theta \cos 2\theta \, d\theta$

2. $\int \cos^3 a\theta \, d\theta$

• **3.** $\int \cos^2 a\theta \, d\theta$

4. $\int \tan^2 \theta \, d\theta$

• **5.** $\int \tan^4 \theta \, d\theta$

6. $\int \frac{d\theta}{\sec\theta}$

• **7.** $\int \sec^2 \theta \, d\theta$

8. $\int \csc\theta \, d\theta$

9. $\int \sin^2 2x \, dx$

10. $\int \frac{dt}{\sin^4 t}$

• **11.** $\int \frac{d\theta}{\sin\theta - \cos\theta}$

12. $\int \frac{d\theta}{\sin^2\theta + \cos^2\theta}$

13. Show that $0 = \int_0^\pi \sin(m\theta)\sin(n\theta)\, d\theta$ [n and m integers, $n \neq m$].

14. Show that $0 = \int_0^\pi \cos(m\theta)\cos(n\theta)\, d\theta$ [n and m integers, $n \neq m$].

15. (a) In Problem 13, take $m = 1$ and $n = 2$. Try to explain graphically why the integral is zero.
(b) Take $m = 1$ and $n = 3$, and work Problem 13 graphically.

16. Obtain the reduction formula

$$(n-1)\int (\sec\theta)^n \, d\theta = (\sec\theta)^{n-2}\tan\theta + (n-2)\int (\sec\theta)^{n-2}\, d\theta.$$

17. Show that $\int (\cot\theta)^n \, d\theta = -(\cot\theta)^{n-1}/(n-1) - \int (\cot\theta)^{n-2}\, d\theta$, and $\int \cot\theta \, d\theta = \log|\sin\theta| + c$.

VOLUME OF REVOLUTION

18. One arch of the sin curve is revolved about the x axis. Compute the resulting volume.

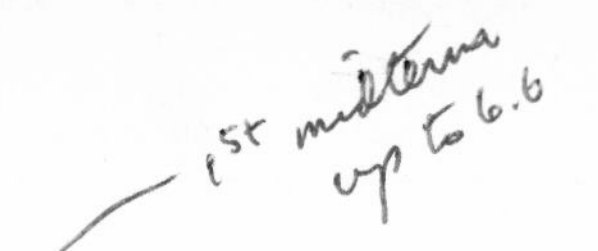

6.6 TRIGONOMETRIC SUBSTITUTION

Each of the identities

$$1 - \sin^2\theta = \cos^2\theta \tag{1}$$

$$1 + \tan^2\theta = \sec^2\theta \tag{2}$$

$$\sec^2\theta - 1 = \tan^2\theta \tag{3}$$

expresses the sum or difference of two squares as a perfect square. This is useful in integrals involving the square root of a quadratic, or even with more general integrals involving sums or differences of squares. The following simple example illustrates the method.

Example 1. $\displaystyle\int \frac{dx}{x\sqrt{x^2-1}} = ?$ *Solution.* The expression $x^2 - 1$ suggests the identity (3). To exploit this identity, make the substitution $x = \sec\theta$; then $\sqrt{x^2-1} = \sqrt{\tan^2\theta} = |\tan\theta|$ and $dx = \sec\theta\tan\theta\,d\theta$, whence

$$\int \frac{dx}{x\sqrt{x^2-1}} = \int \frac{\sec\theta\tan\theta\,d\theta}{\sec\theta\,|\tan\theta|}$$

$$= \int \frac{\tan\theta}{\pm\tan\theta}\,d\theta = \pm\int d\theta = \pm\theta + c = \pm\operatorname{arcsec} x + c. \tag{4}$$

Here, setting $\theta = \operatorname{arcsec} x$, we use "principal values" of $\operatorname{arcsec} x$, i.e. $0 \le \theta < \pi/2$, or $\pi/2 < \theta \le \pi$. The choice of $\pm$ sign depends on the sign of $\tan\theta$, which depends on the sign of $\sec\theta$, which depends on the sign of x. Notice that the integrand $1/x\sqrt{x^2-1}$ is defined only for $x > 1$ and for $x < -1$. If $x > 1$, then $\sec\theta > 1$, so $0 < \theta < \pi/2$, and $\tan\theta > 0$; therefore, $|\tan\theta| = +\tan\theta$, and we choose the $+$ sign in (4). If $x < -1$, on the other hand, then $\sec\theta < -1$, so $\pi/2 < \theta < \pi$, and $\tan\theta < 0$; therefore, $|\tan\theta| = -\tan\theta$, and we choose the $-$ sign in (4).

[The same result (4) can be deduced directly from the formula for the derivative of arcsec,

$$\operatorname{arcsec}'(x) = \frac{1}{|x|\sqrt{x^2-1}}. \tag{5}$$

But the method outlined above avoids this easily forgotten formula. It requires only the familiar trig identity (3), and standard rules for substitution.]

In general, when we have $\sqrt{ax^2+bx+c}$, the first step is to complete the square so that the square root becomes one of the following cases:

Case (i) $\sqrt{a}\sqrt{(x-\alpha)^2+\beta^2}$

[when $a > 0$ and $ax^2 + bx + c$ has no real roots]

Case (ii) $\sqrt{a}\sqrt{(x-\alpha)^2-\beta^2}$

[when $a > 0$ and $ax^2 + bx + c$ has two real roots]

Case (iii) $\sqrt{-a}\sqrt{\beta^2-(x-\alpha)^2}$

[when $a < 0$ and $ax^2 + bx + c$ has two real roots].

We do not list the case of double roots (for then $ax^2 + bx + c$ is a perfect square, and the square root can be eliminated) or the case where $a < 0$ and $ax^2 + bx + c$ has no real roots (for then $ax^2 + bx + c$ is always negative, and we cannot take its square root on any interval).

Case (i) is handled by setting $x - \alpha = \beta \tan\theta$ $(-\pi/2 < \theta < \pi/2)$ so that

$$\sqrt{(x-\alpha)^2+\beta^2} = \beta\sqrt{\tan^2\theta+1} = \beta\sec\theta.$$

Case (ii) is handled by setting $x - \alpha = \beta\sec\theta$ $(0 \le \theta < \pi/2$ or $\pi/2 < \theta \le \pi)$ so that

$$\sqrt{(x-\alpha)^2-\beta^2} = \beta\sqrt{\sec^2\theta-1} = \beta\,|\tan\theta|.$$

Case (iii) is handled by setting $x - \alpha = \beta\sin\theta$ $(-\pi/2 \le \theta \le \pi/2)$, so that

$$\sqrt{\beta^2-(x-\alpha)^2} = \beta\sqrt{1-\sin^2\theta} = \beta\cos\theta.$$

Notice that the range of each substitution is appropriate to the root involved. In case (ii), for example, $x - \alpha$ ranges over $x - \alpha \ge \beta$ when $0 \le \theta < \pi/2$, and over $x - \alpha \le -\beta$ when $\pi/2 < \theta \le \pi$; therefore, the substitution requires implicitly that $|x - \alpha| \ge \beta$. This is appropriate since $\sqrt{(x-\alpha)^2-\beta^2}$ is defined only when $|x - \alpha| \ge \beta$.

Example 2. The upper half of a circle of radius r about the origin has the equation $y = \sqrt{r^2 - x^2}$, so the area of the half circle of radius r is

$$\int_{-r}^{r} \sqrt{r^2-x^2}\,dx.$$

This integral is case (iii), so we set $x = r\sin\theta$; we then have $dx = r\cos\theta\,d\theta$, $\sqrt{r^2 - x^2} = r\cos\theta$; hence

$$\int \sqrt{r^2-x^2}\,dx = \int \sqrt{r^2\cos^2\theta}\;r\cos\theta\,d\theta = r^2\int \cos^2\theta\,d\theta$$

$$= \frac{r^2}{2}\int (\cos 2\theta + 1)\,d\theta \qquad [\S 6.5]$$

$$= \frac{r^2}{2}\left(\frac{\sin 2\theta}{2} + \theta\right) + c.$$

Convert the answer back to x by noting that $\sin 2\theta = 2 \sin\theta \cos\theta$, so

$$\frac{r^2}{2}\sin 2\theta = (r\sin\theta)(r\cos\theta) = x\sqrt{r^2 - x^2}\,.$$

Hence

$$\int \sqrt{r^2 - x^2}\,dx = \frac{1}{2}x\sqrt{r^2 - x^2} + \frac{1}{2}r^2 \arcsin\left(\frac{x}{r}\right) + c,$$

and the area of the half circle is

$$\begin{aligned}\int_{-r}^{r} \sqrt{r^2 - x^2}\,dx &= \left(\frac{1}{2}x\sqrt{r^2 - x^2} + \frac{1}{2}r^2\arcsin\left(\frac{x}{r}\right)\right)\Bigg|_{-r}^{r} \\ &= \tfrac{1}{2}r^2 \arcsin(1) - \tfrac{1}{2}r^2\arcsin(-1) \\ &= \tfrac{1}{2}r^2\left(\frac{\pi}{2}\right) - \tfrac{1}{2}r^2\left(-\frac{\pi}{2}\right) = \tfrac{1}{2}\pi r^2.\end{aligned}$$

Remark 1. In a *definite integral* it is not necessary to convert to the original variable. In Example 2, we found

$$\int \sqrt{r^2 - x^2}\,dx = \frac{r^2}{2}\left(\frac{\sin 2\theta}{2} + \theta\right) + c \qquad [x = r\sin\theta].$$

Now, to evaluate the *definite* integral $\int_{-r}^{r} \sqrt{r^2 - x^2}\,dx$, observe that as θ varies from $-\pi/2$ to $\pi/2$, then $x = r\sin\theta$ varies from $-r$ to r [Fig. 3]. Hence

$$\int_{-r}^{r} \sqrt{r^2 - x^2}\,dx = \tfrac{1}{2}r^2\left(\frac{\sin 2\theta}{2} + \theta\right)\Bigg|_{\theta=-\pi/2}^{\theta=\pi/2} = \tfrac{1}{2}r^2\left(\frac{\pi}{2} + \frac{\pi}{2}\right) = \tfrac{1}{2}\pi r^2.$$

[Formal justification of this method is given in Problem 16 below.]

Example 3. $\displaystyle\int \frac{x\,dx}{\sqrt{x^2 + 2x + 2}} = ?$ *Solution.* Completing the square, we find $x^2 + 2x + 2 = (x+1)^2 + 1$ [case (i)], so we set $x + 1 = \tan\theta$, which makes $dx = \sec^2\theta\,d\theta$, $\sqrt{x^2 + 2x + 2} = \sec\theta$, and we obtain

$$\begin{aligned}\int \frac{x\,dx}{\sqrt{x^2 + 2x + 2}} &= \int \frac{(\tan\theta - 1)\sec^2\theta\,d\theta}{\sec\theta} \\ &= \int \tan\theta\sec\theta\,d\theta - \int \sec\theta\,d\theta = \sec\theta - \log|\sec\theta + \tan\theta| + c.\end{aligned}$$

Since $\sec\theta = \sqrt{x^2 + 2x + 2}$, and $x + 1 = \tan\theta$, we get

$$\int \frac{x\,dx}{\sqrt{x^2 + 2x + 2}} = \sqrt{x^2 + 2x + 2} - \log\left|\sqrt{x^2 + 2x + 2} + x + 1\right| + c.$$

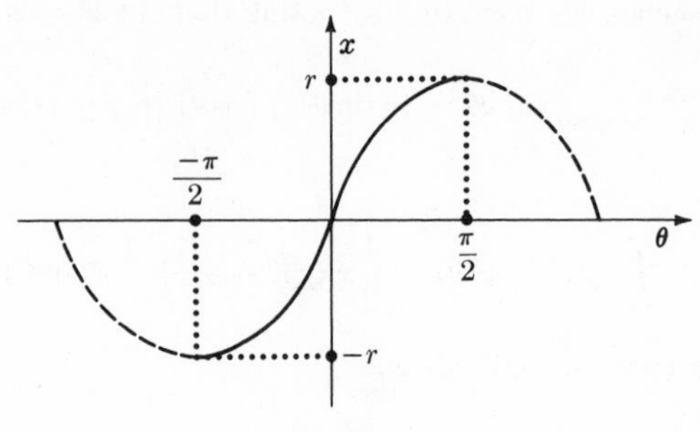

FIGURE 6.3

In Examples 2 and 3 we converted from the "angular variable" θ to the original variable x by using appropriate trigonometric identities. These and other useful identities are suggested by Fig. 4.

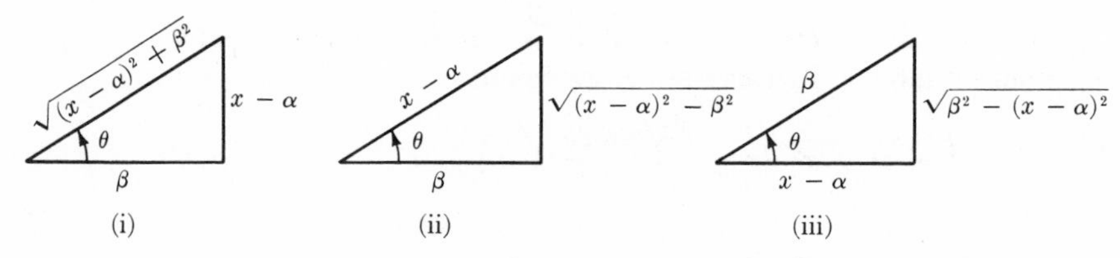

FIGURE 6.4

Example 4. $\int \sqrt{x^2+2x}\,dx = ?$ [Solution below.]

Example 5. The integral $\displaystyle\int_{-1}^{b} \sqrt{\frac{1+x}{1-x}}\,dx$ and the limit $\displaystyle\lim_{b\to 1-} \int_{-1}^{b} \sqrt{\frac{1+x}{1-x}}\,dx$ happen to arise in aerodynamics. The integral can be evaluated by a sequence of substitutions, as follows:

First step. To eliminate the square root, set

$$\sqrt{\frac{1+x}{1-x}} = u, \qquad \text{or} \qquad 1+x = u^2 - xu^2, \qquad \text{or} \qquad x = \frac{u^2-1}{u^2+1},$$

$$dx = \frac{dx}{du}\,du = \frac{(u^2+1)2u - (u^2-1)2u}{(u^2+1)^2}\,du = \frac{4u\,du}{(u^2+1)^2}.$$

Thus

$$\int \sqrt{\frac{1+x}{1-x}}\,dx = \int u \cdot \frac{4u\,du}{(u^2+1)^2}, \tag{6}$$

and we have converted the integrand to a rational function. This could be integrated by partial fractions; but it is easier to make a trigonometric substitution as in the next step.

Second step. To simplify the sum of two squares in the right-hand integral in (6), set $u = \tan\theta$, $du = \sec^2\theta\, d\theta$,

$$\int \frac{4u^2\, du}{(u^2+1)^2} = \int \frac{4\tan^2\theta \sec^2\theta\, d\theta}{\sec^4\theta} = \int 4\sin^2\theta\, d\theta$$

$$= \int 2(1+\cos 2\theta)\, d\theta = 2\theta - \sin 2\theta + C.$$

Third step. To evaluate the definite integral from -1 to b, track down the appropriate limits for θ; as x varies from -1 to b, then

$$u = \sqrt{\frac{1+x}{1-x}} \quad \text{varies from} \quad \sqrt{\frac{1+(-1)}{1-(-1)}} = 0 \quad \text{to} \quad \sqrt{\frac{1+b}{1-b}};$$

hence

$$\theta = \arctan u \quad \text{varies from} \quad \arctan 0 = 0 \quad \text{to} \quad \arctan\sqrt{\frac{1+b}{1-b}}.$$

Thus

$$\int_{-1}^{b} \sqrt{\frac{1+x}{1-x}}\, dx = \Big[2\theta - \sin 2\theta \Big]_0^{\arctan\sqrt{(1+b)/(1-b)}}$$

$$= 2\arctan\sqrt{\frac{1+b}{1-b}} - \sin\left(2\arctan\sqrt{\frac{1+b}{1-b}}\right). \tag{7}$$

Finally, to evaluate the limit as $b \to 1-$, notice that $\lim\limits_{b\to 1-} \sqrt{\frac{1+b}{1-b}} = +\infty$;

therefore,

$$\lim_{b\to 1-} \arctan\sqrt{\frac{1+b}{1-b}} = \frac{\pi}{2}, \quad \text{and} \quad \lim_{b\to 1-} \sin\left(2\arctan\sqrt{\frac{1+b}{1-b}}\right) = \sin\pi = 0,$$

so from (7)

$$\lim_{b\to 1-} \int_{-1}^{b} \sqrt{\frac{1+x}{1-x}}\, dx = \pi.$$

[Notice that this limit is the area of an unbounded region, since the integrand $\sqrt{(1+x)/(1-x)}$ has a vertical asymptote at $x = 1$.]

Solution to Example 4: $x^2 + 2x = (x+1)^2 - 1$. The appropriate identity is $\sec^2\theta - 1 = \tan^2\theta$; we set $x + 1 = \sec\theta$ to obtain $\sqrt{x^2+2x} = \sqrt{\sec^2\theta - 1} = \tan\theta$. (The identity $\sin^2\theta - 1 = -\cos^2\theta$ is no good, because then we would have to take $\sqrt{-\cos^2\theta}$.) Now we have $dx = \sec\theta\tan\theta\, d\theta$, and $\sqrt{x^2+2x} = \tan\theta$; hence

$$\int \sqrt{x^2+2x}\, dx = \int \tan\theta \cdot \sec\theta \cdot \tan\theta\, d\theta = \int (\sec^3\theta - \sec\theta)\, d\theta$$

$$= \tfrac{1}{2}\sec\theta\cdot\tan\theta + \tfrac{1}{2}\int \sec\theta\, d\theta - \log|\sec\theta + \tan\theta|$$

$$= -\tfrac{1}{2}\log|\sec\theta + \tan\theta| + \tfrac{1}{2}\sec\theta\tan\theta + c.$$

PROBLEMS

In Problems 1–11, check your own answers by differentiating them; the derivative of your answer should be the given integrand. Problems 15 and 16 concern change of limits for substitutions in indefinite integrals. The last five problems concern integrals from the various applications in §5.4.

1. $\displaystyle\int \frac{dx}{\sqrt{1-x^2}}$ [Set $x = \sin\theta$.]

2. $\displaystyle\int \frac{dx}{\sqrt{x^2+1}}$ [Set $x = \tan\theta$.]

3. $\displaystyle\int \frac{(x+1)\,dx}{\sqrt{x^2-2x+6}}$

4. $\displaystyle\int \frac{(x-1)\,dx}{\sqrt{x^2-2x+6}}$

5. $\displaystyle\int \frac{(2x+1)\,dx}{\sqrt{2x-x^2}}$

6. $\displaystyle\int \frac{x\,dx}{\sqrt{x^2+4x+5}}$

7. $\displaystyle\int (x^2+x+1)^{3/2}\,dx$

8. $\displaystyle\int \frac{dx}{\sqrt{x^2-2x-15}}$

9. $\displaystyle\int \frac{(2-x)\,dx}{\sqrt{15+2x-x^2}}$

10. $\displaystyle\int \frac{dx}{(x^2+1)^2}$ [Set $x = \tan\theta$.]

11. $\displaystyle\int \frac{x\,dx}{\sqrt{11+4x-x^2}}$

12. (a) The graph of $f(x) = \sqrt{r^2-x^2}$ for $|x| \le r/\sqrt{2}$ is a quarter of a circle of radius r. Using the formula $\int\sqrt{1+f'(x)^2}\,dx$ for arc length, compute the length of a quarter circle.

(b) Explain why $\displaystyle\int_{-r}^{r} \sqrt{1+f'(x)^2}\,dx$ is an improper integral [§5.5] when $f(x) = \sqrt{r^2-x^2}$.

(c) Compute $\displaystyle\lim_{b\to r-}\int_{-b}^{b} \sqrt{1+f'(x)^2}\,dx$ with f as above, and explain the result geometrically.

Problems 13 and 14 concern the *hyperbolic functions* introduced in §2.8: $\sinh\theta = \frac{1}{2}(e^{\theta}-e^{-\theta})$, $\cosh\theta = \frac{1}{2}(e^{\theta}+e^{-\theta})$, etc.

13. Prove that $\sinh 2\theta = 2\sinh\theta\cosh\theta$ and $2\sinh^2\theta = \cosh 2\theta - 1$.

14. (a) Show that the shaded area in Fig. 5(ii) has area

$$\frac{1}{2}\cosh\theta\sinh\theta - \int_{1}^{\cosh\theta} \sqrt{x^2-1}\,dx.$$

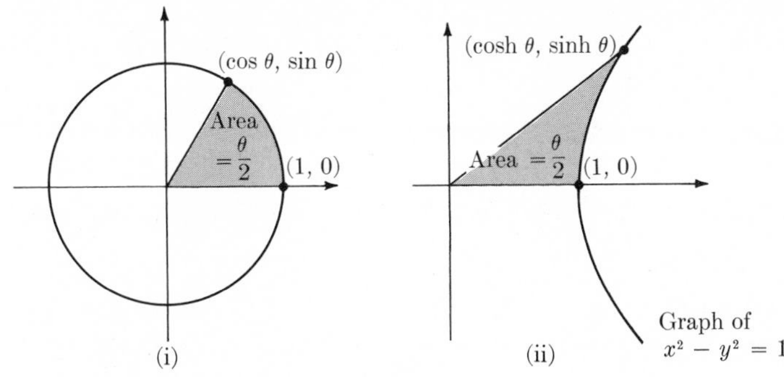

FIGURE 6.5

(b) Evaluate the integral in part (a) by the substitution $x = \cosh t$, together with Problem 13. [Answer in Fig. 5(ii).]

(c) Compute the analogous area in Fig. 5(i).

15. Evaluate each of the following definite integrals $\int_a^b f$ in two ways:

(i) Find an indefinite integral $F(x)$, and compute $F(b) - F(a)$.
(ii) Make the suggested substitution, and the corresponding change in limits of integration, as in Remark 1 above.

(a) $\displaystyle\int_0^1 \frac{x\,dx}{1 + x^2}$ $[u = 1 + x^2]$

(b) $\displaystyle\int_0^1 \sqrt{r^2 - x^2}\,dx$ $[x = r\cos\theta]$

(c) $\displaystyle\int_0^1 \frac{dx}{\sqrt{x^2 + 1}}$ $[x = \tan\theta]$

(d) $\displaystyle\int_1^5 \frac{x + 1}{x^2 + 2x}\,dx$ $[u = x^2 + 2x]$

(e) $\displaystyle\int_{-1}^{-1/2} \frac{x + 1}{x^2 + 2x}\,dx$

(f) $\displaystyle\int_0^{\pi/4} \tan x\,dx$ $[u = \cos x]$

(g) $\displaystyle\int_0^{\pi/2} \sin x \cos x\,dx$ $[u = \sin x]$

(h) $\displaystyle\int_0^b \frac{dt}{\sqrt{a + t}}$ $(a > 0, b > -a)$ $[u = a + t]$

16. (a) Suppose that g has a continuous derivative on $[a,b]$, and that f is continuous on the range of g. Prove the following rule for substitution in a definite integral:

$$\int_a^b f(g(x))g'(x)\,dx = \int_{g(a)}^{g(b)} f(u)\,du. \tag{8}$$

[Hint: Let $F' = f$, and evaluate both sides in terms of F.]

(b) Prove (8) when g is a differentiable *decreasing* function on $[a,b]$, with range $[g(b),g(a)]$.

APPLICATIONS

• **17.** A cylindrical tank of radius a and length b, set on its side, contains water to a depth h. Find the volume of water.

• **18.** Find the length of the curve $\{(x,y): y = x^2\}$ from $(0,0)$ to $(2,4)$.

19. The curve in Problem 18 is revolved about the x axis. Find the surface area generated.

• **20.** Compute the volume of the torus generated by revolving the graph of $(x - a)^2 + y^2 = r^2$ $(r < a)$ about the y axis. [See Problem 11, §5.4.]

21. Find the center of gravity of the area inside the circle

$$\{(x,y): x^2 + y^2 - r^2\}$$

and above the line $\{(x,y): y = b\}$, where $0 < b < r$.

6.7 SEPARABLE DIFFERENTIAL EQUATIONS

In this chapter we have been solving differential equations of a rather simple type: Given a function g, find f such that $f' = g$. Now we consider a slightly more complicated differential equation,

$$f'(x) = g(x)h(f(x)) \tag{1}$$

in which g and h are given functions. This equation is called *separable* for a reason that will soon become apparent.

The Leibniz notation leads almost by magic to a solution of (1). Imagine the desired solution as graphed in the xy plane. Then, on the graph, y and x are related by $y = f(x)$, so we write $f'(x) = dy/dx$. In this notation, (1) becomes

SEPARABLE DIFFERENTIAL EQUATION

$$\frac{dy}{dx} = g(x)h(y). \tag{2}$$

Treating dy and dx as independent entities, we can *separate the variables* x and y in (2) to obtain

$$\frac{dy}{h(y)} = g(x)\,dx. \tag{3}$$

Now, putting an integral sign on both sides of (3), we have

$$\int \frac{dy}{h(y)} = \int g(x)\,dx, \tag{4}$$

and a solution is found by evaluating the integrals.

Example 1. The equation $f'(x) = xf(x)$ becomes, in Leibniz notation,

$$\frac{dy}{dx} = xy.$$

Separate variables:

$$\frac{dy}{y} = x\,dx.$$

Integrate:

$$\int \frac{dy}{y} = \int x\,dx,$$

$$\log|y| = \frac{x^2}{2} + c,$$

$$|y| = e^c e^{x^2/2},$$

$$y = \pm e^c e^{x^2/2} = f(x).$$

Finally, check to see that f really solves the given equation: $f'(x) = \pm e^c x e^{x^2/2} = xf(x)$, as required.

Example 2. Solve $dy/dx = 2y + 1$. *Solution.* Separating variables, we find

$$\frac{dy}{2y+1} = dx,$$

$$\int \frac{2\,dy}{2y+1} = \int 2\,dx,$$

$$\log|2y+1| = 2x + c,$$

$$|2y+1| = e^c e^{2x},$$

$$2y + 1 = \pm e^c e^{2x}.$$

Example 3. Solve $dy/dx = xy^2$. [Solution below.]

Results found (informally) by juggling the Leibniz notation can generally be proved (formally) by using the chain rule, as we did in Example 1. In particular, the method of "separation of variables" can be justified in this way.

Theorem 1. *Let g and h be continuous functions, each defined on some interval. Let $G' = g$ and $H' = 1/h$. Then any differentiable function f such that*

$$H(f(x)) = G(x) + c \tag{5}$$

is a solution of equation (1) above, $f'(x) = g(x)h(f(x))$.

Remark 1. Since $G(x) = \int g(x)\,dx$, $H(y) = \int \frac{dy}{h(y)}$, and $y = f(x)$, equation (5) is the same as (4).

Remark 2. In (1) and (5), it is tacitly assumed that the range of f lies in the domain of h, and also in the domain of H.

Proof of Theorem 1. Differentiate (5) by the chain rule:

$$H'(f(x))f'(x) = G'(x)$$

$$\frac{1}{h(f(x))}f'(x) = g(x) \qquad \left[\text{since } H' = \frac{1}{h} \text{ and } G' = g\right]$$

$$f'(x) = g(x)h(f(x)).$$

Hence f satisfies the given equation. *Q.E.D.*

Remark 3. Theorem 1 does not claim that *every* solution of (1) can be obtained in the form (5) or (4); sometimes, in fact, there are other solutions. For example, in the equation

$$f' = f^{2/3}, \tag{6}$$

the separation method yields the solution $f(x) = (x + c)^3/27$, where c is any constant. But we obtain another solution by setting $f(x) = 0$ for all x. The general method did not include this because the first step, in which we write $dy/(y^{2/3}) = dx$, is not possible when $y = f(x) = 0$.

[The equation (6) has some rather surprising solutions; see Problem 9 below.]

Solution to Example 3

$$y^{-2}\,dy = x\,dx, \qquad \int y^{-2}\,dy = \int x\,dx,$$

$$-y^{-1} = \tfrac{1}{2}x^2 + c, \qquad y = \frac{-1}{c + x^2/2}.$$

Check: $y' = \dfrac{x}{(c + x/2)^2} = xy^2.$

Note that $y = 0$ is also a solution.

PROBLEMS

1. Solve the following equations. Try to include any solutions not given by the separation method.

• (a) $dy/dx = e^{x+y}$

(b) $2xy = dy/dx$

• (c) $dy/dx = y^{1/3}$

(d) $(x \log y) \dfrac{dy}{dx} = y$

• (e) $(xy)\, dy/dx = (1 + x^2) \csc y$. [Leave answer in the form $H(y) = G(x) + c$.]

2. Solve the following equations. Try to include any solutions not given by the separation method.

• (a) $\dfrac{dy}{dx} = y \sin x$ (b) $\dfrac{dy}{dx} = ye^x$ • (c) $\dfrac{dx}{dy} = y \tan x$

3. (a) Solve $dy/dx = y/x$.

(b) Show that the solutions found in part (a) form straight lines through the origin.

(c) Are there any straight lines through the origin *not* obtained as the "separation of variables" solutions in (a)? If so, do these give solutions of the equation?

(d) Use Theorem 1, §0.3, to explain why every nonvertical straight line through the origin solves the equation in part (a).

4. Let a be a continuous function on I. Derive the solution of $f'(x) = f(x)a(x)$. [Note: the equations in Problems 2 and 3 have this form.]

5 • (a) For the equation $dy/dx = e^{x+y}$ in Problem 1(a), find a solution such that $y = 0$ when $x = 0$.

• (b) For the solution $y = f(x)$ in part (a), find $\lim_{x \to -\infty} f(x)$.

• (c) Find a vertical asymptote for the solution in part (a).

(d) Let x_0 and y_0 be given, and let $y = f(x)$ be a solution of $dy/dx = e^{x+y}$ such that $y_0 = f(x_0)$. Find $\lim_{x \to -\infty} f(x)$, and find the vertical asymptote.

• **6.** Suppose a railroad car rolls on a track, and the only appreciable force it encounters is air resistance proportional to its velocity; then the force F is given by $F = -kv$.

AIR RESISTANCE: HYPOTHETICAL MODEL 1

(a) Use Newton's law $F = mv'$ to set up a differential equation for the velocity of the car. Solve the equation.

(b) Does the car ever stop?

(c) If the velocity at $t = 0$ is v_0, does the car roll indefinitely far?

7. Work Problem 6 supposing that the air resistance is proportional to the *square* of the velocity. **MODEL 2**

8. Work Problem 6 supposing that the air resistance is proportional to the *square root* of the velocity. **MODEL 3**

9. Suppose $c_1 < c_2$, and define a function f by

$$f(x) = \begin{cases} \dfrac{(x - c_1)^3}{27} & \text{if } x < c_1 \\ 0 & \text{if } c_1 \le x \le c_2 \\ \dfrac{(x - c_2)^3}{27} & \text{if } x > c_2 . \end{cases}$$

(a) Sketch the graph of f.

(b) Prove that $f'(x)$ exists for every x and that f' is continuous at every point. [Hint: The only points in doubt are $x = c_1$ and $x = c_2$.]

(c) Prove that f solves the equation $f' = f^{2/3}$.

10. Twenty pounds of salt are dumped in a basin containing 120 pounds of pure water at time $t = 0$. The salt proceeds to dissolve. Let $Q(t)$ be the amount of dissolved salt at time $t \ge 0$; thus $Q(0) = 0$, and Q is an increasing function. According to elementary physical chemistry, it dissolves at a rate given by the differential equation

RATE OF DISSOLUTION

$$\frac{dQ}{dt} = k(20 - Q)\left(A - \frac{Q}{120}\right),$$

where k, A are constants; A is the "saturation concentration," given in (pounds of dissolved salt)/(pound of water). Notice that $Q(t)/120$ is the concentration at time t, and $20 - Q(t)$ is the amount of undissolved salt remaining; the rate of dissolution dQ/dt is proportional to the product of these. This is reasonable; the rate of dissolution drops to zero if either (i) the amount of undissolved salt drops to zero or (ii) the concentration reaches its saturation point.

• (a) Find a solution Q of the differential equation with $Q(0) = 0$. Distinguish two cases: $A \ne 1/6$, and $A = 1/6$.

• (b) Show that when $A > 1/6$, then $\lim_{t\to\infty} Q(t) = 20$. Explain why this is reasonable, given the meaning of A.

(c) Show that when $A < 1/6$, then $\lim_{t\to\infty} Q(t) = 120A$. Explain this.

6.8 FIRST ORDER LINEAR DIFFERENTIAL EQUATIONS

FIRST ORDER LINEAR EQUATION

An equation of the special type

$$f' + af = g \tag{1}$$

in which a and g are given continuous functions, is called a *first order linear differential equation.* [The name is explained in the remark below.]

When a is identically zero, equation (1) becomes $f' = g$, where g is a given function; this is precisely the equation discussed in the first six sections of this chapter.

When a is *not* zero, equation (1) can be multiplied by a so-called "integrating factor," which reduces it to the familiar case where $a = 0$.

Example 1. Solve $f'(x) + 2f(x) = \cos x$. *Solution.* If we multiply both sides of the equation by e^{2x}, we get the equivalent equation

$$e^{2x}f'(x) + 2e^{2x}f(x) = e^{2x}\cos x.$$

But notice that $e^{2x}f'(x) + 2e^{2x}f(x) = d(e^{2x}f(x))/dx$, so our equation can be written

$$\frac{de^{2x}f(x)}{dx} = e^{2x}\cos x.$$

This means that

$$e^{2x}f(x) = \int e^{2x}\cos x\,dx$$

$$= \frac{2e^{2x}\cos x + e^{2x}\sin x}{5} + c$$

[integration by parts, §6.3]

so

$$f(x) = \frac{2\cos x + \sin x}{5} + ce^{-2x}.$$

As a final check:

$$f'(x) = -\tfrac{2}{5}\sin x + \tfrac{1}{5}\cos x - 2ce^{-2x},$$
$$2f(x) = \tfrac{2}{5}\sin x + \tfrac{4}{5}\cos x + 2ce^{-2x},$$

hence $f'(x) + 2f(x) = \cos x$, as required.

The trick in Example 1 works for any equation

$$f'(x) + af(x) = g(x)$$

SOLUTION OF $f' + af = g$, a CONSTANT

where a is a *constant.* Multiplying both sides by e^{ax} gives

$$e^{ax}f'(x) + ae^{ax}f(x) = e^{ax}g(x)$$

which can be written

$$\frac{de^{ax}f(x)}{dx} = e^{ax}g(x).$$

Hence

$$e^{ax}f(x) = \int e^{ax}g(x)\,dx,$$

and the solution is completed by evaluating the integral.

More generally, when a is *not* constant, the equation can be solved by multiplying through by the "integrating factor" e^A, where $A = \int a$ denotes any indefinite integral of a. Since e^A is never zero, the equation

GENERAL SOLUTION OF $f' + af = g$

$$f' + af = g \tag{1}$$

is equivalent to

$$(e^A)f' + a(e^A)f = (e^A)g. \tag{2}$$

Differentiating by the product rule, $(e^A f)' = e^A f' + ae^A f$, so equation (2) is equivalent to

$$(e^A f)' = (e^A)g. \tag{3}$$

By the definition of indefinite integral, (3) is equivalent to

$$(e^A)f = \int (e^A)g, \tag{4}$$

and we have found f, provided that the integrals can be evaluated.

This solution of the first order linear case is more complete than the solution of the separable case [§6.7], because we have found *all* solutions. The preceding paragraph actually proves:

Theorem 2. *If a and b are continuous on an interval I, then*

$$f' + af = g \text{ on } I$$

$$\Leftrightarrow$$

$$f = e^{-A}\left(\int ge^A\right) + ce^{-A} \quad \text{for some constant } c,$$

where A denotes any indefinite integral of a.

[Here, contrary to our usual practice, $\int ge^A$ denotes any indefinite integral of ge^A, not including the constant of integration, and $\int ge^A + c$ denotes the general indefinite integral of ge^A.]

If you trust your memory for small details, you can solve first order linear equations by remembering the exact formula in Theorem 2. However, it is probably better to follow the outline in the examples; first write the equation in the standard form $f' + af = g$, then multiply through by e^A, recognize the left-hand side as the derivative of $e^A f$, and evaluate the resulting indefinite integral.

Example 2. $x^3 f'(x) - f(x) = 1$. *Solution.* To put it in the form $f' + af = g$, divide by x^3 to get

$$f'(x) - \frac{1}{x^3}f(x) = \frac{1}{x^3}. \tag{5}$$

Here $a(x) = -\frac{1}{x^3}$, so $A(x) = \int a(x)\,dx = -\int \frac{dx}{x^3} = \frac{1}{2x^2} + c.$

Any indefinite integral of a will do, so we drop the constant c and take the integrating factor $e^A = e^{1/2x^2}$. Multiplying (5) by this, we get

$$e^{1/2x^2}f'(x) - \frac{1}{x^3}e^{1/2x^2}f(x) = \frac{1}{x^3}e^{1/2x^2}$$

$$\Leftrightarrow$$

$$\frac{de^{1/2x^2}f(x)}{dx} = \frac{1}{x^3}e^{1/2x^2}$$

$$\Leftrightarrow$$

$$e^{1/2x^2}f(x) = \int \frac{1}{x^3}e^{1/2x^2}\,dx = -e^{1/2x^2} + c$$

[here we keep the constant c, because we want *all* solutions of the given equation], hence

$$f(x) = -1 + ce^{-1/2x^2}.$$

Example 3. Suppose we have a simple electrical circuit with constant resistance R and capacitance C driven by a voltage E. Then the current I as a function of time satisfies the equation

ELECTRICAL CIRCUIT

$$R\frac{dI}{dt} + C^{-1}I = \frac{dE}{dt}.$$

Suppose further that $E(t) = \sin \omega t$ (alternating voltage of frequency $\omega/2\pi$), so the equation becomes

$$R\frac{dI}{dt} + \frac{1}{C}I = \omega \cos(\omega t).$$

Putting it in the form $f' + af = g$, we get

$$\frac{dI}{dt} + \frac{1}{RC}I = \frac{\omega}{R}\cos(\omega t).$$

The integrating factor is $e^{A(t)} = e^{t/RC}$, so we get

$$e^{t/RC}\frac{dI}{dt} + \frac{1}{RC}e^{t/RC}I = \frac{de^{t/RC}I(t)}{dt} = \frac{\omega}{R}\cos(\omega t)e^{t/RC},$$

and therefore

$$e^{t/RC}I = \frac{\omega}{R}\int e^{t/RC}\cos \omega t\,dt + \text{constant}$$

$$= \frac{(\omega/R)e^{t/RC}(RC\omega \sin \omega t + \cos \omega t)}{1 + (RC\omega)^2} + \text{constant},$$

so

$$I = \frac{\omega}{R}\frac{RC\omega \sin \omega t + \cos \omega t}{1 + (RC\omega)^2} + \text{constant}\cdot e^{-t/RC}.$$

The term $e^{-t/RC}$ grows very small for large t; it is called a *transient*. The other term represents an oscillation of the same frequency $\omega/2\pi$ as the driving voltage E, but the maximum current does not occur at the same time as the maximum voltage; the maximum of $\sin \omega t$ occurs at $t = \pi/2\omega$, whereas the maximum of $RC\omega \sin \omega t + \cos \omega t$ occurs at $t = (1/\omega) \arctan(RC\omega)$.

Remark: Why the equation is called linear

The reason for calling $f' + af = g$ a *first order* equation is obvious; it involves the first derivative f', but no derivative $f'', f''', \ldots$ of higher order. The reason for calling the equation *linear* is not quite so obvious; it can be explained as follows.

Given a function f, we define a new function $L(f)$ by

$$L(f) = f' + af.$$

Then the equation $f' + af = g$ becomes $L(f) = g$. You can check that

$$L(f_1 + f_2) = L(f_1) + L(f_2) \tag{6}$$

and

$$L(cf) = c \cdot L(f). \tag{7}$$

As we pointed out in Chapter II, the properties (6) and (7) of L are shared by the function $l(t) = kt$, whose graph is a straight line through the origin. Thus (6) and (7) are called the properties of *linearity*, and the equation $Lf = g$ is called a *linear* equation.

PROBLEMS

1. Solve the following equations.
- • (a) $f'(x) + xf(x) = 0$ [There are two ways to do this.]
- (b) $f'(x) + xf(x) = x$
- • (c) $xf'(x) + f(x) = \sin x$

2 • (a) Find all solutions of $f'(x) + xf(x) = 0$ such that $f(0) = 0$.
- (b) Find all solutions of $f'(x) + xf(x) = x$ such that $f(0) = 0$.
- • (c) Find all solutions of $xf'(x) + f(x) = \sin x$ such that $f(0) = 0$.
- (d) Find all solutions of $f'(x) + xf(x) = 0$ with $f'(0) = 1$.
- (e) Find all solutions of $f'(x) + xf(x) = x$ with $f(0) = 1$.
- • (f) Find all solutions of $xf'(x) + f(x) = \sin x$ with $f(0) = 1$.

3. Find the general solution of each of the following equations, as in Problem 1; then find the particular solution that satisfies the added condition, as in Problem 2.
- • (a) $xf'(x) = 2f(x) + 3x^4 + x^2$; $\quad f(1) = -2$
- (b) $xf'(x) + (1 + x)f(x) = e^{-x}$; $\quad f(1) = 0$

• (c) $x\dfrac{dy}{dx} = 2y + x^2;$ $y = 0$ when $x = 2$

(d) $(1 - x)\dfrac{dy}{dx} + y = -x(1 - x)^2;$ $y = 3$ when $x = 0$

• (e) $(x + 1)\,dy = (2y + (x + 1)^4)\,dx;$ $y = 4$ when $x = 1$
[This means the same as the equation obtained by dividing through by dx.]

(f) $(1 + xy)\,dx = (1 + x^2)\,dy;$ $y = 0$ when $x = 0$

• (g) $(x - 1)F''(x) + F'(x) = (x - 1)^2;$ $F(2) = 1, F'(2) = 0.$
[Hint: First solve the equation for F', then find F.]

4. (a) Using Theorem 2, prove the following THEOREM:

Let a and g be continuous on an interval I, and let x_0 be any point in I. Given any number y_0, there is one and only one function f defined on I such that

$$f' + af = g, \qquad \textit{and} \qquad f(x_0) = y_0 .$$

(b) Explain why Problem 2(f) does not contradict this theorem.

5. A container holds g gallons of water in which are mixed s_0 pounds of a substance such as salt or soap. At time $t = 0$, pure water is poured in at the constant rate of r gallons per minute. At the same time, the water is constantly mixed, and the mixture flows out at r gallons per minute. Find the amount $s(t)$ of the substance present at time $t > 0$. [Hint: The rate of change of s is due to the mixture flowing out at r gallons per minute, each gallon containing s/g pounds of the substance, so $s' = -rs/g$.]

SLOW FLUSHING

6. In Problem 5, suppose the incoming water is not pure, but contains C pounds per gallon of the substance.
• (a) Write the new differential equation for s.
• (b) Solve the equation.
(c) What happens to $s(t)$ as t grows large and positive? Is this to be expected from an intuitive view of the physical situation of the problem?

7. Let

$$f(x) = \begin{cases} c_1e^{-2/x^2} & x < 0 \\ 0 & x = 0 \\ c_2e^{-2/x^2} & x > 0 \end{cases}$$

EQUATION WITH MANY SOLUTIONS PASSING THROUGH ORIGIN

where c_1 and c_2 are any constants whatsoever. Show that:
(a) f is continuous at 0
(b) $f'(x) \to 0$ as $x \to 0$
(c) $f'(0) = 0$
(d) f is a solution of $x^3f'(x) - f(x) = 0$.
[When the equation is written in standard form $f' + af = 0$, the function $a(x) = x^{-3}$ is unbounded at $x = 0$. This shows up subtly in the solutions; the solution for $x < 0$ is unrelated to that for $x > 0$.]

NAPIER'S ORIGINAL LOGARITHM

8. This problem gives Napier's original definition of logarithms. [See J. F. Scott, *A History of Mathematics* (London: Taylor and Francis, 1958), pp. 129–131.]

Imagine a particle P starting at the point 0 on the line at time $t = 0$ and traveling at the constant velocity of 10^7; thus at time t, P is at the point $p(t) = 10^7 t$. And imagine a second particle Q starting at the point 0, but moving in such a way that its velocity at any time t equals its distance from the point 10^7. Thus if $q(t)$ is the position of Q at time t, then $q(0) = 0$ and $q'(t) = 10^7 - q(t)$. With this set-up, at each time t, $p(t)$ is (Napier's) logarithm of $10^7 - q(t)$.

(a) Find an explicit expression for $q(t)$.
(b) Set $y = 10^7 - q(t)$, and show that (Napier's) logarithm of y is $-10^7 \log(10^{-7}y)$.
(c) Let $\operatorname{Nog}(y)$ denote Napier's logarithm of y. Express $\operatorname{Nog}(ab)$ in terms of $\operatorname{Nog}(a)$ and $\operatorname{Nog}(b)$, and show that Nog reduces multiplication to addition.

OSCILLATING CIRCUIT

• 9. Solve the equation

$$L\frac{dI}{dt} + RI = E \sin \omega t,$$

where L, R, E, and ω are constants, and I is a function of t. [This equation gives the current I in a circuit of resistance R and inductance L, driven by an alternating current generator of frequency $\omega/2\pi$ and maximum voltage E.]

LINEAR OPERATIONS

10. An operation L on functions is called *linear* if

$$L(f+g) = L(f) + L(g) \qquad \text{and} \qquad L(cf) = cL(f);$$

see the Remark preceding these problems. Which of the following operations are linear?

(a) $L(f) = f'$
(b) $L(f) = (f')^2$
(c) $L(f) = f + 1$
(d) $L(f) = \sqrt{1 + (f')^2}$
(e) $L(f)(x) = x^2 f(x) + e^{-x} f'(x)$
(f) $L(f)(x) = \int_0^x f$

11. Suppose that L is a linear operation on functions, and that f_1 and f_2 are two particular functions such that $L(f_1) = 0$ and $L(f_2) = 0$. Prove that $f = c_1 f_1 + c_2 f_2$ (where c_1 and c_2 are any constants) solves the equation $L(f) = 0$.

12. [*Compare Problem 11*] Let $L(f) = (f')^2 + f^2 - 1$.

(a) Show that $f_1(x) = \cos x$ and $f_2(x) = \sin x$ satisfy $L(f_1) = 0$ and $L(f_2) = 0$.
(b) Does $L(f_1 + f_2) = 0$?
(c) Does $L(2f_1) = 0$?

6.9 GENERALITIES ON DIFFERENTIAL EQUATIONS

This chapter compiled tricks and stratagems to find explicit formulas solving differential equations of the following types:

$$f' = g$$

$$f'(x) = g(x)h(f(x))$$

$$f' + af = g.$$

These are all special cases of the general *first order differential equation*

GENERAL FIRST ORDER EQUATION

$$f'(x) = \varphi(x,f(x)) \tag{1}$$

or, in Leibniz notation,

$$\frac{dy}{dx} = \varphi(x,y). \tag{1'}$$

In this general form, φ is a function assigning to each point (x,y) in the plane a number; the equation (1) says that at each point $(x,f(x))$ on the graph of f, the slope $f'(x)$ is to be the one prescribed by φ. For example, consider the equation

$$f'(x) = \frac{1}{x^3}f(x) \qquad \text{or} \qquad \frac{dy}{dx} = \frac{y}{x^3}. \tag{2}$$

Without solving the equation, we can see by graphical means what it says. At many points (x,y), draw short line segments having the required slope $\varphi(x,y) = y/x^3$. Figure 6 shows the result, and sketches a curve which has the prescribed slope at each of these points.

The particular equation (2) can be solved explicitly; the solution is

$$f(x) = \begin{cases} c_1e^{-1/2x^2} & x < 0 \\ 0 & x = 0 \\ c_2e^{-1/2x^2} & x > 0 \end{cases}$$

where c_1 and c_2 are any constants [Problem 7 above]. These solutions behave just as Fig. 6 suggests; each one passes through the origin, and each one approaches a constant value c_1 as $x \to +\infty$, and a (possibly different) constant value c_2 as $x \to -\infty$.

Equations of the special form $f'(x) = g(x)$, where the function $\varphi(x,y) = g(x)$ is actually independent of y, assign the same slope all along each vertical line, as in Fig. 7. This shows clearly that raising or lowering a solution by any constant produces another solution. This is the "law of parallel graphs" in §3.2; it is the reason for writing $+ c$ after evaluating an indefinite integral.

These two examples illustrate the geometric point of view toward differential equations. The equation

$$dy/dx = \varphi(x,y)$$

gives a "family of slopes," assigning a slope at each point where φ is defined [Figs. 6 and 7]. The solution of the equation is a "family of curves"; through each point where φ is defined, there passes one of these solution curves. Figure 8 gives representative members of the family of curves solving the equations illustrated in Figs. 6 and 7.

This geometric point of view comes out in physics in the study of potentials and forces. The *equipotential lines* form a family of curves, and the *lines of force* are curves at right angles to the equipotential lines. For a "central force" [see further §7.9] the equipotentials are concentric circles [Fig. 9] and the lines of force are straight lines through the center of the circles.

For an example more interesting from the geometric point of view, suppose the equipotential lines are the concentric ellipses [Fig. 10]

$$\frac{x^2}{a^2} + \frac{y^2}{b^2} = c, \qquad 0 < c. \tag{3}$$

What, then, are the lines of force? First of all, we find the slope dy/dx on an equipotential line (3) by differentiating (3):

$$\frac{2x}{a^2} + \frac{2y}{b^2}\frac{dy}{dx} = 0, \qquad \frac{dy}{dx} = -\frac{b^2x}{a^2y}.$$

Thus, the equipotential curve (3) passing through a point (x,y) has its slope given by $dy/dx = -b^2x/a^2y$.

The lines of force, being perpendicular to the equipotentials, must have the negative reciprocal slope. Therefore, at each point (x,y) in the plane, the line of force passing through (x,y) has slope

$$\frac{dy}{dx} = \frac{a^2y}{b^2x}.$$

This equation is easily solved by separation of variables:

$$\frac{dy}{y} = \frac{a^2}{b^2}\frac{dx}{x}$$

$$\log y = \frac{a^2}{b^2}\log x + c$$

$$y = \pm C\,|x|^{a^2/b^2} \tag{4}$$

These curves are graphed in Fig. 10 for the case where $a^2/b^2 = 2$; the lines of force in this case are parabolas through the origin. When $a^2/b^2 = 1$, the "ellipses" are circles, and the curves (4) are straight lines through the origin, as in Fig. 9.

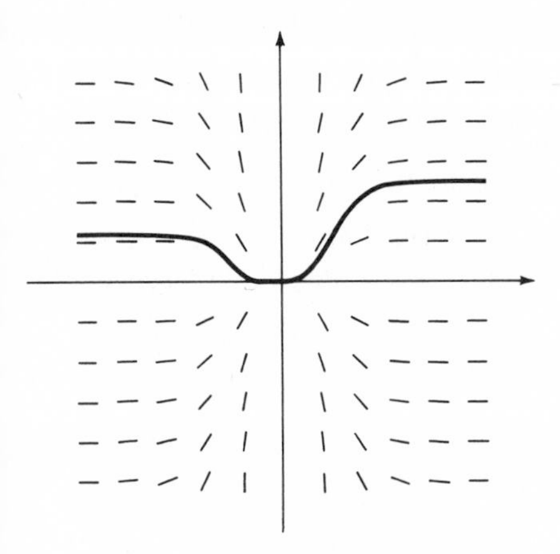

FIGURE 6.6 Slope at (x,y) is $\varphi(x,y) = y/x^3$

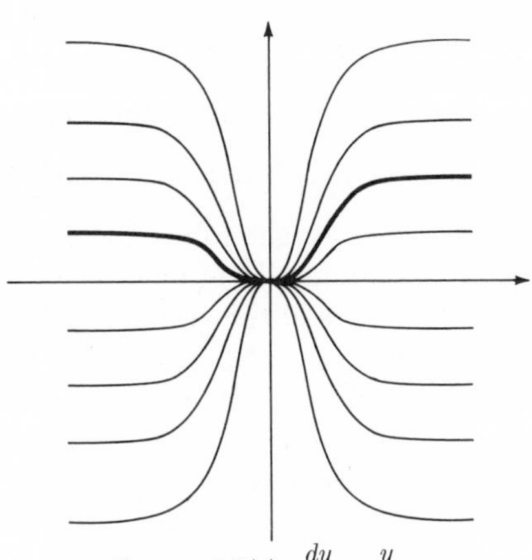

FIGURE 6.8(a) $\dfrac{dy}{dx} = \dfrac{y}{x^3}$

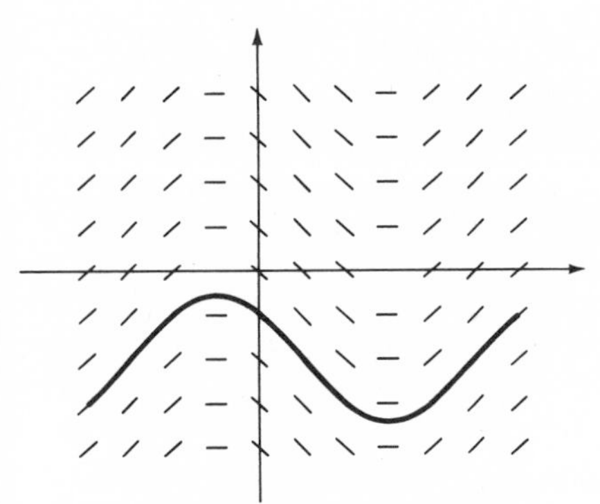

Slope $\varphi(x,y) = g(x)$ does not depend on y

FIGURE 6.7

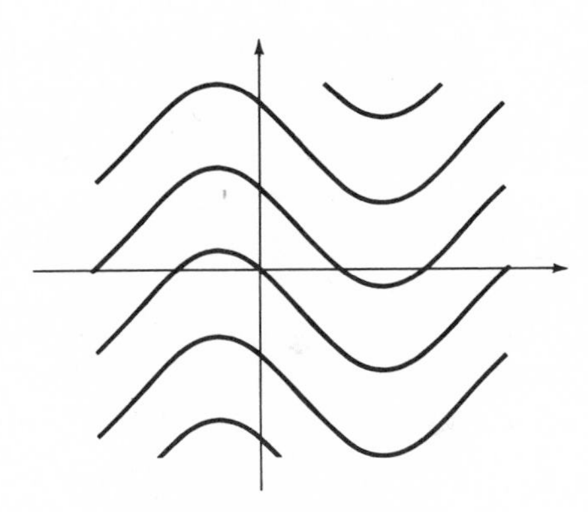

FIGURE 6.8(b) $\dfrac{dy}{dx} = g(x)$

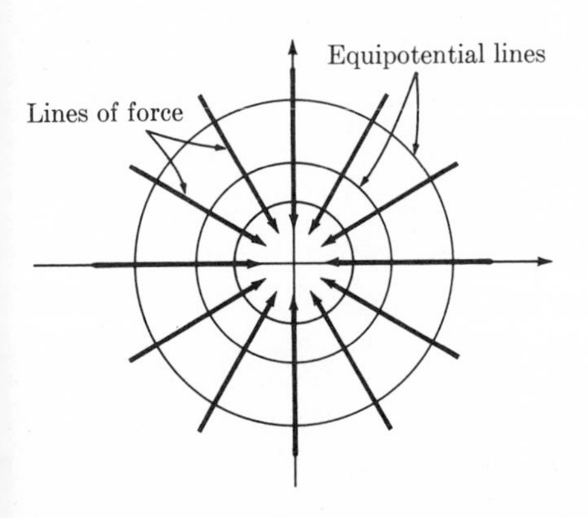

FIGURE 6.9

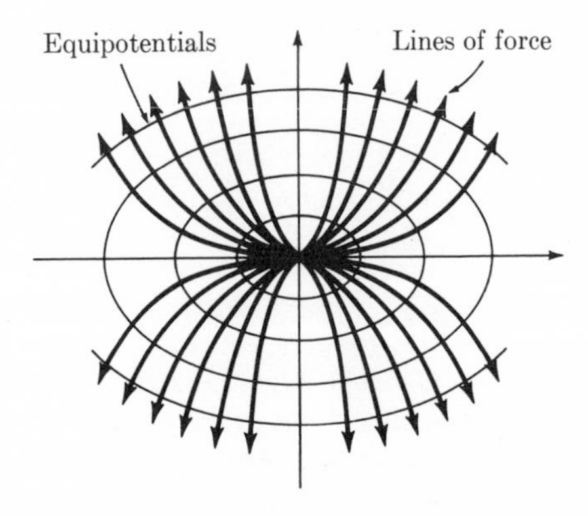

FIGURE 6.10

PROBLEMS

1. For the equation $dy/dx = y/x$ [Problem 3, §6.7], sketch the family of slopes, and sketch representative solution curves.

2. Repeat Problem 1, using each of the equations in Problem 2, §6.7.

3. Repeat Problem 1, using each of the equations in Problem 1, §6.7.

4. For each of the following families of equipotential lines, find the corresponding family of lines of force. [These lines are called the *orthogonal trajectories* of the equipotentials.]
   - • (a) $x^2/2 - y^2 = c$
   - (b) $x^2/a^2 - y^2/b^2 = c$
   - (c) $xy = c$
   - • (d) $y = x^2 + c$
   - • (e) $y = cx^2$ [Hint: Write the slopes of the equipotential lines as a function of x and y, but *not* of c; notice that on these lines, $c = y/x^2$.]

Chapter VII

Vectors and the Laws of Motion

One of the greatest scientific achievements of all time was that of Isaac Newton, who established the connection between Kepler's empirical laws of planetary motion and the inverse-square law of gravitational attraction. In this chapter we present the mathematical ideas relating to the motion of a particle in a plane, and use these ideas to prove those famous theorems of Newton.

7.1 PLANE VECTORS

Until now, we have interpreted an ordered pair of real numbers as a point in the plane. The analysis of motion requires a new, more subtle interpretation, and with it a new notation. We write $[a,b]$ for the ordered pair; and we interpret $[a,b]$ not as a point, but as a *change in position*, a units to the right and b units upward.

Ordered pairs in this interpretation are called *vectors*. The word is quite appropriate to motion, for it is cognate with "convey," derived from the Latin *vehere* (to carry). We think of the vector $[a,b]$ as conveying points in the plane a units to the right and b units upwards.

It is natural to follow one change of position by another; this leads to the notion of *addition of vectors*. And, given a particular change of position, we may consider others in the same direction, but twice, half, or ten times as large, and so on. This leads to the notion of multiplication of a vector by a real number, called *scalar multiplication*.

The interpretation of an ordered pair as a change in position thus leads to certain algebraic operations. We first define the operations, and then discuss the interpretation.

The *sum* of two vectors $[a_1, a_2]$ and $[b_1, b_2]$ is defined by

VECTOR SUM

$$[a_1, a_2] + [b_1, b_2] = [a_1 + a_2, b_1 + b_2], \tag{1}$$

and the *product* of a vector $[a_1, a_2]$ by a real number c. defined by

NUMBER TIMES VECTOR

$$c \cdot [a_1, a_2] = [ca_1, ca_2]. \tag{2}$$

A third operation, *subtraction*, can be constructed out of these two: $[a_1, a_2] - [b_1, b_2]$ is defined to be $[a_1, a_2] + (-1) \cdot [b_1, b_2]$. Since $(-1) \cdot [b_1, b_2] = [(-1)b_1, (-1)b_2] = [-b_1, -b_2]$, this amounts to

VECTOR DIFFERENCE

$$[a_1, a_2] - [b_1, b_2] = [a_1 - b_1, a_2 - b_2].$$

Without these operations, the set of ordered pairs is the plane of plane analytic geometry. *With* these operations, the set of ordered pairs forms an algebraic structure called the *vector space* R^2. The R stands for the real numbers, whose additive and multiplicative structure enters into the defining formulas (1) and (2) above. The superscript in R^2 indicates that we are considering ordered *pairs* of real numbers. If, instead, we considered ordered triples, quadruples, etc., we would denote the resulting structure R^3, R^4, etc.

A member of the plane is called a *point;* a member of the vector space is called a *vector*. For a point (x,y), the numbers x and y are called *coordinates* of the point. For a vector $[a,b]$, a and b are called *components* of the vector.

When we think of an ordered pair as a point, we denote it by any capital letter; when we think of an ordered pair as a vector we denote it by a boldface capital letter such as **A**, **B**, **U**. Finally, real numbers are denoted by italic lower case letters such as a, b, x.

The fundamental rules for calculating with vectors are easily stated and proved. They are the *associative laws*

ELEMENTARY VECTOR LAWS

$$(\mathbf{A} + \mathbf{B}) + \mathbf{C} = \mathbf{A} + (\mathbf{B} + \mathbf{C}), \tag{3}$$

$$c \cdot (d\mathbf{A}) = (c\,d)\mathbf{A} \tag{4}$$

the *commutative law*,

$$\mathbf{A} + \mathbf{B} = \mathbf{B} + \mathbf{A}, \tag{5}$$

the *distributive law*,

$$c \cdot (\mathbf{A} + \mathbf{B}) = c \cdot \mathbf{A} + c \cdot \mathbf{B}, \tag{6}$$

and finally the "obvious" law,

$$1 \cdot \mathbf{A} = \mathbf{A}.$$

Because $(\mathbf{A} + \mathbf{B}) + \mathbf{C} = \mathbf{A} + (\mathbf{B} + \mathbf{C})$, we do not really need the parentheses in a sum of three vectors, and we write simply $\mathbf{A} + \mathbf{B} + \mathbf{C}$.

As an example of the proofs of these rules, take (5). The letter **A** stands for some ordered pair $[a_1, a_2]$, and **B** stands for another ordered pair $[b_1, b_2]$. Then

$$\begin{aligned}
\mathbf{A} + \mathbf{B} &= [a_1\,,\, a_2] + [b_1\,,\, b_2] \\
&= [a_1 + b_1\,,\, a_2 + b_2] && \text{[definition (1)]} \\
&= [b_1 + a_1\,,\, b_2 + a_2] && \text{[property of numbers]} \\
&= [b_1\,,\, b_2] + [a_1\,,\, a_2] && \text{[definition (1)]} \\
&= \mathbf{B} + \mathbf{A}.
\end{aligned}$$

Graphical representation

We interpret the vector $[a,b]$ as a change in position, starting at any point (x,y) and ending at the point $(x + a,\ y + b)$. The natural way to represent this idea is to draw an arrow from (x,y) to $(x + a,\ y + b)$, as in Fig. 1. For the starting point (x,y), we may choose any convenient point. All the arrows in Fig. 2 represent the same vector.

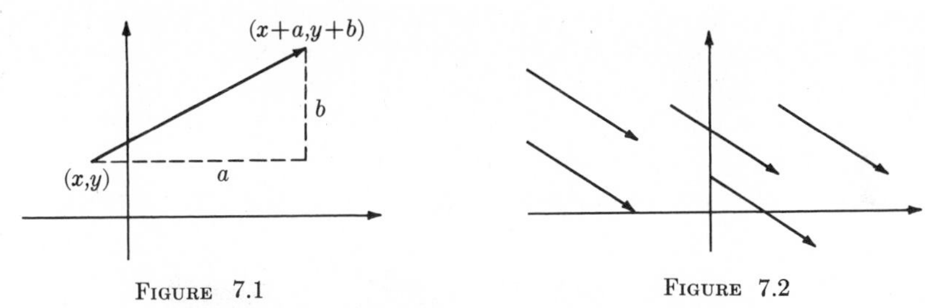

FIGURE 7.1

FIGURE 7.2

The representation as arrows is a great help in thinking about vectors. Figure 3 illustrates addition. We think of a particle moved by **A** from O to P, and then by **B** from P to Q. The result of these two motions is **A** + **B**.

The commutative law **A** + **B** = **B** + **A** is illustrated in Fig. 4. We imagine a particle moved first by **A** from O to P_1, and then by **B** from P_1 to Q. Performing the operations in reverse order (first **B** and then **A**), we arrive at the same point Q by passing along the other side of the parallelogram in Fig. 4, from O to P_1 to Q.

This figure suggests the *parallelogram rule* for addition: When **A**, **B**, and **A** + **B** are all represented as arrows starting at the same point, then the head of the arrow for **A** + **B** is at a vertex of the parallelogram determined by the arrows for **A** and **B**.

PARALLELOGRAM RULE

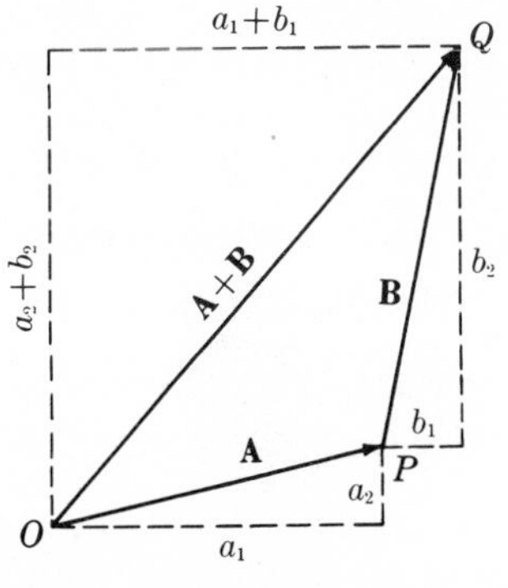

FIGURE 7.3

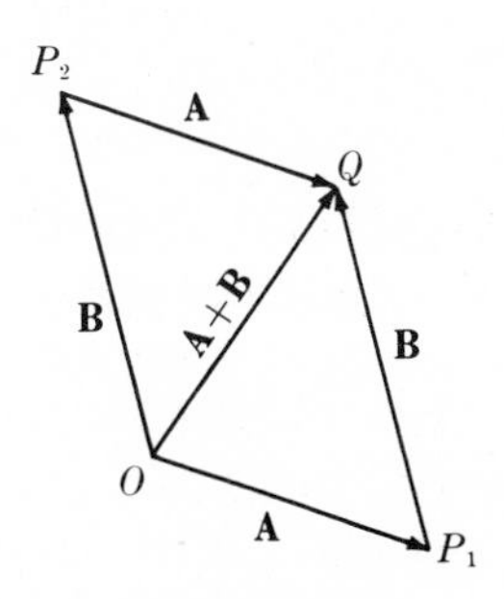

FIGURE 7.4

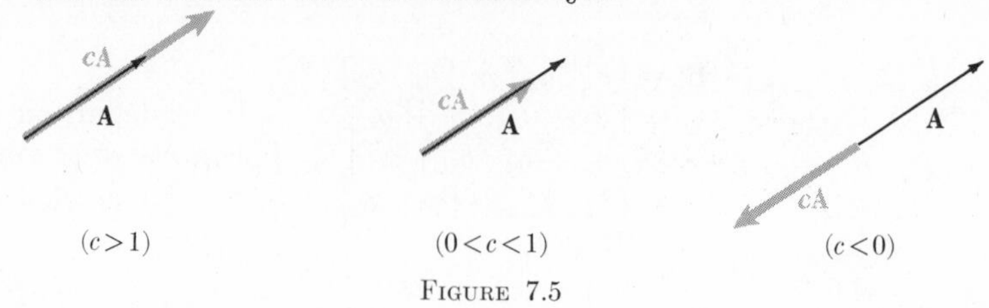

FIGURE 7.5

Figure 5 illustrates scalar multiplication of a vector $\mathbf{A} = [a_1, a_2]$ by a real number c. When $c > 0$, the vector $c\mathbf{A}$ points in the same direction as $\mathbf{A}$, but is c times as long; when $c < 0$, $c\mathbf{A}$ points in the direction *opposite* to $\mathbf{A}$, and is $|c|$ times as long.

B − A INTERPRETED

To interpret subtraction of $\mathbf{A}$ from $\mathbf{B}$, represent both $\mathbf{A}$ and $\mathbf{B}$ by arrows starting at the same point, as in Fig. 6. Then represent $\mathbf{B} - \mathbf{A}$ by an arrow *from* the head of the $\mathbf{A}$ arrow *to* the head of the $\mathbf{B}$ arrow. Notice that the resulting diagram is just what we would draw to show $\mathbf{B}$ as the sum of $\mathbf{A}$ and $\mathbf{B} - \mathbf{A}$; this explains why the arrow for $\mathbf{B} - \mathbf{A}$ is drawn as it is.

Finally, we introduce a notation linking points and vectors. Let $P_1 = (x_1, y_1)$ and $P_2 = (x_2, x_2)$ be any two points in the plane. Then the vector $[x_2 - x_1, y_2 - y_1]$ is denoted $\overrightarrow{P_1P_2}$. The arrow from P_1 to P_2 represents this vector [Fig. 7], and we can think of $\overrightarrow{P_1P_2}$ as a motion from P_1 to P_2.

FIGURE 7.6

FIGURE 7.7

PROBLEMS

• **1.** Represent the vector $[-1,2]$ as an arrow starting
  (a) at the origin
  (b) at the point $(1,1)$.

• **2.** (a) Represent the vectors $\mathbf{A} = [-1,1]$, $\mathbf{B} = [2,3]$, and their sum $\mathbf{A} + \mathbf{B}$, as arrows starting at the origin.
  (b) Sketch the parallelogram illustrating the identity $\mathbf{A} + \mathbf{B} = \mathbf{B} + \mathbf{A}$.

• **3.** Redraw Fig. 6, and add an appropriate arrow representing $\mathbf{A} + \mathbf{B}$. Where does this arrow intersect the arrow for $\mathbf{B} - \mathbf{A}$?

4. Prove:
  • (a) The associative law (3)
  (b) The distributive law (6)
  (c) The associative law (4)

5. Draw sketches illustrating
(a) the associative law (3)
(b) the distributive law (6). [This recalls theorems about similar triangles.]

6. Show that $\overrightarrow{PQ} = -\overrightarrow{QP}$.

7. Let $\mathbf{A} = [1,3]$, $\mathbf{B} = [2,-4]$. In a single diagram, sketch the following vectors as arrows starting at the origin.

(a) $\mathbf{A}$ and $\mathbf{B}$
(b) $\mathbf{A} + \mathbf{B}$
(c) $\frac{1}{2}(\mathbf{A} + \mathbf{B})$
(d) $\frac{1}{3}\mathbf{A} + \frac{2}{3}\mathbf{B}$
(e) $2\mathbf{A} - \mathbf{B}$
(f) $3\mathbf{A} - 2\mathbf{B}$
(g) $4\mathbf{A} - 3\mathbf{B}$
(h) $\frac{3}{2}\mathbf{B} - \frac{1}{2}\mathbf{A}$

8. In the previous problem, the heads of all the arrows except the one for $\mathbf{A} + \mathbf{B}$ should lie on one straight line. Notice that each of these vectors has the form $c\mathbf{A} + d\mathbf{B}$ for some constants c and d. What relation between c and d holds in each of these cases, *except* for the vector $\mathbf{A} + \mathbf{B}$?

7.2 LENGTH AND INNER PRODUCT

The *length* of a vector $\mathbf{V} = [a,b]$ in R^2, denoted $|\mathbf{V}|$, is defined by

$$|\mathbf{V}| = |[a,b]| = \sqrt{a^2 + b^2}\,. \tag{1}$$

VECTOR LENGTH

If we represent $\mathbf{V} = [a,b]$ by an arrow from (x,y) to $(x + a,\, y + b)$, then $|\mathbf{V}|$ is the length of the arrow. If we think of $\mathbf{V}$ as the change of position in moving from P to Q, then $|\mathbf{V}|$ is the distance traveled between P and Q [Fig. 8].

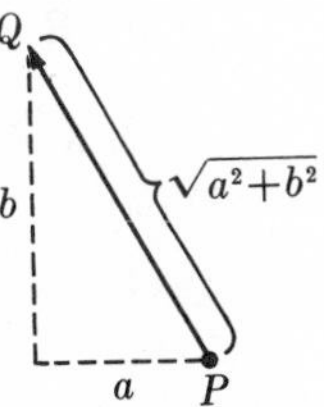

FIGURE 7.8

The *inner product* of two vectors $\mathbf{A} = [a_1, a_2]$ and $\mathbf{B} = [b_1, b_2]$, denoted $\mathbf{A}\cdot\mathbf{B}$, is defined by

INNER (DOT) PRODUCT

$$\mathbf{A}\cdot\mathbf{B} = [a_1, a_2]\cdot[b_1, b_2] = a_1b_1 + a_2b_2\,. \tag{2}$$

(Because of its notation, the inner product $\mathbf{A}\cdot\mathbf{B}$ is also called the *dot product* of **A** and **B**.) The geometric interpretation of the inner product comes out when we compute the angle θ between the arrows representing **A** and **B**. Figure 9 suggests that $\theta = \varphi - \psi$, and

$$\cos\varphi = \frac{a_1}{|\mathbf{A}|}, \qquad \sin\varphi = \frac{a_2}{|\mathbf{A}|},$$

$$\cos\psi = \frac{b_1}{|\mathbf{B}|}, \qquad \sin\psi = \frac{b_2}{|\mathbf{B}|},$$

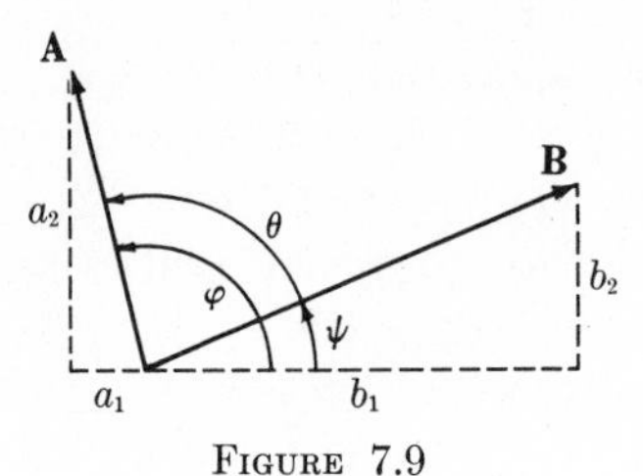

FIGURE 7.9

and hence

$$\cos\theta = \cos(\varphi - \psi) = \cos\varphi\cos\psi + \sin\varphi\sin\psi$$

$$= \frac{a_1b_1 + a_2b_2}{|\mathbf{A}|\cdot|\mathbf{B}|} = \frac{\mathbf{A}\cdot\mathbf{B}}{|\mathbf{A}|\cdot|\mathbf{B}|}.$$

So we interpret the inner product of **A** and **B** by the formula

B·A INTERPRETED

$$\mathbf{A}\cdot\mathbf{B} = |\mathbf{A}|\cdot|\mathbf{B}|\cos\theta, \tag{3}$$

where $|\mathbf{A}|$ and $|\mathbf{B}|$ are the lengths of **A** and **B**, and θ is the angle between **A** and **B** (or rather, the angle between the arrows representing them). The combination of a simple algebraic formula [given by (2)] and a clear geometric interpretation [given by (3)] make the inner product a useful operation, as you will see.

RULES FOR INNER PRODUCT

From (1) and (2), you can easily derive the formula

$$\mathbf{A}\cdot\mathbf{A} = |\mathbf{A}|^2, \tag{4}$$

and the *commutative law*

$$\mathbf{A}\cdot\mathbf{B} = \mathbf{B}\cdot\mathbf{A}, \tag{5}$$

the *distributive laws*

$$\mathbf{A}\cdot(\mathbf{B} + \mathbf{C}) = \mathbf{A}\cdot\mathbf{B} + \mathbf{A}\cdot\mathbf{C}, \tag{6}$$

$$(\mathbf{A} + \mathbf{B})\cdot\mathbf{C} = \mathbf{A}\cdot\mathbf{C} + \mathbf{B}\cdot\mathbf{C}, \tag{7}$$

the *associative law*

$$(c\mathbf{A})\cdot\mathbf{B} = c\cdot(\mathbf{A}\cdot\mathbf{B}) = \mathbf{A}\cdot(c\mathbf{B}), \tag{8}$$

and, finally, a formula connecting length and scalar multiplication,

$$|c\mathbf{A}| = |c|\cdot|\mathbf{A}|. \tag{9}$$

Problem 3 asks you to derive these. In addition, you will prove the *Schwarz inequality* [Problem 4]

SCHWARZ INEQUALITY

$$|\mathbf{A}\cdot\mathbf{B}| \le |\mathbf{A}|\cdot|\mathbf{B}|, \tag{10}$$

Just so you won't think we're leaving *everything* to you, we will do our bit by proving the *triangle inequality*

TRIANGLE INEQUALITY

$$|\mathbf{A} + \mathbf{B}| \leq |\mathbf{A}| + |\mathbf{B}|. \tag{11}$$

(Notice that the meaning of the bars and dots in these formulas is determined by the context. On the left-hand side of (10), for instance, the dot indicates the inner product of vectors, and the vertical bars indicate the absolute value of a number. On the right-hand side, the bars indicate lengths of vectors, and the dot indicates the product of two numbers.)

The geometric interpretation of the triangle inequality explains its name. In Fig. 10, the arrows representing **A**, **B**, and **A** + **B** form the three sides of a triangle, and inequality (11) says that one of these sides is less than or equal to the sum of the other two.

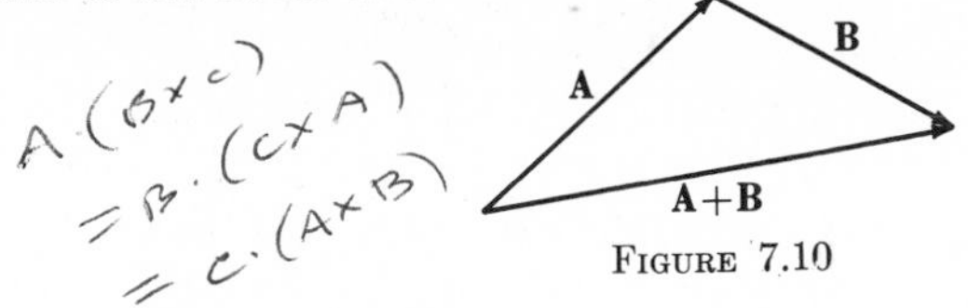

FIGURE 7.10

The interpretation of the Schwarz inequality (10) is found in the formula

$$\mathbf{A}\cdot\mathbf{B} = |\mathbf{A}|\cdot|\mathbf{B}| \cos\theta;$$

(10) follows from the observation that $|\cos\theta| \leq 1$.

Of course, we do not rely on geometric interpretations to prove (10) and (11). Problem 4 below gives a hint for an algebraic proof of (10), and we can easily derive (11) from (10):

$$\begin{aligned}
|\mathbf{A} + \mathbf{B}|^2 &= (\mathbf{A} + \mathbf{B})\cdot(\mathbf{A} + \mathbf{B}) && [\text{by } (4)] \\
&= (\mathbf{A} + \mathbf{B})\cdot\mathbf{A} + (\mathbf{A} + \mathbf{B})\cdot\mathbf{B} && [\text{by } (6)] \\
&= \mathbf{A}\cdot\mathbf{A} + \mathbf{B}\cdot\mathbf{A} + \mathbf{A}\cdot\mathbf{B} + \mathbf{B}\cdot\mathbf{B} && [\text{by } (7)] \\
&= |\mathbf{A}|^2 + \mathbf{B}\cdot\mathbf{A} + \mathbf{A}\cdot\mathbf{B} + |\mathbf{B}|^2 && [\text{by } (4)] \\
&\leq |\mathbf{A}|^2 + |\mathbf{B}|\cdot|\mathbf{A}| + |\mathbf{A}|\cdot|\mathbf{B}| + |\mathbf{B}|^2 && [\text{by } (10)] \\
&= (|\mathbf{A}| + |\mathbf{B}|)^2.
\end{aligned}$$

Taking square roots, we obtain the triangle inequality (11).

Unit vectors

UNIT VECTORS

A vector **U** is called a *unit vector* if $|\mathbf{U}| = 1$. The vectors [1,0], [0,1], $[1/\sqrt{2}, -1/\sqrt{2}]$ are unit vectors, but [1,1] is not.

Recall that if $c > 0$, then we interpret $c\mathbf{V}$ as a vector in the same direction as **V**, but with a different length (unless $c = 1$, in which case the length is of course the same). Given any $\mathbf{V} \neq [0,0]$, there is a unit vector in the same direction as **V**, namely, $(1/|\mathbf{V}|)\mathbf{V}$. This vector is written simply $\mathbf{V}/|\mathbf{V}|$. It has the same direction as **V**, since it was obtained from **V** by multiplying by the positive number $1/|\mathbf{V}|$; and it is a unit vector since $|(1/|\mathbf{V}|)\mathbf{V}| = (1/|\mathbf{V}|)|\mathbf{V}| = 1$, by virtue of formula (9).

The "essential" properties of a vector are traditionally considered to be magnitude and direction. For a vector $\mathbf{V}$, the length $|\mathbf{V}|$ distills out the first essence, and the unit vector $\mathbf{V}/|\mathbf{V}|$ the second. The formula $\mathbf{V} = (|\mathbf{V}|)(\mathbf{V}/|\mathbf{V}|)$ reconstructs $\mathbf{V}$ out of its magnitude and direction.

Example 1. Find a vector bisecting the angle between the vectors $\mathbf{V} = [1,2]$ and $\mathbf{W} = [-1,3]$.

First solution. Let $\mathbf{U} = [a,b]$ be the desired vector, θ the angle between $\mathbf{U}$ and $\mathbf{V}$, and φ the angle between $\mathbf{U}$ and $\mathbf{W}$. We want $\cos\theta = \cos\varphi$. We solve for $\cos\theta$ from the equation $\mathbf{U}\cdot\mathbf{V} = |\mathbf{U}|\cdot|\mathbf{V}|\cos\theta$, and similarly solve for $\cos\varphi$. Equating the results, we find

$$\mathbf{U}\cdot\mathbf{V}/|\mathbf{U}|\,|\mathbf{V}| = \mathbf{U}\cdot\mathbf{W}/|\mathbf{U}|\,|\mathbf{W}|.$$

Substituting $\mathbf{U} = [a,b]$ and the given values of $\mathbf{V}$ and $\mathbf{W}$ leads to the equation

$$a + 2b = \frac{-a + 3b}{\sqrt{2}},$$

which has many solutions; one of them is

$$a = 3 - 2\sqrt{2}, \qquad b = 1 + \sqrt{2}.$$

Second solution. Since the only relevant information in $\mathbf{V}$ and $\mathbf{W}$ is their *direction,* we replace them by unit vectors $\mathbf{V}/|\mathbf{V}|$ and $\mathbf{W}/|\mathbf{W}|$. Since the diagonal of a rhombus bisects the angle at the vertex, we expect the vector

$$\mathbf{U} = (\mathbf{V}/|\mathbf{V}|) + (\mathbf{W}/|\mathbf{W}|)$$

to be a solution. [See Fig. 11.] You can check that it is by showing that $\mathbf{U}\cdot\mathbf{V}/|\mathbf{V}| = \mathbf{U}\cdot\mathbf{W}/|\mathbf{W}|$. Another solution is obtained if we multiply this by $|\mathbf{V}|\cdot|\mathbf{W}|$, obtaining $|\mathbf{W}|\mathbf{V} + |\mathbf{V}|\mathbf{W}$ as a vector bisecting the angle between $\mathbf{V}$ and $\mathbf{W}$.

Orthogonal vectors

Two vectors $\mathbf{V}$ and $\mathbf{W}$ are called *orthogonal* if and only if $\mathbf{V}\cdot\mathbf{W} = 0$. The interpretive formula $\mathbf{V}\cdot\mathbf{W} = |\mathbf{V}|\cdot|\mathbf{W}|\cos\theta$ explains the term; in this formula, θ is the angle between $\mathbf{V}$ and $\mathbf{W}$, and $\cos\theta = 0$ when $\theta = \pi/2$, i.e. when $\mathbf{V}$ and $\mathbf{W}$ are at right angles to each other. ["Orthogonal" means "right-angled."]

It is not difficult to prove that, given a vector $\mathbf{V} = [v_1, v_2] \neq [0,0]$, there are exactly two vectors $\mathbf{A}$ and $\mathbf{B}$ which are orthogonal to $\mathbf{V}$ and have the same length as $\mathbf{V}$. They are [Fig. 12]

$$\mathbf{A} = [-v_2, v_1], \qquad \mathbf{B} = [v_2, -v_1].$$

We think of the first as obtained by rotating $\mathbf{V}$ through an angle of $\pi/2$ radians, and the second by rotating $\mathbf{V}$ through an angle of $-\pi/2$ radians. The orthogonality is easy to check:

$$\mathbf{A}\cdot\mathbf{V} = [-v_2, v_1]\cdot[v_1, v_2] = -v_2v_1 + v_1v_2 = 0,$$

and similarly $\mathbf{B}\cdot\mathbf{V} = 0$. The angle of rotation is discussed in Problem 11 below.

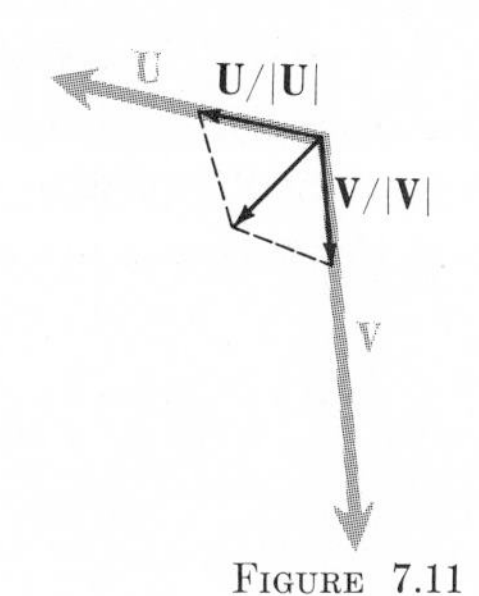

FIGURE 7.11

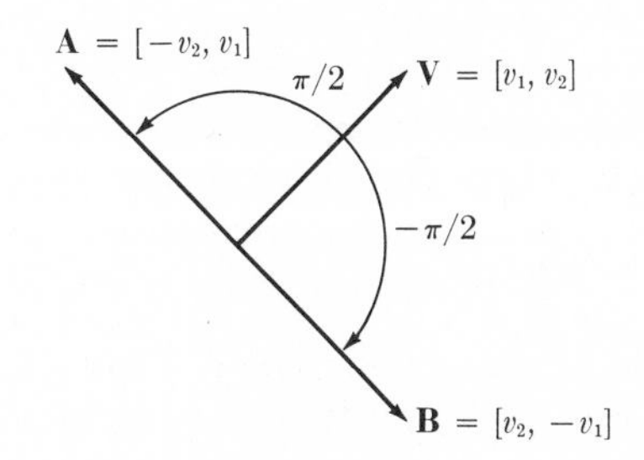

FIGURE 7.12

PROBLEMS

Problem 9 is used in several later sections.

1. Let $\mathbf{U} = [1,0]$, $\mathbf{V} = [1,1]$, $\mathbf{W} = [1,-1]$. Compute the following:
   - • (a) $|\mathbf{V}|$, $|\mathbf{U}|$, $|\mathbf{W}|$
   - (b) $|\mathbf{U} + \mathbf{V}|$, $|\mathbf{V} + \mathbf{W}|$
   - (c) $\mathbf{U}\cdot\mathbf{V}$, $\mathbf{V}\cdot\mathbf{U}$, $\mathbf{V}\cdot\mathbf{W}$, $\mathbf{U}\cdot\mathbf{W}$
   - • (d) The angle between $\mathbf{U}$ and $\mathbf{V}$; the angle between $\mathbf{V}$ and $\mathbf{W}$.
   - (e) $(2\mathbf{U} + 3\mathbf{V})\cdot(7\mathbf{U} - 5\mathbf{V})$

2. Check that the results of Problem 1(b) and 1(c) are consistent with inequalities (10) and (11) of the text.

3 • (a) Prove formulas (4) and (5) of the text.
   (b) Prove formulas (6) and (7) of the text.
   (c) Prove formulas (8) and (9) of the text.

4. Let $\mathbf{A} = [a_1, a_2]$, $\mathbf{B} = [b_1, b_2]$, and prove the Schwarz inequality (10). [Hint: $a_1^2b_2^2 + b_1^2a_2^2 - 2a_1a_2b_1b_2 = (a_1b_2 - a_2b_1)^2$. If you find this proof of the Schwarz inequality too routine, there is a more interesting one in §7.3, Problem 12.]

5. Suppose $\mathbf{V}$, $\mathbf{W}$, and $\frac{1}{2}(\mathbf{V} + \mathbf{W})$ are represented as arrows starting at the origin. Let the endpoints of these arrows be P_1, P_2, and Q respectively. Show that Q bisects the segment from P_1 to P_2, i.e. $|\overrightarrow{P_1Q}| = |\overrightarrow{QP_2}|$, and Q lies on the line through P_1 and P_2. **MIDPOINT**

6. Show that $\mathbf{V}\cdot\mathbf{W} > 0$ when the angle between $\mathbf{V}$ and $\mathbf{W}$ is acute, but $\mathbf{V}\cdot\mathbf{W} < 0$ when the angle is obtuse. [When is $\cos\theta > 0$?]

7. [*Law of cosines*] **LAW OF COSINES**
   (a) Prove that $|\mathbf{W} - \mathbf{V}|^2 = |\mathbf{W}|^2 + |\mathbf{V}|^2 - 2\mathbf{W}\cdot\mathbf{V}$. [Hint: $|\mathbf{W} - \mathbf{V}|^2 = (\mathbf{W} - \mathbf{V})\cdot(\mathbf{W} - \mathbf{V}) = \mathbf{W}\cdot(\mathbf{W} - \mathbf{V}) - \mathbf{V}\cdot(\mathbf{W} - \mathbf{V}) = \cdots$.]
   (b) Draw $\mathbf{V}$, $\mathbf{W}$, and $\mathbf{W} - \mathbf{V}$ as three sides of a triangle, and interpret the formula in (a) as the law of cosines $c^2 = a^2 + b^2 - 2ab\cos\theta$.

8. [*Parallelogram Law*]

(a) Prove that $|\mathbf{V} - \mathbf{W}|^2 + |\mathbf{V} + \mathbf{W}|^2 = 2\,|\mathbf{V}|^2 + 2\,|\mathbf{W}|^2$. [See Problem 7(a).]

(b) Show that $|\mathbf{V} - \mathbf{W}|$ and $|\mathbf{V} + \mathbf{W}|$ are the lengths of two diagonals of a parallelogram with sides $|\mathbf{V}|$ and $|\mathbf{W}|$; the formula in (a) is therefore called the *parallelogram law.*

9. [*Pythagorean theorem*] Prove that $|\mathbf{V} + \mathbf{W}|^2 = |\mathbf{V}|^2 + |\mathbf{W}|^2$ if (and only if) $\mathbf{V}$ and $\mathbf{W}$ are orthogonal. [See Problem 7.]

10. [*Preparation for polar coordinates and* §7.10]
Let $\mathbf{V} = [v_1, v_2]$ be any nonzero vector.

(a) Show that $\left(\dfrac{v_1}{|\mathbf{V}|}\right)^2 + \left(\dfrac{v_2}{|\mathbf{V}|}\right)^2 = 1.$

(b) From part (a), it follows that there is a number θ such that $v_1/|\mathbf{V}| = \cos\theta$ and $v_2/|\mathbf{V}| = \sin\theta$. Show geometrically that θ is the angle from the positive x axis to the vector $\mathbf{V}$.

(c) Show that any vector $\mathbf{V}$ can be written in the form
$$\mathbf{V} = [|\mathbf{V}|\cos\theta, |\mathbf{V}|\sin\theta].$$

(d) The *slope* of the vector $\mathbf{V}$ is defined to be v_2/v_1, if $v_1 \neq 0$. Show that $v_2/v_1 = \tan\theta$, with θ as in (b).

11. [*Rotation of vectors; do Problem* 10 *first*] From Problem 10, any vector $\mathbf{V}$ can be written in the form $\mathbf{V} = [|\mathbf{V}|\cos\theta, |\mathbf{V}|\sin\theta]$. If $\mathbf{V}$ is rotated counterclockwise through an angle φ, the result is defined to be

$$[|\mathbf{V}|\cos(\theta + \varphi), |\mathbf{V}|\sin(\theta + \varphi)],$$

as in Fig. 13.

(a) Show that when $\mathbf{V} = [v_1, v_2]$ is rotated counterclockwise through an angle φ, the result is the vector
$$\mathbf{W} = [v_1\cos\varphi - v_2\sin\varphi, v_1\sin\varphi + v_2\cos\varphi].$$

(b) Show that rotating $\mathbf{V} = [v_1, v_2]$ through $\pi/2$ yields $[-v_2, v_1]$, and rotating through $-\pi/2$ yields $[v_2, -v_1]$, as claimed in the text above.

(c) Taking $\mathbf{W}$ as in (a), show that $\mathbf{W}\cdot\mathbf{V} = |\mathbf{V}|^2\cos\varphi$, and hence that $\mathbf{W}$ and $\mathbf{V}$ are orthogonal iff $\mathbf{W} = \pm[v_2, -v_1]$.

(d) Let $\mathbf{A}$ be any vector such that $|\mathbf{A}| = |\mathbf{V}|$. Show that $\mathbf{A}$ is obtained from $\mathbf{V}$ by rotating through some angle φ. [Hint: $\mathbf{A} = [|\mathbf{V}|\cos\alpha, |\mathbf{V}|\sin\alpha]$ for some α; set $\varphi = \alpha - \theta$, with θ as above.]

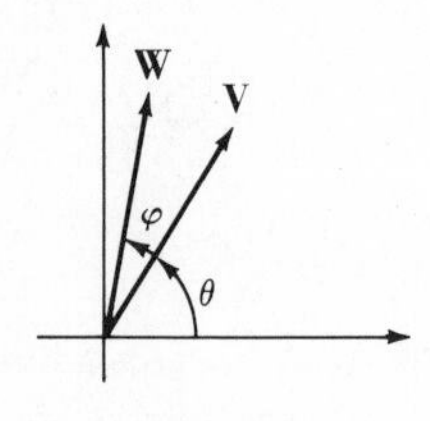

FIGURE 7.13

12. (a) Sketch the vector $\mathbf{V} = [-1/\sqrt{2}, 1/\sqrt{2}]$.
  • (b) Assuming the formula in Problem 11(a), compute the vector $\mathbf{W}$ obtained by rotating the vector $\mathbf{V}$ in part (a) through an angle $\varphi = \pi$, and sketch the result.
  • (c) Repeat part (b) with $\varphi = \pi/4$.
  (d) Repeat part (b) with $\varphi = -\pi/4$.
  • (e) Repeat part (b) with $\varphi = \pi/4$ and $\mathbf{V} = [1,3]$.

13. (a) Show that $(\mathbf{A} + \mathbf{B})\cdot(\mathbf{A} - \mathbf{B}) = |\mathbf{A}|^2 - |\mathbf{B}|^2$.
  (b) Note that $\mathbf{A} + \mathbf{B}$ and $\mathbf{A} - \mathbf{B}$ are the diagonals of a parallelogram of sides $\mathbf{A}$ and $\mathbf{B}$. Show that the parallelogram is a *rhombus* iff the diagonals $\mathbf{A} + \mathbf{B}$ and $\mathbf{A} - \mathbf{B}$ are orthogonal.

14. Suppose $\mathbf{A} \neq \mathbf{0}$. Does $\mathbf{A}\cdot\mathbf{B} = \mathbf{A}\cdot\mathbf{C}$ imply $\mathbf{B} = \mathbf{C}$?

15. Does $(\mathbf{A}\cdot\mathbf{B})\cdot\mathbf{C} = \mathbf{A}\cdot(\mathbf{B}\cdot\mathbf{C})$? Does the equation even make sense?

7.3 VECTORS IN ANALYTIC GEOMETRY

Of the many concepts that are illuminated by the use of vectors, the parametric equations of a line are among the simplest and most important.

Given a point P_0 and a vector $\mathbf{V} \neq \mathbf{0}$, represent $\mathbf{V}$ by an arrow beginning at P_0 [Fig. 14]. Clearly, from the figure, this arrow determines a *line*, the line through P_0 in the direction $\mathbf{V}$. If P is any point on the line, then the vector $\overrightarrow{OP}$ equals $\overrightarrow{OP_0}$ plus some vector in the direction $\pm\mathbf{V}$; i.e. for some number t,

$$\overrightarrow{OP} = \overrightarrow{OP_0} + t\mathbf{V}. \tag{1}$$

Figure 14(b) shows which part of the line we get with $t > 1$, which part with $0 < t < 1$, and which part with $t < 0$.

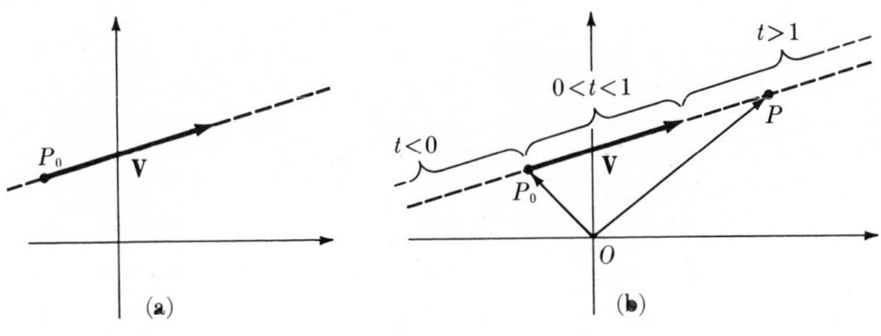

FIGURE 7.14

If we let $\mathbf{V} = [v_1, v_2]$, $P_0 = (x_0, y_0)$, and $P = (x,y)$, then (1) becomes

$$[x,y] = [x_0, y_0] + t[v_1, v_2] = [x_0 + tv_1, y_0 + tv_2].$$

This means the same as the two equations obtained by equating corresponding vector components on the left- and right-hand sides,

PARAMETRIC EQUATIONS OF A LINE

$$x = x_0 + tv_1, \qquad y = y_0 + tv_2. \tag{2}$$

These are called *parametric equations of the line through P_0 in the direction* $\mathbf{V}$, and t is called the *parameter*. Each value of t gives a point on the line, and, conversely, each point (x,y) on the line is obtained by substituting a particular value of t in the equations. In other words, the line is the set

$$\{(x,y) : \text{there is a number } t \text{ such that } x = x_0 + tv_1 \text{ and } y = y_0 + tv_2.\} \tag{3}$$

Our claim that the set (3) is a line was based on Fig. 14, but we will soon prove it is a line, and, further, that *every* line can be written in this parametric form, with an appropriate choice of x_0, y_0, v_1, and v_2.

Example 1. Let $P_0 = (2,\pi)$ and $\mathbf{V} = [1,-2]$. Sketch P_0, and represent $\mathbf{V}$ by an arrow starting at P_0. Find parametric equations of the line through P_0 in the direction $\mathbf{V}$. Find some point on the line other than P_0. [Do it yourself; solution below.]

Example 2. Show that the parametric equations

$$x = 4 + 2t, \qquad y = -1 + t \tag{4}$$

define a line, and find its slope. *Solution.* Let (x,y) be a point represented in the form (4), for some t. Then

$$t = \frac{x-4}{2}, \qquad \text{so} \quad y = -1 + t = -1 + \frac{x-4}{2} = \frac{1}{2}x - 3,$$

so (x,y) lies on the line with equation

$$y = \frac{1}{2}x - 3. \tag{5}$$

The slope is $1/2$.

We have shown that every point (x,y) represented as in (4) lies on the graph of (5). To complete the solution, we must prove conversely that every point (x,y) on the graph of (5) can be represented as in (4). To do this, simply reverse the steps. Given that $y = \frac{1}{2}x - 3$, set $t = (x-4)/2$. Then clearly $x = 4 + 2t$, and

$$y = \frac{1}{2}x - 3 = \frac{1}{2}(4 + 2t) - 3 = -1 + t,$$

so (x,y) is represented as in (4).

Just as in this example, one can prove in general the following:

Theorem 1. *If* $[v_1, v_2] = \mathbf{V} \neq \mathbf{0}$, *then the set* (3) *is a line. Conversely, every line can be represented in the form* (3).

If $v_1 = 0$, *then* $\mathbf{V}$ *points "straight up" and the line is vertical; if* $v_1 \neq 0$, *the line is nonvertical and has slope* $m = v_2/v_1$.

Normal to a line

Let P_0 be a point in the plane, and draw a nonzero vector $\mathbf{N} = [a,b]$ starting at P_0. Consider the set

$$\{P: \overrightarrow{PP_0}\cdot\mathbf{N} = 0\} \tag{6}$$

of all points P such that $\overrightarrow{PP_0}$ is orthogonal to $\mathbf{N}$ [Fig. 15]. This set appears to be a line, and it is easy to prove that it really is. Let $P = (x,y)$ and $P_0 = (x_0, y_0)$. Then

$$\begin{aligned} \overrightarrow{PP_0}\cdot\mathbf{N} = 0 \quad &\Leftrightarrow \quad [x - x_0, y - y_0]\cdot[a,b] = 0 \\ &\Leftrightarrow \quad a(x - x_0) + b(y - y_0) = 0. \end{aligned} \tag{7}$$

If $b \neq 0$, this reduces to

$$y = y_0 - \frac{a}{b}(x - x_0)$$

which is the equation of the line through (x_0, y_0) with slope $m = -a/b$. If $b = 0$, then $a \neq 0$ [because $\mathbf{N} = [a,b] \neq \mathbf{0}$] and (7) reduces to $x = x_0$, the equation of the vertical line through (x_0, y_0).

Thus the set (6) is indeed a line. If we take (7) one step further, we get the equation

LINE THRU (x_0, y_0) $\mathbf{N} = [a,b]$

$$ax + by = ax_0 + by_0 = c, \tag{8}$$

where we have introduced c as an abbreviation for the constant $ax_0 + by_0$. It is easy to see that *every* line, vertical or not, can be written in the form (8); when $b = 0$ the line is vertical, and when $b \neq 0$ it has slope $-a/b$. In either case, the vector $[a,b]$ is perpendicular to the line, as you can see by retracing the steps from (8) to (7) to (6).

The vector $\mathbf{N}$ occurring here is called a *normal vector* to the line (hence the choice of $\mathbf{N}$ as notation). "Normal" means simply "orthogonal" in this context.

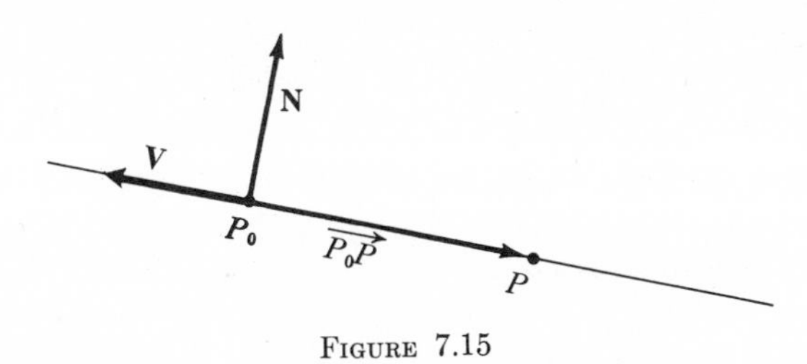

FIGURE 7.15

Example 3. Write the equation of the line through the point (1,5) with normal $\mathbf{N} = [-2,3]$. Sketch the line and the normal. *Solution.* If $P_0 = (1,5)$ is the given point, then a point $P = (x,y)$ is on the line if and only if

$$\overrightarrow{PP_0}\cdot\mathbf{N} = 0$$

$$[x - x_0\,, y - y_0]\cdot[-2,3] = 0$$

$$-2(x - x_0) + 3(y - y_0) = 0$$

$$-2x + 3y = -2x_0 + 3y_0 = -2\cdot 1 + 3\cdot 5 = 13.$$

For the sketch [Fig. 16] we plot (1,5), draw $\mathbf{N} = [-2,3]$, and then draw the line perpendicular to $\mathbf{N}$.

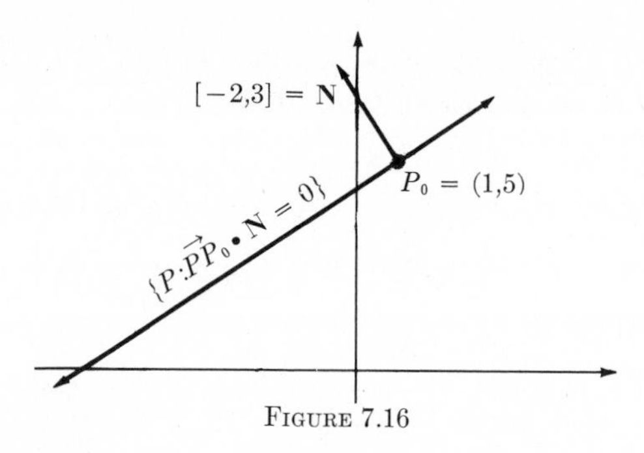

FIGURE 7.16

Orthogonal projection

An important procedure in geometry is to "drop a perpendicular" from a point to a line. (Notice the suggestion of a plumb line in "drop a. . ."; surely, the geometric idea is derived from carpentry.) The corresponding concept for vectors is *orthogonal projection*. By looking carefully at Fig. 17, you should be able to convince yourself that each of the vectors labelled $\mathbf{V_W}$ is given by the formula

ORTHOGONAL PROJECTION OF V ON W

$$\mathbf{V_W} = (|\mathbf{V}|\cos\theta)(\mathbf{W}/|\mathbf{W}|),$$

for this expression has length $|\mathbf{V}|\cos\theta$, and points in the proper direction. In view of the formula $\mathbf{V}\cdot\mathbf{W} = |\mathbf{V}|\,|\mathbf{W}|\cos\theta$, we find for $\mathbf{V_W}$ the simple algebraic expression

$$\mathbf{V_W} = \frac{\mathbf{V}\cdot\mathbf{W}}{\mathbf{W}\cdot\mathbf{W}}\,\mathbf{W}. \tag{9}$$

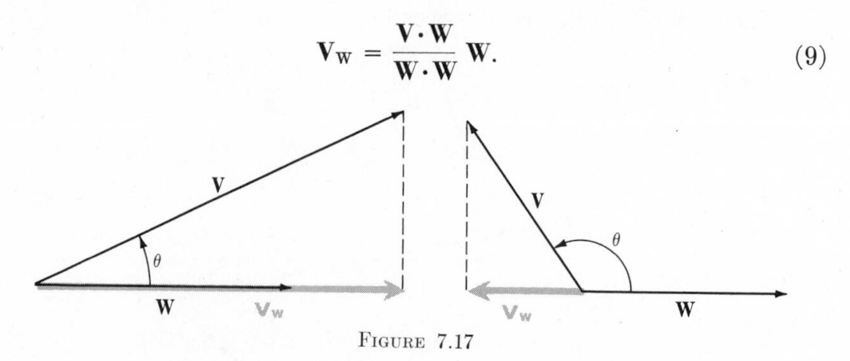

FIGURE 7.17

The previous paragraph is based on a picture, so it proves nothing; but it explains:

Definition 1. If $|\mathbf{W}| \neq 0$, then the *orthogonal projection* of $\mathbf{V}$ on $\mathbf{W}$, denoted $\mathbf{V}_{\mathbf{W}}$, is defined by (9). In particular, if $\mathbf{W}$ is a *unit vector*, then $\mathbf{W} \cdot \mathbf{W} = |\mathbf{W}|^2 = 1$, so

$$\mathbf{V}_{\mathbf{W}} = (\mathbf{V} \cdot \mathbf{W})\mathbf{W}, \qquad \text{if } |\mathbf{W}| = 1.$$

Example 4. Let $\mathbf{V} = [-1,-2]$ and $\mathbf{W} = [3,-4]$. Find the orthogonal projection of $\mathbf{V}$ on $\mathbf{W}$, and of $\mathbf{W}$ on $\mathbf{V}$. Sketch the vectors $\mathbf{V}$, $\mathbf{W}$, $\mathbf{V}_{\mathbf{W}}$, $\mathbf{W}_{\mathbf{V}}$ as arrows from the origin. [Solution below.]

To illustrate the use of orthogonal projections, we compute the distance from a point $P_1 = (x_1, y_1)$ to the line with equation $ax + by = c$. Let P_0 be any point on the line. Then the distance from P_1 to the line is the length of the orthogonal projection of $\overrightarrow{P_0P_1}$ on the normal $\mathbf{N}$ [Fig. 18], which we can easily compute:

$$\left|\frac{\mathbf{N} \cdot \overrightarrow{P_0P_1}}{\mathbf{N} \cdot \mathbf{N}} \mathbf{N}\right| = \frac{|\mathbf{N} \cdot \overrightarrow{P_0P_1}|}{|\mathbf{N}|^2} |\mathbf{N}| = \frac{|\mathbf{N} \cdot \overrightarrow{P_0P_1}|}{|\mathbf{N}|}. \tag{10}$$

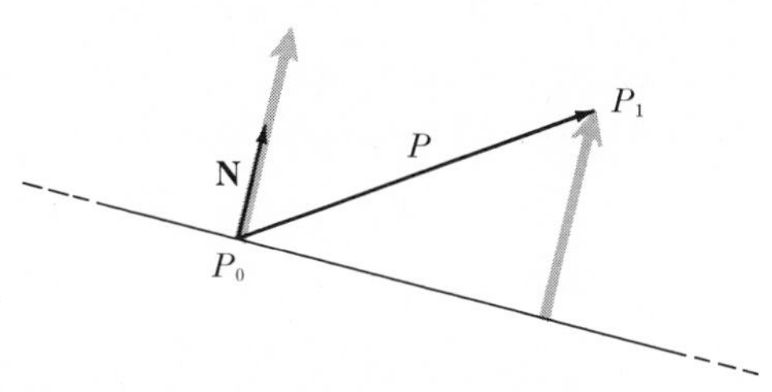

FIGURE 7.18

If we denote the coordinates of P_0 by (x_0, y_0), and note that $\mathbf{N} = [a,b]$, the expression (10) for the distance becomes

$$\frac{|[a,b] \cdot [x_1 - x_0, y_1 - y_0]|}{\sqrt{a^2 + b^2}} = \frac{|ax_1 - ax_0 + by_1 - by_0|}{\sqrt{a^2 + b^2}}.$$

Since (x_0, y_0) is on the line with equation $ax + by = c$, we have $ax_0 + by_0 = c$, and our expression for the distance reduces finally to

$$\frac{|ax_1 + by_1 - c|}{\sqrt{a^2 + b^2}}. \tag{11}$$

Actually, the expression without absolute values,

$$\frac{ax_1 + by_1 - c}{\sqrt{a^2 + b^2}}, \tag{12}$$

is more interesting. This expression gives the distance from (x_1, y_1) to the line if (x_1, y_1) is on the side toward which $\mathbf{N} = [a,b]$ is pointing, and *minus* the distance if (x_1, y_1) is on the side opposite to $\mathbf{N}$.

From (12) we get a very good interpretation of the equation for a line: $(ax + by - c)/\sqrt{a^2 + b^2}$ is plus or minus the distance from (x,y) to the line; when this distance is zero, the point is on the line. The points (x,y) where $ax + by - c > 0$ form one side of the line (the side toward which $\mathbf{N}$ points), and those where $ax + by - c < 0$ form the other side.

Remark. We have strayed from the straight and narrow path of analytic proofs, and based our discussion on Fig. 18; as is so often the case, this illegitimate approach is quicker and easier (but less reliable). An analytic proof of the distance formula (11) is suggested in Problem 9 below.

Solution to Example 1. See Fig. 19 for a sketch of P_0 and $\mathbf{V}$. The parametric equations are $[x,y] = [2,\pi] + t[1,-2]$, or

$$x = 2 + t, \qquad y = \pi - 2t.$$

A second point (different from P_0) on the line can be obtained with any value $t \neq 0$. For example, $t = 1$ gives the point $(3, \pi - 2)$.

Solution to Example 4. $\mathbf{V}\cdot\mathbf{W} = 5$, $|\mathbf{W}|^2 = 25$, $|\mathbf{V}|^2 = 5$, so $\mathbf{V_W} = \frac{5}{25}[3,-4] = [\frac{3}{5},-\frac{4}{5}]$ and $\mathbf{W_V} = \frac{5}{5}[-1,-2] = [-1,-2]$. See Fig. 20.

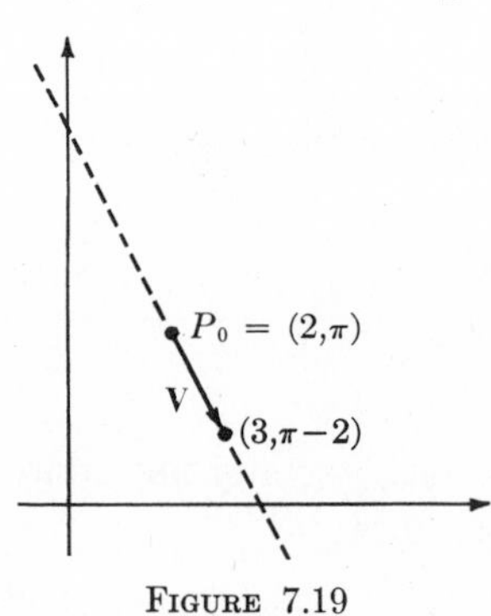

FIGURE 7.19

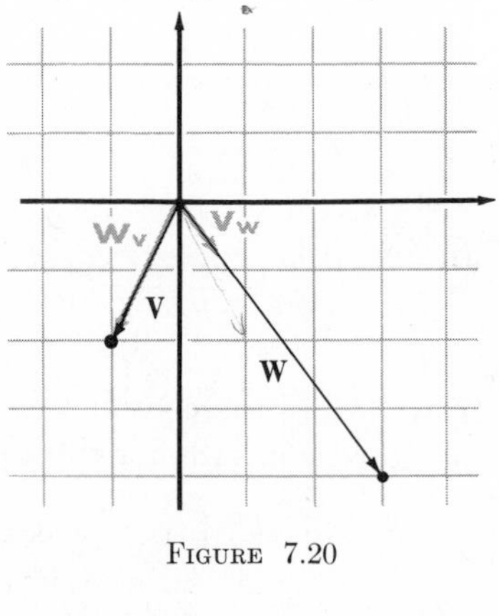

FIGURE 7.20

PROBLEMS

1. Write parametric equations for the line through P_0 in the direction $\mathbf{V}$. Sketch P_0, $\mathbf{V}$, and the line. Write the equation of the line in the form $y = mx + b$ (unless the line is vertical).
   (a) $P_0 = (1,3)$, $\mathbf{V} = [1,-2]$
   (b) $P_0 = (2,-5)$, $\mathbf{V} = [-1,4]$
   (c) $P_0 = (0,0)$, $\mathbf{V} = [0,1]$
   (d) $P_0 = (0,0)$, $\mathbf{V} = [1,0]$
   (e) $P_0 = (1,3)$, $\mathbf{V} = [1,0]$

• **2.** Find the distances from the points $P_1 = (1,1)$, $P_2 = (0,0)$, $P_3 = (1,-1)$, $P_4 = (1,-2)$ to the line $\{(x,y): x - 2y = 3\}$. Decide whether they are on the same side as the normal or on the opposite side. Sketch the line, the normal, and the points.

3. Give parametric equations for the line through each of the following pairs of points. [Hint: The points determine a vector in the direction of the line.]

• (a) $(1,3)$ and $(2,-\frac{1}{2})$
(b) $(-3,10)$ and $(2,10^6)$
(c) $(2,-5)$ and $(5,-2)$

• **4.** Find the cosine of the angle between the lines with equations $x + y = 1$ and $x - 2y = 0$. [The angle between the lines equals the angle between the normal vectors.]

• **5.** Find a line parallel to $x + y = 1$, and passing through the origin $(0,0)$. [Hint: Parallel lines have a common normal.]

6. (a) Find a general equation for the cosine of the angle between the line with equation $a_1x + b_1y = c_1$ and the line with equation $a_2x + b_2y = c_2$.
• (b) When are the lines in (a) parallel? When are they perpendicular?

7 • (a) Find the angle between the line given parametrically by $x = 3 + 10t$, $y = -2 - t$ and the line given by $x = 11t$, $y = 4 - t$.
(b) Do the same for the pairs

$$x = -1 - 5t, \qquad y = 3 + 10t$$

and

$$x = 1 + 5t, \qquad y = 10 - 3t.$$

8. (a) Give a formula for the cosine of the angle between two lines in parametric form.
(b) Give conditions that two such lines be parallel.
(c) Give conditions that two such lines be perpendicular.

9. Let $ax + by = c$ be the equation of a line, and suppose $\mathbf{N} = [a,b]$ is a *unit* normal vector, $a^2 + b^2 = 1$. Let (x_1, y_1) be any point. Let P_2 be the reflection of P_1 in the given line. Show (by a sketch) that P_2 is given by

REFLECTION IN AN ARBITRARY LINE

$$\overrightarrow{P_2P_1} = 2d\mathbf{N},$$

where d is given by (12). [Consider the two possibilities $d > 0$ and $d < 0$.]

10. Let $\mathbf{W}$ be a nonzero vector, and let $\mathbf{V_W}$ be the orthogonal projection of $\mathbf{V}$ on $\mathbf{W}$. Prove algebraically that $\mathbf{V} - \mathbf{V_W}$ is orthogonal to $\mathbf{W}$.

11. Let $x = x_0 + at$, $y = y_0 + bt$ be parametric equations of a line, and let (x_1, y_1) be any point. Let $P = (x,y) = (x_0 + at, y_0 + bt)$, $P_1 = (x_1, y_1)$.

(a) By expanding $|\overrightarrow{PP_1}|^2$, show that the minimum distance from (x_1, y_1) to the line occurs when

$$t = \frac{ax_1 - ax_0 + by_1 - by_0}{a^2 + b^2}.$$

(b) Show that the minimum value of $|\overrightarrow{PP_1}|^2$ is

$$\frac{[b(x_1 - x_0) - a(y_1 - y_0)]^2}{a^2 + b^2}.$$

(c) Prove formula (11).

12. The Schwarz inequality can be proved neatly by noticing the following simple facts:

SCHWARZ INEQUALITY

(i) $|\mathbf{V} + t\mathbf{W}|^2$ is a quadratic in t, and
(ii) $|\mathbf{V} + t\mathbf{W}|^2 \geq 0$ for all t; hence
(iii) the minimum over t of $|\mathbf{V} + t\mathbf{W}|^2$ is ≥ 0.

Check (i)–(iii), and prove the Schwarz inequality $|\mathbf{V}\cdot\mathbf{W}| \leq |\mathbf{V}|\cdot|\mathbf{W}|$ by finding the minimum in (iii).
[Hint: $[\mathbf{V} + t\mathbf{W}, \mathbf{V} + t\mathbf{W}] = [\mathbf{V}, \mathbf{V} + t\mathbf{W}] + t[\mathbf{W}, \mathbf{V} + t\mathbf{W}] = \cdots$.]

13. Suppose $\mathbf{U}$ and $\mathbf{V}$ are *orthogonal unit vectors*, i.e. $\mathbf{U}\cdot\mathbf{V} = 0$, $|\mathbf{U}| = 1$, $|\mathbf{V}| = 1$. Prove that any vector $\mathbf{A}$ equals the sum of its projections on $\mathbf{U}$ and $\mathbf{V}$,

"ORTHOGONAL DECOMPOSITION" OF A

$$\mathbf{A} = (\mathbf{A}\cdot\mathbf{U})\mathbf{U} + (\mathbf{A}\cdot\mathbf{V})\mathbf{V}.$$

[Hint: From §7.2, if $\mathbf{U} = [u_1, u_2]$ then either $\mathbf{V} = [u_2, -u_1]$ or $\mathbf{V} = [-u_2, u_1]$.]

• **14.** Suppose $\mathbf{U}$ and $\mathbf{V}$ are orthogonal unit vectors. Prove that the *only* way to write $\mathbf{A}$ as a linear combination of $\mathbf{U}$ and $\mathbf{V}$,

$$\mathbf{A} = c_1\mathbf{U} + c_2\mathbf{V}$$

is the way given in the previous problem, i.e. $c_1 = \mathbf{A}\cdot\mathbf{U}$ and $c_2 = \mathbf{A}\cdot\mathbf{V}$.
[Hint: In the equation $\mathbf{A} = c_1\mathbf{U} + c_2\mathbf{V}$, take inner products with $\mathbf{U}$ and $\mathbf{V}$.]

7.4 PATHS IN THE PLANE

Imagine a particle moving in the plane. We can always describe the motion by means of two functions f and g; at each time t the particle is at some point (x,y), and we simply let

$$x = f(t) \qquad \text{and} \qquad y = g(t). \tag{1}$$

In other words, at time t the particle is at the point $(f(t),g(t))$.

The equations (1) are called *parametric equations* of the path traced by the particle; the path is the set $\{(x,y): \text{for some } t,\ x = f(t) \text{ and } y = g(t)\}$.

It is convenient to think of the ordered pair $[f,g]$ as a *vector function* $\mathbf{R}$, which assigns to each real number t the vector $\mathbf{R}(t) = [f(t),g(t)]$.

When the vector function $\mathbf{R}$ describes the motion of a particle in the plane, we represent $\mathbf{R}(t)$ by an arrow starting at the origin; then the head of the arrow gives the position of the particle at time t. $\mathbf{R}(t)$ is called the *position vector* [Fig. 21].

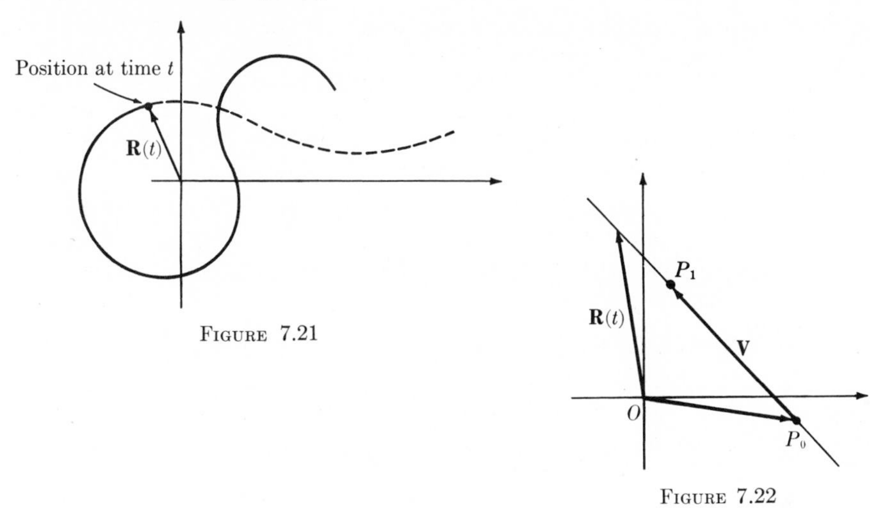

FIGURE 7.21

FIGURE 7.22

Example 1. Let P_0 be a given point and $\mathbf{V}$ a given vector, and set $\mathbf{R}(t) = \overrightarrow{OP_0} + t\mathbf{V}$. Then we have a motion along the straight line through P_0 in the direction $\mathbf{V}$. At time $t = 0$, the particle is at P_0; and at $t = 1$, it is at the point P_1 such that $\overrightarrow{OP} = \overrightarrow{OP_0} + \mathbf{V}$. [See Fig. 22.]

Example 2. Suppose a particle moves counterclockwise around the circle $x^2 + y^2 = r^2$ at a constant speed of one revolution per second, starting at $(r,0)$ when $t = 0$. Then we have $\mathbf{R}(t) = [r \cos 2\pi t, r \sin 2\pi t]$.

We have been thinking of the parameter t as the time. While this is generally the most useful image, others are also common. For example, the equations $x = \cos\theta$, $y = \sin\theta$ are parametric equations of the unit circle with the angle θ as parameter. The cycloid in Example 3 is another path using an angle θ as parameter.

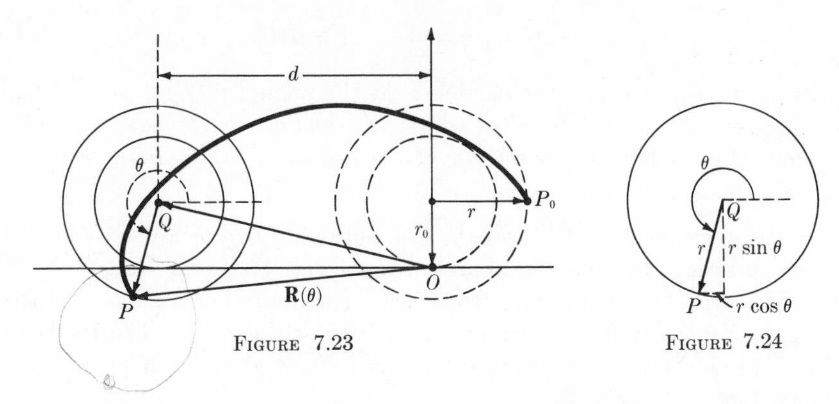

FIGURE 7.23

FIGURE 7.24

CYCLOID

Example 3. Imagine a flanged wheel rolling from right to left on a rail. Let the distance from the center of the wheel to the rail be r_0, and consider a point P at a distance r from the center of the wheel [Fig. 23]. If P is on the flange, then $r > r_0$; otherwise, $r \leq r_0$. Let the wheel start with its center directly above the origin and P at the position P_0 in Fig. 23. Let $\mathbf{R}(\theta)$ be the position vector of P after the wheel has rolled through an angle θ. Then we have $\mathbf{R}(\theta) = \overrightarrow{OQ} + \overrightarrow{QP}$. Figure 24 shows that $\overrightarrow{QP} = [r \cos \theta, r \sin \theta]$. Furthermore, $\overrightarrow{OQ} = [-d, r_0]$, where d is the length of rail that has been in contact with the wheel. To relate d to θ, we assume that d equals the length of circumference of the wheel that has been in contact with the rail; this amounts to assuming that the wheel does not slip. Then $d = r_0\theta$, so $\overrightarrow{OQ} = [-r_0\theta, r_0]$ and

$$\begin{aligned}\mathbf{R}(\theta) &= \overrightarrow{OQ} + \overrightarrow{QP} \\ &= [-r_0\theta, r_0] + [r \cos \theta, r \sin \theta] \\ &= [-r_0\theta + r \cos \theta, r_0 + r \sin \theta].\end{aligned}$$

Thus parametric equations of the path are

$$x = -r_0\theta + r \cos \theta, \qquad y = r_0 + r \sin \theta; \tag{2}$$

the parameter θ is the angle through which the wheel has rolled.

Figure 25 is a time exposure of a light attached to a rolling wheel; the picture graphs the equations (2) with $r = r_0$. The curve is called a *cycloid*.

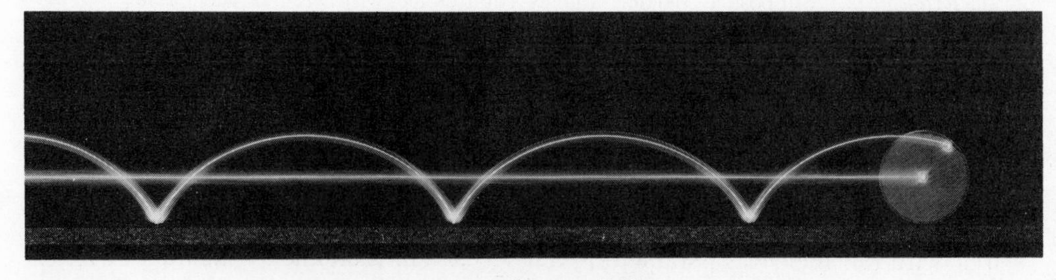

FIGURE 7.25*

*From PSSC PHYSICS, D.C. Heath and Company, Boston, 1965

PROBLEMS

1. Sketch the path given by the parametric equations (2) in the following cases:

(a) $r = 2, r_0 = 1$ (b) $r = 1, r_0 = 2$ (c) $r = 3, r_0 = 2$

2. A bug starts at the center of a very large clock at noon and walks out along the minute hand at 1 cm per sec. Let the coordinate origin be at the center of the clock. Give parametric equations for the position of the bug as follows:

• (a) with time (in seconds) as parameter, and
(b) with parameter θ, the angle through which the minute hand has turned.

RATIONAL POINTS ON THE CIRCLE

• **3.** The position vector for a moving particle is

$$\mathbf{R}(t) = \left[\frac{1-t^2}{1+t^2}, \frac{2t}{1+t^2}\right].$$

(a) Show that the particle moves on a circle centered at the origin.
(b) In what direction does it move as t increases?
(c) Are any points on the circle omitted as t ranges over $(-\infty, +\infty)$?
(d) How many times does the path go through each point?
(e) Show that there are infinitely many points (x,y) on the circle both of whose coordinates are rational numbers.

• **4.** Let $\mathbf{R}(t) = [1 - t^2, t]$. Show that the path is a parabola, and sketch it. Indicate the direction of increasing t.

5. One wheel of radius r_0 is held fixed at the origin of coordinates, and a second wheel, of radius r_1, is rolled around the outside of the circumference of the first wheel. Find parametric equations of the path of a point on the circumference of the second wheel, with parameter θ, the angle through which the second wheel has turned about its center.

INTERSECTING PATHS

6. One particle moves with position vector

$$\mathbf{R}_1(t) = [1 + t, 2 + 3t],$$

and a second particle with position vector

$$\mathbf{R}_2(t) = [1 - t, 3 + t^2].$$

(a) Show that the particles never collide.
(b) Show that the paths traced by the particles intersect at two points. [You can find the intersections of the two paths by solving the simultaneous equations $1 + t = 1 - s$, $2 + 3t = 3 + s^2$.]

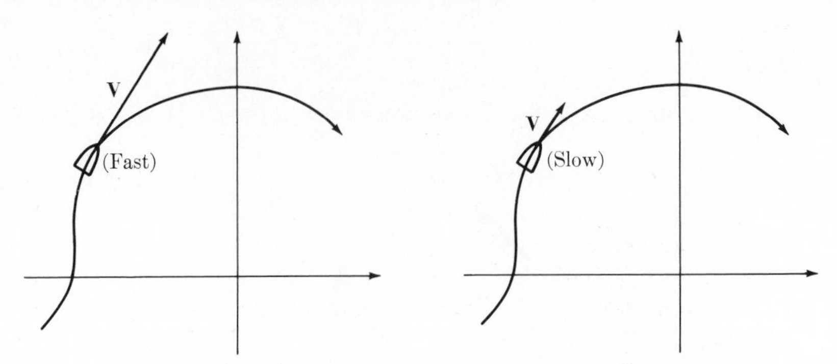

FIGURE 7.26 **V** gives the direction of motion, tangent to the course; $|\mathbf{V}|$ gives the speed

7.5 DIFFERENTIATION OF VECTOR FUNCTIONS; VELOCITY AND ACCELERATION

The motion of a ship or plane at a particular time is described by its *speed* and *direction*, in other words, by a vector **V**, called the *velocity vector.* The direction of **V** is the direction of motion, and the magnitude of **V** is the speed [Fig. 26].

The precise definition of the velocity vector is based on the same idea that led to the definition of derivative. To find the velocity $\mathbf{V}(t)$ at some time t, take the change in position $\mathbf{R}(t + h) - \mathbf{R}(t)$ from time t to time $t + h$, divide by the time h taken to accomplish this change, and take the limit as $h \to 0$ [Fig. 27]. Thus the velocity at time t is the *derivative of the position function* **R**,

VELOCITY IS THE DERIVATIVE OF POSITION

$$\mathbf{V}(t) = \lim_{h\to 0} \frac{1}{h} [\mathbf{R}(t + h) - \mathbf{R}(t)] = \mathbf{R}'(t).$$

This limit relation means that each component of the difference quotient tends to the corresponding component of the derivative. Thus if $\mathbf{R}(t) = [f(t),g(t)]$, then

$$\lim_{h\to 0} \frac{1}{h} [\mathbf{R}(t + h) - \mathbf{R}(t)] = \lim_{h\to 0} \left[\frac{f(t + h) - f(t)}{h}, \frac{g(t + h) - g(t)}{h}\right]$$

$$= \left[\lim_{h\to 0} \frac{f(t + h) - f(t)}{h}, \lim_{h\to 0} \frac{g(t + h) - g(t)}{h}\right]$$

$$= [f'(t),g'(t)].$$

Thus to compute $\mathbf{R}'$, simply differentiate each component of **R**:

DIFFERENTIATE **R** BY DIFFERENTIATING EACH COMPONENT

$$\mathbf{V}(t) = \mathbf{R}'(t) = [f'(t),g'(t)].$$

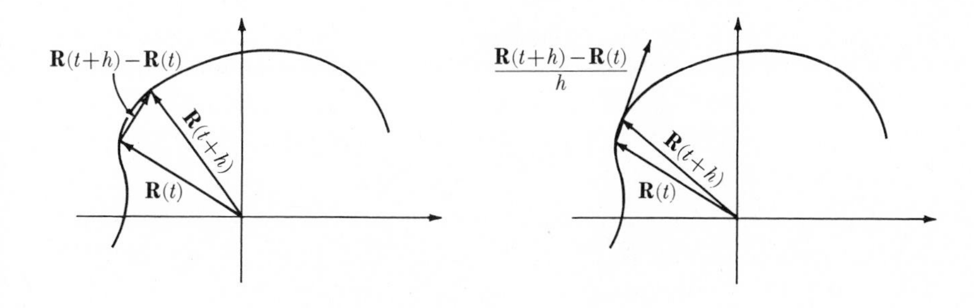

FIGURE 7.27 $\mathbf{V}(t) = \mathbf{R}'(t)$

Example 1. If $\mathbf{R}(t) = [x_0 + at, y_0 + bt]$, then $\mathbf{R}'(t) = [a,b]$. This is motion with constant velocity $\mathbf{V} = [a,b]$. *Conversely,* if $\mathbf{R}$ is any vector function and the derivative $\mathbf{R}'$ is a constant vector $\mathbf{C}$, then $\mathbf{R} = \mathbf{R}(0) + t\mathbf{C}$. To prove this, let the vector function be $\mathbf{R}(t) = [f(t), g(t)]$, and let its constant velocity be $\mathbf{C} = [a,b]$. Then $\mathbf{R}' = [f',g'] = \mathbf{C} = [a,b]$, i.e. $f' = a$ and $g' = b$. It follows that $f(t) = f(0) + at$, $g(t) = g(0) + bt$, and finally that

CONSTANT VELOCITY IMPLIES STRAIGHT LINE MOTION

$$\mathbf{R}(t) = [f(0), g(0)] + t[a, b] = [f(0) + at, g(0) + bt] = \mathbf{R}(0) + t\mathbf{C},$$

as we claimed.

Example 2. Let $\mathbf{R}(t) = [r \cos t, r \sin t]$. Then $\mathbf{V}(t) = \mathbf{R}'(t) = [-r \sin t, r \cos t]$. The *speed* of the motion is the length of the velocity vector, $|\mathbf{V}(t)| = \sqrt{r^2 \cos^2 t + r^2 \sin^2 t} = r$, which is constant. Although the path is traversed at constant speed, the velocity is not constant; its *direction* is always changing. Notice that in this case, $\mathbf{V}$ is obtained from $\mathbf{R}$ by a rotation through $\pi/2$ [Fig. 28]; $\mathbf{R} = [r \cos t, r \sin t]$, and $\mathbf{V} = [-r \sin t, r \cos t]$.

UNIFORM CIRCULAR MOTION

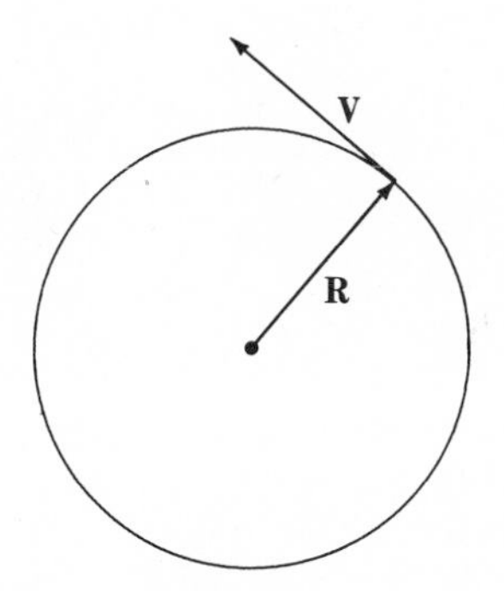

FIGURE 7.28

Differentiation of vector functions

We have imagined the vector function $\mathbf{R}$ to describe the path of a particle, and then $\mathbf{R}'$ was the velocity of the particle. But the definition of $\mathbf{R}'$,

$$\mathbf{R}'(t) = \lim_{h\to 0} \frac{1}{h} [\mathbf{R}(t+h) - \mathbf{R}(t)] \tag{1}$$

is independent of any geometric interpretations. Now we take a more abstract viewpoint: we think of the derivative simply as an operation defined by (1), and deduce its algebraic behavior.

Let $\mathbf{R}$, $\mathbf{R}_1$, $\mathbf{R}_2$, be vector functions, let b be an ordinary function, and let c be a number. Then:

RULES FOR DIFFERENTIATING VECTORS

$$(c\mathbf{R})' = c\mathbf{R}', \tag{2}$$

$$(\mathbf{R}_1 + \mathbf{R}_2)' = \mathbf{R}_1' + \mathbf{R}_2', \tag{3}$$

$$(b\mathbf{R})' = b'\mathbf{R} + b\mathbf{R}', \tag{4}$$

$$(\mathbf{R}_1 \cdot \mathbf{R}_2)' = \mathbf{R}_1' \cdot \mathbf{R}_2 + \mathbf{R}_1 \cdot \mathbf{R}_2', \tag{5}$$

$$(\mathbf{R} \circ b)' = b'(\mathbf{R}' \circ b). \tag{6}$$

Perhaps (6) needs elucidation. The symbol $\circ$ denotes composition, i.e. $(\mathbf{R} \circ b)(t) = \mathbf{R}(b(t))$, and (6) is the chain rule. To prove (6), let $\mathbf{R} = [f,g]$; then $\mathbf{R} \circ b = [f \circ b, g \circ b]$ and

$$\begin{aligned}(\mathbf{R} \circ b)' &= [(f \circ b)', (g \circ b)'] \\ &= [b' \cdot (f' \circ b), b' \cdot (g' \circ b)] \\ &= b'[f' \circ b, g' \circ b] \\ &= b'(\mathbf{R}' \circ b).\end{aligned}$$

Thus the chain rule for vector functions is deduced from the ordinary chain rule, simply by writing out the components of the vector. The same approach leads to the other formulas (2)–(5).

Acceleration and Newton's law

We have already seen [§3.4] how motion in a straight line can be analyzed on the basis of Newton's law. We get much more significant results from the formulation of Newton's law for motion in the plane. Let $\mathbf{F}(t)$ denote the total force acting on a particle at time t; the force acts with a certain strength in a certain direction, so $\mathbf{F}$ is a *vector* function. Newton's law says that

NEWTON'S LAW

$$\mathbf{F} = (m\mathbf{V})', \tag{7}$$

where m is the mass of the particle and $\mathbf{V}$ its velocity. The only difference between (7) and the law given in §3.4 is that now the force $\mathbf{F}$ and the velocity $\mathbf{V}$ are *vector* functions.

In many situations the mass m may be considered independent of t, and by (2) the law (7) simplifies to $\mathbf{F} = m\mathbf{V}'$. The derivative $\mathbf{V}'$ that occurs here is called the *acceleration*, denoted $\mathbf{A}$;

$$\mathbf{A} = \mathbf{V}' = \mathbf{R}''.$$

Thus when the mass is constant, Newton's law takes the simple form

NEWTON'S LAW WITH CONSTANT MASS m

$$\mathbf{F} = m\mathbf{A}. \tag{8}$$

We usually apply (8) to deduce the acceleration $\mathbf{A}$ (and hence, we hope, the position vector $\mathbf{R}$) from known forces. It can also be used the other way around, of course.

In the rest of this chapter, we always assume the mass m is constant, and use Newton's law as given in (8).

UNIFORM STRAIGHT LINE MOTION

Example 3. Let $\mathbf{R}(t) = \mathbf{R}(0) + t\mathbf{C}$ with $\mathbf{C}$ a constant vector. Then $\mathbf{R}' = \mathbf{C}$, $\mathbf{R}'' = 0$. There is no acceleration and hence no force. In the absence of any force (or any change in mass), the particle moves along a straight line at a constant speed.

Example 4. Let $\mathbf{R}(t) = [r \cos t, r \sin t]$, as in Example 2. Then $\mathbf{R}''(t) = [-r \cos t, -r \sin t]$. In this case, we find that $\mathbf{R}'' = -\mathbf{R}$, so the acceleration is directly opposite to the position vector. The position vector aims out from the center of the circle, and the acceleration aims in toward the center; this is called *centripetal* acceleration. [See Fig. 29.]

CENTRIPETAL ACCELERATION

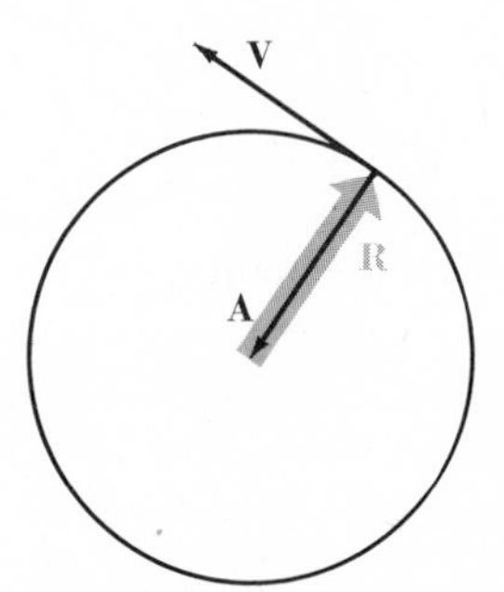

FIGURE 7.29

Example 5. Suppose that a projectile is launched into the air from the origin, with an initial velocity of $\mathbf{V}_0 = [a_0, b_0]$. Neglecting air resistance, the curvature of the Earth, and the variations in gravitational attraction, we may suppose that there is a constant downward force of strength $m\gamma$; thus the force vector is $\mathbf{F} = [0, -m\gamma]$. Then if $\mathbf{R}(t) = [f(t), g(t)]$ is the position vector, we have $\mathbf{V} = \mathbf{R}' = [f', g']$, and $\mathbf{A} = \mathbf{V}' = [f'', g'']$, so Newton's law $\mathbf{F} = m\mathbf{A}$ becomes

$$[0, -m\gamma] = [mf'', mg'']. \tag{9}$$

We will deduce the equation of the position function $\mathbf{R} = [f,g]$ from the differential equation (9), together with the given initial position (0,0) and initial velocity $\mathbf{V}_0 = [a_0, b_0]$.

The vector equation (9) means that corresponding components are equal,

$$0 = mf'' \qquad \text{and} \qquad -m\gamma = mg''.$$

From these equations it follows that f' is constant (since its derivative is zero) and $g' = -m\gamma t + c$ for some constant c. We are also given $[a_0, b_0] = \mathbf{V}_0 = [f'(0), g'(0)]$, so

$$f' = a_0 \qquad \text{and} \qquad g' = b_0 - m\gamma t.$$

Since $f(0) = g(0) = 0$, we conclude finally that

$$f = a_0 t \qquad \text{and} \qquad g = b_0 t - \frac{m\gamma t^2}{2}.$$

Thus if the sun were directly overhead, the shadow of the projectile would move with constant velocity, and the height would vary quadratically.

PROBLEMS

1. For each of the following position vectors $\mathbf{R}$, find $\mathbf{V}$ and $\mathbf{A}$. Sketch the path, and represent $\mathbf{V}(1)$ and $\mathbf{A}(1)$ as arrows starting from the position of the particle at $t = 1$. [The vector $\mathbf{V}(1)$ should be tangent to the path.]

(a) $\mathbf{R}(t) = [t, 1 - t^2]$

• (c) $\mathbf{R}(t) = \left[\dfrac{1 - t^2}{1 + t^2}, \dfrac{2t}{1 + t^2}\right]$

(b) $\mathbf{R}(t) = [t^2, t^3]$

2. At each time t, a particle is at the point (x,y), where $x = t$, $y = \sin t^2$.
(a) Sketch the path traced by the particle from $t = 0$ to $t = 3$.
(b) Find the velocity and acceleration at the times $t_0 = 0$, $t_1 = \sqrt{\pi/2}$, $t_2 = \sqrt{\pi}$.
(c) Represent the velocity and acceleration at the times t_0, t_1, t_2 in part (b) by arrows starting from the position of the particle at the corresponding time. [See Figs. 28 and 29.]

3. Prove the following differentiation rules for vector functions.
(a) $(c\mathbf{R})' = c\mathbf{R}'$ (for any constant c)
(b) $(\mathbf{R}_1 + \mathbf{R}_2)' = \mathbf{R}_1' + \mathbf{R}_2'$
(c) $(b\mathbf{R})' = b'\mathbf{R} + b\mathbf{R}'$
(d) $(\mathbf{R}_1 \cdot \mathbf{R}_2)' = \mathbf{R}_1' \cdot \mathbf{R}_2 + \mathbf{R}_1 \cdot \mathbf{R}_2'$

4. (a) Prove that if $\mathbf{R}$ is of constant length, then $\mathbf{R}'$ is orthogonal to $\mathbf{R}$. [Hint: $\mathbf{R} \cdot \mathbf{R} = c$; use formula (5).]
(b) Prove conversely that if $\mathbf{R}$ and $\mathbf{R}'$ are orthogonal, then $|\mathbf{R}|$ is constant.
(c) Try to interpret (a) and (b) in terms of motion and the velocity vector. [Note that $|\mathbf{R}| =$ constant means that the motion is along a circle of fixed radius about the origin.]

PART OF THE RAILROAD WHEEL MOVES BACKWARD

5. In Example 3, §7.4, we derived the equations

$$x = -r_0\theta + r\cos\theta, \qquad y = r_0 + r\sin\theta$$

for the path of a point on a rolling flanged wheel. Suppose the wheel is rolling at a constant rate so that $\theta = kt$ for some constant k.

(a) Show that $\mathbf{R}(t) = [-r_0kt + r\cos kt, r_0 + r\sin kt]$.

(b) Find the velocity vector $\mathbf{V}$.

(c) Show that the horizontal component of the velocity (that is, f' in $\mathbf{V} = [f', g']$) equals $-ky$. [Since y is the distance above the rail, this means that those points on the wheel *above* the rail are moving to the left, and those below to the right. Part of the flange is always moving backward!]

UNIFORM CIRCULAR MOTION

6. The path of an object moving with constant speed along a circle of radius r is described by $\mathbf{R}(t) = [r\cos kt, r\sin kt]$.

(a) Show that $|\mathbf{V}| = rk$ is the rate at which the perimeter is traversed; that is, rk units of length on the perimeter are traversed in each unit of time. [On a circle, arc length = radius × angle (in radians).]

(b) Show that $|\mathbf{A}| = rk^2 = |\mathbf{V}|^2/r$ and that $\mathbf{A}$ is directed toward the center of the circle. Thus, in uniform circular motion, the magnitude of the centripetal acceleration is $1/r$ times the square of the speed $|\mathbf{V}|$.

7. A motorcycle and rider of combined mass m are going around an unbanked circle of radius r at speed v. According to Problem 6, there is a centripetal acceleration of v^2/r. The circular track must provide the force for this acceleration, as well as the force to overcome gravity [Fig. 30], so the total force provided by the track is

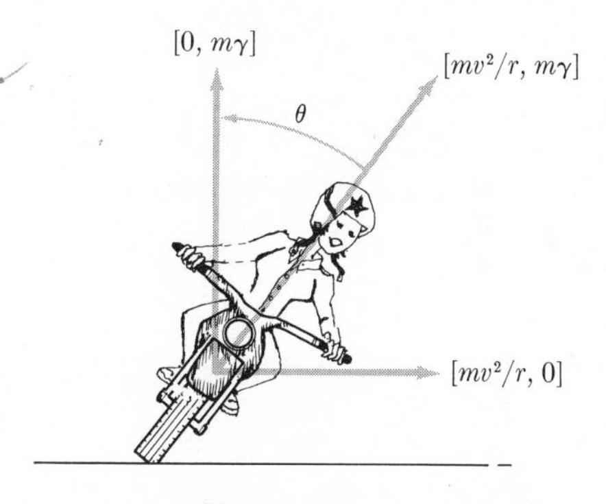

FIGURE 7.30

$$m[v^2/r, 0] + m[0, \gamma] = m[v^2/r, \gamma].$$

The motorcycle must be tipped precisely in the direction of this force, or it will fall over. Show that the inclination θ of the motorcycle is $\arctan(v^2/r\gamma)$.

BANKED CURVE

8. On an icy road, a curve of radius 0.1 mile is banked at an angle θ, where $\tan\theta = 1/4$. Cars going too slowly slip toward the inside of the curve, and going too fast they slip toward the outside. What is the optimal speed to avoid slipping? [See Problem 7 for a very closely related question. The acceleration of gravity is about 78,500 mi/hr^2.]

9. On a race track, a curve of radius 0.2 miles is intended to carry cars at about 150 mph. How steeply should it be banked to keep skidding to a minimum? [Compare with Problem 7.]

10. This problem proves that

$$\lim_{t\to t_0} W(t) = W_0 \quad \text{if and only if} \quad \lim_{t\to t_0} |W(t) - W_0| = 0.$$

Thus, in principle, vector limits can be computed without referring explicitly to the components. Let $W(t) = [f(t), g(t)]$ and $W_0 = [a,b]$.

(a) Show that $|f(t) - a| \le |W(t) - W_0|$ and $|g(t) - b| \le |W(t) - W_0|$.

(b) Conclude from part (a) that

$$\lim_{t\to t_0} |W(t) - W_0| = 0 \Rightarrow \lim_{t\to t_0} f(t) = a \text{ and } \lim_{t\to t_0} g(t) = b,$$

or

$$\lim_{t\to t_0} W(t) = W_0 .$$

(c) Show that $|W(t) - W_0| \le |f(t) - a| + |g(t) - b|$.

(d) Conclude that $\lim_{t\to t_0} W(t) = W_0 \Rightarrow \lim_{t\to t_0} |W(t) - W_0| = 0$. [Here, $\lim W = W_0$ means that each component of W converges to the corresponding component of W_0.]

11. [*The hanging chain*] A perfectly flexible uniform chain or rope supported at two ends assumes a predictable shape; this problem derives the formula for the shape, and justifies the name "catenary" for the graph of the hyperbolic cosine. Imagine the section of the chain between the bottom point P_0 and a variable point $P = (x,y)$. Three forces act on this section [Fig. 31]; a force (call it **H**) pulling to the left at P_0, another force (**T**) pulling up and to the right at P, and the force of gravity (**G**) pulling down all along the length. The force of gravity is proportional to the arc length of the curve from P_0 to P;

CATENARY

thus $\mathbf{G} = \left[0, -\gamma \int_0^x \sqrt{1 + (f')^2}\right]$ for a constant γ. The rest of the derivation rests on two principles:

(i) The force exerted by tension in the chain at any point P must be tangent to the chain at P.

(ii) The vector sum of all forces is 0 [otherwise the chain would accelerate and move.]

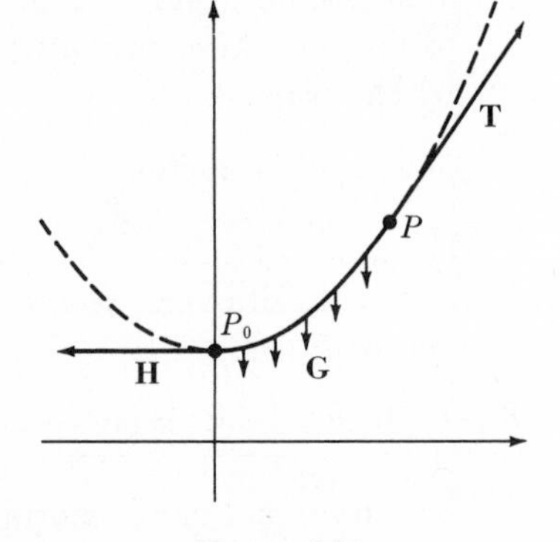

FIGURE 7.31

(a) Show that $h_2 = 0$, and $t_2/t_1 = f'(x)$, where $\mathbf{H} = [h_1, h_2]$ and $\mathbf{T} = [t_1, t_2]$. [Use (i)]

(b) Show that $h_1 = -t_1$ and $t_2 = \int_0^x \sqrt{1 + (f')^2}$. [Use (ii)]

(c) Deduce that $h_1 f' = -\gamma \int_0^x \sqrt{1 + (f')^2}$.

(d) Explain why $f'' = -\dfrac{\gamma}{h_1} \sqrt{1 + (f')^2}$.

(e) Observe that (d) gives a separable differential equation [§6.7] for f', not for f. Denote f' by p, and solve the equation for p.

$\left[\text{In } \int \dfrac{dp}{\sqrt{1+p^2}},\ \text{set } p = \sinh\theta,\ \S 2.8.\right]$ Evaluate the constant that appears in the integration by noting [Fig. 31] that $p(0) = f'(0) = 0$.

(f) Show that $f(x) = c - \dfrac{h_1}{\gamma} \cosh\left(\dfrac{\gamma x}{h_1}\right)$, where c is a constant.

[Note that $h_1 < 0$, from Fig. 31, so the coefficient of cosh is positive.]

7.6 L'HÔPITAL'S RULE *

Our basic Proposition A on limits [Chapter II] handles quotients *if* the denominator has a nonzero limit:

$$\lim_{t \to a} \frac{g(t)}{f(t)} = \frac{\lim g(t)}{\lim f(t)}, \qquad \text{if } \lim f(t) \neq 0.$$

But in many quotients the denominator inconveniently has a zero limit:

$$\lim_{t \to 0} \frac{1}{t}, \qquad \lim_{t \to 0} \frac{\cos t}{t}, \qquad \lim_{t \to 0} \frac{\sin t}{t}, \qquad \lim_{t \to 0} \frac{1 - \cos t}{t^2}, \qquad \lim_{t \to \pi} \frac{t^2 - \pi^2}{\sin t}.$$

In the first two cases, the limit in question does not exist; the denominator goes to zero, while the numerator does not, so the whole quotient "blows up." In the last three cases, however, *both* numerator and denominator tend to zero, and the ultimate behavior of the quotient is not immediately obvious. These are called "limits of the form 0/0"; they can frequently be evaluated by *L'Hôpital's rule*, which replaces the given quotient with a

* This section uses vector ideas, but does not contribute to them. It will be used to evaluate limits in Chapters X and XI.

quotient of derivatives. Figure 32 suggests how. Consider the curve $\mathbf{R}(t) = [f(t), g(t)]$. We are interested in the case where $\lim_{t \to a} f(t) = 0 = \lim_{t \to a} g(t)$, so the curve goes through the origin at $t = a$. Notice [Fig. 32] that the vector $\mathbf{R}'(t) = [f'(t), g'(t)]$ is tangent to the curve at the point $P = ((f(t), g(t))$, so the slope of this tangent line is $g'(t)/f'(t)$. As $t \to a$, the point P approaches the origin, so the slope of the tangent $g'(t)/f'(t)$ at P should approach the slope of the tangent at the origin:

$$\lim_{t \to a} \frac{g'(t)}{f'(t)} = \text{slope at origin, call it } L.$$

On the other hand, $g(t)/f(t)$ is the slope of the line through the origin and the point P. As $t \to a$, P approaches the origin, so the slope $g(t)/f(t)$ should approach the slope at the origin,

$$\lim_{t \to a} \frac{g(t)}{f(t)} = \text{slope at origin} = L = \lim_{t \to a} \frac{g'(t)}{f'(t)}.$$

L'Hôpital's rule says that this is so:

L'HOPITAL'S RULE

Theorem 2. *Suppose that* $\lim_{t \to a} f(t) = 0 = \lim_{t \to a} g(t)$, *and* $\lim_{t \to a} \frac{g'(t)}{f'(t)} = L$. *Then* $\lim_{t \to a} \frac{g(t)}{f(t)} = L$.

Two remarks must be made at once:

Remark 1. The theorem *requires* that $\lim_{t \to a} f(t) = 0$ and $\lim_{t \to a} g(t) = 0$. If either of these limits is *not* zero, the limit can be handled without L'Hôpital's rule, and in fact $\lim \frac{g'(t)}{f'(t)}$ can *not* be expected to agree with $\lim \frac{g(t)}{f(t)}$; see Fig. 33.

Remark 2. When the rule is applied, it is g'/f' that occurs, not $(g/f)'$.

Example 1. $\lim_{t \to 0} \frac{\sin t}{t} = ?$ *Solution.* By L'Hôpital's rule: Since both $f(t) = t$ and $g(t) = \sin t$ have limit 0, we can try L'Hôpital's rule. We get

$$\lim_{t \to 0} \frac{g'(t)}{f'(t)} = \lim_{t \to 0} \frac{\cos t}{1} = 1.$$

Therefore, the original quotient $\frac{\sin t}{t}$ has the same limit, $\lim_{t \to 0} \frac{\sin t}{t} = 1$. [Of

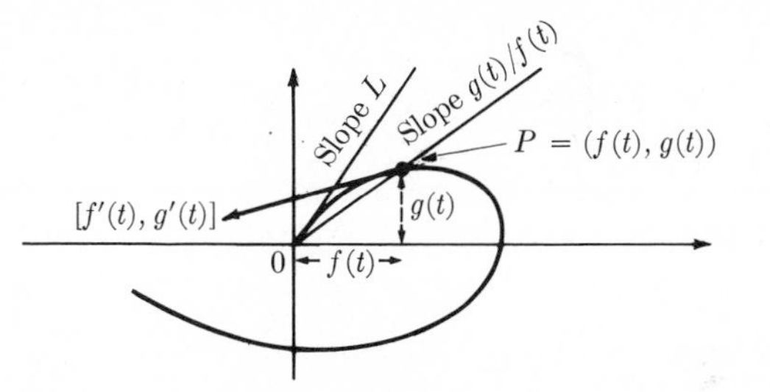

FIGURE 7.32 The tangent vector $[f'(t), g'(t)]$ has slope $g'(t)/f'(t)$; the line through 0 and $(f(t), g(t))$ has slope $g(t)/f(t)$

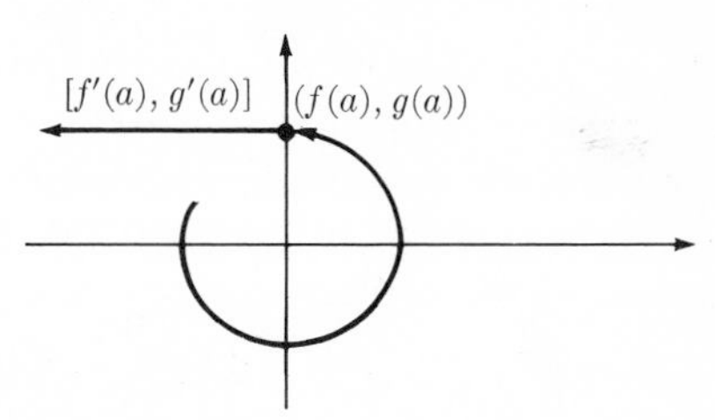

FIGURE 7.33 $\lim_{t\to a} \frac{g}{f} = \infty, \quad \lim_{t\to a} \frac{g'}{f'} = 0;$ L'Hôpital's rule does not apply since $\lim_{t\to a} f(t)$ and $\lim_{t\to a} g(t)$ are not both zero

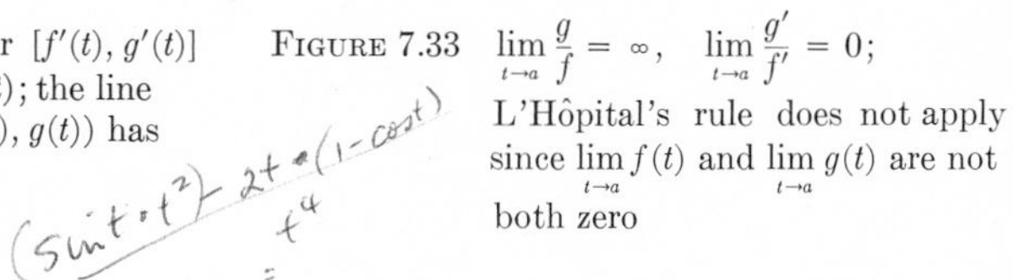

course, we knew this limit already; it is the main step in deriving the formula $\sin' = \cos$.]

Example 2. $\lim_{t\to 0} \frac{1 - \cos t}{t^2} = ?$ *Solution.* The numerator $g(t) = 1 - \cos t$ and the denominator $f(t) = t^2$ have limit 0, so we try

$$\lim_{t\to 0} \frac{g'}{f'} = \lim_{t\to 0} \frac{\sin t}{2t} = \frac{1}{2} \lim_{t\to 0} \frac{\sin t}{t} = \frac{1}{2}.$$

Therefore, the original quotient has the same limit: $\lim_{t\to 0} \frac{1 - \cos t}{t^2} = \frac{1}{2}$.

Note that in this application of L'Hôpital's rule, the new quotient g'/f' was itself a limit of the form 0/0; if we had not known the value of this new limit, we could therefore have tried L'Hôpital's rule on it, as in Example 1.

The proof of L'Hôpital's rule requires a slight generalization of the Mean Value Theorem, illustrated in Figure 34.

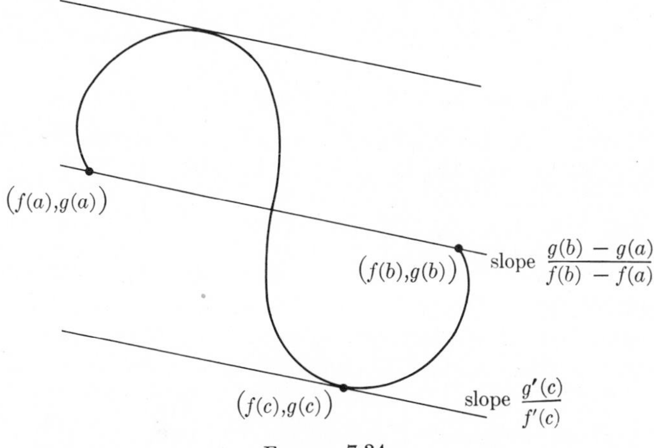

FIGURE 7.34

Lemma 1. *Let f and g be continuous on $[a,b]$ and differentiable on (a,b), and let $g(a) \neq g(b)$. Then there is a number c in (a,b) such that*

$$\frac{f(b) - f(a)}{g(b) - g(a)} = \frac{f'(c)}{g'(c)}. \tag{1}$$

Proof. If we clear the desired formula (1) of fractions, we obtain

$$[f(b) - f(a)]g'(c) = f'(c)[g(b) - g(a)].$$

This suggests that we apply the Mean Value Theorem to the function

$$h(x) = [f(b) - f(a)]g(x) - f(x)[g(b) - g(a)].$$

Indeed, $h(a) = f(b)g(a) - f(a)g(b) = h(b)$, so the Mean Value Theorem provides a point c between a and b where h has zero slope, i.e. where $h'(c) = 0$. And $h'(c) = 0$ is equivalent to (1).

Proof of Theorem 2. Since $\lim f(t) = 0$ and $\lim g(t) = 0$, we may set $f(a) = 0 = g(a)$, making f and g continuous at a. Then

$$\frac{f(t)}{g(t)} = \frac{f(t) - f(a)}{g(t) - g(a)} = \frac{f'(c)}{g'(c)}, \tag{2}$$

where c lies between a and t. It follows that $\lim\limits_{t \to a} c = a$, hence [Proposition C, §2.4.]

$$\lim_{t \to a} \frac{g'(c)}{f'(c)} = L.$$

In view of (2), this proves the theorem.

[We have glossed over the old bugaboo of division by 0, which we will now justify. To say that $\lim\limits_{t \to a} \dfrac{g'(t)}{f'(t)}$ exists implies, among other things, that the fraction is defined for all t sufficiently close to a; thus $f'(t) \neq 0$ for all t sufficiently close to a. It follows that $f(t) \neq 0$ for all t sufficiently close to a; for, $f(a) = 0$, so the Mean Value Theorem gives

$$f(t) = f(a) + (t - a)f'(c) = (t - a)f'(c)$$

for some c between a and t, and the right-hand side is the product of two nonzero terms.]

OTHER TYPES OF LIMIT

L'Hôpital's rule holds if $\lim\limits_{t \to a}$ is replaced by $\lim\limits_{t \to a+}$, $\lim\limits_{t \to a-}$, $\lim\limits_{t \to +\infty}$, or $\lim\limits_{t \to -\infty}$; the proof in all these cases is essentially the same.

Another version of the rule, covering limits of the form ∞/∞, has a more difficult proof, which is postponed to §AII.10. It reads as follows:

Theorem 3. *If* $\lim_{t \to a} f(t) = \infty$ *and* $\lim_{t \to a} g(t) = \infty$, *and* $\lim_{t \to a} \dfrac{f'(t)}{g'(t)} = L$,

then $\lim_{t \to a} \dfrac{f(t)}{g(t)} = L$.

Example 3. $\lim_{t \to 0+} \dfrac{\log t}{\csc t} = ?$ *Solution.* Since both numerator and denominator have limit ∞ as $t \to 0+$, we are in the "∞/∞" case for L'Hôpital's rule. Setting $g(t) = \log(t)$, $f(t) = \csc t$, we have

$$\lim_{t \to 0+} \frac{g'(t)}{f'(t)} = \lim_{t \to 0+} \frac{1/t}{-\csc t \cot t} = -\lim_{t \to 0+} \frac{\sin t \tan t}{t} = -\lim_{t \to 0+} \left(\frac{\sin t}{t}\right) \tan t = 0.$$

Therefore, the original limit is also 0, $\lim_{t \to 0+} \dfrac{\log t}{\csc t} = 0$.

PROBLEMS

1. Evaluate the following by L'Hôpital's rule, wherever it applies.

(a) $\lim_{t \to 0} \dfrac{e^t - 1}{\tan t}$

• (b) $\lim_{t \to 0} \dfrac{e^t}{\tan t}$

(c) $\lim_{x \to \pi} \dfrac{\sin x}{x^2 - \pi^2}$

• (d) $\lim_{x \to +\infty} \dfrac{\sin x}{x}$

(e) $\lim_{x \to +\infty} xe^{-x}$ [$= x/e^x$]

• (f) $\lim_{x \to +\infty} x^2e^{-x}$ [Apply it twice.]

(g) $\lim_{x \to +\infty} x^n e^{-x}$

• (h) $\lim_{x \to +\infty} \dfrac{\log x}{\log(x+1)}$

(i) $\lim_{x \to 0} \dfrac{1 - \cos x}{\sin x - x}$

• (j) $\lim_{x \to 0} \dfrac{1 - \cos x}{\sin x - x + x^2}$

(k) $\lim_{x \to 0+} x^x$ [Take logs.]

2. The following limits are not covered by L'Hôpital's rule. Explain why not, and state in each case what the limit is. [No proof required.]

(a) $\lim_{x \to 0+} \dfrac{\log x}{x}$

(b) $\lim_{x \to 0+} \dfrac{x}{\log x}$

(c) $\lim_{x \to +\infty} xe^x$

[Compare Problem 1(e).]

3. It seems natural to reduce the "∞/∞" rule to the "0/0" rule by writing $\dfrac{f}{g} = \dfrac{1/g}{1/f}$. Can Theorem 3 be deduced from Theorem 2 by this trick? Why, or why not?

7.7 GEOMETRY OF PARAMETRIC CURVES (Optional)

The analysis of curves belongs to differential geometry, one of the most important and attractive branches of mathematics. Here we give a very small sample of this subject by discussing plane curves. We begin with the *length*, then discuss tangents, curvature, and normal vectors, and conclude with an analysis of the acceleration based on these ideas.

Appendix II.9 gives a general definition of arc length (also called curve length), and proves that if the position vector **R** has a continuous derivative $\mathbf{R}' = [f',g']$ on a closed interval $[a,b]$, then the length of curve traversed as t varies from a to b is

ARC LENGTH OF THE CURVE $[f,g]$

$$\int_a^b \sqrt{f'(t)^2 + g'(t)^2}\, dt. \tag{1}$$

Rather than repeat the entire discussion here, we give a hasty argument to suggest why the integral (1) is correct. We divide the interval $[a,b]$ into subintervals by a partition $a = t_0 < t_1 < \cdots < t_n = b$. Then the points $(f(t_j), g(t_j))$ mark the ends of the segments sketched in Fig. 35, and the sum of the lengths of these segments is

$$\sum_1^n ([f(t_j) - f(t_{j-1})]^2 + [g(t_j) - g(t_{j-1})]^2)^{1/2}$$

$$= \sum_1^n \left(\left[\frac{f(t_j) - f(t_{j-1})}{t_j - t_{j-1}}\right]^2 + \left[\frac{g(t_j) - g(t_{j-1})}{t_j - t_{j-1}}\right]^2 \right)^{1/2} (t_j - t_{j-1}). \tag{2}$$

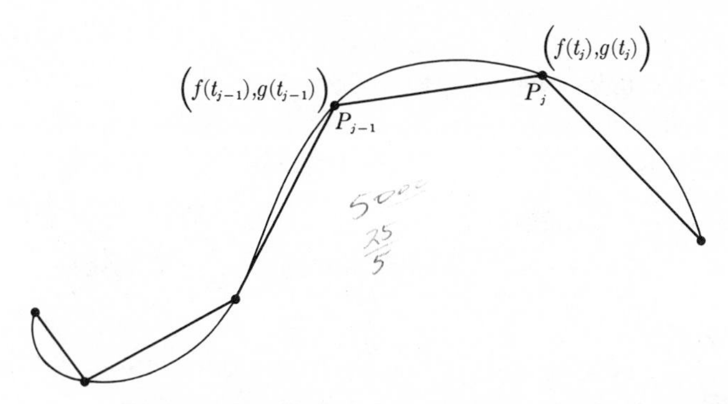

FIGURE 7.35 The length of the segment $P_{j-1}P_j$ is

$$\sqrt{\left(f(t_j)-f(t_{j-1})\right)^2+\left(g(t_j)-g(t_{j-1})\right)^2}$$

As the partition is made finer, the segments grow shorter, and the sum of their lengths tends to the arc length of the curve. At the same time, the denominators $t_j - t_{j-1} \to 0$, so

$$\frac{f(t_j) - f(t_{j-1})}{t_j - t_{j-1}} \to \frac{df}{dt}, \qquad \frac{g(t_j) - g(t_{j-1})}{t_j - t_{j-1}} \to \frac{dg}{dt},$$

the factor $(t_j - t_{j-1}) = \Delta t_j$ at the end of (2) turns into dt, the $\sum_1^n$ turns into $\int_a^b$, and the whole sum (2) turns into the integral

$$\int_a^b \left(\left(\frac{df}{dt}\right)^2 + \left(\frac{dg}{dt}\right)^2\right)^{1/2} dt,$$

which is the same as (1). [It's a very hasty argument, but we warned you!]

Example 1. $x = r\cos t$, $y = r\sin t$ are parametric equations of a circle of radius r. The circle is traversed once as t ranges from 0 to 2π. Use (1) to find the perimeter of the circle. [Do it yourself; solution below.]

The arc length integral (1) helps to interpret the velocity vector $\mathbf{V}$. Suppose $\mathbf{R} = [f,g]$ is a position vector, with f and g differentiable on some closed interval $[a,b]$. We define the *arc length function*

$$s(t) = \int_a^t \sqrt{(f')^2 + (g')^2}. \tag{3}$$

Clearly, from (1), $s(t)$ is the length of the path traversed by the particle from time a to time t. Notice that the integrand, $\sqrt{(f')^2 + (g')^2}$, is the *length* $|\mathbf{V}|$ of the velocity vector $\mathbf{V} = [f', g']$. Thus, by the Fundamental Theorem of Calculus, $|\mathbf{V}|$ is the derivative s' of the arc length function s, the rate of change of distance with respect to time. This is what we call the *speed:*

SPEED IS THE DERIVATIVE OF ARC LENGTH

$$|\mathbf{V}| = \text{speed} = s' = \frac{ds}{dt} = \frac{d(\text{arc length})}{dt}.$$

If $|\mathbf{V}| \neq 0$, we can divide by it and obtain from $\mathbf{V}$ the unit vector $\mathbf{V}/|\mathbf{V}|$. This vector points in the same direction as $\mathbf{V}$, i.e. tangent to the curve, and in the direction of increasing t; it is called the *unit tangent vector*, and usually denoted $\mathbf{T}$. Whereas $\mathbf{V}$ gives the speed and direction of the particle, the unit tangent vector $\mathbf{T} = \mathbf{V}/|\mathbf{V}|$ gives *only* the direction.

UNIT TANGENT VECTOR $\mathbf{T} = \mathbf{V}/|\mathbf{V}|$

Curvature

The curvature of a path is essentially the *rate of change of its direction*, but we take the rate of change with respect to the *curve length*, not with respect to the parameter. (The curvature of a road doesn't depend on how fast you drive—it only seems to. What really does depend on the speed is the sideways force on the car, as we will show a little later in analyzing the acceleration vector with the help of formula (7) below.) Suppose that s stands for the curve length measured from some fixed point. Now, the direction of the curve is given by the unit tangent $\mathbf{T} = \mathbf{V}/|\mathbf{V}|$, so we define the *curvature* κ as the magnitude of $d\mathbf{T}/ds$, the rate of change of $\mathbf{T}$ with respect to arc length:

CURVATURE κ

$$\kappa = \left|\frac{d\mathbf{T}}{ds}\right|. \tag{4}$$

By the chain rule,

$$\frac{d\mathbf{T}}{ds} = \left(\frac{d\mathbf{T}}{dt}\right)\left(\frac{dt}{ds}\right) = \frac{(d\mathbf{T}/dt)}{(ds/dt)}, \qquad \text{so} \qquad \kappa = \left|\frac{d\mathbf{T}/dt}{ds/dt}\right|.$$

Substituting $\mathbf{T} = \mathbf{V}/|\mathbf{V}|$, $|\mathbf{V}| = ds/dt$, and $\mathbf{V} = [f', g']$, you can "reduce" formula (4) to

$$\kappa = \frac{|f'g'' - g'f''|}{[(f')^2 + (g')^2]^{3/2}}. \tag{5}$$

[See Problem 5.] But (4) is easier to remember because of the natural way it was derived.

Example 2. Let $x = r\cos t$, $y = r\sin t$, with $r > 0$. Find $\mathbf{V}(t)$, $|\mathbf{V}(t)|$, $\mathbf{T}(t)$, and ds/dt. Show that $\kappa = 1/r$. Is this reasonable? [Solution below.]

Unit normal vector

Problem 4 of §7.5 showed that the derivative $\mathbf{R}'$ of a vector $\mathbf{R}$ of constant length is orthogonal to $\mathbf{R}$. Now, $\mathbf{T}$ is a unit vector, so $d\mathbf{T}/ds$ is orthogonal to $\mathbf{T}$. Dividing $d\mathbf{T}/ds$ by its length κ, we obtain a unit vector $\mathbf{N}$ orthogonal to $\mathbf{T}$, called the *unit normal:*

UNIT NORMAL $\mathbf{N}$

$$\mathbf{N} = \frac{d\mathbf{T}/ds}{|d\mathbf{T}/ds|} = \frac{d\mathbf{T}/ds}{\kappa}, \qquad d\mathbf{T}/ds = \kappa\mathbf{N}. \tag{6}$$

Since $\mathbf{N}$ indicates the direction in which $\mathbf{T}$ is changing, it always points toward the *inside* of the curve. When $\kappa = 0$, formula (6) does not define $\mathbf{N}$; indeed, when $\kappa = 0$, it may be impossible to distinguish between the "inside" and "outside" of the curve. [See Fig. 36; which way should $\mathbf{N}$ go at the point where $\kappa = 0$?]

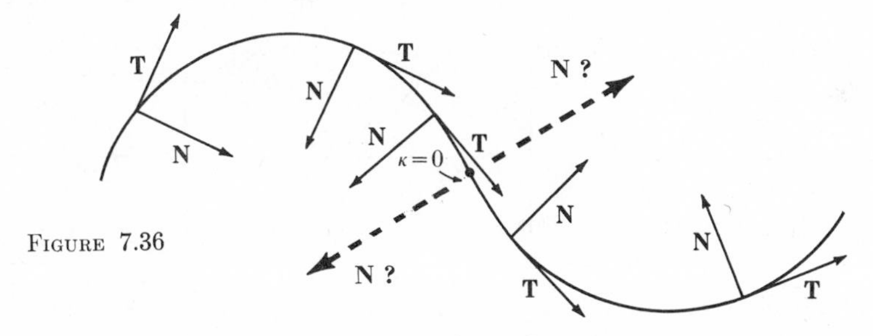

FIGURE 7.36

Tangential and normal components of acceleration

The acceleration vector has an interesting analysis in connection with $\mathbf{T}$, $\mathbf{N}$, κ, and ds/dt. Since $|\mathbf{V}| = \dfrac{ds}{dt}$ and $\mathbf{T} = \dfrac{\mathbf{V}}{|\mathbf{V}|}$, we have $\mathbf{V} = \dfrac{ds}{dt}\mathbf{T}$. Hence

TANGENTIAL, NORMAL COMPONENTS OF ACCELERATION

$$\mathbf{A} = \frac{d\mathbf{V}}{dt} = \frac{d}{dt}\left(\frac{ds}{dt}\mathbf{T}\right) = \frac{d^2s}{dt^2}\mathbf{T} + \frac{ds}{dt}\frac{d\mathbf{T}}{dt}$$

$$= \frac{d^2s}{dt^2}\mathbf{T} + \frac{ds}{dt}\left(\frac{ds}{dt}\frac{d\mathbf{T}}{ds}\right) = \frac{d^2s}{dt^2}\mathbf{T} + \left(\frac{ds}{dt}\right)^2\kappa\mathbf{N}. \tag{7}$$

[Here, d^2s/dt^2 is the Leibniz notation for the derivative of ds/dt; that is, for the second derivative of s.] The coefficients of the orthogonal unit vectors $\mathbf{T}$ and $\mathbf{N}$ in (7) are called respectively the *tangential* and *normal* components of the acceleration.

The tangential component d^2s/dt^2 is the rate of change of the speed ds/dt; thus the tangential component of the acceleration is the part that changes the speed of the particle. The normal component $(ds/dt)^2\kappa$ is the part that changes the direction of the particle, giving rise to the curvature κ. The greater the speed ds/dt, the smaller the curvature κ effected by a given normal component of acceleration.

Now imagine that the "particle" we have been discussing is a car. Then the curvature is imposed by the shape of the road, and the driver must use the forces at his disposal to maintain that curvature. All the forces are transmitted by the friction of the wheels on the road, and when the total force exceeds a certain amount, the car skids. From (7), you can see that in a curve, where $\kappa \neq 0$, there is a limit to how large the speed ds/dt can be without causing a skid. Further, once the driver is in the sharpest part of the curve, he should neither speed up nor slow down, since *either* of these contributes to the tangential component d^2s/dt^2 of the acceleration and hence *increases* the total force. Speeding up is particularly bad, because it increases both d^2s/dt^2 and ds/dt; but braking is not much better, since the increase in d^2s/dt^2 is felt immediately, whereas the decrease in ds/dt occurs only gradually [Fig. 37].

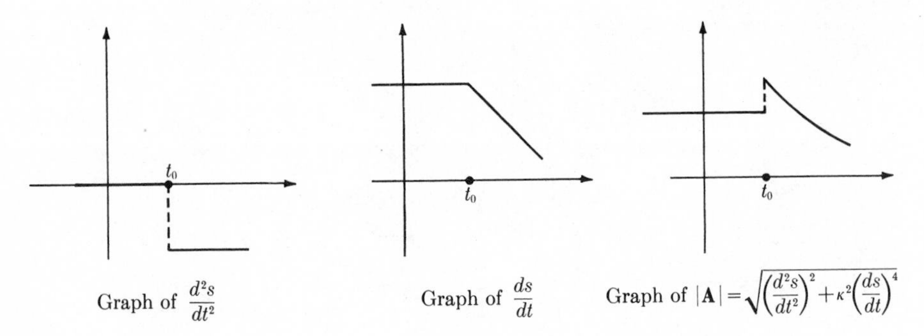

FIGURE 7.37 Effect of braking at time t_0 in a curve of curvature κ

Incidentally, formula (7) provides a handy way to compute the curvature κ. By the Pythagorean theorem for vectors [Problem 9, §7.2],

$$|\mathbf{A}|^2 = \left|\frac{d^2s}{dt^2}\mathbf{T}\right|^2 + \left|\left(\frac{ds}{dt}\right)^2 \kappa\mathbf{N}\right|^2 = \left|\frac{d^2s}{dt^2}\right|^2 + \left(\frac{ds}{dt}\right)^4 \kappa^2.$$

Hence

A FORMULA FOR CURVATURE

$$\kappa^2 = \frac{|\mathbf{A}|^2 - (d^2s/dt^2)^2}{(ds/dt)^4}. \tag{8}$$

Solution to Example 1.

$$\text{Perimeter} = \int_0^{2\pi} \sqrt{f'(t)^2 + g'(t)^2}\, dt = \int_0^{2\pi} \sqrt{(-r \sin t)^2 + (r \cos t)^2}\, dt$$

$$= \int_0^{2\pi} \sqrt{r^2}\, dt = 2\pi r.$$

Solution to Example 2.

$$\mathbf{V}(t) = [-r \sin t, r \cos t]; |\mathbf{V}(t)| = \sqrt{(-r \sin t)^2 + (r \cos t)^2} = r;$$

$$\mathbf{T}(t) = [-\sin t, \cos t],\ ds/dt = r;\ \kappa = \frac{|d\mathbf{T}/dt|}{ds/dt} = \frac{|[-\cos t, -\sin t]|}{r} = \frac{1}{r}.$$

This is reasonable: the smaller the circle, the greater the curvature.

PROBLEMS

1. For each of the following curves, find the velocity, acceleration, speed, and curvature.
   • (a) $\mathbf{R}(t) = [r \cos kt, r \sin kt]$
   (b) $\mathbf{R}(t) = [t, 1 + t^2]$
   (c) $\mathbf{R}(t) = [t^2, t^3]$
   • (d) $\mathbf{R}(t) = [-t + \cos t, 1 + \sin t]$

2. For each of the curves in Problem 1, sketch the unit tangent, unit normal, and acceleration vectors at $t = 1$. Indicate the tangential and normal components of acceleration.

• 3. The position vector $\mathbf{R}(t) = [-t + \cos t, 1 + \sin t]$ describes a *cycloid*, the path traced by a point on a rolling wheel [see Example 3, §7.4, and Fig. 25]. Find the arc length of one arch of the cycloid. [Hint: $1 + \sin t = \cos^2 (t/2) + \sin^2 (t/2) + 2 \cos (t/2) \sin (t/2)$.]

• 4. Let $\mathbf{R}(t) = [-2t + 3 \cos t, 2 + 3 \sin t]$. [See Example 3, §7.4.] Show that the curvature is least at the highest point on the path, and greatest at the lowest point. Find the points of maximum and minimum speed. Sketch the curve, and the velocity and acceleration vectors at the highest and lowest points on the curve.

5. Derive formula (5) of the text from formula (8), by letting $\mathbf{R} = [f,g]$ and computing $|\mathbf{A}|^2$, ds/dt, and d^2s/dt^2 in terms of f and g.

7.8 POLAR COORDINATES

In rectangular coordinates, we locate points in the plane by an ordered pair of numbers (x,y), where x and y are the distances from the point to two perpendicular axes. Sometimes it is more convenient to describe a point by giving its *distance from the origin* and its *direction from the origin.*

The distance is some number r, and the direction can be described by the angle θ, as in Fig. 38. From the figure, you can see the relation between these numbers r and θ, and the rectangular coordinates (x,y):

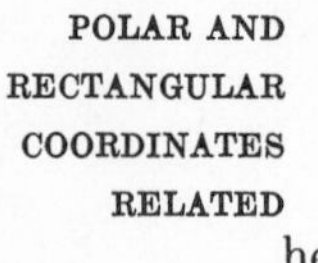
POLAR AND RECTANGULAR COORDINATES RELATED

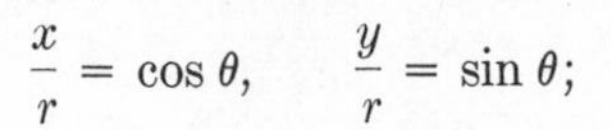

$$\frac{x}{r} = \cos\theta, \qquad \frac{y}{r} = \sin\theta;$$

hence

$$(1) \qquad x = r\cos\theta, \qquad y = r\sin\theta.$$

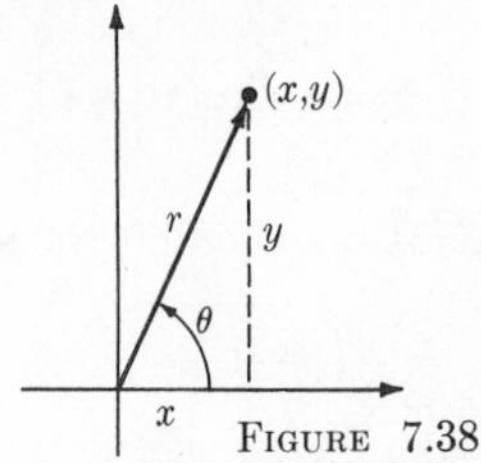

FIGURE 7.38

The pair of numbers r and θ in (1) are called *polar coordinates* of the point (x,y). Figure 38 suggests that $r \geq 0$ and $0 \leq \theta < 2\pi$, but we do not make these restrictions in general; in fact, *any pair of numbers r and θ such that equations* (1) *hold* are called *polar coordinates of the point* (x,y).

Example 1. Each of the following pairs of numbers r and θ are polar coordinates of the point $(0,1)$:

(a) $r = 1, \theta = \pi/2$; (b) $r = -1, \theta = 3\pi/2$; (c) $r = 1, \theta = -3\pi/2$.

With each of these choices of r and θ, we have $0 = r\cos\theta$, $1 = r\sin\theta$. [See Fig. 39.]

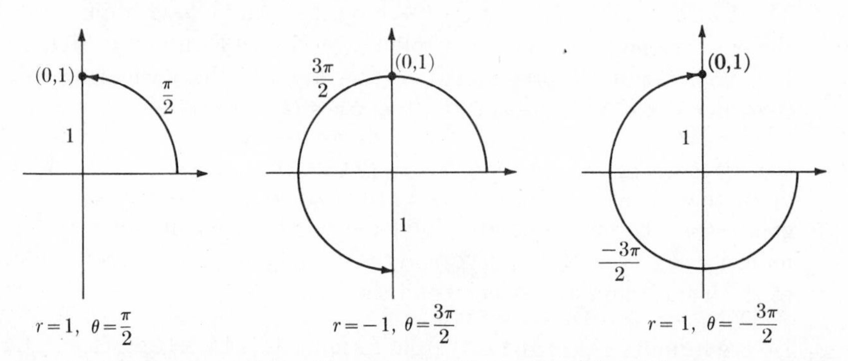

FIGURE 7.39 Three sets of polar coordinates for (0,1)

The origin plays a special role in polar coordinates, which suggests (rightly) that they are used mainly in applications where one particular point has a central importance. We have introduced them here in order to study planetary motion, with the sun as the central point [§7.11].

Each choice of polar coordinates r and θ determines a point in the plane, namely, $(r\cos\theta, r\sin\theta)$. We will denote this point $(r,\theta)_p$. [Warning: Standard practice uses (r,θ) as an alternate notation for the point $(r\cos\theta, r\sin\theta)$, and depends on the context to keep things straight.]

PLOTTING IN POLAR COORDINATES

To locate the point $(r,\theta)_p$, proceed counterclockwise around the unit circle a (signed) distance θ; this determines a certain direction from the origin. Then if $r > 0$, the point is r units from the origin in that direction. If $r < 0$, the point is $|r|$ units from the origin, but in the *opposite* direction. Figure 40 shows how to find $(2, 5\pi/4)_p$ and $(-2, 5\pi/4)_p$.

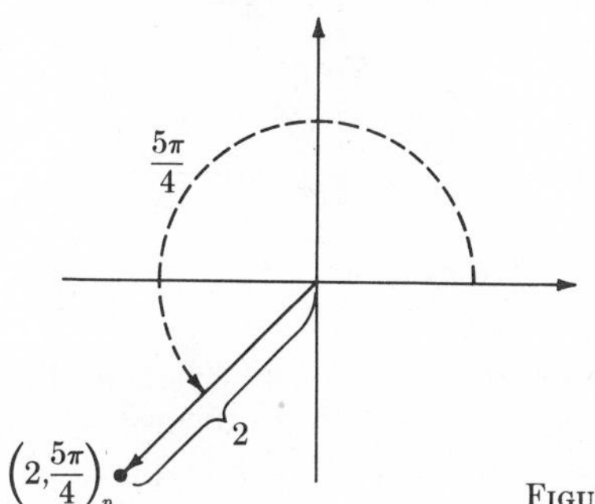

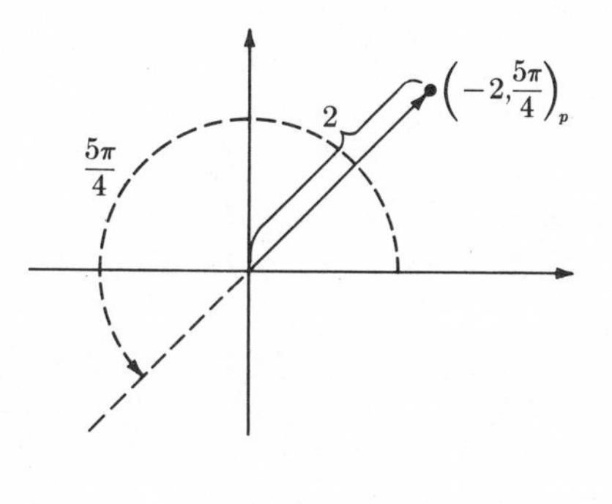

FIGURE 7.40

MANY POLAR COORDINATES FOR ONE GIVEN POINT

It is not true that each point in the plane has a unique set of polar coordinates. [See Example 1.] This is the main disadvantage of the system. You can see from (1) that $r^2 = x^2 + y^2$, so $r = \pm\sqrt{x^2 + y^2}$ is determined up to sign. But there is a great latitude in choosing θ. In fact, when $r = 0$, then *every* value of θ gives you the same point, $(0,\theta)_p = (0,0)$.

Example 2. Plot the points $(1,\pi/2)_p$, $(-1,\pi/2)_p$, $(\sqrt{2},\pi/4)_p$, $(-\sqrt{2},5\pi/4)_p$ directly, without converting to rectangular coordinates. Then find the rectangular coordinates using (1), and check your results. [Solution below.]

Example 3. Find polar coordinates for (0,0), (1,−1), $(-1, \sqrt{3})$. Find at least two sets of polar coordinates for (1,−1). [Solution below.]

We can describe figures in the plane by giving relations between their polar coordinates, as in the next example.

Example 4. Graph the set $\{(r,\theta)_p : r = \sin\theta\}$. *Simple-minded solution:* We tabulate some values and locate the corresponding points in Fig. 41. When $\pi < \theta < 2\pi$, θ indicates a direction *below* the x axis; but since r is negative for these values of θ, the corresponding points $(r,\theta)_p$ are *above* the x axis. The points sketched in Fig. 41 appear to lie on a circle of diameter 1. We can prove that they do by converting the equation $r = \sin\theta$ to rectangular coordinates. Multiply by r to obtain $r^2 = r\sin\theta$, which is the same as $x^2 + y^2 = y$. You can complete the square on y to show that $x^2 + y^2 = y$ is the equation of the circle with radius $\frac{1}{2}$ and center $(0,\frac{1}{2})$.

WARNING

Note, incidentally, that we do not graph $r = \sin\theta$ as a sine curve, with a vertical "r axis" and a horizontal "θ axis." This would be the wrong picture to have in mind when working in polar coordinates!

θ	r	Corresponding point
0	0	P_0
$\pi/6$	$1/2$	P_1
$\pi/3$	$\sqrt{3}/2$	P_2
$\pi/2$	1	P_3
$3\pi/4$	$1/\sqrt{2}$	P_4
π	0	P_5
$5\pi/4$	$-1/\sqrt{2}$	P_6
$3\pi/2$	-1	P_7

FIGURE 7.41

In general, any relation $r = f(\theta)$ yields a graph $\{(r,\theta)_p : r = f(\theta)\}$, as in Example 4. The most important of these graphs are the conic sections. Recall that a parabola is determined by a line L, the *directrix*, and a point F, the *focus:* The parabola is the set of all points equidistant from L and F.

DEFINITION: CONIC SECTION OF ECCENTRICITY e

There is a similar definition that gives *all* the conic sections, the ellipse, parabola, and hyperbola. Given a directrix, a focus F, and a number $e > 0$ (called the *eccentricity*), the corresponding conic section is the set of all points P such that $PF = ePD$, where PF denotes the distance from P to F, and PD denotes the distance from P to the directrix [Fig. 42]. When $e = 1$, we have $PF = PD$, the equation of the parabola. When e is very small, then $PF = ePD$ is small, and the set is nearly a small circle about F. When $0 < e < 1$, the set is an ellipse. When $e > 1$, the set is a hyperbola.

Relying on pictures, we can easily derive the polar equation of a conic section with a focus F at the origin. Let the directrix be at distance d from the origin, in a direction given by the angle φ [Fig. 43]. Take any point $P = (r,\theta)_p$ on the conic section. Suppose $r > 0$; then $PF = r$ and, from Fig. 43,

$$PD = d + r \sin \alpha = d + r \sin (\varphi - \theta - \pi/2) = d - r \cos (\theta - \varphi).$$

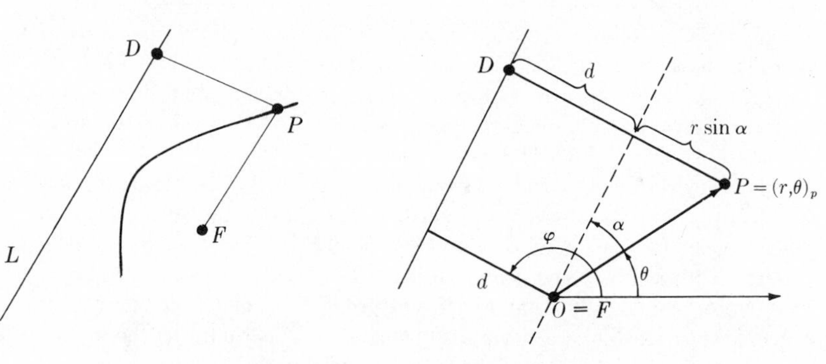

FIGURE 7.42

FIGURE 7.43

Thus the equation $PF = ePD$ becomes

$$r = ed - er\cos(\theta - \varphi).$$

Solving for r, we have

$$r = \frac{ed}{1 + e\cos(\theta - \varphi)}. \tag{2}$$

We leave the graphing of this equation to the problems.

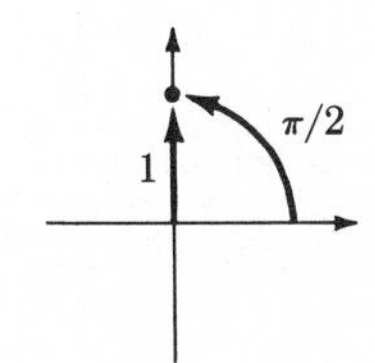

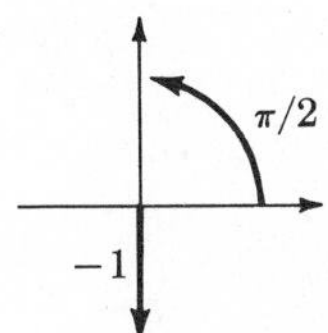

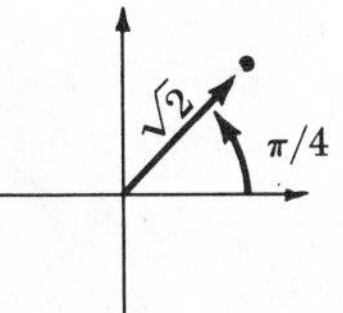

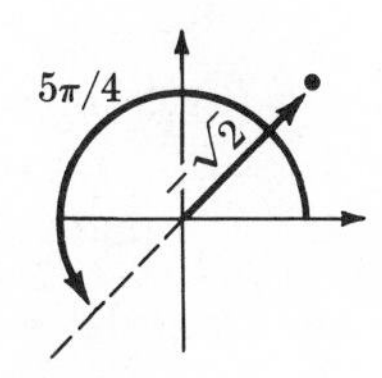

FIGURE 7.44

Solution to Example 2. See Fig. 44. The rectangular coordinates are respectively $(0,1)$, $(0,-1)$, $(1,1)$, $(1,1)$.

Solution to Example 3. The polar coordinates can be guessed from Fig. 45; $(0,0) = (0,0)_p$, $(1,-1) = (\sqrt{2}, -\pi/4)_p = (-\sqrt{2}, 3\pi/4)_p$, $(-1, \sqrt{3}) = (2, 2\pi/3)_p$.

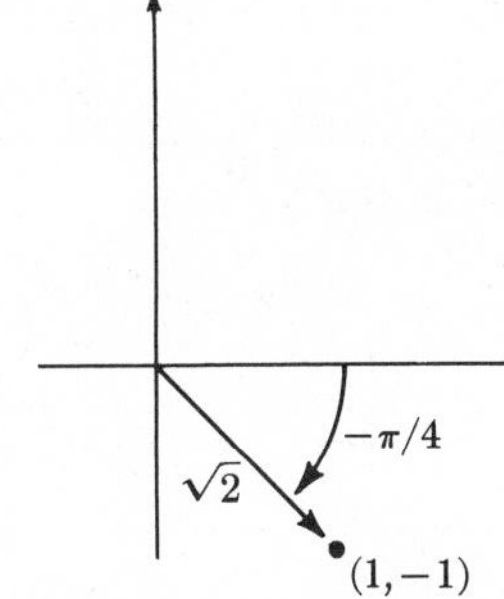

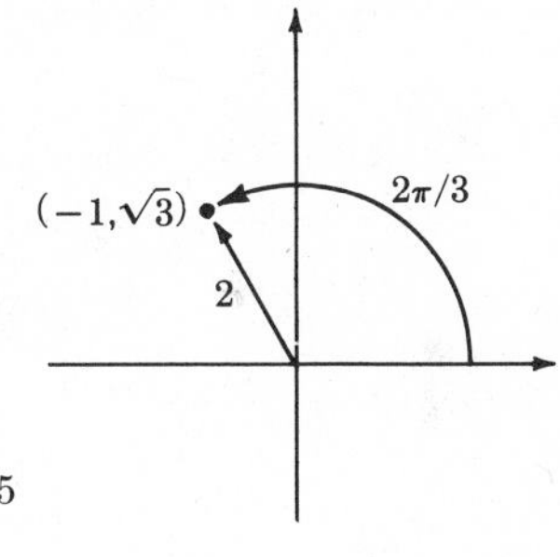

FIGURE 7.45

PROBLEMS

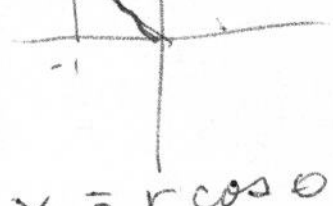

1. Plot the points:
   (a) $(1, \pi/4)_p$
   (b) $(-1, -\pi/4)_p$
   (c) $(\sqrt{2}, 3\pi/4)_p$
   (d) $(\pi,\pi)_p$
   (e) $(1,\pi)_p$
   (f) $(\pi,1)_p$

2. Find *all* the polar coordinates of each of the following points.
   (a) $(0,0)$ (b) $(1,1)$ (c) $(1/2, -1)$

3. Sketch the graphs of the following equations in polar coordinates.
   (a) $r = \cos\theta$
   (b) $r = 1 + \cos\theta$
   (c) $r = \theta$
   (d) $r = \cos 2\theta$
   (e) $r = 2 + \cos\theta$
   (f) $r = 1 + 2\cos\theta$

4. Consider the (polar coordinate) equation $r = 1/(1 + \cos\theta)$.
(a) What choices of e, d, and φ in (2) yield this equation?
(b) Find the minimum value of r. Show that there is no maximum.
(c) Plot the point on the graph of this equation where r is a minimum.
(d) Sketch the directrix and the graph (which is a parabola).

• **5.** Consider the equation $r = 1/(2 + \sin\theta)$.
(a) What choices of e, d, and φ in (2) yield this equation? Is the graph an ellipse, a parabola, or a hyperbola?
(b) Find the minimum and maximum values of r. Plot the corresponding points.
(c) Sketch the directrix and the graph.
(d) Find the length of the major axis of the curve; i.e. find its largest diameter.

6. Consider $r = 1/(1 - 2\cos\theta)$.
(a) What choices of e, d, and φ in (2) yield this equation? Is it an ellipse, a parabola, or a hyperbola?
(b) Find the minimum value of $|r|$. [Notice that r can be negative.]
(c) Find the angles θ at which r becomes infinite. [These give the directions of the *asymptotes* of the curve; however, the asymptotes do not pass through the origin.]
(d) Sketch the directrix and the curve. Notice that the curve consists of two parts: one with $r > 0$ and one with $r < 0$.

7. Convert equation (2) to rectangular coordinates, and show that it has the form of a general quadratic in x and y,

$$Ax^2 + Bxy + Cy^2 + Dx + Ey + F = 0.$$

8. In the derivation of (2), what happens if we take $(r,\theta)_p$ on the other side of the directrix in Fig. 43? What happens if $r < 0$? [Note that (2) has the same graph as $r = -ed/(1 - e\cos(\theta - \varphi))$.]

9. Show that the following are conic sections, and find the eccentricity and the distance from the origin to the directrix.

• (a) $r = \dfrac{1}{1 + \cos\theta + 2\sin\theta}$ (b) $2r + 3r\cos\theta - r\sin\theta - 1 = 0$

10. In rectangular coordinates, the equation of a straight line takes the simple form $ax + by = c$. In polar coordinates it is not quite so simple.
(a) Graph the following polar equations:

$$r\cos\left(\theta - \frac{\pi}{2}\right) = 2; \qquad r = \frac{-1}{\cos\theta + \sin\theta};$$

$$r\sin\left(\theta - \frac{\pi}{2}\right) = 2.$$

(b) Show that the graph of $r \cos(\theta - \varphi) = d$ is a straight line. [Here φ and d are constants. Hint: Convert to rectangular coordinates; $\cos(\theta - \varphi) = ?$]
(c) On the graph in part (b), find the point closest to the origin.
(d) What is the geometric meaning of φ and d in part (b)?
• (e) Show that any equation of the form $ax + by = c$ is equivalent to a polar equation of the form $r \cos(\theta - \varphi) = d$.

11. Show that the square of the distance from point $P_1 = (r_1, \theta_1)_p$ to $P_2 = (r_2, \theta_2)_p$ is $r_1^2 + r_2^2 - 2r_1r_2 \cos(\theta_2 - \theta_1)$. [Use the equations (1) and the distance formula in rectangular coordinates to prove it; explain the result by the law of cosines.]

12. [*Distance from point to line*] Show that the distance from $P_0 = (r_0, \theta_0)_p$ to the polar graph of $r \cos(\theta - \varphi) = d$ is $|d - r \cos(\theta - \theta_0)|$. [Hint: Convert the equation to rectangular coordinates, and see the distance formula in §7.3. Note that this formula confirms the pictorial derivation of the formula for PD from Fig. 42 above.]

7.9 AREA AND ARC LENGTH IN POLAR COORDINATES

The area of the shaded region in Fig. 46 is given by a certain integral. The region lies between the radial lines defined by $\theta = a$ and $\theta = b$, and is bounded on the outer edge by the polar graph of $r = f(\theta)$, where $f > 0$. To find the integral, partition the interval $[a,b]$ into subintervals $[\theta_{j-1}, \theta_j]$, where, as in Chapter V, $a = \theta_0 < \theta_1 < \cdots < \theta_n = b$. Let m_j and M_j be respectively the minimum and maximum of f on the interval $[\theta_{j-1}, \theta_j]$. Denote by A_j the darker shaded region in Fig. 46 corresponding to directions θ with $\theta_{j-1} < \theta < \theta_j$. Then A_j *contains* a sector of radius m_j and angle $\Delta\theta_j = \theta_j - \theta_{j-1}$, and A_j lies *inside* a sector of radius M_j and angle $\Delta\theta_j$.

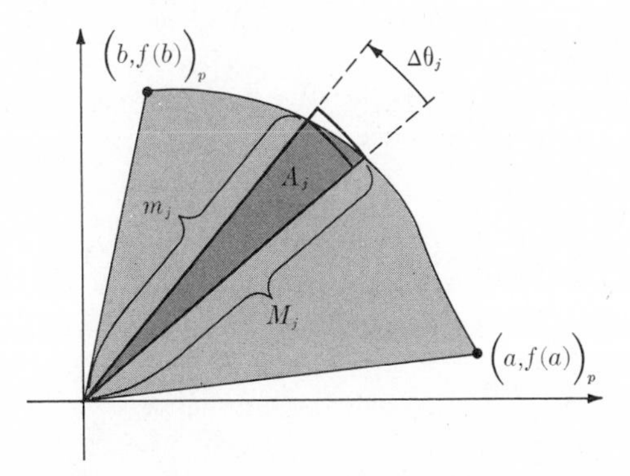

FIGURE 7.46

Now, the area of a sector of angle α and radius r is $\frac{1}{2}r^2\alpha$; thus the inner sector has area $\frac{1}{2}m_j{}^2\Delta\theta_j$ and the outer sector has area $\frac{1}{2}M_j{}^2\Delta\theta_j$, so

$$\frac{1}{2}m_j{}^2\Delta\theta_j \leq A_j \leq \frac{1}{2}M_j{}^2\Delta\theta_j$$

and the entire area $A = \sum_1^n A_j$ satisfies

$$\sum_1^n \frac{1}{2}m_j{}^2\Delta\theta_j \leq A \leq \sum_1^n \frac{1}{2}M_j{}^2\Delta\theta_j\,. \tag{1}$$

The sums appearing in (1) are lower and upper Riemann sums for the integral $\int_a^b \frac{1}{2}f(\theta)^2\,d\theta$; since (1) holds for every partition, we conclude that the area A is given by the integral $\int_a^b \frac{1}{2}f(\theta)^2\,d\theta = \int_a^b \frac{1}{2}r^2\,d\theta.$

Example 1. The circle $r = \sin\theta$ of Example 4, §7.8 is traversed once as θ varies from 0 to π, so the area is

$$\frac{1}{2}\int_0^\pi r^2\,d\theta = \frac{1}{2}\int_0^\pi \sin^2\theta\,d\theta = \frac{1}{4}\int_0^\pi (1 - \cos 2\theta)\,d\theta = \frac{1}{4}\left[\theta - \frac{\sin 2\theta}{2}\right]_0^\pi = \frac{\pi}{4}.$$

This is reassuring, since the region enclosed by this graph is a circle of radius 1/2.

Example 2. Compute the area inside the polar graph of $r = \dfrac{ed}{1 + 3\cos\theta}$, where $e < 1$ [the area of an ellipse]. *Solution.* The evaluation is long, but gives a good review of integration methods; and the final result is useful in applying Kepler's so-called third law of planetary motion. We have

$$A = \frac{1}{2}\int_{-\pi}^{\pi} \frac{e^2d^2}{(1 + e\cos\theta)^2}\,d\theta = \int_0^\pi \frac{e^2d^2}{(1 + e\cos\theta)^2}\,d\theta,$$

since the integrand is an even function. This is a rational function of θ, so [§6.6] we can set $u = \tan(\theta/2)$, which makes $d\theta = \dfrac{2\,du}{1 + u^2}$, $\cos\theta = \dfrac{1 - u^2}{1 + u^2}$.

Moreover, as θ goes from 0 to π, then $u = \tan(\theta/2)$ goes from 0 to ∞. So the area is given by an improper integral [§5.5]

$$A = \int_0^\infty \frac{2e^2d^2\,du}{(1 + u^2)\left[1 + e\dfrac{1 - u^2}{1 + u^2}\right]^2} = \frac{2e^2d^2}{(1 + e)^2}\int_0^\infty \frac{(1 + u^2)\,du}{\left[1 + \dfrac{1 - e}{1 + e}u^2\right]^2}$$

$$= \frac{2e^2d^2}{(1 - e^2)^{3/2}}\int_0^\infty \frac{(1 - e) + (1 + e)v^2}{(1 + v^2)^2}\,dv \qquad \left[v = \sqrt{\frac{1 - e}{1 + e}}\,u\right]$$

$$A = \frac{2e^2d^2}{(1 - e^2)^{3/2}}\left[(1 + e)\int_0^\infty \frac{1 + v^2}{(1 + v^2)^2}\,dv - 2e\int_0^\infty \frac{dv}{(1 + v^2)^2}\right]. \tag{2}$$

But, integrating by parts,

$$\int_0^\infty \frac{dv}{1+v^2} = \frac{v}{1+v^2}\bigg|_0^\infty + \int_0^\infty \frac{2v^2}{(1+v^2)^2}\,dv = 0 + 2\int_0^\infty \frac{dv}{1+v^2} - 2\int_0^\infty \frac{dv}{(1+v^2)^2}. \tag{3}$$

Since $\displaystyle\int_0^\infty \frac{dv}{1+v^2} = \frac{\pi}{2}$, formula (3) shows easily that $\displaystyle\int_0^\infty \frac{dv}{(1+v^2)^2} = \frac{\pi}{4}$,

and thus from (2) we obtain at last $\displaystyle A = \frac{2e^2d^2}{(1-e^2)^{3/2}}\left[(1+e)\frac{\pi}{2} - 2e\frac{\pi}{4}\right]$

$$= \frac{\pi e^2 d^2}{(1-e^2)^{3/2}}.$$

Arc length in polar coordinates

To find the arc length of the graph of $r = f(\theta)$ between $\theta = a$ and $\theta = b$, we can think of θ as a parameter, and form the parametric equations

$$x = r\cos\theta = f(\theta)\cos\theta,$$
$$y = r\sin\theta = f(\theta)\sin\theta.$$

Then, using formula (1), §7.7, we find the arc length between $\theta = a$ and $\theta = b$ to be [after suitable trigonometric simplification]

$$\int_a^b \sqrt{(dx/d\theta)^2 + (dy/d\theta)^2}\,d\theta = \int_a^b \sqrt{f(\theta)^2 + f'(\theta)^2}\,d\theta$$

$$= \int_a^b \sqrt{(r\,d\theta)^2 + (dr)^2}.$$

The arc length formula $s = \int\sqrt{(r\,d\theta)^2 + (dr)^2}$ can be remembered from the small triangle in Fig. 47; $s = \int ds$, and ds is the hypotenuse of a small triangle whose sides are $r\,d\theta$ and dr.

Similarly, the area formula $\int\frac{1}{2}r^2\,d\theta$ can be remembered from Fig. 47 as follows: The region is divided into very thin slices, each of which is a thin triangle of altitude r and angle $d\theta$; the base of the triangle is $r\,d\theta$, and the altitude is r, so the area is $\frac{1}{2}r\cdot r\,d\theta = \frac{1}{2}r^2\,d\theta$.

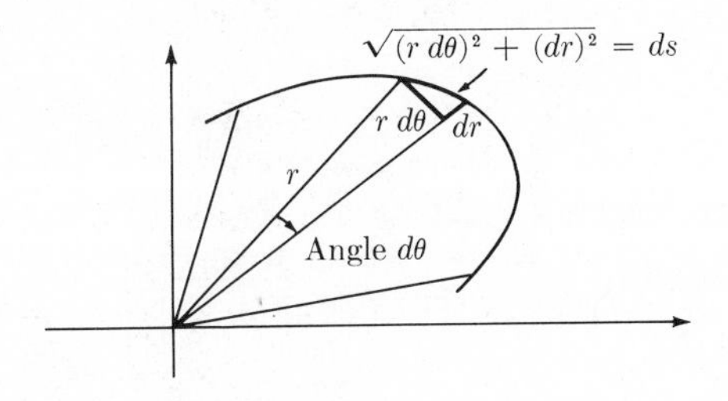

FIGURE 7.47

PROBLEMS

1. Find the area inside the following polar graphs. Consider carefully the appropriate limits for θ.
   - (a) $r = c$, [c is a constant] (b) $r = 2 + \cos\theta$ (c) $r = 1 + \sin\theta$
   - (d) $r = \theta, 0 \le \theta \le 2\pi$ [Part of the boundary of this region consists of a ray emanating from the origin.]

2. Set up an integral for the length of each curve in Problem 1. Evaluate all except (b). [Hint: $1 + \sin\theta = \cos^2\theta/2 + \sin^2\theta/2 + \cos\theta/2 \sin\theta/2$.]

3. As θ increases from a to b in Fig. 46, the ray from the origin to the point $(\theta, f(\theta))_p$ on the graph sweeps out the shaded region in the figure. Let $A(\varphi)$ denote the part of the region swept out from $\theta = a$ to $\theta = \varphi$; thus $A(a) = 0$ and $A(b)$ is the entire shaded region. Show that $dA/d\varphi = \frac{1}{2}f(\varphi)^2$. [Assume f is continuous, and use the Fundamental Theorem.]

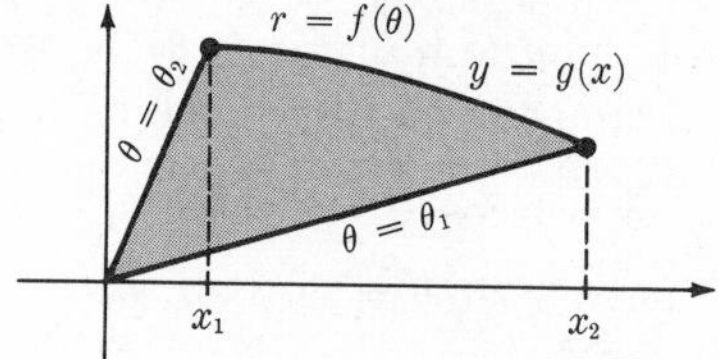

FIGURE 7.48

4. Let the curve in Fig. 48 be the polar coordinate graph of $r = f(\theta)$, and also the rectangular coordinate graph of $y = g(x)$. Then we have two formulas for the shaded area in the figure:

$$\int_{\theta_1}^{\theta_2} \frac{1}{2} f(\theta)^2\, d\theta \qquad \text{and} \qquad \int_{x1}^{x2} g(x)\, dx + \frac{1}{2} x_1 g(x_1) - \frac{1}{2} x_2 g(x_2).$$

This problem explains the second formula, and shows that the two formulas agree.

(a) Explain the second formula as the area under the graph of g, plus a triangle and minus another triangle.

(b) Explain the second formula as $\int_0^{x2} (g - h)$, where g and h are appropriately defined for $0 \le x \le x_2$.

(c) In $\int_{x1}^{x2} g(x)\, dx + \frac{1}{2}x_1 g(x_1) - \frac{1}{2}x_2 g(x_2)$, substitute $x = f(\theta)\cos\theta$, and show that the result is

$$\int_{\theta_1}^{\theta_2} [f^2 \sin^2\theta - ff' \sin\theta\cos\theta]\, d\theta + \frac{1}{2}\Big[f^2 \sin\theta \cos\theta\Big]_{\theta_1}^{\theta_2}.$$

[Note that if $r = f(\theta)$, then $g(r\cos\theta) = r\sin\theta$. Why? Note also that θ_1 corresponds to x_2, and θ_2 to x_1.]

(d) Let $A = \int_{\theta_1}^{\theta_2} f^2 \sin^2 \theta \, d\theta - \int_{\theta_1}^{\theta_2} ff' \sin \theta \cos \theta \, d\theta + \frac{1}{2}\Big[f^2 \sin \theta \cos \theta \Big]_{\theta_1}^{\theta_2}$,

as in part (c). Integrate the second integral by parts [with $dv = ff' \, d\theta$, $u = \sin \theta \cos \theta$] to show that $A = \frac{1}{2}\int_{\theta_1}^{\theta_2} f^2 \, d\theta$. Conclude that the two formulas mentioned at the beginning of this problem agree.

7.10 VECTORS AND POLAR COORDINATES

In §7.7 we analyzed the acceleration **A** by decomposing it into tangential and normal components, $\mathbf{A} = (d^2s/dt^2)\mathbf{T} + (ds/dt)^2\kappa\mathbf{N}$. In the ensuing discussion, this decomposition revealed a lot of information that is not at all obvious from the simpler formula $\mathbf{A} = [f'',g'']$.

This brief reminder† is a pep talk for another decomposition of the acceleration vector, this time into so-called *radial* and *circumferential* components. This decomposition is particularly useful when one or the other of these components is zero. In the study of *central forces*, the circumferential component of the acceleration is zero; in Example 1 below, it is the radial component that is zero. In the study of the *pendulum* [Example 2 below] the fact that the radius r is constant leads easily to the equation of motion.

We introduce at each point $(r,\theta)_p$ two orthogonal vectors, the radial vector $\mathbf{U}_r$ and the circumferential vector $\mathbf{U}_\theta$, as in Fig. 49. You can see that

$$\mathbf{U}_r = [\cos \theta, \sin \theta] \tag{1}$$

and

$$\mathbf{U}_\theta = \left[\cos\left(\theta + \frac{\pi}{2}\right), \sin\left(\theta + \frac{\pi}{2}\right)\right] = [-\sin \theta, \cos \theta]. \tag{2}$$

Notice that $\mathbf{U}_r$ and $\mathbf{U}_\theta$ depend on θ; when it is necessary to exhibit this dependence, we write $\mathbf{U}_r(\theta)$ and $\mathbf{U}_\theta(\theta)$. From (1) and (2), you can see directly that

$$\frac{d\mathbf{U}_r}{d\theta} = [-\sin \theta, \cos \theta],$$

$$\frac{d\mathbf{U}_\theta}{d\theta} = [-\cos \theta, -\sin \theta].$$

FIGURE 7.49

† If you skipped §7.7, never mind; we don't use any of that section here.

Comparing the preceding formulas with (1) and (2), we find

DERIVATIVES OF $\mathbf{U}_r$ AND $\mathbf{U}_\theta$

$$\frac{d\mathbf{U}_r}{d\theta} = \mathbf{U}_\theta\,, \tag{3}$$

$$\frac{d\mathbf{U}_\theta}{d\theta} = -\mathbf{U}_r\,. \tag{4}$$

From Fig. 49, this is reasonable: As θ increases, $\mathbf{U}_r$ turns toward $\mathbf{U}_\theta$, and $\mathbf{U}_\theta$ turns toward $-\mathbf{U}_r$.

Now, suppose we describe the path of a particle by giving polar coordinates r and θ of its position at each time t,

$$r = f(t) \qquad \text{and} \qquad \theta = g(t). \tag{5}$$

Then at time t, the particle is at a distance $f(t)$ from the origin, and in the direction given by $\theta = g(t)$. Since $\mathbf{U}_r(\theta)$ is a unit vector in the direction given by θ, the position vector is

$$\mathbf{R}(t) = f(t)\mathbf{U}_r(g(t)). \tag{6}$$

We write (6) more simply as $\mathbf{R} = r\mathbf{U}_r$, remembering that r depends on t, $\mathbf{U}_r$ depends on θ, and θ depends on t. Then by the rules for differentiating vectors,

$$\mathbf{V} = \frac{d\mathbf{R}}{dt} = \frac{d(r\mathbf{U}_r)}{dt} = \left(\frac{dr}{dt}\right)\mathbf{U}_r + r\frac{d\mathbf{U}_r}{dt}$$

$$= \frac{dr}{dt}\mathbf{U}_r + r\frac{d\mathbf{U}_r}{d\theta}\frac{d\theta}{dt} \qquad \text{[chain rule]}$$

$$= \frac{dr}{dt}\mathbf{U}_r + r\mathbf{U}_\theta\frac{d\theta}{dt}\,. \qquad \text{[by (3)]}$$

Now, it is conventional to abbreviate dr/dt as $\dot{r}$, $d\theta/dt$ as $\dot{\theta}$, and so on, using a dot to indicate a derivative with respect to t. Thus what we have just found is written

$$\mathbf{V} = \dot{r}\mathbf{U}_r + r\dot{\theta}\mathbf{U}_\theta\,.$$

Having found $\mathbf{V}$, we find the acceleration $\mathbf{A}$ by differentiating once more:

$$\mathbf{A} = \dot{\mathbf{V}} = \ddot{r}\mathbf{U}_r + \dot{r}\dot{\mathbf{U}}_r + \dot{r}\dot{\theta}\mathbf{U}_\theta + r\ddot{\theta}\mathbf{U}_\theta + r\dot{\theta}\dot{\mathbf{U}}_\theta\,,$$

where each dot indicates differentiation with respect to t, and double dots indicate second derivatives with respect to t. From the chain rule and equations (3) and (4) we have

CIRCUMFERENTIAL AND RADIAL COMPONENTS OF ACCELERATION

$$\dot{\mathbf{U}}_r = \dot{\theta}\mathbf{U}_\theta, \qquad \dot{\mathbf{U}}_\theta = -\dot{\theta}\mathbf{U}_r,$$

so the last formula for $\mathbf{A}$ becomes

$$\begin{aligned}\mathbf{A} &= \ddot{r}\mathbf{U}_r + \dot{r}\dot{\theta}\mathbf{U}_\theta + \dot{r}\dot{\theta}\mathbf{U}_\theta + r\ddot{\theta}\mathbf{U}_\theta - r\dot{\theta}^2\,\mathbf{U}_r \\ &= (\ddot{r} - r\dot{\theta}^2)\mathbf{U}_r + (2\dot{r}\dot{\theta} + r\ddot{\theta})\mathbf{U}_\theta. \qquad (7)\end{aligned}$$

This is the decomposition we were looking for, decomposing $\mathbf{A}$ into the sum of its "radial" and "circumferential" components; the radial component is $\ddot{r} - r\dot{\theta}^2$, and the circumferential component is $2\dot{r}\dot{\theta} + r\ddot{\theta}$.

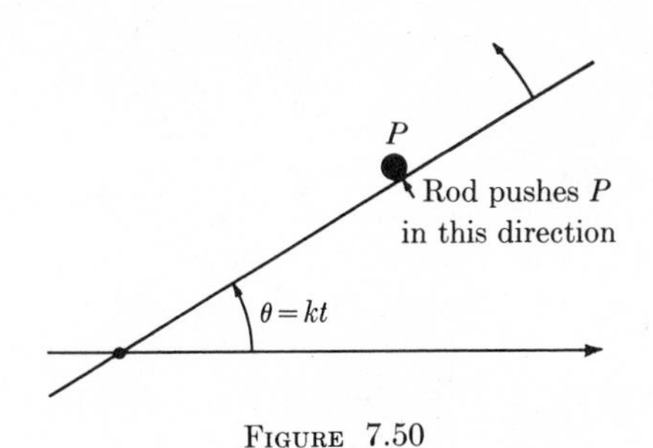

FIGURE 7.50

Example 1. Suppose a rod is rotated about the origin at a constant rate, so that at time t, the angle θ in Fig. 50 is given by $\theta = kt$. Suppose a particle P is pushed along by the rod, and that P can slide freely along the rod; then the force $\mathbf{F}$ exerted on P is always perpendicular to the rod [Fig. 50]. Given all this, we analyze the motion of P. Since $\mathbf{F}$ is perpendicular to the rod, we can write $\mathbf{F} = c\mathbf{U}_\theta$ for some number c (which may depend on r and θ). From Newton's law $\mathbf{F} = m\mathbf{A}$, we find

$$c\mathbf{U}_\theta = m(\ddot{r} - r\dot{\theta}^2)\mathbf{U}_r + m(2\dot{r}\dot{\theta} + r\ddot{\theta})\mathbf{U}_\theta.$$

Since there is no $\mathbf{U}_r$ term on the left this equation suggests that the coefficient $m(\ddot{r} - r\dot{\theta}^2)$ of $\mathbf{U}_r$ on the right must be 0. To *prove* that it's 0, take the inner product with $\mathbf{U}_r$. Noting that $\mathbf{U}_r\cdot\mathbf{U}_r = 1$ and $\mathbf{U}_r\cdot\mathbf{U}_\theta = 0$, we obtain the expected equation

$$0 = \ddot{r} - r\dot{\theta}^2.$$

Since $\theta = kt$, we have $\dot{\theta} = k$, hence

$$\ddot{r} = k^2 r.$$

You can easily check that

$$r = ae^{kt} + be^{-kt} \qquad (8)$$

solves this equation for any choice of a and b. To determine a and b we would need more information, such as the values of $r(0)$ and $\dot{r}(0)$. [In Problem 4 below you show that *every* solution of the equation is obtained from (8) by an appropriate choice of the constants a and b.]

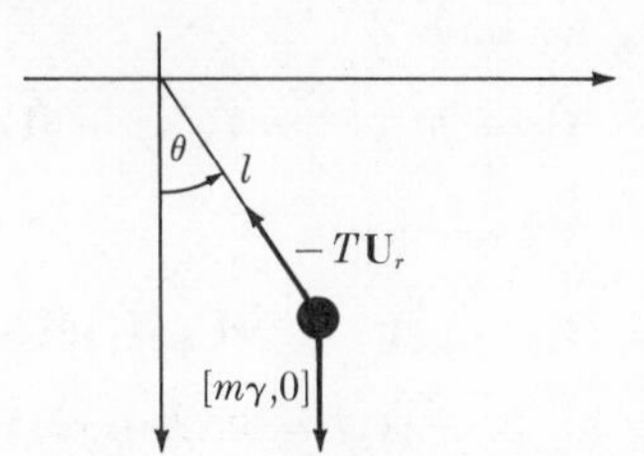

FIGURE 7.51 Forces on a pendulum

Example 2. Figure 51 shows a simple pendulum, consisting of a mass m swinging on a rod of fixed length l. To simplify the problem, the mass of the rod and the friction in the pivot are assumed negligible. Also, for convenience, we have the positive x axis pointing *down*, so that $\theta = 0$ corresponds to the position of the pendulum at rest.

There are two forces on the pendulum: Gravity exerts a downward force $m[\gamma,0]$ where γ is the gravity constant, and the pendulum rod exerts a pull of unknown strength, but in the direction $-\mathbf{U}_r$; call this force $-T\mathbf{U}_r$, where T stands for the tension in the rod.

To analyze the situation, we decompose the force of gravity $m[\gamma,0]$ into radial and circumferential components; it is easy to see that

$$m[\gamma,0] = m\gamma \cos\theta \mathbf{U}_r - m\gamma \sin\theta \mathbf{U}_\theta$$

so the total force acting on the mass m is

$$m\gamma \cos\theta \mathbf{U}_r - m\gamma \sin\theta \mathbf{U}_\theta - T\mathbf{U}_r = (m\gamma \cos\theta - T)\mathbf{U}_r - m\gamma \sin\theta \mathbf{U}_\theta .$$

From Newton's law $\mathbf{F} = m\mathbf{A}$ and formula (7), we get

$$(m\gamma \cos\theta - T)\mathbf{U}_r - m\gamma \sin\theta \mathbf{U}_\theta = m(\ddot{r} - r\dot{\theta}^2)\mathbf{U}_r + m(2\dot{r}\dot{\theta} + r\ddot{\theta})\mathbf{U}_\theta .$$

It follows [by taking inner products with $\mathbf{U}_r$ and $\mathbf{U}_\theta$ as in Example 1] that

$$m\gamma \cos\theta - T = m(\ddot{r} - r\dot{\theta}^2),$$

$$-m\gamma \sin\theta = m(2\dot{r}\dot{\theta} + r\ddot{\theta}).$$

But the pendulum rod has fixed length l, so $r = l$, and $\dot{r} = 0$, $\ddot{r} = 0$. Therefore,

$$\gamma \cos\theta - \frac{T}{m} = -l\dot{\theta}^2,$$

$$-\gamma \sin\theta = l\ddot{\theta}. \tag{9}$$

The last equation does not involve the unknown tension T; it alone describes the motion of the pendulum.

If the oscillations are small, then θ is small, and $\sin\theta$ is nearly equal to θ. Replacing $\sin\theta$ by θ in (9) yields

$$\ddot{\theta} + (\gamma/l)\theta = 0,$$

which is the equation for a simple oscillating motion of frequency $\sqrt{\gamma/l}/2\pi$, as shown in §3.5.

PROBLEMS

Problems 1–4 are based on Example 1.

• **1.** Find the constants a and b in (8) if $r(0) = 1$ and $\dot{r}(0) = 0$. How does the particle move in this case?

• **2.** Find the constants a and b in (8) if $r(0) = 1$ and $\dot{r}(0) = -1$. How does the particle move in this case?

3. Show that the magnitude $|\mathbf{F}|$ of the force exerted by the rod in Example 1 is $2mk^2|ae^{kt} - be^{-kt}|$, when $r(t) = ae^{kt} + be^{-kt}$.

4. Show that $r = ae^{kt} + be^{-kt}$ gives all the solutions of $\ddot{r} = k^2r$. [Hint: Put $r = ue^{kt}$ in the equation $\ddot{r} = k^2r$; you will obtain a differential equation for u', which you can easily solve.]

5. (a) Show that the length of the velocity vector is given in polar coordinates by $|\mathbf{V}|^2 = \dot{r}^2 + r\dot{\theta}^2$. [Use the Pythagorean theorem for vectors, §7.2, Problem 9.]

(b) Conclude that the arc length of a polar coordinate curve is given by $\int \sqrt{\dot{r}^2 + r^2\dot{\theta}^2}\, dt$.

(c) Relate the integral in (b) to the arc length integral in §7.9.

7.11 PLANETARY MOTION

Early in the seventeenth century, Johann Kepler distilled out of the astronomical mass of data compiled by Tycho Brahe the following three laws of planetary motion:

KEPLER'S LAWS

I For each planet, the line segment from the sun to the planet sweeps out equal areas in equal times.

II Each planet moves along an ellipse with the sun as focus.

III The square of the period is proportional to the cube of the major axis of the ellipse, the constant of proportionality being the same for each planet. [The period is the length of time required to complete one orbit.]

In the second half of that century, Isaac Newton used his own law $\mathbf{F} = m\mathbf{A}$ to deduce from Kepler's laws the basic nature of the force that holds each planet in its orbit; the force is directed toward the sun, and is proportional to m/r, where m is the mass of the planet, and r is the distance from planet to Sun.

To begin the deduction, notice that the second law implies that the orbit of each planet lies in a plane, so we can study the problem with the vector methods at hand. We will see that each of Kepler's laws tells a little more about the force.

Theorem 4. *Kepler's first law implies that the force acting on each planet is a central force, with the sun as center.*

In other words, the force is always directed along the line segment between the sun and the planet.

To show this, let the sun be at the origin, and let $A(\theta_0)$ be the area swept out by the segment from the sun to the planet as it moves from $\theta = 0$ to $\theta = \theta_0$. Let $(r,\theta)_p$ be the position of the planet at time t, and assume that r and θ are given by continuous functions of t. By the formula for area in polar coordinates [§7.9],

$$A(\theta_0) = \int_0^{\theta_0} \frac{1}{2} r^2 \, d\theta;$$

hence [by what theorem?]

$$\frac{dA}{d\theta} = \frac{1}{2} r^2$$

and [by what rule?]

$$\frac{dA}{dt} = \frac{1}{2} r^2 \dot{\theta}.$$

Kepler's first law says that dA/dt is constant, so

$$0 = \frac{d^2A}{dt^2} = r\dot{r}\dot{\theta} + \frac{1}{2} r^2 \ddot{\theta}. \tag{1}$$

Now, in polar coordinates the acceleration is

$$\mathbf{A} = (\ddot{r} - r\dot{\theta}^2)\mathbf{U}_r + (2\dot{r}\dot{\theta} + r\ddot{\theta})\mathbf{U}_\theta ,$$

[see §7.10, formula (7)], so (1) shows that

$$\mathbf{A} = (\ddot{r} - r\dot{\theta}^2)\mathbf{U}_r . \tag{2}$$

Formula (2) says that the acceleration is in the direction of $\mathbf{U}_r$, up to sign. Since $\mathbf{F} = m\mathbf{A}$, the force $\mathbf{F}$ likewise has the direction of $\mathbf{U}_r$, up to sign, and Theorem 4 is proved. [See Fig. 52.]

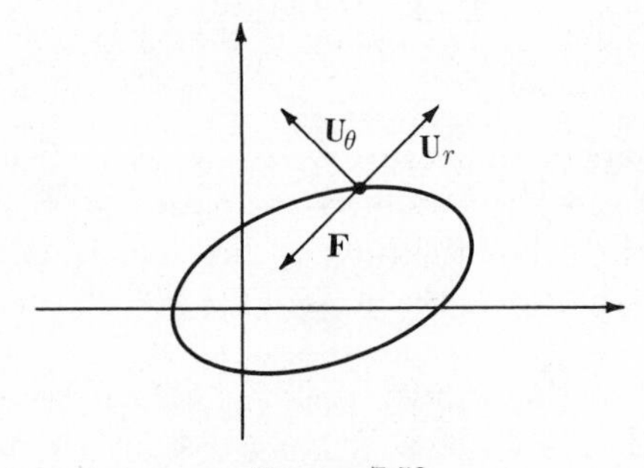

FIGURE 7.52

Theorem 5. *Kepler's first and second laws together imply that the force on the planet is directed toward the Sun and is proportional to* $\dfrac{1}{r^2}$.

The second law can be stated analytically by saying that the planet lies on the graph of an equation of the form

$$r = \frac{ed}{1 + e\cos(\theta - \varphi)}. \tag{3}$$

Since we are assuming the first law, there is a constant h such that

$$r^2\dot{\theta} = h. \tag{4}$$

Moreover, equation (2) holds, and what we have to show is that the radial component of acceleration is

$$\ddot{r} - r\dot{\theta}^2 = -k/r^2 \tag{5}$$

for some constant $k > 0$. Because of the reciprocals in (3) and (5), it is convenient to work with $1/r$ instead of r; we set $v = 1/r$ and $r = 1/v$. Then, by virtue of (4),

$$\frac{dr}{dt} = \frac{dv^{-1}}{dt} = -v^{-2}\frac{dv}{dt} = -v^{-2}\frac{dv}{d\theta}\frac{d\theta}{dt} = -r^2\frac{dv}{d\theta}\dot{\theta} = -h\frac{dv}{d\theta}.$$

Differentiating once more and using (4) again, we have

$$\frac{d^2r}{dt^2} = -h\frac{d^2v}{d\theta^2}\frac{d\theta}{dt} = -h^2v^2\frac{d^2v}{d\theta^2}.$$

Using this and (4) again, we obtain for the acceleration

$$\ddot{r} - r\dot{\theta}^2 = -h^2v^2\frac{d^2v}{d\theta^2} - h^2v^3 = -h^2v^2\left(\frac{d^2v}{d\theta^2} + v\right).$$

Finally, we use Kepler's second law as expressed in (3). Replacing r by $1/v$ in (3), taking reciprocals, and differentiating, we find

$$\frac{d^2v}{d\theta^2} + v = \frac{1}{ed},$$

so

$$\ddot{r} - r\dot{\theta}^2 = \frac{-h^2}{ed}v^2 = -\left(\frac{h^2}{ed}\right)r^{-2}. \tag{6}$$

Thus there is indeed a constant $k > 0$ making (5) valid, and Theorem 5 is proved.

Thus far, we have shown that each planet P is accelerated directly toward the sun, and the magnitude of the acceleration is k_P/r^2, where r is the distance from planet to sun, and k_P is a constant that may depend on which planet P we are discussing. Kepler's third law now shows that this constant k_P is the *same* for every planet P:

Theorem 6. *Kepler's three laws imply that there is a constant k, the same for each planet, such that the magnitude of the acceleration is $-k/r^2$, and the magnitude of the force is therefore $-km/r^2$. Thus the force is proportional to the mass of the planet, and to the inverse of the square of its distance to the sun.*

Proof. The major axis $2a$ of the ellipse given by (3) is the sum of the maximum and minimum values of r,

$$2a = r_{\max} + r_{\min} = \frac{ed}{1-e} + \frac{ed}{1+e} = \frac{2ed}{1-e^2}. \tag{7}$$

The period T can be related to this through the area. By Kepler's first law, area is swept out at the constant rate $\frac{1}{2}h$ where h is the constant introduced in (4) above; and it takes one period T to sweep out the whole ellipse; so the area A is given by $A = \frac{1}{2}hT$. Therefore,

$$T = \frac{2A}{h}.$$

But from §7.9, Example 2, the area of the ellipse is

$$A = \frac{\pi e^2 d^2}{(1-e^2)^{3/2}}, \qquad \text{so} \qquad T = \frac{2\pi e^2 d^2}{h(1-e^2)^{3/2}}.$$

Combining this with (7), we get

$$\frac{a^3}{T^2} = \frac{e^3 d^3}{(1-e^2)^3} \cdot \frac{h^2(1-e^2)^3}{4\pi^2 e^4 d^4} = \frac{h^2}{4\pi^2 ed}.$$

Therefore, from (6), the magnitude of the acceleration of any planet at distance r from the sun is

$$-\frac{h^2}{ed}r^{-2} = -\left(4\pi^2 \frac{a^3}{T^2}\right) r^{-2}.$$

Since a^3/T^2 is the same for each planet [Kepler's third law], Theorem 6 is proved.

We have seen that Kepler's planetary laws, together with Newton's general law of motion $\mathbf{F} = m\mathbf{A}$, lead to the "inverse square" law of gravitational attraction: A planet is attracted to the sun by a force proportional to its mass, and inversely proportional to the square of its distance from the sun. Newton deduced further that the inverse square law, together with $\mathbf{F} = m\mathbf{A}$, imply Kepler's three laws if in the third law we replace "ellipse" by "conic section." This deduction is carried out in the problems below. [Why did Kepler's analysis not reveal any parabolic or hyperbolic orbits?]

Remark. These theorems about the planets are all contained in Newton's great tome *Philosophiae naturalis principia mathematica* (1687). Newton's proofs, however, were based not on calculus, but on Euclidean geometry! This is rather surprising, since calculus makes the proofs much easier, and Newton seems to have developed a good deal of calculus in 1665–1666, twenty years before the appearance of the tome. It is possible that Newton actually did not use calculus in his study of Kepler's laws. It seems more likely, however (although opinions may differ on this), that he did use it, but preferred to base his published proofs on the universally accepted foundation provided by Euclidean geometry rather than on a method that was well understood by him, but practically unknown to most other natural philosophers and mathematicians of his day. Perhaps, too, Newton was aware of the difficulties in the foundations of calculus, and for this reason preferred to give other proofs. There is a precedent for this point of view in Archimedes, who discovered a number of formulas for areas and volumes by a method like the one outlined in §5.1, but then went back and rigorously proved the formulas by accepted principles of Euclidean geometry.

A very readable account of the contributions of Newton, Kepler, Tycho Brahe, and their predecessors is given by A. E. E. McKenzie in *The Major Achievements of Science* (vol. 1, Cambridge University Press, 1960), Chapters 2 and 3.

PROBLEMS

1. (a) Show that if an object is acted upon by a central force, then $2\dot{r}\dot{\theta} + r\ddot{\theta} = 0$.

(b) Multiply $2\dot{r}\dot{\theta} + r\ddot{\theta}$ by r, and notice that the result is the derivative of $r^2\dot{\theta}$. Hence, for a central force, $r^2\dot{\theta}$ is a constant (denote it by h).

(c) Deduce that Kepler's first law holds for *any* central force.

CENTRAL FORCE $\Rightarrow$ KEPLER'S FIRST LAW

2. Assume a central force proportional to m/r^2; that is, add to the conditions in Problem 1 the further condition that $m(\ddot{r} - r\dot{\theta}^2) = -km/r^2$, where k is constant.

KEPLER'S SECOND LAW HOLDS FOR INVERSE SQUARE CENTRAL FORCE

(a) Let $r = 1/v$ as in the text, and show that $(d^2v/d\theta^2) + v = k/h^2$, where $h = r^2\dot{\theta}$ is constant.
(b) Introduce $u = v - k/h^2$, and show that the differential equation in part (a) implies $d^2u/d\theta^2 + u = 0$.
(c) Deduce that $u(\theta) = C \cos(\theta - \varphi)$ for some constants C and φ, and hence that $r = (h^2/k)[1 + e \cos(\theta - \varphi)]^{-1}$, where e is a constant. [See §3.5.]

UNIVERSAL GRAVITATION ⇒ KEPLER'S THIRD LAW

3. Show that if $2a$ is the major axis of the ellipse in Problem 2(c), then $a^3/T^2 = k/4\pi^2$. [Thus, Kepler's third law follows from Newton's *law of universal gravitation*: A planet of mass m at a distance r from a sun of mass M is attracted by a force $\gamma mM/r^2$, where γ is a universal constant, the same for every pair of masses, no matter how large or small. For, this law of gravitation says that the constant k in Problem 2 is given by $k = \gamma M$, which is independent of the mass m of the planet.]

4. The *potential energy* of an object of mass m in a central force field, where the magnitude of the force is km/r^2, is taken to be $-km/r$. Show that km/r is the work required to move the object in a straight line from a position where $r = 1$ to a "position" where "$r = \infty$". [Recall that the work done by a force of magnitude $F(x)$ moving an object along a straight line from $x = a$ to $x = b$ is given by $\int_a^b F(x)\,dx$.]

CONSERVATION OF ENERGY

5. The kinetic energy of an object with velocity $\mathbf{V}$ and mass m is $\frac{1}{2}m|\mathbf{V}|^2$.
(a) Prove that, in polar coordinates, $|\mathbf{V}|^2 = \dot{r}^2 + r^2\dot{\theta}^2$.
(b) Prove that for any planet, the sum of the potential energy and kinetic energy is constant. [The potential energy is km/r; see Problem 4. Hint: Use the formulas $\ddot{r} - r\dot{\theta}^2 = -k/r^2$ and $r^2\dot{\theta} = h =$ constant, developed in the proof of Theorem 5.]

Chapter VIII

Complex Numbers

From time immemorial, numbers have been associated with lengths, and it must have been a terrible shock to discover that some lengths are *not* given by rational numbers. But this fact is inescapable, for the diagonal d of a square of side 1 satisfies $d^2 = 2$, whereas *no* rational number x satisfies $x^2 = 2$. [See Appendix I, Introduction.] The strong intuitive feeling that every length *should* be given by a number led in time to an extension of the rational numbers to form the *real numbers*, which do allow us to solve the equation $x^2 = 2$, and which, in fact, provide all the lengths that geometry requires.

Another extension problem is presented by the quadratic equation $ax^2 + bx + c = 0$, which is solved by the expression

$$x = \frac{-b \pm \sqrt{b^2 - 4ac}}{2a}. \tag{1}$$

Since a negative number has no real square root, formula (1) does not give a real number x when $b^2 - 4ac < 0$. Shall we nevertheless assign a meaning to $\sqrt{b^2 - 4ac}$ if, say, $b^2 - 4ac = -1$? Or shall we simply say that there are no roots in this case? The answer is, once more, to extend the number system and to include objects like $\sqrt{-1}$. This object is called an *imaginary number*, and the extended system containing combinations of this and the real numbers is called the *complex number system*.

The very names "complex number" and "imaginary number" suggest a certain air of mystery, in sharp contrast to the reassuring "real numbers." The mystery is due, most likely, to the fact that the real numbers satisfy a clear geometric need, while imaginary and complex numbers are introduced for purely algebraic reasons.

It would be natural to conclude that complex numbers are not much use in applications of mathematics to real physical situations. However, the fact is that they constitute an extraordinarily powerful tool both in "pure" mathematics and in its applications. This chapter can give only a limited view of these applications and of the results on which they depend.

We begin by defining the complex number system in terms of the real numbers, and we check which of the familiar properties of the real numbers are preserved in the extension. This check is simple, but necessary, even from an intuitive point of view. The $\sqrt{-1}$ is so radically different from the real numbers that our intuition does not suggest which, if any, of the familiar properties of the reals will be retained after they are extended to include $\sqrt{-1}$.

8.1 DEFINITION AND ELEMENTARY ALGEBRAIC PROPERTIES

The complex number system consists of ordered pairs of real numbers, together with certain operations called addition and multiplication. The ordered pair with first member x and second member y is denoted $x + iy$.

REAL AND IMAGINARY PARTS

The number x is called the *real part* of $x + iy$, denoted Re $(x + iy)$, and y is called the *imaginary part*, denoted Im $(x + iy)$:

$$\text{Re}(x + iy) = x \quad \text{and} \quad \text{Im}(x + iy) = y.$$

The addition and multiplication of complex numbers are defined by

ADDITION

$$(a + ib) + (x + iy) = (a + x) + i(b + y), \tag{1}$$

MULTIPLICATION

$$(a + ib)(x + iy) = ax - by + i(bx + ay). \tag{2}$$

The rule (1) for addition looks just like ordinary algebra, regrouping the terms on the left into those *with* an i and those *without*. The "easy way" to multiply is explained below, along with some examples.

REMARKS

This is the third structure we have given based on ordered pairs of real numbers. The first was the plane of plane analytic geometry, the second was the vector space R^2. This third structure has the same definition of addition as R^2, but the multiplication is different from both the inner product of vectors and the product of a vector by a scalar.

The complex number system is denoted by **C**; and any member of **C** is called a complex number.

Notice that we define a complex number as a member of a system with certain rules of operation, *not* as an "expression of the form $x + iy$." Complex numbers are distinguished from vectors and points in the plane not by their labels, but by the rules of addition and multiplication.

Connection with the real numbers

The system **C** contains the real number system, disguised as complex numbers of the form $a + i0$. Such numbers add and multiply like their real counterparts; setting $b = 0$ and $y = 0$ in (1) and (2) yields

$$(a + i0) + (x + i0) = (a + x) + i0,$$

$$(a + i0)(x + i0) = ax + i0.$$

Since a and $a + i0$ behave the same way, we use a as an abbreviation for $a + i0$. Furthermore, we abbreviate $0 + ib$ as ib, and $i1$ as i.

[These abbreviations might conceivably create some confusion: $a + ib$ was defined initially to be the complex number with real part a and imaginary part b; but now a might stand for $a + i0$, i for $0 + i1$, and b for $b + i0$, so $a + ib$ could also stand for $(a + i0) + (0 + i1)(b + i0)$. But in fact, there *is* no danger of confusion since by either interpretation we have exactly the same complex number: From (2) and (1), $(a + i0) + (0 + i1)(b + i0) = (a + i0) + (0 + ib) = a + ib$.]

The fact that **C** contains the real numbers justifies our claim that **C** is an *extension* of the reals. We should also check that the extension has not been made in vain. Does it contain the square roots of negative numbers? The answer is yes, for if $r < 0$, then $-r$ has a real square root $\sqrt{-r}$, and the complex number $i\sqrt{-r}$ is the desired square root of r:

$$(i\sqrt{-r})^2 = (0 + i\sqrt{-r})(0 + i\sqrt{-r}) = -(\sqrt{-r})^2 + i0 = r.$$

In particular, $i^2 = -1$. *This fact makes it unnecessary to remember the complicated formula* (2) *for multiplication;* to multiply $(a + ib)$ by $(x + iy)$, treat all the symbols a, i, b, x, y as if they obeyed all the rules for real numbers, multiply out the product $(a + ib)(x + iy)$, and use the formula $i^2 = -1$:

MULTIPLY AS USUAL, AND SET $i^2 = -1$

$$(a + ib)(x + iy) = ax + ibx + aiy + i^2by = ax - by + i(bx + ay).$$

For example, $(1 + i)(1 - i) = (1 - i) + i(1 - i) = 1 - i + i - i^2 = 1 - (-1) = 2$.

Conjugate and absolute value

With each complex number $x + iy$, we associate the number $x - iy$, called the *conjugate* of $x + iy$, and denoted $\overline{x + iy}$:

CONJUGATE

$$\overline{x + iy} = x - iy.$$

We further define the *absolute value* by

ABSOLUTE VALUE

$$|x + iy| = \sqrt{x^2 + y^2}.$$

You can see that this coincides with the length of the vector $[x,y]$. And for a real number $x = x + i0$, the new sense of absolute value agrees with the old: $|x + i0| = \sqrt{x^2 + 0^2} = |x|$. Moreover, the important properties of the absolute value are preserved:

PROPERTIES OF ABSOLUTE VALUE

$$|z + w| \le |z| + |w|, \tag{3}$$

$$|zw| = |z| \cdot |w|, \tag{4}$$

$$|z| = 0 \iff z = 0. \tag{5}$$

The inequality (3) was proved for vectors in R^2, and the same proof is valid here. It is called the *triangle inequality,* as before. Properties (4) and (5) can be proved directly from the definitions, but (4) is easier to prove after the discussion of multiplication in §8.3.

For the conjugate, we have four simple formulas,

RULES FOR CONJUGATE

$$\overline{z + w} = \bar{z} + \bar{w}, \tag{6}$$

$$\overline{zw} = \bar{z} \cdot \bar{w}, \tag{7}$$

$$z\bar{z} = |z|^2, \tag{8}$$

and, finally, the conjugate of the conjugate of z equals z,

$$\overline{(\bar{z})} = z. \tag{9}$$

All these formulas can be proved directly, but we postpone the proof of (7) to §8.3, where it will be much easier.

Example 1. $|1 + 3i|^2 = 1^2 + 3^2 = 10$ [Example of absolute value]

$\bar{\imath} = -i, \quad \overline{1 + 3i} = 1 - 3i$ [Examples of conjugate]

$\overline{\overline{(1 + 3i)}} = \overline{1 - 3i} = 1 + 3i$ [Illustrating (9)]

$(1 + 3i)\overline{(1 + 3i)} = (1 + 3i)(1 - 3i) = 1 - 9i^2 = 10 = |1 + 3i|^2.$

[Illustrating (8). Warning: Do *not* write $|1 + 3i|^2 = 1^2 + (3i)^2 = 1 - 9 = -8.$]

Basic algebraic properties of complex numbers

The complex numbers share many properties of the reals. Specifically, if z, w, and c stand for any complex numbers, we have the

commutative laws: $$z + w = w + z, \qquad zw = wz, \tag{10}$$

associative laws: $$(z + w) + c = z + (w + c), \qquad (zw)c = z(wc), \tag{11}$$

distributive law: $$c(w + z) = cw + cz. \tag{12}$$

In addition, the identity elements 0 and 1 retain their familiar properties: For every complex number z,

$$z + 0 = 0 + z = z, \qquad 1 \cdot z = z \cdot 1 = z. \tag{13}$$

Finally, each complex number z has a unique negative $-z$ such that $z + (-z) = 0$, and each $z \neq 0$ has a unique inverse z^{-1} such that $z^{-1} \cdot z = 1$.

We don't have to prove the commutative and associative laws for addition; they have already been checked for the vector space R^2, which has the same definition of addition as **C** does. And, using definitions (1) and (2) and the familiar properties of real numbers, it is easy to establish the remaining properties of the complex numbers. We will show how to prove the least simple of these, the existence of z^{-1} when $z \neq 0$.

The key to computing the reciprocal of a complex number $z = x + iy$ is the formula $z\bar{z} = |z|^2$, which shows that multiplying z by $\bar{z}$ produces a real number. We can thus compute the reciprocal informally by "making the denominator real," as follows:

RECIPROCALS
$$\frac{1}{x + iy} = \frac{x - iy}{(x + iy)(x - iy)} = \frac{x - iy}{x^2 + y^2} = \frac{x}{x^2 + y^2} + i\frac{-y}{x^2 + y^2}$$
$$= \frac{1}{x^2 + y^2}(x - iy) = \frac{1}{|z|^2}\bar{z}.$$

If $z \neq 0$, then $|z|^2 \neq 0$, so $\dfrac{1}{|z|^2}\bar{z}$ is well-defined, and it is easy to check formally that it is the desired reciprocal of z, using (11) and (8):

$$\left(\frac{1}{|z|^2}\bar{z}\right)z = \frac{1}{|z|^2}(\bar{z}z) = \frac{|z|^2}{|z|^2} = 1. \tag{14}$$

Similarly, reversing the order of the product still gives 1, $z(|z|^{-2}\bar{z}) = 1$.

Example 2. $\dfrac{1}{2-3i} = \dfrac{2+3i}{(2-3i)(2+3i)} = \dfrac{2+3i}{13} = \dfrac{2}{13} + \dfrac{3}{13}i.$

We have not yet proved all that we stated about z^{-1}, namely, that if $z \neq 0$, there is a *unique* number z^{-1} such that

$$z^{-1}\cdot z = 1;$$

we have only proved that there is *some* number z^{-1} such that $z^{-1}\cdot z = 1$. To prove uniqueness, let w be *any* number such that $wz = 1$. Then by (10), $zw = 1$. Hence by (13), (14), and (11), we get

$$w = 1\cdot w = ((|z|^{-2}\bar{z})z)w = (|z|^{-2}\bar{z})(zw) = (|z|^{-2}\bar{z})\cdot 1 = |z|^{-2}\bar{z}.$$

Thus the only number w such that $wz = 1$ is given by $w = z^{-1}$, where

$$z^{-1} = |z|^{-2}\bar{z} = \frac{\bar{z}}{|z|^2}.$$

COMPLEX NUMBERS NOT ORDERED LIKE REALS

You may well ask whether the complex numbers share *all* the properties of the real numbers. They don't; there is no way to introduce an *ordering* into the complex numbers that satisfies the two basic properties of the ordering of the real numbers:

(I) For every number a, exactly one of the three possibilities holds:

$$a > 0, \qquad \text{or} \qquad a = 0, \qquad \text{or} \qquad -a > 0.$$

(II) If $a > 0$ and $b > 0$, then $a + b > 0$ and $ab > 0$.

In fact, if there were such an ordering, then by property (I) we would have just two possibilities for the number i, either $i > 0$ or $-i > 0$. And each of these leads to a contradiction:

(1) If $i > 0$, then $i^2 = -1 > 0$ by (II), so $1 = (-1)^2 > 0$ by (II) again; but from (I), we cannot have both $-1 > 0$ and $1 > 0$.
(2) If $-i > 0$, then $(-i)^2 = -1 > 0$, and we are led to the same contradiction as before.

Thus when we want to use inequalities with complex numbers, we resort to absolute values; for, $|z|$ and $|w|$ are real numbers, and an inequality like $|z| < |w|$ makes sense.

PROBLEMS

1. Find the conjugates, absolute values, and reciprocals of the following complex numbers.

• (a) 1 • (c) $-i$ • (e) $1 + i$ (g) $-3 + \pi i$
(b) i (d) -1 (f) $-1 - 2i$

2. Compare the sum, and the product, of:
• (a) i and $-i$ • (b) $1 + i$ and $1 + i$ (c) $2 + i$ and $-2 + 2i$.

3. Show that if $|z| = 1$, then $z^{-1} = \bar{z}$.

4 • (a) Prove the commutative laws (10).
(b) Prove the associative laws (11).

5. Prove the distributive law (12).

6. Prove that $1 \cdot z = z$ by writing $1 = 1 + i0$, $z = x + iy$, and using formula (2).

7. Prove that: (a) $\overline{z + w} = \bar{z} + \bar{w}$, and (b) $\overline{zw} = \bar{z} \cdot \bar{w}$.

8. Prove that $|zw| = |z| \cdot |w|$. [Hint: Use Problem 7(b).]

8.2 GEOMETRY OF THE COMPLEX NUMBERS

Because the complex numbers are ordered pairs of real numbers (with a certain structure), we may visualize them as points in the plane; we visualize $x + iy$ as the point (x,y). The real numbers $x + i0$ lie along the horizontal axis, which is accordingly called the *real axis;* and the imaginary numbers $0 + iy$ lie along the vertical axis, called the *imaginary axis.* This leads to the following geometric interpretations, illustrated in Fig. 1.

REAL AXIS, IMAGINARY AXIS

(a) $|z|$ is the distance from 0 to z.

(b) $|z - w|$ is the distance from z to w.

(c) $\bar{z}$ is the reflection of z in the real axis.

(d) $-z$ is the reflection of z in the origin.

(e) $z + w$ is at the vertex opposite 0 in a parallelogram whose other three vertices are 0, z, and w.

(f) z^{-1} is obtained by reflecting z in the real axis, then adjusting the length so that $|z^{-1}| = |z|^{-1}$.

These interpretations are all either familiar, or easy to check. For example, we check (b): if $z = x + iy$ and $w = u + iv$, then $z - w = (x - u) + i(y - v)$ and $|z - w| = \sqrt{(x - u)^2 + (y - v)^2}$, which is indeed the distance between the points (x,y) and (u,v) in the plane. This demonstrates (b), and (a) is a special case of (b). Both (c) and (d) are direct consequences of the sign conventions in connection with up and

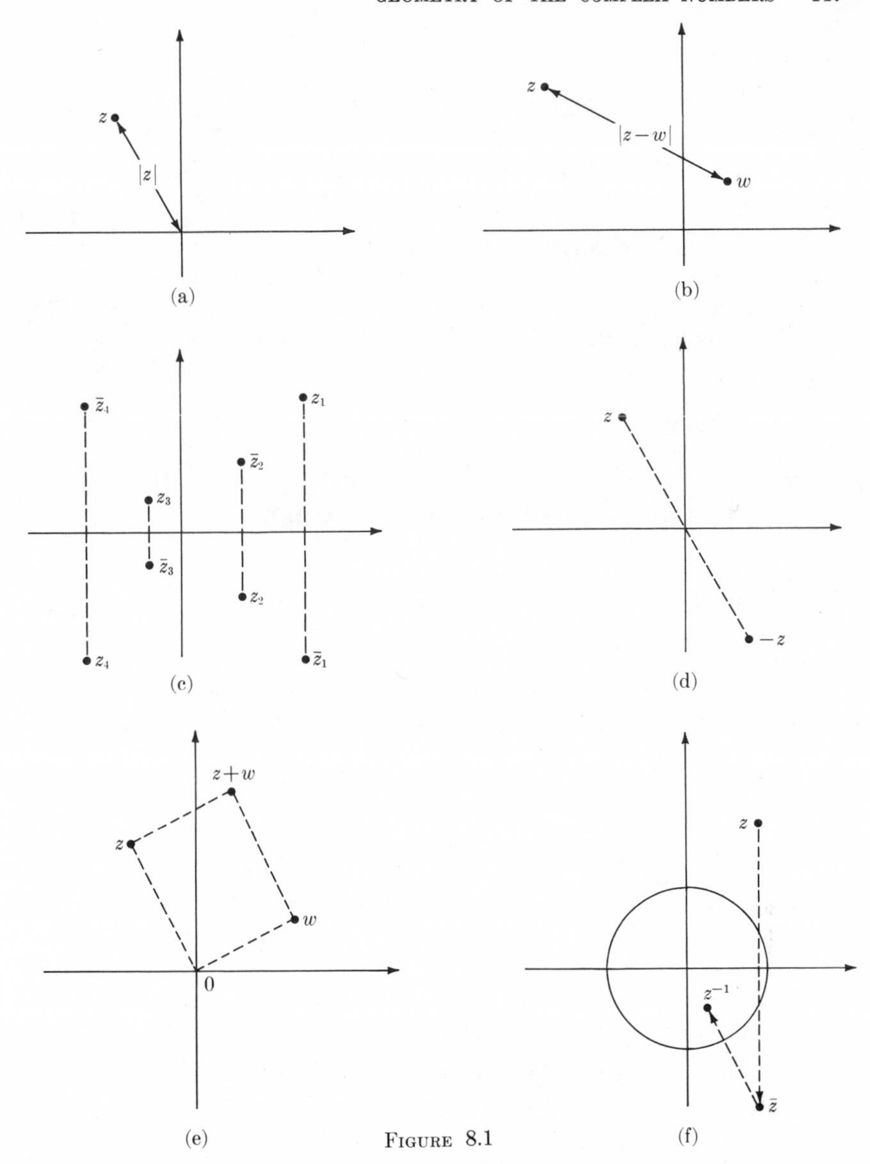

FIGURE 8.1

down and right and left. Figure 1(e) is the familiar parallelogram law for addition in R^2. Finally, we check (f) by noting that

$$(x + iy)^{-1} = \frac{\overline{(x + iy)}}{|x + iy|^2} = \frac{x - iy}{x^2 + y^2} = \frac{x}{x^2 + y^2} + i\,\frac{-y}{x^2 + y^2}\,.$$

Now, $(x,-y)$ is the reflection of (x,y) in the real axis, and dividing both x and $-y$ by the same number $x^2 + y^2$ changes the length of $x - iy$ but not the direction. The length is determined by the equation $|z^{-1}| = |z|^{-1}$, for

$$1 = |1| = |zz^{-1}| = |z| \cdot |z^{-1}|.$$

PROBLEMS

1. For each of the following numbers z, draw a figure showing z, $\bar{z}$, $-z$, and $1/z$.

(a) 1 (b) i (c) $-i$ (d) -1 (e) $1 + i$ (f) $-1 - 2i$ (g) $-3 + \pi i$

2. Describe geometrically the following sets.

(a) $\{z: |z| - 1\}$
(b) $\{z: |z - 1| = 2\}$
• (c) $\{z: |z + 1| = 2\}$
• (d) $\{z: |z - i| = |z - 2|\}$
• (e) $\{z: |z - 1| = \mathrm{Re}(z)\}$
• (f) $\{z: |z - i| = \frac{1}{2}|\mathrm{Re}(z)|\}$ [See §7.8.]
(g) $\{z: |z + i| = 2\,\mathrm{Im}\,(z)\}$.

3. Sketch the set $\{z: |z + 1 - i| < 1 \text{ and } \mathrm{Im}\,(z) > 1/2\}$.

8.3 MULTIPLICATION OF COMPLEX NUMBERS

The definition of multiplication,

$$(x + iy)(u + iv) = (xu - yv) + i(yu + xv),$$

is not always the best way to handle products of complex numbers; a convenient geometric way is based on the idea of *complex exponentials.*

The basic algebraic property of the exponential function is

$$e^{a+b} = e^a e^b. \tag{1}$$

We now define *complex* exponentials by the formula

e^{x+iy} DEFINED

$$e^{x+iy} = e^x(\cos y + i \sin y), \tag{2}$$

and justify this definition by showing that

(A) the new definition agrees with the old when $x + iy$ happens to be a real number, i.e. when $y = 0$; and

(B) the basic property (1) remains valid when a and b are complex numbers.

Now, (A) is obvious, for when $y = 0$, then $\cos y = 1$ and $\sin y = 0$, so $e^{x+i0} = e^x$. To check (B), we let $a = x + iy$ and $b = u + iv$. Then

$$\begin{aligned} e^{a+b} &= e^{(x+u)+i(y+v)} = e^{x+u}[\cos\,(y + v) + i \sin\,(y + v)] \\ &= e^{x+u}(\cos y \cos v - \sin y \sin v + i \sin y \cos v + i \cos y \sin v), \end{aligned}$$

while

$$\begin{aligned} e^a e^b &= e^x(\cos y + i \sin y)e^u(\cos v + i \sin v) \\ &= e^x e^u(\cos y \cos v - \sin y \sin v + i \sin y \cos v + i \cos y \sin v). \end{aligned}$$

The two expressions are indeed the same.

Example 1. Show that $e^{i\pi} = -1$, thus obtaining a mysterious relation between four mysterious numbers. [Do it yourself; solution below.]

To explain complex multiplication from a geometric point of view, we use polar coordinates [Fig. 2]. If $z = x + iy$, we obtain polar coordinates of the corresponding point (x,y) in the plane by choosing r and θ so that

$$r = \sqrt{x^2 + y^2} = |z|$$

and

$$r \cos \theta = x, \qquad r \sin \theta = y. \tag{3}$$

From (3), it follows that

$$x + iy = r \cos \theta + ir \sin \theta = r(\cos \theta + i \sin \theta) = re^{i\theta}.$$

Any number θ chosen as in (3) is traditionally called an *argument* of z, denoted $\arg z$; we think of $\arg z$ as giving the angle from the x axis to the line between the origin and z, as in Fig. 2. Using this notation, and recalling that $r = |z|$, we can write the formula $x + iy = re^{i\theta}$ as

arg $\boldsymbol{z}$

$$z = |z|e^{i \arg z}. \tag{4}$$

Notice that arg is *not* a function, for it assigns to a given z not one, but many numbers; if θ is one of these numbers, then the others are $\theta \pm 2\pi$, $\theta \pm 4\pi, \ldots$.

Example 2. Find values of $\arg 1$, $\arg(-1)$, $\arg i$, $\arg(1 - i)$. Find three different values of $\arg i$. [Solution below.]

The representation in (4) together with formula (1) for products of exponentials leads immediately to the geometric interpretation of complex products. If we write $z = |z|e^{i \arg z}$ and $w = |w|e^{i \arg w}$, then

$$\begin{aligned} zw &= (|z|e^{i \arg z})(|w|e^{i \arg w}) \\ &= |z| \cdot |w| e^{i \arg z} e^{i \arg w} && \text{[commutative law]} \\ &= |z| \cdot |w| e^{i(\arg z + \arg w)}. && \text{[by (1)]} \end{aligned}$$

Thus *to multiply two complex numbers, multiply their absolute values, and add their arguments* [Fig. 3].

$\mathbf{arg(zw) = arg\ z + arg\ w}$

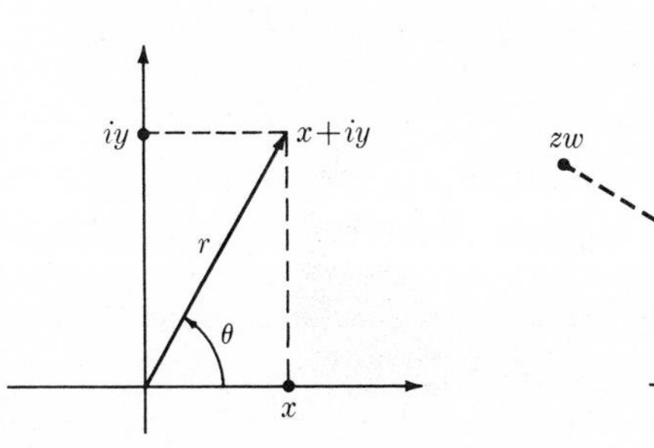

FIGURE 8.2

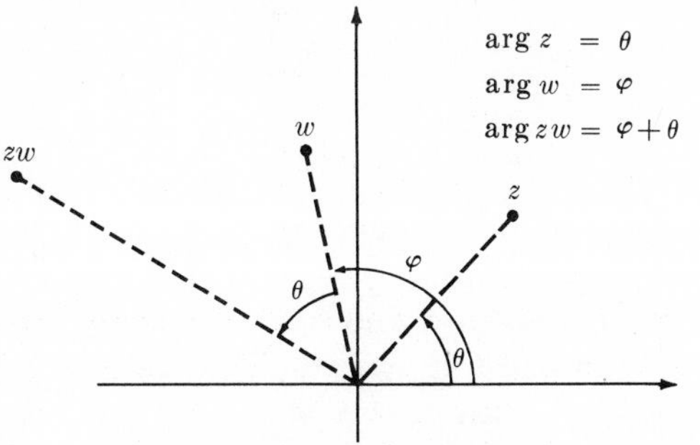

FIGURE 8.3

Example 3. Let $w = i$. Then [Fig. 4] $|i| = 1$ and $\arg i = \pi/2$, so

$$iz = |z| \cdot |i| e^{i(\arg z + \arg i)} = |z| e^{i(\arg z + \pi/2)}.$$

Thus, to multiply z by i, leave the length unchanged, and increase $\arg z$ by $\pi/2$; in other words, rotate z counterclockwise about the origin, through an angle $\pi/2$.

Example 4. Explain the effect of multiplication by -1 by observing that $|-1| = 1$ and $\arg(-1) = \pi$. [Solution below.]

It is easy to take powers of a complex number z written in the form $re^{i\theta}$:

$$(re^{i\theta})^2 = r^2 e^{2i\theta},$$

and generally

$$(re^{i\theta})^n = r^n e^{ni\theta}.$$

This leads, in turn, to a simple way of extracting the roots of a complex number z. Simply write z in the form $re^{i\theta}$, and then

nth ROOTS OF z

$$z^{1/n} = r^{1/n} e^{i\theta/n}. \tag{5}$$

For, as you can see, $(r^{1/n} e^{i\theta/n})^n = re^{i\theta} = z$.

Example 5. To find $\sqrt{-1}$, note [Example 1] that $-1 = e^{i\pi}$, which leads to $\sqrt{-1} = e^{i\pi/2} = i$. Or we can write $-1 = e^{-i\pi}$, leading to $\sqrt{-1} = e^{-i\pi/2} = -i$. Giving both solutions together, $\sqrt{-1} = \pm i$. Thus, i is not the *only* square root of -1.

Example 6. We can find $i^{1/3}$ in any of the following ways [Fig. 5]:

$$i = e^{i\pi/2}, \qquad i^{1/3} = e^{i\pi/6} = \frac{\sqrt{3}}{2} + \frac{i}{2};$$

$$i = e^{i5\pi/2}, \qquad i^{1/3} = e^{i5\pi/6} = -\frac{\sqrt{3}}{2} + \frac{i}{2};$$

$$i = e^{i9\pi/2}, \qquad i^{1/3} = e^{i9\pi/6} = e^{i3\pi/2} = -i;$$

$$i = e^{i13\pi/2}, \qquad i^{1/3} = e^{i13\pi/6} = \frac{\sqrt{3}}{2} + \frac{i}{2};$$

$$i = e^{-i3\pi/2}, \qquad i^{1/3} = e^{-i\pi/2} = -i;$$

and so on.

As Examples 5 and 6 show, the expression (5) produces more than one nth root of a given complex number z, because there is more than one way to choose the angle θ in $z = re^{i\theta}$. The situation is as follows:

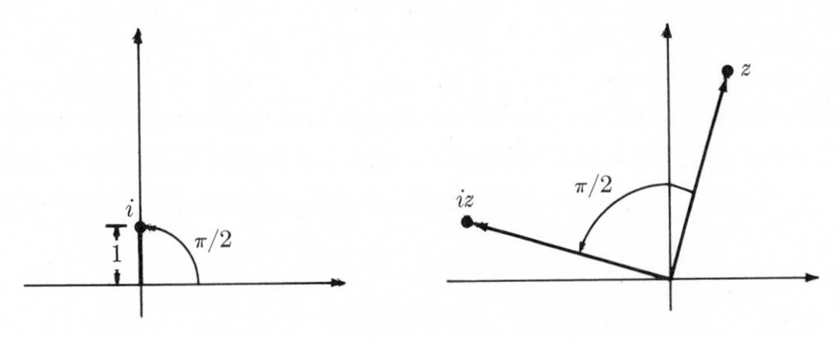

FIGURE 8.4

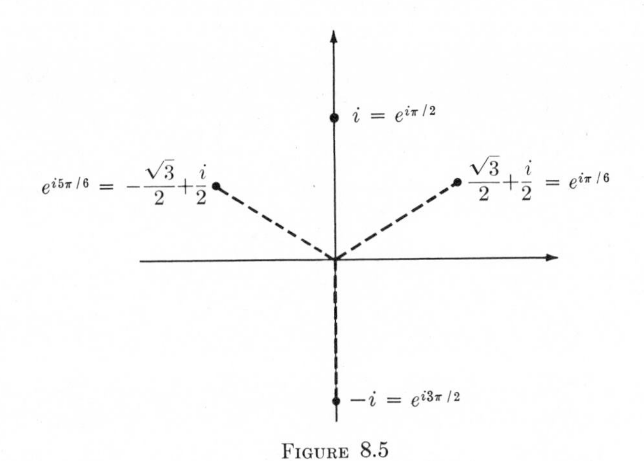

FIGURE 8.5

Theorem 1. *Given any complex number $z \neq 0$, there are exactly n different numbers $z_1, \ldots, z_n$ such that $(z_j)^n = z$. The z_j are uniformly spaced around the circle of radius $|z|^{1/n}$ about the origin.*

We leave the proof as Problem 9 below. Example 6 above provides a hint as to why the theorem is true: the various possible choices of arg z all differ from each other by multiples of 2π, so the nth roots have arguments differing by multiples of $2\pi/n$.

Theorem 1 provides our first example of the simplification brought about by complex numbers. Compare the nice result stated there to the following mess that comes when we deal only with real numbers:

(a) If $x > 0$ and n is even, there are two real nth roots of x, denoted $\sqrt[n]{x}$ and $-\sqrt[n]{x}$.

(b) If $x < 0$ and n is even, there are no real nth roots of x.

(c) If n is odd, then any real number has exactly one real nth root.

Solution to Example 1. $e^{i\pi} = \cos\pi + i\sin\pi = -1.$

Solution to Example 2. From Fig. 6 we read $\arg 1 = 0$, $\arg i = \pi/2$, $\arg(1 - i) = -\pi/4$. The dashed arc and the dotted arc suggest $-3\pi/2$ and $5\pi/2$ as alternate choices of $\arg i$. In general, we get

$$\arg 1 = 0, \pm 2\pi, \pm 4\pi, \ldots$$
$$\arg i = \pi/2, \pi/2 \pm 2\pi, \ldots$$
$$\arg(1 - i) = -\pi/4, -\pi/4 \pm 2\pi, \ldots.$$

Solution to Example 4. Multiplication by -1 amounts to rotation about the origin through an angle of radian measure π, i.e. through a "straight angle."

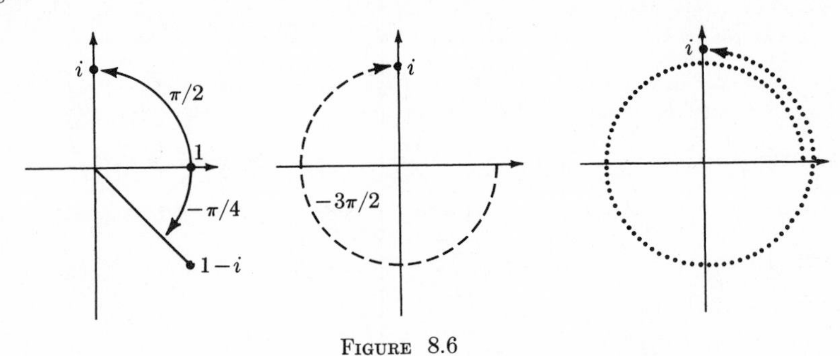

FIGURE 8.6

PROBLEMS

1. Find values of $\arg z$ for the following numbers.
   • (a) $1 + i$
   (b) $1/2 + i/2$
   • (c) $1 + 3i$
   (d) $-i$
   (e) $-1 - i$
   • (f) $-\dfrac{3}{2} + i\dfrac{\sqrt{3}}{2}$

2. Plot the following numbers as points in the plane.
   (a) $e^{i\pi/3}$ (b) $e^{-i\pi/4}$ (c) $2e^{i3\pi/4}$ (d) $\frac{1}{2}e^{i5\pi/6}$

• 3. Show that $e^{z+2\pi i} = e^z$.

4. Plot z, w, and zw in each of the following cases.
   (a) $z = 2i$, $w = 1 + i$
   (b) $z = -i/2$, $w = 3 - i$
   (c) $z = 2i$, $w = e^{\pi i/6}$
   (d) $z = \frac{1}{2}e^{-\pi i/2}$, $w = 3 + i$

• 5. Write each of the four fourth roots of 1 in the form $x + iy$. Plot them.

• 6. Write each of the sixth roots of -2 in the form $x + iy$. Plot them.

7. (a) Find four solutions of $z^4 - 3z^2 + 2 = 0$. [Hint: First solve for z^2.]
   • (b) Find four solutions of $z^4 + 3z^2 + 3 = 0$.

8. Show that $\cos\theta = (e^{i\theta} + e^{-i\theta})/2$. What is the corresponding formula for $\sin\theta$?

9. (a) Find n distinct nth roots of 1.
(b) Show that 1 has at most n distinct nth roots.
(c) Prove Theorem 1 above.

10. Prove that $|zw| = |z|\cdot|w|$.

11. Show that $\overline{re^{i\theta}} = re^{-i\theta}$. Illustrate with an appropriate sketch.

12. Prove that $\overline{zw} = \bar{z}\cdot\bar{w}$. [See Problem 11.]

13. (a) Let r be a complex number. Prove that $(z - r)(z - \bar{r}) = z^2 + bz + c$, where b and c are *real* numbers.

CONJUGATE ROOTS

(b) Suppose that $P(z) = \sum_0^n a_j z^j$, where the a_j are all *real* numbers.

Prove that $P(r) = 0 \Rightarrow P(\bar{r}) = 0$.

14. (a) Prove *De Moivre's Theorem:* $(\cos\theta + i\sin\theta)^n = \cos n\theta + i\sin n\theta$.
(b) Take real and imaginary parts, with $n = 2$, to derive a formula for $\cos 2\theta$ and $\sin 2\theta$ as polynomials in $\cos\theta$ and $\sin\theta$.
(c) Derive formulas for $\cos 3\theta$ and $\sin 3\theta$.
(d) Using De Moivre's Theorem and the binomial theorem

DE MOIVRE'S THEOREM

$$(a + b)^n = a^n + na^{n-1}b + \frac{n(n-1)}{2}a^{n-2}b^2 + \cdots + nab^{n-1} + b^n,$$

derive formulas for $\cos n\theta$ and $\sin n\theta$. [Note that two complex numbers z and w are equal iff $\text{Re}(z) = \text{Re}(w)$ and $\text{Im}(z) = \text{Im}(w)$.]

15. Every nonzero complex number z can be written uniquely in the form $z = re^{i\theta}$, where $r = |z|$ and $-\pi < \theta \leq \pi$. The number θ with these restrictions is called the *principal value* of the argument of z, denoted $\text{Arg}\, z$. Show from appropriate sketches that:

Arg z

(a) $\text{Arg}\,(x + iy) = \arctan\dfrac{y}{x}$ if $x > 0$

(b) $\text{Arg}\,(x + iy) = \text{arccot}\dfrac{x}{y}$ if $y > 0$

(c) $\text{Arg}\,(x + iy) = \text{arccot}\dfrac{x}{y} - \pi$ if $y < 0$

(d) $\text{Arg}\,(x + iy) = \arctan\dfrac{y}{x} + \pi$ if $x < 0$ and $y \geq 0$

(e) $\text{Arg}\,(x + iy) = \arctan\dfrac{y}{x} - \pi$ if $x < 0$ and $y < 0$

[Recall the principal value restrictions $-\pi/2 < \arctan t < \pi/2$ and $0 < \text{arccot}\, t < \pi$.]

16. For any complex z, define the (complex) logarithm by

Log z

$$\text{Log } z = \log |z| + i \text{ Arg } z$$

[see Problem 15 for Arg].

(a) Prove that for all complex z, $e^{\text{Log } z} = z$.

(b) For what values of z is $\text{Log } (e^z) = z$?

(c) Give a general formula for $\text{Log } (e^z)$.

17. The number $z_n = e^{2\pi i/n}$ is called a *primitive* nth *root of unity*.

(a) Show that $z_n, z_n^2, \ldots, z_n^n$ are all the nth roots of 1.

(b) Prove that $z_n + z_n^2 + \cdots + z_n^n = 0$.
[Hint: $(1 - r)(1 + r + \cdots + r^{n-1}) = 1 - r^n$.]

(c) Let $\mathbf{A}_1, \ldots, \mathbf{A}_n$ be vectors pointing from the origin to the vertices of a regular polygon centered at the origin. Prove that $\mathbf{A}_1 + \cdots + \mathbf{A}_n = 0$.

18. Make sense of the equation $i^i = e^{-\pi/2}$.

19. Multiply $x + iy$ by $e^{i\theta}$ to show that rotating the vector $[x,y]$ counterclockwise through an angle θ results in the vector $[x \cos \theta - y \sin \theta, y \cos \theta + x \sin \theta]$.

20. (a) Derive a formula for $(\cosh \theta + i \sinh \theta)^n$, comparable to De Moivre's Theorem [Problem 14]. [See §2.8 for cosh and sinh.]

(b) Relate $\cosh(i\theta)$ and $\sinh(i\theta)$ to $\cos \theta$ and $\sin \theta$.

(c) Make sense of the formula $\cos(i\theta) = \cosh(\theta)$.

8.4 COMPLEX FUNCTIONS OF A REAL VARIABLE

We studied the differential equation for oscillating motion

$$f''(t) + \omega^2 f(t) = 0, \tag{1}$$

and found the general solution [§3.5] as a linear combination of sin and cos,

$$f(t) = A \cos \omega t + B \sin \omega t.$$

The term $\omega^2 f$ in (1) can be thought of as the acceleration due to stretching or compressing a spring. To be more realistic, we should add a frictional force of some kind. If the frictional force is proportional to the velocity f', we get the equation

$$f''(t) + kf'(t) + \omega^2 f(t) = 0, \tag{2}$$

where k is a constant.

Equations (1) and (2) are *linear differential equations with constant coefficients*. The general equation of this type has the form

$$a_n f^{(n)}(t) + a_{n-1} f^{(n-1)}(t) + \cdots + a_1 f'(t) + a_0 f(t) = g, \tag{3}$$

where $a_1, \ldots, a_n, g$ are constants, possibly complex. The solution of such equations depends on the use of complex exponentials, as the following examples suggest.

Example 1. Find solutions of $f'' - 4f = 0$. *Solution.* Try $f(t) = e^{rt}$, where r is a constant to be determined. We get $f' = re^{rt}$, $f'' = r^2e^{rt}$, so

$$f'' - 4f = (r^2 - 4)e^{rt}.$$

Thus, for functions f of this form e^{rt},

$$f'' - 4f = 0 \Leftrightarrow r^2 = 4, \qquad \text{or} \qquad r = \pm 2.$$

We obtain in this way two solutions, e^{2t} and e^{-2t}. You can check further that for any constants A and B, the linear combination

$$f(t) = Ae^{2t} + Be^{-2t}$$

is a solution.

Example 2. Find solutions of $f'' - 3f' + 2f = 0$. *Solution.* Trying $f(t) = e^{rt}$ again, we get easily

$$f'' - 3f' + 2f = (r^2 - 3r + 2)e^{rt}.$$

So e^{rt} is a solution iff $r^2 - 3r + 2 = 0$, which gives $r = 2$ and $r = 1$. You can check that, more generally,

$$f(t) = Ae^{2t} + Be^{t}$$

is a solution for any choice of the constants A and B.

Example 3. Find solutions of $f'' + 9f = 0$. *Solution.* Trying $f(t) = e^{rt}$ as before, we get

$$f'' + 9f = (r^2 + 9)e^{rt},$$

so e^{rt} is a solution iff $r^2 = -9$, or $r = \pm 3i$. We thus get the two functions

$$e^{3it} = \cos 3t + i \sin 3t,$$
$$e^{-3it} = \cos 3t - i \sin 3t.$$

We can easily recognize these as solutions of $f'' + 9f = 0$, *if* complex-valued functions behave as real-valued functions do. This is precisely what we are about to investigate.

A complex function of one real variable has domain on the real line, and assigns to each point of this domain a complex number. The most important examples are the complex polynomials

$$P(t) = a_n t^n + \cdots + a_1 t + a_0$$

[where the numbers $a_0, \ldots, a_n$ may be complex] and the complex exponentials

$$f(t) = e^{rt}$$

[where again the number r may be complex]. Every complex function f can be written as a combination of two real functions u and v:

$$f(t) = u(t) + iv(t)$$

where $u(t) = \operatorname{Re} f(t)$ and $v(t) = \operatorname{Im} f(t)$. For example,

$$f(t) = (1+i)t^2 + t - 3i = (t^2+t) + i(t^2-3) = u(t) + iv(t),$$

or

$$f(t) = e^{2it} = \cos 2t + i \sin 2t = u(t) + iv(t).$$

The *derivative* of a complex function is taken by differentiating the real and imaginary parts separately:

DERIVATIVE OF $u + iv$

$$f' = (u + iv)' = u' + iv'.$$

Thus, if $f(t) = e^{2it} = \cos 2t + i \sin 2t$, then

$$f'(t) = -2 \sin 2t + 2i \cos 2t.$$

You can easily check that the usual rules for differentiation apply to complex functions:

$$(f+g)' = f' + g', \tag{4}$$

$$(fg)' = f'g + fg', \tag{5}$$

$$\left(\frac{f}{g}\right)' = \frac{f'g - fg'}{g^2}, \tag{6}$$

and if h is a real-valued function, then

$$(f \circ h)' = (f' \circ h)h'. \tag{7}$$

Further, the law of parallel graphs remains in force: If f' is identically zero on an interval, then f is constant on that interval. [See Problem 5.]

$\dfrac{de^{rt}}{dt} = re^{rt}$ EVEN WHEN r IS COMPLEX

In addition to these general results, we need a special formula:

$$\text{If} \quad f(t) = e^{rt} \quad (r \text{ complex}), \quad \text{then} \quad f'(t) = re^{rt}. \tag{8}$$

The proof is straightforward. We set $r = a + ib$, and then

$$\begin{aligned} f(t) &= e^{at+ibt} = e^{at}(\cos bt + i \sin bt), \\ f'(t) &= e^{at}(-b \sin bt + ib \cos bt) + ae^{at}(\cos bt + i \sin bt) \\ &= (a+ib)e^{at}(\cos bt + i \sin bt) \\ &= re^{at+ibt} \\ &= re^{rt}. \end{aligned}$$

When f is a complex function of a real variable, we call any function F such that $F' = f$ an *indefinite integral* of f. From the law of parallel graphs, any two indefinite integrals of a given function f differ by a constant. The indefinite integral of f is denoted $\int f$. Thus, from (8), if $r \neq 0$, we have

$$\int e^{rt}\, dt = \frac{1}{r} e^{rt} + c,$$

where c is any complex number.

The net result of the last few paragraphs is that we can calculate with complex-valued functions exactly as we do with real-valued functions. With this assurance, we return to the differential equation with constant coefficients (3), and take the special case where $g = 0$, the homogeneous case:

$$a_n f^{(n)} + \cdots + a_0 f = 0. \tag{9}$$

We try an *exponential solution*,

$$f(t) = e^{rt}.$$

Since $f'(t) = re^{rt}$, $f''(t) = r^2e^{rt}, \ldots$ and $f^{(n)}(t) = r^n e^{rt}$, we are led from (9) to the equation

$$a_n r^n e^{rt} + \cdots + a_0 e^{rt} = 0.$$

Since e^{rt} is never zero, we may divide and obtain a polynomial equation for r,

$$a_n r^n + \cdots + a_0 = 0. \tag{10}$$

Thus *when r satisfies equation* (10), *then e^{rt} is a solution of* (9). The polynomial $\sum_0^n a_j r^j$ occurring in (10) is called the *characteristic polynomial* of equation (9). The solutions of (10) are the *roots* of the characteristic polynomial.

CHARACTERISTIC POLYNOMIAL

Finding these exponential solutions of (9) is the first step in solving linear differential equations with constant coefficients; the solution is completed in the next two sections.

We close with an example illustrating a rather different use of complex functions, in the evaluation of integrals.

Example 4. $\int e^t \cos t\, dt = ?$ *Solution.* $\cos t = \text{Re}\,(e^{it})$, so we consider the integral

$$\int e^t e^{it}\, dt = \int e^{(1+i)t}\, dt = \frac{1}{1+i} e^{(1+i)t} + c = \frac{1-i}{2} e^t e^{it} + c$$

$$= e^t \left(\frac{1-i}{2}\right)(\cos t + i \sin t) + c$$

$$= \frac{e^t}{2} [(\cos t + \sin t) + i(\sin t - \cos t)] + c.$$

Taking real parts, we get

$$\int e^t \cos t\, dt = \frac{e^t}{2} (\cos t + \sin t) + c.$$

Taking imaginary parts, we get a bonus

$$\int e^t \sin t\, dt = \frac{e^t}{2} (\sin t - \cos t) + c.$$

PROBLEMS

1. For each of the following differential equations, find the roots of the characteristic polynomial and the corresponding exponential solutions.

- • (a) $f'' + 4f = 0$
- (b) $f'' - 4f = 0$
- • (c) $f'' + 2f' + f = 0$
- (d) $f'' - 2f' + f = 0$
- • (e) $f'' + 2f' + 2f = 0$
- (f) $f''' + f = 0$

2. Prove formula (4) above.

3. Prove formula (5) above.

4. Prove formula (6) above.

5. Prove that if f' is identically zero throughout an interval I, then f is constant on that interval. [Here, f may be complex, of course.]

6. Prove that $\int (f + g) = \int f + \int g$ and $\int cf = c\int f$, where f and g are any complex functions and c is any complex number.

7. (a) Prove that $\int fg' = fg - \int f'g$, where f and g are differentiable complex functions.

• (b) Find $\int te^{rt}\,dt$, where r is any complex number.

(c) Find $\int t^2e^{rt}\,dt$, where r is any complex number.

8. Let a and b be real numbers.

(a) Show that $\int e^{at} \cos bt\,dt = \operatorname{Re} \int e^{(a+ib)t}\,dt$, and evaluate the integral.

(b) Evaluate $\int e^{at} \sin bt\,dt$ in a similar way.

9. This problem concerns integrals arising in Fourier series, one of the important topics in advanced calculus. Let n and m be integers.

(a) Show that $\displaystyle\int_0^{2\pi} e^{int}e^{-imt}\,dt = 0$ if $n \neq m$, and $= 2\pi$ if $n = m$.

(b) Show that if $n \geq 0$ and $m \geq 0$, then $\displaystyle\int_0^{2\pi} e^{int}e^{imt}\,dt = 0$ unless $n = m = 0$.

(c) Take real and imaginary parts of the integrals in (a) and (b) to show that for integers $n \geq 0$, $m \geq 0$,

$$\int_0^{2\pi} \cos nt \sin mt\,dt = 0 \quad \text{for all } n \text{ and } m$$

$$\int_0^{2\pi} \cos nt \cos mt\,dt = \begin{cases} 0, & n \neq m \\ \pi, & n = m \neq 0 \\ 2\pi, & n = m = 0 \end{cases}$$

$$\int_0^{2\pi} \sin nt \sin mt\,dt = \begin{cases} 0, & n \neq m \\ \pi, & n = m \neq 0 \\ 0, & n = m = 0. \end{cases}$$

10. The complex functions Arg and Log developed in the problems of §8.3 can be used to integrate rational functions.

(a) Define Log z as in Problems 15 and 16, §8.3. Let $r = a + ib$ be a complex number, $b \neq 0$, and define $f(t) = \text{Log}(t - r)$. Show that $f'(t) = \dfrac{1}{t - r}$. [Use Problem 15(c) for $b > 0$, and 15(b) for $b < 0$.]

(b) Note that

$$\int \frac{dt}{t^2 + 1} = \int \frac{dt}{(t - i)(t + i)} = \frac{1}{2i} \int \frac{dt}{t - i} - \frac{1}{2i} \int \frac{dt}{t + i}.$$

Use part (a) to evaluate the last two integrals as Logs, and combine them to produce a formula for arctan t.

(c) Use (a) to evaluate $\displaystyle\int \frac{dt}{t^2 - 2t + 5}$. [Hint: $t^2 - 2t + 5 = 0$ has solutions $t = 1 \pm 2i$, so it factors into $(t - 1 - 2i)(t - 1 + 2i)$. Use a partial fractions decomposition [§6.4], and evaluate the results as Logs. Recombine the Logs.]

(d) Evaluate $\displaystyle\int \frac{dt}{(t^2 + 1)^2}$ using partial fractions and Logs. [The calculations here may be a little longer than those in §6.4, but they are more straightforward and do not require any hard-to-remember reduction formulas.]

11. Evaluate the following improper integrals [§5.5], which are important in the study of heat conduction.

IMPROPER INTEGRALS

• (a) $\displaystyle\int_0^\infty e^{ixt}e^{-t}\,dt$

(b) $\displaystyle\int_{-\infty}^0 e^{ixt}e^{t}\,dt$

• (c) $\displaystyle\int_{-\infty}^\infty e^{ixt}e^{-|t|}\,dt$

(d) $\displaystyle\int_0^\infty \cos(xt)e^{-t}\,dt$

(e) $\displaystyle\int_0^\infty \sin(xt)e^{-t}\,dt$

(f) $\displaystyle\int_{-\infty}^\infty e^{ixt}e^{-a|t|}\,dt, \quad a > 0$

8.5 LINEAR DIFFERENTIAL EQUATIONS WITH CONSTANT COEFFICIENTS; THE HOMOGENEOUS SECOND ORDER CASE

We have at our disposal all the concepts for the solution of equations of the form

LINEAR EQUATION W/CONSTANT COEFFICIENTS

$$a_n f^{(n)} + \cdots + a_1 f' + a_0 f = g, \tag{1}$$

where $a_0, \ldots, a_n$ are complex constants. What we lack is a notation allowing us to manipulate the concepts effectively.

We introduce a new symbol, D, for the operation of differentiation:

$$Df = f'; \qquad Df(t) = f'(t).$$

The familiar rules for differentiation now take the form

$$D(f+g) = Df + Dg$$
$$D(fg) = (Df)g + fDg$$
$$D(cf) = cDf \qquad \text{for any constant } c.$$

Powers of D correspond to higher order derivatives:

$$D^2f = D(Df) = Df' = f''$$
$$D^3f = D(D^2f) = Df'' = f'''$$

and generally

$$D^nf = f^{(n)}.$$

The expression $a_n f^{(n)} + a_{n-1} f^{(n-1)} + \cdots + a_1 f' + a_0 f$ can be written

$$a_nD^nf + a_{n-1}D^{n-1}f + \cdots + a_1Df + a_0 f$$

or even

$$(a_nD^n + a_{n-1}D^{n-1} + \cdots + a_1D + a_0)f.$$

The expression multiplying f in the preceding line is logically denoted $P(D)$, where P is the polynomial

CHARACTERISTIC POLYNOMIAL OF (1)

$$P(r) = a_nr^n + a_{n-1}r^{n-1} + \cdots + a_1r + a_0 .$$

Note that P is the *characteristic polynomial* of the equation (1).

Observe that $De^{rt} = re^{rt}, \ldots, D^ne^{rt} = r^ne^{rt}$; hence for any complex number r,

$$P(D)e^{rt} = P(r)e^{rt}.$$

Thus:

$f(t) = e^{rt}$ solves the differential equation $P(D)f = 0$

$\Leftrightarrow$

r is a root of P, i.e. $P(r) = 0$.

Suppose the characteristic polynomial P can be factored completely,

$$\begin{aligned} P(r) &= a_n r^n + \cdots + a_0 \\ &= a_n(r - r_1)(r - r_2)\cdots(r - r_n). \end{aligned} \tag{2}$$

The roots $r_1, \ldots, r_n$ may be complex numbers, and need not be all distinct; i.e. we may have $r_1 = r_2$, etc. *Corresponding to* (2), *there is a factorization of* $P(D)$: for every differentiable function f,

$$P(D)f = a_n(D - r_1)(D - r_2)\cdots(D - r_n)f. \tag{3}$$

Example 1. If $P(r) = r^2 - 4 = (r - 2)(r + 2)$, we are claiming that $(D^2 - 4)f = (D - 2)(D + 2)f$. This is easily checked: $(D + 2)f = Df + 2f$, so

$$\begin{aligned} (D-2)(D+2)f = (D-2)(Df+2f) &= D(Df + 2f) - 2(DF + 2f) \\ &= (D^2f + 2Df) - 2Df - 4f \\ &= D^2f - 4f = (D^2 - 4)f \\ &= P(D)f. \end{aligned}$$

Example 2. $P(r) = 3r^2 + 9r + 6 = 3(r + 2)(r + 1)$. Then

$$\begin{aligned} 3(D+2)(D+1)f &= 3(D+2)(Df+f) \\ &= 3D(Df+f) + 6(Df+f) \\ &= 3D^2f + 9Df + 6f = (3D^2 + 9D + 6)f \\ &= P(D)f, \end{aligned}$$

checking (3) in this case.

Now we are ready to solve the constant coefficient linear equation. In this section we restrict ourselves to a simple but important case:

Homogeneous second order equations

Equation (1) is of second order if the characteristic polynomial $P(r) = a_n r^n + \cdots + a_0$ has degree 2; it is called *homogeneous* if $g = 0$. Thus, we consider equations of the form

HOMOGENEOUS SECOND ORDER EQUATION

$$a_2 f'' + a_1 f' + a_0 f = (a_2 D^2 + a_1 D + a_0)f = 0. \tag{4}$$

[This includes the equations of elementary circuit theory, and of periodic oscillations with a frictional damping proportional to the velocity.] The characteristic polynomial can always be factored,

$$P(r) = a_2 r^2 + a_1 r + a_0 = a_2(r - r_1)(r - r_2),$$

where the roots r_1 and r_2 of $P(r)$ are given by

$$r_1 = \frac{-a_1 + \sqrt{a_1^2 - 4a_0a_2}}{2a_2}, \qquad r_2 = \frac{-a_1 - \sqrt{a_1^2 - 4a_0a_2}}{2a_2}.$$

Thus (4) can be written

$$a_2(D - r_2)(D - r_1)f = 0. \tag{5}$$

Let f be any twice differentiable function, and set

$$g = (D - r_1)f. \tag{6}$$

Then (5) is equivalent to $a_2(D - r_2)g = 0$, or simply $(D - r_2)g = 0$. This is a familiar equation, solved in §3.3. We multiply through by e^{-r_2t}, obtaining

$$\begin{aligned} e^{-r_2t}(D - r_2)g &= 0 \\ D(e^{-r_2t}g) &= 0 \\ e^{-r_2t}g &= c \\ g(t) &= ce^{r_2t}. \end{aligned}$$

Knowing g, we return to (6) to find f. To solve $(D - r_1)f(t) = g(t) = ce^{r_2t}$, we multiply through by e^{-r_1t}:

$$\begin{aligned} e^{-r_1t}(D - r_1)f &= ce^{r_2t}e^{-r_1t} \\ D(e^{-r_1t}f) &= ce^{(r_2-r_1)t}. \end{aligned}$$

Thus to find $e^{-r_1t}f$, and hence f, we should find the indefinite integral of the right-hand side. The result depends on whether $r_1 = r_2$ or $r_1 \neq r_2$:

$$e^{-r_1t}f(t) = \int ce^{(r_2-r_1)t}\,dt = \begin{cases} \dfrac{c}{r_2 - r_1}\,e^{(r_2-r_1)t} + c_1\,, & \text{if } r_1 \neq r_2 \\[2ex] ct + c_1\,, & \text{if } r_1 = r_2\,. \end{cases}$$

If we relabel $c/(r_2 - r_1)$ as c_2 in the first case $(r_1 \neq r_2)$, and set $c = c_2$ in the second case, and multiply through by e^{r_1t}, we get the solution

$$\begin{aligned} f(t) &= c_1e^{r_1t} + c_2e^{r_2t} && \text{if } r_1 \neq r_2 \\ f(t) &= (c_1 + c_2t)e^{r_1t} && \text{if } r_1 = r_2\,. \end{aligned}$$

Note that the roots r_1 and r_2 may be complex numbers, and the constants c_1 and c_2 may also be complex.

We have proved:

Theorem 2. *Let $a_2f'' + a_1f' + a_0f = 0$ be any second order homogeneous linear equation with constant coefficients, and let r_1 and r_2 be the roots of the characteristic polynomial $P(r) = a_2r^2 + a_1r + a_0$.*

GENERAL SOLUTION OF HOMOGENEOUS SECOND ORDER EQUATION

If $r_1 \neq r_2$, then every solution of the equation has the form

$$f(t) = c_1e^{r_1t} + c_2e^{r_2t}.$$

If $r_1 = r_2$, every solution has the form

$$f(t) = (c_1 + c_2t)e^{r_1t}.$$

Notice that in either case the solution is a linear combination of two functions, with unspecified constants c_1 and c_2. This linear combination is called the *general solution* of the equation, because every solution can be obtained from it by appropriate choice of the constants.

The following table summarizes the various constant coefficient equations whose general solution we have found. Notice in each case that the *number of undetermined constants equals the order of the equation.*

EQUATION	GENERAL SOLUTION
$Df = 0$	$f = c$
$D^2f = 0$	$f(t) = c_1t + c_0$
$D^{n+1}f = 0$	$f(t) = c_nt^n + c_{n-1}t^{n-1} + \cdots + c_1t + c_0$
$(D - r_1)(D - r_2)f = 0 \quad [r_1 \neq r_2]$	$f(t) = c_1e^{r_1t} + c_2e^{r_2t}$
$(D - r_1)^2f = 0$	$f(t) = (c_1 + c_2t)e^{r_1t}$

Example 3. Find the general solution of $f'' + 2f' + 2f = 0$. *Solution.* The characteristic polynomial is $P(r) = r^2 + 2r + 2 = (r + 1)^2 + 1$, and the roots are

$$r_1 = -1 + i, \qquad r_2 = -1 - i.$$

Since $r_1 \neq r_2$, the general solution is

$$\begin{aligned} f(t) &= c_1e^{(-1+i)t} + c_2e^{(-1-i)t} \\ &= e^{-t}(c_1e^{it} + c_2e^{-it}) \\ &= e^{-t}[(c_1 + c_2) \cos t + i(c_1 - c_2) \sin t]. \end{aligned}$$

CONVERTING COMPLEX EXPONENTIALS TO TRIG FUNCTIONS

In the last line of Example 3, we converted the terms with imaginary exponents into sin and cos. This can always be done, using the basic formula

$$e^{ix} = \cos x + i \sin x.$$

In applications it is generally convenient to do this, because we are looking for real-valued solutions. Notice that the constants c_1 and c_2 are in general complex numbers; but it can (and in applications usually does) happen that $c_1 + c_2$ and $i(c_1 - c_2)$ are *real* numbers. These constants are generally to be determined by giving initial conditions, i.e. specifying $f(t_0)$ and $f'(t_0)$ for some particular value t_0. [See Problem 3.]

Example 4. Solve $f'' + 2f' + f = 0$. *Solution.* The characteristic polynomial is $P(r) = r^2 + 2r + 1 = (r + 1)^2$, so $r = -1$ is a *double root.* Thus the solution f requires e^{-t} *and* te^{-t}:

$$f(t) = (c_1 + c_2t)e^{-t}.$$

DAMPED OSCILLATOR *Example 5.* We have found the equation $mf'' = -mk^2f$ for a freely oscillating spring, basing the derivation on the fact that the force mf'' is proportional to the displacement f. Now, suppose there is a frictional damping of some sort exerting a force proportional to the velocity f' and directed opposite to it. Denoting the constant of proportionality $2mb$, we have $mf'' = -mk^2f - 2mbf'$, or $f'' + 2bf' + k^2f = 0$. The characteristic equation is $r^2 + 2br + k^2 = 0$, and the roots are $r_1 = -b + \sqrt{b^2 - k^2}$, $r_2 = -b - \sqrt{b^2 - k^2}$. We consider three cases.

(i) $b^2 > k^2$. Then r_1 and r_2 are real and negative, and $f(t) = c_1e^{r_1t} + c_2e^{r_2t}$ dies down to zero exponentially.

(ii) $b^2 = k^2$. Then $r_1 = r_2 = -b$, and the solution is

$$f(t) = (c_1 + c_2t)e^{-bt} = (c_1 + c_2t)e^{-kt},$$

which dies down faster than e^{-ct} for any $c < k$.

(iii) $b^2 < k^2$. Then

$$r_1 = -b + i\sqrt{k^2 - b^2}, \qquad r_2 = -b - i\sqrt{k^2 - b^2}.$$

Let $\omega = \sqrt{k^2 - b^2}$. Then the solution can be written

$$f(t) = e^{-bt}(c_1e^{i\omega t} + c_2e^{-i\omega t})$$

or, converting to trig functions and relabeling the constants,

$$f(t) = e^{-bt}(C_1 \cos \omega t + C_2 \sin \omega t).$$

The spring oscillates with frequency $\omega = \sqrt{k^2 - b^2}$, damped by the exponential factor e^{-bt}. If $b \to 0$ this damping disappears, and the frequency $\omega = \sqrt{k^2 - b^2} \to k$, the frequency of the undamped oscillation. Thus our earlier solution [§3.5], which neglected damping, gives a good approximation to the situation with small damping, i.e. small b.

PROBLEMS

1. For each of the following polynomials P and functions f, find $P(D)f$.

• (a) $P(r) = r + 1$; $f(t) = e^{-t}$
(b) $P(r) = (r + 1)^2$; $f(t) = e^{-t}$
• (c) $P(r) = (r + 1)^2$; $f(t) = te^{-t}$
(d) $P(r) = (r + 1)^3$; $f(t) = t^2e^{-t}$
• (e) $P(r) = r^2 + i$; $f(t) = e^{at}$

2. For each of the following equations, find the roots of the characteristic polynomial, and the general solution of the equation. Write it without using complex exponentials. [See Example 3.]

• (a) $f'' + 4f = 0$
(b) $f'' + 2if' + f = 0$
• (c) $f'' - 4f' + 5f = 0$
(d) $6f'' + 5f' - 6f = 0$
• (e) $f'' - 4f' = 0$
(f) $f'' + f' = 0$

3. For the equation in the corresponding part of Problem 2, determine the constants in the general solution so that f satisfies the given initial conditions:

• (a) $f(0) = 0,\ f'(0) = 1$ (d) $f(1) = 0,\ f'(1) = 0$
(b) $f(0) = 1,\ f'(0) = 0$ • (e) $f(1) = 1,\ f'(1) = 0$
• (c) $f(0) = 1,\ f'(0) = 1$ (f) $f(-1) = 0,\ f'(-1) = 1$

4. [*Damping*] Suppose you have an instrument whose measuring needle oscillates, governed by the equation $f''(t) + k^2 f(t) = 0$. Suppose the instrument is modified by adding a damping force, so that the needle is governed instead by the equation

$$f''(t) + 2bf'(t) + k^2 f(t) = 0, \tag{7}$$

where $b > 0$. [See Example 5 above.]

(a) Let $f(0) = A$ and $f'(0) = 0$, and suppose f solves (7). Show that

$$f(t) = \begin{cases} \frac{1}{2}A[(1 + b/\sqrt{b^2 - k^2}) \exp(-bt + \sqrt{b^2 - k^2}\,t) \\ \quad + (1 - b/\sqrt{b^2 - k^2}) \exp(-bt - \sqrt{b^2 - k^2}\,t)] & (b > k). \\ A \exp(-bt)[\cos(\sqrt{k^2 - b^2}\,t) \\ \quad + \dfrac{b}{\sqrt{k^2 - b^2}} \sin(\sqrt{k^2 - b^2}\,t)] & (b < k). \end{cases}$$

(b) The rate at which $f(t) \to 0$ as $t \to \infty$ is governed by the number $-b + \sqrt{b^2 - k^2}$ for $b > k$, and by $-b$ in case $b < k$. Explain why this is so.

(c) The value $b = k$ is considered to give the best damping. Explain.

5. (a) Suppose that the roots r_1 and r_2 of $P(r)$ both have negative real part, $\mathrm{Re}(r_1) < 0$ and $\mathrm{Re}(r_2) < 0$. Prove that if f is any solution of $P(D)f = 0$, then $\lim_{t \to +\infty} f(t) = 0$.

(b) What limit is zero when r_1 and r_2 have positive real part?

6. The charge q at a given point in a simple electrical circuit [Fig. 7] is governed by the equation

ELECTRICAL CIRCUIT

$$Lq'' + Rq' + \frac{1}{C}q = 0,$$

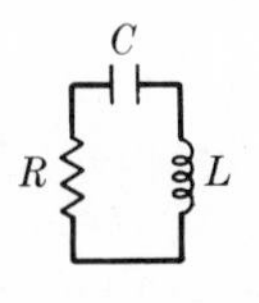

FIGURE 8.7

where the constants L, R, and C are respectively the *inductance*, *resistance*, and *capacitance* of the circuit.

• (a) Find the general solution.

(b) Show that there is an oscillatory solution (involving sines or cosines) if and only if $R^2C < 4L$. Show that the period of the oscillation is $2\pi/\omega_0$, where

$$\omega_0 = \sqrt{(1/LC) - (R^2/4L^2)}.$$

(c) A circuit is normally "tuned" by varying C. Is it possible to obtain an arbitrary preassigned period of oscillation $2\pi/\omega_0$ by varying C while keeping R and L fixed?

8.6 LINEAR DIFFERENTIAL EQUATIONS WITH CONSTANT COEFFICIENTS; THE GENERAL CASE

The general equation

$$P(D)f = g,$$

where $P(r) = \sum_{0}^{n} a_j r^j$ is a polynomial, can always be solved by n successive integrations, provided that P can be completely factored,

$$P(r) = a_n(r - r_1)\cdots(r - r_n).$$

Recall the first order, nonhomogeneous case [§6.8]

$$(D - r)f = g.$$

Multiplying each side by e^{-rt} leads to the following chain of equivalent equations:

$$e^{-rt}(D - r)f = e^{-rt}g$$

$$D(e^{-rt}f) = e^{-rt}g$$

$$e^{-rt}f(t) = \int e^{-rt}g(t)\,dt + c$$

$$f(t) = e^{rt}\int e^{-rt}g(t)\,dt + ce^{rt}.$$

Thus

$$(D - r)f = g \iff f(t) = e^{rt}\int e^{-rt}g(t)\,dt + ce^{rt}. \tag{1}$$

Informally, statement (1) says that we can remove the $(D - r)$ from the left-hand side of $(D - r)f = g$, if we replace the right-hand side g by an integral plus an exponential,

$$e^{rt}\int e^{-rt}g(t)\,dt + ce^{rt}.$$

Now consider the nth order equation

$$(D - r_1)(D - r_2)\cdots(D - r_n)f = g. \tag{2}$$

We can remove $(D - r_1)$ from the left-hand side if we replace the right-hand side by the appropriate integral plus exponential. Thus (2) is equivalent to

$$(D - r_2)\cdots(D - r_n)f = g_1 + c_1e^{r_1t}, \tag{3}$$

where $g_1(t) = e^{r_1t}\int e^{-r_1t}g(t)\,dt$. We can similarly remove $(D - r_2)$ from the left of (3), if we perform the appropriate operations on the right. Thus (3) is equivalent to

$$(D - r_3)\cdots(D - r_n)f = g_2 + c_2e^{r_2t},$$

where $g_2(t) = e^{r_2t}\int e^{-r_2t}[g_1(t) + c_1e^{r_1t}]\,dt$.

Repeating this process n times, we remove all the factors $(D - r_j)$ from f, and thus obtain our solution.

Example 1. Solve $f''(t) + f(t) = 3$. *Solution.* Here $P(D) = D^2 + 1$, $P(r) = r^2 + 1 = (r - i)(r + i)$. Thus $r_1 = i$, $r_2 = -i$, and we have

$$(D - i)(D + i)f = 3.$$

Therefore, removing $(D - i)$ from the left [applying (1) with $r = i$, $g = 3$, and with f replaced by $(D + i)f$], we get

$$(D + i)f(t) = e^{it}\int e^{-it}\cdot 3\,dt + c_1e^{it} = e^{it}\left(\frac{3}{-i}e^{-it}\right) + c_1e^{it} = 3i + c_1e^{it}.$$

To remove $(D + i)$ from the left, we now apply (1) with $r = -i$ and $g = 3i + c_1e^{it}$, obtaining

$$f(t) = e^{-it}\int e^{it}(3i + c_1e^{it})\,dt + c_2e^{-it}$$

$$= e^{-it}\left[3e^{it} + \frac{c_1}{2i}e^{2it}\right] + c_2e^{-it}$$

$$= 3 + \frac{c_1}{2i}e^{it} + c_2e^{-it} = 3 + C_1\cos t + C_2\sin t,$$

where $C_1 = c_1/2i + c_2$ and $C_2 = c_1/2 + ic_2$.

This is the *general* solution; any choice of the arbitrary constants C_1 and C_2 gives a *particular* solution, sometimes denoted f_p. For instance, choosing $C_1 = 0$ and $C_2 = 0$ gives the particular solution $f_p(t) = 3$ of the given equation $(D^2 + 1)f = 3$.

Note that the terms $C_1\cos t + C_2\sin t$ containing the arbitrary constants give the solution of the *homogeneous* equation $(D^2 + 1)f = 0$, obtained by changing the right-hand side of the given equation $(D^2 + 1)f = g$ to 0.

Example 2. Solve $f''(t) + 2f'(t) + f(t) = 3t$. *Solution.* We have $P(D) = D^2 + 2D + 1 = (D + 1)(D + 1)$. We get

$$(D + 1)(D + 1)f(t) = 3t,$$

so from (1), with $r = 1$, $g = 3t$, and f replaced by $(D + 1)f$, we get

$$(D + 1)f(t) = e^{-t}\int e^t3t\,dt + c_1e^{-t} = 3t - 3 + c_1e^{-t}.$$

Removing the remaining factor $(D+1)$ from the left, we get

$$f(t) = e^{-t}\int e^{t}(3t-3+c_1e^{-t})\,dt + c_2e^{-t}$$

$$= e^{-t}(3te^{t}-6e^{t}+c_1t)+c_2e^{-t}$$

$$= 3t-6+c_1te^{-t}+c_2e^{-t}.$$

You can check the calculations by substituting this result back into the equation $f''+2f'+f=3t$. Note again that the terms $c_1te^{-t}+c_2e^{-t}$ with arbitrary constants are precisely the solutions of the homogeneous equation $(D+1)^2f=0$.

The examples suggest some important patterns in the general solution of the equation $P(D)f = g$.

Theorem 3. *Let f_p be any particular solution of $P(D)f = g$,* i.e. *let*

$$P(D)f_p = g,$$

and let f_h be the general solution (with n arbitrary constants) of the corresponding homogeneous equation $P(D)f = 0$. Then the general solution of $P(D)f = g$ *has the form*

$$f = f_p + f_h\,.$$

In Example 1, the particular solution is $f_p(t) = 3$, the general solution of the homogeneous equation is $f_h(t) = C_1\cos t + C_2\sin t$, and the general solution of $(D^2+1)f = 3$ is the sum of these,

$$f(t) = f_p(t) + f_h(t) = 3 + C_1\cos t + C_2\sin t.$$

In Example 2, we have $f_p(t) = 3t-6$, $f_h(t) = c_1te^{-t}+c_2e^{-t}$, and the general solution is the sum

$$f(t) = f_p(t) + f_h(t) = 3t-6+c_1te^{-t}+c_2e^{-t}.$$

The virtue of Theorem 3 is that we can sometimes find a particular solution f_p rather easily, and then it remains only to add on the homogeneous solution f_h.

Example 3. Solve $f''-4f=5$. *Solution.* By inspection, it is clear that $f_p(t) = -5/4$ is a particular solution. From the previous section, the corresponding homogeneous equation $f''-4f=0$ has the general solution $f_h(t) = c_1e^{2t}+c_2e^{-2t}$. Therefore, by Theorem 3, the general solution of the given equation is

$$f(t) = -5/4 + c_1e^{2t}+c_2e^{-2t}.$$

Finding the particular solution f_p is partly a matter of luck. Sometimes it can be seen easily, as in Example 3. And there are various tricks when g has special forms, suggested in the problems below. At worst, it can always be found by successive integration, as in Examples 1 and 2. (The constants of integration $c_1, c_2, \ldots$ can all be dropped when we are looking only for a particular solution; they will reappear when the homogeneous solution f_h is added on.)

The homogeneous solution f_h, on the other hand, can always be written down directly, once the characteristic polynomial has been factored:

Theorem 4. *Let $P(r) = a_n(r - r_1)\cdots(r - r_n)$. Suppose that all the roots $r_1, \ldots, r_n$ are distinct. Then the general solution of $P(D)f = 0$ is*

$$f(t) = c_1e^{r_1t} + c_2e^{r_2t} + \cdots + c_ne^{r_nt}.$$

If any root r is repeated, then in the first repetition, e^{rt} is replaced by te^{rt}; in the second repetition by t^2e^{rt}; and so on.

GENERAL SOLUTION OF HOMOGENEOUS EQUATION $P(D)f = 0$

You should compare this with Theorem 2, which gives the same results in the special case $n = 2$.

Example 4. $D^4f = 0$. The characteristic polynomial is $r^4 = (r - 0)^4$, so 0 is a four-fold root. The general solution is a linear combination of e^{0t}, te^{0t}, t^2e^{0t}, and t^3e^{0t},

$$f(t) = c_1 + c_2t + c_3t^2 + c_4t^3.$$

[This is just a rederivation of a result from §3.4.]

Example 5. $f^{(4)} - 2f'' + f = 0$. *Solution.* The characteristic polynomial is $r^4 - 2r^2 + 1 = (r^2 - 1)^2 = (r - 1)^2(r + 1)^2$. Thus $r = 1$ is repeated once, and $r = -1$ is repeated once, so the general solution is a linear combination of e^t, te^t (for $r = 1$) and e^{-t}, te^{-t} (for $r = -1$):

$$f(t) = c_1e^t + c_2te^t + c_3e^{-t} + c_4te^{-t}.$$

Proofs of Theorems 3 and 4

Theorem 4 is proved by induction on the degree of P, using the same method as Theorem 2 and Examples 1 and 2. We will do the case where all the roots are distinct, and leave the rest to your imagination.

Suppose that Theorem 4 is proved for all polynomials of degree $n - 1$; that is, assume that any equation

$$(D - r_1)\cdots(D - r_{n-1})f = 0$$

in which the roots $r_1, \ldots, r_{n-1}$ are all distinct, has the solution

$$f(t) = c_1e^{r_1t} + \cdots + c_{n-1}e^{r_{n-1}t}.$$

We must then prove the corresponding formula for polynomials of degree n. Suppose that

$$(D - r_1)\cdots(D - r_{n-1})[(D - r_n)f] = 0.$$

Then $(D - r_n)f$ solves an equation of order $n - 1$, with $n - 1$ distinct roots, $r_1, \ldots, r_{n-1}$. Since we are supposing Theorem 4 to be true for such equations, we have

$$(D - r_n)f(t) = c_1e^{r_1t} + c_2e^{r_2t} + \cdots + c_{n-1}e^{r_{n-1}t}$$

for some constants $c_1, \ldots, c_{n-1}$. Therefore, by (1),

$$\begin{aligned} f(t) &= e^{r_nt}\int e^{-r_nt}(c_1e^{r_1t} + \cdots + c_{n-1}e^{r_{n-1}t})\,dt + c_ne^{r_nt} \\ &= e^{r_nt}\left[\frac{c_1}{r_1 - r_n}e^{(r_1-r_n)t} + \cdots + \frac{c_{n-1}}{r_{n-1} - r_n}e^{(r_{n-1}-r_n)t}\right] + c_ne^{r_nt} \\ &= c_1'e^{r_1t} + \cdots + c_{n-1}'e^{r_{n-1}t} + c_ne^{r_nt}, \end{aligned}$$

where $c_1' = \dfrac{c_1}{r_1 - r_n}, \ldots, c_{n-1}' = \dfrac{c_n}{r_{n-1} - r_n}$ are again constants. Thus f has the form given by Theorem 4, and the induction proof is complete.

The case where there are multiple roots is proved similarly; see the case $r_1 = r_2$ in the proof of Theorem 2, §8.5.

We will now return to Theorem 3, which we have not yet proved. The proof relies on the observation that $P(D)$ is *linear:*

$P(D)$ IS LINEAR

$$P(D)(c_1f_1 + c_2f_2) = c_1P(D)f_1 + c_2P(D)f_2. \tag{4}$$

This follows from the familiar properties of D, and of multiplication:

$$\begin{aligned} P(D)(c_1f_1 + c_2f_2) &= (a_nD^n + \cdots + a_0)(c_1f_1 + c_2f_2) \\ &= a_nD^n(c_1f_1 + c_2f_2) + \cdots + a_0(c_1f_1 + c_2f_2) \\ &= a_n(D^nc_1f_1 + D^nc_2f_2) + \cdots + a_0(c_1f_1 + c_2f_2) \\ &= c_1a_nD^nf_1 + c_2a_nD^nf_2 + \cdots + c_1a_0f_1 + c_2a_0f_2 \\ &= c_1a_nD^nf_1 + \cdots + c_1a_0f_1 + c_2a_nD^nf_2 + \cdots + c_2a_0f_2 \\ &= c_1P(D)f_1 + c_2P(D)f_2. \end{aligned}$$

Now we want to prove that the general solution of $P(D)f = g$ has the form $f = f_p + f_h$, where f_p is any particular solution and f_h is the general solution of $P(D) = 0$. We have

PROOF OF THEOREM 3

$$\begin{aligned} P(D)f = g &\Leftrightarrow P(D)f = P(D)f_p \qquad [\text{since } P(D)f_p = g] \\ &\Leftrightarrow P(D)f - P(D)f_p = 0 \\ &\Leftrightarrow P(D)(f - f_p) = 0 \qquad [\text{by (4)}] \\ &\Leftrightarrow f - f_p \text{ is a solution of the homogeneous equation.} \end{aligned}$$

Since f_h stands for the general solution of the homogeneous equation, we have $f - f_p = f_h$, or $f = f_p + f_h$, which was to be proved.

PROBLEMS

1. Find the general solution of the following homogeneous equations, using Theorem 4.

- • (a) $f''' - 27f = 0$
- (b) $f^{(4)} + 5f^{(3)} + 6f^{(2)} = 0$
- • (c) $f''' - 3if'' - 3f' + if = 0$
- (d) $f^{(4)} - 4f^{(2)} + 4f = 0$
- • (e) $f^{(3)} - 3f^{(2)} + 2f' = 0$
- (f) $2f^{(4)} = 32f$.

2. For each of the following equations, find a particular solution f_p and the general solution $f_p + f_h$.

- • (a) $f''' - 27f = 1$
- (b) $f^{(4)} + 5f^{(3)} + 6f^{(2)} = 3$
  [Hint: Try $f_p(t) = at^2 + bt + c$, and determine a, b, c by substituting f_p in the equation.]
- • (c) $f''' - 3if'' - 3f' + if = e^t$
  [Try $f_p(t) = ce^t$; determine c by substituting f_p in the equation.]
- (d) $f''' - 27f = e^{2t}$ [Try $f_p(t) = ce^{2t}$.]
- • (e) $f''' - 27f = 1 + e^{2t}$ [Combine parts (a) and (d).]
- (f) $f'' - f = e^{2it}$ [Try $f_p(t) = ce^{2it}$.]
- • (g) $f'' - f = \cos 2t$
  [Take the real part in the previous equation, or try $f_p(t) = a \cos 2t + b \sin 2t$.]

3. (a) For the equation in Problem 2(f), find the solution f such that $f(0) = 1, f'(0) = 0$.
• (b) Do the same for Problem 2(g).

4 • (a) For the equation in Problem 2(c), find the solution f such that $f(0) = 1, f'(0) = 0, f''(0) = 1$.
(b) Do the same for Problem 2(d).
(c) Do the same for Problem 2(e).

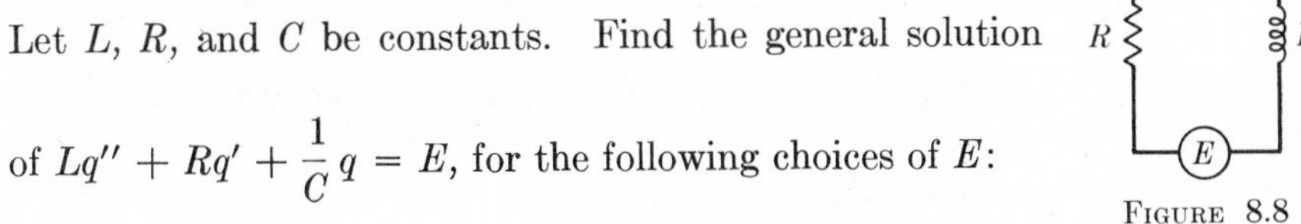

FIGURE 8.8

• **5.** Let L, R, and C be constants. Find the general solution of $Lq'' + Rq' + \frac{1}{C}q = E$, for the following choices of E:

(a) $E =$ constant.

(b) $E(t) = \cos \omega t$. [Assume that $R \neq 0$.]

CIRCUIT EQUATION

Remarks. The homogeneous equation is given in §8.5, Problem 6. The present problem concerns the same circuit as before, but with a voltage E applied to it [Fig. 8]. In part (b), there is a particular solution of the form $A \cos \omega t + B \sin \omega t$.

6. [*Continuation of Problem 5*] The homogeneous part of the solution in

TRANSIENT

Problem 5, i.e. the general solution q_h of $Lq'' + Rq' + \frac{1}{C}q = 0$, satisfies $\lim_{t\to+\infty} q_h = 0$. It is therefore called a *transient*. For very large times, the transient q_h dies out, and all that one sees is the particular solution

STEADY STATE

$$q_p(t) = \frac{R \sin \omega t + (1/C - L\omega^2) \cos \omega t}{[(1/C - L\omega^2)^2 + R^2]^2},$$

which is called the *steady state* solution.

(a) Show that the maximum of q_p is $[(1/C - L\omega^2)^2 + R^2\omega^2]^{-1/2}$. [The maximum of $A \cos \omega t + B \sin \omega t$ is $\sqrt{A^2 + B^2}$.]

(b) The steady state current is the derivative of q_p; show that its maximum is $[R^2 + (1/C\omega - L\omega)^2]^{-1/2}$. This is called the *amplitude* of the current. [The ratio, amplitude of driving voltage over amplitude of current, is called the *impedance* of the circuit at frequency ω, denoted Z. Since our driving voltage $E(t) = \cos \omega t$ has amplitude 1, the impedance is $Z = [R^2 + (1/C\omega - L\omega)^2]^{-1/2}$.]

IMPEDANCE

(c) The impedance Z varies with the frequency $\omega/2\pi$; find the frequency for which the impedance is minimum. [This is clearly $\frac{\omega}{2\pi} = \frac{1}{2\pi\sqrt{LC}}$; it is called the *frequency of resonance*. The circuit responds more strongly to this frequency than to any other; it is the frequency to which the circuit is tuned.]

RESONANCE

7. Suppose all the roots $r_1, \ldots, r_n$ of $P(r)$ have *positive* real part, $\text{Re}(r_j) > 0$. Prove that $f = 0$ is the only solution of $P(D)f = 0$ which is bounded for $t > 0$.

8. Suppose that a and b are functions, and $a' \neq 0$. Does

$$(aD + b)(aD + b)f = (a^2D^2 + 2abD + b^2)f$$

for every twice differentiable function f?

9. This problem concerns ways to find particular solutions.

PARTICULAR SOLUTIONS

(a) Suppose that $P(r_0) \neq 0$. Show that a particular solution of $P(D)f(t) = e^{r_0 t}$ is given by $f_p(t) = e^{r_0 t}/P(r_0)$. [The remaining parts of this problem concern the case where $P(r_0) = 0$.]

(b) Show that for any function g, $D(tg) = tDg + g$, $D^2(tg) = tD^2g + 2Dg, \ldots, D^n(tg) = tD^ng + nD^{n-1}g$.

(c) From part (b), show that $P(D)(tg) = tP(D)g + P'(D)g$.
(d) Let r_0 be a simple root of P, i.e. $P(r_0) = 0$, but $P'(r_0) \neq 0$. Show that $P(D)f = e^{r_0 t}$ has a particular solution $f_p(t) = te^{r_0 t}/P'(r_0)$. [Use part (c).]
(e) From part (c), show that $P(D)t^m g = t^m P(D)g + mt^{m-1}P'(D)g + \frac{m(m-1)}{2} t^{m-2}P''(D)g + \cdots + P^{(m)}(D)g$.
(f) Let r_0 be an m-fold root of P, i.e. $0 = P(r_0) = P'(r_0) = \cdots = P^{(m-1)}(r_0)$, but $P^{(m)}(r_0) \neq 0$. Show that $P(D)f(t) = e^{r_0 t}$ has a particular solution $f_p(t) = t^m e^{r_0 t}/P^{(m)}(r_0)$.
(g) Suppose that $P(r_0) \neq 0$. Show that $P(D)f(t) = te^{r_0 t}$ has a solution of the form $f_p(t) = Ate^{r_0 t} + Be^{r_0 t}$, where A and B are constants.
(h) Discuss the case where r_0 is a root of P, and solve $P(D)f(t) = t^k e^{r_0 t}$ for any k.

10. Let

CAUCHY'S EQUATION

$$Lf(x) = a_n x^n f^{(n)}(x) + \cdots + a_1 x f'(x) + a_0 f(x),$$

in which $a_0, a_1, \ldots, a_n$ are constants. The equation $Lf = g$ is called a *Cauchy equation*, or an *equidimensional equation*.
(a) Show that $Lx^r = Q(r)x^r$ for some polynomial Q.
(b) Show that if $r_1, \ldots, r_s$ are roots of $Q(r)$, then

$$L(c_1 x^{r_1} + \cdots + c_s x^{r_s}) = 0.$$

(c) Let $F(t) = f(e^t)$. Show that the following equations hold when $x = e^t$:

$$\begin{aligned} F(t) &= f(x), \\ F'(t) &= xf'(x), \\ F''(t) &= x^2 f''(x) + xf'(x), \\ F'''(t) &= x^3 f'''(x) + 3x^2 f''(x) + xf'(x). \end{aligned}$$

Find $F^{(4)}(t)$.
(d) Solve the equations in part (c) for $f(x)$, $xf'(x)$, $x^2 f''(x)$, $x^3 f'''(x)$.
(e) Let F and f be related as in parts (c) and (d). Show that $Lf(x) = P(D)F(t)$ for a polynomial P.
(f) Let the polynomial P in part (e) have the factorization $P(r) = c(r - r_1)^{m_1} \cdots (r - r_s)^{m_s}$. Show that when $x > 0$, every solution of $Lf(x) = 0$ is a linear combination of

$$x^{r_1}, x^{r_1}\log x, \ldots, x^{r_1}(\log x)^{m_1 - 1}, \ldots, x^{r_s}, \ldots, x^{r_s}(\log x)^{m_s - 1}.$$

(g) Show that the polynomial P in part (e) is the same as the polynomial Q in part (a). [Hint: In the equation $Lf(x) = P(D)F(t)$, set $f(x) = x^r$ so that $F(t) = e^{rt}$.]

8.7 THE FUNDAMENTAL THEOREM OF ALGEBRA

We bring this chapter on complex numbers to a fitting close by proving:

☆☆☆ **The Fundamental Theorem of Algebra.** *Let P be a polynomial of degree $n > 0$. Then there is a complex number r such that $P(r) = 0$.*

Before giving the proof, we should point out why this theorem is fundamental. Recall that we first introduced complex numbers to provide solutions of quadratic equations such as $z^2 + 1 = 0$, which have no real solutions. The formula $z = (-b \pm \sqrt{b^2 - 4ac})/2a$ for the solutions of $az^2 + bz + c = 0$ shows that every polynomial of degree 2 can be solved by extracting square roots, if complex numbers are allowed. The main contribution of the Italian mathematicians of the fifteenth and sixteenth centuries was to show that equations of degree 3 or 4 could similarly be solved by extracting square and cube roots of complex numbers. It was natural to hope that greater diligence and/or ingenuity would produce similar solutions for equations of degree 5 or more, but nearly three centuries later this was finally proved to be impossible. Neils Abel showed (1824) that there are equations of degree 5 which cannot be solved by extracting square roots, cube roots, fifth roots, or any other roots. We have to accept the unpleasant fact that there is no general way to write down the solutions of polynomial equations of degree 5 or more.

This raises a more serious question: Can we be sure that such equations have any solutions at all? This is precisely the question answered by the Fundamental Theorem of Algebra: Yes, every polynomial equation $P(z) = \sum_0^n a_j z^j = 0$ has at least one solution among the complex numbers. This had been conjectured early in the seventeenth century, and was finally proved by Gauss [three times!].

Once we have proved the hard fact that every polynomial equation has at least one root, it is easy to go further and prove that it can be factored completely. Precisely, the Fundamental Theorem has this consequence:

POLYNOMIALS FACTOR COMPLETELY, IF COMPLEX ROOTS ARE ALLOWED

Corollary 1. *Let P be a polynomial of degree $n > 0$, $P(z) = \sum_0^n a_j z^j$, $a_n \neq 0$, a_j any complex numbers. Then there are complex numbers $r_1, \ldots, r_n$ (not necessarily distinct) such that*

$$P(z) = a_n(z - r_1)\cdots(z - r_n).$$

You can see the significance of the corollary in connection with integration by partial fractions, and the solution of linear differential equations with constant coefficients, both of which depend on factoring a given polynomial.

We prove the corollary by induction on the degree n of the polynomial P. When $n = 1$, we have $P(z) = a_1 z + a_0$, where $a_1 \neq 0$, so P has the root $r_1 = -a_0/a_1$, and can be written $P(z) = a_1(z - r_1)$, as required.

Next, suppose that every polynomial $Q(z) = \sum_0^{n-1} b_j z^j$ of degree $n - 1$ can be factored as $Q(z) = b_{n-1}(z - r_1) \cdots (z - r_{n-1})$, and let $P(z) = \sum_0^n a_j z^j$ be any polynomial of degree n. By the Fundamental Theorem of Algebra, there is a complex number r_n such that $P(r_n) = 0$. It follows from the factor theorem [Problem 1 below] that

$$P(z) = (z - r_n)Q(z) \tag{1}$$

for some polynomial $Q(z) = \sum_0^{n-1} b_j z^j$. Notice that

$$\begin{aligned} a_n z^n + \cdots + a_0 &= (z - r_n)(b_{n-1}z^{n-1} + \cdots + b_0) \\ &= b_{n-1}z^n + (b_{n-2} - r_n b_{n-1})z^{n-1} + \cdots. \end{aligned}$$

Equating the coefficients of z^n, we find $b_{n-1} = a_n \neq 0$. Hence, by the induction assumption, $Q(z) = b_{n-1}(z - r_1) \cdots (z - r_{n-1})$ for some complex numbers $r_1, \ldots, r_{n-1}$; hence $P(z) = b_{n-1}(z - r_1) \cdots (z - r_n)$ from (1): Since $b_{n-1} = a_n$, the corollary is proved.

The Fundamental Theorem itself does not have such an algebraic proof. We derive it from the following lemmas.

Lemma 1. *Let P be a polynomial. Then $|P(z)|^2$ assumes a minimum value; in other words, there is a number z_0 such that $|P(z_0)|^2 \leq |P(z)|^2$ for all z.*

Lemma 2. *If the degree of P is >0 and $P(0) \neq 0$, then 0 is not a minimum point of $|P(z)|^2$,* i.e. *for some z, $|P(z)|^2 < |P(0)|^2$.*

Lemma 3. *If the degree of P is >0 and $P(z_0) \neq 0$, then z_0 is not a minimum point of $|P(z)|^2$.*

The Fundamental Theorem of Algebra follows immediately. From Lemma 1, $|P|^2$ assumes a minimum at some point z_0; from Lemma 3, $P(z_0)$ must equal 0.

Lemma 1 recalls the proposition that a continuous function on a closed bounded interval assumes a minimum value. The proof is similar, and we postpone it to §10.5.

In Lemma 2 we pursue a simple geometric idea. Consider a fixed, real value of θ. Then $f(t) = P(te^{i\theta})$ is a complex function of the real variable t, which can be imagined to trace a path in the plane starting at $P(0) \neq 0$ [Fig. 9]. The idea is to choose θ in such a way that the path leads closer to the origin, and thus to values of $|P(z)|^2$ that are less than $|P(0)|^2$.

To carry out this idea, set

$$P(z) = a_0 + a_k z^k + \cdots + a_n z^n,$$

where $a_0 = P(0) \neq 0$ and k is taken so that a_k is the first nonzero coefficient after a_0. [For example, in the polynomial $1 + 2z - z^2$, $k = 1$ and $a_k = 2$; but in $1 - z^2$, $k = 2$ and $a_k = -1$.] Then, since $|P|^2 = P\bar{P}$, we have

$$\begin{aligned} |P(te^{i\theta})|^2 &= (a_0 + a_k t^k e^{ik\theta} + \cdots)(\bar{a}_0 + \bar{a}_k t^k e^{-ik\theta} + \cdots) \\ &= a_0\bar{a}_0 + t^k(a_0\bar{a}_k e^{-ik\theta} + \bar{a}_0 a_k e^{ik\theta}) + t^{k+1}(\cdots) + \cdots. \end{aligned} \tag{2}$$

Now choose θ so that

$$\arg(-a_0\bar{a}_k) = k\theta, \tag{3}$$

and let $b = -a_0\bar{a}_k e^{-ik\theta}$. Since $a_k \neq 0$ and $a_0 \neq 0$, it follows that $b \neq 0$; and we find from (3) that $\arg(b) = \arg(-a_0\bar{a}_k) + \arg(e^{-ik\theta}) = 0$, so $b > 0$. From this it follows that $\bar{b} = b$, and hence

$$\begin{aligned} |P(te^{i\theta})|^2 &= |a_0|^2 + t^k(-b - \bar{b}) + t^{k+1}(\cdots) + \cdots \\ &= |a_0|^2 - 2bt^k + t^{k+1}(\cdots) + \cdots \qquad [b > 0]. \end{aligned}$$

Hence

$$t^{-k}(|P(te^{i\theta})|^2 - |a_0|^2) = -2b + t(\cdots) + \cdots \tag{4}$$

has the limit $-2b < 0$ as $t \to 0$. It follows that the expression (4) is *negative* for some small positive value of t. For this value of t, we have $t^{-k} > 0$, so $|P(te^{i\theta})|^2 - |a_0|^2 < 0$, in other words, $|P(te^{i\theta})|^2 < |a_0|^2$, and we have proved that $|a_0|^2 = |P(0)|^2$ is not a minimum of $|P(z)|^2$.

Lemma 3 follows immediately from Lemma 2 by writing

$$P(z_0 + z) = Q(z)$$

and applying Lemma 2 to Q.

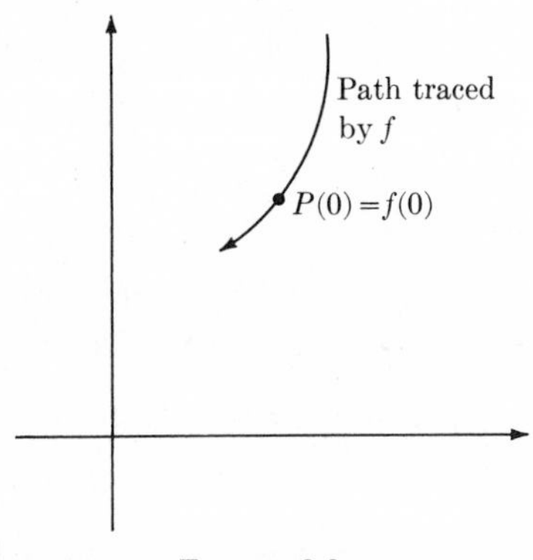

FIGURE 8.9

Equations with real coefficients

When the given polynomial P has real coefficients, it is a little unnatural to factor into complex factors; our discussion of partial fractions required real factors. The Fundamental Theorem of Algebra has the following corollary to cover this case.

FACTORING WITHOUT COMPLEX NUMBERS

Corollary 2. *If* $P(x) = \sum_0^n a_j x^j$ *has real coefficients* $a_0, \ldots, a_n$, *then* P *can be factored completely as a product of factors of the type* $x - r$ *and* $x^2 + px + q$, *where* r, p, *and* q *are real, and* $x^2 + px + q = 0$ *has no real roots.*

This corollary follows from the fact that when P is real, then any complex roots occur in conjugate pairs; if r is a root, so is $\bar{r}$. This fact is easy to see in the quadratic case, where the roots are

$$r_1 = \frac{-b + \sqrt{b^2 - 4ac}}{2a} \quad \text{and} \quad r_2 = \frac{-b - \sqrt{b^2 - 4ac}}{2a}.$$

When $b^2 - 4ac \geq 0$, then r_1 and r_2 are real; and when $b^2 - 4ac < 0$, then

$$r_2 = \frac{-b - i\sqrt{4ac - b^2}}{2a} = \bar{r}_1.$$

In the general case, let r be a root; $P(r) = 0$. Then

$$\begin{aligned} P(\bar{r}) &= a_n(\bar{r})^n + a_{n-1}(\bar{r})^{n-1} + \cdots + a_0 \\ &= a_n(\overline{r^n}) + a_{n-1}(\overline{r^{n-1}}) + \cdots + a_0 \\ &= \overline{a_n}(\overline{r^n}) + \overline{a_{n-1}}(\overline{r^{n-1}}) + \cdots + \overline{a_0} \qquad [\text{since } a_0, \ldots, a_n \text{ are real}] \\ &= \overline{P(r)} = 0. \end{aligned}$$

Thus $\bar{r}$ is a root whenever r is. It follows that in the factorization $P(r) = a_n(r - r_1) \cdots (r - r_n)$, any complex root r_j is matched by an $\overline{r_j}$; the product

$$\begin{aligned} (r - r_j)(r - \overline{r_j}) &= r^2 - (r_j + \overline{r_j})r + r_j\overline{r_j} \\ &= r^2 - 2\,\mathrm{Re}\,(r_j) + |r_j|^2 \end{aligned}$$

is a quadratic with real coefficients and no real roots. When all pairs of conjugate factors have been combined like this, we are left with a factorization of the type given in Corollary 2.

PROBLEMS

FACTOR THEOREM

1. In this problem you prove the Factor Theorem: *If P is a polynomial of degree n and $P(r) = 0$, then $z - r$ is a factor of P;* i.e. *there is a polynomial Q such that $P(z) = (z - r)Q(z)$.*

(a) Show that $P(w + r) = \sum_0^n b_j w^j$ for some complex numbers $b_0, \ldots, b_n$, (which may depend on r).

(b) Show that $b_0 = 0$. [Remember that $P(r) = 0$.]

(c) Conclude that $P(w + r) = \sum_{j=1}^{n} b_j w^j = w \sum_{k=0}^{n-1} b_{k+1} w^k$.

(d) Conclude that $P(z) = (z - r)Q(z)$ for some polynomial Q.

CARDANO'S FORMULA FOR A CUBIC

2. The complicated formula found by the Italians for solving $x^3 + 3px + 2q = 0$ is

$$x = w - \frac{p}{w}, \qquad \text{where } w = (-q \pm \sqrt{q^2 + p^3})^{1/3}$$

["Cardano's formula"]. Use this to solve the following.

(a) $x^3 - 15x - 4 = 0$

(b) $x^3 - 2x - 4 = 0$

3. The general cubic $r^3 + br^2 + cr + d = 0$ can be reduced to the form $x^3 + 3px + 2q$ in Problem 2 by "completing the cube," i.e. by setting $r = x - b/3$. Use this to solve:

(a) $x^3 - x^2 - x - 2 = 0$

(b) $x^3 - x^2 - 4x - 6 = 0$.

Chapter IX

Approximations

The preceding five chapters dealt mainly with equalities of one sort or another; for instance, some area $A = \pi$, or some slope $m = \sqrt{10}$. Such equalities provide the satisfying sense of perfection and completeness that is expected of mathematics.

This chapter, in contrast, deals with *approximations*. Instead of showing $a = b$, we show that a equals b *with an error less than* ϵ.* For example, $\pi = 3.14$ *with an error less than* .005, or $3.135 < \pi < 3.145$. Instead of a single equality $a = b$, we accept *two inequalities:*

$$b - \epsilon < a < b + \epsilon \qquad \text{or equivalently} \qquad |b - a| < \epsilon$$

[sometimes written $a = b \pm \epsilon$, e.g. $\pi = 3.14 \pm .005$]. If the error ϵ is small, then the approximation is a good one.

---

* ϵ is the Greek letter epsilon, corresponding to e, for *error*.

Why consider approximations, when equalities are so much nicer? There are many reasons.

(i) Approximations are available where equalities are not. For example, the integral $\int_0^1 e^{-x^2}\,dx$ is not equal to any familiar number that we can write down, but we can write down very good approximations to this integral by rational numbers.

(ii) To write the equalities $A = \pi$ and $m = \sqrt{10}$ does not tell us everything about A and m unless we have good *approximations* to $\sqrt{10}$ and to π by rational numbers. For example, it is not obvious at a glance whether $\pi > \sqrt{10}$, or $\pi < \sqrt{10}$. But from the decimal approximations

$$\pi = 3.14 \pm 0.005,$$

$$\sqrt{10} = 3.16 \pm 0.005,$$

it is clear that $\pi < \sqrt{10}$.

(iii) Approximations arise in the application of numerical results. A machinist, say, wants his measurements expressed to the nearest .001 cm, and a carpenter wants his measurements to the nearest 1/16, 1/32, or 1/64 of an inch. The machinist thus uses *decimal* approximations $a_n 10^n + a_{n-1}10^{n-1} + \cdots + a_{-m}10^{-m}$, in which each a_j is an integer from 0 through 9, while the carpenter uses *binary* approximations $b_n 2^n + b_{n-1}2^{n-1} + \cdots + b_{-m}2^{-m}$, in which each b_j is either 0 or 1. And digital computers [like carpenters] use binary expansions, accurate up to 2^{-100} or better.

(iv) Approximations enter into the proofs of equalities, including some of those that we have derived. For example, to say that $\lim_{x \to x_0} \dfrac{f(x) - f(x_0)}{x - x_0} = f'(x_0)$ amounts to saying that $\dfrac{f(x) - f(x_0)}{x - x_0}$ is a good approximation to $f'(x_0)$ when x is near x_0. Thus approximations enter prominently in the analytic proofs that have been postponed to Appendix II.

(v) Subtle mathematical questions are frequently settled by appropriate approximations. For example, the question whether the base e of the natural logarithms is rational or not is settled by producing good approximations to e by rational numbers. [See Problem 16, §9.2.]

The approximations most familiar to you are probably those provided by tables or slide rules. These have a serious shortcoming: their accuracy goes so far and no further; a 5-place table of logarithms is useless if 8-place accuracy is required. While the methods in this chapter are not so simple as a table or a slide rule, they have the virtue that whatever accuracy is desired can be obtained *by carrying the methods sufficiently far.* Although the approximations themselves lack the finality of an equality, the methods of obtaining them do not.

Chapter X and Appendix II show how this notion of approximations with arbitrarily great accuracy provides the key to understanding—even more—to creating the underlying theory of calculus. This is a major reason why we have postponed the theoretical discussion of calculus for so long. Sections 1 and 2 below provide enough background for this theoretical discussion. Sections 3 and 4 give further methods of approximation for the purposes suggested in (i), (ii), and (iii) above.

9.1 APPROXIMATION BY THE TANGENT LINE

In Chapter I, we approximated the derivative $f'(a)$ by the slope of a line through $(a,f(a))$ and a nearby point $(b,f(b))$; when b is near a, $[f(b) - f(a)]/(b - a)$ is nearly equal to $f'(a)$ [Fig. 1]. Now we reverse the process: When $f(a)$ and $f'(a)$ are known, we solve the "equation"

$$\frac{f(b) - f(a)}{b - a} =_{\text{(approx)}} f'(a)$$

for $f(b)$, and take $f(a) + (b - a)f'(a)$ as an approximation to $f(b)$ [Fig. 2],

$$f(b) =_{\text{(approx)}} f(a) + (b - a)f'(a).$$

TANGENT LINE APPROXIMATION

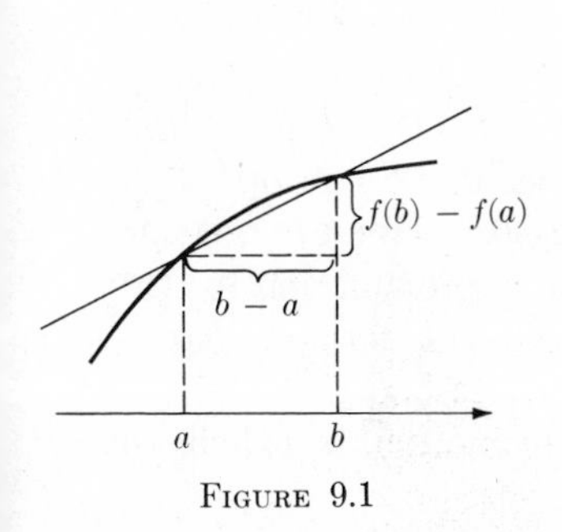

FIGURE 9.1

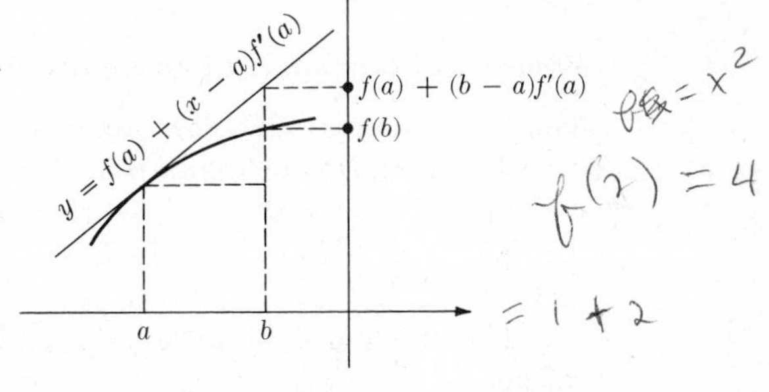

FIGURE 9.2

Example 1. We approximate $(2.1)^3$ by setting $f(x) = x^3$, $a = 2$, $b = 2.1$. Then $b - a = 0.1$, $f'(a) = 12$, and so the tangent line approximation gives

$$(2.1)^3 =_{\text{(approx)}} f(a) + (b - a)f'(a) = 2^3 + (0.1)\cdot 12 = 9.2.$$

Example 2. To estimate $(1.98)^3$, we take f and a as before, and $b = 1.98$. Then $f(a) = 8$, $b - a = -0.02$, $f'(a) = 12$, so we obtain

$$(1.98)^3 =_{\text{(approx)}} 8 + (-0.02)(12) = 7.76.$$

Of course, you could compute $(1.98)^3$ directly (but laboriously) to find $(1.98)^3 = 7.762392$. But if we only need this result to two decimal places, the tangent line approximation is adequate, and much easier.

Example 3. Take $f(x) = \sin x$ and $a = 0$. Then $f(a) = \sin 0 = 0$, and $f'(a) = \cos(0) = 1$. Hence $\sin b =_{\text{(approx)}} 0 + (b - 0)\cdot 1 = b$. In other words, when b is small, $\sin b$ is approximately equal to b. [Compare TABLE 1.]

Example 4. Let $f(x) = \tan x$. Show that when b is small, $\tan b$ is approximately equal to b. [Compare TABLE 1. Solution below.]

TABLE 1

b	$\sin b$	$\tan b$
.00	.00000	.00000
.01	.01000	.01000
.02	.02000	.02000
.03	.03000	.03001
.04	.03999	.04002
.05	.04998	.05004
.06	.05996	.06007
.07	.06994	.07011
.08	.07991	.08017
.09	.08988	.09024
.10	.09983	.10033

Accuracy of the tangent line approximation

You may be grumbling that our first two examples don't live up to the expectations raised in the introduction to this chapter. We are not supposed to say simply that $f(a) + (b - a)f'(a)$ is an approximation to $f(b)$; we should say $f(b) = f(a) + (b - a)f'(a)$ *with an error no more than* ϵ, and presumably we should give you an effective way to find ϵ.

The difference between $f(b)$ and the approximation is called the remainder R,

$$R = f(b) - [f(a) + (b - a)f'(a)].$$

This can be estimated by the Mean Value Theorem [§4.2]:

$$R = [f(b) - f(a)] - (b - a)f'(a) = (b - a)f'(c_1) - (b - a)f'(a)$$
$$= (b - a)[f'(c_1) - f'(a)], \qquad \text{with } c_1 \text{ between } a \text{ and } b.$$

Applying the Mean Value Theorem to f' on the interval $[a,c_1]$, we find that

$$R = (b - a)(c_1 - a)f''(c_2), \qquad \text{with } c_2 \text{ between } a \text{ and } c_1 .$$

We don't know c_1 and c_2 exactly, but in spite of this we can usually tell pretty well how large the remainder R is. Suppose $|f''(x)| \leq M$ for all x between a and b; then $|f''(c_2)| \leq M$. Further, since c_1 lies between a and b, we have $|c_1 - a| < |b - a|$, so

$$|R| = |b - a| \cdot |c_1 - a| \cdot |f''(c_2)| \leq |b - a|^2 M. \tag{1}$$

Example 5. Taking $a = 0$ and $f(x) = \tan x$, you obtain [Example 4] the approximation $\tan b =$ (approx) b. Now we investigate the accuracy of this for b in the range $0 \leq b \leq \pi/6$. We have $f''(x) = 2 \sin x/\cos^3 x$, and for the given values of x, we have $\sin x \leq x$ and $\cos^3 x \geq \left(\cos \frac{\pi}{6}\right)^3 = (\sqrt{3}/2)^3 > \frac{1}{2}$, so

$$|f''(x)| \leq 2x \leq 2b, \qquad \text{for } 0 \leq x \leq b.$$

Thus in (1) we can take $M = 2b$, and find

$$|R| \leq |b - 0|^2 2b = 2b^3.$$

For instance, when $b = .03$, the difference between $\tan b$ and b is less than $2(.03)^3 = .000054$. [TABLE 1 above gives a difference of .00001.] When $b = .01$, the difference between $\tan b$ and b is less than $2(.01)^3 = .000002$. [TABLE 1 does not show any difference at all, since it is only a five-place table.]

Remark. If f' were constant, then the graph of f would be a straight line, and the tangent line "approximation" would be perfect. In general, however, f' changes, and its rate of change is given by f''; in this sense, f'' measures how far f deviates from its tangent line, and it is reasonable that the accuracy of the approximation is controlled by $M = \max f''$.

Solution to Example 4. Take $a = 0$, so $f(a) = f(0) = \tan(0) = 0$, $f'(a) = \sec^2 a = \sec^2 0 = 1$, hence

$$\tan b = \text{(approx) } f(0) + (b - 0)f'(0) = 0 + b = b.$$

PROBLEMS

Note: Review Problem 13, §3.4, in preparation for the polynomial approximations in the next section.

1. Find tangent line approximations to
- •(a) $(1.02)^8$
- (b) $(1 + \frac{1}{10})^{10}$
- •(c) $\sin(5\pi/24)$ [Use the tangent line at $(\pi/6, 1/2)$.]

2. Find approximations to
- (a) $(1.01)^{10} - (1.01)^5 - 5$
- •(b) $(0.99)^{100} + (0.99)^{10}$
- (c) $\cos(21\pi/10)$

3. A pendulum of length l and period T determines the constant of gravity γ by the equation $\gamma = 4\pi^2 l/T^2$ [§7.10, Example 2].
- •(a) Assuming T is known exactly, find the exact error in γ due to an error h in measuring l.
- (b) Assuming l is known exactly, find the approximate error in γ due to an error k in measuring T.

4. Notice in TABLE 1 that $\sin b$ tends to be slightly less than b, while $\tan b$ tends to be slightly more. Relate this to the second derivatives $\sin''$ and $\tan''$, noting that $\sin''(0) < 0$ and $\tan''(0) > 0$.

5. Use the inequality (1) in the text to estimate the error in approximating $(1.98)^3$ by $2^3 - (0.02) \cdot 12$, as in Example 2.

6 •(a) Show that the error in approximating $\sin b$ by b is less than b^3.
(b) Estimate this error when $b = 0.01$.

9.2 THE TAYLOR EXPANSION

The function

$$f(a) + (x - a)f'(a) \tag{1}$$

used in the tangent line approximation [Fig. 9.2] is characterized by three properties:

it is a polynomial of degree 1

it agrees with f at the point $x = a$

its first derivative agrees with f' at the point $x = a$.

We call this function the *first degree Taylor polynomial* of f at a, and denote it $P_{1,f,a}$,

$$P_{1,f,a}(x) = f(a) + (x - a)f'(a).$$

One might naturally expect better approximations from polynomials of higher degree. The second degree Taylor polynomial $P_{2,f,a}$ should agree with f at a, and its first *and second* derivatives should agree as well:

$$P_{2,f,a}(a) = f(a), \qquad (P_{2,f,a})'(a) = f'(a), \qquad (P_{2,f,a})''(a) = f''(a). \quad (2)$$

It is easy to check [§3.4] that there is just one second degree polynomial satisfying these conditions, namely

$$P_{2,f,a}(x) = f(a) + (x - a)f'(a) + \tfrac{1}{2}(x - a)^2 f''(a). \quad (3)$$

Example 1. Show that the three equations (2) are satisfied by the function $P_{2,f,a}$ defined above. [Solution below.]

More generally, *the* nth *degree Taylor polynomial of* f *at* a is defined by

TAYLOR POLYNOMIAL OF f AT THE POINT a

$$\begin{aligned} P_{n,f,a}(x) &= f(a) + (x - a)f'(a) + \tfrac{1}{2}(x - a)^2 f''(a) + \cdots \\ &\qquad + \frac{1}{n!}(x - a)^n f^{(n)}(a) \\ &= \sum_{j=0}^{n} \frac{f^{(j)}(a)}{j!}(x - a)^j. \end{aligned} \quad (4)$$

It agrees with f and all its derivatives up to order n, at $x = a$;

$$P_{n,f,a}(a) = f(a),\ (P_{n,f,a})'(a) = f'(a), \ldots, (P_{n,f,a})^{(n)}(a) = f^{(n)}(a). \quad (5)$$

[We define $f^{(0)} = f$, and $0! = 1$, just so that we can write convenient formulas like (4). In this formula we also define $(x - a)^0 = 1$ even when $x - a = 0$, although in most other contexts 0^0 *is not defined*, any more than $0/0$ is.]

Example 2. Take $a = 0$ and $f(x) = \sin x$. The successive derivatives are:

$f(0)$	$f'(0)$	$f''(0)$	$f'''(0)$	$f^{(4)}(0)$	$f^{(5)}(0)$	...
sin (0)	cos (0)	−sin (0)	−cos (0)	sin (0)	cos (0)	...
0	1	0	−1	0	1	...

You can recognize the pattern 0, 1, 0, −1, 0, 1, 0, −1, ..., and so the Taylor polynomial is

$$P_{n,f,a}(x) = 0 + x + 0 - \frac{x^3}{3!} + 0 + \frac{x^5}{5!} + \cdots + \begin{cases} 0 & \text{if } n \text{ is even} \\ x^n/n! & \text{if } n = 4k + 1 \\ -x^n/n! & \text{if } n = 4k + 3. \end{cases}$$

If the Taylor polynomial $P_{n,f,a}$ is to tell us anything about the given function f, we need a handy expression for the difference $f - P_{n,f,a}$. In the simplest case ($n = 0$) we already have it. The zero degree Taylor polynomial is $P_{0,f,a}(x) = f(a)$, and the difference in this case is

$$f(x) - P_{0,f,a}(x) = f(x) - f(a) = (x - a)f'(c) \quad (6)$$

by the Mean Value Theorem. Here, c is some point lying between x and a; we generally don't know exactly which point, but the formula is valuable in spite of this indeterminacy, as we saw in §4.2 and §9.1.

To extend formula (6) from the case $n = 0$, rewrite it as

$$f(x) = f(a) + (x - a)f'(c),$$

and notice that the right-hand side is precisely the *first* degree Taylor polynomial (1), except that in the last term $f'(a)$ is replaced by $f'(c)$. The corresponding formula resembling the second degree polynomial (3) is

$$f(x) = f(a) + (x - a)f'(a) + \tfrac{1}{2}(x - a^2)f''(c),$$

and generally, for any n,

TAYLOR'S EXPANSION –LAGRANGE REMAINDER

$$f(x) = f(a) + (x - a)f'(a) + \cdots + \frac{(x-a)^n}{n!} f^{(n)}(a) + \frac{(x-a)^{n+1}}{(n+1)!} f^{(n)}(c),$$

$$= P_{n,f,a}(x) + \frac{(x-a)^{n+1}}{(n+1)!} f^{(n)}(c) \qquad (7)$$

for some c between a and x. This remarkable formula is called *Taylor's expansion,* with the *Lagrange form* of the remainder. The "remainder" is the last term, generally denoted R_{n+1},

$$R_{n+1} = f(x) - P_{n,f,a}(x) = \frac{(x-a)^{n+1}}{(n+1)!} f^{(n+1)}(c).$$

This is what remains if we subtract $P_{n,f,a}$ from f.

Although the particular formula (7) is due to Lagrange (1736–1813), all formulas involving the polynomials $P_{n,f,a}$ bear the name of the British mathematician Brook Taylor (1685–1731), who used the "complete" expansion $f(x) = f(a) + (x - a)f'(a) + \cdots = \sum_{j=0}^{\infty} \frac{(x-a)^j}{j!} f^{(j)}(a)$ with no remainder. [This is an *infinite series,* to be discussed in Chapter XI.] Formula (7) gives the Taylor expansion with Lagrange's form of the remainder, which is generally the easiest to remember and to apply. In the problems you will find both *Cauchy's form* of the remainder and the *integral form* of the remainder.

In the proof of (7), it is convenient to replace the letter x by b, and to assume that $b > a$; when we are done, it will be clear that the proof also works for $b < a$.

As for the hypotheses required to prove (7), recall [§4.2] that the Mean Value Theorem $f(b) - f(a) = (b - a)f'(c)$ is valid when f is continuous on $[a,b]$, and $f'(x)$ exists for $a < x < b$. This suggests:

Theorem 1. *Let $f, f', \ldots, f^{(n)}$ be continuous on $[a,b]$, and $f^{(n+1)}(x)$ exist for $a < x < b$. Then the remainder R_{n+1} has the form*

$$R_{n+1} = f(b) - P_{n,f,a}(b) = \frac{(b-a)^{n+1}}{(n+1)!} f^{(n+1)}(c), \qquad (8)$$

where c is some point between a and b.

Proof. Referring again to the Mean Value Theorem, the proof in §4.2 introduced the function

$$d(x) = f(x) - f(a) - (x - a)\frac{f(b) - f(a)}{b - a}$$

which can be rewritten

$$d(x) = f(x) - P_{0,f,a}(x) - \frac{x - a}{b - a}[f(b) - P_{0,f,a}(b)].$$

In the more general situation at hand, we set

$$d(x) = f(x) - P(x) - \left(\frac{x - a}{b - a}\right)^{n+1}[f(b) - P(b)]; \tag{9}$$

we have written P instead of $P_{n,f,a}$ for simplicity of notation. Notice that $(x - a)^{n+1}$ is zero at $x = a$, and so do all its derivatives of order $\leq n$, while the $(n + 1)$st derivative of $(x - a)^{n+1}$ is the constant $(n + 1)!$. Hence, differentiating both sides of (9) and noting (5), we see that

$$d^{(j)}(a) = f^{(j)}(a) - P^{(j)}(a) - 0 = 0 \quad \text{for } j = 0, 1, 2, \ldots, n; \tag{10}$$

and $P^{(n+1)}(x) = 0$ [since P is a polynomial of degree $\leq n$], so

$$d^{(n+1)}(x) = f^{(n+1)}(x) - 0 - \frac{(n + 1)!}{(b - a)^{n+1}}[f(b) - P(b)]. \tag{11}$$

Further, by direct substitution in (9), $d(b) = 0 = d(a)$. Hence, by the Mean Value Theorem, there is a point c_0 such that $a < c_0 < b$ and

$$d'(c_0) = \frac{d(b) - d(a)}{b - a} = \frac{0 - 0}{b - a} = 0.$$

Now apply the Mean Value Theorem again, this time to the function d' on the interval $[a,c_0]$, to find a point c_1 such that $a < c_1 < c_0$ and

$$d''(c_1) = \frac{d'(c_0) - d'(a)}{b - a} = \frac{0 - 0}{b - a} = 0.$$

Continuing this process $n + 1$ times, each time applying (10), we obtain points $c_0, c_1, \ldots, c_n$ such that $b > c_0 > \cdots > c_n > a$ and

$$d^{(n+1)}(c_n) = 0. \tag{12}$$

Now, c_n is the number c required in the statement of Theorem 1; for if we set $c_n = c$ in (12), use formula (11), and solve for $f(b) - P(b)$, we arrive at (8). Thus Theorem 1 is proved.

We have stated Theorem 1 only for $a < b$; it remains true when $b < a$, as you are asked to show in Problem 19 below.

Applications of the Taylor expansion

Example 3. Take $a = 0$ and $f(x) = \sin x$ in Taylor's formula. Then Taylor's expansion up to the term in $(b - a)^4$ is

$$\sin(b) = 0 + b - 0 - b^3/3! + 0 + R_5 .$$

[Compare this to the Taylor polynomial found in Example 2.] Suppose we are interested in values of b in the interval $0 \leq b \leq \pi/4$. Then we can estimate the remainder R_5 as follows.

$$|f^{(5)}(x)| = |\cos x| \leq 1, \qquad 0 \leq x \leq \frac{\pi}{4},$$

so using (6′), and noting that $\pi < 3.2$, we find $(\pi/4)^5 < 1/3$, so

$$|R_5| \leq \frac{b^5}{5!} \leq \left(\frac{\pi}{4}\right)^5 \cdot \frac{1}{5!} < \left(\frac{1}{3}\right)\left(\frac{1}{5!}\right) = \frac{1}{360} < .003, \qquad 0 \leq b \leq \frac{\pi}{4}.$$

Hence the simple polynomial approximation $b - b^3/3!$ is accurate enough to construct a table of $\sin b$ for $0 \leq b \leq \pi/4$, accurate to two decimals.

Example 4. Show that $b - b^3/3!$ approximates $\sin b$ with an error less than 0.0001 on the interval $0 \leq b \leq \pi/8$. [Solution below.]

When we take the Taylor expansion (7) with $a \neq 0$ and $n > 1$, it is convenient to replace $x - a$ by a single letter h. We also introduce $\theta = \dfrac{c - a}{x - a}$; since c lies between a and x, we have $0 < \theta < 1$. With these changes, the Taylor expansion becomes

$$f(a + h) = f(a) + hf'(a) + \cdots + \frac{h^n}{n!} f^{(n)}(a) + \frac{h^{n+1}}{(n + 1)!} f^{(n+1)}(a + \theta h)$$

where $0 < \theta < 1$.

Example 5. A table of sines can be constructed by using Taylor's expansion about each of the points $a = 0$, $a = \pi/4$, $a = \pi/2$. We use the expansion about $a = 0$ for the interval $0 \leq x \leq \pi/6$, about $a = \pi/4$ for $\pi/6 < x < \pi/3$, and about $a = \pi/2$ for $\pi/3 \leq x \leq \pi/2$. [See Fig. 3.] How many terms of the expansion about $\pi/4$ will provide accuracy to five decimals in the required interval $[\pi/6, \pi/3]$?

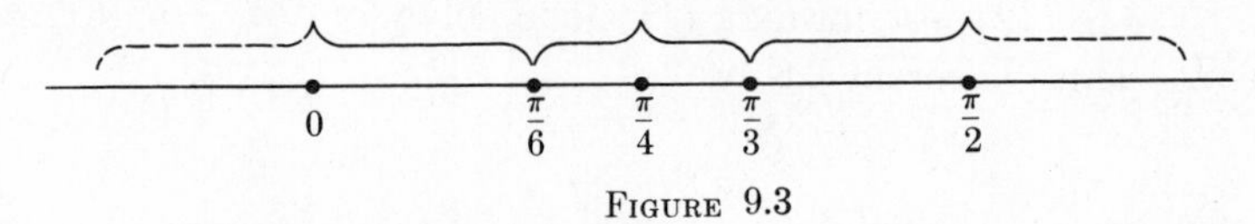

FIGURE 9.3

Solution. The derivatives at $a = \pi/4$ are

$$\sin\frac{\pi}{4} \qquad \cos\frac{\pi}{4} \qquad -\sin\frac{\pi}{4} \qquad -\cos\frac{\pi}{4} \qquad \ldots.$$

$$\frac{1}{\sqrt{2}} \qquad \frac{1}{\sqrt{2}} \qquad \frac{-1}{\sqrt{2}} \qquad \frac{-1}{\sqrt{2}} \qquad \ldots.$$

Thus

$$\sin\left(\frac{\pi}{4} + h\right) = \frac{1}{\sqrt{2}} + \frac{h}{\sqrt{2}} - \frac{h^2}{\sqrt{2}\,2!} - \frac{h^3}{\sqrt{2}\,3!} + \cdots + R_{n+1}$$

and $R_{n+1} = h^{n+1}\sin^{(n+1)}(c)/(n+1)!$, where $\sin^{(n+1)} = (n+1)$st derivative of sin. Since $\pi/6 < \pi/4 + h < \pi/3$, we have $|h| < \pi/12 < 0.3$ and $|\sin^{(n+1)}(c)| < 1$, so $|R_{n+1}| \leq (0.3)^{n+1}/(n+1)!$.

To see when $|R_{n+1}|$ will be small enough to provide accuracy to five decimals, we simply try various possibilities for n in the expression $(0.3)^{n+1}/(n+1)!$. We find that $(0.3)^6/6! < 2\cdot 10^{-6}$, so the expansion

$$\sin\left(\frac{\pi}{4} + h\right) = \frac{1}{\sqrt{2}}\left(1 + h - \frac{h^2}{2!} - \frac{h^3}{3!} + \frac{h^4}{4!} + \frac{h^5}{5!}\right) + R_6$$

gives five-place accuracy for $|h| < \pi/12$.

Remark. We now have two different estimates of the remainder R_2 in the tangent line approximation. Formula (8) gives

$$\begin{aligned}|f(b) - f(a) - (b - a)f'(a)| = |R_2| &= |\tfrac{1}{2}(b - a)^2 f''(c)| \\ &\leq \tfrac{1}{2}(b - a)^2 \max_{[a,b]} |f''|,\end{aligned}$$

while in §9.1 we found

$$|f(b) - f(a) - (b - a)f'(a)| \leq (b - a)^2 \max_{[a,b]} |f''|.$$

Thus formula (8) is better, and the less sophisticated estimate from §9.1 should be conveniently forgotten.

Solution to Example 1. $P_2(a) = f(a) + 0f'(a) + 0f''(a)$;

$P_2'(x) = 1\cdot f'(a) + (x - a)f''(a)$; $P_2'(a) = f'(a)$;

$P_2''(x) = f''(a)$ for all x.

Solution to Example 4. As in Example 2, $|f^{(5)}(x)| = |\cos x| \leq \cos \pi/8 < 1$, so

$$|\sin b - (b - b^3/3!)| \leq \frac{b^5}{5!} \leq \frac{(\pi/8)^5}{120} < \frac{(3.2/8)^5}{120} = \frac{(0.4)^5}{120} = \frac{0.01024}{120} < 0.0001.$$

PROBLEMS

1. Obtain the Taylor expansion of $f(a + h)$ for the given function f. Obtain an estimate of Lagrange's form of the remainder R_{n+1} for the given restrictions on h.

•(a) $\cos x$ $(a = 0, \text{ all } h)$

(b) e^x $(a = 0, h < 0)$

•(c) e^x $(a = 0, |h| < 1)$

(d) $\log x$ $(a = 1, |h| < \frac{1}{2})$

•(e) $\log x$ $(a = 1, |h| < 1)$

(f) $\cos x$ $(a = \pi/4, \text{ all } h)$

•(g) $(1 + x)^s$ $(a = 0, |h| < 1, s \text{ any real number})$
[This is called the *binomial expansion.*]

(h) $\dfrac{1}{1 - x}$ $(a = 0, |h| < 1)$

TAYLOR'S EXPANSION: GRAPHICAL VIEW

2. Let $f(x) = \sin x$. Graph carefully, on a single piece of graph paper, the functions $f, P_{0,f,0}, \ldots, P_{5,f,0}$, on the interval $[-4,4]$.

3. Repeat Problem 2 for $f(x) = \cos x$.

4. Repeat Problem 2 for $f(x) = 1/(1 - x)$ on $(-1,1)$.

•**5.** Compute $\cos(0.1)$ to four decimals.

6. Compute e to three decimals. Prove your answer is this accurate.

7. Write out the Taylor expansion of $\tan x$ about 0 to the term in h^3. How accurately does this expansion determine $\tan(0.1)$?

8. Write down the Taylor expansions of the following functions about $a = 0$, to the term in h^4.

•(a) $\sin^2 x$, $(a = 0)$
(b) $\cos^2 x$, $(a = 0)$
(c) $(\log x)^2$, $(a = 1)$

•**9.** Square the expansion

$$\sin h = h - h^3/3! + R_5,$$

and compare the result with part (a) in Problem 8.

10. Square the expansion

$$\cos h = 1 - h^2/2 + R_4,$$

and compare the result with part (b) in Problem 8.

11. Multiply the expansion of $\tan x$ to the remainder R_4 [Problem 7] by the expansion of $\cos x$ to R_4, and compare the result with the expansion of $\sin x$. [They should agree up through the terms in h^3.]

12. The Taylor expansion can be used to show that e, like $\sqrt{2}$, is not a rational number.

THE NUMBER *e* IS IRRATIONAL

(a) Use the Taylor expansion of e^x about 0 to show that

$$e = 1 + \frac{1}{1!} + \frac{1}{2!} + \cdots + \frac{1}{n!} + R_{n+1},$$

with $0 \leq R_{n+1} \leq e/(n+1)! < 3/(n+1)!$.

(b) Show that if $e = p/q$ with p and q integers, then $q \geq 2$, and

$$q!\left(\frac{p}{q} - 1 - \frac{1}{1!} - \frac{1}{2!} - \cdots - \frac{1}{q!}\right) \text{ is an integer between 0 and 1.}$$

13. (a) Use the Taylor expansion up to the term in h^n to show that $e^h \geq h^n/n!$ for $h \geq 0$.

(b) For each integer $n > 0$, show that $h^{n-1}e^{-h}$ tends to 0 as h grows large and positive.

• **14.** Find the Taylor expansion of $f(x) = \int_0^x e^{-t^2}\,dt$ about $x = 0$. Take the expansion to the term in h^3, and estimate the remainder for $|h| \leq 1$.

15. Go through the proof of Theorem 1 in the case $b < a$, and list the changes that must be made.

16. Prove the formula giving the Taylor expansion with *Cauchy's form* of the remainder, $R_{n+1} = \dfrac{(b-a)(b-c)^n}{n!} f^{(n+1)}(c)$. [Hint: Apply the Mean Value Theorem just once to the function

CAUCHY FORM OF REMAINDER

$$F(x) = f(x) + (b-x)f'(x) + \frac{(b-x)^2}{2!} f''(x) + \cdots + \frac{(b-x)^n}{n!} f^{(n)}(x)$$

on the interval $[a,b]$. In computing F', you will find that most of the terms cancel each other out.]

17. Find the *integral form* of the remainder from the formula $f(b) - f(a) = \int_a^b f'(x)\,dx$, by making n integrations by parts. [Hint: In $\int u\,dv = uv - \int v\,du$, set $u = f'$, $dv = dx$, and $v = x - b$. You should find

INTEGRAL FORM OF REMAINDER

$$R_{n+1} = \int_a^b f^{(n+1)}(t)\,(b-t)^n/n!\,dt.]$$

Remark. Each form of the remainder is obtained by some algebraic trick. For the integral form [Problem 17], it is the choice of $v = x - b$ instead of $v = x$; for Cauchy's form [Problem 16], it is the choice of F, and for Lagrange's form [Theorem 1], the choice of d.

18. Try to obtain the Taylor expansion up to R_{n+1} for $f(x) = \arctan x$ about $a = 0$. [You will find this quite difficult beyond the first few terms, and it is still more difficult to estimate the remainder. This expansion, like many others, is easier to obtain by the methods of infinite series, Chapter XI.]

9.3 NEWTON'S METHOD

When an equation such as $2x^3 + 3x^2 - 2 = 0$ cannot be solved exactly by simple algebraic methods, Newton's method gives an efficient sequence of approximations, $x_1, x_2, x_3, \ldots$, particularly for polynomial equations.

We discussed the simple equation $x^2 - c = 0$ in §1.6, and found effective estimates of the accuracy; now, using the Taylor expansion, we discuss the general case. Figure 4 recalls the method.

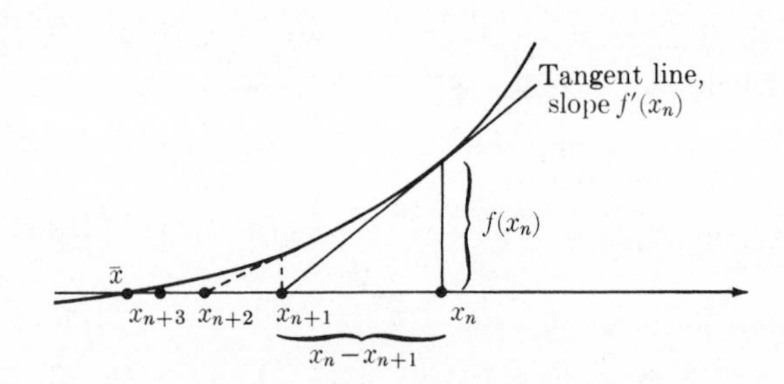

FIGURE 9.4

Suppose we have found the nth approximation x_n to a solution $\bar{x}$ of the equation $f(x) = 0$. Then we construct the tangent line to the graph of f at $(x_n, f(x_n))$, and determine our next approximation x_{n+1} by the intersection of this tangent line with the x axis. The slope of the tangent line is $f'(x_n)$, so from Fig. 4 we read $f(x_n)/(x_n - x_{n+1}) = f'(x_n)$, and this leads to the formula for x_{n+1}:

NEWTON'S APPROXIMATION TO A SOLUTION OF $f(x) = 0$

$$x_{n+1} = x_n - \frac{f(x_n)}{f'(x_n)}. \tag{1}$$

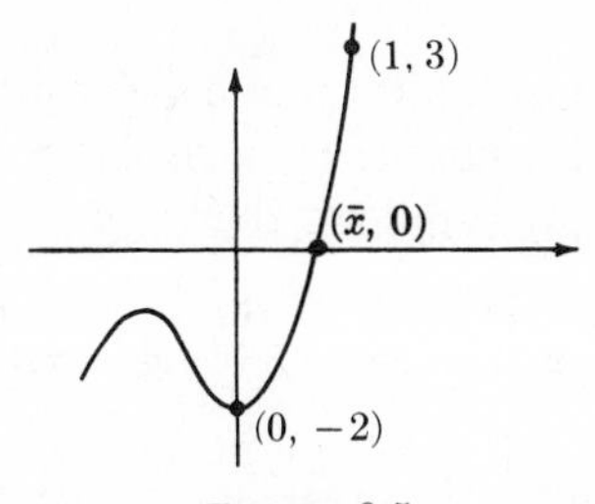

FIGURE 9.5

Example 1. In §1.4 we sketched the graph of $f(x) = 2x^3 + 3x^2 - 2$ [Fig. 5] and found that it has a root $\bar{x}$ somewhere in the interval $(0,1)$, probably closer to 1 than to 0. To start things off, we therefore take $x_0 = 1$ as our first approximation, and find from (1) that

$$x_1 = x_0 - \frac{f(x_0)}{f'(x_0)} = 1 - \frac{3}{12} = \frac{3}{4},$$

$$x_2 = x_1 - \frac{f(x_1)}{f'(x_1)} = \frac{3}{4} - \frac{17/32}{63/8} = .6825\ldots.$$

The next approximation would require a desk calculator; a computer could repeat the process many times in less than a second!

As with all approximations, we want an estimate of the accuracy. Let $\bar{x}$ be the actual solution of the equation $f(\bar{x}) = 0$. Then by the Mean Value Theorem,

$$f(\bar{x}) = 0 = f(x_n) + (\bar{x} - x_n)f'(c)$$

for some c between $\bar{x}$ and x_n, so

$$\bar{x} - x_n = -f(x_n)/f'(c). \tag{2}$$

This shows that the accuracy depends on having $f(x_n)$ not too large, and $f'(c)$ not too near zero. Precisely, it follows from (2) that

NEWTON'S METHOD: 1st ERROR ESTIMATE

$$\textit{if} \quad |f'(x)| \geq m_1 \textit{ between } x_n \textit{ and } \bar{x},$$

$$\textit{then} \quad |\bar{x} - x_n| \leq \frac{|f(x_n)|}{m_1}. \tag{3}$$

The next approximation x_{n+1} will be much better than x_n if the second derivative f'' is not too large. To prove this, use Taylor's expansion about $\bar{x}$:

$$f(\bar{x}) = 0 = f(x_n) + (\bar{x} - x_n)f'(x_n) + \tfrac{1}{2}(\bar{x} - x_n)^2 f''(c)$$

for some c between x_n and $\bar{x}$. Eliminating $f(x_n)$ from this and from (1), we find

$$x_{n+1} - \bar{x} = \frac{f''(c)}{2f'(x_n)}(\bar{x} - x_n)^2. \tag{4}$$

To use this formula, we look for a number M_2 such that $|f''(x)| \leq M_2$ for x between x_n and $\bar{x}$, and a number m_1 such that $|f'(x)| \geq m_1$ for x between x_n and $\bar{x}$. Then from (4) it follows that

2d ERROR ESTIMATE

$$|x_{n+1} - \bar{x}| \leq \frac{M_2}{2m_1}(\bar{x} - x_n)^2. \tag{5}$$

Formula (5) estimates the error $|x_{n+1} - \bar{x}|$ if we already know about the previous error $|\bar{x} - x_n|$. In fact, we do know something about $|\bar{x} - x_n|$ from (3); combining (3) and (5), we obtain

3d ERROR ESTIMATE

$$|x_{n+1} - \bar{x}| \leq \frac{M_2}{2m_1^3}|f(x_n)|^2. \tag{6}$$

The estimate (6) is particularly convenient because we have already computed $f(x_n)$ in finding x_{n+1}.

Example 2. Estimate the accuracy of $x_2 = .6825\ldots$, found in Example 1, approximating the solution $\bar{x}$ of $f(\bar{x}) = 2\bar{x}^3 + 6\bar{x}^2 - 2 = 0$.
Solution. Observe that all our approximations lie in the interval $[\frac{1}{2},1]$, so we take for the terms in (6)

$$m_1 = \min_{[\frac{1}{2},1]} |f'(x)| = f'(\tfrac{1}{2}) = 9/2$$

[note that f' is increasing on $[\frac{1}{2},1]$]

$$M_2 = \max_{[\frac{1}{2},1]} |f''(x)| = f''(1) = 18$$

[since f'' is increasing on $[\frac{1}{2},1]$].

$$|f(x_1)|^2 = (17/32)^2 < 0.3.$$

Therefore, from (6),

$$|x_2 - \bar{x}| \leq \frac{18}{2\cdot(9/2)^3}(0.3) = \left(\frac{8}{81}\right)(0.3) < .03.$$

Note that $x_2 = .6825\ldots$ is roughly accurate to two decimals, so in any further calculations we would drop the last two digits and proceed from $x_2 = .68$. If we computed the next term x_3, the remaining error could be estimated from (5):

$$|x_3 - \bar{x}| \leq \frac{M_2}{2m_1}(x_2 - \bar{x})^2 = \frac{18}{9}(.03)^2 < .002.$$

PROBLEMS

• **1.** In Example 2, show graphically why all successive approximations $x_2, x_3, \ldots$ lie in the interval $[1/2, 1]$.

2. Given the result of Problem 1, we can take $m_1 = 9/2$ and $M_2 = 18$ for all the successive approximations in Example 2. Show that x_4 is good to five decimals: $|x_4 - \bar{x}| \leq 10^{-5}$. [Use the inequality (5).]

3. Continue with Example 2. For which n can you be sure that $|x_n - \bar{x}| < 10^{-20}$?

4. Show graphically why Newton's method works poorly, or not at all, if f' is very small or f'' very large.

• **5.** Decide graphically whether Newton's method will provide successively better approximations to a zero of f in each of the following cases. [Compare Problem 4.]

(a) $f(x) = x^{1/3}, \qquad x_1 = 1$

(b) $f(x) = \sin x, \qquad x_1 = \dfrac{\pi}{2}$

(c) $f(x) = \dfrac{x}{1 + x^2}, \qquad x_1 = 2$

(d) $f(x) = \dfrac{x}{1 + x^2}, \qquad x_1 = \dfrac{1}{2}$

6. Let $f(x) = x^4 + x - 4$.

(a) Show that f is convex [§3.4]; hence there are at most two solutions of $f(x) = 0$.

(b) Sketch a graph of f, and observe that there is exactly one positive number $\bar{x}$ with $f(\bar{x}) = 0$.

(c) Compute $\bar{x}$ to one decimal place, and prove the accuracy.

(d) Compute $\bar{x}$ to three decimal places, and prove the accuracy.

9.4 THE TRAPEZOID RULE AND SIMPSON'S RULE

When it appears too difficult (or impossible) to evaluate an integral by the Fundamental Theorem of Calculus [Chapter V], approximations are called for. In principal, the Riemann sums give arbitrarily accurate approximations when the subdivisions are small enough, but this is generally inefficient; for example, approximating $\int_0^1 e^{-x^2}\,dx$ to two decimals by the Riemann sum $\sum_{j=1}^{n} f(x_j)\,\Delta x_j$ would require a subdivision into at least fifty intervals!

There are numerous improvements on the Riemann sums; we discuss here the *trapezoid rule* and *Simpson's rule.* Problem 7 below gives the *midpoint rule,* and finally §11.5 shows how to evaluate integrals by infinite series.

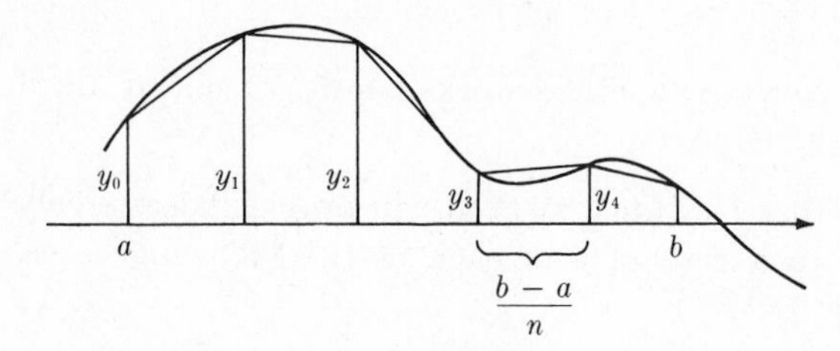

FIGURE 9.6

The trapezoid rule improves on the Riemann sums by substituting trapezoids for rectangles [Fig. 6]. Suppose the interval $[a,b]$ is divided into n equal subintervals, so that each subinterval has length $(b-a)/n$. Then the sum of the areas of the trapezoids in Fig. 6 is

$$\frac{b-a}{n} \times \frac{y_0+y_1}{2} + \frac{b-a}{n} \times \frac{y_1+y_2}{2} + \cdots + \frac{b-a}{n} \times \frac{y_{n-1}+y_n}{2}$$

$$= \frac{b-a}{2n}(y_0 + 2y_1 + 2y_2 + \cdots + 2y_{n-1} + y_n).$$

The trapezoid rule says simply that

TRAPEZOID RULE

$$\int_a^b f(x)\,dx = _{\text{(approx)}} \frac{b-a}{2n}(y_0 + 2y_1 + 2y_2 + \cdots + 2y_{n-1} + y_n), \quad (1)$$

where $y_k = f\left(a + \frac{k}{n}(b-a)\right)$.

We will show that the error in this approximation is no larger than

ERROR ESTIMATE

$$\frac{(b-a)^3}{12n^2} \max_{[a,b]} |f''(x)|. \quad (2)$$

Notice that if f is a first degree polynomial, then its graph is a straight line and the trapezoid rule obviously commits no error [Fig. 6]. Thus it is reasonable that the error is estimated by $|f''|$, which measures the deviation from a straight line.

Example 1. Estimate $\log 3 = \int_1^3 dx/x$, accurate to one decimal, by the trapezoid rule. *Solution.* We will have to choose n so that the expression (2) is less than 0.1. The function in this case is $f(x) = 1/x$, so $f''(x) = 2x^{-3}$, and $\max_{[1,3]} |f''| = 2$. Since $a = 1$, $b = 3$, we have $(b - a)^3 = 8$, so we need

$$\frac{8}{12n^2}\cdot 2 < 0.1 = \frac{1}{10} \qquad \text{or} \qquad n^2 > \frac{40}{3} = 13\tfrac{1}{3}\,.$$

[Note the reversal of the inequality in taking reciprocals of positive numbers.] This is satisfied for $n = 4$, so we apply (1) with $n = 4$. The points $x_0 = 1$, $x_1 = 3/2$, $x_2 = 2$, $x_3 = 5/2$, $x_4 = 3$ divide the given interval of integration [1,3] into 4 equal subintervals; the corresponding values $y_j = f(x_j) = 1/x_j$ are

$$y_0 = 1, \quad y_1 = 2/3, \quad y_2 = 1/2, \quad y_3 = 2/5, \quad y_4 = 1/3,$$

so the trapezoid rule (1) gives

$$\log 3 = \int_1^3 \frac{1}{x}\,dx = {\scriptstyle(\text{approx})}\ \frac{2}{2\cdot 4}\left(1 + \frac{4}{3} + 1 + \frac{4}{5} + \frac{1}{3}\right) = \frac{67}{60} = 1.116\ldots.$$

The error in this approximation is, according to (2), less than $\dfrac{2^3}{12\cdot 4^2}\cdot 2 = \dfrac{1}{12}$,

so $\log 3 = 1.1$, to one decimal.

Simpson's rule

The approximation of definite integrals that is generally most convenient and accurate is Simpson's rule. Here we replace sections of the graph by parabolas, whereas in the trapezoid rule we replace them by line segments.

First we find the approximation to $\int_{-h}^{h} f$ in which the graph of f is replaced by a parabola through the three points $(-h,f(-h))$, $(0,f(0))$, $(h,f(h))$. [See Fig. 7.] Let this approximating parabola be the graph of

$$y = Ax^2 + Bx + C = P(x).$$

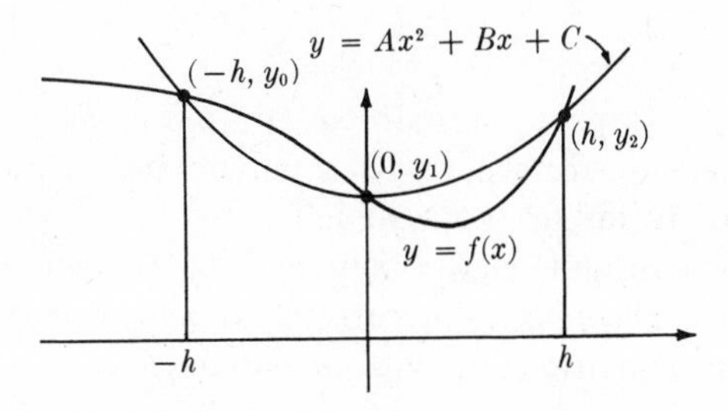

FIGURE 9.7 $y_0 = f(-h)$, $y_1 = f(0)$, $y_2 = f(h)$

If we set $y_0 = f(-h)$, $y_1 = f(0)$, and $y_2 = f(h)$, then the parabola [Fig. 7] passes through the points $(-h, y_0)$, $(0, y_1)$, and (h, y_2), so we have

$$\left.\begin{aligned} P(-h) &= Ah^2 - Bh + C = y_0\,, \\ P(0) &= C = y_1\,, \\ P(h) &= Ah^2 + Bh + C = y_2\,. \end{aligned}\right\} \tag{3}$$

The integral of the approximating parabola is

$$\int_{-h}^{h} (Ax^2 + Bx + C)\, dx = \frac{2Ah^3}{3} + 2Ch.$$

Equations (3) give $C = y_1$, and eliminating Bh from (3) gives

$$2Ah^2 = y_0 + y_2 - 2C = y_0 + y_2 - 2y_1\,.$$

Hence the parabolic approximation to $\int_{-h}^{h} f(x)\, dx$ is

$$\frac{2Ah^3}{3} + 2Ch = \frac{1}{3} h(y_0 + 4y_1 + y_2). \tag{4}$$

To approximate $\int_a^b f(x)\, dx$ for a general interval $[a,b]$, we take $2n$ subintervals of length $h = (b - a)/2n$, letting $x_j = a + j(a - b/2n)$ and $y_j = f(x_j)$. [See Fig. 8.] We then obtain Simpson's rule,

SIMPSON'S RULE

$$\int_a^b f(x)\, dx = \text{(approx)}\ \frac{1}{3} h(y_0 + 4y_1 + y_2) + \frac{1}{3} h(y_2 + 4y_3 + y_4)$$

$$+ \cdots + \frac{1}{3} h(y_{2n-2} + 4y_{2n-1} + y_{2n})$$

$$= \frac{b-a}{6n} (y_0 + 4y_1 + 2y_2 + 4y_3 + 2y_4$$

$$+ \cdots + 4y_{2n-1} + y_{2n}). \tag{5}$$

Since Simpson's rule approximates by parabolas, it seems clear that there should be *no* error when f is a polynomial of second degree. We might therefore estimate the remainder by $|f'''|$, which measures the deviation from a second degree polynomial. In fact, however, the best estimate involves the *fourth* derivative; Problem 6 below shows that the error in the approximation (5) with $2n$ subdivisions is no more than

ERROR ESTIMATE

$$\left(\frac{b-a}{2}\right)^5 \frac{\max |f^{(4)}(x)|}{90n^4}\,. \tag{6}$$

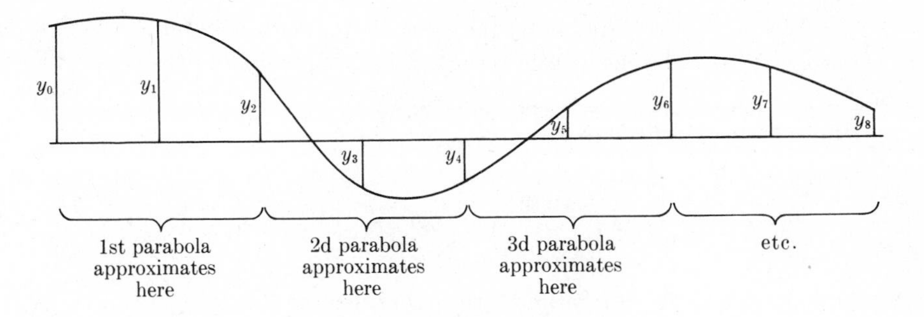

FIGURE 9.8

Example 2. Try Simpson's rule on the integral $\int_1^3 \frac{dx}{x}$ considered in Example 1.

Solution. We have $f^{(4)}(x) = 24x^{-5}$, so $\max_{[1,3]} |f^{(4)}(x)| = 24$. With $2n$ subdivisions of the interval $[a,b] = [1,3]$, formula (6) shows the error to be less than

$$\left(\frac{b-a}{2}\right)^5 \times \frac{24}{90n^4} = \frac{4}{15n^4}.$$

To compare this with the trapezoid rule method in Example 1, we should take four subintervals, i.e. $n = 2$; the error in Simpson's rule is then $\frac{4}{15 \times 2^4} = \frac{1}{60}$, five times as small as in the trapezoid rule. With $n = 3$ (six subintervals) the error is less than

$$\frac{4}{15 \times 3^4} = \frac{4}{15 \cdot 81} < \frac{1}{300},$$

which is good for a two-decimal approximation. Carrying out (5) with this case, we get

$$x_0 = 1, \quad x_1 = 4/3, \quad x_2 = 5/3, \quad x_3 = 2, \quad x_4 = 7/3, \quad x_5 = 8/3, \quad x_6 = 3,$$

so $y_0 = 1$, $y_1 = 3/4$, $y_2 = 3/5$, etc., and (5) yields

$$\int_1^3 \frac{1}{x}\,dx = \text{(approx)}\ \frac{1}{9}\left(1 + 3 + \frac{6}{5} + 2 + \frac{6}{7} + \frac{3}{2} + \frac{1}{3}\right) = 1.0989\ldots$$

with an error known [by (6)] to be $< .0033$. [Actually, the error is less than .0004, but it is more difficult to prove this.]

Estimating the error

The error estimates (2) and (5) are both proved by the same technique; we will prove (2) and leave (5) as a problem.

We begin by estimating the error in the first subdivision $[x_0, x_1]$. Let $h = x_1 - x_0$ be the length of the subdivision, and $E(h)$ the error. The trapezoid approximation is $h\dfrac{f(x_0) + f(x_1)}{2}$, so the error is

$$E(h) = \left(\int_{x_0}^{x_0+h} f\right) - \frac{1}{2}h[f(x_0) + f(x_0 + h)],$$

since $x_1 = x_0 + h$. To estimate this, consider the function

$$E(t) = \left(\int_{x_0}^{x_0+t} f\right) - \frac{1}{2}t[f(x_0) + f(x_0 + t)], \qquad 0 \le t \le h. \tag{7}$$

We will compute $E''(t)$, and then reconstruct E. From the Fundamental Theorem of Calculus, the indefinite integral $\int_{x_0}^{x_0+t} f$ has the derivative $f(x_0 + t)$, so

$$\begin{aligned} E'(t) &= f(x_0 + t) - \frac{1}{2}[f(x_0) + f(x_0 + t)] - \frac{1}{2}tf'(x_0 + t) \\ &= \frac{1}{2}[f(x_0 + t) - f(x_0)] - \frac{1}{2}tf'(x_0 + t), \end{aligned} \tag{8}$$

and therefore, differentiating once more and combining terms,

$$E''(t) = -\frac{1}{2}tf''(x_0 + t). \tag{9}$$

Suppose that $|f''| \le M$ on the interval $[x_0, x_0 + h]$. Then (9) shows that $|E''(t)| \le \frac{1}{2}tM$, or

$$-\frac{1}{2}tM \le E''(t) \le \frac{1}{2}tM.$$

Integrating each term of the inequality from 0 to x gives

$$-\frac{1}{4}x^2M \le \int_0^x E''(t)\,dt \le \frac{1}{4}x^2M.$$

But [by what theorem?] $\int_0^x E''(t)\,dt = E'(x) - E'(0) = E'(x)$ [since $E'(0) = 0$, from (8)], so

$$-\frac{1}{4}x^2M \le E'(x) \le \frac{1}{4}x^2M.$$

Now integrate each term from 0 to h, obtaining in the same way as before

$$-\frac{1}{12}h^3M \leq E(h) \leq \frac{1}{12}h^3M. \tag{10}$$

Thus in each subinterval of length h, the error can be estimated by the inequality

$$|E(h)| \leq \frac{1}{12}h^3M.$$

Now, suppose that the original interval $[a,b]$ is divided into n equal subintervals of length $h = \dfrac{b-a}{n}$, and that $|F''| \leq M$ on the entire interval $[a,b]$. Then the error in each subinterval is no more than

$$\frac{1}{12}h^3M = \frac{1}{12}\left(\frac{b-a}{n}\right)^3 M,$$

and the combined error for all n intervals is no more than n times this, which gives the number $\dfrac{1}{12}\dfrac{(b-a)^3}{n^2}M$ appearing in (2).

PROBLEMS

1. (a) Use the trapezoid rule with four subdivisions ($n = 4$) to approximate log 2, and estimate the error.

• (b) Use Simpson's rule with four subdivisions ($n = 2$) to approximate log 2, and estimate the error.

2. • (a) The graph of $y = \dfrac{b}{a}\sqrt{a^2 - x^2}$ on $0 \leq x \leq a$ is a quarter ellipse.

Show that its arc length [§5.5] is $\displaystyle\int_0^{\pi/2} \sqrt{a^2 \sin^2\theta + b^2\cos^2\theta}\, d\theta.$

[Hint: Set $x = a\cos\theta$.]

(b) Use Simpson's rule with two subdivisions ($n = 1$) to approximate the integral.

(c) Use Simpson's rule with four subdivisions to approximate the integral.

(d) Use the trapezoid rule with four subdivisions, and estimate the error. [Hint: Assume that $b < a$. Then I think you can show that $|f''| \leq \dfrac{a^2}{b^3}(a^2 - b^2)$.]

3. (a) Show that

$$\frac{\pi}{4} = \int_0^1 \frac{dx}{1+x^2}.$$

• (b) How many subdivisions would allow you to compute π to four decimals, using Simpson's rule on the integral in part (a)?

• **4.** Show that the error in the trapezoid rule with one subinterval is *exactly* $[(b-a)^3/12] \max_{[a,b]} |f''(x)|$, when f is an appropriate polynomial of the second degree. [Thus, in a sense, (2) gives the best possible error estimate.]

5. Show that the error in Simpson's rule with two subintervals is *exactly* $[h^4(b-a)/180] \max_{[a,b]} |f^{(4)}(x)|$, when f is an appropriate polynomial of the fourth degree. [Thus, in a sense, (6) gives the best possible estimate of the error.]

6. Estimate the error in Simpson's rule. [Hints: Estimate the error $E(h) = \left(\int_{-h}^{h} f\right) - \frac{1}{3}h[f(-h) + 4f(0) + f(h)]$ in the basic interval illustrated in Fig. 7. Compute E''', and use the Mean Value Theorem to show that $|E'''(t)| \leq \frac{2t^2}{3}M$, if $|f^{(4)}| \leq M$ on $[-h,h]$. In differentiating $E(t)$, note that $\int_{-t}^{t} f = \int_0^t f - \int_0^{-t} f.$]

7. The *Midpoint rule*

MIDPOINT RULE

$$\int_a^b f =_{(\text{approx})} \frac{b-a}{n}[f(\xi_1) + f(\xi_2) + \cdots + f(\xi_n)]$$

$\left[\text{where } \xi_j = \frac{x_{j-1}+x_j}{2}\right.$ is the midpoint of the jth out of n equal sub-$\left.\text{intervals}\right]$ approximates integrals better than the trapezoid rule, but is a little less convenient to compute.

(a) Use the midpoint rule with $n = 4$ to approximate $\int_1^2 \frac{1}{x}\,dx$.

(b) Supposing that $|f''| \leq M$ on the interval $[-h/2, h/2]$, show that

$\left| \int_{-h/2}^{h/2} f - hf(0) \right| \leq \frac{h^3}{24} M$. [Hint: From the Taylor expansion,

$-\frac{x^2}{2} M \leq f(x) - f(0) - xf'(0) \leq \frac{x^2}{2} M$; integrate these inequal-

ities from $-h/2$ to $h/2$.]

(c) Show that the error in the midpoint rule with n equal subdivisions is no more than $\frac{(b-a)^3}{24n^2} \max_{[a,b]} |f''|$.

MIDPOINT RULE: ERROR ESTIMATE

(d) Estimate the error in part (a); compare with Example 1 above.

(e) Use the midpoint rule with $n = 4$ to approximate the integral in Problem 2(a), and estimate the error. [See hint in Problem 2(d).]

Chapter X

Infinite Sequences

Working with real numbers leads, sooner or later, to *infinite sequences.* First of all, there are the infinite sequences of *positive integers*

$$1, 2, 3, \ldots,$$

even integers

$$2, 4, 6, \ldots,$$

prime numbers

$$2, 3, 5, 7, 11, 13, 17, \ldots,$$

and so on.

Examples closer to the spirit of this chapter arise in the approximation of irrational numbers by decimals. For instance, the infinite sequence [Fig. 1]

$$1, \quad 1.7, \quad 1.73, \quad 1.732, \quad 1.7320, \quad 1.73205, \ldots$$

gives successively better decimal approximations to $\sqrt{3}$. Another sequence of approximations to $\sqrt{3}$ can be obtained by Newton's method [§1.6]; taking $x_1 = 2$ as the first approximation, and applying repeatedly the Newton formula

$$x_{n+1} = \frac{x_n^2 + c}{2x_n} \qquad \text{with } c = 3,$$

we generate the infinite sequence [Fig. 1]

$$x_1 = 2, \quad x_2 = \frac{x_1^2 + 3}{2x_1} = \frac{7}{4}, \quad x_3 = \frac{x_2^2 + 3}{2x_2} = \frac{97}{56}, \ldots, \quad x_{n+1} = \frac{x_n^2 + 3}{2x_n}, \ldots.$$

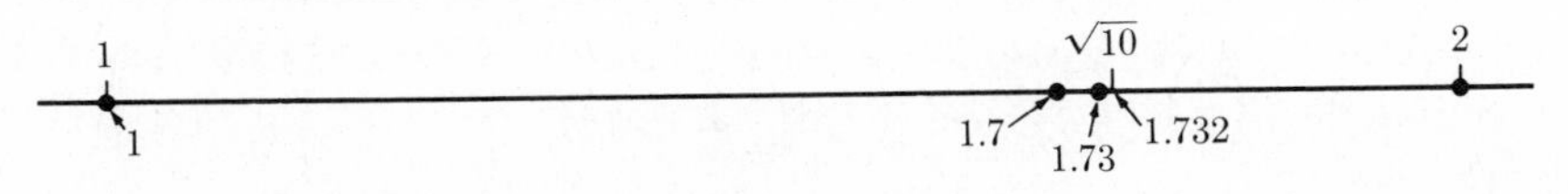

FIGURE 10.1

With either of these two sequences, you can approximate $\sqrt{3}$ as closely as you like simply by going far enough along in the sequence. The sequences are said to *converge to* $\sqrt{3}$, or to *have* $\sqrt{3}$ *as limit.*

The concept of limit suggested here is the fundamental idea distinguishing calculus from other branches of mathematics. This chapter gives the formal analytic definition of the *limit of an infinite sequence,* and proves the basic facts about these limits.

The definition and the proofs are a little more subtle than those in elementary algebra and geometry; but not really difficult, once you get used to them. The first prerequisite is some confidence with inequalities [§AI.3]. Later on, in §10.3 and §10.5, we also need the least upper bound axiom [§AI.5].

10.1 LIMIT OF A SEQUENCE

An infinite sequence of real numbers

$$s_1\,,\ s_2\,,\ s_3\,,\ldots,\ s_n\,,\ldots \tag{1}$$

is, formally speaking, a rule assigning a number s_n to each positive integer n. In other words, it is a *real-valued function whose domain is the positive integers.* The value of the function at the point n is s_n. [The standard notation for the value of a function s at a point n would be $s(n)$ rather than s_n; but in the context of sequences, s_n is traditional, and simpler to write.] The sequence as a whole can be indicated using "dots" as in (1), or within brackets, as $\{s_n\}_{n=1}^{\infty}$ or more simply as $\{s_n\}$.

Like any function, a sequence $\{s_n\}$ has a *graph,* consisting of the points $(1,s_1)$, $(2,s_2)$, $(3,s_3)$,.... Figure 2 graphs the six sequences

$$\{a_n\} = 1, \frac{1}{2}, \frac{1}{3}, \ldots, \frac{1}{n}, \ldots; \quad \left(a_n = \frac{1}{n}\right)$$

$$\{b_n\} = -1, \frac{1}{4}, -\frac{1}{9}, \frac{1}{18}, \ldots, \frac{(-1)^n}{n^2}, \ldots; \quad \left(b_n = \frac{(-1)^n}{n^2}\right)$$

$$\{c_n\} = 0, \frac{3}{2}, -\frac{2}{3}, \frac{5}{4}, \ldots, (-1)^n + \frac{1}{n}, \ldots$$

$$\{d_n\} = 1,\ 1.7,\ 1.73,\ 1.732, \ldots;$$
$$(d_n = \text{decimal expansion of } \sqrt{3} \text{ to } n \text{ decimal places})$$

$$\{e_n\} = 2, \left(\frac{3}{2}\right)^2, \left(\frac{4}{3}\right)^3, \ldots, \left(1 + \frac{1}{n}\right)^n, \ldots$$

$$\{f_n\} = 4, \frac{4^2}{2!}, \frac{4^3}{3!}, \ldots, \frac{4^n}{n!}, \ldots.$$

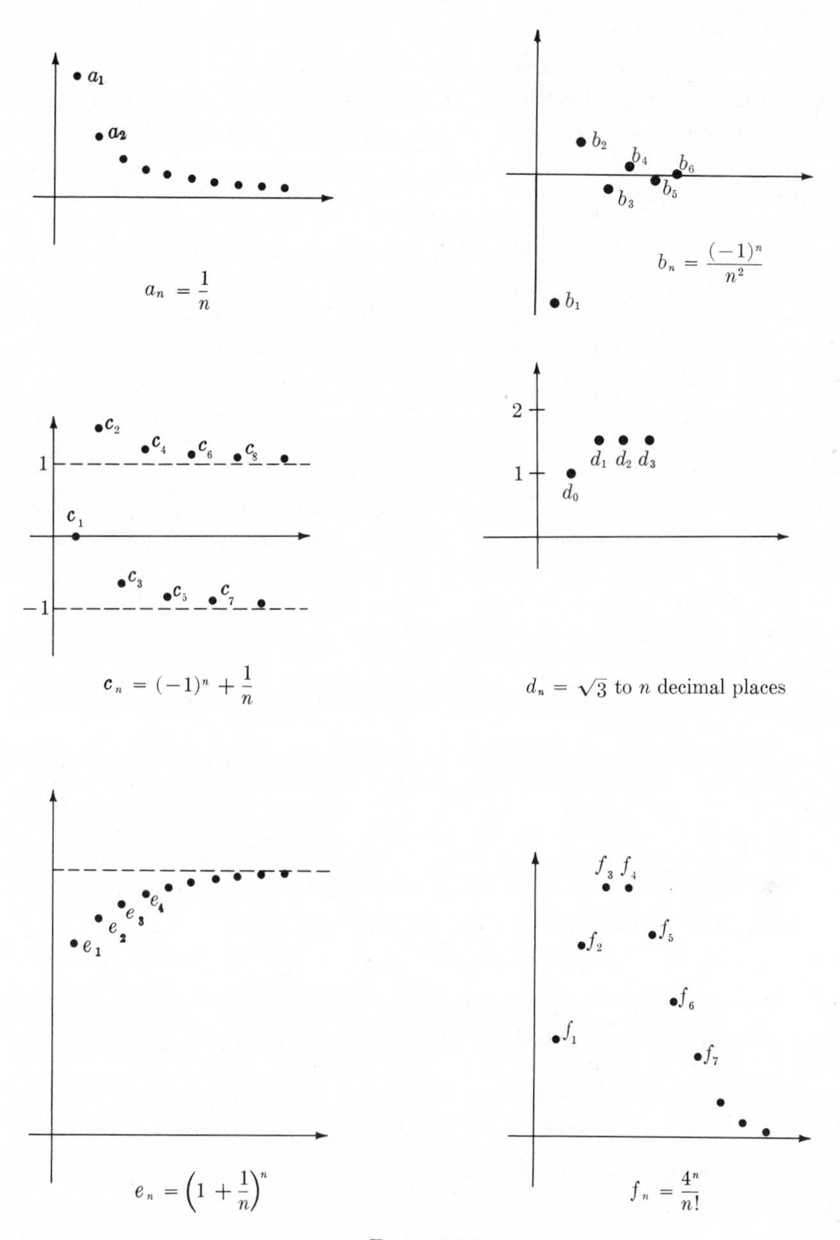

FIGURE 10.2

[The scales on the horizontal and vertical axes in Fig. 2 are different, so that the graphs can show more points in a limited space.] From these graphs, the sequences $\{a_n\}$, $\{b_n\}$, and $\{f_n\}$ appear to converge toward zero as n grows large; each of the sequences $\{d_n\}$ and $\{e_n\}$ appears to converge toward some positive number as n grows large; but the sequence $\{c_n\}$ does not appear to converge to any single number as n grows large.

Another way to picture a sequence $\{s_n\}$ is simply to plot the points $s_1, s_2, \ldots$ on a line. Figure 1 plots the sequence of decimals $\{d_n\}$. You can "see" that the numbers d_n converge toward $\sqrt{3}$.

The formal definition of convergence is based on the idea of *approximation.* In the decimal sequence $\{d_n\}$, the nth term approximates $\sqrt{3}$ with an error less than 10^{-n}:

$$d_0 = 1, \qquad |d_0 - \sqrt{3}| < 1 \qquad (\text{error} < 10^0)$$

$$d_1 = 1.7, \qquad |d_1 - \sqrt{3}| < .1 \qquad (\text{error} < 10^{-1})$$

$$d_2 = 1.73, \qquad |d_2 - \sqrt{3}| < .01 \qquad (\text{error} < 10^{-2})$$

$$d_3 = 1.732, \qquad |d_3 - \sqrt{3}| < .001 \qquad (\text{error} < 10^{-3})$$

and so on.

Different applications will generally require different degrees of accuracy in the approximation; say in one case the error must be no more than 10^{-2}, or in another it must be no more than 10^{-5}. But in any given case, no matter how small the maximum allowable error (call it ϵ), you can approximate $\sqrt{3}$ with an error $<\epsilon$ simply by taking a term d_n with n sufficiently large.

Just how large n must be depends, of course, on ϵ; when ϵ is very small, then n may have to be very large.

The definition of limit of a sequence is a precise analytic statement of this idea.

Definition 1. A sequence $\{s_n\}$ *converges to* S if and only if for every positive number ϵ, there is a number N_ϵ such that

$$n > N_\epsilon \;\Rightarrow\; |s_n - S| < \epsilon. \tag{2}$$

LIMIT OF A SEQUENCE

The number S is called the *limit* of the sequence $\{s_n\}$, and we write

$$\lim_{n\to\infty} s_n = S, \qquad \text{or} \qquad s_n \to S.$$

CONVERGENT SEQUENCE

If $\lim_{n\to\infty} s_n = S$ for some number S, we say the sequence $\{s_n\}$ *has a limit*, or *is convergent.* If there is no such limit S, the sequence $\{s_n\}$ is called *divergent.*

In Fig. 2, the sequences $\{a_n\}$, $\{b_n\}$, $\{d_n\}$, $\{e_n\}$, and $\{f_n\}$ are convergent, and $\{c_n\}$ is divergent.

The following examples show how the definition works.

BASIC EXAMPLE

Example 1. Prove (the obvious fact) that $\lim_{n\to\infty} 1/n = 0$. *Solution.* To prove that $\lim_{n\to\infty} s_n = S$, let ϵ stand for an arbitrary positive number; we must then fill in the question mark so that the following implication becomes true:

$$n > \textbf{(?)} \Rightarrow |s_n - S| < \epsilon.$$

In our case, where $s_n = 1/n$ and $S = 0$, we have $|s_n - S| = 1/n$, so the implication to be completed is

$$n > \textbf{(?)} \Rightarrow \frac{1}{n} < \epsilon.$$

Clearly, this is true is if we replace **(?)** by $1/\epsilon$:

$$n > \frac{1}{\epsilon} \Rightarrow \frac{1}{n} < \epsilon.$$

Thus Definition 1 is satisfied by taking $N_\epsilon = 1/\epsilon$. Hence $\lim_{n\to\infty} 1/n = 0$.

Example 2. Prove (the even more obvious fact) that if $s_n = c$ for all n (i.e. if $\{s_n\}$ is a constant sequence) then $\lim_{n\to\infty} s_n = c$. *Solution.* Given $\epsilon > 0$, we have to complete

$$n > \textbf{(?)} \Rightarrow |s_n - c| < \epsilon.$$

But $s_n = c$, so $|s_n - c| = 0$, hence $|s_n - c| < \epsilon$ for *every* n. We can therefore replace **(?)** by 0:

$$n > 0 \Rightarrow |s_n - c| < \epsilon.$$

Hence Definition 1 is satisfied with $N_\epsilon = 0$, which proves that $\lim_{n\to\infty} s_n = c$.

[There is no special reason for $N_\epsilon = 0$ in this example; we could just as well take $N_\epsilon = 1$ or $N_\epsilon = 100$.]

Example 3. Prove that $\lim_{n\to\infty} \dfrac{1}{n-2} = 0$. *Solution.* This example does not exactly fit our definition of a sequence, since $s_n = 1/(n-2)$ is not defined for $n = 2$. However, it *is* defined for all $n > 2$; moreover, for any $\epsilon > 0$ we have

$$n > 2 + \frac{1}{\epsilon} \Rightarrow n - 2 > \frac{1}{\epsilon} \Rightarrow 0 < \frac{1}{n-2} < \epsilon \Rightarrow \left|\frac{1}{n-2} - 0\right| < \epsilon.$$

Thus Definition 1 is satisfied with $N_\epsilon = 2 + 1/\epsilon$, and we conclude that $\lim_{n\to\infty} \dfrac{1}{n-2} = 0$.

Generally, we may consider sequences $\{s_n\}$ which are defined for all but finitely many n; in all such cases, Definition 1 can be applied to determine whether $\{s_n\}$ converges or not, just as in Example 3.

ANOTHER BASIC EXAMPLE

Example 4. Let $|r| < 1$. Prove that $\lim_{n\to\infty} r^n = 0$. *Solution.* For every $\epsilon > 0$, we have to fill in

$$n > (?) \Rightarrow |r^n - 0| < \epsilon.$$

But $|r^n - 0| = |r|^n$, so the problem becomes

$$n > (?) \Rightarrow |r|^n < \epsilon.$$

The solution here is not so obvious as in the previous examples; but it can be found by solving the inequality $|r|^n < \epsilon$ for n:

$$\begin{aligned} |r|^n < \epsilon &\Leftrightarrow \log(|r|^n) < \log(\epsilon) && \text{[property of log]} \\ &\Leftrightarrow n \log |r| < \log \epsilon \\ &\Leftrightarrow n > \frac{\log \epsilon}{\log |r|}. && \text{[because } \log|r| < 0 \text{ when } |r| < 1\text{]} \end{aligned}$$

Therefore,

$$n > \frac{\log \epsilon}{\log |r|} \Rightarrow |r^n - 0| < \epsilon,$$

and Definition 1 is satisfied with $N_\epsilon = \log \epsilon/\log |r|$. [Notice that when ϵ is very small, then $\log \epsilon$ is a large negative number, and $N_\epsilon = \log \epsilon/\log |r|$ is a large *positive* number. Thus to come very close to the limit 0 (to come within ϵ of zero), you must go far out in the sequence (beyond N_ϵ), as would be expected.]

Remark. Since infinite sequences entered in our informal proof of the properties of the logarithm [§2.6], objections can be raised against the use of logarithms in proving elementary results about sequences. An alternate solution of Example 4, not using logarithms, is outlined in Problem 7 below. However, when we renounce the use of logarithms, the solution is less direct.

The two particular limits $\lim_{n\to\infty} 1/n = 0$ and $\lim_{n\to\infty} r^n = 0$ ($|r| < 1$) yield many further results, with the aid of the following "comparison theorem."

COMPARISON THEOREM

Theorem 1. *If* $|a_n| \leq b_n$ *for all* n, *and* $\lim b_n = 0$, *then* $\lim a_n = 0$.

Proof. In view of the hypothesis $\lim b_n = 0$, for every $\epsilon > 0$ there is a number N_ϵ such that

$$n > N_\epsilon \Rightarrow |b_n - 0| < \epsilon. \tag{3}$$

We will prove that the N_ϵ that works for $\{b_n\}$ also works for $\{a_n\}$, i.e. with the same N_ϵ as in (3),

$$n > N_\epsilon \quad \Rightarrow \quad |a_n - 0| < \epsilon. \tag{4}$$

To prove (4), suppose that $n > N_\epsilon$; then

$|a_n - 0| = |a_n| \leq b_n = |b_n - 0|$ [given hypothesis; note that $b_n > 0$, so $b_n = |b_n - 0|$]

$< \epsilon.$ [by (3), since $n > N_\epsilon$]

Hence (4) is true, and $\lim\limits_{n\to\infty} a_n = 0$. *Q.E.D.*

Example 5. $\lim \dfrac{(-1)^n}{n^2} = ?$ *Solution.* Figure 2 suggests that $\lim \dfrac{(-1)^n}{n^2} = 0$. This is easy to prove; in Theorem 1 take $a_n = (-1)^n/n^2$ and $b_n = 1/n$. Then $|a_n| \leq b_n$ and $\lim\limits_{n\to\infty} b_n = 0$, so it follows that $\lim\limits_{n\to\infty} a_n = 0$.

As it stands, the comparison method [Theorem 1] can only prove that the limit of something is zero. The following simple result allows it to be applied more generally.

Theorem 2. *Let $\{s_n\}$ be any sequence. Then*

$$\lim_{n\to\infty} s_n = S \quad \Leftrightarrow \quad \lim_{n\to\infty} |s_n - S| = 0.$$

Proof. The statement "$\lim |s_n - S| = 0$" means "for every $\epsilon > 0$, there is an N_ϵ such that

$$n > N_\epsilon \quad \Rightarrow \quad \Big| |s_n - S| - 0 \Big| < \epsilon."$$

But this is exactly the same as Definition 1 for $\lim\limits_{n\to\infty} s_n = S$, since

$$\big| |s_n - S| - 0 \big| = |s_n - S|. \qquad Q.E.D.$$

Example 6. Let $\{d_n\}$ be the sequence of decimal approximations of $\sqrt{3}$; $d_0 = 1$, $d_1 = 1.7$, $d_2 = 1.73, \ldots$. Prove that $\lim\limits_{n\to\infty} d_n = \sqrt{3}$. *Solution.* By Theorem 2, $\lim\limits_{n\to\infty} d_n = \sqrt{3} \Leftrightarrow \lim\limits_{n\to\infty} |d_n - \sqrt{3}| = 0$. The second limit can be proved by Theorem 1. We have $|d_n - \sqrt{3}| < 10^{-n}$, and $\lim\limits_{n\to\infty} 10^{-n} = \lim (1/10)^n = 0$, by Example 4. Hence, applying Theorem 1 with $a_n = |d_n - \sqrt{3}|$ and $b_n = 10^{-n}$, we obtain $\lim\limits_{n\to\infty} |d_n - \sqrt{3}| = 0$.

PROBLEMS

[Note that $\lim\limits_{n\to\infty}$ has been abbreviated to lim.]

1. Prove the following, using only Definition 1 and elementary algebra and inequalities.

(a) $\lim \dfrac{(-1)^n}{n} = 0$

•(b) $\lim n^{-1/3} = 0$

(c) $\lim \dfrac{n+1}{n} = 1$

•(d) $\lim \sqrt{\dfrac{n+1}{n}} = 1$

2. Use the comparison method [Theorem 1] with Examples 1 and 4 to prove the following:

•(a) $\lim \dfrac{1}{n(n+1)} = 0$

(b) $\lim \dfrac{1}{n!} = 0$, where $n! = n(n-1)(n-2)\cdots(2)(1)$

•(c) $\lim \dfrac{r^n}{n} = 0$, where $|r| \leq 1$

(d) $\lim n^{-n} = 0$

•(e) $\lim 10^{-2^n} = 0$. [Hint: $2^n > n$; this can be proved by induction, or by expanding $(1+1)^n$. *Remark.* Problem 4 below uses the relation $\lim 10^{-2^n} = 0$.]

3. Let $d_n = a.a_1a_2\ldots a_n$ be the decimal expansion of a number A to n places; the nth digit a_n is determined by the pair of inequalities $d_n \leq A < d_n + 10^{-n}$.
(a) Prove that $|d_n - A| < 10^{-n}$.
(b) Evaluate $\lim 10^{-n}$. [See Example 3.]
(c) Prove that $\lim d_n = A$.

4. In Newton's method for extracting the square root of c [§1.6] we obtain a sequence of approximations $x_0, x_1, \ldots$ which satisfy

NEWTON'S METHOD CONVERGES

$$|x_{n+1} - \sqrt{c}| \leq \frac{|x_n - \sqrt{c}|^2}{2x_n}.$$

Suppose that $2x_n \geq 1$ for all n, and $|x_0 - \sqrt{c}| < 1/10$.
(a) Prove that $|x_1 - \sqrt{c}| < 10^{-2}$.
(b) Prove that $|x_2 - \sqrt{c}| < 10^{-4}$.
(c) Prove that $|x_3 - \sqrt{c}| < 10^{-8}$.
(d) Prove that $|x_n - \sqrt{c}| < 10^{-2^n}$.
(e) Prove that $\lim |x_n - \sqrt{c}| = 0$, and hence that $\lim x_n = \sqrt{c}$. [See Problem 2(e).]

5. Prove that $\lim a_n = 0 \iff \lim |a_n| = 0$.

6. Suppose that $\lim s_n = S$. Let $t_n = s_{k+n}$. [For example, if

IMPORTANT PROBLEM

$$\{s_n\} = 1, \frac{1}{2}, \frac{1}{3}, \ldots, \frac{1}{n}, \ldots,$$

then

$$\{t_n\} = \frac{1}{k+1}, \frac{1}{k+2}, \frac{1}{k+3}, \ldots, \frac{1}{k+n}, \ldots]$$

Prove that $\lim t_n = S$. [Hint: If N_ϵ works for $\{s_n\}$, then $N_\epsilon - k$ works for $\{t_n\}$.]

7. This problem proves that $\lim r^n = 0$ for $|r| < 1$, without using logarithms.

(a) If $R > 1$, prove that $\dfrac{R^n - 1}{R - 1} > n$. [Hint: Carry out the division.]

(b) If $0 < r < 1$, prove that $\left(\dfrac{1}{r}\right)^n > n\left(\dfrac{1}{r} - 1\right) + 1 > n\dfrac{1-r}{r}$.

[Use part (a) with $R = 1/r$.]

(c) If $0 < r < 1$, prove that $\lim r^n = 0$.

(d) If $|r| < 1$, prove that $\lim r^n = 0$. [Use Theorem 1, with $b_n = |r|^n$.]

10.2 THE ALGEBRA OF LIMITS

Our ability to evaluate limits is extended still further by the following two theorems. For simplicity of notation, we write lim instead of $\lim\limits_{n\to\infty}$.

Theorem 3. *Suppose that $a_n \le b_n$ for all n, and that $\lim a_n = A$ and $\lim b_n = B$. Then $A \le B$.*

INEQUALITIES FOR LIMITS

Theorem 4. *Let $\{a_n\}$ and $\{b_n\}$ be convergent sequences. Then*

EQUALITIES FOR LIMITS

(1) $\lim (ca_n) = c \lim a_n$ *for any constant* c

(2) $\lim (a_n + b_n) = (\lim a_n) + (\lim b_n)$

(3) $\lim (a_n b_n) = (\lim a_n)(\lim b_n)$

(4) $\lim \left(\dfrac{a_n}{b_n}\right) = \dfrac{\lim a_n}{\lim b_n}$ *if* $\lim b_n \neq 0$.

Before proving the theorems, let's see what they're good for. Theorem 3 can be used to answer a question which may or may not have occurred to you: Can a given sequence converge to more than one limit? If the answer were "yes," then a symbol such as $\lim_{n\to\infty} \frac{n+1}{n}$ might stand for several different numbers, a rather confusing situation. Fortunately, the answer is "no":

LIMITS ARE UNIQUELY DEFINED

A given sequence $\{a_n\}$ cannot converge to two different limits.

The proof is by contradiction: Suppose that $\{a_n\}$ does converge to two different limits; call the larger one A, and the smaller one B. Now apply Theorem 3, taking $\{b_n\}$ to be the same as $\{a_n\}$, i.e. define $b_n = a_n$ for all n. Obviously $a_n \le b_n$ for all n, hence by Theorem 3, $A \le B$. This contradicts our selection of A as the larger of the two supposedly different limits. Hence there could not have been two different limits $A > B$.

With this fundamental question resolved, we give some computational examples.

Example 1. The area under the graph of $f(x) = mx$ from $x = 0$ to $x = b$ is a right triangle of base b and height $h = mb$ [Fig. 3]. The Riemann sum

$$S_n = \sum_{j=0}^{n} \frac{b}{n} f\left(\frac{jb}{n}\right) = \sum_{j=0}^{n} \frac{b}{n} \frac{mjb}{n} = \frac{mb^2}{n^2} \sum_{j=0}^{n} j = \frac{mb^2}{n^2} \frac{n(n+1)}{2}$$

$$= \frac{bh}{n^2} \frac{n(n+1)}{2} \qquad [\text{recall that } h = mb]$$

approximates the triangular area by the sum of areas of rectangles. Compute $\lim S_n$.

Solution. $\lim S_n = \lim \left(bh \frac{n(n+1)}{2n^2} \right)$

$$= bh \lim \frac{n(n+1)}{2n^2} \qquad [\text{by } (1)]$$

$$= bh \lim \left(\frac{1}{2} + \frac{1}{2n} \right) \qquad [\text{simple algebra}]$$

$$= bh \left[\lim \frac{1}{2} + \lim \frac{1}{2n} \right] \qquad [\text{by } (2)]$$

$$= bh \left[\frac{1}{2} + \frac{1}{2} \lim \frac{1}{n} \right] \qquad [\text{by (1) and Example 2, §10.1}]$$

$$= bh \left(\frac{1}{2} + 0 \right) = \frac{1}{2} bh.$$

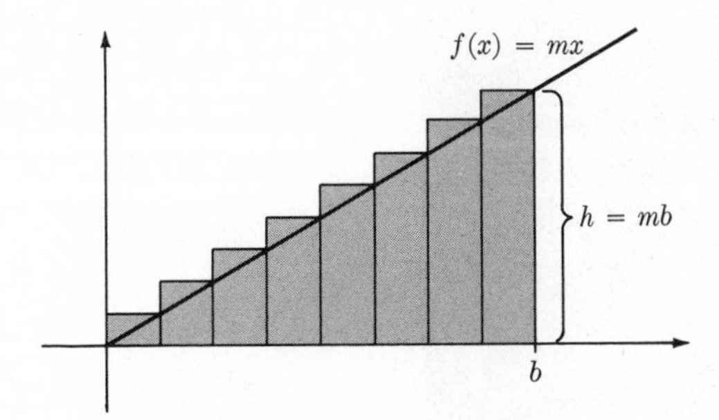

FIGURE 10.3 S_n = sum of rectangular areas

[From a careful reading of Theorem 4, formulas (1)–(4) apply *only* if the sequences on the right-hand side are known to converge; thus the above chain of equalities appears unjustified. However, once completed, the chain can be justified by starting at the *end* and working back, noting at each step that the sequences in question actually do converge. For example, $\lim (1/n)$ exists, so we can legitimately conclude from (1) that $\lim \frac{1}{2n} = \frac{1}{2} \lim \frac{1}{n} = 0$.]

Example 2. $\lim\limits_{n\to\infty} \dfrac{n^2 + n}{2n^2 + 1} = ?$ *Solution.* When n is large, then the dominant term in the numerator is n^2, and in the denominator is $2n^2$, so we can expect a limit equal to $n^2/2n^2 = 1/2$. A simple trick allows us to prove this, using Theorem 3. Divide numerator and denominator of the given sequence by n^2, and proceed as follows:

$$\lim \frac{n^2 + n}{2n^2 + 1} = \lim \frac{1 + 1/n}{2 + 1/n^2} \qquad \text{[by algebra]}$$

$$= \frac{\lim (1 + 1/n)}{\lim (2 + 1/n^2)} \qquad \text{[by (4)]}$$

$$= \frac{\lim 1 + \lim 1/n}{\lim 2 + (\lim 1/n)^2} \qquad \text{[by (2) and (3)]}$$

$$= \frac{1 + 0}{2 + 0} \qquad \text{[Examples 1 and 2, §10.1]}$$

$$= \frac{1}{2}.$$

Example 3. The resilience of a ball is measured by its *coefficient of restitution r;* dropped from height h onto a very hard surface, the ball comes back up to height rh [Fig. 4]. The livelier the ball, the larger r; but in any case, it cannot bounce back to the height from which it was dropped, so $rh < h$, i.e. $r < 1$.

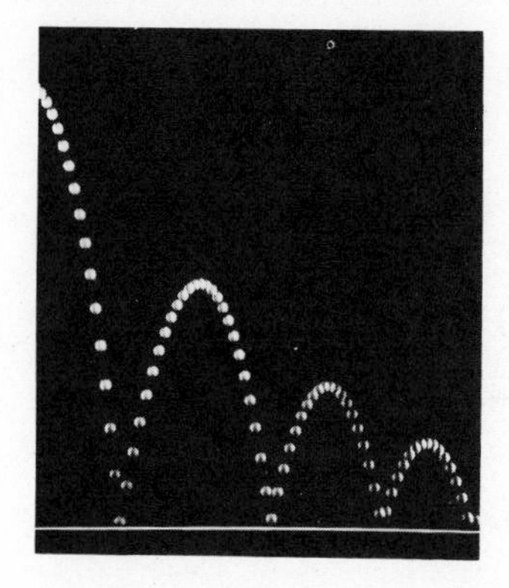

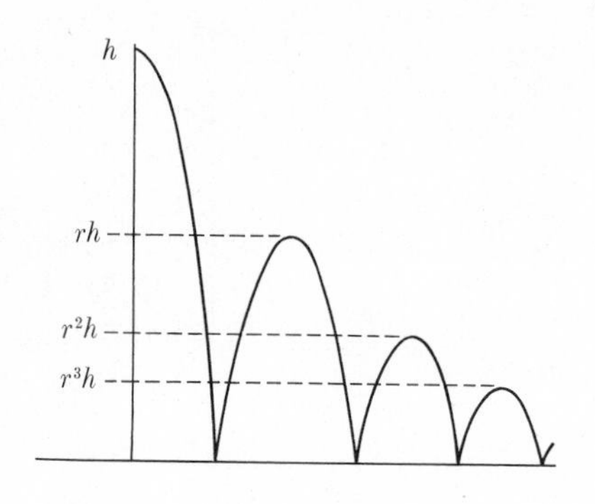

FIGURE 10.4*

If the ball takes two bounces, then the second bounce amounts to a drop from height rh, so it comes back up to height $r \cdot rh = r^2h$. After n bounces, it comes back up to height r^nh. Thus no matter how lively the ball, in the long run the height decreases to zero, because

$$\begin{aligned} \lim (r^n h) &= h \lim r^n \qquad &[\text{by } (1)] \\ &= 0. &[\text{Example 4, §10.1}] \end{aligned}$$

Example 4. Define a sequence $\{s_n\}$ by setting $s_0 = 1$, and $s_{n+1} = \dfrac{s_n{}^2 + 10}{2s_n}$.

$\left(\text{Thus, } s_1 = \dfrac{s_0{}^2 + 10}{2s_0} = \dfrac{11}{2},\quad s_2 = \dfrac{s_1{}^2 + 10}{2s_1} = \dfrac{161}{41}, \text{ and so on.}\right)$ *Supposing that s_n converges,* find its limit. *Solution.* Suppose that $\{s_n\}$ converges to S, i.e. $\lim s_n = S$. Then also $\lim s_{n+1} = S$ [see Problem 6, §10.1]. Hence, taking limits in the defining relation $s_{n+1} = (s_n{}^2 + 10)/2s_n$, we obtain

$$S = \lim s_{n+1} = \lim \frac{s_n{}^2 + 10}{2s_n} = \frac{\lim (s_n)^2 + 10}{2 \lim s_n} = \frac{S^2 + 10}{2S}.$$

This gives an equation for S,

$$S = \frac{S^2 + 10}{2S}, \quad \text{or} \quad 2S^2 = S^2 + 10, \quad \text{or} \quad S^2 = 10, \quad \text{or} \quad S = \pm\sqrt{10}.$$

We determine the sign as follows: Since $s_0 = 1$, it is easy to see that $s_n > 0$ for every n. Hence, by Theorem 3, $S = \lim s_n \geq 0$, so $S = \sqrt{10}$, not $-\sqrt{10}$.

This last example brings out the restricted nature of Theorem 4; granted that certain limits exist, others can be deduced from them. In Example 4, however, we have as yet no proof that $\{s_n\}$ does in fact converge, so the calculations are conditional on a proof of convergence. This gap is filled in the next section.

**From PSSC* PHYSICS, *D.C. Heath and Company, Boston, 1965*

We return now to the postponed proofs of Theorems 3 and 4. The statement of Theorem 3 seems perfectly obvious intuitively:

If $a_n \leq b_n$ *for all* n, *and* $\lim a_n = A$ *and* $\lim b_n = B$, *then* $A \leq B$.

The trick is to construct a proof, based on the "ϵ, N_ϵ" condition in Definition 1. Suppose that A is *not* $\leq B$. Then $A > B$, and Fig. 5 suggests how to obtain a contradiction; go so far out in both sequences that $b_n < \frac{1}{2}(A + B)$ and $\frac{1}{2}(A + B) < a_n$, and thus contradict $a_n \leq b_n$. The following proof carries out this idea in detail.

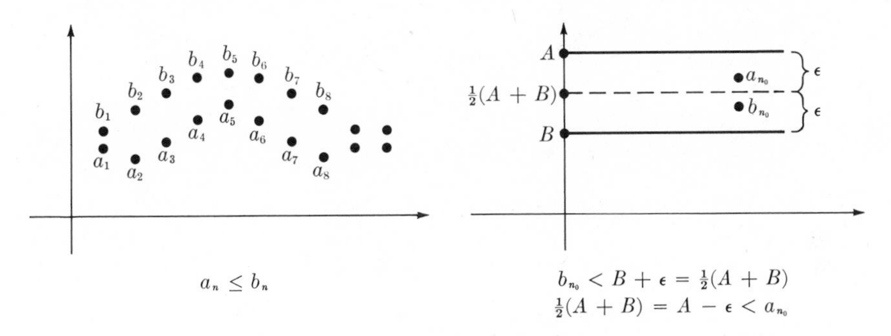

FIGURE 10.5

Proof of Theorem 3

Suppose that $A > B$. Then $\epsilon = \frac{1}{2}(A - B) > 0$. Since $\lim a_n = A$ and $\lim b_n = B$, there are numbers n_1 and n_2 such that

$$n > n_1 \quad \Rightarrow \quad |a_n - A| < \tfrac{1}{2}(A - B) \tag{5}$$

$$n > n_2 \quad \Rightarrow \quad |b_n - B| < \tfrac{1}{2}(A - B). \tag{6}$$

Let n_0 be any integer greater than both n_1 and n_2.† Then from (5) and (6),

$$|a_{n_0} - A| < \tfrac{1}{2}(A - B) \qquad \text{and} \qquad |b_{n_0} - B| < \tfrac{1}{2}(A - B).$$

The first inequality entails

$$a_{n_0} > A - \tfrac{1}{2}(A - B) = \tfrac{1}{2}(A + B),$$

and the second entails

$$b_{n_0} < B + \tfrac{1}{2}(A - B) = \tfrac{1}{2}(A + B).$$

Hence, by the transitive law, $b_{n_0} < a_{n_0}$. This contradicts the hypothesis that $a_n \leq b_n$ for all n. Since the supposition that $A > B$ led to a contradiction, it must be false; hence $A \leq B$. *Q.E.D.*

---

† That such an integer n_0 exists is intuitively obvious; the formal justification is the *Archimedean property:* The integers have no upper bound [§AI.6].

Proof of Theorem 4

We prove the equations (2), (3), (1), and (4), in that order.

Sketchy proof of (2). $\lim (a_n + b_n) = \lim a_n + \lim b_n$. Let $\lim a_n = A$ and $\lim b_n = B$; then we must prove that $\lim (a_n + b_n) = A + B$. In other words, given $\epsilon > 0$, we must make $a_n + b_n$ approximate $A + B$ with an error $<\epsilon$. This is easy; let a_n approximate A and b_n approximate B, each with an error $<(\epsilon/2)$.

Formal proof of (2). Since $\lim a_n = A$, and $\epsilon/2$ is a positive number, there is, by Definition 1, a number n_1 such that

$$n > n_1 \quad \Rightarrow \quad |a_n - A| < \epsilon/2. \tag{7}$$

Similarly, there is a number n_2 such that

$$n > n_2 \quad \Rightarrow \quad |b_n - B| < \epsilon/2. \tag{8}$$

Let N_ϵ be the larger of n_1 and n_2.

Suppose that $n > N_\epsilon$. Then $n > n_1$, so $|a_n - A| < \epsilon/2$, by (7). Further, $n > n_2$, so $|b_n - B| < \epsilon/2$, by (8). Hence

$$\begin{aligned} |(a_n + b_n) - (A + B)| &= |(a_n - A) + (b_n - B)| \\ &\leq |a_n - A| + |b_n - B| \qquad \text{[Triangle inequality]} \\ &< \epsilon/2 + \epsilon/2 = \epsilon. \end{aligned}$$

Thus

$$n > N_\epsilon \quad \Rightarrow \quad |(a_n + b_n) - (A + B)| < \epsilon.$$

Hence, by Definition 1, $\lim (a_n + b_n) = A + B = (\lim a_n) + (\lim b_n)$.

Sketchy proof of (3). $\lim (a_n b_n) = (\lim a_n)(\lim b_n)$. Let $A = \lim a_n$ and $B = \lim b_n$; we want to show that for any $\epsilon > 0$, we can make $|a_n b_n - AB| < \epsilon$. But

$$a_n b_n - AB = (a_n - A)b_n + A(b_n - B),$$

so the problem is solved if we can make

$$|(a_n - A)b_n| < \epsilon/2 \qquad \text{and} \qquad |A(b_n - B)| < \epsilon/2.$$

To do this, we must make $|a_n - A|$ small [done in (11) below], $|b_n|$ not too large [done in (10) below], and $|b_n - B|$ small [done in (9) below]. Exactly how small $|a_n - A|$ and $|b_n - B|$ are made is determined by the way all the pieces fit together at the end of the proof.

Formal proof of (3). Given $\epsilon > 0$, there is a number n_1 such that

$$n > n_1 \Rightarrow |b_n - B| < \frac{\epsilon}{2|A| + 1} \tag{9}$$

[definition of $\lim b_n = B$]. Since $|b_n| \leq |B| + |b_n - B|$, it follows from (9) that

$$n > n_1 \Rightarrow |b_n| < |B| + \frac{\epsilon}{2|A| + 1} = M, \tag{10}$$

where, for brevity, we have set $|B| + \epsilon/(2|A| + 1) = M$. Notice that $M > 0$, so $\epsilon/2M > 0$. Hence there is a number n_2 such that

$$n > n_2 \Rightarrow |a_n - A| < \epsilon/2M. \tag{11}$$

Let N_ϵ be the larger of n_1 and n_2.

Suppose that $n > N_\epsilon$. Then $n > n_1$ and $n > n_2$, so

$$\begin{aligned}
|a_n b_n - AB| &= |(a_n - A)b_n + A(b_n - B)| && \text{[algebra]} \\
&\leq |a_n - A| \cdot |b_n| + |A| \cdot |b_n - B| && \text{[Triangle inequality]} \\
&< \frac{\epsilon}{2M} \cdot M + |A| \frac{\epsilon}{2|A| + 1} && \text{[by (9), (10), and (11)]} \\
&= \epsilon \left(\frac{1}{2} + \frac{|A|}{2|A| + 1} \right) \\
&< \epsilon.
\end{aligned}$$

Thus $n > N_\epsilon \Rightarrow |a_n b_n - AB| < \epsilon$, so $\lim a_n b_n = AB$. *Q.E.D.*

Remark. If $A \neq 0$, the right-hand side in (9) could be simply $\epsilon/2|A|$; we make it $\epsilon/(2|A| + 1)$ to avoid a possible division by zero.

Proof of (1). $\lim ca_n = c \lim a_n$. This follows immediately from (3), on setting $b_n = c$.

Sketchy proof of (4). $\lim \dfrac{a_n}{b_n} = \dfrac{\lim a_n}{\lim b_n}$ if $\lim b_n \neq 0$. We will prove the simpler relation

$$\lim \frac{1}{b_n} = \frac{1}{\lim b_n} \quad \text{if } \lim b_n \neq 0, \tag{12}$$

and then (4) will follow from (3) and (12):

$$\lim \frac{a_n}{b_n} = \lim \left(a_n \cdot \frac{1}{b_n} \right) = (\lim a_n) \left(\lim \frac{1}{b_n} \right) = (\lim a_n) \frac{1}{\lim b_n}.$$

To prove (12), we must make $|1/b_n - 1/B| < \epsilon$, where $B = \lim b_n$. Since

$$\frac{1}{b_n} - \frac{1}{B} = \frac{B - b_n}{b_n B},$$

we must make $|B - b_n|$ small, *and* $|b_n B|$ *not too small.* We begin by controlling $|b_n B|$.

Formal proof of (12). Since $|B| > 0$ by assumption, there is a number n_1 such that

$$n > n_1 \quad \Rightarrow \quad |b_n - B| < \tfrac{1}{2}|B|.$$

Further, it is easy to see [Problem 6 below] that

$$|b_n - B| < \tfrac{1}{2}|B| \quad \Rightarrow \quad |b_n| > \tfrac{1}{2}|B|,$$

so

$$n > n_1 \quad \Rightarrow \quad |b_n| > \tfrac{1}{2}|B| \quad \Rightarrow \quad |b_n B| > \tfrac{1}{2}|B|^2. \tag{13}$$

This will keep $|b_n B|$ from getting too small.

Now, let $\epsilon > 0$ be given. There is a number n_2 such that

$$n > n_2 \quad \Rightarrow \quad |B - b_n| < \tfrac{1}{2}B^2\epsilon. \tag{14}$$

Let N_ϵ be the larger of n_1 and n_2. Suppose that $n > N_\epsilon$. Then by (13) and (14),

$$\left|\frac{1}{b_n} - \frac{1}{B}\right| = \frac{|B - b_n|}{|b_n B|} < \frac{\frac{1}{2}B^2\epsilon}{\frac{1}{2}B^2} = \epsilon.$$

This completes the proof of (12); and (12) and (3) together imply (4); so Theorem 4 is proved.

PROBLEMS

The first nine problems use only elementary algebra and inequalities, together with the theory above. The last six refer to material on calculus from Chapters I–IX.

1. Evaluate the following limits as $n \to \infty$, using results above.

•(a) $\lim \dfrac{3n^2 + 2n + 1}{n^2 - 5}$ (b) $\lim \dfrac{n(n+1)}{(n+2)(n+3)}$

• (c) $\lim \dfrac{1}{n(n+1)}$

(d) $\lim 2^{-n}$

• (e) $\lim \dfrac{1-r^n}{1-r}, \quad |r| < 1$

(f) $\lim r^{2n}, \quad |r| < 1$

2. In §5.1, we obtained

$$S_n = \frac{b}{n}\sum_{k=1}^{n}\left(\frac{bk}{n}\right)^2 = \frac{b^3}{n^3}\frac{n(n+1)(2n+1)}{6}$$

as an approximation to the area under a parabola. Prove that $\lim S_n = \frac{1}{3}b^3$.

3. The ball in Example 3 travels a distance h on its first drop, $2 \times rh$ on the rise and fall during the first bounce, $2 \times r^2h$ on the rise and fall of the second bounce, and so on. This problem computes the total distance traveled if the ball continues bouncing like this forever.

(a) Show that $(1-r)(1+r+r^2+\cdots+r^n) = 1-r^{n+1}$, and hence

$$1+r+\cdots+r^n = \frac{1-r^{n+1}}{1-r} \quad \text{if } r \neq 1.$$

(b) Show that $r+r^2+\cdots+r^n = \dfrac{r-r^{n+2}}{1-r}$ if $r \neq 1$.

(c) Show that with n bounces, the distance traveled is

$$h(1+r-2r^{n+1})/(1-r).$$

• (d) Take $\lim\limits_{n\to\infty}$, and find the total distance traveled if the ball bounces forever.

4. Suppose that $\lim\limits_{n\to\infty} s_n = S$.

(a) Prove that $\lim\limits_{n\to\infty} (s_n - s_{n+k}) = 0$. [See Problem 6, §10.1.]

(b) Prove that $\lim\limits_{n\to\infty} |s_n - s_{n+k}| = 0$. [See Problem 5, §10.1.]

5. Prove that the sequence $c_n = (-1)^n + 1/n$ is not convergent. $\left[\text{Hint: Use Problem 4(b), noting that } |c_n - c_{n+1}| = 2 + \dfrac{(-1)^n}{n(n+1)}.\right]$

• **6.** Prove that $|b-B| < \frac{1}{2}|B| \Rightarrow |b| > \frac{1}{2}|B|$.

POE PRINCIPLE

7. This problem proves the *Poe Principle* for sequences: *If* $a_n \le b_n \le c_n$ *for all n, and* $\lim a_n = L$, *and* $\lim c_n = L$, *then* $\lim b_n = L$. [§2.6 gives the Poe Principle for functions.]

(a) Show that $|b_n - L| \le |a_n - L|$ if $L \ge b_n$.

(b) Show that $|b_n - L| \le |c_n - L|$ if $L \le b_n$.

(c) From (a) and (b), show that $|b_n - L| \le |a_n - L| + |c_n - L|$ for all n.

(d) From the hypotheses on $\{a_n\}$ and $\{c_n\}$, show that $\lim |b_n - L| = 0$, hence $\lim b_n = L$. [See Theorems 2 and 4.]

8. Define a sequence by setting $a_1 = 1$, and $a_{n+1} = \dfrac{1}{1 + a_n}$.

(a) Express a_2, a_3, a_4, a_5 as fractions.

(b) Show that *if* $\{a_n\}$ is convergent, then $\lim a_n = (\sqrt{5} - 1)/2$. [Hint: Use Problem 6, §10.1, and the definition of a_{n+1}.]

CONTINUED FRACTION

Remark. The sequence $\{a_n\}$ is called a *continued fraction* sequence, since

$$a_{n+1} = \frac{1}{1 + a_n} = \frac{1}{1 + \dfrac{1}{1 + a_{n-1}}} = \frac{1}{1 + \dfrac{1}{1 + \dfrac{1}{1 + \cdots}}}.$$

CAUTION

9. Set $a_1 = 1$, and $a_{n+1} = 2/a_n$.

(a) Express a_2, a_3, a_4, a_5 as fractions.

(b) Show that *if* $\{a_n\}$ is convergent, then $\lim a_n = \sqrt{2}$.

(c) Is $\{a_n\}$ convergent or not? Does $\lim a_n = \sqrt{2}$?

10. Let T_n be the approximation to $\int_a^b f$ obtained by the trapezoid rule with n subdivisions; then from §9.4,

$$\left| T_n - \int_a^b f \right| \le \frac{(b - a)^3}{12n^2} \max_{[a,b]} |f''(x)|,$$

if f'' is continuous on $[a,b]$. Prove that $\lim T_n = \int_a^b f$.

11. Let $s_n(x) = x - \dfrac{x^3}{3!} + \dfrac{x^5}{5!} - \cdots + (-1)^n \dfrac{x^{2n+1}}{(2n + 1)!}$. The Taylor expansion [§9.2] shows that $|\sin x - s_n(x)| \le \dfrac{|x|^{2n+3}}{(2n + 3)!}$. Prove that $\lim s_n(x) = \sin x$, at least for $|x| \le 1$. [See Problem 13 for $|x| > 1$.]

12. Let $e_n(x) = 1 + \frac{x}{1!} + \cdots + \frac{x^n}{n!}$. The Taylor expansion [§9.2] shows that $|e_n(x) - e^x| \le e^{|x|} \cdot \frac{|x|^{n+1}}{(n+1)!}$. Prove that $\lim e_n(x) = e^x$, at least for $|x| \le 1$. [See Problem 13 for $|x| > 1$.]

13. This problem proves that $\lim_{n\to\infty} \frac{x^n}{n!} = 0$ for any $x > 0$, and applies this to complete Problems 11 and 12.

(a) Let $a_n = x^n/n!$. Show that $a_{n_0+1}/a_{n_0} = x/(n_0 + 1)$.

(b) Show that $a_{n_0+2}/a_{n_0} < (x/n_0)^2, \ldots, a_{n_0+k}/a_{n_0} < (x/n_0)^k$.

(c) Show that for any $n > n_0$, $a_n < a_{n_0}\left(\frac{x}{n_0}\right)^{n-n_0} = a_{n_0}\left(\frac{n_0}{x}\right)^{n_0}\left(\frac{x}{n_0}\right)^n$.

(d) Show that $\lim_{n\to\infty} a_n = 0$. [Choose $n_0 > x$ in part (c).]

(e) Show that in Problem 13, $\lim_{n\to\infty} e_n(x) = e^x$ for *every* real x.

(f) Show that in Problem 12, $\lim_{n\to\infty} s_n(x) = \sin x$ for *every* real x.

14. Establish (a)–(e) below, and then prove *Wallis's formula:*

$$\frac{\pi}{2} = \lim_{n\to\infty}\left(\frac{2}{1}\cdot\frac{2}{3}\cdot\frac{4}{3}\cdot\frac{4}{5}\cdot\frac{6}{5}\cdot\frac{6}{7}\cdots\frac{2n}{2n-1}\cdot\frac{2n}{2n+1}\right).$$

WALLIS'S FORMULA

(a) $\int_0^{\pi/2} \sin^k x\, dx = \frac{k}{2k+1}\int_0^{\pi/2} \sin^{k-2} x\, dx.$ [See §6.5.]

(b) $\int_0^{\pi/2} \sin^{2n} x\, dx = \frac{(2n-1)(2n-3)\cdots 1}{(2n)(2n-2)\cdots 2}\cdot\frac{\pi}{2}.$

(c) $\int_0^{\pi/2} \sin^{2n+1} x\, dx = \frac{2n(2n-2)\cdots 2}{(2n+1)(2n-1)\cdots 3}.$

(d) $0 < \int_0^{\pi/2} \sin^{2n+1} x\, dx \le \int_0^{\pi/2} \sin^{2n} x\, dx \le \int_0^{\pi/2} \sin^{2n-1} x\, dx.$

[Hint: Do *not* use parts (a)–(c) for this; use your head.]

(e) $1 \le \frac{\int_0^{\pi/2} \sin^{2n} x\, dx}{\int_0^{\pi/2} \sin^{2n+1} x\, dx} \le 1 + \frac{1}{2n}.$

(continued)

Now prove Wallis's formula. [John Wallis, thirty years older than Newton, was among the first to work with infinite sums and infinite products. The infinite product formula proved here is more famous for antiquity than for utility. However, we will use it in the next problem to evaluate

$$\int_0^\infty e^{-x^2}\,dx = \lim_{m\to\infty}\int_0^m e^{-x^2}\,dx.]$$

15. This problem shows that $\int_0^\infty e^{-x^2}\,dx = \frac{1}{2}\sqrt{\pi}$. The proof outlined here is complicated, but easier proofs require more advanced mathematics.

$\int_0^\infty e^{-x^2}\,dx$

(a) Show that $1 - x^2 \le e^{-x^2}$ if $0 \le x \le 1$, and $e^{-x^2} \le 1/(1 + x^2)$ if $0 \le x$. [Note that e^{-x^2}, $1 - x^2$, and $1/(1 + x^2)$ all agree for $x = 0$; compare their derivatives.]

(b) Show that

$$\int_0^1 (1 - x^2)^n\,dx = \frac{2}{3}\cdot\frac{4}{5}\cdots\frac{2n}{2n+1}$$

and

$$\lim_{k\to\infty}\int_0^k (1 + x^2)^{-n}\,dx = \frac{\pi}{2}\cdot\frac{1}{2}\cdot\frac{3}{4}\cdots\frac{2n-3}{2n-2}.$$

[Use trigonometric substitutions [§6.6], the previous problem, and §6.5.]

(c) Show that $\int_0^1 e^{-(mx)^2}\,dx = \frac{1}{m}\int_0^m e^{-u^2}\,du$, and

$$\lim_{k\to\infty}\int_0^k e^{-(mx)^2}\,dx = \frac{1}{m}\lim_{k\to\infty}\int_0^{km} e^{-u^2}\,du = \frac{1}{m}\int_0^\infty e^{-u^2}\,du.$$

(d) Raise the inequalities in part (a) to the power m^2, and show that

$$m\cdot\frac{2}{3}\cdot\frac{4}{5}\cdots\frac{2m^2}{2m^2+1} \le \int_0^m e^{-u^2}\,du \le \lim_{k\to\infty}\int_0^{km} e^{-u^2}\,du$$

$$\le m\cdot\frac{\pi}{2}\cdot\frac{1}{2}\cdot\frac{3}{4}\cdots\frac{2m^2-3}{2m^2-2}.$$

(e) Show, using Wallis's formula [previous problem], that

$$\lim_{n\to\infty}\left(\frac{1}{\sqrt{n}}\cdot\frac{2\cdot4\cdot6\cdots(2n)}{1\cdot3\cdot5\cdots(2n-1)}\right) = \sqrt{\pi};$$

hence

$$\lim_{m\to\infty}\left(\frac{1}{m}\cdot\frac{2\cdot4\cdot6\cdots(2m^2)}{1\cdot3\cdot5\cdots(2m^2-1)}\right) = \sqrt{\pi}.$$

(f) From (d) and (e), deduce that $\lim_{m\to\infty} \int_0^m e^{-u^2}\,du = \dfrac{\sqrt{\pi}}{2}$.

10.3 BOUNDED AND MONOTONE SEQUENCES

In Theorems 1–4, sequences were *assumed* to converge, and conclusions were drawn about relations involving their limits. This section takes up the equally important question of determining which sequences are, in fact, convergent.

A simple example of a divergent sequence is the positive integers, 1, 2, 3,.... It seems obvious that this sequence, whose terms grow arbitrarily large, cannot converge to any number. ("Infinity," of course, is not a number.) As usual, the obvious is true, but it has to be proved. The divergence of this sequence, and many others, follows from

CONVERGENT SEQUENCES ARE BOUNDED

Theorem 5. *Every convergent sequence* $\{s_n\}$ *is bounded,* i.e. *there is a number* M *such that* $|s_n| \le M$ *for all* n.

Proof. Let $\lim s_n = S$. We apply Definition 1 with $\epsilon = 1$; thus there is a number N_1 such that

$$n > N_1 \quad \Rightarrow \quad |s_n - S| < 1. \tag{1}$$

By the Triangle Inequality, $|s_n| = |S + (s_n - S)| \le |S| + |s_n - S|$, hence from (1),

$$n > N_1 \quad \Rightarrow \quad |s_n| < |S| + 1. \tag{2}$$

This bounds the terms s_n for $n > N_1$, and the rest, being finite in number, are also bounded [Fig. 6]. To get a bound M for the whole sequence, we simply take the *largest* of the numbers

$$|s_1|,\ |s_2|, \ldots,\ |s_{N_1}|,\ |S| + 1.$$

This automatically makes $|s_n| \le M$ for $1 \le n \le N_1$; and (2) makes $|s_n| < |S| + 1 \le M$ for $n > N_1$. *Q.E.D.*

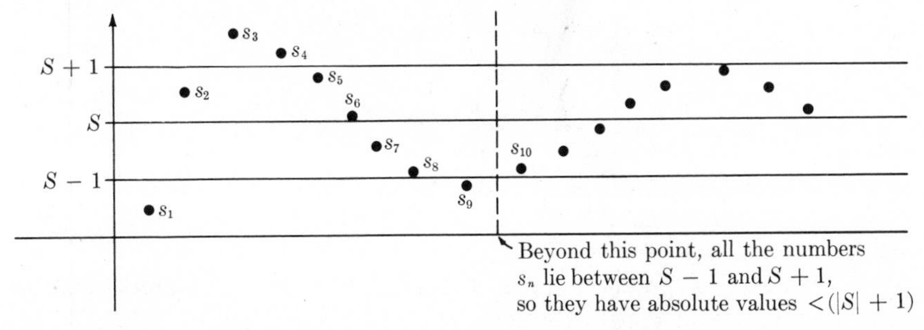

FIGURE 10.6 Every convergent sequence is bounded. In this picture, $|s_n| \le |s_3|$ for every n.

The converse of Theorem 5 is obviously false; "bounded" does not imply "convergent," as you can see from the sequence $c_n = (-1)^n + 1/n$, which is bounded ($|c_n| \leq 3/2$) but not convergent (Problem 5, §10.2). The fundamental theorem guaranteeing convergence of a sequence $\{s_n\}$ requires that it be bounded *and monotone*, i.e. either $s_1 \leq s_2 \leq s_3 \leq \cdots$ (monotone increasing) or $s_1 \geq s_2 \geq \cdots$ (monotone decreasing).

MONOTONE CONVERGENCE THEOREM

Theorem 6. *If $\{s_n\}$ is bounded and monotone, then $\{s_n\}$ converges.*

Proof. We use the *least upper bound axiom of the real numbers:* Any nonempty set of real numbers having an upper bound has a least upper bound [§AI.6]. As Fig. 7 suggests, when $\{s_n\}$ is increasing and bounded, then it converges to its least upper bound. To prove this, suppose that $s_1 \leq s_2 \leq \cdots \leq s_n \leq \cdots \leq M$. Then the set of numbers $\{s_1, s_2, s_3, \ldots\}$ is not empty, and has an upper bound M, so it has a *least* upper bound L. We prove that $\lim_{n\to\infty} s_n = L$. Suppose that $\epsilon > 0$ is given. Since L is an upper bound for the s_n, we have

$$s_n \leq L \qquad \text{for every } n. \tag{3}$$

Since L is the *least* upper bound, $L - \epsilon$ is *not* an upper bound; therefore, for some s_N, we must have

$$L - \epsilon < s_N. \tag{4}$$

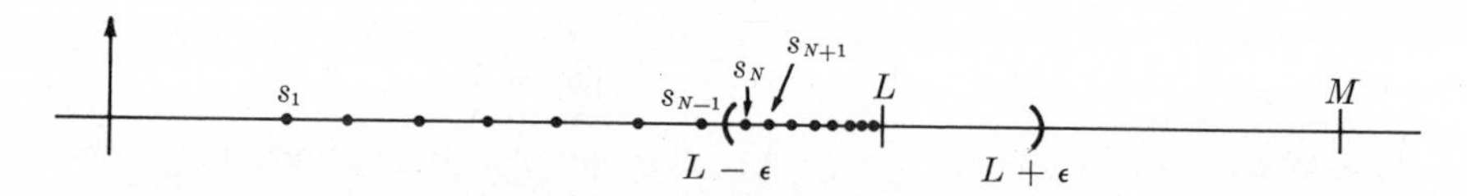

FIGURE 10.7 M is an upper bound for the numbers s_n, and L is the least upper bound

Since $\{s_n\}$ is *increasing*, $n > N \Rightarrow s_n \geq s_{n-1} \geq \cdots \geq s_N$. Hence, from (3) and (4),

$$n > N \Rightarrow L - \epsilon < s_N \leq s_n \leq L.$$

But

$$L - \epsilon < s_n < L \Rightarrow |s_n - L| < \epsilon;$$

hence

$$n > N \Rightarrow |s_n - L| < \epsilon,$$

which proves that $\lim s_n = L$.

When $\{s_n\}$ is *decreasing* and bounded, then $\{-s_n\}$ is *increasing* and bounded, so $\{-s_n\}$ converges; hence $\{s_n\}$ converges, too. *Q.E.D.*

Example 1. Let $a_1, a_2, \ldots$ be any sequence of "digits," i.e. integers from 0 through 9. Define the corresponding sequence of finite decimals

$$d_n = 0.a_1a_2\ldots a_n.$$

Prove that $\{d_n\}$ converges. *Solution.* $0 \leq d_n \leq 1$ for every n, and $d_n \leq d_{n+1}$ for every n. Thus $\{d_n\}$ is bounded and monotone increasing, so it converges. The limit of this sequence is generally denoted $0.a_1a_2a_3\ldots$, and called an *infinite decimal*. Thus every infinite decimal is a real number, no matter how the digits are chosen. For example,

$$0.10110011100011110000\ldots$$

is a real number.

Example 2. Consider the sequence defined by

$$s_0 = 1, \quad s_{n+1} = \frac{s_n{}^2 + 10}{2s_n}.$$

With Theorems 3 and 4, it is easy to show that *if* $\{s_n\}$ converges, then $\lim s_n = \sqrt{10}$. The Monotone Convergence Theorem can be used to complete the argument, by proving that $\{s_n\}$ is in fact convergent; to apply it, we will show that $s_1 > s_2 > s_3 > \cdots > 0$.

First, it is clear that $s_n > 0$ for all n. Next, observe that $s_n{}^2 > 10$ for $n = 1, 2, \ldots$, since

$$s_n{}^2 - 10 = \left(\frac{s_{n-1}^2 + 10}{2s_{n-1}}\right)^2 - 10 = \frac{s_{n-1}^4 - 20s_{n-1}^2 + 100}{(2s_{n-1})^2} = \left(\frac{s_{n-1} - 10}{2s_{n-1}}\right)^2 > 0. \quad (5)$$

Finally, $s_n > s_{n+1}$ for $n = 1, 2, \ldots$, since

$$s_n - s_{n+1} = s_n - \frac{s_n{}^2 + 10}{2s_n} = \frac{s_n{}^2 - 10}{2s_n} > 0 \qquad [\text{by } (5)].$$

Thus $s_1, s_2, \ldots$ is monotone decreasing, and bounded, so $\lim s_n$ exists.

Remark 1. The argument in Example 2 can be used to prove that *every positive number has a square root,* a cube root, and in general a kth root, $k = 0, 1, 2, \ldots$. [See Problems 2–4 below.]

PHILOSOPHICAL REMARKS

Remark 2. We used the least upper bound axiom to prove the Monotone Convergence Theorem, but the roles could be reversed. We could replace the least upper bound axiom with a "monotone convergence axiom": *every bounded monotone sequence converges;* and then prove a "least upper bound theorem":

Every nonempty set, bounded above, has a least upper bound.

Proof. Let M be an upper bound for S, and s_0 any member of S. We will define a decreasing sequence $a_1 \geq a_2 \geq \cdots \geq s_0$, such that $\lim a_n$ is the least upper bound of S.

Let $d = M - s_0$ be the distance between M and s_0, and let $c_1 = M - \frac{1}{2}d$ be the midpoint between M and s_0. If c_1 is an upper bound for S, set $a_1 = c_1$; if not, set $a_1 = M$. In either case,

$$a_1 \text{ is an upper bound for } S, \quad \text{and}$$
$$\text{there is a member } s_1 \text{ of } S \text{ such that } a_1 - \tfrac{1}{2}d < s_1 .$$

Now let $c_2 = a_1 - \frac{1}{4}d$. If c_2 is an upper bound for S, set $a_2 = c_2$; if not, set $a_2 = c_1$. In either case,

$$a_2 \text{ is an upper bound for } S,$$
$$a_1 \geq a_2 \geq s_0, \quad \text{and}$$
$$\text{there is a member } s_2 \text{ of } S \text{ such that } a_2 - \tfrac{1}{4}d < s_2 .$$

Continuing thus, we define a sequence $\{a_n\}$ such that $a_1 \geq a_2 \geq a_3 \geq \cdots \geq s_0$, and

(i) a_n is an upper bound for S, and

(ii) there is a member s_n of S such that $a_n - 2^{-n}d < s_n$.

By the "monotone convergence axiom", $\{a_n\}$ converges; let $A = \lim a_n$. Then *A is an upper bound for S;* because if s is any member of S, then $s \leq a_n$ for all n [by (i)], so $s \leq A = \lim a_n$. Finally, *A is the least upper bound for S;* because if $B < A$, then $B < A - 2^{-n}d$ for some n, hence $B < a_n - 2^{-n}d < s_n$ by (ii); hence B is not an upper bound. *Q.E.D.*

Thus the "least upper bound" property and the "monotone convergence" property are equivalent; each implies the other. And both have a similar intuitive validity. The reason for making the first of these properties an axiom and the second a theorem is, roughly, that the first can be applied directly in more cases than the second can.

PROBLEMS

1. Which of the following sequences are monotone? Which are bounded? Which are convergent? [No proof required. Tools that make the last two questions routine are given in the next chapter; for now, make as educated a guess as you can.]

(a) $D_n = n$th digit in the decimal expansion of $1/3$

•(b) $D_n = n$th digit in the decimal expansion of $3/7$ B MI C

(c) $s_n = \sin(\pi n/2)$

•(d) $s_n = \log n$

(e) $s_n = \sin(\pi n/2) \log n$

• (f) $s_n = 1 - \frac{1}{2} + \frac{1}{3} - \frac{1}{4} + \cdots + \frac{(-1)^{n+1}}{n}$

(g) $s_n = 1 + \frac{1}{2} + \frac{1}{3} + \frac{1}{4} + \cdots + \frac{1}{n}$

• (h) $s_n = 1 + \frac{1}{4} + \frac{1}{9} + \cdots + \frac{1}{n^2}$

(i) s_n is the perimeter of an equilateral polygon of 2^n sides inscribed in a circle of radius 1.

• (j) $M \geq f \geq 0$, and s_n is the lower Riemann sum for $\int_a^b f$ obtained by partitioning the interval $[a,b]$ into 2^n equal subintervals.

(k) s_n is the lower Riemann sum for $\int_0^1 \frac{1}{x}\,dx$ obtained by partitioning $[0,1]$ into 2^n equal subintervals.

• (l) Same as part (j), but s_n uses only n equal subintervals. [Don't be too hasty.]

2. Define a sequence $\{a_n\}$ by $a_0 = 1$, $a_{n+1} = \frac{a_n{}^2 + c}{2a_n}$, where $c > 0$. Prove the following.

PROOF THAT SQUARE ROOTS EXIST

(a) $a_n{}^2 \geq c$ for $n = 1, 2, \ldots$ [See Example 2.]
(b) $a_1 \geq a_2 \geq \cdots > 0$
(c) $\{a_n\}$ converges
(d) $(\lim a_n)^2 = c$
(e) $\lim a_n > 0$
(f) c has a positive square root

3. Define a sequence $\{b_n\}$ by $b_0 = 1$, $b_{n+1} = \frac{2b_n{}^3 + c}{3b_n{}^2}$, where $c > 0$. Prove the following.

PROOF THAT CUBE ROOTS EXIST

(a) $b_n > 0$ for all n
(b) $b_n{}^3 \geq c$ for $n = 1, 2, \ldots$

$\left[\text{Hint: Show that } b_n{}^3 - c = \frac{(b_{n-1}^3 - c)^2(8b_{n-1}^3 + c)}{27b_{n-1}^6}.\right]$

(c) $b_1 \geq b_2 \geq \cdots > 0$
(d) $\{b_n\}$ converges
(e) $(\lim b_n)^3 = c$
(f) $\lim b_n > 0$
(g) c has a positive cube root

Remark. The formulas for the sequences in Problems 2 and 3 were derived by Newton's method [§1.6]. As applied here, they show that every positive number has a positive square root, and a positive cube root. There are simpler ways to prove this; see the *Intermediate Value Theorem* [§AII.5].

4. Let $c > 0$.
(a) Prove that c has a positive fourth root.
(b) Prove that c has a positive fifth root.

5. In the first generation of a certain population, a proportion λ_1 carries a fatal gene. Assuming that all "double carriers" die without reproducing, and that no mutations produce new carriers, and that the effects of the gene are independent of all other vital factors, then the proportion of carriers in the second generation is $\lambda_1(1 + \frac{1}{2}\lambda_1)^{-1}$, and in the nth generation it is $\lambda_n = \lambda_{n-1}(1 + \frac{1}{2}\lambda_{n-1})^{-1}$.
(a) Prove that $\lim \lambda_n$ exists.
(b) Prove that $\lim \lambda_n = 0$.

6. This problem proves the *Archimedean property:* The positive integers have no upper bound.
(a) Let $s_n = n$. Prove that $\{s_n\}$ does not converge. [See Problem 4, §10.2.]
(b) Prove that $\{s_n\}$ is not bounded. [Hint: $\{s_n\}$ is monotone.]

10.4 SEQUENCE LIMITS AND FUNCTION LIMITS

Results about sequence limits translate easily into results on function limits, for the definition of $\lim_{x\to x_0} f(x) = L$ is modeled on that of $\lim_{n\to\infty} s_n = L$.

$\lim_{x\to x_0} f(x) = L$

Definition 2. Let f be defined in an open interval containing x_0, except perhaps at x_0 itself. Then $\lim_{x\to x_0} f(x) = L$ iff, for every $\epsilon > 0$, there is a $\delta_\epsilon > 0$ such that

$$0 < |x - x_0| < \delta_\epsilon \Rightarrow |f(x) - L| < \epsilon. \tag{1}$$

Compare (1) to the corresponding condition for convergence of a sequence $\{s_n\}$ to a limit L: There is an N_ϵ such that

$$n > N_\epsilon \Rightarrow |s_n - L| < \epsilon. \tag{2}$$

In (2), $n > N_\epsilon$ means that s_n is "far out in the sequence"; when s_n is this far out, then $|s_n - L| < \epsilon$. In (1), $0 < |x - x_0| < \delta_\epsilon$ means that x is "very close to x_0, but not equal to x_0"; when x is this close to x_0, but not equal to x_0, then $|f(x) - L| < \epsilon$.

In the definition of $\lim_{x \to x_0} f(x)$, the value $f(x_0)$ need not be defined; and even if $f(x_0)$ *is* defined, we don't take it into account in computing the limit [Fig. 8]. (This is not a deep point, merely a standard convention.)

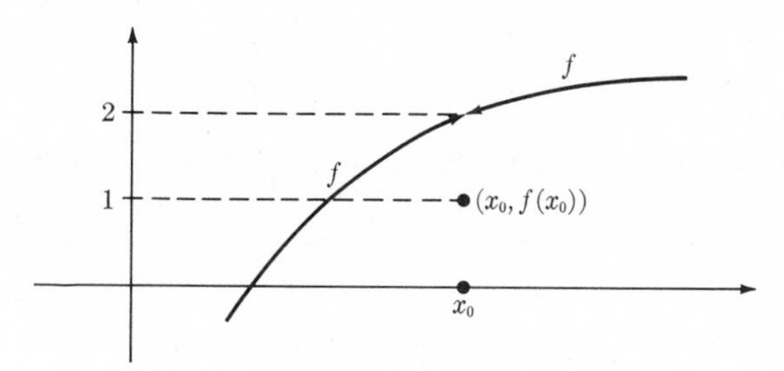

FIGURE 10.8 f is peculiarly defined at x_0, since $f(x_0)$ is less than we would expect. However, we still say that $\lim_{x \to x_0} f(x) = 2$.

It should be fairly easy for you to prove the basic theorems about function limits simply by referring to the corresponding theorems for sequences, replacing every inequality of the form "$n > N$" by an inequality of the form "$0 < |x - x_0| < \delta$". The full details are given in §AII.1. Our aim here is to develop some of the interplay between sequences and functions, based on the following theorem.

Theorem 7. *Suppose that*

(i) $\lim_{x \to x_0} f(x)$ *exists,*

(ii) $\lim_{n \to \infty} x_n = x_0$, *and*

(iii) *for all* n, $x_n \neq x_0$.

Then $\lim_{n \to \infty} f(x_n) = \lim_{x \to x_0} f(x)$.

Proof. Let $\lim_{x \to x_0} f(x) = L$. Given $\epsilon > 0$, there is a number $\delta > 0$ such that

$$0 < |x - x_0| < \delta \quad \Rightarrow \quad |f(x) - L| < \epsilon \tag{3}$$

[definition of $\lim_{x \to x_0} f(x) = L$].

Further, there is a number N_δ such that

$$n > N_\delta \quad \Rightarrow \quad |x_n - x_0| < \delta \tag{4}$$

[definition of $\lim_{n\to\infty} x_n = x_0$].

Since $x_n \neq x_0$ is assumed, we automatically have $0 < |x_n - x_0|$ for every n; hence the implication (4) can be replaced by a stronger one:

$$n > N_\delta \quad \Rightarrow \quad 0 < |x_n - x_0| < \delta. \tag{5}$$

Combining (5) and (3), we get

$$n > N_\delta \quad \Rightarrow \quad 0 < |x_n - x_0| < \delta \quad \Rightarrow \quad |f(x_n) - L| < \epsilon.$$

Since this can be obtained for every $\epsilon > 0$, it follows that $\lim_{n\to\infty} f(x_n) = L$. Since we began the proof by setting $L = \lim_{x\to x_0} f(x)$, the theorem is complete.

The theorem also applies when $x_0 = +\infty$ or $x_0 = -\infty$; the proof is practically the same [Problem 5].

Theorem 7 can be used in a positive way, to evaluate sequence limits when function limits are known; or in a negative way, to prove that certain function limits do not exist. Our first application is this "negative" type.

Example 1. Find, if possible, $\lim_{x\to 0} f(x)$ for the "step function"

$$f(x) = \begin{cases} 1, & x \geq 0 \\ 0, & x < 0. \end{cases}$$

Solution. Suppose that $\lim_{x\to 0} f(x) = L$ for some number L. Consider the sequence $\{1/n\}$, $n = 1, 2, \ldots$. We have $f(1/n) = 1$, and $\lim (1/n) = 0$, and $1/n \neq 0$ for all n. Therefore, Theorem 7 implies that

$$L = \lim_{n\to\infty} f(1/n) = \lim_{n\to\infty} 1 = 1. \tag{6}$$

On the other hand, $f(-1/n) = 0$, and $\lim (-1/n) = 0$, and $-1/n \neq 0$, for all n; therefore Theorem 7 implies that

$$L = \lim_{n\to\infty} f\left(-\frac{1}{n}\right) = \lim_{n\to\infty} 0 = 0.$$

But this contradicts (6), so $\lim_{x\to 0} f(x) = L$ is not possible for any L whatsoever. In other words, in this case $\lim_{x\to 0} f(x)$ does not exist.

To apply Theorem 7 in a positive way, we assume all the results in Chapter II, particularly the following: *If $f'(x_0)$ exists, then* $\lim_{x\to x_0} f(x) = f(x_0)$.

In these positive applications, it is convenient to restate Theorem 7 as follows:

Theorem 8. *If* (i) $\lim_{x \to x_0} f(x) = f(x_0)$ *and*

(ii) $\lim_{n \to \infty} x_n = x_0$,

then

$$\lim_{n \to \infty} f(x_n) = f\left(\lim_{n \to \infty} x_n\right).$$

$\lim f(x_n) = f(\lim x_n)$

Proof. Given $\epsilon > 0$, there is a number $\delta > 0$ such that

$$0 < |x - x_0| < \delta \Rightarrow |f(x) - f(x_0)| < \epsilon \tag{7}$$

$$\left[\text{because } \lim_{x \to x_0} f(x) = f(x_0)\right].$$

This implication can actually be replaced by a stronger one,

$$|x - x_0| < \delta \Rightarrow |f(x) - f(x_0)| < \epsilon. \tag{8}$$

For, the difference between (7) and (8) is that (8) draws a conclusion for $|x - x_0| = 0$, whereas (7) does not. But when $0 = |x - x_0|$, then $x = x_0$, so $|f(x) - f(x_0)| = 0$, and the conclusion of (8) is valid.

Next, since $\lim_{n \to \infty} x_n = x_0$, there is an N_δ such that

$$n > N_\delta \Rightarrow |x_n - x_0| < \delta. \tag{9}$$

Combining (8) and (9), we find that

$$n > N_\delta \Rightarrow |x_n - x_0| < \delta \Rightarrow |f(x_n) - f(x_0)| < \epsilon,$$

which proves that $\lim_{n \to \infty} f(x_n) = f(x_0) = f\left(\lim_{n \to \infty} x_n\right)$. *Q.E.D.*

Example 2. Evaluate $\lim_{n \to \infty} \left(1 + \frac{1}{n}\right)^n$. *Solution.* Take logs:

$$\log \left(1 + \frac{1}{n}\right)^n = n \log \left(1 + \frac{1}{n}\right).$$

This suggests the sequence $x_n = \frac{1}{n}$, the function $f(x) = \frac{1}{x} \log (1 + x)$, and the limit

$$\lim_{x \to 0} \left(\frac{1}{x} \log (1 + x)\right) = \lim_{x \to 0} \frac{\log (1 + x) - \log 1}{x} \qquad [\text{since } \log 1 = 0]$$

$$= \log' (1) \qquad [\text{definition of derivative}]$$

$$= 1. \qquad [\log' (x) = 1/x]$$

By Theorem 7, with $x_n = 1/n$, we thus find

$$\lim_{n\to\infty}\left(n\log\left(1+\frac{1}{n}\right)\right) = \lim_{x\to 0}\left(\frac{1}{x}\log(1+x)\right) = 1.$$

To recover the original limit $\lim\left(1+\frac{1}{n}\right)^n$, "undo" the logarithm by exponentiating:

$$\begin{aligned}\lim_{n\to\infty}\left(1+\frac{1}{n}\right)^n &= \lim_{n\to\infty}\exp\left(n\log\left(1+\frac{1}{n}\right)\right) && \text{[property of exp]}\\ &= \exp\left(\lim_{n\to\infty} n\log\left(1+\frac{1}{n}\right)\right) && \text{[Theorem 8]}\\ &= \exp(1) = e.\end{aligned}$$

PROBLEMS

1. Show that the following limits do not exist.

(a) $\lim_{x\to 0}\frac{x}{|x|}$

• (b) $\lim_{x\to 0} 1/x$

(c) $\lim_{x\to 0}\sin(1/x)$

• (d) $\lim_{x\to 0}\log|x|$

(e) $\lim_{x\to 0} e^{-1/x}$

2. Evaluate $\lim_{n\to\infty}\left(1+\frac{a}{n}\right)^n$. [This limit arises as follows. If a bank gives simple interest of r percent per year, then the principal P is increased by $\frac{r}{100}P$ at the end of the year, giving a new principal $P + \frac{r}{100}P = \left(1+\frac{r}{100}\right)P$. If the interest is compounded twice yearly, then at the end of 1/2 year the principal is increased by $\frac{1}{2}\frac{r}{100}P$, giving a new principal of $\left(1+\frac{1}{2}\frac{r}{100}\right)P$; and this new principal is similarly increased after the second half-year, giving $\left(1+\frac{1}{2}\frac{r}{100}\right)\left(\left(1+\frac{1}{2}\frac{r}{100}\right)P\right)$ at the end of the year. When interest is compounded n times a year, the principal at the end of the year is $\left(1+\frac{1}{n}\frac{r}{100}\right)^n P$. What happens

COMPOUND INTEREST

as interest is compounded more and more often? That is, find $\lim_{n\to\infty}\left(1+\frac{a}{n}\right)^n$, where we have set $a = r/100$.]

3. Evaluate the following limits.

• (a) $\lim_{n\to\infty} n \sin(1/n)$

(b) $\lim_{n\to\infty} n^2(\cos(1/n) - 1)$ [Use L'Hôpital's rule twice.]

• (c) $\lim_{n\to\infty} n(e^{1/n} - 1)$

(d) $\lim_{n\to\infty} \frac{\log n}{n}$

• (e) $\lim_{n\to\infty} e^{-n}$

(f) $\lim_{n\to\infty} e^{-n^2}$

• (g) $\lim_{n\to\infty} \frac{(n+1)e^{-(n+1)^2}}{ne^{-n^2}}$ [Rewrite the given expression.]

(h) $\lim_{n\to\infty} \left(\frac{n}{n+1}\right)^n$

• (i) $\lim_{n\to\infty} \frac{\log n}{\log(n+1)}$

4. Set $f(x) = 1$ if x is rational, $f(x) = -1$ if x is irrational.
(a) Try to sketch the graph of f.
(b) For what numbers x_0 does $\lim_{x\to x_0} f(x)$ not exist? Prove your answer.

5. (a) Define "$\lim s_n = +\infty$."
(b) Prove that if $\lim_{x\to+\infty} f(x) = L$, and $\lim_{n\to+\infty} x_n = +\infty$, then $\lim_{n\to+\infty} f(x_n) = L$.

[*Note:* The definition of $\lim_{x\to+\infty} f(x) = L$ is the following:

For every $\epsilon > 0$, there is a number X_ϵ such that
$$x > X_\epsilon \Rightarrow |f(x) - L| < \epsilon.]$$

6. Prove the following theorem for bounded monotone functions [Proposition K, §5.4]: *Let F be increasing on $[a, +\infty)$, and bounded above by a constant M. Then F has a limit as $x \to +\infty$, and* $\lim_{x\to+\infty} F(x) \le M$. [See the definition of $\lim_{x\to+\infty}$ in Problem 5, and recall the proof of Theorem 6.]

PROPOSITION K

10.5 THE BOLZANO-WEIERSTRASS THEOREM

In proving the deeper facts about continuous functions, we are led to construct sequences $\{x_n\}$ which are bounded, but *not necessarily convergent* [Fig. 9]. Sequences of this type are not as useless as they may seem. We analyze them as follows.

Let S_k be the set of numbers occurring in the sequence with subscripts $\geq k$,

$$S_k = \{x_k, x_{k+1}, x_{k+2}, \ldots\}.$$

Since $\{x_n\}$ is a bounded sequence, S_k is a bounded set; hence it has a least upper bound (call it U_k) and a greatest lower bound (call it L_k):

$$U_k = \sup \{x_k, x_{k+1}, \ldots\} \tag{1}$$

$$L_k = \inf \{x_k, x_{k+1}, \ldots\} \tag{2}$$

[Fig. 10]. Here, sup (short for "supremum") denotes the least upper bound, and inf (short for "infimum") denotes the greatest lower bound. Obviously, $U_1 \geq U_2 \geq \cdots$; for U_k, being an upper bound for $S_k = \{x_k, x_{k+1}, x_{k+2}, \ldots\}$, is also an upper bound for $S_{k+1} = \{x_{k+1}, x_{k+2}, \ldots\}$; hence U_k is $\geq U_{k+1}$, the *least* upper bound of S_{k+1}. Thus the sequence $\{U_k\}$ of least upper bounds is a decreasing sequence. Similarly, $L_1 \leq L_2 \leq \cdots$. Further, both sequences are bounded (since $\{x_k\}$ is), so they are convergent. Their limits are called the "lim sup" and "lim inf" of the sequence $\{x_k\}$:

$$\limsup x_k = \lim_{k\to\infty} (\sup \{x_k, x_{k+1}, \ldots\}) = \lim U_k$$

$$\liminf x_k = \lim_{k\to\infty} (\inf \{x_k, x_{k+1}, \ldots\}) = \lim L_k.$$

Obviously, $\sup \{x_k, x_{k+1}, \ldots\} \geq \inf \{x_k, x_{k+1}, \ldots\}$, so $\limsup x_k \geq \liminf x_k$. Moreover, the lim sup and lim inf are equal if and only if $\{x_k\}$ is convergent [Problems 5 and 7]. When they differ, then "in the long run" the sequence oscillates back and forth between $\liminf x_k$ and $\limsup x_k$ [Fig. 11]. This oscillation can be described more precisely in two ways:

(i) No matter how far out you go in the sequence $\{x_n\}$, there are points arbitrarily close to $\limsup x_k$, and other points arbitrarily close to $\liminf x_k$; or alternatively

(ii) by selecting infinitely many of the terms from the given sequence $\{x_n\}$, you can form a sequence which converges to $\limsup x_k$; by selecting other terms, you can form a sequence which converges to $\liminf x_k$.

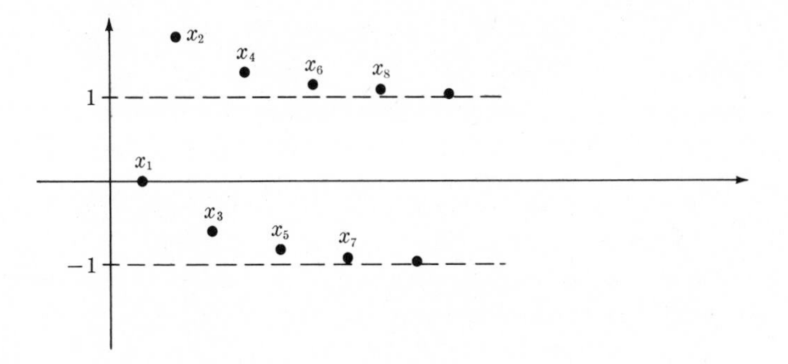

FIGURE 10.9

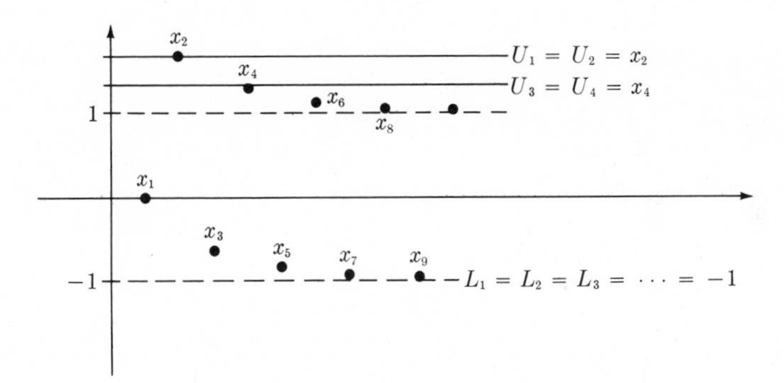

FIGURE 10.10

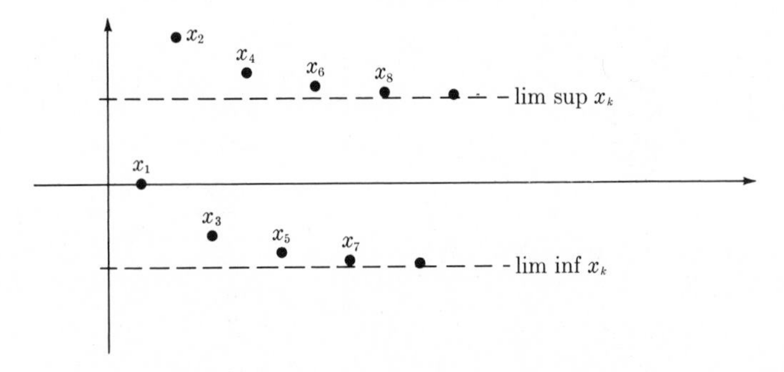

FIGURE 10.11

Statement (ii) contains the *Bolzano-Weierstrass Theorem*, a powerful tool in the study of continuous functions. Statement (i) is the main step [Lemma 1 below] in proving the Bolzano-Weierstrass Theorem.

Example 1. Let $x_n = (-1)^n + 1/n$ [Fig. 11]. Then $\limsup x_n = 1$, and $\liminf x_n = -1$. No matter how far out you go in the sequence, there are *even* terms $x_{2k} = 1 + 1/2k$ arbitrarily close to $\limsup x_n = 1$, and *odd* terms $x_{2k+1} = -1 + 1/(2k + 1)$ arbitrarily close to $\liminf x_n = -1$.

In proving the Bolzano-Weierstrass Theorem, we begin with a precise statement of (i) above.

Lemma 1. *Let* $\{x_n\}$ *be a bounded sequence. Given any integer* N *(no matter how large), and any positive number* ϵ *(no matter how small), there is a term* x_n *such that* $n > N$, *and* $|x_n - \limsup x_k| < \epsilon$. [A similar statement holds for $\liminf x_n$.]

Proof. Let $\bar{x} = \limsup x_k$. Define U_k as in (1). Since $\lim U_k = \bar{x}$, there is an integer N such that

$$|U_N - \bar{x}| < \epsilon/2. \tag{3}$$

But $U_N = \sup \{x_N, x_{N+1}, \ldots\}$, so all of the terms $x_N, x_{N+1}, \ldots$ are $\leq U_N$, and at least one of them, call it x_n, is $> (U_N - \epsilon/2)$ [since $U_N - \epsilon/2$ is *not* an upper bound for $\{x_N, x_{N+1}, \ldots\}$]. Hence there is a term x_n with $n \geq N$ and $U_N - \epsilon/2 < x_n < U_N$. Combining this with (3) we get $|x_n - \bar{x}| < \epsilon$. *Q.E.D.*

BOLZANO-WEIERSTRASS THEOREM

Theorem 9. *Let* $\{x_n\}$ *be a bounded sequence. Then there is an increasing sequence of integers* $n_1 < n_2 < \cdots$ *such that*

$$\lim_{k \to \infty} x_{n_k} = \limsup x_n .$$

Further, $n_k \geq k$. [A similar statement holds for $\liminf x_n$.]

Proof. Let $\bar{x} = \limsup x_n$. Apply Lemma 1 with $N = 1$ and $\epsilon = 1$ to find an integer n_1 such that

$$n_1 \geq 1, \qquad \text{and} \qquad |x_{n_1} - \bar{x}| < 1.$$

Next, apply Lemma 1 with $N = n_1 + 1 \geq 2$ and $\epsilon = 1/2$, to find an integer n_2 such that

$$n_2 \geq n_1 + 1 \geq 2, \qquad \text{and} \qquad |x_{n_2} - \bar{x}| < 1/2.$$

Continuing like this, we obtain for each k an integer n_k such that

$$n_k \geq n_{k-1} + 1 \geq k, \qquad \text{and} \qquad |x_{n_k} - \bar{x}| < 1/k.$$

Thus $n_1 < n_2 < \cdots$, and $n_k \geq k$, and $\lim x_{n_k} = \bar{x}$. *Q.E.D.*

Any sequence $x_{n_1}, x_{n_2}, x_{n_3}, \ldots$, where $1 \leq n_1 < n_2 < \cdots$, is called a *subsequence* of $x_1, x_2, \ldots$. The Bolzano-Weierstrass Theorem thus proves that *every bounded sequence contains a convergent subsequence.* This is the main point of the theorem; in most applications, it is not really important *what* the actual limit is. The lim sup was introduced here merely to show *why* there is a convergent subsequence.

As an application, we prove the basic lemma behind the Fundamental Theorem of Algebra [about the complex roots of a polynomial], as given in §8.7.

Lemma 2. *Let* $P(z) = \sum_0^n a_j z^j$ *be a complex polynomial of degree* $n > 0$. *Then* $|P(z)|^2$ *assumes a minimum value* m.

Proof. We must find a number z_0 such that $|P(z_0)|^2 \leq |P(z)|^2$ for every complex z. Our strategy is (i) to find the value $m = |P(z_0)|^2$, and (ii) to find z_0.

The first step is easy. Clearly, *if* $|P(z)|^2$ has a minimum m, then m is the g.l.b. of all the values $|P(z)|^2$. Although we are not yet sure that $|P(z)|^2$ really assumes a minimum, we can nevertheless define the number

$$m = \text{g.l.b. } \{|P(z)|^2 : z \text{ is complex}\}.$$

For, the set of numbers $|P(z)|^2$ is not empty, and has the lower bound zero, so it has a greatest lower bound m.

The next step is to find z_0. By definition of m, for every integer j there is a number $|P(z_j)|^2$ such that $m \leq |P(z_j)|^2 < m + 1/j$; clearly

$$\lim_{j\to\infty} |P(z_j)|^2 = m. \tag{4}$$

We will extract from $\{z_j\}$ a convergent subsequence; the limit of this will be called z_0, and (4) will show that $|P(z_0)|^2 = m$.

First, we show [by contradiction] that the sequence $\{z_j\}$ is bounded. If it were not, there would be a subsequence $\{z_{j_k}\}$ such that $\lim_{k\to\infty} |z_{j_k}| = \infty$, and hence

$$\lim_{k\to\infty} \frac{1}{|P(z_{j_k})|^2} = \lim_{k\to\infty} \frac{|z_{j_k}|^{-2n}}{|a_n + (a_{n-1}/z_{j_k}) + \cdots + (a_0/z_{jk}^n)|^2} = \frac{0}{|a_n|} = 0. \tag{5}$$

On the other hand, $|P(z_{j_k})|^2$ is a convergent sequence, so it is bounded, $|P(z_{j_k})|^2 \leq M_0$; therefore, $\dfrac{1}{|P(z_{j_k})|^2} \geq \dfrac{1}{M_0}$.

This contradiction with (5) shows that $\{z_j\}$ cannot have a subsequence $\{z_{j_k}\}$ tending to ∞, so $\{z_j\}$ is bounded: $|z_j| \leq M$, for some M independent of j.

Two applications of Bolzano-Weierstrass now extract from $\{z_j\}$ a convergent subsequence $\{w_k\}$. Let $z_j = x_j + iy_j$. Since $|z_j| \leq M$, the same is true of $|x_j|$ and $|y_j|$. Therefore, $\{x_j\}$ contains a convergent subsequence $\{x_{j_k}\}$, whose limit we shall call x_0. But $|y_{j_k}| \leq M$, so $\{y_{j_k}\}$ contains a convergent subsequence $\{y_{j_{k_l}}\}$, whose limit we shall call y_0. To simplify the notation at this point, set

$$x_{j_{k_l}} + iy_{j_{k_l}} = u_l + iv_l = w_l .$$

Then:

(6) $\{w_l\}$ is a subsequence of $\{z_j\}$, so $\lim |P(w_l)|^2 = \lim |P(z_j)|^2 = m$.

(7) $\{u_l\}$ is a subsequence of $\{x_j\}$, so $\lim u_l = x_0$; similarly $\lim v_l = y_0$.

(8) $|P(w_l)|^2$ is a polynomial in u_1 and v_1 , so

$$\lim |P(w_l)|^2 = |P(\lim w_l)|^2 = |P(x_0 + iy_0)|^2.$$

Combining statements (6)–(8) and (4) gives $|P(x_0 + iy_0)|^2 = m =$ g.l.b. $\{|P(z)|^2\colon z \text{ is complex}\}$, and Lemma 2 is proved; the minimum of $|P(z)|^2$ is achieved at $z_0 = x_0 + iy_0$.

PROBLEMS

1. For each of the following sequences, find U_k and L_k for $1 \leq k \leq 5$; find the lim sup and the lim inf; and find a convergent subsequence.

(a) $x_n = (-1)^n - 1/n$

•(b) $\{x_n\} = 1, \frac{1}{2}, \frac{1}{3}, \frac{2}{3}, \frac{1}{4}, \frac{2}{4}, \frac{3}{4}, \frac{1}{5}, \frac{2}{5}, \frac{3}{5}, \frac{4}{5}, \frac{1}{6}, \frac{2}{6}, \frac{3}{6}, \frac{4}{6}, \frac{5}{6}, \frac{1}{7}, \ldots$

[This sequence contains all the rational numbers between 0 and 1.]

(c) $a_n = (-1)^{3n} \dfrac{n-1}{n}$

•(d) $a_n =$ the nth digit in the decimal expansion of 3/7

Chapter XI

Infinite Series

11.1 SOME USES AND ABUSES OF INFINITE SERIES

With Taylor expansions [§9.2], we generally found that including more terms gives approximations with greater accuracy. Suppose we could somehow add up *all* the infinitely many terms in the Taylor expansion of a function f; would this give $f(x)$ exactly? The idea of adding up infinitely many terms, crazy as it may sound, is worth pursuing.

Thus, from the Taylor expansions

$$\frac{1}{1-x} = 1 + x + x^2 + \cdots + x^n + R_{n+1}\,,$$

$$e^x = 1 + x + \frac{x^2}{2} + \cdots + \frac{x^n}{n!} + R_{n+1}$$

let us postulate the equations

$$\frac{1}{1-x} = 1 + x + \cdots + x^n + \cdots = \sum_{n=0}^{\infty} x^n, \tag{1}$$

$$e^x = 1 + x + \cdots + \frac{x^n}{n!} + \cdots = \sum_{n=0}^{\infty} \frac{x^n}{n!}. \tag{2}$$

The "infinite sums" on the right in (1) and (2), obtained by "adding up all terms" in the Taylor expansion, are called the *Taylor series* of the functions on the left. The two series displayed here are so important that they have been given special names: (1) is called the *geometric series*, and (2) the *exponential series*.

Let us postulate further that these infinite sums may be treated in every respect as though they were polynomials. Any difficulties that arise from these postulations will be dealt with later. Our first interest is in the possible use of Taylor series in various applications, ranging from idle philosophy to economics to physics; and in mathematical problems that are beyond the methods developed up to this point in the book.

A. *Zeno's paradox explained*

Nearly two hundred years before Euclid, Zeno of Elea offered his famous paradoxical arguments as a challenge to philosophers and geometers. The simplest of Zeno's paradoxes is the dichotomy, which takes the following form. In order to go a mile, you must first go a half mile; in order to go that half mile, you must first go a quarter mile; in order to go that quarter mile, you must first go one-eighth; and so on *ad infinitum*. Thus by repeated division into halves, the whole length to be traversed can be presented as the sum of an infinite number of lengths; and to go any distance at all, you must traverse infinitely many of these lengths. [See Fig. 1.] It thus appears impossible ever to begin the mile, so that motion in itself is impossible!

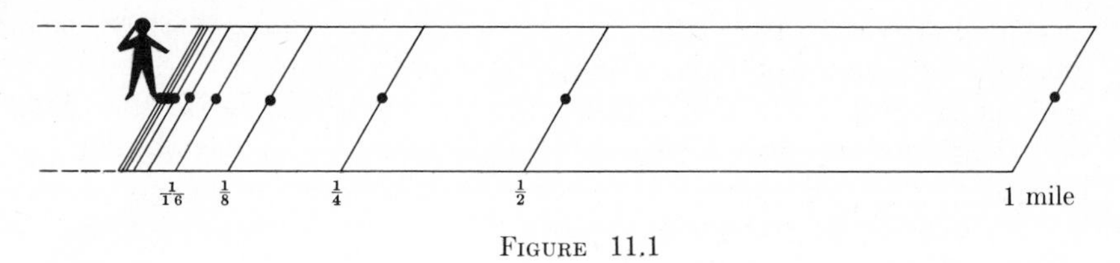

FIGURE 11.1

This conclusion is obviously a paradox, but it is not so obvious how the paradox arose. Perhaps the source of the trouble is our feeling that it is impossible to do infinitely many things in a finite length of time. In mathematical terms, we feel that it is impossible for infinitely many positive numbers to add up to a finite sum.

However, Zeno's construction actually shows that this feeling is not justified. Since the mile is unchanged by our conceptual division of it into half a mile plus a quarter mile plus an eighth of a mile and so on, we must have the equality

$$1 = \frac{1}{2} + \frac{1}{4} + \frac{1}{8} + \frac{1}{16} + \cdots.$$

The dots indicate that if you had *all* those infinitely many pieces, you would have the whole mile. This formula is a special case of the geometric series (1) above; set $x = \frac{1}{2}$ and then divide by 2:

$$2 = \frac{1}{1-\frac{1}{2}} = 1 + \frac{1}{2} + \frac{1}{4} + \frac{1}{8} + \cdots, \qquad \text{[formula (1)]}$$

$$1 = \frac{1}{2} + \frac{1}{4} + \frac{1}{8} + \frac{1}{16} + \cdots. \qquad \text{[dividing by 2]}$$

If Zeno can imagine cutting up the mile into infinitely many pieces, then we can imagine putting the pieces back together again.

B. *The "spending multiplier"*

The many conventions held in Chicago are obviously a source of income to the city. In fact, in a certain sense, they contribute to the city's economic activity much more than just the amount spent there by the conventioners. To show why, let us make the following simple assumptions about a hypothetical convention: (a) the out-of-town conventioners spend $1,000,000 in Chicago, and (b) on the average, each Chicagoan spends 3/4 of his income in Chicago; the other 1/4 is spent elsewhere, or saved. We now compute the total amount of money spent in Chicago as a result of the convention.

First of all, one million dollars is spent in Chicago by the conventioners themselves; this is the "direct effect." Then the Chicagoans spend 3/4 of this money in the city, causing a secondary effect of 3/4 million. Then, 3/4 of *this* 3/4 million is spent in the city, causing a tertiary effect of $(3/4)^2$ million. Continuing like this n times, we find that

$$1 + \frac{3}{4} + \left(\frac{3}{4}\right)^2 + \cdots + \left(\frac{3}{4}\right)^n$$

million dollars is spent. Thus, ultimately, the total spending† is

$$1 + \frac{3}{4} + \left(\frac{3}{4}\right)^2 + \cdots = \sum_0^\infty \left(\frac{3}{4}\right)^n$$

million dollars. Assuming the geometric series formula (1) is true, we find

$$\sum_{n=0}^\infty \left(\frac{3}{4}\right)^n = \frac{1}{1-\frac{3}{4}} = 4;$$

so the one million dollars brought in led to the spending of four million dollars. This is the "multiplier" effect; the multiplier is four, in this case.

---

† "...late and soon, / Getting and spending, we lay waste our powers..."
—William Wordsworth, SONNET, 1806

C. *The bouncing ball*

Ideally, a ball dropped from a height h returns after one bounce to a height rh, where $r < 1$ and r is independent of h [Fig. 2]. If the ball is never stopped, will it go on bouncing forever? Since it bounces infinitely many times, we expect that it will actually go on forever; but Zeno's paradox indicates that this argument is not reliable; we need a more careful analysis.

To find the time elapsed during the drop from height h, recall that the distance traversed in time t by a falling body starting at rest is $s(t) = gt^2/2$; hence it falls the distance h in a time t given by $gt^2/2 = h$, or $t = \sqrt{2h/g}$. Similarly, the time taken to rise to height rh and fall again to height 0 is $2\sqrt{2rh/g}$. Computing the time for each successive bounce in this way, we find the time that elapses during infinitely many bounces to be

$$\begin{aligned} T &= \sqrt{2h/g} + 2\sqrt{2rh/g} + 2\sqrt{2r^2h/g} + \cdots \\ &= \sqrt{2h/g} + 2\sqrt{2h/g}\,(r^{1/2} + r^{2/2} + r^{3/2} + \cdots) \\ &= \sqrt{2h/g}\,[1 + 2\sum_1^\infty (\sqrt{r})^n]. \end{aligned}$$

To evaluate $\sum_1^\infty (\sqrt{r})^n = r^{1/2} + r^{2/2} + r^{3/2} + \cdots$, set $x = r^{1/2}$ in the geometric series (1) and multiply by $r^{1/2}$, obtaining $\dfrac{r^{1/2}}{1 - r^{1/2}}$. Thus, in spite of any expectations to the contrary, the ball makes all its bounces in the finite amount of time

$$T = \sqrt{2h/g}\left[1 + \frac{2\sqrt{r}}{1 - \sqrt{r}}\right] = \sqrt{2h/g}\,\frac{1 + \sqrt{r}}{1 - \sqrt{r}}.$$

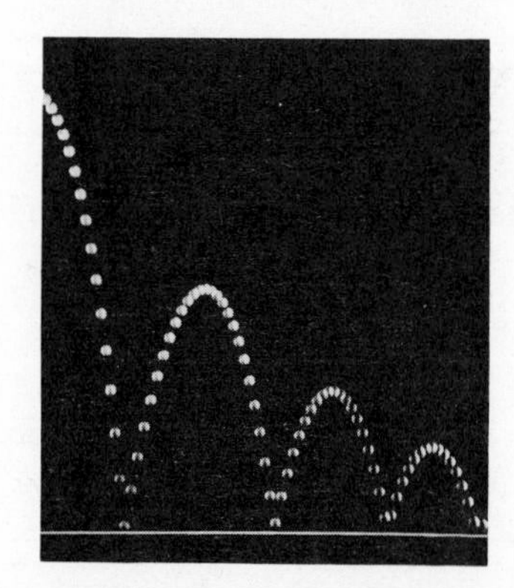

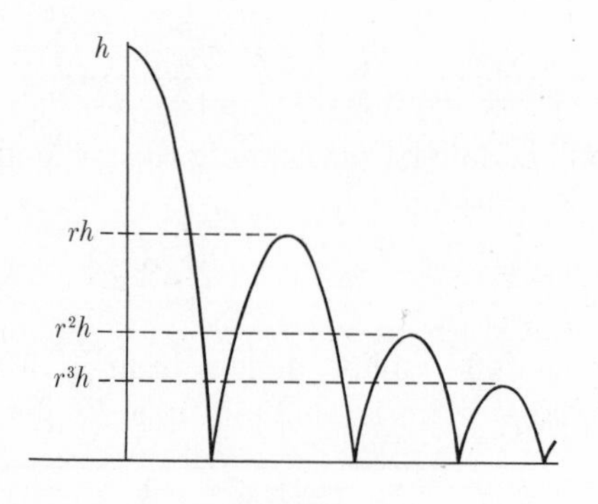

FIGURE 11.2*

*From PSSC PHYSICS, D.C. Heath and Company, Boston, 1965

D. *Integrals*

The error function $\operatorname{erf}(x) = \dfrac{2}{\sqrt{\pi}} \displaystyle\int_0^x e^{-t^2}\, dt$ is one of those stubborn integrals that cannot be expressed as an elementary function. However, we *can* obtain a Taylor series for it as follows. Substitute $-t^2$ for x in the exponential series (2), obtaining

$$e^{-t^2} = 1 - t^2 + \frac{t^4}{2!} - \cdots = \sum_0^\infty \frac{(-t^2)^n}{n!}.$$

Then integrate this as if it were a polynomial, obtaining

$$\operatorname{erf}(x) = \frac{2}{\sqrt{\pi}} \int_0^x e^{-t^2}\, dt = \frac{2}{\sqrt{\pi}} \int_0^x \left(1 - t^2 + \frac{t^4}{2!} - \cdots + \frac{(-t^2)^n}{n!} + \cdots\right) dt$$

$$= \frac{2}{\sqrt{\pi}} \left(x - \frac{x^3}{3} + \frac{x^5}{5 \cdot 2!} - \cdots + \frac{(-1)^n x^{2n+1}}{(2n+1) \cdot n!} + \cdots\right).$$

Thus, although the integral cannot be expressed as an elementary function, it *can* be expressed as an infinite Taylor series.

E. *A simple differential equation*

Consider the simple differential equation

$$f'(x) = x + f(x). \tag{3}$$

We can solve this equation for the unknown function f by obtaining f in the form of a Taylor series of the type in (1) and (2) above,

$$f(x) = a_0 + a_1 x + a_2 x^2 + a_3 x^3 + \cdots = \sum_0^\infty a_n x^n. \tag{4}$$

We operate on this sum as if it were a polynomial; thus the left-hand side of (3) is

$$f'(x) = a_1 + 2a_2 x + 3a_3 x^2 + \cdots \tag{5}$$

and the right-hand side is

$$x + f(x) = x + (a_0 + a_1 x + a_2 x^2 + \cdots)$$
$$= a_0 + (a_1 + 1)x + a_2 x^2 + \cdots. \tag{6}$$

To satisfy the given equation $f' = x + f$, we need only equate the coefficients of like powers of x in (5) and (6), obtaining

$$\begin{aligned} a_1 &= a_0 && [\text{coefficients of } x^0] \\ 2a_2 &= 1 + a_1 && [\text{coefficients of } x] \\ 3a_3 &= a_2 && [\text{coefficients of } x^2] \\ 4a_4 &= a_3 && [\text{etc.}] \\ \cdots&\cdots\cdots \end{aligned}$$

These equations allow us to express each coefficient successively in terms of a_0:

$$a_1 = a_0$$

$$a_2 = \frac{1 + a_1}{2} = \frac{1 + a_0}{2}$$

$$a_3 = \frac{a_2}{3} = \frac{1 + a_0}{3 \cdot 2}$$

$$a_4 = \frac{a_3}{4} = \frac{1 + a_0}{4!}$$

$$\vdots$$

$$a_n = \frac{a_{n-1}}{n} = \frac{1 + a_0}{n!}.$$

Substituting these values in the series (4) for $f(x)$, we arrive at the solution

$$\begin{aligned} f(x) &= a_0 + a_0 x + \frac{1 + a_0}{2} x^2 + \frac{1 + a_0}{3!} x^3 + \cdots \\ &= -1 - x + (1 + a_0)\left(1 + x + \frac{x^2}{2} + \frac{x^3}{3!} + \cdots\right) \\ &= -1 - x + (1 + a_0)e^x. \end{aligned}$$

It is easy to check this solution:

$$f'(x) = -1 + (1 + a_0)e^x = f(x) + x.$$

F. *Bessel's equation*

The analysis of the vibrations of a circular drum head, and of the radiation from certain types of radio antennas, leads to the differential equation

$$x^2 f'' + x f' + x^2 f = 0, \tag{7}$$

called *Bessel's equation.* As before, we try to solve this with a Taylor series

$$f(x) = a_0 + a_1 x + a_2 x^2 + \cdots.$$

Carrying out the appropriate differentiations and multiplications, you can easily check that

$$\begin{aligned} x^2 f(x) &= a_0 x^2 + a_1 x^3 + a_2 x^4 + \cdots \\ x f'(x) &= a_1 x + 2a_2 x^2 + 3a_3 x^3 + 4a_4 x^4 + \cdots \\ x^2 f''(x) &= 2a_2 x^2 + 3 \cdot 2a_3 x^3 + 4 \cdot 3a_4 x^4 + \cdots \end{aligned}$$

and hence from (7) we find, after collecting like powers of x, that

$$a_1 x + (a_0 + 2^2 a_2)x^2 + (a_1 + 3^2 a_3)x^3 + (a_2 + 4^2 a_4)x^4 + \cdots = 0.$$

To satisfy this equation, we set the coefficient of each power of x equal to zero:

$$\begin{aligned} a_1 &= 0 \\ a_0 + 2^2 a_2 &= 0 \\ a_1 + 3^2 a_3 &= 0 \\ a_2 + 4^2 a_4 &= 0 \\ a_3 + 5^2 a_5 &= 0 \\ &\cdots\cdots \end{aligned}$$

Take first the equations with odd subscripts, which yield

$$\begin{aligned} a_1 &= 0 \\ a_3 &= \frac{-a_1}{9} = 0 \\ a_5 &= \frac{-a_3}{25} = 0 \\ &\cdots\cdots \end{aligned}$$

The remaining equations yield

$$\begin{aligned} a_2 &= \frac{-a_0}{2^2} \\ a_4 &= \frac{-a_2}{4^2} = \frac{a_0}{2^2 \cdot 4^2} \\ a_6 &= \frac{-a_4}{6^2} = \frac{-a_0}{2^2 \cdot 4^2 \cdot 6^2} \\ &\vdots \\ a_{2j} &= \frac{(-1)^j a_0}{2^2 \cdot 4^2 \cdots (2j)^2}. \end{aligned} \tag{8}$$

The denominator of a_{2j} is

$$(2 \cdot 4 \cdots 2j)^2 = (2^j \cdot 1 \cdot 2 \cdots j)^2 = 2^{2j}(j!)^2.$$

Since a_{2j} is the coefficient of x^{2j} in $\sum_0^\infty a_n x^n$, and all the *odd* coefficients are zero, we find from (8) that

$$f(x) = \sum_{j=0}^{\infty} a_{2j} x^{2j} = a_0 \sum_{j=0}^{\infty} \frac{(-1)^j (x/2)^{2j}}{(j!)^2}. \tag{9}$$

In contrast to the simple case considered before, the infinite sum here is *not* the Taylor series of any familiar function. It is, however, the Taylor series of a new function, called a *Bessel function,* in honor of Friedrich Wilhelm Bessel, who was the first to study this equation effectively. The Bessel function is *defined* by the series (9).

So far, each application of infinite sums has been at least partially successful. We obtained quite tangible results in A, B, C, and E; and in D and F, we could at least express otherwise intractible functions as Taylor series. The next two applications, however, do not turn out so well.

G. *Euler's funny formula*

From the geometric series (1) we deduce, on multiplying by x, the formula

$$\frac{x}{1-x} = x + x^2 + x^3 + \cdots = \sum_1^{\infty} x^n. \tag{10}$$

Again from (1), replacing x by $1/x$, we get

$$\frac{1}{1-1/x} = \frac{x}{x-1} = 1 + \frac{1}{x} + \frac{1}{x^2} + \cdots.$$

Adding this to (10), we obtain a formula ascribed to the great mathematician Leonhard Euler:

$$\cdots + \frac{1}{x^2} + \frac{1}{x} + 1 + x + x^2 + \cdots = \frac{x}{1-x} + \frac{x}{x-1} = 0.$$

But this is absurd! For when x is positive, then the left side is 1 plus an infinite number of *positive* terms, while the right side is 0.

H. *The harmonic series*

The Taylor series of $\log(1-x)$ is easily found, either directly or by integrating the geometric series (1), to be

$$\log(1-x) = -x - \frac{x^2}{2} - \frac{x^3}{3} - \cdots.$$

Taking $x = -1$ yields

$$\log 2 = 1 - \frac{1}{2} + \frac{1}{3} - \frac{1}{4} + \cdots.$$

From this, we can try to find the sum of the so-called *harmonic* series, $\sum_1^{\infty} 1/n = 1 + \frac{1}{2} + \frac{1}{3} + \cdots$. Let the unknown sum be L, $1 + \frac{1}{2} + \frac{1}{3} + \cdots = L$. This sum is related to our formula for log 2, as follows:

$$\log 2 = 1 - \frac{1}{2} + \frac{1}{3} - \frac{1}{4} + \cdots$$

$$= 1 + \frac{1}{2} + \frac{1}{3} + \frac{1}{4} + \frac{1}{5} + \frac{1}{6} + \frac{1}{7} + \frac{1}{8} + \cdots$$

$$\updownarrow \qquad \updownarrow \qquad \updownarrow \qquad \updownarrow$$

$$-2\left(\frac{1}{2}\right) - 2\left(\frac{1}{4}\right) - 2\left(\frac{1}{6}\right) - 2\left(\frac{1}{8}\right) - \cdots,$$

where the entries in the last row are to be added to those in the previous row, as indicated. Regrouping terms, we get

$$\log 2 = \left(1 + \frac{1}{2} + \frac{1}{3} + \frac{1}{4} + \frac{1}{5} + \cdots\right) - 2\left(\frac{1}{2} + \frac{1}{4} + \frac{1}{6} + \frac{1}{8} + \cdots\right)$$

$$= \qquad L \qquad - 2\cdot\frac{1}{2}\left(1 + \frac{1}{2} + \frac{1}{3} + \frac{1}{4} + \cdots\right)$$

$$= \qquad L \qquad - 2\cdot\frac{1}{2} L = 0.$$

But this is impossible, for $\log 2 > 0$. Our attempt to determine $\sum_{1}^{\infty} 1/n$ has not merely failed, it has led to a contradiction.

What is the meaning of all this? The six successful examples suggest (correctly) that infinite sums are a powerful, but possibly mysterious, mathematical tool. The two paradoxical results show beyond a doubt that series cannot be handled blindly.

This chapter aims to dispel the mystery and demonstrate the power of infinite series. The basic questions to be answered are:

What does a formula such as $\dfrac{1}{1-x} = \sum_{n=0}^{\infty} x^n$ mean? [§2]

When does an expression such as $f(x) = \sum_{n=0}^{\infty} a_n x^n$ really define a function f? [§§3–4]

What manipulations (adding, differentiating, etc.) can be performed on infinite sums? [§§2, 6, 7]

The answers to these questions will justify the successful applications above, and explain what is wrong in the unsuccessful ones.

11.2 THE SUM OF AN INFINITE SERIES

The meaning of an "infinite sum" $\sum_0^\infty a_n$ is determined by the finite sums $\sum_0^N a_n$, called the *partial sums* of $\sum_0^\infty a_n$.

Definition 1. Let $a_0, a_1, \ldots$ be any sequence. Then

SUM OF AN INFINITE SERIES

(i) $\sum_0^\infty a_n = L$ means $\lim_{N\to\infty} (\sum_0^N a_n) = L$; L is called the *sum of the infinite series* $\sum_0^\infty a_n$. The number a_n is called the nth *term* of the series.

CONVERGENT AND DIVERGENT SERIES

(ii) $\sum_0^\infty a_n$ is called *convergent* if the sequence of partial sums $\sum_0^N a_n$ converges; otherwise $\sum_0^N a_n$ is called *divergent.*

Example 1. Is the geometric series $\sum_0^\infty x^n$ convergent? If so, what is its sum?

Solution. The partial sums $\sum_0^N x^n$ can be expressed quite simply. From the factorization

$$1 - x^{N+1} = (1 - x)(1 + x + \cdots + x^N)$$

we find

$$\sum_0^N x^n = \frac{1 - x^{N+1}}{1 - x}, \qquad x \neq 1. \tag{1}$$

When $|x| < 1$, then $\lim_{N\to\infty} x^{N+1} = 0$, hence $\lim \dfrac{1 - x^{N+1}}{1 - x} = \dfrac{1}{1 - x}$. Thus, in view of (i) and (1),

$$\sum_0^\infty x^n = \lim_{N\to\infty} \sum_0^N x^n = \frac{1}{1 - x}, \quad \text{if } |x| < 1. \tag{2}$$

When $|x| > 1$, then x^{N+1} diverges (since it is unbounded); thus $\sum_0^\infty x^n$ diverges for $|x| > 1$. Finally, we must consider the case $|x| = 1$, i.e. $x = -1$ or $x = 1$.

When $x = -1$, the partial sum is $\sum_0^N (-1)^n = 1 + (-1)^N$, which diverges; and when $x = 1$ the partial sum is $\sum_0^N 1^n = N + 1$, which also diverges. Thus

$$\sum_0^\infty x^n \quad \text{diverges if} \quad |x| \geq 1.$$

Definition 1 does not really introduce a new concept; it is essentially a glossary, listing the *idiomatic uses* of the symbol $\sum_0^\infty a_n$. The basic mathematical concept involved is the limit of a sequence, which we already know. Using the glossary, it is easy to translate certain results about sequences into the corresponding results for series.

Theorem 1. *Let $\sum_0^\infty a_n$ and $\sum_0^\infty b_n$ be convergent. Then*

$$\sum_0^\infty (a_n + b_n) = (\sum_0^\infty a_n) + (\sum_0^\infty b_n),$$

and

$$\sum_0^\infty ca_n = c\sum_0^\infty a_n \quad \textit{for any number } c.$$

Proof. By elementary algebra, the partial sums of $\sum_0^\infty (a_n + b_n)$ can be written

$$\sum_0^N (a_n + b_n) = \sum_0^N a_n + \sum_0^N b_n\ ;$$

hence

$$\lim_{N\to\infty} \sum_0^N (a_n + b_n) = \lim_{N\to\infty} (\sum_0^N a_n + \sum_0^N b_n)$$

$$= \lim_{N\to\infty} \sum_0^N a_n + \lim_{N\to\infty} \sum_0^N b_n\,.$$

In view of (i), this means

$$\sum_0^\infty (a_n + b_n) = \sum_0^\infty a_n + \sum_0^\infty b_n\,.$$

The proof that $\sum_0^\infty ca_n = c\sum_0^\infty a_n$ is similar.

TERMS OF CONVERGENT SERIES FORM SEQUENCE CONVERGING TO 0

Theorem 2. *If* $\sum_0^\infty a_n$ *converges, then* $\lim_{n\to\infty} a_n = 0$.

Proof. For any sequence $\{s_N\}$, if $\lim s_N = S$, then $\lim_{N\to\infty} s_{N-1} = S$ [Problem 6, §10.1]. We apply this observation to the partial sums of $\sum_0^\infty a_n$. Let $A = \sum_0^\infty a_n$. Notice that $a_N = \sum_0^N a_n - \sum_0^{N-1} a_n$; therefore,

$$\lim_{N\to\infty} a_N = \lim_{N\to\infty}\left(\sum_0^N a_n - \sum_0^{N-1} a_n\right)$$

$$= \lim_{N\to\infty}\sum_0^N a_n - \lim_{N\to\infty}\sum_0^{N-1} a_n = A - A = 0. \quad Q.E.D.$$

Example 2. Discuss the convergence of $\sum_1^\infty n^a$, for different values of a, on the basis of Theorem 2. *Solution.* When $a \geq 0$, then $\lim_{n\to\infty} n^a \neq 0$, so $\sum_1^\infty n^a$ *cannot* converge. When $a < 0$, then $\lim_{n\to\infty} n^a = 0$; in this case, Theorem 2 gives no information. The theorem says

$$\sum_0^\infty a_n \text{ converges} \Rightarrow \lim_{n\to\infty} a_n = 0,$$

but not the converse

$$\lim_{n\to\infty} a_n = 0 \Rightarrow \sum_0^\infty a_n \text{ converges}.$$

The complete answer to this question is deferred to §11.3 below.

The simple formulas (1) and (2) above, for the geometric series, lead to many useful results.

REPEATING DECIMALS GIVE RATIONAL NUMBERS & VICE VERSA

Example 3. The infinite decimal 1.11111... stands for the limit of the sequence of numbers

$$1, \quad 1.1, \quad 1.11, \quad 1.111, \quad \ldots,$$

which are the successive partial sums of the geometric series [Example 1]

$$1 + \frac{1}{10} + \frac{1}{100} + \frac{1}{1000} + \cdots = \sum_0^\infty \left(\frac{1}{10}\right)^n = \frac{1}{1-\frac{1}{10}} = \frac{10}{9} = 1\tfrac{1}{9}.$$

Thus $1.11111\ldots = 1\frac{1}{9}$. Similarly, $3.1646464\ldots$ stands for the limit of

$$3.1, \quad 3.164, \quad 3.16464, \quad 3.1646464, \quad \ldots,$$

which is the sequence of partial sums of the series

$$3.1 + 64 \times 10^{-3} + 64 \times 10^{-5} + 64 \times 10^{-7} + \cdots$$

$$= 3.1 + \sum_0^\infty \frac{64}{1000}(10^{-2})^n = 3.1 + \frac{64}{1000}\sum_0^\infty (10^{-2})^n \qquad \text{[Theorem 1]}$$

$$= 3.1 + \frac{64}{1000} \cdot \frac{1}{1 - 10^{-2}} = \frac{3133}{990}.$$

Similarly, *every repeating decimal is a rational number.* Conversely, *every rational number p/q has a repeating decimal expansion.* For, in the division

$$\begin{array}{l} \quad d.d_1d_2\ldots \\ q)\overline{p\ldots} \\ \quad \underline{\ldots.} \\ \qquad r_1 \\ \qquad \underline{\ldots.} \\ \qquad\quad r_2 \\ \qquad\quad \ldots \quad \text{etc.} \end{array}$$

the remainders $r_1, r_2, \ldots$ are all less than q; hence, by the time you reach the qth remainder r_q, you must have repeated an earlier remainder at least once; and, having the same remainder as before, you must repeat the same digits as before.

Example 4. Obtain the formula

SERIES FOR log(1 − a)

$$\sum_0^\infty \frac{a^{n+1}}{n+1} = -\log(1-a), \qquad -1 \le a < 1. \tag{3}$$

Solution. Integrate each term on the left of (2) from 0 to a, and you obtain $\sum_0^\infty \frac{a^{n+1}}{n+1}$; integrate the right-hand side of (2) from 0 to a, and you obtain $-\log(1-a)$. These two observations "lead to" the desired formula, *but they do not prove it;* they do not show that $\lim_{N\to\infty} \sum_0^N \frac{x^{n+1}}{n+1} = -\log(1-x)$, which is what must be done. To prove this limit, we integrate (1), not (2):

$$\int_0^a \sum_0^N x^n\,dx = \sum_0^N \frac{a^{n+1}}{n+1} = \int_0^a \frac{1-x^{N+1}}{1-x}\,dx \qquad \text{[by (1)]}$$

$$= \int_0^a \frac{dx}{1-x} - \int_0^a \frac{x^{N+1}}{1-x}\,dx$$

$$= -\log(1-a) - \int_0^a \frac{x^{N+1}}{1-x}\,dx.$$

Hence

$$\sum_{0}^{N} \frac{a^{n+1}}{n+1} + \log(1-a) = -\int_0^a \frac{x^{N+1}}{1-x}\,dx. \tag{4}$$

We will show that when $0 \leq a < 1$, then

$$0 \leq \int_0^a \frac{x^{N+1}}{1-x}\,dx \leq \frac{1}{(N+2)(1-a)}, \tag{5}$$

and this together with (4) shows that

$$\lim_{N\to\infty} \left(\sum_{0}^{N} \frac{a^{n+1}}{n+1} + \log(1-a) \right) = 0,$$

which is equivalent to the formula in (3). [Problem 4 will outline the case where $-1 \leq a < 0$, and this completes the proof of (3).]

To prove (5), *do not* evaluate the integral exactly; this is unnecessary, and in fact useless. Instead, "estimate" it, as follows. The maximum value of $\dfrac{1}{1-x}$ occurs when the denominator is smallest; thus when $a < 1$,

$$1 \leq \frac{1}{1-x} \leq \frac{1}{1-a} \quad \text{for } 0 \leq x \leq a.$$

Since $x^{N+1} \geq 0$ for $0 \leq x$, it follows that we can multiply the preceding inequality by x^{N+1}, obtaining

$$x^{N+1} \leq \frac{x^{N+1}}{1-x} \leq \frac{x^{N+1}}{1-a}, \quad \text{for } 0 \leq x \leq a.$$

We can integrate the preceding inequality to obtain

$$\int_0^a x^{N+1}\,dx \leq \int_0^a \frac{x^{N+1}}{1-x}\,dx \leq \int_0^a \frac{a^{N+1}}{1-a}\,dx.$$

Evaluating the integrals on the far left and far right, we obtain

$$\frac{a^{N+2}}{N+2} \leq \int_0^a \frac{x^{N+1}}{1-x}\,dx \leq \frac{a^{N+2}}{(N+2)(1-a)},$$

hence (since $0 \leq a < 1$)

$$0 \leq \int_0^a \frac{x^{N+1}}{1-x}\,dx \leq \frac{a^{N+2}}{(N+2)(1-a)} \leq \frac{1}{(N+2)(1-a)}.$$

This proves (5), and the desired relation (3) follows.

Example 5. Obtain a formula for $\log\left(\dfrac{1+a}{1-a}\right)$. *Solution.* $\log\left(\dfrac{1+a}{1-a}\right) =$ $\log(1+a) - \log(1-a)$, so formula (3) is obviously relevant. Replacing a by $-a$ in (3) is legitimate, *if* $-1 < a \leq 1$, for then $-1 \leq -a < 1$. We thus find

$$\log(1+a) = -\sum_{0}^{\infty} \frac{(-a)^{n+1}}{n+1}, \qquad -1 < a \leq 1. \tag{6}$$

When $-1 < a < 1$, both (3) and (6) are valid; so for these values of a,

$$\begin{aligned}
\log(1+a) - \log(1-a) &= -\sum_{0}^{\infty} \frac{(-a)^{n+1}}{n+1} + \sum_{0}^{\infty} \frac{a^{n+1}}{n+1} \\
&= \sum_{0}^{\infty}\left(-\frac{(-a)^{n+1}}{n+1} + \frac{a^{n+1}}{n+1}\right) \qquad \text{[Theorem 1]} \\
&= \sum_{0}^{\infty}((-1)^n + 1)\frac{a^{n+1}}{n+1} \\
&= 2a + 0\frac{a^2}{2} + 2\frac{a^3}{3} + 0\frac{a^4}{4} + 2\frac{a^5}{5} + \cdots \\
&= 2a + 2\frac{a^3}{3} + 2\frac{a^5}{5} + \cdots \\
&= 2\left(a + \frac{a^3}{3} + \frac{a^5}{5} + \cdots\right) \qquad \text{[Theorem 1]}
\end{aligned}$$

and we obtain the following result:

$$\log\left(\frac{1+a}{1-a}\right) = 2\sum_{k=0}^{\infty} \frac{a^{2k+1}}{2k+1}, \qquad -1 < a < 1. \tag{7}$$

Remark 1. Formula (7) gives a series for the log of any positive number, since $\dfrac{1+a}{1-a}$ assumes every positive value as a varies in the range $-1 < a < 1$.

Remark 2. In Example 5, the equation

$$2a + 0\frac{a^2}{2} + 2\frac{a^3}{3} + \cdots = 2a + 2\frac{a^3}{3} + 2\frac{a^5}{5} + \cdots \tag{8}$$

needs to be justified. The left-hand side is the limit of the sequence of partial sums

$$(2a),\ (2a+0),\ \left(2a + 2\frac{a^3}{3}\right),\ \left(2a + 2\frac{a^3}{3} + 0\right), \tag{9}$$

and the right-hand side is the limit of

$$(2a),\ \left(2a + 2\frac{a^3}{3}\right),\ \left(2a + 2\frac{a^3}{3} + 2\frac{a^5}{5}\right),\ \left(2a + 2\frac{a^3}{3} + 2\frac{a^5}{5} + 2\frac{a^7}{7}\right), \ldots \tag{10}$$

which is not quite the same. In fact, (10) is a *subsequence* of (9); if we denote (9) by $s_1, s_2, s_3, \ldots, s_n, \ldots$, then (10) is $s_1, s_3, s_5, \ldots$. Hence, (8) is justified by the following THEOREM: *If* $\lim_{n\to\infty} s_n = S$, *then* $\lim_{k\to\infty} s_{2k+1} = S$. More generally, $\lim_{k\to\infty} s_{n_k} = S$ *for any subsequence* $\{s_{n_k}\}$ *of the given sequence* $\{s_n\}$. [See §10.5, Problem 6.]

PROBLEMS

The formulas deduced in Problems 4, 6, and 11 will be much easier to prove when we get to §5.

1. From formulas (2), (3), and (7), evaluate the following series:

• (a) $1 + \frac{1}{3} + \frac{1}{9} + \frac{1}{27} + \cdots$

(b) $\frac{1}{3} + \frac{1}{9} + \frac{1}{27} + \cdots$

• (c) $0.999\ldots$

(d) $\frac{1}{1\cdot 2} + \frac{1}{2\cdot 2^2} + \frac{1}{3\cdot 2^3} + \frac{1}{4\cdot 2^4} + \cdots$

• (e) $1 - \frac{1}{2} + \frac{1}{3} - \frac{1}{4} + \cdots$

(f) $2\cdot 10^{-1} + \frac{2}{3}10^{-3} + \frac{2}{5}10^{-5} + \cdots$

• (g) $1 + \frac{1}{12} + \frac{1}{80} + \frac{1}{384} + \cdots + \frac{2^{-2k}}{2k+1} + \cdots$

2. Express the following repeating decimals as rational numbers (not necessarily in simplest form: $p + q/r$ is a rational number when p, q, and r are rational).

• (a) .4999...

(b) .12121212...

• (c) .123123123...

(d) 3.1414...

• (e) 2.718281828...

3. Is $\sum_0^\infty 10^{-n^2}$ a rational or an irrational number?

4. This problem completes the proof of formula (3) in the text. Throughout the problem, *assume that* $-1 \leq a \leq 0$.

(a) Show that $\int_0^a \frac{x^{N+1}}{1-x}\,dx = (-1)^N \int_0^{-a} \frac{t^{N+1}}{1+t}\,dt \qquad [x = -t]$.

(b) Show that $\frac{1}{2} \leq \frac{1}{1+t} \leq 1$ for $0 \leq t \leq -a$.

(c) Show that $\left|\int_0^{-a} \frac{t^{N+1}}{1+t}\,dt\right| < \frac{1}{N+2}$.

(d) Prove (3), for $-1 \le a \le 0$.

5. Prove that for $|t| < 1$, $\frac{1}{1+t^2} = \sum_0^\infty (-t^2)^N$. [Hint: Use (2).]

6. (a) Show that

SERIES FOR **arctan *a***

$$\frac{1}{1+t^2} - \sum_0^N (-t^2)^n = \frac{(-t^2)^{N+1}}{1+t^2}.$$

(b) Show that

$$\arctan x - \sum_0^N (-1)^n \frac{x^{2n+1}}{2n+1} = (-1)^{N+1} \int_0^x \frac{t^{2N+2}}{1+t^2}\,dt.$$

(c) Show that when $0 \le x \le a$,

$$\left|\arctan x - \sum_0^N (-1)^n \frac{x^{2n+1}}{2n+1}\right| \le \frac{a^{2N+3}}{2N+3}.$$

(d) Show that when $0 \le x \le 1$,

$$\arctan x = \sum_0^\infty (-1)^n \frac{x^{2n+1}}{2n+1}$$

(e) Show that when $-1 \le x \le 0$, the formula in (d) still holds.

7. Show that when $|x| \le \frac{1}{5}$,

$$\left|\arctan x - \sum_0^2 (-1)^n \frac{x^{2n+1}}{2n+1}\right| < 2\cdot 10^{-6}.$$

[See Problem 6.]

8. From the addition formula for tangent, show that

$$4 \arctan\left(\frac{1}{5}\right) - \arctan\left(\frac{1}{239}\right) = \frac{\pi}{4}.$$

[Hint: Let $\tan\theta = \frac{1}{5}$, and find $\tan 2\theta$, $\tan 4\theta$, $\tan(4\theta - \pi/4)$.]

9. Use Problems (7) and (8) to compute π to four decimals.

APPROXIMATING π

10. Prove that $c \sum_0^\infty a_n = \sum_0^\infty c a_n$, if $\sum_0^\infty a_n$ is convergent.

11. Prove that $\sum_0^\infty nx^{n-1} = \dfrac{1}{(1-x)^2}$, if $|x| < 1$. [Hint: This is what you would get by differentiating (2). To *prove* that it is true, differentiate (1). You may also need an inequality such as $Nr^N \leq \dfrac{1}{N}\left(\dfrac{2}{e \log r}\right)^2$ for $N > 0$, which follows from $t^2e^{-t} \leq \left(\dfrac{2}{e}\right)^2$, $t \geq 0$, [§3.1] on setting $t = -N \log r$.]

12. [*Probability*] A parts factory tests its product, and finds that the proportion of good parts is p; hence p is the probability that a tested part is good, and $1 - p$ is the probability that it is not good. If p is nearly equal to 1, then most of the parts are good, and the tester finds long strings of good parts, interrupted by an occasional bad part or two:

ggbggggbbgggggggggggbgggg....

We ask, *What is the average length of a string of good parts?*

Suppose a string of g's (good parts) has just ended, i.e. we have arrived at a b (bad part). Consider the string of g's following this b; the string has length *zero* if and only if the next part is a b (bad part), and the probability of this is $(1 - p)$. The string of g's has length 1 if the next parts come in the order g b, and the probability of this is $p(1 - p)$. [The probability that each of *two* independent events occurs is the product of the probabilities of each individual event.] Generally, the next string of good parts has length n if the parts test like this,

$$\ldots \text{b}\,\underbrace{\text{g}\ldots\ldots\text{g}}_{n \text{ times}}\,\text{b}$$

and the probability of this event is $p^n(1 - p)$. Hence in all the strings of g's that occur,

$(1 - p)$ is the proportion of strings of length 0

$p(1 - p)$ is the proportion of strings of length 1

$\vdots$

$p^n(1 - p)$ is the proportion of strings of length n.

Now prove the following:

(a) If $p(n)$ denotes the proportion of strings of length n, show that $\sum_0^\infty p(n) = 1$. Why is this reasonable?

(b) To obtain the average length of the "good strings," multiply each string length by the proportion of times it occurs, and add:

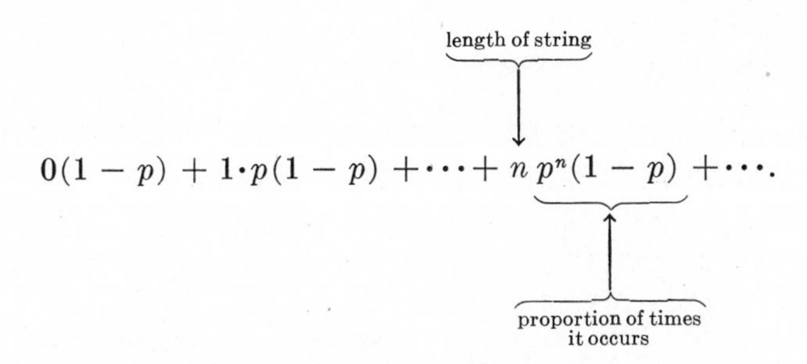

Show that the average "good string" length is $\dfrac{p}{1-p}$. [See the previous problem.]

(c) Show that the average "good string" length is an *increasing* function of p. Why is this reasonable?

Remark. The same analysis applies to many other situations. *Example 1:* p is the probability that a basketball player sinks his foul shots; then $\dfrac{p}{1-p}$ is the average length of a string of good shots.

Example 2: At a certain stop sign, cars travel on the main street with random gaps between them. Let p be the probability that a gap is long enough for a car on the stop street to cross the main street; then $\dfrac{p}{1-p}$ is the average number of cars he has to wait for before crossing.

11.3 POSITIVE SERIES

A series $\sum_1^\infty a_n$ is called *positive* if its terms a_n are all ≥ 0. In this case, the partial sums $s_N = \sum_1^N a_n$ form an increasing sequence; $s_{N+1} = \sum_1^{N+1} a_n$ is obtained from s_N by adding the term $a_{N+1} \geq 0$, so $s_{N+1} = s_N + a_{N+1} \geq s_N$. For such series there are two possibilities.

(I) There is a bound M for the partial sums. Then, according to the Monotone Convergence Theorem [§10.3], $s_n \to S$, and $\sum_1^\infty a_n = S$.

(II) There is *no* bound M for the partial sums. Then $\sum_1^\infty a_n$ does not converge.

We indicate the second case symbolically by writing $\sum_1^\infty a_n = \infty$, and the first case by writing $\sum_1^\infty a_n < \infty$. These expressions are generally used only for positive series; they do not introduce a new number ∞, but serve only to indicate whether we are in case (I) or case (II).

All the tests for convergence of positive series consist in finding an upper bound M for the partial sums, guaranteeing that we are in case (I). Theorem 3 gives a particularly simple and obvious test.

COMPARISON TEST

Theorem 3. *Suppose* $0 \leq a_n \leq b_n$. *Then*

$$\sum_1^\infty b_n < \infty \quad \Rightarrow \quad \sum_1^\infty a_n < \infty, \tag{1}$$

and

$$\sum_1^\infty a_n = \infty \quad \Rightarrow \quad \sum_1^\infty b_n = \infty. \tag{2}$$

Proof. Suppose $\sum_1^\infty b_n < \infty$, so this series converges, $\sum_1^\infty b_n = S$. Then

$$\sum_1^N a_n \leq \sum_1^N b_n \leq S,$$

so the partial sums of $\sum_1^\infty a_n$ are bounded, and this series converges, proving (1). The implication (2) follows by contradiction; *if* $\sum_1^\infty b_n$ were convergent, then by (1), $\sum_1^\infty a_n$ would also converge.

Example 1. The geometric series $\sum_1^\infty (1/2)^n$ is known to converge. Therefore $\sum_1^\infty \dfrac{1 - 1/n}{1 + 2^n}$ converges, since $0 \leq \dfrac{1 - 1/n}{1 + 2^n} \leq (1/2)^n$.

Example 2. The series $\sum_1^\infty \frac{1}{n^n}$ converges, since $\frac{1}{n^n} \le \frac{1}{2^{n-1}}$, and

$$\sum_1^\infty \frac{1}{2^{n-1}} = \sum_0^\infty \frac{1}{2^n} = \frac{1}{1-\frac{1}{2}} = 2.$$

Our next test compares a series to an integral.

Theorem 4. *Let f be a positive decreasing function for $x \ge 1$. Then $\sum_1^\infty f(n)$ converges if and only if $\lim_{N\to\infty} \int_1^N f$ exists.*

INTEGRAL TEST

Recall [§5.5] that when $\lim_{N\to\infty} \int_1^N f$ exists, it is denoted by $\int_1^\infty f$, and called an *improper integral,* because the "interval of integration" is not finite.

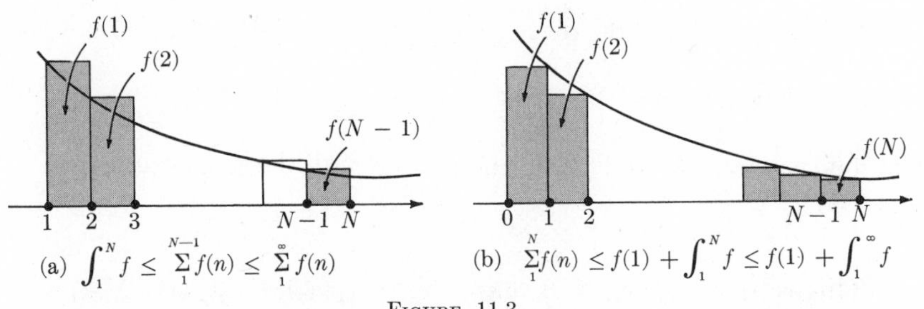

FIGURE 11.3

Figure 3 illustrates the simple proof of the integral test. Suppose first that $\sum_1^\infty f(n) < \infty$. Since $f \ge 0$, the sequence $\int_1^N f$ is increasing, and [Fig. 3(a)] it has the upper bound $\sum_1^\infty f(n)$, so it converges. Conversely, if $\int_1^N f$ converges to a finite value $\int_1^\infty f$, then $f(1) + \int_1^\infty f$ is an upper bound for the sequence s_N of partial sums [Fig. 3(b)], so they converge.

Warning: As Fig. 3 shows, it is not generally true that $\int_1^\infty f = \sum_1^\infty f(n)$.

Generally, $\int_1^\infty f < \sum_1^\infty f(n) < f(1) + \int_1^\infty f$.

IMPORTANT EXAMPLE

Example 3. Prove $\sum_1^\infty n^{-a} < \infty$ if $a > 1$. *Solution.* The function $f(x) = x^{-a}$ is positive and decreasing for $x > 0$, so the integral test applies. We find

$$\int_1^N x^{-a}\,dx = \frac{x^{1-a}}{1-a}\bigg|_1^N = \frac{1 - N^{1-a}}{a-1}.$$

Since $1 - a < 0$, $\lim N^{1-a} = 0$, and

$$\int_1^\infty x^{-a}\,dx = \lim \int_1^N x^{-a}\,dx = 1/(a-1) < \infty.$$

Hence $\sum_1^\infty n^{-a} < \infty$ if $a > 1$.

Example 4. Discuss the convergence of $\sum_1^\infty n^{-a} = \infty$ if $a \leq 1$. [Do it yourself; solution below.]

Our final convergence test uses a comparison with the geometric series.

RATIO TEST

Theorem 5. *If* $a_n > 0$ *and* $\lim_{n\to\infty} (a_{n+1}/a_n) < 1$, *then* $\sum_1^\infty a_n < \infty$.

If $\lim_{n\to\infty} (a_{n+1}/a_n) > 1$, *then* $\sum_1^\infty a_n = \infty$.

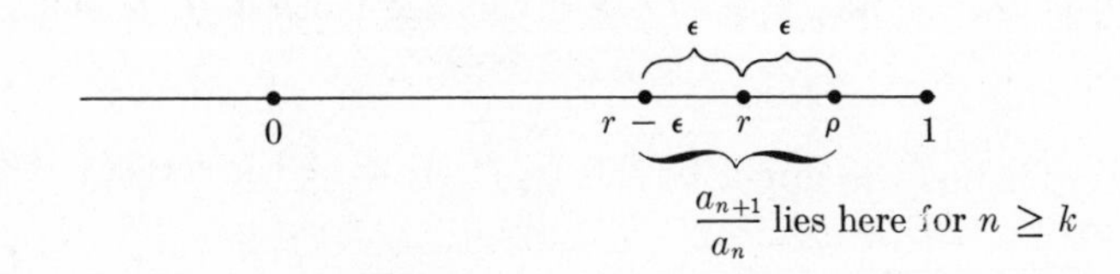

FIGURE 11.4

Proof. Suppose $\lim_{n\to\infty} (a_{n+1}/a_n) = r < 1$. Pick a number ρ between r and 1, and let $\epsilon = \rho - r$ [Fig. 4]. By the definition of limit, there is an integer K such that $|a_{n+1}/a_n - r| < \epsilon$ whenever $n \geq K$. In particular, $a_{n+1}/a_n < \rho$ when $n \geq K$ [Fig. 4], and so $a_{n+1} < \rho a_n$ whenever $n \geq K$. From this we find successively

$$
\begin{aligned}
&a_{K+1} < \rho a_K \\
&a_{K+2} < \rho a_{K+1} < \rho^2 a_K \\
&a_{K+3} < \rho a_{K+2} < \rho^3 a_K \\
&\cdots\cdots\cdots\cdots\cdots\cdots \\
&a_{K+j} < \rho^j a_K .
\end{aligned}
$$

Hence, for $N \geq K$,

$$
\begin{aligned}
\sum_1^N a_n &= \left(\sum_1^{K-1} a_n\right) + (a_K + a_{K+1} + \cdots + a_N) \\
&< \left(\sum_1^{K-1} a_n\right) + a_K(1 + \rho + \cdots + \rho^{N-K}) = \left(\sum_1^{K-1} a_n\right) + \frac{a_K(1 - \rho^{N-K+1})}{1-\rho} \\
&\leq \left(\sum_1^{K-1} a_n\right) + \frac{a_K}{1-\rho} .
\end{aligned}
$$

The last term in this chain of inequalities, being independent of N, provides the desired upper bound for all the partial sums; hence $\sum_1^\infty a_n < \infty$.

In case $\lim_{n\to\infty} (a_{n+1}/a_n) = r > 1$, there is a number ρ greater than 1 and a number K such that $n \geq K \Rightarrow a_{n+1} \geq \rho a_n$. It follows that $a_{K+1} > a_K \geq 0$, and $a_{K+N} \geq a_{K+1}$, so we do *not* have $\lim_{n\to\infty} a_n = 0$, and the series diverges, by Theorem 2. *Q.E.D.*

Example 5. Consider $\sum_1^\infty n^2 x^n$, which is a positive series for $x \geq 0$. Applying the ratio test, we find

$$
\frac{a_{n+1}}{a_n} = \frac{(n+1)^2 x^{n+1}}{n^2 x^n} = x\left(\frac{n+1}{n}\right)^2 \to x.
$$

Thus the ratio test says that the series converges if $0 < x < 1$, and diverges if $x > 1$. If $x = 1$, the ratio test gives no decision. But when $x = 1$, the series is $\sum_1^\infty n^2$, which obviously diverges.

Example 6. For what positive values of x does $\sum_{1}^{\infty} nx^n$ converge?

[Solution below.]

Example 7. For what positive values of x does $\sum_{0}^{\infty} x^n/n!$ converge? *Solution.*

The function $f(n) = x^n/n!$ is not suitable for integration, so we reject the integral test. The ratio test works well, however;

$$\frac{a_{n+1}}{a_n} = \frac{x^{n+1}/(n+1)!}{x^n/n!} = x\,\frac{n(n-1)\cdots(1)}{(n+1)(n)(n-1)\cdots(1)} = \frac{x}{n+1} \to 0.$$

Hence $\sum_{0}^{\infty} x^n/n!$ *converges for every* $x > 0$.

Incidentally, this shows that

$$\lim_{n\to\infty} \frac{x^n}{n!} = 0 \tag{3}$$

no matter how large x may be, for the terms of a convergent series form a sequence converging to 0 [Theorem 2].

Example 8. Prove that $\sum_{0}^{\infty} \frac{t^n}{n!} = e^t$ for every real t. *Solution.* The partial sum

$\sum_{0}^{N} \frac{t^n}{n!}$ is precisely the Taylor polynomial of e^t about $t = 0$. Hence, by the

Taylor expansion [§9.2],

$$e^t = \sum_{0}^{N} \frac{t^n}{n!} + \frac{t^{N+1}}{(N+1)!} e^c,$$

where c is between 0 and t. It follows that $e^c \leq e^{|t|}$, hence

$$\left| e^t - \sum_{0}^{N} \frac{t^n}{n!} \right| = \left| \frac{t^{N+1}}{(N+1)!} e^c \right| \leq e^{|t|} \frac{|t|^{N+1}}{(N+1)!}.$$

By (3), $\lim_{N\to\infty} |t|^{N+1}/(N+1)! = 0$; therefore

$$\lim_{n\to\infty} \left(e^t - \sum_{0}^{N} \frac{t^n}{n!} \right) = 0 \quad \text{and} \quad \sum_{0}^{\infty} \frac{t^n}{n!} = e^t.$$

Remark 1. The ratio test is very convenient when it gives a decision, i.e. when $\lim (a_{n+1}/a_n) = r \neq 1$. Unfortunately, there are many series where it does *not* decide the issue, for example, $\sum_{1}^{\infty} 1/n$, where we find

$$\lim (a_{n+1}/a_n) = \lim (n/(n+1)) = 1.$$

The same situation arises with

$$\sum_{1}^{\infty} 1/n^2, \quad \sum_{2}^{\infty} 1/(n \log n), \quad \sum_{1}^{\infty} (\log n)/n,$$

and so on. Such series lie close to the boundary line between convergence and divergence, and the ratio test is not precise enough to tell which side of the line they are on. In this respect, the integral test is far more precise, and often gives a quick and easy answer in those cases left undecided by the ratio test. For example, the integral test shows that $\sum_{1}^{\infty} 1/n^2$ converges, but that $\sum_{1}^{\infty} 1/n$, $\sum_{2}^{\infty} 1/(n \log n)$, and $\sum_{1}^{\infty} (\log n)/n$ all diverge.

Remark 2. The convergence question for $\sum_{1}^{\infty} (\log n)/n$ can be answered by a comparison to $\sum_{1}^{\infty} 1/n$; since $\sum_{1}^{\infty} 1/n$ diverges and $(\log n)/n > 1/n$ [at least when $n \geq 3$], $\sum_{1}^{\infty} (\log n)/n$ must also diverge. This line of reasoning is probably clear enough but, strictly speaking, it is *not* justified immediately by Theorem 3; that theorem requires $a_n \leq b_n$ for *all* n, and we only have $a_n \leq b_n$ for $n \geq 3$. The comparison test can be strengthened to cover examples such as this one, along with others less obvious. Here are three versions; the proofs are left to you. In each version, we assume that $a_n \geq 0$ and $b_n \geq 0$.

STRONGER VERSIONS OF COMPARISON TEST

(i) If there are a number N and a constant c such that $0 \leq a_n \leq cb_n$ for $n \geq N$, and if $\sum_{1}^{\infty} b_n < \infty$, then $\sum_{1}^{\infty} a_n < \infty$. [Problem 14 below]

(ii) If $\lim (a_n/b_n)$ exists, and $\sum_{1}^{\infty} b_n < \infty$, then $\sum_{1}^{\infty} a_n < \infty$. [Problem 15]

(iii) If $\lim (a_n/b_n) = L \neq 0$, then $\sum_{1}^{\infty} a_n$ and $\sum_{1}^{\infty} b_n$ are either both convergent or both divergent. [Problem 16]

Example 9. Test $\sum_1^\infty 1/(2^n - 1)$ for convergence. The obvious comparison series is $\sum_1^\infty 1/2^n$. Theorem 3 does not apply directly, but each of the above versions does. To apply version (i), set $a_n = 1/(2^n - 1)$ and $b_n = 1/2^n$, and notice that $\frac{1}{2^n - 1} \leq \frac{2}{2^n}$; to apply versions (ii) and (iii), compute the limit

$$\lim \frac{a_n}{b_n} = \lim \frac{1/(2^n - 1)}{1/2^n} = \lim \frac{1}{1 - 2^{-n}} = 1.$$

Example 10. Does $\sum_1^\infty 4/(3^n - 2)$ converge? [Solution below.]

Example 11. Does $\sum_1^\infty 1/(n + 10^6)$ converge? [Solution below.]

Remark 3. The divergence of $\sum_1^\infty 1/n$ explains the contradiction that arose in trying to "compute" this sum [§11.1]. The trouble comes at the very beginning, where we said

"Let the unknown sum be L, $1 + \frac{1}{2} + \frac{1}{3} + \frac{1}{4} + \cdots = L$".

Actually, there *is* no number L such that $\sum_1^\infty 1/n = L$. This example illustrates the importance of knowing which things exist, and which do not!

Solution to Example 4. $\int_1^N x^{-a}\,dx = \left.\frac{x^{1-a}}{1-a}\right|_1^N = \frac{1}{1-a}(N^{1-a} - 1)$, if $a < 1$. When $a < 1$, the exponent $1 - a$ is positive, so $N^{1-a} \to +\infty$, and $\int_1^N x^{-a}\,dx$ is unbounded. Therefore $\sum_1^\infty n^{-a} = \infty$, by the integral test. In the remaining case where $a = 1$, we have $\int_1^N x^{-1}\,dx = \log N$, which is unbounded, so $\sum_1^\infty n^{-1} = \infty$. Combining this with Example 3 shows that

$\sum_1^\infty n^{-a} < \infty$ IFF $a > 1$

$$\sum_1^\infty n^{-a} \text{ converges if and only if } a > 1.$$

Solution to Example 6. Applying the ratio test, we find

$$\frac{a_{n+1}}{a_n} = \frac{(n+1)x^{n+1}}{nx^n} = \frac{n+1}{n}x \to x,$$

so the series $\sum_1^\infty nx^n$ converges when $0 \le x < 1$. For $x \ge 1$, it diverges, since the terms $a_n = nx^n$ do not tend to 0.

Solution to Example 10. Compare the given term $a_n = 4/(3^n - 2)$ with the simpler term $b_n = 1/3^n$. We get

$$\lim \frac{a_n}{b_n} = \lim \frac{4 \cdot 3^n}{3^n - 2} = \lim \frac{4}{1 - 2 \cdot 3^{-n}} = 4 \ne 0.$$

Therefore by version (iii) above, $\sum a_n$ converges iff $\sum b_n$ does. But $\sum b_n = (1/3)^n$ is a convergent geometric series, so $\sum 4/(3^n - 2)$ converges.

Solution to Example 11. Compare the given term $a_n = 1/(n + 10^6)$ with the simpler term $b_n = 1/n$. We get $\lim a_n/b_n = 1$. Since $\sum b_n = \sum 1/n = \infty$, we conclude from (iii) that the given series also diverges, $\sum a_n = \sum 1/(n + 10^6) = \infty$.

PROBLEMS

1. Decide the convergence of the following series by comparison with $\sum_1^\infty n^{-a}$ (convergent if and only if $a > 1$), with $\sum_0^\infty r^n$ (convergent if and only if $|r| < 1$), or with $\sum_0^\infty 1/n!$ (convergent to e).

• (a) $\sum_1^\infty \frac{n}{10^6 + n^2}$

(b) $\sum_0^\infty \frac{2^n + n}{3^n - n}$

• (c) $\sum_1^\infty n^{-n^2}$

(d) $\sum_0^\infty n(1 + n^7)^{-1/3}$

2. Decide the convergence of the following series *by the integral test.*

(a) $\sum_2^\infty \frac{1}{n \log n}$

• (b) $\sum_0^\infty ne^{-n^2}$

(c) $\sum_0^\infty r^n \quad (0 < r < 1)$

• (d) $\sum_1^\infty \frac{n}{1 + n^2}$

(e) $\sum_0^\infty \frac{\log n}{n}$

(f) $\sum_0^\infty r^n \quad (r \ge 1)$

3. Decide the convergence of the following series *by the ratio test* whenever possible. If the ratio test fails, try something else.

• (a) $\sum_0^\infty ne^{-n^2}$

(b) $\sum_0^\infty r^n, \quad r > 0$

(c) $\sum_1^\infty \dfrac{n}{1+n^2}$

• (d) $\sum_1^\infty \dfrac{n!}{n^n}$ $\left[\text{Hint: } \lim \left(1+\dfrac{1}{n}\right)^n = e.\right]$

Decide the convergence of the following series [Problems 4–10]:

• **4.** $\sum_1^\infty \dfrac{n^3}{(n+1)!}$

5. $\sum_1^\infty \dfrac{n^3}{n^4+n^3+n+1}$

• **6.** $\sum_2^\infty \dfrac{1}{n(\log n)^2}$

7. $\sum_1^\infty ne^{-n}$

• **8.** $\sum_1^\infty \dfrac{1}{n(n+1)}$. [A "partial fractions" decomposition of $\dfrac{1}{n(n+1)}$ will allow you to express the partial sums concisely and evaluate the series exactly.]

9. $\sum_1^\infty n^2 2^{n+1} 3^{-n}$

• **10.** $\sum_0^\infty n!\,2^{-n}$

• **11.** Decide for which positive values of x the following series converge.

(a) $\sum_2^\infty \dfrac{1}{n(\log n)^x}$

(b) $\sum_0^\infty \dfrac{(x/2)^{2n}}{(n!)^2}$

(c) $\sum_1^\infty \dfrac{\cos^2(xn)}{1+n^2}$

(d) $\sum_3^\infty \dfrac{1}{n(\log n)(\log(\log n))^x}$

(e) $\sum_1^\infty \left(\dfrac{1}{n} - e^{-nx}\right)$

(f) $\sum_0^\infty x^n n!\, n^{-n}$

(g) $\sum_0^\infty \dfrac{x^n (n!)^2}{(2n)!}$

(h) $\sum_0^\infty x^n n!\, 2^{-n^2}$

(i) $\sum_2^\infty \dfrac{x^n}{n(\log n)^{2x}}$

$\left[\text{Hint: By L'Hôpital's rule, §7.6, } \lim_{n\to\infty} \dfrac{\log(n+1)}{\log n} = 1.\right]$

12. (a) Prove that $\sin x = \sum_0^\infty (-1)^n \frac{x^{2n+1}}{(2n+1)!}$. [See Example 8.]

(b) Prove that $\cos x = \sum_0^\infty (-1)^n \frac{x^{2n}}{(2n)!}$.

(c) Obtain a series for $(\sin x)/x$, valid for $x \neq 0$. To what does the series converge for $x = 0$? Is this reasonable?

13. (a) Show that $e^{-x} = \sum_0^\infty (-1)^n \frac{x^n}{n!}$. [Use Example 8.]

(b) Obtain a series for $\sinh x = \frac{1}{2}(e^x - e^{-x})$. [This series bears a striking resemblance to the series for $\sin x$, Problem 12(a).]

(c) Obtain a series for $\cosh x = \frac{1}{2}(e^x + e^{-x})$. [Compare the series for $\cos x$, Problem 12(b).]

14. Suppose $\sum_1^\infty b_n < \infty$ and $0 \leq a_n \leq cb_n$ for $n \geq N$. Prove that $\sum_1^\infty a_n < \infty$.

15. Suppose $a_n \geq 0$, $b_n \geq 0$, $\sum_1^\infty b_n < \infty$, and $\lim_{n\to\infty} (a_n/b_n)$ exists. Prove that $\sum_1^\infty a_n$ converges.

16. Suppose $a_n \geq 0$, $b_n \geq 0$, and $\lim_{n\to\infty} a_n/b_n$ exists and is different from 0. Prove that $\sum_1^\infty a_n$ and $\sum_1^\infty b_n$ either both converge or both diverge. [Hint: Use Problem 15.]

17. [*Variations on the Ratio Test*]

(a) Suppose that $\frac{a_{n+1}}{a_n} \geq 1$ for all n. Prove that $\sum_1^\infty a_n = \infty$.

• (b) Suppose that $\frac{a_{n+1}}{a_n} < 1$ for all n. Is $\sum_1^\infty a_n < \infty$?

18. [*Variation on the Integral Test*] Suppose f is positive and decreasing, not on $[1, \infty)$ but on the smaller interval $[k, \infty)$, where $k > 0$. Show that $\sum_1^\infty f(n)$ and $\lim_{N\to\infty} \int_1^N f(x)\,dx$ either both converge or both diverge.

• **19.** Let P and Q be polynomials, and suppose $P(x)/Q(x) > 0$ for $x > 0$. Under what conditions does $\sum_1^\infty \frac{P(n)}{Q(n)}$ converge?

20. Suppose that $\sum_1^\infty a_n$ converges *by the ratio test.* Prove that $\sum_1^\infty n^k a_n < \infty$ for every k.

21. (a) $\int_n^{n+1} \frac{1}{x}\,dx$ represents a certain area. Sketch the area. Show that

$$\frac{1}{2}\left(\frac{1}{n} - \frac{1}{n+1}\right) < \frac{1}{n} - \int_n^{n+1} \frac{1}{x}\,dx < \left(\frac{1}{n} - \frac{1}{n+1}\right).$$

(b) Show that $\sum_1^\infty \left(\frac{1}{n} - \frac{1}{n+1}\right) = 1$. [Hint: Evaluate the partial sums explicitly.]

(c) Show that $\sum_1^\infty \left(\frac{1}{n} - \int_n^{n+1} \frac{1}{x}\,dx\right)$ converges to a sum γ, $\frac{1}{2} < \gamma < 1$.

EULER'S CONSTANT

(d) Show that $\gamma = \lim_{N\to\infty} \left(\sum_1^N \frac{1}{n} - \log N\right)$. This number γ is called the *Euler constant.*

APPENDIX: ERROR ESTIMATES (optional)

The integral, ratio, and comparison tests all implicitly include estimates of the error in replacing the infinite sum $S = \sum_k^\infty a_n$ by the finite sum $s_N = \sum_k^N a_n$. This is best shown by examples.

Example 1. By the integral test, $\sum_1^\infty n^{-2}$ converges. Moreover, the error E_N in replacing $\sum_1^\infty nx^n$ by $\sum_1^N nx^n$ is simply $\sum_{N+1}^\infty n^{-2}$, and

$$\int_{N+1}^\infty x^{-2}\,dx < \sum_{N+1}^\infty n^{-2} < \int_N^\infty x^{-2}\,dx.$$

[Draw a sketch like Fig. 3 above to verify this.] By evaluating the improper integrals, we find

$$\frac{1}{N+1} < \sum_{N+1}^\infty n^{-2} < \frac{1}{N}.$$

Hence [see Fig. 5]

$$\left|\sum_{N+1}^\infty n^{-2} - \frac{2N+1}{2N(N+1)}\right| < \frac{1}{2}\left(\frac{1}{N} - \frac{1}{N+1}\right) = \frac{1}{2N(N+1)},$$

and we have

$$\sum_1^\infty n^{-2} = \sum_1^N n^{-2} + \sum_{N+1}^\infty n^{-2} = \sum_1^N n^{-2} + \frac{2N+1}{2N(N+1)} \pm \frac{1}{2N(N+1)}.$$

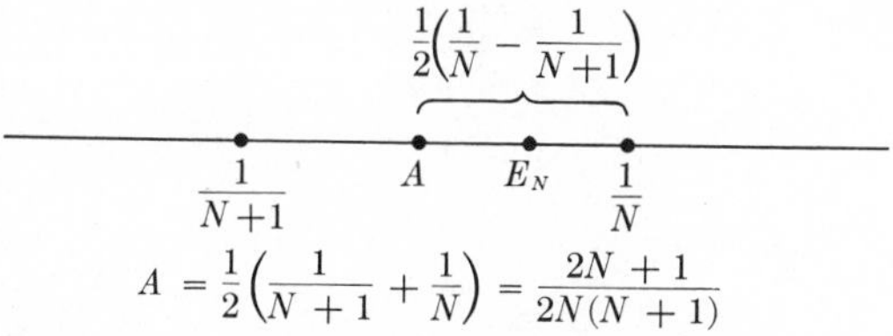

$\frac{1}{2}\left(\frac{1}{N} - \frac{1}{N+1}\right)$

$\frac{1}{N+1}$ $\quad A \quad E_N \quad$ $\frac{1}{N}$

$$A = \frac{1}{2}\left(\frac{1}{N+1} + \frac{1}{N}\right) = \frac{2N+1}{2N(N+1)}$$

FIGURE 11.5

In Example 1, notice that the error after N terms is roughly $1/(N+1)$, *much larger* than the first neglected term $1/(N+1)^2$. Notice also that when we have both upper *and* lower bounds for the error, we can compensate for the error after N terms and approximate the whole sum within $\pm 1/2N^2$.

Example 2. $\sum_1^\infty nx^n$ converges when $0 \le x < 1$, by the ratio test. The error after N terms is

$$\sum_{N+1}^\infty nx^n = (N+1)x^{N+1} + (N+2)x^{N+2} + \cdots$$

$$= (N+1)x^{N+1}\left(1 + \frac{N+2}{N+1}x + \frac{N+3}{N+1}x^2 + \frac{N+4}{N+1}x^3 + \cdots\right). \quad (1)$$

Now,

$$1 \le \frac{N+k+1}{N+1} = \frac{N+k+1}{N+k} \times \frac{N+k}{N+k-1} \times \cdots \times \frac{N+2}{N+1} \le \left(\frac{N+2}{N+1}\right)^k,$$

since $\dfrac{n+1}{n} \le \dfrac{m+1}{m}$ when $n \ge m$. Hence, from (1), we have the estimate

$$\frac{(N+1)x^{N+1}}{1-x} = (N+1)x^{N+1}\sum_0^\infty x^k \le \sum_{N+1}^\infty nx^n$$

$$\le (N+1)x^{N+1}\sum_0^\infty \left(\frac{N+2}{N+1}\right)^k x^k$$

$$= \frac{(N+1)x^{N+1}}{1 - x(N+2)/(N+1)},$$

when $\dfrac{N+2}{N+1}x < 1$.

Specifically, when $x = \frac{1}{2}$, the error $E_N = \sum_{N+1}^\infty n2^{-n}$ satisfies

$$(N+1)2^{-N} \le E_N \le \frac{(N+1)2^{-N-1}}{1 - \dfrac{1}{2}\left(\dfrac{N+2}{N+1}\right)},$$

and

$$\sum_1^N n2^{-n} + (N+1)2^{-N} < \sum_1^N n2^{-n} < \sum_1^N n2^{-n} + (N+1)2^{-N} + \frac{N+1}{N}2^{-N}.$$

Notice that the error after N terms, $\sum_1^N n2^{-n}$, is about $2^{-N}(N+1)$; but we can obtain an approximation with N terms whose error is merely $2^{-N}(N+1)/N$ by finding upper *and* lower bounds for the error after N terms.

PROBLEMS

1. We have proved that $e = \sum_0^\infty 1/n!$. Show that

$$\sum_0^N \frac{1}{n!} + \frac{1}{(N+1)!} < e < \sum_0^N \frac{1}{n!} + \frac{N+2}{N+1}\cdot\frac{1}{(N+1)!}.$$

Thus when N is large, the error after adding $1/N!$ is very nearly equal to the first neglected term, $1/(N+1)!$. [Compare with Example 1.]

2. Find the Taylor series for $-\log(1-x)$. Following Example 2, give a careful estimate of the error after N terms when $0 < x < 1$, and compute $\log 2 = -\log(1/2)$ to two decimals. By your method, how many terms are necessary for accuracy to four decimals?

3. Use the integral test to estimate the error after N terms in $\sum_1^\infty nx^n$.

Compare the results with Example 2.

11.4 ABSOLUTE CONVERGENCE; ALTERNATING SERIES

The convergence tests for *positive* series can be applied to prove the convergence of more general series, on the strength of the following theorem.

ABSOLUTE CONVERGENCE IMPLIES ORDINARY CONVERGENCE

Theorem 6. *If* $\sum_1^\infty |a_n| < \infty$, *then* $\sum_1^\infty a_n$ *converges.*

Thus the convergence of $\sum_1^\infty a_n$ can be proved by considering the positive series $\sum |a_n|$.

When $\sum_1^\infty |a_n| < \infty$, we call $\sum_1^\infty a_n$ *absolutely convergent.* Theorem 6 says that every absolutely convergent series is also convergent in the ordinary sense. The proof is based on the comparison test and the addition of two series. We define two *positive* series $\sum a_n{}^+$ and $\sum a_n{}^-$ as follows:

$$a_n{}^+ = \begin{cases} a_n & \text{if } a_n \geq 0 \\ 0 & \text{if } a_n < 0, \end{cases} \qquad a_n{}^- = \begin{cases} 0 & \text{if } a_n \geq 0 \\ -a_n & \text{if } a_n < 0. \end{cases}$$

It is easy to see that $0 \leq a_n{}^+ \leq |a_n|$, so $\sum a_n{}^+$ converges, by comparison with $\sum |a_n|$, which is convergent by hypothesis. Similarly, $0 \leq a_n{}^- \leq |a_n|$, so $\sum a_n{}^-$ converges. Finally, $a_n = a_n{}^+ - a_n{}^-$, so $\sum_1^\infty a_n$ converges to $\sum_1^\infty a_n{}^+ - \sum_1^\infty a_n{}^-$, and Theorem 6 is proved.

Sample application: $\sum_1^\infty (-1)^n/n^2$ converges, since $\sum_1^\infty 1/n^2$ does.

Theorem 6 leads immediately to more general versions of the comparison and ratio tests:

COMPARISON TEST

Theorem 7. *If* $|a_n| \leq b_n$ *and* $\sum_1^\infty b_n < \infty$, *then* $\sum_1^\infty a_n$ *converges absolutely.*

RATIO TEST

Theorem 8. *If* $\lim \frac{|a_{n+1}|}{|a_n|} = r < 1$, *then* $\sum_1^\infty a_n$ *converges absolutely.*

If $\lim \frac{|a_{n+1}|}{|a_n|} = r > 1$, *then* $\sum_1^\infty a_n$ *diverges.*

Example 1. To apply the ratio test to $\sum_1^\infty nx^n$, consider the limit

$$\lim \frac{|a_{n+1}|}{|a_n|} = \lim \frac{|(n+1)x^{n+1}|}{|nx^n|} = \lim \frac{n+1}{n}|x| = |x|.$$

This series therefore converges absolutely for $|x| < 1$, and diverges for $|x| > 1$. For $|x| = 1$, the ratio test gives no result, but you can see easily that the series diverges in that case.

Example 2. $\sum_{n=1}^\infty 3^{-n}n^{-5}x^n$ converges absolutely for any $|x| \leq 3$. For,

$$\lim_{n\to\infty} \frac{|a_{n+1}|}{|a_n|} = \lim_{n\to\infty} \frac{|3^{-n-1}(n+1)^{-5}x^{n+1}|}{|3^{-n}n^{-5}x^n|} = \frac{|x|}{3}\lim_{n\to\infty}\left(\frac{n+1}{n}\right)^{-5} = \frac{|x|}{3},$$

which is <1 when $|x| < 3$. When $|x| > 3$, the series diverges, by the ratio test. When $|x| = 3$, it converges absolutely because then the general term satisfies $|3^{-n}n^{-5}x^n| = n^{-5}$, and $\sum_1^\infty n^{-5} < \infty$.

Example 3. The series $\sum_1^\infty (-1)^n/n$ converges to $-\log 2$ [§11.2], but does not converge absolutely. For, this is $\sum a_n$ with $a_n = (-1)^n/n$, so we have $|a_n| = \frac{1}{n}$, and $\sum_1^\infty \frac{1}{n} = \infty$. This example shows that the condition $\sum |a_n| < \infty$, though *sufficient* for convergence, is *not necessary*.

Series that converge, but not absolutely, are called *conditionally convergent.* Their convergence is conditional on having the terms arranged so that appropriate cancellation occurs. The simplest form of this cancellation is found in *alternating series,* where the terms are alternately positive and negative, i.e. where

$$\sum_0^\infty a_n = \pm \sum_0^\infty (-1)^n b_n \quad \text{with } b_n = |a_n| \geq 0.$$

ALTERNATING SERIES

Theorem 9. *An alternating series converges if the absolute values b_n tend monotonically to zero. The sum of the series lies between any two successive partial sums s_N and s_{N+1}.*

To say that the numbers b_n tend monotonically to zero means that $b_n \geq b_{n+1}$ for every n and that $b_n \to 0$.

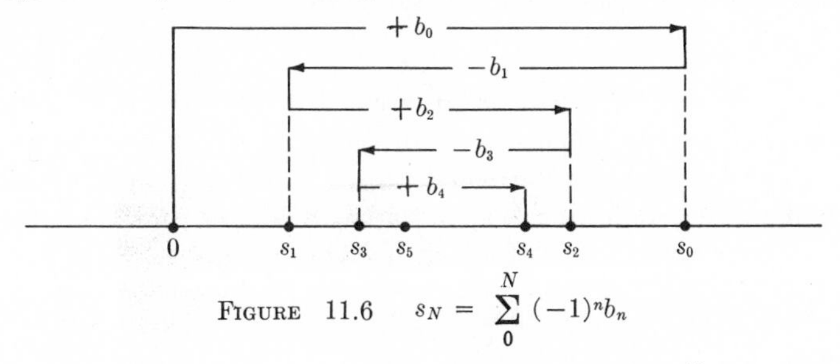

FIGURE 11.6 $s_N = \sum_0^N (-1)^n b_n$

Proof. Suppose we have the alternating series $\sum_0^\infty (-1)^n b_n$, where $b_1 \geq b_2 \geq b_3 \geq \cdots \geq 0$, and $b_n \to 0$. The convergence is illustrated in Fig. 6, which shows that the *odd* partial sums form an increasing sequence bounded above by $b_0 = s_0$. Thus the odd partial sums converge to some limit S. Since $b_n \to 0$, the difference between the even and odd partial sums tends to zero, so the even partial sums also converge to S.

This can all be demonstrated formally. For every m, $b_m - b_{m+1} \geq 0$, so

$$\begin{aligned} s_{2n+1} &= (b_0 - b_1) + (b_2 - b_3) + \cdots + (b_{2n} - b_{2n+1}) \\ &\geq (b_0 - b_1) + \cdots + (b_{2n-2} - b_{2n-1}) = s_{2n-1} \end{aligned}$$

and, grouping terms differently,

$$s_{2n+1} = b_0 - (b_1 - b_2) - \cdots - (b_{2n-1} - b_{2n}) - b_{2n+1} \leq b_0 .$$

Thus the odd partial sums do indeed form an increasing sequence bounded by b_0, and it follows that

$$\lim_{n\to\infty} s_{2n+1} = S \tag{1}$$

for some S.

We can now show that S, which was defined as the limit of the odd partial sums s_{2n+1}, is also the limit of the even partial sums s_{2n}. Since $s_{2n+1} = s_{2n} - b_{2n+1}$, and $\lim_{n\to\infty} b_{2n+1} = 0$, we obtain

$$\lim_{n\to\infty} s_{2n} = \lim_{n\to\infty} (s_{2n+1} + b_{2n+1}) = (\lim_{n\to\infty} s_{2n+1}) + 0 = S.$$

Next, to prove that S lies between any two successive partial sums, observe that

(i) $\{s_{2n+1}\}$ is an increasing sequence, so $s_{2n+1} \leq \lim_{n\to\infty} s_{2n+1} = S$, and similarly

(ii) $\{s_{2n}\}$ is a decreasing sequence, so $s_{2n} \geq \lim_{n\to\infty} s_{2n} = S$.

Since S lies above each odd partial sum, and below each even partial sum, it lies between any two successive partial sums.

Finally, although it must be obvious by now, we have not really proved that the entire sequence $\{s_k\}$ converges to S; we have only proved it separately for the odd and the even terms. This is now easy to complete. Given $\epsilon > 0$, there is a number N such that $k > N \Rightarrow b_k < \epsilon$. It follows that $|s_k - S| < \epsilon$ for $k > N$. Indeed, if k is odd, then

$$s_k \leq S \leq s_{k+1} = s_k + b_{k+1}, \quad \text{so} \quad |s_k - S| \leq b_{k+1} < \epsilon.$$

Similarly, if k is even, then

$$s_k \geq S \geq s_{k+1} = s_k - b_{k+1}, \quad \text{so again} \quad |s_k - S| \leq b_{k+1} < \epsilon. \quad Q.E.D.$$

Example 4. Does $\sum_{3}^{\infty} (-1)^n/\log(\log n)$ converge? *Solution.* This is an obvious candidate for Theorem 9, so we check the hypotheses with $b_n = 1/\log(\log n)$. The sequence $\{b_n\}$ converges monotonically to zero; for the function $\log(\log x)$ is positive and monotone increasing for $x > e$, hence $b_n = 1/\log(\log n)$ is monotone decreasing for $n \geq 3$; and $\lim_{n\to\infty} b_n = 0$. Therefore, by Theorem 9, $\sum_{3}^{\infty} (-1)^n/\log(\log n)$ converges. But notice how slowly the series converges to its sum. To get a partial sum accurate within $1/10$ would take about e^{e^5} terms, more than 10^{64} terms! For, $b_n < 1/5$ iff $n > e^{e^5}$.

Example 5. For what values of x does the series $\sum_{1}^{\infty} (x - 2)^n/\sqrt{n}$ converge?

Solution. The usual application of the ratio test shows absolute convergence for $|x - 2| < 1$, and divergence for $|x - 2| > 1$. When $|x - 2| = 1$ the ratio test gives no result, so we try other tests. When $x - 2 = 1$ our series becomes

$\sum_1^\infty 1/\sqrt{n}$, which diverges by the integral test. When $x - 2 = -1$, we get

$\sum_1^\infty (-1)^n/\sqrt{n}$, which is a convergent alternating series, by Theorem 9.

PROBLEMS

For which values of x do the following series converge absolutely?

• **1.** $\sum_0^\infty (-1)^n \dfrac{(x/2)^{2n}}{(n!)^2}$.

[This series defines the Bessel function $J_0(x)$; see §11.1, Example F.]

2. $\sum_0^\infty \dfrac{\cos(xn)}{1 + n^2}$.

• **3.** $\sum_0^\infty \dfrac{\sin(xn)}{1 + n}$.

[Make a guess; a complete discussion of this example is *very* difficult.]

4. $\sum_1^\infty n!\left(\dfrac{x}{n}\right)^n$

• **5.** $\sum_0^\infty \dfrac{x^n(n!)^2}{(2n)!}$

6. $\sum_1^\infty n^2(x + 1)^n$

• **7.** $\sum_1^\infty n^2(x + 1)^{n^4-n}$

8. $\sum_1^\infty \dfrac{(x - 1)^n}{n}$

9. Let b_n decrease monotonically to zero and $s_N = \sum_0^N (-1)^n b_n$. Show that $s_0, s_2, s_4, \ldots$ is a monotone sequence.

10. Let s_N be as in Problem 9. Show that $s_{2n} \geq 0$.

11. Base a proof of the test for alternating series on Problems 9 and 10.

12. Which of the following series are alternating? Of those that are alternating, which are convergent?

• (a) $\sum_0^\infty (-1)^n$ (b) $\sum_1^\infty (\cos \pi n)/n$ • (c) $\sum_1^\infty (\cos n)^n/n$

(d) $\sum_5^\infty (-1)^n \log(\log(\log(\log n)))^{-1/100}$

• (e) $\sum_1^\infty a_n$, where $a_n = n^{-2}$ if n is even and $a_n = -n^{-1}$ if n is odd.

13. For what values of x do the following series converge?

• (a) $\sum_1^\infty \frac{n}{n+1} x^n$ (d) $\sum_2^\infty \frac{(2x+1)^n}{(\log n)^{1/2}}$

(b) $\sum_1^\infty \frac{1}{n+1} x^n$ • (e) $\sum_1^\infty x^n \sin\left(\frac{1}{n}\right)$

• (c) $\sum_1^\infty x^n \log\left(\frac{n+1}{n}\right)$ • (f) $\sum_1^\infty \frac{(\sin x)^n}{n}$

14. Let $\{b_n\}$ be any sequence converging to 0, however slowly or irregularly. Prove that there is a convergent series $\sum_1^\infty a_n$ with $|a_n| \geq |b_n|$ for all n.

15. Prove that $\sum_1^\infty \frac{1}{2n(2n-1)} = \log 2.$

$\left[\text{Recall the series } \log 2 = 1 - \frac{1}{2} + \frac{1}{3} - \frac{1}{4} + \cdots.\right]$

• **16.** Compute $J_0(1)$ to three decimals, where $J_0(x) = \sum_{j=0}^\infty \frac{(-1)^j}{(j!)^2}\left(\frac{x}{2}\right)^{2j}$.

[Use Theorem 9 to determine how many terms are required.]

11.5 POWER SERIES

Series like $\sum_0^\infty x^n/n!$ and $\sum_0^\infty x^n$ are called *power* series, for the obvious reason that their terms involve powers of x. They arise naturally either as Taylor series or as solutions of differential equations. In either case, it is important to know for what values of x the series converge.

In the two examples above, simple applications of the ratio test show that $\sum_0^\infty x^n/n!$ converges for all x, while $\sum_0^\infty x^n$ converges only for $|x| < 1$. Thus each of these series converges in an open interval symmetric about the origin, and diverges outside that interval.

Practically the same thing is true for *every* power series $\sum_0^\infty a_n x^n$, no matter what the coefficients a_n may be. We will soon prove:

EVERY POWER SERIES HAS RADIUS OF CONVERGENCE

Theorem 10. *Let $\sum_0^\infty a_n x^n$ be any power series. Then either*

(i) *the series converges absolutely for all x, or*

(ii) *there is a number R (called the* radius of convergence*) such that the series converges absolutely for $|x| < R$, and diverges for $|x| > R$.*

Notice that nothing is said about convergence or divergence at the endpoints of the interval, where $x = \pm R$; the series may converge at both, or one, or neither of these two points. [See Fig. 7 and Examples 1–3 below.]

FIGURE 11.7 R = "radius of convergence" of $\sum a_n x^n$

The radius of convergence can often be found by the ratio test; the convergence for $|x| = R$ must then be investigated by more delicate tests, such as the integral, comparison, or alternating series tests.

Example 1. Find the radius of convergence of $\sum_0^\infty n^4(2x)^n$. *Solution.* To apply the ratio test, take the limit of the $(n+1)$st term over the nth,

$$\lim_{n\to\infty} \frac{|(n+1)^4(2x)^{n+1}|}{|n^4(2x)^n|} = \lim_{n\to\infty} \left(\frac{n+1}{n}\right)^4 2|x| = 2|x|.$$

The series converges when the limit is <1, i.e. when $2|x| < 1$, or $|x| < 1/2$; and it diverges when the limit is >1, i.e. when $|x| > 1/2$. Thus $R = 1/2$ is the radius of convergence. It is easy to see that the series *diverges* when $|x| = 1/2$, for then the general term $n^4(2x)^n$ does not tend to 0 as $n \to \infty$.

Example 2. From §11.2, $\sum_1^\infty x^n/n = -\log(1-x)$, $-1 \le x < 1$. In this case, the radius of convergence is obviously $R = 1$. The series converges at one endpoint ($x = -1$) by the "alternating" test, and diverges at the other endpoint ($x = 1$) by the integral test.

Example 3. For what values of x does $\sum_1^\infty x^n/n^2$ converge? *Solution.* By the ratio test, it diverges for $|x| > 1$. It converges absolutely for $|x| \le 1$, by the comparison test: When $|x| \le 1$, then $|x^n/n^2| \le 1/n^2$; and $\sum 1/n^2 < \infty$.

Remark. Theorem 10 extends easily to more general power series $\sum_0^\infty a_n(x-b)^n$ in powers of $(x-b)$. Either

(i) the series converges for all x, or

(ii) there is a radius of convergence R, the series converging for $|x-b| < R$ and diverging for $|x-b| > R$.

To prove this, substitute $y = x - b$: By Theorem 10, $\sum_0^\infty a_n y^n$ converges absolutely for $|y| < R$, so $\sum a_n(x-b)^n$ converges absolutely for $|x-b| < R$, i.e. for all x in an interval of radius R centered at b.

Example 4. $\sum_0^\infty n(x-1)^n$ converges absolutely for $|x-1| < 1$, i.e. for $0 < x < 2$, as you can see by the usual application of the ratio test.

The proof of Theorem 10 requires two lemmas, and an application of the least upper bound axiom for the real numbers [§AI.6].

Lemma 1. *Suppose that for some particular number $\bar{x}$, $\sum a_n\bar{x}^n$ converges. Then $\sum a_n x^n$ converges absolutely for all x closer to zero,* i.e. [Fig. 8] *for $|x| < |\bar{x}|$.*

MAIN STEP IN PROOF OF THEOREM 10

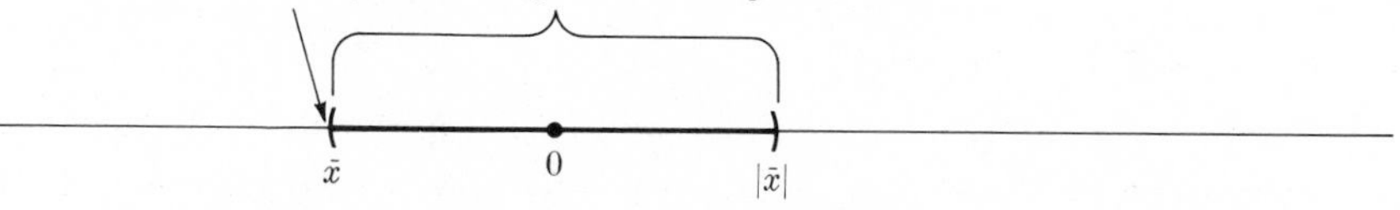

FIGURE 11.8

Proof. If $\sum a_n\bar{x}^n$ converges, then $\lim_{n\to\infty} a_n\bar{x}^n = 0$ [Theorem 2]. Since the sequence $\{a_n\bar{x}^n\}$ is convergent, it is *bounded;* there is a number M such that $|a_n\bar{x}^n| \le M$ for all n. It follows that

$$|a_nx^n| = |a_n\bar{x}^n|\cdot|x/\bar{x}|^n \le M|x/\bar{x}|^n. \tag{1}$$

When $|x| < |\bar{x}|$, then $|x/\bar{x}| < 1$, so $\sum_0^\infty M|x/\bar{x}|^n$ converges. By comparison, using (1), $\sum a_nx^n$ converges absolutely. *Q.E.D.*

Lemma 2. *Suppose that for some particular number $\bar{\bar{x}}$, $\sum a_n\bar{\bar{x}}^n$ diverges. Then $\sum a_nx^n$ diverges for all x farther from zero,* i.e. [Fig. 9] *for $|x| > |\bar{\bar{x}}|$.*

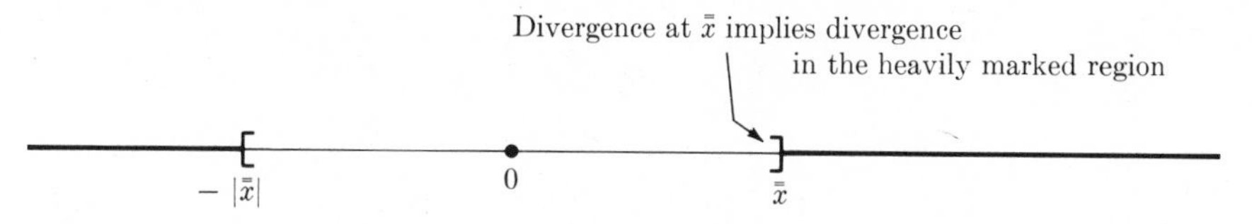

FIGURE 11.9

Proof by contradiction. If $\sum a_nx^n$ converged for some $|x| > |\bar{\bar{x}}|$, then $\sum a_n\bar{\bar{x}}^n$ would converge [Lemma 1], contradicting the given hypothesis. *Q.E.D.*

Lemmas 1 and 2 say that convergence at any point $\bar{x}$ implies convergence at all points closer to zero, and divergence at any point $\bar{\bar{x}}$ implies divergence at all points farther from zero. This suggests that there must be some distance R from the origin that distinguishes the points of convergence from the points of divergence, the series converging when $|x| < R$, and diverging when $|x| > R$, as stated in Theorem 10.

Proof of Theorem 10. In case (i) of the statement of the theorem, there is nothing to prove; the series converges for all x.

In case (ii), the series does *not* converge for some $\bar{\bar{x}}$. Consider, then, the set of points x for which the series *does* converge, the set

$$C = \{x: \sum_0^\infty a_n x^n \text{ converges}\}.$$

We want to prove that C is an interval $(-R,R)$, perhaps with one or both of its endpoints adjoined. Our problem is to define the appropriate number R, and then to prove that it works.

To define R, notice that *if* the theorem is true, then C is an interval with right endpoint R; thus it is natural to *define* R as the least upper bound of C,

$$R = \text{l.u.b. } C.$$

To justify this, we must prove that C is nonempty and bounded. It *is* nonempty, because it contains 0; $\sum a_n 0^n$ converges. It is also *bounded*, because of Lemma 2; since $\sum a_n \bar{\bar{x}}^n$ diverges, Lemma 2 asserts that $\sum a_n x^n$ also diverges for $x > |\bar{\bar{x}}|$, showing that $|\bar{\bar{x}}|$ is an upper bound for the set C.

Now we have to prove that $\sum a_n x^n$ converges for $|x| < R$, and diverges for $|x| > R$. Suppose first that $|x| < R$. Then $|x|$ is not an upper bound for C [since R is the *least* upper bound]. This means there is a number $\bar{x}$ in C such that $|x| < \bar{x}$. Since $\bar{x}$ is in C, $\sum_0^\infty a_n \bar{x}^n$ converges. It follows [Lemma 1] that $\sum_0^\infty a_n x^n$ converges absolutely, as was to be proved.

Suppose next that $|x| > R$. Then there is a number $\bar{\bar{x}}$ such that $|x| > \bar{\bar{x}} > R$. Since $\bar{\bar{x}} > R$, and R is an upper bound for C, it follows that $\bar{\bar{x}}$ is *not* in C; thus the series $\sum_0^\infty a_n \bar{\bar{x}}^n$ does not converge; by Lemma 2, $\sum_0^\infty a_n x^n$ does not converge either. This completes the proof of Theorem 10.

Derivatives of power series

When a power series $\sum a_n x^n$ has a positive radius of convergence $R > 0$, its sum defines a function f:

$$f(x) = \sum_0^\infty a_n x^n, \qquad |x| < R. \tag{2}$$

For example, the series $\sum_{0}^{\infty} (-1)^n \frac{(x/2)^{2n}}{(n!)^2}$ defines the Bessel function J_0,

$$J_0(x) = \sum_{0}^{\infty} \frac{(-1)^n}{(n!)^2} \left(\frac{x}{2}\right)^{2n}, \qquad |x| < \infty.$$

An important question arises: Does the function f in (2) have a derivative? And if so, how can it be found? Fortunately the derivative exists and can be taken in the easiest possible way; differentiate the infinite series term by term, as if it were a polynomial:

TERM-BY-TERM DIFFERENTIATION OF (2)

$$f'(x) = \sum_{0}^{\infty} n a_n x^{n-1}, \qquad |x| < R. \tag{3}$$

The differentiated series has precisely the same radius of convergence as the original.

Example 5. The Bessel function

$$J_0(x) = \sum_{0}^{\infty} \frac{(-1)^j}{2^{2j}(j!)^2} x^{2j}, \qquad |x| < \infty,$$

has the derivative

$$J_0'(x) = \sum_{0}^{\infty} \frac{(-1)^j(2j)}{2^{2j}(j!)^2} x^{2j-1}.$$

The function

$$-\log(1-x) = \sum_{1}^{\infty} \frac{1}{n} x^n, \qquad |x| < 1,$$

has the derivative

$$\frac{1}{1-x} = \sum_{1}^{\infty} \frac{n}{n} x^{n-1}, \qquad |x| < 1,$$

and this in turn has the derivative

$$\frac{1}{(1-x)^2} = \sum_{1}^{\infty} \frac{n(n-1)}{n} x^{n-2}, \qquad |x| < 1.$$

[Notice that the series for $1/(1-x)$ can be rewritten $\sum_{j=0}^{\infty} x^j$, and the series for $1/(1-x)^2$ can be rewritten $\sum_{j=0}^{\infty} (j+1)x^j$.]

Remember, the elementary theorem about the derivative of a linear combination justifies term-by-term differentiation for *finite* sums $\sum_0^N a_n x^n$, but not for infinite power series as in (2). To prove formula (3), we must go back to the definition of derivative and work with difference quotients. The following result about the difference quotient for x^n is crucial:

Lemma 3. *For any real x and h, and any positive integer n,*

$$\left| \frac{(x+h)^n - x^n}{h} - nx^{n-1} \right| \leq \frac{n(n-1)}{2} |h|(|x| + |h|)^{n-2}. \qquad (4)$$

Proof. This could easily be proved from the binomial expansion of $(x+h)^n$. However, it is even easier to use a Taylor expansion with two terms and remainder,

$$(x+h)^n = x^n + nhx^{n-1} + \frac{n(n-1)}{2} h^2 \cdot (x + \theta h)^{n-2}, \qquad 0 < \theta < 1.$$

Hence

$$\begin{aligned} \left| \frac{(x+h)^n - x^n}{h} - nx^{n-1} \right| &= \frac{n(n-1)}{2} |h| \cdot |x + \theta h|^{n-2} \\ &\leq \frac{n(n-1)}{2} |h| \cdot (|x| + |\theta h|)^{n-2} \\ &\leq \frac{n(n-1)}{2} |h| \cdot (|x| + |h|)^{n-2}. \quad \textit{Q.E.D.} \end{aligned}$$

Now we are ready for the main result.

Theorem 11. *Suppose that $\sum_0^\infty a_n x^n$ converges for $|x| < R$, where $R > 0$. Then:*

(i) $\sum_0^\infty n a_n x^{n-1}$ *converges absolutely for $|x| < R$.*

(ii) *The function defined by (2) has the derivative (3).*

Proof of (i). [Compare the proof of the ratio test.] Let $|x| < R$. Choose a number t between $|x|$ and R; then $\sum_0^\infty a_n t^n$ converges, so the sequence $\{a_n t^n\}$ is bounded, $|a_n t^n| \le M$. Furthermore,

$$|na_n x^{n-1}| = n|a_n t^{n-1}| \cdot \left|\frac{x}{t}\right|^{n-1} = \frac{|a_n t^n|}{t} n \left|\frac{x}{t}\right|^{n-1} \le \left(\frac{M}{t}\right) n \left|\frac{x}{t}\right|^{n-1}.$$

Since $|x/t| < 1$, it follows from the ratio test that $\sum n|x/t|^{n-1} < \infty$. By the comparison test, $\sum |na_n x^{n-1}|$ converges.

Applying (i) with $\sum_0^\infty na_n x^{n-1}$ in place of $\sum_0^\infty a_n x^n$, we find that $\sum n(n-1)a_n x^{n-2}$ converges absolutely for $|x| < R$. In fact, no matter how many times the series is differentiated, term by term, the result converges absolutely for $|x| < R$.

Proof of (ii). We have to show that

$$\lim_{h\to 0} \frac{f(x+h) - f(x)}{h} = \sum_0^\infty na_n x^{n-1}.$$

Since $f(x) = \sum_0^\infty a_n x^n$ and $f(x+h) = \sum_0^\infty a_n (x+h)^n$, we have

$$\frac{f(x+h) - f(x)}{h} - \sum_0^\infty na_n x^{n-1} = \sum_0^\infty a_n \left[\frac{(x+h)^n - x^n}{h} - nx^{n-1}\right]. \quad (5)$$

Putting (4) and (5) together, we obtain the inequality

$$\left|\frac{f(x+h) - f(x)}{h} - \sum_0^\infty na_n x^{n-1}\right| \le \left(\frac{|h|}{2}\right) \sum_0^\infty n(n-1)|a_n|(|x| + |h|)^{n-2}. \quad (6)$$

As in part (i), choose a number t such that $|x| < t < R$. When $|h| < t - |x|$, then $|x| + |h| < t$, hence from (5),

$$\left|\frac{f(x+h) - f(x)}{h} - \sum_0^\infty na_n x^{n-1}\right| \le \left(\frac{|h|}{2}\right) \sum_0^\infty n(n-1)|a_n| t^{n-2}.$$

Since $t < R$, the series on the far right converges. Thus the right-hand side of this inequality clearly tends to zero as $h \to 0$, so

$$f'(x) = \lim_{h\to 0} \frac{f(x+h) - f(x)}{h} = \sum_0^\infty na_n x^{n-1},$$

and Theorem 11 is proved.

We now have all the tools to solve differential equations by power series [Example 7 below]. But before working out these practical examples, we round out our discussion of power series with some theoretical applications of Theorem 11.

Integration of power series

Theorem 12. *If* $f(x) = \sum_0^\infty a_n x^n$, $|x| < R$ *(where* $R > 0$*), then* f *has the indefinite integral*

$$\int f(x)\,dx = \sum_0^\infty \frac{a_n x^{n+1}}{n+1} + c, \qquad |x| < R \tag{7}$$

and in particular

$$\int_0^x f(t)\,dt = \sum_0^\infty \frac{a_n}{n+1}\,x^{n+1}, \qquad |x| < R. \tag{8}$$

Proof. We begin by showing that the series in (8) converges for $|x| < R$. Note that $\left|\frac{a_n x^{n+1}}{n+1}\right| \le R|a_n x^n|$ and $\sum_0^\infty |a_n x^n|$ converges for $|x| < R$, [Theorem 10], so the series

$$F(x) = \sum_0^\infty \frac{a_n x^{n+1}}{n+1}$$

converges for $|x| < R$. Differentiating this last series term by term [Theorem 11] yields $F' = f$, which proves (7).

To prove (8), notice that $x^{n+1} = 0$ when $x = 0$, for $n = 0, 1, 2, \ldots$. Hence $F(0) = 0$, and $\int_0^x f(t)\,dt = F(x) - F(0) = F(x)$, proving (8).

Power series versus Taylor series

The *Taylor series* of a function f is constructed by the same general scheme as the Taylor polynomials; in particular, the Taylor series of f at 0 is

$$\sum_0^\infty \frac{f^{(n)}(0)}{n!}\,x^n. \tag{9}$$

Now, suppose our function f is defined by a power series, $f(x) = \sum_0^\infty a_n x^n$. It is fair to ask: Is the Taylor series of f exactly the same as the series $\sum_0^\infty a_n x^n$ defining f? Fortunately it is.

Theorem 13. *If* $\sum_0^\infty a_n x^n$ *has a positive radius of convergence* R, *then* $\sum_0^\infty a_n x^n$ *is the Taylor series (9) of the function* $f(x) = \sum_0^\infty a_n x^n$, $|x| < R$. *That is,* $\dfrac{f^{(n)}(0)}{n!} = a_n$.

Proof. Successive applications of Theorem 11 (term-by-term differentiation) yield the following results, for $|x| < R$:

$$f(x) = \sum_0^\infty a_n x^n = a_0 + a_1 x + a_2 x^2 + \cdots,$$

$$f'(x) = \sum_1^\infty n a_n x^{n-1} = 1 \cdot a_1 + 2 \cdot a_2 x + 3 \cdot a_3 x^2 + \cdots,$$

$$f''(x) = \sum_2^\infty n(n-1) a_n x^{n-1} = 2 \cdot 1 \cdot a_2 + 3 \cdot 2 \cdot a_3 x + 4 \cdot 3 \cdot a_4 x^2 + \cdots,$$

$$\vdots$$

$$f^{(k)}(x) = \sum_k^\infty n(n-1) \cdots (n-k+1) a_n x^{n-k}$$

$$= k!\, a_k + \frac{(k+1)!}{1} a_{k+1} x + \frac{(k+2)!}{2!} a_{k+2} x^2 + \cdots.$$

Setting $x = 0$ gives $f(0) = a_0$, $f'(0) = a_1, \ldots, f^{(k)}(0) = k!\, a_k$, so $a_k = \dfrac{f^{(k)}(0)}{k!}$. *Q.E.D.*

Theorem 13 implies

COEFFICIENTS IN CONVERGENT POWER SERIES—DETERMINED BY THE SUM

Theorem 14. *If two power series* $\sum_0^\infty a_n x^n$ *and* $\sum_0^\infty b_n x^n$ *both converge to the same values* $f(x)$ *for all* $|x| < R$, *where* $R > 0$, *then the coefficients are identical;* $a_0 = b_0$, $a_1 = b_1, \ldots$.

Proof. By Theorem 13, $a_n = f^{(n)}(0)/n! = b_n$. *Q.E.D.*

There is one potentially troublesome aspect of Taylor series:

SURPRISING FACT *A function f and all its derivatives may exist for all x, yet the Taylor series (9) may converge only for $x = 0$; and even if it converges everywhere, it may not converge to the given function f.*

In such cases, the Taylor series is not much use in studying f. Fortunately, such functions are "shy"; they do not come forward naturally, but must be sought out. Nevertheless, their existence reminds us that in using the Taylor series (9), we must prove not only that (9) converges, but that it *converges to f*. The examples in §11.2 and §11.3 did this; notice particularly Example 8, §11.3, proving that $e^x = \sum x^n/n!$.

TERM-BY-TERM DIFFERENTIATION: MORE EXAMPLES

Example 6. Suppose that we did not know the derivative of exp, but knew only the formula

$$\exp(x) = \sum_0^\infty \frac{x^n}{n!}.$$

We could then prove that $\exp' = \exp$ as follows:

$$\exp'(x) = \sum_0^\infty \frac{nx^{n-1}}{n!} \qquad \text{[Theorem 11]}$$

$$= \sum_1^\infty \frac{nx^{n-1}}{n!} \qquad \left[\text{the term with } n = 0 \text{ drops out, since } \frac{0}{0!} = \frac{0}{1} = 0\right]$$

$$= \sum_1^\infty \frac{x^{n-1}}{(n-1)!} \qquad \left[\frac{n}{n\cdot(n-1)(n-2)\cdots} = \frac{1}{(n-1)(n-2)\cdots}\right]$$

$$= \sum_0^\infty \frac{x^j}{j!} \qquad \text{[substitute } n - 1 = j\text{, with the appropriate change in limits of summation]}$$

$$= \exp(x).$$

[We changed from n to j to obtain a direct comparison, term by term, of the original and the differentiated series. This is often necessary in dealing with differentiated series, since after differentiation the nth term has x^{n-1} instead of x^n.]

Example 7. In §11.1, we tried to solve Bessel's equation

$$x^2f'' + xf' + x^2f = 0 \tag{10}$$

with a series $f(x) = \sum_0^\infty a_nx^n$. After various manipulations we found the series

$$J_0(x) = \sum_{j=0}^\infty \frac{(-1)^jx^{2j}}{2^{2j}(j!)^2} = 1 - \frac{x^2}{2^2\cdot 1^2} + \frac{x^4}{2^4\cdot(2!)^2} - \frac{x^6}{2^6\cdot(3!)^2} + \cdots. \tag{11}$$

It is now possible to prove that this series really does solve the equation. By the ratio test, the series converges for all x. Hence, by Theorem 11,

$$J_0'(x) = \sum_{j=1}^{\infty} \frac{(-1)^j \, 2j \, x^{2j-1}}{2^{2j}(j!)^2}$$

$$J_0''(x) = \sum_{j=1}^{\infty} (-1)^j \frac{2j(2j-1)x^{2j-2}}{2^{2j}(j!)^2},$$

so the terms in (10), with $f = J_0$, are

$$x^2 J_0''(x) = \sum_{j=1}^{\infty} (-1)^j \frac{2j(2j-1)x^{2j}}{2^{2j}(j!)^2}$$

$$x J_0'(x) = \sum_{j=1}^{\infty} (-1)^j \frac{2j \, x^{2j}}{2^{2j}(j!)^2}$$

$$x^2 J_0(x) = \sum_{k=0}^{\infty} (-1)^k \frac{x^{2k+2}}{2^{2k}(k!)^2} = \sum_{j=1}^{\infty} (-1)^{j-1} \frac{x^{2j}}{2^{2j-2}((j-1)!)^2}.$$

[The substitution $k = j - 1$ in the last line was made so that now, in each series, the jth term has x^{2j}.] Adding, we get

$$x^2 J_0''(x) + x J_0'(x) + x^2 J_0(x)$$

$$= \sum_{j=1}^{\infty} \left((-1)^j \frac{2j(2j-1)}{2^{2j}(j!)^2} + (-1)^j \frac{2j}{2^{2j}(j!)^2} + \frac{(-1)^{j-1}}{2^{2j-2}((j-1)!)^2} \right) x^{2j}$$

$$= \sum_{j=1}^{\infty} (-1)^j \frac{2j(2j-1) + j - 2^2 j^2}{2^{2j}(j!)^2} x^{2j}$$

$$= 0. \qquad Q.E.D.$$

Remark 1. The Bessel function J_0 is *defined* by the series (11); there is no less analytic, more geometric way to define it, as there is with $\sin x$ and $\log x$.

Remark 2. Power series provide rigorous analytic definitions of the trigonometric and exponential functions, to replace the intuitive geometric definitions in Chapter II. The next section shows how to derive the familiar basic properties from the definitions by power series.

Our final two examples concern the integration of power series [Theorem 12].

TERM-BY-TERM INTEGRATIONS

Example 8. $\int \frac{1}{1-x}\, dx = \int \left(\sum_0^{\infty} x^n\right) dx = \sum_0^{\infty} \frac{x^{n+1}}{n+1} + c.$ If we take $c = 0$, we have

$$\sum_0^\infty \frac{x^{n+1}}{n+1} = \int_0^x \frac{1}{1-t}\,dt = -\log(1-x).$$

Theorem 12 gives this result for $|x| < 1$. [Actually, we saw in §11.2 that it remains true for $x = -1$ as well.]

Example 9. Using the series in Example 8, we find

$$-\int \frac{1}{x}\log(1-x)\,dx = \int \frac{1}{x}\sum_0^\infty \frac{x^{n+1}}{n+1}\,dx$$

$$= \int \sum_0^\infty \frac{x^n}{n+1}\,dx \qquad \text{[Theorem 1]}$$

$$= \sum_0^\infty \frac{x^{n+1}}{(n+1)^2} + c. \qquad \text{[Theorem 12]}$$

Example 10. The integral $\int (1+x^2)^{1/3}\,dx$ does not yield to any elementary method. However, the Taylor series for $(1+t)^s$, which is

$$1 + st + \frac{s(s-1)}{2}t^2 + \frac{s(s-1)(s-2)}{3!}t^3 + \cdots,$$

actually converges to $(1+t)^s$ for $|t| < 1$ [Problem 5 below]. Hence

$$(1+x^2)^{1/3}$$

$$= 1 + \tfrac{1}{3}x^2 + \frac{\frac{1}{3}(\frac{1}{3}-1)}{2}(x^2)^2 + \frac{\frac{1}{3}(\frac{1}{3}-1)(\frac{1}{3}-2)}{3!}x^6 + \cdots, \qquad |x| < 1,$$

and by Theorem 12,

$$\int (1+x^2)^{1/3}\,dx = c + x + \frac{1}{3\cdot 3}x^3 + \frac{\frac{1}{3}(\frac{1}{3}-1)}{5\cdot 2}x^5 + \cdots, \qquad |x| < 1.$$

Notice that the series is valid only for $|x| < 1$. If we required an indefinite integral for some other interval, say for $|x-2| < 1$, we could integrate the Taylor expansion of $(1+x^2)^{1/3}$ about $x = 2$, obtaining for the integral a series in powers of $x - 2$.

PROBLEMS

1. Verify, by differentiating the series for sin and cos, that
(a) $\sin' x = \cos x$,
(b) $\cos' x = -\sin x$.

2. Verify that the Bessel function of order 1,

$$J_1(x) = \sum_0^\infty \frac{(-1)^n (x/2)^{2n+1}}{(n!)(n+1)!},$$

satisfies Bessel's equation of order 1, $x^2f'' + xf' + (x^2 - 1)f = 0$.

3. (a) Show that if $f(x) = \sum_0^\infty a_n x^n$ satisfies $f'' - f = 0$, then $n(n-1)a_n - a_{n-2} = 0$ for $n = 2, 3, \ldots$. [Hint: Differentiate twice, term by term, and collect like powers of x in the resulting expression for $f'' - f$.]

(b) Show that $a_{2k} = (2k)!\,a_0$ and $a_{2k+1} = (2k+1)!\,a_1$.

• (c) Find the radius of convergence of f. [Hint: Break f into an "even" and an "odd" series, and apply the ratio test to each.]

(d) *Prove* that $f'' - f = 0$, no matter what a_0 and a_1 are. [This amounts to checking your work in part (a), in reverse.]

(e) Show that $f(x) = a_0 \cosh x + a_1 \sinh x$, where $\cosh x = \frac{1}{2}(e^x + e^{-x})$ and $\sinh x = \frac{1}{2}(e^x - e^{-x})$.

4. The general form of Bessel's equation is

$$x^2f'' + xf' + (x^2 - p^2)f = 0,$$

and solutions are called *Bessel functions of order p*.

• (a) Find J_2, the Bessel function of order 2 such that $J_2''(0) = \frac{1}{4}$.

(b) Find the radius of convergence of J_2.

(c) Find $J_p(x)$, given that $J_p^{(p)}(0) = 2^{-p}$.

5. This problem proves that the *binomial series*

$$1 + sx + \frac{s(s-1)}{2}x^2 + \frac{s(s-1)(s-2)}{3!}x^3 + \cdots$$

BINOMIAL SERIES

converges to $(1+x)^s$, for $|x| < 1$. [This fact was used to evaluate an integral in Example 10.]

(a) Prove that the Taylor series of $(1+x)^s$ is the binomial series given above.

(b) Prove that the binomial series has radius of convergence 1, i.e. converges for $|x| < 1$ and diverges for $|x| > 1$.

(c) Let $f(x)$ be the sum of the binomial series for $|x| < 1$. Prove that $(1+x)f'(x) = sf(x)$. [Note that you may *not yet assume* that $f(x) = (1+x)^s$, only that $f(x)$ is given by the binomial series.]

(d) Prove that $(1+x)^{-s}f(x)$ is constant, for $|x| < 1$. [This is the sort of thing we did in §3.2.]

(e) Prove that $f(x) = (1+x)^s$, $|x| < 1$.

6 • (a) Find the Taylor series for $f(x) = (1+x)^{1/3}$ about $x = b > 0$,

$$\sum_{0}^{\infty} \frac{f^{(n)}(b)}{n!}(x-b)^n.$$

(b) Show that the series converges for $|x - b| < 1 + b$.
(c) Obtain a series for $\int (1+x^2)^{1/3}\,dx$, valid in an open interval containing $x = 1$. [Compare Example 10.]

7. Obtain the Taylor series for arcsin x as follows.

SERIES FOR arcsin

(a) From the binomial series [Problem 5] show that

$$\frac{1}{\sqrt{1-x^2}} = \sum_{n=0}^{\infty} \frac{\frac{1}{2}\cdot\frac{3}{2}\cdot\frac{5}{2}\cdots(\frac{1}{2}+n-1)}{n!}x^{2n}, \qquad |x| < 1.$$

(b) Show that $\arcsin x = \sum_{n=0}^{\infty} \frac{1\cdot 3\cdot 5\cdots(2n-1)}{2\cdot 4\cdot 6\cdots(2n)}\cdot\frac{x^{2n+1}}{2n+1}$.

(c) Determine the radius of convergence of the series in (b).

8. Obtain a power series for $\int_0^x \frac{\sin t}{t}\,dt$. For what values of x is the series valid?

9. Write the power series for $\operatorname{erf}(x) = \frac{2}{\sqrt{\pi}}\int_0^x e^{-t^2}\,dt$. [Hint: The power series for e^{-t^2} is obtained simply by putting $x = -t^2$ in the series for e^x.]

10. Obtain the series

$$\log\frac{1+x}{1-x} = 2\sum_{n=1}^{\infty}\frac{x^{2n-1}}{2n-1}$$

in the following two ways:
(a) by subtracting the series for $\log(1-x)$ from that for $\log(1+x)$;
(b) by integrating the series for $1/(1-x^2)$.

11. The *Bessel function of order p* is defined by

BESSEL FUNCTIONS

$$J_p(x) = \sum_{k=0}^{\infty}\frac{(-1)^k(x/2)^{2k+p}}{k!(k+p)!}, \qquad x \text{ real}, p \ge 0.$$

Prove the following identities:
(a) $[xJ_1(x)]' = xJ_0(x)$
(b) $J_0' = J_1$
(c) $[xJ_p(x)]' = xJ_{p-1}(x), \quad p \ge 1$
(d) $(x^{-p}J_p)' = -x^{-p}J_{p+1}$

• **12.** Compute $J_p(2)$ [Problem 11] with an error $< \dfrac{1}{2(p+2)!}$. [Use Theorem 9 on alternating series.]

13. Bessel's equation is a *second order* equation, and we have found only one solution to it, the function J_0. A second solution can be found by the following trick. Let

$$f(x) = J_0(x) \log x + g(x).$$

(a) Using the fact that J_0 solves Bessel's equation, show that $f = J_0 \log x + g$ satisfies that same equation iff

$$x^2 g''(x) + x g'(x) + x^2 g(x) = 2x J_0'(x).$$

• (b) Obtain a power series for g, and hence obtain *Neumann's solution* of Bessel's equation,

$$Y_{(0)}(x) = J_0(x) \log x + g(x).$$

14. We know that $\sum_0^\infty x^n = \dfrac{1}{1-x}$ for $|x| < 1$. By appropriate manipulation of this series, obtain analogous formulas for the following series. [See Example 9 above.]

(a) $\sum_1^\infty n x^{n-1} = \sum_0^\infty (n+1) x^n = ?$

(b) $\sum_1^\infty n x^n = ?$

(c) $\sum_1^\infty n(n+1) x^{n-1} = \sum_0^\infty (n+2)(n+1) x^n = ?$

(d) $\sum_1^\infty n^2 x^n = ?$

(e) $\sum_1^\infty n^3 x^n = ?$

15. Starting with the series $e^x = \sum_0^\infty x^n/n!$, obtain, by operating on power series, the sums of the following series.

• (a) $\sum_0^\infty x^n/(n+1)!$

(b) $\sum_1^\infty n x^{n-1}/(n+1)!$

(c) $\sum_0^\infty (n+1)/(n+2)!$

(d) $\sum_2^\infty (n+1) x^n/n!$

STRANGE EXAMPLE **16.** This problem exhibits a function f whose Taylor series $\sum_{0}^{\infty} f^{(n)}(0)x^n/n!$ converges for every x, but *does not converge to* f. Define $f(0) = 0$ and $f(x) = e^{-1/x^2}$ if $x \neq 0$, and prove the following.

(a) For any integer n, $x^{-n}e^{-1/x^2} \to 0$ as $x \to 0$.
(b) $f'(0) = 0$.
(c) If $x \neq 0$, then $f'(x) = P_1(1/x)e^{-1/x^2}$ for some polynomial P_1.
(d) Prove, by induction, that if $x \neq 0$, then $f^{(n)}(x) = P_n(1/x)e^{-1/x^2}$ for some polynomial P_n.
(e) Prove, by induction, that $f^{(n)}(0) = 0$.
(f) Prove that $\sum_{0}^{\infty} f^{(n)}(0)x^n/n! = f(x)$ only when $x = 0$.

FIBONACCI NUMBERS **17.** (a) Show that *if* $f(x) = \dfrac{1}{1 - x - x^2}$ has a power series $\sum_{0}^{\infty} a_n x^n$ converging for $|x| < R$ (where $R > 0$), then $a_0 = 1$, $a_1 = 1$, and $a_n = a_{n-1} + a_{n-2}$ for $n \geq 2$. [Hint: $f(x) - xf(x) - x^2f(x) = 1$.]

(b) Find a_2, a_3, a_4, a_5, a_6.
(c) Show that $a_{n+1}/a_n \leq 2$.
(d) Show that $\sum_{0}^{\infty} a_n x^n$ converges at least for $|x| < 1/2$.
(e) Show that $\sum_{0}^{\infty} a_n x^n = \dfrac{1}{1 - x - x^2}$ wherever the series converges.

[Hint: If $g(x)$ is the sum of the series, what is $g - xg - x^2g$?]

(f) Obtain another power series for f, by noting that

$$\frac{1}{1 - x - x^2} = \frac{A}{r_1 - x} + \frac{B}{r_2 - x}$$

where r_1 and r_2 are the roots of $1 - x - x^2 = 0$, and A and B are appropriate constants. Note further that

$$\frac{1}{r - x} = \frac{1}{r}\left(\frac{1}{1 - x/r}\right) = \frac{1}{r}\sum_{0}^{\infty} (x/r)^n.$$

(g) By Theorem 13, show that

$$a_{n-1} = \frac{(1 + \sqrt{5})^n - (1 - \sqrt{5})^n}{2^n\sqrt{5}}.$$

[The numbers $a_0, a_1, \ldots$ are called *Fibonacci numbers,* in honor of Fibonacci of Pisa.* Fibonacci hypothesized that a pair of newborn

---

* Leonardo of Piso (ca. 1170–1250), also called Fibonacci ["son of Bonaccio"].

rabbits would bear no offspring the first month, but would bear a new pair of rabbits every month thereafter. Thus, beginning with one pair of bearing rabbits, the population looks like this month by month:

Month number	1	2	3	4	5	6	
Pairs of bearing rabbits	1 →	1 →	2 →	3 →	5 →	8	
		↗	↗	↗	↗	↗	
Pairs of nonbearing rabbits	0	1	1	2	3	5	etc.
		↗	↗	↗	↗	↗	
Pairs of births	1	1	2	3	5	8	

Hence the formula $a_n = a_{n-1} + a_{n-2}$, where a_n is the number of pairs of bearing rabbits in month n.]

11.6 ANALYTIC DEFINITIONS OF TRIGONOMETRIC AND EXPONENTIAL FUNCTIONS

Power series offer the most convenient way to replace the intuitive geometric definition of the elementary functions by rigorous analytic ones.

Definition 1. For any real number x,

$$\sin x = \sum_0^\infty (-1)^n \frac{x^{2n+1}}{(2n+1)!},$$

$$\cos x = \sum_0^\infty (-1)^n \frac{x^{2n}}{(2n)!},$$

$$\exp(x) = \sum_0^\infty \frac{x^n}{n!}.$$

Of course, just giving the definitions is not enough; we must derive the important properties associated with these functions. The text does this for sine and cosine, and the problems take care of exp and its inverse, log.

Properties of sin and cos

First of all, putting $x = 0$ in the definition gives

$$\sin(0) = 0, \qquad \cos(0) = 1. \tag{1}$$

Further, a glance at the series shows that sin is an odd function, and cos is even.

Term-by-term differentiation gives

$$\sin' = \cos, \qquad \cos' = \sin. \tag{2}$$

From (2) it follows that $(\sin^2 + \cos^2)' = 0$, so $\sin^2 x + \cos^2 x$ is constant. Setting $x = 0$ and using (1), we find that the constant must be 1. Thus

$$\sin^2 x + \cos^2 x = 1. \tag{3}$$

Now consider a general linear combination of sin and cos,

$$f(x) = A \cos x + B \sin x. \tag{4}$$

From (2) and (1), straightforward calculations show that f satisfies the three conditions:

$$f'' + f = 0; \qquad f(0) = A; \qquad f'(0) = B. \tag{5}$$

The rest of our discussion depends on the following fact:

The *only* function satisfying (5) is the function f in (4). (6)

This is proved in §3.5. The proof uses no information about sin and cos except formulas (1) and (2), and is therefore valid in the present context.

It is now easy to prove the addition formulas

$$\sin(a + x) = \sin a \cos x + \cos a \sin x \tag{7}$$

$$\cos(a + x) = \cos a \cos x - \sin a \sin x. \tag{8}$$

Consider the function $f(x) = \sin(a + x)$ on the left-hand side of (7). From (2) and (1), we find that

$$f'' + f = 0, \qquad f(0) = \sin a, \qquad f'(0) = \cos a.$$

By (6), the *only* function satisfying these conditions is the function $\sin a \cos x + \cos a \sin x$ given on the right-hand side of (7); so (7) is proved. Now (8) follows by differentiating each side of (7).

Next, we construct the graph of cos. An elementary argument using (1) and (2) shows that

$$1 - \frac{x^2}{2} \leq \cos x \leq 1 - \frac{x^2}{2} + \frac{x^4}{24} \tag{9}$$

$$x - \frac{x^3}{6} \leq \sin x \leq x \tag{10}$$

for all x [§3.1, Problem 19]. The inequalities (9) control the graph of $\cos x$ reasonably well for $|x| < \sqrt{6 - 2\sqrt{3}} =$ (approx) 1.6 [Fig. 10]. In particular, they show that $\cos x > 0$ for $|x| < \sqrt{2}$, and that $\cos(2) < 0$. Therefore, $\cos x$ has at least one zero between $\sqrt{2}$ and 2, by the Intermediate Value Theorem [§AII.5].

The inequalities (10) control the derivative $\cos'(x)$ to some extent; in particular

$$\cos' x = -\sin x \leq \frac{x^3}{6} - x = x\left(\frac{x^2}{6} - 1\right) < 0 \quad \text{if } 0 < |x| < \sqrt{6}. \tag{11}$$

Thus cos is strictly decreasing on $(0, \sqrt{6})$, so $\cos x$ has *only one* zero between 0 and $\sqrt{6}$; obviously, it is the one we found between $\sqrt{2}$ and 2.

In the traditional development of sin and cos, the first zero of $\cos x$ occurs at $x = \pi/2$; therefore, we now *define* $\pi/2$ *to be the first positive zero of* $\cos x$. We have seen that $\sqrt{2} < \pi/2 < 2$ [Fig. 10], and therefore (11) shows that $\sin \pi/2 > 0$. It then follows from (3) that

$$\sin \pi/2 = +\sqrt{1 - \cos^2 \pi/2} = 1,$$

so we obtain

$$\sin \pi/2 = 1, \qquad \cos \pi/2 = 0. \tag{12}$$

We now show that the graphs of sin and cos consist of arcs congruent to the graph of $\cos x$ for $0 \leq x \leq \pi/2$, which we have already sketched [Fig. 11].

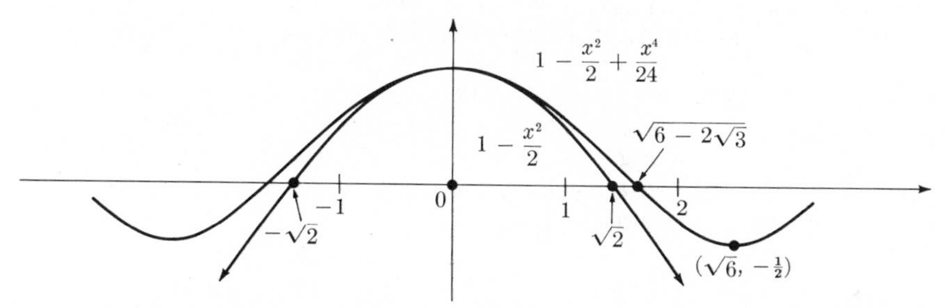

FIGURE 11.10 The graph of cos lies between the two graphs shown, so $\cos(\sqrt{2}) > 0$ and $\cos(2) < 0$

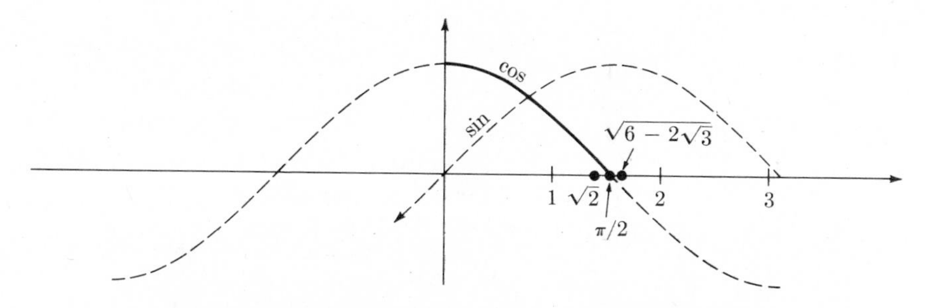

FIGURE 11.11 The solid arc is practically determined by Fig. 10, and the rest of the graph consists of congruent arcs

Since cos is even, this arc is reflected in the y axis, giving the graph for $-\pi/2 \le x \le \pi/2$. Next, from (12) and the addition formulas (7) and (8) we get

$$\sin(x + \pi/2) = \cos x, \qquad \cos(x + \pi/2) = -\sin x.$$

The first of these gives the graph of $\sin x$ for $0 \le x \le \pi$, and then the second gives the graph of $\cos x$ for $\pi/2 \le x \le 3\pi/2$. Continuing like this, we obtain the familiar periodic graphs.

These graphs lead us back at last to the geometric definition of sin and cos, for they show that *as θ increases from 0 to 2π, ranging in the interval $[0,2\pi)$, the point $(\cos\theta, \sin\theta)$ passes through each point on the unit circle exactly once.* Moreover, θ is arc length on the circle. For, in general, a differentiable parametric curve $x = f(\theta)$, $y = g(\theta)$ has arc length given by

$s(\theta) = \int_0^\theta \sqrt{(f')^2 + (g')^2}$. In this case, $(f')^2 + (g')^2 = (-\sin)^2 + \cos^2 = 1$,

so $s(\theta) = \int_0^\theta 1 = \theta$, which proves that θ is the length of arc along the circle from $(1,0)$ to $(\cos\theta, \sin\theta)$.

Finally, the problems below suggest how the addition formulas and (12) can be exploited to obtain the values $\sin(\pi/4) = 1/\sqrt{2}$, $\sin(\pi/6) = 1/2$, and so on. Thus we recover all the familiar properties from the formal analytic definition by means of power series.

PROBLEMS

Obtain the following results by using (i) Definition 1, (ii) the facts deduced in this section, and (iii) general theorems about derivatives and series, but *not* any material deduced in earlier sections about sin, cos, log, or exp.

1. (a) $\cos 2x = \cos^2 x - \sin^2 x = 2\cos^2 x - 1, \quad \sin 2x = 2\cos x \sin x$

(b) $\cos(x/2) = \pm\sqrt{\dfrac{1 + \cos x}{2}}$ ("half-angle" formula)

(c) $\cos(\pi/4) = 1/\sqrt{2}$ [Use (12), and explain your choice of sign.]

(d) $\sin(\pi/4) = 1/\sqrt{2}$

2. (a) $\cos 3x = \cos x(4\cos^2 x - 3)$

(b) $\cos(\pi/6) = \sqrt{3}/2$ and $\sin(\pi/6) = 1/2$

(c) $\cos(\pi/3) = 1/2$ and $\sin(\pi/3) = \sqrt{3}/2$

3. (a) $\cos(x + \pi/2) = -\sin x$

(b) $\sin(x + \pi/2) = \cos x$

(c) $\cos(x + \pi) = -\cos x$

(d) $\sin(x + \pi) = -\sin x$

(e) $\cos(x + 2\pi) = \cos x$

(f) $\sin(x + 2\pi) = \sin x$

4. $\sin'(x) = -\cos x$

5. $\exp' = \exp$

6. $\exp(x)\exp(-x) = 1$ for all x.

7. There is one and only one differentiable function f such that $f' = f$ and $f(0) = A$; it is given by $f(x) = Ae^x$. [Suppose $f' = f$, and show that $e^{-x}f(x)$ is constant.]

8. $e^{a+x} = e^a e^x$. [Use Problem 7.]

9. $e^x > 0$ for all x. [Use Problem 8 to show that e^x is a square, and Problem 6 to show that $e^x \neq 0$.]

10. exp is a strictly increasing function.

11. $\lim_{x\to+\infty} e^x = +\infty$. [Hint: When $x \geq 0$, then $e^x \geq 1 + x$, from the power series.]

12. $\lim_{x\to-\infty} e^x = 0$ [See Problems 6 and 11.]

13. When $x \geq 0$, $e^x \geq x^{n+1}/(n+1)!$.

14. For every integer n, $\lim_{x\to+\infty} x^n e^{-x} = 0$.

15. For every integer n, $\lim_{x\to+\infty} x^n e^{-x^2} = 0$ and $\lim_{x\to-\infty} x^n e^{-x^2} = 0$.

16. For every integer n, $\lim_{x\to 0} x^{-n} e^{-1/x^2} = 0$. [For a surprising application of this, see §11.5, Problem 16.]

17. Let $f(x) = x^2 \sin(1/x)$. Show that f is differentiable at zero, i.e. that the limit defining $f'(0)$ exists.

18. If k is any constant, and $f'(x) = kf(x)$ for all x, then there is a constant A such that $f(x) = Ae^{kx}$. [See Problem 7. For applications of this result, see §3.3.]

19. Prove that exp has range $(0, +\infty)$. [See Problems 10–12.]

20. By Problem 19, exp has an inverse (call it L), defined on $(0, +\infty)$. Prove the following:
(a) $L(1) = 0$
(b) $L'(x) = 1/x$
(c) $L(xy) = L(x) + L(y)$
[Hint: The arguments used in §2.7 for exp can be run backwards.]

11.7 GROUPING, REORDERING, AND PRODUCTS OF SERIES

We conclude this chapter by considering how three of the rules for finite sums carry over to infinite sums.

Grouping

A familiar law of arithmetic, the associative law, says that the terms in a finite sum may be grouped in any way; for instance,

$$((a + b) + c) + d = (a + b) + (c + d) = a + ((b + c) + d).$$

This procedure is also legitimate for *convergent* infinite series $\sum_1^\infty a_n$. Writing $(a_1 + \cdots + a_{n_1}) + (a_{n_1+1} + \cdots + a_{n_2}) + \cdots$ means that instead of considering *all* the partial sums $s_N = \sum_1^N a_n$, we consider only $s_{n_1}, s_{n_2}, \ldots$. If $\lim_{N\to\infty} s_N = S$, then $\lim_{j\to\infty} s_{n_j} = S$ also.

If the original series is divergent, however, the grouping may make it convergent, and different groupings may make it converge to different sums. Example 2 below gives a convincing illustration of this phenomenon.

Example 1. $\log 2 = \sum_1^\infty \frac{(-1)^{n+1}}{n} = 1 - \frac{1}{2} + \frac{1}{3} - \frac{1}{4} + \cdots$

$$= \left(1 - \frac{1}{2}\right) + \left(\frac{1}{3} - \frac{1}{4}\right) + \left(\frac{1}{5} - \frac{1}{6}\right) + \cdots$$

$$= \frac{1}{1\cdot 2} + \frac{1}{3\cdot 4} + \frac{1}{5\cdot 6} + \cdots$$

$$= \sum_1^\infty \frac{1}{2n(2n-1)}.$$

The grouping of terms is legitimate, since the series converges.

Example 2. $\sum_0^\infty (-1)^n = 1 - 1 + 1 - 1 + 1 - \cdots$ does not converge, but

$$(1 - 1) + (1 - 1) + (1 - 1) + \cdots = 0$$

and

$$1 + (-1 + 1) + (-1 + 1) + \cdots = 1.$$

Reordering

From the commutative law $a + b = b + a$, we deduce that the terms in a finite sum can be rearranged in any order without affecting the sum. In contrast, the terms of an infinite series may be arbitrarily reordered only when the series is *absolutely convergent.*

We should say a word or two about what we mean by reordering the terms of an infinite series. A series $\sum_1^\infty b_m$ is called a *reordering* of the series $\sum_1^\infty a_n$ if each a_n occurs exactly once in $\sum_1^\infty b_m$ and if no other terms occur in $\sum_1^\infty b_k$. More precisely, we require a sequence $\{n_k\}$ which takes each positive integer value exactly once, and such that $b_k = a_{n_k}$.

Example 3. $1 + \frac{1}{3} - \frac{1}{2} + \frac{1}{5} + \frac{1}{7} - \frac{1}{4} + \frac{1}{9} + \frac{1}{11} - \frac{1}{6} + \frac{1}{13} + \frac{1}{15} - \frac{1}{8} + \cdots$

is a reordering of

$$\sum_1^\infty \frac{(-1)^{n+1}}{n} = 1 - \frac{1}{2} + \frac{1}{3} - \frac{1}{4} + \frac{1}{5} - \frac{1}{6} + \cdots.$$

In this case the sequence $n_1, n_2, n_3, \ldots$ is

$$1, 3, 2, 5, 7, 4, 9, 11, 6, 13, 15, 8, 17, 19, \ldots.$$

You can see that in the sequence $\{n_k\}$ the odd integers come twice as fast as the even integers, but each positive integer comes exactly once.

By Example 3, we can show that reordering may change the sum of a *conditionally* convergent series. Group the reordered series in blocks of 3,

$$\left(1 + \frac{1}{3} - \frac{1}{2}\right) + \left(\frac{1}{5} + \frac{1}{7} - \frac{1}{4}\right) + \cdots.$$

Notice that each block is positive, and that the first block is $\frac{5}{6}$, which is greater than $\log 2 = \sum_1^\infty (-1)^{n+1}/n$. Thus the sum of the reordered series is *larger* than the sum of the original series. (It is not too difficult to show that the reordered series does, in fact, converge.)

On the other hand, reordering does not affect *absolutely* convergent series. We begin the proof of this by discussing positive series.

Theorem 15. *If* $\sum_1^\infty b_m$ *is a reordering of a convergent positive series* $\sum_1^\infty a_n$, *then* $\sum_1^\infty b_m = \sum_1^\infty a_n$.

Proof. Let $s_M = \sum_1^M b_m$. Since all the terms b_m in $\sum_1^M b_m$ come from the terms a_n, there is an integer N such that each of the terms $b_1, b_2, \ldots, b_M$ comes from one of $a_1, a_2, \ldots, a_N$. Then $\sum_1^M b_m \le \sum_1^N a_n \le \sum_1^\infty a_n$. It follows that the reordered series converges to a sum no larger than $\sum_1^\infty a_n$,

$$\sum_1^\infty b_m \le \sum_1^\infty a_n . \tag{1}$$

Now we know that $\sum_1^\infty b_m$ is a convergent series, and $\sum_1^\infty a_n$ is a reordering of it, so we may reverse the roles of a_n and b_m in the argument above to prove that

$$\sum_1^\infty a_n \le \sum_1^\infty b_m . \tag{2}$$

The inequalities (1) and (2) together prove Theorem 15.

Theorem 16. *If* $\sum_1^\infty b_m$ *is a reordering of an absolutely convergent series* $\sum_1^\infty a_n$, *then* $\sum_1^\infty b_m$, *too, is absolutely convergent, and*

$$\sum_1^\infty b_m = \sum_1^\infty a_n .$$

Proof. $\sum_1^\infty |b_m|$ is a reordering of the convergent positive series $\sum_1^\infty |a_n|$, so it converges, and $\sum_1^\infty b_m$ is absolutely convergent. Further, if we set

$$a_n^+ = \begin{cases} a_n & \text{if } a_n \ge 0 \\ 0 & \text{if } a_n < 0, \end{cases} \qquad a_n^- = \begin{cases} 0 & \text{if } a_n \ge 0 \\ -a_n & \text{if } a_n < 0, \end{cases}$$

and similarly define b_n^+ and b_n^-, we find that $\sum b_n^+$ and $\sum b_n^-$ are reorderings of $\sum a_n^+$ and $\sum a_n^-$, so they have the same sums as these latter series. Hence

$$\sum_1^\infty b_n = \sum_1^\infty b_n^+ - \sum_1^\infty b_n^- = \sum_1^\infty a_n^+ - \sum_1^\infty a_n^- = \sum_1^\infty a_n .$$

Remark. In §11.1, where we attempted to find the sum of the harmonic series $\sum_1^\infty 1/n$, one of our errors lay in the reordering of the series $1 - 1/2 + 1/3 - 1/4 + \cdots$ as the difference of two divergent series,

$$\left(1 + \frac{1}{3} + \frac{1}{5} + \cdots\right) - \left(\frac{1}{2} + \frac{1}{4} + \frac{1}{6} + \cdots\right).$$

This is obviously not legitimate.

Products of series

An important application of reordering occurs in the *multiplication of series.* Let $A_N = \sum_0^N a_n$ and $B_N = \sum_0^N b_n$ be the partial sums of two series with sums A and B respectively. Since $\lim A_N = A$ and $\lim B_N = B$, we have $\lim A_N B_N = AB$. The connection of the sequence $A_N B_N$ to the original series can be seen in Fig. 12: $A_N B_N$ is the sum of all the products in the indicated square array. Grouping the terms as indicated by the arrows in Fig. 12, we can write

$$A_N B_N = \sum_{k=0}^N \left(b_k \sum_0^k a_n + a_k \sum_0^{k-1} b_n\right).$$

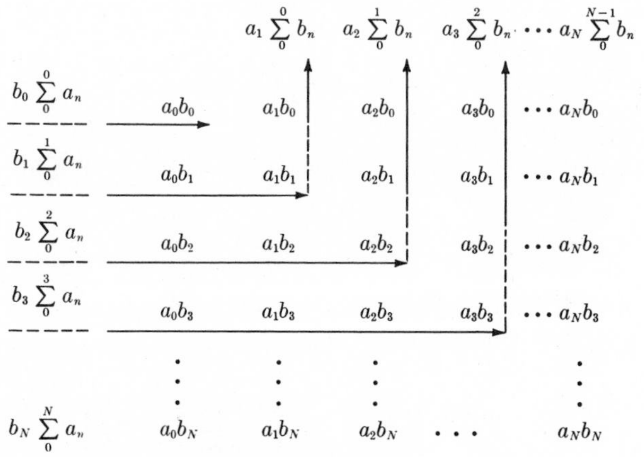

		$a_1 \sum_0^0 b_n$	$a_2 \sum_0^1 b_n$	$a_3 \sum_0^2 b_n$	$\cdots\ a_N \sum_0^{N-1} b_n$
$b_0 \sum_0^0 a_n$	a_0b_0	a_1b_0	a_2b_0	a_3b_0	$\cdots\ a_Nb_0$
$b_1 \sum_0^1 a_n$	a_0b_1	a_1b_1	a_2b_1	a_3b_1	$\cdots\ a_Nb_1$
$b_2 \sum_0^2 a_n$	a_0b_2	a_1b_2	a_2b_2	a_3b_2	$\cdots\ a_Nb_2$
$b_3 \sum_0^3 a_n$	a_0b_3	a_1b_3	a_2b_3	a_3b_3	$\cdots\ a_Nb_3$
	$\vdots$	$\vdots$	$\vdots$		$\vdots$
$b_N \sum_0^N a_n$	a_0b_N	a_1b_N	a_2b_N	$\cdots$	a_Nb_N

FIGURE 11.12

Since $A_N B_N \to (\sum_0^\infty a_n)(\sum_0^\infty b_n)$, we find

$$(\sum_0^\infty a_n)(\sum_0^\infty b_n) = \sum_{k=0}^\infty \left(b_k \sum_0^k a_n + a_k \sum_0^{k-1} b_n\right). \tag{3}$$

Thus the product of the two series is expressed as a sum of all the products $a_n b_n$ arranged with a certain order and a certain grouping.

This formula is not convenient in dealing with power series; the product of $\sum_0^\infty a_n x^n$ and $\sum_0^\infty b_n x^n$ as given by (3) is

$$\sum_{k=0}^\infty x^k \left(b_k \sum_{n=0}^k a_n x^n + a_k \sum_{n=0}^{k-1} b_n x^n\right),$$

which is *not* a power series. The appropriate order and grouping from this point of view is the *Cauchy product*

$$\begin{aligned}(\sum_0^\infty a_n)(\sum_0^\infty b_m) &= \sum_{k=0}^\infty \left(\sum_{n=0}^k a_n b_{k-n}\right) \\ &= a_0 b_0 + (a_0 b_1 + a_1 b_0) + (a_0 b_2 + a_1 b_1 + a_2 b_0) \\ &\qquad + (a_0 b_3 + \cdots) + \cdots,\end{aligned}$$

illustrated in Fig. 13.

We will show that this rearrangement is justified for *absolutely convergent* series. Since power series converge absolutely inside the interval of convergence, we can therefore use the Cauchy product to write

$$\begin{aligned}(\sum_0^\infty a_n x^n)(\sum_0^\infty b_n x^n) &= \sum_{k=0}^\infty \left(\sum_{n=0}^k a_n x^n b_{k-n} x^{k-n}\right) \\ &= \sum_{k=0}^\infty \left(\sum_{n=0}^k a_n b_{k-n}\right) x^k,\end{aligned}$$

thus showing that the product of two power series is itself a power series.

Theorem 17. *When* $\sum_0^\infty a_n$ *and* $\sum_0^\infty b_n$ *are absolutely convergent, then*

$$(\sum_0^\infty a_n)(\sum_0^\infty b_n) = \sum_{k=0}^\infty \left(\sum_{n=0}^k a_n b_{k-n}\right).$$

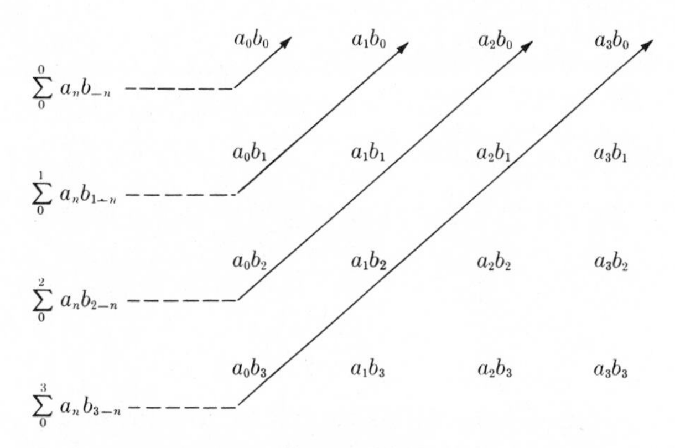

FIGURE 11.13

Proof. Let $c_0 = a_0b_0$, $c_1 = a_0b_1$, $c_2 = a_1b_1$, $c_3 = a_1b_0$, $c_4 = a_0b_2$, and in general let c_j be the jth term in the sequence of products of a_m's and b_n's indicated by the arrows in Fig. 13. Then we have

$$\sum_0^N |c_j| < \sum_0^{N^2} |c_j| = \left(\sum_0^N |a_n|\right)\left(\sum_0^N |b_n|\right) < \left(\sum_0^\infty |a_n|\right)\left(\sum_0^\infty |b_n|\right).$$

It follows that $\sum_0^\infty c_j$ is absolutely convergent and yields the same sum with any order and any grouping. Since the arrangement in (3) yields the sum $(\sum_0^\infty a_n)(\sum_0^\infty b_n)$, so does the arrangement in the Cauchy product. Theorem 17 is proved.

Example 4. We have from Problem 5, §11.5, the binomial series

$$(1 + x)^s = \sum_0^\infty \binom{s}{n} x^n, \tag{4}$$

where $\binom{s}{n}$ is a standard abbreviation for $s(s - 1)\cdots(s - n + 1)/n!$. Multiply (4) by the series for $(1 + x)^{-1}$ to obtain

$$(1 + x)^{s-1} = \left(\sum_0^\infty \binom{s}{n} x^n\right)\left(\sum_0^\infty (-1)^n x^n\right) = \sum_{k=0}^\infty \left[\sum_{n=0}^k \binom{s}{n}(-1)^{k-n}\right] x^k.$$

The binomial series provides another power series for the same function:

$$(1+x)^{s-1} = \sum_{k=0}^{\infty} \binom{s-1}{k} x^k.$$

According to Theorem 14, §11.5, the coefficients in these series must be equal, and we obtain the formula

$$\binom{s-1}{k} = \sum_{n=0}^{k} \binom{s}{n} (-1)^{k-n}.$$

PROBLEMS

1. Obtain the power series for $(1-x)^{-2}$ by differentiating the series for $(1-x)^{-1}$, and again by squaring the series for $(1-x)^{-1}$.

• 2. Obtain the series for $(1-x)^{-3}$ by differentiating the series found in Problem 1, and again by multiplying the series for $(1-x)^{-1}$ and $(1-x)^{-2}$. By comparing the two results, deduce the formula

$$\sum_{n=0}^{k} n = \frac{k(k+1)}{2}.$$

3. With the technique of Problem 2, develop a formula for $\sum_{n=0}^{k} n(n+1)$.

   Combine this with the result of Problem 2 to find a formula for $\sum_{n=0}^{k} n^2$.

4. Multiply two power series to show that $e^{x+y} = e^x e^y$.

• 5. Suppose there is a series $f(x) = \sum_{0}^{\infty} a_n x^n$ such that $f'(x) = x^2 + f^2(x)$ and $f(0) = 1$. Find a_0, a_1, a_2, a_3, a_4.

6. Do Problem 5 for the equations

   (a) $f'(x) = 1 + xf^2(x), \qquad f(0) = 0,$

   (b) $f'(x) = x + f^3(x), \qquad f(0) = -1.$

Part Two

Calculus of Several Variables

Part II is essentially the author's *Calculus of Several Variables*. It can be taken up at any time after §8.2 of the first part.

The basic minimum for an introduction to functions of several variables is given in §§1.1–1.4, §2.1, §3.1–3.4, §3.6, §4.1–4.3, §5.1 and §5.4. The rest of the material reaches out in various directions: linear algebra, differential geometry of curves, physics, extensions of the Fundamental Theorem of Calculus, and differential forms. The use of these other sections depends, of course, on the time available.

Complete proofs are given for the theorems on differentiation. In integration, the basic result that $\int\int f\,dx\,dy = \int\int f\,dy\,dx$ is clearly formulated but not proved; see §4.1 for the treatment of this point. Assuming this result, we continue giving complete proofs up to Gauss' theorem and differential forms; the presentation of these two topics is very informal but, we hope, suggestive.

Chapter I

Vectors

This chapter provides the setting for the rest of the book. We assume as background a few geometric concepts (such as parallelism, orthogonality, and the Pythagorean theorem, and elementary trigonometry) and develop an abstract algebraic system, called the vector space R^3, in which these concepts have simple algebraic definitions.

§1.1 presents the properties that R^3 shares with the simpler two-dimensional space R^2; familiarity with R^2, though not required, would be useful in reading this section.

§1.2 introduces the cross product.

§1.3 applies vector space methods to analytic geometry.

§1.4 introduces R^n, carrying over the definitions and terminology from R^2 and R^3.

§1.5 introduces the rather abstract idea of *linear independence*. This section, though not essential to the rest of the book, is included partly as preparation for linear algebra (which many students will study sooner or later), and partly because it seems the best way to prove certain basic facts (for example, that **A**, **B**, and **C** form a basis if $\mathbf{A} \cdot (\mathbf{B} \times \mathbf{C}) \neq 0$). Such facts are used only rarely in the text, so it is possible to omit §1.5 and give intuitive geometric arguments on those few occasions when a reference to §1.5 is made.

We observe a strict separation of powers between "geometry" and "algebra"; geometry suggests and interprets many results, but definitions and proofs are based on elementary algebra. In keeping with this, we are not obliged to prove any statements about the geometric interpretations; they are given only to guide the intuition, not to provide proofs.

1.1 THE VECTOR SPACE R^3

When the Greeks studied solid geometry, they laid the foundations of a remarkably accurate mathematical model of the "physical space" in which stars, planets, rockets, baseballs, electrons, and so on all move about. Solid geometry is still the underlying model for a large part of natural science, but its study has been tremendously simplified by shifting from the purely geometric point of view taken by the Greeks to an algebraic one. From the modern (algebraic) point of view, we define the *vector space* R^3, and establish its various properties by elementary algebraic calculations. To relate the vector space to our intuitive understanding of "physical space" we introduce a coordinate system. Once this is done, all the algebraic results have a more or less obvious geometric meaning.

Definition 1.* The vector space R^3 consists of all ordered triples of real numbers (usually denoted $\mathbf{A} = (a_1,a_2,a_3)$, $\mathbf{B} = (b_1,b_2,b_3)$, etc.), together with the following algebraic operations:

$$\mathbf{A} + \mathbf{B} = (a_1 + b_1\,,\, a_2 + b_2\,,\, a_3 + b_3) \qquad \text{(addition)}$$

$$t\mathbf{A} = (ta_1,ta_2,ta_3), \quad t \text{ any real number} \qquad \text{(scalar multiplication)}$$

$$\mathbf{A}\cdot\mathbf{B} = a_1b_1 + a_2b_2 + a_3b_3 \qquad \text{(inner product, or dot product)}$$

The members of R^3 are called *points* or *vectors*. The vector (0,0,0) is denoted **0**. The *length* of a vector **A** is

$$|\mathbf{A}| = (\mathbf{A}\cdot\mathbf{A})^{1/2} = \sqrt{a_1^2 + a_2^2 + a_3^2}\,.$$

The numbers a_1, a_2, and a_3 are called *components* or *coordinates* of the vector (a_1,a_2,a_3).

From this definition follow many simple but useful identities, which for convenience we sum up in Theorem 1. You will not find it necessary to memorize all these, since they all reflect familiar properties of numbers.

---

* There is a slight discrepancy in notation and terminology between this definition and Chapter VII in Part I (One Variable), which distinguishes between points and vectors, denoting points with parentheses (x_0,y_0) and vectors with brackets $[a,b]$. Here the distinction is dropped, and we feel free to think of an ordered triple (a,b,c) either as a point or as a vector. This dual point of view is explained in the discussion following Theorem 1.

Theorem 1. *Let* **A**, **B**, *and* **C** *be any members of* R^3, *and let* t *and* s *be any real numbers. Then*

$$\left.\begin{array}{ll} (1) & t(s\mathbf{A}) = (ts)\mathbf{A} \\ (2) & (\mathbf{A} + \mathbf{B}) + \mathbf{C} = \mathbf{A} + (\mathbf{B} + \mathbf{C}) \\ (3) & t(\mathbf{A}\cdot\mathbf{B}) = (t\mathbf{A})\cdot\mathbf{B} = \mathbf{A}\cdot(t\mathbf{B}) \end{array}\right\} \quad (\textit{associative laws})$$

$$\left.\begin{array}{ll} (4) & \mathbf{A} + \mathbf{B} = \mathbf{B} + \mathbf{A} \\ (5) & \mathbf{A}\cdot\mathbf{B} = \mathbf{B}\cdot\mathbf{A} \end{array}\right\} \quad (\textit{commutative laws})$$

$$\left.\begin{array}{ll} (6) & (t + s)\mathbf{A} = t\mathbf{A} + s\mathbf{A} \\ (7) & t(\mathbf{A} + \mathbf{B}) = t\mathbf{A} + t\mathbf{B} \\ (8) & (\mathbf{A} + \mathbf{B})\cdot\mathbf{C} = (\mathbf{A}\cdot\mathbf{C}) + (\mathbf{B}\cdot\mathbf{C}) \\ (9) & \mathbf{A}\cdot(\mathbf{B} + \mathbf{C}) = \mathbf{A}\cdot\mathbf{B} + \mathbf{A}\cdot\mathbf{C} \end{array}\right\} \quad (\textit{distributive laws})$$

$$(10) \quad 1\cdot\mathbf{A} = \mathbf{A}, \quad 0\mathbf{A} = \mathbf{0}$$

$$\left.\begin{array}{ll} (11) & \mathbf{A} + \mathbf{0} = \mathbf{A} \\ (12) & |\mathbf{A}| = 0 \Leftrightarrow \mathbf{A} = \mathbf{0} \end{array}\right\} \quad (\textit{laws for the zero vector})$$

$$(13) \quad |t\mathbf{A}| = |t|\cdot|\mathbf{A}|.$$

Proof. Formula (1) follows directly from the commutativity of real numbers; since $a_1 + b_1 = b_1 + a_1$, etc., we have

$$\begin{aligned} \mathbf{A} + \mathbf{B} &= (a_1 + b_1, a_2 + b_2, a_3 + b_3) \\ &= (b_1 + a_1, b_2 + a_2, b_3 + a_3) = \mathbf{B} + \mathbf{A}. \end{aligned}$$

The proofs of (2)–(11) follow the same basic pattern; you compute each side, and observe that the corresponding components are equal. Finally, (12) and (13) follow from the formula for the length $|\mathbf{A}| = \sqrt{a_1^2 + a_2^2 + a_3^2}$. For example, to prove (12), observe that a sum of squares of real numbers $a_1^2 + a_2^2 + a_3^2$ is zero if and only if each term is zero.

In view of the associative law (2), we can let $\mathbf{A} + \mathbf{B} + \mathbf{C}$ stand for both $(\mathbf{A} + \mathbf{B}) + \mathbf{C}$ and $\mathbf{A} + (\mathbf{B} + \mathbf{C})$. More generally, we omit parentheses from any sum of three or more vectors; for example, we simplify $(\mathbf{A} + \mathbf{B}) + (\mathbf{C} + \mathbf{D})$ to $\mathbf{A} + \mathbf{B} + \mathbf{C} + \mathbf{D}$.

Any mathematical system in which formulas (1), (2), (4), (6), (7), (10), and (11) hold is called a *real vector space.* When all of formulas (1)–(13) hold, it is called a *vector space with inner product.* Thus, Theorem 1 states that R^3 *is a vector space with inner product.*

The geometric interpretation of R^3 is based on a rectangular coordinate system. Picture three mutually perpendicular lines intersecting at a given point 0 (the origin), as in Fig. 1. Call these lines the x axis, the y axis, and the z axis, and space the real numbers uniformly along each axis, with zero at the origin. With this picture, to every ordered triple

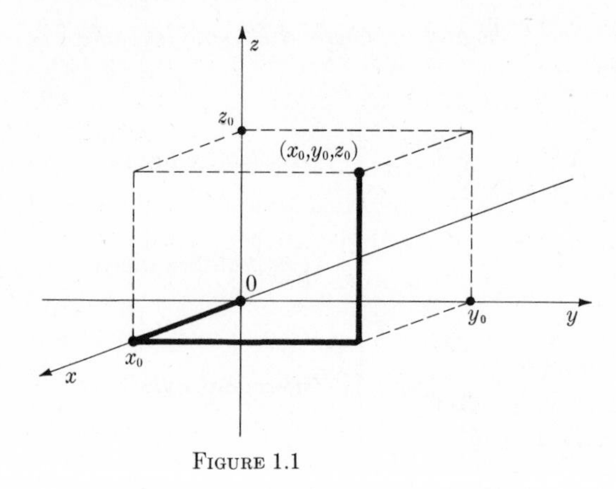

FIGURE 1.1

of real numbers (x_0,y_0,z_0) there corresponds a particular point in space, as shown in Fig. 1. The basic assumption underlying the applications of algebraic methods to concrete geometric and physical problems is that this correspondence is *reversible*; given the coordinate system, every ordered triple corresponds to a unique point in "physical space," and conversely every point corresponds to a unique ordered triple.

Since the members of R^3 are ordered triples, we can think of R^3 geometrically as the points in space. When we have this image in mind, we call the members of R^3 *points*, and generally label them $\mathbf{P} = (x,y,z)$, $\mathbf{P}_0 = (x_0,y_0,z_0)$, and so on. Figure 2 shows that the length $|\mathbf{P}_0|$ is the distance from the origin to the point representing $\mathbf{P}_0$.

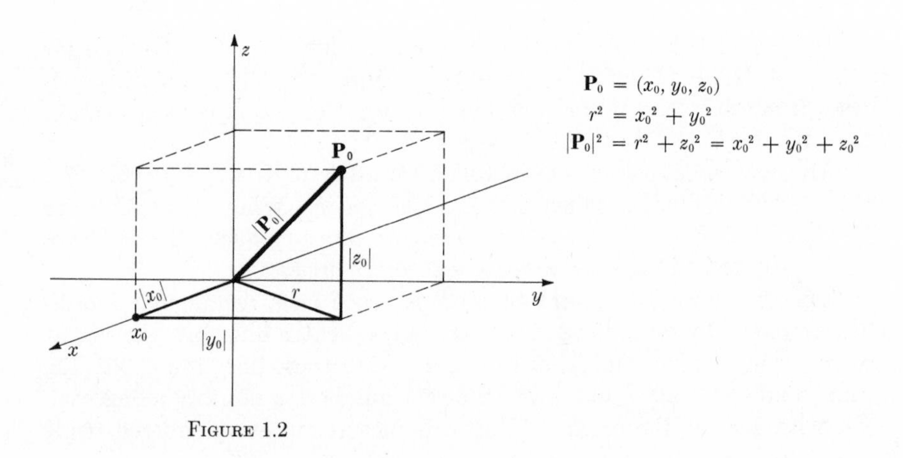

FIGURE 1.2

A second way to represent R^3 is by arrows. If $\mathbf{A} = (a_1,a_2,a_3)$ is any member of R^3, we represent it by an arrow from the origin, as in Fig. 3(a), or more generally by an arrow beginning at any point (x,y,z) and ending at the point $(x + a_1, y + a_2, z + a_3)$, as in Fig. 3(b). (You can think of $\mathbf{A}$ as giving a change in position from the initial point (x,y,z) to the terminal point $(x + a_1, y + a_2, z + a_3)$.) When we have this image in mind, we call the members of R^3 *vectors*. The arrow representing $\mathbf{A}$ can start at any point (and this turns out to be very helpful in visualizing the applications of vector theory), but no matter where it is drawn, it always has the same direction, and the length $|\mathbf{A}| = \sqrt{a_1^2 + a_2^2 + a_3^2}$. (We are overlooking the distortions of perspective entailed in drawing lines in three-space on two-dimensional paper.)

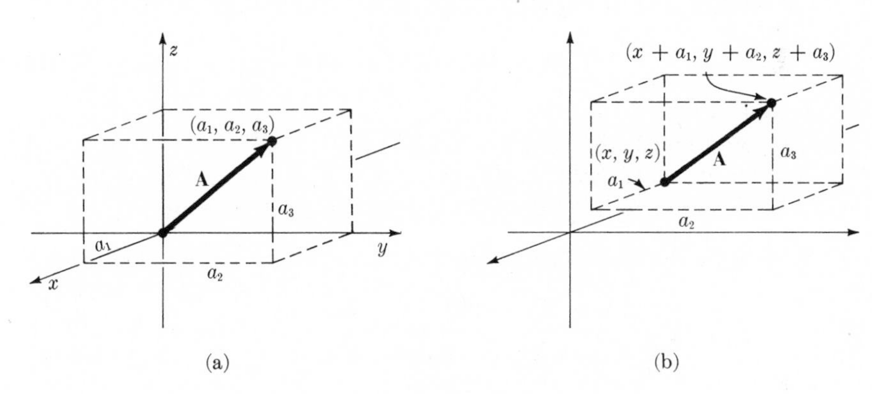

FIGURE 1.3

These two alternate interpretations (point in space, or arrow between two points in space) are closely related. When $\mathbf{A} = (a_1,a_2,a_3)$ is represented by an *arrow beginning at the origin*, as in Fig. 3(a), then the tip of the arrow coincides with the *point* representing (a_1,a_2,a_3). When in doubt as to which interpretation to use, use both, but let the arrow begin at the origin.

Addition has the effect of following one change of position, $\mathbf{A}$, by another, $\mathbf{B}$. Thus $\mathbf{A} + \mathbf{B}$ is represented by the third side of a triangle whose other two sides represent $\mathbf{A}$ and $\mathbf{B}$, as in Fig. 4. Combining this picture with the corresponding one for $\mathbf{B} + \mathbf{A}$ (as in Fig. 5), we find that the commutative law (1) expresses an "obvious fact": $\mathbf{A} + \mathbf{B}$ and $\mathbf{B} + \mathbf{A}$ are both represented by the same diagonal of the parallelogram in Fig. 5.

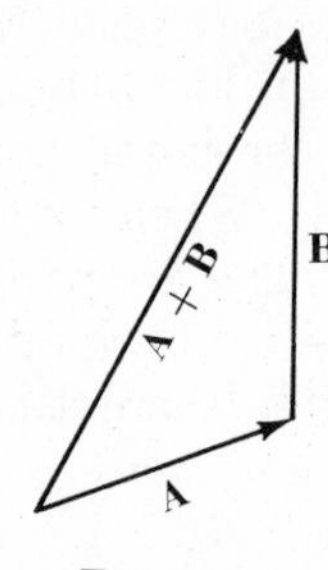

FIGURE 1.4

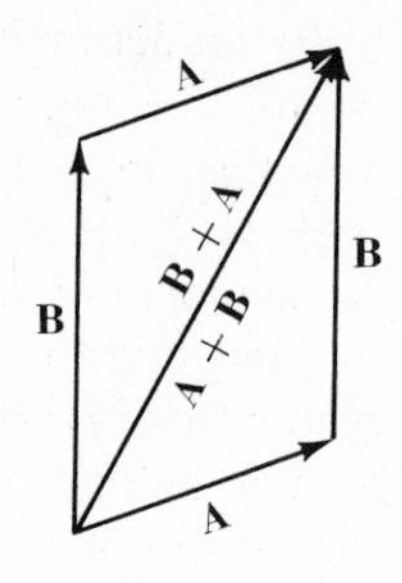

FIGURE 1.5

A familiar theorem of classical geometry states that

$$|\mathbf{A} + \mathbf{B}|^2 = |\mathbf{A}|^2 + |\mathbf{B}|^2 \tag{14}$$

if and only if the arrows representing **A** and **B** in Fig. 4 are orthogonal; thus it would be natural to define the vectors themselves to be orthogonal when equation (14) holds. However, this equation can be reduced to a much simpler form if we expand $|\mathbf{A} + \mathbf{B}|^2$ as a dot product:

$$\begin{aligned}
|\mathbf{A} + \mathbf{B}|^2 &= (\mathbf{A} + \mathbf{B})\cdot(\mathbf{A} + \mathbf{B}) && \text{(by Definition 1)} \\
&= (\mathbf{A}\cdot(\mathbf{A} + \mathbf{B})) + (\mathbf{B}\cdot(\mathbf{A} + \mathbf{B})) && \text{(by (8))} \\
&= (\mathbf{A}\cdot\mathbf{A} + \mathbf{A}\cdot\mathbf{B}) + (\mathbf{B}\cdot\mathbf{A} + \mathbf{B}\cdot\mathbf{B}) && \text{(by (9))} \\
&= |\mathbf{A}|^2 + 2(\mathbf{A}\cdot\mathbf{B}) + |\mathbf{B}|^2 && \text{(by (2) and (5).)}
\end{aligned}$$

Hence $|\mathbf{A} + \mathbf{B}|^2 = |\mathbf{A}|^2 + |\mathbf{B}|^2$ if and only if $\mathbf{A}\cdot\mathbf{B} = 0$. This motivates

Definition 2. **A** and **B** are called *orthogonal* if and only if $\mathbf{A}\cdot\mathbf{B} = 0$.

Turning next to the *scalar product* $t\mathbf{A}$ of a real number t and a vector **A**, we find that if $t > 0$, then $t\mathbf{A}$ is a change of position in the same direction as **A** but t times as far, while if $t < 0$, then $t\mathbf{A}$ is in the opposite direction from **A** but $|t|$ times as far (Fig. 6). In any case, the vector $t\mathbf{A}$ appears to be *parallel* to **A**; this motivates

Definition 3. Two vectors are *parallel* if and only if one is a scalar multiple of the other. In other words, **A** and **B** are parallel if and only if either $\mathbf{A} = t\mathbf{B}$ for some real t or $\mathbf{B} = t\mathbf{A}$ for some real t.

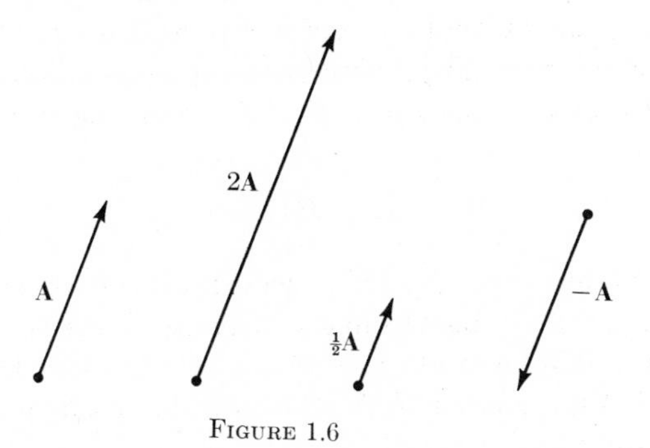

FIGURE 1.6

Combining addition and scalar multiplication leads to interesting results. Given two vectors **A** and **B**, with $\mathbf{B} \neq \mathbf{0}$, look at the vectors of the form

$$\mathbf{A} + t\mathbf{B}, \tag{15}$$

where t varies over all real numbers. With $t = 0$ we get **A**; with $t = 1$ we get $\mathbf{A} + \mathbf{B}$; with $t = -1$ we get $\mathbf{A} + (-1)\mathbf{B}$, which is usually written simply $\mathbf{A} - \mathbf{B}$ and called the *difference* of **A** and **B**. In general, as t varies over the real numbers, the vectors (15) generate a *line* as sketched in Fig. 7, called the line through **A** in the direction **B**. (It must be assumed that $\mathbf{B} \neq \mathbf{0}$, for otherwise (15) gives only the point **A**, not a whole line.) Visualize the line as consisting of all the points that represent $\mathbf{A} + t\mathbf{B}$ for various t or, equivalently, as the tips of the arrows representing $\mathbf{A} + t\mathbf{B}$, when the arrows begin at the origin.

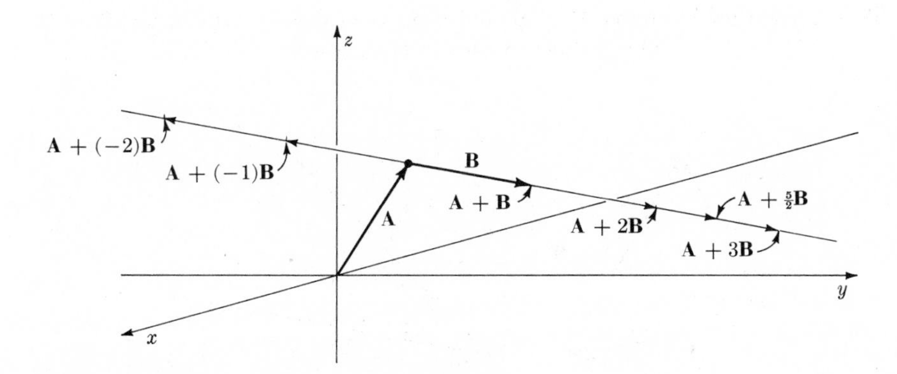

FIGURE 1.7

Figure 7 shows the difference $\mathbf{A} - \mathbf{B}$ as the sum of the vectors $\mathbf{A}$ and $(-1)\mathbf{B}$. Figure 8 shows another useful representation; when $\mathbf{A}$ and $\mathbf{B}$ are drawn from a common initial point, then $\mathbf{A} - \mathbf{B}$ is represented by an arrow from the tip of $\mathbf{B}$ to the tip of $\mathbf{A}$. The resulting triangle illustrates the identity

$$\mathbf{B} + (\mathbf{A} - \mathbf{B}) = \mathbf{A}.$$

Returning to the line given by (15), we can collect an unexpected dividend by computing the distance from the origin to the line. By definition, this distance is the minimum of $|\mathbf{A} + t\mathbf{B}|$ as t varies over all real numbers. To find the minimum, expand $|\mathbf{A} + t\mathbf{B}|^2$ as a dot product (just like the calculation preceding Definition 2):

$$|\mathbf{A} + t\mathbf{B}|^2 = \mathbf{A}\cdot\mathbf{A} + 2t(\mathbf{A}\cdot\mathbf{B}) + t^2\mathbf{B}\cdot\mathbf{B}. \tag{16}$$

On the right in (16) is a quadratic in t (since $|\mathbf{B}| \neq 0$), and its minimum is easily found to occur at $t = -\mathbf{A}\cdot\mathbf{B}/\mathbf{B}\cdot\mathbf{B}$. Putting this value of t in each side of (16), we find the square of the distance from the line to the origin to be

$$\begin{aligned} \left|\mathbf{A} - \frac{\mathbf{A}\cdot\mathbf{B}}{\mathbf{B}\cdot\mathbf{B}}\mathbf{B}\right|^2 &= \mathbf{A}\cdot\mathbf{A} - 2\frac{(\mathbf{A}\cdot\mathbf{B})^2}{\mathbf{B}\cdot\mathbf{B}} + \frac{(\mathbf{A}\cdot\mathbf{B})^2}{\mathbf{B}\cdot\mathbf{B}} \\ &= |\mathbf{A}|^2 - \frac{(\mathbf{A}\cdot\mathbf{B})^2}{|\mathbf{B}|^2} \\ &= \frac{1}{|\mathbf{B}|^2}\left(|\mathbf{A}|^2|\mathbf{B}|^2 - (\mathbf{A}\cdot\mathbf{B})^2\right). \end{aligned} \tag{17}$$

Since the number in (17) is a square, it cannot be negative; hence $(\mathbf{A}\cdot\mathbf{B})^2 \leq |\mathbf{A}|^2\cdot|\mathbf{B}|^2$. Taking square roots, we collect our dividend, the *Schwarz inequality*:

$$|(\mathbf{A}\cdot\mathbf{B})| \leq |\mathbf{A}|\cdot|\mathbf{B}|. \tag{18}$$

This derivation assumed $\mathbf{B} \neq \mathbf{0}$, but (18) is obviously true when $\mathbf{B} = \mathbf{0}$ as well, since both sides reduce to zero in that case.

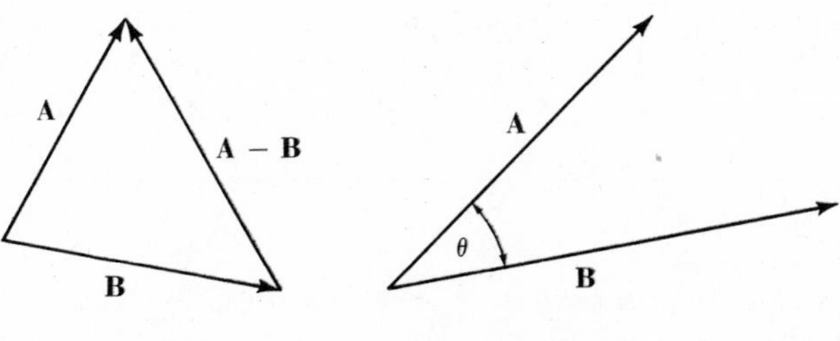

FIGURE 1.8 FIGURE 1.9

The Schwarz inequality implies, in turn, the *triangle inequality*:

$$|\mathbf{A} + \mathbf{B}| \leq |\mathbf{A}| + |\mathbf{B}|. \tag{19}$$

The proof of (19) is straightforward; from (16) we have already found that

$$|\mathbf{A} + \mathbf{B}|^2 = |\mathbf{A}|^2 + 2(\mathbf{A}\cdot\mathbf{B}) + |\mathbf{B}|^2; \tag{20}$$

hence

$$\begin{aligned} |\mathbf{A} + \mathbf{B}|^2 &\leq |\mathbf{A}|^2 + 2|(\mathbf{A}\cdot\mathbf{B})| + |\mathbf{B}|^2 \\ &\leq |\mathbf{A}|^2 + 2|\mathbf{A}|\cdot|\mathbf{B}| + |\mathbf{B}|^2 \qquad \text{(by (18))} \\ &= (|\mathbf{A}| + |\mathbf{B}|)^2, \end{aligned}$$

and the triangle inequality (19) follows by taking square roots. Figure 4 interprets this inequality as a familiar principle of Euclidean geometry, namely, any side of a triangle is less than or equal to the sum of the other two sides. (Hence the name "triangle inequality.")

The geometric interpretation of the dot product is based on the Schwarz inequality. When $\mathbf{A} \neq \mathbf{0}$ and $\mathbf{B} \neq \mathbf{0}$, we can rewrite (18) as

$$-1 \leq \frac{\mathbf{A}\cdot\mathbf{B}}{|\mathbf{A}|\cdot|\mathbf{B}|} \leq 1.$$

Hence $\arccos\left(\dfrac{\mathbf{A}\cdot\mathbf{B}}{|\mathbf{A}|\cdot|\mathbf{B}|}\right)$ is defined; namely, it is the number θ such that $0 \leq \theta \leq \pi$ and

$$\cos\theta = \frac{\mathbf{A}\cdot\mathbf{B}}{|\mathbf{A}|\cdot|\mathbf{B}|}\,. \tag{21}$$

This is called the *angle between* **A** *and* **B** (see Fig. 9). Notice that (21) is consistent with the definition of orthogonal vectors, since for $0 \leq \theta \leq \pi$

$$\theta = \frac{\pi}{2} \quad\Leftrightarrow\quad \cos\theta = 0 \quad\Leftrightarrow\quad \mathbf{A}\cdot\mathbf{B} = 0.$$

It is also consistent with the definition of parallel vectors, and with the law of cosines, as shown in the problems below. When (21) is multiplied out, it gives

$$\mathbf{A}\cdot\mathbf{B} = |\mathbf{A}|\cdot|\mathbf{B}|\cos\theta,$$

which is the geometric interpretation we were looking for: the dot product of two vectors is the product of the lengths times the cosine of the angle between the vectors.

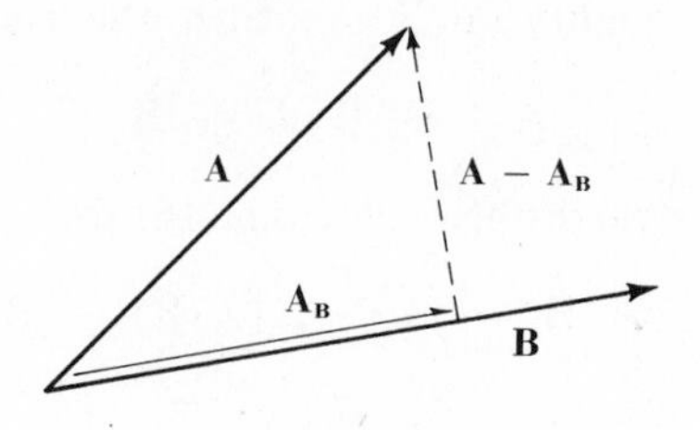

FIGURE 1.10

Figure 10 illustrates *orthogonal projection*, which is the main geometric use of the dot product. The orthogonal projection $\mathbf{A}_\mathbf{B}$ of $\mathbf{A}$ on a nonzero vector $\mathbf{B}$ is characterized by two properties:

(i) $\mathbf{A}_\mathbf{B}$ is parallel to $\mathbf{B}$

(ii) $\mathbf{A} - \mathbf{A}_\mathbf{B}$ is orthogonal to $\mathbf{B}$.

Condition (i) implies $\mathbf{A}_\mathbf{B} = t\mathbf{B}$ for some t, and combining this with condition (ii) gives

$$0 = (\mathbf{A} - \mathbf{A}_\mathbf{B})\cdot\mathbf{B} = (\mathbf{A} - t\mathbf{B})\cdot\mathbf{B} = \mathbf{A}\cdot\mathbf{B} - t(\mathbf{B}\cdot\mathbf{B}).$$

Hence $t = \mathbf{A}\cdot\mathbf{B}/\mathbf{B}\cdot\mathbf{B}$, and the projection is

$$\mathbf{A}_\mathbf{B} = \frac{\mathbf{A}\cdot\mathbf{B}}{\mathbf{B}\cdot\mathbf{B}}\mathbf{B}. \tag{22}$$

When $\mathbf{B}$ happens to be a unit vector, i.e. when $|\mathbf{B}| = 1$, the projection formula simplifies to

$$\mathbf{A}_\mathbf{B} = (\mathbf{A}\cdot\mathbf{B})\mathbf{B} \qquad (\mathbf{B} \text{ a unit vector}).$$

For example, the projection of $\mathbf{A} = (a_1,a_2,a_3)$ on the unit vector $\mathbf{E}_1 = (1, 0, 0)$ is

$$(\mathbf{A}\cdot\mathbf{E}_1)\mathbf{E}_1 = a_1(1,0,0) = (a_1,0,0).$$

Similarly, the projections on $\mathbf{E}_2 = (0,1,0)$ and $\mathbf{E}_3 = (0,0,1)$ are

$$(\mathbf{A}\cdot\mathbf{E}_2)\mathbf{E}_2 = (0,a_2,0), \qquad (\mathbf{A}\cdot\mathbf{E}_3)\mathbf{E}_3 = (0,0,a_3).$$

From this you can see immediately that $\mathbf{A}$ equals the sum of its projections on $\mathbf{E}_1$, $\mathbf{E}_2$, and $\mathbf{E}_3$ (see Fig. 11):

$$\mathbf{A} = (\mathbf{A}\cdot\mathbf{E}_1)\mathbf{E}_1 + (\mathbf{A}\cdot\mathbf{E}_2)\mathbf{E}_2 + (\mathbf{A}\cdot\mathbf{E}_3)\mathbf{E}_3 .$$

Toward the end of this chapter we will be able to show that this same formula holds when $\mathbf{E}_1$, $\mathbf{E}_2$, and $\mathbf{E}_3$ are replaced by any three unit vectors $\mathbf{U}_1$, $\mathbf{U}_2$, $\mathbf{U}_3$ which are orthogonal to each other.

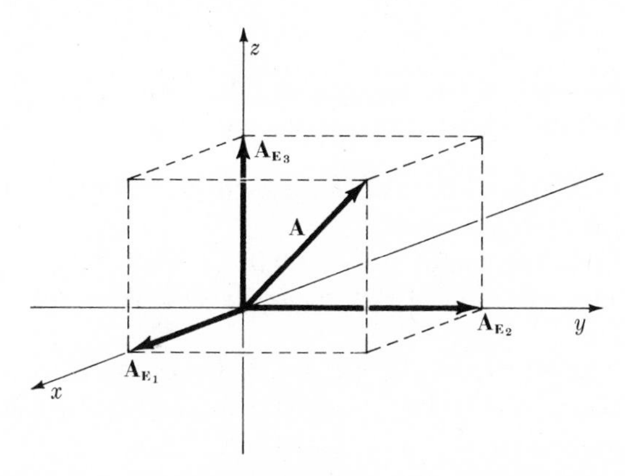

FIGURE 1.11

PROBLEMS

1. Represent the following vectors as arrows starting at the origin **0**.
(a) (0,1,3) (b) (−1,2,5) (c) (−1,−3,−4)
(d) (1,−5,3) (e) (1,−3,−5) (f) (1,3,−5)

2. Represent the vectors in Problem 1 as arrows starting at the point (1,1,1).

3. Let $\mathbf{A} = (0,1,5)$, $\mathbf{B} = (-\sqrt{14},5,1)$.
(a) Find $|\mathbf{A}|$ and $|\mathbf{B}|$.
(b) Find $\mathbf{A}\cdot\mathbf{B}$.
(c) Find the angle between **A** and **B**.

4. (a) Find the angle between the vectors in Problem 1(a) and (b).
(b) Find the angle between the vectors in Problem 1(e) and (f).

5. Which of the following vectors are parallel to (1,1,−1)?
(a) (2,2,−2) (b) (2,−2,2) (c) (−2,2,2)
(d) (1,2,−2) (e) $(-\frac{1}{2},-\frac{1}{2},\frac{1}{2})$ (f) $(\pi,\pi,-\pi)$

6. (a) Find *all* vectors that are orthogonal to $\mathbf{E}_1 = (1,0,0)$.
(b) Find *all* vectors that are orthogonal both to $\mathbf{E}_1$ and to $\mathbf{E}_2 = (0,1,0)$.
(c) Find *all* vectors that are orthogonal to $\mathbf{E}_1$, to $\mathbf{E}_2$, and to $\mathbf{E}_3 = (0,0,1)$. (There is exactly one.)

7. (a) Find a nonzero vector orthogonal to (1,2,−1). (There are many.)
(b) Find a nonzero vector orthogonal to (1,2,−1) and to (−1,0,3). (There are still quite a few.)

8. Let

$$\mathbf{U}_1 = \left(\frac{1}{\sqrt{3}}, \frac{1}{\sqrt{3}}, \frac{-1}{\sqrt{3}}\right), \quad \mathbf{U}_2 = \left(\frac{1}{\sqrt{6}}, \frac{1}{\sqrt{6}}, \frac{2}{\sqrt{6}}\right), \quad \mathbf{U}_3 = \left(\frac{1}{\sqrt{2}}, \frac{-1}{\sqrt{2}}, 0\right).$$

(a) Show that each of $\mathbf{U}_1$, $\mathbf{U}_2$, $\mathbf{U}_3$ is orthogonal to the other two, and that each is a unit vector.
(b) Find the projection of $\mathbf{E}_1 = (1,0,0)$ on each of $\mathbf{U}_1$, $\mathbf{U}_2$, $\mathbf{U}_3$.
(c) Find the projection of the general vector $\mathbf{A} = (a_1,a_2,a_3)$ on $\mathbf{U}_1$, $\mathbf{U}_2$, $\mathbf{U}_3$.
(d) Prove that $\mathbf{A} = (\mathbf{A}\cdot\mathbf{U}_1)\mathbf{U}_1 + (\mathbf{A}\cdot\mathbf{U}_2)\mathbf{U}_2 + (\mathbf{A}\cdot\mathbf{U}_3)\mathbf{U}_3$. (This is a tedious calculation. Later on, we will develop an easier but more sophisticated way to prove such formulas.)

9. (a) Prove that $|\mathbf{A} - \mathbf{B}|^2 = |\mathbf{A}|^2 + |\mathbf{B}|^2 - 2(\mathbf{A}\cdot\mathbf{B})$.
(b) Referring to Figs. 8 and 9, show that the formula in part (a) is the *law of cosines*: $c^2 = a^2 + b^2 - 2ab\cos\theta$.

10. Suppose that $\mathbf{A}$ and $\mathbf{B}$ are orthogonal, and let θ be the angle between $\mathbf{A}$ and $\mathbf{A} + \mathbf{B}$. Prove (algebraically) that $|\mathbf{A}| = |\mathbf{A} + \mathbf{B}|\cos\theta$ and $|\mathbf{B}| = |\mathbf{A} + \mathbf{B}|\sin\theta$. Draw a sketch relating these formulas to the definitions of the sine and the cosine of an angle in a right triangle.

11. Prove the *parallelogram law*: $|\mathbf{A} + \mathbf{B}|^2 + |\mathbf{A} - \mathbf{B}|^2 = 2|\mathbf{A}|^2 + 2|\mathbf{B}|^2$. Draw a figure showing why this is called the parallelogram law. (Hint: Fig. 5 above shows $\mathbf{A} + \mathbf{B}$ as the diagonal of a parallelogram. What is the other diagonal?)

12. Suppose that $t\mathbf{A} = \mathbf{0}$. Prove that either $t = 0$ or $\mathbf{A} = \mathbf{0}$.

13. Suppose that $\mathbf{A}$ and $\mathbf{B}$ are parallel, and that $\mathbf{B} \neq \mathbf{0}$. Prove that $\mathbf{A} = t\mathbf{B}$ for some real t. (Hint: Parallelism means *either* $\mathbf{A} = t\mathbf{B}$ or $\mathbf{B} = s\mathbf{A}$. In the first case, there is nothing to prove; in the second case, use the previous problem.)

14. Suppose that $\mathbf{B} \neq \mathbf{0}$. Prove that a vector $\mathbf{A}$ is parallel to $\mathbf{B}$ if and only if $\mathbf{A}$ equals $\mathbf{A}_\mathbf{B}$, its projection on $\mathbf{B}$. (Hint: In proving $\mathbf{A} = \mathbf{A}_\mathbf{B}$, use formula (22) and the previous problem.)

15. Suppose that $\mathbf{A}$ and $\mathbf{B}$ are both nonzero. Prove that $\mathbf{A}$ and $\mathbf{B}$ are parallel if and only if the angle θ between them is 0 or π. (Hint: In proving the "if" part, show that $|\mathbf{A} - \mathbf{A}_\mathbf{B}|^2 = \mathbf{0}$ when $|\cos\theta| = 1$.)

16. (a) Obtain a formula for the distance d from a point $\mathbf{P}_0$ to the line with parametric formula $\mathbf{A} + t\mathbf{B}$. (When $\mathbf{P}_0 = \mathbf{0}$, your formula should reduce to the expression in the text for the distance from the line to the origin.)

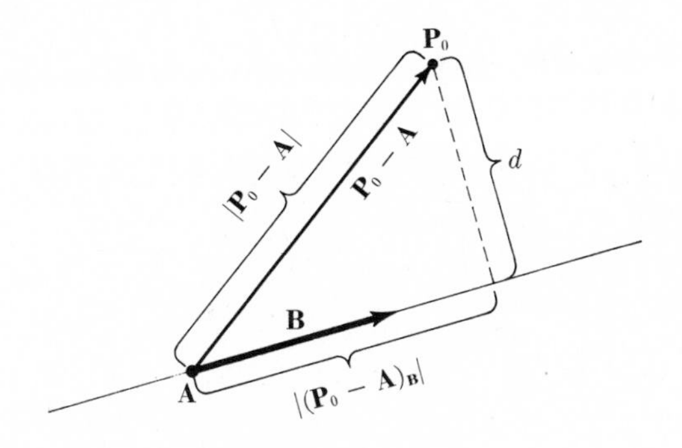

FIGURE 1.12 $d^2 = |\mathbf{P}_0 - \mathbf{A}|^2 - |(\mathbf{P}_0 - \mathbf{A})_{\mathbf{B}}|^2$

(b) Show that your formula for the distance d can be described in terms of orthogonal projection as $d^2 = |\mathbf{P}_0 - \mathbf{A}|^2 - |(\mathbf{P}_0 - \mathbf{A})_{\mathbf{B}}|^2$. (This is an easy geometric way to remember the formula; see Fig. 12.)

17. Prove at least some of the following formulas in the text above: (2), (3), (4), (5), (6), (7), (8), (9), (10), (11), (13).

1.2 THE CROSS PRODUCT

Many geometric constructions in R^3 reduce ultimately to the same algebraic problem: Given two nonparallel vectors **A** and **B**, find a nonzero vector **C** which is orthogonal to both **A** and **B**. (See Fig. 13.) This problem has a standard solution called the *cross product* of **A** and **B**, denoted $\mathbf{A} \times \mathbf{B}$.

Definition 2. If $\mathbf{A} = (a_1,a_2,a_3)$ and $\mathbf{B} = (b_1,b_2,b_3)$, then the *cross product* of **A** and **B** is defined by the formula

$$\mathbf{A} \times \mathbf{B} = (a_2b_3 - a_3b_2\,,\, a_3b_1 - a_1b_3\,,\, a_1b_2 - a_2b_1). \tag{1}$$

The rest of this section develops the algebraic properties and geometric meaning of the strange expression (1).

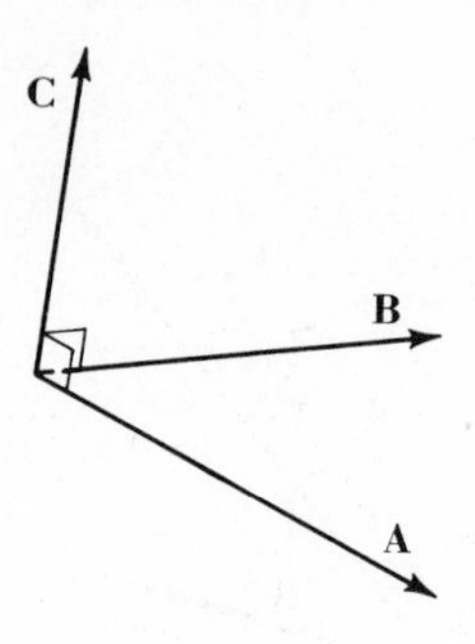

FIGURE 1.13

Theorem 2. **A** × **B** *is orthogonal to both* **A** *and* **B**. *Further,*

$$\mathbf{A} \times \mathbf{B} = -(\mathbf{B} \times \mathbf{A}) \tag{2}$$

$$\mathbf{A} \times \mathbf{A} = \mathbf{0} \tag{3}$$

$$(t\mathbf{A}) \times \mathbf{B} = t(\mathbf{A} \times \mathbf{B}) = \mathbf{A} \times (t\mathbf{B}) \tag{4}$$

$$|\mathbf{A} \times \mathbf{B}|^2 = |\mathbf{A}|^2 \cdot |\mathbf{B}|^2 - (\mathbf{A} \cdot \mathbf{B})^2 \tag{5}$$

$$\mathbf{A} \cdot (\mathbf{B} \times \mathbf{C}) = \mathbf{B} \cdot (\mathbf{C} \times \mathbf{A}) = \mathbf{C} \cdot (\mathbf{A} \times \mathbf{B}) \tag{6}$$

$$(\mathbf{A} + \mathbf{B}) \times \mathbf{C} = (\mathbf{A} \times \mathbf{C}) + (\mathbf{B} \times \mathbf{C}) \tag{7}$$

$$\mathbf{A} \times (\mathbf{B} + \mathbf{C}) = (\mathbf{A} \times \mathbf{B}) + (\mathbf{A} \times \mathbf{C}). \tag{8}$$

Proof. We begin with formulas (2)–(8). Interchanging **A** and **B** in Definition 2 produces

$$\mathbf{B} \times \mathbf{A} = (b_2a_3 - b_3a_2\,,\, b_3a_1 - b_1a_3\,,\, b_1a_2 - b_2a_1)\,;$$

comparing this to the expression (1) for **A** × **B**, you can see that **B** × **A** = −(**A** × **B**), so (2) is proved. Formula (3) can be seen immediately by setting **B** = **A** in (1). Formula (4) is an easy calculation which is left to you. Formulas (5)–(8) are not quite so easy, but they can be proved simply by expanding both sides; this, too, is left to you (Problem 5).

Finally, we prove that **A** × **B** is orthogonal to **A**. Setting **C** = **A** in (6) yields

$$\mathbf{A} \cdot (\mathbf{A} \times \mathbf{B}) = \mathbf{B} \cdot (\mathbf{A} \times \mathbf{A})\,;$$

since **A** × **A** = **0**, we find that

$$\mathbf{A} \cdot (\mathbf{A} \times \mathbf{B}) = \mathbf{B} \cdot \mathbf{0} = 0,$$

which says precisely that **A** × **B** is orthogonal to **A**. Similarly, **A** × **B** is orthogonal to **B**, and Theorem 2 is proved.

Notice that we have distributive laws (7) and (8) and an associative law (4), all of which resemble the usual "laws of algebra." But formula (2) is *not* the usual commutative law; generally, $\mathbf{A} \times \mathbf{B} \neq \mathbf{B} \times \mathbf{A}$ unless the product is $\mathbf{0}$. Moreover, the associative "law" $\mathbf{A} \times (\mathbf{B} \times \mathbf{C}) = (\mathbf{A} \times \mathbf{B}) \times \mathbf{C}$ is *not* valid. (See Problem 7 below.)

Formula (5) gives

$$|\mathbf{A} \times \mathbf{B}|^2 = |\mathbf{A}|^2 \cdot |\mathbf{B}|^2 - |\mathbf{A}|^2 \cdot |\mathbf{B}|^2 \cos^2 \theta$$
$$= |\mathbf{A}|^2 \cdot |\mathbf{B}|^2 (1 - \cos^2 \theta);$$

hence

$$|\mathbf{A} \times \mathbf{B}| = |\mathbf{A}| \cdot |\mathbf{B}| \sin \theta, \tag{9}$$

where θ is the angle between $\mathbf{A}$ and $\mathbf{B}$. (Since $0 \leq \theta \leq \pi$, we have $\sin \theta \geq 0$; hence $\sqrt{1 - \cos^2 \theta}$ equals $\sin \theta$, not $-\sin \theta$.) Geometrically, (9) says that the length of $\mathbf{A} \times \mathbf{B}$ is the area of the parallelogram spanned by $\mathbf{A}$ and $\mathbf{B}$ (Fig. 14).

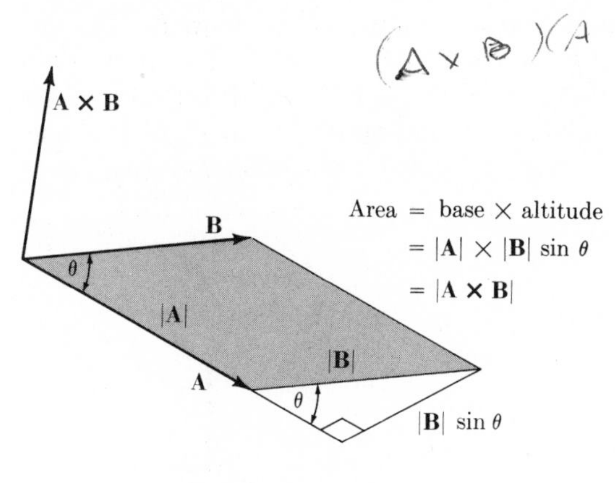

FIGURE 1.14

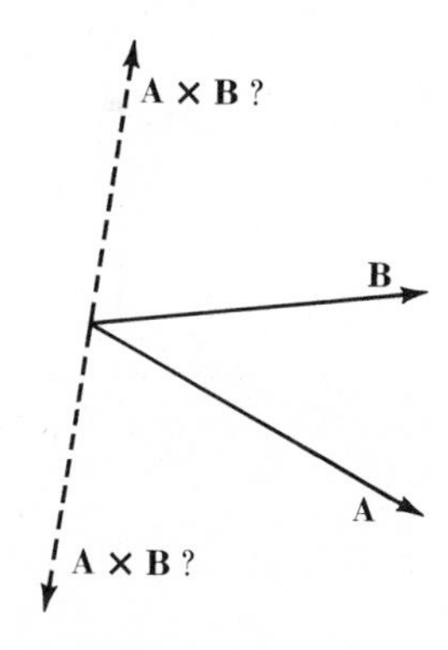

FIGURE 1.15

Now we know that the cross product $\mathbf{A} \times \mathbf{B}$ is perpendicular to both $\mathbf{A}$ and $\mathbf{B}$, and we know its length $|\mathbf{A} \times \mathbf{B}|$. Intuitively, this leaves only two possibilities for $\mathbf{A} \times \mathbf{B}$, as suggested in Fig. 15. Which arrow represents $\mathbf{A} \times \mathbf{B}$ can be determined by a convenient "rule of thumb," whose exact nature depends on the labeling of the coordinate axes.

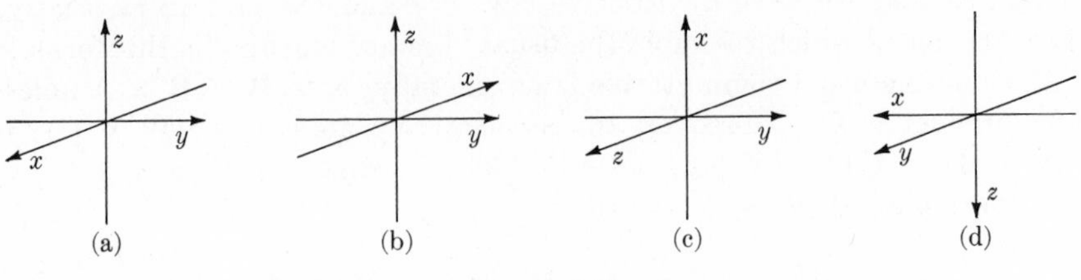

FIGURE 1.16

There are many ways to label the axes (see Fig. 16), but they fall into two main types called *right-hand* and *left-hand.* To tell which type you have, sketch the three basic unit vectors

$$\mathbf{E}_1 = (1,0,0), \qquad \mathbf{E}_2 = (0,1,0), \qquad \mathbf{E}_3 = (0,0,1),$$

which point respectively in the positive direction along the first, second, and third axes (i.e. the x, y, and z axes). Figure 17(a) shows $\mathbf{E}_1$, $\mathbf{E}_2$, and $\mathbf{E}_3$ in what is called a *right-hand* coordinate system. The mysterious hand in the figure explains this name; if you hold the third unit vector $\mathbf{E}_3 = (0,0,1)$ with your *right* hand, fingers pointing from $\mathbf{E}_1$ to $\mathbf{E}_2$, then the thumb points in the same direction as $\mathbf{E}_3$. A *left-hand* system works the same way, but (naturally) with the left hand (see Fig. 17(b)).

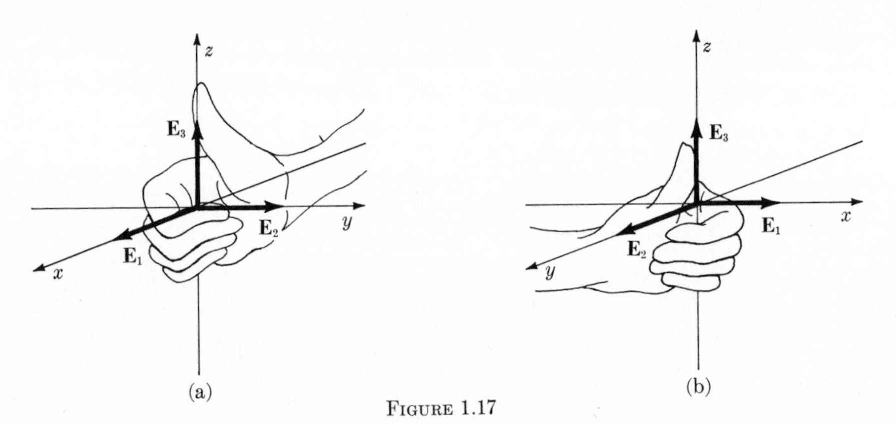

FIGURE 1.17

The rule for visualizing the cross product is that the three vectors **A**, **B**, and $\mathbf{A} \times \mathbf{B}$ are oriented like the three vectors $\mathbf{E}_1$, $\mathbf{E}_2$, $\mathbf{E}_3$. Specifically: In a right-hand coordinate system the cross product is directed so that *if you hold* $\mathbf{A} \times \mathbf{B}$ *with the right hand, with the fingers pointing from* **A** *toward* **B**, *then the thumb points in the direction of* $\mathbf{A} \times \mathbf{B}$ (see Fig. 18). In a left-hand coordinate system, the same method works if you use the left hand.

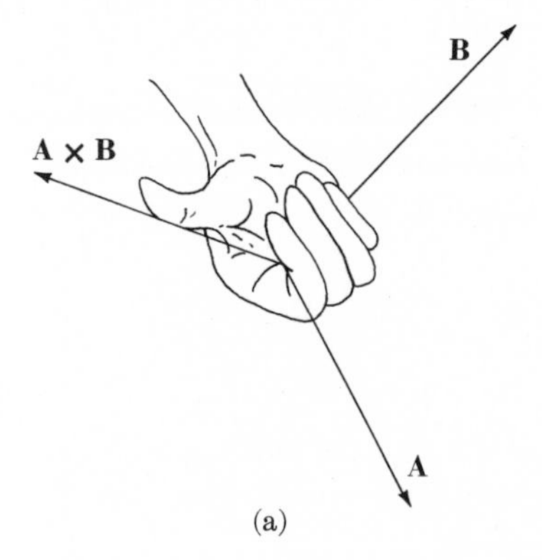

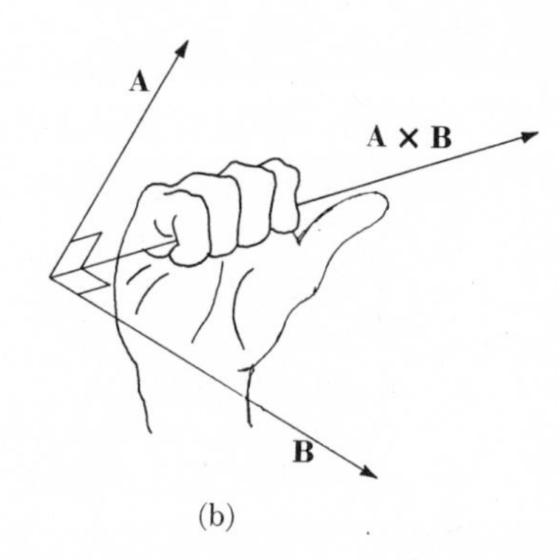

FIGURE 1.18

Please do not expect a mathematical proof of the "right-hand rule." It is just an experimental fact that when we think of vectors as arrows, the cross product works out this way.

Since $|\mathbf{B} \times \mathbf{C}|$ is the area of the parallelogram spanned by $\mathbf{B}$ and $\mathbf{C}$, it follows that $|\mathbf{A}\cdot(\mathbf{B} \times \mathbf{C})|$ is the volume of the parallelepiped spanned by $\mathbf{A}$, $\mathbf{B}$, and $\mathbf{C}$; for, as Fig. 19 shows,

$$\begin{aligned}\mathbf{A}\cdot(\mathbf{B} \times \mathbf{C}) &= |\mathbf{A}|\cdot|\mathbf{B} \times \mathbf{C}| \cos\varphi \\ &= (|\mathbf{B} \times \mathbf{C}|)(|\mathbf{A}| \cos\varphi) \\ &= \pm(\text{area of base})\cdot(\text{altitude}).\end{aligned}$$

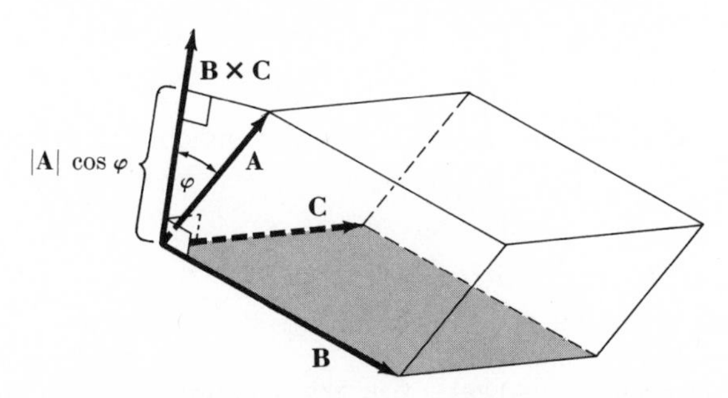

FIGURE 1.19 $|\mathbf{A}\cdot(\mathbf{B} \times \mathbf{C})|$ = area of parallelepiped

We have shown how to visualize everything about the cross product except the complicated formula (1) that defines it. This is easy to remember as a 3×3 determinant

$$\mathbf{A} \times \mathbf{B} = \begin{vmatrix} \mathbf{E}_1 & \mathbf{E}_2 & \mathbf{E}_3 \\ a_1 & a_2 & a_3 \\ b_1 & b_2 & b_3 \end{vmatrix}.$$

In the first row are the basic unit vectors

$$\mathbf{E}_1 = (1,0,0), \qquad \mathbf{E}_2 = (0,1,0), \qquad \mathbf{E}_3 = (0,0,1),$$

and in the second and third rows are the components of **A** and **B** respectively. Expanding by minors of the first row gives

$$\mathbf{A} \times \mathbf{B} = \begin{vmatrix} a_2 & a_3 \\ b_2 & b_3 \end{vmatrix} \mathbf{E}_1 - \begin{vmatrix} a_1 & a_3 \\ b_1 & b_3 \end{vmatrix} \mathbf{E}_2 + \begin{vmatrix} a_1 & a_2 \\ b_1 & b_2 \end{vmatrix} \mathbf{E}_3,$$

which reduces to (1).

In case you are not familiar with determinants, we can easily fill in the necessary background. A 2×2 determinant assigns numbers to square arrays of four numbers. If the square array is

$$\begin{matrix} a & b \\ c & d \end{matrix}$$

then the determinant is

$$\begin{vmatrix} a & b \\ c & d \end{vmatrix} = ad - bc,$$

that is, the product of the entries a and d on one diagonal minus the product of the entries b and c on the other diagonal. For example,

$$\begin{vmatrix} 0 & 1 \\ 1 & 0 \end{vmatrix} = -1, \qquad \begin{vmatrix} 1 & 0 \\ 0 & 1 \end{vmatrix} = 1, \qquad \begin{vmatrix} 1 & 2 \\ 3 & 4 \end{vmatrix} = -2, \qquad \begin{vmatrix} -1 & 5 \\ 0 & 2 \end{vmatrix} = -2.$$

Using 2×2 determinants, the cross product (1) is written

$$\mathbf{A} \times \mathbf{B} = \left(\begin{vmatrix} a_2 & a_3 \\ b_2 & b_3 \end{vmatrix}, -\begin{vmatrix} a_1 & a_3 \\ b_1 & b_3 \end{vmatrix}, \begin{vmatrix} a_1 & a_2 \\ b_1 & b_2 \end{vmatrix} \right).$$

To remember this expression, write the components of **A** and **B** in two rows,

$$\begin{matrix} a_1 & a_2 & a_3 \\ b_1 & b_2 & b_3 \end{matrix} . \tag{10}$$

Then the components of $\mathbf{A} \times \mathbf{B}$ are obtained as follows:

First component: Disregard the *first* column in (10), and take the determinant of the remaining entries:

$$\not{a_1} \begin{vmatrix} a_2 & a_3 \\ b_2 & b_3 \end{vmatrix}$$
$$\not{b_1}$$

Second component: Disregard the *second* column in (10), and take *minus* the determinant of the remaining entries:

$$-\begin{vmatrix} a_1 & \not{a_2} & a_3 \\ b_1 & \not{b_2} & b_3 \end{vmatrix}$$

Third component: Disregard the *third* column in (10), and take the determinant of the remaining entries:

$$\begin{vmatrix} a_1 & a_2 \\ b_1 & b_2 \end{vmatrix} \not{a_3} \; \not{b_3}$$

For example, if **A** = (1,2,3) and **B** = (4,5,6), write the components in two rows,

$$\begin{matrix} 1 & 2 & 3 \\ \\ 4 & 5 & 6 \end{matrix}$$

and following the prescription above, find

$$\begin{aligned} \mathbf{A} \times \mathbf{B} &= \left(\begin{vmatrix} 2 & 3 \\ 5 & 6 \end{vmatrix}, -\begin{vmatrix} 1 & 3 \\ 4 & 6 \end{vmatrix}, \begin{vmatrix} 1 & 2 \\ 4 & 5 \end{vmatrix} \right) \\ &= ((12 - 15), -(6 - 12), (5 - 8)) \\ &= (-3, 6, -3). \end{aligned}$$

As a check, you can verify that $(-3, 6, -3)$ is orthogonal to both **A** and **B**.

This method provides a useful expansion of the "triple product" $\mathbf{A}\cdot(\mathbf{B} \times \mathbf{C})$. We have

$$\mathbf{B} \times \mathbf{C} = \left(\begin{vmatrix} b_2 & b_3 \\ c_2 & c_3 \end{vmatrix}, -\begin{vmatrix} b_1 & b_3 \\ c_1 & c_3 \end{vmatrix}, \begin{vmatrix} b_1 & b_2 \\ c_1 & c_2 \end{vmatrix} \right);$$

hence

$$\mathbf{A}\cdot(\mathbf{B} \times \mathbf{C}) = a_1\begin{vmatrix} b_2 & b_3 \\ c_2 & c_3 \end{vmatrix} - a_2\begin{vmatrix} b_1 & b_3 \\ c_1 & c_3 \end{vmatrix} + a_3\begin{vmatrix} b_1 & b_2 \\ c_1 & c_2 \end{vmatrix}.$$

The expression on the right-hand side is exactly the definition of the 3×3 *determinant*

$$\begin{vmatrix} a_1 & a_2 & a_3 \\ b_1 & b_2 & b_3 \\ c_1 & c_2 & c_3 \end{vmatrix} = a_1\begin{vmatrix} b_2 & b_3 \\ c_2 & c_3 \end{vmatrix} - a_2\begin{vmatrix} b_1 & b_3 \\ c_1 & c_3 \end{vmatrix} + a_3\begin{vmatrix} b_1 & b_2 \\ c_1 & c_2 \end{vmatrix}.$$

Hence the formulas in Theorem 2 can be used to deduce some of the properties of 3×3 determinants. (See Problem 10 below.)

Example 1. Find $(0,-2,-1) \times (-1,-1,1)$. *Solution.* From the array

$$\begin{matrix} 0 & -2 & -1 \\ \\ -1 & -1 & 1 \end{matrix}$$

we obtain the cross product

$$\begin{aligned} (-2\cdot 1 - (-1)\cdot(-1), \; -1\cdot(-1) - 0\cdot 1, \; 0\cdot(-1) - (-2)(-1)) \\ = (-3,1,-2). \end{aligned}$$

Example 2. Show that $\mathbf{P}_1 = (1,2,3)$, $\mathbf{P}_2 = (1,0,2)$, $\mathbf{P}_3 = (0,-1,3)$ and $\mathbf{P}_4 = (0,1,4)$ form the four vertices of a parallelogram, and find its area. *Solution.* The four points are the vertices of a parallelogram, since (Fig. 20) $\mathbf{P}_2 - \mathbf{P}_1 = \mathbf{P}_3 - \mathbf{P}_4 = (0,-2,-1)$. To compute the area, notice that $\mathbf{P}_2 - \mathbf{P}_1$ and $\mathbf{P}_4 - \mathbf{P}_1$ form two adjacent sides of the parallelogram. Thus, referring to Fig. 20, the area of the parallelogram is

$$\begin{aligned}|(\mathbf{P}_2 - \mathbf{P}_1) \times (\mathbf{P}_4 - \mathbf{P}_1)| &= |(0,-2,-1) \times (-1,-1,1)| \\ &= |(-3,1,-2)| = \sqrt{14}\,.\end{aligned}$$

Example 3. Compute the area of the triangle with vertices $\mathbf{P}_1 = (1,5,2)$, $\mathbf{P}_2 = (-1,3,0)$ and $\mathbf{P}_3 = (0,1,4)$. *Solution.* This is half the area of the parallelogram spanned by arrows representing $\mathbf{P}_2 - \mathbf{P}_1$ and $\mathbf{P}_3 - \mathbf{P}_1$, so the area is

$$\begin{aligned}\tfrac{1}{2}|(\mathbf{P}_2 - \mathbf{P}_1) \times (\mathbf{P}_3 - \mathbf{P}_1)| &= \tfrac{1}{2}|(-2,-2,-2) \times (-1,-4,2)| \\ &= \tfrac{1}{2}|(-12,6,6)| = 3|(-2,1,1)| = 3\sqrt{6}\,.\end{aligned}$$

You can check this result by interchanging the vertices; you will find, for instance, that

$$\tfrac{1}{2}|(\mathbf{P}_1 - \mathbf{P}_3) \times (\mathbf{P}_2 - \mathbf{P}_3)| = \tfrac{1}{2}|(-12,6,6)| = 3\sqrt{6}\,.$$

PROBLEMS

The first four problems are routine calculations with the cross product. Problems 5–8 ask you to prove some general formulas for the cross product. Problem 9 proves that $\mathbf{A} \times \mathbf{B} = \mathbf{0}$ if and only if $\mathbf{A}$ and $\mathbf{B}$ are parallel. Problems 10 and 11 concern trigonometry; 12 and 13 concern Cramer's rule and determinants.

1. (a) Check algebraically that

$$\mathbf{E}_1 \times \mathbf{E}_2 = \mathbf{E}_3\,, \qquad \mathbf{E}_2 \times \mathbf{E}_3 = \mathbf{E}_1\,, \qquad \mathbf{E}_3 \times \mathbf{E}_1 = \mathbf{E}_2\,.$$

(b) Find $\mathbf{E}_1 \times \mathbf{E}_3$.

(c) Sketch a right-hand coordinate system, and check the products in (a) and (b) visually by the right-hand rule.

2. In the following cases, compute $\mathbf{A} \times \mathbf{B}$ and $(\mathbf{A} \times \mathbf{B})\cdot\mathbf{C}$.

(a) $\mathbf{A} = (1,2,0)$, $\mathbf{B} = (-3,1,0)$, $\mathbf{C} = (4,9,-3)$. (Here $(\mathbf{A} \times \mathbf{B})\cdot\mathbf{C} = -21$.)

(b) $\mathbf{A} = (-3,1,-2)$, $\mathbf{B} = (2,0,4)$, $\mathbf{C} = (1,1,1)$. (Here $(\mathbf{A} \times \mathbf{B})\cdot\mathbf{C} = 10$.)

3. (a) Find the area of the parallelogram whose vertices are at the points $\mathbf{P}_1 = (0,0,0)$, $\mathbf{P}_2 = (1,1,1)$, $\mathbf{P}_3 = (2,0,0)$, and $\mathbf{P}_4 = (1,-1,-1)$. (Notice that $\mathbf{P}_3 = \mathbf{P}_2 + \mathbf{P}_4$; hence this parallelogram is spanned by arrows representing $\mathbf{P}_2 = (1,1,1)$ and $\mathbf{P}_4 = (1,-1,-1)$.)

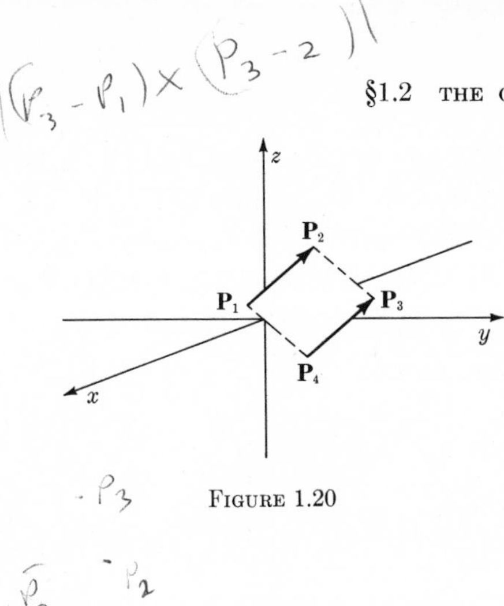

FIGURE 1.20

(b) Find the area of the triangle with vertices (0,0,0), (1,1,1), and (1,−1,−1).

4. Four points $\mathbf{P}_1$, $\mathbf{P}_2$, $\mathbf{P}_3$, $\mathbf{P}_4$, taken in order, form the vertices of a parallelogram if and only if $\mathbf{P}_2 - \mathbf{P}_1 = \mathbf{P}_3 - \mathbf{P}_4$ (see Fig. 20). Prove that $\mathbf{P}_1 = (1,1,1)$, $\mathbf{P}_2 = (1,2,2)$, $\mathbf{P}_3 = (2,3,2)$, and $\mathbf{P}_4 = (2,2,1)$ are the vertices of a parallelogram, and compute its area. (Hint: The vectors $\mathbf{P}_2 - \mathbf{P}_1$, $\mathbf{P}_3 - \mathbf{P}_2$, $\mathbf{P}_4 - \mathbf{P}_3$, $\mathbf{P}_1 - \mathbf{P}_4$ correspond to the sides of the parallelogram, as in Fig. 20.)

5. In Theorem 2 prove:
(a) Formula (4)
(b) Formula (5)
(c) Formula (6)
(d) Formula (7)
(e) Formula (8) (Hint: Instead of expanding both sides, you can use (2) and (7).)

6. Prove the following equalities, noticing particularly those with minus signs. (None of the proofs require expansion; they should be based on formulas (2)–(8), plus properties of the inner product.)
(a) $(\mathbf{A} \times \mathbf{B})\cdot\mathbf{C} = \mathbf{C}\cdot(\mathbf{A} \times \mathbf{B})$
(b) $(\mathbf{A} \times \mathbf{B})\cdot\mathbf{C} = \mathbf{A}\cdot(\mathbf{B} \times \mathbf{C})$
(c) $(\mathbf{A} \times \mathbf{B})\cdot\mathbf{C} = -(\mathbf{A} \times \mathbf{C})\cdot\mathbf{B}$
(d) $(\mathbf{A} \times \mathbf{B})\cdot\mathbf{C} = -\mathbf{A}\cdot(\mathbf{C} \times \mathbf{B})$

7. Prove that $\mathbf{E}_1 \times (\mathbf{E}_1 \times \mathbf{E}_2) \neq (\mathbf{E}_1 \times \mathbf{E}_1) \times \mathbf{E}_2$.

8. (a) Prove the *Lagrange identity:*

$$(\mathbf{A} \times \mathbf{B})\cdot(\mathbf{C} \times \mathbf{D}) = (\mathbf{A}\cdot\mathbf{C})(\mathbf{B}\cdot\mathbf{D}) - (\mathbf{A}\cdot\mathbf{D})(\mathbf{B}\cdot\mathbf{C}).$$

(b) Deduce the formula for $|\mathbf{A} \times \mathbf{B}|^2$ from part (a).

9. (a) Prove that if $\mathbf{A}$ and $\mathbf{B}$ are parallel, then $\mathbf{A} \times \mathbf{B} = \mathbf{0}$.
(b) Suppose that $\mathbf{B} \neq \mathbf{0}$. Prove that if $\mathbf{A} \times \mathbf{B} = \mathbf{0}$, then $\mathbf{A} = t\mathbf{B}$ for some constant t.
(c) Prove that $\mathbf{A} \times \mathbf{B} = \mathbf{0}$ if and only if $\mathbf{A}$ and $\mathbf{B}$ are parallel.

10. (a) Prove that $\mathbf{A} \times (\mathbf{A} + \mathbf{B}) = \mathbf{A} \times \mathbf{B}$.
(b) Prove the *law of sines:*

$$\frac{|\mathbf{A}|}{\sin \alpha} = \frac{|\mathbf{B}|}{\sin \beta} = \frac{|\mathbf{C}|}{\sin \gamma}.$$

(See Fig. 21; use formula (9) to compute $\sin \alpha$, etc.)

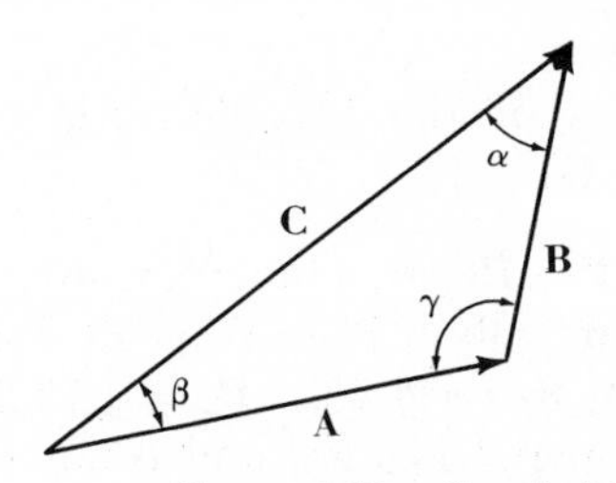

FIGURE 1.21 $\mathbf{C} = \mathbf{A} + \mathbf{B}$

11. Let $\mathbf{A}$ and $\mathbf{B}$ be any vectors in R^3, and let α, β, and γ be the angles indicated in Fig. 21. Prove that $\cos(\alpha + \beta + \gamma) = -1$. (Hint: Prove that $\cos(\alpha + \beta) = -\cos\gamma$ and $\sin(\alpha + \beta) = \sin\gamma$. Use the addition formula for $\cos(\alpha + \beta)$, etc., and formulas (9) and (5) above.)

12. This problem deduces half of *Cramer's rule* for three simultaneous linear equations. Given $a_1, b_1, c_1, d_1, a_2, \ldots, d_3$, the system of equations

$$ra_1 + sb_1 + tc_1 = d_1$$

$$ra_2 + sb_2 + tc_2 = d_2$$

$$ra_3 + sb_3 + tc_3 = d_3 \tag{11}$$

can be written as a single vector equation:

$$r\mathbf{A} + s\mathbf{B} + t\mathbf{C} = \mathbf{D}. \tag{12}$$

(a) Take the inner product of both sides of (12) with $\mathbf{B} \times \mathbf{C}$, and show that

$$r(\mathbf{A}\cdot(\mathbf{B} \times \mathbf{C})) = \mathbf{D}\cdot(\mathbf{B} \times \mathbf{C}).$$

(b) Show that

$$s(\mathbf{B}\cdot(\mathbf{A}\times\mathbf{C})) = \mathbf{D}\cdot(\mathbf{A}\times\mathbf{C})$$

and

$$t(\mathbf{C}\cdot(\mathbf{A}\times\mathbf{B})) = \mathbf{D}\cdot(\mathbf{A}\times\mathbf{B}).$$

(c) Deduce that if $\mathbf{A}\cdot(\mathbf{B}\times\mathbf{C}) \neq 0$, then the only possible solution of (11) for r, s, and t is

$$r = \frac{\mathbf{D}\cdot(\mathbf{B}\times\mathbf{C})}{\mathbf{A}\cdot(\mathbf{B}\times\mathbf{C})}$$

$$s = \frac{\mathbf{A}\cdot(\mathbf{D}\times\mathbf{C})}{\mathbf{A}\cdot(\mathbf{B}\times\mathbf{C})}$$

$$t = \frac{\mathbf{A}\cdot(\mathbf{B}\times\mathbf{D})}{\mathbf{A}\cdot(\mathbf{B}\times\mathbf{C})}.$$

(In this problem you have *not* proved that these values of r, s, and t actually solve the equation (11); that is the other half of Cramer's rule. Its proof is outlined in §1.5, Problem 21.)

13. Let D stand for the determinant

$$D = \begin{vmatrix} a_1 & a_2 & a_3 \\ b_1 & b_2 & b_3 \\ c_1 & c_2 & c_3 \end{vmatrix} = \mathbf{A}\cdot(\mathbf{B}\times\mathbf{C}).$$

Use Theorem 2 to prove the following:

(a) $$\begin{vmatrix} b_1 & b_2 & b_3 \\ c_1 & c_2 & c_3 \\ a_1 & a_2 & a_3 \end{vmatrix} = D$$

(b) $$\begin{vmatrix} b_1 & b_2 & b_3 \\ a_1 & a_2 & a_3 \\ c_1 & c_2 & c_3 \end{vmatrix} = -D$$

(c) $$\begin{vmatrix} a_1 & a_2 & a_3 \\ b_1+ta_1 & b_2+ta_2 & b_3+ta_3 \\ c_1 & c_2 & c_3 \end{vmatrix} = D$$

(d) $$\begin{vmatrix} a_1+tc_1 & a_2+tc_2 & a_3+tc_3 \\ b_1 & b_2 & b_3 \\ c_1 & c_2 & c_3 \end{vmatrix} = D$$

1.3 SPHERES, PLANES, AND LINES

We are ready to define the elementary geometric figures in R^3.

The *sphere of radius r with center at a given point* $\mathbf{P}_0$ is the set of all points $\mathbf{P}$ such that $|\mathbf{P} - \mathbf{P}_0| = r$ (Fig. 22). This set is denoted

$$\{\mathbf{P}: |\mathbf{P} - \mathbf{P}_0| = r\}$$

or equivalently

$$\{(x,y,z): (x - x_0)^2 + (y - y_0)^2 + (z - z_0)^2 = r^2\}.$$

The *closed ball* of radius r about $\mathbf{P}_0$ is the set

$$\{\mathbf{P}: |\mathbf{P} - \mathbf{P}_0| \leq r\},$$

and the set

$$\{\mathbf{P}: |\mathbf{P} - \mathbf{P}_0| < r\}$$

is the *open ball* of radius r about $\mathbf{P}_0$. A closed ball (like a closed interval) contains all its boundary points, and an open ball contains none of them.

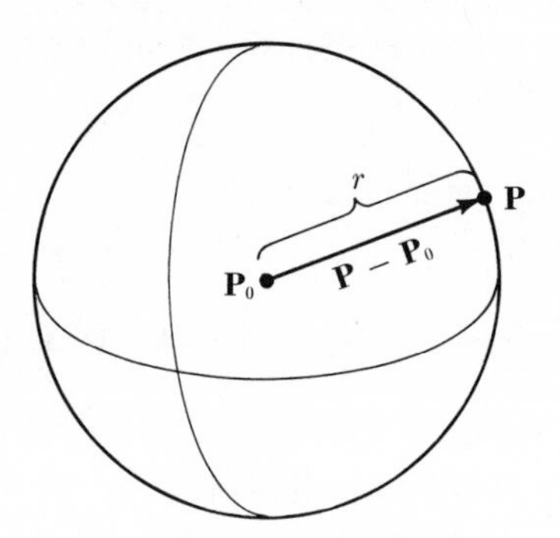

FIGURE 1.22

A *plane* is by definition any set of the form

$$\{\mathbf{P}: \mathbf{N}\cdot\mathbf{P} = d\} \tag{1}$$

where $\mathbf{N}$ is a given nonzero vector and d is a given number. If $\mathbf{N} = (a,b,c)$, then the condition $\mathbf{N}\cdot\mathbf{P} = d$ takes the form

$$ax + by + cz = d; \tag{2}$$

this is the *equation* of the plane (1). For example,

$\{(x,y,z): x = 0\}$ is the yz plane (Fig. 23)

$\{(x,y,z): x = 1\}$ is a plane parallel to the yz plane (Fig. 24)

$\{(x,y,z): z = 0\}$ is the xy plane (Fig. 25)

$\{(x,y,z): y = z\}$ is a plane containing the x axis and bisecting the y and z axes. (Fig. 26)

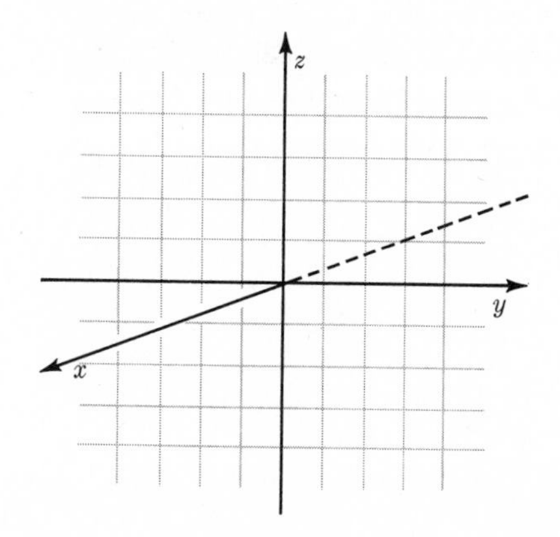

FIGURE 1.23

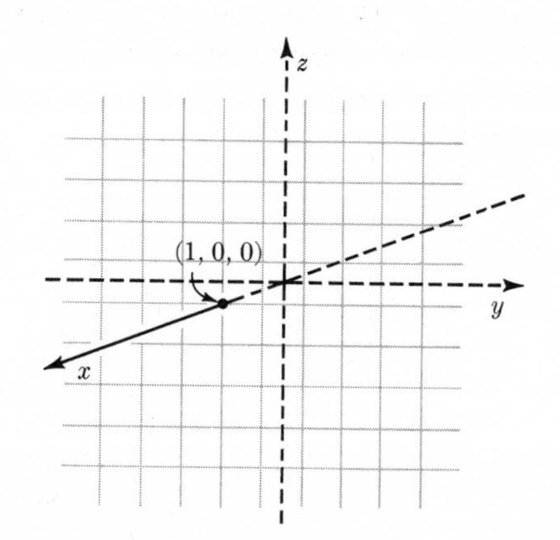

FIGURE 1.24

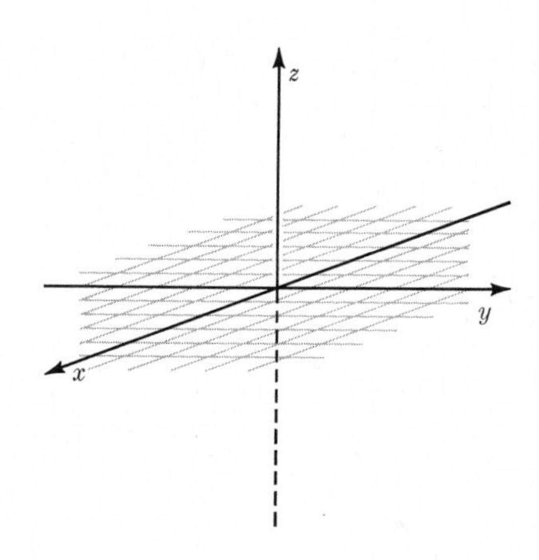

FIGURE 1.25

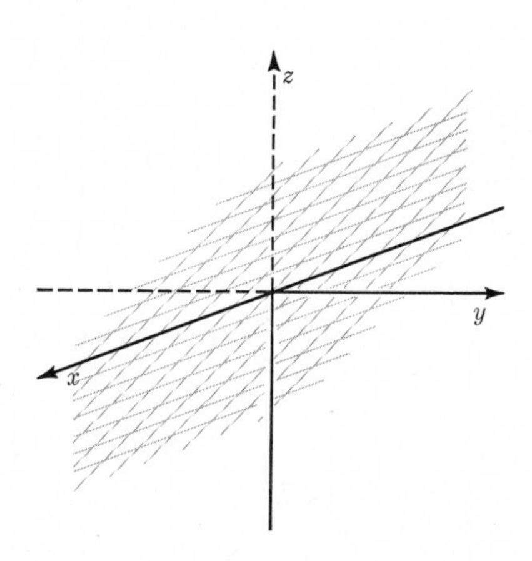

FIGURE 1.26

Let $\mathbf{P}_0$ be any point on the plane (1). Then $\mathbf{N}\cdot\mathbf{P}_0 = d$, so the equation $\mathbf{N}\cdot\mathbf{P} = d$ can be written

$$\mathbf{N}\cdot\mathbf{P} = \mathbf{N}\cdot\mathbf{P}_0$$

or

$$\mathbf{N}\cdot(\mathbf{P} - \mathbf{P}_0) = 0. \tag{3}$$

Thus the plane consists of all points $\mathbf{P}$ such that $\mathbf{P} - \mathbf{P}_0$ is orthogonal to $\mathbf{N}$ (see Fig. 27). The vector $\mathbf{N}$ is called a *normal vector* to the plane. Equation (3) is an algebraic version of a traditional Euclidean characterization: the set of all lines through a given point $\mathbf{P}_0$ and orthogonal to a given line through $\mathbf{P}_0$ is a plane.

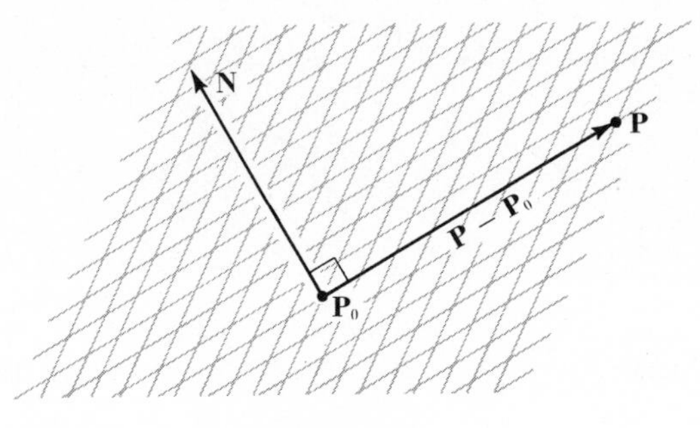

FIGURE 1.27

Lines were introduced in §1.1; by definition, a line is any set of the form

$$\{\mathbf{P}: \mathbf{P} = \mathbf{A} + t\mathbf{B} \text{ for some real number } t\}, \tag{4}$$

where $\mathbf{B}$ is assumed to be nonzero. The equation $\mathbf{P} = \mathbf{A} + t\mathbf{B}$ is called a *parametric equation of the line,* with t as parameter. You can think of t as time, and imagine the line as traced by a point moving with constant speed and direction; at time $t = 0$ the point is at $\mathbf{A}$, at $t = 1$ it is at $\mathbf{A} + \mathbf{B}$, at $t = -1$ it was at $\mathbf{A} - \mathbf{B}$, and so on. (See Fig. 28.)

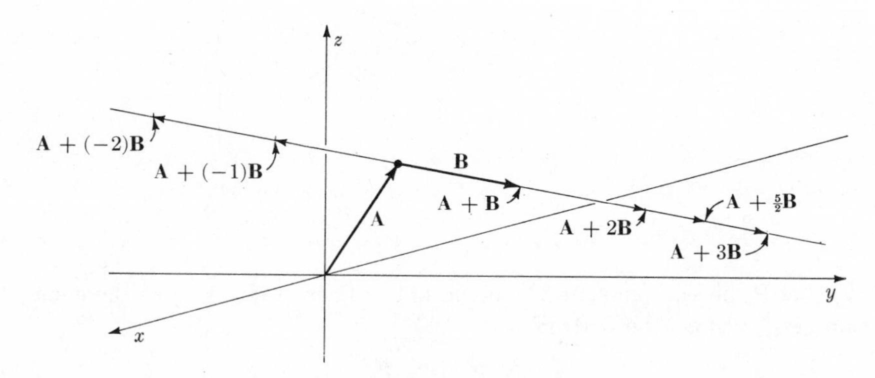

FIGURE 1.28

Given any two distinct points $\mathbf{P}_1$ and $\mathbf{P}_2$, there is a line

$$\{\mathbf{P}: \mathbf{P} = \mathbf{P}_1 + t(\mathbf{P}_2 - \mathbf{P}_1) \text{ for some } t\}$$

which contains both $\mathbf{P}_1$ (set $t = 0$) and $\mathbf{P}_2$ (set $t = 1$). (See Fig. 29.) The points obtained with $0 \leq t \leq 1$ form the *line segment* from $\mathbf{P}_1$ to $\mathbf{P}_2$; the *midpoint* of the segment, obtained with $t = \frac{1}{2}$, is the point

$$\mathbf{M} = \mathbf{P}_1 + \tfrac{1}{2}(\mathbf{P}_2 - \mathbf{P}_1) = \tfrac{1}{2}(\mathbf{P}_1 + \mathbf{P}_2).$$

The term "midpoint" is justified since $|\mathbf{P}_1 - \mathbf{M}| = |\mathbf{P}_2 - \mathbf{M}|$, as you can easily check.

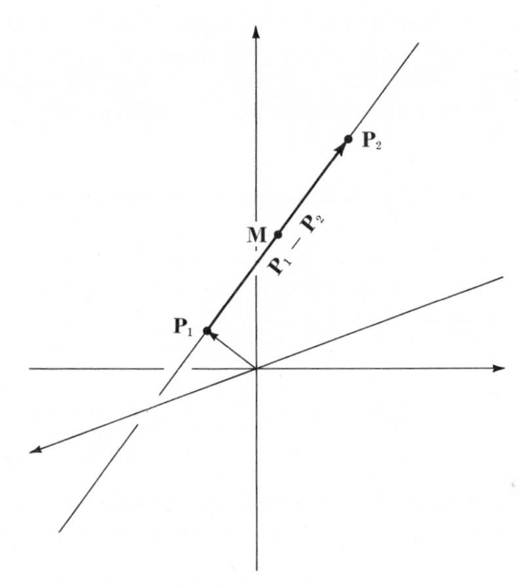

FIGURE 1.29

Now that we have defined spheres, planes, and lines, we could prove algebraically all the classical axioms and theorems of Euclidean solid geometry, but that would be a long and somewhat irrelevant detour. The main objective is to understand the translation of geometric statements into vector language and vice versa, and to be familiar with algebraically based proofs. This is the point of the following examples and problems.

Example 1. The y axis is defined as the set $\{(x,y,z): x = 0 \text{ and } z = 0\}$. Prove that the plane $\{(x,y,z): x + 3z = 0\}$ contains the y axis. *Solution.* If (x,y,z) is any point on the y axis, then $x = 0$ and $z = 0$; hence $x + 3z = 0$, and (x,y,z) is in the given plane.

Example 2. Do the following two lines intersect?

$$\{\mathbf{P}: \mathbf{P} = (0,1,-1) + t(-3,4,5)\} \quad \text{and} \quad \{\mathbf{Q}: \mathbf{Q} = (0,-1,1) + t(4,3,2)\}.$$

Solution. The lines might intersect in a point **R** even though **P** and **Q** do not "arrive at **R** at the same time," so we have to use different parameters t and s for the two lines. The question is, are there numbers t and s such that

$$(0,1,-1) + t(-3,4,5) = (0,-1,1) + s(4,3,2)\ ?$$

This vector equation is equivalent to the three simultaneous equations obtained by equating components:

$$\begin{aligned} 0 - 3t &= 0 + 4s \\ 1 + 4t &= -1 + 3s \\ -1 + 5t &= 1 + 2s. \end{aligned} \tag{5}$$

The first equation says that $s = -3t/4$, and then the second gives $1 + 4t = -1 - 9t/4$, or $t = -8/25$. Thus, if the equations (5) have any solution at all, it must be $t = -8/25$ and $s = -3t/4 = 6/25$. But these values do *not* satisfy the third equation, so there is no solution. Hence the lines do not intersect.

Example 3. Find the equation of a plane through the three points $\mathbf{P}_0 = (0,0,0)$, $\mathbf{P}_1 = (1,1,1)$, $\mathbf{P}_2 = (-1,1,1)$.

First Method. We use the definition of a plane as the set with equation $ax + by + cz = d$. The constants a, b, c, d must be chosen so that the three given points lie on the plane: thus

$$\begin{aligned} a\cdot 0 + b\cdot 0 + c\cdot 0 &= d && (\mathbf{P}_0 \text{ is on the plane}) \\ a + b + c &= d && (\mathbf{P}_1 \text{ is on the plane}) \\ a\cdot(-1) + b + c &= d && (\mathbf{P}_2 \text{ is on the plane}). \end{aligned}$$

These are three homogeneous equations in four unknowns, so they have lots of solutions. One solution is $a = 0$, $b = 1$, $c = -1$, $d = 0$; hence

$$\{(x,y,z): y - z = 0\} \tag{6}$$

is a plane containing $\mathbf{P}_0$, $\mathbf{P}_1$, and $\mathbf{P}_2$.

Second Method. We use equation (3). A point $\mathbf{P}_0$ is already given, so we only have to find a normal **N**. Since $\mathbf{P}_1 - \mathbf{P}_0$ and $\mathbf{P}_2 - \mathbf{P}_0$ should both be perpendicular to **N**, we take **N** to be their cross product:

$$\mathbf{N} = (\mathbf{P}_1 - \mathbf{P}_0) \times (\mathbf{P}_2 - \mathbf{P}_0) = (1,1,1) \times (-1,1,1) = (0,-2,2).$$

Thus the desired plane has the equation $\mathbf{N}\cdot(\mathbf{P} - \mathbf{P}_0) = 0$, which with $\mathbf{N} = (0,-2,2)$, $\mathbf{P} = (x,y,z)$, $\mathbf{P}_0 = (0,0,0)$ reduces to $-2y + 2z = 0$. This is not exactly the same as the equation we found in (6), $y - z - 0$; but the two equations have the same solutions, so they define the same plane. It is sketched in Fig. 26.

Example 4. The line $\{\mathbf{P}: \mathbf{P} = \mathbf{A} + t\mathbf{B}\}$ is called *orthogonal* to the plane $\{\mathbf{P}: \mathbf{P}\cdot\mathbf{N} = d\}$ if and only if **B** and **N** are parallel. Find a line passing through $(-1,3,2)$ and orthogonal to the plane with equation

$$x + 2y - z = 5. \tag{7}$$

Solution. The normal to the plane can be read off from the coefficients of x, y, and z in equation (7); it is $\mathbf{N} = (1,2-1)$. Hence the desired line is

$$\{\mathbf{P}: \mathbf{P} = (-1,3,2) + t(1,2,-1)\}.$$

PROBLEMS

The first twelve problems give plenty of routine practice with lines and planes. The remaining eleven problems illustrate the vector approach to solid geometry: tangents to a sphere, distance from a point to a plane, a line as the intersection of two planes, two points determine a line, direction cosines, triangles, tetrahedra, cones.

1. Find parametric equations of lines through the following pairs of points, and find the midpoint of the segment between each pair.
(a) $(-5,-6,8)$ and $(1,3,7)$
(b) $(2,4,6)$ and $(1,2,3)$
(c) $(1,3,10)$ and $(-3,6,-2)$
(d) $(10,3,1)$ and $(6,-2,-3)$

2. For each of the following equations, find a normal vector to the corresponding plane, and find any point $\mathbf{P}_0$ on the plane.
(a) $x + y + z = 1$
(b) $2x + 3y - z = 2$
(c) $(x - 2) + 3(y - 5) - 4(z + 1) = 0$

3. Find an equation of a plane through the three given points $\mathbf{P}_0$, $\mathbf{P}_1$, $\mathbf{P}_2$.
(a) $\mathbf{P}_0 = (1,0,0)$, $\mathbf{P}_1 = (0,1,0)$, $\mathbf{P}_2 = (0,0,1)$
(b) $\mathbf{P}_0 = (1,0,0)$, $\mathbf{P}_1 = (-1,0,0)$, $\mathbf{P}_2 = (0,1,0)$
(c) $\mathbf{P}_0 = (0,1,0)$, $\mathbf{P}_1 = (0,2,0)$, $\mathbf{P}_2 = (0,0,-1)$

4. (a) Prove (algebraically) that the plane in Problem 3(b) contains the x axis, the set $\{(x,y,z): y = 0 \text{ and } z = 0\}$. Sketch the three points in 3(b), and sketch the plane.
(b) Prove (algebraically) that the plane in Problem 3(c) contains the y axis. Sketch the three points in 3(c), and sketch the plane.
(c) Prove that the plane in Problem 3(a) does not contain the origin.

5. (a) Prove that the x axis is a line, by finding parametric equations for it.
(b) Do the same for the y axis and z axis.

6. Find a point of intersection of the following two lines:

$$\{\mathbf{P}: \mathbf{P} = (1,-5,2) + t(-3,4,0)\} \quad \text{and} \quad \{\mathbf{P}: \mathbf{P} = (3,-13,1) + t(4,0,1)\}.$$

7. Prove that the line $\{\mathbf{P}: \mathbf{P} = (1,3,-1) + t(0,3,5)\}$ lies entirely in the plane $\{(x,y,z): 2x - 5y + 3z = -16\}$.

8. Suppose that $\mathbf{P}_1$ and $\mathbf{P}_2$ lie on the plane $\{\mathbf{P}: \mathbf{P}\cdot\mathbf{N} = d\}$. Prove that every point of the line $\{\mathbf{P}: \mathbf{P} = \mathbf{P}_1 + t(\mathbf{P}_2 - \mathbf{P}_1)\}$ lies in the given plane.

9. (a) The two lines $\{\mathbf{P}: \mathbf{P} = (-2,4,6) + t(1,2,3)\}$ and $\{\mathbf{P}: \mathbf{P} = (-2,4,6) + t(3,2,1)\}$ intersect in the point $\mathbf{P}_0 = (-2,4,6)$. Find the equation of a plane containing the two lines. (Since you have $\mathbf{P}_0$, all you need is the normal $\mathbf{N}$.)

(b) Suppose that $\mathbf{B}$ and $\mathbf{C}$ are nonzero and not parallel. Prove that there is a plane containing the two lines $\{\mathbf{P}: \mathbf{P} = \mathbf{A} + t\mathbf{B}\}$ and $\{\mathbf{P}: \mathbf{P} = \mathbf{A} + t\mathbf{C}\}$.

10. Find all points of intersection of the given line and the given plane.

(a) $\{\mathbf{P}: \mathbf{P} = t(1,-3,6)\}$ and $\{\mathbf{P}: x + 3y + z = 2\}$

(b) $\{\mathbf{P}: \mathbf{P} = (1,-3,6) + t(1,0,0)\}$ and $\{\mathbf{P}: z = 6\}$

(c) $\{\mathbf{P}: \mathbf{P} = (1,-3,6) + t(1,0,0)\}$ and $\{\mathbf{P}: z = 0\}$

11. (a) Prove that if $\mathbf{B}\cdot\mathbf{N} \neq 0$, then the line $\{\mathbf{P}: \mathbf{P} = \mathbf{A} + t\mathbf{B}\}$ intersects the plane $\{\mathbf{P}: (\mathbf{P} - \mathbf{P}_0)\cdot\mathbf{N} = 0\}$ in exactly one point. (See Problem 10(a).)

(b) Prove that if $\mathbf{B}\cdot\mathbf{N} = 0$ and $\mathbf{A}$ is on the plane $\{\mathbf{P}: (\mathbf{P} - \mathbf{P}_0)\cdot\mathbf{N} = 0\}$, then the entire line $\{\mathbf{P}: \mathbf{P} = \mathbf{A} + t\mathbf{B}\}$ lies in the plane. (See Problem 10(b).)

(c) Prove that if $\mathbf{B}\cdot\mathbf{N} = 0$ and $\mathbf{A}$ is *not* on the plane

$$\{\mathbf{P}: (\mathbf{P} - \mathbf{P}_0)\cdot\mathbf{N} = 0\},$$

then the line $\{\mathbf{P}: \mathbf{P} = \mathbf{A} + t\mathbf{B}\}$ does not intersect the plane in any point. (See Problem 10(c).)

(d) Draw sketches illustrating parts (a), (b), and (c). (The line is called *parallel* to the plane when $\mathbf{B}\cdot\mathbf{N} = 0$. You have proved a *theorem*: A line L and a plane Π intersect in a single point if and only if they are not parallel.)

12. The line $\{\mathbf{P}: \mathbf{P} = \mathbf{A} + t\mathbf{B}\}$ is called *normal* to the plane

$$\{\mathbf{P}: (\mathbf{P} - \mathbf{P}_0)\cdot\mathbf{N} = 0\}$$

if and only if $\mathbf{B}$ is parallel to $\mathbf{N}$.

(a) Find a line through the point $\mathbf{P}_1 = (-5,2,1)$ and normal to the plane $\{(x,y,z): x = y\}$.

(b) Find the intersection $\mathbf{P}_2$ of the line and plane in part (a).

(c) Show that $|\mathbf{P}_2 - \mathbf{P}_1| = (\mathbf{P}_1 - \mathbf{P}_0)_{\mathbf{N}}$, where $\mathbf{P}_0 = \mathbf{0}$.

(d) Sketch the plane, $\mathbf{N}$, $\mathbf{P}_0$, $\mathbf{P}_1$, and $\mathbf{P}_2$.

13. (a) Find a line through a given point $\mathbf{P}_1 = (x_1,y_1,z_1)$ and normal to the plane $\{(x,y,z): ax + by + cz = d\}$.

(b) Find the point $\mathbf{P}_0$ in which the line and plane in part (a) intersect.

(c) Prove that if $\mathbf{P}$ is any point on the plane in part (a), and $\mathbf{P}_0$ is the point found in (b), then $\mathbf{P}_1 - \mathbf{P}_0$ is orthogonal to $\mathbf{P} - \mathbf{P}_0$. Sketch this situation.

(d) Continuing part (c), prove that if $\mathbf{P}$ is any point on the plane, then $|\mathbf{P}_1 - \mathbf{P}|^2 \geq |\mathbf{P}_1 - \mathbf{P}_0|^2$. (Hint: Use the Pythagorean theorem.)

(e) Part (d) shows that $|\mathbf{P}_1 - \mathbf{P}_0|$ is the distance from the point $\mathbf{P}_1$ to the plane $\{(x,y,z): ax + by + cz = d\}$. Show that this distance is

$$\frac{|ax_1 + by_1 + cz_1 - d|}{\sqrt{a^2 + b^2 + c^2}}.$$

14. Let $\mathbf{P}_0$ be a point on the sphere $\{\mathbf{P}: |\mathbf{P}| = r\}$. Prove that the line $\{\mathbf{P}: \mathbf{P} = \mathbf{P}_0 + t\mathbf{B}\}$ intersects the sphere in two distinct points unless $\mathbf{B} \cdot \mathbf{P}_0 = 0$.

15. (a) Suppose that $\mathbf{P}_0$ lies on the sphere

$$\{\mathbf{P}: |\mathbf{P}| = r\}. \tag{8}$$

Prove that the plane $\{\mathbf{P}: (\mathbf{P} - \mathbf{P}_0) \cdot \mathbf{P}_0 = 0\}$ intersects the sphere *only* at $\mathbf{P}_0$. (Hint: Assuming $\mathbf{P}$ is on the plane, you can compute $|\mathbf{P}|$ by setting $\mathbf{P} = \mathbf{P}_0 + (\mathbf{P} - \mathbf{P}_0)$ and using the Pythagorean theorem.)

(b) Suppose that $\mathbf{P}_0$ lies on the sphere (8), and

$$\{\mathbf{P}: (\mathbf{P} - \mathbf{P}_0) \cdot \mathbf{N} = 0\} \tag{9}$$

is a plane through $\mathbf{P}_0$. Prove that there is a point $\mathbf{P}_1$ which lies on the sphere (8), on the plane (9), and on the line

$$\{\mathbf{P}: \mathbf{P} = -\mathbf{P}_0 + t\mathbf{N}\}. \qquad \text{(See Fig. 30.)}$$

(c) In part (b), prove that $\mathbf{P}_1 = \mathbf{P}_0$ if and only if $\mathbf{N}$ is parallel to $\mathbf{P}_0$.

(d) Conclude that the plane (9) intersects the sphere (8) in more than one point, unless $\mathbf{P}_0$ is normal to the plane (9). (When $\mathbf{P}_0$ is normal to it, the plane is called *tangent* to the sphere.)

16. Let $\mathbf{A} = (1,8,2)$ and $\mathbf{B} = (-3,1,1)$.

(a) Write out the equation $\mathbf{P} = \mathbf{A} + t\mathbf{B}$ by components, setting $\mathbf{P} = (x,y,z)$.

(b) Show that

$$\mathbf{P} = \mathbf{A} + t\mathbf{B} \quad \text{for some } t$$

if and only if

$$x = 7 - 3z \qquad \text{and} \qquad y = 6 + z.$$

(Hint: Eliminate t from the equation in part (a). Part (b) proves that the line $\{\mathbf{P}: \mathbf{P} = \mathbf{A} + t\mathbf{B}\}$ is the intersection of two planes.)

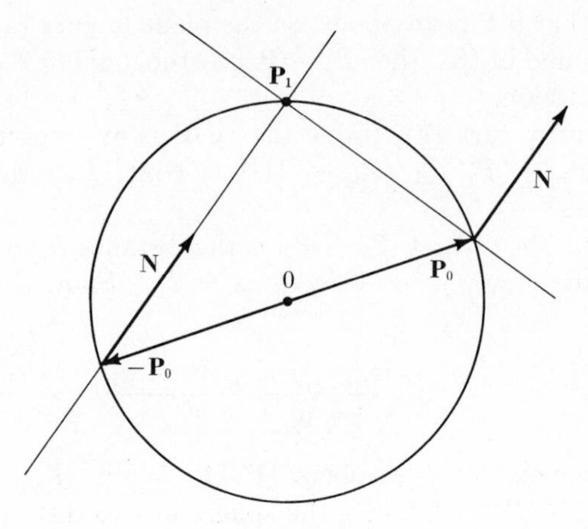

FIGURE 1.30

17. Prove that every line is the intersection of two planes. (See the previous problem.)

18. Given two distinct points $\mathbf{P}_1$ and $\mathbf{P}_2$, prove that the set

$$\{\mathbf{P}: |\mathbf{P} - \mathbf{P}_1| = |\mathbf{P} - \mathbf{P}_2|\}$$

is a plane containing $\frac{1}{2}(\mathbf{P}_1 + \mathbf{P}_2)$, with normal $\mathbf{P}_2 - \mathbf{P}_1$. Draw an appropriate sketch.

19. The *direction cosines* of the line $\{\mathbf{P}: \mathbf{P} = \mathbf{A} + t\mathbf{B}\}$ are the cosines of the angles between $\mathbf{B}$ and the three basic unit vectors $\mathbf{E}_1$, $\mathbf{E}_2$, $\mathbf{E}_3$, as shown in Fig. 31.

(a) Show that

$$\cos\alpha = \frac{b_1}{|\mathbf{B}|}, \qquad \cos\beta = \frac{b_2}{|\mathbf{B}|}, \qquad \cos\gamma = \frac{b_3}{|\mathbf{B}|}.$$

(b) Show that if $\mathbf{P}_1 = (x_1,y_1,z_1)$ and $\mathbf{P}_2 = (x_2,y_2,z_2)$ are any two points on the line $\{\mathbf{P}: \mathbf{P} = \mathbf{A} + t\mathbf{B}\}$, then the direction cosines are either

$$\cos\alpha = \frac{x_1 - x_2}{|\mathbf{P}_1 - \mathbf{P}_2|}, \qquad \cos\beta = \frac{y_1 - y_2}{|\mathbf{P}_1 - \mathbf{P}_2|}, \qquad \cos\gamma = \frac{z_1 - z_2}{|\mathbf{P}_1 - \mathbf{P}_2|},$$

or

$$\cos\alpha = \frac{x_2 - x_1}{|\mathbf{P}_2 - \mathbf{P}_1|}, \qquad \cos\beta = \frac{y_2 - y_1}{|\mathbf{P}_2 - \mathbf{P}_1|}, \qquad \cos\gamma = \frac{z_2 - z_1}{|\mathbf{P}_2 - \mathbf{P}_1|}.$$

(c) Find direction cosines of a line through the points $(-8,3,5)$ and $(6,1,0)$.

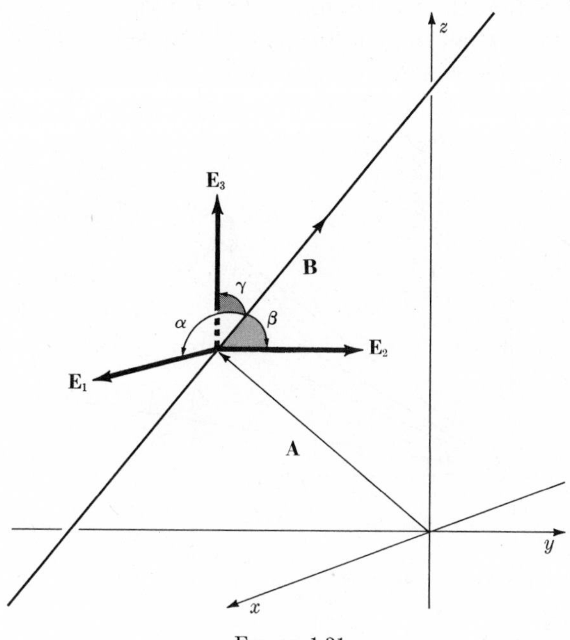

FIGURE 1.31

20. The *triangle* with three given vertices $\mathbf{P}_1$, $\mathbf{P}_2$, and $\mathbf{P}_3$ is the set

$$\{\mathbf{P}: \mathbf{P} = t_1\mathbf{P}_1 + t_2\mathbf{P}_2 + t_3\mathbf{P}_3, \ t_j \geq 0, \ \sum_{j=1}^{3} t_j = 1\}. \tag{10}$$

For example, taking $t_1 = 1$, $t_2 = 0$, $t_3 = 0$, we find that $1 \cdot \mathbf{P}_1 + 0 \cdot \mathbf{P}_2 + 0 \cdot \mathbf{P}_3 = \mathbf{P}_1$ is in the triangle. Taking $t_1 = t_2 = t_3 = \frac{1}{3}$, we obtain the *barycenter* $\mathbf{C}$ of the triangle:

$$\mathbf{C} = \tfrac{1}{3}\mathbf{P}_1 + \tfrac{1}{3}\mathbf{P}_2 + \tfrac{1}{3}\mathbf{P}_3 .$$

(See Fig. 32.)

(a) Prove that the vertex $\mathbf{P}_2$ is in the triangle (10); i.e. find nonnegative numbers t_1, t_2, t_3 such that $\sum_{j=1}^{3} t_j = 1$ and

$$\mathbf{P}_2 = t_1\mathbf{P}_1 + t_2\mathbf{P}_2 + t_3\mathbf{P}_3 .$$

(b) Show that $\mathbf{P}_3$ is in the triangle; show also that the midpoint of the segment from $\mathbf{P}_1$ to $\mathbf{P}_2$ is in the triangle.

(c) Let $\mathbf{M}_3$ denote the midpoint found in part (b). Prove that the segment from $\mathbf{P}_3$ to $\mathbf{M}_3$ lies in the triangle. (This segment is called a *median* of the triangle.)

(d) Prove that the median in part (c) contains the barycenter **C**.
(e) Prove that all three medians of the triangle interesect at **C**.

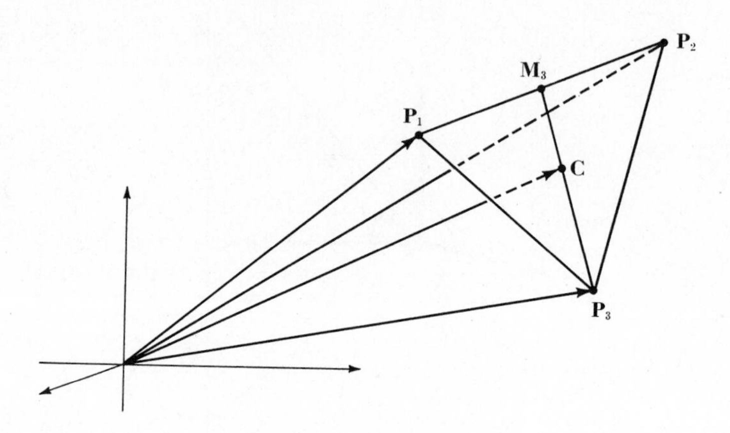

FIGURE 1.32

21. Four points $\mathbf{P}_1$, $\mathbf{P}_2$, $\mathbf{P}_3$, $\mathbf{P}_4$ determine a *tetrahedron*

$$T = \{\mathbf{P}: \mathbf{P} = \sum_1^4 t_j\mathbf{P}_j, \ t_j \geq 0, \ \sum_1^4 t_j = 1\},$$

a polygonal solid with four triangular faces and the four vertices $\mathbf{P}_1$, $\mathbf{P}_2$, $\mathbf{P}_3$, $\mathbf{P}_4$. The *barycenter* of the tetrahedron is the point

$$\mathbf{C} = \sum_1^4 \tfrac{1}{4}\mathbf{P}_j .$$

(a) Prove that T contains the triangle with vertices $\mathbf{P}_1$, $\mathbf{P}_2$, $\mathbf{P}_3$. (See the previous problem.)
(b) Prove that **P** contains the segment from $\mathbf{P}_4$ to the barycenter of the triangle in part (a). (This segment is a *median* of the tetrahedron.)
(c) Prove that the barycenter **C** lies on the median in part (b).
(d) Prove that the barycenter **C** lies on the intersection of all four medians of T.
(e) Prove that if $\mathbf{Q}_1$ and $\mathbf{Q}_2$ are any two points in T, then T contains the entire segment between $\mathbf{Q}_1$ and $\mathbf{Q}_2$.

22. This problem proves that "two points determine a line" by showing that if $\mathbf{A} + t\mathbf{B}$ is any line containing the two distinct points $\mathbf{P}_1$ and $\mathbf{P}_2$, then

$$\{\mathbf{P}: \mathbf{P} = \mathbf{A} + t\mathbf{B} \text{ for some } t\} \qquad (11)$$

is exactly the same set as

$$\{\mathbf{P}: \mathbf{P} = \mathbf{P}_1 + s(\mathbf{P}_2 - \mathbf{P}_1)\} \text{ for some } s\}. \tag{12}$$

(a) Suppose that $\mathbf{P}_1$ and $\mathbf{P}_2$ lie on the line (11). Prove that there are two *distinct* numbers t_1 and t_2 such that $\mathbf{P}_1 = \mathbf{A} + t_1\mathbf{B}$ and $\mathbf{P}_2 = \mathbf{A} + t_2\mathbf{B}$.

(b) Prove that

$$\mathbf{B} = \frac{1}{t_2 - t_1}(\mathbf{P}_2 - \mathbf{P}_1) \quad \text{and} \quad \mathbf{A} = \mathbf{P}_1 - \frac{t_1}{t_2 - t_1}(\mathbf{P}_2 - \mathbf{P}_1).$$

(c) Prove that

$$\mathbf{A} + t\mathbf{B} = \mathbf{P}_1 + \frac{t - t_1}{t_2 - t_1}(\mathbf{P}_2 - \mathbf{P}_1),$$

and conclude that every point on the line (11) lies on the line (12).

(d) Prove, conversely, that every point on (12) lies on (11).

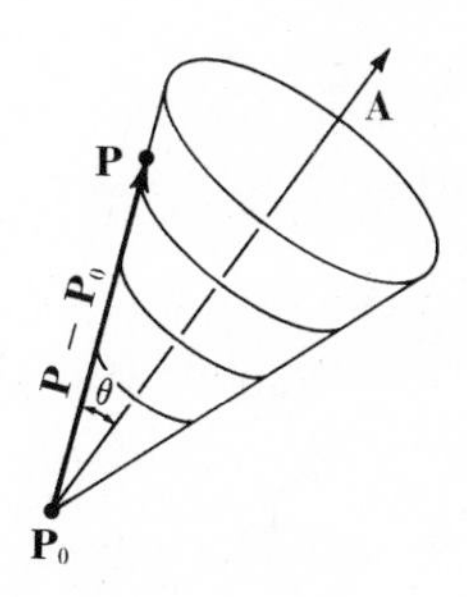

FIGURE 1.33

23. Given a point $\mathbf{P}_0$, a vector $\mathbf{A}$, and a number θ, the equation

$$(\mathbf{P} - \mathbf{P}_0) \cdot \mathbf{A} = |\mathbf{A}| \cos\theta \, |\mathbf{P} - \mathbf{P}_0|$$

defines a *single cone* with *vertex* $\mathbf{P}_0$, *axial direction* $\mathbf{A}$, and *angle* θ. (See Fig. 33.) The equation

$$|(\mathbf{P} - \mathbf{P}_0) \cdot \mathbf{A}| = |\mathbf{A}| \, |\cos\theta| \, |\mathbf{P} - \mathbf{P}_0|$$

defines the corresponding *double cone*.

(a) Prove that if $\theta = \pi/2$, the "cone" is actually a plane.

(b) Prove that if $\theta = 0$, the "double cone" is actually the line $\{\mathbf{P}: \mathbf{P} = \mathbf{P}_0 + t\mathbf{A}\}$.

(c) Let $\mathbf{P}_0 = (0,0,2)$, $\mathbf{A} = (0,1,-1)$, $\theta = \pi/4$, and find the intersection of the corresponding cone with the xy plane. (This is a familiar parabola.)

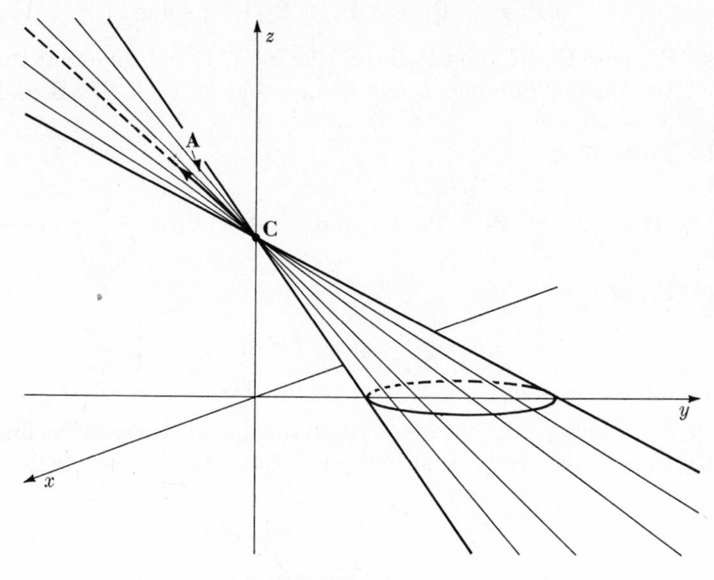

FIGURE 1.34

(d) Let $\mathbf{P}_0 = (0,0,2)$, $\mathbf{A} = (0,-1,1)$, and θ be arbitrary; find the intersection of the resulting double cone with the xy plane. (For $\theta < \pi/4$, this is an ellipse

$$\frac{x^2}{a^2} + \frac{(y - y_0)^2}{b^2} = 1,$$

and for $\theta > \pi/4$, it is a hyperbola

$$-\frac{x^2}{a^2} + \frac{(y - y_0)^2}{b^2} = 1.$$

See Fig. 34.)

(e) Prove that every double cone is defined by a quadratic equation

$$ax^2 + bxy + cxz + dy^2 + eyz + fz^2 + gx + hy + iz + j = 0.$$

(However, not all quadratic equations define cones; for example, a sphere is defined by a quadratic equation, but a sphere is not a cone. From part (e) it follows that the intersection of any double cone with the xy plane is determined by the equation $z = 0$ and a quadratic in x and y,

$$ax^2 + bxy + dy^2 + gx + hy + j = 0.)$$

1.4 THE VECTOR SPACE R^n

R^3 is surely the most interesting vector space because it reflects faithfully the three-dimensional space around us. But vector analysis neither starts nor stops there; we need the plane R^2, the four-dimensional space R^4, and in fact we need R^n for every positive integer n.

Fortunately, almost everything can be done in R^n just as in R^3. A point, or vector, is an ordered n-tuple of real numbers, $\mathbf{A} = (a_1, \ldots, a_n)$, $\mathbf{X} = (x_1, \ldots, x_n)$, etc. *Addition* is defined by

$$\mathbf{A} + \mathbf{B} = (a_1 + b_1, a_2 + b_2, \ldots, a_n + b_n),$$

subtraction by

$$\mathbf{A} - \mathbf{B} = (a_1 - b_1, \ldots, a_n - b_n),$$

scalar multiplication by

$$t\mathbf{A} = (ta_1, ta_2, \ldots, ta_n),$$

the *inner product* by

$$\mathbf{A}\cdot\mathbf{B} = a_1b_1 + \cdots + a_nb_n = \sum_1^n a_jb_j,$$

and the *length* by

$$|\mathbf{A}| = (\mathbf{A}\cdot\mathbf{A})^{1/2} = \left(\sum_1^n a_j^2\right)^{1/2}$$

The *zero vector* is $(0, 0, \ldots, 0)$, denoted $\mathbf{0}$.

The facts about R^3 established in §1.1 hold equally well for R^n. The proofs are exactly the same as before; we merely list the most important results for reference.

Theorem 4. *Let* $\mathbf{A}$, $\mathbf{B}$, *and* $\mathbf{C}$ *be any vectors in* R^n, *and let* t *and* s *be any real numbers. Then*

$$t(s\mathbf{A}) = (ts)\mathbf{A} \tag{1}$$

$$(\mathbf{A} + \mathbf{B}) + \mathbf{C} = \mathbf{A} + (\mathbf{B} + \mathbf{C}) \tag{2}$$

$$t(\mathbf{A}\cdot\mathbf{B}) = (t\mathbf{A})\cdot\mathbf{B} = \mathbf{A}\cdot(t\mathbf{B}) \tag{3}$$

$$\mathbf{A} + \mathbf{B} = \mathbf{B} + \mathbf{A} \tag{4}$$

$$\mathbf{A}\cdot\mathbf{B} = \mathbf{B}\cdot\mathbf{A} \tag{5}$$

$$(\mathbf{A} + \mathbf{B})\cdot\mathbf{C} = \mathbf{A}\cdot\mathbf{C} + \mathbf{B}\cdot\mathbf{C} \tag{6}$$

$$\mathbf{A} + \mathbf{0} = \mathbf{A} \tag{7}$$

$$1\mathbf{A} = \mathbf{A} \tag{8}$$

$$|\mathbf{A}| = 0 \iff \mathbf{A} = \mathbf{0} \tag{9}$$

$$|t\mathbf{A}| = |t|\cdot|\mathbf{A}| \tag{10}$$

$$|\mathbf{A}\cdot\mathbf{B}| \le |\mathbf{A}|\cdot|\mathbf{B}| \tag{11}$$

$$|\mathbf{A} + \mathbf{B}| \le |\mathbf{A}| + |\mathbf{B}|. \tag{12}$$

The Schwarz inequality (11) shows that $\mathbf{A}\cdot\mathbf{B}/|\mathbf{A}|\cdot|\mathbf{B}|$ lies between -1 and 1, so we can define the angle between **A** and **B** by

$$\cos\theta = \frac{\mathbf{A}\cdot\mathbf{B}}{|\mathbf{A}|\cdot|\mathbf{B}|};$$

hence

$$\mathbf{A}\cdot\mathbf{B} = |\mathbf{A}|\cdot|\mathbf{B}|\cos\theta. \tag{13}$$

A and **B** are *orthogonal* if $\mathbf{A}\cdot\mathbf{B} = 0$, and *parallel* if one vector is a scalar multiple of the other.

We visualize vectors in R^2 as arrows or points in the plane, and vectors in R^3 as arrows or points in three-space, but beginning with R^4 there is no such concrete geometric image. We either forget about visualization, or cheat a little and continue to imagine n-dimensional vectors as if they were in R^2 or R^3. The terminology used for R^n reflects this. For example, the set $\{\mathbf{X}: |\mathbf{X} - \mathbf{A}| < r\}$ is called the *open ball* (or open disk) *of radius r with center* **A**, and its boundary $\{\mathbf{X}: |\mathbf{X} - \mathbf{A}| = r\}$ is called the *sphere* of radius r with center **A**. (In R^2, such a "sphere" is really a circle.)

A *line* in R^n is any set of the form

$$\{\mathbf{X}: \mathbf{X} = \mathbf{A} + t\mathbf{B} \text{ for some } t\}$$

where **A** and **B** are fixed vectors in R^n and $\mathbf{B} \neq \mathbf{0}$.

Any set of the form $\{\mathbf{X}: \mathbf{X}\cdot\mathbf{N} = d\}$, where **N** is a given nonzero vector and d is a given constant, is called a *hyperplane.* In R^2 a hyperplane is a line, and in R^3 it is an ordinary plane. In general, we think of a hyperplane in R^n as something that looks like R^{n-1}. For example, in R^4 the set of all vectors $(x,y,z,0)$ is a hyperplane, since it is obtained by taking $N = (0,0,0,1)$ and $d = 0$ in the definition $\{\mathbf{X}: \mathbf{X}\cdot\mathbf{N} = d\}$; this set looks very much like R^3, i.e., like the set of all vectors (x,y,z).

The cross product has been defined only in R^3, where it solved the problem of finding a nonzero vector orthogonal to two given (nonparallel) vectors. The corresponding problem in R^n is to find a nonzero vector orthogonal to $n - 1$ given vectors $\mathbf{A}_1, \ldots, \mathbf{A}_{n-1}$. It is not necessary to have an explicit solution like the cross product, but it *is* important to know that the problem can always be solved.

Theorem 5. *Given any k vectors $\mathbf{A}_1, \ldots, \mathbf{A}_k$ in R^n, $k < n$, there is a nonzero vector* **X** *orthogonal to each of $\mathbf{A}_1, \ldots, \mathbf{A}_k$.*

To prove the theorem, we rephrase the result in terms of simultaneous linear equations. Denote the components of the jth given vector $\mathbf{A}_j$ by $a_{j1}, \ldots, a_{jn}$. Then the required orthogonality conditions

$$\begin{aligned} \mathbf{A}_1\cdot\mathbf{X} &= 0, \\ \vdots\quad & \quad\vdots \\ \mathbf{A}_k\cdot\mathbf{X} &= 0 \end{aligned}$$

are equivalent to the system of k simultaneous linear homogeneous equations

$$\begin{aligned} a_{11}x_1 + \cdots + a_{1n}x_n &= 0 \\ \vdots \qquad\qquad \vdots \quad & \;\; \vdots \\ a_{k1}x_1 + \cdots + a_{kn}x_n &= 0. \end{aligned}$$

Thus Theorem 5 is equivalent to the following basic result:

Theorem 6. *Any system of k simultaneous linear homogeneous equations in n unknowns $x_1, \ldots, x_n$, with $k < n$, has a nonzero solution.*

Proof. We use induction on k, the number of equations.

(i) When $k = 1$, we are solving a single equation

$$a_{11}x_1 + a_{12}x_2 + \cdots + a_{1n}x_n = 0 \tag{14}$$

with $n > 1$. If $a_{11} = 0$, there is a nonzero solution $x_1 = 1$, $x_2 = 0, \ldots, x_n = 0$. If $a_{11} \neq 0$, we can set $x_2 = 1$, $x_3 = 0, \ldots, x_n = 0$, and $x_1 = -a_{12}/a_{11}$.

(ii) Suppose that Theorem 6 is true for systems of $k - 1$ equations in m unknowns whenever $m > k - 1$. We then want to prove it for an arbitrary system of k equations in n unknowns

$$\begin{aligned} a_{11}x_1 + \cdots + a_{1n}x_n &= 0 & (15_1) \\ a_{21}x_1 + \cdots + a_{2n}x_n &= 0 & (15_2) \\ \vdots \qquad\qquad \vdots \quad & \;\; \vdots & \vdots \\ a_{k1}x_1 + \cdots + a_{kn}x_n &= 0, & (15_k) \end{aligned}$$

where $n > k$. If the coefficients $a_{11}, \ldots, a_{k1}$ in the first column are all zero, we obtain a nonzero solution of (15) simply by setting

$$x_1 = 1,\ x_2 = 0,\ \ldots,\ x_n = 0.$$

Suppose, then, that one of the coefficients in the first column is different from zero. By relabeling the equations, we can assume that a_{11} is different from zero. Now eliminate x_1 from the last $k - 1$ equations in (15) by subtracting appropriate multiples of the first equation, obtaining

$$\begin{aligned} \left(a_{22} - \frac{a_{21}a_{12}}{a_{11}}\right)x_2 + \cdots + \left(a_{2n} - \frac{a_{21}a_{1n}}{a_{11}}\right)x_n &= 0 & (16_2) \\ \vdots \qquad\qquad\qquad\qquad \vdots \qquad\qquad\qquad & \;\; \vdots \\ \left(a_{k2} - \frac{a_{k1}a_{12}}{a_{11}}\right)x_2 + \cdots + \left(a_{kn} - \frac{a_{k1}a_{1n}}{a_{11}}\right)x_n &= 0. & (16_k) \end{aligned}$$

This is a set of $k - 1$ simultaneous linear homogeneous equations in the $n - 1$ unknowns $x_2, \ldots, x_n$, and $n - 1 > k - 1$. Since we are supposing Theorem 6 to be true for systems of $k - 1$ equations, the system (16) has a nonzero solution. Let $x_2, \ldots, x_n$ stand for this solution, and set

$$x_1 = -\frac{1}{a_{11}}(a_{12}x_2 + \cdots + a_{1n}x_n). \tag{17}$$

Then $(x_1, \ldots, x_n)$ obviously satisfies equation (15_1), and together (16) and (17) show that the other equations in (15) are satisfied too. For example, (15_2) follows from (16_2) and (17), since

$$\begin{aligned} a_{21}x_1 + a_{22}x_2 + \cdots + a_{2n}x_n \\ &= -\frac{a_{21}}{a_{11}}(a_{12}x_2 + \cdots + a_{1n}x_n) + a_{22}x_2 + \cdots + a_{2n}x_n \quad \text{(by (17))} \\ &= \left(a_{22} - \frac{a_{21}a_{12}}{a_{11}}\right)x_2 + \cdots + \left(a_{2n} - \frac{a_{21}a_{1n}}{a_{11}}\right)x_n = 0. \quad \text{(by } (16_2)\text{)} \end{aligned}$$

Example 1. Find a nonzero vector **X** orthogonal to $(-1,2)$. *Solution.* Following the first step in the proof of Theorem 6, we find $\mathbf{X} = (2,1)$.

Example 2. Find a nonzero vector **X** orthogonal to (1,2,3,4), (2,3,4,5), and (3,4,5,6). *Solution.* If $\mathbf{X} = (x_1, \ldots, x_4)$, we have to solve

$$x_1 + 2x_2 + 3x_3 + 4x_4 = 0 \tag{18$_1$}$$

$$2x_1 + 3x_2 + 4x_3 + 5x_4 = 0 \tag{18$_2$}$$

$$3x_1 + 4x_2 + 5x_3 + 6x_4 = 0. \tag{18$_3$}$$

Eliminating x_1 from the last two equations (as in the proof of Theorem 6), we get

$$-x_2 - 2x_3 - 3x_4 = 0 \tag{19$_1$}$$

$$-2x_2 - 4x_3 - 6x_4 = 0. \tag{19$_2$}$$

These two equations are obviously equivalent; a nonzero solution is found by taking $x_2 = 1$, $x_3 = -\frac{1}{2}$, $x_4 = 0$. Returning to (18_1), we find $x_1 = -2x_2 - 3x_3 - 4x_4 = -\frac{1}{2}$. Thus $\mathbf{X} = (-\frac{1}{2},1,-\frac{1}{2},0)$ is orthogonal to the given vectors. So is $-2\mathbf{X} = (1,-2,1,0)$.

PROBLEMS

1. Find the angle between the given pair of vectors.
(a) (1,0,0,0) and (0,1,1,3)
(b) (1,1,1,1) and $(1,-1,1,1)$
(c) (1,1,1,1) and $(1,-1,-1,1)$
(d) (4,3,2,1) and (8,6,4,2)
(e) (4,3,2,1) and $(-8,-6,-4,-2)$
(f) $(2,-3,4,6,1)$ and $(1,3,1,0,-3)$

2. Find the intersection of the line and the hyperplane with the given equations. (The intersection may be a single point, or the whole line, or it may be empty.)
(a) $\mathbf{X} = (1,3) + t(-2,-1)$; $\mathbf{X}\cdot(1,2) = 4$
(b) $\mathbf{X} = (0,3,8,-1) + t(2,-1,3,0)$; $\mathbf{X}\cdot(1,3,2,0) = 1$
(c) $\mathbf{X} = t(3,5)$; $\mathbf{X}\cdot(5,-3) = 0$
(d) $\mathbf{X} = t(3,5)$; $\mathbf{X}\cdot(5,-3) = 1$
(e) $\mathbf{X} = t(3,5,7,2)$; $\mathbf{X}\cdot(2,4,-2,1) = 0$
(f) $\mathbf{X} = t(3,5,7,2)$; $\mathbf{X}\cdot(2,4,-2,1) = 5$

3. Find the points of intersection of the given line and the given sphere.
(a) $\mathbf{X} = (1,3) + t(-2,-1)$; $|\mathbf{X} - (1,3)| = 2$
(b) $\mathbf{X} = (1,3) + t(-2,-1)$; $|\mathbf{X}| = r$
(c) $\mathbf{X} = (0,3,8,-1) + t(2,-1,3,0)$; $|\mathbf{X}| = r$

4. (a) In Problem 3(b), choose r so that there is exactly one point of intersection. (This gives the distance from the line to the origin.)
(b) In Problem 3(c), choose r so that there is exactly one point of intersection.

5. Prove Theorem 4. (This problem has twelve parts. For hints, you can refer to the proofs in §1.1.)

6. Prove that $|\mathbf{A}| \leq \sum_1^n |a_j|$

(a) by squaring both sides; or
(b) by writing

$$\mathbf{A} = (a_1,0,\ldots,0) + (0,a_2,0,\ldots,0) + \cdots + (0,\ldots,0,a_n)$$

and applying the triangle inequality.

7. Prove that $-|\mathbf{A} - \mathbf{B}| \leq |\mathbf{A}| - |\mathbf{B}| \leq |\mathbf{A} - \mathbf{B}|$. (Hint: To get the first inequality, notice that $|\mathbf{B}| = |\mathbf{A} + (\mathbf{B} - \mathbf{A})| \leq |\mathbf{A}| + |\mathbf{B} - \mathbf{A}|$.)

8. In each case, find a nonzero vector orthogonal to the given set of vectors $\mathbf{A}_1, \ldots, \mathbf{A}_k$.
(a) $\mathbf{A}_1 = (2,-5)$
(b) $\mathbf{A}_1 = (2,3,4)$, $\mathbf{A}_2 = (-1,3,-4)$
(c) $\mathbf{A}_1 = (1,2,3)$, $\mathbf{A}_2 = (-2,-4,-6)$
(d) $\mathbf{A}_1 = (1,2,3,4)$, $\mathbf{A}_2 = (2,3,4,5)$, $\mathbf{A}_3 = (3,4,5,0)$

9. (a) Although Theorem 5 does not guarantee that there is one, try to find a nonzero vector in R^3 that is orthogonal to $\mathbf{A}_1 = (1,0,-1)$, $\mathbf{A}_2 = (0,1,3)$, and $\mathbf{A}_3 = (1,1,2)$.
(b) Prove that $\mathbf{A}_1$, $\mathbf{A}_2$, and $\mathbf{A}_3$ lie on a common plane through the origin.

10. Suppose that $\mathbf{A}_1, \ldots, \mathbf{A}_k$ are any vectors in R^n, where $k > n$. Prove that there are numbers $t_1, \ldots, t_k$, not all zero, such that

$$t_1\mathbf{A}_1 + \cdots + t_k\mathbf{A}_k = 0.$$

(Hint: Write out this relation as a system of simultaneous linear equations.)

1.5 LINEAR DEPENDENCE AND BASES

To avoid misunderstandings, we advise you that this final section is a little more subtle than the rest of the chapter, and does not form part of the main line of development of the text. The material is included in spite of these considerations because it solves many problems that are inaccessible by less subtle methods, for example the existence of a partial fractions decomposition for every rational function (see the problems below); and because it provides a brief introduction to the important subject of linear algebra.

A sum $t_1\mathbf{A}_1 + \cdots + t_k\mathbf{A}_k$, where the t_j are numbers and the $\mathbf{A}_j$ are vectors, is called a *linear combination* of $\mathbf{A}_1, \ldots, \mathbf{A}_k$. For example, $(5,-1)$ is a linear combination of $(1,1)$ and $(1,-1)$, since

$$(5,-1) = 2(1,1) + 3(1,-1).$$

Given any vectors $\mathbf{A}_1, \ldots, \mathbf{A}_k$ $(k \geq 1)$, we can write the zero vector in a rather simple-minded way as a linear combination

$$\mathbf{0} = 0 \cdot \mathbf{A}_1 + \cdots + 0 \cdot \mathbf{A}_k.$$

If there is any other way to write $\mathbf{0}$ as a linear combination of these vectors, they are called *linearly dependent*. Precisely:

Definition 3. A set of vectors $\mathbf{A}_1, \ldots, \mathbf{A}_k$ is called *linearly dependent* if and only if there are numbers $t_1, \ldots, t_k$, not all zero, such that

$$t_1\mathbf{A}_1 + t_2\mathbf{A}_2 + \cdots + t_k\mathbf{A}_k = \mathbf{0}. \tag{1}$$

This imposing definition has a rather simple geometric meaning, which shows up when we examine the cases $k = 1, 2, 3$.

A single vector $\mathbf{A}_1$ *forms a linearly dependent set if and only if* $\mathbf{A}_1 = \mathbf{0}$.

Proof. $t_1\mathbf{A}_1 = \mathbf{0}$ for some $t_1 \neq 0$ if and only if $\mathbf{A}_1 = \mathbf{0}$.

Two vectors $\mathbf{A}_1$ *and* $\mathbf{A}_2$ *form a linearly dependent set if and only if they are parallel* (see Fig. 35).

Proof. Suppose that $\mathbf{A}_1$ and $\mathbf{A}_2$ are linearly dependent, i.e.

$$t_1\mathbf{A}_1 + t_2\mathbf{A}_2 = \mathbf{0},$$

where t_1 and t_2 are not both zero. If $t_1 \neq 0$, then $\mathbf{A}_1 = -\frac{t_2}{t_1}\mathbf{A}_2$, and if $t_2 \neq 0$, then $\mathbf{A}_2 = -\frac{t_1}{t_2}\mathbf{A}_1$; in either case, the vectors are parallel.

Conversely, if $\mathbf{A}$ and $\mathbf{B}$ are parallel, then either $\mathbf{A} = t\mathbf{B}$ or $\mathbf{B} = t\mathbf{A}$. In the first case, $1\cdot\mathbf{A} + (-t)\mathbf{B} = \mathbf{0}$; in the second case, $t\mathbf{A} + (-1)\mathbf{B} = \mathbf{0}$; in either case, the pair is linearly dependent.

Three vectors $\mathbf{A}_1$, $\mathbf{A}_2$, $\mathbf{A}_3$ in R^3 form a linearly dependent set if and only if they lie on a common plane through the origin (see Fig. 36).

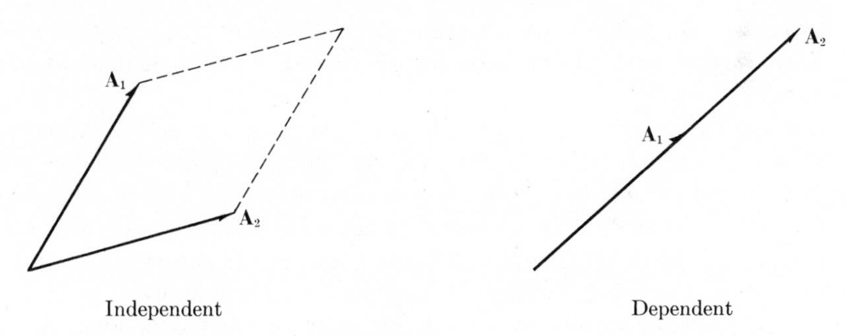

FIGURE 1.35

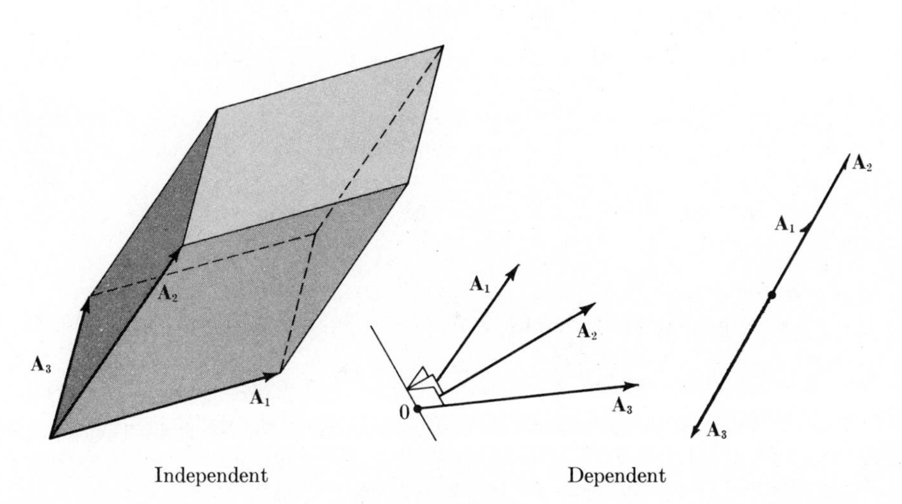

FIGURE 1.36

Proof. If $\mathbf{A}_1$, $\mathbf{A}_2$, $\mathbf{A}_3$ are linearly dependent, then

$$t_1\mathbf{A}_1 + t_2\mathbf{A}_2 + t_3\mathbf{A}_3 = \mathbf{0},$$

where at least one of the t_j is not zero, say $t_3 \neq 0$. By Theorem 5, there is a nonzero vector $\mathbf{N}$ orthogonal to both $\mathbf{A}_1$ and $\mathbf{A}_2$, so these two vectors lie on the plane through the origin

$$\{\mathbf{P}: \mathbf{P}\cdot\mathbf{N} = 0\}. \tag{2}$$

Since $t_3 \neq 0$, we have

$$\mathbf{A}_3 = -\frac{t_1}{t_3}\mathbf{A}_1 - \frac{t_2}{t_3}\mathbf{A}_2,$$

so

$$\mathbf{A}_3\cdot\mathbf{N} = -\frac{t_1}{t_3}\mathbf{A}_1\cdot\mathbf{N} - \frac{t_2}{t_3}\mathbf{A}_2\cdot\mathbf{N} = 0;$$

hence $\mathbf{A}_1$, $\mathbf{A}_2$, $\mathbf{A}_3$ all lie on the plane (2). The converse implication (that three vectors coplanar with the origin form a linearly dependent set) is Problem 6 below.

These three cases show that linear dependence is a sort of degeneracy. A pair of vectors generally spans a parallelogram, but when the pair is linearly dependent, the parallelogram degenerates to a line segment (see Fig. 35) or even to a single point (when both vectors are zero). Three vectors generally span a parallelepiped, but when the three are linearly dependent, the parallelepiped degenerates to a plane figure.

A nondegenerate set, i.e. a set that is not linearly dependent, is called *linearly independent*. Thus $\mathbf{A}_1, \ldots, \mathbf{A}_k$ is a linearly independent set if no relation (1) exists except the trivial one with $t_1 = 0, \ldots, t_k = 0$. In other words, $\mathbf{A}_1, \ldots, \mathbf{A}_k$ *is a linearly independent set if and only if*

$$t_1\mathbf{A}_1 + \cdots + t_k\mathbf{A}_k = \mathbf{0} \;\Rightarrow\; t_1 = t_2 = \cdots = t_k = 0.$$

It can be rather tedious to determine whether a given set of vectors in R^n is linearly dependent or not, especially when $n > 3$; but two important cases can be decided at a glance. The first is the case where $\mathbf{A}_1, \ldots, \mathbf{A}_k$ are mutually orthogonal, i.e. where $\mathbf{A}_j\cdot\mathbf{A}_m = 0$ for $m \neq j$.

Theorem 7. *If $\mathbf{A}_1, \ldots, \mathbf{A}_k$ are all nonzero and mutually orthogonal, then they are linearly independent.*

Proof. Suppose that

$$t_1\mathbf{A}_1 + \cdots + t_k\mathbf{A}_k = \mathbf{0}.$$

Taking the inner product of each side with $\mathbf{A}_j$, we get

$$t_1\mathbf{A}_1\cdot\mathbf{A}_j + \cdots + t_j\mathbf{A}_j\cdot\mathbf{A}_j + \cdots + t_k\mathbf{A}_k\cdot\mathbf{A}_j = \mathbf{0}\cdot\mathbf{A}_j = 0. \tag{3}$$

Since $\mathbf{A}_m \cdot \mathbf{A}_j = 0$ for $m \neq j$, equation (3) reduces to

$$t_j \mathbf{A}_j \cdot \mathbf{A}_j = 0.$$

Since $\mathbf{A}_j \cdot \mathbf{A}_j = |\mathbf{A}_j|^2 \neq 0$, it follows that $t_j = 0$. Thus $t_1\mathbf{A}_1 + \cdots + t_k\mathbf{A}_k = \mathbf{0}$ implies $t_j = 0$ for every j. *Q.E.D.*

The second important case arises when there are "more vectors than dimensions"; in this case the set is always *linearly dependent.*

Theorem 8. *Let $\mathbf{A}_1, \ldots, \mathbf{A}_k$ be vectors in R^n. If $k > n$, then the set of vectors is linearly dependent. (If $k \leq n$, the set may or may not be linearly dependent.)*

Proof. We have to find $t_1, \ldots, t_k$, not all zero, such that

$$t_1\mathbf{A}_1 + \cdots + t_k\mathbf{A}_k = \mathbf{0}. \tag{4}$$

If we denote $\mathbf{A}_j$ by $(a_{j1}, \ldots, a_{jn})$ and write out (4) by components, we get

$$\begin{aligned} t_1a_{11} + \cdots + t_ka_{k1} &= 0 \\ t_1a_{12} + \cdots + t_ka_{k2} &= 0 \\ \vdots \qquad \qquad \vdots \quad &\quad \vdots \\ t_1a_{1n} + \cdots + t_ka_{kn} &= 0. \end{aligned}$$

This is a set of n simultaneous linear homogeneous equations in k unknowns, with $k > n$. By Theorem 6, there is a nonzero solution, and the proof is complete.

The most important consequence of Theorem 8 is

Theorem 9. *If $\mathbf{A}_1, \ldots, \mathbf{A}_n$ is a linearly independent set of vectors in R^n, then every vector $\mathbf{X}$ in R^n can be written in one and only one way as a linear combination of $\mathbf{A}_1, \ldots, \mathbf{A}_n$,*

$$\mathbf{X} = t_1\mathbf{A}_1 + \cdots + t_n\mathbf{A}_n .$$

Proof. By Theorem 8, the set of $n + 1$ vectors $\mathbf{X}, \mathbf{A}_1, \ldots, \mathbf{A}_n$ is linearly dependent; hence there is a linear combination

$$t_0\mathbf{X} + t_1\mathbf{A}_1 + \cdots + t_n\mathbf{A}_n = \mathbf{0}, \tag{5}$$

with at least one of the t_j different from zero. It follows that $t_0 \neq 0$; for if $t_0 = 0$, then $t_1, \ldots, t_n$ are not all zero, and $t_1\mathbf{A}_1 + \cdots + t_n\mathbf{A}_n = \mathbf{0}$, contradicting the assumption that $\mathbf{A}_1, \ldots, \mathbf{A}_n$ are linearly independent. Since $t_0 \neq 0$, we can solve (5) and obtain $\mathbf{X}$ as the linear combination

$$\mathbf{X} = \left(-\frac{t_1}{t_0}\right)\mathbf{A}_1 + \cdots + \left(-\frac{t_n}{t_0}\right)\mathbf{A}_n .$$

It remains to prove that there is *only one* such combination giving **X**. Suppose that

$$\mathbf{X} = t_1\mathbf{A}_1 + \cdots + t_n\mathbf{A}_n , \tag{6}$$

and also

$$\mathbf{X} = s_1\mathbf{A}_1 + \cdots + s_n\mathbf{A}_n . \tag{7}$$

Subtracting (7) from (6), we obtain

$$\mathbf{0} = (t_1 - s_1)\mathbf{A}_1 + \cdots + (t_n - s_n)\mathbf{A}_n . \tag{8}$$

Since $\mathbf{A}_1, \ldots, \mathbf{A}_n$ are linearly independent, equation (8) implies that $t_1 - s_1 = 0, \ldots, t_n - s_n = 0$; hence the representations (6) and (7) are identical, and the proof is complete.

Any linearly independent set of vectors whose linear combinations produce the whole space is called a *basis*; in this language, Theorem 9 says that *in R^n any linearly independent set of n vectors forms a basis.* This fundamental result is the reason for having introduced the concept of linear independence.

According to Theorem 9, any pair of nonparallel vectors **A**, **B** in R^2 forms a basis, so every vector **P** can be written uniquely as a linear combination $\mathbf{P} = s\mathbf{A} + t\mathbf{B}$, as suggested in Fig. 37.

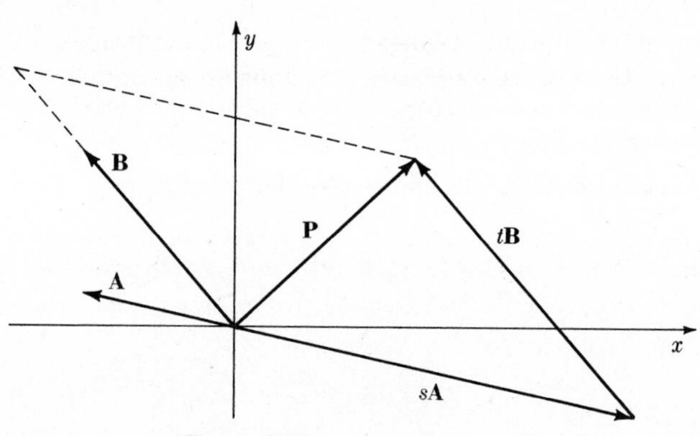

FIGURE 1.37

In R^n the unit vectors

$$\mathbf{E}_1 = (1,0,\ldots,0), \mathbf{E}_2 = (0,1,0,\ldots,0), \ldots, \mathbf{E}_n = (0,\ldots,0,1)$$

are linearly independent, and every vector $\mathbf{X} = (x_1, \ldots, x_n)$ in R^n can be written as

$$\mathbf{X} = x_1\mathbf{E}_1 + \cdots + x_n\mathbf{E}_n ;$$

hence $\mathbf{E}_1, \ldots, \mathbf{E}_n$ form a basis, called the "standard" or "canonical" basis of R^n. It is not, however, the best basis to use in every case. Problems in vector analysis can be simplified by a good choice of basis, just as problems in physics or traditional analytic geometry are simplified by a good choice of coordinates. We conclude with two theorems (10 and 12) about bases, and two theorems (11 and 13) that depend on a good choice of basis.

Theorem 10. *If* $\mathbf{A}_1, \ldots, \mathbf{A}_n$ *are mutually orthogonal unit vectors in* R^n, *then they form a basis, and any vector* $\mathbf{X}$ *in* R^n *can be written*

$$\mathbf{X} = (\mathbf{X}\cdot\mathbf{A}_1)\mathbf{A}_1 + \cdots + (\mathbf{X}\cdot\mathbf{A}_n)\mathbf{A}_n . \tag{9}$$

Proof. By Theorem 7, the vectors $\mathbf{A}_1, \ldots, \mathbf{A}_n$ are linearly independent. By Theorem 9, $\mathbf{X}$ can be written as a linear combination

$$\mathbf{X} = t_1\mathbf{A}_1 + \cdots + t_n\mathbf{A}_n . \tag{10}$$

Taking the inner product of each side with $\mathbf{A}_j$, we get

$$\mathbf{X}\cdot\mathbf{A}_j = t_1\mathbf{A}_1\cdot\mathbf{A}_j + \cdots + t_j\mathbf{A}_j\cdot\mathbf{A}_j + \cdots + t_n\mathbf{A}_n\cdot\mathbf{A}_j = t_j$$

since $\mathbf{A}_m\cdot\mathbf{A}_j = 0$ for $m \neq j$ and $\mathbf{A}_j\cdot\mathbf{A}_j = |\mathbf{A}_j|^2 = 1$. The proof is completed by substituting $\mathbf{X}\cdot\mathbf{A}_j$ for t_j in (10).

A basis of the type in Theorem 10 is called "orthonormal"; "ortho-" refers to the orthogonality condition $\mathbf{A}_m\cdot\mathbf{A}_j = 0$ for $m \neq j$, and "-normal" refers to the condition $|\mathbf{A}_j| = 1$, which standardizes or "normalizes" the lengths of the vectors in the basis. Formula (9) expresses $\mathbf{X}$ as the sum of its orthogonal projections on $\mathbf{A}_1, \ldots, \mathbf{A}_n$.

As an application of Theorem 10, we prove some interesting results about the cross product in R^3.

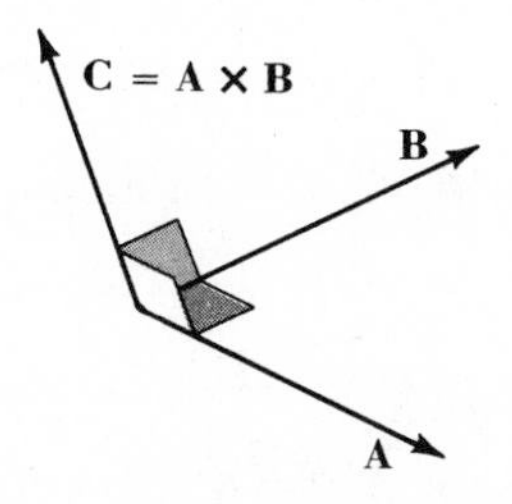

FIGURE 1.38

Theorem 11. *Suppose that* $\mathbf{A}$ *and* $\mathbf{B}$ *are orthogonal unit vectors in* R^3, *and set* $\mathbf{C} = \mathbf{A} \times \mathbf{B}$. *Then*

(i) $\mathbf{A}, \mathbf{B}, \mathbf{C}$ *is an orthonormal basis;*

(ii) $\mathbf{A} = \mathbf{B} \times \mathbf{C}$ *and* $\mathbf{B} = \mathbf{C} \times \mathbf{A}$.

(The formulas in (ii) are suggested by the right-hand rule; see Fig. 38.)

Proof. (i) By assumption, **A** and **B** are orthogonal unit vectors, and by Theorem 2, $\mathbf{C} = \mathbf{A} \times \mathbf{B}$ is orthogonal to **A** and **B**, and

$$|\mathbf{C}|^2 = |\mathbf{A}|^2 \cdot |\mathbf{B}|^2 - |\mathbf{A} \cdot \mathbf{B}|^2 = 1 \cdot 1 - 0 = 1.$$

Hence (i) is proved.

(ii) Applying part (i) to **B** and **C** instead of **A** and **B** shows that **B**, **C**, and $\mathbf{B} \times \mathbf{C}$ are mutually orthogonal unit vectors. Hence by Theorem 10 and Theorem 2, formulas (4) and (6),

$$\begin{aligned} \mathbf{A} &= (\mathbf{A} \cdot \mathbf{B})\mathbf{B} + (\mathbf{A} \cdot \mathbf{C})\mathbf{C} + [\mathbf{A} \cdot (\mathbf{B} \times \mathbf{C})](\mathbf{B} \times \mathbf{C}) \\ &= \quad \mathbf{0} \quad + \quad \mathbf{0} \quad + ([\mathbf{C} \cdot (\mathbf{A} \times \mathbf{B})]\mathbf{B}) \times \mathbf{C} \\ &= ([\mathbf{C} \cdot \mathbf{C}]\mathbf{B}) \times \mathbf{C} = \mathbf{B} \times \mathbf{C}. \end{aligned}$$

Similarly, **C**, **A**, and $\mathbf{C} \times \mathbf{A}$ are mutually orthogonal unit vectors, so

$$\begin{aligned} \mathbf{B} &= (\mathbf{B} \cdot \mathbf{C})\mathbf{C} + (\mathbf{B} \cdot \mathbf{A})\mathbf{A} + [\mathbf{B} \cdot (\mathbf{C} \times \mathbf{A})]\mathbf{C} \times \mathbf{A} \\ &= [\mathbf{C} \cdot (\mathbf{A} \times \mathbf{B})]\mathbf{C} \times \mathbf{A} = \mathbf{C} \times \mathbf{A}. \end{aligned} \qquad \textit{Q.E.D.}$$

So far we have only two simple ways to identify a basis: (1) in R^2, any pair of nonparallel vectors forms a basis by Theorem 9; (2) in R^n, an orthonormal set of n vectors forms a basis. We add a third way, valid in R^3:

Theorem 12. *Three vectors* **A**, **B**, *and* **C** *in* R^3 *form a basis if and only if* $\mathbf{A} \cdot (\mathbf{B} \times \mathbf{C}) \neq 0$. (*Notice that* $\mathbf{A} \cdot (\mathbf{B} \times \mathbf{C})$ *is* $\pm$ *the volume of the parallelepiped in* Fig. 36.)

Proof. Suppose that $\mathbf{A} \cdot (\mathbf{B} \times \mathbf{C}) \neq 0$, and

$$r\mathbf{A} + s\mathbf{B} + t\mathbf{C} = \mathbf{0}. \tag{11}$$

Taking the dot product with $\mathbf{B} \times \mathbf{C}$ on both sides gives

$$r\mathbf{A} \cdot (\mathbf{B} \times \mathbf{C}) = 0;$$

since $\mathbf{A} \cdot (\mathbf{B} \times \mathbf{C}) \neq 0$, we have $r = 0$. Further, by Theorem 2,

$$\mathbf{B} \cdot (\mathbf{C} \times \mathbf{A}) = \mathbf{A} \cdot (\mathbf{B} \times \mathbf{C}) \neq 0,$$

so taking dot products with $\mathbf{C} \times \mathbf{A}$ on both sides of (11) gives $s = 0$, and similarly taking dot products with $\mathbf{A} \times \mathbf{B}$ gives $t = 0$. Hence if

$$\mathbf{A} \cdot (\mathbf{B} \times \mathbf{C}) \neq 0,$$

then **A**, **B**, **C** is a linearly independent set, so by Theorem 9 it forms a basis.

Conversely, suppose **A**, **B**, **C** is a basis. Then, by definition, these three vectors are linearly independent; it follows that **B**, **C** is a linearly independent set (see Problem 10), so **B** and **C** are not parallel, and $\mathbf{B} \times \mathbf{C} \neq \mathbf{0}$.

Since **A**, **B**, **C** form a basis, we can express $\mathbf{B} \times \mathbf{C}$ as a linear combination

$$\mathbf{B} \times \mathbf{C} = r\mathbf{A} + s\mathbf{B} + t\mathbf{C}.$$

Taking the inner product on both sides with $\mathbf{B} \times \mathbf{C}$, we get

$$r\mathbf{A}\cdot(\mathbf{B} \times \mathbf{C}) = |\mathbf{B} \times \mathbf{C}|^2 \neq 0;$$

hence $\mathbf{A}\cdot(\mathbf{B} \times \mathbf{C}) \neq 0$. *Q.E.D.*

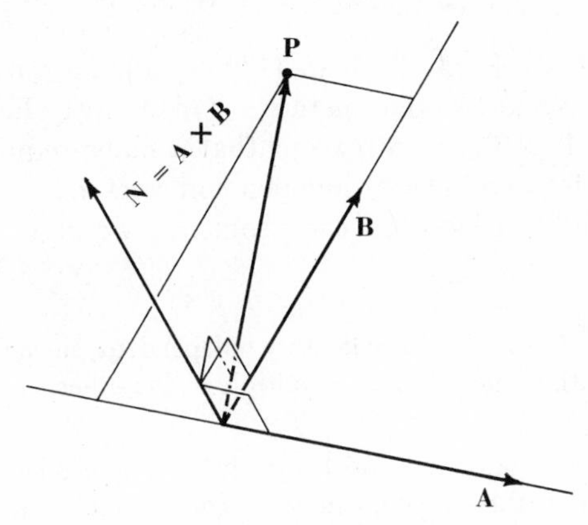

FIGURE 1.39

A typical application of Theorem 12 occurs in the proof of

Theorem 13. *If* **A** *and* **B** *are linearly independent vectors in* R^3, *then their linear combinations form a plane through the origin.*

Proof. From Fig. 39 it seems that the linear combinations of **A** and **B** should coincide with the plane

$$\{\mathbf{P}: \mathbf{P}\cdot\mathbf{N} = 0\}, \tag{12}$$

where $\mathbf{N} = \mathbf{A} \times \mathbf{B}$. The appropriate basis to use in proving this is **N**, **A**, **B**; the first step, then, is to prove that this really is a basis.

Since **A** and **B** are linearly independent, they are not parallel; hence $\mathbf{A} \times \mathbf{B} \neq \mathbf{0}$, and $\mathbf{N}\cdot(\mathbf{A} \times \mathbf{B}) = |\mathbf{A} \times \mathbf{B}|^2 \neq \mathbf{0}$, so by Theorem 12, **N**, **A**, **B** is a basis. Thus every vector **P** can be written uniquely as

$$\mathbf{P} = r\mathbf{N} + s\mathbf{A} + t\mathbf{B}; \tag{13}$$

using this representation of **P**, we find that

$$\begin{aligned} \mathbf{P}\cdot\mathbf{N} = 0 \quad &\Leftrightarrow \quad r|\mathbf{N}|^2 = 0 && (\text{since } \mathbf{A}\cdot\mathbf{N} = 0 = \mathbf{B}\cdot\mathbf{N}) \\ &\Leftrightarrow \quad r = 0 && (\text{since } \mathbf{N} \neq \mathbf{0}) \\ &\Leftrightarrow \quad \mathbf{P} = s\mathbf{A} + t\mathbf{B} && (\text{in view of } (13)) \\ &\Leftrightarrow \quad \mathbf{P} \text{ is a linear combination of } \mathbf{A} \text{ and } \mathbf{B}. \end{aligned}$$

Hence the linear combinations of **A** and **B** coincide with the points of the plane (12). *Q.E.D.*

By a very similar proof, you can show that if **A** and **B** are linearly independent, and $\mathbf{P}_0$ is any given point in R^3, then

$$\{\mathbf{P}: \mathbf{P} = \mathbf{P}_0 + s\mathbf{A} + t\mathbf{B} \text{ for some real } s \text{ and } t\} \tag{14}$$

coincides with the plane

$$\{\mathbf{P}: (\mathbf{P} - \mathbf{P}_0)\cdot(\mathbf{A} \times \mathbf{B}) = 0\}. \tag{15}$$

The equation $\mathbf{P} = \mathbf{P}_0 + s\mathbf{A} + t\mathbf{B}$ in (14) is called a *parametric equation* of the plane, with s and t as parameters. This is just like the parametric equation of a line $\mathbf{P} = \mathbf{P}_0 + s\mathbf{A}$, except that a plane requires not one but two parameters, and two linearly independent vectors.

Before turning you loose on the problems, we give some simple examples.

Example 1. (2,3) and (1,0) are linearly independent, since they are obviously not parallel; neither one is a scalar multiple of the other.

Example 2. (2,3), (1,0), and (0,1) are linearly dependent, by Theorem 8; they form a set of three vectors in R^2. You can also see directly from the definition that they are dependent, since

$$(2,3) + (-2)(1,0) + (-3)(0,1) = \mathbf{0}.$$

Example 3. (0,0) and (1,0) are linearly *dependent*, since

$$1\cdot(0,0) + 0(1,0) = \mathbf{0}.$$

Example 4. (0,1,0), (0,1,1), and (2,0,1) are linearly *independent*. This can be proved by Theorem 12, but it can also be proved directly:

$$r(0,1,0) + s(0,1,1) + t(2,0,1) = \mathbf{0} \tag{16}$$

is equivalent to

$$\begin{aligned} 2t &= 0 \quad &&\text{(equating first components)} \\ r + s &= 0 \quad &&\text{(equating second components)} \\ s + t &= 0 \quad &&\text{(equating third components).} \end{aligned}$$

It follows that $t = 0$, hence $s = 0$ (from the third equation), hence $r = 0$ (from the second equation). Thus (16) implies that $r = s = t = 0$, so the three vectors are linearly independent.

Example 5. Since (2,3) and (1,0) are two linearly independent vectors in R^2, they form a basis; hence any vector (x,y) can be written as a linear combination of these two. We can find explicitly what the combination is as follows:

$$(x,y) = s(2,3) + t(1,0)$$

means

$$\begin{aligned} x &= 2s + t \\ y &= 3s; \end{aligned}$$

hence $s = y/3$ and $t = x - 2y/3$. Thus

$$(x,y) = \frac{y}{3}(2,3) + \left(x - \frac{2y}{3}\right)(1,0)$$

is the desired linear combination.

Example 6'. $\mathbf{A} = (1,1,1)$ and $\mathbf{B} = (2,-1,-1)$ are orthogonal and nonzero, but they do *not* form a basis of R^3. Theorem 9 suggests this, but does not prove it (why not?). To prove it, notice that $\mathbf{C} = \mathbf{A} \times \mathbf{B}$ cannot be written as a linear combination of $\mathbf{A}$ and $\mathbf{B}$; for if $\mathbf{C} = s\mathbf{A} + t\mathbf{B}$, then $\mathbf{C}\cdot\mathbf{C} = s\mathbf{A}\cdot\mathbf{C} + t\mathbf{B}\cdot\mathbf{C} = 0$, which contradicts $|\mathbf{C}| = |\mathbf{A}|\cdot|\mathbf{B}| \sin\theta = |\mathbf{A}|\cdot|\mathbf{B}| \neq 0$. (Problem 8 below shows that in R^n, every basis must have exactly n members; using that result, this example could be done simply by observing that no two vectors can form a basis for R^3.)

PROBLEMS

Problems 1–17 are for practice with the ideas of linear independence, basis, and orthogonal basis. The first five are fairly routine, while Problems 6–17 ask more theoretical questions. Problems 18–20 are geometric (intersection of two planes, conic sections), and 21–28 are algebraic (Cramer's rule for a 3×3 system, partial fractions decomposition).

1. Which of the following sets are linearly dependent? Which are bases in the appropriate space R^n?
(a) $(1,-2)$ and $(-1,2)$
(b) $(1,0)$ and $(1,1)$
(c) $(1,0)$, $(1,1)$, and $(1,-1)$
(d) $(1,-1)$
(e) $(0,0)$
(f) $(0,0,0)$, $(0,1,0)$, and $(0,0,1)$
(g) $(0,1,0)$ and $(0,0,1)$
(h) $(1,0,0)$, $(1,1,0)$, and $(0,0,1)$
(i) $(1,1,0)$, $(1,-1,0)$, and $(0,0,1)$
(j) $(1,1,1)$, $(1,1,0)$, and $(1,0,0)$
(k) $(1,1,1)$, $(1,0,0)$, $(0,1,0)$, and $(0,0,1)$
(l) $(1,1,1,0)$, $(1,0,0,0)$, $(0,1,0,0)$, and $(0,0,1,0)$
(m) $(1,1,1,1)$, $(1,1,1,0)$, $(1,1,0,0)$, and $(1,0,0,0)$. (Hint: Write out $t_1\mathbf{A}_1 + t_2\mathbf{A}_2 + t_3\mathbf{A}_3 + t_4\mathbf{A}_4 = 0$ by components.)

2. (a) Prove that $\mathbf{A}_1 = (1,1)$ and $\mathbf{A}_2 = (1,-1)$ form a basis for R^2.
(b) Express the vector $(2,3)$ as a linear combination of $\mathbf{A}_1$ and $\mathbf{A}_2$. (Write out $(2,3) = t_1\mathbf{A}_1 + t_2\mathbf{A}_2$ by components, and solve for t_1 and t_2.)
(c) Express the general vector (x,y) as a linear combination of $\mathbf{A}_1$ and $\mathbf{A}_2$.

3. (a) Prove that $\mathbf{A}_1 = (1,1,1)$, $\mathbf{A}_2 = (1,-1,0)$, and $\mathbf{A}_3 = (1,1,-2)$ are mutually orthogonal.
(b) Prove that the vectors in part (a) form a basis for R^3.
(c) Express $(1,1,-1)$ as a linear combination of $\mathbf{A}_1, \mathbf{A}_2, \mathbf{A}_3$. (*Warning:* This basis is not orthonormal; it's "ortho-" but not "-normal." You can use brute force as in Problem 2(b), or be clever and take inner products as in the proof of Theorem 10.)

4. Let $$\mathbf{A} = \left(\frac{1}{\sqrt{2}}, \frac{1}{\sqrt{2}}, 0\right) \quad \text{and} \quad \mathbf{B} = \left(\frac{1}{\sqrt{2}}, \frac{-1}{\sqrt{2}}, 0\right).$$
(a) Prove that $\mathbf{A}$ and $\mathbf{B}$ are orthogonal unit vectors.
(b) Prove that $\mathbf{A}$ and $\mathbf{B}$ do *not* form a basis of R^3, by finding a vector $\mathbf{C}$ that is not a linear combination of $\mathbf{A}$ and $\mathbf{B}$.
(c) Find a vector $\mathbf{C}$ such that $\mathbf{A}, \mathbf{B}, \mathbf{C}$ is an orthonormal basis of R^3.
(d) Express $(1,1,1)$ as a linear combination of $\mathbf{A}$, $\mathbf{B}$, and $\mathbf{C}$.

5. Let $\mathbf{A} = (1,1,0)$ and $\mathbf{B} = (1,-1,0)$.
(a) Prove that $\mathbf{A}, \mathbf{B}$ is not a basis of R^3 (see Problem 4(b)).
(b) Find a vector $\mathbf{C}$ such that $\mathbf{A}, \mathbf{B}, \mathbf{C}$ is a basis of R^3.

6. Prove that if $\mathbf{A}, \mathbf{B}, \mathbf{C}$ are vectors in R^3, all lying in the plane
$$\{\mathbf{P}: \mathbf{P}\cdot\mathbf{N} = 0\},$$
then $\mathbf{A}, \mathbf{B}, \mathbf{C}$ are linearly dependent. (Hint: If they were independent, they would form a basis, hence $\mathbf{N} = r\mathbf{A} + s\mathbf{B} + t\mathbf{C}$; obtain a contradiction by taking the dot product with $\mathbf{N}$ on both sides of this equation.)

7. Suppose that $\mathbf{B}$ is orthogonal to each of the vectors $\mathbf{A}_1, \ldots, \mathbf{A}_k$. Prove that $\mathbf{B}$ is orthogonal to every linear combination $t_1\mathbf{A}_1 + \cdots + t_k\mathbf{A}_k$.

8. Prove that in R^n no set of k vectors $\mathbf{A}_1, \ldots, \mathbf{A}_k$ forms a basis if $k < n$; i.e. show that there is a vector $\mathbf{B}$ which cannot be written as a linear combination $t_1\mathbf{A}_1 + \cdots + t_k\mathbf{A}_k$. (Use Theorem 5 and the previous problem.)

9. Prove that if *any* of the vectors $\mathbf{A}_1, \ldots, \mathbf{A}_k$ is zero, then the set is linearly dependent. (Hint: For linear dependence, it suffices to have just *one* of $t_1, \ldots, t_k$ different from zero in the sum $t_1\mathbf{A}_1 + \cdots + t_k\mathbf{A}_k = \mathbf{0}$.)

10. Suppose that $\mathbf{A}_1, \ldots, \mathbf{A}_k$ is a linearly dependent set. Prove that any set $\mathbf{A}_1, \ldots, \mathbf{A}_k, \mathbf{A}_{k+1}, \ldots, \mathbf{A}_{k+m}$ containing $\mathbf{A}_1, \ldots, \mathbf{A}_k$ is also linearly dependent.

11. Suppose that $\mathbf{A}_1, \ldots, \mathbf{A}_k$ is an orthonormal set in R^n, i.e. $\mathbf{A}_j\cdot\mathbf{A}_m = 0$

if $j \neq m$, and $|\mathbf{A}_j| = 1$ for every j. Prove that there is an orthonormal basis $\mathbf{A}_1, \ldots, \mathbf{A}_n$ containing the given orthonormal set. (Use Theorem 5.)

12. The *span* of a given set of vectors $\mathbf{A}_1, \ldots, \mathbf{A}_k$ is the set of all linear combinations $t_1\mathbf{A}_1 + \cdots + t_k\mathbf{A}_k$. For example, if $\mathbf{A}_1, \ldots, \mathbf{A}_n$ is a basis of R^n, then the span of this set is all of R^n. Theorem 13 proves that the span of two nonparallel vectors in R^3 is a plane through $\mathbf{0}$.

(a) Choose $\mathbf{A}, \mathbf{B}, \mathbf{C}$ so that they span all of R^3.

(b) Prove that the span of (1,0,0), (0,1,0), and (1,1,0) is the xy plane.

(c) Choose $\mathbf{A}, \mathbf{B}, \mathbf{C}$ in R^3 so that they span only a line.

(d) Choose $\mathbf{A}, \mathbf{B}, \mathbf{C}$ in R^3 so that they span only the origin.

(e) Prove that the span of any three vectors $\mathbf{A}, \mathbf{B}, \mathbf{C}$ in R^3 is either the *origin*, a *line* through the origin, a *plane* through the origin, or all of R^3. (This is hard, starting from scratch.)

13. Let $\mathbf{A}_1, \ldots, \mathbf{A}_n$ be an orthonormal basis in R^n. Let

$$\mathbf{X} = \sum_1^n t_j\mathbf{A}_j \qquad \text{and} \qquad \mathbf{Y} = \sum_1^n s_k\mathbf{A}_k .$$

(a) Prove that $\mathbf{X}\cdot\mathbf{Y} = \sum_1^n t_j s_j$.

(b) Prove that $|\mathbf{X}| = \sqrt{\sum_1^n t_j^2}$.

14. Let $\mathbf{A}_1, \ldots, \mathbf{A}_n$ be mutually orthogonal nonzero vectors in R^n. Prove that every vector $\mathbf{X}$ can be written as

$$\mathbf{X} = \sum_1^n \frac{\mathbf{X}\cdot\mathbf{A}_j}{|\mathbf{A}_j|^2}\mathbf{A}_j = \sum \mathbf{X}_{\mathbf{A}_j},$$

where $\mathbf{X}_{\mathbf{A}_j}$ is the orthogonal projection of $\mathbf{X}$ on $\mathbf{A}_j$. (Hint: Review the proof of Theorem 10.)

15. Suppose that $\mathbf{A}, \mathbf{B}, \mathbf{C}$ is a basis of R^3. Prove that $\mathbf{B} \times \mathbf{C}, \mathbf{C} \times \mathbf{A}, \mathbf{A} \times \mathbf{B}$ is also a basis. (Hint: In $r(\mathbf{B} \times \mathbf{C}) + s(\mathbf{C} \times \mathbf{A}) + t(\mathbf{A} \times \mathbf{B}) = \mathbf{0}$, take dot products with $\mathbf{A}$, $\mathbf{B}$, and $\mathbf{C}$.)

16. Two sets of vectors $\mathbf{A}_1, \ldots, \mathbf{A}_k$ and $\mathbf{B}_1, \ldots, \mathbf{B}_k$ in R^n are called *biorthogonal* if

$$\mathbf{A}_j\cdot\mathbf{B}_m \begin{cases} = 0 & \text{for } j \neq m \\ \neq 0 & \text{for } j = m. \end{cases}$$

(a) Prove that if the two sets are biorthogonal, then each set is linearly independent.

(b) Prove that if the two sets are biorthogonal and $k = n$, then each set is a basis.

(c) With the assumptions in part (b), prove that

$$\mathbf{X} = \sum \frac{\mathbf{X}\cdot\mathbf{B}_j}{\mathbf{A}_j\cdot\mathbf{B}_j}\mathbf{A}_j$$

for every $\mathbf{X}$ in R^n. (This generalizes the formula in Theorem 10. When $\mathbf{A}_j = \mathbf{B}_j$, we have an orthogonal basis, and the formula in part (c) reduces to the one in Problem 14. If we assume further that $|\mathbf{A}_j| = 1$, we have an orthonormal basis, as in Theorem 10.)

17. Suppose that $\mathbf{A}_1, \ldots, \mathbf{A}_k$ is an orthonormal set in R^n, i.e.

$$\mathbf{A}_j\cdot\mathbf{A}_m = \begin{cases} 0 & \text{if } j \neq m \\ 1 & \text{if } j = m. \end{cases}$$

Prove *Bessel's inequality:*

$$|\mathbf{X}|^2 \geq \sum_1^k (\mathbf{X}\cdot\mathbf{A}_j)^2,$$

i.e. $|\mathbf{X}|^2$ is greater than or equal to the sum of the squares of the lengths of the projections of $\mathbf{X}$ on $\mathbf{A}_1, \ldots, \mathbf{A}_k$. (Hint: Show that

$$\left|\mathbf{X} - \sum_1^k (\mathbf{X}\cdot\mathbf{A}_j)\mathbf{A}_j\right|^2 = |\mathbf{X}|^2 - \sum_1^k (\mathbf{X}\cdot\mathbf{A}_j)^2.)$$

18. Suppose that $\mathbf{A}$ and $\mathbf{B}$ are not parallel. Prove that the set

$$\{\mathbf{P}: \mathbf{P} = \mathbf{P}_0 + s\mathbf{A} + t\mathbf{B} \text{ for some } s \text{ and } t\}$$

is the plane $\{\mathbf{P}: (\mathbf{P} - \mathbf{P}_0)\cdot(\mathbf{A} \times \mathbf{B}) = 0\}$.

19. Suppose that $\mathbf{A}$ and $\mathbf{B}$ are not parallel.

(a) Prove that the two planes $\{\mathbf{P}: \mathbf{P}\cdot\mathbf{A} = 0\}$ and $\{\mathbf{P}: \mathbf{P}\cdot\mathbf{B} = 0\}$ intersect in the line $\{\mathbf{P}: \mathbf{P} = t\mathbf{A} \times \mathbf{B} \text{ for some } t\}$. (Hint: Look at the proof of Theorem 13.)

(b) Let $\mathbf{P}_0$ be any point. Prove that the two planes

$$\{\mathbf{P}: (\mathbf{P} - \mathbf{P}_0)\cdot\mathbf{A} = 0\} \quad \text{and} \quad \{\mathbf{P}: (\mathbf{P} - \mathbf{P}_0)\cdot\mathbf{B} = 0\}$$

intersect in the line $\{\mathbf{P}: \mathbf{P} = \mathbf{P}_0 + t(\mathbf{A} \times \mathbf{B})\}$. (Hint: Write $\mathbf{P} - \mathbf{P}_0$ as a linear combination of appropriate basis vectors.)

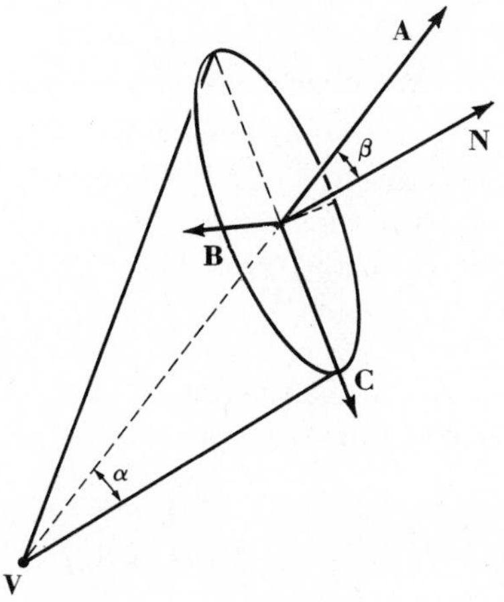

FIGURE 1.40

20. This problem analyzes the intersection of a plane $\{\mathbf{P}: \mathbf{P}\cdot\mathbf{N} = d\}$ and a double cone $\{\mathbf{P}: |(\mathbf{P} - \mathbf{V})\cdot\mathbf{A}| = |\mathbf{P} - \mathbf{V}| \cos\alpha\}$. We assume (without loss of generality) that $|\mathbf{N}| = 1$ and $|\mathbf{A}| = 1$. We also assume that $\mathbf{N}$ and $\mathbf{A}$ are not parallel. Let β be the angle between $\mathbf{N}$ and $\mathbf{A}$.

(a) If $\mathbf{B} = \dfrac{1}{\sin\beta}\mathbf{N} \times \mathbf{A}$ and $\mathbf{C} = \mathbf{N} \times \mathbf{B}$, then $\mathbf{N}, \mathbf{B}, \mathbf{C}$ is an orthonormal basis of R^3 (see Fig. 40).

(b) Prove that $\mathbf{A}\cdot\mathbf{B} = 0$ and $\mathbf{A}\cdot\mathbf{C} = -\sin\beta$. (Use Theorem 2.)

(c) Let $\mathbf{P}_0 = \mathbf{V} + t_0\mathbf{A}$ be the intersection of the given plane and the axis of the cone (the axis is the line $\{\mathbf{P}: \mathbf{P} = \mathbf{V} + t\mathbf{A}$ for some $t\}$). Prove that a point $\mathbf{P}$ is on the plane if and only if $\mathbf{P} = \mathbf{V} + t_0\mathbf{A} + s\mathbf{B} + t\mathbf{C}$ for some real s and t. (See Problem 19.)

(d) Prove that the distance between two points $\mathbf{P} = \mathbf{P}_0 + s\mathbf{B} + t\mathbf{C}$ and $\mathbf{P}' = \mathbf{P}_0 + s'\mathbf{B} + t'\mathbf{C}$ on the plane is given by

$$|\mathbf{P} - \mathbf{P}'|^2 = (s - s')^2 + (t - t')^2.$$

(e) Prove that $\mathbf{P}$ is on both the plane and the cone if and only if

$\mathbf{P} = \mathbf{P}_0 + s\mathbf{B} + t\mathbf{C}$ and

$$(t_0 - t\sin\beta)^2 - (t_0^2 + s^2 + t^2 + 2t_0t\sin\beta)\cos^2\alpha. \tag{17}$$

(f) Let $e = (\sin\beta)/(\cos\alpha)$. Prove that if $e \neq 1$, then equation (17) is equivalent to

$$\frac{s^2}{1 - e^2} + \left(t + \frac{t_0 \sin\beta \tan^2\alpha}{1 - e^2}\right)^2 = \frac{t_0^2 \tan^2\alpha \cos^2\beta}{(1 - e^2)^2};$$

and if $e = 1$, it is equivalent to

$$s^2 = t_0 \tan^2 \alpha (t_0 - 2t \sin \beta).$$

(In view of part (d), distances on the plane $\{\mathbf{P}: \mathbf{P} \cdot \mathbf{N} = d\}$ are computed in terms of s and t just as distances in the plane R^2 are computed in terms of x and y. Hence, by the methods of elementary analytic geometry, part (f) shows that the curve is an ellipse when $e < 1$, a parabola when $e = 1$, and a hyperbola when $e > 1$.)

21. (Cramer's rule.) Suppose that $\mathbf{A} \cdot (\mathbf{B} \times \mathbf{C}) \neq 0$, and let $\mathbf{D}$ be any vector in R^3. Prove that $\mathbf{D} = r\mathbf{A} + s\mathbf{B} + t\mathbf{C}$ if and only if

$$r = \frac{\mathbf{D} \cdot (\mathbf{B} \times \mathbf{C})}{\mathbf{A} \cdot (\mathbf{B} \times \mathbf{C})}, \qquad s = \frac{\mathbf{D} \cdot (\mathbf{A} \times \mathbf{C})}{\mathbf{B} \cdot (\mathbf{A} \times \mathbf{C})}, \qquad t = \frac{\mathbf{D} \cdot (\mathbf{A} \times \mathbf{B})}{\mathbf{C} \cdot (\mathbf{A} \times \mathbf{B})}.$$

(Hint: Theorem 12 plays an important part in the proof.)

22. Prove that if $a_1 b_2 \neq b_1 a_2$, then the simultaneous equations

$$\begin{aligned} ra_1 + sb_1 &= d_1 \\ ra_2 + sb_2 &= d_2 \end{aligned}$$

have the unique solution

$$r = \frac{d_1 b_2 - d_2 b_1}{a_1 b_2 - a_2 b_1}, \qquad s = \frac{a_1 d_2 - a_2 d_1}{a_1 b_2 - a_2 b_1}.$$

(Hint: In the previous problem, set $\mathbf{A} = (a_1, a_2, 0)$, $\mathbf{B} = (b_1, b_2, 0)$, $\mathbf{C} = (0,0,1)$, and $\mathbf{D} = (d_1, d_2, 0)$.

23. Show that Theorem 9 is equivalent to the following theorem: *Suppose that the $n \times n$ system of equations*

$$\begin{aligned} t_1 a_{11} + \cdots + t_n a_{n1} &= 0 \\ t_1 a_{12} + \cdots + t_n a_{n2} &= 0 \\ \vdots \qquad\qquad & \vdots \\ t_1 a_{1n} + \cdots + t_n a_{nn} &= 0 \end{aligned}$$

has no solution other than $t_1 = t_2 = \cdots = t_n = 0$. Then for each choice of the constants $b_1, \ldots, b_n$ the system

$$\begin{aligned} t_1 a_{11} + \cdots + t_n a_{n1} &= b_1 \\ \vdots \qquad\qquad & \vdots \\ t_1 a_{1n} + \cdots + t_n a_{nn} &= b_n \end{aligned}$$

has a unique solution $t_1, \ldots, t_n$.

The next five problems prove that partial fraction decompositions are always possible.

24. Prove Theorem 6 for systems of equations where the coefficients (and the solutions) are *complex* numbers.

25. $\mathbf{C}^n$ is the vector space of n-tuples of *complex* numbers; addition and scalar multiplication are defined exactly as in §1.4, except that now the numbers t, a_1, b_1, ... are complex. The inner product is defined as $\sum a_j\bar{b}_j$, and the length as $\sqrt{\sum |a_j|^2}$. Prove the analog of Theorem 1 with R^3 replaced by $\mathbf{C}^n$. (The formula $\mathbf{A}\cdot\mathbf{B} = \mathbf{B}\cdot\mathbf{A}$ must be replaced by $\mathbf{A}\cdot\mathbf{B} = \overline{\mathbf{B}\cdot\mathbf{A}}$, and $t(\mathbf{A}\cdot\mathbf{B}) = \mathbf{A}\cdot t\mathbf{B}$ by $t\mathbf{A}\cdot\mathbf{B} = \mathbf{A}\cdot\bar{t}\mathbf{B}$.)

26. A linear combination in $\mathbf{C}^n$ is a sum $t_1\mathbf{A}_1 + \cdots + t_k\mathbf{A}_k$, where the t_j are complex numbers. *Linear dependence* and *basis* are defined for $\mathbf{C}^n$ exactly as for R^n. Prove the analogs of Theorems 8 and 9 for $\mathbf{C}^n$. (See Problem 24.)

27. Prove the analog of Problem 23 for a *complex* system of equations.

28. Let $r_1, \ldots, r_k$ be distinct complex numbers, let $n_1, \ldots, n_k$ be positive integers, and set $n = n_1 + \cdots + n_k$.

(a) Prove that when z is not equal to any of $r_1, \ldots, r_k$, then any linear combination

$$\frac{t_1}{z - r_1} + \frac{t_1}{(z - r_1)^2} + \cdots + \frac{t_{n1}}{(z - r_1)^{n1}} \qquad (n_1 \text{ terms})$$

$$+ \frac{t_{n1+1}}{z - r_2} + \frac{t_{n1+2}}{(z - r_2)^2} + \cdots \qquad (n_2 \text{ terms})$$

$$\vdots \qquad\qquad \vdots$$

$$+ \frac{t_{n-nk+1}}{z - z_k} + \cdots + \frac{t_n}{(z - r_k)^{nk}} \qquad (n_k \text{ terms})$$

can be written in the form

$$\frac{\left(\sum_1^n a_{j1}t_j\right)z^{n-1} + \left(\sum_1^n a_{j2}t_j\right)z^{n-2} + \cdots + \left(\sum_1^n a_{jn}t_j\right)}{(z - r_1)^{n1} \cdots (z - r_k)^{nk}}, \qquad (18)$$

where the a_{jk} are complex numbers depending on $r_1, \ldots, r_k$ and $n_1, \ldots, n_k$, but not on $t_1, \ldots, t_n$.

(b) Prove that if the numerator on the right in (18) is the zero polynomial, then $t_1 = \cdots = t_n = 0$. (Hint: Notice that the numerator N in (18) equals $t_{n_1}(z - r_2)^{n2} \cdots (z - r_k)^{nk}$ + [a polynomial that vanishes when $z = r_1$]; hence if $N \equiv 0$, then $t_{n_1} = 0$. With this established, show that $t_{n_1-1} = 0$, etc. Similarly, all the other coefficients are zero.)

(c) Conclude from Problem 27 that for *any* given complex numbers $b_1, \ldots, b_n$, the constants $t_1, \ldots, t_n$ can be chosen uniquely so that the numerator in (18) equals $b_1 z^{n-1} + \cdots + b_{n-1} z + b_n$.

(d) Prove that if $r_1 = \bar{r}_2$ and $n_1 = n_2$, then

$$\frac{t_1}{z - r_1} + \cdots + \frac{t_{n_1}}{(z - r_1)^{n_1}} + \frac{t_{n_1+1}}{(z - \bar{r}_1)} + \cdots + \frac{t_{2n_1}}{(z - \bar{r}_1)^{n_1}}$$

$$= \frac{t_1' + t_1'' Z}{z^2 - 2\,\mathrm{Re}(r_1) z + |r_1|^2} + \cdots + \frac{t_{n_1}' + t_{n_1}'' Z}{(z^2 - 2\,\mathrm{Re}(r_1) z + |r_1|^2)^{n_1}}$$

for uniquely determined constants $t_1', t_1'', \ldots, t_{n_1}', t_{n_1}''$.

(Parts (c) and (d) prove that if

$$Q = (x - a_1)^{r_1} \cdots (x - a_k)^{r_k} (x^2 + b_1 x + c_1)^{s_1} \cdots (x^2 + b_m x + c_m)^{s_m},$$

where $x^2 + b_j x + c_j$ has no real roots, and if R is a polynomial of degree $< n = r_k + 2s_m$, then R/Q can be written uniquely as a linear combination of

$$\frac{1}{x - a_1}, \ldots, \frac{1}{(x - a_1)^{r_1}}, \ldots,$$

$$\frac{1}{(x^2 + b_m x + c_m)^{s_m}}, \frac{x}{(x^2 + b_m x + c_m)^{s_m}}.)$$

29. (a) Prove the Schwarz inequality in $\mathbf{C}^n$:

$$|\mathbf{A} \cdot \mathbf{B}| \leq |\mathbf{A}| \cdot |\mathbf{B}|.$$

(Hint: Minimize $|\theta \mathbf{A} + t\mathbf{B}|^2$, where θ is fixed so that $|\theta| = 1$ and $\theta \mathbf{A} \cdot \mathbf{B}$ is real.)

(b) Prove the triangle inequality in $\mathbf{C}^n$:

$$|\mathbf{A} + \mathbf{B}| \leq |\mathbf{A}| + |\mathbf{B}|.$$

Chapter II

Curves in R^n

The study of curves is interesting in its own right, and moreover provides a useful tool in the study of functions of several variables.

§2.1 gives the basic theory of curves.

§2.2 applies it to some problems of motion (primarily, the motion of an electron) in order to show how certain curves (the helix and the parabola, for example) arise in nature.

§2.3 studies curves from a geometric point of view.

None of §2.2 is used in the rest of the book, nor is much of §2.3, but these sections put some meat on the skeleton given in §2.1.

2.1 DEFINITIONS AND ELEMENTARY PROPERTIES

A function $\mathbf{F}$ that assigns a vector $\mathbf{F}(t)$ to each point t in some interval of the real line is called a *vector function of one real variable.* We write

$$\mathbf{F}: [a,b] \to R^n$$

to show that $\mathbf{F}$ is a function from the interval $[a,b]$ to R^n.

The vector $\mathbf{F}(t)$ has n components $f_1(t), \ldots, f_n(t)$ such that $\mathbf{F}(t) = (f_1(t), \ldots, f_n(t))$; for example, if $\mathbf{F}(t) = (t, e^t, \sin t)$, the components are

$$f_1(t) = t, \qquad f_2(t) = e^t, \qquad f_3(t) = \sin t.$$

If the components of $\mathbf{F}$ are all continuous functions, then $\mathbf{F}$ is called continuous. A continuous vector-valued function defined on an interval is called simply a *curve*, and is usually denoted by the letter $\boldsymbol{\gamma}$, instead of $\mathbf{F}$.

This definition of curve as a function conflicts slightly with the general idea of a curve as a figure in space, but there are good reasons for our terminology. First, we need a reasonable abbreviation for "continuous vector-valued function of one variable." Second, the easiest way to describe analytically a curved line in space is by means of a vector-valued function of one variable, as you will see in the following examples.

The simplest example of a curve is

$$\boldsymbol{\gamma}(t) = \mathbf{A} + t\mathbf{B},$$

where $\mathbf{A}$ and $\mathbf{B}$ are fixed vectors in R^n, $\mathbf{B} \neq \mathbf{0}$. The range of $\boldsymbol{\gamma}$, i.e. the set

$$\{\mathbf{X}: \mathbf{X} = \boldsymbol{\gamma}(t) \text{ for some } t\},$$

is a straight line, by definition (see Fig. 1 and §1.4). When $n = 3$, this curve describes the motion of a particle in space, with constant speed and direction; at time t, the particle is at $\boldsymbol{\gamma}(t)$. If $\mathbf{A} = (a_1, \ldots, a_n)$ and $\mathbf{B} = (b_1, \ldots, b_n)$, then

$$\boldsymbol{\gamma}(t) = (a_1 + tb_1, \ldots, a_n + tb_n);$$

hence the components of $\boldsymbol{\gamma}$ are

$$\begin{aligned} \gamma_1(t) &= a_1 + tb_1 \\ &\vdots \\ \gamma_n(t) &= a_n + tb_n. \end{aligned}$$

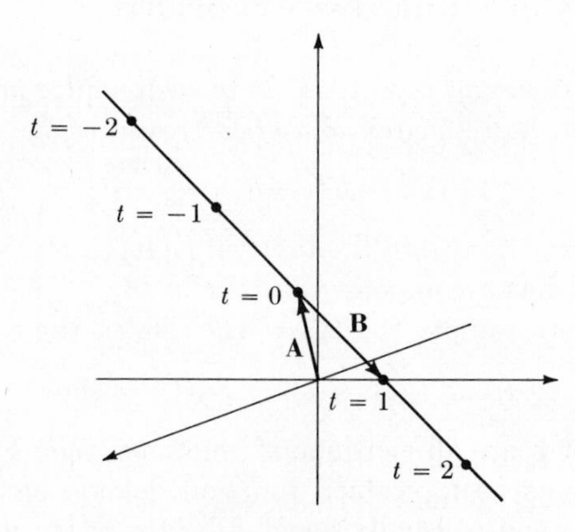

FIGURE 2.1

A more interesting example is the helix

$$\boldsymbol{\gamma}(t) = (a \cos t, a \sin t, bt) \tag{1}$$

which spirals steadily around a circular cylinder of radius a (see Fig. 2). This describes the motion of an electron in a constant magnetic field (see §2.2 below). The components of $\boldsymbol{\gamma}$ are

$$\begin{aligned} \gamma_1(t) &= a \cos t \\ \gamma_2(t) &= a \sin t \\ \gamma_3(t) &= bt. \end{aligned}$$

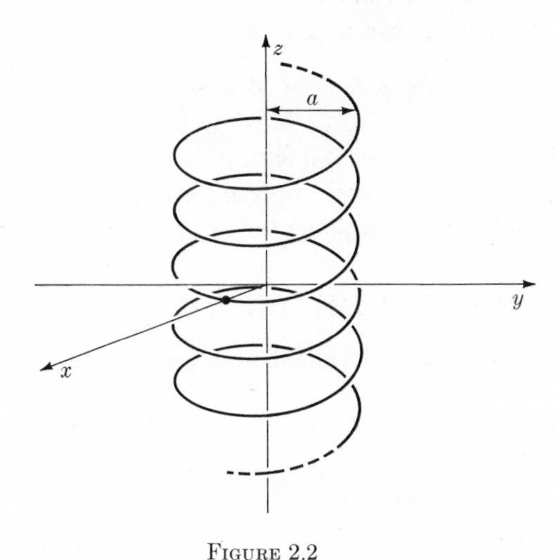

FIGURE 2.2

Notice that Figs. 1 and 2 do not show the *graph* of $\boldsymbol{\gamma}$; this would be the set in R^{n+1}

$$\{(t, x_1, \ldots, x_n) : x_1 = \gamma_1(t), \ldots, x_n = \gamma_n(t)\}.$$

What the figures actually show is the *range* of $\boldsymbol{\gamma}$, the set of points $\{\mathbf{X}: \mathbf{X} = \boldsymbol{\gamma}(t)$ for some $t\}$, sometimes called the *trace* of the curve $\boldsymbol{\gamma}$. Any point $\boldsymbol{\gamma}(t)$ in the range of $\boldsymbol{\gamma}$ is called a *point on the curve.*

The *derivative* of a curve $\boldsymbol{\gamma} = (\gamma_1, \ldots, \gamma_n)$ is defined by

$$\boldsymbol{\gamma}'(t) = (\gamma_1'(t), \ldots, \gamma_n'(t)).$$

For example, if $\boldsymbol{\gamma}$ is the helix (1), then

$$\boldsymbol{\gamma}'(t) = (-a \sin t, a \cos t, b).$$

If $\boldsymbol{\gamma}'(t)$ exists at each point where $\boldsymbol{\gamma}(t)$ is defined, then $\boldsymbol{\gamma}$ is called a *differentiable* curve.

The usual formulas for differentiation carry over to curves, as in the following theorem.

Theorem 1. *Let* $\boldsymbol{\gamma}$ *and* $\boldsymbol{\delta}$ *be curves, let* h *be a real-valued function, and suppose that* $\boldsymbol{\gamma}'(c)$, $\boldsymbol{\delta}'(c)$, *and* $h'(c)$ *exist. Then*

$$(\boldsymbol{\gamma} + \boldsymbol{\delta})'(c) = \boldsymbol{\gamma}'(c) + \boldsymbol{\delta}'(c) \tag{2}$$

$$(\boldsymbol{\gamma} \cdot \boldsymbol{\delta})'(c) = \boldsymbol{\gamma}'(c) \cdot \boldsymbol{\delta}(c) + \boldsymbol{\gamma}(c) \cdot \boldsymbol{\delta}'(c) \tag{3}$$

$$(h\boldsymbol{\gamma})'(c) = h'(c)\,\boldsymbol{\gamma}(c) + h(c)\,\boldsymbol{\gamma}'(c). \tag{4}$$

In the case of curves in R^3,

$$(\boldsymbol{\gamma} \times \boldsymbol{\delta})'(c) = \boldsymbol{\gamma}'(c) \times \boldsymbol{\delta}(c) + \boldsymbol{\gamma}(c) \times \boldsymbol{\delta}'(c). \tag{5}$$

Finally, if φ *is a real-valued function such that* $\varphi'(c)$ *exists, and* $\boldsymbol{\gamma}'(\varphi(c))$ *exists, then the chain rule holds:*

$$(\boldsymbol{\gamma} \circ \varphi)'(c) = \varphi'(c)\,\boldsymbol{\gamma}'(\varphi(c)). \tag{6}$$

Each of these formulas is proved simply by writing out the components on both sides of the equation and applying known results for real-valued functions to the individual components. For example, to prove (6), we have

$$(\boldsymbol{\gamma} \circ \varphi)(t) = \boldsymbol{\gamma}(\varphi(t)) = \big(\gamma_1(\varphi(t)), \ldots, \gamma_n(\varphi(t))\big);$$

hence by the chain rule for real-valued functions

$$\begin{aligned}(\boldsymbol{\gamma} \circ \varphi)'(c) &= \big(\gamma_1'(\varphi(c))\varphi'(c), \ldots, \gamma_n'(\varphi(c))\varphi'(c)\big) \\ &= \varphi'(c)\big(\gamma_1'(\varphi(c)), \ldots, \gamma_n'(\varphi(c))\big) \\ &= \varphi'(c)\,\boldsymbol{\gamma}'(\varphi(c)).\end{aligned}$$

Notice that the order of the factors in formula (5) for the derivative of the cross product $\boldsymbol{\gamma} \times \boldsymbol{\delta}$ has to be preserved, since generally

$$\boldsymbol{\gamma}' \times \boldsymbol{\delta} \neq \boldsymbol{\delta} \times \boldsymbol{\gamma}'.$$

Example 1. If $\boldsymbol{\gamma}(t) \equiv \mathbf{B}$ (a constant vector), then each component of $\boldsymbol{\gamma}$ is constant; hence $\boldsymbol{\gamma}'(t) = \mathbf{0}$ for all t.

Example 2. Applying formula (4) with $\boldsymbol{\gamma}(t) = \mathbf{B}$ and $h(t) = t$, we find that $t\mathbf{B}$ has the derivative $1 \cdot \mathbf{B} + t \cdot \mathbf{0} = \mathbf{B}$.

Example 3. Applying formula (2) with $\boldsymbol{\gamma}(t) = \mathbf{A}$ and $\boldsymbol{\delta}(t) = t\mathbf{B}$, we find $\boldsymbol{\gamma}' \equiv \mathbf{0}$ (by Example 1) and $\boldsymbol{\delta}' \equiv \mathbf{B}$ (by Example 2), hence by (2) the sum $\mathbf{A} + t\mathbf{B}$ has the derivative

$$\mathbf{0} + \mathbf{B} = \mathbf{B}.$$

We have defined continuity and differentiability of a vector function in terms of its components, but these properties can be characterized more intuitively without referring to components.

Theorem 2. *A vector function* $\mathbf{F}\colon [a,b] \to R^n$ *is continuous at a point* c, $a < c < b$, *if and only if*

$$\lim_{t\to c} |\mathbf{F}(t) - \mathbf{F}(c)| = 0. \tag{7}$$

Similarly, $\mathbf{F}$ *is continuous at the left endpoint* a *if and only if*

$$\lim_{t\to a+} |\mathbf{F}(t) - \mathbf{F}(a)| = 0,$$

and $\mathbf{F}$ *is continuous at the right endpoint* b *if and only if*

$$\lim_{t\to b-} |\mathbf{F}(t) - \mathbf{F}(b)| = 0.$$

Formula (7) says that the distance between $\mathbf{F}(t)$ and $\mathbf{F}(c)$ tends to zero as t tends to c. Intuitively, when $\mathbf{F}$ is continuous, the moving particle does not jump around, but moves continuously from point to point.

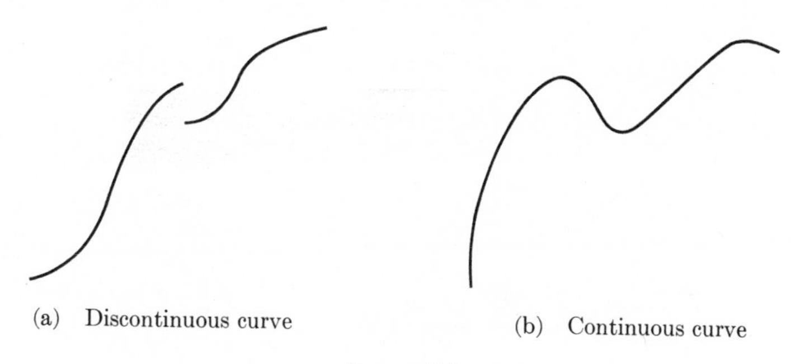

(a) Discontinuous curve (b) Continuous curve

FIGURE 2.3

Proof of Theorem 2. Suppose that $\mathbf{F}$ is continuous at c; i.e. each component f_j is continuous at c. By Problem 6, §1.4,

$$0 \le |\mathbf{F}(t) - \mathbf{F}(c)| \le \sum_{j=1}^{n} |f_j(t) - f_j(c)|. \tag{8}$$

Since f_j is continuous at c, we have $\lim_{t\to c} |f_j(t) - f_j(c)| = 0$ for each j; hence the sum on the right in (8) has limit zero, and (7) follows.

Conversely, suppose (7) holds. Then for each j,

$$0 \le |f_j(t) - f_j(c)| = \sqrt{[f_j(t) - f_j(c)]^2}$$
$$\le \sqrt{\sum_{k=1}^{n} [f_k(t) - f_k(c)]^2} = |\mathbf{F}(t) - \mathbf{F}(c)|. \tag{9}$$

By (7), the last term on the right has limit zero as $t \to c$; hence $|f_j(t) - f_j(c)|$ has limit zero, so f_j is continuous at c.

The corresponding result when c is replaced by an endpoint a or b is proved in the same way.

Theorem 2 characterizes continuity at c by comparing $\mathbf{F}(t)$ to the *constant* vector $\mathbf{F}(c)$. We get an analogous characterization of the derivative $\boldsymbol{\gamma}'(c)$ of a curve $\boldsymbol{\gamma}$ by comparing $\boldsymbol{\gamma}(t)$ to the *straight line* curve $\boldsymbol{\gamma}(c) + (t - c)\boldsymbol{\gamma}'(c)$.

Theorem 3. $\boldsymbol{\gamma}'(c)$ *exists and equals* $\mathbf{D}$ *if and only if*

$$\lim_{t \to c} \frac{1}{|t - c|} |\boldsymbol{\gamma}(t) - \boldsymbol{\gamma}(c) - (t - c)\mathbf{D}| = 0. \tag{10}$$

Proof. The theorem follows from a chain of equivalent statements, beginning with $\boldsymbol{\gamma}'(c) = \mathbf{D}$. Let $\mathbf{D} = (d_1, \ldots, d_n)$. By definition, $\boldsymbol{\gamma}'(c) = \mathbf{D}$ if and only if $\gamma_j'(c) = d_j$ for every j. Further, $\gamma_j'(c) = d_j$ if and only if the function

$$f_j(t) = \begin{cases} \dfrac{1}{t - c}[\gamma_j(t) - \gamma_j(c)] & t \neq c \\ d_j & t = c \end{cases} \tag{11}$$

is continuous at $t = c$. Let $\mathbf{F} = (f_1, \ldots, f_n)$. By Theorem 2, the functions f_j are all continuous at c if and only if

$$\lim_{t \to c} |\mathbf{F}(t) - \mathbf{F}(c)| = 0. \tag{12}$$

But looking at (11), you can see that

$$|\mathbf{F}(t) - \mathbf{F}(c)| = \frac{1}{|t - c|} |\boldsymbol{\gamma}(t) - \boldsymbol{\gamma}(c) - (t - c)\mathbf{D}|;$$

hence (12) is equivalent to (10). *Q.E.D.*

Notice that if $\boldsymbol{\gamma}$ is continuous at c, then for *any* choice of $\mathbf{D}$ we have the limit relation

$$\lim_{t \to c} |\boldsymbol{\gamma}(t) - \boldsymbol{\gamma}(c) - (t - c)\mathbf{D}| = 0; \tag{13}$$

for

$$\begin{aligned} 0 &\leq |\boldsymbol{\gamma}(t) - \boldsymbol{\gamma}(c) - (t - c)\mathbf{D}| \\ &\leq |\boldsymbol{\gamma}(t) - \boldsymbol{\gamma}(c)| + |t - c|\,|\mathbf{D}|, \end{aligned} \tag{14}$$

and the right-hand side of (14) tends to zero as $t \to c$, by Theorem 2. However, dividing the expression in (13) by $|t - c|$ leads to the much stronger relation (10) which, by Theorem 3, can be satisfied for *only one* choice of $\mathbf{D}$, namely, for $\mathbf{D} = \boldsymbol{\gamma}'(c)$. The resulting straight line

$$\{\mathbf{X}: \mathbf{X} = \boldsymbol{\gamma}(c) + (t - c)\boldsymbol{\gamma}'(c) \text{ for some } t\}$$

is by definition the *tangent line* to $\boldsymbol{\gamma}$ at the point $\boldsymbol{\gamma}(c)$.

A simple rewriting of (10) yields

$$\lim_{t \to c} \left| \frac{\boldsymbol{\gamma}(t) - \boldsymbol{\gamma}(c)}{t - c} - \boldsymbol{\gamma}'(c) \right| = 0; \tag{15}$$

i.e. the difference quotient $\dfrac{\boldsymbol{\gamma}(t) - \boldsymbol{\gamma}(c)}{t - c}$ tends to $\boldsymbol{\gamma}'(c)$ as $t \to c$ (see Fig. 4).

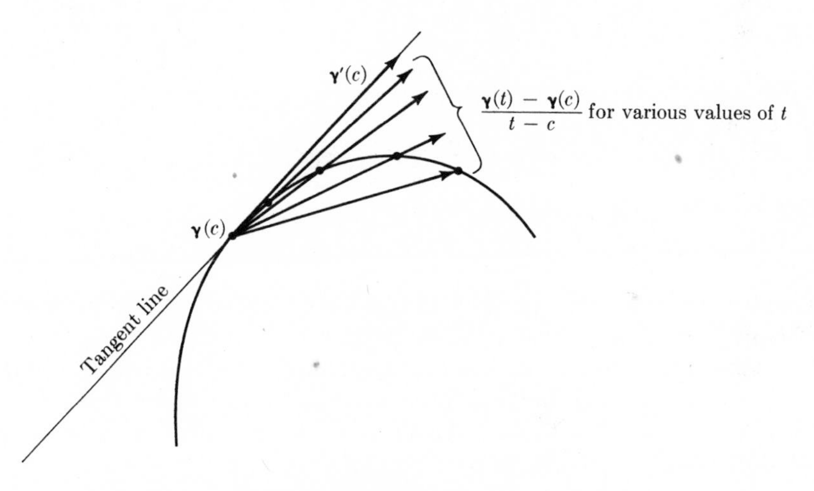

Figure 2.4

When $\boldsymbol{\gamma}$ describes the motion of a particle, the ratio

$$\frac{\boldsymbol{\gamma}(t) - \boldsymbol{\gamma}(c)}{t - c}$$

is the average velocity of the particle from time c to time t; hence, in view of (15), $\boldsymbol{\gamma}'(c)$ is called the *instantaneous velocity* at time c. The derivative $\boldsymbol{\gamma}'(c)$ is represented by an arrow beginning at the point $\boldsymbol{\gamma}(c)$ and tangent to the curve at that point, as in Fig. 4. For example, with the helix

$$\boldsymbol{\gamma}(t) = (2 \cos t, 2 \sin t, t) \tag{16}$$

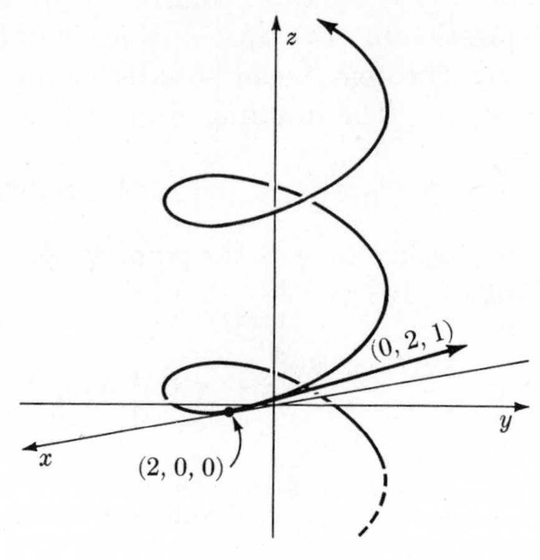

FIGURE 2.5

we have

$$\gamma'(t) = (-2 \sin t, 2 \cos t, 1) \tag{17}$$

and in particular

$$\gamma'(0) = (0,2,1).$$

Figure 5 shows $\gamma'(0)$ as an arrow tangent to the curve at the point $\gamma(0) = (2,0,0)$.

This same helix illustrates an interesting difference between curves in R^1 (i.e. real-valued functions of one variable) and curves in R^n, $n > 1$. In R^1 the mean value theorem holds: *If f: $[a,b] \to R^1$ is continuous, and f is differentiable on (a,b), then*

$$\frac{f(b) - f(a)}{b - a} = f'(t_0)$$

for some t_0 in (a,b). For curves in R^n, $n > 1$, this is no longer true. For example, taking the helix (16), we have

$$\frac{\gamma(2\pi) - \gamma(0)}{2\pi} = (0,0,1),$$

and you can see from (17) that the derivative $\gamma'(t)$ is never equal to $(0,0,1)$; in fact, $\gamma'(t)$ is never even parallel to $(0,0,1)$ (see Fig. 6). In terms of velocity, this shows that the average velocity from time a to time b may *not* equal the instantaneous velocity at any intermediate time.

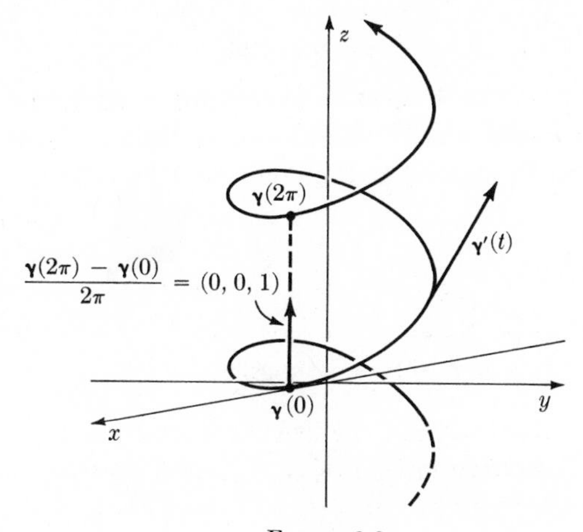

FIGURE 2.6

PROBLEMS

1. (a) Find the derivative of the helix

$$\gamma(t) = \left(\cos\frac{t}{2}, \sin\frac{t}{2}, t\right).$$

(b) Sketch the curve for $-\pi \leq t \leq \pi$, and sketch $\gamma'(0)$ as an arrow beginning at the corresponding point $\gamma(0) = (1,0,0)$ on the curve.

2. Repeat Problem 1 for the "elliptical spiral"

$$\gamma(t) = (\cos t, 2\sin t, 3t).$$

3. Repeat Problem 1 for the "expanding spiral"

$$\gamma(t) = (t\cos t, t\sin t, t).$$

4. (a) At what points does the curve $\gamma(t) = (t - 1, 3 + t^2, 2t^2)$ intersect the plane with equation $14x + y + 3z = 10$?

(b) Find the angle of intersection in part (a), i.e. the complement of the angle between γ' and the normal to the plane.

5. (a) Find a point of intersection of the two curves γ and $\bar{\gamma}$, where $\gamma(t) = (\cos t, \sin t, 0)$ and $\bar{\gamma}(t) = (0, \cos t, \sin t)$. (Hint: Do not simply set $\cos t = 0$, $\sin t = \cos t$, $0 = \sin t$; the curves may go through the same point at different times.)

(b) Find a parametric equation of a line that is perpendicular to both γ and $\bar{\gamma}$ at the point of intersection found in part (a).

6. For the helix $\boldsymbol{\gamma}(t) = (a \cos t, a \sin t, bt)$, prove that $\boldsymbol{\gamma}'(t)$ is orthogonal to $(a \cos t, a \sin t, 0)$. Draw a sketch illustrating this situation.

7. The *normal plane* to a curve $\boldsymbol{\gamma}$ at a point $\boldsymbol{\gamma}(t_0)$ is the plane through $\boldsymbol{\gamma}(t_0)$ with normal $\boldsymbol{\gamma}'(t_0)$. Find the equation of the normal plane to the helix in Problem 6 at the point $(a,0,0)$.

8. Let $\boldsymbol{\gamma}(t) = \left(\dfrac{2t}{1+t^2}, \dfrac{1-t^2}{1+t^2}\right)$.
   (a) Prove that $|\boldsymbol{\gamma}(t)| \equiv 1$.
   (b) Prove that $\boldsymbol{\gamma}$ and $\boldsymbol{\gamma}'$ are orthogonal.

9. (a) Find the derivative of the "twisted cubic" $\boldsymbol{\gamma}(t) = (t,t^2,t^3)$.
   (b) Sketch the curve for $-1 \leq y \leq 1$, and sketch $\boldsymbol{\gamma}'(0)$.
   (c) Is there any point t_0, $0 < t_0 < 1$, such that

$$\boldsymbol{\gamma}'(t_0) = \frac{1}{1-0}[\boldsymbol{\gamma}(1) - \boldsymbol{\gamma}(0)]?$$

   (Compare this equation to the mean value theorem.)
   (d) Is there any point t_0, $0 < t_0 < 1$, such that $\boldsymbol{\gamma}'(t_0)$ is parallel to $\boldsymbol{\gamma}(1) - \boldsymbol{\gamma}(0)$?

10. Repeat Problem 9, but this time use the curve $\boldsymbol{\gamma}(t) = (t,t^3)$ in R^2.

11. Suppose that $\boldsymbol{\gamma}$ has a derivative $\boldsymbol{\gamma}'$, that $\boldsymbol{\gamma}'$ has a derivative $\boldsymbol{\gamma}''$, and that $\boldsymbol{\gamma}''$ has a derivative $\boldsymbol{\gamma}'''$. Prove that

$$[\boldsymbol{\gamma}\cdot(\boldsymbol{\gamma}' \times \boldsymbol{\gamma}'')]' = \boldsymbol{\gamma}\cdot(\boldsymbol{\gamma}' \times \boldsymbol{\gamma}''').$$

12. Suppose that $|\boldsymbol{\gamma}(t)|^2$ is constant. Prove that for every t, $\boldsymbol{\gamma}(t)$ is orthogonal to $\boldsymbol{\gamma}'(t)$. Draw an illustrating sketch when $\boldsymbol{\gamma}$ is a curve in R^2, and another sketch for the case of a curve in R^3. (Hint: Use formula (3) with $\boldsymbol{\gamma} = \boldsymbol{\delta}$.)

13. Let $\mathbf{P}_0 = (0,1,0)$.
   (a) Let $\boldsymbol{\gamma}(t) = \mathbf{P}_1$ be the point on the helix in Problem 6 closest to $\mathbf{P}_0$. Show that $a \cos t = b^2 t$.
   (b) Prove that $\mathbf{P}_0 - \mathbf{P}_1$ is orthogonal to the velocity vector of the helix at $\mathbf{P}_1$.

14. Let $\boldsymbol{\gamma}$ be any differentiable curve and $\mathbf{P}_0$ be any point *not* on the curve. Suppose that there is a point $\mathbf{P}_1$ on the curve, not an endpoint, such that $\mathbf{P}_0$ is closer to $\mathbf{P}_1$ than to any other point on the curve. Prove that $\mathbf{P}_1 - \mathbf{P}_0$ is orthogonal to the curve at $\mathbf{P}_1$. (Hint: If $\mathbf{P}_1 = \boldsymbol{\gamma}(t_1)$, then $|\boldsymbol{\gamma}(t) - \mathbf{P}_0|^2$ has a minimum at $t = t_1$.)

15. Prove the derivative formulas (2), (3), (4), and (5).

16. Prove that if $\boldsymbol{\gamma}(t_0)$ is not zero, then

$$\left(\frac{1}{|\boldsymbol{\gamma}|}\boldsymbol{\gamma}\right)'(t_0) = \frac{1}{|\boldsymbol{\gamma}(t_0)|}\boldsymbol{\gamma}'(t_0) - \frac{\boldsymbol{\gamma}(t_0)\cdot\boldsymbol{\gamma}'(t_0)}{|\boldsymbol{\gamma}(t_0)|^3}\boldsymbol{\gamma}(t_0).$$

17. Suppose that $\boldsymbol{\gamma}$ is a curve in R^3 such that $\boldsymbol{\gamma}(t)$ is never zero and $\boldsymbol{\gamma}'(t)$ is parallel to $\boldsymbol{\gamma}(t)$ for every t.

(a) Prove that $\dfrac{1}{|\boldsymbol{\gamma}|}\boldsymbol{\gamma}$ is constant. (Hint: Use Problem 16 above and the fact that $\boldsymbol{\gamma}'(t) = c(t)\,\boldsymbol{\gamma}(t)$.)

(b) Prove that there is a single straight line through the origin which contains $\boldsymbol{\gamma}(t)$ for every t.

2.2 NEWTON'S LAW OF MOTION

Vector functions are perfectly adapted to the study of particles (electrons, for example) moving in space. If $\boldsymbol{\gamma}(t)$ denotes the position of the particle at time t, then the first derivative $\boldsymbol{\gamma}'(t)$ is called the *velocity*, its absolute value $|\boldsymbol{\gamma}'(t)|$ is the *speed*, and the second derivative $\boldsymbol{\gamma}''(t)$ is the *acceleration*. Newton's law of motion says that if a particle of mass m is moved by a force $\mathbf{F}$, then

$$\mathbf{F} = (m\boldsymbol{\gamma}')'.$$

If the mass m is constant, this reduces to the familiar rule that force equals mass times acceleration:

$$\mathbf{F} = m\boldsymbol{\gamma}''.$$

Two examples will suggest the impressive consequences of this simple equation.

First, suppose that the force $\mathbf{F}$ is constant. (This happens, for example, when an electron is moved by an electric field of constant strength and direction.) We write out Newton's law by components, setting $\boldsymbol{\gamma} = (x,y,z)$ and $\mathbf{F} = (a,b,c)$, and find

$$\begin{aligned} mx''(t) &= a \\ my''(t) &= b \\ mz''(t) &= c. \end{aligned}$$

The solutions of these three equations are well known to be

$$x = \frac{a}{2m} t^2 + x_1 t + x_0$$

$$y = \frac{b}{2m} t^2 + y_1 t + y_0$$

$$z = \frac{c}{2m} t^2 + z_1 t + z_0$$

where $x_0, x_1, y_0, y_1, z_0, z_1$ are constants. These three equations are summed up in the vector equation

$$\boldsymbol{\gamma}(t) = \frac{t^2}{2m} \mathbf{F} + t\mathbf{V}_0 + \mathbf{P}_0$$

where $\mathbf{P}_0 = (x_0, y_0, z_0)$ and $\mathbf{V}_0 = (x_1, y_1, z_1)$ are constant vectors (see Fig. 7). Setting $t = 0$, we find that $\mathbf{P}_0 = \boldsymbol{\gamma}(0)$; differentiating and setting $t = 0$ shows that $\mathbf{V}_0 = \boldsymbol{\gamma}'(0)$. ($\mathbf{P}_0$ is called the *initial position,* and $\mathbf{V}_0$ is called the *initial velocity.* Warning: Remember that the above formula for $\boldsymbol{\gamma}$ is valid *only when the force* $\mathbf{F}$ *is constant.*)

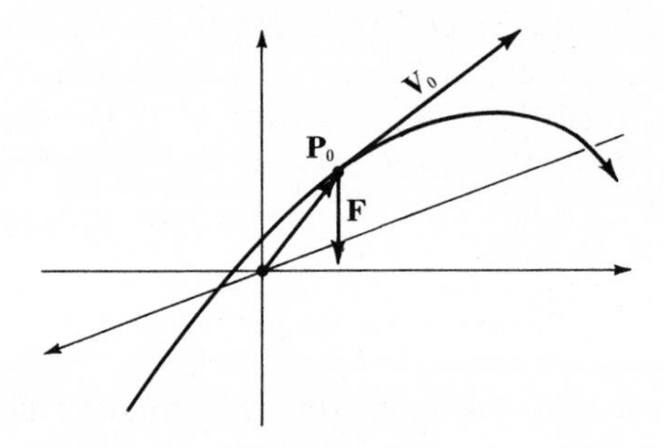

FIGURE 2.7

In particular, when there is no force, i.e. when $\mathbf{F} = \mathbf{0}$, we have

$$\boldsymbol{\gamma}(t) = t\mathbf{V}_0 + \mathbf{P}_0$$

so the particle moves with constant speed and direction along a straight line. This is another of Newton's laws of motion, sometimes called the law of inertia.

Our second example is a little more complicated. Suppose that $\boldsymbol{\gamma}(t)$ is the position at time t and that the force $\mathbf{F}$ has the form $q\mathbf{H} \times \boldsymbol{\gamma}'$, where q is a constant number and $\mathbf{H}$ is a constant vector. (This is the force acting on a particle of charge q in a constant magnetic field $\mathbf{H}$.) We will show that $\boldsymbol{\gamma}$ is either a *straight line,* a *circle,* or a *helix.*

To simplify matters, suppose that $\mathbf{H} = (0,0,h)$, where h is a numerical constant. Then

$$\mathbf{H} \times \boldsymbol{\gamma}' = \left(\begin{vmatrix} 0 & h \\ y' & z' \end{vmatrix}, -\begin{vmatrix} 0 & h \\ x' & z' \end{vmatrix}, \begin{vmatrix} 0 & 0 \\ x' & y' \end{vmatrix} \right)$$
$$= (-hy', kx', 0),$$

so Newton's law $\mathbf{F} = m\boldsymbol{\gamma}''$ becomes, with $\mathbf{F} = q\mathbf{H} \times \boldsymbol{\gamma}'$,

$$mx'' = -qhy', \qquad my'' = qhx', \qquad mz'' = 0,$$

or

$$\begin{aligned} x'' &= -cy' \\ y'' &= cx' \\ z'' &= 0 \end{aligned}$$

where we have set $c = qh/m$. The equation for z is easy to solve:

$$z = z_0 + z_1 t, \qquad z_0 \text{ and } z_1 \text{ constants.}$$

The other two equations look harder; they are simultaneous linear differential equations in the two unknown functions x and y. Actually, they are not so hard; we simply imitate the solution of simultaneous linear equations in algebra. Write them in the form

$$x'' + cy' = 0 \tag{1}$$

$$cx' - y'' = 0. \tag{2}$$

To eliminate y, differentiate equation (1), multiply equation (2) by c, and add; the result is

$$x''' + c^2 x' = 0.$$

This can be regarded as an equation in x', and, as such, its solution is

$$x'(t) = a\cos(ct) + b\sin(ct) \tag{3}$$

for some constants a and b. We integrate to find x, and evaluate the constant of integration by setting $t = 0$ to find

$$x(t) = \frac{a}{c}\sin(ct) - \frac{b}{c}\cos(ct) + \left(x_0 + \frac{b}{c}\right) \tag{4}$$

where $x_0 = x(0)$. Going back to solve for y, we have from equations (1) and (3) that

$$cy' = ac\sin ct - bc\cos ct; \tag{5}$$

hence

$$y = -\frac{a}{c}\cos ct - \frac{b}{c}\sin ct + \left(y_0 + \frac{a}{c}\right) \tag{6}$$

where $y_0 = y(0)$. Finally, relating the constants a and b to the initial velocity, i.e. to $x_1 = x'(0)$ and $y_1 = y'(0)$, we find from (3) that $a = x_1$ and from (5) that $b = -y_1$; hence

$$\begin{aligned} x(t) &= \frac{x_1}{c}\sin(ct) + \frac{y_1}{c}\cos(ct) + \left(x_0 - \frac{y_1}{c}\right), \\ y(t) &= -\frac{x_1}{c}\cos(ct) + \frac{y_1}{c}\sin(ct) + \left(y_0 + \frac{x_1}{c}\right), \\ z(t) &= z_0 + z_1 t. \end{aligned} \tag{7}$$

The three equations in (7) can be written as a single vector equation:

$$\boldsymbol{\gamma}(t) = \mathbf{A} + \overline{\boldsymbol{\gamma}}(t)$$

where $\mathbf{A} = \left(x_0 - \frac{y_1}{c}, y_0 + \frac{x_1}{c}, z_0\right)$ is a constant vector and

$$\overline{\boldsymbol{\gamma}}(t) = (r\cos(ct - \varphi), r\sin(ct - \varphi), z_1 t)$$

with

$$r^2 = \frac{x_1^2 + y_1^2}{c^2}, \qquad \frac{x_1}{rc} = \sin\varphi, \qquad \frac{y_1}{rc} = \cos\varphi.$$

The exact shape of the curve depends on the initial velocity $\mathbf{V}_0$.

(a) If $\mathbf{V}_0 = \mathbf{0}$, then $\boldsymbol{\gamma}(t) = \mathbf{P}_0$, and the particle does not move.

(b) If $x_1 = y_1 = 0$, but $z_1 \neq 0$, then $\boldsymbol{\gamma}(t) = \mathbf{P}_0 + t(0,0,z_1)$. The particle moves with constant speed and direction; there is no force acting since $\mathbf{H} \times \boldsymbol{\gamma}' = \mathbf{0}$. (Fig. 8)

(c) If $z_1 = 0$, but $x_1^2 + y_1^2 = r^2c^2 > 0$, then

$$\boldsymbol{\gamma}(t) = \mathbf{A} + (r\cos(ct - \varphi), r\sin(ct - \varphi), 0)$$

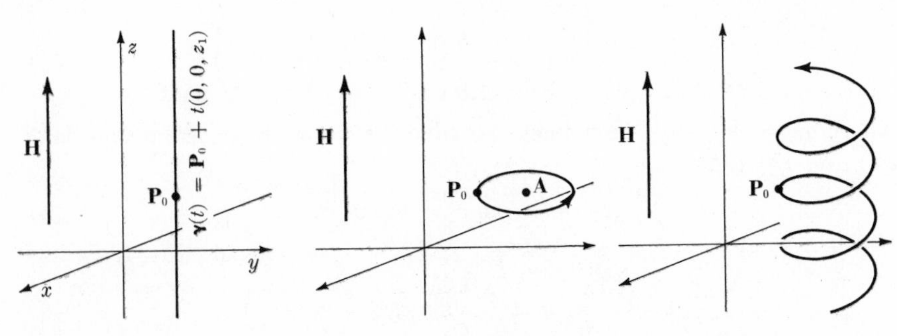

FIGURE 2.8 FIGURE 2.9 FIGURE 2.10

and the particle moves in a circle of radius r, parallel to the xy plane, with center $\mathbf{A}$. (Fig. 9)

(d) In the remaining case, the particle moves in a helix spiraling around the line through $\mathbf{A}$ and parallel to the z axis. (Fig. 10)

PROBLEMS

1. In each of the following problems, find the velocity, the acceleration, and the force.

(a) $\boldsymbol{\gamma}(t) = (2 \sin t, 3 \cos t, 4t)$, $m(t) = 5$

(b) $\boldsymbol{\gamma}(t) = (2e^t, 2e^{-t}, 5 \cos t)$; $m(t) = e^{-t}$

(c) $\boldsymbol{\gamma}(t) = (t^2, 3 + t^3, t^4)$, $m(t) = 1$

(d) $\boldsymbol{\gamma}(t) = \mathbf{A} + t\mathbf{B}$ (where $\mathbf{A}$ and $\mathbf{B}$ are constant vectors in R^n), $m(t) = 1$

(e) $\boldsymbol{\gamma}(t) = \mathbf{A} + t\mathbf{B}$, $m(t) = t$

2. Describe the shape of the curve in Problem 1(a).

3. The paths in 1(a) and (c) intersect at $t = 0$. Find the angle between the forces at $t = 0$.

4. Let the constant force $\mathbf{F} = (0,0,e)$ move a particle of constant mass m on a curve $\boldsymbol{\gamma}$, and assume that $\boldsymbol{\gamma}(0) = (1,2,3)$, $\boldsymbol{\gamma}'(0) = (-1,-3,0)$. Find $\boldsymbol{\gamma}(t)$ for all t.

5. Let the variable force $\mathbf{F}(t) = (\cos t,0,1)$ move a particle of constant mass m in a curve $\boldsymbol{\gamma}$ with $\boldsymbol{\gamma}(0) = \boldsymbol{\gamma}'(0) = \mathbf{0}$. Find $\boldsymbol{\gamma}(t)$, and sketch the path.

6. A particle of constant mass m is moved in a curve $\boldsymbol{\gamma}(t) = (x(t),y(t),z(t))$ by a force $\mathbf{F} = (x(t),y(t),z(t))$, with $\boldsymbol{\gamma}(0) = \mathbf{P}_0$ and $\boldsymbol{\gamma}'(0) = \mathbf{V}_0$. Find $\boldsymbol{\gamma}(t)$ for all t. (Hint: The solution of $f'' = cf$ is $f(t) = Ae^{\sqrt{c}t} + Be^{-\sqrt{c}t}$, where A and B are constants.)

7. (a) Prove that if the acceleration $\boldsymbol{\gamma}''$ is always orthogonal to the direction of motion $\boldsymbol{\gamma}'$, then the speed $|\boldsymbol{\gamma}'|$ is constant. (Hint: $|\boldsymbol{\gamma}'|^2 = \boldsymbol{\gamma}' \cdot \boldsymbol{\gamma}'$.)

(b) Prove the converse of part (a).

8. The acceleration of a curve $\boldsymbol{\gamma}$ is called *central* if $\boldsymbol{\gamma}''$ is always parallel to $\boldsymbol{\gamma}$, i.e. if $\boldsymbol{\gamma}'' \times \boldsymbol{\gamma} \equiv \mathbf{0}$. Suppose that $\boldsymbol{\gamma}$ has central acceleration.

(a) Prove that $\mathbf{N} = \boldsymbol{\gamma} \times \boldsymbol{\gamma}'$ is constant.

(b) Assuming $\mathbf{N} \neq \mathbf{0}$, prove that $\boldsymbol{\gamma}(t)$ always lies on the plane through the origin with normal $\mathbf{N}$. (See Fig. 11.)

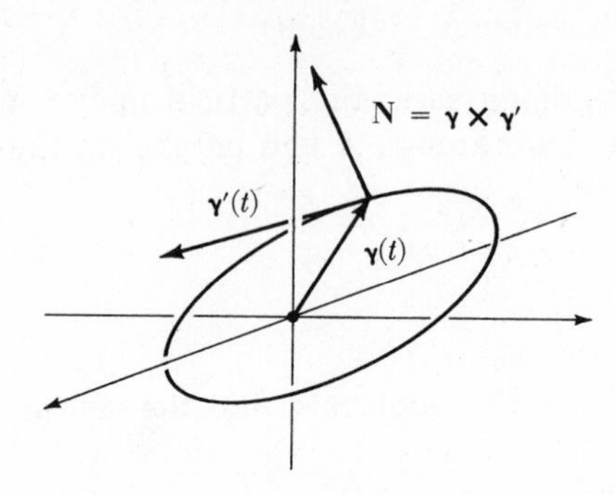

FIGURE 2.11

9. A moving point $\gamma(t)$ has a constant acceleration $\mathbf{A}$, initial velocity $\mathbf{V}_0 = \gamma'(0)$, and initial position $\mathbf{P}_0 = \gamma(0)$.

(a) Prove that

$$\gamma(t) = \frac{t^2}{2}\mathbf{A} + t\mathbf{V}_0 + \mathbf{P}_0 .$$

(b) If $\mathbf{A}$ and $\mathbf{V}_0$ are parallel, prove that $\gamma(t)$ moves in a straight line (though not necessarily with constant speed).

(c) Assuming $\mathbf{A}$ and $\mathbf{V}_0$ are not parallel, prove that $\gamma(t)$ lies in the plane through $\mathbf{P}_0$ with normal $\mathbf{A} \times \mathbf{V}_0$. (See Fig. 12.)

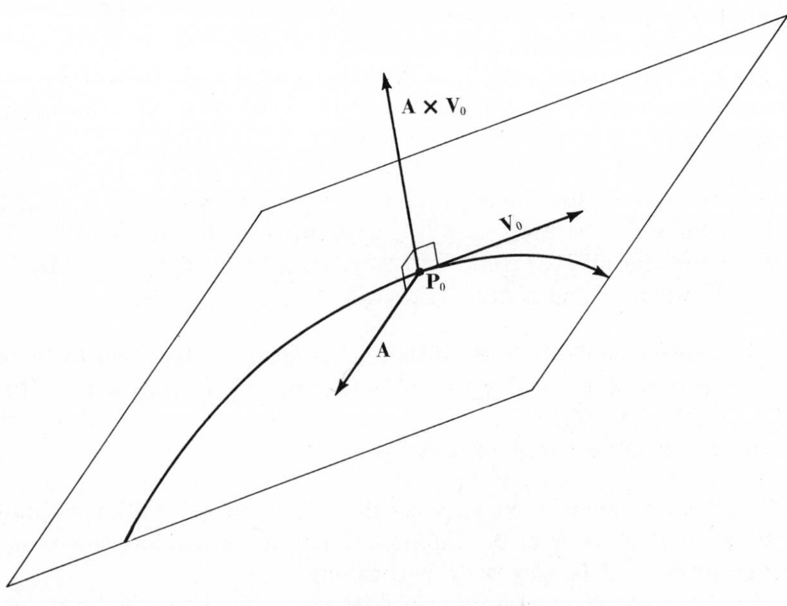

FIGURE 2.12

(d) The *speed* of the particle at time t is $|\boldsymbol{\gamma}'(t)|$. Assuming that $\mathbf{A} \neq \mathbf{0}$, find the time t at which the speed is minimum.

(e) Show that the minimum speed is zero if and only if the acceleration $\mathbf{A}$ is parallel to the initial velocity $\mathbf{V}_0$.

(f) Show that the velocity $\boldsymbol{\gamma}'(t)$ is orthogonal to the acceleration $\mathbf{A}$ when $|\boldsymbol{\gamma}'(t)|$ is minimum.

(g) Assume that the acceleration $\mathbf{A}$ and the initial velocity $\mathbf{V}_0$ are orthogonal, and let $\mathbf{F}$ be the point

$$\mathbf{F} = \mathbf{P}_0 + \frac{|\mathbf{V}_0|^2}{2|\mathbf{A}|^2}\mathbf{A}.$$

Prove that for every t, $|\mathbf{F} - \boldsymbol{\gamma}(t)|$ equals the distance from $\boldsymbol{\gamma}(t)$ to the line $\{\mathbf{Q}\colon \mathbf{Q} = \mathbf{B} + s\mathbf{V}_0 \text{ for some } s\}$ where

$$\mathbf{B} = \mathbf{P}_0 - \frac{|\mathbf{V}_0|^2}{2|\mathbf{A}|^2}\mathbf{A}.$$

(See Fig. 13. Hint: Recall that the distance d from a point $\mathbf{P}$ to the line $\{\mathbf{Q}\colon \mathbf{Q} = \mathbf{B} + s\mathbf{V}_0\}$ is given by

$$d^2 = |\mathbf{Q} - \mathbf{B}|^2 - \frac{1}{|\mathbf{V}_0|^2}((\mathbf{Q} - \mathbf{B})\cdot\mathbf{V}_0)^2,$$

and recall the Pythagorean theorem for orthogonal vectors.)

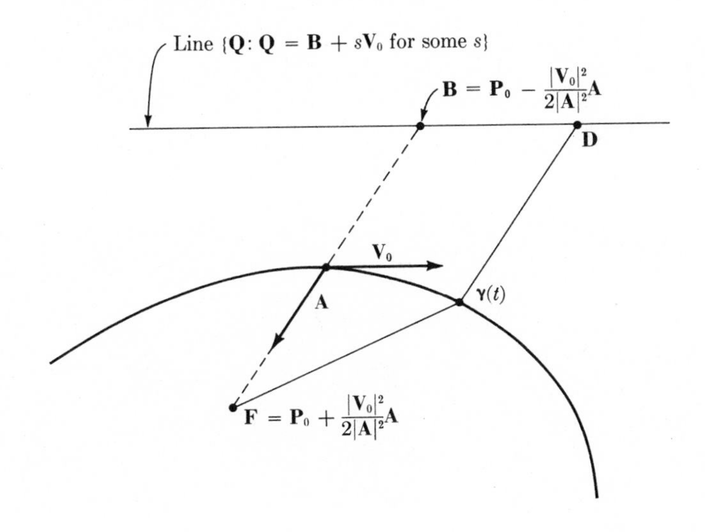

FIGURE 2.13 $|\boldsymbol{\gamma}(t) - \mathbf{F}| = |\boldsymbol{\gamma}(t) - \mathbf{D}|$

(*Remarks:* Part (f) shows that by shifting the time scale (i.e. by considering $\bar{\boldsymbol{\gamma}}(t) = \boldsymbol{\gamma}(t - t_0)$, where t_0 is the time at which $|\boldsymbol{\gamma}'(t_0)|$ is minimum) we can assume that $\mathbf{V}_0$ and $\mathbf{A}$ are orthogonal. Parts (c) and (g) show that $\boldsymbol{\gamma}(t)$ lies on a parabola whose focus is

$$\mathbf{F} = \mathbf{P}_0 + \frac{|\mathbf{V}_0|^2}{2|\mathbf{A}|^2}\mathbf{A}$$

and whose directrix is the line $\{\mathbf{Q}: \mathbf{Q} = \mathbf{B} + s\mathbf{V}_0\}$, where

$$\mathbf{B} = \mathbf{P}_0 - \frac{|\mathbf{V}_0|^2}{2|\mathbf{A}|^2}\mathbf{A}.)$$

10. A cathode ray tube lies in a constant magnetic field $\mathbf{H} = (0,0,h)$ parallel to the axis of the tube. (See Fig. 14.) At time $t = 0$ an electron of charge q and mass m leaves the origin 0 with velocity $(x_1, 0, z_1)$. The electron is intercepted by the face of the tube, which lies in the plane $\{(x,y,z) : z = d\}$.

(a) When, and at what point, is the electron intercepted? (Hint: In equation (7), recall that $c = qh/m$.)

(b) Show that the electron hits the face of the tube along the line through $(0,0,d)$ in the direction

$$\left(\sin\frac{dqh}{mz_1}, 1 - \cos\frac{dqh}{mz_1}, 0\right). \qquad (*)$$

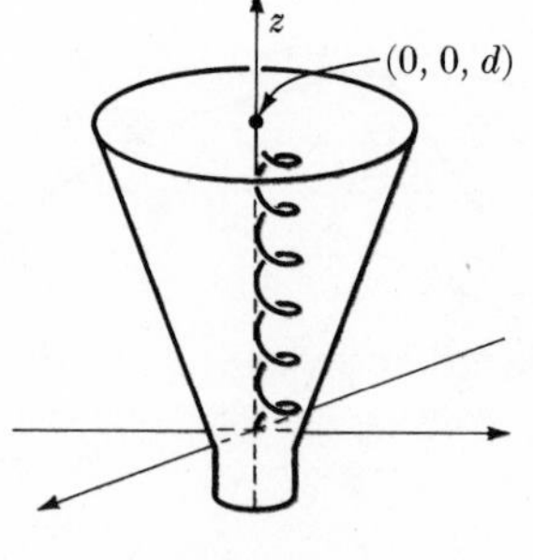

FIGURE 2.14

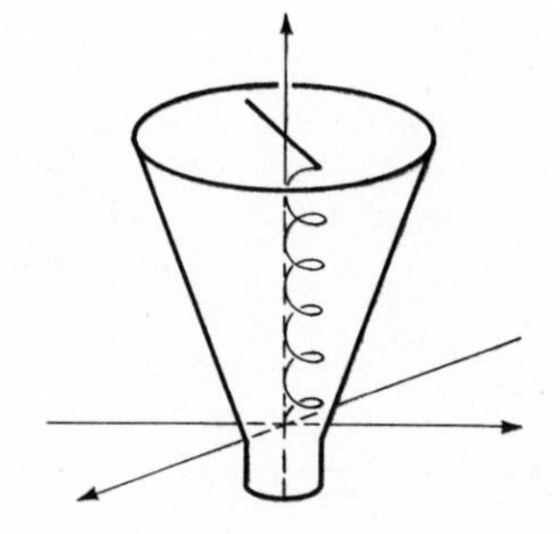

FIGURE 2.15

(*Remark:* When the electron hits the face of the tube, a luminous spot appears briefly at the point of impact. In practice, electrons are sent in a steady stream, all having the same value of z_1 but varying values of x_1; thus as the stream hits the face of the tube a segment of the line in part (b) is illuminated. (See Fig. 15.) The ratio q/m of charge to mass for the electron can be determined by the direction of this line segment because:

(i) the distance d from the origin to the face of the tube is known;
(ii) the constant z_1 equals $a\sqrt{q/m}$, where a is a constant determined by the construction of the tube.

Thus the number dqh/mz_1 in (*) has the form $Ah\sqrt{q/m}$, where A is a known constant. Suppose now that a certain value h_1 of the magnetic field h produces a line of slope m on the cathode ray tube. If h is increased gradually to the next larger value h_2 that produces the same slope m, then it follows from the equation (*) that

$$Ah_2\sqrt{q/m} = Ah_1\sqrt{q/m} + 2\pi,$$

and this equation determines $\sqrt{q/m}$.)

11. Solve the following systems of differential equations. (Note: x' and y' denote dx/dt and dy/dt.)

(a) $$\begin{aligned} y' &= x \\ x' &= y \end{aligned}$$

(b) $$\begin{aligned} x' - y &= 0 \\ x + y' &= 0 \end{aligned}$$

(c) $$\begin{aligned} x' + 2y + 3y' &= 0 \\ x + 3x' - y &= 0 \end{aligned}$$

(Hint: Write the equations (c) in the form

$$\begin{aligned} \mathbf{D}x + (2 + 3\mathbf{D})y &= 0 \\ (1 + 3\mathbf{D})x \qquad\qquad - y &= 0.) \end{aligned}$$

2.3 THE GEOMETRY OF CURVES IN R^3

In the study of motion, the first and second derivatives with respect to time (the velocity and acceleration) are crucial. From a geometric point of view, however, the time is irrelevant; what matters is the figure traced out, and particularly the features listed below (see Fig. 16):

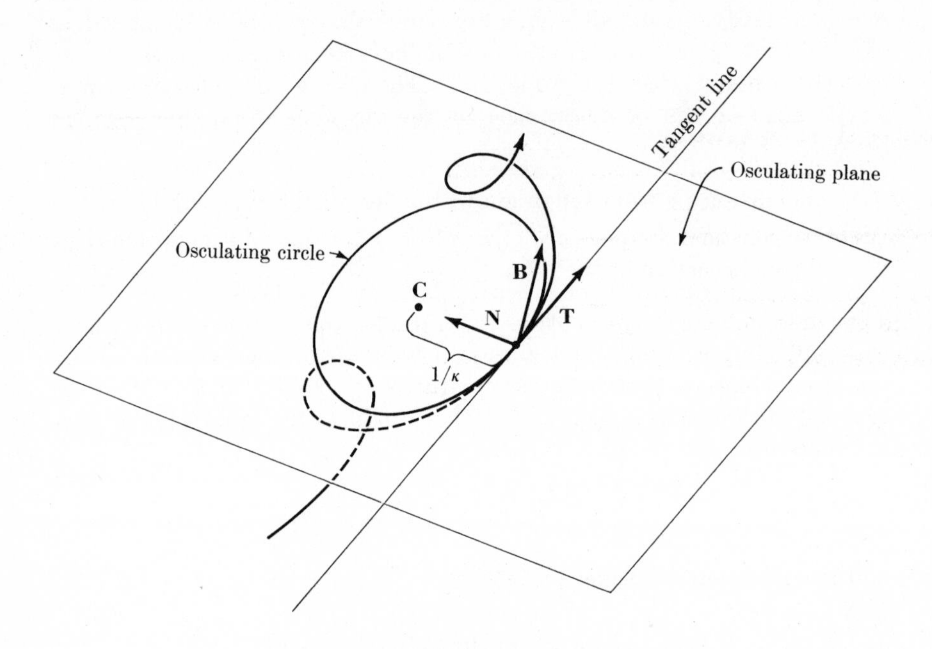

Center of osculating circle $= \mathbf{C}$

Radius of osculating circle $= \dfrac{1}{\kappa} = \dfrac{1}{\text{curvature}}$

Normal to osculating plane $= \mathbf{B} = \mathbf{T} \times \mathbf{N}$

FIGURE 2.16

Arc length
Tangent line (line of best fit)
Osculating* plane (plane of best fit)
Osculating circle (circle of best fit)
Unit tangent vector **T** (direction the curve is going)
Principal normal vector **N** (direction toward which the curve is turning)
Curvature (rate of change of direction per unit length)
Torsion (rate of "twisting" of the curve per unit length)

---

* Look that up in your Funk and Wagnalls.

The development of these ideas begins with the tangent line.

Let $\boldsymbol{\gamma}$ be a curve. If $\boldsymbol{\gamma}'(t_0) \neq 0$, the *tangent line* at $\boldsymbol{\gamma}(t_0)$ is given parametrically as the set

$$\{\mathbf{P}: \mathbf{P} = \boldsymbol{\gamma}(t_0) + (t - t_0)\boldsymbol{\gamma}'(t_0)\}.$$

When $\boldsymbol{\gamma}'(t_0) = 0$, on the other hand, there is generally no way to define the tangent, and Fig. 17 suggests why. Both curves shown there are differentiable everywhere, but neither curve has a tangent line at $\boldsymbol{\gamma}(0)$; the moving point comes in to the origin, stops there ($\boldsymbol{\gamma}'(0) = 0$), and goes off in a new direction. In the rest of this section we rule out such "corners" by requiring that $\boldsymbol{\gamma}' \neq 0$. We also rule out the irregularities which arise when $\boldsymbol{\gamma}'(t)$ does not exist or is not continuous; in fact, we shall assume that $\boldsymbol{\gamma}'$, $\boldsymbol{\gamma}''$, and $\boldsymbol{\gamma}'''$ all exist and that $\boldsymbol{\gamma}'$ is never zero.

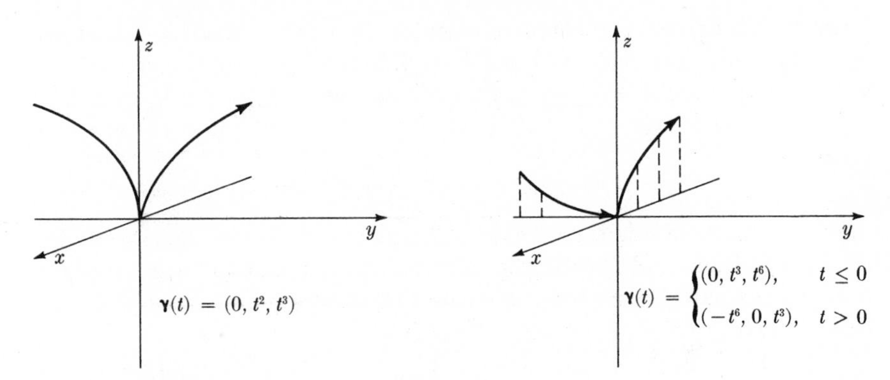

FIGURE 2.17

The length of curve, or *arc length* traversed by the curve $\boldsymbol{\gamma}$ from $t = t_0$ to $t = t_1$, is given by the integral

$$\int_{t_0}^{t_1} |\boldsymbol{\gamma}'|.$$

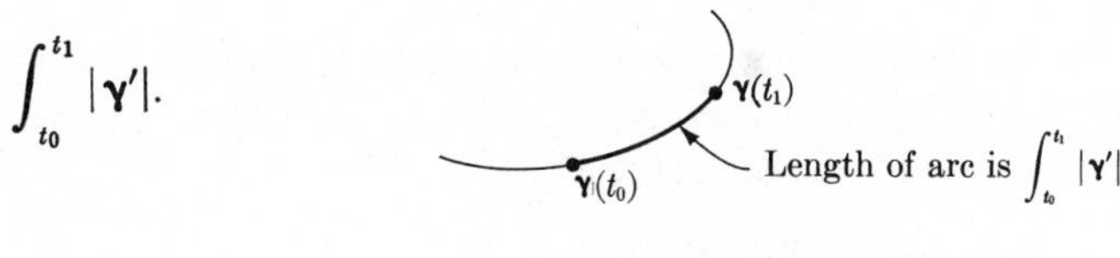

FIGURE 2.18

(See Fig. 18 above. We use this integral as a definition of the arc length, although it can be derived from a more intuitive definition based on "inscribed polygons," as in §AII.9; see Fig. 24 there.) Since we assume that γ'' exists, it follows that γ' is continuous, so $|\gamma'|$ is continuous, and the integral exists. Since

$$\int_{t_0}^{t_1} |\gamma'|$$

gives the length of path traversed, it follows that $|\gamma'(t)|$ (the speed) is the rate at which the path is traversed. The speed is precisely what is irrelevant from the geometric point of view, though essential in the study of motion. By an appropriate change of variable, we can replace the given curve γ by a new curve $\mathbf{G}$ which traces out the same points but does not contain this geometrically irrelevant information.

Taking any "base point" t_0, we define an *arc length function*

$$\ell(t) = \int_{t_0}^{t} |\gamma'|.$$

From the fundamental theorem of calculus, $\ell' = |\gamma'|$. Since γ' is assumed to be different from zero, we have $\ell' > 0$, hence ℓ is strictly increasing. It follows that there is a differentiable inverse function, call it g, such that $\ell(g(s)) = s$ for every s in the range of ℓ. We now define a new vector function

$$\mathbf{G}(s) = \gamma(g(s)).$$

Since the range of g is the domain of ℓ, and g is monotone increasing, the function $\mathbf{G}$ traces out exactly the same points as γ and in the same order; but now the parameter s has a geometric significance, namely, it is the arc length. To prove this, notice that $\mathbf{G}'(s) = \gamma'(g(s))g'(s)$ and $g'(s) = 1/\ell'(g(s)) = 1/|\gamma'(g(s))|$; hence $\mathbf{G}'(s) = \gamma'(g(s))/|\gamma'(g(s))|$. It follows that $|\mathbf{G}'(s)| = 1$; hence the arc length traversed by $\mathbf{G}$ as s varies from 0 to s_1 is

$$\int_0^{s_1} 1 = s_1;$$

in other words, s gives the arc length on the curve $\mathbf{G}$ measured from the point $\mathbf{G}(0) = \gamma(t_0)$.

In the rest of this section we study $\mathbf{G}$ instead of γ. The vector $\mathbf{G}'(s)$ is a unit vector (since $|\mathbf{G}'(s)| = 1$) giving the direction of the curve; it is generally denoted $\mathbf{T}(s)$.*

Example 1. Take the helix $\gamma(t) = (\cos t, \sin t, t)$, and the base point $t_0 = 0$. Compute $\ell(t)$, $g(s)$, $\mathbf{G}(s)$, and $\mathbf{T}(s)$. *Solution*:

$$\gamma'(t) = (-\sin t, \cos t, 1), \qquad |\gamma'(t)| = \sqrt{2}$$

$$\ell(t) = \int_0^t \sqrt{2} = t\sqrt{2}.$$

* The rest of this section is not required in any other part of the text.

To find g, use the defining equation $\ell(g(s)) = s$, i.e.

$$g(s)\sqrt{2} = s, \qquad g(s) = \frac{s}{\sqrt{2}}.$$

Hence

$$\mathbf{G}(s) = \boldsymbol{\gamma}(g(s)) = \boldsymbol{\gamma}\left(\frac{s}{\sqrt{2}}\right) = \left(\cos\frac{s}{\sqrt{2}}, \sin\frac{s}{\sqrt{2}}, \frac{s}{\sqrt{2}}\right).$$

It is obvious that **G** traces the same path as $\boldsymbol{\gamma}$. Finally,

$$\mathbf{T}(s) = \mathbf{G}'(s) = \left(-\frac{1}{\sqrt{2}}\sin\frac{s}{\sqrt{2}}, \frac{1}{\sqrt{2}}\cos\frac{s}{\sqrt{2}}, \frac{1}{\sqrt{2}}\right)$$

$$= \frac{1}{\sqrt{2}}\left(-\sin\frac{s}{\sqrt{2}}, \cos\frac{s}{\sqrt{2}}, 1\right).$$

Curvature. Since $\mathbf{T}(s)$ is a unit vector giving the direction of the curve, its derivative $\mathbf{T}'(s)$ gives the *rate of change of direction with respect to arc length.* We define the length of this derivative to be the *curvature* κ,

$$\kappa(s) = |\mathbf{T}'(s)|.$$

Moreover, since **T** has constant length, the derivative $\mathbf{T}'$ is orthogonal to **T** (see Problem 12, §2.1). When $\mathbf{T}'(s) \neq \mathbf{0}$, we can divide by $|\mathbf{T}'(s)|$ to obtain a unit vector orthogonal to $\mathbf{T}(s)$, called the *principal normal vector* **N**:

$$\mathbf{N}(s) = \frac{1}{|\mathbf{T}'(s)|}\mathbf{T}'(s) = \frac{1}{\kappa}\mathbf{T}'. \tag{1}$$

(See Fig. 16.) To simplify the rest of the discussion, we restrict our attention to points where $\mathbf{T}'(s) \neq \mathbf{0}$. Geometrically, $\mathbf{T} = \mathbf{G}'$ gives the direction of the curve, and $\mathbf{N} = (1/\kappa)\mathbf{T}'$ gives the direction toward which the curve is turning.

Suppose that **G** describes a circle of radius r. Then it is easy to show that the curvature $\kappa = 1/r$; the smaller the radius, the greater the curvature. Moreover, the principal normal **N** points toward the center of the circle (see Fig. 19), so the center is located at $\mathbf{C} = \mathbf{G} + (1/\kappa)\mathbf{N}$. (See Problems 9 and 11.)

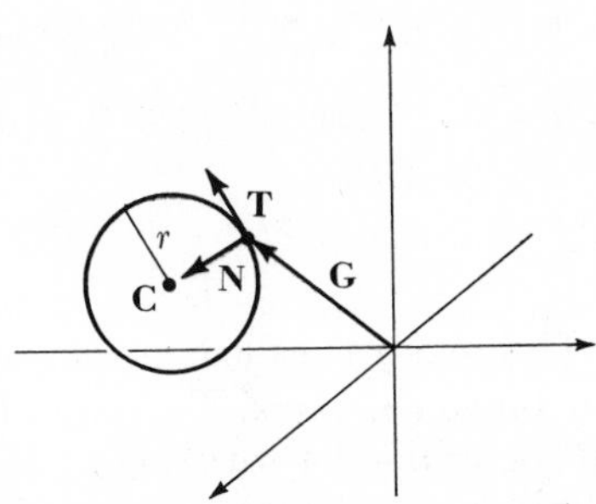

FIGURE 2.19 $\mathbf{C} = \mathbf{G} + r\mathbf{N} + \mathbf{G} + \frac{1}{\kappa}\mathbf{N}$

For a general curve, the point

$$\mathbf{C}(s_0) = \mathbf{G}(s_0) + \frac{1}{\kappa(s_0)}\mathbf{N}(s_0)$$

is called the *center of curvature* at $\mathbf{G}(s_0)$. This point together with the tangent line determines a plane, the *osculating plane* (see Fig. 16). The circle lying in this plane with center $\mathbf{C}(s_0)$ and radius $1/\kappa(s_0)$ is the *osculating circle*; it matches the curve both in direction and in curvature at the point $\mathbf{G}(s_0)$.

The cross product of $\mathbf{T}$ and $\mathbf{N}$ is a normal to the osculating plane, called the *binormal* $\mathbf{B}$:

$$\mathbf{B}(s_0) = \mathbf{T}(s_0) \times \mathbf{N}(s_0).$$

Since $\mathbf{T}$ and $\mathbf{N}$ are orthogonal unit vectors, $\mathbf{B} = \mathbf{T} \times \mathbf{N}$ is a unit vector orthogonal to both $\mathbf{T}$ and $\mathbf{N}$. Just as the line to which the curve adheres most closely at $\mathbf{G}(s_0)$ is the tangent line, the plane to which the curve adheres most closely at $\mathbf{G}(s_0)$ is the osculating plane. (See Fig. 16 and Problem 16 below.)

Torsion. Change in the binormal $\mathbf{B}$ amounts geometrically to a "twist" in the curve. If your intuition about space curves is very good, it may occur to you that this change in $\mathbf{B}$ must be directed toward $\pm\mathbf{N}$ and is perpendicular to $\mathbf{T}$ and $\mathbf{B}$, as the following theorem proves:

Theorem 5. *For each s where* $\mathbf{T}'(s) \neq \mathbf{0}$, *there is a number* $\tau(s)$ *such that* $\mathbf{B}'(s) = \tau(s)\mathbf{N}(s)$.

Proof. Since $\mathbf{T}$, $\mathbf{N}$, and $\mathbf{B}$ are mutually orthogonal unit vectors in R^3, we have

$$\mathbf{B}' = (\mathbf{B}'\cdot\mathbf{T})\mathbf{T} + (\mathbf{B}'\cdot\mathbf{N})\mathbf{N} + (\mathbf{B}'\cdot\mathbf{B})\mathbf{B},$$

by Theorem 10, §1.5. But $\mathbf{B}$ is a unit vector, so $\mathbf{B}'\cdot\mathbf{B} = 0$. And

$$\mathbf{B}' = (\mathbf{T} \times \mathbf{N})' = \mathbf{T}' \times \mathbf{N} + \mathbf{T} \times \mathbf{N}' = \mathbf{T} \times \mathbf{N}',$$

since $\mathbf{T}' \times \mathbf{N} = \kappa\mathbf{N} \times \mathbf{N} = \mathbf{0}$. Hence $\mathbf{B}'\cdot\mathbf{T} = 0$, and we have proved that $\mathbf{B}' = \tau\mathbf{N}$, with $\tau = \mathbf{B}'\cdot\mathbf{N}$. *Q.E.D.*

The number τ in Theorem 5 is called the *torsion* of the curve at the point in question. The torsion τ, like κ, $\mathbf{T}$, $\mathbf{N}$, and $\mathbf{B}$, is a function of the arc length s. Since $\mathbf{B}'$ describes the twisting of the curve and $|\mathbf{B}'| = |\tau\mathbf{N}| = |\tau|$, the torsion τ measures the amount of twisting; the sign of τ determines whether the curve is twisted toward $\mathbf{N}$ or toward $-\mathbf{N}$.

We now have two important derivative formulas. By the definition of $\mathbf{N}$,

$$\mathbf{T}' = \kappa\mathbf{N}, \qquad \kappa = \text{curvature}, \tag{2}$$

and by Theorem 5,

$$\mathbf{B}' = \tau\mathbf{N}, \qquad \tau = \text{torsion}. \tag{3}$$

There is a similar formula for $\mathbf{N}'$, namely,

$$\mathbf{N}' = -\kappa\mathbf{T} - \tau\mathbf{B}. \tag{4}$$

Formulas (2), (3), and (4) are the three *Frenet-Serret formulas.* The last one, (4), can be easily derived from the other two, as outlined in Problem 8 below.

We conclude this brief exposition with a few examples of the general theorems that can be proved about space curves.

Theorem 6. *If the curvature* $\kappa = |\mathbf{T}'|$ *is identically zero, then the curve* $\mathbf{G}$ *is a straight line.*

Proof. Since $\mathbf{T}' \equiv \mathbf{0}$, $\mathbf{T}$ is a constant vector. Since $\mathbf{G}'(s) = \mathbf{T}$ and $\mathbf{T}$ is constant, we have $\mathbf{G}(s) = s\mathbf{T} + \mathbf{C}$ for some constant vector $\mathbf{C}$, and the theorem is proved.

Theorem 7. *If the torsion* τ *is identically zero, then the curve lies in a plane, and this is the osculating plane at every point of the curve.*

Proof. If $\tau \equiv 0$, then $\mathbf{B}' \equiv \mathbf{0}$ by (3); hence $\mathbf{B}$ is constant. Let $\mathbf{G}(s_0)$ be any point on the curve; we want to prove that the curve lies in the osculating plane

$$\{\mathbf{P}: [\mathbf{P} - \mathbf{G}(s_0)]\cdot\mathbf{B} = 0\},$$

in other words, that the function $d(s) = [\mathbf{G}(s) - \mathbf{G}(s_0)]\cdot\mathbf{B}$ is zero for every s. But $d(s_0) = 0$, and

$$\begin{aligned} d'(s) &= \mathbf{G}'(s)\cdot\mathbf{B} && \text{(since } \mathbf{B} \text{ and } \mathbf{G}(s_0) \text{ are constant)} \\ &= \mathbf{T}(s)\cdot\mathbf{B} = 0 && \text{(since } \mathbf{T} \text{ and } \mathbf{B} \text{ are orthogonal)}; \end{aligned}$$

hence d is identically zero, as was to be proved.

Theorem 8. *If the torsion* $\tau \equiv 0$ *and the curvature* κ *is a nonzero constant, then the curve is a circle of radius* $1/\kappa$.

Proof. Since $\tau = 0$, the curve lies in a plane. Thus, to prove that it is a circle of radius $1/\kappa$, all we have to do is find a fixed point $\mathbf{C}$ such that

$$|\mathbf{C} - \mathbf{G}(s)| = \frac{1}{\kappa} \quad \text{for all } s.$$

Clearly, $\mathbf{C}$ should be the center of curvature, so it is natural to look at the function

$$\mathbf{C}(s) = \mathbf{G}(s) + \frac{1}{\kappa}\mathbf{N}(s).$$

Differentiate this equation, and apply the Frenet-Serret formula (4). Since κ is constant and $\tau = 0$, the result is

$$\mathbf{C}'(s) = \mathbf{G}'(s) + \frac{1}{\kappa}\mathbf{N}'(s) = \mathbf{T}(s) - \mathbf{T}(s) = \mathbf{0}.$$

Hence $\mathbf{C}$ is constant; and from its definition,

$$|\mathbf{C} - \mathbf{G}(s)| = \left|\frac{1}{\kappa}\mathbf{N}(s)\right| = \frac{1}{\kappa},$$

so the theorem is proved.

PROBLEMS

1. Write parametric equations for the tangent line to the given curve at the given point.
(a) $\boldsymbol{\gamma}(t) = (t, \sin 4t, \cos 4t)$ at the point $\boldsymbol{\gamma}(\pi/8)$
(b) $\boldsymbol{\gamma}(t) = (2t,t^2,t)$ at the point $(2,1,1)$
(c) $\boldsymbol{\gamma}(t) = (e^{3t},e^{-3t},3\sqrt{2}t)$ at $\boldsymbol{\gamma}(1)$
(d) $\boldsymbol{\gamma}(t) = (t,t^4,t^5)$ at $(1,1,1)$

2. For each curve in Problem 1, find the unit tangent vector $\mathbf{T} = \boldsymbol{\gamma}'/|\boldsymbol{\gamma}'|$ at the given point.

3. Find the length of the given curve over the given interval:
(a) The curve in Problem 1(a) from $t = 0$ to $t = \pi$
(b) The curve in Problem 1(b) from $t = 0$ to $t = 1$
(c) The curve in Problem 1(c) from $t = -1$ to $t = 2$

4. For the curve in Problem 1(a):
(a) Find the arc length function $\ell(t)$.
(b) Solve the equation $s = \ell(t)$ for t, and thus find the inverse g to ℓ.
(c) Write the curve with arc length as parameter, $\mathbf{G}(s) = \boldsymbol{\gamma}(g(s))$.

5. Repeat Problem 4 for the curve in Problem 1(c).

6. (a) For the helix $\boldsymbol{\gamma}(t) = (a\cos(kt), a\sin(kt), bkt)$, where a, b, and k are constants, find the arc length function $\ell(t) = \int_0^t |\boldsymbol{\gamma}'|$.

(b) Show that the inverse function g is

$$g(s) = \frac{s}{k\sqrt{a^2 + b^2}}.$$

(c) Show that the equation of the helix with arc length as parameter is

$$\mathbf{G}(s) = \left(a \cos \frac{s}{c}, a \sin \frac{s}{c}, \frac{bs}{c}\right),$$

where $c = \sqrt{a^2 + b^2}$.

7. For the helix $\mathbf{G}$ in Problem 6(c):
(a) find $\mathbf{T}(s)$
(b) find $\kappa(s)$ and $\mathbf{N}(s)$
(c) find $\mathbf{B}(s)$
(d) find $\tau(s)$
(e) Show that the Frenet-Serret formulas (2)–(4) are true for this particular curve.

8. For the helix in Problem 6, prove that $\mathbf{T}$, $\mathbf{N}$, and $\mathbf{B}$ all make constant angles with the vector (0,0,1).

9. $\boldsymbol{\gamma}(t) = (r \cos t, r \sin t, 0)$ describes a circle of radius r in the xy plane. Show that its curvature is $1/r$.

10. Suppose that $\boldsymbol{\gamma}$ is a curve, ℓ the arc length function, g its inverse function, and $\mathbf{G}$ the corresponding curve with arc length as parameter. Suppose further that $s = \ell(t)$. Prove that

(a) $g'(s) = \dfrac{1}{\ell'(t)}$

(b) $\mathbf{T}(s) = \dfrac{1}{|\boldsymbol{\gamma}'(t)|} \boldsymbol{\gamma}'(t)$

(c) $\mathbf{T}'(s) = \dfrac{|\boldsymbol{\gamma}'(t)|^2 \boldsymbol{\gamma}''(t) - [\boldsymbol{\gamma}'(t) \cdot \boldsymbol{\gamma}''(t)]\boldsymbol{\gamma}'(t)}{|\boldsymbol{\gamma}'(t)|^4}$ $\left(\text{Hint: } \dfrac{d\mathbf{T}}{ds} = \dfrac{d\mathbf{T}/dt}{ds/dt}\right)$

(d) $\kappa(s) = \dfrac{|\boldsymbol{\gamma}'(t) \times \boldsymbol{\gamma}''(t)|}{|\boldsymbol{\gamma}'(t)|^3}$

(Hint: Compute the length of the numerator in part (c) by taking its inner product with itself.)

11. The general circle of radius r about a given center $\mathbf{C}$ and perpendicular to a given unit vector $\mathbf{B}$ has parametric equations

$$\boldsymbol{\gamma}(t) = \mathbf{C} + (r \cos t)\mathbf{D} + (r \sin t)\mathbf{E},$$

where $\mathbf{D}$ is a unit vector orthogonal to $\mathbf{B}$, and $\mathbf{E} = \mathbf{B} \times \mathbf{D}$. Prove that the curvature of this circle is $1/r$. (There are two ways: (i) rewrite the circle with arc length as parameter, or (ii) use Problem 10(d). Note that $\mathbf{D}$ and $\mathbf{E}$ are orthogonal unit vectors.)

12. Recall that the center of curvature at $\mathbf{G}(s_0)$ is

$$\mathbf{C}(s_0) = \mathbf{G}(s_0) + \frac{1}{\kappa(s_0)} \mathbf{N}(s_0).$$

(a) For the helix $\mathbf{G}$ in Problem 6(c) prove that the center of curvature $\mathbf{C}(s)$ describes a helix. (See Fig. 20.)
(b) Prove that the curvature of $\mathbf{C}$ is the same as that of $\mathbf{G}$. (*Caution:* s is *not* the arc length on the curve $\mathbf{C}$.)

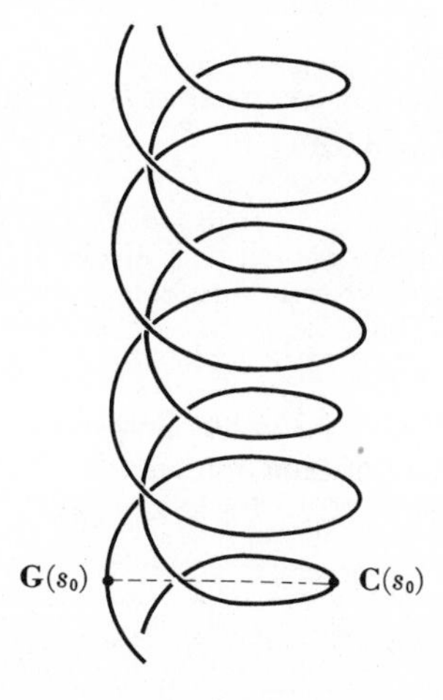

FIGURE 2.20

13. Derive the Frenet-Serret formula (4). (Hint: Use Theorem 11, §1.5, together with (2) and (3).)

14. Suppose that the curve $\mathbf{G}$ lies in a plane $\{\mathbf{P}: (\mathbf{P} - \mathbf{P}_0) \cdot \mathbf{A} = 0\}$.
(a) Prove that the tangent vector $\mathbf{T}$ is orthogonal to $\mathbf{A}$.
(b) Prove that the principal normal $\mathbf{N}$ is orthogonal to $\mathbf{A}$.
(c) Prove that $\mathbf{A} = (\mathbf{A} \cdot \mathbf{B})\mathbf{B}$, where $\mathbf{B}$ is the binormal. (See Theorem 10, §1.5.)

(d) Prove that

$$\mathbf{B} = \frac{\pm 1}{|\mathbf{A}|}\,\mathbf{A}.$$

(e) Prove that where $\mathbf{B}'$ exists, $\mathbf{B}' = \mathbf{0}$. (Hint: $\mathbf{B}(s) = b(s)\mathbf{A}$, where $b(s) = \pm 1/|\mathbf{A}|$; and $\mathbf{B}'(s_0)$ exists $\Leftrightarrow$ $b'(s_0)$ exists $\Leftrightarrow$ b has constant sign in an open interval about s_0.)

(*Remark:* The difficulty with the $\pm$ sign here can be eliminated only by using a more subtle definition of the principal normal $\mathbf{N}$. To see what the problem is, look at the curve $\boldsymbol{\gamma}(t) = (t,t^3,0)$ in Fig. 21, and compare the vectors $\mathbf{N}$ and $\mathbf{B}$ for $t < 0$ with the corresponding vectors for $t > 0$; at $t = 0$, they both switch signs.)

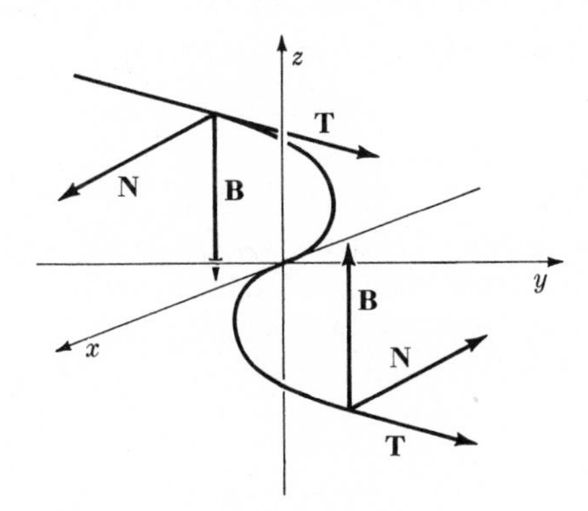

FIGURE 2.21

15. Suppose that $\mathbf{G}$, $\mathbf{G}'$, and $\mathbf{G}''$ are continuous in an open interval containing s_0. Prove that

$$\mathbf{G}(s) = \mathbf{G}(s_0) + (s - s_0)\mathbf{T}(s_0) + \tfrac{1}{2}(s - s_0)^2\kappa(s_0)\mathbf{N}(s_0) + (s - s_0)^2\mathbf{R}(s),$$

where $\lim_{s\to s_0} |\mathbf{R}(s)| = 0$. (Hint: Replace each component of $\mathbf{G}$ by a Taylor expansion of the form

$$f(s) = f(s_0) + (s - s_0)f'(s_0) + \tfrac{1}{2}(s - s_0)^2 f''(s_0) + (s - s_0)^2\frac{f''(\bar{s}) - f''(s_0)}{2},$$

with $\bar{s}$ between s and s_0.)

16. The signed distance from a point $\mathbf{P}_1$ to a plane

$$\pi = \{\mathbf{P}: (\mathbf{P} - \mathbf{P}_0) \cdot \mathbf{A} = 0\}$$

is given by

$$d(\mathbf{P}_1, \pi) = \frac{(\mathbf{P}_1 - \mathbf{P}_0) \cdot \mathbf{A}}{|\mathbf{A}|}.$$

(a) Prove that if π is the osculating plane to $\mathbf{G}$ at $\mathbf{G}(s_0)$, then

$$\lim_{s \to s_0} \frac{d(\mathbf{G}(s), \pi)}{(s - s_0)^2} = 0. \tag{5}$$

(Hint: Use Problem 15.)

(b) Prove that if $\pi = \{\mathbf{P}: (\mathbf{P} - \mathbf{P}_0) \cdot \mathbf{A} = 0\}$ is any plane and $\lim_{s \to s_0} d(\mathbf{G}(s), \pi) = 0$, then $\mathbf{G}(s_0)$ lies on π.

(c) Prove that if $\pi = \{\mathbf{P}: (\mathbf{P} - \mathbf{G}(s_0)) \cdot \mathbf{A} = 0\}$ is any plane containing $\mathbf{G}(s_0)$, and

$$\lim_{s \to s_0} \frac{d(\mathbf{G}(s), \pi)}{|s - s_0|} = 0, \tag{6}$$

then $\mathbf{A} \cdot \mathbf{T}(s_0) = 0$.

(d) Prove that if π is any plane and (5) holds, and $\mathbf{G}''(s_0) \neq \mathbf{0}$, then π is the osculating plane. (Parts (a) and (d) together show that the osculating plane is the "plane of best fit"; it is the only plane π for which (5) is valid.)

Differentiation of Functions of Two Variables

Functions of several variables present some new and interesting questions that do not arise in the one-variable theory, not even in the theory of curves in R^n. Since most of these new questions are rather easy to visualize in the case of two variables, we begin by studying that case in detail, proceeding quickly through the parts that are close to the one-variable theory and dwelling on the new features. This chapter covers differentiation, and the next one combines the derivative theory with integration.

3.1 DEFINITIONS, EXAMPLES, AND ELEMENTARY THEOREMS

The *domain* of a function of two real variables is a set in the plane R^2, that is, a set of ordered pairs of real numbers. To each point (x,y) in its domain, the function f assigns a real number, denoted $f(x,y)$, for example,

$f(x,y) = \sqrt{1 - x^2 - y^2}$ the domain of f is $\{(x,y) : x^2 + y^2 \leq 1\}$

$g(x,y) = xy/(x^2 + y^2)$ the domain of g is all points except $(0,0)$

$h(x,y) = e^x \cos y$ the domain of h is all points in the plane

$\varphi(x,y) = 1/\sqrt{xy}$ the domain of φ is $\{(x,y) : xy > 0\}$.

Nonmathematical examples are just as easy to find. For instance, the latitudes x and longitudes y of all points on the island of Bali form a set of ordered pairs (x,y). If the height above sea level at the point (x,y) is denoted $A(x,y)$, then A is a function of two variables, which we call the altitude function.

The *range* of a function f is the set of values assumed by f, the set

$$\{z: z = f(x,y) \text{ for some } (x,y) \text{ in the domain of } f\}.$$

In the examples listed above, the range of the first function f is the interval $[0,1]$ on the real line, the range of g is $[-\frac{1}{2},\frac{1}{2}]$, the range of h is $(-\infty,+\infty)$, and the range of φ is $(0,+\infty)$. The range of the Balinese altitude function is $[0,3142]$, if altitude is measured in meters.

The *graph* of a function f of two variables is defined to be the set of points in R^3 given by

$$\{(x,y,z): z = f(x,y) \text{ for some } (x,y) \text{ in the domain of } f\}. \tag{1}$$

This definition is virtually the same as for one variable, but sketching the graph is something else again. In general, the graph is a surface in R^3; for instance, the graph of the altitude function A is part of the surface of the earth. (This is a slight oversimplification neglecting the curvature of the earth, but the general idea is right.) The graph of the function

$$f(x,y) = \sqrt{1 - x^2 - y^2} \tag{2}$$

is the upper half of a sphere of radius 1, sketched in Fig. 1(a). (Notice that the z axis points up; this is a standard convention in sketching a graph (1), just as the y axis generally points up for the graph of a function of one variable.) The graph of (2) is a hemisphere because

$$\begin{aligned} z = \sqrt{1 - x^2 - y^2} &\Leftrightarrow z^2 = 1 - x^2 - y^2 \text{ and } z \geq 0, \\ &\Leftrightarrow x^2 + y^2 + z^2 = 1 \text{ and } z \geq 0, \\ &\Leftrightarrow |\mathbf{P}| = 1 \text{ and } z \geq 0, \end{aligned}$$

where $\mathbf{P} = (x,y,z)$. The condition $|\mathbf{P}| = 1$ puts $\mathbf{P}$ on a sphere, and $z \geq 0$ puts $\mathbf{P}$ on or above the xy plane, i.e. on the upper half of the sphere.

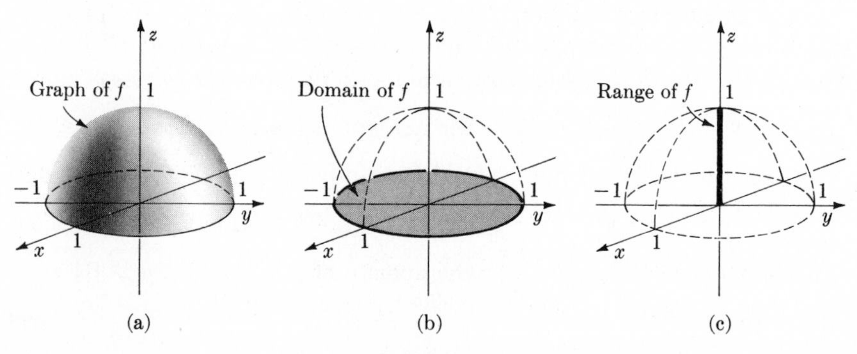

FIGURE 3.1

The *domain* of f is obtained by projecting the graph onto the xy plane; in Fig. 1(b) this projection is the unit disk $\{(x,y): x^2 + y^2 \leq 1\}$. The *range* of f is obtained by projecting the graph onto the z axis; in Fig. 1(c) this projection is the interval $[0,1]$.

The graph of a function can be illustrated on paper by a perspective drawing like Fig. 1, but this is a severe artistic challenge to most of us. Another way to present the graph, calling for less artistic ability in the grapher but more visual imagination in the beholder, is to draw a "contour map"; that is, draw an xy plane, and show the curves along which f assumes various equally spaced values. Figure 2(a) does this for the Balinese altitude function, and Fig. 2(b) for the function f in (2). The result is called a "contour graph." Figure 3 gives three more contour graphs, and Fig. 4 gives the corresponding perspective drawings. The lines in Fig. 3 along which the function is constant are called *contour lines*, or *level lines*. Notice that the graph is steeper where the contour lines are closer together. The contour line where f assumes the value c shows exactly where the graph of f intersects the plane $\{(x,y,z): z = c\}$ (see Fig. 5).

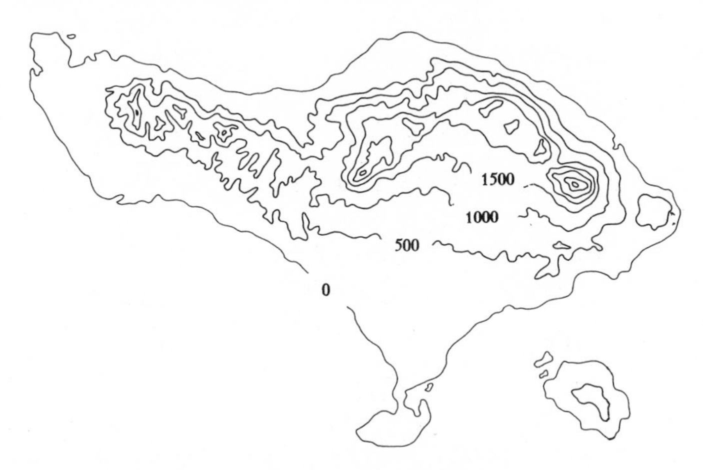

FIGURE 3.2 (a) Topographic map of Bali

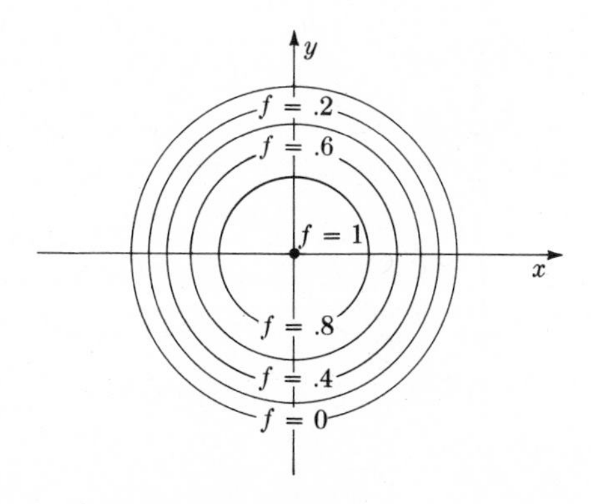

FIGURE 3.2 (b)

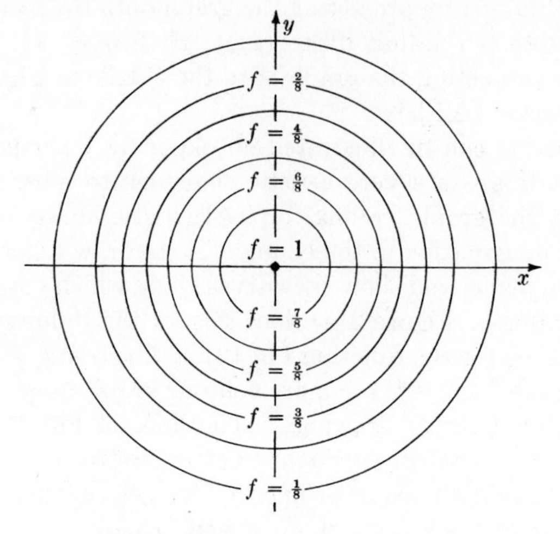

3.3(a) $f(x,y) = \exp\left[-\frac{1}{2}(x^2 + y^2)\right]$

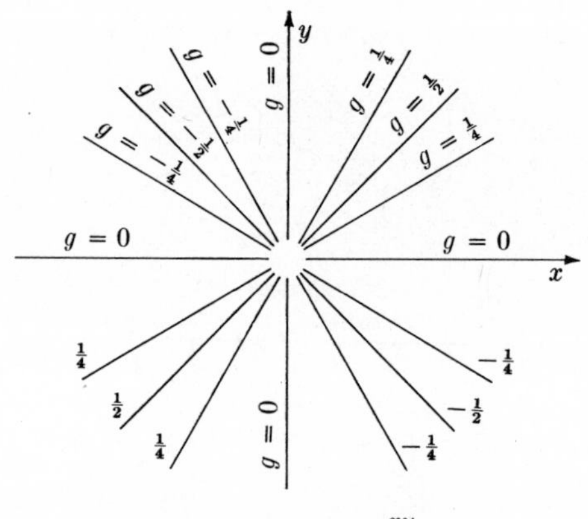

3.3(b) $g(x,y) = \dfrac{xy}{x^2 + y^2}$

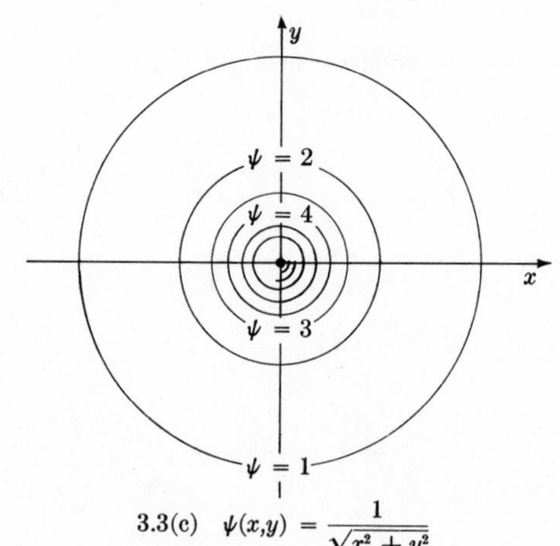

3.3(c) $\psi(x,y) = \dfrac{1}{\sqrt{x^2 + y^2}}$

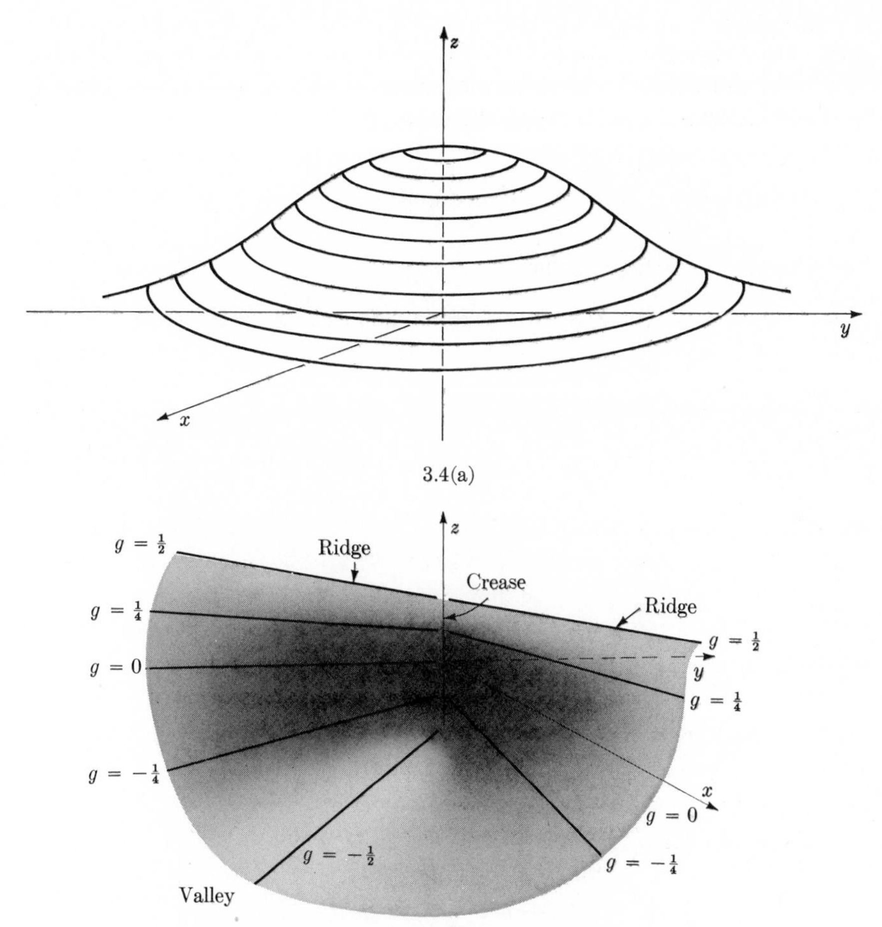

3.4(a)

3.4(b) The perspective is changed (x axis to the right, not to the left) in order to show a "valley" of the graph.

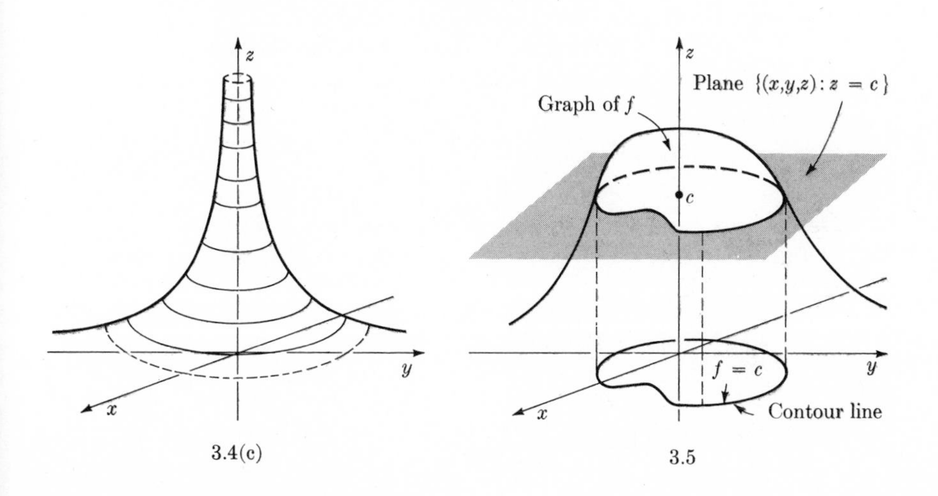

3.4(c)

3.5

The ideas of limit and continuity carry over immediately from functions of one variable. We will state the definitions and theorems, but give no proofs, since they are exactly the same as before. We use the letter $\mathbf{P}$ to stand for (x,y), and $\mathbf{P}_0$ to stand for (x_0, y_0).

Definition 1. $\lim_{\mathbf{P}\to\mathbf{P}_0} f(\mathbf{P}) = L$ if and only if, for every $\epsilon > 0$, there is a number $\delta > 0$ such that

$$0 < |\mathbf{P} - \mathbf{P}_0| < \delta \Rightarrow f(\mathbf{P}) \text{ is defined and } |f(\mathbf{P}) - L| < \epsilon;$$

that is,

$$0 < \sqrt{(x - x_0)^2 + (y - y_0)^2} < \delta \Rightarrow$$
$$f(x,y) \text{ is defined and } |f(x,y) - L| < \epsilon.$$

An alternate notation for $\lim_{\mathbf{P}\to\mathbf{P}_0} f(\mathbf{P})$ is

$$\lim_{(x,y)\to(x_0,y_0)} f(x,y),$$

which is clumsy but sometimes convenient.

A given function f can have at most one limit at a given point $\mathbf{P}_0$. Moreover, limits obey the usual algebraic rules:

Theorem 1. *If* $\lim_{\mathbf{P}\to\mathbf{P}_0} f(\mathbf{P}) = a$ *and* $\lim_{\mathbf{P}\to\mathbf{P}_0} g(\mathbf{P}) = b$, *then*

$$\lim_{\mathbf{P}\to\mathbf{P}_0} (f(\mathbf{P}) + g(\mathbf{P})) = a + b,$$

$$\lim_{\mathbf{P}\to\mathbf{P}_0} f(\mathbf{P})g(\mathbf{P}) = ab,$$

$$\lim_{\mathbf{P}\to\mathbf{P}_0} cf(\mathbf{P}) = ca \qquad (\textit{where } c \textit{ is any constant}),$$

$$\lim_{\mathbf{P}\to\mathbf{P}_0} \frac{f(\mathbf{P})}{g(\mathbf{P})} = \frac{a}{b} \qquad (\textit{if } b \neq 0).$$

Definition 2. f is *continuous* at $\mathbf{P}_0$ if and only if $\mathbf{P}_0$ is in the domain of f and, for every $\epsilon > 0$, there is $\delta > 0$ such that for all points $\mathbf{P}$ in the domain of f,

$$|\mathbf{P} - \mathbf{P}_0| < \delta \Rightarrow |f(\mathbf{P}) - f(\mathbf{P}_0)| < \epsilon;$$

that is,

$$\sqrt{(x - x_0)^2 + (y - y_0)^2} < \delta \Rightarrow |f(x,y) - f(x_0, y_0)| < \epsilon.$$

Theorem 2. *If f and g are continuous at $\mathbf{P}_0$, then so are $f + g$, fg, and cf (where c is any constant). If, in addition, $g(\mathbf{P}_0) \neq 0$, then f/g is continuous at $\mathbf{P}_0$.*

Theorem 3. Composition of Continuous Functions. *Let f be continuous at $\mathbf{P}_0$.*

(i) *If g is continuous at $f(\mathbf{P}_0)$, then $g \circ f$ is continuous at $\mathbf{P}_0$. (Here, g is a function of* one *real variable, and $g \circ f$ is the composite function defined by $(g \circ f)(\mathbf{P}) = g(f(\mathbf{P}))$.)*

(ii) *If $\boldsymbol{\gamma}$ is a curve in the plane which is continuous at t_0, and $\boldsymbol{\gamma}(t_0) = \mathbf{P}_0$, then $f \circ \boldsymbol{\gamma}$ is continuous at t_0.*

The functions we ordinarily encounter are continuous at "most" of the points where they are defined; the exceptional points of discontinuity show up on the contour graph where contour lines for different values of f come together (*example:* the point $(0,0)$ in Fig. 3(b)) or where contour lines pile up (*example:* the point $(0,0)$ in Fig. 3(c)). The first possibility corresponds to a "crinkly" part of the graph (Fig. 4(b)), and the second to a place where the graph "blows up" (Fig. 4(c)).

The domains of functions of one variable are generally taken to be intervals, but in several variables they may be more complicated. The plane sets most nearly analogous to open intervals are *open rectangles*

$$R = \{(x,y) : a < x < b,\ c < y < d\}$$

(Fig. 6) and *open disks*

$$D = \{\mathbf{P} : |\mathbf{P} - \mathbf{A}| < r\}.$$

But these are too special. The general concept we need is that of an *open set.*

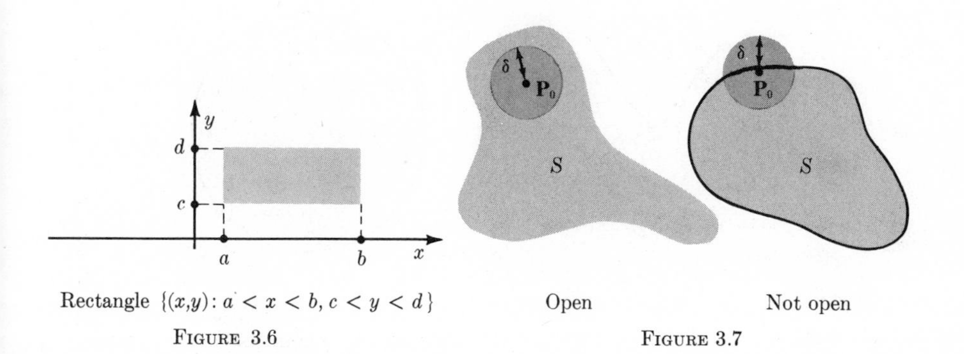

Rectangle $\{(x,y) : a < x < b,\ c < y < d\}$ Open Not open

FIGURE 3.6 FIGURE 3.7

A set S in R^2 is called *open* if every point $\mathbf{P}_0$ in S is the center of an open disk of positive radius contained entirely in S. (See Fig. 7.) More formally:

Definition 3. A set S in R^2 is *open* if for every point $\mathbf{P}_0$ in S there is a number $\delta > 0$ such that

$$|\mathbf{P} - \mathbf{P}_0| < \delta \quad \Rightarrow \quad \mathbf{P} \text{ is in } S.$$

Intuitively, a set S is open if it contains none of the points on its boundary. Every open disk $S = \{\mathbf{P}: |\mathbf{P} - \mathbf{A}| < r\}$ is an open set in the sense of Definition 3. To prove it, let $\mathbf{P}_0$ be any point in S. Then

$$\delta = r - |\mathbf{P}_0 - \mathbf{A}|$$

is a positive number, and the disk

$$\{\mathbf{P}: |\mathbf{P} - \mathbf{P}_0| < \delta\}$$

is contained entirely in S, since by the triangle inequality

$$|\mathbf{P} - \mathbf{P}_0| < \delta \quad \Rightarrow \quad |\mathbf{P} - \mathbf{A}| \leq |\mathbf{P} - \mathbf{P}_0| + |\mathbf{P}_0 - \mathbf{A}| < \delta + |\mathbf{P}_0 - \mathbf{A}| = r$$
$$\Rightarrow \quad \mathbf{P} \quad \text{is in } S.$$

(See Fig. 8.) As you might expect, the *closed* disk $\{\mathbf{P}: |\mathbf{P}| \leq 1\}$ is not an open set. Any boundary point $\mathbf{P}_0$ violates the condition in Definition 3; no matter how small $\delta > 0$ is chosen, the disk

$$\{\mathbf{P}: |\mathbf{P} - \mathbf{P}_0| < \delta\}$$

contains points $\mathbf{P}$ that are *not* in the closed unit disk. (See Fig. 9.)

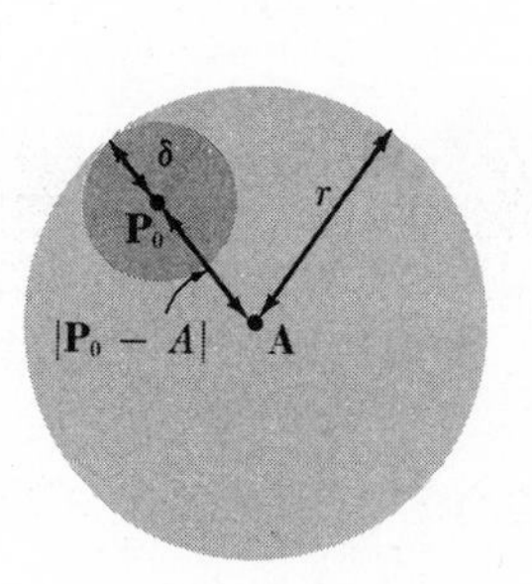

FIGURE 3.8

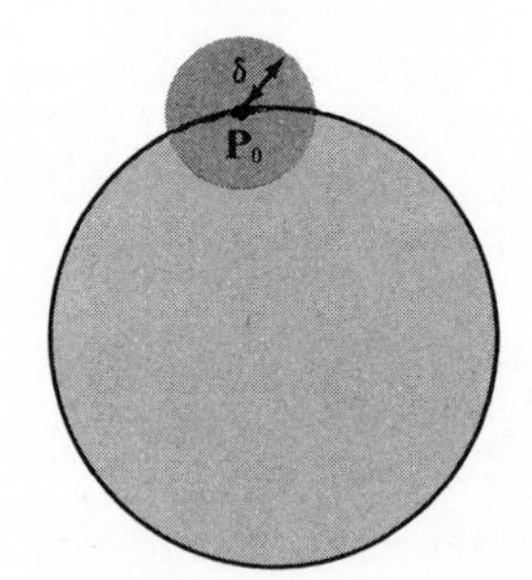

FIGURE 3.9

Similarly, the set $\{(x,y): xy > 0\}$ is open, but the set $\{(x,y): xy \geq 0\}$ is not. Roughly speaking, sets defined by strict inequalities ($<$ or $>$) are open, and sets defined by $\leq$, $\geq$, or $=$ are generally not open.

One advantage of considering functions defined on open sets shows up in

Theorem 4. *If the domain of f is an open set, then f is continuous at $\mathbf{P}_0$ if and only if* $\lim\limits_{\mathbf{P}\to\mathbf{P}_0} f(\mathbf{P}) = f(\mathbf{P}_0)$.

Proof. The only difference between the two statements

$$f \text{ is continuous at } \mathbf{P}_0$$

and

$$\lim_{\mathbf{P}\to\mathbf{P}_0} f(\mathbf{P}) = f(\mathbf{P}_0)$$

is that the second statement requires f to be defined in some disk of radius $\delta > 0$ about $\mathbf{P}_0$ (see Definitions 1 and 2). Since the domain of f is an open set, f is in fact defined in such a disk, so the two statements about f are equivalent.

We conclude with some examples of limits, continuity, and open sets.

Example 1. If $f(\mathbf{P}) = c$ (a constant), then $\lim\limits_{\mathbf{P}\to\mathbf{P}_0} f(\mathbf{P}) = c$.

Proof. Given $\epsilon > 0$, *every* $\delta > 0$ satisfies the conditions of Definition 1, since

$$0 < |\mathbf{P} - \mathbf{P}_0| < \delta \Rightarrow |f(\mathbf{P}) - f(\mathbf{P}_0)| = |c - c| = 0 < \epsilon.$$

Example 2. $\lim\limits_{(x,y)\to(x_0,y_0)} x = x_0$.

Proof. Given $\epsilon > 0$, choose $\delta = \epsilon$. Since $|x - x_0| \leq \sqrt{|x - x_0|^2 + |y - y_0|^2}$, we have

$$0 < \sqrt{|x - x_0|^2 + |y - y_0|^2} < \delta \Rightarrow |x - x_0| < \delta \Rightarrow |x - x_0| < \epsilon;$$

hence Definition 1 is satisfied.

A similar proof shows that $\lim\limits_{(x,y)\to(x_0,y_0)} y = y_0$. These results combine with Theorems 2, 3, and 4 to prove the continuity of functions given by simple expressions such as $x^2 + y^2$, $e^{x^2+y^2}$, $\sin(xe^y)$, and so on.

Example 3. The set $S = \{(x,y): |y| > 1\}$ is open.

Proof. If $|y_0| > 1$, then $\delta = |y_0| - 1 > 0$, and the disk of radius δ lies in S (see Fig. 10) since

$$\sqrt{(x - x_0)^2 + |y - y_0|^2} < \delta \Rightarrow |y - y_0| < \delta \Rightarrow |y| > |y_0| - \delta = 1.$$

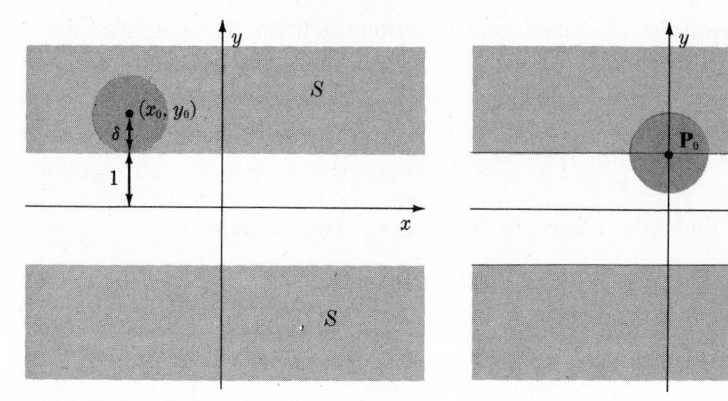

FIGURE 3.10 $S = \{(x,y): |y| > 1\}$, $\delta = |y_0| - 1$ 3.11 $\bar{S} = \{(x,y): |y| \geq 1\}$, $\mathbf{P}_0 = (0,1)$

Example 4. The set $\bar{S} = \{(x,y): |y| \geq 1\}$ is *not* open. Intuitively, the reason for this is that it contains boundary points along the lines $y = \pm 1$. To prove that $\bar{S}$ is not open, we take the point $\mathbf{P}_0$ in Definition 3 to be one of these boundary points, say $\mathbf{P}_0 = (0,1)$. It is clear from Fig. 11 that *every* open disk about $\mathbf{P}_0$ contains points which are not in $\bar{S}$; hence, by Definition 3, $\bar{S}$ is not open. (To make a rigorous analytic proof, suppose there *is* $\delta > 0$ such that

$$|\mathbf{P} - \mathbf{P}_0| < \delta \implies \mathbf{P} \text{ is in } \bar{S}. \tag{*}$$

Choose a number h such that $0 < h < 2$ and $h < \delta$, and let $\mathbf{P} = (0,1 - h)$. Since $0 < h < 2$, we get $|1 - h| < 1$, so $\mathbf{P}$ is *not* in $\bar{S}$. Since $0 < h < \delta$, we have $|\mathbf{P} - \mathbf{P}_0| = |(0,1 - h) - (0,1)| = h < \delta$; hence, by (*), $\mathbf{P}$ *is* in $\bar{S}$. This contradiction shows that $\bar{S}$ is not open, after all.)

Example 5. If $g(x,y) = xy/(x^2 + y^2)$ for $(x,y) \neq \mathbf{0}$ (see Figs. 3(b) and 4(b)), then $\lim_{\mathbf{P}\to\mathbf{0}} g(\mathbf{P})$ does not exist. The reason for this, intuitively, is that $g(\mathbf{P})$ approaches *different* values as $\mathbf{P} \to \mathbf{0}$ from different directions. For example, if $\mathbf{P} \to \mathbf{0}$ along either axis, then $g(\mathbf{P}) \to 0$, but if $\mathbf{P} \to \mathbf{0}$ along the main diagonal (where $x = y$), then $g(\mathbf{P}) \to 1/2$, since $g(x,x) = x^2/2x^2 = 1/2$. To prove analytically that $\lim_{\mathbf{P}\to\mathbf{0}} g(\mathbf{P})$ does not exist, suppose on the contrary that $\lim_{\mathbf{P}\to\mathbf{0}} g(\mathbf{P}) = L$ for some L. Then the function

$$\bar{g}(\mathbf{P}) = \begin{cases} \dfrac{xy}{x^2 + y^2}, & (x,y) \neq \mathbf{0} \\ L, & (x,y) = \mathbf{0} \end{cases}$$

is continuous at $\mathbf{0}$, by Theorem 4. Now let $\boldsymbol{\gamma}(t) = (0,t)$. Then $\boldsymbol{\gamma}(0) = \mathbf{0}$, so

$$L = \bar{g}(\boldsymbol{\gamma}(0)) \qquad \text{(by definition of } \bar{g})$$

$$= \lim_{t\to 0} \bar{g}(\boldsymbol{\gamma}(t)) \qquad \text{(by Theorem 3, since } \boldsymbol{\gamma} \text{ is continuous at 0 and } \bar{g} \text{ is continuous at } \boldsymbol{\gamma}(0))$$

$$= 0 \qquad \left(\text{since } \bar{g}(\boldsymbol{\gamma}(t)) = \frac{0\cdot t}{0+t^2} = 0 \text{ for } t \neq \mathbf{0}\right);$$

hence $L = 0$. On the other hand, if we take $\boldsymbol{\gamma}(t) = (t,t)$, the same argument shows that $L = \frac{1}{2}$, and this contradiction shows that $\lim_{\mathbf{P}\to\mathbf{0}} g(\mathbf{P}) = L$ is impossible.

This example shows that when $\lim_{\mathbf{P}\to\mathbf{P}_0} f(\mathbf{P}) = L$, then $f(\mathbf{P})$ must tend to L as $\mathbf{P}$ approaches $\mathbf{P}_0$ *from any direction.* Moreover, it could happen that $f(\mathbf{P})$ tends to L along every straight line through $\mathbf{P}_0$, and yet $\lim_{\mathbf{P}\to\mathbf{P}_0} f(\mathbf{P}) \neq L$ because $f(\mathbf{P})$ does not tend to L along certain curves through $\mathbf{P}_0$. (See Problem 14 below.)

PROBLEMS

1. Sketch contour graphs (and, when practical, perspective graphs) of the following functions:

(a) $f(x,y) = x^2 + y^2$. (Sketch the contour lines for $f = 0, \frac{1}{2}, 1, \frac{3}{2}, 2$.)

(b) $f(x,y) = \dfrac{1}{x^2 + y^2}$. (Sketch the contour lines for $f = 1, 2, \ldots$.)

(c) $f(x,y) = xy$. (Sketch the contour lines for $f = 0, \pm\frac{1}{2}, \pm 1, \pm\frac{3}{2}, \pm 2$.)

(d) $f(x,y) = \dfrac{x}{x^2 + y^2}$. (Sketch the contour lines for $f = 0, \pm 1, \pm 2$. Notice that the contour lines are circles.)

(e) $f(x,y) = x^2 + 3y^2$.

(f) $f(x,y) = e^x \cos y$. (Sketch the contour lines for $f = 0, \pm\frac{1}{2}, \pm 1, \pm\frac{3}{2}$.)

2. From the graphs in Problem 1, which of the functions appear to be continuous at $(0,0)$? not (b) (d)

3. (a) Prove that $\lim_{(x,y)\to(x_0,y_0)} y = y_0$.

(b) Prove that if $Q(x,y)$ is a polynomial in x and y, then Q is continuous at every point (x_0,y_0). (Use part (a), Examples 1 and 2, and Theorems 2 and 4.)

(c) Prove that if f is a rational function, i.e. if $f = Q/R$ where Q and R are polynomials, then f is continuous at every point where $R \neq 0$.

4. Prove that the following functions are continuous at every point. (Use Problem 3 and the theorems given above.)

(a) $f(x,y) = e^{xy}$

(b) $g(x,y) = \cos(x^2 + y^2)$

(c) $h(x,y) = xe^{xy}$

For Problems 5–7 review the corresponding proof for functions of one variable, if necessary.

5. Prove Theorem 1.

6. Prove Theorem 2.

7. Prove Theorem 3.

8. Prove that $S = \{\mathbf{P}: |\mathbf{P}| > 0\}$ is open. (Hint: Prove that if $\mathbf{P}_0$ is in S, then the open disk about $\mathbf{P}_0$ of radius $|\mathbf{P}_0|$ is contained entirely in S.)

9. Is the set $\{(x,y): |y| > 0\}$ open or not? Prove your answer.

10. Prove that the x axis is not open.

11. Suppose that f is continuous and its domain is an open set. Prove that the set $\{\mathbf{P}: f(\mathbf{P}) > 0\}$ is open. (Hint: Use Theorem 4 and the definition of limit with $\epsilon = f(\mathbf{P}_0)$.)

12. Suppose that f is continuous and its domain is an open set. Prove that for any numbers a and b, the set $\{\mathbf{P}: a < f(\mathbf{P}) < b\}$ is open. (Hint: $a < z < b \;\Leftrightarrow\; (b - z)(z - a) > 0$.)

13. Use Problem 12 to show that the following sets are open:

(a) $\{\mathbf{P}: |\mathbf{P}| < r\}$ (r a constant)

(b) $\{\mathbf{P}: 1 < |\mathbf{P}| < 2\}$

(c) $\{(x,y): x^2 + y^2 < 1 \text{ and } y > 0\}$. (Hint: Consider the function $f(x,y) = y$ defined only on the open unit disk.)

(d) $\{(x,y): x^2 + y^2 < 1 \text{ and } 0 < y < \frac{1}{2}\}$

(e) $\{(x,y): \sin^2 x < \frac{1}{2}\}$

14. Let

$$f(x,y) = \begin{cases} 1 & \text{if } x^2 < y < 2x^2 \\ 0 & \text{otherwise.} \end{cases}$$

Figure 12(a) shows a "contour graph"; $f = 1$ in the shaded area, and $f = 0$ elsewhere. Prove that this function has the following properties:

(a) $f(\mathbf{P})$ tends to zero along every straight line through the origin; that is, $\lim_{t\to 0} f(t\cos\theta, t\sin\theta) = 0$. (See Fig. 12(b).)

(b) $f(\mathbf{P})$ tends to 1 along certain curves through the origin; for example, $\lim_{t\to 0} f(2t, 6t^2) = 1$. (See Fig. 12(b).)

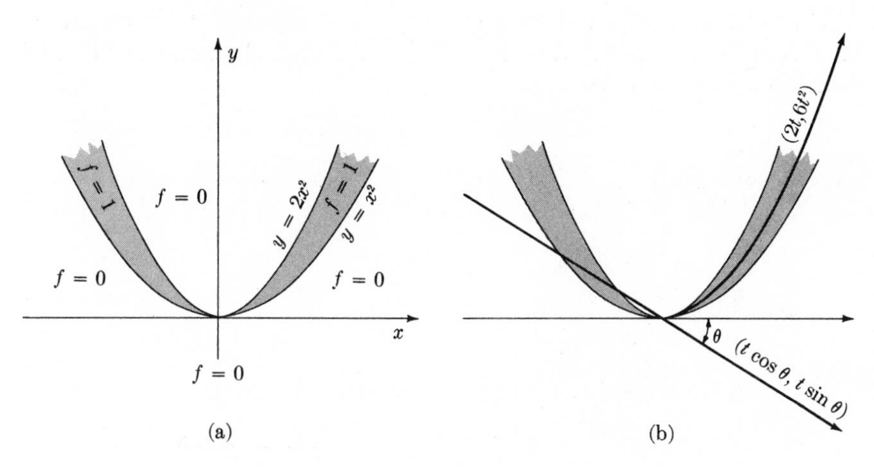

FIGURE 3.12

3.2 POLYNOMIALS OF DEGREE ONE

The study of functions of one variable begins, as you know, with the general first degree polynomial $f(x) = mx + b$. The graph of f, the set $\{(x,y) : y = mx + b\}$, is a straight line in R^2. The number m, called the *slope*, tells how steeply the line is tipped, and in which direction.

Our starting point for two variables is exactly the same. A first degree polynomial is a function of the form $f(x,y) = ax + by + c$, where a, b, and c are constants. More concisely, $f(\mathbf{P}) = \mathbf{M}\cdot\mathbf{P} + c$, where $\mathbf{P} = (x,y)$ and $\mathbf{M} = (a,b)$. The graph of f is the set $\{(x,y,z) : z = ax + by + c\}$, which is a plane in R^3. The vector $\mathbf{M} = (a,b)$ is called the *slope* of the plane. To see the geometric meaning of the slope, think of the plane as a hillside with the z axis pointing straight up, as in Fig. 13. Then *the slope* $\mathbf{M}$ *points in the direction of steepest ascent, and its length gives the rate of ascent in this direction.* This slightly vague statement can be made precise as follows: moving from point $\mathbf{P}_0$ to point $\mathbf{P}$ causes f to increase by $f(\mathbf{P}) - f(\mathbf{P}_0)$, and the *rate* of increase is

$$\frac{f(\mathbf{P}) - f(\mathbf{P}_0)}{|\mathbf{P} - \mathbf{P}_0|} = \frac{\mathbf{M}\cdot\mathbf{P} + c - (\mathbf{M}\cdot\mathbf{P}_0 + c)}{|\mathbf{P} - \mathbf{P}_0|}$$

$$= \frac{\mathbf{M}\cdot(\mathbf{P} - \mathbf{P}_0)}{|\mathbf{P} - \mathbf{P}_0|} = |\mathbf{M}| \cos\theta, \tag{1}$$

where θ is the angle between $\mathbf{P} - \mathbf{P}_0$ and $\mathbf{M}$. Our italicized claim about the slope follows immediately from (1): the rate of increase of f is maximum when $\cos\theta = 1$ (that is, when $\mathbf{P} - \mathbf{P}_0$ has the same direction as $\mathbf{M}$), and the maximum rate is $|\mathbf{M}|$.

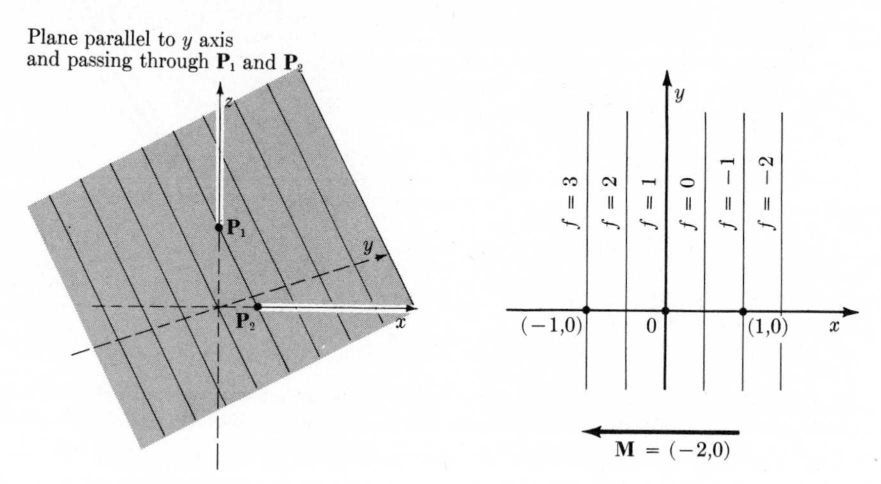

FIGURE 3.13 $f(x,y) = -2x + 1$

FIGURE 3.14 $f(x,y) = -2x + 1$

Figure 13 illustrates this for the function $f(x,y) = -2x + 1$. The slope $\mathbf{M} = (-2,0)$ points directly to the left, which is clearly the direction of steepest ascent; and the length of $\mathbf{M}$ is 2, which corresponds to the fact that moving one unit to the left causes f to increase by 2 units (see contour graph in Fig. 14).

Another important fact shows up in Fig. 14, namely, *the slope is orthogonal to the contour lines.* This is nothing new; it is basic in analytic geometry that the vector (a,b) is a normal to the line $\{(x,y): ax + by + c = d\}$. In other words, the slope vector (a,b) is orthogonal to the level curve $f(x,y) = ax + by + c = d$.

So far we have discussed the slope vector as a whole, but each component has its own interpretation as well. In the polynomial $f(x,y) = ax + by + c$, suppose we hold y fixed, but increase x by some amount h. Then f is increased by

$$f(x + h,y) - f(x,y) = a(x + h) + by + c - ax - by - c = ah,$$

which shows that a is the rate of increase of f per unit increase in x, with y fixed. Geometrically, this is the slope in the x direction. (See Fig. 15.) Similarly, b is the slope in the y direction.

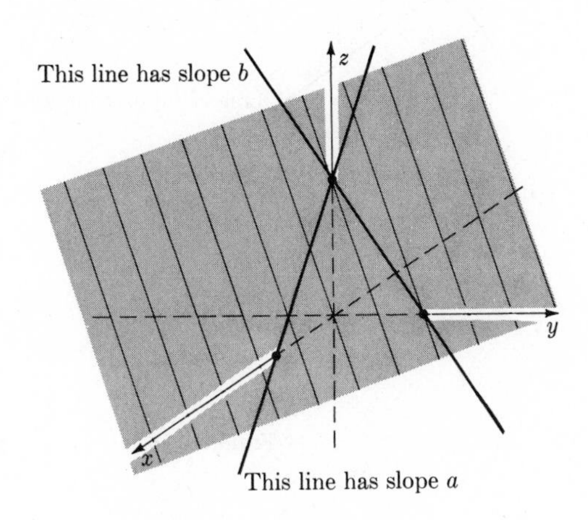

FIGURE 3.15

To summarize: The graph of the first degree polynomial $f(x,y) = ax + by + c$ is a plane. The vector $\mathbf{M} = (a,b)$ is its slope; the first component, a, is the slope in the x direction, and the second component, b, is the slope in the y direction. The slope vector $\mathbf{M}$ points in the direction of maximum increase of f. The contour lines of f are precisely the lines orthogonal to $\mathbf{M}$.

This connection between contour lines and slope can be used to solve maximum problems for any first degree polynomial f. Let R be a given region in the plane; then $\max_R f = d$ means that

(i) $f(\mathbf{P}) \leq d$ for every point $\mathbf{P}$ in R

and

(ii) $f(\mathbf{P}) = d$ for *some* point $\mathbf{P}$ in R.

To see the geometric meaning of (i), notice that the contour line

$$\{\mathbf{P}: f(\mathbf{P}) = d\} \tag{2}$$

divides the plane into two open half-planes, namely,

$$\{\mathbf{P}: f(\mathbf{P}) > d\} \qquad \text{and} \qquad \{\mathbf{P}: f(\mathbf{P}) < d\}.$$

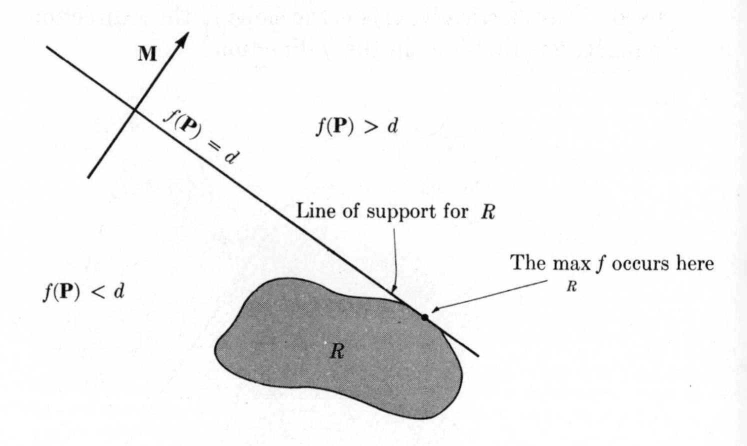

FIGURE 3.16

The first, called the *positive* half-plane, is the one toward which the slope **M** points (see Fig. 16). The second, naturally, is called the *negative* half-plane. Thus condition (i) above means that R lies on the *negative* side of the contour line (2), and (ii) means that at least one point of R lies *on* the contour line itself. A line (2) with these two properties is called a *line of support for R orthogonal to* **M**.

We have shown that finding the maximum $\max_R f$ is equivalent to finding such a line of support; this can often be done simply by drawing an accurate picture of R and **M**. The appendix following this section gives many examples of simple "managerial" problems (maximizing profit, for example) that can be solved in this way.

Example 1. Find the maximum of $f(x,y) = 3x + 4y - 10$ over the disk $\{(x,y): x^2 + y^2 \leq 1\}$. *Solution.* Figure 17 shows the region R, the slope vector $\mathbf{M} = (3,4)$, and a line of support orthogonal to **M**. By elementary geometry, this line of support must be tangent to the unit circle at $(\frac{3}{5},\frac{4}{5})$. Hence the maximum of f over the unit disk is $f(\frac{3}{5},\frac{4}{5}) = -5$. (This result, found by graphical means, can be proved analytically by observing that the expression $3x + 4y$ is a dot product. Thus

$$3x + 4y = (3,4)\cdot(x,y) = |(3,4)|\cdot|(x,y)|\cos\theta$$
$$= 5\sqrt{x^2 + y^2}\cos\theta, \tag{3}$$

where θ is the angle between (3,4) and (x,y). Hence the maximum of the expression (3) under the restriction $x^2 + y^2 \leq 1$ is achieved by taking $x^2 + y^2 = 1$ and $\cos\theta = 1$, and this maximum is 5. Since $3x + 4y$ has a maximum of 5, it follows that $3x + 4y - 10$ has a maximum of -5.)

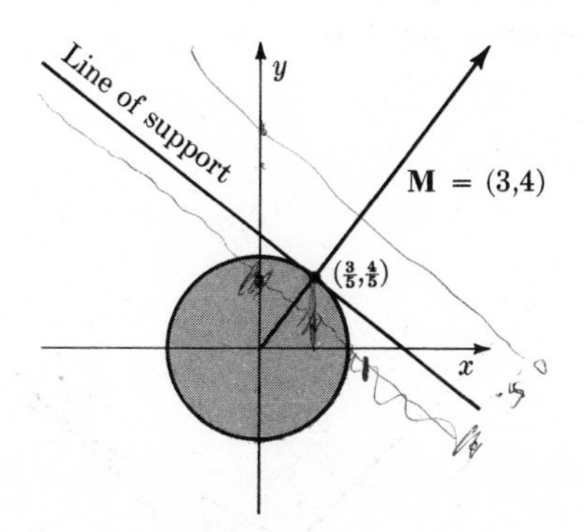

FIGURE 3.17

PROBLEMS

1. Show that the equation of a plane through a given point (x_0, y_0, z_0) with given slope $\mathbf{M} = (a,b)$ is

$$f(\mathbf{P}) = z_0 + (\mathbf{P} - \mathbf{P}_0)\cdot\mathbf{M},$$

where $\mathbf{P}_0 = (x_0, y_0)$.

2. Find the slope of the plane defined by the given equation:

(a) $z = x - y + 100$
(b) $z = 2$
(c) $z = x - y - 100$
(d) $4z + 5x = 1$
(e) $x + 2y + 3z = 0$
(f) $x + y = 1$

(Notice that in part (b) the slope is (0,0); the graph is horizontal and there is no "uphill" direction. Notice that in part (f) *no slope is defined*; the plane is vertical.)

3. For each of the following functions f, sketch a contour graph (as in Fig. 14), and sketch the slope vector.

(a) $f(x,y) = 2x + 3y - 1$
(b) $f(x,y) = 10x - 5y + 3$
(c) $f(x,y) = -5x + 4y$
(d) $f(x,y) = -x - y + 10$

4. Let $f(x,y) = ax + by + c$. Show that $\mathbf{N} = (a,b,-1)$ is a normal vector to the graph of f.

5. For each of the functions f in Problem 3, find (by graphical means) the maximum value of f over the unit circle $\{(x,y) : x^2 + y^2 \leq 1\}$.

6. Repeat Problem 5, but find the minimum instead of the maximum.

7. Find the maximum of f over the rectangle $\{(x,y): |x| \leq 1, |y| \leq 2\}$. Use graphical methods.
(a) $f(x,y) = 10x - 3y$
(b) $f(x,y) = -4x + 5y + 6$

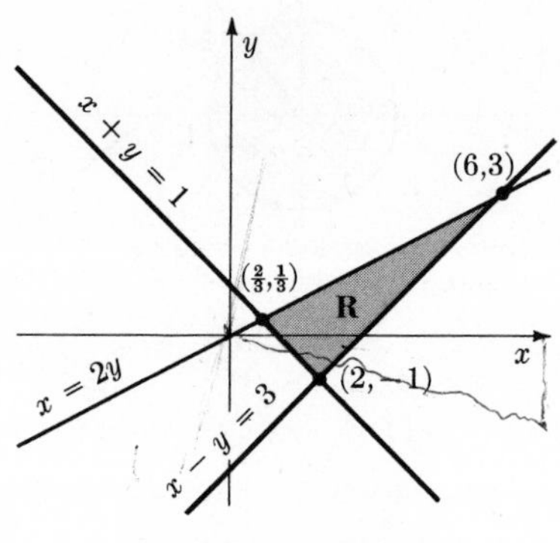

FIGURE 3.18

8. Figure 18 shows the region R in the (x,y) plane determined by the inequalities $x \geq 2y$, $x + y \geq 1$, $x - y \leq 3$. For each of the functions in Problem 7, find (by graphical means) the maximum value over R.

9. What happens in Problem 8 if R is defined by the strict inequalities $x > 2y$, $x + y > 1$, $x - y < 3$? Is there still a maximum?

10. Find (graphically) the maximum of $f(x,y) = x + y$ over the ellipse $\{(x,y): x^2 + 2y^2 \leq 2\}$.

11. Suppose a hillside is sloped like the plane $\{(x,y,z): z = x\}$, and a man is going "northeast" up the hill, i.e. in the direction of the vector (1,1). Show that he rises 1 unit whenever he moves $\sqrt{2}$ units toward the northeast; i.e. the upward slope of his path is $1/\sqrt{2}$.

12. Suppose a hillside is sloped like the plane $\{(x,y,z): z = ax + by + c\}$, and a man is walking on the hillside in the direction of the unit vector $\mathbf{V} = (v_1,v_2)$. Find the upward slope of his path, i.e. the increase of altitude due to his going one unit in the direction $\mathbf{V}$. (The answer can be expressed as a dot product.)

13. Under what conditions on the slope $\mathbf{M}$ does the first degree polynomial $f(\mathbf{P}) = \mathbf{M}\cdot\mathbf{P} + c$ achieve a maximum in the region $R = \{(x,y): y \geq x^2\}$?

APPENDIX: TWO-DIMENSIONAL LINEAR PROGRAMMING

In the maximum problems solved by "linear programming," we find the maximum of a first degree polynomial over a region R, where R itself is defined by inequalities involving first degree polynomials. The following example shows how these problems arise, and how to solve them in the simple case of two variables.

Example 1. A trucker going from farm to city has room for only 100 bushels of fruit in his truck. The farm will sell up to 75 bushels of apples at \$1 a bushel, and up to 65 bushels of peaches at \$2 a bushel. The trucker has \$150 to spend on the fruit, and he can make a profit of \$1 a bushel on peaches, and \$.75 a bushel on apples. Exactly how should he divide the load between peaches and apples to make the maximum profit?

We have to sort out the information and express it mathematically. Let p be the number of bushels of peaches he takes, and a the number of bushels of apples. Then the restrictions on what he can take are

$$p + a \leq 100 \tag{1}$$

$$0 \leq a \leq 75 \tag{2}$$

$$0 \leq p \leq 65 \tag{3}$$

$$2p + a \leq 150. \tag{4}$$

The profit which he wants to maximize is

$$P(p,a) = p + \tfrac{3}{4}a. \tag{5}$$

The conditions (1)–(4) define a certain region R in the (p, a) plane, and we want to maximize the profit P on that region. So we sketch the region R, the slope vector of P, and a typical contour line, and then locate the maximum.

To see what the restriction (1) means geometrically, notice that the graph of $a + p = 100$ is the straight line in Fig. 19, and the inequality $a + p \leq 100$ says that (p,a) lies *on or below* this line; thus $\{(p,a)\colon (1) \text{ is satisfied}\}$ is the shaded area in Fig. 19. Taking account of the other restrictions (2)–(4) as well, we find the shaded region R in Fig. 20 representing the trucker's buying possibilities. Finally, we sketch a vector parallel to the slope $\mathbf{M} = (1,\frac{3}{4})$ of the profit function P in (5), sketch a line of support orthogonal to $\mathbf{M}$, and thus find the point marked with a heavy dot in Fig. 20 to yield the maximum profit, given the trucker's possibilities. This point lies at the intersection of the two lines

$$p + a = 100$$

$$2p + a = 150;$$

subtracting the first from the second, we find $p = 50$, hence $a = 50$. Hence the trucker should buy 50 bushels each of apples and peaches, thus spending all his money, filling his truck, and maximizing his profit.

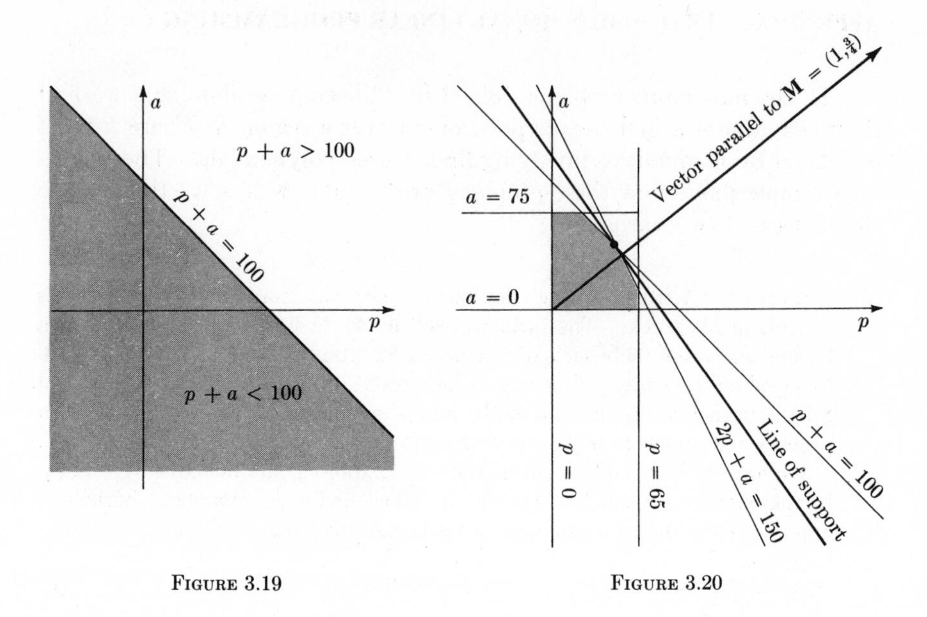

FIGURE 3.19

FIGURE 3.20

Remark 1. It is worth comparing our graphical solution to a plausible "common sense" solution. The trucker might reason as follows: "The profit is highest on peaches, so I'll buy as many peaches as possible; 65 bushels at \$2 a bushel comes to \$130, so with the \$20 left over I'll buy 20 bushels of apples. But look, now I have only 85 bushels, and the truck has room for 15 more. Maybe I should try to fill up the truck, but with as high a proportion of peaches as possible. That means I spend \$150 to buy 100 bushels. Let's see, ..." Thus, eventually, he may arrive at our solution; then he has to compare the profit made by buying 65 bushels of peaches to the profit made by buying 50 bushels of each. The virtue of Fig. 20 is that it shows *all* the possibilities at a glance, and you can immediately pick out the best one without weighing each possibility.

Remark 2. Graphical solutions of mathematical problems are convenient, but not completely satisfactory. Fortunately, there is an analytic solution to the problems of linear programming, suitable for use on computers. In the example above, notice that the maximum was found at a vertex of the polygonal region R; this is also true in all the problems below, and in fact it can be proved that the maximum is generally found at a vertex. More important yet, it can always be found by solving systems of linear equations, something computers do rather well. We will describe the method in detail for the case of two variables.

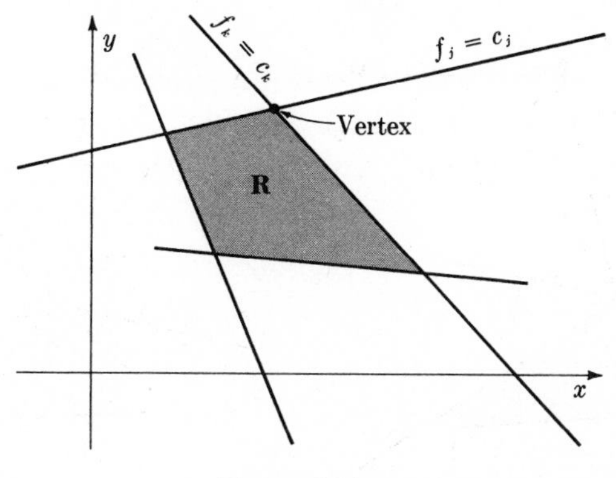

FIGURE 3.21

Let $f(x,y) = ax + by + c$ be the function to be maximized, and let the region R be defined by the inequalities $f_1 \le c_1, \ldots, f_n \le c_n$, where $f_j(x,y) = a_j x + b_j y$. Notice that $f_j \le c_j$ defines a half-plane and $f_j = c_j$ defines its boundary line; the region R consists of the points in common to all these half-planes. A point (x,y) in R is called a *boundary point* if at least one of the equations $f_j(x,y) = c_j$ is satisfied, and a *vertex* if two of these equations are satisfied. Thus, a vertex is a point in R lying on two boundary lines (Fig. 21).

Let $\mathbf{M} = (a,b)$ denote the slope vector for f, and $\mathbf{M}_j = (a_j, b_j)$ the slope for f_j. To avoid uninteresting "degenerate" cases, we assume that none of the vectors $\mathbf{M}, \mathbf{M}_1, \ldots, \mathbf{M}_n$ is zero.

Proposition 1.* *Suppose that none of* $\mathbf{M}, \mathbf{M}_1, \ldots, \mathbf{M}_n$ *is zero, and that* $\mathbf{V}$ *is a point in* R *lying on the two lines* $f_j = c_j$ *and* $f_k = c_k$, *and that*

$$\mathbf{M} = t_j\mathbf{M}_j + t_k\mathbf{M}_k \quad \text{with constants } t_j \ge 0,\ t_k \ge 0. \tag{6}$$

Then $f(\mathbf{P}) \le f(\mathbf{V})$ *for every point* $\mathbf{P}$ *in* R.

The proof of this is outlined in the problems below. Condition (6) is easy to interpret geometrically; it means that $\mathbf{M}$ lies between $\mathbf{M}_j$ and $\mathbf{M}_k$, as in Fig. 22(a). If (6) held with $t_k < 0$ (as in Fig. 22(b)), we could find larger values of f to one side of $\mathbf{V}$ along the line $f_j = c_j$, and if t_j were < 0 there would be larger values of f to the other side of $\mathbf{V}$ along the line $f_k = c_k$; in either case, $f(\mathbf{V})$ would not be a maximum.

* This fact is labeled "proposition" instead of "theorem" because it is not part of the main line of development of the chapter.

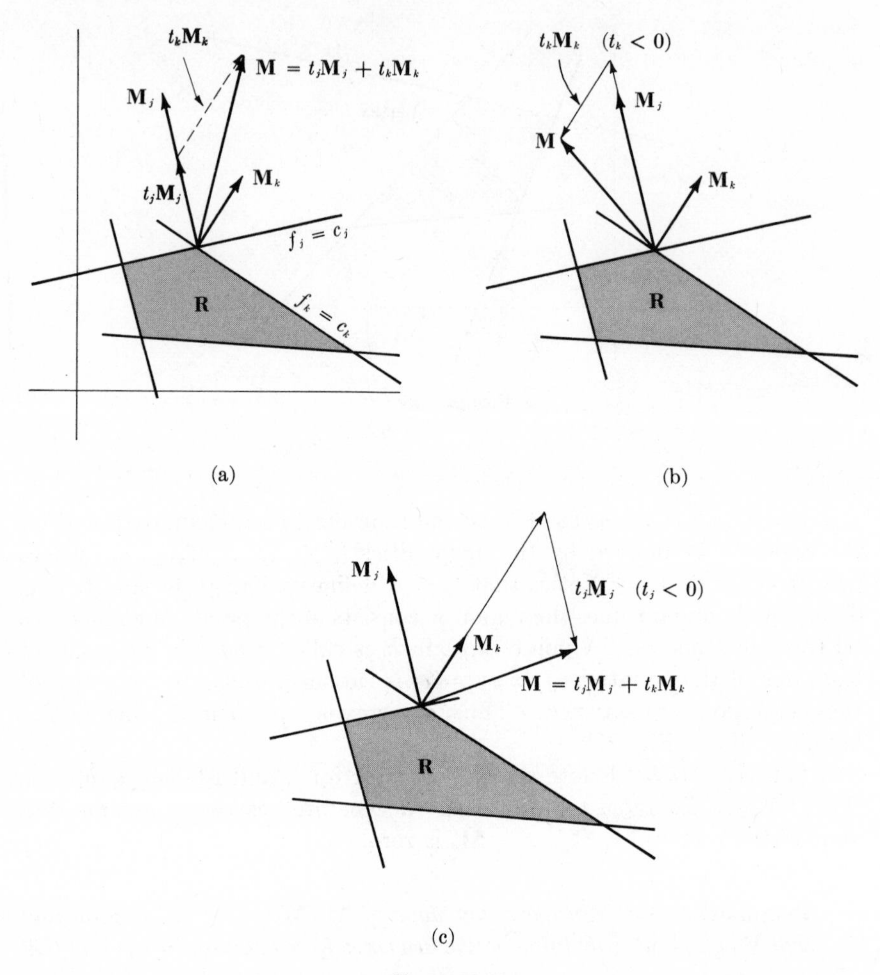

FIGURE 3.22

It may not be clear in a given case how to decide analytically whether (6) holds or not. Fortunately, (6) can be rewritten in a form that is easily decided:

Proposition 2. *Let* $\mathbf{M} = (a,b)$, $\mathbf{M}_j = (a_j, b_j)$, *and* $\mathbf{M}_k = (a_k, b_k)$. *Then condition* (6) *is equivalent to:*

$$ab_j - ba_j = t_k(a_kb_j - b_ka_j) \text{ and } a_kb - b_ka = t_j(a_kb_j - b_ka_j) \quad \text{for some constants } t_k \geq 0 \text{ and } t_j \geq 0. \tag{6'}$$

The proof of this is left as a problem. Propositions 1 and 2 complement the graphical method of finding maxima; once you have found $\mathbf{V}$ graphically, you can prove that $f(\mathbf{V})$ is maximum by verifying (6′).

Example 2. Prove that the point found in the solution of Example 1 actually does give maximum profit. *Solution.* We found graphically that the function

$$f(x,y) = x + \tfrac{3}{4}y$$

has a maximum in R at the intersection of the two lines

$$\begin{aligned} x + y &= 100 \\ 2x + y &= 150. \end{aligned} \tag{7}$$

Thus the various slopes we have to consider are

$$\begin{aligned} \mathbf{M} &= (a,b) = (1,\tfrac{3}{4}) \\ \mathbf{M}_1 &= (a_1,b_1) = (1,1) \\ \mathbf{M}_2 &= (a_2,b_2) = (2,1). \end{aligned}$$

We find

$$ab_1 - ba_1 = 1 - \tfrac{3}{4} = \tfrac{1}{4}, \qquad ab_2 - ba_2 = 1 - \tfrac{3}{2} = -\tfrac{1}{2}$$

$$a_1b_2 - b_1a_2 = 1 - 2 = -1, \qquad a_2b_1 - b_2a_1 = 1;$$

hence

$$ab_1 - ba_1 = \tfrac{1}{4}(a_2b_1 - b_2a_1) \qquad \text{and} \qquad ab_2 - ba_2 = \tfrac{1}{2}(a_1b_2 - b_1a_2),$$

so (6′) is satisfied by the nonnegative constants $t_1 = \frac{1}{2}$ and $t_2 = \frac{1}{4}$. Thus by Propositions 1 and 2, the point (x,y) solving (7) does give a maximum of f over the region R.

One question remains: What if *no* vertex of R satisfies the conditions (6) or (6′)? In most cases, this implies that f has *no* maximum in R. The precise statement in this situation is a little complicated, and we will not give the (tedious and uninspiring) proof, but we will at least state the result:

Proposition 3. *If the vectors $\mathbf{M}$, $\mathbf{M}_1, \ldots, \mathbf{M}_n$ are nonzero and not all parallel, and if no vertex of R satisfies* (6) *or* (6′), *then f has no maximum in R. If $\mathbf{M}$, $\mathbf{M}_1, \ldots, \mathbf{M}_n$* are *all parallel, and R is not empty, then there are two cases:*

Case (i) *For some k, $\mathbf{M} = t_k\mathbf{M}_k$ with $t_k > 0$, and the line*

$$\{\mathbf{P} : \mathbf{P} \cdot \mathbf{M}_k = c_k\} \tag{8}$$

lies in R; in this case, f is constant on the line (8) *and every point on this line gives a maximum value of f.*

Case (ii) *For every* k, $\mathbf{M} = t_k\mathbf{M}_k$ *with* $t_k < 0$. *In this case,* R *is one of the half-planes* $\{\mathbf{P}: \mathbf{P}\cdot\mathbf{M}_j \leq c_j\}$, *and* $\mathbf{M}$ *points "into"* R; f *has no maximum in* R, *but it has a minimum along the line* $\{\mathbf{P}: \mathbf{P}\cdot\mathbf{M}_j = c_j\}$.

PROBLEMS

1. In Example 1 in the text, suppose that the profit on peaches were \$.75 and the profit on apples were \$1. How should the trucker maximize his profit then? (You will find that he fills the truck, but does *not* spend all his money.)

2. Superduper Farms, Inc. has a square feet of livestock storage, and B cubic feet of feed storage. Each cow requires c square feet of room, and C cubic feet of feed storage, while each sheep requires s square feet of room and S cubic feet of feed storage. The net income per year from a cow is I_c, and the net income from a sheep is I_s. Superduper's farm is to have x cows and y sheep.

(a) Set up the inequalities governing Superduper's possibilities, and the function giving their income.

(b) Determine x and y to give maximum income, when $a = 1{,}000$, $B = 12{,}500$, $c = 20$, $C = 200$, $s = 10$, $S = 150$, $I_c = \$200$, and $I_s = \$120$.

3. A dealer has space for 50 appliances: washing machines, stoves, or refrigerators. He is sure to sell more washers than refrigerators, and more refrigerators than stoves; and he must fill orders for 3 stoves, 5 refrigerators, and 10 washers. Beyond these facts, he has no idea how many of each item he will sell in a given time. However, he does know that his profit is \$50 per stove, \$60 per refrigerator, and \$30 per washer. He wants to order 50 appliances in such a way that his profit will be maximum if they are all sold.

(a) Suppose he orders w washers, s stoves, and r refrigerators. Write down all the conditions that w, s, and r must satisfy, and write the profit P as a function of w, s, and r. (There are five inequalities, one equality, and one equation giving the profit P.)

(b) Use the equality found in part (a) to express w in terms of r and s, and rewrite all the other expressions in terms of r and s.

(c) Sketch the region in the (r,s) plane determined by the inequalities found in part (b).

(d) Find the maximum profit for the region found in part (c). How many of each appliance does the dealer order?

4. A manufacturer has 52 workers, each provided with adequate machines. Twenty are skilled, earning \$5/hour, and 32 are unskilled, earning

\$2/hour. The manufacturer can make either of two products. Product X takes 10 man-hours of skilled labor, 5 man-hours of unskilled labor, and brings a net profit of \$120, *not counting* labor costs. Product Y takes 4 man-hours of skilled and 8 man-hours of unskilled labor, and brings a net profit of \$80, again *not counting* labor costs. How much of each product should be made per hour to maximize the profit, taking into account labor costs? Assume that the skilled laborers will not do unskilled work.

(a) Suppose the manufacturer produces x items per hour of product X, and y items per hour of product Y. Taking into account the fixed labor cost

$$20 \times \$5/\text{hour} + 32 \times \$2/\text{hour} = \$164/\text{hour},$$

show that the net hourly profit in dollars is

$$P(x,y) = 120x + 80y - 164.$$

(b) Show that the total number of skilled man-hours required is $10x + 4y$, and deduce the restriction $10x + 4y \leq 20$.

(c) Taking into account the supply of unskilled labor, show that $5x + 8y \leq 32$.

(d) Sketch the region R determined by the inequalities in parts (b) and (c), and by the obvious restrictions $x \geq 0$, $y \geq 0$.

(e) Find the maximum in R of the function P given in part (a). (Hint: The slope vector itself is much too long to sketch. All you need is a vector in the right direction, so scale down the slope vector; for instance, divide it by 40.)

5. Rework Problem 4 above, but now assume that there are 20 unskilled workers and 32 skilled workers.

6. Rework Problem 4 above, but now assume that the skilled laborers will do unskilled work, as long as they are still paid \$5/hour. Show that the inequality in part (c) of Problem 4 should be replaced by $5x + 8y + 10x + 4y \leq 52$.

7. Find three first-degree polynomials f, f_1, and f_2 such that f does not assume a maximum in the region $R = \{\mathbf{P}: f_1(\mathbf{P}) \leq 0 \text{ and } f_2(\mathbf{P}) \leq 0\}$.

8. Prove Proposition 2. (Treat separately the two cases $a_kb_j - b_ka_j \neq 0$ and $a_kb_j - b_ka_j = 0$.)

9. This problem outlines the proof of Proposition 1 under the assumption that $\mathbf{M}_j$ and $\mathbf{M}_k$ are not parallel. Without loss of generality, we assume that $j = 1$ and $k = 2$.

(a) Let $\mathbf{A}_1$ and $\mathbf{A}_2$ be nonzero, and $\mathbf{A}_1 \cdot \mathbf{M}_2 = 0$, $\mathbf{A}_2 \cdot \mathbf{M}_1 = 0$. Prove that $\mathbf{A}_1 \cdot \mathbf{M}_1 \neq 0$ and $\mathbf{A}_2 \cdot \mathbf{M}_2 \neq 0$. (Hint: $\mathbf{M}_1$ and $\mathbf{M}_2$ form a basis of R^2: see §1.5).

(b) Let $\mathbf{B}_1 = \dfrac{1}{\mathbf{A}_1 \cdot \mathbf{M}_1} \mathbf{A}_1$ and $\mathbf{B}_2 = \dfrac{1}{\mathbf{A}_2 \cdot \mathbf{M}_2} \mathbf{A}_2$. Prove that

$$\mathbf{B}_l \cdot \mathbf{M}_m = \begin{cases} 0 & \text{if } l \neq m \\ 1 & \text{if } l = m \end{cases}$$

and that $\mathbf{B}_1$ and $\mathbf{B}_2$ form a basis of R^2. (See Fig. 23).

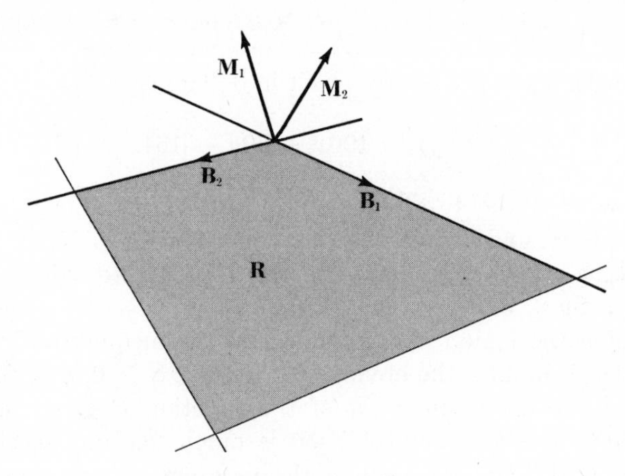

FIGURE 3.23

(c) Let $\mathbf{M}_1 \cdot \mathbf{V} = c_1$ and $\mathbf{M}_2 \cdot \mathbf{V} = c_2$ (i.e. $\mathbf{V}$ is the vertex in the statement of Proposition 1.) By part (b), any point $\mathbf{P}$ in R^2 can be written in the form $\mathbf{P} = \mathbf{V} + s_1\mathbf{B}_1 + s_2\mathbf{B}_2$. Prove that if $\mathbf{P}$ is in the given region R, then $s_1 \leq 0$ and $s_2 \leq 0$. (Compute $f_1(\mathbf{P}) = \mathbf{M}_1 \cdot \mathbf{P}$ and $f_2(\mathbf{P}) = \mathbf{M}_2 \cdot \mathbf{P}$.)

(d) Using parts (b) and (c), prove that if $f(\mathbf{P}) = \mathbf{M} \cdot \mathbf{P} + c$, and $\mathbf{M}$ satisfies (6), then $f(\mathbf{P}) \leq f(\mathbf{V})$ for every point $\mathbf{P}$ in R.

10. This problem proves Proposition 1 in the remaining case, where $\mathbf{M}_j$ and $\mathbf{M}_k$ are assumed to be parallel.

(a) Suppose that $\mathbf{M}_j$ and $\mathbf{M}_k$ are parallel and nonzero, and

$$\mathbf{M} = t_j\mathbf{M}_j + t_k\mathbf{M}_k$$

with $t_j \geq 0$ and $t_k \geq 0$. Prove that either $\mathbf{M} = t\mathbf{M}_j$ or $\mathbf{M} = t\mathbf{M}_k$, with $t \geq 0$.

(b) Suppose that $\mathbf{M} = t\mathbf{M}_j$, $t \geq 0$, and let $\mathbf{V} \cdot \mathbf{M}_j = c_j$. Prove that $\mathbf{P} \cdot \mathbf{M}_j \leq c_j \Rightarrow f(\mathbf{P}) \leq f(\mathbf{V})$, where $f(\mathbf{P}) = \mathbf{M} \cdot \mathbf{P} + c$.

(c) Conclude that $f(\mathbf{P}) \leq f(\mathbf{V})$ for $\mathbf{P}$ in R.

3.3 PARTIAL DERIVATIVES, THE GRADIENT, AND THE CHAIN RULE

We come at last to the question of differentiating functions of two variables. In general, a derivative is a rate of change. For a function $f(x,y)$ there are two easily computed rates of change, one with respect to x, and one with respect to y. The first of these is the *partial derivative with respect to x*, whose value at a point (x_0, y_0) is

$$\lim_{x \to x_0} \frac{f(x,y_0) - f(x_0,y_0)}{x - x_0}, \tag{1}$$

and the second is the partial derivative with respect to y, whose value at (x_0, y_0) is

$$\lim_{y \to y_0} \frac{f(x_0,y) - f(x_0,y_0)}{y - y_0}. \tag{2}$$

There are at least three current notations for partial derivatives: for the x derivative (1) they are

$$\frac{\partial f}{\partial x}(x_0,y_0), \qquad f_x(x_0,y_0), \qquad \text{or} \qquad D_1 f(x_0,y_0),$$

and for the y derivative (2)

$$\frac{\partial f}{\partial y}(x_0,y_0), \qquad f_y(x_0,y_0), \qquad \text{or} \qquad D_2 f(x_0,y_0).$$

The $\partial f/\partial x$ version (read "*dfdx*" or "partial f partial x") depends on the conventional use of x for the first variable, and shares some (but not all!) of the advantages of the Leibniz notation df/dx for functions of one variable. The f_x version is the easiest to write, and $D_1 f$ is the safest when the conventions about x and y become ambiguous. (What does $f_x(y_0, x_0)$ mean?) Figure 24 shows the geometric meaning of partial derivatives. The vertical plane $\{(x,y,z): y = y_0\}$ intersects the graph of f in a curve which is essentially the graph of $f(x,y_0)$ considered as a function of the one variable x. The limit (1) is precisely the derivative of this function of one variable, and thus $f_x(x_0,y_0)$ is the slope of the line tangent to this curve

at $Q_0 = (x_0, y_0, f(x_0,y_0))$, as shown in Fig. 24(a). Similarly, $\dfrac{\partial f}{\partial y}(x_0,y_0)$ is the slope of the line in Fig. 24(b).

To compute $f_x(x_0,y_0)$ simply think of $y = y_0$ as a constant, and take the derivative with respect to x. For example, if $f(x,y) = x^2 + y^2$, then $f_x(x_0, y_0) = 2x_0$; if $g(x,y) = 2x + 3xy$, then $g_x(x_0, y_0) = 2 + 3y_0$. Similarly, to compute f_y, think of x as constant and differentiate with respect to y. The functions f and g just given have the derivatives

$$f_y(x_0,y_0) = \frac{\partial f}{\partial y}(x_0,y_0) = 2y_0, \qquad \frac{\partial g}{\partial y}(x_0,y_0) = 3x_0.$$

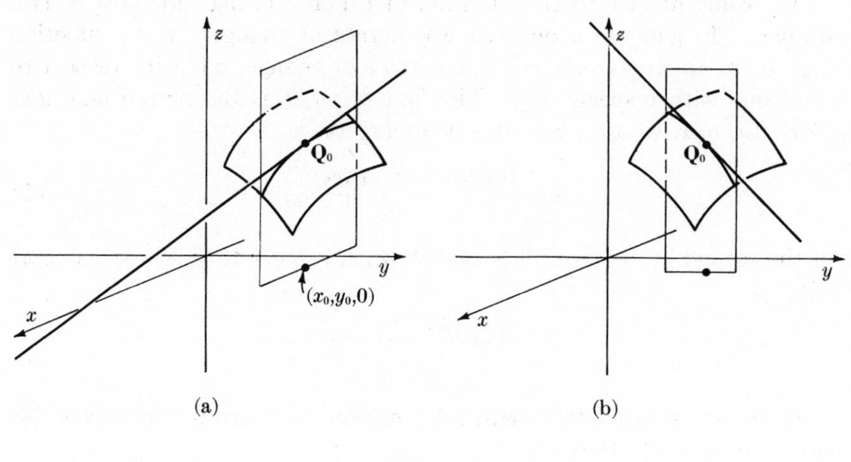

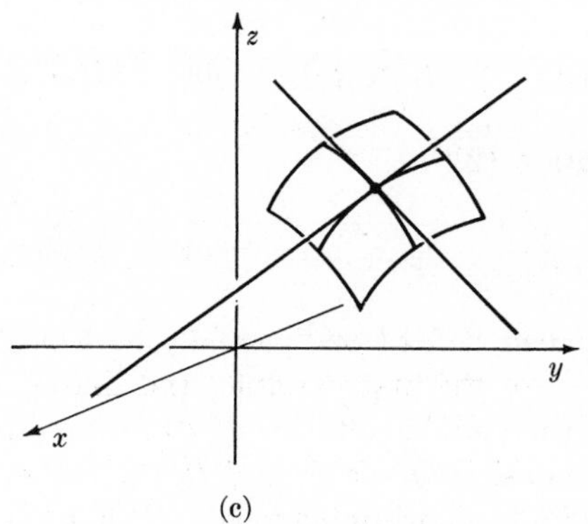

FIGURE 3.24

A less trivial example is the function graphed in Fig. 25,

$$h(x,y) = \begin{cases} \dfrac{xy}{x^2 + y^2}, & x^2 + y^2 \neq 0 \\ 0, & x^2 + y^2 = 0. \end{cases}$$

When $(x,y) \neq (0,0)$, you can easily find the partial derivatives

$$D_1h(x,y) = \frac{y^3 - x^2y}{(x^2 + y^2)^2} \quad \text{and} \quad D_2h(x,y) = \frac{x^3 - y^2x}{(x^2 + y^2)^2}.$$

At the origin these formulas do not apply, but a direct application

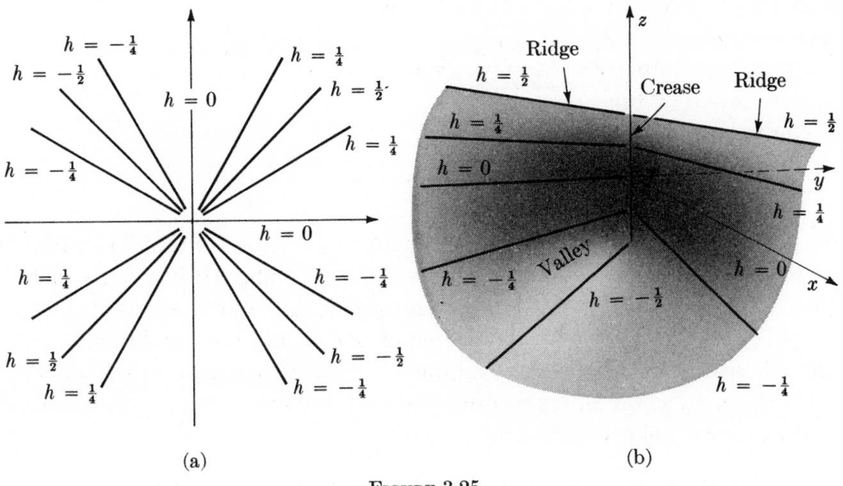

FIGURE 3.25

of the definition of D_1 yields

$$D_1h(0,0) = \lim_{x\to 0} \frac{h(x,0) - h(0,0)}{x - 0} = \lim_{x\to 0} \frac{0 - 0}{x} = 0.$$

This reflects the fact that when h is restricted to the x axis (this is the effect of setting $y = 0$), then it is a "nice" function at $x = 0$; in fact, h is identically zero along the x axis. Similarly, h is identically zero along the y axis, so $D_2h(0,0) = 0$. Thus the *partial derivatives of h exist at every point, even though h itself is not continuous at* $(0,0)$. (See Fig. 25 and Example 5, §3.1.)

Thus the mere existence of the partial derivatives does not tell very much about the function. For this reason, the term "differentiable" is *not* taken to mean "having partial derivatives", but is given a more sophisticated definition. The existence of $f_x(x_0, y_0)$ and $f_y(x_0, y_0)$ depends only on the behaviour of f along the lines $x = x_0$ and $y = y_0$. By contrast, differentiability takes into account *all* points near (x_0, y_0).

Definition 4. A function f is *differentiable at* $\mathbf{P}_0$ if and only if there is a first degree polynomial $g(\mathbf{P}) = f(\mathbf{P}_0) + (\mathbf{P} - \mathbf{P}_0)\cdot\mathbf{M}$ such that

$$\lim_{\mathbf{P}\to\mathbf{P}_0} \frac{f(\mathbf{P}) - g(\mathbf{P})}{|\mathbf{P} - \mathbf{P}_0|} = 0, \tag{3}$$

that is,

$$\lim_{(x,y)\to(x_0,y_0)} \frac{f(x,y) - f(x_0,y_0) - a(x - x_0) - b(y - y_0)}{\sqrt{(x - x_0)^2 + (y - y_0)^2}} = 0. \tag{3'}$$

This definition is modeled on a simple result about functions of one variable. Elementary algebra shows that the relation

$$\lim_{x\to x_0} \frac{F(x) - F(x_0)}{x - x_0} = a \qquad (\text{i.e. } F'(x_0) = a)$$

is equivalent to

$$\lim_{x\to x_0} \frac{F(x) - F(x_0) - a(x - x_0)}{x - x_0} = 0.$$

The similarity between this last limit and (3′) suggests that Definition 4 is at least a reasonable one. However, the real importance of the definition lies in its consequences, and the most efficient course now is to develop these consequences, point out their geometric significance, and finally apply the whole theory to specific examples. (The theoretical development is a little heavy, so on a first reading you may prefer to skip to the summary and examples below, beginning on page 735.)

First of all we prove that differentiability implies continuity, and also implies the existence of the partial derivatives.

Theorem 5. *Suppose that* $g(\mathbf{P}) = f(\mathbf{P}_0) + (\mathbf{P} - \mathbf{P}_0)\cdot\mathbf{M}$ *is a first degree polynomial satisfying* (3). *Then* f *is continuous at* $\mathbf{P}_0$, *the partial derivatives* $D_1 f$ *and* $D_2 f$ *exist at* $\mathbf{P}_0$, *and* g *has the slope*

$$\mathbf{M} = (D_1 f(\mathbf{P}_0), D_2 f(\mathbf{P}_0)).$$

The proof takes three easy steps.

(i) $\lim_{\mathbf{P}\to\mathbf{P}_0} f(\mathbf{P}) = f(\mathbf{P}_0)$. For

$$f(\mathbf{P}) = g(\mathbf{P}) + |\mathbf{P} - \mathbf{P}_0| \frac{f(\mathbf{P}) - g(\mathbf{P})}{|\mathbf{P} - \mathbf{P}_0|}; \tag{4}$$

since g is a polynomial and $g(\mathbf{P}_0) = f(\mathbf{P}_0)$, we have $\lim_{\mathbf{P}\to\mathbf{P}_0} g(\mathbf{P}) = f(\mathbf{P}_0)$, so (3) shows that the right-hand side of (4) tends to $f(\mathbf{P}_0) + 0\cdot 0$ as $\mathbf{P} \to \mathbf{P}_0$.

(ii) If $\mathbf{M} = (a,b)$, then

$$\lim_{x\to x_0} \left| \frac{f(x,y_0) - f(x_0,y_0)}{x - x_0} - a \right| = 0. \tag{5}$$

To prove this, let $\mathbf{P}_0 = (x_0,y_0)$ and $\mathbf{P} = (x,y_0)$. Then $g(\mathbf{P}) = f(x_0,y_0) + a(x - x_0)$, and $|\mathbf{P} - \mathbf{P}_0| = |x - x_0|$, so

$$\left| \frac{f(x,y_0) - f(x_0,y_0)}{x - x_0} - a \right| = \frac{|f(x,y_0) - f(x_0,y_0) - a(x - x_0)|}{|x - x_0|}$$

$$= \frac{|f(\mathbf{P}) - g(\mathbf{P})|}{|\mathbf{P} - \mathbf{P}_0|}. \tag{6}$$

As x tends to x_0, then $\mathbf{P}$ tends to $\mathbf{P}_0$, so the right-hand side of (6) tends to zero, and (5) is proved.

(iii) If $\mathbf{M} = (a,b)$, then

$$\lim_{y\to y_0}\left|\frac{f(x_0,y) - f(x_0,y_0)}{y - y_0} - b\right| = 0.$$

The proof of this is just like (ii).

This completes the theorem, for (i) says that f is continuous at $\mathbf{P}_0$, (ii) says $D_1f(\mathbf{P}_0) = a$, and (iii) says $D_2f(\mathbf{P}_0) = b$, so $\mathbf{M} = (a,b) = (D_1f(\mathbf{P}_0),D_2f(\mathbf{P}_0))$.

An immediate consequence of Theorem 5 is that *there is only one vector* $\mathbf{M}$ such that $g(\mathbf{P}) = \mathbf{P}_0 + (\mathbf{P} - \mathbf{P}_0)\cdot\mathbf{M}$ satisfies (3), namely

$$\mathbf{M} = (f_x(\mathbf{P}_0), f_y(\mathbf{P}_0)).$$

This vector is called the *gradient* of f at $\mathbf{P}_0$, denoted $\nabla f(\mathbf{P}_0)$ (read "del f of $\mathbf{P}_0$"). Thus, when f is differentiable at $\mathbf{P}_0$ we have the limit relation

$$\lim_{\mathbf{P}\to\mathbf{P}_0}\frac{f(\mathbf{P}) - f(\mathbf{P}_0) - (\mathbf{P} - \mathbf{P}_0)\cdot\nabla f(\mathbf{P}_0)}{|\mathbf{P} - \mathbf{P}_0|} = 0, \tag{7}$$

where $\nabla f(\mathbf{P}_0) = (f_x(\mathbf{P}_0), f_y(\mathbf{P}_0))$.

In many respects the gradient $\nabla f(\mathbf{P}_0)$ replaces the ordinary derivative $F'(x_0)$ of a function F of one variable. For example, the graph of

$$G(x) = F(x_0) + (x - x_0)F'(x_0)$$

is defined to be the tangent line to the graph of F at the point $(x_0,F(x_0))$. Analogously, we define the graph of

$$g(\mathbf{P}) = f(\mathbf{P}_0) + (\mathbf{P} - \mathbf{P}_0)\cdot\nabla f(\mathbf{P}_0)$$

to be the *tangent plane* to the graph of f at the point $(\mathbf{P}_0, f(\mathbf{P}_0))$. Notice that $\nabla f(\mathbf{P}_0)$ is the *slope* of the tangent plane, and the *equation* of the tangent plane is

$$z = f(\mathbf{P}_0) + (\mathbf{P} - \mathbf{P}_0)\cdot\nabla f(\mathbf{P}_0)$$

or

$$z = f(x_0,y_0) + (x - x_0)f_x(x_0,y_0) + (y - y_0)f_y(x_0,y_0).$$

Another analogy between ∇f and F' shows up in the following important *chain rule.*

Theorem 6. *Let $\boldsymbol{\gamma}$ be a curve in R^2 and f a function of two variables. If f is differentiable at $\mathbf{P}_0 = \boldsymbol{\gamma}(t_0)$, and if $\boldsymbol{\gamma}'(t_0)$ exists, then $f\circ\boldsymbol{\gamma}$ is differentiable at t_0 and*

$$(f\circ\boldsymbol{\gamma})'(t_0) = \nabla f(\mathbf{P}_0)\cdot\boldsymbol{\gamma}'(t_0). \tag{8}$$

Proof. Let

$$Q(\mathbf{P}) = \begin{cases} \dfrac{f(\mathbf{P}) - f(\mathbf{P}_0) - (\mathbf{P} - \mathbf{P}_0) \cdot \nabla f(\mathbf{P}_0)}{|\mathbf{P} - \mathbf{P}_0|} & \text{if } \mathbf{P} \neq \mathbf{P}_0 \\ 0 & \text{if } \mathbf{P} = \mathbf{P}_0 . \end{cases}$$

Then (7) says that Q is continuous at $\mathbf{P}_0$. Now set $\mathbf{P} = \boldsymbol{\gamma}(t)$ and $\mathbf{P}_0 = \boldsymbol{\gamma}(t_0)$. By simple algebra we have, for every $t \neq t_0$,

$$\frac{f(\boldsymbol{\gamma}(t)) - f(\boldsymbol{\gamma}(t_0))}{t - t_0} = \frac{\boldsymbol{\gamma}(t) - \boldsymbol{\gamma}(t_0)}{t - t_0} \cdot \nabla f(\mathbf{P}_0) + \frac{|\boldsymbol{\gamma}(t) - \boldsymbol{\gamma}(t_0)|}{t - t_0} Q(\boldsymbol{\gamma}(t)). \quad (9)$$

The first term on the right-hand side of (9) converges to $\boldsymbol{\gamma}'(t_0) \cdot \nabla f(\mathbf{P}_0)$ as $t \to t_0$, so we have to prove that the remaining term converges to zero, or what is the same, that its absolute value

$$\left| \frac{\boldsymbol{\gamma}(t) - \boldsymbol{\gamma}(t_0)}{t - t_0} \right| |Q(\boldsymbol{\gamma}(t))| \qquad (10)$$

converges to zero. But Q, $\boldsymbol{\gamma}$, and the absolute value are all continuous functions, so by Theorem 3 on composite functions

$$\lim_{t \to t_0} |Q(\boldsymbol{\gamma}(t))| = |Q(\boldsymbol{\gamma}(t_0))| = |Q(\mathbf{P}_0)| = 0.$$

Similarly,

$$\lim_{t \to t_0} \left| \frac{\boldsymbol{\gamma}(t) - \boldsymbol{\gamma}(t_0)}{t - t_0} \right| = |\boldsymbol{\gamma}'(t_0)|,$$

so the expression in (10) converges to $|\boldsymbol{\gamma}'(t_0)| \cdot 0 = 0$, and the theorem is proved.

Various special cases of the chain rule formula give some insight into the gradient.

(i) If $\boldsymbol{\gamma}(t) = (x_0 + t, y_0)$, then $\boldsymbol{\gamma}$ traces a straight line $\{(x,y) : y = y_0\}$ parallel to the x axis, with speed 1, and $f \circ \boldsymbol{\gamma}$ gives the values of f along this line. (See Fig. 26.) We find $\boldsymbol{\gamma}'(t) = (1,0)$, hence

$$(f \circ \boldsymbol{\gamma})'(0) = \boldsymbol{\gamma}'(0) \cdot \nabla f(\mathbf{P}_0) = (1,0) \cdot \nabla f(\mathbf{P}_0) = D_1 f(\mathbf{P}_0).$$

Thus, the derivative of f along this curve $\boldsymbol{\gamma}$ equals the partial derivative of f with respect to x.

(ii) If $\mathbf{U}$ is any unit vector and $\boldsymbol{\gamma}(t) = \mathbf{P}_0 + t\mathbf{U}$, then $\boldsymbol{\gamma}$ is the line passing through $\mathbf{P}_0$ with unit speed in the direction $\mathbf{U}$. The composite

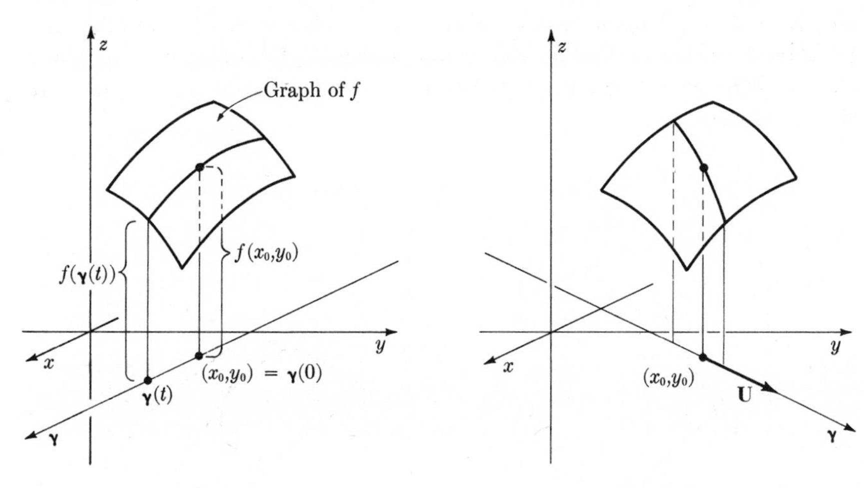

FIGURE 3.26

FIGURE 3.27

function $f\circ\boldsymbol{\gamma}$ gives the values of f along this line, and its derivative at $t = 0$,

$$(f\circ\boldsymbol{\gamma})'(0) = \boldsymbol{\gamma}'(0)\cdot\nabla f(\mathbf{P}_0) = \mathbf{U}\cdot\nabla f(\mathbf{P}_0),$$

gives the rate of change of f with respect to distance in the direction **U**. This is called the *directional derivative of f in the direction* **U** (see Fig. 27). When $\mathbf{U} = (1,0)$, the directional derivative is $\mathbf{U}\cdot\nabla f(\mathbf{P}_0) = D_1 f(\mathbf{P}_0) =$ the partial derivative of f with respect to x, as we found in case (i) above; when $\mathbf{U} = (0,1)$, we get the directional derivative $\mathbf{U}\cdot\nabla f(\mathbf{P}_0) = D_2 f(\mathbf{P}_0)$, the partial derivative with respect to y. In general, the derivative in the direction **U** is a combination of $D_1 f$ and $D_2 f$.

One sometimes speaks of a directional derivative in a direction **V**, where $|\mathbf{V}| \neq 1$; in this book, that will be understood as

$$\frac{1}{|\mathbf{V}|}\mathbf{V}\cdot\nabla f.$$

(iii) Suppose $\boldsymbol{\gamma}$ is a curve lying entirely in a contour line $\{\mathbf{P}: f(\mathbf{P}) = c\}$. Then

$$f(\boldsymbol{\gamma}(t)) = c \tag{11}$$

for every t. Differentiating each side of (11) with respect to t yields

$$\nabla f(\boldsymbol{\gamma}(t))\cdot\boldsymbol{\gamma}'(t) = 0.$$

Thus, for every point $\mathbf{P}_0$ on the curve $\boldsymbol{\gamma}$, $\nabla f(\mathbf{P}_0)$ is orthogonal to the curve at $\mathbf{P}_0$ (see Fig. 28). We describe this situation by saying simply that *the gradient is orthogonal to the contour line.* (This phrase has a fairly clear intuitive meaning (as in Fig. 28); the precise analytic meaning is, of course, that $\nabla f(\mathbf{P}_0)$ is orthogonal at $\mathbf{P}_0$ to every curve lying in the contour line through $\mathbf{P}_0$.)

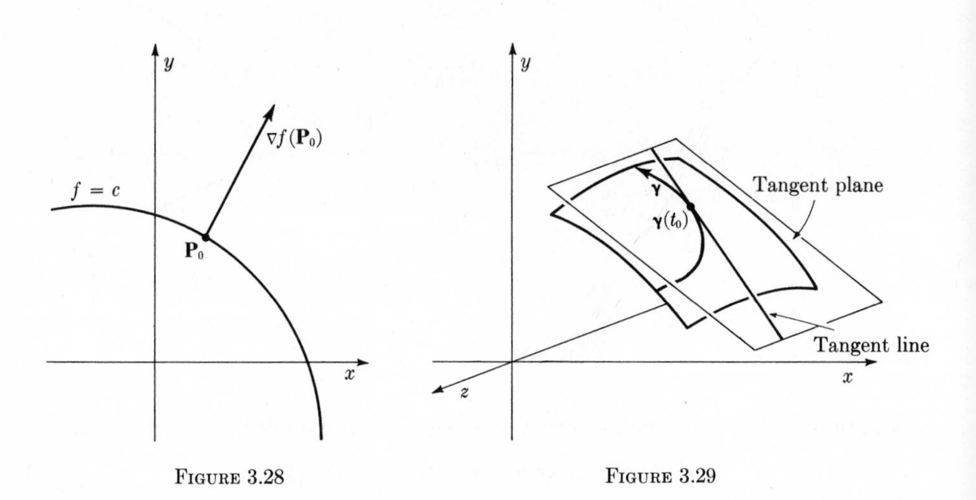

FIGURE 3.28

FIGURE 3.29

(iv) Suppose $\boldsymbol{\gamma}$ is a curve in R^3 lying in the graph of f. Then, as Fig. 29 suggests, *the tangent line to* $\boldsymbol{\gamma}$ *at* $\boldsymbol{\gamma}(t_0)$ *lies in the tangent plane to* f *at* $\boldsymbol{\gamma}(t_0)$. This fact (whose proof is left as a problem) lends geometric support to the rather abstract definition of tangent plane that we gave, based on Definition 5 and Theorem 5. It shows that our definition of tangent plane is intuitively consistent with the definition of tangent line to a curve.

This whole discussion about tangent planes, the chain rule, and the directional derivative is valid when f is differentiable at the point $\mathbf{P}_0$ in question, in other words, when the limit relation (7) holds. What we need now is an effective way to establish differentiability. Here it is:

Theorem 7. *If $D_1 f$ and $D_2 f$ exist at all points in some disk centered at $\mathbf{P}_0$, and are continuous at $\mathbf{P}_0$, then f is differentiable at $\mathbf{P}_0$.*

Proof. We have to prove that

$$\lim_{\mathbf{P}\to\mathbf{P}_0} \frac{|f(\mathbf{P}) - g(\mathbf{P})|}{|\mathbf{P} - \mathbf{P}_0|} = 0, \tag{12}$$

where

$$g(\mathbf{P}) = f(\mathbf{P}_0) + (\mathbf{P} - \mathbf{P}_0) \cdot \nabla f(\mathbf{P}_0). \tag{13}$$

Referring to Fig. 30, we write

$$\begin{aligned} f(\mathbf{P}) - f(\mathbf{P}_0) &= f(x,y) - f(x_0,y_0) \\ &= f(x,y) - f(x_0,y) + f(x_0,y) - f(x_0,y_0). \end{aligned} \tag{14}$$

If we think of y as fixed, and apply the mean value theorem to f as a function of the one variable x, we get (see Fig. 30)

$$f(x,y) - f(x_0,y) = (x - x_0)f_x(\xi,y) \quad \text{with } \xi \text{ between } x \text{ and } x_0.$$

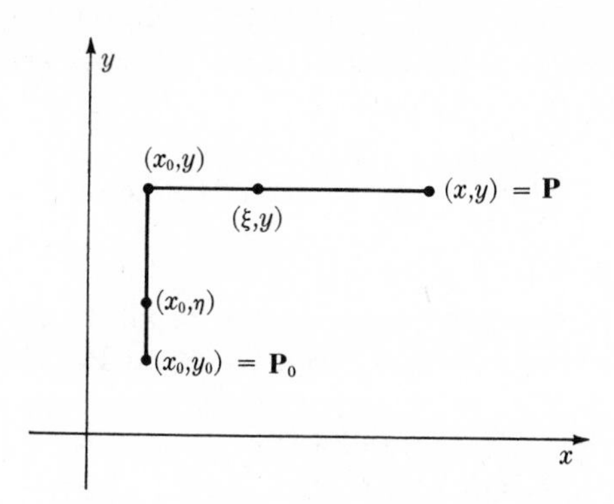

FIGURE 3.30

Similarly, fixing x_0 and considering f as a function only of y,

$$f(x_0,y) - f(x_0,y_0) = (y - y_0)f_y(x_0,\eta) \quad \text{with } \eta \text{ between } y \text{ and } y_0.$$

Hence, from (14),

$$\begin{aligned} f(\mathbf{P}) - f(\mathbf{P}_0) &= (x - x_0)f_x(\xi,y) + (y - y_0)f_y(x_0,\eta) \\ &= (x - x_0)f_x(x_0,y_0) + (y - y_0)f_y(x_0,y_0) + R(\mathbf{P}) \\ &= (\mathbf{P} - \mathbf{P}_0) \cdot \nabla f(\mathbf{P}_0) + R(\mathbf{P}) \end{aligned} \tag{15}$$

where R makes up for changing (ξ,y) and (x_0,η) into (x_0,y_0), i.e

$$\begin{aligned} R(\mathbf{P}) = (x - x_0)[f_x(\xi,y) - f_x(x_0,y_0)] \\ + (y - y_0)[f_y(x_0,\eta) - f_y(x_0,y_0)]. \end{aligned} \tag{16}$$

From (13), (15), and (16),

$$\frac{|f(\mathbf{P}) - g(\mathbf{P})|}{|\mathbf{P} - \mathbf{P}_0|} = \frac{|R(\mathbf{P})|}{|\mathbf{P} - \mathbf{P}_0|}$$

$$\leq \frac{|x - x_0|}{|\mathbf{P} - \mathbf{P}_0|} |f_x(\xi,y) - f_x(x_0,y_0)|$$

$$+ \frac{|y - y_0|}{|\mathbf{P} - \mathbf{P}_0|} |f_y(x_0,\eta) - f_y(x_0,y_0)|$$

$$\leq |f_x(\xi,y) - f_x(x_0,y_0)| + |f_x(x_0,\eta) - f_y(x_0,y_0)|. \tag{17}$$

In arriving at the last line we used $|x - x_0| \leq \sqrt{(x - x_0)^2 + (y - y_0)^2} = |\mathbf{P} - \mathbf{P}_0|$, and similarly $|y - y_0| \leq |\mathbf{P} - \mathbf{P}_0|$. Now the end of the proof is in sight: since f_x and f_y are continuous at $\mathbf{P}_0$, the two terms in the last line of (17) tend to zero as $\mathbf{P} \to \mathbf{P}_0$, so

$$\frac{|f(\mathbf{P}) - g(\mathbf{P})|}{|\mathbf{P} - \mathbf{P}_0|} \quad \text{tends to zero}$$

and (12) is proved.

If you want to complete the argument in all its gory detail, suppose $\epsilon > 0$ is given. Since f_x and f_y are continuous at $\mathbf{P}_0$, there is $\delta > 0$ such that

$$|\mathbf{Q} - \mathbf{P}_0| < \delta \quad \Rightarrow \quad |f_x(\mathbf{Q}) - f_x(\mathbf{P}_0)| < \frac{\epsilon}{2} \text{ and } |f_y(\mathbf{Q}) - f_y(\mathbf{P}_0)| < \frac{\epsilon}{2}. \tag{18}$$

Further, if $\mathbf{Q}_1 = (\xi,y)$ and $\mathbf{Q}_2 = (x_0,n)$ then

$$|\mathbf{Q}_1 - \mathbf{P}_0| = \sqrt{(\xi - x_0)^2 + (y - y_0)^2} \leq \sqrt{(x - x_0)^2 + (y - y_0)^2}$$
$$\leq |\mathbf{P} - \mathbf{P}_0|$$

since ξ is between x and x_0; and similarly $|\mathbf{Q}_2 - \mathbf{P}_0| \leq |\mathbf{P} - \mathbf{P}_0|$. Hence from (18)

$$|\mathbf{P} - \mathbf{P}_0| < \delta \quad \Rightarrow \quad |\mathbf{Q}_1 - \mathbf{P}_0| < \delta \quad \Rightarrow \quad |f_x(\mathbf{Q}_1) - f_x(\mathbf{P}_0)| < \frac{\epsilon}{2}$$

and

$$|\mathbf{P} - \mathbf{P}_0| < \delta \quad \Rightarrow \quad |\mathbf{Q}_2 - \mathbf{P}_0| < \delta \quad \Rightarrow \quad |f_y(\mathbf{Q}_2) - f_y(\mathbf{P}_0)| < \frac{\epsilon}{2}.$$

Combining this with (17) and recalling the definitions of $\mathbf{Q}_1$ and $\mathbf{Q}_2$, we obtain

$$|\mathbf{P} - \mathbf{P}_0| < \delta \quad \Rightarrow \quad \frac{|f(\mathbf{P}) - g(\mathbf{P})|}{|\mathbf{P} - \mathbf{P}_0|} < \frac{\epsilon}{2} + \frac{\epsilon}{2} = \epsilon,$$

and the proof of (12) is complete.

Summary. This has been a long section with some difficult proofs, but the end results can be summed up rather briefly. If the first partials $D_1 f$ and $D_2 f$ exist throughout a disk centered at $\mathbf{P}_0$ and are continuous at $\mathbf{P}_0$, then f is *differentiable at* $\mathbf{P}_0$, that is, f satisfies the condition

$$\lim_{\mathbf{P}\to\mathbf{P}_0} \frac{f(\mathbf{P}) - f(\mathbf{P}_0) - (\mathbf{P} - \mathbf{P}_0)\cdot\nabla f(\mathbf{P}_0)}{|\mathbf{P} - \mathbf{P}_0|} = 0,$$

where $\nabla f(\mathbf{P}_0) = (D_1 f(\mathbf{P}_0), D_2 f(\mathbf{P}_0))$ is the *gradient* of f at $\mathbf{P}_0$. The gradient is orthogonal to the contour lines, and gives the slope of the tangent plane. The equation of the tangent plane at a point (x_0, y_0, z_0) on the graph of f is

$$z = z_0 + (\mathbf{P} - \mathbf{P}_0)\cdot\nabla f(\mathbf{P}_0),$$

where $\mathbf{P} = (x,y)$, $\mathbf{P}_0 = (x_0, y_0)$, and $z_0 = f(\mathbf{P}_0)$.

If $\boldsymbol{\gamma}$ is any differentiable curve such that $\boldsymbol{\gamma}(t_0) = \mathbf{P}_0$, then the *chain rule* holds,

$$(f\circ\boldsymbol{\gamma})'(t_0) = \boldsymbol{\gamma}'(t_0)\cdot\nabla f(\mathbf{P}_0).$$

The directional derivative of f in the direction $\mathbf{U}$ is $\nabla f\cdot\mathbf{U}$, where $\mathbf{U}$ is any unit vector.

Example 1. Consider the function $f(x,y) = x^2 + 2y^2 - 1$. The partial derivatives $f_x = 2x$ and $f_y = 4y$ are continuous everywhere, so the general theory applies. Figure 31 shows the contour lines and the gradient $(2x,4y)$ at several points; notice that the gradient $\nabla f(\mathbf{P}_0)$ really is orthogonal to the contour line through $\mathbf{P}_0$. At the origin we find $\nabla f = (0,0)$, so the directional derivative $\nabla f\cdot\mathbf{U}$ is zero in every direction $\mathbf{U}$. At the point $(1,1)$ we have $\nabla f = (2,4)$. The directional derivative at $(1,1)$ in the direction $(1/\sqrt{2}, 1/\sqrt{2})$ is $(2,4)\cdot(1/\sqrt{2}, 1/\sqrt{2}) = 3\sqrt{2}$.

The point $(0,0,-1)$ is on the graph of f, and the tangent plane at this point has slope $\nabla f(0,0) = (0,0)$, so its equation is

$$z = -1 + [\mathbf{P} - (0,0)]\cdot\nabla f(0,0), \qquad \text{or} \qquad z = -1.$$

This is a horizontal plane, as shown in Fig. 31. The tangent plane at $(1,1,2)$ has slope $\nabla f(1,1) = (2,4)$, so its equation is

$$z = 2 + [\mathbf{P} - (1,1)]\cdot(2,4).$$

Setting $\mathbf{P} = (x,y)$ and multiplying out the dot product, we get the equation of the tangent plane in the form

$$z = 2 + (2x + 4y) - 2 - 4 = -4 + 2x + 4y.$$

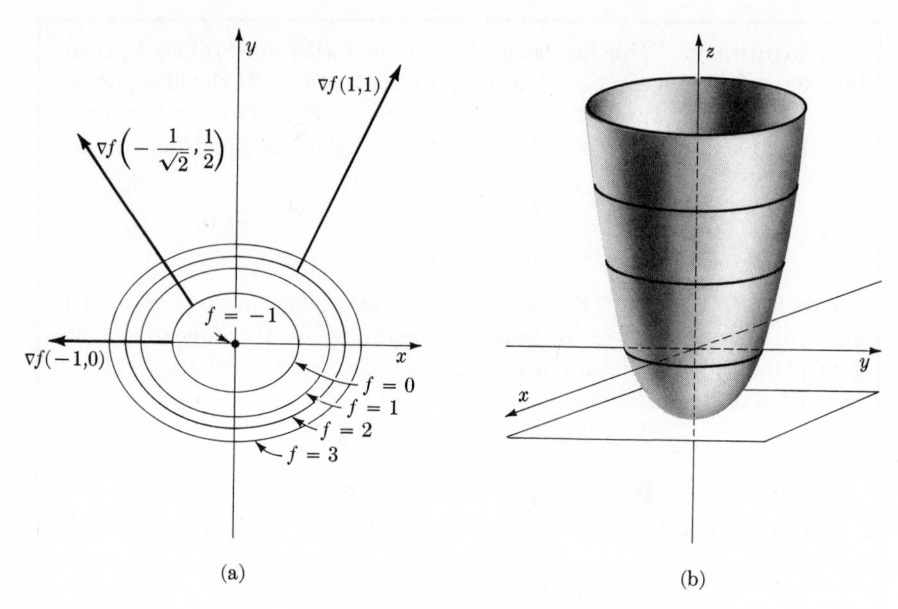

FIGURE 3.31

Example 2. Consider the function $f(x,y) = xy$. The partial derivatives $f_x = y$ and $f_y = x$ are continuous everywhere, so the general theory applies. Figure 32 shows the contour lines and the gradient $\nabla f(x,y) = (y,x)$ at several points. At the point $(-2,3)$, for example, we have $\nabla f(-2,3) = (3,-2)$. Starting at $(-2,3)$, the unit vector $\left(\frac{2}{\sqrt{13}},\frac{-3}{\sqrt{13}}\right)$ points toward the origin. Thus the directional derivative "toward the origin" at this point is

$$\nabla f(-2,3)\cdot\left(\frac{2}{\sqrt{13}},\frac{-3}{\sqrt{13}}\right) = (3,-2)\cdot\left(\frac{2}{\sqrt{13}},\frac{-3}{\sqrt{13}}\right) = \frac{12}{\sqrt{13}}$$

and the directional derivative "away from the origin" is

$$\nabla f(-2,3)\cdot\left(\frac{-2}{\sqrt{13}},\frac{3}{\sqrt{13}}\right) = \frac{-12}{\sqrt{13}}.$$

Thus f increases as you move from $(-2,3)$ toward the origin, and decreases as you move away; this is consistent with Fig. 32. The equation of the tangent plane to the graph at $(-2,3,-6)$ is

$$z = -6 + (\mathbf{P} - (-2,3))\cdot\nabla f(-2,3),$$

or

$$\begin{aligned} z &= -6 + ((x,y) - (-2,3))\cdot(3,-2) \\ &= -6 + 3x - 2y + 6 + 6 = 6 + 3x - 2y. \end{aligned}$$

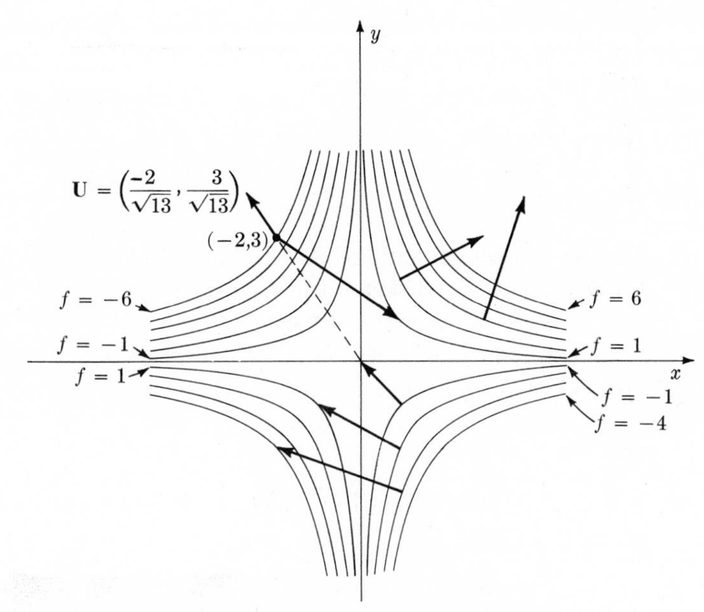

U is a unit vector pointing away from the origin, *not* a gradient; all other arrows represent gradients

FIGURE 3.32

Example 3. Find a line orthogonal to the graph of $f(x,y) = xy$ at the point $(-2,3,-6)$. *Solution.* "Orthogonal to the graph" means "orthogonal to the tangent plane to the graph." Since the equation of the tangent plane is

$$z = 6 + 3x - 2y,$$

or

$$3x - 2y - z = -6,$$

a vector normal to this plane is $\mathbf{N} = (3,-2,-1)$. Thus the line orthogonal to the plane (and hence to the graph) at $(-2,3,-6)$ is given parametrically by

$$\mathbf{Q} = (-2,3,-6) + t\mathbf{N} = (-2,3,-6) + t(3,-2,-1),$$

or

$$x = -2 + 3t, \qquad y = 3 - 2t, \qquad z = -6 - t.$$

PROBLEMS

The first nine problems give routine practice with gradients, directional derivatives, tangent planes, and normals. Problems 10–14 concern the theory developed above. The final five problems develop the concept of an *open connected set.*

1. Find the gradients of the following functions. At what points are the partial derivatives continuous?
(a) e^{xy}
(b) $\sqrt{x^2 + y^2}$
(c) $\cos\left(\sqrt{x^2 + y^2}\right)$
(d) y/x
(e) $x \int_0^y e^{t^2}\,dt$
(f) $\arctan(x/y)$

2. For the following functions, sketch the contour line that passes through the point $(1,2)$, and sketch the gradient at that point.
(a) $f(x,y) = x^2 + y^2$ (Since $f(1,2) = 5$, the required contour line is $\{(x,y): f(x,y) = 5\}$.)
(b) $f(x,y) = xy$
(c) $f(x,y) = 4x^2 - y^2$
(d) $f(x,y) = x^2 + y$

3. For each of the functions in Problem 2, find the equation of the tangent plane at $(1,1,f(1,1))$.

4. Find the directional derivative of the given function at the given point in the given direction.
(a) $x \arctan(x/y)$ at $(1,1)$ in the direction $(0,-1)$
(b) $x \arctan(x/y)$ at $(1,1)$ in the direction $(1,0)$
(c) $e^{-x^2-y^2}$ at $(0,0)$ in the direction $(\frac{3}{5}, \frac{4}{5})$
(d) $e^{-x^2-y^2}$ at $(1,-1)$ in the direction $(\frac{3}{5}, \frac{4}{5})$

5. If two surfaces intersect at a point (x_0,y_0,z_0), their *angle of intersection* at (x_0,y_0,z_0) is the angle between the normals to their tangent planes at that point. Find the angle of intersection of the graphs of $f(x,y) = xy$ and $g(x,y) = 1 - x^2 - 2y^2$ at the point $(1,0,0)$. (You should find the angle $\arccos(1/\sqrt{10})$.)

6. (a) Find all points at which the graph of $f(x,y) = 1 - x^2 - y^2$ intersects the xy plane.
(b) Show that at each of these points the angle of intersection is $\arccos(-1/\sqrt{5})$. (See Problem 5. The answer $\arccos(1/\sqrt{5})$ is equally good; it corresponds to a different choice of normals.)

7. Let $f(x,y) = x^2 + y^2$ and $\gamma(t) = (\cos t, \sin t)$.
(a) Check that $\gamma(t)$ lies in the contour line $\{(x,y): f(x,y) = 1\}$.
(b) Show that the velocity vector $\gamma'(0)$ is orthogonal to the gradient ∇f at the point $\gamma(0) = (1,0)$. Sketch the curve γ, the velocity $\gamma'(0)$, and the gradient $\nabla f(1,0)$.
(c) Check that for every t, $\gamma'(t)$ is orthogonal to the gradient ∇f at the point $\gamma(t)$.

8. Let $f(x,y) = y - x^2$ and $\gamma(t) = (t,t^2)$.
(a) Check that $\gamma(t)$ lies in the contour line $\{(x,y): f(x,y) = 0\}$.
(b) Show that the velocity vector $\gamma'(1)$ is orthogonal to the gradient ∇f at the point $\gamma(1) = (1,1)$. Sketch the curve γ, the velocity $\gamma'(1)$, and the gradient $\nabla f(0,0)$.
(c) Check that, for every t, $\gamma'(t)$ is orthogonal to the gradient ∇f at the point $\gamma(t)$.

9. Let $f(x,y) = x^2y^3$.
(a) Find the equation of the tangent plane at the point $(1,1,1)$.
(b) Show that the vector $(2,3,-1)$ is a normal to the tangent plane at the point $(1,1,1)$.
(c) Show that the curve $\gamma(t) = (\sqrt{2}\cos t, \sqrt{2}\sin t, 4\sqrt{2}\cos^2 t \sin^3 t)$ lies entirely in the graph of f.
(d) Show that the curve in part (c) passes through the point $(1,1,1)$, and that its velocity vector at that point is $(-1,1,1)$.
(e) Prove that the line tangent to the curve γ at the point $(1,1,1)$ lies in the tangent plane to the graph of f at that point.

10. Suppose $\gamma(t) = (g_1(t),g_2(t),g_3(t))$ is a curve lying entirely in the graph of f.
(a) Show that $g_3(t) = f(g_1(t),g_2(t))$.
(b) Show that $g_3'(t) = D_1f(g_1(t),g_2(t))g_1'(t) + D_2f(g_1(t),g_2(t))g_2'(t)$.
(c) Show that at any point $(x_0,y_0,z_0) = \gamma(t_0)$, the tangent vector $\gamma'(t_0)$ is orthogonal to $\mathbf{N}$, the normal of the tangent plane to the graph of f at (x_0,y_0,z_0). (Hence, the line tangent to γ lies in the plane tangent to the graph.)

11. (a) Let f and g be any functions. Prove that if

$$\lim_{\mathbf{P}\to\mathbf{P}_0} |\mathbf{P} - \mathbf{P}_0|^{-1}\, |f(\mathbf{P}) - g(\mathbf{P})| = 0,$$

then $\lim_{\mathbf{P}\to\mathbf{P}_0} |f(\mathbf{P}) - g(\mathbf{P})| = 0$.

(b) If $f(\mathbf{P})$ is defined for all $\mathbf{P}$ near $\mathbf{P}_0$, and is continuous at $\mathbf{P}_0$, and if g is *any* polynomial of degree 1 such that $g(\mathbf{P}_0) = f(\mathbf{P}_0)$, prove that

$$\lim_{\mathbf{P}\to\mathbf{P}_0} |g(\mathbf{P}) - f(\mathbf{P})| = 0. \tag{19}$$

(c) Suppose that g is a first degree polynomial with $g(\mathbf{P}_0) = f(\mathbf{P}_0)$, and suppose that $\lim_{\mathbf{P}\to\mathbf{P}_0} |g(\mathbf{P}) - f(\mathbf{P})| = 0$. Prove that f is continuous at $\mathbf{P}_0$. (This problem should be compared to the definition of differentiability. Part (a) shows that condition (19) is weaker than the condition for differentiability. Parts (b) and (c) show that (19) corresponds closely to the continuity of f at $\mathbf{P}_0$, whereas in the text we showed that the stronger condition

$$\lim_{\mathbf{P}\to\mathbf{P}_0} |\mathbf{P} - \mathbf{P}_0|^{-1}\, |f(\mathbf{P}) - g(\mathbf{P})| = 0$$

corresponds closely to the behavior of the partial derivatives of f at $\mathbf{P}_0$.)

12. Suppose that $f(\mathbf{P}_0) \geq f(\mathbf{P})$ for all $\mathbf{P}$, and that the partial derivatives of f exist at $\mathbf{P}_0$. Prove that $D_1 f(\mathbf{P}_0) = 0$ and $D_2 f(\mathbf{P}_0) = 0$. (Hint: Look at the functions $f_1(x) = f(x,y_0)$ and $f_2(y) = f(x_0,y)$, and apply a familiar theorem about functions of one variable.)

13. This problem concerns the geometric meaning of the condition

$$\lim_{\mathbf{P}\to\mathbf{P}_0} \frac{f(\mathbf{P}) - f(\mathbf{P}_0) - (\mathbf{P} - \mathbf{P}_0)\cdot\nabla f(\mathbf{P}_0)}{|\mathbf{P} - \mathbf{P}_0|} = 0. \qquad (*)$$

Referring to Fig. 33, let $\mathbf{P} = (x,y)$, $\mathbf{P}_0 = (x_0,y_0)$, $\mathbf{Q} = (x,y,f(x,y))$, $\mathbf{Q}_0 = (x_0,y_0,f(x_0,y_0))$, $\nabla f(\mathbf{P}_0) = (a,b)$, $\mathbf{N} = (-a,-b,1)$, and let θ be the angle between $\mathbf{Q} - \mathbf{Q}_0$ and $\mathbf{N}$. You will show that $\mathbf{N}$ is normal to the tangent plane at $\mathbf{Q}_0$, and that $\theta \to \pi/2$ as $\mathbf{P} \to \mathbf{P}_0$.

(a) Prove that $\mathbf{N}$ is normal to the tangent plane at $\mathbf{Q}_0$.

(b) Prove that

$$\cos\theta = \frac{f(\mathbf{P}_0) + (\mathbf{P} - \mathbf{P}_0)\cdot\nabla f(\mathbf{P}_0) - f(\mathbf{P})}{|\mathbf{Q} - \mathbf{Q}_0|\cdot|\mathbf{N}|}.$$

(c) Prove that $|\mathbf{N}| \geq 1$ and $|\mathbf{Q} - \mathbf{Q}_0| \geq |\mathbf{P} - \mathbf{P}_0|$.

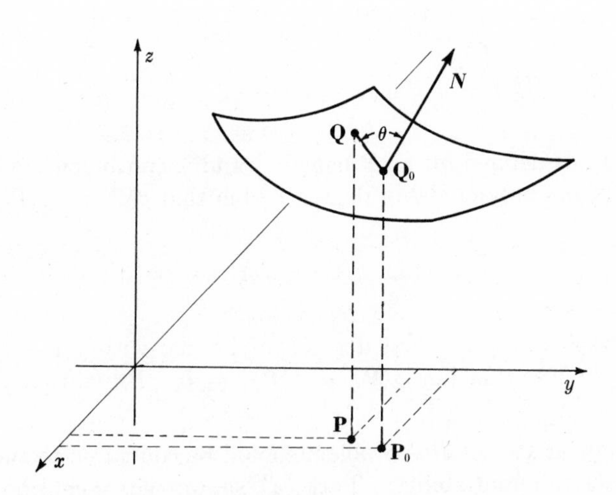

FIGURE 3.33

(d) Prove that if (*) holds, then $\lim_{\mathbf{P} \to \mathbf{P}_0} \cos \theta = 0$.

(e) Prove that if (*) holds, then $\lim_{\mathbf{P} \to \mathbf{P}_0} \theta = \pi/2$. (Hint: arccos is a continuous function.)

14. Suppose that f and h are differentiable at $\mathbf{P}_0$. Prove that $f + h$ is differentiable at $\mathbf{P}_0$, and $(\nabla(f + h))(\mathbf{P}_0) = \nabla f(\mathbf{P}_0) + \nabla h(\mathbf{P}_0)$.

Problems 15–19 concern *connected open sets.* An open set S in which every pair of points $\mathbf{P}_1$ and $\mathbf{P}_2$ can be joined by a differentiable curve γ lying in S is called *connected.* Figure 34(a) shows a connected open set, and Fig. 34(b) shows a disconnected one. Every open disk D is connected, since any two points in D can be joined by the line segment between them. For exactly the same reason, every open rectangle is connected.

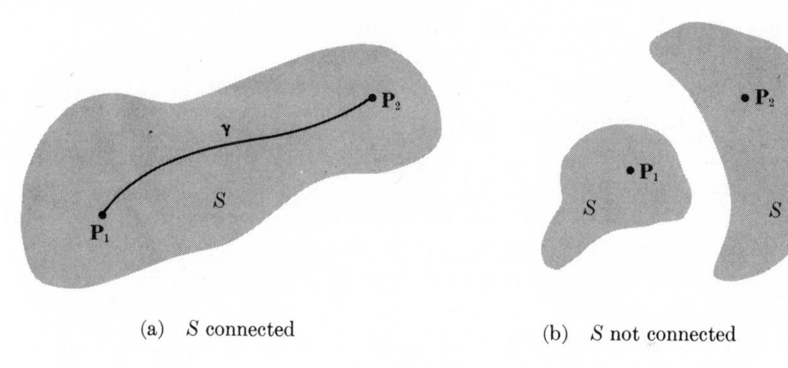

(a) S connected (b) S not connected

FIGURE 3.34

15. Suppose that S is a connected open set, that f is differentiable on S, and that $\nabla f \equiv \mathbf{0}$. Prove that f is constant on S. (Hint: Take any fixed point $\mathbf{P}_0$ in S. For every $\mathbf{P}_1$ in S there is a curve $\boldsymbol{\gamma}: [a,b] \to S$ such that $\boldsymbol{\gamma}(a) = \mathbf{P}_0$ and $\boldsymbol{\gamma}(b) = \mathbf{P}_1$. Use the chain rule (8) to prove that $f(\boldsymbol{\gamma}(b)) = f(\boldsymbol{\gamma}(a))$.)

16. Suppose S is a connected open set, f and g are differentiable, and $\nabla f \equiv \nabla g$ on S. What can you conclude about the difference between f and g? (See the two preceding problems.)

17. Suppose that S is a connected open set, that f is continuous on S, that $\mathbf{P}_1$ and $\mathbf{P}_2$ are two points in S. Prove the *Intermediate Value theorem:* If v is any number between $f(\mathbf{P}_1)$ and $f(\mathbf{P}_2)$, then there is a point $\mathbf{P}$ in S such that $f(\mathbf{P}) = v$. (Hint: Use the definition of connectedness given above, and apply the intermediate value theorem for functions of one variable; see Fig. 35.)

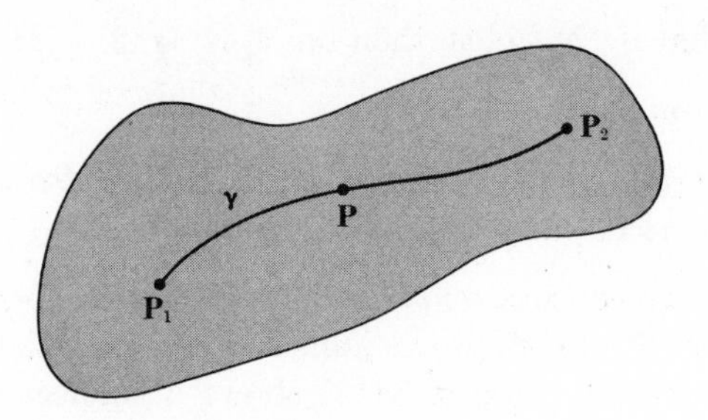

FIGURE 3.35

18. Prove that the following sets are connected.

(a) R^2

(b) $U = \{(x,y): y > 0\}$ (Hint: Given $\mathbf{P}_1$ and $\mathbf{P}_2$ in U, prove that the line segment given by $\boldsymbol{\gamma}(t) = (1-t)\mathbf{P}_1 + t\mathbf{P}_2$ lies in U for $0 \leq t \leq 1$, while $\boldsymbol{\gamma}(0) = \mathbf{P}_1$ and $\boldsymbol{\gamma}(1) = \mathbf{P}_2$.)

(c) $\{(x,y): x > 0\}$

(d) $R = \{(x,y): a < x < b \text{ and } c < y < d\}$

(e) $D = \{(x,y): x^2 + y^2 < 1\}$
(Hint: Use the same definition of $\boldsymbol{\gamma}$ as in part (b).)

(f) $\{\mathbf{P}: |\mathbf{P}| > 0\}$
(Hint: If $\mathbf{P}_1 = (r_1 \cos\theta_1,\ r_1 \sin\theta_1)$ and $\mathbf{P}_2 = (r_2\cos\theta_2,\ r_2\sin\theta_2)$, let $r(t) = (1-t)r_1 + tr_2$, $\theta(t) = (1-t)\theta_1 + t\theta_2$, and $\boldsymbol{\gamma}(t) = \big(r(t)\cos\theta(t),\ r(t)\sin\theta(t)\big)$. See Fig. 36.)

(g) $\{\mathbf{P}: 1 < |\mathbf{P}| < 2\}$

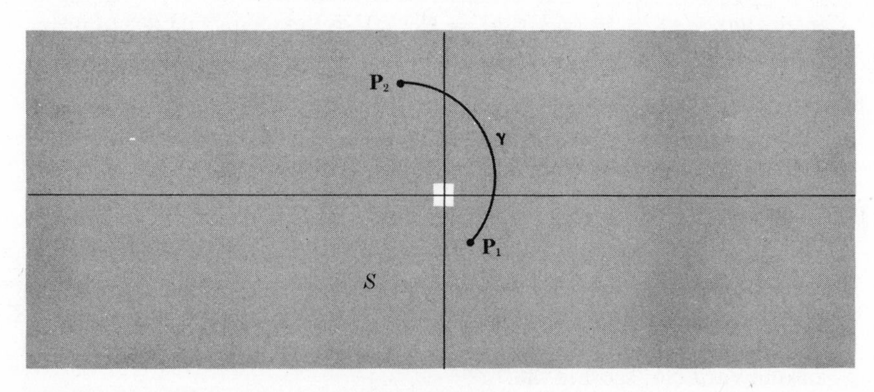

FIGURE 3.36

19. Prove that the following sets are *not* connected.
(a) $\{(x,y): y \neq 0\}$ (Hint: Apply Problem 17 with $\mathbf{P}_1 = (0,-1)$, $\mathbf{P}_2 = (0,1)$, $f(x,y) = y$, and $v = 0$.)
(b) $\{(x,y): x \neq 0\}$
(c) $\{\mathbf{P}: |\mathbf{P}| < 1 \text{ or } |\mathbf{P}| > 2\}$
(Hint: Apply Problem 17 with $f(\mathbf{P}) = |\mathbf{P}|^2$.)

3.4 COMPUTATIONS WITH THE CHAIN RULE

In geometric questions about contour lines, directional derivatives, and tangent planes, the chain rule can be applied best in the form that we have given,

$$(f\circ\boldsymbol{\gamma})'(t) = \nabla f(\boldsymbol{\gamma}(t))\cdot\boldsymbol{\gamma}'(t). \tag{1}$$

However, in many computational problems Leibniz notation is more convenient. To convert (1) into this notation we write

$$z = f(x,y), \qquad \text{so} \qquad \nabla f = \left(\frac{\partial z}{\partial x}, \frac{\partial z}{\partial y}\right),$$

and we set

$$x = \gamma_1(t), \qquad y = \gamma_2(t), \qquad \text{so} \qquad \boldsymbol{\gamma}'(t) = \left(\frac{dx}{dt}, \frac{dy}{dt}\right).$$

Then the chain rule (1) takes the form

$$\frac{dz}{dt} = \frac{\partial z}{\partial x}\frac{dx}{dt} + \frac{\partial z}{\partial y}\frac{dy}{dt}. \tag{2}$$

Notice that dx does *not* cancel ∂x, and dy does not cancel ∂y; with two or more variables, Leibniz notation no longer makes the chain rule look like simple algebra.

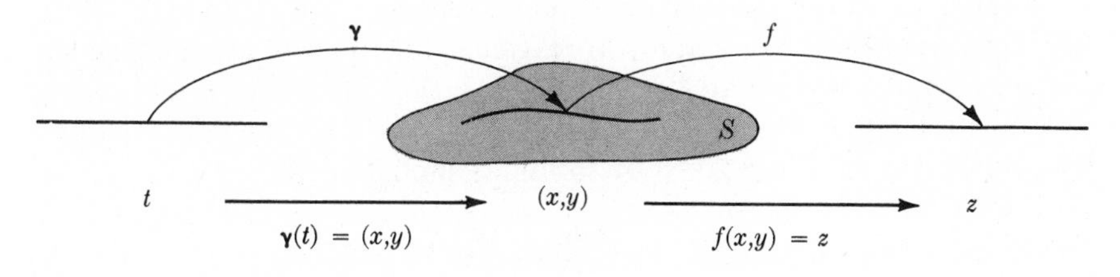

FIGURE 3.37

Figure 37 shows the situation schematically. The function $\boldsymbol{\gamma}$ carries a real number t into a point $(x,y) = \boldsymbol{\gamma}(t)$ in the domain of f, and f carries (x,y) into a real number $z = f(x,y)$. The composite function thus carries t into z, and its derivative is given by (2). In Fig. 37, z is called the "dependent variable," t is the "independent variable," and x and y are the "intermediate variables."

The chain rule computes the derivative of the dependent variable z with respect to the independent variable t; the formula involves the derivative of z with respect to each intermediate variable, multiplied by the derivative of the intermediate variable with respect to the independent variable (Fig. 38).

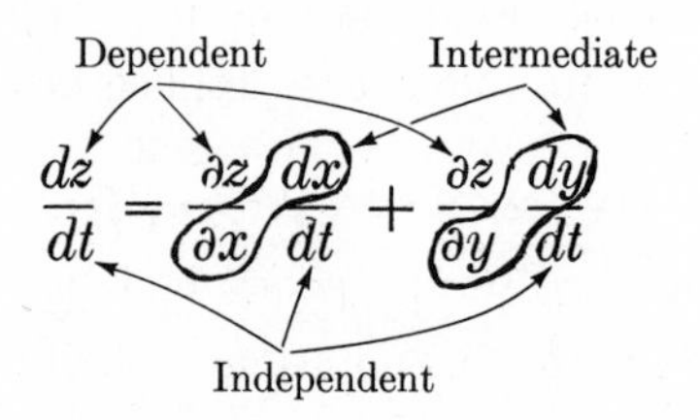

FIGURE 3.38

Example 1. If $z = f(x,y) = e^{xy}$, $x = t^2$, $y = 1/t$, we get $\partial z/\partial x = ye^{xy}$, $\partial z/\partial y = xe^{xy}$, hence from (2)

$$\frac{dz}{dt} = ye^{xy}2t + xe^{xy}(-1/t^2).$$

Substituting for x and y their expressions in terms of t, we find

$$\frac{dz}{dt} = \frac{1}{t}e^t \cdot 2t + t^2e^t\left(-\frac{1}{t^2}\right) = e^t.$$

You can easily check this computation by noting that $z = f(t^2,1/t) = e^t$, hence $dz/dt = e^t$.

Figure 39 shows another important version of the chain rule. Here $z = f(x,y)$ as before, but x and y are given as functions of *two* independent variables s and t, say $x = \varphi(s, t)$ and $y = \psi(s, t)$. Thus z can be considered as a function of the two variables s and t. In this case the partial derivatives $\partial z/\partial s$ and $\partial z/\partial t$ are given by formulas similar to (2),

$$\frac{\partial z}{\partial s} = \frac{\partial z}{\partial x}\frac{\partial x}{\partial s} + \frac{\partial z}{\partial y}\frac{\partial y}{\partial s} \tag{3}$$

$$\frac{\partial z}{\partial t} = \frac{\partial z}{\partial x}\frac{\partial x}{\partial t} + \frac{\partial z}{\partial y}\frac{\partial y}{\partial t}, \tag{4}$$

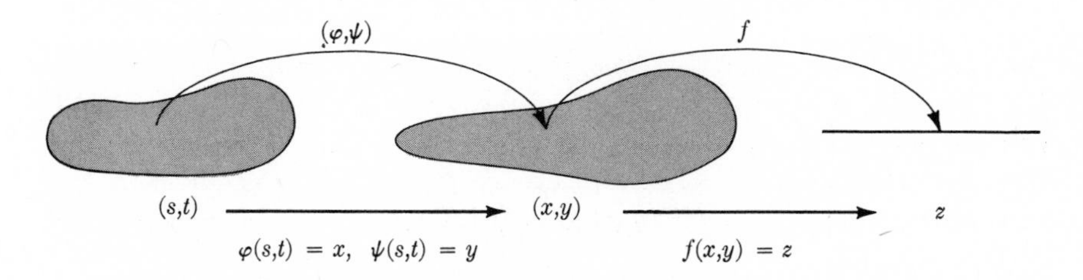

FIGURE 3.39

and for a very simple reason. To compute $\partial z/\partial t$, $\partial x/\partial t$, and $\partial y/\partial t$ we hold s fixed and differentiate as if we had functions of the one variable t; thus formula (2) implies (4). Similarly, to compute $\partial z/\partial s$ we hold t constant, so the rule (2) for ordinary derivatives implies (3).

Notice in (3) and (4) that cancellation of ∂x and ∂y from numerator and denominator in each term leads to complete nonsense.

Example 2. Let $z = xe^y$, $x = s + t$, $y = s^2 - t$. Then

$$\frac{\partial z}{\partial s} = \frac{\partial z}{\partial x}\frac{\partial x}{\partial s} + \frac{\partial z}{\partial y}\frac{\partial y}{\partial s} = e^y \cdot 1 + xe^y \cdot 2s = e^{s^2-t}(1 + (s + t)2s)$$

$$\frac{\partial z}{\partial t} = \frac{\partial z}{\partial x}\frac{\partial x}{\partial t} + \frac{\partial z}{\partial y}\frac{\partial y}{\partial t} = e^y \cdot 1 + xe^y(-1) = e^{s^2-t}(1 - s - t).$$

You can check these results by differentiating $(s + t)e^{s^2-t}$ directly.

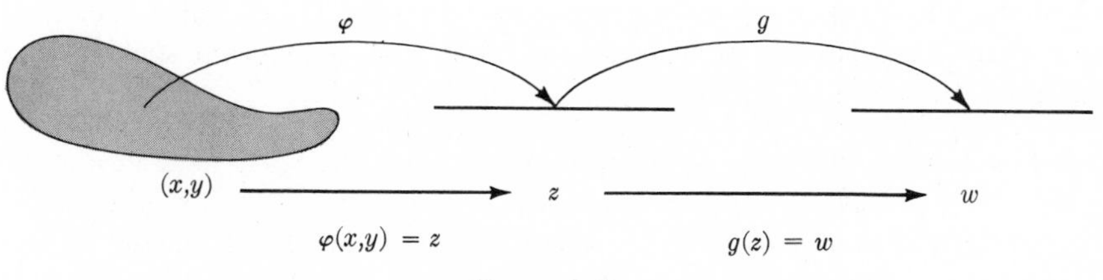

FIGURE 3.40

Figure 40 shows yet another version of the chain rule:

$$\frac{\partial w}{\partial x} = \frac{dw}{dz}\frac{\partial z}{\partial x} \tag{5}$$

$$\frac{\partial w}{\partial y} = \frac{dw}{dz}\frac{\partial z}{\partial y} \tag{6}$$

or

$$\nabla(g\circ\varphi) = (g'\circ\varphi)\nabla\varphi. \tag{7}$$

Since y is held constant in taking $\partial/\partial x$, formula (5) follows from the one-variable chain rule, and (6) follows analogously. Finally, (7) is simply a different notation for (5) and (6) together.

Example 3. Suppose g is a differentiable function of one variable, and let $f(\mathbf{P}) = g(|\mathbf{P}|)$, i.e., $f(x,y) = g(\sqrt{x^2+y^2})$. Find ∇f. *Solution.* We have the situation in Fig. 40 with $\varphi(x,y) = \sqrt{x^2+y^2}$, and $f = g\circ\varphi$. Here

$$\nabla\varphi(x,y) = \left(\frac{x}{\sqrt{x^2+y^2}}, \frac{y}{\sqrt{x^2+y^2}}\right),$$

so by (7)

$$\begin{aligned}(\nabla f)(x,y) &= g'(\varphi(x,y))\nabla\varphi(x,y)\\ &= g'(\sqrt{x^2+y^2})\left(\frac{x}{\sqrt{x^2+y^2}}, \frac{y}{\sqrt{x^2+y^2}}\right).\end{aligned} \tag{8}$$

Setting $\mathbf{P} = (x,y)$ gives $|\mathbf{P}| = \sqrt{x^2+y^2}$, hence from (8)

$$\nabla f(\mathbf{P}) = \frac{1}{|\mathbf{P}|}g'(|\mathbf{P}|)\mathbf{P}.$$

In applying the chain rule it is essential to know which variables are respectively independent, intermediate, and dependent. There is no standard convention governing which letters play which roles, but you can easily keep the given situation in mind with a diagram such as, for example, the one in Fig. 40,

$$(x,y) \to z \to w.$$

Example 4. Let $u = f(x+y)$, where f is any differentiable function of one variable. Show that $\partial u/\partial x = \partial u/\partial y$. *Solution.* We have the situation

$$(x,y) \to z \to u$$

where

$$z = x+y, \qquad u = f(z).$$

Hence, as in (5) and (6),

$$\frac{\partial u}{\partial x} = \frac{du}{dz}\frac{\partial z}{\partial x} = f'(z)\cdot 1 = f'(x+y)$$

$$\frac{\partial u}{\partial y} = \frac{du}{dz}\frac{\partial z}{\partial y} = f'(z)\cdot 1 = f'(x+y),$$

so $\partial u/\partial x = \partial u/\partial y$. •

Example 5. Show that if F is a function defined on R^2 and has continuous derivatives satisfying $\partial F/\partial x = \partial F/\partial y$, then $F(x,y) = f(x+y)$ for some differentiable function f of one variable. (Notice that this is the converse of Example 4.) *Solution.* We use a tricky substitution. Set

$$s = x + y, \qquad t = x - y, \tag{9}$$

so

$$x = \tfrac{1}{2}(s+t), \qquad y = \tfrac{1}{2}(s-t). \tag{10}$$

Thus we are in the situation $(s,t) \to (x,y) \to u$ where x and y are given by (10), and $u = F(x,y)$. Then the hypothesis that $\partial F/\partial x = \partial F/\partial y$ becomes $\partial u/\partial x = \partial u/\partial y$. Hence

$$\begin{aligned}\frac{\partial u}{\partial t} &= \frac{\partial u}{\partial x}\frac{\partial x}{\partial t} + \frac{\partial u}{\partial y}\frac{\partial y}{\partial t}\\ &= \frac{\partial u}{\partial x}\cdot\frac{1}{2} + \frac{\partial u}{\partial y}\cdot\left(-\frac{1}{2}\right) \qquad \text{(by (10))}\\ &= 0.\end{aligned}$$

Since $\partial u/\partial t = 0$, it follows that for any fixed s, u is a constant independent of t. Call this constant $f(s)$; then $f(s) = u = F(x,y)$. Thus, by (9), $F(x,y) = f(x+y)$, and it remains only to show that f is differentiable. But, setting $y = 0$, we get $f(x) = F(x,0)$, so the existence of $\partial F/\partial x$ implies the existence of $f'(x)$, and the problem is solved.

PROBLEMS

The first six problems are for practice in computations. The others concern *homogeneous functions* (defined in Problem 7) and *thermodynamics* (Problem 10).

1. Find the required derivatives in two ways: (i) by the chain rule, (ii) by substituting and computing directly. Reconcile the two results, as in Example 1.

(a) $w = x^2 + 2y$, $x = e^r$, $y = e^s \cos r$; find $\partial w/\partial r$ and $\partial w/\partial s$. (The scheme here is $(r, s) \to (x, y) \to w$.)

(b) $z = \log(x^2 + y^2)$, $x = r\cos\theta$, $y = r\sin\theta$; find $\partial z/\partial r$ and $\partial z/\partial\theta$.
(c) $z = r^2\cos 2\theta$, $r = \sqrt{x^2 + y^2}$, $\theta = \arctan(y/x)$; find $\partial z/\partial x$ and $\partial z/\partial y$.
(d) $w = u^3 + \sin u$, $u = ax + y$; find $\partial w/\partial x$ and $\partial w/\partial y$.
(e) $u = z^2 + e^z$, $z = ax - y$; find $\partial u/\partial x$ and $\partial u/\partial y$.

2. Suppose $F(x,t) = f(x + ct)$, where f is a differentiable function of one variable. Prove that

$$c\frac{\partial F}{\partial x} = \frac{\partial F}{\partial t}.$$

3. Suppose that F is differentiable at every point of R^2, and $cD_1F = D_2F$, where $c \neq 0$. Prove that $F(x,t) = f(x + ct)$ for some differentiable function of one variable. (Hint: Let $z = F(x,t)$, $x = (u + v)/2$, $t = (u - v)/2c$, and show that $\partial z/\partial v = 0$.)

4. Let $f(\mathbf{P}) = |\mathbf{P}|^a$, where a is a real constant. Show that $\nabla f(\mathbf{P}) = a|\mathbf{P}|^{a-2}\mathbf{P}$.

5. Suppose $f(u,v)$ has continuous partial derivatives, and let

$$z = f(x + y, x - y).$$

Prove that

$$\frac{\partial z}{\partial x}\frac{\partial z}{\partial y} = \left(\frac{\partial f}{\partial u}\right)^2 - \left(\frac{\partial f}{\partial v}\right)^2.$$

6. Let $z = f(x,y)$, $x = r\cos\theta$, $y = r\sin\theta$. Show that

(a) $\dfrac{\partial z}{\partial r} = f_x\cos\theta + f_y\sin\theta$ and $\dfrac{\partial z}{\partial\theta} = -rf_x\sin\theta + rf_y\cos\theta$.

(b) $(f_x)^2 + (f_y)^2 = \left(\dfrac{\partial z}{\partial r}\right)^2 + \dfrac{1}{r^2}\left(\dfrac{\partial z}{\partial\theta}\right)^2$. (This gives the formula for $|\nabla f|^2$ in polar coordinates.)

7. A function $f(\mathbf{P})$ defined for $|\mathbf{P}| \neq 0$ is called *homogeneous of degree n* if $f(t\mathbf{P}) = t^nf(\mathbf{P})$ for every $\mathbf{P} \neq \mathbf{0}$ and every $t > 0$.
(a) Prove that every polynomial of the form

$$f(x,y) = a_0x^n + a_1x^{n-1}y + \cdots + a_ny^n$$

is homogeneous of degree n.
(b) Prove that $h(x,y) = xy/(x^2 + y^2)$ is homogeneous of degree zero, and that $f(\mathbf{P}) = |\mathbf{P}|^n$ is homogeneous of degree n.
(c) Prove that if f is homogeneous of degree zero, and f is continuous at $(0,0)$, then f is constant. (Hint: $f(0,0) = \lim_{t\to 0+} f(t\mathbf{P})$ for every $\mathbf{P} \neq (0,0)$.)

8. Suppose that f is homogeneous of degree n in the sense of Problem 7, and that f is differentiable at every point $\mathbf{P}_0 \neq (0,0)$. Prove that $\mathbf{P}_0 \cdot \nabla f(\mathbf{P}_0) = nf(\mathbf{P}_0)$, i.e. $xf_x + yf_y = nf$. (Hint: Differentiate the identity $f(tx_0, ty_0) = t^n f(x_0, y_0)$ with respect to t. The scheme here is $t \to (x,y) \to z$, where $x = tx_0$, $y = ty_0$, and $z = f(x,y)$. In more geometric notation, you are differentiating $f(\boldsymbol{\gamma}(t)) = t^n f(\mathbf{P}_0)$, where $\boldsymbol{\gamma}(t) = t\mathbf{P}_0$.)

9. Prove that if f is homogeneous of degree n and has continuous partial derivatives, then $D_1 f$ and $D_2 f$ are homogeneous of degree $n - 1$. (Hint: For $t > 0$,

$$D_1 f(x,y) = \lim_{h \to 0} \frac{f(x + th, y) - f(x,y)}{th}.)$$

10. In *thermodynamics* there are five basic quantities: T (temperature), V (volume), P (pressure), E (energy), and S (an esoteric quantity called *entropy*). It is generally assumed that *any two* of these may be considered as independent variables; the other three are then functions of those two. A symbol such as $\partial S/\partial T$ would imply that T is considered an independent variable, but it would not tell which is the *other* independent variable. This ambiguity is eliminated by writing $(\partial S/\partial T)_V$ for the derivative of S with respect to T when T and V are considered independent, and $(\partial S/\partial T)_P$ for the derivative of S with respect to T when T and P are independent, and so on. With this convention, certain identities are obvious; for example, when T and V are independent we have

$$\left(\frac{\partial T}{\partial T}\right)_V = 1, \qquad \left(\frac{\partial T}{\partial V}\right)_T = 0$$

$$\left(\frac{\partial V}{\partial T}\right)_V = 0, \qquad \left(\frac{\partial V}{\partial V}\right)_T = 1.$$

Beyond these formal identities, the theory of thermodynamics assumes two basic equations:

$$T\left(\frac{\partial S}{\partial T}\right)_V = \left(\frac{\partial E}{\partial T}\right)_V \tag{*}$$

and

$$T\left(\frac{\partial S}{\partial V}\right)_T = \left(\frac{\partial E}{\partial V}\right)_T + P. \tag{**}$$

Corresponding relations when other independent variables are chosen can then be deduced by the chain rule, as outlined below.†

† Part (a) requires §3.6. Assume the formula in (a) and do (b)–(g). Sorry!

(a) Considering T and V as independent variables, differentiate (*) with respect to V and (**) with respect to T, and conclude that

$$\left(\frac{\partial P}{\partial T}\right)_V = \left(\frac{\partial S}{\partial V}\right)_T.$$

(Assume that all first and second derivatives are continuous.)

(b) Suppose that T and P are chosen as independent. Using the scheme $(T,P) \to (T,V) \to P$, explain why

$$\left(\frac{\partial P}{\partial P}\right)_T = \left(\frac{\partial P}{\partial T}\right)_V \left(\frac{\partial T}{\partial P}\right)_T + \left(\frac{\partial P}{\partial V}\right)_T \left(\frac{\partial V}{\partial P}\right)_T.$$

(c) From part (b), deduce that

$$1 = \left(\frac{\partial P}{\partial V}\right)_T \left(\frac{\partial V}{\partial P}\right)_T.$$

(d) Compute $(\partial P/\partial T)_P$ by the scheme in part (b), and deduce that

$$\left(\frac{\partial P}{\partial T}\right)_V = -\left(\frac{\partial P}{\partial V}\right)_T \left(\frac{\partial V}{\partial T}\right)_P.$$

(e) Compute $(\partial S/\partial P)_T$ by the scheme $(T,P) \to (T,V) \to S$. Using parts (a), (c), and (d), show that

$$\left(\frac{\partial S}{\partial P}\right)_T = -\left(\frac{\partial V}{\partial T}\right)_P.$$

(f) Prove that

$$\left(\frac{\partial E}{\partial P}\right)_T = -T\left(\frac{\partial V}{\partial T}\right)_P - P\left(\frac{\partial V}{\partial P}\right)_T,$$

using (*) or (**) and any methods or results from parts (a)–(e).

(g) Show that $T = \left(\frac{\partial E}{\partial S}\right)_V$.

3.5 THE IMPLICIT FUNCTION THEOREM*

For the mathematician (like the philosopher) existence is a fundamental question. Does there exist a square root of 2? Does a function have a maximum value? Does a system of simultaneous linear equations have a nonzero solution? Given f, is there a function F satisfying the equation $F'(x) = f(x)$?

---

* This section is not required in the rest of the text.

The implicit function theorem concerns another existence question: When does an equation $f(x,y) = c$ define y as a function of x? The answer is a little complicated, and has to be digested slowly.

Theorem 8. Implicit Function Theorem. *Suppose that $f(x_0, y_0) = c$, $f_y(x_0, y_0) \neq 0$, and that f has continuous first derivatives in an open set containing (x_0, y_0). Then there is a rectangle*

$$R = \{(x,y): |x - x_0| < \delta, |y - y_0| < \epsilon\}, \qquad \delta > 0, \epsilon > 0,$$

and there is a differentiable function φ defined on the open interval $(x_0 - \delta, x_0 + \delta)$ such that

$$\text{for } (x,y) \text{ in } R, \quad f(x,y) = c \iff y = \varphi(x).$$

In other words, if we start at a point (x_0, y_0) on the contour line

$$\{(x,y): f(x,y) = c\}, \tag{1}$$

and if $f_y \neq 0$ at that point, then near (x_0, y_0) the contour line (1) is precisely the graph of φ (see Fig. 41). The theorem guarantees that such a function φ exists, even though it might be impossible to write an explicit formula for it. The equation $f(x,y) = c$ is said to determine y *implicitly* as a function $\varphi(x)$, hence the name of the theorem.

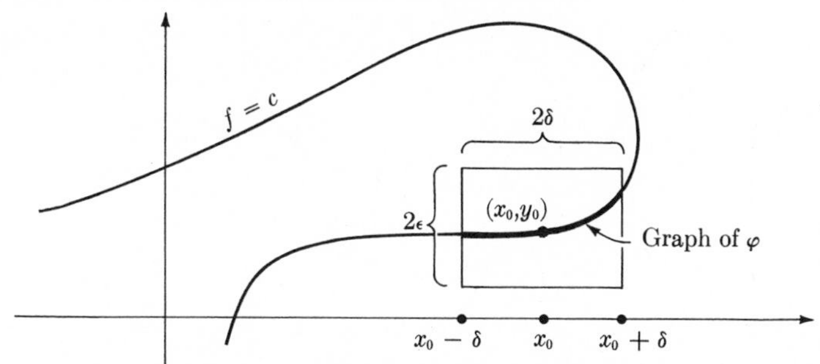

FIGURE 3.41

This is a *local* existence theorem; we assume things about f only near the point (x_0, y_0) and the resulting function φ is guaranteed to exist only near x_0 (precisely, for $x_0 - \delta < x < x_0 + \delta$), and its graph gives only that part of the contour line $f = c$ which is near x_0 (precisely, in the rectangle R).

The simple equation

$$x^2 + y^2 = 1 \tag{2}$$

shows why it is important that $f_y(x_0, y_0) \neq 0$, and why φ may be defined only in a small interval about x_0. If we take $f(x,y) = x^2 + y^2$, $c = 1$, and $(x_0, y_0) = (0,1)$, then the implicit function theorem applies; for f has continuous derivatives everywhere, $f(x_0, y_0) = 1$, and $f_y(x_0, y_0) = 2 \neq 0$.

Hence there is a rectangle R such that the part of the contour line (2) lying in R is the graph of a function φ. In fact, as Fig. 42(a) shows, we can take

$$R = \{(x,y): |x - 0| < 1, |y - 1| < 1\}, \qquad \varphi(x) = \sqrt{1 - x^2}\,.$$

Taking another point $(x_0, y_0) = (3/5, -4/5)$, Fig. 42(b), we find the rectangle

$$R = \{(x,y): |x - \tfrac{3}{5}| < \tfrac{2}{5}, |y + \tfrac{4}{5}| < \tfrac{4}{5}\}$$

and the implicit function $\varphi(x) = -\sqrt{1 - x^2}$ (note the minus sign); φ is defined for $|x - 3/5| < 2/5$, but it does not give *all* of the contour line (1) lying over that interval, only the part near $(3/5, -4/5)$. Generally, if (x_0, y_0) lies on the unit circle, then equation (2) defines y as a function of x on the interval $(x_0 - \delta, x_0 + \delta)$ with $\delta = 1 - |x_0|$ (Fig. 42(c)); when x_0 is near ± 1, this interval is very small, and when $x_0 = \pm 1$ (the points where $f_y = 0$), it disappears completely.

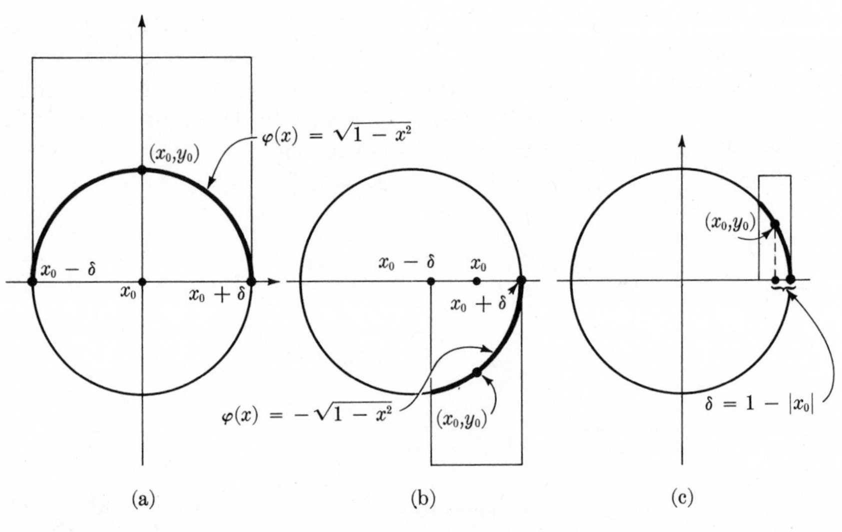

FIGURE 3.42

The proof of the implicit function theorem is divided into two lemmas.

Lemma 1. *Suppose that f has continuous first derivatives in the rectangle*

$$R = \{(x,y): |x - x_0| < \delta, |y - y_0| < m\delta\},$$

and $f_y > 0$, $|f_x/f_y| \le m$, $f(x_0, y_0) = c$. Then there is a unique function φ defined on the interval $(x_0 - \delta, x_0 + \delta)$ such that

$$\textit{for } (x,y) \textit{ in } R, \quad f(x,y) = c \iff y = \varphi(x).$$

Further,

$$\varphi'(x_0) = \frac{-f_x(x_0, y_0)}{f_y(x_0, y_0)}.$$

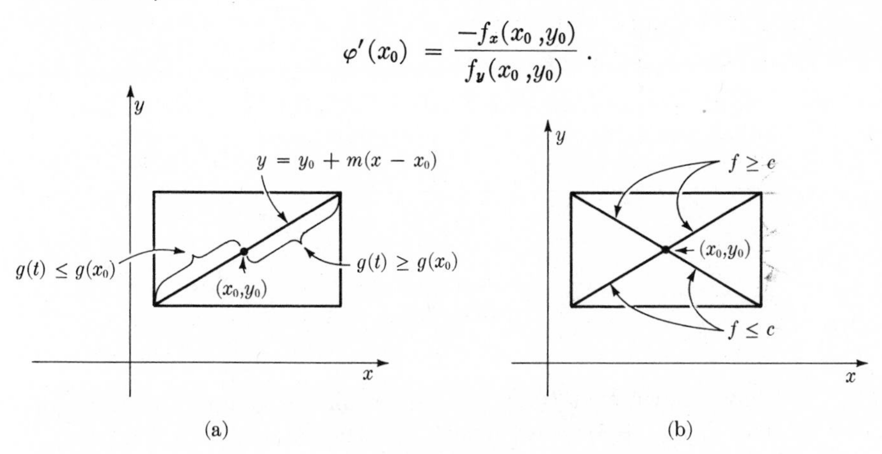

FIGURE 3.43

Proof. The values of f along the line of slope m in Fig. 43(a) are given by $g(t) = f(t, y_0 + m(t - x_0))$. By the chain rule,

$$g'(t) = f_x(t, y_0 + m(t - x_0)) + mf_y(t, y_0 + m(t - x_0)).$$

By assumption, $mf_y \geq |f_x|$, so $g' \geq 0$. Hence for $t \geq x_0$, we have $g(t) \geq g(x_0) = c$, and for $t \leq x_0$, we have $g(t) \leq g(x_0) = c$.

In other words, along the line of slope m, $f \geq c$ for $x \geq x_0$ and $f \leq c$ for $x \leq x_0$, as shown in Fig. 43(b). Similarly, along the line of slope $-m$, $f \leq c$ for $x \geq x_0$ and $f \geq c$ for $x \leq x_0$.

Now fix x in the interval $[x_0, x_0 + \delta)$, and look at the function

$$h(y) = f(x,y)$$

which gives the values of f along the vertical line in Fig. 44(a). We have just shown that

$$h(y_0 + m(x - x_0)) \geq c \qquad \text{and} \qquad h(y_0 - m(x - x_0)) \leq c;$$

by the Intermediate Value theorem, there is a number y_1 such that

$$y_0 - m(x - x_0) \leq y_1 \leq y_0 + m(x - x_0) \qquad \text{and} \qquad h(y_1) = c. \tag{3}$$

Moreover, $h'(y) = f_y(x,y) > 0$, so h is strictly increasing on the interval $(y_0 - m\epsilon, y_0 + m\epsilon)$; thus there is *only one* y_1 in that interval for which $h(y_1) = c$. Denote this unique value of y_1 by $\varphi(x)$. From (3),

$$y_0 - m(x - x_0) \leq \varphi(x) \leq y_0 + m(x - x_0) \tag{4}$$

(see Fig. 44(b)).

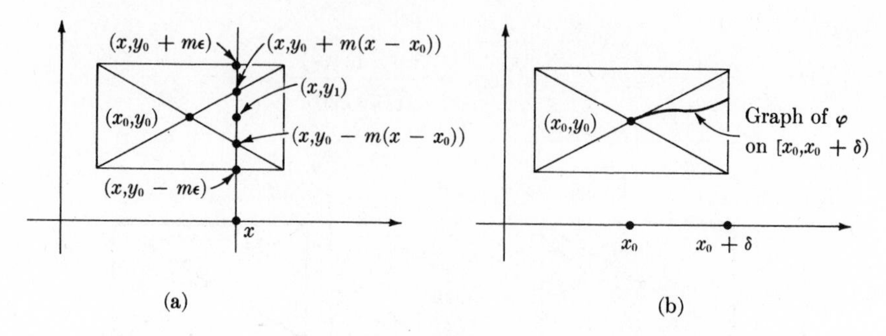

FIGURE 3.44

We have shown that $f(x,y) = c$ determines a unique function φ on the interval $[x_0, y_0 + \delta)$; φ is similarly determined on $(x_0 - \delta, x_0]$ and

$$y_0 + m(x - x_0) \leq \varphi(x) \leq y_0 - m(x - x_0), \qquad x \leq x_0. \tag{4'}$$

Hence φ is uniquely defined on $(x_0 - \delta, x_0 + \delta)$, and we see from (4) and (4′) that

$$|\varphi(x) - y_0| \leq m|x - x_0|. \tag{5}$$

It remains to take the derivative $\varphi'(x_0)$. Because f has continuous derivatives, it is differentiable at (x_0, y_0); thus (as in the proof of Theorem 6)

$$f(x,y) = f(x_0,y_0) + (x - x_0)f_x(x_0,y_0) + (y - y_0)f_y(x_0,y_0) + \sqrt{(x - x_0)^2 + (y - y_0)^2}\,Q(x,y),$$

where Q is continuous and $Q(x_0,y_0) = 0$. Setting $y = \varphi(x)$, we have $f(x,\varphi(x)) = c$, so

$$c = c + (x - x_0)f_x(x_0,y_0) + (\varphi(x) - y_0)f_y(x_0,y_0) + \sqrt{(x - x_0)^2 + (\varphi(x) - y_0)^2}\,Q(x,\varphi(x)),$$

hence

$$\frac{\varphi(x) - y_0}{x - x_0} = -\frac{f_x(x_0,y_0)}{f_y(x_0,y_0)} - \frac{\sqrt{(x - x_0)^2 + (\varphi(x) - y_0)^2}}{x - x_0}\,\frac{Q(x,\varphi(x))}{f_y(x_0,y_0)}. \tag{6}$$

We complete the proof by using (5) to show that

$$\lim_{x\to x_0}\left(\frac{\sqrt{(x - x_0)^2 + (\varphi(x) - y_0)^2}}{|x - x_0|}\,\frac{Q(x,\varphi(x))}{f_y(x_0,y_0)}\right) = 0. \tag{7}$$

First of all, (5) shows that φ is continuous at x_0 and $\varphi(x_0) = y_0$, so by composition $Q(x,\varphi(x))$ is continuous at x_0 and

$$\lim_{x\to x_0} Q(x,\varphi(x)) = Q(x_0,\varphi(x_0)) = 0. \tag{8}$$

Again by (5),

$$\frac{\sqrt{(x-x_0)^2+(\varphi(x)-y_0)^2}}{|x-x_0|} \le \frac{\sqrt{(x-x_0)^2+m^2(x-x_0)^2}}{|x-x_0|}$$

$$\le \sqrt{1+m^2}\,; \tag{9}$$

and (8) and (9) imply (7). Since $\varphi(x_0) = y_0$, (7) and (6) imply $\varphi'(x_0) = -f_x(x_0,y_0)/f_y(x_0,y_0)$, and Lemma 1 is proved.

Lemma 1 looks a lot like the implicit function theorem itself, but it makes more specific hypotheses about f. Lemma 2 bridges this gap.

Lemma 2. *If f has continuous derivatives in a disk of radius r about $\mathbf{P}_0 = (x_0,y_0)$, and $f_y(\mathbf{P}_0) > 0$, then there are positive numbers m and δ such that*

$$f_y > 0 \qquad \textit{and} \qquad \left|\frac{f_x}{f_y}\right| \le m \tag{10}$$

in the rectangle

$$R = \{(x,y) : |x-x_0| < \delta,\ |y-y_0| < m\delta\}. \tag{11}$$

Proof. Since f_y is continuous, there is a number $\delta_1 > 0$ such that $\delta_1 \le r$ and

$$|\mathbf{P}-\mathbf{P}_0| < \delta_1 \ \Rightarrow\ |f_y(\mathbf{P}) - f_y(\mathbf{P}_0)| < f_y(\mathbf{P}_0)\,;$$

hence

$$f_y(\mathbf{P}) > 0 \quad \text{for } |\mathbf{P}-\mathbf{P}_0| < \delta_1\,. \tag{12}$$

It follows that $|f_x/f_y|$ is continuous for $|\mathbf{P}-\mathbf{P}_0| < \delta_1$, so there is a number $\delta_2 \le \delta_1$ such that

$$|\mathbf{P}-\mathbf{P}_0| < \delta_2 \ \Rightarrow\ \left|\frac{f_x(\mathbf{P})}{f_y(\mathbf{P})}\right| < \left|\frac{f_x(\mathbf{P}_0)}{f_y(\mathbf{P}_0)}\right| + 1. \tag{13}$$

Now define the numbers m and δ of the lemma by

$$m = \left|\frac{f_x(\mathbf{P}_0)}{f_y(\mathbf{P}_0)}\right| + 1, \qquad (1+m^2)\delta^2 = \delta_2{}^2.$$

Then the rectangle (11) lies in the disk of radius δ_2 about $\mathbf{P}_0$ (see Fig. 45); hence (10) follows from (12) and (13), and Lemma 2 is proved.

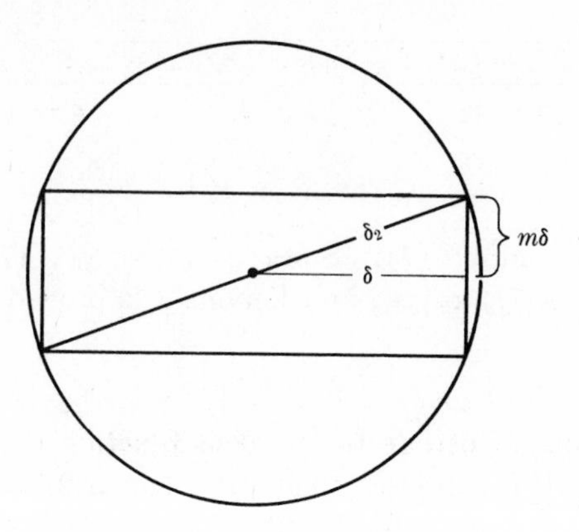

FIGURE 3.45 $\delta^2 + (m\delta)^2 = \delta_2{}^2$

Returning at last to Theorem 8 itself, we can suppose that $f_y(x_0\,,y_0) > 0$; for if $f_y(x_0\,,y_0) < 0$, simply replace f by $-f$ and c by $-c$, and all the hypotheses will still be satisfied. Supposing $f_y(x_0\,,y_0) > 0$, Lemma 2 provides a rectangle R in which Lemma 1 applies; then Lemma 1 provides a function φ such that

$$\text{for } (x,y) \text{ in } R, \quad f(x,y) = c \iff y = \varphi(x),$$

and φ is differentiable *at* x_0. But the hypotheses of Theorem 8 apply at *every* point of the graph of φ, so by what we have already shown φ is differentiable everywhere. *Q.E.D.*

We can, of course, interchange the roles of x and y. If $f_x(x_0\,,y_0) \neq 0$, then $f(x,y) = c$ determines x as a function ψ of y; i.e. near $(x_0\,,y_0)$ the contour line $f = c$ is the set $\{(x,y): |y - y_0| < \delta,\ x = \psi(y)\}$.

Example 1. Represent the contour line

$$y - xe^y = 1 \tag{14}$$

near the point $(-1,0)$ as a function. *Solution.* Here $f(x,y) = y - xe^y$, so $\nabla f(x,y) = (-e^y, 1 - xe^y)$, and $\nabla f(-1,0) = (-1,2)$. Since $f_x \neq 0$ and $f_y \neq 0$, we can represent the contour line either as $y = \varphi(x)$, or as $x = \psi(y)$. In fact, in the second case you can see easily that $\psi(y) = (y - 1)e^{-y}$ (simply solve (14) for x). The function φ, on the other hand, cannot be given explicitly; we only know that it exists by Theorem 8.

Example 2. Let φ be the function in Example 1. Compute $\varphi'(-1)$, and $\varphi'(x)$ for x near -1. *Solution.* By Theorem 8, φ is differentiable. Since

$$\varphi(x) - xe^{\varphi(x)} = 1,$$

we find

$$\varphi'(x) - e^{\varphi(x)} - x\varphi'(x)e^{\varphi(x)} = 0;$$

hence

$$\varphi'(x) = \frac{e^{\varphi(x)}}{1 - xe^{\varphi(x)}}. \tag{15}$$

In particular, since $(-1,0)$ is on the graph of φ (see Example 1), we have $\varphi(-1) = 0$; hence

$$\varphi'(-1) = \frac{1}{1-(-1)} = \frac{1}{2}.$$

The next example illustrates how to find a specific interval in which φ is defined, even though φ cannot be found explicitly.

Example 3. Find a rectangle R as guaranteed by Theorem 8 for the equation $y - xe^y = 1$, taking $(x_0,y_0) = (-1,0)$. (This is the equation considered in Example 1.) *Solution.* From the proof of Theorem 8, we only need to find a rectangle R satisfying the hypotheses in Lemma 1 for some number m. We have

$$\frac{f_x(x,y)}{f_y(x,y)} = \frac{-e^y}{1 - xe^y} = \frac{-1}{e^{-y} - x}.$$

If $|x - (-1)| < \frac{1}{2}$ and $|y| < 1$, then $-x > \frac{1}{2}$, so $e^{-y} - x > \frac{1}{2}$; hence

$$\left|\frac{f_x(x,y)}{f_y(x,y)}\right| = \frac{1}{e^{-y} - x} < 2.$$

Thus Lemma 1 is satisfied with $\delta = \frac{1}{2}$ and $m = 2$. It follows that φ is defined for $|x - (-1)| < \frac{1}{2}$.

The implicit function theorem justifies the technique of "implicit differentiation," as exemplified by the following calculation:

$$x \cos xy = 1 \tag{16}$$

$$\cos xy - xy \sin xy - x^2 \sin xy \frac{dy}{dx} = 0$$

$$\frac{dy}{dx} = \frac{\cos xy - xy \sin xy}{x^2 \sin xy}.$$

Here we differentiated (16) *as if* y were a function of x, then solved for dy/dx. The general scheme is

$$f(x,y) = c, \tag{17}$$

$$f_x + f_y \frac{dy}{dx} = 0 \tag{18}$$

$$\frac{dy}{dx} = -\frac{f_x}{f_y}. \tag{19}$$

Here (18) follow from (17) by the chain rule, *if* y is a function of x. However, in (19) we see that the final calculation works only when $f_y \neq 0$; and in precisely this case, the implicit function theorem guarantees that y *is* a function of x. Thus, whenever (19) is algebraically legitimate (i.e. does not require division by zero), the whole calculation is legitimate (i.e. y really is a differentiable function of x).

Our final application of the implicit function theorem concerns the "Lagrange multiplier" condition for finding maxima and minima of one function g along the contour line of another function f.

Theorem 9. *Suppose that f and g have continuous derivatives throughout a disk centered at $\mathbf{P}_0$, that $f(\mathbf{P}_0) = c$, and that $g(\mathbf{P}_0) \geq g(\mathbf{P})$ for every $\mathbf{P}$ on the contour line $\{\mathbf{P}: f(\mathbf{P}) = c\}$. If $\nabla f(\mathbf{P}_0) \neq \mathbf{0}$, then $\nabla g(\mathbf{P}_0)$ is a multiple of $\nabla f(\mathbf{P}_0)$,*

$$\nabla g(\mathbf{P}_0) = \lambda \nabla f(\mathbf{P}_0).$$

(The number λ is the "Lagrange multiplier" referred to above. Since ∇f is orthogonal to the contour line $f = c$, the criterion $\nabla g = \lambda \nabla f$ says that ∇g is orthogonal to the contour line; see Fig. 46, and compare §3.2.)

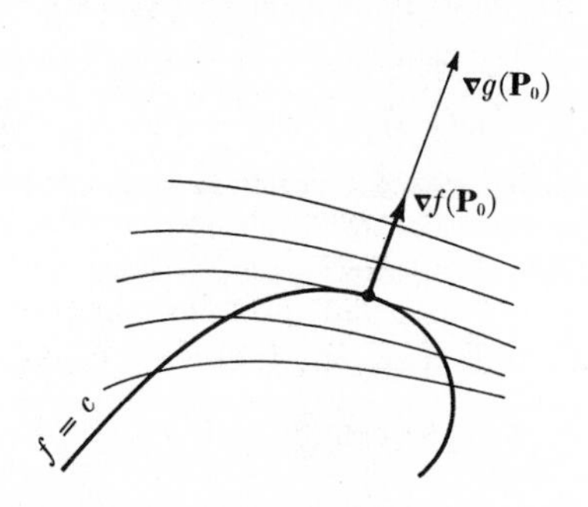

FIGURE 3.46

Proof. Since the vector $\nabla f(\mathbf{P}_0) \neq \mathbf{0}$, at least one of its components is not zero; say $f_x(\mathbf{P}_0) \neq 0$. By the implicit function theorem, the contour line near $\mathbf{P}_0 = (x_0, y_0)$ has the form

$$\{(x,y): x = \psi(y)\}$$

where ψ is differentiable on an open interval about y_0, and $\psi(y_0) = x_0$. By assumption, the function

$$h(y) = g(\psi(y),y))$$

has a maximum at $y = y_0$, hence

$$0 = h'(y_0) = g_x(\mathbf{P}_0)\psi'(y_0) + g_y(\mathbf{P}_0). \tag{20}$$

To eliminate $\psi'(y_0)$, differentiate the identity $f(\psi(y),y) \equiv c$ by the chain rule, obtaining

$$f_x(\mathbf{P}_0)\psi'(y_0) + f_y(\mathbf{P}_0) = 0;$$

solve for ψ' and substitute in (20), obtaining

$$0 = -g_x(\mathbf{P}_0)\,\frac{f_y(\mathbf{P}_0)}{f_x(\mathbf{P}_0)} + g_y(\mathbf{P}_0). \tag{21}$$

If we set $\lambda = g_x(\mathbf{P}_0)/f_x(\mathbf{P}_0)$, then obviously

$$g_x(\mathbf{P}_0) = \lambda f_x(\mathbf{P}_0),$$

and from (21)

$$g_y(\mathbf{P}_0) = \lambda f_y(\mathbf{P}_0),$$

so $\nabla g(\mathbf{P}_0) = \lambda\nabla f(\mathbf{P}_0)$. *Q.E.D.*

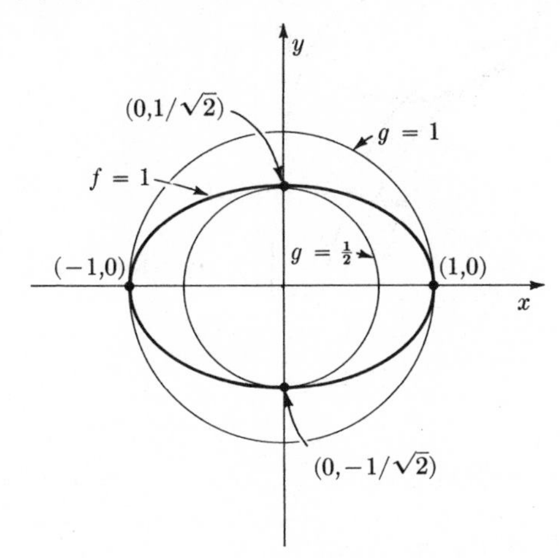

FIGURE 3.47

Example 4. Find the maxima and minima of $g(x,y) = x^2 + y^2$ on the ellipse $f(x,y) = x^2 + 2y^2 = 1$ in Fig. 47. *Solution.* Here

$$\nabla g = (2x,2y), \qquad \nabla f = (2x,4y).$$

If $\nabla g = \lambda\nabla f$, then $2x = \lambda\cdot 2x$ and $2y = \lambda\cdot 4y$, so either

$$x = 0, \qquad \lambda = \tfrac{1}{2}$$

or

$$y = 0, \qquad \lambda = 1.$$

Since we consider only the points on $x^2 + 2y^2 = 1$, we find for $x = 0$ that $y = \pm 1/\sqrt{2}$, and for $y = 0$ that $x = \pm 1$. Since $g(0, \pm 1/\sqrt{2}) = \frac{1}{2}$, and $g(\pm 1,0) = 1$, we conclude that *if* there is any maximum it occurs at $(\pm 1,0)$, and *if* there is a minimum it occurs at $(0,\pm 1/\sqrt{2})$. (In fact, since $x^2 + y^2$ is the square of the distance from (x,y) to the origin, it is intuitively obvious that 1 really is the maximum and $\frac{1}{2}$ really is the minimum.)

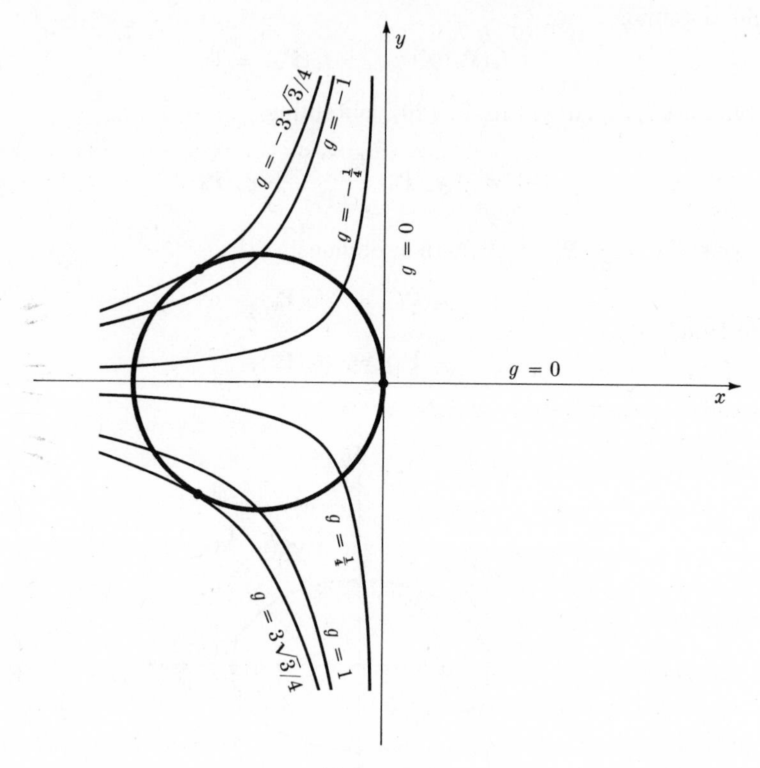

FIGURE 3.48

Example 5. Find the maximum of $g(x,y) = xy$ on the curve $f(x,y) = (x + 1)^2 + y^2 = 1$, *assuming* that such a maximum exists. *Solution.* $\nabla f(x,y) = (2x + 2,2y)$. Thus $\nabla f(\mathbf{P}) = \mathbf{0}$ only at $\mathbf{P} = (-1,0)$. Since $f(-1,0) = 0$, it follows that $\nabla f(\mathbf{P}) \neq \mathbf{0}$ at every point on the contour $f = 1$, so the Lagrange condition $\nabla g = \lambda \nabla f$ must be satisfied at the maximum point. Since $\nabla g(x,y) = (y,x)$, the two conditions $\nabla g = \lambda \nabla f$, $f = 1$ become

$$
\begin{aligned}
y &= \lambda \cdot (2x + 2) \\
x &= \lambda \cdot (2y) \\
(x + 1)^2 + y^2 &= 1.
\end{aligned}
\tag{22}
$$

In these equations, $y = 0 \Rightarrow x = 0 \Rightarrow \lambda = 0$; we thus find that the point $(0,0)$ satisfies the Lagrange condition. For any other solution of (22), $y \neq 0$; thus $\lambda = x/2y$, and

$$
y = \frac{x}{2y}(2x + 2), \qquad \text{or} \qquad y^2 = x^2 + x.
$$

Solving this simultaneously with $(x + 1)^2 + y^2 = 1$ gives

$$(x + 1)^2 + x^2 + x = 1 \quad \text{or} \quad 2x^2 + 3x = 0,$$

$$x = 0 \quad \text{or} \quad x = -3/2.$$

For $x = 0$ we get the point $(0,0)$ on the given curve $f = 1$, and for $x = 1$, and for $x = -3/2$ we get the two points $(-3/2, \pm\sqrt{3}/2)$. From the contour lines of g given in Fig. 48 it is clear that $(-3/2, -\sqrt{3}/2)$ gives a maximum, $(-3/2, \sqrt{3}/2)$ gives a minimum, and $(0,0)$ gives neither.

PROBLEMS

1. Find a function φ as guaranteed by the implicit function theorem, in the following cases:
(a) $xy = 1$, $(x_0, y_0) = (1,1)$
(b) $x^2 + 2y^2 = 3$, $(x_0, y_0) = (-1,1)$
(c) $x - y^2 = 0$, $(x_0, y_0) = (1,-1)$
(d) $xe^{xy} = 2$, $(x_0, y_0) = (1, \log 2)$

2. For each part of Problem 1, find a rectangle R as guaranteed by the implicit function theorem. (Hint: Sketch the contour line first.)

3. In each of the following problems, find dy/dx in general, and evaluate dy/dx as an explicit number when $x = x_0$.
(a) $(x + y)e^{xy} = 2e$, $(x_0, y_0) = (1,1)$
(b) $x^3 + 5y^5 + 7xy + y = 1$, $(x_0, y_0) = (1,0)$
(c) $x^2 - xy + y^2 = 3$, $(x_0, y_0) = (1,2)$
(d) $x \cos xy = 0$, $(x_0, y_0) = (1, \pi/2)$
(e) $x^5 + y^5 + xy = -4$, $(x_0, y_0) = (2,-2)$

4. Find a rectangle R as guaranteed by Theorem 8 for the equation in Problem 3(a). (Hint: Show that $|f_x/f_y| \le 8$ when $|x - 1| < 1/8$ and $|y - 1| < 1$, and apply Lemma 1.)

5. Find a rectangle R as guaranteed by Theorem 8 for the equation in Problem 3(b). (Hint: You can show that $|f_x/f_y| \le 20$ when $|x - 1| < 1$ and $|y| < 1$.)

6. Show that in the following cases there is *no* rectangle about the origin

$$R = \{(x,y) : |x| < \delta,\ |y| < m\delta\} \quad (\text{with } \delta > 0)$$

such that the part of the contour line $f = 0$ lying in R is the graph of a function of x on $(-\delta,\delta)$, or the graph of a function of y on $(-m\delta, m\delta)$.
(a) $f(x,y) = xy$
(b) $f(x,y) = x^2 + y^2$

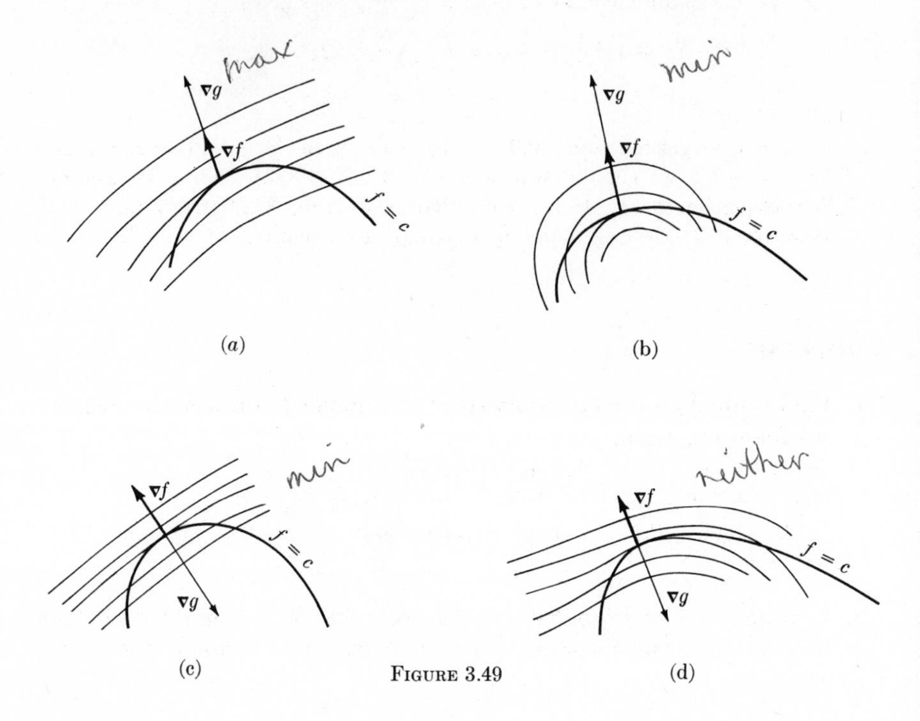

FIGURE 3.49

7. Suppose that $\nabla f(\mathbf{P}_0) \neq \mathbf{0}$ and $f(\mathbf{P}_0) = c$; thus, near $\mathbf{P}_0$, the contour line $f = c$ is the graph of a function $\varphi(x)$ or a function $\psi(y)$. Show that in either case the tangent line to the graph at (x_0, y_0) is orthogonal to $\nabla f(\mathbf{P}_0)$. (Thus, the gradient is orthogonal to the tangent line.)

8. Let g be a function of one variable. Show that near any point $x_0 = g(y_0)$ where $g'(y_0) \neq 0$ there is an inverse function φ such that $g(\varphi(x)) = x$, and $\varphi'(x) = 1/g(\varphi(x))$. (Hint: Let $f(x,y) = x - g(y)$. *Remark:* In this problem you are deducing part of the *inverse function* theorem for one variable from the *implicit function* theorem for two variables.)

9. Use Lagrange multipliers to find maxima of the function g on the given contour line of f. (Assume in each case that a maximum exists.)

(a) $g(x,y) = 3x^2 + 2\sqrt{2}\,xy + 4y^2$ on $x^2 + y^2 = 9$

(b) $g(x,y) = x^2 + y^2$ on $5x^2 + 6xy + 5y^2 = 16$

(c) $g(x,y) = xy$ on $x + y = 1$

(Notice that in part (b) you are finding the points on the given contour line which are farthest from the origin.)

10. Let F be a differentiable function of one variable, and set

$$f(x,y) = F(x) - y \qquad \text{and} \qquad g(x,y) = y.$$

Show that the Lagrange condition

$$\nabla g(x,y) = \lambda \nabla f(x,y)$$

is equivalent to $F'(x) = 0$. (Thus the familiar necessary condition for a local maximum or minimum of a function of one variable is a special case of the Lagrange multiplier condition.)

11. Figure 49 shows a variety of points where the Lagrange condition $\nabla g = \lambda \nabla f$ is satisfied. Which (if any) are maxima of g on $f = c$, which are minima, and which are neither? (The light lines are contour lines of g.)

3.6 DERIVATIVES OF HIGHER ORDER

Our study of a function f, based up to now on the first derivatives $D_1 f$ and $D_2 f$, can be extended by taking derivatives of higher order. The first derivative

$$D_1 f(a,b) = \lim_{h \to 0} \frac{f(a + h,b) - f(a,b)}{h} \tag{1}$$

defines a function $D_1 f$ whose domain is the set of points (a,b) where the limit (1) exists. The derivatives of this function $D_1 f$ are denoted $D_1 D_1 f$ and $D_2 D_1 f$, and by a variety of other notations as well:

$$D_1 D_1 f = D_1^2 f = \frac{\partial}{\partial x}\left(\frac{\partial f}{\partial x}\right) = \frac{\partial^2 f}{\partial x^2} = f_{xx},$$

$$D_2 D_1 f = \frac{\partial}{\partial y}\left(\frac{\partial f}{\partial x}\right) = \frac{\partial^2 f}{\partial y \partial x} = f_{xy}.$$

Similarly, we can differentiate $D_2 f$, obtaining

$$D_1 D_2 f = \frac{\partial^2 f}{\partial x \partial y} = f_{yx}$$

and

$$D_2 D_2 f = D_2^2 f = \frac{\partial^2 f}{\partial y^2} = f_{yy}.$$

These are the *second order* partial derivatives of f. For example, if $f(x,y) = e^{xy}$, we find

$$f_x = ye^{xy}, \qquad f_y = xe^{xy},$$

so the derivatives of f_x are

$$f_{xx} = y^2e^{xy}, \qquad f_{xy} = e^{xy} + xye^{xy},$$

and the derivatives of f_y are

$$f_{yx} = e^{xy} + xye^{xy}, \qquad f_{yy} = x^2e^{xy}.$$

If $g(x,y) = x \cos y$, then

$$\begin{aligned} g_x &= \cos y, & g_y &= -x \sin y \\ g_{xx} &= 0, & g_{yx} &= -\sin y \\ g_{xy} &= -\sin y, & g_{yy} &= -x \cos y \\ g_{xyx} &= 0, & g_{yxx} &= 0. \end{aligned}$$

Notice that $f_{xy} = f_{yx}$, and $g_{xy} = g_{yx}$. This is not just a coincidence.

Theorem 10. *If f_x, f_y, f_{xy}, and f_{yx} all exist in an open set containing $\mathbf{P}_0$, and if f_{xy} and f_{yx} are continuous at $\mathbf{P}_0$, then $f_{xy}(\mathbf{P}_0) = f_{yx}(\mathbf{P}_0)$.*

The proof depends on obtaining f_{xy} and f_{yx} as limits of difference quotients; it turns out that both are limits of the *same* quotient, hence they are equal. We begin by investigating f_{yx}.

By definition (1), $f_{yx}(a,b)$ is approximated by the difference quotient

$$\frac{f_y(a+h,b) - f_y(a,b)}{h}. \tag{2}$$

Going further, in (2) we can approximate $f_y(a,b)$ by

$$\frac{f(a,b+h) - f(a,b)}{h}$$

and $f_y(a+h,b)$ by

$$\frac{f(a+h,b+h) - f(a+h,b)}{h}.$$

Putting these approximations in (2), we get the "mixed second difference quotient"

$$Q(h) = \frac{f(a+h,b+h) - f(a+h,b) - f(a,b+h) + f(a,b)}{h^2}. \tag{3}$$

After all these approximations, it is not entirely clear that Q bears any relation to the second derivative $f_{yx}(a,b)$. However:

Lemma 1. *If f_y and f_{yx} exist in a disk centered at (a,b), and f_{yx} is continuous at (a,b), then*

$$\lim_{h\to 0} Q(h) = f_{yx}(a,b).$$

Proof. If we define the function of one variable

$$F(y) = f(a + h,y) - f(a,y),$$

then

$$Q(h) = \frac{F(b + h) - F(b)}{h^2}.$$

By the mean value theorem,

$$F(b + h) - F(b) = hF'(b + \theta h) \qquad 0 < \theta < 1$$
$$= h[f_y(a + h,b + \theta h) - f_y(a,b + \theta h)];$$

hence

$$Q(h) = \frac{f_y(a + h,b + \theta h) - f_y(a,b + \theta h)}{h}.$$

Applying the Mean Value theorem now to the function $G(x) = f_y(x,b + \theta h)$, we get $G(a + h) - G(a) = hG'(a + \theta' h)$, $0 < \theta' < 1$; hence

$$Q(h) = f_{yx}(a + \theta' h,b + \theta h), \qquad 0 < \theta < 1, \quad 0 < \theta' < 1.$$

Since f_{yx} is continuous at (a,b) we get $\lim_{h\to 0} Q(h) = f_{yx}(a,b)$. *Q.E.D.*

Now if you construct a mixed second difference quotient to approximate f_{xy} instead of f_{yx}, you will get exactly the same expression Q as in (3). Hence, reversing the roles of x and y, you can prove:

Lemma 2. *If f_x and f_{xy} exist in a disk centered at (a,b), and f_{xy} is continuous at (a,b), then*

$$\lim_{h\to 0} Q(h) = f_{xy}(a,b).$$

Returning to Theorem 10, the hypotheses there guarantee that Lemmas 1 and 2 are both valid; hence $f_{yx}(a,b) = \lim_{h\to 0} Q(h) = f_{xy}(a,b)$. *Q.E.D.*

Problem 18(b) below shows that $f_{xy}(a,b)$ may actually be different from $f_{yx}(a,b)$ if these partial derivatives are *not* continuous at (a,b); thus we cannot drop the hypothesis of continuity at (a,b).

What we have proved is essentially a commutative law: $D_1D_2 = D_2D_1$. Like most commutative laws, it extends to products of more than two terms. Precisely: *In any mixed partial derivative, the order of differentiation is irrelevant as long as all the derivatives that occur are continuous in an open set containing the point where the derivative is taken.* For example, if f_{xx}, f_{xy}, f_{yx}, f_{xxy}, f_{xyx}, and f_{yxx} are all continuous in an open set U, then Theorem 10 applies at every point in U, so $f_{xy} = f_{yx}$ throughout U; differentiating this with respect to x shows that $f_{xyx} = f_{yxx}$ throughout U. Further, applying

Theorem 10 to f_x instead of f, we get $f_{xyx} = f_{xxy}$; thus

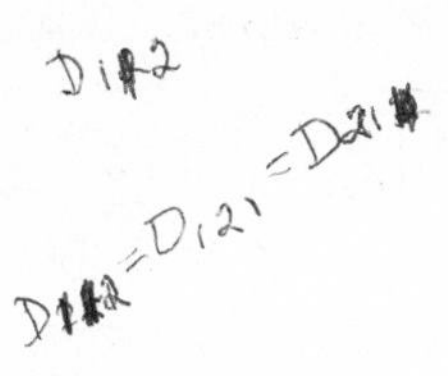

$$f_{yxx} = f_{xyx} = f_{xxy} ,$$

or

$$D_1{}^2D_2f = D_1D_2D_1f = D_2D_1{}^2f.$$

PROBLEMS

Problems 1 and 2 are just routine differentiation. The others fall into three groups: Problems 3–8 concern the *wave equation* $f_{xx} = c^2f_{yy}$, which governs the motion of a vibrating string (as in a violin, guitar, piano, etc.). Problems 9–14 concern the *Laplace equation* $f_{xx} + f_{yy} = 0$, which arises in the study of temperature distribution. The last four are related directly to Theorem 10 and its proof.*

1. Let $f(x,y) = x \cos (x + y)$. Compute

(a) f_{xy} (c) $D_1{}^2D_2 f$ (e) $\dfrac{\partial^3 f}{\partial x \partial y \partial x}$

(b) f_{yx} (d) $D_2D_1{}^2f$

2. Define a function f of your choice, and show by direct calculation that $f_{xy} = f_{yx}$, $f_{xyy} = f_{yxy}$ wherever these derivatives are all continuous.

3. Prove that the following functions satisfy $\dfrac{\partial^2 f}{\partial x^2} = \dfrac{\partial^2 f}{\partial y^2}$.

(a) $\cos (x - y)$ (d) $\cos x \sin y$
(b) $\sin(x + y)$ (e) $\cos 3x \sin 3y$
(c) $e^{x+y} + e^{(x-y)^2}$ (f) $\cos 2x \sin 2y$

4. Prove that the following functions satisfy $f_{xx} = c^2f_{yy}$ where c is a constant.
(a) $f(x,y) = \cos (cx - y)$
(b) $f(x,y) = \log (cx + y)$
(c) $f(x,y) = \cos (cx + y) + e^{cx-y}$

5. (a) Prove that if u and v are functions of one variable such that u'' and v'' exist, then $f(x,y) = u(x + y) + v(x - y)$ satisfies

$$\frac{\partial^2 f}{\partial x^2} = \frac{\partial^2 f}{\partial y^2}.$$

(b) With u and v as in part (a), prove that $f(x,y) = u(cx + y) + v(cx - y)$ satisfies

$$\frac{\partial^2 f}{\partial x^2} = c^2 \frac{\partial^2 f}{\partial y^2}.$$

---

* You can also do Problem 10(a), page 750.

6. (a) Show that every first degree polynomial f satisfies

$$\frac{\partial^2 f}{\partial x^2} = \frac{\partial^2 f}{\partial y^2}.$$

(b) Show that a homogeneous second degree polynomial $f(x,y) = Ax^2 + Bxy + Cy^2$ satisfies

$$\frac{\partial^2 f}{\partial x^2} = \frac{\partial^2 f}{\partial y^2}$$

if and only if $A = C$.

7. Define an operator $D_1 + D_2$ by $(D_1 + D_2)f = f_x + f_y$, and similarly $(D_1 - D_2)f = f_x - f_y$. Prove that $(D_1 - D_2)(D_1 + D_2) = D_1^2 - D_2^2$, or in other words, $(D_1 - D_2)((D_1 + D_2)f) = D_1^2 f - D_2^2 f$ for every function f having continuous second derivatives.

8. Suppose that f has continuous partial derivatives of order ≤ 2, and that $f_{xx} = f_{yy}$ at every point in R^2. This problem shows that there are functions u and v such that $f(x,y) = u(x - y) + v(x + y)$. (This is a converse of Problem 5(a).)

(a) Define $g(s,t) = f(s + t, s - t)$. Prove that

$$D_2\, g(s,t) = D_1 f(s + t, s - t) - D_2 f(s + t, s - t)$$

or, in slightly sloppy notation, $g_t = f_x - f_y$. (Apply the chain rule, §3.4.)

(b) Prove that $D_1 g(s,t) = D_1 f(s + t, s - t) + D_2 f(s + t, s - t)$, or $g_s = f_x + f_y$.

(c) Prove that $g_{st} = f_{xx} - f_{yy}$.

(d) Suppose $h(s,t)$ is defined for all s and t, and $h_t \equiv 0$. Prove that there is a function $H(s)$ such that $h(s,t) \equiv H(s)$. Prove that H is continuous if h is.

(e) Suppose that $f_{xx} = f_{yy}$, and g is defined as in part (a). Prove that $g_s = H(s)$ for some continuous function H. (Use parts (c) and (d).)

(f) Continuing part (e), prove that $g(s,t) = g(0,t) + \int_0^s H(r)\, dr$.

(g) Continuing part (f), prove that there are functions u and v such that $f(x,y) = u(x - y) + v(x + y)$.

$$\left(\text{Hint:} \quad \left.\begin{matrix} x = s + t \\ y = s - t \end{matrix}\right\} \Leftrightarrow \left\{\begin{matrix} s = \frac{1}{2}(x + y) \\ t = \frac{1}{2}(x - y) \end{matrix}\right.\right)$$

9. (a) Show that every first degree polynomial f satisfies Laplace's equation

$$\frac{\partial^2 f}{\partial x^2} + \frac{\partial^2 f}{\partial y^2} = 0.$$

(b) Show that a homogeneous second degree polynomial $f(x,y) = Ax^2 + Bxy + Cy^2$ satisfies Laplace's equation if and only if $A + C = 0$.

(c) Prove that every second degree polynomial f satisfying

$$\frac{\partial^2 f}{\partial x^2} + \frac{\partial^2 f}{\partial y^2} = 0$$

is a linear combination of the functions 1, x, y, $x^2 - y^2$, and xy, i.e. $f(x,y) = c_1 + c_2x + c_3y + c_4(x^2 - y^2) + c_5xy$ for some constants $c_1, \ldots, c_5$.

10. Which of the following functions satisfy Laplace's equation? (Three do and two do not.)

(a) $e^{-y} \cos x$ (b) $e^{-2x} \cos 2y$

(c) $\sin y \cos x$ (d) $e^{-2x} \cos 3y$

(e) $\log (x^2 + y^2)$ (Disregard the point $(x,y) = \mathbf{0}$, where the function is not defined.)

11. This problem concerns Laplace's equation in polar coordinates. Suppose that all the second derivatives of f are continuous, and define a function g by $g(r,\theta) = f(r \cos \theta, r \sin \theta)$.

(a) Prove that

$$g_r(r,\theta) = \cos \theta D_1 f(r \cos \theta, r \sin \theta) + \sin \theta D_2 f(r \cos \theta, r \sin \theta).$$

(Hint: Apply the chain rule, §3.4.)

(b) Prove that

$$g_\theta(r,\theta) = -r \sin \theta D_1 f(r \cos \theta, r \sin \theta) + r \cos \theta D_2 f(r \cos \theta, r \sin \theta).$$

(c) Find corresponding expressions for g_{rr} and $g_{\theta\theta}$, and prove that for $r \neq 0$,

$$g_{rr} + \frac{1}{r} g_r + \frac{1}{r^2} g_{\theta\theta} = f_{xx} + f_{yy}.$$

(You will need Theorem 10.)

12. Suppose that $f(x,y)$ is a function of $r = \sqrt{x^2 + y^2}$ only, that is, $f(r \cos \theta, r \sin \theta) = G(r)$ for some function G. Suppose that G has a continuous second derivative for $r > 0$.

(a) Using Problem 11(c), prove that if $f_{xx} + f_{yy} = 0$, then $rG'' = -G'$.

(b) Prove that $G' = A/r$ for some constant A.

(c) Prove that $G = A \log r + B$ for constants A and B.

13. Suppose that $f(r \cos \theta, r \sin \theta) = G(r) \cos \theta$.
(a) Using Problem 11(c), prove that

$$f_{xx} + f_{yy} = \left(G'' + \frac{1}{r} G' - \frac{1}{r^2} G\right) \cos \theta.$$

(b) Prove that $(Ar^{-1} + Br) \cos \theta$ satisfies Laplace's equation.
(c) Prove that

$$A \frac{\cos \theta}{r} = D_1 F(r \cos \theta, r \sin \theta),$$

where $F(x,y) = A \log \sqrt{x^2 + y^2}$.

14. Suppose that $f(r \cos \theta, r \sin \theta) = G(r)(a \cos n\theta + b \sin n\theta)$. Prove that $f_{xx} + f_{yy} = 0$ if $G(r) = Ar^{-n} + Br^n$ for some constants A and B.

15. Derive a second difference quotient approximating $f_{xy}(a,b)$, and show that it is the same as the quotient $Q(h)$ in (3) above that approximates $f_{yx}(a,b)$.

16. Prove Lemma 2. (Hint: Begin by considering the function $F(x) = f(x, b + h) - f(x,b)$.)

17. (a) Obtain the second difference quotient

$$Q_{hh} = \frac{f(a + 2h,b) - 2f(a + h,b) + f(a,b)}{h^2}$$

as an approximation to $f_{xx}(a,b)$.
(b) Suppose that f_{xx} is continuous at (a,b); prove that

$$\lim_{h \to 0} Q_{hh} = f_{xx}(a,b).$$

18. Let

$$f(x,y) = \frac{xy(x^2 - y^2)}{x^2 + y^2}$$

if $x^2 + y^2 \neq 0$, and $f(0,0) = 0$.
(a) Compute f_x and f_y, and show that they are continuous. (To compute $f_x(0,0)$, you will have to use the defining formula (1); similarly for $f_y(0,0)$.)
(b) Compute $f_{xy}(0,0)$ and $f_{yx}(0,0)$; note that they are *not* equal, but that this does not contradict Theorem 10.
(Notice that f is homogeneous of degree 2, in the sense explained in Problem 7, §3.4. Homogeneous functions provide many other curious examples, for instance the function $h(x,y) = xy(x^2 + y^2)^{-1}$ in §3.3, which has derivatives everywhere, but is not continuous at the origin.)

3.7 THE TAYLOR EXPANSION

The Taylor expansion is a way of approximating a given function near a given point by polynomials. Thanks to the chain rule and Theorem 10 on mixed partials, the expansion for functions of two variables is obtained easily from the one-variable version.

Suppose that f is defined throughout a disk $D = \{\mathbf{P}: |\mathbf{P} - \mathbf{A}| < r\}$ centered at the point $\mathbf{A} = (a,b)$, and that all the partial derivatives we write down are continuous in D. Given a vector $\mathbf{H} = (h,k)$ with $|\mathbf{H}| < r$, the point $\mathbf{A} + \mathbf{H}$ lies in D; we want to expand $f(\mathbf{A} + \mathbf{H})$ as a polynomial in h and k, plus a remainder term. To do this, look at the function

$$g(t) = f(\mathbf{A} + t\mathbf{H}), \qquad 0 \le t \le 1, \tag{1}$$

which gives the values of f along the straight line segment from $\mathbf{A}$ to $\mathbf{A} + \mathbf{H}$. By the mean value theorem, $g(1) - g(0) = (1 - 0)g'(\theta)$ with $0 < \theta < 1$, or

$$g(1) = g(0) + g'(\theta), \qquad 0 < \theta < 1. \tag{2}$$

By (1), $g(0) = f(\mathbf{A})$ and $g(1) = f(\mathbf{A} + \mathbf{H})$, and by the chain rule, $g'(t) = \mathbf{H}\cdot\nabla f(\mathbf{A} + t\mathbf{H})$, so (2) becomes

$$f(\mathbf{A} + \mathbf{H}) = f(\mathbf{A}) + \mathbf{H}\cdot\nabla f(\mathbf{A} + \theta\mathbf{H}), \qquad 0 < \theta < 1. \tag{3}$$

This is the *Mean Value theorem for functions of two variables.*

Formula (3) is the beginning of the Taylor expansion of f. To expand further, take another term in the expansion of g,

$$g(t) = g(0) + tg'(0) + \frac{t^2}{2!}g''(\theta t), \qquad 0 < \theta < 1,$$

then set $t = 1$, obtaining

$$g(1) = g(0) + g'(0) + \frac{1}{2!}g''(\theta), \qquad 0 < \theta < 1. \tag{4}$$

To rewrite this result in terms of f, we have to differentiate the function

$$\begin{aligned} g'(t) &= \mathbf{H}\cdot\nabla f(\mathbf{A} + t\mathbf{H}) \\ &= hD_1 f(a + th, b + tk) + kD_2 f(a + th, b + tk). \end{aligned} \tag{5}$$

By the chain rule,

$$\frac{d[D_1 f(a + th,b + tk)]}{dt} = hD_1D_1 f(a + th,b + tk) + kD_2D_1 f(a + th,b + tk),$$

and similarly

$$\frac{d[D_2 f]}{dt} = hD_1D_2 f + kD_2D_2 f.$$

Thus, differentiating (5) and recalling that $D_1D_2 = D_2D_1$, we find

$$g''(t) = h^2D_1^2f(\mathbf{A} + t\mathbf{H}) + 2hkD_1D_2f(\mathbf{A} + t\mathbf{H}) + k^2D_2^2f(\mathbf{A} + t\mathbf{H}).$$

Putting this in (4), we obtain the *first degree Taylor expansion of f with remainder:*

$$f(\mathbf{A} + \mathbf{H}) = f(\mathbf{A}) + hD_1f(\mathbf{A}) + kD_2f(\mathbf{A})$$
$$+ \frac{1}{2}h^2D_1^2f(\mathbf{A} + \theta\mathbf{H}) + hkD_1D_2f(\mathbf{A} + \theta\mathbf{H}) + \frac{1}{2}k^2D_2^2f(\mathbf{A} + \theta\mathbf{H}). \quad (6)$$

Example 1. If $f(x,y) = e^x \cos y$, then $f_x = e^x \cos y$, $f_y = -e^x \sin y$, $f_{xx} = e^x \cos y$, $f_{xy} = -e^x \sin y$, $f_{yy} = -e^x \cos y$. Hence (6) becomes

$$e^{a+h} \cos (b + k) = e^a \cos b + he^a \cos b + k(-e^a \sin b)$$
$$+ \tfrac{1}{2}h^2e^{a+\theta h} \cos (b + \theta k) - hke^{a+\theta h} \sin (b + \theta k)$$
$$- \tfrac{1}{2}k^2e^{a+\theta h} \cos (b + \theta k).$$

Formula (6) shows how to approximate f by a second degree polynomial, and how good the approximation is. Precisely:

Theorem 11. *If f has continuous partial derivatives of order ≤ 2 in the disk $D = \{\mathbf{P}: |\mathbf{P} - \mathbf{A}| < r\}$, then*

$$\lim_{\mathbf{H}\to 0} \frac{f(\mathbf{A} + \mathbf{H}) - f_2(\mathbf{H})}{|\mathbf{H}|^2} = 0, \quad (7)$$

where $f_2(\mathbf{H}) = f_2(h,k)$ is the second degree polynomial in h and k given by

$$f(\mathbf{A}) + hf_x(\mathbf{A}) + kf_y(\mathbf{A}) + \frac{h^2}{2}f_{xx}(\mathbf{A}) + hkf_{xy}(\mathbf{A}) + \frac{k^2}{2}f_{yy}(\mathbf{A}).$$

Proof. By (6),

$$\frac{f(\mathbf{A} + \mathbf{H}) - f_2(\mathbf{H})}{|\mathbf{H}|^2} = \frac{h^2}{2|\mathbf{H}|^2}[f_{xx}(\mathbf{A} + \theta\mathbf{H}) - f_{xx}(\mathbf{A})]$$
$$+ \frac{hk}{|\mathbf{H}|^2}[f_{xy}(\mathbf{A} + \theta\mathbf{H}) - f_{xy}(\mathbf{A})]$$
$$+ \frac{k^2}{2|\mathbf{H}|^2}[f_{yy}(\mathbf{A} + \theta\mathbf{H}) - f_{yy}(\mathbf{A})]. \quad (8)$$

Because the second derivatives of f are continuous, the three terms in square brackets have limit 0 as $\mathbf{H} \to \mathbf{0}$. Further,

$$\frac{h^2}{|\mathbf{H}|^2} \leq 1, \qquad \frac{k^2}{|\mathbf{H}|^2} \leq 1, \qquad \text{and} \qquad \frac{|hk|}{|\mathbf{H}|^2} \leq 1,$$

so the right-hand side of (8) has limit zero as $\mathbf{H} \to \mathbf{0}$, and the proof is done.

Formula (7) should be compared to the condition for differentiability of f at $\mathbf{A}$,

$$\lim_{\mathbf{H}\to 0} \frac{f(\mathbf{A}+\mathbf{H}) - f_1(\mathbf{H})}{|\mathbf{H}|} = 0,$$

where $f_1(\mathbf{H}) = f(\mathbf{A}) + \mathbf{H}\cdot\nabla f(\mathbf{A})$. In (7) there is a second degree polynomial f_2 instead of the first degree polynomial f_1, and a correspondingly closer approximation, since the denominator is $|\mathbf{H}|^2$ instead of $|\mathbf{H}|$.

Theorem 11 and its extension to polynomials of degree n is fundamental in the study of functions of several variables. The simplest application, concerning maxima and minima, is given in the next section.

It is easy, in principle, to extend Theorem 11 and obtain approximations by polynomials of degree higher than 2; the only difficulty is in the notation. We begin with the expansion

$$g(1) = \sum_{m=0}^{n-1} \frac{1}{m!} g^{(m)}(0) + \frac{1}{n!} g^{(n)}(\theta), \qquad 0 < \theta < 1, \tag{9}$$

where $g(t) = f(\mathbf{A} + t\mathbf{H})$ as before. To express the derivatives of g in terms of f and $\mathbf{H}$, we use the symbol

$$\mathbf{H}\cdot\nabla = (hD_1 + kD_2).$$

(This is the inner product of $\mathbf{H} = (h,k)$ with the "vector" $\nabla = (D_1, D_2)$.) By formula (5), differentiating a function $f(\mathbf{A} + t\mathbf{H})$ with respect to t is the same as "multiplying" f by $\mathbf{H}\cdot\nabla$; hence differentiating n times is the same as multiplying n times by $\mathbf{H}\cdot\nabla$, and (9) yields the nth degree Taylor expansion of f in the form

$$f(\mathbf{A}+\mathbf{H}) = \sum_{m=0}^{n-1} \frac{1}{m!}((\mathbf{H}\cdot\nabla)^m f)(\mathbf{A}) + \frac{1}{n!}((\mathbf{H}\cdot\nabla)^n f)(\mathbf{A}+\theta\mathbf{H}). \tag{10}$$

This looks less mysterious when we work out the expressions $(\mathbf{H}\cdot\nabla)^m f$:

$$(\mathbf{H}\cdot\nabla)f = hD_1 f + kD_2 f \tag{11}$$

$$(\mathbf{H}\cdot\nabla)^2 f = (\mathbf{H}\cdot\nabla)((\mathbf{H}\cdot\nabla)f) = hD_1((\mathbf{H}\cdot\nabla)f) + kD_2((\mathbf{H}\cdot\nabla)f);$$

substituting (11) for $(\mathbf{H}\cdot\nabla)f$, and noting that $D_1D_2 = D_2D_1$, we get

$$(\mathbf{H}\cdot\nabla)^2 f = h^2D_1^2 f + 2hkD_1D_2 f + k^2D_2^2 f.$$

Similarly,

$$\begin{aligned}(\mathbf{H}\cdot\nabla)^3 f &= hD_1((\mathbf{H}\cdot\nabla)^2 f) + kD_2((\mathbf{H}\cdot\nabla)^2 f)\\ &= h^3D_1^3 f + 3h^2kD_1^2D_2 f + 3hk^2D_1D_2^2 f + k^3D_2^3 f,\end{aligned}$$

and generally we obtain the binomial expansion of $(hD_1 + kD_2)^m f$:

$$\begin{aligned}(\mathbf{H}\cdot\nabla)^m f &= h^mD_1^m f + mh^{m-1}kD_1^{m-1}D_2 f + \cdots + k^mD_2^m f\\ &= \sum_{l=0}^{m} \frac{m!}{l!(m-l)!} h^l k^{m-l} D_1^l D_2^{m-l} f.\end{aligned} \tag{12}$$

Putting this in (10) gives an alternate form of the Taylor expansion,

$$f(\mathbf{A} + \mathbf{H}) = \sum_{m=0}^{n-1} \sum_{l=0}^{m} \frac{1}{l!(m-l)!} h^l k^{m-l} (D_1{}^l D_2{}^{m-l} f)(\mathbf{A})$$

$$+ \sum_{l=0}^{n} \frac{1}{l!(n-l)!} h^l k^{n-l} (D_1{}^l D_2{}^{n-l} f)(\mathbf{A} + \theta\mathbf{H}) \qquad 0 < \theta < 1. \qquad (13)$$

From the Taylor expansion (10) or (13) follows

Theorem 12. *Suppose that f is defined in the disk $D = \{\mathbf{P}: |\mathbf{P} - \mathbf{A}| < r\}$, and that all the partial derivatives of f of order $\leq n$ are continuous in D. Let*

$$f_n(\mathbf{H}) = \sum_{m=0}^{n} \frac{1}{m!} (\mathbf{H} \cdot \nabla)^m f(\mathbf{A}). \qquad (14)$$

Then

$$\lim_{|\mathbf{H}| \to 0} \frac{f(\mathbf{A} + \mathbf{H}) - f_n(\mathbf{H})}{|\mathbf{H}|^n} = 0. \qquad (15)$$

The proof is just like Theorem 11; the details are left to you.

Example 2. Taking $f(x,y) = e^x \cos y$ as in Example 1, we find the third degree polynomial approximation

$$\begin{aligned} f_3(h,k) &= e^a \cos b + he^a \cos b - ke^a \sin b \\ &\quad + \tfrac{1}{2}(h^2 e^a \cos b - 2hke^a \sin b - k^2 e^a \cos b) \\ &\quad + \tfrac{1}{6}(h^3 e^a \cos b - 3h^2 k e^a \sin b - 3hk^2 e^a \cos b + k^3 e^a \sin b). \end{aligned}$$

According to Theorem 12, then,

$$\lim_{(h,k) \to 0} \frac{e^{a+h} \cos (b + k) - f_3(h,k)}{(h^2 + k^2)^{3/2}} = 0.$$

In particular, taking $\mathbf{A} = (a,b)$ to be the origin, we get for our approximation to $e^h \cos k$ the polynomial

$$f_3(h,k) = 1 + h + \tfrac{1}{2}h^2 - \tfrac{1}{2}k^2 + \tfrac{1}{6}h^3 - \tfrac{1}{2}hk^2.$$

(You would obtain the same result by multiplying the one-variable expansions of e^h and $\cos k$, and neglecting all terms of degree > 3.)

PROBLEMS

1. Find the second degree Taylor polynomial f_2 for the given function f at the given point $\mathbf{A}$.

(a) $f(x,y) = \sin(xy)$, $\mathbf{A} = (0,0)$
(b) $f(x,y) = \sin(xy)$, $\mathbf{A} = (\pi,1)$
(c) $f(x,y) = e^{-x^2-y^2}$, $\mathbf{A} = (0,0)$

2. Find the third degree Taylor polynomial

$$\sum_{m=0}^{3} \frac{1}{m!} (\mathbf{H}\cdot\nabla)^m f(\mathbf{A})$$

for the functions f and points $\mathbf{A}$ given in Problem 1.

3. (a) Find the second degree Taylor polynomial at $\mathbf{A} = \mathbf{0}$ for the function

$$f(x,y) = \begin{cases} \dfrac{\sin xy}{x}, & x \neq 0 \\ y, & x = 0. \end{cases}$$

(b) Take the third degree Taylor polynomial $t - t^3/6$ for $\sin t$, substitute $t = hk$, divide by x, and compare the result with the answer in part (a); notice that the difference is a polynomial containing no terms of degree < 2.

4. Prove that when f is a second degree polynomial and $\mathbf{A}$ is any given point, then $f = f_2$, where f_2 is the second degree Taylor polynomial of f at the point $\mathbf{A}$. (Hint 1: You can use formula (6), and notice that the second derivatives of f are all constant. Hint 2: Alternatively, you can use formula (10) with $n = 3$, and notice that the third derivatives of f are all zero.)

5. Suppose that the second derivatives of f are all constant on R^2. Prove that f is a second degree polynomial of degree < 2. (Hint: Use (6).)

6. Suppose that the nth derivatives of f are all identically zero on R^2. Prove that f is a polynomial of degree $< n$.

7. Prove that formula (3) in the text is valid whenever f is differentiable at every point of the closed segment from $\mathbf{A}$ to $\mathbf{A} + \mathbf{H}$.

8. (a) Suppose f is differentiable at every point of a disk D, and that $|\nabla f| \leq M$ in D, where M is a constant. Prove that when $\mathbf{A}$ and $\mathbf{A} + \mathbf{H}$ are in D, then $|f(\mathbf{A} + \mathbf{H}) - f(\mathbf{A})| \leq M|\mathbf{H}|$. (Hint: The segment from $\mathbf{A}$ to $\mathbf{A} + \mathbf{H}$ lies in D, so Problem 7 applies.)

(b) A set S is *convex* if, whenever it contains two points $\mathbf{P}$ and $\mathbf{Q}$, it contains the segment between them. Suppose f is differentiable in a convex set S, and $|\nabla f| \leq M$ there. Prove that for all points $\mathbf{P}$ and $\mathbf{Q}$ in S,

$$|f(\mathbf{P}) - f(\mathbf{Q})| \leq M\cdot|\mathbf{P} - \mathbf{Q}|.$$

9. (a) Suppose f is differentiable at $\mathbf{A}$ and $|f(\mathbf{A}+\mathbf{H}) - f(\mathbf{A})| \leq M|\mathbf{H}|$ for some constant M. Prove that $\nabla f(\mathbf{A}) \leq M$. (Hint: By the chain rule,

$$\nabla f(\mathbf{A}) \cdot \mathbf{U} = \lim_{t \to 0} \frac{f(\mathbf{A} + t\mathbf{U}) - f(\mathbf{A})}{t}$$

for every constant vector $\mathbf{U}$. Show that $|\nabla f(\mathbf{A}) \cdot \mathbf{U}| \leq M|\mathbf{U}|$, then set $\mathbf{U} = \nabla f(\mathbf{A})$.)

(b) From part (a) and Problem 8(b), show that *if f is differentiable at every point of an open convex set S, then the following conditions* (i) *and* (ii) *are equivalent:*

(i) $|\nabla f(\mathbf{P})| \leq M$ at every point $\mathbf{P}$ in S

(ii) $|f(\mathbf{P}) - f(\mathbf{Q})| \leq M|\mathbf{P} - \mathbf{Q}|$ for every pair of points $\mathbf{P}$ and $\mathbf{Q}$ in S.

10. Prove Theorem 12.

3.8 MAXIMA AND MINIMA

Suppose that f is defined in an open set containing $\mathbf{A}$. Then $\mathbf{A}$ is called a *local maximum of f* if $f(\mathbf{A}) \geq f(\mathbf{P})$ for all points $\mathbf{P}$ in some disk centered at $\mathbf{A}$.

Theorem 13. *If $\mathbf{A}$ is a local maximum, then the gradient* $\nabla f(\mathbf{A}) = \mathbf{0}$.

Proof. For any constant $\mathbf{H}$, the function

$$F(t) = f(\mathbf{A} + t\mathbf{H})$$

has a local maximum at $t = 0$, hence

$$0 = F'(0) = \mathbf{H} \cdot \nabla f(\mathbf{A}).$$

In particular, setting $\mathbf{H} = \nabla f(\mathbf{A})$, we get $0 = |\nabla f(\mathbf{A})|^2$. *Q.E.D.*

When $\nabla f(\mathbf{A}) = \mathbf{0}$, then $\mathbf{A}$ is called a *critical point* of f. We have shown that every local maximum occurs at a critical point; and it is easy to see that every local minimum also occurs at a critical point. Some critical points, however, are neither maxima nor minima. Consider the three examples

$$f(x,y) = -x^2 - y^2$$
$$g(x,y) = x^2 + y^2$$
$$h(x,y) = x^2 - y^2.$$

All three have a vanishing gradient at the point $\mathbf{A} = \mathbf{0}$, but f has a local *maximum* there, g has a local *minimum*, and h has neither; $h(0,0) = 0$, but in any disk about $\mathbf{0}$ there are points along the x axis where $h(x,y) = x^2 > 0$, and points along the y axis where $h(x,y) = -y^2 < 0$. The graph of h at this point looks like a saddle (see Fig. 50). Moving from "front to back," (0,0) is a minimum, but moving from "side to side" it is a maximum.

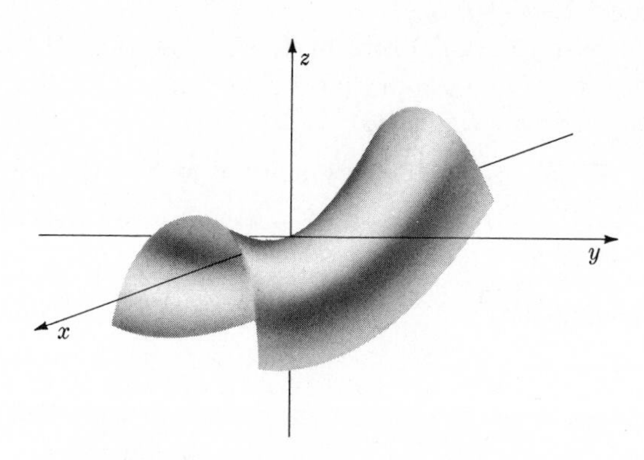

FIGURE 3.50 Graph of $h(x,y) = x^2 - y^2$ near the origin

The second derivatives of f usually distinguish between maxima, minima, and saddle points.

Theorem 14. *Suppose f has continuous derivatives of order ≤ 2 in a disk centered at* $\mathbf{A}$, *and that* $\nabla f(\mathbf{A}) = \mathbf{0}$. *Then* $\mathbf{A}$ *is*

(i) *a local maximum point of f if $f_{xx}f_{yy} > (f_{xy})^2$ and $f_{xx} + f_{yy} < 0$;*
(ii) *a local minimum point of f if $f_{xx}f_{yy} > (f_{xy})^2$ and $f_{xx} + f_{yy} > 0$;*
(iii) *a saddle point of f if $f_{xx}f_{yy} < (f_{xy})^2$.*

In the remaining case, where $f_{xx}f_{yy} = (f_{xy})^2$, the test fails. The proof, based on the Taylor expansion, is outlined in the problems below.

The three examples f, g, h given above are useful mnemonic aids for the three conditions in Theorem 14. It is obvious that f has a maximum, g a minimum, h a saddle point; and it is easy to check that f is in case (i), g in case (ii), and h in case (iii),

$$h_{xx}h_{yy} = -4 < (h_{xy})^2.$$

The proof of Theorem 14 will show that in the special case where f is a polynomial of degree 2, the criteria in (i) and (ii) give *absolute* maxima and minima; i.e. in case (i), $f(\mathbf{A}) \geq f(\mathbf{P})$ for all $\mathbf{P}$ in R^2, and in case (ii), $f(\mathbf{A}) \leq f(\mathbf{P})$ for all $\mathbf{P}$ in R^2.

Example 1. Let $f(x,y) = x^2 - 2xy + 2y^2 + x - 5$. We have $\nabla f(x,y) = (2x - 2y + 1, -2x + 4y)$, so $\nabla f = \mathbf{0}$ when

$$2x - 2y = -1$$
$$-2x + 4y = 0,$$

which gives $x = -1$, $y = -\frac{1}{2}$. Thus we apply Theorem 14 at the critical point $\mathbf{A} = (-1,-\frac{1}{2})$. We find

$$f_{xx} = 2, \qquad f_{xy} = -2, \qquad f_{yy} = 4,$$

so

$$f_{xx}f_{yy} = 8 > (f_{xy})^2 = 4, \qquad f_{xx} + f_{yy} = 6 > 0,$$

and we are in case (ii), the minimum-point case. This is easy to see from the contour graph in Fig. 51.

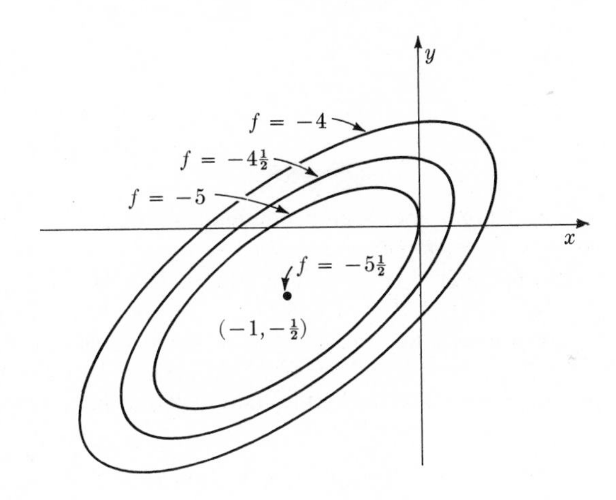

FIGURE 3.51 Contour graph of $f(x,y) = x^2 - 2xy + 2y^2 + x - 5$

Incidentally, the criteria in Theorem 14 can help in plotting the contour lines of quadratics; when there is a maximum or minimum, the contour lines are *ellipses* with center at the critical point $\mathbf{A}$ (as in Fig. 51). When there is a saddle point, the contours are *hyperbolas* with center at the critical point $\mathbf{A}$; the particular contour $\{\mathbf{P}: f(\mathbf{P}) = f(\mathbf{A})\}$ is a degenerate hyperbola, consisting of two lines that form the *asymptotes* of the other hyperbolas. (See Fig. 52.) When there are *no* critical points, the contour lines are parabolas, and their vertices all lie on a line where $|\nabla f|^2$ is minimum (See Fig. 53.) All this is proved in analytic geometry (without referring to critical points and second derivatives, of course).

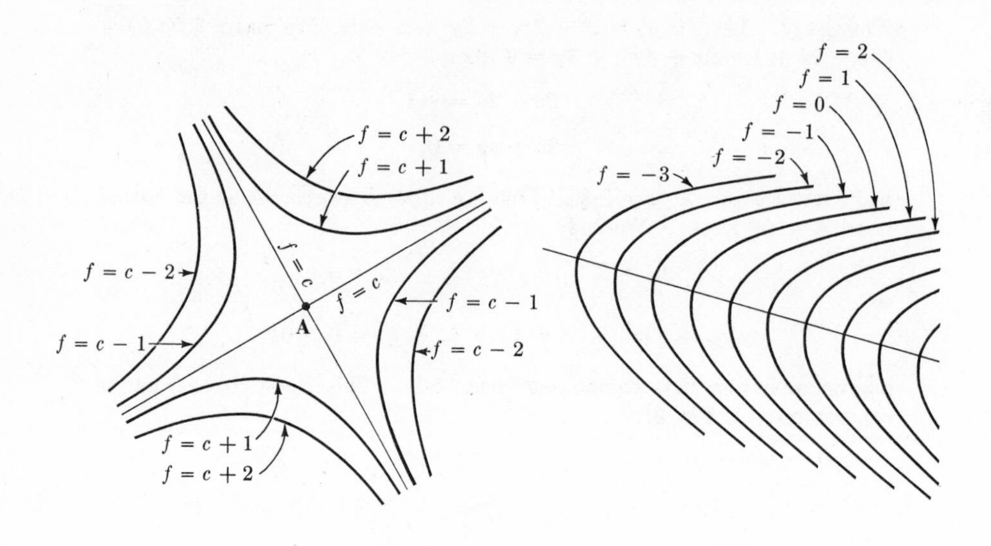

FIGURE 3.52 $c = f(\mathbf{A})$

FIGURE 3.53

Example 2. Find the maximum and minimum of $f(x,y) = xy - y + x - 1$ in the closed disk $D = \{\mathbf{P}: |\mathbf{P}| \leq 2\}$. *Solution.* We first look for a *local* maximum in the open disk $\{\mathbf{P}: |\mathbf{P}| < 2\}$. Since $\nabla f(x,y) = (y + 1, x - 1)$, the only critical point is at $x = 1$, $y = -1$. By applying Theorem 14 you can show that this is a saddle point. Hence there can be no maximum or minimum in the open disk; if they exist at all, the maximum and minimum must occur on the boundary. To check this, represent the boundary of D parametrically by

$$x = 2\cos\theta, \qquad y = 2\sin\theta,$$

and seek the maximum and minimum of

$$f(2\cos\theta, 2\sin\theta) = 4\cos\theta\sin\theta - 2\sin\theta + 2\cos\theta - 1.$$

By the usual one-variable methods we find that among all points on the circle $\{\mathbf{P}: |\mathbf{P}| = 2\}$, f assumes a *maximum* at

$$\left(\frac{1+\sqrt{3}}{\sqrt{2}}, \frac{\sqrt{3}-1}{\sqrt{2}}\right) \quad \text{and at} \quad \left(\frac{1-\sqrt{3}}{\sqrt{2}}, -\frac{\sqrt{3}+1}{\sqrt{2}}\right),$$

and a *minimum* at $(-\sqrt{2},\sqrt{2})$. Now we have shown that there is no maximum or minimum in the open disk, but that considering only the points on the boundary circle, f assumes a maximum of

$$f\left(\frac{1+\sqrt{3}}{\sqrt{2}}, \frac{\sqrt{3}-1}{\sqrt{2}}\right) = \sqrt{2},$$

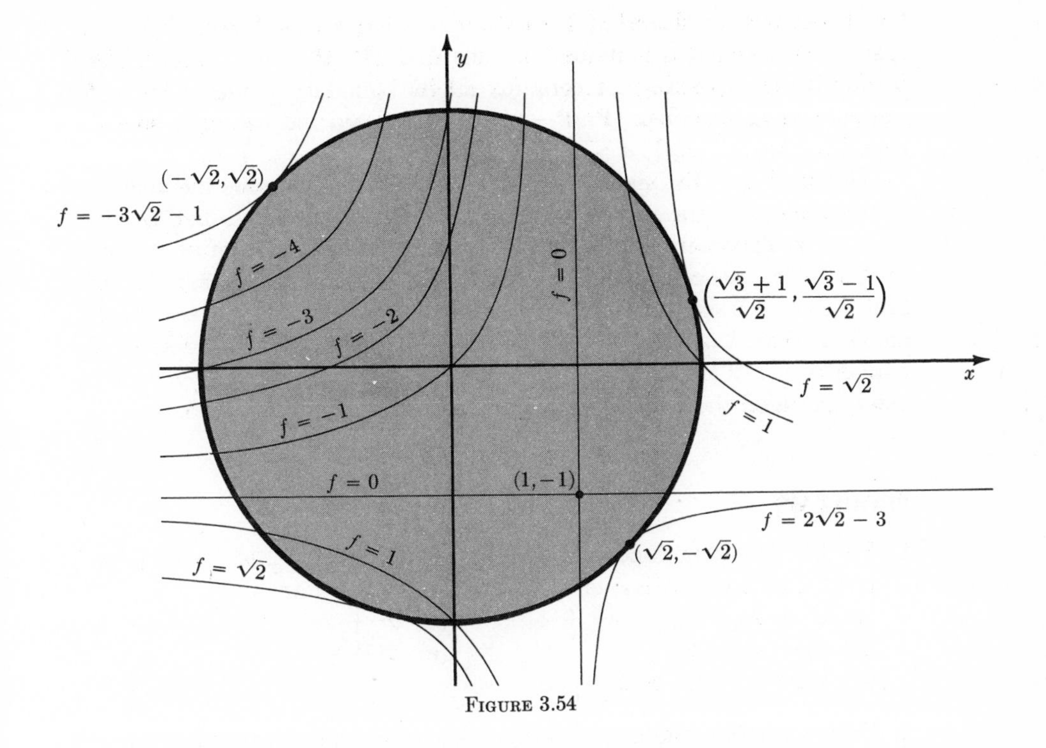

FIGURE 3.54

and a minimum of $f(-\sqrt{2},\sqrt{2}) = -3\sqrt{2} - 1$. It follows from the special nature of f that $\max_D f = \sqrt{2}$ and $\min_D f = -3\sqrt{2} - 1$, because every value assumed by f in D is actually assumed on the boundary. This is clear from Fig. 54, since every contour line having a point in D must intersect the boundary of D. (We omit the analytic proof.)

Remark 1. The function f in Example 2 satisfies *Laplace's equation* $f_{xx} + f_{yy} = 0$ for steady state temperature distribution. The property that the maximum and minimum of f are achieved on the boundary of D holds for *every* function f satisfying $f_{xx} + f_{yy} = 0$ in D.

Remark 2. The maximum and minimum on the boundary can be found by the method of Lagrange multipliers given in §3.5, instead of by parametrizing the boundary.

Remark 3. The discussion of maximum and minimum problems is simplified when it is known beforehand that a maximum and minimum exist. For functions of one variable, a function f that is continuous on a closed bounded interval I assumes a maximum and a minimum on I. For functions of two variables, there is an analogous *Maximum Value theorem: A function f that is continuous on a closed bounded set S in the plane assumes a maximum and a minimum on S.* The proof goes beyond the scope of this book, but we should at least describe what a closed, bounded set is. A set is *bounded* if it is contained in some disk $\{\mathbf{P}: |\mathbf{P}| \le r\}$. A set is *closed* if (intuitively speaking) it contains all its boundary points. Thus, for example, any closed disk $\{\mathbf{P}: |\mathbf{P} - \mathbf{A}| \le r\}$ is closed and bounded; so is any closed rectangle $\{(x,y): a \le x \le b, c \le y \le d\}$.

Returning to Example 2, we see by the maximum value theorem that f *must* have a maximum and minimum on the given closed disk D; since there is no maximum or minimum in the open disk, they must occur on the boundary, hence they are the points we found by considering the function $f(2\cos\theta, 2\sin\theta)$. Thus, the special argument at the end of Example 2 (based on the observation that the contour lines intersect the boundary of D if they intersect D at all) is replaced by a very general argument, based on the fact that a maximum and minimum must exist.

PROBLEMS

1. For the following functions, find all critical points, and for each such point decide which case of Theorem 14 applies (if any).

(a) $x^2 + 2x + 3y^2 + 4y + 1$
(b) $x^3 + y^3 - 3xy$
(c) $\log(3x^2 + y^2 + 2x + 10)$
(d) $x^2 + xy$
(e) $x^2 + y$

2. Decide whether the given contour line is an ellipse, a hyperbola, or a parabola. If it is an ellipse, find the center; if a hyperbola, find the center and asymptotes; if a parabola, find the axis. (See the remarks after Example 1.)

(a) $x^2 + xy = c$
(b) $x^2 + 2xy + y^2 + x - y = c$
(c) $y^2 - 2xy + x - y = c$
(d) $2x^2 + 2xy + 5y^2 + x - 2y = c$

3. Find the maximum and minimum of the following functions over the unit disk $D = \{\mathbf{P}: |\mathbf{P}| \le 1\}$.

(a) The function in Problem 1(d)
(b) The function in Problem 1(e)
(c) The function in Problem 1(c)

4. Suppose that f solves Laplace's equation $f_{xx} + f_{yy} = 0$, and that $\mathbf{A}$ is a critical point of f, i.e. $\nabla f(\mathbf{A}) = \mathbf{0}$. Prove that either (i) f_{xx}, f_{yy}, and f_{xy} all vanish at $\mathbf{A}$ or (ii) $\mathbf{A}$ is a *saddle point.*

5. (Method of least squares.) Given a number of points $\mathbf{P}_1 = (x_1, y_1), \ldots, \mathbf{P}_n = (x_n, y_n)$ in the plane, the problem arises of finding the equation $y = mx + b$ of a straight line which "most closely" passes through the given points. The usual solution is to take the equation $y = mx + b$ for which the following sum of squares is minimum:

$$S(m,b) = \sum_1^n (y - y_j)^2 = \sum_1^n (mx_j + b - y_j)^2.$$

(a) Show that if there are just two points $\mathbf{P}_1$ and $\mathbf{P}_2$, and $x_1 \neq x_2$, then $S(m,b)$ has a minimum of zero, and the corresponding line actually passes through the two given points.

(b) Find m and b to minimize $S(m,b)$ if the given points are $(0,2)$, $(1,1)$, and $(4,-1)$.

(c) In the case of n points $(x_1, y_1), \ldots, (x_n, y_n)$, show that $\partial S/\partial m = 0$ and $\partial S/\partial b = 0$ if and only if

$$m\left(\sum_1^n x_j\right) + nb = \sum_1^n y_j \tag{16}$$

and

$$m\left(\sum_1^n x_j^2\right) + b\left(\sum_1^n x_j\right) = \sum_1^n x_j y_j \,. \tag{17}$$

(d) Show that the simultaneous equations (16) and (17) have a unique solution for m and b if

$$(\sum_1^n x_j)^2 \neq n \sum_1^n x_j^2.$$

(See Problem 22, §1.5.)

(e) Show that $(\sum_1^n x_j)^2 < n \sum_1^n x_j^2$ unless the x_j are all equal, in which case $(\sum_1^n x_j)^2 = n \sum_1^n x_j^2$. (Hint: Notice that $\sum_1^n x_j$ is the dot product of $\mathbf{X} = (x_1, \ldots, x_n)$ and $\mathbf{A} = (1, \ldots, 1)$. Use the fact that $|\mathbf{X}\cdot\mathbf{A}| < |\mathbf{X}|\cdot|\mathbf{A}|$ if $\mathbf{X}$ and $\mathbf{A}$ are not parallel, and $|\mathbf{X}\cdot\mathbf{A}| = |\mathbf{X}|\cdot|\mathbf{A}|$ if $\mathbf{X}$ and $\mathbf{A}$ are parallel.)

(f) Use part (e) and Theorem 14 to show that, unless the x_j are all equal, the function $S(m,b)$ has a unique minimum given by (1) and (2). (Note that $S(m,b)$ is a polynomial of degree 2 in m and b, so Theorem 14 gives absolute maxima and minima.)

6. This problem proves Theorem 14 in the simplest interesting case, that of a homogeneous quadratic polynomial $f(x,y) = \alpha x^2 + 2\beta xy + \gamma y^2$. Prove the following:

(a) $f_{xx} = 2\alpha$, $f_{xy} = 2\beta$, $f_{yy} = 2\gamma$.

(b) $f(r\cos\theta, r\sin\theta) = \frac{1}{2}r^2[(\alpha+\gamma) + 2\beta\sin 2\theta + (\alpha-\gamma)\cos 2\theta]$.

(c) $f(r\cos\theta, r\sin\theta) = \frac{1}{2}r^2[(\alpha+\gamma) + \sqrt{(\alpha-\gamma)^2 + 4\beta^2}\cos(2\theta-\varphi)]$, where

$$\cos\varphi = \frac{\alpha-\gamma}{\sqrt{(\alpha-\gamma)^2+4\beta^2}} \quad \text{and} \quad \sin\varphi = \frac{2\beta}{\sqrt{(\alpha-\gamma)^2+4\beta^2}}\,.$$

(d) On the circle $\{\mathbf{P}: |\mathbf{P}| = r\}$, f has the maximum value

$$\tfrac{1}{2}r^2(\alpha+\gamma+\sqrt{(\alpha-\gamma)^2+4\beta^2})$$

and minimum value $\frac{1}{2}r^2(\alpha+\gamma-\sqrt{(\alpha-\gamma)^2+4\beta^2})$.

(e) The maximum and minimum in part (d) have opposite signs if and only if $4\alpha\gamma < 4\beta^2$, or $f_{xx} f_{yy} < (f_{xy})^2$.

(f) If $4\alpha\gamma > 4\beta^2$, then the maximum and minimum are both positive if $\alpha + \gamma > 0$, and both negative if $\alpha + \gamma < 0$.

(g) From (d)–(f), deduce that (0,0) is a *saddle point* for f if $f_{xx} f_{yy} < (f_{xy})^2$; a *maximum* if $f_{xx} f_{yy} > (f_{xy})^2$ and $f_{xx} + f_{yy} < 0$; and a *minimum* if $f_{xx} f_{yy} > (f_{xy})^2$ and $f_{xx} + f_{yy} > 0$.

(h) If $4\alpha\gamma = 4\beta^2$, prove that α and γ have the same sign, and $f(x,y) = \pm(\sqrt{|\alpha|}x \pm \sqrt{|\gamma|}y)^2$. What can be said about maxima or minima of f in this case? (This is the only case not covered in parts (e) and (f).)

7. This problem proves Theorem 14 in general. Suppose f has continuous second partial derivatives in a disk centered at $\mathbf{A} = (a,b)$, and that $\nabla f(\mathbf{A}) = \mathbf{0}$. Set $2\alpha = f_{xx}(\mathbf{A})$, $2\beta = f_{xy}(\mathbf{A})$, $2\gamma = f_{yy}(\mathbf{A})$.

(a) Prove that $f(a + h, b + k) = f(a,b) + \alpha h^2 + 2\beta hk + \gamma k^2 + R(h,k)$ where, setting $\mathbf{H} = (h,k)$, we have

$$\lim_{\mathbf{H}\to 0} \frac{R(\mathbf{H})}{|\mathbf{H}|^2} = 0.$$

(b) Prove that if f is a polynomial of degree two, then $R(\mathbf{H}) \equiv 0$. (Hint: In formula (8) of §3.7, note that f_{xx}, f_{yy}, and f_{xy} are constant.)

(c) Using (a) and (b), prove the remark after Theorem 14, about polynomials of degree two. (See Problem 6(g).)

(d) Suppose $\alpha + \gamma > 0$ and $\alpha\gamma > \beta^2$. Prove that $\mathbf{A}$ gives a local minimum. (Hint: Let m be the minimum of

$$\frac{\alpha h^2 + 2\beta hk + \gamma k^2}{h^2 + k^2};$$

by Problem 6(f), $m > 0$. Prove that when $|\mathbf{H}|$ is so small that $|R(\mathbf{H})|/|\mathbf{H}|^2 < m$, then $f(\mathbf{A} + \mathbf{H}) \geq f(\mathbf{A})$; in fact $f(\mathbf{A} + \mathbf{H}) > f(\mathbf{A})$ unless $\mathbf{H} = \mathbf{0}$.)

(e) Suppose $\alpha + \gamma < 0$ and $\alpha\gamma > \beta^2$. Prove that $\mathbf{A}$ gives a local maximum.

(f) Prove that if $\alpha\gamma < \beta^2$, then $\mathbf{A}$ is a saddle point.

Chapter IV

Double Integrals, Vector Fields, and Line Integrals

The title of this chapter suggests a collection of random topics; in fact, however, the topics are united in *Green's theorem,* an important two-dimensional version of the fundamental theorem of calculus, relating the line integral of a vector field to a double integral.

§4.1 Double integrals and Leibniz' rule for differentiating integrals. (Here, for the first time, we give a theorem without the proof.)

§4.2 Vector fields

§4.3 Line integrals

§4.4 Green's theorem

§4.5 Green's theorem applied to the problem of change of variable in double integrals.

4.1 DOUBLE INTEGRALS

The main theorems about integrating functions of two variables are important and conceptually clear, but difficult to formulate and prove. In fact, the first really good theory was not developed until recently, by Henri Lebesgue (1902) and Guido Fubini (1910), to mention two of the most famous names. We will begin with the general ideas, and make only a few steps toward formulating and proving the theorems.

Suppose f is a function defined in a set S of the plane, and $f \geq 0$ on S. Then the double integral $\iint_S f$ means, intuitively, the volume of the region lying over S and under the graph of f.

Example 1. S is the unit disk $\{P: |P| \leq 1\}$, and $f(x,y) = \sqrt{1 - x^2 - y^2}$. The region lying over S and under the graph of f is the upper half of the unit ball $\{(x,y,z): x^2 + y^2 + z^2 \leq 1\}$; hence

$$\iint_S f = \tfrac{1}{2}\left(\tfrac{4}{3}\pi 1^3\right) = \tfrac{2}{3}\pi.$$

(See Fig. 1.)

Example 2. S is the square $\{(x,y): 0 \leq x \leq 1, 0 \leq y \leq 1\}$, and $f(x,y) = y$. The region lying over S and under the graph of f is a wedge, a unit cube sliced in half, so $\iint_S f = \frac{1}{2}$. (See Fig. 2.)

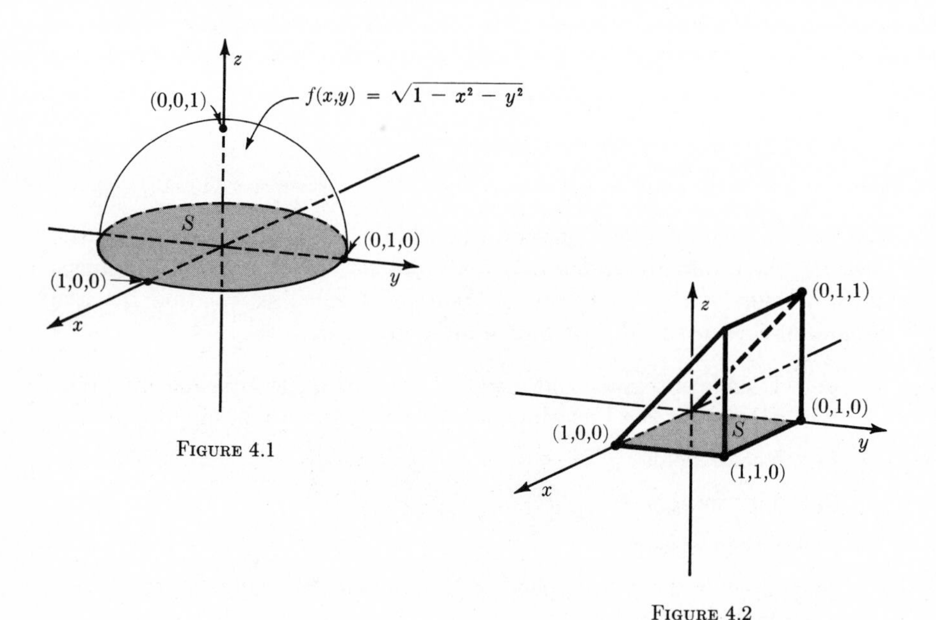

FIGURE 4.1

FIGURE 4.2

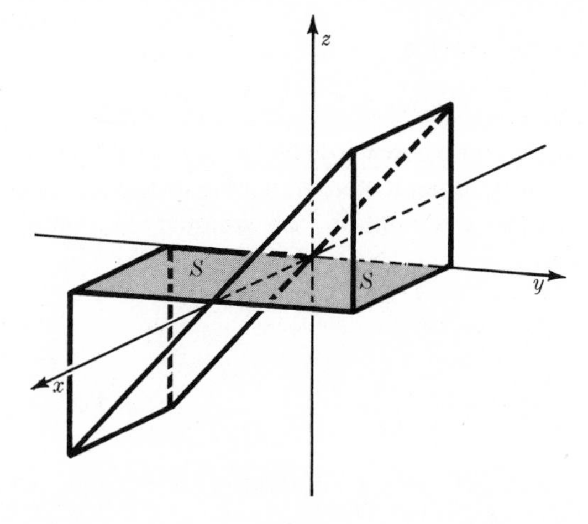

FIGURE 4.3

When f takes negative values in part of S, then $\iint_S f$ means, intuitively, the volume lying under the graph of f and over S, *minus* the volume lying over the graph and under S. In other words, volumes above the xy plane are taken as positive, and those below the xy plane as negative.

For example, if S is the rectangle $\{(x,y): -1 \leq x \leq 1, 0 \leq y \leq 1\}$, and $f(x,y) = y$, then $\iint_S f$ is the difference in the volumes of two congruent wedges, so $\iint_S f = 0$. (See Fig. 3.)

Obviously, simple pictures like these can evaluate only very special integrals. A much more general method is suggested by the formula for a volume of revolution. Recall that if we take the graph of a function g on an interval $[a,b]$ and revolve it about the x axis, we enclose a volume

$$V = \int_a^b \pi g(x)^2 \, dx.$$

Here, $\pi g(x)^2$ is the area of a cross section of the enclosed volume, and the volume itself is the integral of these cross-sectional areas (Fig. 4).

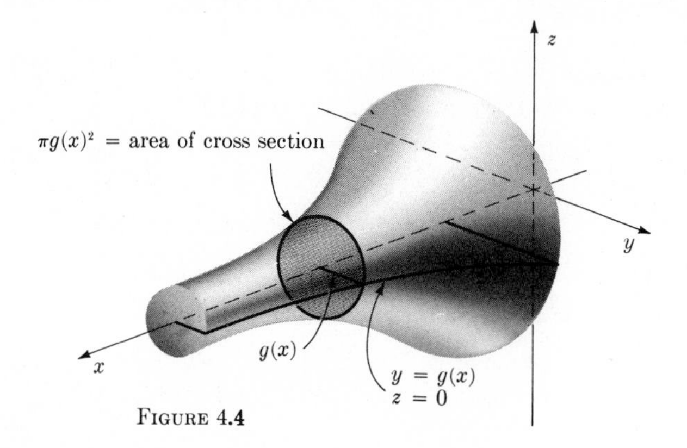

FIGURE 4.4

Now consider a function $f \geq 0$ defined on a rectangle

$$S = \{(x,y): a \leq x \leq b,\ c \leq y \leq d\};$$

then $\iint_S f$ is the volume shown in Fig. 5. Let $A(x)$ denote the area of a typical cross section by a plane perpendicular to the x axis at $(x,0,0)$. By analogy with volumes of revolution, the volume $\iint_S f$ should be

$$\iint_S f = \int_a^b A(x)\, dx. \tag{1}$$

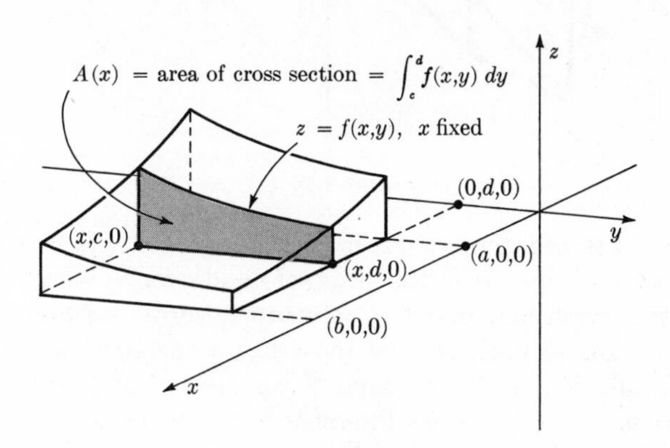

FIGURE 4.5

Further, for any fixed x, the cross-sectional area $A(x)$ is the area under the graph of a function of one variable, $g(y) = f(x,y)$, so

$$A(x) = \int_c^d f(x,y)\, dy. \tag{2}$$

Putting (2) into (1), we get

$$\iint_S f = \int_a^b \left[\int_c^d f(x,y)\, dy \right] dx. \tag{3}$$

This formula reduces the evaluation of $\iint_S f$ to the evaluation of two ordinary integrals, a question we have already studied.

Example 3. $S = \{(x,y): 0 \leq x \leq 1, 1 \leq y \leq 2\}$

$$f(x,y) = xe^{xy}$$

$$\iint_S f = \int_0^1 \left[\int_1^2 xe^{xy}\, dy \right] dx = \int_0^1 [e^{xy}]_{y=1}^{y=2}\, dx$$

$$= \int_0^1 (e^{2x} - e^x)\, dx = [\tfrac{1}{2}e^{2x} - e^x]_0^1 = \tfrac{1}{2}e^2 - e + \tfrac{1}{2}.$$

The same method can be applied when S is bounded by the graphs of two functions $\varphi_1(x)$ and $\varphi_2(x)$ on an interval $[a,b]$,

$$S = \{(x,y) : a \leq x \leq b,\ \varphi_1(x) \leq y \leq \varphi_2(x)\}, \tag{4}$$

where we assume that $\varphi_1 \leq \varphi_2$ on the interval $[a,b]$. As shown in Fig. 6, a typical cross section perpendicular to the x axis has area

$$A(x) = \int_{\varphi_1(x)}^{\varphi_2(x)} f(x,y)\, dy,$$

so the total volume is

$$\iint_S f = \int_a^b A(x)\, dx = \int_a^b \left[\int_{\varphi_1(x)}^{\varphi_2(x)} f(x,y)\, dy\right] dx. \tag{5}$$

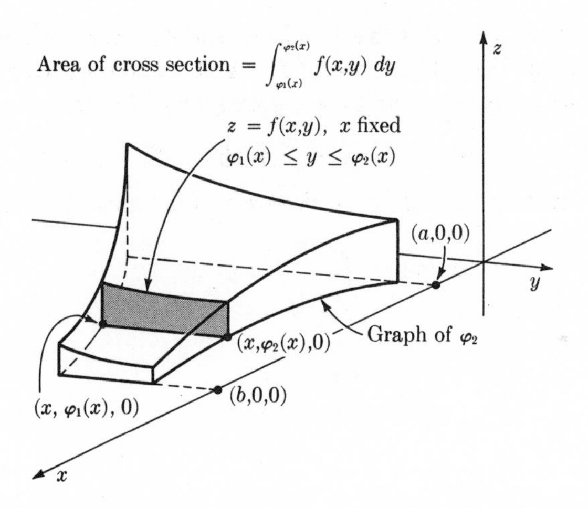

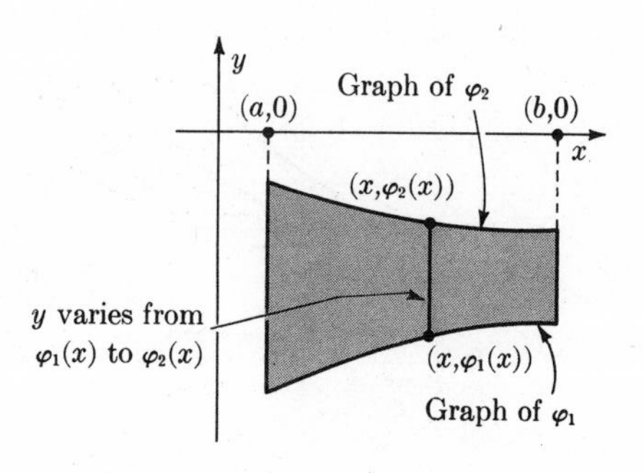

FIGURE 4.6

Example 4. For the double integral illustrated in Fig. 1 we have

$$S = \{(x,y): -1 \le x \le 1, -\sqrt{1-x^2} \le y \le \sqrt{1-x^2}\} \quad \text{(the unit disk)}$$

$$f(x,y) = \sqrt{1-x^2-y^2},$$

$$\iint_S f = \int_{-1}^{1} \left[\int_{-\sqrt{1-x^2}}^{\sqrt{1-x^2}} \sqrt{1-x^2-y^2}\, dy \right] dx$$

$$= \int_{-1}^{1} \left[\frac{1-x^2}{2} \arcsin \frac{y}{\sqrt{1-x^2}} + \frac{1}{2} y\sqrt{1-x^2-y^2} \right]_{y=-\sqrt{1-x^2}}^{y=\sqrt{1-x^2}} dx$$

$$= \int_{-1}^{1} \frac{\pi}{2}(1-x^2)\, dx = \frac{\pi}{2}\left[x - \frac{x^3}{3}\right]_{-1}^{1} = \frac{\pi}{2}\cdot\frac{4}{3} = \frac{2\pi}{3}.$$

(Here we used the integral $\int \sqrt{a^2-y^2}\, dy = \frac{a^2}{2} \arcsin \frac{y}{a} + \frac{1}{2} y\sqrt{a^2-y^2}$.)

The restriction $f \ge 0$ was made only to simplify the picture; the same method works, regardless of the sign of f. And, of course, we could just as well take sections perpendicular to the y axis (Fig. 7): if

$$S = \{(x,y): \psi_1(y) \le x \le \psi_2(y),\ c \le y \le d\}, \tag{6}$$

where $\psi_1 \le \psi_2$ on the interval $[c,d]$, then a typical cross section has area

$$A(y) = \int_{\psi_1(y)}^{\psi_2(y)} f(x,y)\, dx,$$

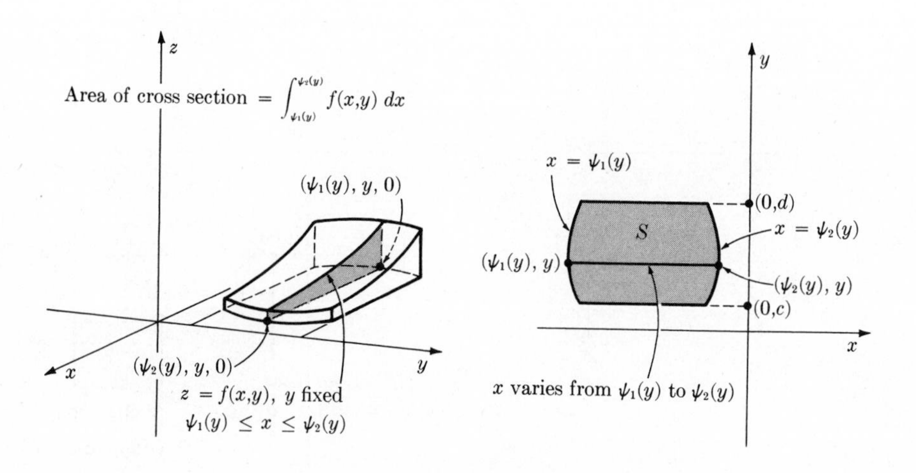

FIGURE 4.7

and

$$\iint_S = \int_c^d A(y)\,dy = \int_c^d \left[\int_{\psi_1(y)}^{\psi_2(y)} f(x,y)\,dx\right] dy. \tag{7}$$

In both (5) and (7) there is an "inner integral" (the one in square brackets) and an "outer integral"; the inner integral has variable limits, and the outer integral has constant limits. When the inner integral is with respect to x (as in (7)), its limits depend on y, and in the evaluation of $\int f(x,y)\,dx$ we keep y constant. When the inner integral is with respect to y (as in (5)), its limits depend on x, and in $\int f(x,y)\,dy$ we keep x constant.

Example 5. $S = \{(x,y): 0 \le y \le x^2, 0 \le x \le 2\}$ (see Fig. 8(a))

$$f(x,y) = x + y.$$

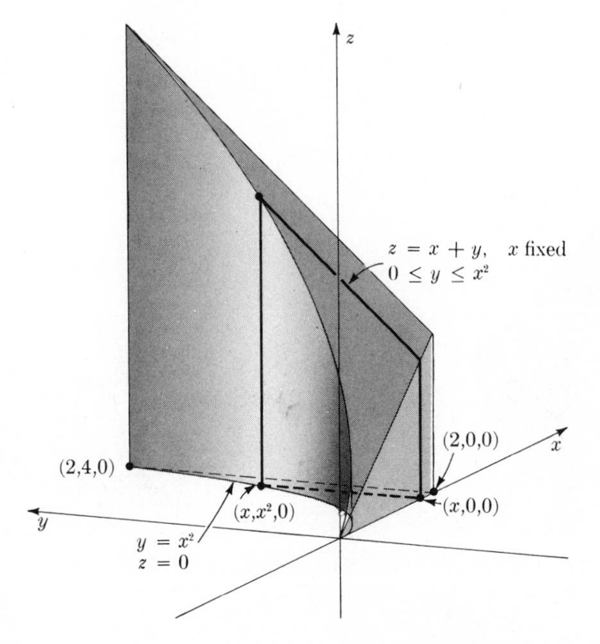

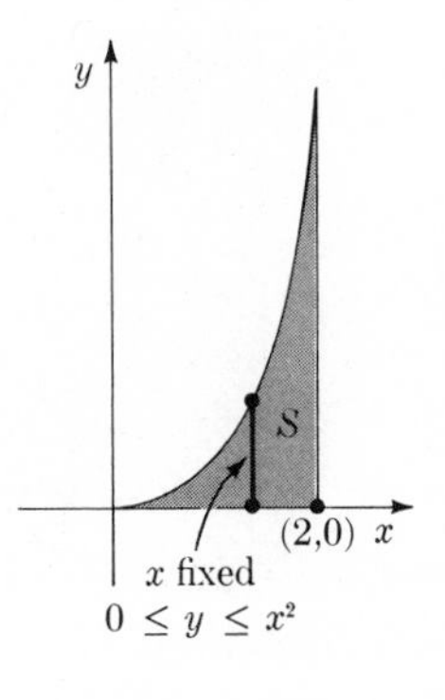

FIGURE 4.8 (a)

The set S is given in the form (4), with $\varphi_1 = 0$ and $\varphi_2 = x^2$, so we have, as in (5),

$$\iint_S f = \int_0^2 \left[\int_0^{x^2} (x+y)\,dy\right] dx = \int_0^2 \left[xy + \frac{y^2}{2}\right]_{y=0}^{y=x^2} dx$$

$$= \int_0^2 \left[x^3 + \frac{x^4}{2}\right] dx = \left[\frac{x^4}{4} + \frac{x^5}{10}\right]_0^2 = \frac{36}{5}.$$

The same set can also be described in the form (6),

$$S = \{\,(x,y)\colon \sqrt{y} \le x \le 2,\, 0 \le y \le 4\} \qquad \text{(see Fig. 8(b))},$$

so

$$\iint_S f = \int_0^4 \left[\int_{\sqrt{y}}^2 (x+y)\,dx\right] dy = \int_0^4 \left[\frac{x^2}{2} + xy\right]_{x=\sqrt{y}}^{x=2} dy$$

$$= \int_0^4 \left(2 + 2y - \frac{y}{2} - y^{3/2}\right) dy$$

$$= \left[2y + y^2 - \frac{y^2}{4} - \frac{2}{5}y^{5/2}\right]_0^4 = \frac{36}{5}.$$

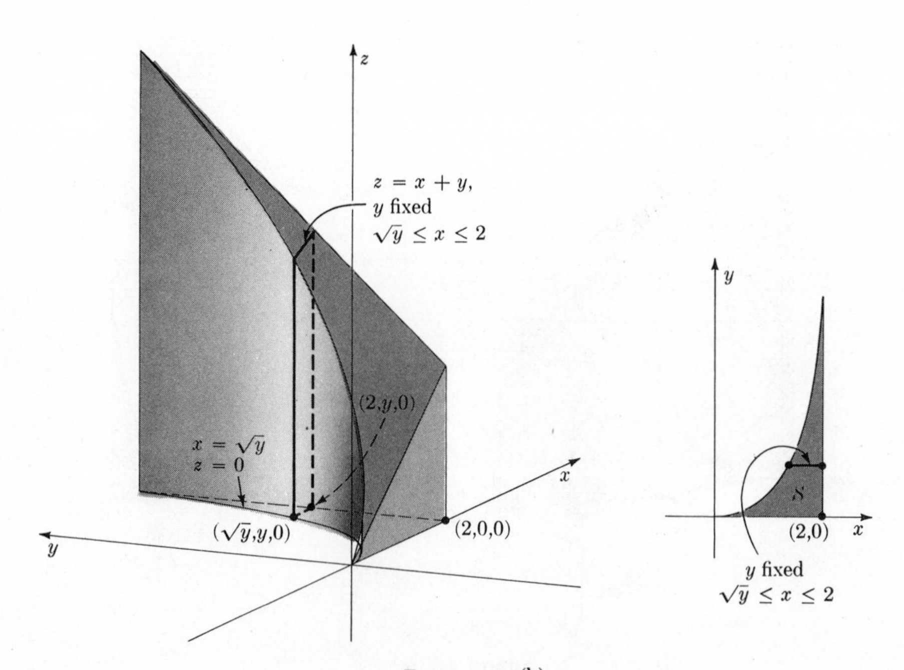

FIGURE 4.8 (b)

We have written the "repeated integral"

$$\int_a^b \left[\int_{\varphi_1(x)}^{\varphi_2(x)} f(x,y)\, dy \right] dx$$

with square brackets to show clearly how it is to be evaluated; first take

$$\int_{\varphi_1(x)}^{\varphi_2(x)} f(x,y)\, dy,$$

and then integrate the result with respect to x. Usually it is written simply

$$\int_a^b \int_{\varphi_1(x)}^{\varphi_2(x)} f(x,y)\, dy\, dx,$$

with the understanding that the "inner integral" is to be evaluated first.

This whole discussion has been based on an analogy and a few pictures. Two important theorems give these pictures the necessary rigorous basis.

Theorem 1. *Let f be continuous on the set*

$$S = \{(x,y) : a \leq x \leq b,\ \varphi_1(x) \leq y \leq \varphi_2(y)\}, \tag{4}$$

where φ_1 and φ_2 are continuous on $[a, b]$ and $\varphi_1 \leq \varphi_2$. Let

$$G(x) = \int_{\varphi_1(x)}^{\varphi_2(x)} f(x,y)\, dy.$$

Then G is continuous on $[a,b]$.

Similarly, if S is the set

$$S = \{(x,y) : \psi_1(y) \leq x \leq \psi_2(y),\ c \leq y \leq d\}, \tag{6}$$

where ψ_1 and ψ_2 are continuous and $\psi_1 \leq \psi_2$, then

$$H(y) = \int_{\psi_1(y)}^{\psi_2(y)} f(x,y)\, dx$$

is continuous on $[c,d]$.

Theorem 2. *If the same set S is described both by (4) and by (6), then*

$$\int_a^b \left[\int_{\varphi_1(x)}^{\varphi_2(x)} f(x,y)\, dy \right] dx = \int_c^d \left[\int_{\psi_1(y)}^{\psi_2(y)} f(x,y)\, dx \right] dy. \tag{8}$$

If these theorems could be proved intelligibly in a page or two, we would do it, but in fact they are by-products of a thorough investigation of double integration, as well as the deeper properties of continuous functions of two variables. Rather than give "ad hoc" direct proofs of such special results, we refer you to the general theory as expounded in the references listed at the end of this section.

The common value of the two integrals in (8) is called the *double integral* of f over S, denoted

$$\iint_S f \qquad \text{or} \qquad \iint_S f(x,y)\,dx\,dy.$$

Since we are defining the double integral by (8), the linearity property

$$\iint_S (\alpha f + \beta g) = \alpha \iint_S f + \beta \iint g \qquad (\alpha \text{ and } \beta \text{ are constants})$$

follows from the linearity of the single integral (Problem 7). Similarly,

$$f \leq g \Rightarrow \iint_S f \leq \iint_S g.$$

Theorems 1 and 2 have an important corollary, *Leibniz' rule* for differentiating under the integral sign:

Theorem 3. *Suppose $D_1 f(t,y)$ is continuous on the rectangle*

$$R = \{(t,y) : a \leq t \leq b,\ c \leq y \leq d\}.$$

Then the function

$$G(t) = \int_c^d f(t,y)\,dy$$

has the derivative

$$G'(t) = \int_c^d D_1 f(t,y)\,dy.$$

Proof. Consider $f(t,y)$ as a function of t, with y held fixed. By the fundamental theorem of calculus

$$f(t,y) = f(a,y) + \int_a^t D_1 f(x,y)\,dx;$$

hence

$$G(t) = \int_c^d f(t,y)\,dy = \int_c^d f(a,y)\,dy + \int_c^d \left[\int_a^t D_1 f(x,y)\,dx\right] dy$$

$$= \int_c^d f(a,y)\,dy + \int_a^t \left[\int_c^d D_1 f(x,y)\,dy\right] dx,$$

by Theorem 2. The first term in the last line is constant, so its derivative is zero. The second term is

$$\int_a^t H(x)\,dx, \quad \text{where} \quad H(x) = \int_c^d D_1 f(x,y)\,dy.$$

By Theorem 1, H is continuous, so by the fundamental theorem of calculus $G(t) = \int_a^t H(x)\,dx$ has the derivative

$$G'(t) = H(t) = \int_c^d D_1 f(t,y)\,dy. \quad Q.E.D.$$

Perhaps it is fitting to write Leibniz' rule in Leibniz notation:

$$\frac{d}{dt}\int_c^d f(t,y)\,dy = \int_c^d \frac{\partial f}{\partial t}(t,y)\,dy.$$

Thus, under appropriate conditions, the derivative d/dt can be moved from the outside of the integral sign to the inside. (But once inside, it has to be rewritten as $\partial/\partial t$, since it now applies to a function of two variables.) In this guise, it is easy to see how the rule works when other letters are used for the variables; for example,

$$\frac{d}{dx}\int_a^b f(x,s)\,ds = \int_a^b \frac{\partial f}{\partial x}(x,s)\,ds,$$

$$\frac{d}{dt}\int_a^b f(t,x)\,dx = \int_a^b \frac{\partial f}{\partial t}(t,x)\,dx.$$

Further, since s is held constant in computing $\partial f(t,s,x)/\partial t$, we have

$$\frac{\partial}{\partial t}\int_a^b f(t,s,x)\,dx = \int_a^b \frac{\partial f}{\partial t}(t,s,x)\,dx.$$

Example 6. Let

$$G(t) = \int_0^1 \sin(t + y)\, dy.$$

Applying Theorem 3 with $f(t, y) = \sin(t + y)$, we find $D_1 f(t,y) = f_t = \cos(t + y)$; hence

$$G'(t) = \int_0^1 \cos(t + y)\, dy = [\sin(t + y)]_0^1$$

$$= \sin(t + 1) - \sin t.$$

We can check this by first evaluating the integral for $G(t)$, obtaining $G(t) = -\cos(t + 1) + \cos t$, which yields $G'(t) = \sin(t + 1) - \sin t$.

Example 7. Let

$$G(t) = \int_1^2 \frac{1}{y} e^{ty}\, dy.$$

Applying Theorem 3 with

$$f(t,y) = \frac{1}{y} e^{ty},$$

we find $D_1 f(t,y) = f_t = e^{ty}$; hence

$$G'(t) = \int_1^2 e^{ty}\, dy = \left[\frac{1}{t} e^{ty}\right]_1^2 = \frac{e^{2t} - e^t}{t}.$$

We cannot check this result by evaluating $G(t)$ explicitly, since none of the standard methods evaluate the integral

$$\int \frac{1}{y} e^{ty}\, dy.$$

In a case like this, Theorem 3 is the only practical way to obtain a useful expression for $G'(t)$.

Example 8. Let

$$G(x,y) = \int_0^x f(s,y)\, ds,$$

where f and f_y are continuous. Find G_x and G_y. *Solution.* By the fundamental theorem of calculus, $G_x(x,y) = f(x,y)$, and by Leibniz' rule,

$$G_y(x,y) = \int_0^x f_y(s,y)\, ds.$$

Example 9. Let

$$H(x,y) = \int_0^{\varphi(x,y)} f(s,y)\, ds,$$

where f, f_y, φ, φ_x, and φ_y are all continuous. Find H_x and H_y.
Solution. We can use the chain rule, with the scheme

$$(x,y) \;\longrightarrow\; (\varphi(x,y),y) \;\longrightarrow\; \int_0^{\varphi(x,y)} f(s,y)\, ds = H(x,y)$$

or

$$(x,y) \;\longrightarrow\; (u,v) \;\longrightarrow\; z$$

where $u = \varphi(x,y)$, $v = y$, and $z = \int_0^u f(s,v)\, ds$. We find

$$\begin{aligned} H_x &= \frac{\partial z}{\partial x} = \frac{\partial z}{\partial u}\frac{\partial u}{\partial x} + \frac{\partial z}{\partial v}\frac{\partial v}{\partial x} \\ &= f(u,v)\cdot\varphi_x + \left(\int_0^u D_2 f(s,v)\, ds\right)\cdot 0 \\ &= f(\varphi(x,y),y)\varphi_x(x,y), \end{aligned}$$

and

$$\begin{aligned} H_y &= \frac{\partial z}{\partial y} = \frac{\partial z}{\partial u}\frac{\partial u}{\partial y} + \frac{\partial z}{\partial v}\frac{\partial v}{\partial y} \\ &= f(u,v)\cdot\varphi_y + \left(\int_0^u D_2 f(s,v)\, ds\right)\cdot 1 \\ &= f(\varphi(x,y),y)\varphi_y(x,y) + \int_0^{\varphi(x,y)} D_2 f(s,y)\, ds. \end{aligned}$$

PROBLEMS

1. Evaluate the following repeated integrals.

(a) $\int_0^{\pi}\int_0^{y} y \sin x\, dx\, dy$ (d) $\int_0^1\int_{\sqrt{x}}^1 dy\, dx$

(b) $\int_1^2\int_x^{x^2} dy\, dx$ (e) $\int_0^2\int_1^{e^y} dx\, dy$

(c) $\int_1^3\int_0^{\log y} e^{x+y}\, dx\, dy$ (f) $\int_0^1\int_0^{x^2} \frac{1}{1+x^2}\, dy\, dx.$

2. Describe each of the sets in Fig. 9 in the form (4), and in the form (6).

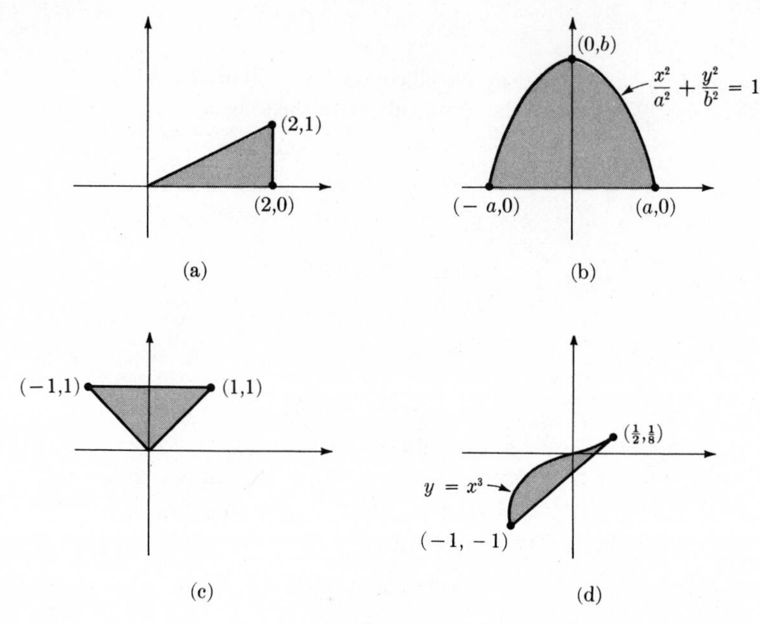

FIGURE 4.9

3. For each of the integrals in Problem 1, sketch the set S over which you are integrating, and rewrite the integral with dx and dy interchanged. If possible, evaluate the new repeated integral, and compare it to the result found in Problem 1. (*Warning:* It is difficult to obtain the limits for the new repeated integral just by looking at the original limits; you need a clear picture of the set S over which you are integrating.)

4. Find the partial derivatives of the following functions by applying the fundamental theorem of calculus and Leibniz' rule.

(a) $f(x,y) = \int_1^x e^{sy}\, ds$

(b) $g(x,y) = \int_1^{x^2} e^{sy}\, ds$

(Hint: $g(x,y) = f(x^2,y)$, with f as in (a); use the chain rule).

(c) $h(x,y) = \int_0^x \sin (sy)\, ds$

(d) $i(x,y) = \int_0^{x^3} \sin(sy)\, ds$

(e) $j(x,y) = \int_0^{x+y} \sin(sy)\, ds$

(Hint: $j(x,y) = h(x + y, y)$; use the chain rule.)

(f) $k(x,y) = \int_0^{\varphi(x)} \sin(sy)\, ds$, where φ is differentiable

(g) $l(x,y) = \int_0^{\varphi(x,y)} \sin(sy)\, ds$, where φ is differentiable

(h) $m(x,y) = \int_{\varphi_1(x,y)}^{\varphi_2(x,y)} \sin(sy)\, ds$ (Hint: $\int_{\varphi_1}^{\varphi_2} = \int_0^{\varphi_2} - \int_0^{\varphi_1}$)

5. Evaluate the integrals in Problem 4(a)–(e), and check the results you found in Problem 4.

6. Suppose that f and f_y are continuous on the rectangle

$$R = \{(x,y) : a \leq x \leq b,\ c \leq y \leq d\},$$

that φ_1 and φ_2 are differentiable on R, that $a \leq \varphi_1 \leq b$, and that $a \leq \varphi_2 \leq b$. Prove that if

$$F(x,y) = \int_{\varphi_1(x,y)}^{\varphi_2(x,y)} f(s,y)\, ds,$$

then

$$F_x(x,y) = f(\varphi_2(x,y),y)\frac{\partial \varphi_2}{\partial x} - f(\varphi_1(x,y),y)\frac{\partial \varphi_1}{\partial x},$$

and

$$F_y(x,y) = \int_{\varphi_1}^{\varphi_2} f_y(s,y)\, ds + f(\varphi_2(x,y),y)\frac{\partial \varphi_2}{\partial y} - f(\varphi_1(x,y),y)\frac{\partial \varphi_1}{\partial y}.$$

(Hint: Do Problem 4 first.)

7. Suppose that S is a region of the type in Theorem 2, and f and g are continuous on S.

(a) Prove that $\iint_S (\alpha f + \beta g) = \alpha \iint_S f + \beta \iint_S g$ for all constants α and β.

(b) Prove that if $f \geq g$, then $\iint_S f \geq \iint_S g$.

8. (a) Suppose that U is an open set, f is continuous on U, and $\iint_D f = 0$ for every disk D contained in U. Prove that $f \equiv 0$ in U. (Hint: If $f \not\equiv 0$, then $f(\mathbf{P}_0) \neq 0$ for some point $\mathbf{P}_0$ in U; say $f(\mathbf{P}_0) > 0$.

Since f is continuous and U is open, there is a disk D contained in U such that $f(\mathbf{P}) > \frac{1}{2}f(\mathbf{P}_0)$ for every point $\mathbf{P}$ in R. Now apply Problem 7(b).)

(b) Suppose that S is a region of the type in Theorem 2, f is continuous on S, $f \geq 0$, and $\iint_S f = 0$. Let U be any open set contained in S. Prove that $f \equiv 0$ on U. (Hint: Show that $0 \leq \iint_D f \leq \iint_S f = 0$ for every disk D contained in U.)

References. The following books give a thorough development of double integrals, including the proofs omitted above. We will refer to them again later in connection with Green's theorem and its extensions.

T. M. Apostol, *Mathematical Analysis*, Addison-Wesley, 1957

W. H. Fleming, *Functions of Several Variables*, Addison-Wesley, 1965

M. Spivak, *Calculus on Manifolds*, Benjamin, 1965

4.2 VECTOR FIELDS

A vector field $\mathbf{F}$ over a set S is a function assigning a vector $\mathbf{F}(\mathbf{P})$ to each point $\mathbf{P}$ in S. Here the set S will be in the plane, and the vectors will be in R^2. The two components of $\mathbf{F}(\mathbf{P})$ are often denoted $M(\mathbf{P})$ and $N(\mathbf{P})$; thus $\mathbf{F}(\mathbf{P}) = (M(\mathbf{P}), N(\mathbf{P}))$, where M and N are ordinary real-valued functions on S. To visualize $\mathbf{F}$, picture at each point $\mathbf{P}$ in S an arrow representing $\mathbf{F}(\mathbf{P})$, as in Fig. 10.

The two main physical examples are *force fields* and *velocity fields.* A particle of mass M at the origin attracts a particle of mass m at the point $\mathbf{P}$ by a force $-\gamma mM\mathbf{P}/|\mathbf{P}|^3$, where γ is a gravity constant. The function

$$\mathbf{F}(\mathbf{P}) = -\gamma mM\, |\mathbf{P}|^{-3}\mathbf{P} \qquad (*)$$

is a vector field defined on the set where $|\mathbf{P}| \neq 0$ (Fig. 10(a)); since it describes a force, it is called a *force field.*

For a simple example of a velocity field, imagine that a plate lying on the plane is rotated counterclockwise about the origin at a rate of c radians per minute. Then for any (x,y) in the plane, the point on the plate lying over (x,y) moves with velocity $\mathbf{F}(x,y) = (-cy,cx)$, as you

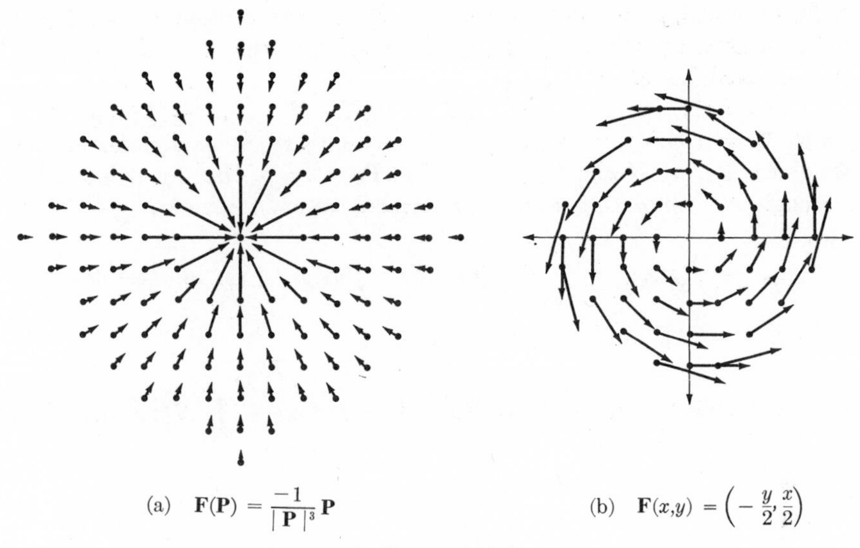

(a) $\mathbf{F}(\mathbf{P}) = \dfrac{-1}{|\mathbf{P}|^3}\mathbf{P}$ (b) $\mathbf{F}(x,y) = \left(-\dfrac{y}{2}, \dfrac{x}{2}\right)$

FIGURE 4.10

can easily check (see Problem 4). This vector field $\mathbf{F}$ is called the *velocity field* of the rotation (Fig. 10(b)).

In a purely mathematical context, vector fields arise from gradients; precisely, if f is differentiable at every point in S, then $\mathbf{F}(\mathbf{P}) = \nabla f(\mathbf{P})$ defines a vector field $\mathbf{F}$ on S. For example, if $f(\mathbf{P}) = \gamma mM/|\mathbf{P}|$, you can check that $\nabla f(\mathbf{P}) = -\gamma mM\ |\mathbf{P}|^{-3}\,\mathbf{P}$, and we get the force field (*) as a gradient. The function f is called a *potential function* of the force $\mathbf{F} = \nabla f$.

Not every vector field is a gradient, though. For example, the velocity field $\mathbf{F}(x,y) = (-cy,cx)$ is not the gradient of any function f. Suppose, on the contrary, that $\mathbf{F}$ were a gradient ∇f; then we would have $f_x = -cy$, $f_y = cx$; hence $f_{xy} = -c$, $f_{yx} = c$, and the mixed partials of f would be continuous but unequal, contradicting Theorem 10 of the previous chapter. Hence there is no function f such that $\nabla f(x,y) = (-cy,cx)$.

This raises the question: Which vector fields are gradients? Or, in physical terms, which force fields have potentials? The answer is easy for a vector field defined over a rectangle:

Theorem 4. *Suppose that* $\mathbf{F} = (M,N)$ *is defined on an open rectangle*

$$R = \{(x,y) : a < x < b,\ c < y < d\},$$

and the partial derivatives M_y *and* N_x *are continuous on* R. *Then* $\mathbf{F}$ *is a gradient if and only if* $M_y \equiv N_x$.

Proof. If $\mathbf{F}$ is the gradient of a function f, then $(M,N) = (D_1 f, D_2 f)$, so $D_2M = D_2D_1 f$ and $D_1N = D_1D_2 f$. By the assumptions on D_2M and D_1N, the mixed partials of f are continuous, hence they must be equal, so $D_2M = D_1N$, i.e. $M_y = N_x$.

Conversely, suppose that $D_2M = D_1N$. We must find a function f with $\nabla f = (M,N)$, in other words a function f whose partial derivatives are given, $D_1 f = M$ and $D_2 f = N$. Let $\mathbf{P}_0 = (x_0,y_0)$ be any point in the rectangle R. Since $D_1 f = M$ is prescribed, we must have

$$f(x,y_0) = f(x_0,y_0) + \int_{x_0}^{x} D_1 f(s,y_0)\, ds = f(x_0,y_0) + \int_{x_0}^{x} M(s,y_0)\, ds. \quad (1)$$

Since $D_2 f = N$ is prescribed, we must have

$$f(x,y) = f(x,y_0) + \int_{y_0}^{y} D_2 f(x,t)\, dt = f(x,y_0) + \int_{y_0}^{y} N(x,t)\, dt;$$

hence, from (1),

$$f(x,y) = f(x_0,y_0) + \int_{x_0}^{x} M(s,y_0)\, ds + \int_{y_0}^{y} N(x,t)\, dt. \quad (2)$$

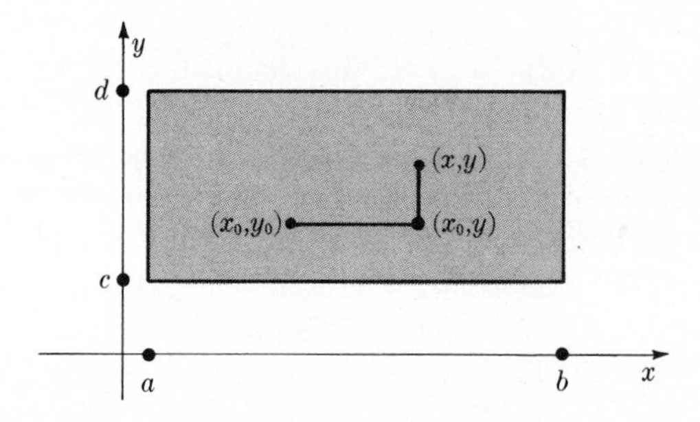

FIGURE 4.11

(Conceptually, the integrals in (2) are taken along the two line segments shown in Fig. 11. When (x,y) is in R, these segments lie entirely in R, so the integrals are defined.) What we have proved is this: If there is a function f with $\nabla f = (M,N)$, it must be given by (2). It remains to be seen whether (2) actually works. By the fundamental theorem of calculus we see immediately that $f_y = N(x,y)$, as desired. Using Leibniz' rule (Theorem 3), we get

$$f_x = M(x,y_0) + \int_{y_0}^{y} N_x(x,t)\, dt,$$

which does not look like M. But $N_x = M_y$, so

$$f_x = M(x,y_0) + \int_{y_0}^{y} M_y(x,t)\,dt$$
$$= M(x,y_0) + M(x,y) - M(x,y_0) = M(x,y). \quad Q.E.D.$$

Theorem 4 is so important that its main hypothesis and its conclusion have special names. A vector field $\mathbf{F} = (M,N)$ is called *closed* if $M_y = N_x$, and *exact* if $\mathbf{F} = \nabla f$ for some function f. In these terms, the theorem says that *a vector field whose domain is a rectangle is closed if and only if it is exact.* When the domain is more general the situation becomes more complicated, but more interesting. Every exact field is still closed; but if the domain S has any "holes" (e.g. if $S = \{\mathbf{P}: |\mathbf{P}| > 0\}$, which has a hole at the origin, Fig. 12), then there are vector fields which are closed (i.e. $M_y = N_x$) but are *not* exact (i.e. (M,N) is not a gradient). Problem 3 gives one example. Problem 7, on the other hand, gives examples of sets "without holes" in which it is easy to prove (as in Theorem 4) that "closed" implies "exact" (Fig. 13).

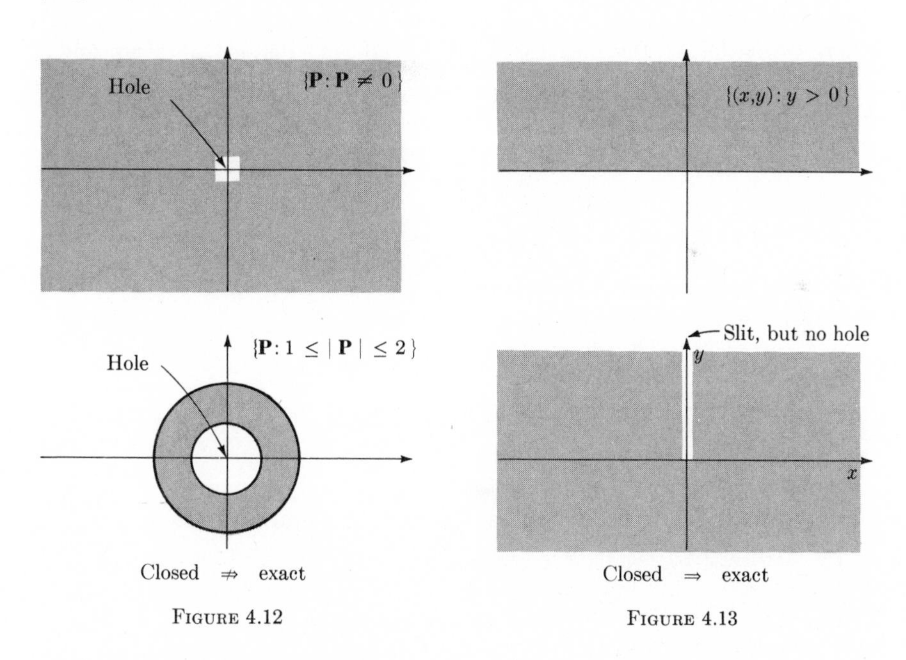

FIGURE 4.12

FIGURE 4.13

The proof of Theorem 4 does not merely say that a closed vector field has a potential f; it gives the explicit formula (2). Given a specific M and N and a little luck in evaluating the integrals, you can write down exactly what f is, up to the undetermined constant $f(x_0,y_0)$. This is as much as could be expected, since changing f by a constant does not affect ∇f.

Example 1. $M(x,y) = y, N(x,y) = x$. This vector field is defined in all of R^2, and $M_y = 1$, $N_x = 1$, so it is closed; hence formula (2) defines a potential f. Take $x_0 = 0$, $y_0 = 0$. Then $M(s,y_0) = y_0 = 0$, and $N(x,t) = x$, so formula (2) becomes

$$f(x,y) = f(0,0) + \int_0^x 0\,ds + \int_0^y x\,dt = f(0,0) + xy.$$

You can check that $\nabla f = (M,N)$, as required.

A certain special class of differential equations can be solved by Theorem 4. Suppose we want to solve

$$\varphi'(t) = G(t,\varphi(t)), \tag{3}$$

where $G(x,y)$ is a given function defined on a rectangle, and φ is to be determined by (3). Suppose that G can be written in the special form

$$G(x,y) = -\frac{M(x,y)}{N(x,y)}, \tag{4}$$

where $M_y = N_x$. Because (M,N) is closed, and defined on a rectangle, it is exact; there is an f with $\nabla f = (M,N)$. Then *any differentiable function* φ *such that*

$$f(t,\varphi(t)) = \text{constant} \tag{5}$$

is a solution of (3). For, differentiating (5) by the chain rule gives

$$D_1 f(t,\varphi(t)) + D_2 f(t,\varphi(t))\varphi'(t) = 0.$$

Since $D_1 f = M$ and $D_2 f = N$, we get

$$M(t,\varphi(t)) + N(t,\varphi(t))\varphi'(t) = 0,$$

which on account of (4) reduces to (3). (Notice that if $N = f_y \neq 0$, the implicit function theorem guarantees the existence of φ, at least on a small interval.)

Equation (3) is often written in Leibniz notation with $x = t$ and $y = \varphi(t)$. It then looks like

$$\frac{dy}{dx} = G(x,y),$$

and if $G = -M/N$ this becomes

$$\frac{dy}{dx} = -\frac{M}{N}, \qquad \text{or} \qquad N\frac{dy}{dx} + M = 0,$$

or

$$M\,dx + N\,dy = 0. \tag{6}$$

We have found that φ is a solution if $f(t,\varphi(t)) = \text{constant}$; in other words, φ is a solution if its graph lies along a contour line of f. Summing up briefly: *To solve the equation* (6) *when* $M_y = N_x$, *find a potential* f *with* $\nabla f = (M,N)$. *Then the contour lines of* f *are the graphs of solutions of* (6). Differential equations of this type are called *exact*.

Example 2. In the equation

$$y\,dx + x\,dy = 0 \tag{7}$$

we have $M = y$, $N = x$, hence $M_y = 1$ and $N_x = 1$, so (M,N) is closed. This vector field is the gradient of $f(x,y) = xy$, so the solutions of (7) are the contour lines $xy = c$, or $y = c/x$, where c is a constant. As a check: when $y = c/x$, then $dy/dx = -c/x^2$, hence $x^2\,dy + c\,dx = 0$; but $c = xy$, so $x^2\,dy + xy\,dx = 0$, or $x\,dy + y\,dx = 0$, as required by (7).

PROBLEMS

1. Decide which of the following vector fields are closed. For those **F** which are closed, find a potential function f such that $\nabla f = \mathbf{F}$.

(a) $(2xy,x^2)$ (i.e. $M(x,y) = 2xy$, $N(x,y) = x^2$.)

(b) $(2xy,-x^2)$

(c) (e^{x+y},e^{x+y})

(d) $(\sin(xy),\cos(xy))$

(e) (e^y,xe^y)

(f) $(\cos(x+y),\cos(x+y))$

(g) $\left(\dfrac{x}{x^2+y^2},\dfrac{y}{x^2+y^2}\right)$ (Notice that Theorem 4 does not apply; nevertheless, the vector field *is* exact. Find the potential f such that $f(1,0) = 0$.)

2. Of the following differential equations, solve the exact ones. (You need not express y explicitly as a function of x in every case.)

(a) $xe^{xy}\dfrac{dy}{dx} + ye^{xy} = 0$

(b) $ye^{xy}\dfrac{dy}{dx} + xe^{xy} = 0$

(c) $x\,dx + y\,dy = 0$

(d) $x^2\,dx + y^2\,dy = 0$

(e) $y^2\,dx + x^2\,dy = 0$

(f) $(x-y)\,dx + (y^2-x)\,dy = 0$

3. Let

$$\mathbf{F}(x,y) = \left(\frac{-y}{x^2+y^2},\frac{x}{x^2+y^2}\right)$$

be defined on the set $\{\mathbf{P}: |\mathbf{P}| > 0\}$.

(a) Prove that $\mathbf{F}$ is closed.

(b) Prove that if $\mathbf{F} = \nabla f$, where f is differentiable for $|\mathbf{P}| > 0$, then there are constants c_+ and c_- such that

$$f(x,y) = \arctan\frac{y}{x} + c_+ \quad \text{for } x > 0$$

$$f(x,y) = \arctan\frac{y}{x} + c_- \quad \text{for } x < 0.$$

(Recall that two functions defined in an open connected set and having the same gradient there can differ by at most a constant. See Problem 16, §3.3.)

(c) Prove that if the function f in part (b) is continuous at points on the positive half of the y axis, then $\pi/2 + c_+ = -\pi/2 + c_-$.

(d) Obtain a similar condition for continuity of f along the negative y axis.

(e) Show that the conditions in parts (c) and (d) are incompatible. (This proves that the function f postulated in part (b) cannot exist. Thus, $\mathbf{F}$ is closed, but not exact.)

4. The curve $\boldsymbol{\gamma}(t) = (r\cos(ct), r\sin(ct))$ describes a motion of rotation clockwise about the origin, c radians per unit time. Prove that when the moving point is at (x,y), then its velocity is $(-cy,cx)$. (Thus, $\mathbf{F}(x,y) = (-cy,cx)$ is the velocity field for a rotation of the plane about the origin.)

5. Suppose $\mathbf{F} = (M,N)$ is exact in an arbitrary open set S, and M_y and N_x are continuous in S. Prove that $\mathbf{F}$ is closed in S.

6. Problem 3 shows that a closed vector field need not be exact if it is defined on the set $S = \{\mathbf{P}: |\mathbf{P}| > 0\}$. What goes wrong in the proof of Theorem 4 given above if we replace the rectangle R by this set S?

7. Rectangles are not the only sets for which "closed" implies "exact." Show that the proof of Theorem 4 remains valid when the rectangle R is replaced by

(a) a disk

(b) the whole plane R^2

(c) the upper half plane $\{(x,y): y > 0\}$

(d) the set $S = \{(x,y): x \neq 0 \text{ or } y < 0\}$, which consists of the plane with the upper half of the y axis removed, Fig. 13. (Hint: Take $(x_0,y_0) = (0,-1)$).

(e) Define $\mathbf{F}$ as in Problem 3, and find a function f such that $\nabla f = \mathbf{F}$ for points in the set S given in part (d).

8. It can happen that the equation

$$M\,dx + N\,dy = 0 \tag{8}$$

is not exact, but that the equivalent equation obtained by multiplying by some nonzero function ρ, $\rho M\,dx + \rho N\,dy = 0$, *is* exact, i.e.

$$\frac{\partial \rho M}{\partial y} = \frac{\partial \rho N}{\partial x}.$$

Such a function ρ is called an *integrating factor*. The usual way to find an integrating factor is simply to guess. This doesn't succeed very often, so hints are provided with the following problems. Solve the given equations.

(a) $(1 + xy)\,dx + x^2\,dy = 0$ (Try $\rho(x,y) = e^{xy}$)
(b) $(1 - xy)\,dx - x^2\,dy = 0$
(c) $(xy^2 + y)\,dx - x\,dy = 0$ (Try $\rho(x,y) = y^{-2}$)
(d) $(y^2 + 1)\,dx + (xy - y\sqrt{y^2 + 1})\,dy = 0$ (Try $\rho = (y^2 + 1)^{-1/2}$)
(e) $\dfrac{dy}{dx} = \dfrac{e^x - y}{x}$

4.3 LINE INTEGRALS

In both of the main physical examples of vector fields there arises an important concept, the *line integral* of a vector field over a curve. In a force field, the work done by the force in moving a particle along a curve is given by a line integral; and in a velocity field, the rate of flow across a curve is given by a line integral.

The mathematical definition of line integrals is simple. Given a differentiable curve $\boldsymbol{\gamma}: [a,b] \to R^2$ and a vector field $\mathbf{F}$ which is continuous at every point on $\boldsymbol{\gamma}$, the line integral of $\mathbf{F}$ along $\boldsymbol{\gamma}$ is defined by

$$\int_\gamma \mathbf{F} = \int_a^b (\mathbf{F}\circ\boldsymbol{\gamma})\cdot\boldsymbol{\gamma}'. \tag{1}$$

We will give some examples and remarks on the evaluation and interpretation of (1), and finally prove a theorem relating line integrals and exact vector fields.

Example 1. If $\mathbf{F}(x,y) = (y,x)$ and $\boldsymbol{\gamma}(t) = (r\cos t, r\sin t)$ for $0 \le t \le \pi/4$, then $\boldsymbol{\gamma}'(t) = (-r\sin t, r\cos t)$, so

$$\begin{aligned}\mathbf{F}(\boldsymbol{\gamma}(t))\cdot\boldsymbol{\gamma}'(t) &= (r\sin t, r\cos t)\cdot(-r\sin t, r\cos t)\\ &= r^2(\cos^2 t - \sin^2 t) = r^2\cos(2t);\end{aligned}$$

hence

$$\int_\gamma \mathbf{F} = \int_0^{\pi/4} r^2\cos 2t\,dt = \frac{1}{2}r^2\sin 2t\Big]_0^{\pi/4} = \frac{r^2}{2}.$$

The Leibniz notation for the line integral (1) is

$$\int_\gamma M(x,y)\,dx + N(x,y)\,dy, \tag{2}$$

where M and N are the components of $\mathbf{F}$. To evaluate (2), simply substitute for x and y their values on the given curve, set

$$dx = \frac{dx}{dt}\,dt, \qquad dy = \frac{dy}{dt}\,dt,$$

and evaluate the integral between the given limits on t. Thus if $x = \gamma_1(t)$ and $y = \gamma_2(t)$, $a \le t \le b$, the expression (2) becomes

$$\int_a^b M(\gamma_1(t),\gamma_2(t))\gamma_1'(t)\,dt + N\,(\gamma_1(t),\gamma_2(t))\gamma_2'(t)\,dt$$

$$= \int_a^b (M(\boldsymbol{\gamma}(t))\gamma_1'(t) + N(\boldsymbol{\gamma}(t))\gamma_2'(t))\,dt$$

$$= \int_a^b \mathbf{F}(\boldsymbol{\gamma}(t))\cdot\boldsymbol{\gamma}'(t)\,dt,$$

which is equivalent to the definition (1).

Example 2. In Leibniz notation, the line integral in Example 1 is given as

$$\int_\gamma y\,dx + x\,dy, \qquad x = r\cos t, \quad y = r\sin t, \quad 0 \le t \le \pi/4.$$

Making the indicated substitutions, we get

$$\int_0^{\pi/4} (r\sin t(-r\sin t\,dt) + r\cos t(r\cos t\,dt)) = \int_0^{\pi/4} r^2(\cos^2 t - \sin^2 t)\,dt = \frac{r^2}{2}.$$

Another common notation for line integrals is derived from the formulas

$$\frac{ds}{dt} = |\boldsymbol{\gamma}'(t)|, \qquad \mathbf{T} = \frac{1}{|\boldsymbol{\gamma}'|}\boldsymbol{\gamma}',$$

where s is arc length on the curve and $\mathbf{T}$ is the unit tangent vector (§2.3). Substituting these in (1) suggests

$$\int_\gamma \mathbf{F} = \int_a^b (\mathbf{F}\circ\boldsymbol{\gamma})\cdot\frac{\boldsymbol{\gamma}'}{|\boldsymbol{\gamma}'|}|\boldsymbol{\gamma}'|\,dt = \int_\gamma \mathbf{F}\cdot\mathbf{T}\,ds.$$

Since $\mathbf{T}$ is a unit vector, $\mathbf{F}\cdot\mathbf{T}$ is the component of $\mathbf{F}$ in the tangent direction $\mathbf{T}$, called the *tangential component* of $\mathbf{F}$ (see Fig. 14); the line integral $\int_\gamma \mathbf{F}\cdot\mathbf{T}\,ds$ is the integral of this tangential component with respect to arc length on the curve.

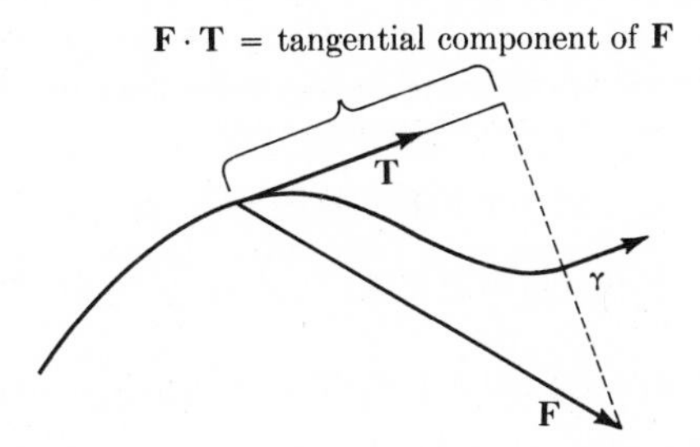

FIGURE 4.14

When **F** is a force, then $\int_\gamma \mathbf{F} = \int_\gamma \mathbf{F}\cdot\mathbf{T}\,ds$ is defined to be the *work* done by **F** in moving a particle along $\boldsymbol{\gamma}$ from beginning to end.

When $\mathbf{F} = (M,N)$ is the velocity field of a flow, there is a line integral giving the rate of flow across $\boldsymbol{\gamma}$ from left to right; it is not $\int_\gamma \mathbf{F} = \int_\gamma M\,dx + N\,dy$, but

$$\int_\gamma (-N\,dx + M\,dy). \tag{3}$$

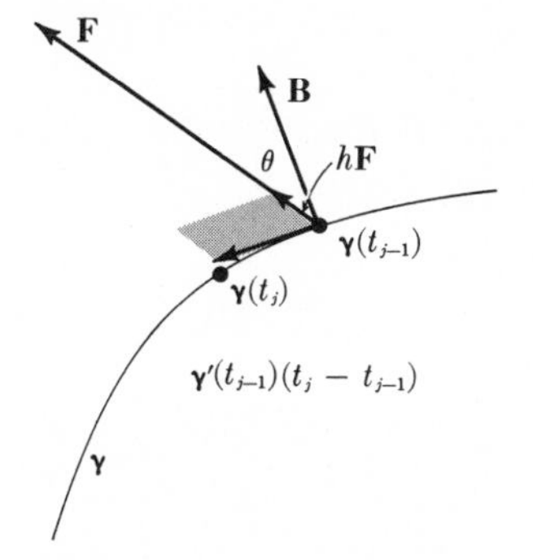

FIGURE 4.15

To see why this integral is interpreted as the rate of flow, consider a small section of the curve from $\boldsymbol{\gamma}(t_{j-1})$ to $\boldsymbol{\gamma}(t_j)$. In a short time interval of length h, the fluid near this part of the curve has a displacement of approximately $h\mathbf{F}(\boldsymbol{\gamma}(t_{j-1}))$. Hence the fluid that has crossed this section of curve in the given time h is contained approximately in the parallelogram in Fig. 15. The parallelogram has sides $h\mathbf{F}(\boldsymbol{\gamma}(t_{j-1}))$ and $\boldsymbol{\gamma}'(t_{j-1})(t_j - t_{j-1})$, so its area is

$$A = h \cdot |\mathbf{F}| \cdot |\boldsymbol{\gamma}'| \cdot (t_j - t_{j-1}) \cos\theta,$$

where θ is the angle between $\mathbf{F}$ and the normal $\mathbf{B}$ to the curve, obtained by rotating $\boldsymbol{\gamma}'(t_{j-1})$ clockwise through an angle $\pi/2$. (Rotating through $\pi/2$ makes the area positive for a flow from left to right, and negative from right to left; rotating through $-\pi/2$ would reverse this.) Since $\mathbf{B} = (\gamma_2'(t_{j-1}), -\gamma_1'(t_{j-1}))$, we have $|\mathbf{B}| = |\boldsymbol{\gamma}'|$, and $|\mathbf{F}|\,|\boldsymbol{\gamma}'| \cos\theta = \mathbf{F}\cdot\mathbf{B}$, so the parallelogram has area

$$A = h(\mathbf{F}\cdot\mathbf{B})(t_j - t_{j-1}).$$

Evaluating the inner product of $\mathbf{B}$ with $\mathbf{F} = (M, N)$, we get $\mathbf{F}\cdot\mathbf{B} = M\gamma_2' - N\gamma_1'$, hence

$$A = h[M(\boldsymbol{\gamma}(t_{j-1}))\gamma_2'(t_{j-1}) - N(\boldsymbol{\gamma}(t_{j-1}))\gamma_1'(t_{j-1})].$$

Adding up these areas for each section of the curve, we get a Riemann sum for

$$h\int_a^b [M(\boldsymbol{\gamma})\gamma_2' - N(\boldsymbol{\gamma})\gamma_1'] = h\int_{\boldsymbol{\gamma}} M\,dy - N\,dx.$$

This is the fluid that crosses in time h, so the *rate* of flow is given by (3).

These physical applications, and the mathematical theory as well, require line integrals not only over smooth differentiable curves (ellipses, parabolas, exponentials, etc.), but over "curves with corners" (rectangles and triangles, for example). To handle this situation, we introduce the concept of a *chain of curves*, namely, a finite sequence of differentiable curves $\boldsymbol{\gamma}^1, \ldots, \boldsymbol{\gamma}^n$, each curve $\boldsymbol{\gamma}^j$ defined on some interval $[a_j, b_j]$. The chain is called *connected* if the endpoint of each curve coincides with the beginning of the next, i.e. $\boldsymbol{\gamma}^{j-1}(b_{j-1}) = \boldsymbol{\gamma}^j(a_j)$. (See Fig. 16.) The *endpoints* of a connected chain are the beginning, $\boldsymbol{\gamma}^1(a_1)$, and the end, $\boldsymbol{\gamma}^n(b_n)$. A connected chain is called *closed* if the beginning and end coincide, that is, if $\boldsymbol{\gamma}^1(a_1) = \boldsymbol{\gamma}^n(b_n)$. (See Fig. 17.)

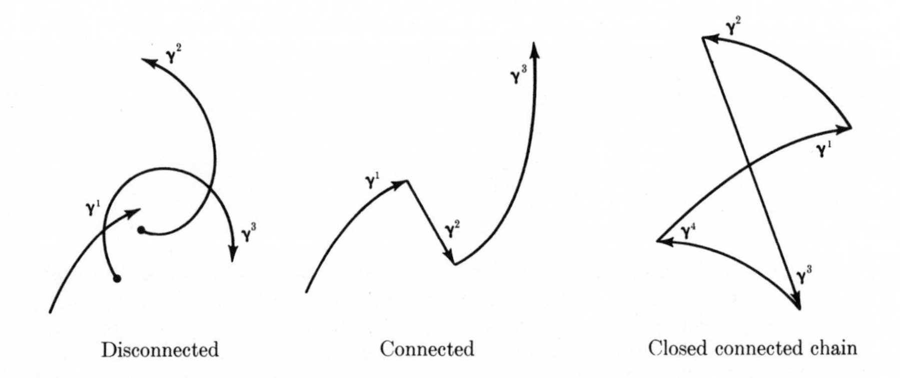

FIGURE 4.16

FIGURE 4.17

A chain $\boldsymbol{\gamma}^1, \ldots, \boldsymbol{\gamma}^n$ can be denoted by a single letter $\boldsymbol{\gamma}$. The line integral of $\mathbf{F}$ over such a chain is simply the sum of the integrals over all the pieces,

$$\int_\gamma \mathbf{F} = \sum_{j=1}^{n} \int_{\gamma_j} \mathbf{F}.$$

Now we are ready to take up the connection between exact vector fields and line integrals. The starting point is a simple expression for the integral of a gradient ∇f over a differentiable curve $\boldsymbol{\gamma}$. Using the chain rule and the fundamental theorem of calculus, we find

$$\int_\gamma \nabla f = \int_a^b (\nabla f \circ \boldsymbol{\gamma}) \cdot \boldsymbol{\gamma}' = \int_a^b (f \circ \boldsymbol{\gamma})' = (f \circ \boldsymbol{\gamma})(b) - (f \circ \boldsymbol{\gamma})(a),$$

or

$$\int_\gamma \nabla f = f(\boldsymbol{\gamma}(b)) - f(\boldsymbol{\gamma}(a)). \tag{4}$$

This handy formula bypasses the work of evaluating an integral; but beyond that, it is interesting because, given f, *the value of the integral depends only on the endpoints of the curve* $\boldsymbol{\gamma}$, i.e. it depends only on $\boldsymbol{\gamma}(b)$ and $\boldsymbol{\gamma}(a)$. This is obvious from the right-hand side of (4). The same phenomenon occurs, more generally, if $\boldsymbol{\gamma}$ is a *connected chain* consisting of differentiable curves $\boldsymbol{\gamma}^1, \ldots, \boldsymbol{\gamma}^n$; in this case we have

$$\begin{aligned}\int_\gamma \nabla f = \sum_{j=1}^{n} \int_{\gamma^j} \nabla f = f(\boldsymbol{\gamma}^1(b_1)) &- f(\gamma^1(a_1)) \\ + f(\boldsymbol{\gamma}^2(b_2)) &- f(\boldsymbol{\gamma}^2(a_2)) \\ + f(\boldsymbol{\gamma}^3(b_3)) &- f(\boldsymbol{\gamma}^3(a_3)) \\ + \cdots & \\ + f(\boldsymbol{\gamma}^{n-1}(b_{n-1})) &- f(\boldsymbol{\gamma}^{n-1}(a_{n-1})) \\ + f(\boldsymbol{\gamma}^n(b_n)) &- f(\boldsymbol{\gamma}^n(a_n)).\end{aligned}$$

On the right-hand side, all but two terms cancel out (as suggested by the dotted lines), since $\boldsymbol{\gamma}^{j-1}(b_{j-1}) = \boldsymbol{\gamma}^j(a_j)$. Thus the sum reduces to $f(\boldsymbol{\gamma}^n(b_n)) - f(\boldsymbol{\gamma}^1(a_1))$, which depends only on the endpoints of the chain.

This phenomenon, that $\int_\gamma \mathbf{F}$ is determined by the endpoints of $\boldsymbol{\gamma}$, characterizes the exact vector fields:

Theorem 5. *Let* $\mathbf{F}$ *be a continuous vector field defined in an open connected set* S. *Then* $\mathbf{F}$ *is exact if and only if for every connected chain* $\boldsymbol{\gamma}$ *in* S, $\int_\gamma \mathbf{F}$ *is determined by the endpoints of* $\boldsymbol{\gamma}$.

Proof. If $\mathbf{F}$ is exact, then $\mathbf{F} = \nabla f$ for some f, so $\int_\gamma \mathbf{F} = \int_\gamma \nabla f$ is determined by the endpoints, as we have already seen. Conversely, suppose that $\int_\gamma \mathbf{F}$ is determined by the endpoints of $\boldsymbol{\gamma}$. Pick a fixed point $\mathbf{P}_0$ in S. Since S is open and connected, each point $\mathbf{P}$ in S can be joined to $\mathbf{P}_0$ by a differentiable curve $\boldsymbol{\gamma}^{\mathbf{P}}$. (This is precisely the *definition* of a connected open set; see the problems in §3.3.) We define a function f by

$$f(\mathbf{P}) = \int_{\gamma \mathbf{P}} \mathbf{F}; \tag{5}$$

because of our assumption about $\mathbf{F}$, *the integral* (5) *gives the same value for* $f(\mathbf{P})$ *no matter what chain of curves* $\boldsymbol{\gamma}^{\mathbf{P}}$ *is chosen from* $\mathbf{P}_0$ *to* $\mathbf{P}$. It remains to show that $\nabla f = \mathbf{F}$; in other words, if $\mathbf{F} = (M,N)$, then $D_1 f = M$ and $D_2 f = N$. We will prove the first equality,

$$D_1 f(x,y) = \lim_{h \to 0} \frac{f(x + h,y) - f(x,y)}{h} = M(x,y), \tag{6}$$

and leave the other to you.

Let $\boldsymbol{\gamma}^{\mathbf{P}}$ be any curve from $\mathbf{P}_0$ to $\mathbf{P} = (x,y)$, and set

$$\boldsymbol{\gamma}^h(t) = (x + th,y), \qquad 0 \le t \le 1.$$

Then $f(x + h,y)$ can be evaluated by integrating over the chain $\boldsymbol{\gamma}$ consisting of $\boldsymbol{\gamma}^{\mathbf{P}}$ and $\boldsymbol{\gamma}^h$ (Fig. 18). Thus, by (5),

$$f(x + h,y) = \int_{\gamma \mathbf{P}} \mathbf{F} + \int_{\gamma h} \mathbf{F} = f(x,y) + \int_{\gamma h} \mathbf{F},$$

so $f(x + h,y) - f(x,y) = \int_{\gamma h} \mathbf{F}$, and

$$\frac{f(x + h,y) - f(x,y)}{h} = \frac{1}{h}\int_{\gamma h} \mathbf{F} = \frac{1}{h}\int_0^1 \mathbf{F}(x + th,y) \cdot (h,0)\, dt$$

$$\text{(since } (\boldsymbol{\gamma}^h)'(t) = (h,0))$$

$$= \frac{1}{h}\int_0^1 M(x + th,y)h\, dt \qquad \text{(since } \mathbf{F} = (M,N))$$

$$= \frac{1}{h}\int_0^h M(x + s,y)\, ds \qquad \text{(substituting } s = th).$$

Now, as $h \to 0$, this last expression tends to $M(x,y)$, by the fundamental theorem of calculus, so (6) is proved. Similarly, $D_2 f = N$, and Theorem 5 is proved.

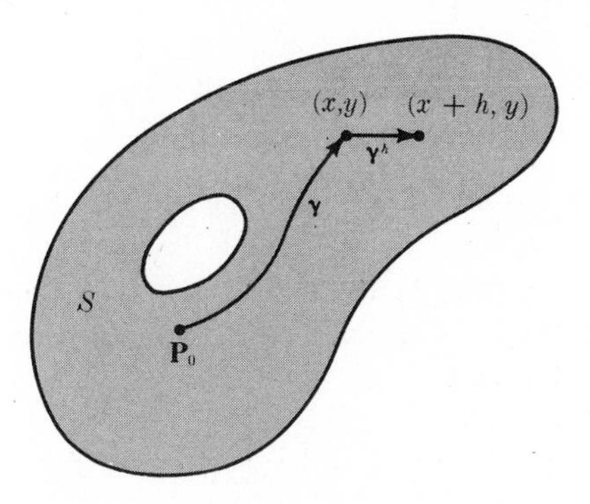

FIGURE 4.18

Corollary 1. *If* $\boldsymbol{\gamma}$ *is a closed chain and* **F** *is exact, then* $\int_\gamma \mathbf{F} = 0$.

Proof. Since **F** is exact, we have $\mathbf{F} = \nabla f$. Since $\boldsymbol{\gamma}$ is connected, we have

$$\int_\gamma \mathbf{F} = \int_\gamma \nabla f = f(\mathbf{P}_2) - f(\mathbf{P}_1) \tag{7}$$

where $\mathbf{P}_1$ and $\mathbf{P}_2$ are the endpoints of the chain. But $\mathbf{P}_2 = \mathbf{P}_1$ (since $\boldsymbol{\gamma}$ is a closed chain), so (7) shows that $\int_\gamma \mathbf{F} = 0$. *Q.E.D.*

We now have two ways to prove that a vector field $\mathbf{F} = (M,N)$ is *not* exact: (i) show that it is not closed, i.e. $M_y \neq N_x$, or (ii) find a closed chain $\boldsymbol{\gamma}$ such that $\int_\gamma \mathbf{F} \neq 0$ (which, by the corollary above, implies that **F** is not exact.)

It is a little harder to prove that a vector field **F** *is* exact. We can (i) find explicitly a function f such that $\nabla f = \mathbf{F}$, or (ii) apply Theorem 4: **F** is exact in a rectangle R if **F** is closed in R. The second half of Theorem 5 (which says that **F** is exact if $\int_\gamma \mathbf{F}$ depends only on the endpoints of $\boldsymbol{\gamma}$) is hard to apply, since it has to be checked for *all* chains $\boldsymbol{\gamma}$.

Example 3. Problem 3 of the preceding section showed that

$$\mathbf{F}(x,y) = \left(\frac{-y}{x^2+y^2}, \frac{x}{x^2+y^2}\right)$$

is closed, but not exact. Now we can prove the nonexactness very simply by integrating **F** around a circle $\boldsymbol{\gamma}(t) = (r\cos t, r\sin t)$, $0 \le t \le 2\pi$:

$$\int_\gamma \mathbf{F} = \int_0^{2\pi} (\mathbf{F}\circ\boldsymbol{\gamma})\cdot\boldsymbol{\gamma}' = \int_0^{2\pi} \left(\frac{-r\sin t}{r^2}, \frac{r\cos t}{r^2}\right)\cdot(-r\sin t, r\cos t)\,dt$$

$$= \int_0^{2\pi} (\sin^2 t + \cos^2 t)\,dt = 2\pi.$$

Since $\boldsymbol{\gamma}$ is a closed curve and the integral is not zero, $\mathbf{F}$ is not exact, by Corollary 1.

Example 4. The line integral in Examples 1 and 2,

$$\int_\gamma y\,dx + x\,dy, \qquad x = r\cos t,\ y = r\sin t,\ 0 \le t \le \pi/4,$$

can be evaluated by formula (4). Since the vector field $\mathbf{F} = (y,x)$ is the gradient of $f(x,y) = xy$, and since the curve begins at $(r,0)$ and ends at $(r/\sqrt{2}, r/\sqrt{2})$ we have

$$\int_\gamma y\,dx + x\,dy = f\left(\frac{r}{\sqrt{2}}, \frac{r}{\sqrt{2}}\right) - f(r,0) = \frac{r}{\sqrt{2}}\cdot\frac{r}{\sqrt{2}} - 0\cdot r = \frac{r^2}{2}.$$

Example 5. Evaluate $\int_\gamma x^2\,dx + xy\,dy$, where $\boldsymbol{\gamma}$ is the straight line segment from (1,0) to (2,3). *Solution.* Strictly speaking, this is not a well-defined problem, since $\int_\gamma \mathbf{F}$ is defined only when $\boldsymbol{\gamma}$ is a curve, i.e. a function from an interval $[a,b]$ to the plane. However, "the straight line from (1,0) to (2,3)" has a natural parametrization of the form $\mathbf{P} = \mathbf{P}_1 + t(\mathbf{P}_2 - \mathbf{P}_1)$, which gives

$$x = 1 + t, \qquad y = 3t, \qquad 0 \le t \le 1, \tag{8}$$

and we use this to evaluate the integral:

$$\int_\gamma x^2\,dx + xy\,dy = \int_0^1 (1+t)^2\,dt + (1+t)3t\cdot 3\,dt$$

$$= \int_0^1 (1 + 11t + 10t^2)\,dt = 1 + \frac{11}{2} + \frac{10}{3}.$$

Example 5 raises an obvious question: What if you use some other parametrization for the given line segment, for example,

$$y = \bar{t}, \qquad x = 1 + \tfrac{1}{3}\bar{t}, \qquad 0 \le \bar{t} \le 3\,? \tag{9}$$

You would find, in fact, the same value for the line integral. The curves given in (8) and (9) are related by a change of parameter, and this does not affect the line integral (see Problem 14 below). It is important, however, that in each parametrization the curve runs *from* (1,0) *to* (2,3), not vice versa; reversing the endpoints would change the sign of the integral, as you can easily check by evaluating $\int x^2\,dx + xy\,dy$ with a parametrization of the segment from (2,3) to (1,0), for example,

$$x = 1 - t, \qquad y = -3t, \qquad -1 \le t \le 0.$$

Example 6. A force $\mathbf{F}$ is called *conservative* if it is exact. Show that the force $\mathbf{F}(x,y) = (y\cos xy, x\cos xy)$ is conservative, and find the work done by this force in moving a particle from the origin to the point (3,8).

Solution. The force **F** is *closed*, since

$$\frac{\partial M}{\partial y} = \frac{\partial}{\partial y}(y \cos xy) = \cos xy - xy \sin xy = \frac{\partial N}{\partial x}.$$

Since **F** is defined in all of R^2, it is exact; you can easily find that **F** is the gradient of $f(xy) = \sin xy$. Hence the work done by **F**, the line integral $\int_\gamma \mathbf{F}$, depends only on the endpoints of $\boldsymbol{\gamma}$. In particular, the work done in moving from the origin to the point (3,8) is $f(3,8) - f(0,0) = \sin 24$.

Example 7. The *divergence* of a flow $\mathbf{F} = (M,N)$ is defined to be $M_x + N_y$. Show that if the flow **F** is defined in a rectangle R, and has zero divergence at every point of R, then the rate of flow across every closed chain in R is zero. *Solution.* The rate of flow across $\boldsymbol{\gamma}$ is $\int_\gamma (-N\,dx + M\,dy)$. From the given equation $M_x + N_y \equiv 0$, it follows that the vector field $(-N, M)$ is closed in R; hence (by Theorem 4) it is exact in R; hence (by the Corollary of Theorem 5) $\int_\gamma (-N\,dx + M\,dy) = 0$ for every closed chain $\boldsymbol{\gamma}$ in R.

PROBLEMS

Problems 10, 12, and 13 concern *change of parameter*, and 17–19 concern the geometric concept of *winding number of a curve.*

1. Compute the following line integrals $\int_\gamma \mathbf{F}$:
(a) $\mathbf{F}(x,y) = (x^2y, xy^2)$, $\boldsymbol{\gamma}(t) = (t,t^2)$, $0 \le t \le 1$
(b) $\int x^2y\,dx + xy^2\,dy$, $x = c_1t$, $y = c_2t$, $0 \le t \le 1$
(c) $\mathbf{F}(x,y) = (x \sin y, \cos y)$, $\boldsymbol{\gamma}$ as in part (b)
(d) $\mathbf{F}(x,y) = (y,x)$, $\boldsymbol{\gamma}$ is the chain with

$$\boldsymbol{\gamma}^1(t) = (t^2,t), \qquad 0 \le t \le 1$$

$$\boldsymbol{\gamma}^2(t) = (1-t, 1-t), \qquad 0 \le t \le 1$$

(e) $\mathbf{F}(x,y) = (-y,0)$, $\boldsymbol{\gamma}$ is the straight line path from (x_1,y_1) to (x_2,y_2) (*Answer:* $\frac{1}{2}(x_1 - x_2)(y_1 + y_2)$.)

2. Obtain a parametric representation for each of the following curves $\boldsymbol{\gamma}$.
(a) $\boldsymbol{\gamma}$ is the graph of $y = x^3$ from (0,0) to (1,1).
(b) $\boldsymbol{\gamma}$ is the straight line from a given point (x_1,y_1) to (x_2,y_2).
(c) $\boldsymbol{\gamma}$ is the upper half of the unit circle, taken from left to right.
(d) $\boldsymbol{\gamma}$ is a closed chain running counterclockwise around the boundary of the triangle in Fig. 19.

3. Evaluate $\int (dx - dy)$ over each of the curves (or chains) in Problem 2. (This can actually be done without answering Problem 2.)

4. Find the work done by the following forces **F** over the given curves $\boldsymbol{\gamma}$. (Work $= \int_\gamma \mathbf{F}$.)
(a) $\mathbf{F} = (0,-1)$, $\boldsymbol{\gamma}(t) = (t, 1 - t^2)$, $-1 \le t \le 0$

(b) Same $\mathbf{F}$ and γ, but $0 \leq t \leq 1$

(c) $\mathbf{F}(\mathbf{P}) = -\mathbf{P}/|\mathbf{P}|^3, \quad \gamma(t) = (\cos t, \sin t), \quad 0 \leq t \leq \pi$

(d) $\mathbf{F}(\mathbf{P}) = -\mathbf{P}/|\mathbf{P}|^3, \quad \gamma(t) = \left(\frac{t}{\sqrt{2}}, \frac{t}{\sqrt{2}}\right), \quad a \leq t \leq b$

(e) $\mathbf{F}(\mathbf{P}) = -\mathbf{P}/|\mathbf{P}|^3, \quad \gamma(t) = (t,0), \quad a \leq t \leq b$

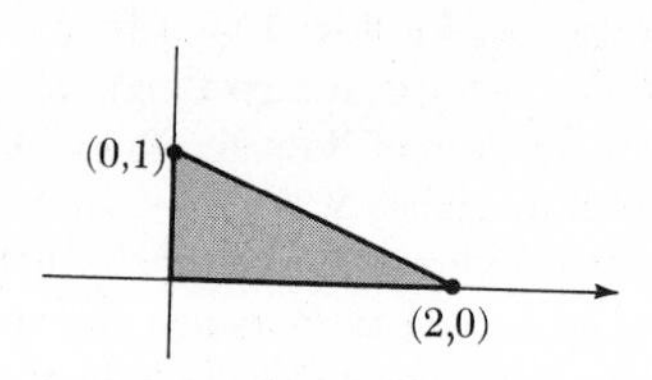

FIGURE 4.19

5. Let $f(\mathbf{P}) = c/|\mathbf{P}|$ for $|\mathbf{P}| \neq 0$, and $\mathbf{F} = \nabla f$.

(a) A particle originally at a point $\mathbf{P}_1$ at distance r_1 from the origin is moved by $\mathbf{F}$ to a point $\mathbf{P}_2$ at distance r_2 from the origin. Find the work done. (Note: By Theorem 5, the answer depends only on the starting point $\mathbf{P}_1$ and ending point $\mathbf{P}_2$, but in this case you will find that it actually depends only on r_1 and r_2.)

(b) Find the work done in moving from a point r_1 units from the origin "all the way to ∞." (When the constant c is properly chosen, and r_1 is taken as the radius of the earth, your answer gives the work required to move the "particle" from the surface of the earth to a point "completely out of the earth's gravity.")

(c) Repeat parts (a) and (b), but with $f(\mathbf{P}) = -\log|\mathbf{P}|$.

6. Let $\mathbf{F}(x,y) = (1,0)$ be the velocity field for a constant flow parallel to the x axis. Find the rate of flow from left to right across each of the five curves shown in Fig. 20. (To interpret "left to right" in each figure, face in the direction of the given curve.)

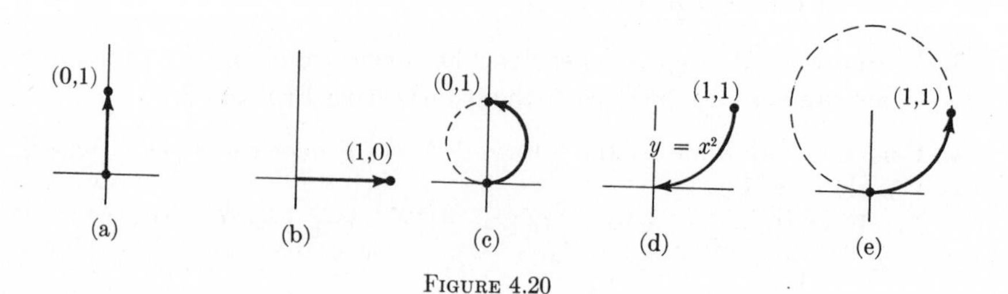

FIGURE 4.20

7. Repeat Problem 6 with $\mathbf{F}(x,y) = (x,0)$.

8. Show that the function $f(x,y)$ constructed in Theorem 4, page 799, is actually $f(x_0,y_0) + \int_\gamma \mathbf{F}$ for a certain chain $\boldsymbol{\gamma}$ from (x_0,y_0) to (x,y).

9. (a) Show that if $\boldsymbol{\gamma}$ is a *constant* curve, then $\int_\gamma \mathbf{F} = 0$ for every $\mathbf{F}$.
(b) Show that if $\boldsymbol{\gamma}$ is defined on a "one-point interval" $[a,a]$, then $\int_\gamma \mathbf{F} = 0$ for every $\mathbf{F}$.

10. If $\boldsymbol{\gamma}$ is defined on $[a,b]$, its *inverse* $\boldsymbol{\gamma}^{-1}$ is given by

$$\boldsymbol{\gamma}^{-1}(t) = \boldsymbol{\gamma}(-t), \qquad -b \le t \le -a.$$

(a) Prove that $\boldsymbol{\gamma}^{-1}$ runs through exactly the same points as $\boldsymbol{\gamma}$, but starting at $\boldsymbol{\gamma}(b)$ and ending at $\boldsymbol{\gamma}(a)$, i.e. γ^{-1} runs backward over $\boldsymbol{\gamma}$. (See Fig. 21.)
(b) Prove that for any continuous $\mathbf{F}$, $\int_{\gamma^{-1}} \mathbf{F} = -\int_\gamma \mathbf{F}$.

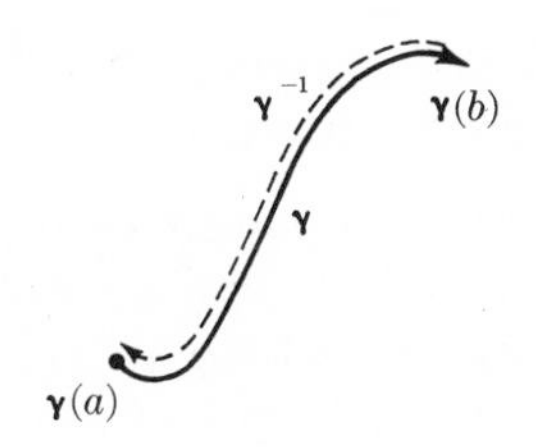

FIGURE 4.21

11. In the second half of Theorem 5, prove that $D_2 f = N$.

12. Suppose that $\boldsymbol{\gamma}$ is a differentiable curve lying in the graph of a function φ defined on $c \le x \le d$, beginning at $(c,\varphi(c))$ and ending at $(d,\varphi(d))$. Prove that

$$\int_\gamma M(x,y)\, dx = \int_c^d M(x,\varphi(x))\, dx.$$

(Hint: If $\boldsymbol{\gamma} = (\gamma_1,\gamma_2)$, then $c \le \gamma_1 \le d$ and $\gamma_2 = \varphi \circ \gamma_1$; make a change of variable in the integral.)

13. (*Change of parameter.*) If $\boldsymbol{\gamma}$ is defined on the interval $[a,b]$ and g is a differentiable function from some interval $[\bar{a},\bar{b}]$ into $[a,b]$, then we can form a new curve $\bar{\boldsymbol{\gamma}} = \boldsymbol{\gamma} \circ g$ defined on $[\bar{a},\bar{b}]$ by the formula

$$\bar{\boldsymbol{\gamma}}(\bar{t}) = \boldsymbol{\gamma}(g(\bar{t})), \qquad \bar{a} \le \bar{t} \le \bar{b}.$$

(a) If $g' \ge 0$, $g(\bar{a}) = a$ and $g(\bar{b}) = b$, then g is called an *orientation-preserving change of parameter.* Prove that for such a g, $\boldsymbol{\gamma} \circ g$ runs through the same points as $\boldsymbol{\gamma}$ and in the same order (although the speed is generally different, since $\bar{\gamma}' = g'\gamma'$ by the chain rule). (Hint: Use the one-variable Intermediate Value theorem.)

(b) If $g' \leq 0$, $g(\bar{a}) = b$ and $g(\bar{b}) = a$, then g is an *orientation-reversing* change of parameter. Prove that for such a g, $\gamma \circ g$ runs through the same points as γ but in the *opposite* order.

(c) Prove that if g is an orientation-preserving change of parameter, and $\mathbf{F}$ is continuous along γ, then $\int_{\gamma \circ g} \mathbf{F} = \int_\gamma \mathbf{F}$.

(d) Prove that if g is orientation-reversing, then $\int_{\gamma \circ g} \mathbf{F} = -\int_\gamma \mathbf{F}$.

(*Remark 1:* The moral of (c) and (d) is that $\int_\gamma \mathbf{F}$ depends on $\mathbf{F}$, on the points through which γ passes, and on the order (reversing the order reverses the sign of the integral), but *not* on the speed of γ. *Remark 2:* If the new curve $\gamma \circ g$ is parametrized by arc length s, we have the curve $\mathbf{G}$ discussed in §2.3. Here, $g' > 0$, so by part (c)

$$\int_\gamma \mathbf{F} = \int_{\mathbf{G}} \mathbf{F} = \int_{\bar{a}}^{\bar{b}} \mathbf{F}(\mathbf{G}(s)) \cdot \mathbf{G}'(s)\, ds.$$

Since $\mathbf{G}'(s) = \mathbf{T}(s) =$ the unit tangent vector, we arrive again at the notation $\int_\gamma \mathbf{F} \cdot \mathbf{T}\, ds$ suggested in the text.)

14. (*Conservation of energy.*) The *kinetic energy* of a particle of mass m is defined to be $\frac{1}{2}m\,|\gamma'|^2$, where γ' is the velocity. Assuming Newton's law $\mathbf{F} = m\gamma''$, prove that the work done by the force $\mathbf{F}$ in moving the particle along the curve γ equals the increase in kinetic energy. (Hint: $(|\gamma'|^2)' = 2\gamma' \cdot \gamma''$.)

15. Suppose that $\mathbf{F} = (M,N)$ is defined on a set S, and $\int_\gamma \mathbf{F} = 0$ whenever γ is a *closed* connected chain lying in S.

(a) Prove that when γ is any connected chain in S, then $\int_\gamma \mathbf{F}$ depends only on the endpoints of γ. (Hint: See Problem 10.)

(b) Prove that $\mathbf{F}$ is exact in S.

16. The *divergence* of a flow $\mathbf{F} = (M,N)$ is defined to be $M_x + N_y$. Show that if the rate of flow across every closed chain γ is zero, then the divergence is zero. (Hint: If $\int_\gamma -N\, dx + M\, dy = 0$ for every closed chain γ, then $(-N,M)$ is exact, by Problem 15.)

17. (*Winding numbers.*) Suppose that γ is a closed connected chain that does not pass through the origin, i.e. for every t, $\gamma(t) \neq (0,0)$. The *winding number of* γ *about the origin* is defined to be

$$\frac{1}{2\pi} \int_\gamma \frac{-y}{x^2 + y^2}\, dx + \frac{x}{x^2 + y^2}\, dy.$$

Compute the winding numbers of the following chains about the origin. In each case sketch the chain, and note that the winding number gives the net number of times that the chain encircles the origin counterclockwise.

(a) $\boldsymbol{\gamma}^1(t) = (r\cos t, r\sin t), \quad 0 \le t \le 2\pi$
(b) $\boldsymbol{\gamma}^2(t) = (r\cos(-t), r\sin(-t)), \quad 0 \le t \le 2\pi$
(c) $\boldsymbol{\gamma}^3(t) = (r\cos t, r\sin t), \quad -2\pi \le t \le 2\pi$
(d) $\boldsymbol{\gamma}^4(t) = (2r - r\cos t, r\sin t), \quad 0 \le t \le 2\pi$
(Hint: Prove that

$$\left(\frac{-y}{x^2+y^2}, \frac{x}{x^2+y^2}\right) = \nabla\left(\arctan\frac{y}{x}\right)$$

in a rectangle containing $\boldsymbol{\gamma}^4$.)
(e) $\boldsymbol{\gamma}$ is the chain consisting of $\boldsymbol{\gamma}^3$ and $\boldsymbol{\gamma}^4$
(f) $\boldsymbol{\gamma}$ is a chain running clockwise around the square

$$\{(x,y) : |x| \le 1 \text{ and } |y| \le 1\}.$$

18. (a) Prove that the vector field

$$\left(\frac{-y}{x^2+y^2}, \frac{x}{x^2+y^2}\right)$$

is exact in the set $S = \{(x,y) : y \ne 0, \text{ or } x > 0\}$, consisting of the plane with the left half of the x axis removed. In other words, there is a differentiable function f defined in S with

$$\nabla f = \left(\frac{-y}{x^2+y^2}, \frac{x}{x^2+y^2}\right).$$

(Hint: Imitate the proof of Theorem 4, as in §4.2, Problem 7(e).)
(b) Prove that if $\boldsymbol{\gamma}$ is a closed chain not passing through the left half of the x axis (the set $\{(x,y) : y = 0 \text{ and } x \le 0\}$), then the winding number of $\boldsymbol{\gamma}$ about the origin is zero. (See the previous problem for the definition of winding number.)

19. The vector field

$$\left(\frac{-y}{x^2+y^2}, \frac{x}{x^2+y^2}\right)$$

is *not* the gradient of any function defined in all of $\{\mathbf{P} : |\mathbf{P}| > 0\}$; nevertheless, it can reasonably be denoted $\nabla\theta$, where θ is the angular polar coordinate. The angle θ in polar coordinates is traditionally called the *argument*, and the integral $\int_{\boldsymbol{\gamma}} \nabla\theta$ is called the *change in argument along* $\boldsymbol{\gamma}$. This problem justifies the name "change in argument," and draws some conclusions about winding numbers. (See also Problem 17(d), and note that $\tan\theta = y/x$.)
(a) If $\boldsymbol{\gamma}$ is a differentiable curve defined on the interval $[a,b]$, set

$$\varphi(s) = \int_a^s [(\nabla\theta)\circ\boldsymbol{\gamma}]\cdot\boldsymbol{\gamma}',$$

that is, $\varphi(s)$ is the integral of $\nabla\theta$ along the part of the curve where $a \leq t \leq s$. Prove that the derivative of

$$\exp\left[-i\varphi(s)\right]\left(\frac{\gamma_1(s) + i\gamma_2(s)}{|\gamma(s)|}\right)$$

is identically zero. (This is a long but straightforward calculation, so do it carefully. The notation is simplified if you set $x = \gamma_1(s)$, $y = \gamma_2(s)$, $r = \sqrt{x^2 + y^2}$, denote the derivatives by $x' = \gamma_1'(s)$, $y' = \gamma_2'(s)$, and notice that $r' = xx'/r + yy'/r$.)

(b) With $\varphi(s)$ as above, let θ_1 denote an angular coordinate for $\boldsymbol{\gamma}(a)$, that is, $\gamma_1(a) + i\gamma_2(a) = |\boldsymbol{\gamma}(a)|e^{i\theta_1}$. Prove that

$$\exp\left\{-i[\varphi(s) + \theta_1]\right\}\left(\frac{\gamma_1(s) + i\gamma_2(s)}{|\boldsymbol{\gamma}(s)|}\right) \equiv 1,$$

hence $\theta_1 + \varphi(s)$ is an angular coordinate for $\boldsymbol{\gamma}(s)$. (This has the interpretation that $\varphi(s)$ gives the net change in angular coordinate (argument) as t varies from a to s, hence in particular that $\varphi(b) = \int_\gamma \nabla\theta$ is the change in argument along $\boldsymbol{\gamma}$.)

(c) Let $\boldsymbol{\gamma}$ be any connected chain of differentiable curves $\boldsymbol{\gamma}^1, \ldots, \boldsymbol{\gamma}^n$, and let θ_1 be an argument for the beginning, $\boldsymbol{\gamma}^1(a_1)$. Prove that $\theta_1 + \int_\gamma \nabla\theta$ is an argument for the end of the chain $\boldsymbol{\gamma}^n(b_n)$.

(d) Prove that if $\boldsymbol{\gamma}$ is a *closed* connected chain, then

$$\frac{1}{2\pi}\int_\gamma \nabla\theta$$

is an integer. (This integer is called the *winding number* of $\boldsymbol{\gamma}$ about the origin.)

4.4 GREEN'S THEOREM

The formula

$$\int_\gamma \nabla f = f(\boldsymbol{\gamma}(b)) - f(\boldsymbol{\gamma}(a)) \tag{1}$$

is a sort of fundamental theorem of calculus for line integrals, expressing the integral of the "derivative" of f as the difference of the values of f at the endpoints of the curve of integration.

There is an analogous formula for double integrals, called *Green's theorem:*

$$\iint_S (N_x - M_y)\,dx\,dy = \int_\gamma M\,dx + N\,dy. \tag{2}$$

In formula (2), S is a set in the plane, and $\boldsymbol{\gamma}$ is a closed connected chain running counterclockwise around the boundary of S, as in Fig. 22.

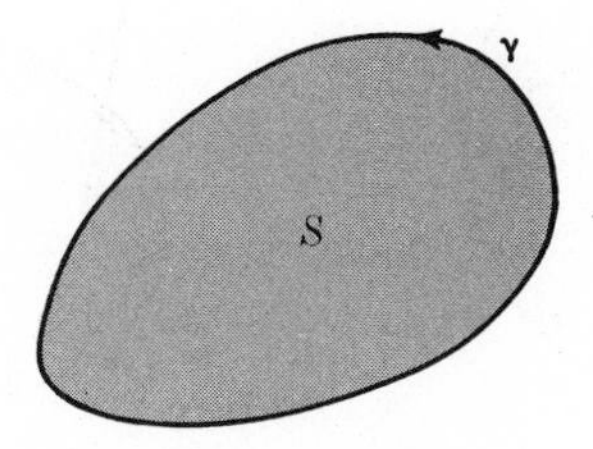

FIGURE 4.22

The following table gives the precise analogy between (1), (2), and the fundamental theorem of calculus; each formula says that the integral of the "derivative" of some function over some domain equals the integral of the function itself over the "oriented boundary" of the domain.

Function	Derivative	Domain of Integration	Oriented Boundary	Formula
f	f'	interval $[a, b]$	b taken "+" a taken "−"	$\int_a^b f' = f(b) - f(a)$
f	∇f	curve $\boldsymbol{\gamma}$	$\boldsymbol{\gamma}(b)$ taken "+" $\boldsymbol{\gamma}(a)$ taken "−"	(1) above
(M, N)	$N_x - M_y$	plane set S	a curve $\boldsymbol{\gamma}$ running counterclockwise around S	(2) above

Before proving Green's theorem, we have to translate "$\boldsymbol{\gamma}$ runs counterclockwise around the boundary of S" into a precise mathematical statement. This is deceptively difficult, and has given rise to concepts which, alas, require a longer development than our time and space allow here. (See the references listed in §4.1.) In view of this, we will work a few examples in which the connection between $\boldsymbol{\gamma}$ and S is intuitively clear, and then give Green's theorem in a form which, though inelegant, has two virtues: it covers a reasonable range of examples, and can be proved with the tools at our disposal.

Example 1. $M(x,y) = -y$, $N(x,y) = x$, and $S = \{\mathbf{P}: |\mathbf{P}| \leq r\}$. Clearly (Fig. 23), the curve

$$\boldsymbol{\gamma}(t) = (r \cos t, r \sin t), \qquad -\pi \leq t \leq \pi,$$

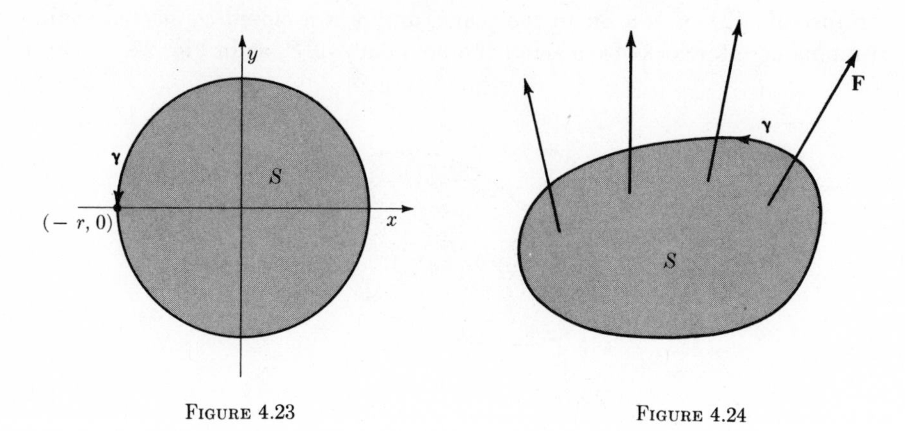

FIGURE 4.23

FIGURE 4.24

runs counterclockwise around the boundary of S. We have

$$\iint_S (N_x - M_y) = \iint_S 2 = 2\cdot\text{(area of } S) = 2\pi r^2,$$

since $M_y = -1$, $N_x = 1$, and S is a circle of radius r. On the other hand,

$$\int_\gamma (-y)\,dx + x\,dy = \int_{-\pi}^{\pi} (-r\sin t)(-r\sin t)\,dt + (r\cos t)(r\cos t)\,dt$$

$$= \int_{-\pi}^{\pi} r^2\,dt = 2\pi r^2.$$

Thus Green's theorem (2) is verified in this example.

Example 2. Applying Green's theorem to a plane flow, we can show that the rate of flow out of S equals the integral over S of a certain function, called the *divergence* of the flow. Recall that for a velocity field $\mathbf{F} = (M,N)$ the rate of flow from left to right across $\boldsymbol{\gamma}$ is $\int_\gamma (-N\,dx + M\,dy)$. When $\boldsymbol{\gamma}$ runs counterclockwise around the boundary of S, as in Fig. 24, then S "lies to the left of gamma"; hence this line integral gives the rate of flow *out* of S. According to Green's theorem, the rate can be expressed by an integral over S,

$$\int_\gamma (-N\,dx + M\,dy) = \iint_S (M_x + N_y).$$

The expression $M_x + N_y$ is the divergence of the flow $\mathbf{F} = (M,N)$, denoted div $\mathbf{F}$ or $\nabla\cdot\mathbf{F}$. (Think of ∇ as the vector $(\partial/\partial x, \partial/\partial y)$, so

$$\nabla\cdot\mathbf{F} = \frac{\partial}{\partial x}M + \frac{\partial}{\partial y}N = M_x + N_y\,.)$$

The divergence $(\nabla\cdot\mathbf{F})(\mathbf{P})$ measures the rate of flow out from the point $\mathbf{P}$; by Green's theorem, integrating the divergence over S gives the total rate of flow out of S.

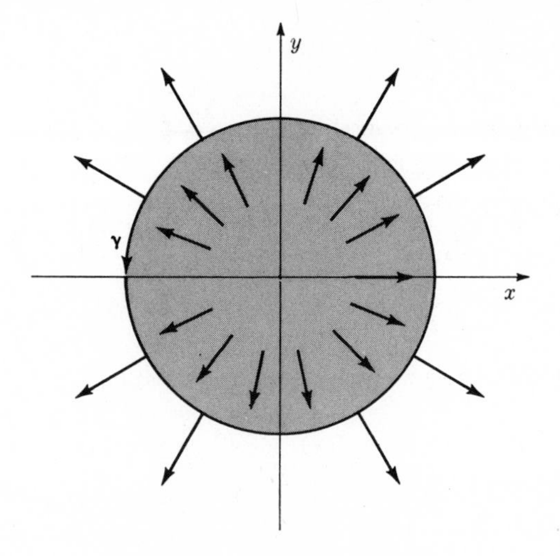

FIGURE 4.25 $\mathbf{F}(x,y) = (x,y)$

We can interpret Example 1 in connection with the flow $\mathbf{F}(x,y) = (x,y)$ in Fig. 25. Obviously, there is a positive rate of flow out of the disk of radius r, which is given by the integral $\int_\gamma (-y\,dx + x\,dy)$, where $\boldsymbol{\gamma}$ is the curve in Fig. 25 and Example 1; thus the outward rate of flow, computed by this boundary integral, is $2\pi r^2$. On the other hand, the *divergence* of the flow is $\partial x/\partial x + \partial y/\partial y = 2$ (recall that the divergence of $\mathbf{F} = (M,N)$ is defined as $M_x + N_y$.) Computing the outward rate of flow as the integral of the divergence thus gives $\int\int_D 2 = 2\pi r^2$, in agreement with the boundary integral.

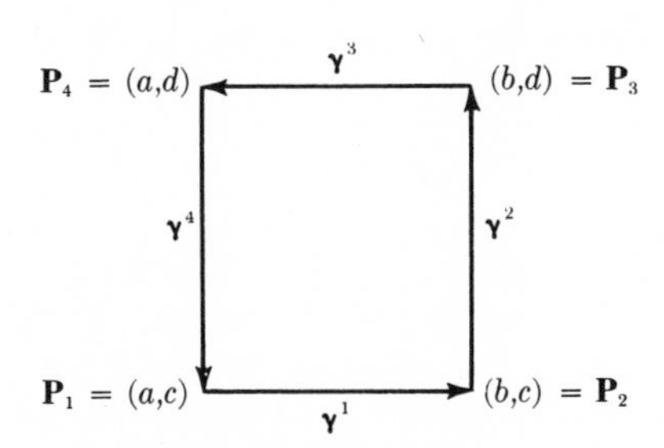

FIGURE 4.26

Example 3. $M(x,y) = xy, N(x,y) = x^2, S = \{(x,y): a \leq x \leq b, c \leq y \leq d\}$. Here the boundary of S is a chain consisting of the four straight line segments from $\mathbf{P}_1$ to $\mathbf{P}_2$, $\mathbf{P}_2$ to $\mathbf{P}_3$, $\mathbf{P}_3$ to $\mathbf{P}_4$, and $\mathbf{P}_4$ to $\mathbf{P}_1$ (see Fig. 26), with any convenient parametrization. The integral over S is

$$\iint_S (N_x - M_y) = \int_a^b \int_c^d x\,dy\,dx = \frac{1}{2}(d-c)(b^2 - a^2), \tag{3}$$

and the integral over the boundary of S is

$$\int_\gamma xy\,dx + x^2\,dy = \int_{\gamma^1} (xy\,dx + x^2\,dy) + \cdots + \int_{\gamma^4} (xy\,dx + x^2\,dy)$$

$$= \int_a^b tc\,dt + \int_c^d b^2\,dt + \int_{-b}^{-a} [(-t)d]\,dt + \int_{-d}^{-c} a^2\,dt$$

$$= \tfrac{1}{2}c(b^2 - a^2) + (d - c)b^2 + \tfrac{1}{2}d(a^2 - b^2) + a^2(d - c). \quad (4)$$

In the next to last line we used the parametrizations

$$\gamma^1(t) = (t,c), \qquad a \leq t \leq b$$

$$\gamma^2(t) = (b,t), \qquad c \leq t \leq d$$

$$\gamma^3(t) = (-t,d), \qquad -b \leq t \leq -a$$

$$\gamma^4(t) = (a,-t), \qquad -d \leq t \leq -c.$$

At first glance, the results in (3) and (4) appear different, but a little algebra shows them to be the same; thus Green's theorem is verified in this example.

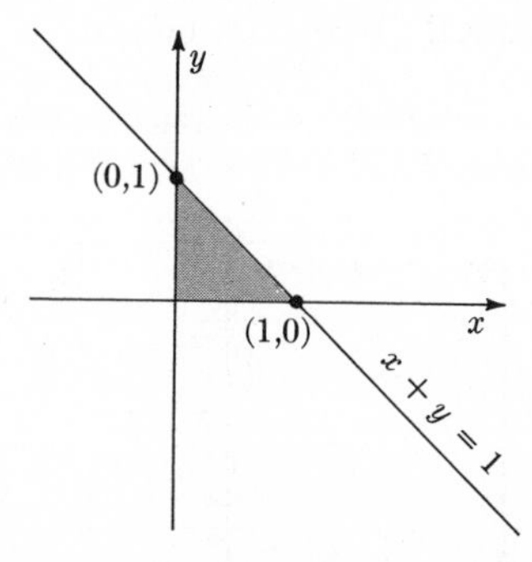

FIGURE 4.27

Example 4. $M(x,y) = y$, $N(x,y) = x$, S is the triangle

$$\{(x,y): 0 \leq x \leq 1, 0 \leq y \leq 1, x + y \leq 1\}$$

in Fig. 27. In this case, $N_x - M_y = 1 - 1 = 0$, so

$$\iint_S (N_x - M_y) = 0. \quad (5)$$

On the other hand, it is clear that the boundary of S is a closed connected chain, no matter how it is parametrized; and by Theorem 4, the vector field (M,N) is exact, so

$$\int_{\gamma} M\,dx + N\,dy = 0, \tag{6}$$

by the Corollary of Theorem 5. Since (5) and (6) agree, Green's theorem is verified in this example.

We come now to the formulation and proof of Green's theorem. Any formulation must state (i) what functions M and N are allowed, (ii) what regions S are allowed, and (iii) when a connected chain runs counterclockwise around S. We shall make the following hypotheses:

(i) M, N, M_y, and N_x are assumed continuous at all points of S. (This is a very natural restriction.)

(ii) S can be described both in the form

$$S = \{(x,y) : a \le x \le b,\ \varphi_1(x) \le y \le \varphi_2(x)\}$$

and

$$S = \{(x,y) : \psi_1(y) \le x \le \psi_2(y),\ c \le y \le d\},$$

where the φ's and ψ's are continuous, $\varphi_1 \le \varphi_2$, and $\psi_1 \le \psi_2$. (This is a rather special sort of domain, but our limited development of double integrals allows nothing more general.)

(iii) The question of the boundary is more complicated. For this, we introduce the following concept: Given a real-valued function φ defined on an interval $[a,b]$ and a connected chain $\boldsymbol{\gamma}$, we say that $\boldsymbol{\gamma}$ *runs from left to right along the graph of* φ if each point on $\boldsymbol{\gamma}$ lies in the graph of φ, and $\boldsymbol{\gamma}$ begins at $(a, \varphi(a))$ and ends at $(b, \varphi(b))$. (See Fig. 28.) Under these conditions a simple change of variables (Problem 12, §4.3) proves that

$$\int_{\gamma} M\,dx = \int_a^b M(x,\varphi(x))\,dx. \tag{7}$$

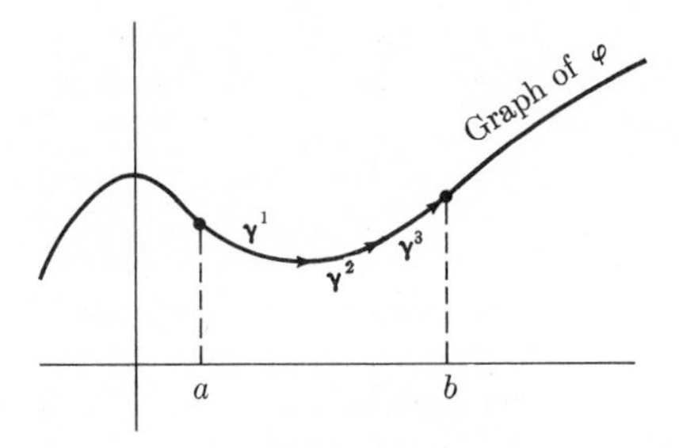

FIGURE 4.28

On the other hand, if $\boldsymbol{\gamma}$ runs from right to left, i.e. $\boldsymbol{\gamma}$ begins at $(b,\varphi(b))$ and ends at $(a,\varphi(a))$, we have

$$\int_{\gamma} M\,dx = -\int_{a}^{b} M(x,\varphi(x))\,dx. \tag{8}$$

Analogously, $\boldsymbol{\gamma}$ *runs from top to bottom along the set* $\{(x,y): x = \psi(y),\ c \leq y \leq d\}$ if each point on $\boldsymbol{\gamma}$ lies in that set, and $\boldsymbol{\gamma}$ begins at $(d,\psi(d))$ and ends at $(c,\psi(c))$. In this case,

$$\int_{\gamma} N\,dy = -\int_{c}^{d} N(\psi(y),y)\,dy; \tag{9}$$

and if $\boldsymbol{\gamma}$ runs from bottom to top, then

$$\int_{\gamma} N\,dy = \int_{c}^{d} N(\psi(y),y)\,dy. \tag{10}$$

Finally (Fig. 29), we say that a *connected chain* $\boldsymbol{\gamma} = \boldsymbol{\gamma}^1, \ldots, \boldsymbol{\gamma}^8$ *runs counterclockwise around the boundary of the set* S in (ii) above if the chain $\boldsymbol{\gamma}^1, \boldsymbol{\gamma}^2, \boldsymbol{\gamma}^3$ runs from left to right along the graph of φ_1; $\boldsymbol{\gamma}^3, \boldsymbol{\gamma}^4, \boldsymbol{\gamma}^5$ runs from bottom to top along $\{(x,y): x = \psi_2(y)\}$; $\boldsymbol{\gamma}^5, \boldsymbol{\gamma}^6, \boldsymbol{\gamma}^7$ runs from right to left along the graph of φ_2, and $\boldsymbol{\gamma}^7, \boldsymbol{\gamma}^8, \boldsymbol{\gamma}^1$ runs from top to bottom along $\{(x,y): x = \psi_1(y)\}$.

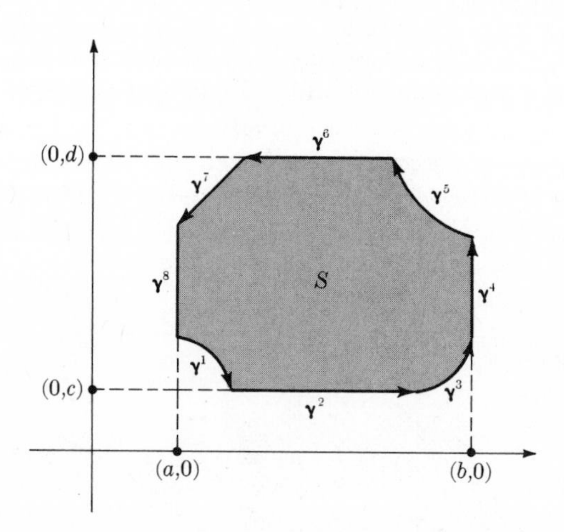

FIGURE 4.29

With the above notions defined, we can at last prove a version of Green's theorem:

Theorem 6. *If M, N, S, and $\boldsymbol{\gamma}$ satisfy conditions* (i)–(iii) *above, then*

$$\iint_S N_x - M_y \, dx \, dy = \int_\gamma M \, dx + N \, dy.$$

Proof. By Theorem 2 and the fundamental theorem of calculus,

$$-\iint_S M_y \, dx \, dy = -\int_a^b \left[\int_{\varphi_1(x)}^{\varphi_2(x)} M_y(x,y) \, dy \right] dx = -\int_a^b [M(x,y)]_{\varphi_1(x)}^{\varphi_2(x)} \, dx$$

$$= -\int_a^b M(x,\varphi_2(x)) \, dx + \int_a^b M(x,\varphi_1(x)) \, dx.$$

Using the assumptions (iii) on the chain $\boldsymbol{\gamma}^1, \ldots, \boldsymbol{\gamma}^8$, and formulas (7) and (8), we get

$$-\iint_S M_y \, dx \, dy = \int_{\gamma^5} M \, dx + \int_{\gamma^6} M \, dx + \int_{\gamma^7} M \, dx$$

$$+ \int_{\gamma^1} M \, dx + \int_{\gamma^2} M \, dx + \int_{\gamma^3} M \, dx. \tag{11}$$

Further, since $\boldsymbol{\gamma}^4$ lies in the line where $x = b$, the first component $\gamma_1{}^4$ of $\boldsymbol{\gamma}^4$ is constant, so its derivative $(\gamma_1{}^4)'$ is zero, and

$$\int_{\gamma^4} M \, dx = \int_{a_4}^{b_4} M(\gamma_1{}^4(t), \gamma_2{}^4(t))(\gamma_1{}^4)'(t) \, dt = 0.$$

Similarly, $\int_{\gamma^8} M \, dx = 0$, and adding these results to (11) yields

$$-\iint_S M_y \, dx \, dy = \sum_{j=1}^{8} \int_{\gamma^j} M \, dx = \int_\gamma M \, dx. \tag{12}$$

In a completely analogous way, beginning with the formula

$$\iint_S N_x \, dx \, dy = \int_c^d \left[\int_{\psi_1(y)}^{\psi_2(y)} N_x(x,y) \, dx \right] dy$$

$$= \int_c^d N(\psi_2(y),y) \, dy - \int_c^d N(\psi_1(y),y) \, dy,$$

we get $\iint_S N_x \, dx \, dy = \int_\gamma N \, dy$. Combining this with (12) proves the theorem.

We conclude with some examples showing how to pick the curves $\boldsymbol{\gamma}^1, \ldots, \boldsymbol{\gamma}^8$ in Theorem 6.

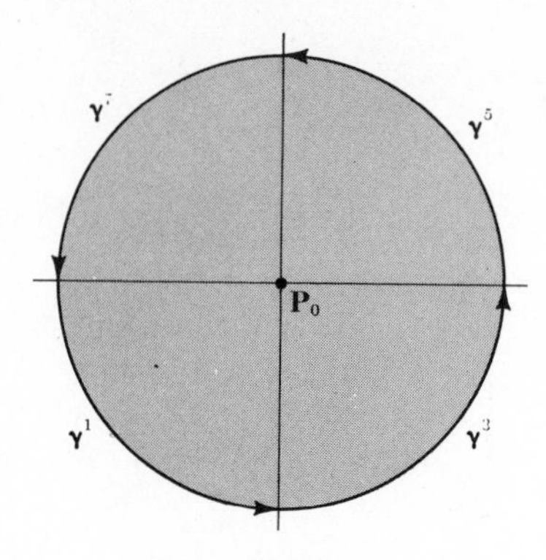

FIGURE 4.30

Example 5. To apply Theorem 6 to a disk $D = \{\mathbf{P}: |\mathbf{P} - \mathbf{P}_0| \leq r\}$ we begin with the "intuitively obvious" boundary curve

$$\boldsymbol{\gamma}(t) = \mathbf{P}_0 + (r \cos t, r \sin t), \qquad -\pi \leq t \leq \pi,$$

and break it up into four pieces, as in Fig. 30. Thus $\boldsymbol{\gamma}^1(t)$ is defined for $-\pi \leq t \leq -\pi/2$, $\boldsymbol{\gamma}^3(t)$ for $-\pi/2 \leq t \leq 0$, etc. The curves $\boldsymbol{\gamma}^2$, $\boldsymbol{\gamma}^4$, $\boldsymbol{\gamma}^6$, $\boldsymbol{\gamma}^8$ referred to in Theorem 6 reduce to a single point, and contribute nothing to the boundary integral. Thus if M, N, M_y, and N_x are continuous in D, we have

$$\iint_D (N_x - M_y)\,dx\,dy = \int_{\gamma} M\,dx + N\,dy.$$

Example 6. In applying Theorem 6 to a rectangle

$$R = \{(x,y): a \leq x \leq b, c \leq y \leq d\},$$

the boundary $\boldsymbol{\gamma}$ consists of the four straight line curves $\boldsymbol{\gamma}^2$, $\boldsymbol{\gamma}^4$, $\boldsymbol{\gamma}^6$, $\boldsymbol{\gamma}^8$ (Fig. 31); each of the others reduces to a single point, and contributes nothing to the boundary integral. Thus, with the usual hypotheses on $\mathbf{F} = (M,N)$,

$$\iint_R (N_x - M_y)\,dx\,dy = \int_{\gamma 2} \mathbf{F} + \int_{\gamma 4} \mathbf{F} + \int_{\gamma 6} \mathbf{F} + \int_{\gamma 8} \mathbf{F}.$$

Example 7. For the set S in Fig. 32, $\boldsymbol{\gamma}^1$, $\boldsymbol{\gamma}^2$, $\boldsymbol{\gamma}^5$, and $\boldsymbol{\gamma}^8$ can be easily parametrized, and each of the other four curves reduces to a point; thus with the usual hypotheses on $\mathbf{F} = (M,N)$, we find

$$\iint_S (N_x - M_y)\,dx\,dy = \int_{\gamma 1} \mathbf{F} + \int_{\gamma 2} \mathbf{F} + \int_{\gamma 5} \mathbf{F} + \int_{\gamma 8} \mathbf{F}.$$

Example 8. For the triangle T in Fig. 33 we can take

$$\boldsymbol{\gamma}^1(t) = \mathbf{P}_1 + t(\mathbf{P}_2 - \mathbf{P}_1), \qquad 0 \le t \le 1$$

$$\boldsymbol{\gamma}^5(t) = \mathbf{P}_2 + t(\mathbf{P}_3 - \mathbf{P}_2)$$

$$\boldsymbol{\gamma}^7(t) = \mathbf{P}_3 + t(\mathbf{P}_1 - \mathbf{P}_3).$$

Each of the other five curves reduces to a point, and contributes nothing to the boundary integral; thus, with the usual hypotheses on $\mathbf{F} = (M, N)$,

$$\iint_T (N_x - M_y) = \int_{\gamma 1} \mathbf{F} + \int_{\gamma 5} \mathbf{F} + \int_{\gamma 7} \mathbf{F}.$$

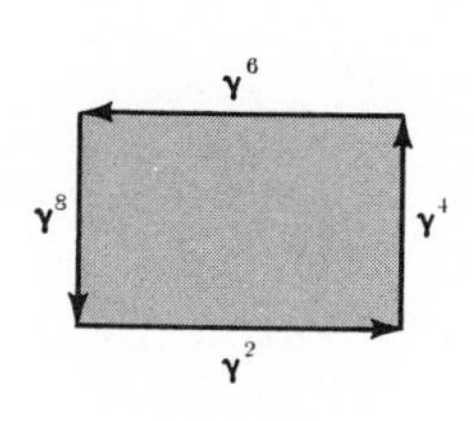

FIGURE 4.31

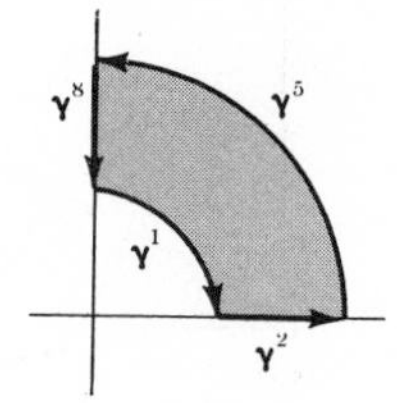

FIGURE 4.32

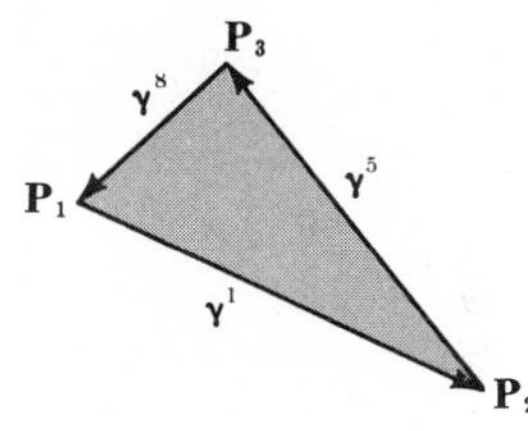

FIGURE 4.33

PROBLEMS

The first three problems are routine applications of Green's theorem. Problem 4 suggests an extension to a new type of region S, and Problem 5 fills in a detail of the proof of Theorem 6. Problems 6–11 concern applications of Green's theorem to the study of Laplace's equation $f_{xx} + f_{yy} = 0$. Problem 12 develops the theory of the planimeter.

1. Compute the rate of flow out of S both by a line integral

$$\int_\gamma -N\,dx + M\,dy$$

and by the double integral of the divergence $\iint_S (M_x + N_y)\,dx\,dy$. Sketch the flow $\mathbf{F} = (M,N)$ to see if your answer seems reasonable.

(a) $M = -y,\quad N = x,\quad S = \{\mathbf{P}: |\mathbf{P}| < r\}$ ($\mathbf{F}$ is a rotation.)
(b) $M = x,\quad N = y,\quad S = \{(x,y): a \le x \le b,\ c \le y \le d\}$
(c) $M = 1,\quad N = 0,\quad S = \{(x,y): x \ge 0,\ y \ge 0,\ x + y \le 1\}$
(d) $M = x,\quad N = 0,\quad S = \{(x,y): |x| \le 1,\ |y| \le 1\}$
(e) $M = x,\quad N = 0,\quad S = \{(x,y): 0 \le x \le 1, 0 \le y \le 1\}$
(f) $M = 0,\quad N = x,\quad S = \{(x,y): 0 \le x \le 1, 0 \le y \le 1\}$

2. Show that the area of S is given by any of the line integrals

$$\int_\gamma x\,dy, \qquad -\int_\gamma y\,dx, \qquad \frac{1}{2}\int_\gamma (x\,dy - y\,dx),$$

where $\boldsymbol{\gamma}$ denotes a chain running counterclockwise around the boundary of S.

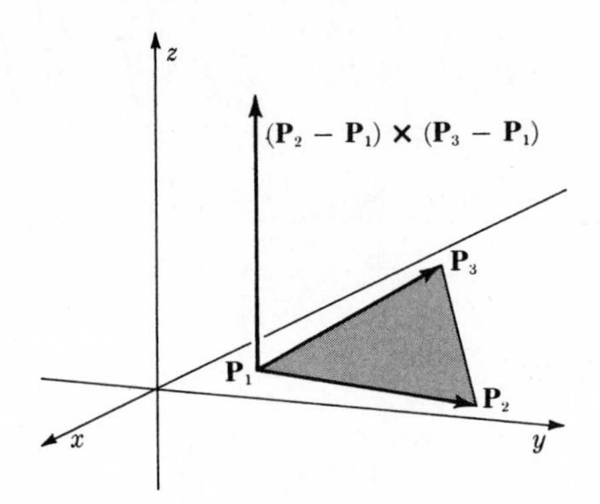

FIGURE 4.34

3. (a) Use $-\int_\gamma y\,dx$ to show that the area of the triangle S with vertices (x_1,y_1), (x_2,y_2), (x_3,y_3) is

$$\pm\frac{(x_2 - x_1)(y_2 + y_1) + (x_3 - x_2)(y_3 + y_2) + (x_1 - x_3)(y_1 + y_3)}{2}$$

(See Problem 1(e), §4.3, to evaluate the integrals on line segments.)

(b) Check the answer in part (a) by computing half the length of the cross product $|(\mathbf{P}_2 - \mathbf{P}_1) \times (\mathbf{P}_3 - \mathbf{P}_1)|$, where $\mathbf{P}_1 = (x_1,y_1,0)$, $\mathbf{P}_2 = (x_2,y_2,0)$, $\mathbf{P}_3 = (x_3,y_3,0)$. (See Fig. 34).

4. Figure 35 shows an annulus S divided into four regions S_1, S_2, S_3, S_4, to each of which Theorem 6 applies. Prove Green's theorem for the annulus

$$\sum_{j=1}^{8} \int_{\gamma^j} (M\,dx + N\,dy) = \sum_{j=1}^{4} \iint_{S_j} (N_x - M_y)\,dx\,dy.$$

(Notice that the sum on the right amounts to an integral over S, and the sum on the left adds up to $\int_\gamma (M\,dx + N\,dy)$, where $\boldsymbol{\gamma}$ runs around the boundary of S (intuitively speaking). In treating the radial curves $\boldsymbol{\gamma}^9, \ldots, \boldsymbol{\gamma}^{12}$ it may help to review Problem 10, §4.3.)

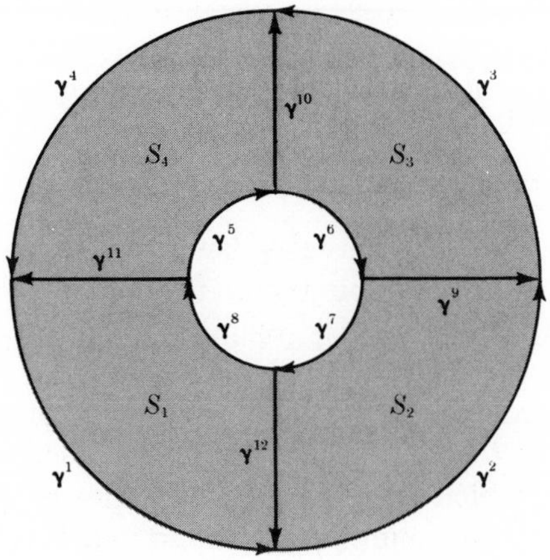

FIGURE 4.35

5. (a) Prove formula (7). (Hint: $a = \gamma_1{}^1(a_1)$, $\gamma_1{}^1(b_1) = \gamma_1{}^2(a_2)$, $\gamma_1{}^2(b_2) = \gamma_1{}^3(a_3)$, $\ldots$, $\gamma_1{}^n(b_n) = b$, so for any continuous function f,

$$\int_a^b f = \int_{\gamma_1{}^1(a_1)}^{\gamma_1{}^1(b_1)} f + \int_{\gamma_1{}^2(a_2)}^{\gamma_1{}^2(b_2)} f + \cdots + \int_{\gamma_1{}^n(a_n)}^{\gamma_1{}^n(b_n)} f.)$$

(b) Prove formula (8).

6. Suppose that $\mathbf{F} = (M,N)$ is an "incompressible flow," in the sense that $\int_\gamma (-N\,dx + M\,dy) = 0$ for every closed chain $\boldsymbol{\gamma}$. Problem 16 of the previous section proved that $\mathbf{F}$ has zero divergence, $M_x + N_y \equiv 0$.

(a) Assume now the weaker condition that $\int_\gamma (-N\,dx + M\,dy) = 0$ *for every* $\boldsymbol{\gamma}$ *which is the oriented boundary of a disk* D. Prove again that $M_x + N_y = 0$. (Use Green's theorem and Problem 7, §4.1, and assume that M_x and N_y are continuous.)

(b) Suppose that $\mathbf{F} = \nabla f$, and $\mathbf{F}$ is an incompressible flow. Prove that f satisfies Laplace's equation $f_{xx} + f_{yy} = 0$. (When $f(x,y)$ is the temperature at (x,y), then ∇f is parallel to the velocity field for the flow of heat; when the temperature does not change with time (so-called "steady state temperature distribution"), then heat cannot flow out of any rectangle, hence the flow is "incompressible" in the sense of part (a), hence $f_{xx} + f_{yy} = 0$.)

7. Suppose that S and γ satisfy the conditions of Theorem (6), and that $f, g, f_x, f_{xx}, f_y, f_{yy}, g_x$, and g_y are all continuous on S. Prove *Green's formula*

$$\iint_S (g\Delta f + \nabla g \cdot \nabla f) = \int_\gamma (-gf_y\,dx + gf_x\,dy), \tag{13}$$

where Δf denotes the *Laplacian* $f_{xx} + f_{yy}$.

8. (a) Suppose that the curve γ has arc length as parameter; thus $\gamma'(s) = \mathbf{T}(s)$ is a unit vector tangent to the curve. If $\mathbf{T}(s) = (\gamma_1'(s), \gamma_2'(s))$, let $\mathbf{N}(s) = (\gamma_2'(s), -\gamma_1'(s))$. Prove that $\mathbf{N}(s)$ is a unit vector orthogonal to the curve at $\gamma(s)$. (See §2.3.)

(b) With $\mathbf{T}$ and $\mathbf{N}$ as above, choose θ so that $\mathbf{T} = (\cos\theta, \sin\theta)$; show that $\mathbf{N} = (\cos(\theta - \pi/2), \sin(\theta - \pi/2))$, i.e. $\mathbf{N}$ is obtained by rotating $\mathbf{T}$ clockwise through an angle of $\pi/2$. (Hence, intuitively, if γ runs counterclockwise around S, then $\mathbf{N}$ is a unit normal pointing out of S.)

(c) Show that with γ as above, $\int_\gamma - gf_y\,dx + gf_x\,dy = \int_a^b g(\nabla f \cdot \mathbf{N})\,ds$. (The expression $\nabla f \cdot \mathbf{N}$ is the directional derivative of f in the direction of the normal $\mathbf{N}$ pointing out of S, called the "outer normal derivative of f," denoted $\partial f/\partial n$. Thus Green's formula (13) says that

$$\iint_S (g\Delta f + \nabla g \cdot \nabla f) = \int_\gamma g\frac{\partial f}{\partial n}\,ds.)$$

9. Prove the following. (See Problems 7 and 8.)

(a) $$\iint_S f\,\Delta g + \nabla f \cdot \nabla g = \int_\gamma f\frac{\partial g}{\partial n}\,ds \tag{14}$$

(b) $$\iint_S (f\,\Delta g - g\,\Delta f) = \int_\gamma \left(f\frac{\partial g}{\partial n} - g\frac{\partial f}{\partial n}\right) ds \tag{15}$$

10. Suppose that f, S, and γ satisfy the conditions in Problem 7, that $\Delta f \equiv 0$, and that $f = 0$ at each point on the boundary γ of S.

(a) Prove that $\nabla f \equiv 0$ in S. (Hint: Apply (13) with $f = g$, and apply Problem 8, §4.1.)

(b) Prove that $f \equiv 0$ in S. (In the language of temperature distributions: *If a steady state temperature distribution is zero on the boundary of S, then it is zero at every point in S.* Hint: See Problem 15, §3.3.)

11. Suppose that f, S, and γ satisfy the conditions in Problem 10, except that "$f \equiv 0$ on γ" is replaced by "$\partial f/\partial n \equiv 0$ on γ" (see Problem 8). Prove that f is constant on S.

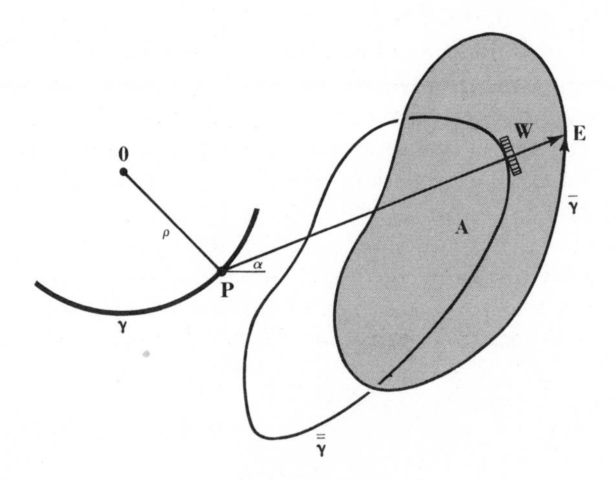

FIGURE 4.36

12. Figure 36 shows a *planimeter*, an ingenious device for measuring plane areas. An arm of radius ρ is pivoted at a fixed point **0**, and on the free end **P** of this arm is pivoted a second arm, of radius R. On the second arm, at distance r from the pivot **P**, there is a little knurled wheel W. The planimeter measures an area A by recording the net turning of the wheel W as the free end **E** of the second arm is moved around the boundary of A. This problem shows why the turning of W is proportional to the area of A.

As **E** is moved around the boundary of A, it traces a closed curve $\bar{\boldsymbol{\gamma}}$, and by Problem 2, $A = \frac{1}{2}\int_{\bar{\gamma}} (x\,dy - y\,dx)$. At the same time, the pivot **P** traces a closed curve $\boldsymbol{\gamma}$ lying entirely on the circle of radius ρ about **0**, and the wheel W is dragged over a closed curve $\bar{\bar{\boldsymbol{\gamma}}}$, as in Fig. 36. Let α be the angle between the arm **PE** and the x axis. Then $\mathbf{U}_1 = (\cos\alpha, \sin\alpha)$ is parallel to the arm **PE**, and $\mathbf{U}_2 = (-\sin\alpha, \cos\alpha)$ is perpendicular to it. Since the little knurled wheel is unaffected by motion parallel to the arm **PE**, but responds to motion perpendicular to **PE**, it turns at a rate proportional to $\mathbf{U}_2 \cdot \bar{\bar{\boldsymbol{\gamma}}}'$. Thus if $[a,b]$ is the interval on which $\boldsymbol{\gamma}$ is defined, the net turning of the wheel is

$$c \int_a^b \mathbf{U}_2 \cdot \bar{\bar{\boldsymbol{\gamma}}}' \tag{16}$$

where c is a constant of proportionality. The steps below show that if the end **E** is drawn around the boundary A so that the planimeter returns to its original position, and neither arm swings all the way around in a full circle, then $A = R\int_a^b \mathbf{U}_2 \cdot \bar{\bar{\boldsymbol{\gamma}}}'$. Thus A is proportional to the turning of the wheel, and the planimeter is justified.

(a) Suppose that $\boldsymbol{\gamma}(t)$, $\overline{\boldsymbol{\gamma}}(t)$, $\overline{\overline{\boldsymbol{\gamma}}}(t)$, and $\alpha(t)$ are respectively the position of **P**, position of **E**, position of W, and angle α at time t, and that t varies from a to b as **E** is drawn around the area A. Show that $\overline{\boldsymbol{\gamma}} = \boldsymbol{\gamma} + (R \cos \alpha, R \sin \alpha)$, $\overline{\overline{\boldsymbol{\gamma}}} = \boldsymbol{\gamma} + (r \cos \alpha, r \sin \alpha)$.

(b) If the second arm returns to its original position, and $-\pi < \alpha < \pi$ throughout the motion, and α is a differentiable function of t, then $\int_a^b \alpha' = 0$. Why?

(c) Under the conditions of part (b), show that the integral (16) giving the turning of the wheel reduces to $c\int_a^b (-\sin \alpha \gamma_1' + \cos \alpha \gamma_2')$.

(d) Show that $A = \frac{1}{2}\int_{\overline{\gamma}} (x\,dy - y\,dx)$ can be "reduced" to

$$\frac{1}{2}\int_{\gamma} (x\,dy - y\,dx) + \frac{R^2}{2}\int_a^b \alpha'$$

$$+ \frac{R}{2}\int_a^b [\gamma_1(\sin \alpha)' - \gamma_1' \sin \alpha - \gamma_2(\cos \alpha)' + \gamma_2' \cos \alpha].$$

(e) If the first arm returns to its original position without swinging all the way around **0**, then $\int_\gamma (x\,dy - y\,dx) = 0$. Why?

(f) In the expression in part (d), show that

$$\int_a^b \gamma_1(\sin \alpha)' = -\int_a^b \gamma_1' \sin \alpha$$

and

$$\int_a^b -\gamma_2(\cos \alpha)' = \int_a^b \gamma_2' \cos \alpha.$$

(g) Combining parts (b)–(f), show that (16) equals $\dfrac{c}{R} A$.

4.5 CHANGE OF VARIABLE

Green's theorem can be ingeniously applied to the problem of change of variable in double integrals. The most important case involves polar coordinates.* Suppose a region S of the plane is described in terms of polar coordinates as

$$S = \{(x,y) : x = r \cos \theta,\ y = r \sin \theta,\ r_1 \le r \le r_2,\ \theta_1 \le \theta \le \theta_2\}, \quad (1)$$

where $0 \le r_1 < r_2$ and $\theta_1 \le \theta_2$ (Fig. 37(a)). Then we have the formula

* Recall that r and θ are polar coordinates of a point $\mathbf{P} = (x,y)$ if and only if $x = r \cos \theta$ and $y = r \sin \theta$. When $r > 0$, then $r = |\mathbf{P}|$ and θ is the angle between **P** and the positive x axis.

$$\iint_S f = \int_{\theta_1}^{\theta_2} \left[\int_{r_1}^{r_2} f(r\cos\theta, r\sin\theta) r\, dr \right] d\theta \tag{2}$$

for every function f continuous on S. The point is that it is often easier to compute the double integral on the right in (2) than to compute the integral on the left in the usual way as

$$\int \left[\int_{\varphi_1}^{\varphi_2} f\, dy \right] dx \qquad \text{or} \qquad \int \left[\int_{\psi_1}^{\psi_2} f\, dx \right] dy. \tag{3}$$

For example, taking $f = 1$ in (2), we find easily that

$$\text{Area of } S = \iint_S 1 = \int_{\theta_1}^{\theta_2} \left[\int_{r_1}^{r_2} r\, dr \right] d\theta$$

$$= \int_{\theta_1}^{\theta_2} \frac{1}{2} (r_2^2 - r_1^2)\, d\theta = \frac{1}{2} (\theta_2 - \theta_1)(r_2^2 - r_1^2)$$

which is the formula for the area of an annular sector S of radii r_1 and r_2 and angle $\theta_2 - \theta_1$. By contrast, it is a chore to set up and evaluate $\iint_S 1$ in either of the forms (3).

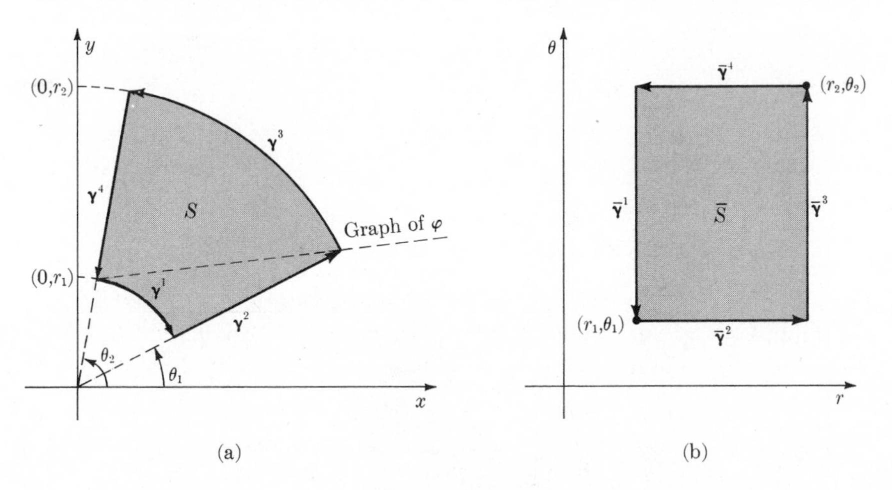

FIGURE 4.37

To understand formula (2), visualize two planes, an xy plane and an $r\theta$ plane. The function f is defined in a region S of the xy plane; according to (2), its integral can be found by integrating a *different* function, call it $\bar{f}$, over the rectangle $\bar{S}$ in the $r\theta$ plane shown in Fig. 37(b). Precisely,

$$\bar{f}(r,\theta) = rf(r\cos\theta, r\sin\theta),$$
$$\bar{S} = \{(r,\theta) : r_1 \le r \le r_2\,,\ \theta_1 \le \theta \le \theta_2\}.$$

Suppose that $f \ge 0$. Then, intuitively, $\iint_S f$ is the volume between S and the graph of f, as shown in Fig. 38. This volume can be approximated by dividing S into small sections, using radial and circular lines, as in Fig. 39. Denote the radial dimension of such a section by Δr, and the

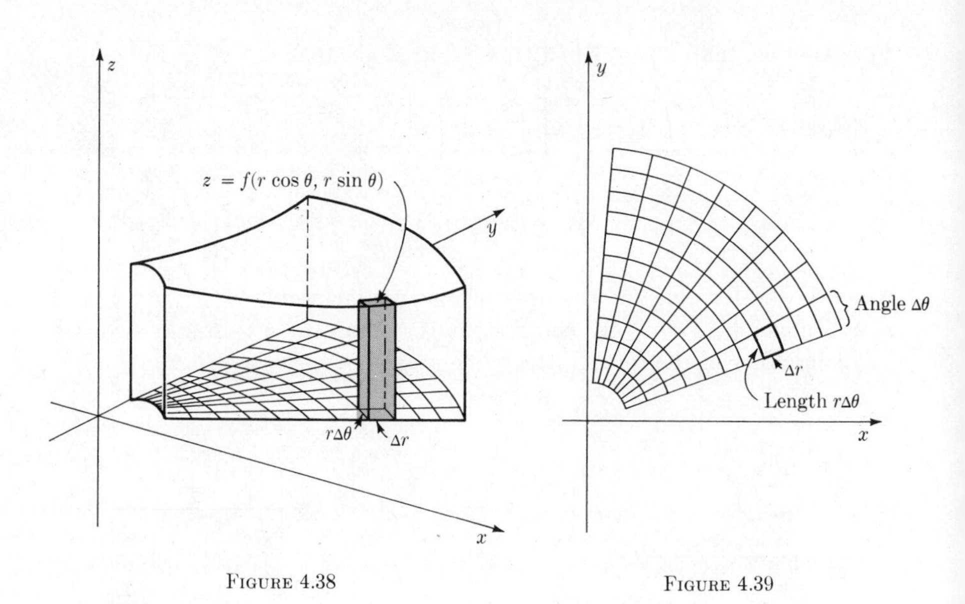

FIGURE 4.38

FIGURE 4.39

angular dimension by $\Delta\theta$. The area of such a section is $(r\Delta\theta)(\Delta r)$, so the volume lying over that section is approximately $f(r\cos\theta, r\sin\theta)r\Delta\theta\,\Delta r$. Since the whole volume is the sum of these parts, we find

$$\iint_S f \underset{\text{approx}}{=} \sum \overbrace{\underbrace{f(r\cos\theta, r\sin\theta)}_{\substack{\text{height of}\\ \text{column over } \bar{S}}}}^{\substack{\text{height of}\\ \text{column over } S}} \overbrace{r\,\underbrace{\Delta\theta\,\Delta r}_{\substack{\text{area of}\\ \text{base in } \bar{S}}}}^{\substack{\text{area of}\\ \text{base in } S}}. \tag{4}$$

Now, the radial and circular lines subdividing S correspond to a rectangular grid of lines subdividing $\bar{S}$ into small rectangles of area $\Delta r\,\Delta\theta$ (Fig. 40). Thus the sum (4) approximates the volume between $\bar{S}$ and the graph of $f(r\cos\theta, r\sin\theta)r$, and this volume is precisely the integral on the right in (2). Thus the factor r by which f is multiplied here is, intuitively, the

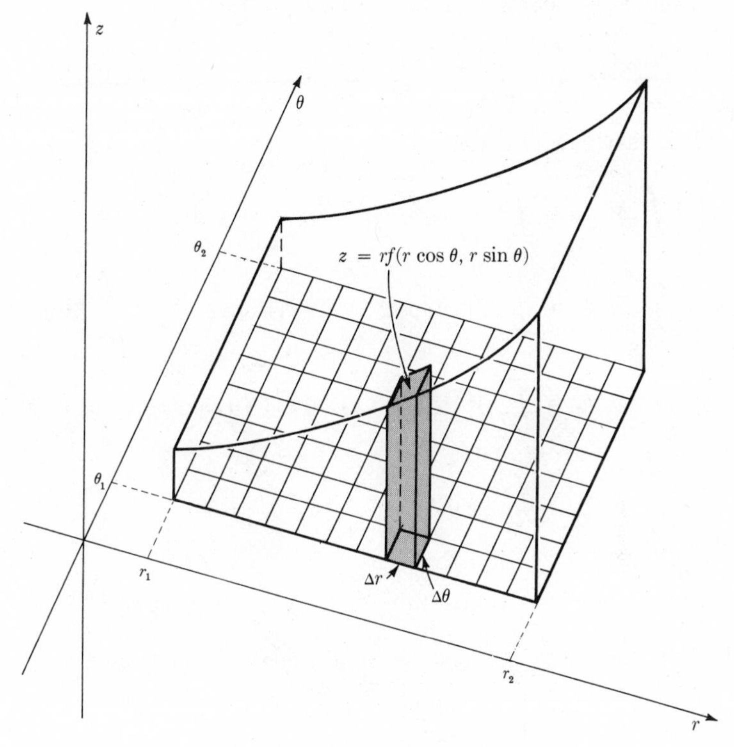

FIGURE 4.40

ratio of the area of a small piece of the S region to the corresponding small piece of the $\bar{S}$ region. This ratio is called the "magnification factor" in transforming from $\bar{S}$ to S.

In proving (2) we will apply Green's theorem to write $\iint_S f$ as a line integral around the boundary of S, transform this into a line integral around the boundary of $\bar{S}$ (using the change of variable formula for single integrals), and finally apply Green's theorem again to obtain an integral over $\bar{S}$. Suppose for simplicity that $0 \leq \theta_1 < \theta_2 \leq \pi/2$. Then Theorem 6 applies if we take the four curves $\boldsymbol{\gamma}^1$, $\boldsymbol{\gamma}^2$, $\boldsymbol{\gamma}^3$, $\boldsymbol{\gamma}^4$ in Fig. 37 as the boundary of S, and the corresponding curves $\overline{\boldsymbol{\gamma}}^1$, $\overline{\boldsymbol{\gamma}}^2$, $\overline{\boldsymbol{\gamma}}^3$, $\overline{\boldsymbol{\gamma}}^4$ as the boundary of $\bar{S}$. (This fits the terminology in (iii), §4.4, except that the curves are numbered differently, and those that reduce to a point are left out.) Let φ denote the function whose graph is the dotted line in Fig. 37(a), and define

$$M(x,y) = -\int_{\varphi(x)}^{y} f(x,t)\,dt, \qquad N(x,y) = 0.$$

Then $N_x - M_y = f$, so by Green's theorem,

$$\iint_S f = \iint_S (N_x - M_y) = \sum_{j=1}^{4} \int_{\gamma^j} M\,dx. \tag{5}$$

It is easy to check that

$$\int_{\gamma^1} M\,dx = \int_{\theta_2}^{\theta_1} M(r_1 \cos\theta, r_1 \sin\theta)(-r_1 \sin\theta)\,d\theta = \int_{\bar{\gamma}^1} \bar{M}\,dr + \bar{N}\,d\theta$$

$$\int_{\gamma^2} M\,dx = \int_{r_1}^{r_2} M(r\cos\theta_1, r\sin\theta_1)\cos\theta_1\,dr = \int_{\bar{\gamma}^2} \bar{M}\,dr + \bar{N}\,d\theta$$

$$\int_{\gamma^3} M\,dx = \int_{\theta_1}^{\theta_2} M(r_2 \cos\theta, r_2 \sin\theta)(-r_2 \sin\theta)\,d\theta = \int_{\bar{\gamma}^3} \bar{M}\,dr + \bar{N}\,d\theta$$

$$\int_{\gamma^4} M\,dx = \int_{r_2}^{r_1} M(r\cos\theta_2, r\sin\theta_2)(\cos\theta_2)\,dr = \int_{\bar{\gamma}^4} \bar{M}\,dr + \bar{N}\,d\theta$$

where

$$\bar{M}(r,\theta) = M(r\cos\theta, r\sin\theta)\cos\theta \tag{6}$$

$$\bar{N}(r,\theta) = -M(r\cos\theta, r\sin\theta)r\sin\theta. \tag{7}$$

(See Problem 2.) Applying (5) and Green's theorem, we get

$$\iint_S f = \sum_{j=1}^{4} \int_{\gamma^j} M\,dx = \sum_{j=1}^{4} \int_{\bar{\gamma}^j} \bar{M}\,dr + \bar{N}\,d\theta = \iint_{\bar{S}} (\bar{N}_r - \bar{M}_\theta). \tag{8}$$

Applying the chain rule in (6) and (7),

$$\bar{M}_\theta = -M\sin\theta + [M_x\cdot(-r\sin\theta) + M_y\cdot(r\cos\theta)](\cos\theta)$$

$$\bar{N}_r = -M\sin\theta - [M_x\cdot(\cos\theta) + M_y\cdot(\sin\theta)](r\sin\theta);$$

hence $\bar{N}_r - \bar{M}_\theta = -M_y\cdot r = f\cdot r$. (It is understood that f, like M, M_x, and M_y, is to be evaluated at the point $(r\cos\theta, r\sin\theta)$.) Thus from (8)

$$\iint_S f = \iint_{\bar{S}} f\cdot r = \int_c^d \left[\int_a^b f(r\cos\theta, r\sin\theta)r\,dr\right] d\theta,$$

and formula (2) is proved.

If you check the details of this proof, you will find that f has to be differentiable, not just continuous (see Problem 12). This is a weakness in the method of proof; actually, formula (2) remains valid for every continuous function f.

As we have already suggested, Theorem 6 is just a special case of a much more general result, various forms of which can be found in the references in §4.1. To illustrate what is gained by proving the general version, we show how it leads to a general rule for change of variables in a double integral. Many of the terms we use will, of necessity, not be precisely defined, but only suggested by pictures.

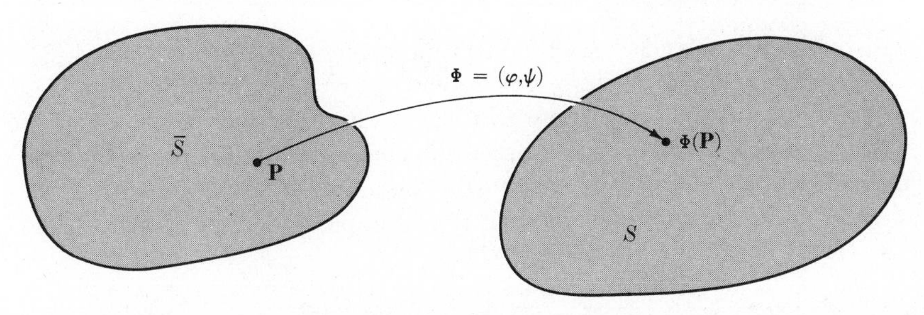

FIGURE 4.41

Suppose $\bar{S}$ is a region in the plane, and φ and ψ are two differentiable real-valued functions on $\bar{S}$. Then the *pair* of functions (φ,ψ) assigns to each point $\mathbf{P}$ in $\bar{S}$ a point $(\varphi(\mathbf{P}),\psi(\mathbf{P}))$ in the plane, in other words, (φ,ψ) is a function from $\bar{S}$ to R^2, as suggested in Fig. 41. Denote this function by $\mathbf{\Phi}$; thus $\mathbf{\Phi}(\mathbf{P}) = (\varphi(\mathbf{P}),\psi(\mathbf{P}))$. Let S be the set of all points in the range of $\mathbf{\Phi}$,

$$S = \{\mathbf{Q}\colon \mathbf{Q} = \mathbf{\Phi}(\mathbf{P}) \text{ for some } \mathbf{P} \text{ in } \bar{S}\}.$$

The function $\mathbf{\Phi}$ is called *one-to-one* if each point $\mathbf{Q}$ in S comes from just one point $\mathbf{P}$ in $\bar{S}$. We denote points $\mathbf{P}$ in $\bar{S}$ by (u,v) (in other words, "$\bar{S}$ lies in the uv plane"), and the derivatives of φ and ψ are denoted $\partial\varphi/\partial u$, etc. The rule for change of variable states that *if (φ,ψ) is a differentiable one-to-one map of $\bar{S}$ onto S, then*

$$\iint_S f = \iint_{\bar{S}} f(\varphi(u,v),\psi(u,v)) \left|\frac{\partial(\varphi,\psi)}{\partial(u,v)}\right| du\, dv, \tag{9}$$

where $\partial(\varphi,\psi)/\partial(u,v)$ is the *Jacobian* of (φ,ψ), defined by

$$\frac{\partial(\varphi,\psi)}{\partial(u,v)} = \begin{vmatrix} \varphi_u & \psi_u \\ \varphi_v & \psi_v \end{vmatrix} = \frac{\partial\varphi}{\partial u}\frac{\partial\psi}{\partial v} - \frac{\partial\psi}{\partial u}\frac{\partial\varphi}{\partial v}. \tag{10}$$

Thus the absolute value of the Jacobian (10) is the "magnification factor" in transforming from $\bar{S}$ to S. For example, if

$$\varphi(u,v) = u \cos v$$

$$\psi(u,v) = u \sin v,$$

then

$$\frac{\partial(\varphi,\psi)}{\partial(u,v)} = \begin{vmatrix} \cos v & \sin v \\ -u \sin u & u \cos v \end{vmatrix} = u;$$

hence (9) gives the same result as (2),

$$\iint_S f = \iint_{\bar{S}} f(u \cos v, u \sin v)\ |u|\ du\, dv,$$

except that now S and $\bar{S}$ can be much more general than they are in (2).

We can outline a proof of (9) following the same reasoning that led to (2). For simplicity, suppose that the positively oriented boundary of $\bar{S}$ is formed by a single closed differentiable curve $\bar{\boldsymbol{\gamma}}$ defined on an interval $[a,b]$. Then the curve $\boldsymbol{\gamma} = \boldsymbol{\Phi}\circ\bar{\boldsymbol{\gamma}}$ forms the boundary of S, but it may be oriented either positively or negatively. Suppose there is a vector field $\mathbf{F} = (M,0)$ such that $-M_y = f$. Then

$$\iint_S f = -\iint_S M_y = \pm\int_\gamma M\, dx = \pm\int_a^b (M\circ\boldsymbol{\gamma})\gamma_1{}'.$$

(The $\pm$ sign arises because $\boldsymbol{\gamma}$ may be oriented either positively or negatively.) Since $\boldsymbol{\gamma} = \boldsymbol{\Phi}\circ\bar{\boldsymbol{\gamma}} = (\varphi\circ\bar{\boldsymbol{\gamma}}, \psi\circ\bar{\boldsymbol{\gamma}})$, we have $\gamma_1 = \varphi\circ\bar{\boldsymbol{\gamma}}$, hence $\gamma_1{}' = (\boldsymbol{\nabla}\varphi\circ\bar{\boldsymbol{\gamma}})\cdot\bar{\boldsymbol{\gamma}}'$, and

$$\iint_S f = \pm\int_a^b [(M\circ\boldsymbol{\Phi}\circ\bar{\boldsymbol{\gamma}})\ \boldsymbol{\nabla}\varphi\circ\bar{\boldsymbol{\gamma}}]\cdot\bar{\boldsymbol{\gamma}}'$$

$$= \pm\int_{\bar{\gamma}} (M\circ\boldsymbol{\Phi})\ \boldsymbol{\nabla}\varphi\circ\bar{\boldsymbol{\gamma}}$$

$$= \pm\int_{\bar{\gamma}} M(\varphi,\psi)\frac{\partial\varphi}{\partial u}\, du + M(\varphi,\psi)\frac{\partial\varphi}{\partial v}\, dv$$

$$= \pm\iint_{\bar{S}} \left[\frac{\partial}{\partial u}\left(M(\varphi,\psi)\frac{\partial\varphi}{\partial v}\right) - \frac{\partial}{\partial v}\left(M(\varphi,\psi)\frac{\partial\varphi}{\partial u}\right)\right] du\, dv$$

$$= \pm\iint_{\bar{S}} \left\{\left[\left(M_x(\varphi,\psi)\frac{\partial\varphi}{\partial u} + M_y(\varphi,\psi)\frac{\partial\psi}{\partial u}\right)\frac{\partial\varphi}{\partial v} + M(\varphi,\psi)\frac{\partial^2\varphi}{\partial u\partial v}\right]\right.$$

$$\left. -\left[\left(M_x(\varphi,\psi)\frac{\partial\varphi}{\partial v} + M_y(\varphi,\psi)\frac{\partial\psi}{\partial v}\right)\frac{\partial\varphi}{\partial u} + M(\varphi,\psi)\frac{\partial^2\varphi}{\partial v\partial u}\right]\right\} du\, dv$$

$$= \pm\iint_{\bar{S}} \left\{-M_y(\varphi,\psi)\left(\frac{\partial\varphi}{\partial u}\frac{\partial\psi}{\partial v} - \frac{\partial\varphi}{\partial v}\frac{\partial\psi}{\partial u}\right)\right\} du\, dv$$

$$= \pm\iint_{\bar{S}} f(\varphi,\psi)\frac{\partial(\varphi,\psi)}{\partial(u,v)}\, du\, dv.$$

Now we have to choose the $+$ or the $-$ sign. If $f = 1$, we get

$$\iint_S 1 = \iint_{\bar{S}} \pm \frac{\partial(\varphi,\psi)}{\partial(u,v)}\, du\, dv.$$

Since the left-hand side is not negative, neither is the right, so we must have the sign that makes

$$\pm \frac{\partial(\varphi,\psi)}{\partial(u,v)} \geq 0,$$

in other words, we have $|\partial(\varphi,\psi)/\partial(u,v)|$, and we arrive at (9).

Formula (9) should be compared to the rule for change of variable in a single integral,

$$\int_{\varphi(a)}^{\varphi(b)} f(x)\, dx = \int_a^b f(\varphi(u))\varphi'(u)\, du. \tag{10}$$

Imagine φ mapping the u axis into the x axis, as in Fig. 42. If $\varphi' \geq 0$, then φ maps the interval $\bar{S} = [a,b]$ onto $S = [\varphi(a),\varphi(b)]$; but if $\varphi' \leq 0$, it maps onto $S = [\varphi(b),\varphi(a)]$, as in Fig. 42(b). In the first case, $\varphi' = |\varphi'|$, so by (10)

$$\int_S f = \int_{\varphi(a)}^{\varphi(b)} f = \int_a^b f(\varphi(u))\varphi'(u)\, du = \int_a^b f(\varphi(u))\, |\varphi'(u)|\, du.$$

In the second case, $\varphi' = -|\varphi'|$, so

$$\int_S f = \int_{\varphi(b)}^{\varphi(a)} f = -\int_{\varphi(a)}^{\varphi(b)} f = -\int_a^b f(\varphi(u))\varphi'(u)\, du$$

$$= \int_a^b f(\varphi(u))\, |\varphi'(u)|\, du.$$

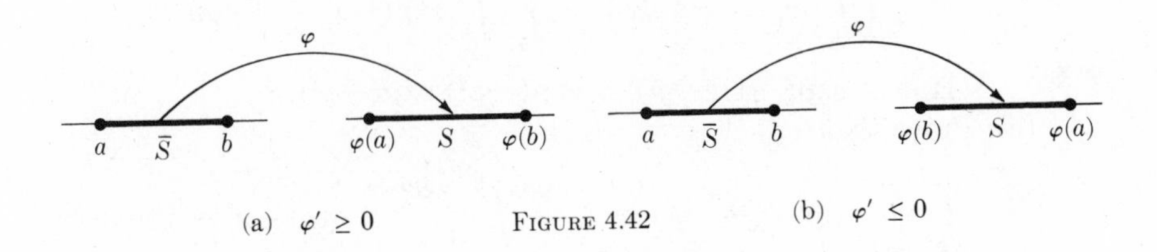

(a) $\varphi' \geq 0$ FIGURE 4.42 (b) $\varphi' \leq 0$

Hence in either case,

$$\int_S f = \int_{\bar{S}} f(\varphi(u))\, |\varphi'(u)|\, du.$$

This formula is the one-variable version of (9); the derivative $\varphi' = d\varphi/du$ corresponds to the Jacobian $\partial(\varphi,\psi)/\partial(u,v)$ in two variables. Just as $\varphi' \leq 0$ corresponds to the fact that φ reverses the endpoints of the interval, $\partial(\varphi,\psi)/\partial(u,v) \leq 0$ means that (φ,ψ) reverses the direction of the boundary curve.

PROBLEMS

1. Evaluate the following integrals in polar coordinates, using formula (2).

(a) $\displaystyle\int_0^1 \int_0^{\sqrt{1-y^2}} x\, dx\, dy$ (Sketch the set S over which you integrate.)

(b) $\displaystyle\iint_S \frac{xy}{x^2+y^2}\, dx\, dy$, where S is the annulus $\{\mathbf{P}: r_1 \leq |\mathbf{P}| \leq r_2\}$

(c) $\displaystyle\int_0^R \int_0^{\sqrt{R^2-x^2}} \exp(x^2+y^2)\, dy\, dx$ (It is not practical to evaluate the integral explicitly *without* changing to polar coordinates.)

2. This problem evaluates an important "improper integral,"

$$\int_0^\infty \exp(-x^2)\, dx = \lim_{R\to\infty} \int_0^R \exp(-x^2)\, dx = \frac{\sqrt{\pi}}{2}.$$

(a) Prove that $\displaystyle\int_0^R \int_0^{\sqrt{R^2-x^2}} \exp(-x^2-y^2)\, dy\, dx = \frac{\pi}{4}(1-e^{-R^2}).$

(b) Show that

$$\left(\int_0^R \exp(-x^2)\, dx\right)^2 = \int_0^R \int_0^R \exp(-x^2-y^2)\, dy\, dx.$$

(Hint: $\exp(-x^2-y^2) = \exp(-x^2)\exp(-y^2)$.)

(c) Prove that $0 \leq$

$$\int_0^R \int_0^R \exp(-x^2-y^2)\, dy\, dx - \int_0^R \int_0^{\sqrt{R^2-x^2}} \exp(-x^2-y^2)\, dy\, dx$$
$$\leq R^2 \exp(-R^2).$$

(Notice that the difference in question is the integral of $e^{-x^2-y^2}$ over the set S in Fig. 43 and that $\exp(-x^2-y^2) \leq \exp(-R^2)$ on S.)

(d) Using parts (a)–(c), show that

$$\lim_{R\to\infty}\left(\int_0^R e^{-x^2}\,dx\right)^2 = \frac{\pi}{4}.$$

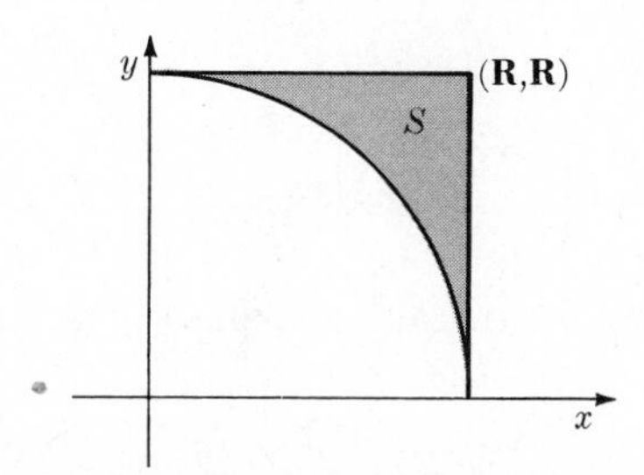

FIGURE 4.43

3. Let $\varphi(u,v) = u + v$, $\psi(u,v) = u - v$, and $\mathbf{\Phi} = (\varphi,\psi)$.

(a) Show that $\dfrac{\partial(\varphi,\psi)}{\partial(u,v)} = -2$.

(b) Let $\bar{S} = \{(u,v) : u^2 + v^2 \leq 1\}$. Show that $\mathbf{\Phi}$ carries $\bar{S}$ one-to-one onto the set $S = \{(x,y) : x^2 + y^2 \leq 2\}$.

(c) The curve $\boldsymbol{\gamma}(t) = (\cos t, \sin t)$, $0 \leq t \leq 2\pi$, forms a positively oriented boundary of $\bar{S}$. Show that $\mathbf{\Phi}\circ\boldsymbol{\gamma}$ is the *negatively* oriented boundary of S. (*Hint:* $\cos t + \sin t = \sqrt{2}\cos(\pi/4 - t)$, and there is a similar formula for $\cos t - \sin t$. *Note:* The switch in orientation is expected, since the Jacobian in part (a) is negative.)

(d) Without using (9), check by comparing areas that $\int\int_S dx\,dy = 2\int\int_{\bar{S}} du\,dv$, with S and $\bar{S}$ as in part (b).

(e) Show that the result in part (d) agrees with (9).

4. (a) The formula

$$\int_0^1 \int_{v-1}^{1-v} 2f(u + v, u - v)\,du\,dv = \int_{-1}^1 \int_{-1}^x f(x,y)\,dy\,dx$$

is a special case of (9). What choices of φ, ψ, S, and $\bar{S}$ reduce (9) to this?

(b) Use Part (a) to evaluate

$$\int_{-1}^1 \int_{-1}^x \frac{1}{x - y - 2}\,dy\,dx.$$

5. In the proof of the change of variable formula (2), show that

$$\int_{\gamma^1} M\,dx = \int_{\bar{\gamma}^1} \bar{M}\,dr + \bar{N}\,d\theta.$$

6. In proving (2), we neglected to show that the functions M, $\bar{M}$, $\bar{N}$, M_x, etc. are continuous. This problem fills the gap, on the assumption that f, f_x, and f_y are continuous and bounded on an open set containing S.

(a) Prove that $M(x,y) = \int_y^{\varphi(x)} f(x,t)\,dt$ is continuous. (Note that

$$M(x,y) - M(x_0,y_0)$$

$$= \left[\int_{y_0}^{\varphi(x)} f(x,t)\,dt - \int_{y_0}^{\varphi(x_0)} f(x_0,t)\,dt\right] + \int_y^{y_0} f(x,t)\,dt;$$

Use Theorem 1 for the term in square brackets, and use the boundedness of f for the other term.)

(b) Prove that $M_x(x,y) = \int_y^{\varphi(x)} f_x(x,t)\,dt + \varphi'(x)f(x,\varphi(x))$ and is continuous. (See §4.1, Problem 6.)

(c) Conclude from parts (a), (b) and a theorem on the composition of continuous functions that $\bar{M}$ and $\bar{N}$ are continuous and have continuous first partial derivatives.

7. The planimeter in Problem 12 of the previous section can be analyzed in a slightly different way. Let β denote the angle made by the first arm and the x axis. Then if $\boldsymbol{\gamma}$ denotes the same curve as in Problem 12 above, we have $\boldsymbol{\gamma}(t) = (\rho \cos \beta(t), \rho \sin \beta(t))$, and the turning of the wheel is proportional to

$$\int_a^b (-\gamma_1' \sin \alpha + \gamma_2' \cos \alpha) = \rho \int_a^b (\sin \alpha \sin \beta + \cos \alpha \cos \beta)\beta'. \qquad (*)$$

(See part (c) of Problem 12, §4.4.)

(a) Since α and β are functions of t, we have a curve $\mathbf{\Gamma}(t) = (\alpha(t),\beta(t))$ in the $\alpha\beta$ plane. Show that (*) is $\int_{\Gamma} \cos(\beta - \alpha)\,d\beta$.

(b) Suppose $\mathbf{\Gamma}$ forms the positively oriented boundary of a region $\bar{S}$ in the $\alpha\beta$ plane; show that (*) equals $\rho \int\int_{\bar{S}} \sin(\beta - \alpha)\,d\alpha\,d\beta$.

(c) For any given α and β, the coordinates of the endpoint $\mathbf{E}$ of the planimeter are

$$x = \rho \cos \beta + R \cos \alpha = \varphi(\alpha,\beta)$$

$$y = \rho \sin \beta + R \sin \alpha = \psi(\alpha,\beta).$$

Suppose that $\mathbf{\Phi} = (\varphi,\psi)$ defines a one-to-one function from $\bar{S}$ onto the region S in the xy plane whose area A we are computing. Suppose further that $\sin(\beta - \alpha) \geq 0$ in $\bar{S}$. Show that

$$\rho \iint_{\bar{S}} \sin(\beta - \alpha)\,d\alpha\,d\beta = \frac{1}{R}\iint_S dx\,dy = \frac{1}{R}A.$$

Chapter V

Functions of n Variables

The elementary theory of functions of n variables is so similar to the two-variable theory that the first few sections of this chapter are practically a review of Chapter IV:

These four sections give the basic outline, omitting the many proofs that are virtually the same as those already given in the case of two variables. The last four sections introduce new material:

Here, again, the topics are covered in outline, partly because of the similarity to Chapter IV, but primarily because a thorough discussion is beyond the scope of this book. For a thorough discussion, see the references listed at the end of §4.1.

5.1 CONTINUITY, PARTIAL DERIVATIVES, AND GRADIENTS

The *domain* of a function f of n variables is a set in R^n, and f assigns to each point $\mathbf{X}$ in its domain a real number $f(\mathbf{X})$. The *range* of f is the set of values assumed by f, that is, the set $\{y: y = f(\mathbf{X}) \text{ for some } \mathbf{X}\}$. The graph of f is the set

$$\{(\mathbf{X},y): \mathbf{X} \text{ is in the domain of } f \text{ and } y = f(\mathbf{X})\}.$$

Here, $\mathbf{X}$ denotes a point $(x_1, \ldots, x_n)$ in R^n, y is a real number, and $(\mathbf{X},y)$ is the point $(x_1, \ldots, x_n, y)$ in R^{n+1}. When $n = 3$, the graph lies in R^4 and cannot be visualized very well. However, the *level surfaces*

$$\{\mathbf{X}: f(\mathbf{X}) = c\}$$

are sets in R^3, and are sometimes rather easy to visualize. For example, if

$$f(x_1,x_2,x_3) = x_1^2 + x_2^2 + 2x_3^2,$$

then the level surface where $f = c$ is an *ellipsoid* for $c > 0$ (see Fig. 1).

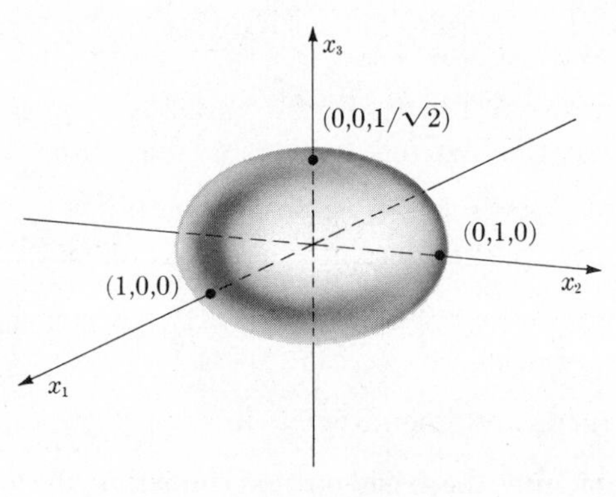

FIGURE 5.1

A function f is *continuous* at $\mathbf{X}_0$ if (i) $f(\mathbf{X}_0)$ is defined, and (ii) for every $\epsilon > 0$, there is $\delta > 0$ such that for all $\mathbf{X}$ in the domain of f,

$$|\mathbf{X} - \mathbf{X}_0| < \delta \Rightarrow |f(\mathbf{X}) - f(\mathbf{X}_0)| < \epsilon.$$

A function f has *limit* L at $\mathbf{X}_0$ if, for every $\epsilon > 0$, there is $\delta > 0$ such that

$$0 < |\mathbf{X} - \mathbf{X}_0| < \delta \Rightarrow f(\mathbf{X}) \text{ is defined and } |f(\mathbf{X}) - L| < \epsilon.$$

In treating continuity it is useful to consider the more general concept of *vector function,* that is, a function $\mathbf{F}$ whose domain is a set S in R^n, and which assigns to each point $\mathbf{X}$ in S a point $\mathbf{F}(\mathbf{X})$ in R^m. We have already studied various special cases in some detail:

$n = 1, m = 1$ (real-valued functions of one variable)

$n = 1, m$ general (m-dimensional vector functions of one variable, curves, §2.1)

$n = 2, m = 1$ (real functions of two variables, Chapter III)

$n = 2, m = 2$ (vector fields in the plane, §4.2; change of variable in double integrals, §4.5).

A vector function $\mathbf{F}$ is *continuous* at $\mathbf{X}_0$ if (i) $\mathbf{F}(\mathbf{X}_0)$ is defined and (ii) for every $\epsilon > 0$, there is $\delta > 0$ such that for all $\mathbf{X}$ in the domain of $\mathbf{F}$,

$$|\mathbf{X} - \mathbf{X}_0| < \delta \quad \Rightarrow \quad |\mathbf{F}(\mathbf{X}) - \mathbf{F}(\mathbf{X}_0)| < \epsilon.$$

A vector function $\mathbf{F}$ has m components $f_1, \ldots, f_m$, real-valued functions of n variables, such that $\mathbf{F}(\mathbf{X}) = (f_1(\mathbf{X}), \ldots, f_m(\mathbf{X}))$. It is easy to prove that $\mathbf{F}$ is continuous at $\mathbf{X}_0$ if and only if each component f_j is continuous at $\mathbf{X}_0$ (Problem 13).

Let S be a set in R^n, and T be a set in R^m. Given a vector-valued function $\mathbf{F}$: $S \to R^m$ and another function $\mathbf{G}$: $T \to R^l$, we can form the composition $\mathbf{G}\circ\mathbf{F}(\mathbf{X}) = \mathbf{G}(\mathbf{F}(\mathbf{X}))$, Fig. 2. It can be proved (exactly as in the case of one variable) that *if* $\mathbf{F}$ *is continuous at* $\mathbf{X}_0$ *and* $\mathbf{G}$ *is continuous at* $\mathbf{F}(\mathbf{X}_0)$, *then* $\mathbf{G}\circ\mathbf{F}$ *is continuous at* $\mathbf{X}_0$.

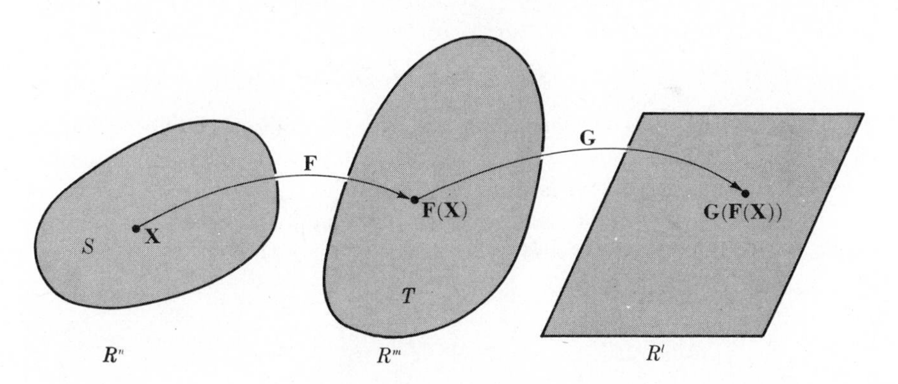

FIGURE 5.2

A set S in R^n is *open* if for every point $\mathbf{X}_0$ in S there is an open ball of positive radius centered at $\mathbf{X}_0$ and contained entirely in S. Intuitively, an open set (like an open interval) is one that contains none of its boundary points. When the domain of a function $\mathbf{F}$ is an open set, then continuity can be expressed in terms of limits: $\mathbf{F}$ is continuous at $\mathbf{X}_0$ if and only if

$$\lim_{\mathbf{X}\to\mathbf{X}_0} |\mathbf{F}(\mathbf{X}) - \mathbf{F}(\mathbf{X}_0)| = 0.$$

Let f be a real-valued function of n variables. The *partial derivatives* of f, denoted by $D_1 f, \ldots, D_n f$, are defined by

$$D_j f(\mathbf{X}) = \lim_{h \to 0} \frac{f(x_1, \ldots, x_j + h, x_{j+1}, \ldots, x_n) - f(x_1, \ldots, x_j, \ldots, x_n)}{h}.$$

Thus, to compute $D_j f$, hold all variables except x_j fixed, and differentiate with respect to x_j in the usual way. This derivative is also denoted

$$\frac{\partial f}{\partial x_j} \quad \text{or} \quad f_{x_j}.$$

For example, if

$$f(x_1,x_2,x_3) = x_1^2 + x_2^2 + 2x_3^2,$$

then

$$D_1 f(x_1,x_2,x_3) = f_{x_1} = \frac{\partial f}{\partial x_1} = 2x_1$$

$$D_2 f(x_1,x_2,x_3) = f_{x_2} = \frac{\partial f}{\partial x_2} = 2x_2$$

$$D_3 f(x_1,x_2,x_3) = f_{x_3} = \frac{\partial f}{\partial x_3} = 4x_3.$$

When the partial derivatives are continuous in an open set, then "mixed partials" may be taken in any order; for example,

$$D_j D_k f = D_k D_j f, \qquad D_1^2 D_3 f = D_1 D_3 D_1 f = D_3 D_1^2 f, \qquad \text{etc.}$$

This follows from the corresponding result for functions of two variables (Theorem 10, §3.6), since $D_j D_k f$ and $D_k D_j f$ are computed by holding constant all but the two variables x_j and x_k.

In the case of three variables, it is customary to write $\mathbf{P} = (x,y,z)$ instead of $\mathbf{X} = (x_1,x_2,x_3)$, and f_y or $\partial f/\partial y$ instead of $D_2 f$, etc. Thus if $g(x,y,z) = xe^{yz}$, then

$$g_x = e^{yz}, \qquad g_y = xze^{yz}, \qquad \frac{\partial g}{\partial z} = xye^{yz}.$$

The concept of *differentiability* of f requires more than the mere existence of the partial derivatives; it is based on the relation between f and an appropriate polynomial of first degree, i.e. a function g of the form

$$g(\mathbf{X}) = \mathbf{X} \cdot \mathbf{M} + c;$$

here, c is a constant and $\mathbf{M}$ is a given vector in R^n, called the *slope* of g. Given a point $\mathbf{X}_0$, a slope vector $\mathbf{M}$, and a constant d, there is a unique first degree polynomial g with slope $\mathbf{M}$ such that $g(\mathbf{X}_0) = d$, namely

$$g(\mathbf{X}) = d + (\mathbf{X} - \mathbf{X}_0) \cdot \mathbf{M}. \tag{1}$$

The graph of g is a hyperplane in R^{n+1}; equation (1) gives the hyperplane of prescribed slope $\mathbf{M}$ passing through a given point $(\mathbf{X}_0, d)$ in R^{n+1}.

From this brief digression on first degree polynomials, we return to the concept of differentiability. A function f is *differentiable at* $\mathbf{X}_0$ if and only if there is a first degree polynomial

$$g(\mathbf{X}) = f(\mathbf{X}_0) + (\mathbf{X} - \mathbf{X}_0)\cdot\mathbf{M}$$

such that

$$\lim_{\mathbf{X}\to\mathbf{X}_0} \frac{f(\mathbf{X}) - g(\mathbf{X})}{|\mathbf{X} - \mathbf{X}_0|} = 0. \tag{3}$$

When f is differentiable at $\mathbf{X}_0$, then the partial derivatives $D_1 f(\mathbf{X}_0), \ldots, D_n f(\mathbf{X}_0)$ all exist, and the vector $\mathbf{M}$ in (2) is the *gradient*

$$\nabla f(\mathbf{X}_0) = (D_1 f(\mathbf{X}_0), \ldots, D_n f(\mathbf{X}_0)).$$

Conversely, when the first partials all exist in an open ball about $\mathbf{X}_0$ *and are continuous* at $\mathbf{X}_0$, then f is differentiable at $\mathbf{X}_0$. When the first partials exist and are continuous, f is called *continuously differentiable.*

If f is differentiable at $\mathbf{X}_0$, and $\boldsymbol{\gamma}$ is a curve in R^n with $\boldsymbol{\gamma}(t_0) = \mathbf{X}_0$, and $\boldsymbol{\gamma}'(t_0)$ exists, then the *chain rule* holds:

$$(f\circ\boldsymbol{\gamma})'(t_0) = \nabla f(\boldsymbol{\gamma}(t_0))\cdot\boldsymbol{\gamma}'(t_0), \tag{4}$$

or in Leibniz notation

$$\frac{df\circ\boldsymbol{\gamma}}{dt} = \sum_1^n \frac{\partial f}{\partial x_j}\frac{dx_j}{dt}. \tag{5}$$

If we take the particular curve $\boldsymbol{\gamma}(t) = \mathbf{X}_0 + t\mathbf{U}$, with $\mathbf{U}$ a unit vector, then at time zero $\boldsymbol{\gamma}$ moves through the point $\mathbf{X}_0$ in the direction $\mathbf{U}$ (Fig. 3); the derivative of $f\circ\boldsymbol{\gamma}$,

$$(f\circ\boldsymbol{\gamma})'(0) = \nabla f(\mathbf{X}_0)\cdot\mathbf{U},$$

is called the *directional derivative* of f in the direction $\mathbf{U}$. This is maximum when $\mathbf{U}$ points in the direction $\nabla f(\mathbf{X}_0)$, and zero when $\mathbf{U}$ is orthogonal to $\nabla f(\mathbf{X}_0)$.

If $\boldsymbol{\gamma}$ is any curve lying in a level surface

$$\{\mathbf{X}: f(\mathbf{X}) = c\}, \tag{6}$$

then $f\circ\boldsymbol{\gamma} \equiv c$, so $(f\circ\boldsymbol{\gamma})' \equiv 0$. Since $(f\circ\boldsymbol{\gamma})' = (\nabla f\circ\boldsymbol{\gamma})\cdot\boldsymbol{\gamma}'$, it follows that ∇f is orthogonal to $\boldsymbol{\gamma}'$, in other words, *the gradient at each point of the level surface* (6) *is orthogonal to every curve lying in that surface* (Fig. 4). Because of this, we say simply that the gradient is orthogonal to the level surface (just as in two dimensions the gradient is orthogonal to the level line, or contour line). Further, when $\mathbf{X}_0$ is on the level surface (6) and $\nabla f(\mathbf{X}_0) \neq \mathbf{0}$,

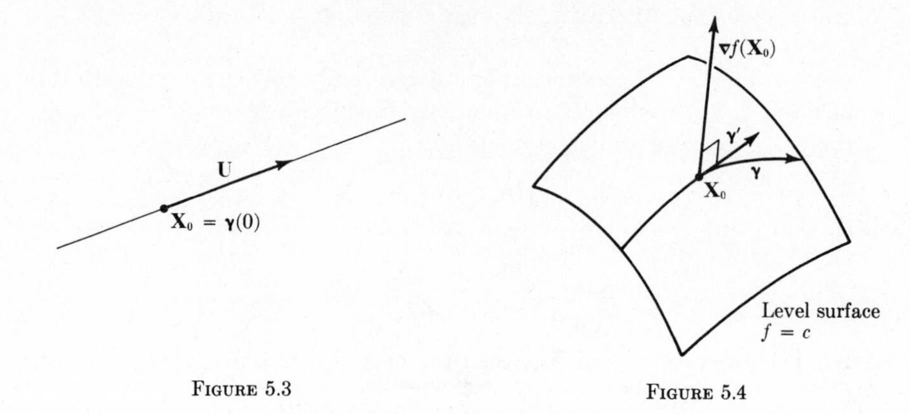

FIGURE 5.3

FIGURE 5.4

we define the *tangent plane* to the surface at $\mathbf{X}_0$ to be the plane with equation

$$(\mathbf{X} - \mathbf{X}_0) \cdot \nabla f(\mathbf{X}_0) = 0.$$

When $n = 2$, the level "surface" is actually a curve, and the tangent "plane" is actually a line; when $n = 3$, the tangent plane is a genuine two-dimensional plane in R^3, and when $n > 3$, it is an $(n - 1)$-dimensional object called a "hyperplane."

In the case $n = 3$, a possible conflict appears between the definition of tangent plane to the level surface of a function of three variables given here and the definition of tangent plane to the graph of a function of two variables given in the previous chapter. What if a level surface is also a graph? Do the definitions then agree? They do, and it is easy to prove this by the chain rule. (See Example 2 below.)

The chain rule (4) or (5) applies to partial derivatives as well. Suppose that f is a continuously differentiable function of n variables, and that $\varphi_1, \ldots, \varphi_n$ are continuously differentiable functions of m variables. Then the composite function

$$F(x_1, \ldots, x_m) = f(\varphi_1(x_1, \ldots, x_m), \ldots, \varphi_n(x_1, \ldots, x_m))$$

has continuous partial derivatives given by

$$D_1F = D_1 f(\varphi_1, \ldots, \varphi_n) D_1\varphi_1 + D_2 f(\varphi_1, \ldots, \varphi_n) D_1\varphi_2 + \cdots$$
$$+ D_n f(\varphi_1, \ldots, \varphi_n) D_1\varphi_n$$

$$\vdots$$

$$D_mF = D_1 f(\varphi_1, \ldots, \varphi_n) D_m\varphi_1 + \cdots + D_n f(\varphi_1, \ldots, \varphi_n) D_m\varphi_n .$$

These formulas may look more familiar in Leibniz notation, as in §3.4.

If we represent the situation by a scheme such as

$$\begin{array}{ccccc} (x_1, \ldots, x_m) & \rightarrow & (y_1, \ldots, y_n) & \rightarrow & z \\ y_j = \varphi_j(x_1, \ldots, x_m), & & z = f(y_1, \ldots, y_n) & = & F(x_1, \ldots, x_m) \end{array} \tag{7}$$

then the derivatives of the composite function F are given by

$$\frac{\partial z}{\partial x_1} = \frac{\partial z}{\partial y_1}\frac{\partial y_1}{\partial x_1} + \frac{\partial z}{\partial y_2}\frac{\partial y_2}{\partial x_1} + \cdots + \frac{\partial z}{\partial y_n}\frac{\partial y_n}{\partial x_1}$$

$$\vdots$$

$$\frac{\partial z}{\partial x_m} = \frac{\partial z}{\partial y_1}\frac{\partial y_1}{\partial x_m} + \cdots + \frac{\partial z}{\partial y_n}\frac{\partial y_n}{\partial x_m}.$$

Example 1. Find the equation of the tangent plane to the hyperboloid

$$x^2 + y^2 - z^2 = 1 \tag{8}$$

at the point $\mathbf{P}_0 = (1,1,-1)$. (See Fig. 5.)

First solution. Equation (8) defines a level surface for the function $f(x,y,z) = x^2 + y^2 - z^2$, so the gradient $\nabla f = (2x,2y,-2z)$ is normal to the tangent plane. Substituting $(1,1,-1)$ for (x,y,z) gives the normal $\mathbf{N} = (2,2,2)$, so the tangent plane has the equation $\mathbf{N}\cdot(\mathbf{P}-\mathbf{P}_0) = 0$, or

$$2(x-1) + 2(y-1) + 2(z+1) = 0,$$

or

$$x + y + z = 1.$$

Second solution. Near the point $(1,1,-1)$, equation (8) determines z as a function of x and y,

$$z = \varphi(x,y) = -\sqrt{x^2 + y^2 - 1}$$

(*not* $z = \sqrt{x^2 + y^2 - 1}$). According to the previous chapter, the tangent plane has the equation

$$\begin{aligned} z &= \varphi(\mathbf{P}_0) + (\mathbf{P} - \mathbf{P}_0)\cdot\nabla\varphi(\mathbf{P}_0) \\ &= \varphi(1,1) + (x-1)\varphi_x(1,1) + (y-1)\varphi_y(1,1) \\ &= -1 + (x-1)(-1) + (y-1)(-1), \end{aligned}$$

or

$$(z+1) + (x-1) + (y-1) = 0.$$

This is the same plane found in the first solution.

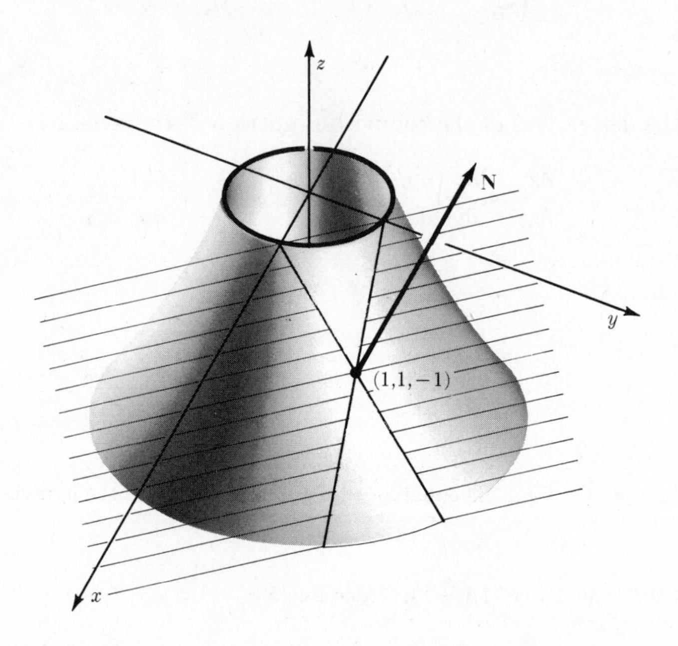

FIGURE 5.5 Graph and tangent plane shown only for $z \leq 0$

Example 2. Suppose that the level surface $\{(x,y,z): f(x,y,z) = c\}$ contains the graph of a differentiable function φ. Let $\mathbf{P}_0 = (x_0,y_0,z_0)$ be a point on the graph, and suppose that f has continuous derivatives and $\nabla f(\mathbf{P}_0) \neq \mathbf{0}$. Prove that the tangent plane to the graph of φ at $\mathbf{P}_0$ has the equation

$$(\mathbf{P} - \mathbf{P}_0)\cdot\nabla f(\mathbf{P}_0) = 0.$$

Solution. Since the graph of φ lies in the level surface of f, we have

$$f(x,y,\varphi(x,y)) \equiv c.$$

Differentiating this with respect to x by the chain rule,

$$D_1 f(x,y,\varphi(x,y)) + D_3 f(x,y,\varphi(x,y)) \frac{\partial \varphi}{\partial x} \equiv 0, \tag{9}$$

and differentiating with respect to y,

$$D_2 f(x,y,\varphi(x,y)) + D_3 f(x,y,\varphi(x,y)) \frac{\partial \varphi}{\partial y} \equiv 0. \tag{10}$$

Since $\nabla f(\mathbf{P}_0) \neq \mathbf{0}$, it follows that $D_3 f(\mathbf{P}_0) \neq 0$; (for if $D_3 f(\mathbf{P}_0) = 0$, then

$D_1 f(\mathbf{P}_0) = 0$ by (9), and $D_2 f(\mathbf{P}_0) = 0$ by (10)). Hence we can solve (9) and (10) to find

$$\frac{\partial \varphi}{\partial x} = -\frac{D_1 f}{D_3 f}, \qquad \frac{\partial \varphi}{\partial y} = -\frac{D_2 f}{D_3 f}. \tag{11}$$

By §3.3, the tangent plane to the graph of φ at the point $\mathbf{P}_0 = (x_0,y_0,z_0)$ has the equation

$$(z - z_0) - (x - x_0)\frac{\partial \varphi}{\partial x}(x_0,y_0) - (y - y_0)\frac{\partial \varphi}{\partial y}(x_0,y_0) = 0. \tag{12}$$

Substituting the expressions from (11) and multiplying by $D_3 f$ reduces (12) to the form $(\mathbf{P} - \mathbf{P}_0)\cdot\nabla f(\mathbf{P}_0) = 0$, and the problem is solved.

Example 3. Suppose that the temperature at the point (x,y,z) is

$$T(x,y,z) = xy + yz + zx.$$

A particle travels along the curve

$$\boldsymbol{\gamma}(t) = (t, 3t^2, 2\cos t).$$

What is the rate of change of temperature observed by the particle at time t?

First solution. The temperature observed at the point $\boldsymbol{\gamma}(t)$ is $T(\boldsymbol{\gamma}(t))$, so we compute the derivative $(T\circ\boldsymbol{\gamma})'(t) = \nabla T(\boldsymbol{\gamma}(t))\cdot\boldsymbol{\gamma}'(t)$. We have

$$\nabla T(x,y,z) = (T_x,T_y,T_z) = (y + z, x + z, x + y),$$

so

$$\nabla T(\boldsymbol{\gamma}(t)) = (3t^2 + 2\cos t, t + 2\cos t, t + 3t^2).$$

Further,

$$\boldsymbol{\gamma}'(t) = (1, 6t, -2\sin t),$$

so

$$\begin{aligned}(T\circ\boldsymbol{\gamma})'(t) &= (3t^2 + 2\cos t)\cdot 1 + (t + 2\cos t)\cdot 6t + (t + 3t^2)(-2\sin t)\\ &= 9t^2 + (12t + 2)\cos t - 2(t + 3t^2)\sin t.\end{aligned}$$

Second solution. At time t the particle is at the point with coordinates

$$x = t, \qquad y = 3t^2, \qquad z = 2\cos t,$$

so the temperature observed is

$$xy + yz + zx = 3t^3 + 6t^2\cos t + 2t\cos t,$$

and the rate of change is

$$9t^2 + (12t + 2)\cos t - (6t^2 + 2t)\sin t.$$

Example 4. Suppose that the "electrical potential" at the point (x,y,z) is $E(x,y,z) = x^2 + y^2 - 2z^2$. According to the principles of electrodynamics, a positively charged particle is accelerated in the direction of maximum decrease of potential. What is the direction of acceleration at the point (1,3,2)?

Solution. The maximum decrease of E is the maximum increase of $-E$, so the charged particle moves in the direction $\nabla(-E) = -\nabla E$. We find that

$$-\nabla E = (-2x,-2y,4z),$$

and the value at (1,3,2) is

$$(-2,-6,8);$$

hence the direction of acceleration is parallel to $(-2,-6,8)$, or parallel to $(-1,-3,4)$.

PROBLEMS

1. Let $f(x,y,z) = e^{xy}\cos(yz)$, and compute the following derivatives.
(a) f_x
(b) $D_1^2 f$
(c) f_{xy}
(d) $D_2D_1 f$
(e) $D_2D_1D_3 f$
(f) $D_1D_2D_3 f$

2. Show that each of the following functions satisfies Laplace's equation $f_{xx} + f_{yy} + f_{zz} = 0$. (This equation is satisfied by electrical potentials, and by steady-state temperature distributions.)
(a) $x^2 - 2y^2 + z^2$
(b) $(x^2 + y^2 + z^2)^{-1/2}$
(c) $2x^3 - 3(y^2 + z^2)x$
(d) $e^{5x+12y}\sin 13z$
(e) $e^{3y+4z}\cos 5x$
(f) $\cos 3x \sin 4y e^{5z}$

3. Find an equation of the plane tangent to the given surface at the given point.
(a) $x^2 + y^2 + 2z^2 = 1$, at (1,0,0)
(b) $e^{xy}\cos(yz) = 1$, at (1,0,1)
(c) $x^2 + y^3 - z^4 = -16$, at $(1,-1,2)$

4. Find an equation of the hyperplane tangent to the given "surface" at the given point.
(a) $x_1^2 + x_2^2 = 2$, at (1,1)
(b) $x_1^2 + x_2^2 + x_3^2 + x_4^2 = 4$, at (1,1,1,1)
(c) $x_1^2 + x_2^2 - x_3^2 - x_4^2 = 0$, at $(1,1,1,-1)$

5. Suppose that the temperature T at the point (x,y,z) is given by $T(x,y,z) = xyz$.
(a) Starting at (1,2,3), in what direction should you move to achieve the maximum rate of increase of temperature?
(b) Starting at $(1,-3,1)$, in what direction should you move to achieve the maximum rate of cooling?

6. Suppose that ∇f is orthogonal to a differentiable curve γ at every point on γ. Prove that f is *constant* on γ, in other words, that γ lies in a single level surface of f.

7. (a) Prove that the graph of any function f of two variables is automatically a level surface of a function F of three variables. (Hint: All you have to do is write $z = f(x,y)$ in the form $F(x,y,z) = 0$.)
(b) Show that not every level surface is a graph.

8. Suppose that f is a function of three variables having continuous partial derivatives, and define a new function

$$F(r,\theta,z) = f(r\cos\theta, r\sin\theta, z).$$

(a) Find F_r, F_θ, and F_z in terms of $D_1 f$, $D_2 f$, $D_3 f$.
(b) Show that

$$D_1{}^2f + D_2{}^2f + D_3{}^2f = F_{rr} + \frac{1}{r}F_r + \frac{1}{r^2}F_{\theta\theta} + F_{zz}.$$

(The variables r, θ, and z are called *cylindrical coordinates* of (x,y,z) when $x = r\cos\theta$, $y = r\sin\theta$, $z = z$. See Fig. 6. The formula in part (b) shows how to write Laplace's equation in cylindrical coordinates.)

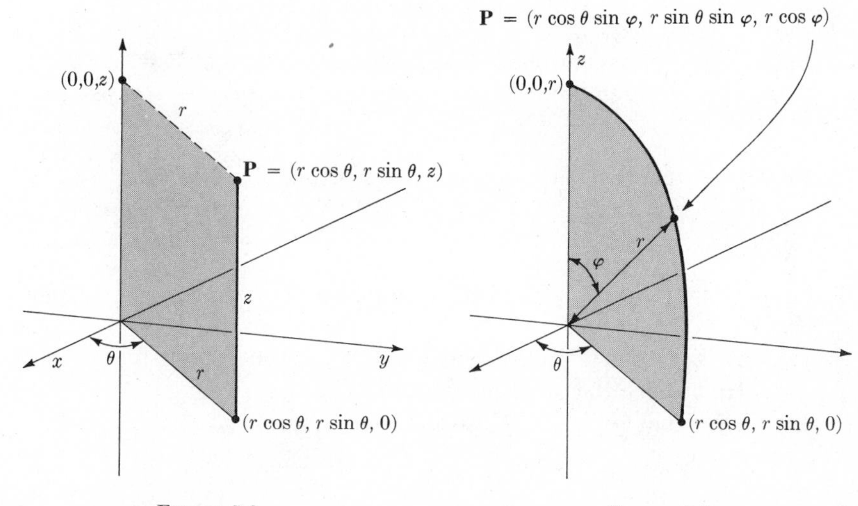

FIGURE 5.6

FIGURE 5.7

9. Suppose that f is a function of three variables having continuous partial derivatives, and set

$$F(r,\theta,\varphi) = f(r\cos\theta\sin\varphi, r\sin\theta\sin\varphi, r\cos\varphi).$$

(See Fig. 7; r, θ, and φ are the *spherical coordinates* of (x,y,z) when $x = r\cos\theta\sin\varphi$, $y = r\sin\theta\sin\varphi$, $z = r\cos\varphi$.)

(a) Show that for $r \sin \varphi \neq 0$,

$$D_1^2f + D_2^2f + D_3^2f = F_{rr} + \frac{2}{r}F_r + \frac{1}{r^2}F_{\varphi\varphi} + \frac{\cot\varphi}{r^2}F_\varphi + \frac{1}{r^2\sin^2\varphi}F_{\theta\theta}.$$

(b) Show that if $f(x,y,z) = g(\sqrt{x^2+y^2+z^2})$, where $g(r)$ has continuous second derivatives for $r > 0$, then

$$D_1^2f + D_2^2f + D_3^2f = g''(r) + \frac{2}{r}g'(r) \quad \text{with } r = \sqrt{x^2+y^2+z^2}.$$

(c) Show that $g(\sqrt{x^2+y^2+z^2})$ satisfies $f_{xx} + f_{yy} + f_{zz} = 0$ (Laplace's equation) if and only if $g(r) = A/r + B$ for some constants A and B.

10. Prove the chain rule

$$(f\circ\boldsymbol{\gamma})' = \sum_1^n \gamma_j'(D_jf)\circ\boldsymbol{\gamma}$$

or

$$\frac{df}{dt} = \sum_1^n \frac{df}{dx_j}\frac{dx_j}{dt}.$$

11. Prove that if f has continuous first partial derivatives at $\mathbf{P}_0 = (x_0,y_0,z_0)$, then f is differentiable at $\mathbf{P}_0$. (Hint: $f(x,y,z) - f(x_0,y_0,z_0) = f(x,y,z) - f(x_0,y,z) + f(x_0,y,z) - f(x_0,y_0,z) + f(x_0,y_0,z) - f(x_0,y_0,z_0)$.)

12. Prove the theorem on continuity of a composite function: *If* $\mathbf{F}$ *is continuous at* $\mathbf{X}_0$ *and* $\mathbf{G}$ *is continuous at* $\mathbf{F}(\mathbf{X}_0)$, *then* $\mathbf{G}\circ\mathbf{F}$ *is continuous at* $\mathbf{X}_0$.

13. Suppose that $f_1, \ldots, f_m$ are the components of a vector-valued function $\mathbf{F}$.

(a) Prove that if $\mathbf{F}$ is continuous at $\mathbf{X}_0$, then so are $f_1, \ldots, f_m$. (Hint: See the proof of Theorem 2, §2.1.)

(b) Prove that if $f_1, \ldots, f_m$ are all continuous at $\mathbf{X}_0$, then so is $\mathbf{F}$.

5.2 THE IMPLICIT FUNCTION THEOREM

Section 3.5 proved that under certain natural hypotheses an equation

$$f(x,y) = c \tag{1}$$

determines y as a function of x, locally, near any point (x_0,y_0) on the contour line defined by (1). In other words, near (x_0,y_0) the contour line can be represented as the graph of a function of one variable.

There is a similar *implicit function theorem* which states that under certain natural hypotheses, the equation

$$f(x_1, \ldots, x_n) = c \tag{2}$$

determines x_n as a function of $x_1, \ldots, x_{n-1}$, locally, near any point $\mathbf{X}_0$ on the level surface (2). To simplify the notation, we take $\mathbf{X}_0 = \mathbf{0}$; then the theorem reads as follows:

Suppose that $f(\mathbf{0}) = c$, *and* $D_n f(\mathbf{0}) \neq 0$, *and* f *has continuous first partial derivatives in an open set containing the origin. Then there is a cylinder*

$$C = \{\mathbf{X}: x_1^2 + \cdots + x_{n-1}^2 < \delta^2 \text{ and } |x_n| < \epsilon\}$$

and there is a function φ *with continuous first derivatives defined in the ball*

$$B = \{(x_1, \ldots, x_{n-1}): \sum_1^{n-1} x_j^2 < \delta^2\}$$

such that for $\mathbf{X}$ *in* C,

$$f(\mathbf{X}) = c \iff x_n = \varphi(x_1, \ldots, x_{n-1}).$$

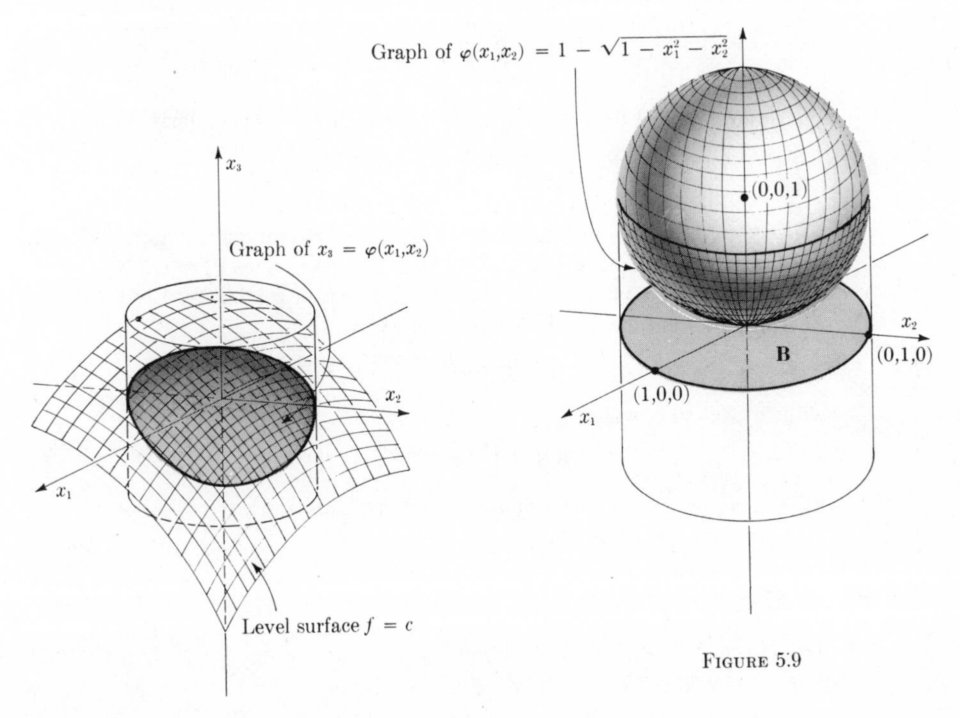

FIGURE 5.8 The cylinder has height 2ϵ and radius δ

FIGURE 5.9

Figure 8 illustrates the situation with $n = 3$. A simple example is the equation

$$f(x_1,x_2,x_3) = x_1^2 + x_2^2 + (x_3 - 1)^2 = 1,$$

which is satisfied with $x_1 = x_2 = x_3 = 0$. Since $D_3 f(\mathbf{0}) = -2 \neq 0$, the implicit function theorem applies. It is easy to see (from Fig. 9) that the biggest possible B in this case is the unit ball

$$B = \{(x_1,x_2) : x_1^2 + x_2^2 < 1\},$$

that $\varphi(x_1,x_2) = 1 - \sqrt{1 - x_1^2 - x_2^2}$, and that

$$C = \{(x_1,x_2,x_3) : x_1^2 + x_2^2 < 1 \text{ and } |x_3| < 1\}.$$

Of course, we don't usually bother to apply the implicit function theorem when the function φ can be written down so easily; the real point is that an equation like

$$z + \cos(xyz)e^{x^2+y^2+z^2} = 1 \tag{3}$$

defines z as a function of x and y (at least near the origin), although it is not possible to solve (3) explicitly for z.

Example 1. Let $f(x,y,z) = xe^{x+y+z}$. The point $\mathbf{P}_0 = (1,0,-1)$ lies on the level surface $f = 1$. The gradient is $\nabla f = e^{x+y+z}(1 + x, x, x)$; hence $\nabla f(\mathbf{P}_0) = (2,1,1)$. Since

$$\frac{\partial f}{\partial z}(\mathbf{P}_0) = 1 \neq 0,$$

the equation $f = 1$ defines z locally as a function $\varphi(x,y)$; similarly,

$$\frac{\partial f}{\partial x}(\mathbf{P}_0) \neq 0 \quad \text{and} \quad \frac{\partial f}{\partial y}(\mathbf{P}_0) \neq 0,$$

so $f = 1$ defines x locally as a function $\psi(y,z)$, or y as a function $\eta(x,z)$. We will deduce some of the properties of $\psi(y,z)$. First of all, since $\mathbf{P}_0 = (1,0,-1)$ is on the level surface $f = 1$, it must be on the graph of ψ, that is, $1 = \psi(0,-1)$. Further, differentiating the identity

$$\psi(y,z)e^{\psi(y,z)+y+z} = 1$$

with respect to y, we get

$$\psi_y e^{\psi+y+z} + \psi e^{\psi+y+z}(\psi_y + 1) = 0,$$

whence $\psi_y = -\psi/(1 + \psi)$, and in particular $\psi_y(0,-1) = -1/2$.

PROBLEMS

1. Let $f(x,y,z) = z + \cos(xyz)e^{x^2+y^2+z^2}$.

(a) Prove that there is a cylinder

$$C = \{(x,y,z) : x^2 + y^2 < \delta^2 \text{ and } |z| < \epsilon\}$$

and there is a function $\varphi(x,y)$ defined for $x^2 + y^2 < \delta^2$ such that for (x,y,z) in C,

$$f(x,y,z) = 1 \iff z = \varphi(x,y).$$

(Simply check that the implicit function theorem applies.)

(b) Show that $\varphi(0,0) = 0$.

(c) Show that $\varphi_x(0,0) = 0$ and $\varphi_y(0,0) = 0$. (Hint: Differentiate $\varphi + \cos(xy\varphi)e^{x^2+y^2+\varphi^2} = 1$.)

(d) Show that $\varphi(x,y) = \varphi(y,x)$.

2. (a) Show that the equation

$$x \sin(x + y + z) = 0$$

determines x as a function $\psi(y,z)$, locally, near the point $(2,-1,-1)$.

(b) Show that $\psi(-1,-1) = 2$.

(c) Show that $\psi_z(-1,-1) = -1$.

(d) Find $\psi_y(-1,-1)$.

(e) Show that there is a function φ of one variable such that $\psi(y,z) = \varphi(y + z)$ when y and z are in the domain of ψ.

3. Let φ be a differentiable function of two variables, and set $f(x,y,z) = [z - \varphi(x,y)]^2$.

(a) Prove that $f = 0 \iff z = \varphi(x,y)$.

(b) Prove that at every point $\mathbf{P}_0$ where $f(\mathbf{P}_0) = 0$ we have $\nabla f(\mathbf{P}_0) = \mathbf{0}$.

(c) Reconcile parts (a) and (b) with the implicit function theorem.

4. Prove the implicit function theorem in R^3: *If f has continuous first partial derivatives at $\mathbf{P}_0 = (x_0,y_0,z_0)$, and $D_3 f(\mathbf{P}_0) \neq 0$, then there is a cylinder*

$$C = \{(x,y,z): (x - x_0)^2 + (y - y_0)^2 < \delta^2 \text{ and } |z - z_0| < \epsilon\}$$

and a differentiable function φ such that for (x,y,z) in C,

$$f(x,y,z) = f(x_0,y_0,z_0) \iff z = \varphi(x,y).$$

(Hint: Review the proof of the implicit function theorem in §3.5.)

5. The implicit function theorem shows when *one* equation $f(x,y,z) = c$ can determine z as a function of x and y. This problem shows when *two* equations

$$f(x,y,z) = c$$

$$g(x,y,z) = d$$

can determine y and z both as functions of x; thus the two equations together define a curve (Fig. 10).

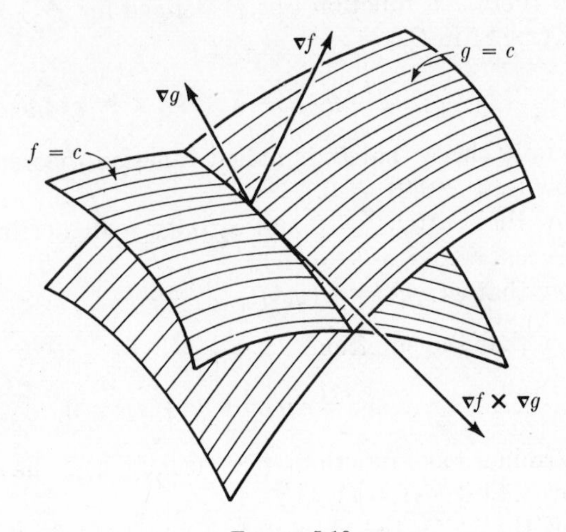

FIGURE 5.10

Suppose that f and g have continuous first partial derivatives at $\mathbf{P}_0 = (x_0,y_0,z_0)$, and the first component of $\nabla f(\mathbf{P}_0) \times \nabla g(\mathbf{P}_0)$ is not zero.

(a) Prove that either $f_z(\mathbf{P}_0) \neq 0$ or $g_z(\mathbf{P}_0) \neq 0$.

(b) Supposing $f_z(\mathbf{P}_0) \neq 0$, show that there is a differentiable function φ and a cylinder C of height 2ϵ centered at $\mathbf{P}_0$ in which

$$f(x,y,z) = f(\mathbf{P}_0) \quad \Leftrightarrow \quad z = \varphi(x,y).$$

(c) Using the result in part (b), show that for points in the cylinder C, the two equations

$$\begin{aligned} f(x,y,z) &= f(\mathbf{P}_0) \\ g(x,y,z) &= g(\mathbf{P}_0) \end{aligned} \tag{4}$$

are equivalent to the equations

$$G(x,y) = g(\mathbf{P}_0), \qquad z = \varphi(x,y)$$

where $G(x,y) = g(x,y,\varphi(x,y))$.

(d) Assuming $f_z(\mathbf{P}_0) \neq 0$ and the first component of $\nabla f(\mathbf{P}_0) \times \nabla g(\mathbf{P}_0)$ is not zero, prove that $G_y(x_0,y_0) \neq 0$.

(e) Prove that there is a rectangle

$$R = \{(x,y,z): |x - x_0| < \delta_1 \text{ and } |y - y_0| < \epsilon_1\}$$

in which the equation $G(x,y) = g(\mathbf{P}_0)$ is equivalent to $y = \psi(x)$, where ψ is a differentiable function.

(f) Prove that for $|x - x_0| < \delta_1$, $|y - y_0| < \epsilon_1$, and $|z - z_0| < \epsilon$, the two equations (4) are equivalent to

$$y = \psi(x)$$

$$z = \varphi(x,\psi(x))$$

or

$$y = \psi_1(x), \qquad z = \psi_2(x). \tag{5}$$

(g) The two equations (5) define a curve, given parametrically by

$$x = t, \qquad y = \psi_1(t), \qquad z = \psi_2(t).$$

Prove that $\nabla f(\mathbf{P}_0) \times \nabla g(\mathbf{P}_0)$ is tangent to this curve at time $t = x_0$.

(h) Suppose that the *third* component of $\nabla f(\mathbf{P}_0) \times \nabla g(\mathbf{P}_0)$ is not zero; draw a conclusion analogous to the one in part (f) above.

6. Suppose $\mathbf{P}_0 = (x_0,y_0,z_0)$ lies on the level surface $S = \{\mathbf{P}: f(\mathbf{P}) = c\}$, and $g(\mathbf{P}_0) \geq g(\mathbf{P})$ for all $\mathbf{P}$ on the surface S. Assuming that f and g have continuous derivatives, and that $\nabla f(\mathbf{P}_0) \neq \mathbf{0}$, prove that there is a "Lagrange multiplier" λ such that $\nabla g(\mathbf{P}_0) = \lambda \nabla f(\mathbf{P}_0)$. (Hint: See the discussion of Lagrange multipliers in §3.5.)

7. Find the minimum of $g(x,y,z) = x^2 + y^2 + z^2$ on the surface defined by $xyz = 1$. (You are finding the point on the surface that is closest to the origin. Use Lagrange multipliers, Problem 6, and assume that a minimum exists.)

8. Suppose that $\mathbf{P}_0 = (x_0,y_0,z_0)$ lies in the set

$$C = \{\mathbf{P}: f(\mathbf{P}) = c \text{ and } g(\mathbf{P}) = d\},$$

and suppose that h is a function such that $h(\mathbf{P}_0) \geq h(\mathbf{P})$ for every $\mathbf{P}$ in C. Prove that $\nabla h(\mathbf{P}_0) \cdot (\nabla f(\mathbf{P}_0) \times \nabla g(\mathbf{P}_0)) = 0$. (Hint: Use Problem 5. Geometrically, the condition $\nabla h \cdot (\nabla f \times \nabla g) = 0$ says that ∇h is orthogonal to the curve C at the given point.)

9. Find the maximum of $h(x,y,z) = x^2 + y^2 + z^2$ on the intersection of the two surfaces $xyz = 1$, $x^2 + y^2 + 2z^2 = 4$. (Hint 1: Assume that a maximum exists. Hint 2: The intersection of the two surfaces is the set of points (x,y,z) such that

$$f(x,y,z) = xyz = 1 \tag{6}$$

and

$$g(x,y,z) = x^2 + y^2 + 2z^2 = 4. \tag{7}$$

By Problem 8, you are looking for points $\mathbf{P}_0$ that satisfy (6), (7), and the equation $\nabla h \cdot (\nabla f \times \nabla g) = 0$. There are four such points, and two of them maximize h.)

5.3 TAYLOR EXPANSIONS

The Taylor expansion in n variables approximates $f(\mathbf{A} + \mathbf{H})$ by a polynomial in $\mathbf{H}$; the approximation is best when $\mathbf{H}$ is small, or when the degree of the polynomial is large. To obtain a simple expression for the approximating polynomial, we introduce the "vector operator"

$$\nabla = (D_1, \ldots, D_n)$$

and the dot product $\mathbf{H} \cdot \nabla$, defined by

$$(\mathbf{H} \cdot \nabla f)(\mathbf{X}) = \sum_{j=1}^{n} h_j D_j f(\mathbf{X}). \tag{1}$$

Then the kth *degree Taylor polynomial* of f at the point $\mathbf{A}$ is defined to be

$$f_k(\mathbf{A},\mathbf{H}) = \sum_{j=0}^{k} \frac{1}{j!} ((\mathbf{H} \cdot \nabla)^j f)(\mathbf{A}).$$

The first term, $\frac{1}{0!}((H \cdot \nabla)^0 f)(\mathbf{A})$, stands for $f(\mathbf{A})$, and the second term is simply (1) with $\mathbf{X} = \mathbf{A}$. In the remaining terms, the derivatives in $(\mathbf{H} \cdot \nabla)^j$ are applied to f, with $\mathbf{H}$ considered constant; thus if the second derivatives of f are continuous at $\mathbf{A}$, we have

$$\begin{aligned}
((\mathbf{H} \cdot \nabla)^2 f)(\mathbf{A}) &= (h_1 D_1 + \cdots + h_n D_n)^2 f(\mathbf{A}) \qquad (2) \\
&= \sum_{l=1}^{n} \sum_{m=1}^{n} h_l h_m D_l D_m f(\mathbf{A}) \\
&= h_1^2 D_1^2 f(\mathbf{A}) + \cdots + h_n^2 D_n^2 f(\mathbf{A}) \\
&\quad + 2[h_1 h_2 D_1 D_2 f(\mathbf{A}) + h_1 h_3 D_1 D_3 f(\mathbf{A}) + \cdots \\
&\qquad + h_{n-1} h_n D_{n-1} D_n f(\mathbf{A})],
\end{aligned}$$

where the square brackets above contain all the "cross" terms that come from expanding the square on the right-hand side of (2).

Taylor's theorem says that if all the derivatives of f of order $\leq k$ are continuous in a ball of radius ϵ about $\mathbf{A}$, then for every $|\mathbf{H}| < \epsilon$ we can write

$$f(\mathbf{A} + \mathbf{H}) = f_{k-1}(\mathbf{A},\mathbf{H}) + \frac{1}{k!} (\mathbf{H} \cdot \nabla)^k f(\mathbf{A} + \theta \mathbf{H})$$

for some number θ between 0 and 1. It follows that

$$\lim_{\mathbf{H} \to 0} \frac{f(\mathbf{A} + \mathbf{H}) - f_k(\mathbf{A},\mathbf{H})}{|\mathbf{H}|^k} = 0. \tag{3}$$

In particular, taking $k = 1$, we have the first degree Taylor polynomial

$$f_1(\mathbf{A},\mathbf{H}) = f(\mathbf{A}) + \mathbf{H}\cdot\nabla f(\mathbf{A}),$$

and (3) becomes

$$\lim_{\mathbf{H}\to\mathbf{0}} \frac{f(\mathbf{A}+\mathbf{H}) - f(\mathbf{A}) - \mathbf{H}\cdot\nabla f(\mathbf{A})}{|\mathbf{H}|} = 0. \tag{4}$$

This is precisely the condition for differentiability of f at $\mathbf{A}$.

A useful tool in the study of maxima and minima is the second degree Taylor polynomial

$$\begin{aligned} f_2(\mathbf{A},\mathbf{H}) &= f(\mathbf{A}) + \mathbf{H}\cdot\nabla f(\mathbf{A}) + \tfrac{1}{2}(\mathbf{H}\cdot\nabla)^2 f(\mathbf{A}) \\ &= f(\mathbf{A}) + \sum_{j=1}^{n} h_j D_j f(\mathbf{A}) + \frac{1}{2}\sum_{l=1}^{n}\sum_{m=1}^{n} h_l h_m D_l D_m f(\mathbf{A}). \end{aligned} \tag{5}$$

$\mathbf{A}$ is a *critical point* for f if $\nabla f(\mathbf{A}) = \mathbf{0}$; at a critical point, we have

$$f_2(\mathbf{A},\mathbf{H}) = f(\mathbf{A}) + \frac{1}{2}\sum_{l=1}^{n}\sum_{m=1}^{n} h_l h_m D_l D_m f(\mathbf{A}).$$

Just as with two variables, it is generally possible to determine whether a critical point $\mathbf{A}$ is a local maximum, minimum, or saddle point by studying the "quadratic form"

$$\sum_{l=1}^{n}\sum_{m=1}^{n} h_l h_m D_l D_m f(\mathbf{A});$$

however, when $n > 2$, the general study of quadratic forms requires concepts from linear algebra, and we will have to restrict ourselves to particularly simple cases.

Example 1. Let

$$f(x_1,x_2,x_3) = x_1^2 + 2x_2^2 + 3x_3^2 - \cos(x_1 + x_3) + \sin(x_1x_2x_3).$$

Find the Taylor polynomial $f_2(\mathbf{0},\mathbf{H})$ about the point $\mathbf{A} = \mathbf{0}$. *Solution.* You can check that $f(\mathbf{0}) = -1$, $\nabla f(\mathbf{0}) = \mathbf{0}$, $D_1^2 f(\mathbf{0}) = 3$, $D_2^2 f(\mathbf{0}) = 4$, $D_3^2 f(\mathbf{0}) = 7$, $D_1D_2 f(\mathbf{0}) = 0$, $D_1D_3 f(\mathbf{0}) = 1$, $D_2D_3 f(\mathbf{0}) = 0$. Hence

$$\begin{aligned} f_2(\mathbf{H}) &= f(\mathbf{0}) + \mathbf{H}\cdot\nabla f(\mathbf{0}) + \tfrac{1}{2}((h_1D_1 + h_2D_2 + h_3D_3)^2 f)(\mathbf{0}) \\ &= f(\mathbf{0}) + 0 + \tfrac{1}{2}(h_1^2D_1^2 f + h_2^2D_2^2 f + h_3^3D_3^2 f)(\mathbf{0}) \\ &\qquad + \tfrac{1}{2}\cdot 2(h_1h_2D_1D_2 f + h_1h_3D_1D_3 f + h_2h_3D_2D_3 f)(\mathbf{0}) \\ &= -1 + \tfrac{1}{2}(3h_1^2 + 4h_2^2 + 7h_3^2) + 1(h_1h_3) \\ &= -1 + \tfrac{3}{2}h_1^2 + 2h_2^2 + \tfrac{7}{2}h_3^2 + h_1h_3. \end{aligned} \tag{6}$$

Example 2. The function f in Example 1 has $\mathbf{0}$ as a critical point. Examine this point to see whether it is a local maximum or minimum.

Solution. From (6) it seems likely that $f_2(\mathbf{0},\mathbf{H}) > -1$ if $\mathbf{H} \neq \mathbf{0}$. This can be proved by writing $f_2(\mathbf{0},\mathbf{H})$ in the form

$$-1 + h_1^2 + 2h_2^2 + 3h_3^2 + \tfrac{1}{2}(h_1 + h_3)^2$$

(which is suggested by the definition of f in Example 1). From this it is clear that

$$f_2(\mathbf{0},\mathbf{H}) \geq -1 + h_1^2 + h_2^2 + h_3^2 = -1 + |\mathbf{H}|^2; \tag{7}$$

hence the *Taylor polynomial* f_2 has an absolute minimum at $\mathbf{0}$.

Using (3) and (7), we will show that f itself has a *local* minimum at $\mathbf{0}$. First of all, (3) implies that there is a number $\delta > 0$ such that

$$0 < |\mathbf{H}| < \delta \Rightarrow \frac{|f(\mathbf{H}) - f_2(\mathbf{0},\mathbf{H})|}{|\mathbf{H}|^2} < 1 \Rightarrow |f(\mathbf{H}) - f_2(\mathbf{0},\mathbf{H})| < |\mathbf{H}|^2. \tag{8}$$

Hence for $0 < |\mathbf{H}| < \delta$ we have

$$\begin{aligned} f(\mathbf{H}) &\geq f_2(\mathbf{0},\mathbf{H}) - |\mathbf{H}| \\ &> -1 + |\mathbf{H}|^2 - |\mathbf{H}|^2 = -1 \qquad \text{(by (7))}; \end{aligned}$$

thus

$$0 < |\mathbf{H}| < \delta \Rightarrow f(\mathbf{H}) > -1 = f(\mathbf{0}),$$

and $\mathbf{0}$ is a local minimum point of f.

PROBLEMS

1. Let $f(x_1,x_2,x_3) = x_1e^{x_2+x_3}$, and find the Taylor polynomial $f_k(\mathbf{A},\mathbf{H})$ when
(a) $\mathbf{A} = \mathbf{0}$, $k = 1$;
(b) $\mathbf{A} = (1,-1,1)$, $k = 2$;
(c) $\mathbf{A} = (0,0,1)$, $k = 3$.

2. (a) Suppose that f is differentiable at $\mathbf{A}$, and has a local maximum at $\mathbf{A}$; i.e. there is a number $\delta > 0$ such that

$$|\mathbf{H}| < \delta \Rightarrow f(\mathbf{A} + \mathbf{H}) \leq f(\mathbf{A}).$$

Prove that $\nabla f(\mathbf{A}) = \mathbf{0}$.

(b) Prove that the function in Problem 1 has no local maximum or minimum.

3. Prove that if f has continuous partial derivatives of order $\leq k$ in the ball of radius ϵ about $\mathbf{A}$, then for $|\mathbf{H}| < \epsilon$

$$f(\mathbf{A} + \mathbf{H}) = \sum_0^{k-1} \frac{1}{j!}(\mathbf{H}\cdot\nabla)^j f(\mathbf{A}) + \frac{1}{k!}(\mathbf{H}\cdot\nabla)^k f(\mathbf{A} + \theta\mathbf{H}), \qquad 0 < \theta < 1,$$

and alternatively

$$f(\mathbf{A}+\mathbf{H}) = \sum_{0}^{k-1} \frac{1}{j!} (\mathbf{H}\cdot\nabla)^j f(\mathbf{A})$$

$$+ \frac{1}{(k-1)!} \int_0^1 (\mathbf{H}\cdot\nabla)^k f(\mathbf{A}+t\mathbf{H})\,(1-t)^{k-1}\,dt.$$

4. Prove that f is a polynomial of degree $< k$ if and only if all the kth order derivatives of f are identically zero.

5. Prove that if f is a homogeneous polynomial of degree k (i.e. f is a sum of monomial terms each of degree k), then

$$f(\mathbf{H}) = \frac{1}{k!} (\mathbf{H}\cdot\nabla)^k f(\mathbf{0}).$$

6. (a) Prove that if f is homogeneous of degree k (i.e. $f(t\mathbf{X}) = t^k f(\mathbf{X})$ for every $\mathbf{X} \neq \mathbf{0}$ and every $t > 0$), and f is differentiable for $\mathbf{X} \neq \mathbf{0}$, then $D_1 f, \ldots, D_n f$ are all homogeneous of degree $k - 1$.

(b) Suppose that f is homogeneous of degree k, and has kth order partial derivatives which are continuous at $\mathbf{0}$. Prove that the kth order derivatives are constant. (See Problem 7(c), §3.4.)

(c) For the f in part (b), prove that f is a *homogeneous polynomial* of degree k.

5.4 VECTOR FIELDS AND LINE INTEGRALS IN R^3

A vector field $\mathbf{F}$ over a set V in R^3 assigns to each point $\mathbf{X}$ in V a vector $\mathbf{F}(\mathbf{X})$ in R^3. The three components of this vector are denoted $f_1(\mathbf{X})$, $f_2(\mathbf{X})$, $f_3(\mathbf{X})$. Obviously, a single vector function $\mathbf{F}$ is equivalent to three real-valued functions f_1, f_2, f_3.

In physics, vector fields generally stand for *forces* or *flows*. Mathematically, vector fields can arise as gradients. If the vector field $\mathbf{F} = (f_1, f_2, f_3)$ happens to be the gradient of a function f, i.e. if $f_1 = f_x$, $f_2 = f_y$, and $f_3 = f_z$, then $\mathbf{F}$ is called *exact*. If $\mathbf{F}$ is exact and f_1, f_2, f_3 have continuous partial derivatives, we find that these components must satisfy certain relations, for example,

$$\frac{\partial f_1}{\partial y} = \frac{\partial}{\partial y}\left(\frac{\partial f}{\partial x}\right) = \frac{\partial}{\partial x}\left(\frac{\partial f}{\partial y}\right) = \frac{\partial f_2}{\partial x}.$$

Working out similar expressions for the other components and other partial derivatives, we find that

if **F** *is exact and has continuous derivatives, then*

$$\frac{\partial f_1}{\partial y} = \frac{\partial f_2}{\partial x}, \qquad \frac{\partial f_1}{\partial z} = \frac{\partial f_3}{\partial x}, \qquad \frac{\partial f_2}{\partial z} = \frac{\partial f_3}{\partial y}. \tag{1}$$

A vector field **F** which satisfies the conditions (1) is called *closed.* We have just shown that "exact" implies "closed"; on the other hand, the example in (2) below shows that a vector field can be closed but *not* exact.

Given a vector field **F** and a differentiable curve

$$\boldsymbol{\gamma}: [a,b] \to R^3,$$

the line integral of **F** over $\boldsymbol{\gamma}$ is defined by

$$\int_\gamma \mathbf{F} = \int_a^b (\mathbf{F}\circ\boldsymbol{\gamma})\cdot\boldsymbol{\gamma}'.$$

If $\boldsymbol{\gamma}$ is a chain of curves $\boldsymbol{\gamma}^1, \ldots, \boldsymbol{\gamma}^n$, then

$$\int_\gamma \mathbf{F} = \sum_{j=1}^{n} \int_{\gamma^j} \mathbf{F}.$$

In physics, when **F** is a force, then $\int_\gamma \mathbf{F}$ is the work done by **F** in moving a particle along the curve $\boldsymbol{\gamma}$. When **F** is a flow, $\int_\gamma \mathbf{F}$ has no special meaning; for flows, we integrate not over curves but over surfaces, as discussed in the next section.

Line integrals help to clarify the difference between closed and exact vector fields. By the same reasoning as in §4.3, you can prove:

(I) If **F** is exact in V, i.e. if there is a function f in V such that $\mathbf{F} = \nabla f$, then $\int_\gamma \mathbf{F} = f(\mathbf{P}_2) - f(\mathbf{P}_1)$ for every connected chain $\boldsymbol{\gamma}$ lying in V, beginning at $\mathbf{P}_1$ and ending at $\mathbf{P}_2$. In particular, when $\boldsymbol{\gamma}$ is a closed chain (i.e. when $\mathbf{P}_2 = \mathbf{P}_1$), then $\int_\gamma \mathbf{F} = 0$.

(II) Conversely, if $\int_\gamma \mathbf{F}$ depends only on the endpoints of $\boldsymbol{\gamma}$ for every connected chain $\boldsymbol{\gamma}$ lying in V, then **F** is exact in V.

From (I) we find, for example, that the vector field

$$\mathbf{F}(x,y,z) = \left(\frac{-y}{x^2+y^2}, \frac{x}{x^2+y^2}, z\right), \qquad x^2+y^2 \neq 0, \tag{2}$$

is *not* exact, since $\int_\gamma \mathbf{F} = 2\pi$ when $\boldsymbol{\gamma}$ is the closed curve

$$\boldsymbol{\gamma}(t) = (\cos t, \sin t, 0), \qquad 0 \le t \le 2\pi.$$

This vector field *is* closed, however, since the equations (1) are satisfied.

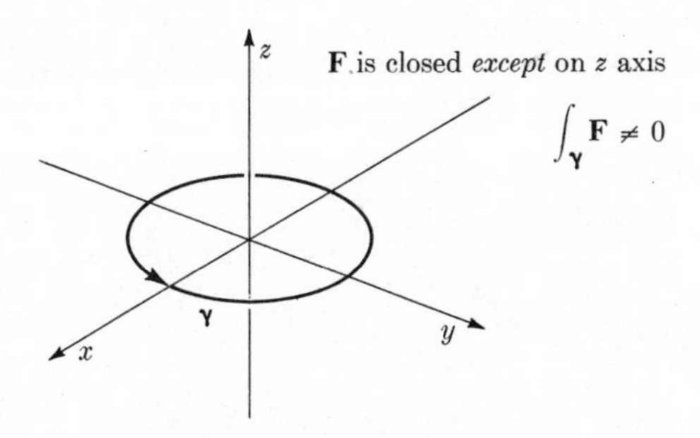

FIGURE 5.11

The reason for this discrepancy between "closed" and "exact" lies in the shape of the region V where the vector field $\mathbf{F}$ is defined. The vector field (2) is defined everywhere *except along the z axis*; it is no coincidence that the curve $\boldsymbol{\gamma}$ on which $\int_\gamma \mathbf{F} \neq 0$ encircles the set where $\mathbf{F}$ is *not* defined (Fig. 11). By contrast:

(III) When a closed vector field $\mathbf{F}$ is defined everywhere in R^3, then $\mathbf{F}$ is exact in R^3.

Proof. For any point $\mathbf{P}$ in R^3, let $\boldsymbol{\gamma}^{\mathbf{P}}$ be the straight line path from $\mathbf{0}$ to $\mathbf{P}$,

$$\boldsymbol{\gamma}^{\mathbf{P}}(t) = t\mathbf{P}, \qquad 0 \leq t \leq 1.$$

Define a function f by

$$f(\mathbf{P}) = \int_{\gamma^{\mathbf{P}}} \mathbf{F}. \tag{3}$$

(This definition is practically forced by (1) above. If we are to have $\mathbf{F} = \boldsymbol{\nabla} f$, then $f(\mathbf{P}) = f(\mathbf{0}) + \int_{\gamma^{\mathbf{P}}} \mathbf{F}$, and the constant $f(\mathbf{0})$ does not affect the gradient. Thus, if $\mathbf{F}$ is the gradient of anything at all, it is the gradient of the function f defined by (3).) For each point $\mathbf{P} = (x,y,z)$, we have $(\boldsymbol{\gamma}^{\mathbf{P}})'(t) = \mathbf{P} = (x,y,z)$; hence

$$f(x,y,z) = \int_0^1 \mathbf{F}(t\mathbf{P}) \cdot \mathbf{P}\, dt$$

$$= \int_0^1 (xf_1(tx,ty,tz) + yf_2(tx,ty,tz) + zf_3(tx,ty,tz))\, dt, \tag{4}$$

where $\mathbf{F} = (f_1,f_2,f_3)$. Differentiate (4) with respect to x, applying Leibniz' rule and the chain rule, to obtain

$$\frac{\partial f}{\partial x} = \int_0^1 [f_1(t\mathbf{P}) + txD_1 f_1(t\mathbf{P}) + tyD_1 f_2(t\mathbf{P}) + tzD_1 f_3(t\mathbf{P})]\, dt.$$

Since $\mathbf{F}$ is closed, we can set $D_1 f_2 = D_2 f_1$ and $D_1 f_3 = D_3 f_1$; hence

$$\frac{\partial f}{\partial x} = \int_0^1 [f_1(t\mathbf{P}) + txD_1 f_1(t\mathbf{P}) + tyD_2 f_1(t\mathbf{P}) + tzD_3 f_1(t\mathbf{P})]\, dt$$

$$= \int_0^1 \frac{d}{dt}[tf_1(t\mathbf{P})]\, dt = [tf_1(t\mathbf{P})]_0^1 = f_1(\mathbf{P}),$$

as desired. Similarly, $\partial f/\partial y = f_2$ and $\partial f/\partial z = f_3$, so the claim (III) is established.

Exactly the same proof works when $\mathbf{F}$ is defined not in all of R^3 but in a more restricted set, such as a ball of radius r centered at $\mathbf{0}$, or an open rectangular solid containing $\mathbf{0}$. In fact, it works for every open set V with the following property: *Whenever* $\mathbf{P}$ *is in* V, *then* V *contains the straight line segment between* $\mathbf{0}$ *and* $\mathbf{P}$. Such a set V is called "star-shaped with respect to the origin" (Fig. 12). More generally, V is called star-shaped with respect to a point $\mathbf{P}_0$ if, whenever $\mathbf{P}$ is in V, then V contains the straight line segment from $\mathbf{P}_0$ to $\mathbf{P}$. You can easily prove:

(III′) If $\mathbf{F}$ is a closed vector field defined in a region V which is star-shaped with respect to some point $\mathbf{P}_0$, then $\mathbf{F}$ is exact in V.

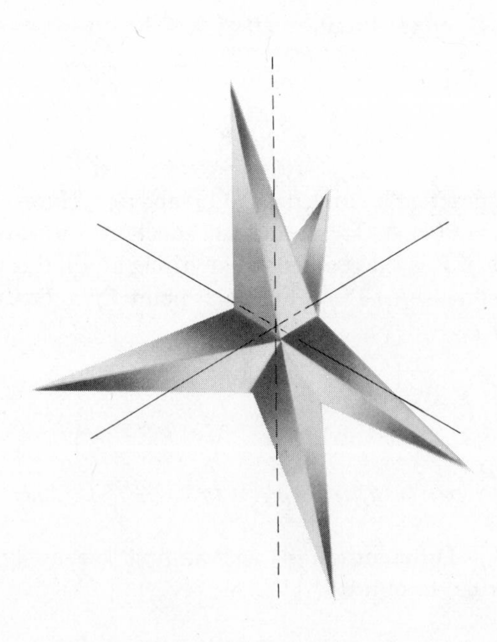

FIGURE 5.12 V is star-shaped with respect to $\mathbf{0}$

For example, the "half space"

$$H = \{(x,y,z) : z > 0\}$$

is star-shaped with respect to the point $\mathbf{P}_0 = (0,0,1)$; but the set

$$S = \{(x,y,z) : x^2 + y^2 > 0\},$$

in which the vector field (2) is closed but not exact, is not star-shaped with respect to any point $\mathbf{P}_0$ (Fig. 13).

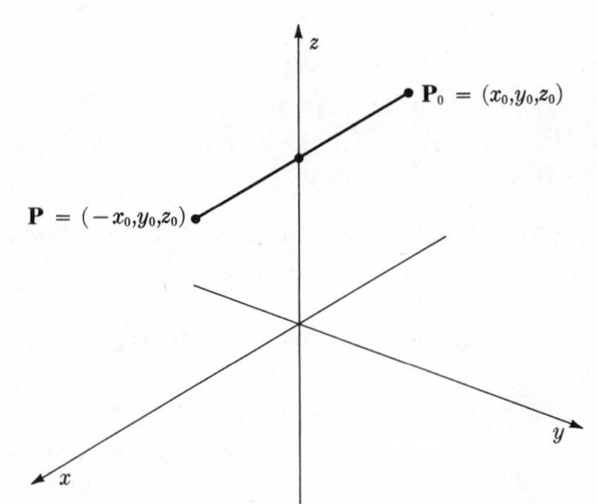

FIGURE 5.13 The segment from $\mathbf{P}_0$ to $\mathbf{P}$ does not lie entirely in $S = \{(x,y,z): x^2 + y^2 > 0\}$

PROBLEMS

1. Compute the line integral $\int_\gamma \mathbf{F}$ in the following cases:

(a) $\mathbf{F}(x,y,z) = (e^x,e^y,e^z)$, $\boldsymbol{\gamma}(t) = (ta,tb,tc)$, $0 \leq t \leq 1$, a, b, and c constants.

(b) $\mathbf{F}(x,y,z) = (y + z, z + x, x + y)$, $\boldsymbol{\gamma}(t) = (t^2,t^3,t^4)$, $0 \leq t \leq 1$.

(c) $\mathbf{F}(\mathbf{P}) = \dfrac{\mathbf{P}}{|\mathbf{P}|^3}$, $\boldsymbol{\gamma}(t) = (ta,tb,tc)$, $r \leq t \leq R$.

(Here, a, b, c, r, R are constants and $0 < r < R$.)

2. Let $\mathbf{F}(\mathbf{P}) = \dfrac{\mathbf{P}}{|\mathbf{P}|^3}$.

(a) Show that $\mathbf{F}$ is closed, using the formulas (1).

(b) Show that $\mathbf{F} = \nabla f$ when $f(\mathbf{P}) = -|\mathbf{P}|^{-1}$.

(c) Show that $\displaystyle\int_\gamma \frac{\mathbf{P}}{|\mathbf{P}|^3} = \frac{1}{|\boldsymbol{\gamma}(a)|} - \frac{1}{|\boldsymbol{\gamma}(b)|}$.

(d) Compute the work done by $\mathbf{F}$ in moving a particle from $(1,3,5)$ to $(100, 50, -80)$.

(e) Compute the work done by $\mathbf{F}$ in moving a particle from $(1,3,5)$ "all the way to infinity."

3. Prove statement (I) in the text, namely, $\int_\gamma \nabla f = f(\mathbf{P}_2) - f(\mathbf{P}_1)$, where $\boldsymbol{\gamma}$ is a connected chain beginning at $\mathbf{P}_1$ and ending at $\mathbf{P}_2$.

4. Prove statement (II) in the text namely: If $\mathbf{F}$ is a continuous vector field in an open connected set V, and for every curve $\boldsymbol{\gamma}$ in V, $\int_\gamma \mathbf{F}$ depends only on the endpoints of $\boldsymbol{\gamma}$, then there is a function f defined in V such that $\nabla f = \mathbf{F}$.

5. Prove statement (III′) in the text. (Hint: Use the paths $\boldsymbol{\gamma}_{\mathbf{P}_0}{}^{\mathbf{P}}(t) = \mathbf{P}_0 + t(\mathbf{P} - \mathbf{P}_0)$. Alternate Hint: Consider the vector field $\mathbf{G}$ defined by $\mathbf{G}(\mathbf{P}) = \mathbf{F}(\mathbf{P} - \mathbf{P}_0)$; $\mathbf{G}$ is defined in a region which is star-shaped with respect to the origin.)

6. The *curl* of a vector field $\mathbf{F}$, denoted $\nabla \times \mathbf{F}$, is defined by

$$\nabla \times \mathbf{F} = (D_2 f_3 - D_3 f_2, D_3 f_1 - D_1 f_3, D_1 f_2 - D_2 f_1).$$

This is the cross product of the "vector" $\nabla = (D_1, D_2, D_3)$ and the vector $\mathbf{F} = (f_1, f_2, f_3)$.

(a) Show that $\mathbf{F}$ is closed if and only if $\nabla \times \mathbf{F} = \mathbf{0}$.

(b) Show that if f has continuous second derivatives, then

$$\nabla \times (\nabla f) = \mathbf{0}.$$

7. A particle of mass m moves along a curve $\boldsymbol{\gamma}$ from time a to time b, propelled by a force $\mathbf{F}$ according to Newton's law $\mathbf{F} = m\boldsymbol{\gamma}''$. At each time t, its *kinetic energy* is defined to be $\frac{1}{2}m\,|\boldsymbol{\gamma}'(t)|^2$, and its *potential energy* to be $-\int_a^{t_0} (\mathbf{F}\circ\boldsymbol{\gamma})\cdot\boldsymbol{\gamma}'$. Prove the "conservation of energy" law, namely, the sum of the kinetic energy and the potential energy is constant.

8. Suppose that $|\mathbf{F}(\mathbf{X})| \leq M$ for every point on the differentiable curve $\boldsymbol{\gamma}: [a,b] \to R^3$. Prove that $|\int_\gamma \mathbf{F}| \leq M\cdot(\text{length of } \boldsymbol{\gamma})$. (Hint: Use the Schwarz inequality.)

5.5 SURFACE INTEGRALS AND STOKES' THEOREM

A *parametric surface* is a continuous vector-valued function $\boldsymbol{\sigma}$ defined on some set S in R^2, and assigning to each point (u,v) in S a point $\boldsymbol{\sigma}(u,v)$ in R^3. The set S is called the *parameter domain*, and u and v are the *parameters*.

For example, the function

$$\boldsymbol{\sigma}(u,v) = (\cos u \sin v, \sin u \sin v, \cos v) \tag{1}$$

defined on the set

$$S = \{(u,v) : 0 < u < \pi, 0 < v < \pi\}$$

is a parametric surface representing one half of the unit sphere (Fig. 14). The lines where u is constant and v varies are the "meridians" (the great-circle lines from the North Pole to the South Pole), and the lines where v is constant but u varies are the "parallels" (circles parallel to the equator).

If we retain formula (1) for $\boldsymbol{\sigma}$, but enlarge the parameter domain S to the rectangle

$$\bar{S} = \{(u,v) : 0 \leq u \leq 2\pi, 0 \leq v \leq \pi\},$$

we get the entire sphere in Fig. 15. The points along the heavily drawn meridian are all obtained twice (once with $u = 0$ and once with $u = 2\pi$), and the North and South Poles are obtained infinitely often (with every value of u, when $v = 0$ or $v = \pi$).

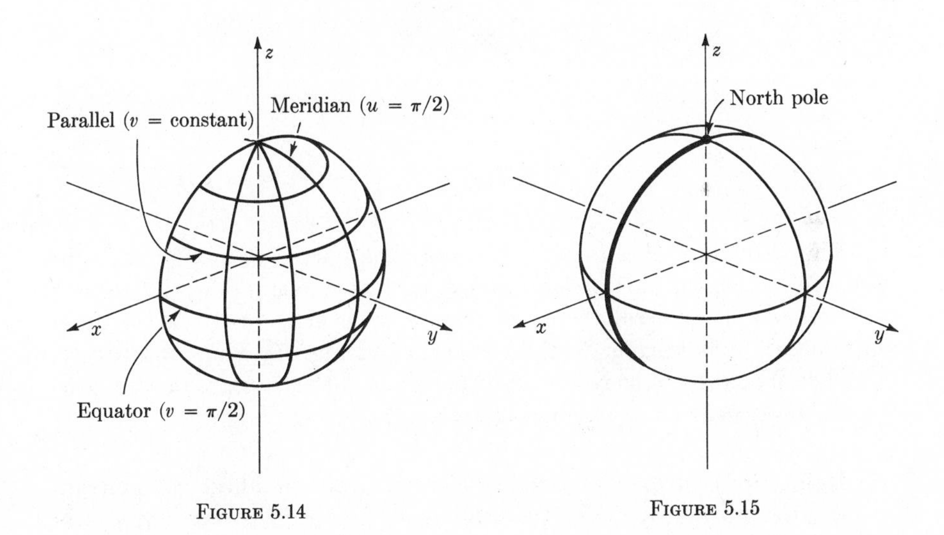

FIGURE 5.14 FIGURE 5.15

The graph of any function f of two variables can be considered as a parametric surface. We take the parameter domain S to be the same as the domain of f, and set

$$\boldsymbol{\sigma}(u,v) = (u,v,f(u,v)).$$

Obviously, every point $\boldsymbol{\sigma}(u,v)$ on this surface $\boldsymbol{\sigma}$ lies in the graph of f and, conversely, every point $(x,y,f(x,y))$ on the graph of f has the form $\boldsymbol{\sigma}(u,v)$

with $u = x$ and $v = y$. For example, if $f(x,y) = \sqrt{1 - x^2 - y^2}$ with domain $\{(x,y); x^2 + y^2 < 1\}$, we take $S = \{(u,v): u^2 + v^2 < 1\}$ and $\boldsymbol{\sigma}(u,v) = (u,v,\sqrt{1 - u^2 - v^2})$. This gives the upper half of the unit sphere as a parametric surface.

In any parametric surface, we can obtain a curve by fixing v and allowing u to vary; this is called a "u-curve." Similarly, fixing u and allowing v to vary produces a "v-curve." In the surface (1), the v-curves are the meridians and the u-curves are the parallels.

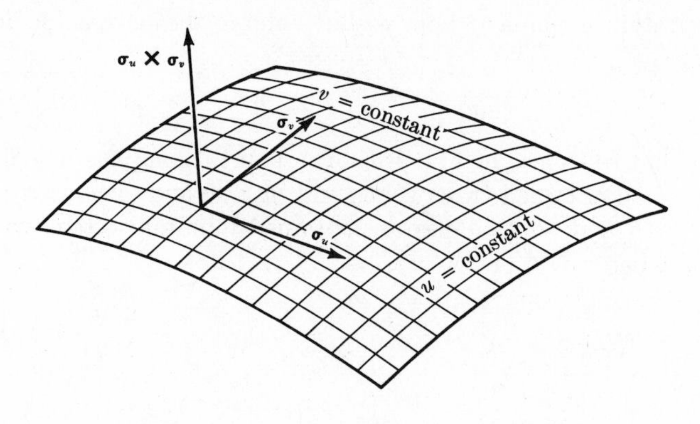

FIGURE 5.16

The derivative of a u-curve is simply the partial derivative $\boldsymbol{\sigma}_u$, obtained by taking the partial derivative of each component of $\boldsymbol{\sigma}$ with respect to u. From the general theory of curves, $\boldsymbol{\sigma}_u$ is tangent to the u-curve, and similarly $\boldsymbol{\sigma}_v$ is tangent to the v-curve, as in Fig. 16. Thus, intuitively, $\boldsymbol{\sigma}_u$ and $\boldsymbol{\sigma}_v$ should be tangent to the surface $\boldsymbol{\sigma}$, so their cross product $\boldsymbol{\sigma}_u \times \boldsymbol{\sigma}_v$ should be *normal* to $\boldsymbol{\sigma}$. This motivates the following

Definition. If $\boldsymbol{\sigma}$ is a parametric surface each of whose components is differentiable, then the vector $\boldsymbol{\sigma}_u \times \boldsymbol{\sigma}_v$ is called the *standard normal vector* to $\boldsymbol{\sigma}$, denoted $\mathbf{N}_\sigma$.

When $\boldsymbol{\sigma}$ happens to be the graph of a function f,

$$\boldsymbol{\sigma}(u,v) = (u,v,f(u,v)),$$

then $\boldsymbol{\sigma}_u = (1,0,f_u)$, $\boldsymbol{\sigma}_v = (0,1,f_v)$, and $\boldsymbol{\sigma}_u \times \boldsymbol{\sigma}_v = (-f_u,-f_v,1)$, which we recognize as a normal to the tangent plane to the graph of f; thus, the new definition of normal agrees with the old. Further, when the surface $\boldsymbol{\sigma}$ lies in the level surface of a function g of three variables, then $\mathbf{N}_\sigma$ is parallel to ∇g at every point of the surface, so the two notions of "normal vector" agree in this case too. (See Problem 12 below.)

There is a far-reaching analogy between surfaces and curves in which the normal $\mathbf{N}_\sigma = \boldsymbol{\sigma}_u \times \boldsymbol{\sigma}_v$ of a surface $\boldsymbol{\sigma}$ corresponds to the derivative $\boldsymbol{\gamma}'$ of a curve $\boldsymbol{\gamma}$. The following table lists the points of analogy, and the rest of this section explains them.

	Curve $\boldsymbol{\gamma}$: $[a, b] \to R^3$	Surface $\boldsymbol{\sigma}$: $S \to R^3$
	Derivative $\boldsymbol{\gamma}'$	Normal $\mathbf{N}_\sigma = \boldsymbol{\sigma}_u \times \boldsymbol{\sigma}_v$
(I)	$\boldsymbol{\gamma}'$ points in positive direction along $\boldsymbol{\gamma}$	$\mathbf{N}_\sigma$ points toward positive side of surface
(II)	$\int_a^b \lvert\boldsymbol{\gamma}'\rvert$ = length of curve	$\iint_S \lvert\mathbf{N}_\sigma\rvert$ = area of surface
(III)	$\int_a^b (\mathbf{F}\circ\boldsymbol{\gamma}) \cdot \boldsymbol{\gamma}'$ is the *line integral* $\int_\gamma \mathbf{F}$	$\iint_S (\mathbf{F}\circ\boldsymbol{\sigma}) \cdot \mathbf{N}_\sigma$ is the *surface integral* $\iint_\sigma \mathbf{F}$
(IV)	$\int_\gamma \nabla f = f(\boldsymbol{\gamma}(b)) - f(\boldsymbol{\gamma}(a))$	$\iint_\sigma \nabla \times \mathbf{F} = \int_{\bar{\gamma}} \mathbf{F}$, where $\bar{\boldsymbol{\gamma}}$ is the positively oriented boundary of $\boldsymbol{\sigma}$

(I) *Oriented normal.* A curved line in space can be "oriented" in either of two ways, in the sense that a particle can move in either of two directions along the line. A parametric curve $\boldsymbol{\gamma}$ chooses a particular so-called "positive" direction (the direction of increasing values of t), and the derivative $\boldsymbol{\gamma}'$ points in that direction.

Similarly, a curved surface in space generally has two sides. A parametric surface chooses a particular side, the one toward which the normal $\mathbf{N}_\sigma$ points, and this is called the positive side of the surface (Fig. 17).

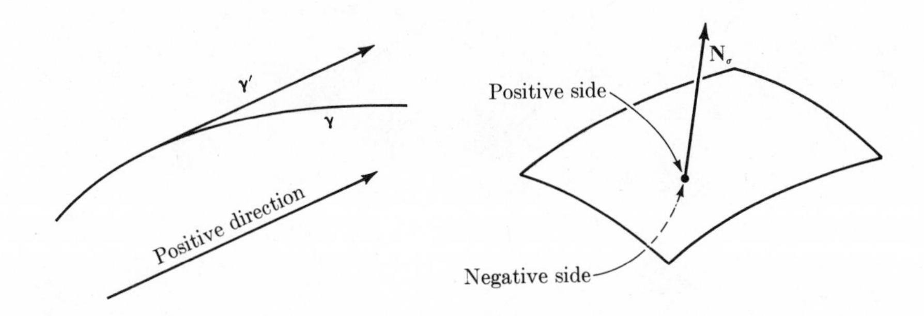

FIGURE 5.17

(II) *Surface area.* It is very difficult to give a rigorous and geometrically motivated development of the theory of surface area, so we take the easy way out and simply make a

Definition. If $\boldsymbol{\sigma}$ is a surface with parameter domain S, and the components of $\boldsymbol{\sigma}$ have continuous first partial derivatives, and $\int\int_S |\boldsymbol{\sigma}_u \times \boldsymbol{\sigma}_v|$ is defined, then this integral is called the *area* of the surface $\boldsymbol{\sigma}$.

We have to show that this definition is consistent with the well-known formula in an important special case, the area of a surface of revolution. Suppose f is a real-valued function defined on an interval $[a,b]$, and f' is continuous on $[a,b]$. Then the area of the surface generated by revolving the graph of f about the x axis is given in one-variable calculus as

$$2\pi \int_a^b |f| \sqrt{1 + (f')^2}\,. \tag{2}$$

Now, we can represent the surface of revolution parametrically by

$$\boldsymbol{\sigma}(u,v) = (u, f(u) \cos v, f(u) \sin v)$$

on the parameter domain

$$S = \{(u,v) : a \leq u \leq b, 0 \leq v \leq 2\pi\}.$$

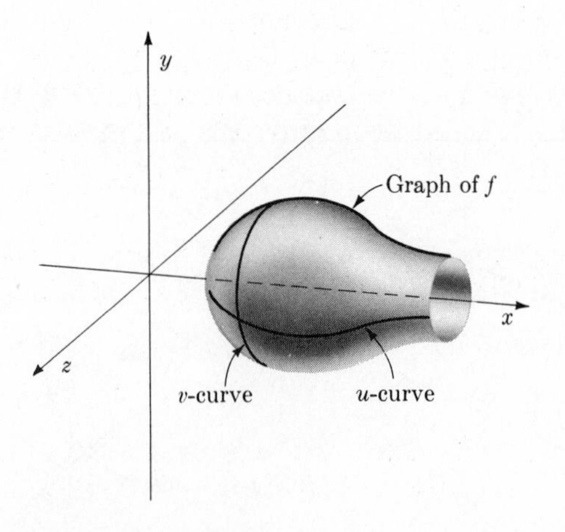

FIGURE 5.18

Each v-curve is a circle around the x axis, and each u-curve looks like the graph of f, but rotated through some angle (Fig. 18). We find

$$\begin{aligned}
\boldsymbol{\sigma}_u &= (1, f'(u)\cos v, f'(u)\sin v)\\
\boldsymbol{\sigma}_v &= (0, -f(u)\sin v, f(u)\cos v)\\
\mathbf{N}_\sigma = \boldsymbol{\sigma}_u \times \boldsymbol{\sigma}_v &= (f'(u)f(u), -f(u)\cos v, -f(u)\sin v)\\
&= f(u)\,(f'(u), -\cos v, -\sin v)
\end{aligned}$$

$$|\boldsymbol{\sigma}_u \times \boldsymbol{\sigma}_v| = |f(u)|\sqrt{1+f'(u)^2}\,.$$

Hence the surface area is

$$\iint_S |\boldsymbol{\sigma}_u \times \boldsymbol{\sigma}_v| = \int_a^b \int_0^{2\pi} |f(u)|\sqrt{1+f'(u)^2}\,dv\,du$$

$$= 2\pi \int_a^b |f(u)|\sqrt{1+f'(u)^2}\,du,$$

which agrees with (2).

Another case we should check, to see whether the above definition of surface area is consistent with our earlier work, is that of a surface lying in the xy plane, i.e. the case where $\boldsymbol{\sigma}$ has the form

$$\boldsymbol{\sigma}(u,v) = (\varphi(u,v), \psi(u,v), 0)$$

for some functions φ and ψ. Here we find

$$\boldsymbol{\sigma}_u = (\varphi_u, \psi_u, 0), \qquad \boldsymbol{\sigma}_v = (\varphi_v, \psi_v, 0)$$

$$\boldsymbol{\sigma}_u \times \boldsymbol{\sigma}_v = (0, 0, \varphi_u\psi_v - \psi_u\varphi_v),$$

$$\iint_S |\boldsymbol{\sigma}_u \times \boldsymbol{\sigma}_v| = \iint_S |\varphi_u\psi_v - \psi_u\varphi_v|. \tag{3}$$

If the function $\boldsymbol{\sigma}$ is one-to-one (i.e. if each point on the surface comes from exactly one point in S), then the formula for change of variable in a double integral shows that (3) is in fact the area of the region onto which $\boldsymbol{\sigma}$ maps S. If $\boldsymbol{\sigma}$ is not one-to-one, then (3) will generally be larger than the area of the region onto which $\boldsymbol{\sigma}$ maps S, reflecting the fact that the parametric surface "covers" some points several times. (See Problem 2 below for examples of "multiple covering" of certain points by a parametric surface.)

(III) *Surface integrals.* We simply *define* surface integrals by the formula

$$\iint_\sigma \mathbf{F} = \iint_S (\mathbf{F}\circ\boldsymbol{\sigma})\cdot\mathbf{N}_\sigma\,.$$

When $\mathbf{F}$ is a flow, then $\iint_\sigma \mathbf{F}$ is interpreted as the rate of flow across $\boldsymbol{\sigma}$ in the positive direction, i.e. from the negative side toward the positive side.

There is an analogy here with the line integral of a plane flow along a plane curve, and the explanation is similar. $\iint_S (\mathbf{F}\circ\boldsymbol{\sigma})\cdot\mathbf{N}_\sigma = \iint_S |\mathbf{F}| \cos\theta\, |\mathbf{N}_\sigma|$, where θ is the angle between the flow $\mathbf{F}$ and the normal $\mathbf{N}_\sigma$. If we look at a small part $\bar{S}$ of the parameter domain S, then $|\mathbf{F}| \cos\theta$ is approximately constant on the small part $\bar{\boldsymbol{\sigma}}$ of the surface corresponding to $\bar{S}$, so $\iint_{\bar{S}} |\mathbf{F}| \cos\theta\, |\mathbf{N}_\sigma|$ is nearly the same as $|\mathbf{F}| \cos\theta \iint_{\bar{S}} |\mathbf{N}_\sigma|$. This last expression is the *normal component of the flow* across $\bar{\boldsymbol{\sigma}}$ times the *area of* $\bar{\boldsymbol{\sigma}}$, thus giving the *rate of flow* across $\bar{\boldsymbol{\sigma}}$ (Fig. 19). Adding up these small pieces $\bar{S}$ over all of S, we get the total rate of flow through the surface $\boldsymbol{\sigma}$.

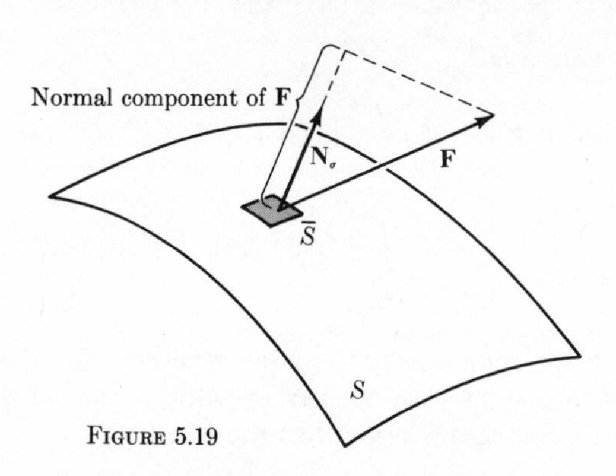

FIGURE 5.19

Example 1. Take $\boldsymbol{\sigma}(u,v) = (\cos u \sin v, \sin u \sin v, \cos v)$ and $S = \{(u,v): 0 < u < \pi, 0 < v < \pi\}$ (Fig. 14). Find the normal $\mathbf{N}_\sigma$, the area of $\boldsymbol{\sigma}$, and the surface integral $\int_\sigma \mathbf{F}$, where $\mathbf{F}(x,y,z) = (0,2,0)$. *Solution.* We have

$$\boldsymbol{\sigma}_u = \frac{\partial\boldsymbol{\sigma}}{\partial u} = (-\sin u \sin v, \cos u \sin v, 0)$$

$$\boldsymbol{\sigma}_v = (\cos u \cos v, \sin u \cos v, -\sin v);$$

hence

$$\mathbf{N}_\sigma = \boldsymbol{\sigma}_u \times \boldsymbol{\sigma}_v = (-\cos u \sin^2 v, -\sin u \sin^2 v, -\sin v \cos v)$$
$$= (-\sin v)\boldsymbol{\sigma}.$$

Since $\sin v > 0$, this shows that at each point $\boldsymbol{\sigma}(u,v)$ on the surface, the vector $\mathbf{N}_\sigma$ points in the direction opposite to $\boldsymbol{\sigma}$, i.e. towards the origin. It is thus "intuitively obvious" that $\mathbf{N}_\sigma$ really is orthogonal to the hemisphere at $\boldsymbol{\sigma}(u,v)$.

The area of the hemisphere is

$$\iint_S |\mathbf{N}_\sigma| = \int_0^\pi\int_0^\pi |(-\sin v)\boldsymbol{\sigma}|\,du\,dv = \int_0^\pi\int_0^\pi \sin v\,du\,dv,$$

since $|\boldsymbol{\sigma}(u,v)| \equiv 1$ and $\sin v \geq 0$ for $0 \leq v \leq \pi$. Hence the area is

$$\pi[-\cos v]_0^\pi = 2\pi,$$

which agrees (as it should) with the traditional formula for the area of a hemisphere of radius 1.

Finally, the surface integral $\int_\sigma \mathbf{F}$, where $\mathbf{F} = (0,2,0)$, is by definition

$$\iint_S \mathbf{N}_\sigma \cdot \mathbf{F} = \int_0^\pi \int_0^\pi 2(-\sin u \sin^2 v)du\, dv = -2\pi.$$

You can see that this is reasonable by thinking of **F** as a flow, flowing constantly from left to right at the speed of two units of distance per unit of time; say 2 feet per second. In one second, the amount that has flowed through S fills the volume in Fig. 20; since the bulging cap on the right has exactly the same volume as the hollow on the left, the total volume flowing through $\boldsymbol{\sigma}$ in one second should be 2π, the volume of a cylinder of height 2 and radius 1. Finally, since the flow actually passes from the positive side of $\boldsymbol{\sigma}$ to the negative side, the rate of flow *toward* the positive side should be -2π.

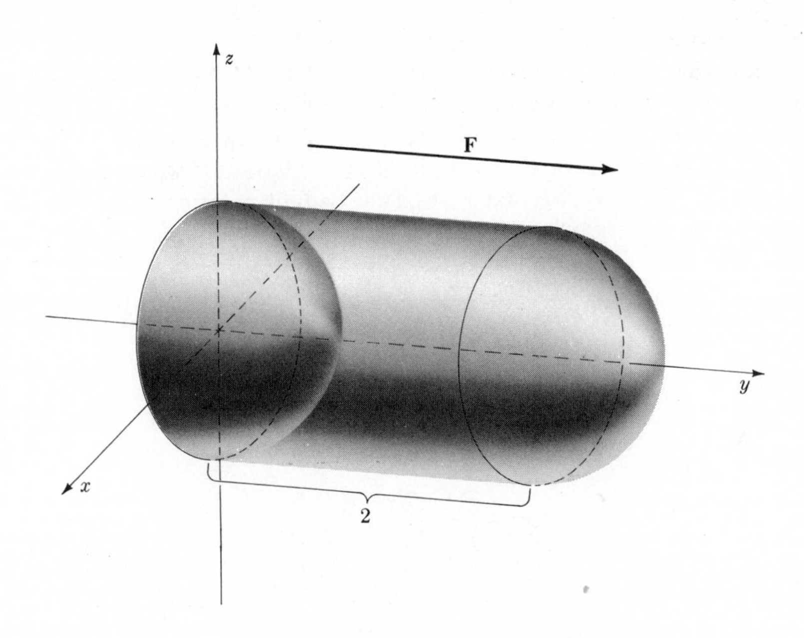

FIGURE 5.20

(IV) *Stokes' theorem.* In the preceding chapter we proved Green's theorem for certain special regions S: if the chain of curves $\boldsymbol{\gamma}$ forms the positively oriented boundary of S, then every vector field (M,N) having continuous partial derivatives in S satisfies the equation

$$\iint_S (N_x - M_y)\, dx\, dy = \int_\gamma M\, dx + N\, dy. \tag{4}$$

Now suppose that such a region S with positively oriented boundary $\boldsymbol{\gamma}$ is the parameter domain of a surface $\boldsymbol{\sigma}$. Then the composite function $\bar{\boldsymbol{\gamma}} = \boldsymbol{\sigma}\circ\boldsymbol{\gamma}$ is a curve in R^3 (or, more generally, a chain of curves); we call this composite function the *positively oriented boundary* of $\boldsymbol{\sigma}$. Stokes' theorem, like Green's theorem (4), relates the integral of a certain "derivative" of $\mathbf{F}$ over a surface $\boldsymbol{\sigma}$ to the integral of $\mathbf{F}$ itself over the boundary of $\boldsymbol{\sigma}$. The derivative in this case is the *curl* of $\mathbf{F}$,

$$\nabla \times \mathbf{F} = (D_2 f_3 - D_3 f_2\,, D_3 f_1 - D_1 f_3\,, D_1 f_2 - D_2 f_1) \tag{5}$$

which is simply the cross product of $\nabla = (D_1,D_2,D_3)$ and $\mathbf{F} = (f_1, f_2, f_3)$. Stokes' theorem says that under appropriate conditions on $\boldsymbol{\sigma}$ and $\mathbf{F}$,

$$\iint_{\boldsymbol{\sigma}} \nabla \times \mathbf{F} = \int_{\bar{\boldsymbol{\gamma}}} \mathbf{F}, \tag{6}$$

where $\bar{\boldsymbol{\gamma}}$ is the positively oriented boundary of $\boldsymbol{\sigma}$.

Now that all the terms have been defined, Stokes' theorem (6) follows from Green's theorem (4) by a straightforward (but long!) calculation, using the chain rule. To simplify the writing, we assume that the second and third components of $\mathbf{F}$ vanish, i.e. $\mathbf{F} = (f,0,0)$. If we write the three components of $\boldsymbol{\sigma}$ as σ_1, σ_2, and σ_3, and use Leibniz notation with the scheme

$$\begin{array}{cccc} t \longrightarrow & (u,v) \longrightarrow & (x,y,z) \longrightarrow & w \\ u = \gamma_1(t) & x = \sigma_1(u,v) & w = f(x,y,z) & \\ v = \gamma_2(t) & y = \sigma_2(u,v) & & \\ & z = \sigma_3(u,v) & & \end{array}$$

we get $\mathbf{F}\circ\boldsymbol{\sigma}\circ\boldsymbol{\gamma} = (w,0,0)$ and $\boldsymbol{\sigma}\circ\boldsymbol{\gamma} = (x,y,z)$; hence

$$(\boldsymbol{\sigma}\circ\boldsymbol{\gamma})' = \left(\frac{dx}{dt}, \frac{dy}{dt}, \frac{dz}{dt}\right),$$

and

$$\int_{\bar{\boldsymbol{\gamma}}} \mathbf{F} = \int_a^b (\mathbf{F}\circ\boldsymbol{\sigma}\circ\boldsymbol{\gamma}) \cdot (\boldsymbol{\sigma}\circ\boldsymbol{\gamma})' = \int_a^b w\frac{dx}{dt}\,dt$$

$$= \int_a^b w\left(\frac{\partial x}{\partial u}\frac{du}{dt} + \frac{\partial x}{\partial v}\frac{dv}{dt}\right) dt = \int_{\boldsymbol{\gamma}} \left(w\frac{\partial x}{\partial u}\,du + w\frac{\partial x}{\partial v}\,dv\right)$$

$$= \iint_S \left(\frac{\partial}{\partial u}\left(w\frac{\partial x}{\partial v}\right) - \frac{\partial}{\partial v}\left(w\frac{\partial x}{\partial u}\right)\right) du\,dv \quad \text{(by Green's theorem)} \tag{7}$$

$$= \iint_S \left(\frac{\partial w}{\partial u}\frac{\partial x}{\partial v} - \frac{\partial w}{\partial v}\frac{\partial x}{\partial u}\right) du\,dv \quad \left(\text{since } \frac{\partial^2 x}{\partial u \partial v} = \frac{\partial^2 x}{\partial v \partial u}\right) \tag{8}$$

$$= \iint_S \left(\left(\frac{\partial w}{\partial x}\frac{\partial x}{\partial u} + \frac{\partial w}{\partial y}\frac{\partial y}{\partial u} + \frac{\partial w}{\partial z}\frac{\partial z}{\partial u} \right) \frac{\partial x}{\partial v} \right.$$

$$\left. - \left(\frac{\partial w}{\partial x}\frac{\partial x}{\partial v} + \frac{\partial w}{\partial y}\frac{\partial y}{\partial v} + \frac{\partial w}{\partial z}\frac{\partial z}{\partial v} \right) \frac{\partial x}{\partial u} \right) du\, dv \quad \text{(chain rule)}$$

$$= \iint_S \left(\frac{\partial w}{\partial z} \left(\frac{\partial z}{\partial u}\frac{\partial x}{\partial v} - \frac{\partial z}{\partial v}\frac{\partial x}{\partial u} \right) - \frac{\partial w}{\partial y} \left(\frac{\partial y}{\partial v}\frac{\partial x}{\partial u} - \frac{\partial y}{\partial u}\frac{\partial x}{\partial v} \right) \right) du\, dv. \qquad (9)$$

But $\mathbf{F} = (f,0,0)$ and $f(x,y,z) = w$, so

$$\nabla \times \mathbf{F} = \nabla \times (f,0,0) = \left(0, \frac{\partial w}{\partial z}, -\frac{\partial w}{\partial y} \right);$$

and

$$\sigma_u \times \sigma_v = \frac{\partial \sigma}{\partial u} \times \frac{\partial \sigma}{\partial v}$$

$$= \left(\frac{\partial x}{\partial u}, \frac{\partial y}{\partial u}, \frac{\partial z}{\partial u} \right) \times \left(\frac{\partial x}{\partial v}, \frac{\partial y}{\partial v}, \frac{\partial z}{\partial v} \right)$$

$$= \left(\frac{\partial y}{\partial u}\frac{\partial z}{\partial v} - \frac{\partial y}{\partial v}\frac{\partial z}{\partial u}, \frac{\partial z}{\partial u}\frac{\partial x}{\partial v} - \frac{\partial z}{\partial v}\frac{\partial x}{\partial u}, \frac{\partial x}{\partial u}\frac{\partial y}{\partial v} - \frac{\partial x}{\partial v}\frac{\partial y}{\partial u} \right),$$

so from (9) we obtain

$$\int_{\tilde{\gamma}} \mathbf{F} = \iint_S (\nabla \times \mathbf{F}) \cdot (\sigma_u \times \sigma_v) = \iint_\sigma \nabla \times \mathbf{F},$$

which proves Stokes' theorem in the case $\mathbf{F} = (f,0,0)$. In the general case $\mathbf{F} = (f_1, f_2, f_3)$, the calculation is exactly the same but requires three times as much space, so we leave it to you.

By looking back at the calculations above, it is easy to see the conditions under which we have proved Stokes' theorem:

(i) step (7) requires that the parameter domain S with its boundary $\boldsymbol{\gamma}$ be such that Green's theorem (4) holds for every pair of functions (M,N) having continuous partial derivatives in S;

(ii) step (8) requires that the components σ_1, σ_2, σ_3 of the surface $\boldsymbol{\sigma}$ have continuous second partial derivatives, and that the components f_1, f_2, f_3 of $\mathbf{F}$ have continuous first partial derivatives at every point $\boldsymbol{\sigma}(u,v)$ on the surface $\boldsymbol{\sigma}$.

Stokes' theorem sheds some light on the line integral of a closed vector field. Since "**F** is closed" means that

$$D_1 f_2 = D_2 f_1 , \qquad D_1 f_3 = D_3 f_1 , \qquad D_2 f_3 = D_3 f_2 ,$$

it follows immediately from the definition of the curl $\nabla \times \mathbf{F}$ in (5) that **F** *is closed if and only if* $\nabla \times \mathbf{F} = \mathbf{0}$. Hence, *if* **F** *is closed, and the curve* $\bar{\boldsymbol{\gamma}}$ *forms the boundary of a surface* $\boldsymbol{\sigma}$ *satisfying conditions* (i) *and* (ii) *above, then the integral of* **F** *over* $\bar{\boldsymbol{\gamma}}$ *is zero:*

$$\int_{\bar{\gamma}} \mathbf{F} = \int_{\sigma} \nabla \times \mathbf{F} = \int_{\sigma} \mathbf{0} = 0.$$

Example 2. If

$$\mathbf{F}(x,y,z) = \left(\frac{-y}{x^2 + y^2}, \frac{x}{x^2 + y^2}, z\right), \tag{10}$$

then $\nabla \times \mathbf{F}$ exists and equals zero everywhere except along the z axis. If $\boldsymbol{\gamma}^1$ is the curve in Fig. 21 (a),

$$\boldsymbol{\gamma}^1(t) = (\sin t, 1, \cos t), \qquad 0 \le t \le 2\pi,$$

then $\boldsymbol{\gamma}^1$ forms the boundary of a disk-shaped surface

$$\boldsymbol{\sigma}(u,v) = (u,1,v)$$

on the parameter domain

$$S = \{(u,v): u^2 + v^2 \le 1\} \tag{11}$$

with boundary

$$\boldsymbol{\gamma}(t) = (\sin t, \cos t), \qquad 0 \le t \le 2\pi. \tag{12}$$

The conditions (i) and (ii) for Stokes' theorem are satisfied, so

$$\int_{\gamma^1} \mathbf{F} = \iint_{\sigma} \nabla \times \mathbf{F} = 0.$$

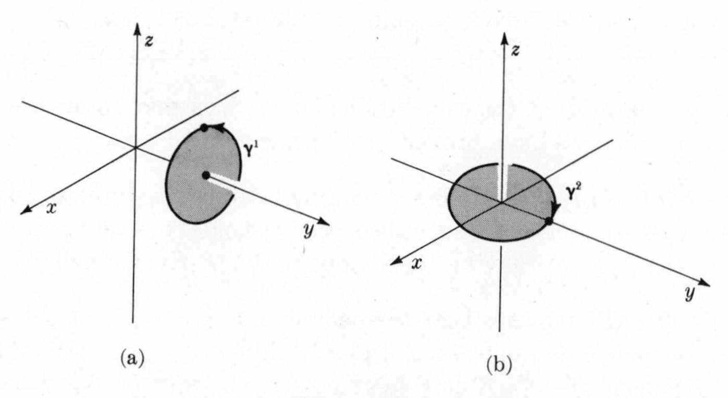

FIGURE 5.21

Example 3. Take $\mathbf{F}$ as in (10), and let $\boldsymbol{\gamma}^2$ be the curve in Fig. 21 (b),

$$\boldsymbol{\gamma}^2(t) = (\sin t, \cos t, 0), \qquad 0 \le t \le 2\pi.$$

Again, $\boldsymbol{\gamma}^2$ is the boundary of a surface

$$\boldsymbol{\sigma}(u,v) = (u,v,0),$$

with the parameter domain S and its boundary $\boldsymbol{\gamma}$ the same as in (11) and (12). However, Stokes' theorem does *not* apply for this $\mathbf{F}$ and $\boldsymbol{\sigma}$, since $\boldsymbol{\sigma}$ contains a point (the origin) where the derivatives of $\mathbf{F}$ do not exist. Thus, we cannot conclude from Stokes' theorem that $\int_{\gamma^2} \mathbf{F} = 0$. In fact, in §5.4 we computed $\int_{\gamma^2} \mathbf{F} = 2\pi$.

PROBLEMS

1. Let $\boldsymbol{\sigma}(u,v) = (\cos u \sin v,\ \sin u \sin v,\ \cos v)$ for $0 \le u \le 2\pi$ and $0 \le v \le \pi$. (This is the sphere in Fig. 15.)
   (a) Find $\mathbf{N}_\sigma = \boldsymbol{\sigma}_u \times \boldsymbol{\sigma}_v$.
   (b) Show that $\boldsymbol{\sigma}$ and $\mathbf{N}_\sigma$ are parallel, in fact $\mathbf{N}_\sigma = (-\sin v)\boldsymbol{\sigma}$. Explain geometrically why $\boldsymbol{\sigma}$ and $\mathbf{N}_\sigma$ should be parallel.
   (c) Compute the area of the unit sphere as $\int\int |\boldsymbol{\sigma}_u \times \boldsymbol{\sigma}_v|$.
   (d) Compute $\int\int_\sigma \mathbf{F}$ where $\mathbf{F}(x,y,z) = (1,0,0)$. Interpret $\mathbf{F}$ as a flow, and thus explain why the integral should be zero.
   (e) Compute $\int\int_\sigma \mathbf{F}$ where $\mathbf{F}(x,y,z) = (x,y,z)$. Interpret $\mathbf{F}$ as a flow, and thus explain why the integral should be -4π.

2. This problem concerns parametric representations of the sphere that do not "cover" the sphere exactly once. In each case, the formula for $\boldsymbol{\sigma}$ is the same as in Problem 1, but the parameter domain S is different. Compute the area of the surface if the parameter domain S is
   (a) $S = \{(u,v) : 0 \le u \le 4\pi,\ 0 \le v \le \pi\}$ (This surface covers the sphere twice, so you should find the area 8π.)
   (b) $S = \{(u,v) : 0 \le u \le 2\pi,\ 0 \le v \le 2\pi\}$
   (c) $S = \{(u,v) : 0 \le u \le \pi,\ 0 \le v \le \pi\}$
   (d) $S = \{(u,v) : 0 \le u \le 4\pi,\ 0 \le v \le 2\pi\}$
   In each part, try to visualize how the parametric surface $\boldsymbol{\sigma}$ wraps around the sphere.

3. Let $\mathbf{A}$ and $\mathbf{B}$ be vectors in R_3, $\mathbf{A} \times \mathbf{B} \ne \mathbf{0}$, and let $\boldsymbol{\sigma}(u,v) = \mathbf{P}_0 + u\mathbf{A} + v\mathbf{B}$, where $\mathbf{P}_0$ is some fixed vector in R^3.
   (a) Taking the parameter domain $S = \{(u,v) : 0 \le u \le 1, 0 \le v \le 1\}$, $\boldsymbol{\sigma}$ defines a parallelogram (Fig. 22). Show that the area of the parallelogram is $|\mathbf{A} \times \mathbf{B}|$. (Thus our definition of area is consistent with the geometric interpretation of $|\mathbf{A} \times \mathbf{B}|$ in §1.2.)

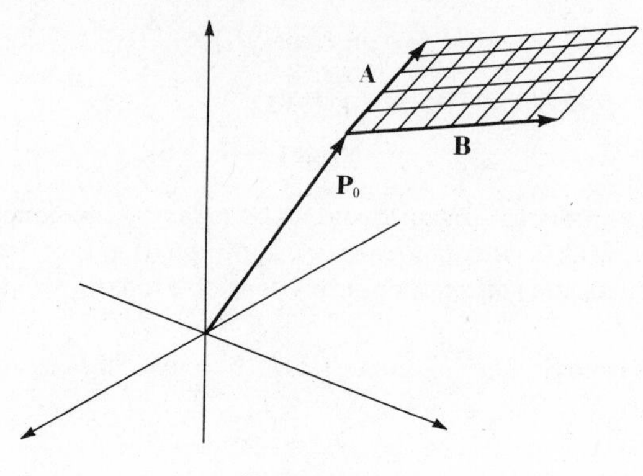

FIGURE 5.22

(b) Take the parameter domain

$$S = \{(u,v) : 0 \leq u \leq 1, 0 \leq v \leq 1 - u\}.$$

What geometric figure does $\boldsymbol{\sigma}$ produce, and what is its area?

4. (a) Let $\boldsymbol{\sigma}$ be the parallelogram in Problem 3(a), and let $\mathbf{F}$ be any vector field which is constant along the parallelogram, that is, $\mathbf{F}(\boldsymbol{\sigma}(u,v)) \equiv \mathbf{C}$ for some constant vector $\mathbf{C}$. Show that

$$\iint_{\sigma} \mathbf{F} = (\mathbf{A} \times \mathbf{B}) \cdot \mathbf{C}.$$

(b) Show that the rate of flow through the parallelogram in part (a) is $(\mathbf{N} \cdot \mathbf{C}) \times$ (area of parallelogram), where

$$\mathbf{N} = \frac{1}{|\mathbf{A} \times \mathbf{B}|} \mathbf{A} \times \mathbf{B}$$

is the positive unit normal to the parallelogram.

5. A cone with vertex $\mathbf{V}$, altitude $\mathbf{A}$, and radius r is given by

$$\boldsymbol{\sigma}(u,v) = \mathbf{V} + u\mathbf{A} + ru(\cos v\mathbf{B} + \sin v\mathbf{C}), \quad 0 \leq u \leq 1, -\pi \leq v \leq \pi,$$

where $\mathbf{B}$ and $\mathbf{C}$ are unit vectors orthogonal to $\mathbf{A}$, and $\mathbf{B} \cdot \mathbf{C} = 0$ (Fig. 23).

(a) Compute the area of the cone. (Hint: $\mathbf{A} \times \mathbf{B}$, $\mathbf{A} \times \mathbf{C}$, and $\mathbf{B} \times \mathbf{C}$ are mutually orthogonal, by §1.5; use the Pythagorean formula.)

(b) Compute $\iint_{\sigma} \mathbf{A}$ and $\iint_{\sigma} \mathbf{B}$, where $\mathbf{A}$ and $\mathbf{B}$ are the constant vectors in the definition of $\boldsymbol{\sigma}$. Explain your results by interpreting $\mathbf{A}$ and $\mathbf{B}$ as constant flows.

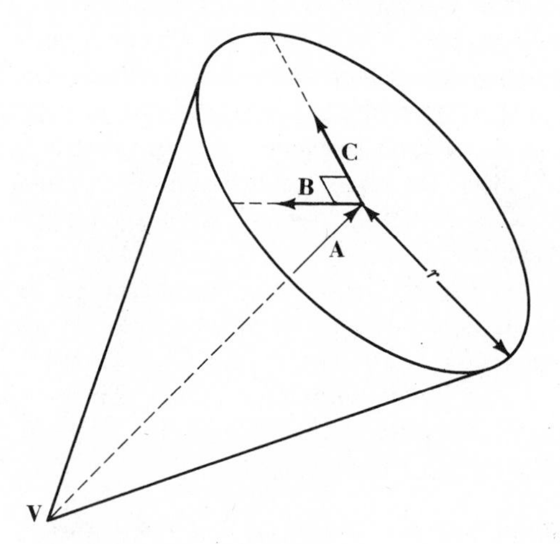

FIGURE 5.23

6. In each of the following cases, compute $\iint_\sigma \nabla \times \mathbf{F}$ by the definition of surface integral (III), and check that $\iint_\sigma \nabla \times \mathbf{F} = \int_{\bar{\gamma}} \mathbf{F}$, where $\bar{\boldsymbol{\gamma}} = \boldsymbol{\sigma} \circ \boldsymbol{\gamma}$ is the positively oriented boundary of $\boldsymbol{\sigma}$.
   (a) $\mathbf{F}(x,y,z) = (x,y,z)$, $\boldsymbol{\sigma}(u,v) = (u,v,0)$, $S = \{(u,v): 0 \le u \le 1, 0 \le v \le 1\}$, $\boldsymbol{\gamma}$ consists of the four obvious straight line curves.
   (b) $\mathbf{F}(x,y,z) = (y,z,x)$ with $\boldsymbol{\sigma}$, S, and $\boldsymbol{\gamma}$ as in part (a).
   (c) $\mathbf{F}(x,y,z) = (z,x,y)$, $\boldsymbol{\sigma}(u,v) = (u,v,u^2 + v^2)$, $S = \{(u,v): u^2 + v^2 \le 1\}$, $\boldsymbol{\gamma}(t) = (\cos t, \sin t)$, $0 \le t \le 2\pi$.

7. An *ellipsoid* with semi-axes a, b, c can be represented parametrically by $\boldsymbol{\sigma}(u,v) = (a \cos u \sin v, b \sin u \sin v, c \cos v)$, $\quad 0 \le u \le 2\pi, 0 \le v \le \pi$. Obtain an integral for the surface area of the ellipsoid. (The integral is complicated; do not try to evaluate it.)

8. Suppose f is a function of two variables (x,y) with domain S. Represent the graph of f as a parametric surface (as in the text above), and thus deduce a formula for the surface area A of the graph,

$$A = \iint_S \sqrt{1 + |\nabla f|^2}.$$

(Notice the resemblance to the integral for the *length* of the graph of a function f of one variable, $\int_a^b \sqrt{1 + (f')^2}$.)

9. Use the formula in the previous problem to compute the area of the following surfaces:
   (a) The graph of $f(x,y) = 3x$ with domain $\{(x,y): 0 \le x \le 1, 0 \le y \le 1\}$. (Your integral gives the area of a rectangle; check the result by computing the sides of the rectangle.)

(b) The part of the plane $\{(x,y,z): z = 3x + 4y\}$ lying over the rectangle $0 \leq x \leq 1$, $0 \leq y \leq 1$. (Your integral gives the area of a parallelogram; check the result by using a cross product.)
(c) The part of the plane $\{(x,y,z): z = 3x\}$ lying over the circle $x^2 + y^2 \leq 1$ in the xy plane. (Your integral gives the area of an ellipse; check the result against the formula πab for the area of an ellipse whose longest diameter is $2a$ and shortest diameter is $2b$.)
(d) The part of the plane $\{(x,y,z): z = 3x + 4y\}$ lying over the disk $x^2 + y^2 \leq 1$ in the xy plane. (Again, check the result against the simple formula for the area of an ellipse.)
(e) The part of the "saddle" $\{(x,y,z): z = xy\}$ lying over the disk $x^2 + y^2 \leq 1$ in the xy plane. (Use polar coordinates, §4.5.)
(f) The part of the paraboloid $\{(x,y,z): z = x^2 + y^2\}$ lying over the disk $x^2 + y^2 \leq 1$ in the xy plane.

10. Show that the graphs of $f(x,y) = 2xy$ and $g(x,y) = x^2 + y^2$, when restricted to lie over the same domain S in the xy plane, always have equal surface area.

11. Suppose that $\boldsymbol{\sigma}$ is a surface lying on the unit sphere, i.e. $|\boldsymbol{\sigma}(u,v)| = 1$ for every u and v. Prove that there is a real-valued function $\lambda(u,v)$ such that $\mathbf{N}_\sigma(u,v) = \lambda(u,v)\boldsymbol{\sigma}(u,v)$. (Hint: Show that $\boldsymbol{\sigma}_u$ and $\boldsymbol{\sigma}_v$ are orthogonal to $\boldsymbol{\sigma}$, and apply §1.5. Problem 1 was a special case, and there you found $\lambda(u,v) = -\sin v$.)

12. Suppose that a surface $\boldsymbol{\sigma}$ lies in a level surface $\{\mathbf{P}: f(\mathbf{P}) = c\}$ of a differentiable function f. Prove that $\mathbf{N}_\sigma$ and ∇f are parallel. (Hint: If the level surface is defined by $f = c$, then $f \circ \boldsymbol{\sigma} \equiv c$. Differentiate this to show that ∇f is orthogonal to $\boldsymbol{\sigma}_u$ and $\boldsymbol{\sigma}_v$. It is then "geometrically obvious" that ∇f is parallel to $\boldsymbol{\sigma}_u \times \boldsymbol{\sigma}_v$; the proof uses §1.5.)

13. Suppose that $\boldsymbol{\gamma}$ is any differentiable curve lying in the parameter domain S of a surface $\boldsymbol{\sigma}$. Then $\boldsymbol{\sigma} \circ \boldsymbol{\gamma}$ is a curve lying in the surface. Show that $(\boldsymbol{\sigma} \circ \boldsymbol{\gamma})'$ is orthogonal to $\mathbf{N}_\sigma$ at every point. (Hint: By the chain rule, write $(\boldsymbol{\sigma} \circ \boldsymbol{\gamma})'$ as a combination of $\boldsymbol{\sigma}_u$ and $\boldsymbol{\sigma}_v$.)

14. Suppose that $\boldsymbol{\sigma}$ is a surface not passing through the origin $\mathbf{0}$. The *solid angle subtended by* $\boldsymbol{\sigma}$ *at* $\mathbf{0}$ is defined to be $\iint_\sigma \boldsymbol{\Omega}$, where $\boldsymbol{\Omega}$ is the vector field $\boldsymbol{\Omega}(\mathbf{P}) = \mathbf{P}/|\mathbf{P}|^3$.
(a) Compute $\iint_\sigma \boldsymbol{\Omega}$ when $\boldsymbol{\sigma}$ is the disk $\boldsymbol{\sigma}(u,v) = (1,u,v)$ with $u^2 + v^2 \leq 1$.
(b) Same question, with $\boldsymbol{\sigma}(u,v) = (a,u,v)$ and $u^2 + v^2 \leq R^2$. What happens as the constant $a \to 0$? $a \to \infty$? $R \to \infty$? Explain geometrically.
(c) Compute $\iint_\sigma \boldsymbol{\Omega}$ when $\boldsymbol{\sigma}$ is the rectangle $\boldsymbol{\sigma}(u,v) = (a,u,v)$ with $|u| \leq b$ and $|v| \leq c$. What happens as $a \to \infty$? (Use a table of integrals.)

(d) Compute $\iint_\sigma \Omega$ when $\boldsymbol{\sigma}$ is the sphere in Problem 1.

(e) Compute $\iint_\sigma \Omega$ when $\boldsymbol{\sigma}$ is the sphere in Problem 2(c).

(f) Suppose that $\boldsymbol{\sigma}$ is any surface lying in a sphere of radius r about the origin, and that $\mathbf{N}_\sigma \cdot \boldsymbol{\sigma} \geq 0$ at every point. Show that $r^2 \iint_\sigma \Omega$ is the area of $\boldsymbol{\sigma}$. (Hint: $\mathbf{N}_\sigma \cdot \boldsymbol{\sigma} = r\,|\mathbf{N}_\sigma|$; see Problem 11.)

(g) Let $\boldsymbol{\sigma}$ be a given surface not passing through the origin, and define a new surface $\bar{\boldsymbol{\sigma}}$ by

$$\bar{\boldsymbol{\sigma}}(u,v) = \frac{1}{|\boldsymbol{\sigma}(u,v)|}\,\boldsymbol{\sigma}(u,v).$$

(This is the "projection" of $\boldsymbol{\sigma}$ onto the unit sphere.) Show that $\iint_\sigma \Omega = \iint_{\bar{\sigma}} \Omega$. (Hint:

$$\left(\frac{\boldsymbol{\sigma}}{|\boldsymbol{\sigma}|}\right)_u = \frac{\boldsymbol{\sigma}_u}{|\boldsymbol{\sigma}|} - \frac{\boldsymbol{\sigma}_u \cdot \boldsymbol{\sigma}}{|\boldsymbol{\sigma}|^3}\,\boldsymbol{\sigma}.)$$

5.6 TRIPLE INTEGRALS

The set V in Fig. 24 is bounded by the graphs of two continuous functions φ_1 and φ_2 with domain S in R^2; i.e.

$$V = \{(x,y,z)\colon (x,y) \text{ is in } S,\ \varphi_1(x,y) \leq z \leq \varphi_2(x,y)\}, \tag{1}$$

where $\varphi_1 \leq \varphi_2$ on S. Suppose that f is continuous on V. Then it can be proved that

$$F(x,y) = \int_{\varphi_1(x,y)}^{\varphi_2(x,y)} f(x,y,z)\,dz$$

is a continuous function on S. If, further, the set S is of the type we considered in studying double integrals in the preceding chapter, then the *triple integral* of f over V is defined by

$$\iiint_V f = \iint_S \left[\int_{\sigma_1(x,y)}^{\varphi_2(x,y)} f(x,y,z)\,dz\right] dx\,dy.$$

For example, if V is the unit ball, then we can take S to be the unit disk, and

$$\iiint_V f = \iint_S \left[\int_{-\sqrt{1-x^2-y^2}}^{\sqrt{1-x^2-y^2}} f\,dz\right] dx\,dy$$

$$= \int_{-1}^{1}\left[\int_{-\sqrt{1-y^2}}^{\sqrt{1-y^2}}\left[\int_{-\sqrt{1-x^2-y^2}}^{\sqrt{1-x^2-y^2}} f\,dz\right] dx\right] dy.$$

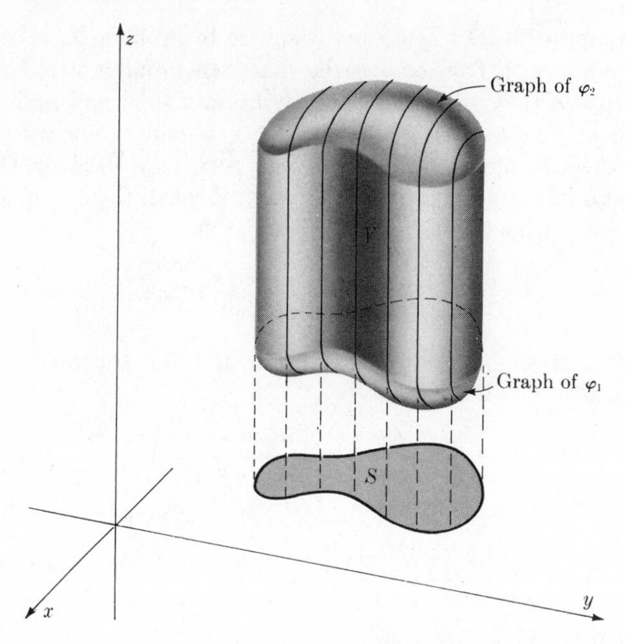

FIGURE 5.24

V is determined in (1) by functions φ_1 and φ_2 of x and y. If this same set V can also be determined by functions of x and z,

$$V = \{(x,y,z): (x,z) \text{ in } S', \psi_1(x,z) \le y \le \psi_2(x,z)\},$$

or by functions of y and z,

$$V = \{(x,y,z): (y,z) \text{ in } S'', \eta_1(y,z) \le x \le \eta_2(y,z)\},$$

then it can be proved that the corresponding integrals all give the same result,

$$\iint_S \left[\int_{\varphi_1}^{\varphi_2} f\,dz\right] dx\,dy = \iint_{S'} \left[\int_{\psi_1}^{\psi_2} f\,dy\right] dx\,dz = \iint_{S''} \left[\int_{\eta_1}^{\eta_2} f\,dx\right] dy\,dz.$$

For example, if V is the rectangular solid

$$V = \{(x,y,z): x_1 \le x \le x_2\,,\ y_1 \le y \le y_2\,,\ z_1 \le z \le z_2\},$$

then the integral over V can be evaluated in numerous ways:

$$\iiint_V f = \int_{z_1}^{z_2} \left[\int_{y_1}^{y_2} \left[\int_{x_1}^{x_2} f\,dx\right] dy\right] dz = \int_{x_1}^{x_2} \left[\int_{y_1}^{y_2} \left[\int_{z_1}^{z_2} f\,dz\right] dy\right] dx$$

$$= \int_{y_1}^{y_2} \left[\int_{x_1}^{x_2} \left[\int_{z_1}^{z_2} f\,dz\right] dx\right] dy = \cdots.$$

The square brackets show how the integral is to be computed, i.e. as three successive single integrals, beginning with the "inner" one. In practice, these brackets are left out.

Example 1. Evaluate $\int_0^1 \int_0^x \int_0^y x\,dz\,dy\,dx$. *Solution.* Since x and y are held constant in taking the integral with respect to z,

$$\int_0^y x\,dz = x \int_0^y dz = xy.$$

Thus

$$\int_0^x \int_0^y x\,dz\,dy = \int_0^x xy\,dy = \frac{1}{2}xy^2\Big|_{y=0}^{y=x} = \frac{1}{2}x^3$$

and

$$\int_0^1 \int_0^x \int_0^y x\,dz\,dy\,dx = \int_0^1 \frac{1}{2}x^3\,dx = \frac{1}{8}x^4\Big|_0^1 = \frac{1}{8}.$$

Example 2. Sketch the region of integration in Example 1, and rewrite the integral in the order $\int\int\int \cdots dx\,dy\,dz$. *Solution.* The two "outer limits" $\int_0^1 \int_0^x \cdots dy\,dx$ indicate an integral over the triangle S in Fig. 25(a); thus V lies somewhere in the prism-shaped region in Fig. 25(a) generated by lines parallel to the z axis and passing through S. (This region is called the *cylinder* parallel to the z axis with S as cross section.) Next, the "inner limits" $\int_0^y \cdots dz$ show that for each (x,y) in S, z varies from 0 to y, so V lies above the plane defined by $z = 0$ and below the plane defined by $z = y$; thus V is the tetrahedron in Fig. 25(b).

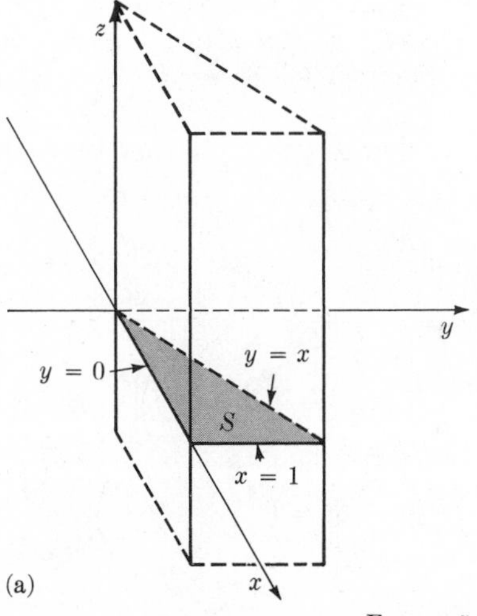

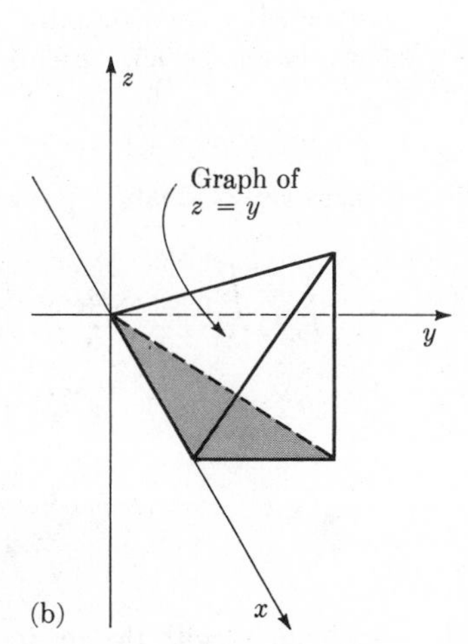

FIGURE 5.25

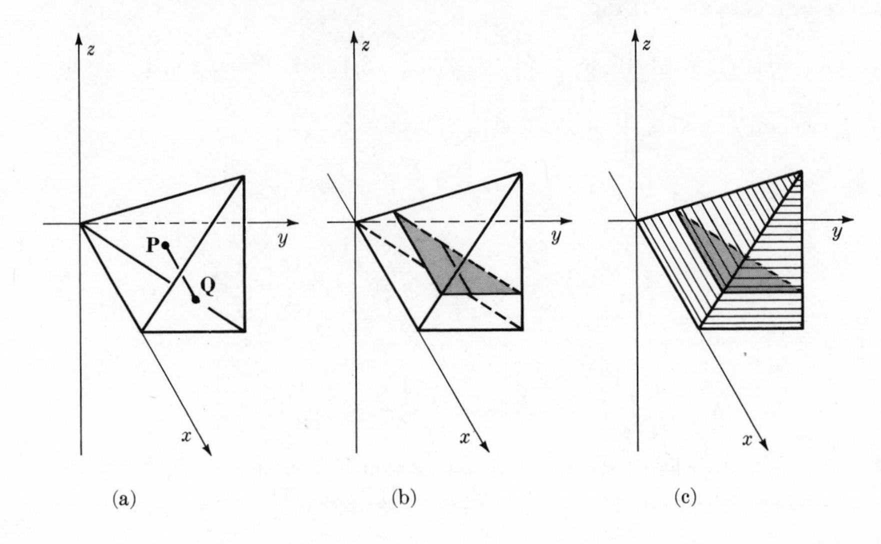

FIGURE 5.26

To write the integral in the order $\iiint \cdots dx\,dy\,dz$, begin with the inner limits. For each fixed y and z, x varies along the heavy segment in Fig. 26 (a), from the point **P** lying in the plane in back (whose equation is $y = x$) to the point **Q** lying in the plane in front (whose equation is $x = 1$). Thus, for each fixed y and z, x varies from y to 1, generating a line segment representing a typical "linear section" of V. Next, for each fixed z, y varies from z to 1, and the corresponding segments in Fig. 26 (a) sweep out the planar cross section of V shown in Fig. 26 (b). Finally, as z varies from 0 to 1 these plane cross sections sweep out all of V (Fig. 26 (c)). Thus the integral is $\int_0^1 \int_z^1 \int_y^1 x\,dx\,dy\,dz$. As a check, we evaluate the integral in this order:

$$\int_0^1 \int_z^1 \int_y^1 x\,dx\,dy\,dz = \int_0^1 \int_z^1 \frac{x^2}{2}\Big|_{x=y}^{x=1} dy\,dz$$

$$= \int_0^1 \int_z^1 \frac{1-y^2}{2}\,dy\,dz = \int_0^1 \left(\frac{y}{2} - \frac{y^3}{6}\right)\Big|_{y=z}^{y=1} dz$$

$$= \int_0^1 \left(\frac{1}{3} - \frac{z}{2} + \frac{z^3}{6}\right) dz = \left(\frac{z}{3} - \frac{z^2}{4} + \frac{z^4}{24}\right)\Big|_0^1 = \frac{1}{8},$$

which agrees with the answer in Example 1.

Example 3. Evaluate $\iiint_V f$ when $f(x,y,z) = x^2 + y^2 + z^2$ and V is the unit ball $\{(x,y,z): x^2 + y^2 + z^2 \leq 1\}$.
Solution.

$$\iiint_V f = \int_{-1}^{1}\int_{-\sqrt{1-y^2}}^{\sqrt{1-y^2}}\int_{-\sqrt{1-x^2-y^2}}^{\sqrt{1-x^2-y^2}} (x^2 + y^2 + z^2)\, dz\, dx\, dy$$

$$= \int_{-1}^{1}\int_{-\sqrt{1-y^2}}^{\sqrt{1-y^2}} \left[x^2z + y^2z + \frac{z^3}{3}\right]_{-\sqrt{1-x^2-y^2}}^{\sqrt{1-x^2-y^2}} dx\, dy$$

$$= \int_{-1}^{1}\int_{-\sqrt{1-y^2}}^{\sqrt{1-y^2}} \frac{2}{3}(1 + 2x^2 + 2y^2)\sqrt{1 - x^2 - y^2}\, dx\, dy.$$

This last integral is an obvious candidate for polar coordinates; it becomes

$$\int_0^{2\pi}\int_0^1 \frac{2}{3}(1 + 2r^2)\sqrt{1 - r^2}\, r\, dr\, d\theta$$

$$= \frac{4\pi}{3}\int_0^1 (1 + 2r^2)\sqrt{1 - r^2}\, r\, dr$$

$$= \frac{4\pi}{3}\int_0^1 [1 + 2(1 - u^2)]u^2 du \qquad (\sqrt{1 - r^2} = u; r\, dr = -u\, du)$$

$$= \frac{4\pi}{3}\left[u^3 - \frac{2}{5}u^5\right]_0^1 = \frac{4\pi}{5},$$

and $\iiint_V f$ is evaluated.

The various interpretations of triple integrals are natural extensions of the one- and two-dimensional cases. When $f \equiv 1$, then $\iiint_V f$ is the *volume* of V. When V is occupied by a substance of variable density f, then $\iiint_V f$ is the *total mass* of the substance in V. When $T(P)$ is the temperature at the point P in V, and $S(P)$ the specific heat of the substance at the point P, then TS is the "heat density," and $\iiint_V TS$ is the total heat in V.

PROBLEMS

1. Evaluate the following integrals, and sketch the region over which the integral is taken.

(a) $\displaystyle\int_2^3\int_1^2\int_0^1 x\, dx\, dy\, dz$

(b) $\displaystyle\int_{-1}^0\int_x^0\int_0^{\sqrt{3-x^2-2y^2}} z\, dz\, dy\, dx$

(c) $\displaystyle\int_1^5\int_0^4\int_1^{y+z} y\, dx\, dz\, dy$

(d) $\displaystyle\int_{\pi/2}^{\pi}\int_1^{2x}\int_0^{\cos(x+z)} 1\, dy\, dz\, dx$

2. Evaluate $\iiint_V f$ in the following cases.

(a) $f(x,y,z) = xy, \quad V = \{(x,y,z)\colon 0 \le z \le x + 2,\ x^2 + 4y^2 \le 4\}$

(b) $f(x,y,z) \equiv 1, \quad V = \{(x,y,z)\colon x^2 + y^2 \le a^2,\ y^2 + z^2 \le a^2\}$

(c) $f \equiv 1, \quad V = \{(x,y,z)\colon x \ge 0,\ y \ge 0,\ z \ge 0,\ x + y + z \le 1\}$

(d) $f \equiv 1, \quad V = \{(x,y,z)\colon x^2 + y^2 + z^2 \le R^2,\ x^2 + y^2 \ge r^2\}$, where $0 < r < R$. (This is the volume of a "cored apple.")

(e) $f \equiv 1, \quad V = \{(x,y,z)\colon x^2 + y^2 + z^2 \le R^2,\ (x - r)^2 + y^2 \ge r^2\}$, where $2r = R$. (This is the volume of an "eccentrically cored apple.")

3. Let V be the cylinder $\{(x,y,z)\colon (x - a)^2 + y^2 \le a^2,\ 0 \le z \le b\}$, and suppose V is occupied by material of constant density ρ. Compute

(a) $\iiint_V \rho$ (the mass of V).

(b) $\iiint_V \rho x\, dx\, dy\, dz$ (the *first moment of* V *with respect to the* yz *plane,* denoted M_{yz}.)

(c) $\iiint_V \rho y\, dx\, dy\, dz$ (the *first moment of* V *with respect to the* xz *plane,* denoted M_{xz}.)

(d) $\iiint_V \rho z\, dx\, dy\, dz$ (the *first moment of* V *with respect to the* xy *plane,* denoted M_{xy}.)

(e) The *center of gravity* of V, the point $(\bar{x},\bar{y},\bar{z})$ where

$$\bar{x} = \frac{M_{yz}}{\text{mass}}, \qquad \bar{y} = \frac{M_{xz}}{\text{mass}}, \qquad \bar{z} = \frac{M_{xy}}{\text{mass}}.$$

(You should find $(\bar{x},\bar{y},\bar{z}) = (a,0,\tfrac{1}{2}b) =$ center of the cylinder.)

4. Let V be the slice of the sphere $\{\mathbf{P}\colon |\mathbf{P}| \le b\}$ lying above the plane defined by $z = a$, where $0 < a < b$ (Fig. 27). Let S be the disk $\{(x,y)\colon x^2 + y^2 \le b^2 - a^2\}$. Assume V is occupied by material of constant density 1. Find the center of gravity by computing

(a) the volume of V,

$$\iiint_V 1 = \iint_S \int_a^{\sqrt{b^2-x^2-y^2}} dz\, dx\, dy.$$

(Use polar coordinates.)

(b) M_{xy}, M_{yz}, M_{xz} (See Problem 3. You should find $M_{xz} = 0$ and $M_{yz} = 0$ with very little work; use polar coordinates on M_{xy}.)

(c) Show that the center of gravity is

$$\left(0, 0, \frac{3(b - a)(b + a)^2}{4(2b^2 - ab - a^2)}\right).$$

(d) As an interesting exercise in calculus of one variable, you can show that if $a < b$, then

$$a < \frac{3(b - a)(b + a)^2}{4(2b^2 - ab - a^2)} < b, \tag{*}$$

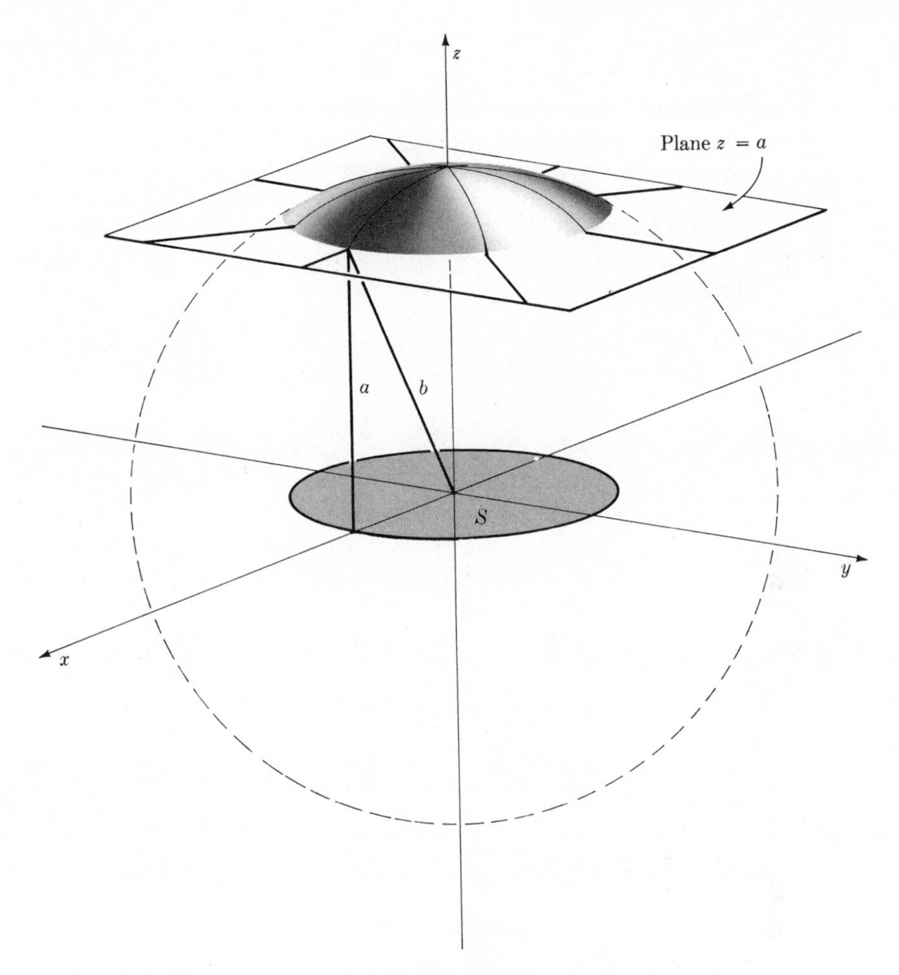

FIGURE 5.27

i.e. the center of gravity lies between the top and the bottom of the slice in Fig. 27. (Hint: Since the denominator in (*) is positive for $a < b$, you can multiply through by it. Then divide through by b^3, set $x = a/b$, and study the resulting functions on the interval $0 \leq x \leq 1$.)

5. Evaluate $\iiint_V f$ when $f \equiv 1$ and

$$V = \{(x,y,z) : 0 \leq x \leq 4 - y^2, 0 \leq z \leq y\}.$$

(Hint: Sketch V, and set up the integral in the order $dz\,dx\,dy$.)

6. Let $V = \{(x,y,z) : 8 - x^2 - y^2 \leq z \leq x^2 + y^2\}$.

(a) Evaluate $\iiint_V (x^2 + y^2)\,dx\,dy\,dz$. (This is the *moment of inertia* of V about the z axis. Set up the integral in the order $dz\,dx\,dy$. It helps if you can sketch, or at least visualize, the boundary surfaces $z = x^2 + y^2$ and $z = 8 - x^2 - y^2$. The first is a paraboloid of revolution opening upward, and the second is a similar paraboloid opening downward. The region for the $dx\,dy$ integration is determined by the intersection of these two surfaces.)

(b) Evaluate $\iiint (x^2 + z^2)\,dx\,dy\,dz$ (the moment of inertia about the y axis).

7. Let $V = \{(x,y,z): a \le x \le b, \varphi_1(x) \le y \le \varphi_2(z), \psi_1(x,y) \le z \le \psi_2(x,y)\}$. Suppose that $f(x,y,z,t)$ and its first derivative $f_t(x,y,z,t)$ are continuous for (x,y,z) in V and $c < t < d$. Prove that

$$\frac{d}{dt}\iiint_V f(x,y,z,t)\,dx\,dy\,dz = \iiint_V f_t(x,y,z,t)\,dx\,dy\,dz$$

for $c < t < d$. (Hint: Begin with the right-hand side, supply the limits of integration, and apply Leibniz' rule for single integrals (Theorem 3, §4.1) three times.)

8. Take V as in Problem 7, and let f be continuous on V. Prove that the function

$$g(x,y,z,t) = t^{-3/2}\iiint_V \exp\left(\frac{(x-\xi)^2 + (y-\eta)^2 + (z-\zeta)^2}{-4t}\right) f(\xi,\eta,\zeta)\,d\xi\,d\eta\,d\zeta$$

satisfies the equation for (non-steady state) heat conduction,

$$g_t = g_{xx} + g_{yy} + g_{zz}\,.$$

9. (a) Take V as in Problem 7. Suppose that f and g are continuous in V, and $f \le g$. Prove that $\iiint_V f \le \iiint_V g$. (Hint: By a known result for single integrals,

$$\int_{\psi_1(x,y)}^{\psi_2(x,y)} f(x,y,z)\,dz \le \int_{\psi_1(x,y)}^{\psi_2(x,y)} g(x,y,z)\,dz;$$

obtain from this a similar inequality for $\displaystyle\int_{\varphi_1(x)}^{\varphi_2(x)} \cdots dx$, and finally obtain the desired inequality for $\iiint_V$.)

(b) Suppose that f is continuous and $m \le f \le M$ in V. Prove that $m\,|V| \le \iiint_V f \le M\,|V|$, where $|V|$ denotes the volume of V.

10. (a) Suppose that f is continuous in an open set V, and that $\iiint_B f = 0$ for every ball B contained in V. Prove that $f \equiv 0$ in V. (Hint: If $f(\mathbf{P}_0) > 0$, there is a ball B of radius δ contained in V, such that $f(\mathbf{P}) > \frac{1}{2}f(\mathbf{P}_0)$ at every point $\mathbf{P}$ in B.)

(b) Suppose that V and V' are sets of the form in Problem 7, and that V contains V', i.e. if V is described by functions φ_j and ψ_j, and V' by φ_j' and ψ_j', then $\varphi_1 \leq \varphi_1' \leq \varphi_2' \leq \varphi_2$ and $\psi_1 \leq \psi_1' \leq \psi_2' \leq \psi_2$. Prove that if $f \geq 0$ in V, then $\iiint_{V'} f \leq \iiint_V f$.

(c) Suppose that f is continuous and $f \geq 0$ in a set V of the form in Problem 7, and that $\iiint_V f = 0$. Let U be any open set contained in V. Prove that $f \equiv 0$ in U. (Hint: If B is any ball contained in U, then B is in V, so $0 \leq \iiint_B f \leq \iiint_V f = 0$.)

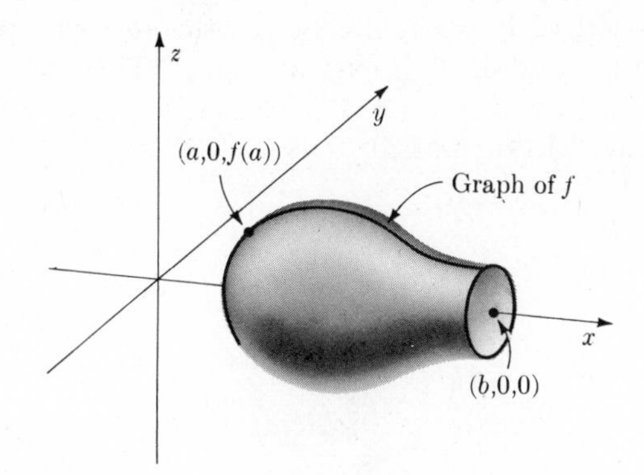

FIGURE 5.28

11. Let V be the solid of revolution in Fig. 28, where $f(x) \geq 0$ for $a \leq x \leq b$.

(a) Show that the volume of V is

$$\int_a^b \int_{-f(x)}^{f(x)} \int_{-\sqrt{f(x)^2-y^2}}^{\sqrt{f(x)^2-y^2}} dz\, dy\, dx.$$

(b) Show that

$$\int_{-f(x)}^{f(x)} \int_{-\sqrt{f(x)^2-y^2}}^{\sqrt{f(x)^2-y^2}} dz\, dy = \pi f(x)^2.$$

(c) Deduce that the volume of V is $\int_a^b \pi f(x)^2\, dx$.

5.7 THE DIVERGENCE THEOREM

Green's theorem in R^2, relating line integrals to double integrals, has an analog in R^3, called *Gauss' theorem* or the *divergence theorem*, relating surface integrals to triple integrals. The *divergence* of a vector field $\mathbf{F} = (f_1, f_2, f_3)$ is defined to be

$$\nabla \cdot \mathbf{F} = D_1 f_1 + D_2 f_2 + D_3 f_3 = \frac{\partial f_1}{\partial x} + \frac{\partial f_2}{\partial y} + \frac{\partial f_3}{\partial z}.$$

The divergence theorem says that if $\mathbf{F}$ has continuous derivatives in V, then

$$\iiint_V \nabla\cdot\mathbf{F} = \iint_\sigma \mathbf{F}, \tag{1}$$

where the surface $\boldsymbol{\sigma}$ forms the positively oriented boundary of V. In this case, "positively oriented" means "with the normal vector pointing out of V," as in Fig. 29. Thus if $\mathbf{F}$ is a flow, then the right-hand side of (1) is the rate of flow through the surface from inside to outside, in other words, the rate of flow out of V. The divergence theorem (1) says that this is the same as the integral of $\nabla\cdot\mathbf{F}$ over all of V. Thus, for any point $\mathbf{P}$ in V, we interpret the divergence $(\nabla\cdot\mathbf{F})(\mathbf{P})$ as the rate at which fluid leaves $\mathbf{P}$; hence the name "divergence."

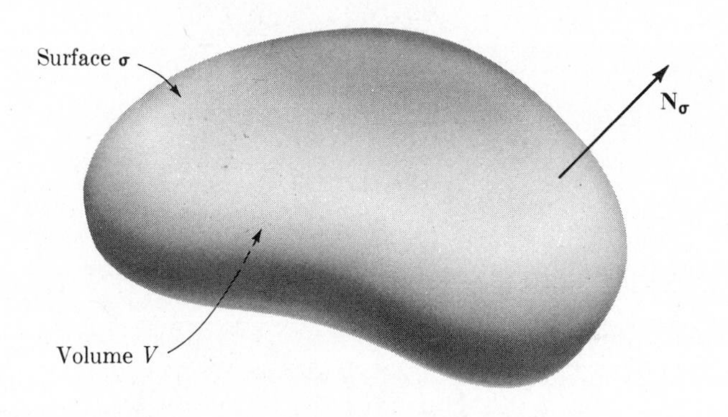

FIGURE 5.29

The divergence theorem, like Green's theorem, is much harder to formulate properly than to prove. In fact, with the divergence theorem the discrepancy is much greater than with Green's, and there is no half-decent elementary formulation. The jig is up; we leave it to you to interpret formula (1) and Fig. 29 in the examples and problems below. (However, Problems 3 and 4 below prove the two most important special cases, where V is either a rectangle or a ball.)

If you want to see a real proof of (1), you should study *differentiable manifolds* and *differential forms*. The next section gives a very brief introduction to differential forms in R^3, and shows how they unify the theorems named for Green, Stokes, and Gauss. The references in §4.1 develop the theory rather thoroughly. Another good reference is J. Woll, *Functions of Several Variables* (Harcourt, Brace, 1966).

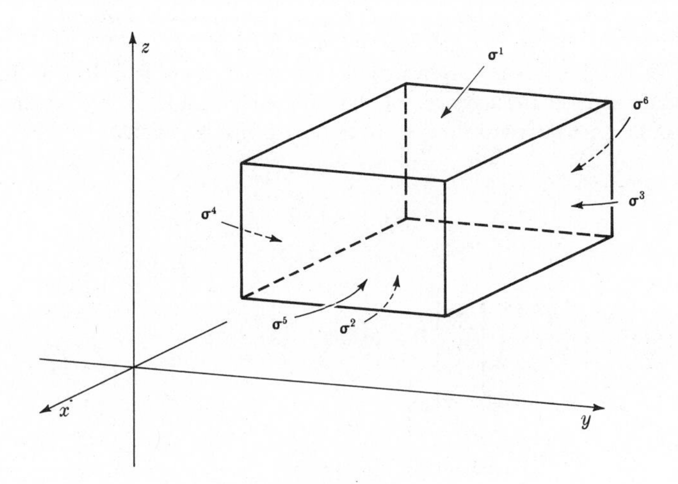

FIGURE 5.30

Example 1. Let $\mathbf{F}(x,y,z) = (0,0,z)$, and let

$$V = \{(x,y,z): a \leq x \leq b, c \leq y \leq d, e \leq z \leq f\}.$$

The oriented boundary of V consists of the six rectangles $\boldsymbol{\sigma}^1, \ldots, \boldsymbol{\sigma}^6$ in Fig. 30. Since $\mathbf{N}_{\sigma^j} \cdot \mathbf{F} = 0$ for each surface except $\boldsymbol{\sigma}^1$ and $\boldsymbol{\sigma}^2$, the integral over the boundary of V is $\iint_{\sigma^1}\mathbf{F} + \iint_{\sigma^2}\mathbf{F}$. Along $\boldsymbol{\sigma}^2$ we have $z = e$, so $\mathbf{F}$ has the constant value $(0,0,e)$ on $\boldsymbol{\sigma}^2$. Further, the unit normal on $\boldsymbol{\sigma}^2$ is $\mathbf{N} = (0,0,-1)$; hence (see Problem 4, §5.5) $\iint_{\sigma^2}\mathbf{F} = \mathbf{N} \cdot (0,0,e) \times (\text{area of } \boldsymbol{\sigma}^2)$. Evaluating $\iint_{\sigma^1}$ analogously, we find

$$\iint_{\sigma^1} \mathbf{F} + \int_{\sigma^2} \mathbf{F} = f \cdot (\text{area of } \boldsymbol{\sigma}^1) - e \cdot (\text{area of } \boldsymbol{\sigma}^2)$$

$$= -e(b-a)(d-c) + f(b-a)(d-c)$$

$$= (f-e)(b-a)(d-c) = \text{volume of } V.$$

On the other hand, the divergence of $\mathbf{F}$ is $\nabla \cdot \mathbf{F} = 1$, so

$$\iiint_V \nabla \cdot \mathbf{F} = \text{volume of } V.$$

Thus the divergence theorem checks in this case.

Figure 31 (a) shows the situation when $a = c = e = -1$ and $b = d = f = 1$, or

$$V = \{ (x,y,z) : |x| \leq 1, |y| \leq 1, |z| \leq 1 \}.$$

There is a flow out of the top of V, and a flow out of the bottom of V; thus it is obvious that there is a net outward flow, and $\mathbf{F}$ has positive divergence in V. In Fig. 31 (b) this is only a little less obvious; there is a flow in through the bottom and out through the top, but the outward flow is greater, so once again there is a net outward flow, and the divergence is positive.

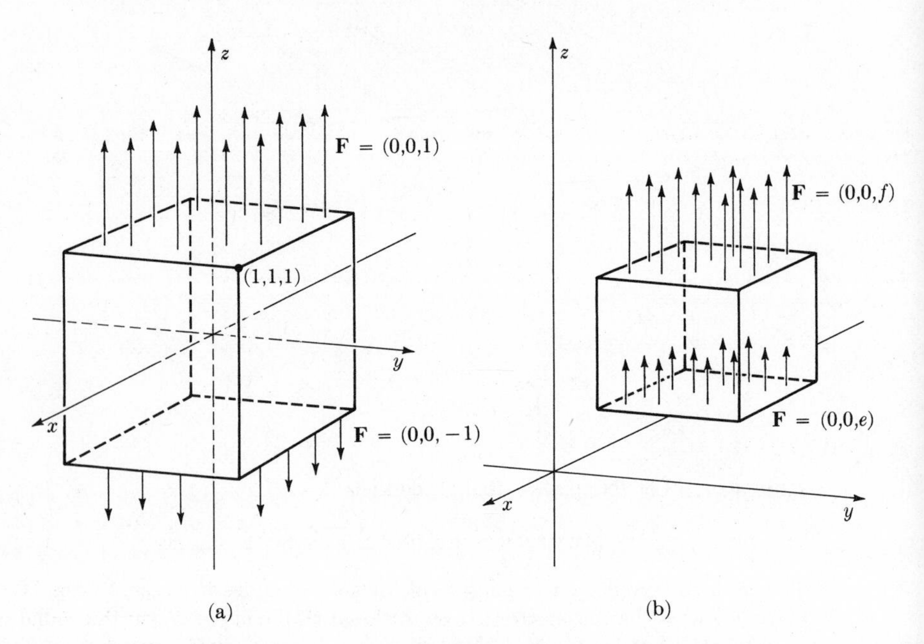

FIGURE 5.31

Example 2. We derive the so-called *equation of continuity* for fluid flows. Let $\mathbf{F}(x,y,z,t)$ denote the velocity of a fluid at the point (x,y,z) and time t, and let $\rho(x,y,z,t)$ denote the *density* of the fluid at the point (x,y,z) and time t. If the fluid is incompressible (as water is, very nearly) then ρ is constant; but if the fluid is compressible (as air is) then ρ can vary with time and position.

Consider a ball B with boundary $\boldsymbol{\sigma}$. The total mass of fluid in B at time t is $\iiint_B \rho(x,y,z,t)dx\,dy\,dz$. By Leibniz' rule (Problem 7, §5.6), the rate of increase of mass in B is

$$\frac{d}{dt}\iiint_B \rho(x,y,z,t)dx\,dy\,dz = \iiint_B \frac{\partial\rho}{\partial t}. \tag{2}$$

On the other hand, unless mass is being produced in B by some mysterious process, then the change of mass in B is due solely to the flow in and out through the boundary $\boldsymbol{\sigma}$. The integral $\iint_\sigma \mathbf{F}$ gives the volume rate of flow out of B; the rate of flow of *mass*, on the other hand, is given by $\iint_\sigma \rho\mathbf{F}$. (You can simply accept this as a proper definition from the point of view of physics, or you can explain it by modifying the discussion based on Fig. 19 in §5.5 so as to take the density into account.) Since the rate of flow of mass *out* of B is $\iint_\sigma \rho\mathbf{F}$, the rate of *increase* of mass must be $-\iint_\sigma \rho\mathbf{F}$. By the divergence theorem, then, the rate of increase of mass in B is given by

$$-\iint_\sigma \rho\mathbf{F} = -\iiint_B \nabla\cdot(\rho\mathbf{F}).$$

Comparing this to (2), we get

$$\iiint_B \frac{\partial\rho}{\partial t} = -\iiint_B \nabla\cdot(\rho\mathbf{F}).$$

Since this equation must hold for *every* ball B, it follows (see Problem 10(a), §5.6) that

$$\frac{\partial\rho}{\partial t} = -\nabla\cdot(\rho\mathbf{F}).$$

This is the equation of continuity. In particular, if the fluid is incompressible (i.e. if ρ is constant) we get $\partial\rho/\partial t = 0$ and $\nabla\cdot(\rho\mathbf{F}) = \rho(\nabla\cdot\mathbf{F})$; hence $\nabla\cdot\mathbf{F} = 0$. Thus, for an incompressible fluid, the divergence is always zero.

PROBLEMS

1. Compute the divergence of the following vector fields in R^3. Try to visualize the flow, and reconcile your answers physically.
(a) $\mathbf{F}(x,y,z) \equiv (1,0,0)$
(b) $\mathbf{F}(x,y,z) = (x,0,0)$
(c) $\mathbf{F}(x,y,z) = (0,x,0)$
(d) $\mathbf{F}(\mathbf{P}) = \mathbf{P}$
(e) $\mathbf{F}(\mathbf{P}) = \dfrac{1}{|\mathbf{P}|^2}\mathbf{P}$
(f) $\mathbf{F}(\mathbf{P}) = |\mathbf{P}|^r\mathbf{P}$, r a real number. For which value of r is $\nabla\cdot\mathbf{F} \equiv 0$?

2. Let V be the ball of radius r, $V = \{\mathbf{P}: |\mathbf{P}| \le r\}$, and

$$\boldsymbol{\sigma}(u,v) = (r\sin u\cos v,\ r\sin u\sin v,\ r\cos u),\quad 0 \le u \le \pi,\ 0 \le v \le 2\pi.$$

For each of the following vector fields, compute $\iiint_V \nabla\cdot\mathbf{F}$ and $\iint_\sigma \mathbf{F}$, and compare the results. Try to visualize the flow, and reconcile the value of $\iint_\sigma \mathbf{F}$ physically.
(a) $\mathbf{F}(x,y,z) \equiv (1,0,0)$
(b) $\mathbf{F}(x,y,z) = (x,0,0)$
(c) $\mathbf{F}(x,y,z) = (0,x,0)$
(d) $\mathbf{F}(\mathbf{P}) = \mathbf{P}$

3. In this problem you prove the divergence theorem for the rectangular solid

$$V = \{(x,y,z): x_1 \leq x \leq x_2 , y_1 \leq y \leq y_2 , z_1 \leq z \leq z_2\},$$

where x_1, x_2, etc. are constants. The *positively oriented* boundary of V consists of the following six faces $\boldsymbol{\sigma}^1, \ldots, \boldsymbol{\sigma}^6$ (Fig. 30):

$$\begin{aligned}
\boldsymbol{\sigma}^1(u,v) &= (u,v,z_2), & x_1 \leq u \leq x_2 , y_1 \leq v \leq y_2 \\
\boldsymbol{\sigma}^2(u,v) &= (v,u,z_1), & y_1 \leq u \leq y_2 , x_1 \leq v \leq x_2 \\
\boldsymbol{\sigma}^3(u,v) &= (v,y_2,u), & z_1 \leq u \leq z_2 , x_1 \leq v \leq x_2 \\
\boldsymbol{\sigma}^4(u,v) &= (u,y_1,v), & x_1 \leq u \leq x_2 , z_1 \leq v \leq z_2 \\
\boldsymbol{\sigma}^5(u,v) &= (x_2,u,v), & y_1 \leq u \leq y_2 , z_1 \leq v \leq z_2 \\
\boldsymbol{\sigma}^6(u,v) &= (x_1,v,u), & z_1 \leq u \leq z_2 , y_1 \leq v \leq y_2 .
\end{aligned}$$

(a) Check that the "standard normal" on $\boldsymbol{\sigma}^1$ points up, and on $\boldsymbol{\sigma}^2$ points down; on $\boldsymbol{\sigma}^3$ points toward positive y, and on $\boldsymbol{\sigma}^4$ toward negative y; on $\boldsymbol{\sigma}^5$ toward positive x, and on $\boldsymbol{\sigma}^6$ toward negative x. In other words, all the normals point out of V. (Hint: Sketch the cube, and on each face sketch $\boldsymbol{\sigma}_u$ and $\boldsymbol{\sigma}_v$.)

(b) Check that

$$\iiint_V \frac{\partial f_1}{\partial x}\, dx\, dy\, dz = \iint_{\sigma^5} (f_1,0,0) + \iint_{\sigma^6} (f_1,0,0)$$

$$= \sum_{j=1}^{6} \iint_{\sigma^j} (f_1,0,0).$$

(In the last step you have to show that $\iint_{\sigma^j} (f_1,0,0) = 0$ for $j = 1, 2, 3, 4$.)

(c) Check that

$$\iiint_V \frac{\partial f_2}{dy}\, dx\, dy\, dz = \sum_{j=1}^{6} \iint_{\sigma^j} (0, f_2,0).$$

(d) Check that

$$\iiint_V \frac{\partial f_3}{\partial z}\, dx\, dy\, dz = \sum_{j=1}^{6} \iint_{\sigma^j} (0,0, f_3).$$

(Adding up the results in (b)–(d) gives

$$\iiint_V \nabla \cdot \mathbf{F} = \sum_{j=1}^{6} \iint_{\sigma^j} \mathbf{F}.)$$

4. This problem proves the divergence theorem for a ball

$$V = \{\mathbf{P}: |\mathbf{P} - \mathbf{P}_0| \leq r\}.$$

The positively oriented boundary of V can be given parametrically by

$$\boldsymbol{\sigma}(u,v) = (x_0 + r \sin u \cos v,\ y_0 + r \sin u \sin v,\ z_0 + r \cos u),$$
$$0 \leq u \leq \pi,\ 0 \leq v \leq 2\pi,$$

where $\mathbf{P}_0 = (x_0,y_0,z_0)$.

(a) Check that $(\boldsymbol{\sigma}_u \times \boldsymbol{\sigma}_v) \cdot (\boldsymbol{\sigma} - \mathbf{P}_0) \geq 0$, in other words, the standard normal points out of V.

(b) Check that

$$\iiint_V \frac{\partial f_3}{\partial z}\, dz\, dy\, dx = \iint_S f_3(x,y,\sqrt{r^2 - x^2 - y^2})\, dy\, dx$$
$$- \iint_S f_3(x,y,-\sqrt{r^2 - x^2 - y^2})\, dy\, dx$$
$$= \iint_{\sigma^1} (0,0,f_3) + \iint_{\sigma^2} (0,0,f_3)$$
$$= \iint_{\sigma} (0,0,f_3),$$

where S is the disk $\{(x,y): (x - x_0)^2 + (y - y_0)^2 \leq r^2\}$, and $\boldsymbol{\sigma}^1$ is the part of the surface $\boldsymbol{\sigma}$ where u is restricted to $0 \leq u \leq \pi/2$, and $\boldsymbol{\sigma}^2$ the part where $\pi/2 \leq u \leq \pi$. (Use formula (9), §4.5.)

(c) Check that $\displaystyle\iiint_V \frac{\partial f_1}{\partial x}\, dx\, dy\, dz = \iint_\sigma (f_1, 0, 0)$.

(d) Check that $\displaystyle\iiint_V \frac{\partial f_2}{\partial y}\, dy\, dz\, dx = \iint_\sigma (0, f_2, 0)$.

(Adding up parts (b)–(d) gives $\displaystyle\iiint_V \nabla \cdot \mathbf{F} = \iint_\sigma \mathbf{F}$.)

(*Remark:* The preceding two problems show that for an explicitly given V, it is often possible to say explicitly what we mean by the positively oriented boundary of V, and then to prove the divergence theorem. In fact, if we allow "proof by picture," then it is possible to "prove" the divergence theorem for a general class of regions V, following the lines of the preceding two problems. These proofs generally rely on a picture to show you what the boundary of V is.)

5. Suppose that $\iint_\sigma \mathbf{F} = 0$ for every sphere $\boldsymbol{\sigma}$ as in Problem 4. Prove that $\nabla \cdot \mathbf{F} \equiv 0$.

6. (a) Suppose that the components of $\mathbf{F}$ have continuous first derivatives, and that ρ has continuous first derivatives. Prove that $\nabla \cdot (\rho \mathbf{F}) = (\nabla \rho) \cdot \mathbf{F} + \rho(\nabla \cdot \mathbf{F})$.

(b) Suppose that f has continuous second derivatives, and g has continuous first derivatives. Prove that $\nabla \cdot (g \nabla f) = \nabla g \cdot \nabla f + g \Delta f$, where $\Delta f = f_{xx} + f_{yy} + f_{zz}$ is the *Laplacian* of f.

(c) Let $\boldsymbol{\sigma}$ be the oriented boundary of V, and prove *Green's formula* $\iiint_V (g\Delta f + \nabla g \cdot \nabla f) = \iint_\sigma g \nabla f$. (Assume the divergence theorem for V.)

(d) Let S be the parameter domain of a surface $\boldsymbol{\sigma}$, and show that

$$\iint_\sigma g \nabla f = \iint_S g \frac{\partial f}{\partial n} |\mathbf{N}_\sigma|,$$

where $\partial f / \partial n$ is the *outer normal derivative* of f, the directional derivative of f in the direction of the unit outer normal

$$\frac{1}{|\mathbf{N}_\sigma|} \mathbf{N}_\sigma .$$

(e) Suppose that $\Delta f \equiv 0$ in V, and $f \equiv 0$ on $\boldsymbol{\sigma}$. Prove that $\nabla f \equiv 0$ in V, f is constant in V, and in fact $f \equiv 0$ in V. (Assume V connected.)

(f) Suppose that $\Delta f \equiv 0$ in V, and the normal derivative $\partial f / \partial n \equiv 0$ on $\boldsymbol{\sigma}$. Prove that f is constant in V. (Assume V connected.)

(g) Assuming that f and g have continuous second derivatives, prove that

$$\iiint_V (g\Delta f - f\Delta g) = \iint_\sigma (g \nabla f - f \nabla g)$$

$$= \iint_S \left(g \frac{\partial f}{\partial n} - f \frac{\partial g}{\partial n} \right) |\mathbf{N}_\sigma|.$$

7. (a) Suppose that the components of $\mathbf{F}$ have continuous second derivatives. Prove that $\nabla \cdot (\nabla \times \mathbf{F}) = 0$.

(b) Show that there is *no* vector field $\mathbf{F}$ such that $(\nabla \times \mathbf{F})(x,y,z) = (x,0,0)$.

(c) Find, if you can, a vector field $\mathbf{F}$ such that

$$(\nabla \times \mathbf{F})(x,y,z) = (y,0,0).$$

8. Suppose that $\nabla \cdot \mathbf{G} = 0$ in an open rectangle R containing the origin $\mathbf{0}$. This problem shows that there is a vector field $\mathbf{F}$ defined in R such that $\nabla \times \mathbf{F} = \mathbf{G}$.

(a) Show that $\nabla \times \mathbf{F} = \mathbf{G}$ is equivalent to the system of equations

$$-D_3 f_2 + D_2 f_3 = g_1 \tag{3_1}$$

$$D_3 f_1 - D_1 f_3 = g_2 \tag{3_2}$$

$$-D_2 f_1 + D_1 f_2 = g_3 . \tag{3_3}$$

(b) Set $f_3 = 0$, and show that equations (3_1) and (3_2) are equivalent to

$$\begin{aligned} f_1(x,y,z) &= \int_0^z g_2(x,y,t)\,dt + \varphi_1(x,y) \\ f_2(x,y,z) &= -\int_0^z g_1(x,y,t)\,dt + \varphi_2(x,y) \end{aligned} \tag{4}$$

for some functions φ_1 and φ_2 having continuous partial derivatives.

(c) Let f_1 and f_2 be defined by (4), and show that

$$\begin{aligned} -D_2 f_1(x,y,z) + D_1 f_2(x,y,z) &= g_3(x,y,z) - g_3(x,y,0) \\ &\quad + D_1\varphi_2(x,y) - D_2\varphi_1(x,y). \end{aligned}$$

(Recall that $\nabla \cdot \mathbf{G} = 0$.)

(d) Show that the functions φ_1 and φ_2 in (4) can be chosen so that $D_1\varphi_2(x,y) - D_2\varphi_1(x,y) = g_3(x,y,0)$. (Actually, you can take $\varphi_1 = 0$.)

(e) Conclude that there is a vector field $\mathbf{F}$ such that $\nabla \times \mathbf{F} = \mathbf{G}$.

5.8 A VERY BRIEF INTRODUCTION TO DIFFERENTIAL FORMS

Differential forms are the ultimate in Leibniz notation; they reduce to a single formula the theorems of Green, Stokes, and Gauss, as well as the fact that $\int_\gamma \nabla f = f(\boldsymbol{\gamma}(b)) - f(\boldsymbol{\gamma}(a))$; and they make the formula for change of variable look as "obvious" in n dimensions as in 1.

A *differential* 1-*form* looks like

$$f_1\,dx + f_2\,dy + f_3\,dz,$$

where f_1, f_2, f_3 are continuous real-valued functions on R^3. A differential 1-form can be integrated over a differentiable curve $\boldsymbol{\gamma}$, by the formula

$$\int_\gamma (f_1\,dx + f_2\,dy + f_3\,dz) = \int_a^b \left(f_1 \frac{d\gamma_1}{dt} + f_2 \frac{d\gamma_2}{dt} + f_3 \frac{d\gamma_3}{dt} \right) dt.$$

(This is the same as the line integral $\int_\gamma \mathbf{F}$ of the vector field $\mathbf{F} = (f_1, f_2, f_3)$.)

A *differential* 2-*form* looks like

$$f_1\,dy\,dz + f_2\,dz\,dx + f_3\,dx\,dy;$$

it can be integrated over a differentiable surface $\boldsymbol{\sigma}$ with standard normal vector $\mathbf{N}_\sigma$, by the formula

$$\iint_\sigma f_1\,dy\,dz + f_2\,dz\,dx + f_3\,dx\,dy = \iint_S \mathbf{F}\cdot\mathbf{N}_\sigma = \int_\sigma \mathbf{F}$$

where S is the parameter domain of $\boldsymbol{\sigma}$, and $\mathbf{F} = (f_1, f_2, f_3)$. A *differential 3-form* looks like $f\,dx\,dy\,dz$, where f is a function on R^3. If f is continuous, the 3-form can be integrated over a set V in R^3, simply as $\iiint_V f\,dx\,dy\,dz$. Finally, a 0-form is just a function; it is not to be integrated at all.

In other words, if a vector field $\mathbf{F} = (f_1, f_2, f_3)$ is destined to be integrated over a curve, we write it in the form $f_1\,dx + f_2\,dy + f_3\,dz$ and call it a 1-form. Similarly, if $\mathbf{F}$ is destined to be integrated over a surface, we write it $f_1\,dy\,dz + f_2\,dz\,dx + f_3\,dx\,dy$ and call it a 2-form. This alone is not a very exciting advance beyond the older notation, but now the fun begins. We define a *multiplication* of the symbols dx, dy, and dz, satisfying the strange rules

$$\left.\begin{aligned} dx\,dy &= -dy\,dx \\ dx\,dz &= -dz\,dx \\ dy\,dz &= -dz\,dy \end{aligned}\right\} \tag{1}$$

$$dx\,dx = dy\,dy = dz\,dz = 0. \tag{2}$$

(Notice the similarity with cross products.) We then extend this multiplication to differential forms by decreeing that the usual associative and distributive laws shall hold, and that *functions* commute with the symbols dx, dy, and dz, i.e. $f\,dx = (dx)f$. Thus for example, we multiply the 1-form $f_1\,dx + f_2\,dy + f_3\,dz$ by the 2-form $g\,dx\,dy$ by applying (1) and (2) as follows:

$$\begin{aligned} (f_1\,dx + f_2\,dy + f_3\,dz)\cdot(g\,dx\,dy) & \\ &= f_1g\,dx\,dx\,dy + f_2g\,dy\,dx\,dy + f_3g\,dz\,dx\,dy \\ &= 0 - f_2g\,dx\,dy\,dy - f_3g\,dx\,dz\,dy \\ &= 0 + 0 + f_3g\,dx\,dy\,dz. \end{aligned}$$

Further, we define *exterior differentiation d*, beginning with the exterior derivative of a function:

$$df = \frac{\partial f}{\partial x}\,dx + \frac{\partial f}{\partial y}\,dy + \frac{\partial f}{\partial z}\,dz. \tag{3}$$

The exterior derivative of a 1-form

$$f_1\,dx + f_2\,dy + f_3\,dz$$

is taken by applying d to each of the components f_1, f_2, f_3, then combining the terms by using the rules (1) and (2):

$$
\begin{aligned}
d(f_1\,dx + f_2\,dy + f_3\,dz) &= d(f_1\,dx) + d(f_2\,dy) + d(f_3\,dz)\\
&= \left(\frac{\partial f_1}{\partial x}\,dx + \frac{\partial f_1}{\partial y}\,dy + \frac{\partial f_1}{\partial z}\,dz\right) dx\\
&\quad + \left(\frac{\partial f_2}{\partial x}\,dx + \frac{\partial f_2}{\partial y}\,dy + \frac{\partial f_2}{\partial z}\,dz\right) dy\\
&\quad + \left(\frac{\partial f_3}{\partial x}\,dx + \frac{\partial f_3}{\partial y}\,dy + \frac{\partial f_3}{\partial z}\,dz\right) dz\\
&= 0 - \frac{\partial f_1}{\partial y}\,dx\,dy + \frac{\partial f_1}{\partial z}\,dz\,dx\\
&\quad + \frac{\partial f_2}{\partial x}\,dx\,dy + 0 - \frac{\partial f_2}{\partial z}\,dy\,dz\\
&\quad - \frac{\partial f_3}{\partial x}\,dz\,dx + \frac{\partial f_3}{\partial y}\,dy\,dz + 0.
\end{aligned}
$$

Hence $d(f_1\,dx + f_2\,dy + f_3\,dz)$

$$
= \left(\frac{\partial f_3}{\partial y} - \frac{\partial f_2}{\partial z}\right) dy\,dz + \left(\frac{\partial f_1}{\partial z} - \frac{\partial f_3}{\partial x}\right) dz\,dx + \left(\frac{\partial f_2}{\partial x} - \frac{\partial f_1}{\partial y}\right) dx\,dy. \tag{4}
$$

Working in the same way, we get the exterior derivative of a 2-form,

$$
\begin{aligned}
&d(f_1\,dy\,dz + f_2\,dz\,dx + f_3\,dx\,dy)\\
&\qquad = \left(\frac{\partial f_1}{\partial x}\,dx + \frac{\partial f_1}{\partial y}\,dy + \frac{\partial f_1}{\partial z}\,dz\right) dy\,dz + \cdots\\
&\qquad = \frac{\partial f_1}{\partial x}\,dx\,dy\,dz + 0 + 0\\
&\qquad\quad + 0 + \frac{\partial f_2}{\partial y}\,dy\,dz\,dx + 0\\
&\qquad\quad + 0 + 0 + \frac{\partial f_3}{\partial z}\,dz\,dx\,dy.
\end{aligned}
$$

By applying the rules (1) twice to each of the last two terms, this reduces to

$$
d(f_1\,dy\,dz + f_2\,dz\,dx + f_3\,dx\,dy) = \left(\frac{\partial f_1}{\partial x} + \frac{\partial f_2}{\partial y} + \frac{\partial f_3}{\partial z}\right) dx\,dy\,dz. \tag{5}
$$

Now the expressions on the right in (3), (4), and (5) are quite familiar; (3) gives the components of ∇f, (4) gives the components of $\nabla \times \mathbf{F}$, and (5) gives $\nabla \cdot \mathbf{F}$. Thus the various formulas we developed take the form:

$$\int_{\gamma} df = f(\boldsymbol{\gamma}(b)) - f(\boldsymbol{\gamma}(a))$$

$$\iint_{\sigma} d\boldsymbol{\omega}^1 = \int_{\gamma} \boldsymbol{\omega}^1$$ (Stokes' theorem; $\boldsymbol{\omega}^1 = f_1\,dx + f_2\,dy + f_3\,dz$ is a 1-form, and $\boldsymbol{\gamma}$ is the oriented boundary of $\boldsymbol{\sigma}$.)

$$\iiint_{V} d\boldsymbol{\omega}^2 = \iint_{\sigma} \boldsymbol{\omega}^2$$ (Divergence theorem; $\boldsymbol{\omega}^2 = f_1\,dy\,dz + f_2\,dz\,dx + f_3\,dx\,dy$ is a 2-form, and $\boldsymbol{\sigma}$ is the oriented boundary of V.)

These formulas can be further unified by letting $\boldsymbol{\sigma}^k$ stand for an oriented k-dimensional "surface," and dropping the notational distinction between $\int$, $\iint$, and $\iiint$. Then everything is given in the single equation

$$\int_{\sigma^k} d\boldsymbol{\omega}^{k-1} = \int_{\sigma^{k-1}} \boldsymbol{\omega}^{k-1},$$

where $\boldsymbol{\sigma}^{k-1}$ forms the oriented boundary of $\boldsymbol{\sigma}$. In the case $k = 1$, the 0-form $\boldsymbol{\omega}^{k-1}$ is a function f, $\boldsymbol{\sigma}^k$ is a curve $\boldsymbol{\gamma}$, the oriented boundary $\boldsymbol{\sigma}^{k-1}$ consists of the two points $\boldsymbol{\gamma}(b)$ taken as "+" and $\boldsymbol{\gamma}(a)$ taken as "−", and the integral

$$\int_{\sigma^{k-1}} \boldsymbol{\omega}^{k-1} \quad \text{stands for} \quad f(\boldsymbol{\gamma}(b)) - f(\boldsymbol{\gamma}(a)).$$

We hope that this sketchy outline, together with the problems below, will suggest the great value of differential forms in unifying old results and suggesting new ones. A real understanding, however, requires a thorough development of the subject. This is given, for example, in the references mentioned at the bottom of page 892.

PROBLEMS

1. This problem shows how the rules for multiplying differentials lead directly to the formulas for change of variable in double and triple integrals.
   (a) Let $x = \varphi(u,v)$, $y = \psi(u,v)$, and use (1), (2), and (3) to show that $dx\,dy = (\varphi_u\psi_v - \varphi_v\psi_u)\,du\,dv$.
   (b) Let $x = \varphi(u,v,w)$, $y = \psi(u,v,w)$, $z = \eta(u,v,w)$, and show that $dx\,dy\,dz = \dfrac{\partial(\varphi,\psi,\eta)}{\partial(u,v,w)}\,du\,dv\,dw$, where $\partial(\varphi,\psi,\eta)/\partial(u,v,w)$ is the *Jacobian determinant*

$$\begin{vmatrix} \varphi_u & \varphi_v & \varphi_w \\ \psi_u & \psi_v & \psi_w \\ \eta_u & \eta_v & \eta_w \end{vmatrix}.$$

(*Caution:* The rules for differentials involve a certain orientation of R^3, which can be reversed by changing the order of the variables; for example, by the rules for multiplying differentials, $dy\,dx\,dz = -dx\,dy\,dz$. Hence, in problems where the orientation is not involved (e.g. computing a volume or a moment of inertia) the change of variable formula uses the *absolute value* of the Jacobian determinant.)

(c) Let $x = \rho \sin \varphi \cos \theta$, $y = \rho \sin \varphi \sin \theta$, $z = \rho \cos \varphi$; here, ρ, φ, and θ are *spherical coördinates* of (x,y,z). Show that $dx\,dy\,dz = \rho^2 \sin \varphi\, d\rho\, d\varphi\, d\theta$.

(d) Let V be a sphere of radius R about the origin, and let $f(\mathbf{P}) = |\mathbf{P}|^2$. Evaluate $\int\int\int_V f$ by using spnerical coordinates. (Compare this to the calculation of $\int\int\int_V (x^2 + y^2 + z^2)\, dx\, dy\, dz$ in §5.6.)

2. (a) Suppose that f has continuous second derivatives. Show that $d(df) = 0$.

(b) Suppose that f_1, f_2, f_3 have continuous second derivatives. Show that $d(d(f_1\, dx + f_2\, dy + f_3\, dz)) = 0$.
(These formulas are equivalent to

$$\nabla \times (\nabla f) = \mathbf{0} \text{ and } \nabla \cdot (\nabla \times \mathbf{F}) = 0.$$

3. This problem shows how the notation $\int\int f_1\, dy\, dz + f_2\, dz\, dx + f_3\, dx\, dy$ for a surface integral, together with the rules for multiplying differentials and the formula

$$df(u,v) = \frac{\partial f}{\partial u} du + \frac{\partial f}{\partial v} dv,$$

leads to the expression $\int\int_S (\mathbf{F} \circ \boldsymbol{\sigma}) \cdot (\boldsymbol{\sigma}_u \times \boldsymbol{\sigma}_v)$ for surface integrals.

(a) Suppose $\boldsymbol{\sigma} = (\sigma_1, \sigma_2, \sigma_3)$ has parameter domain S. Then the surface is given by the parametric equations

$$x = \sigma_1(u,v), \qquad y = \sigma_2(u,v), \qquad z = \sigma_3(u,v),$$

so

$$dx = \frac{\partial \sigma_1}{\partial u} du + \frac{\partial \sigma_1}{\partial v} dv.$$

Find analogous expressions for dy and dz.

(b) Multiply the expressions for dx, dy, and dz, using the rules

$$dv\, du = -du\, dv, \qquad du\, du = 0, \qquad dv\, dv = 0,$$

and thus obtain $f_1\, dy\, dz + f_2\, dz\, dx + f_3\, dx\, dy$ as $\mathbf{F} \cdot (\boldsymbol{\sigma}_u \times \boldsymbol{\sigma}_v)\, du\, dv$.

Appendices

Numbers

The first numbers were the positive integers (for counting prehistoric sheep, perhaps). But necessity is the mother of invention, and as civilization developed, the needs of surveying and commerce gave birth to the *rational numbers*, the basis of all numerical computations since the days of ancient Babylon and Egypt, if not earlier.

The development of geometry was based, naturally, on the rational numbers. By 500 B.C. the two fundamental facts of plane geometry [illustrated in Fig. 1] were well known.

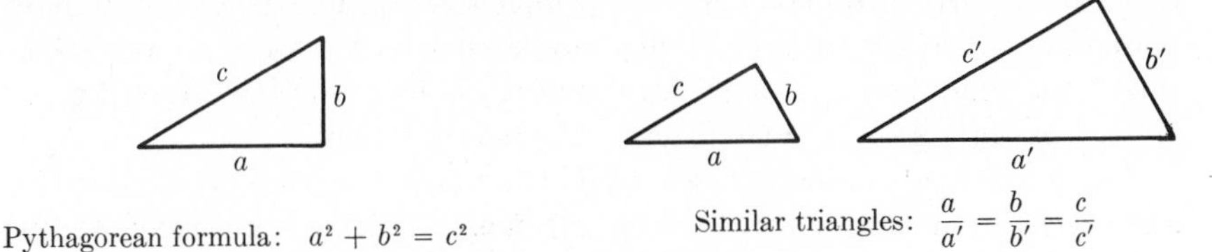

Pythagorean formula: $a^2 + b^2 = c^2$ Similar triangles: $\frac{a}{a'} = \frac{b}{b'} = \frac{c}{c'}$

FIGURE AI.1

The "Pythagorean formula" $a^2 + b^2 = c^2$ was known even at the time of Hammurabi (1950 B.C.), but is named for the legendary founder of a school of mathematical Greek mystics; perhaps Pythagoras explained the formula on geometric grounds, rather than merely accepting it as a fact of life. These same Pythagoreans discovered that *no rational number p/q satisfies the relation* $(p/q)^2 = 2$. To prove it, consider the decomposition of p and q into primes. Let $p = p_1^{r_1} \cdots p_m^{r_m}$ and $q = q_1^{s_1} \cdots q_n^{s_n}$. If $(p/q)^2 = 2$, then $p^2 = 2q^2$, so

$$p_1^{2r_1} \cdots p_m^{2r_m} = 2q_1^{2s_1} \cdots q_n^{2s_n}.$$

But this is impossible, since each prime on the left occurs an even number of times, and on the right, the prime 2 occurs an odd number of times.

This was a shocking discovery, since the Pythagorean theorem *requires* a square root of 2; a 1×1 square has a diagonal d satisfying $d^2 = 1^2 + 1^2 = 2$, so its length d would *have* to be a square root of 2.

The reactions to this crisis were quite varied. Some mystics took vows of secrecy, attempting in vain to prevent the spread of the dismal discovery. Surveyors could afford to ignore it, since it is easy to find rational numbers p/q with $(p/q)^2$ satisfactorily close to 2. But those who were developing a theory of geometry could not ignore it; they abandoned the "naive" idea that lengths are given by numbers. The law of similar triangles could no longer be viewed as a simple equality between numbers, but was explained by an elegant (and rather complicated) theory of "proportions." The elegance of this theory appealed to philosophers, but the complications must have divorced theory from practice, possibly inhibiting the development of science on a mathematical basis for a long time.

Gradually, however, a less restricted idea of number grew up, somewhat informally at first. In the nineteenth century this development was codified and it was shown that the resulting system, the *real numbers*, could be viewed as a natural extension of the rational numbers, much as the rationals extend the integers; and that the real numbers could indeed serve the needs of geometry, as the rational numbers could not.

To see how the reals extend the rationals, think of numbers as points on a line. The integers leave obvious gaps of unit length, which appear to be filled in by the rationals. However, on close examination, we find many points *not* covered by the rationals, points such as $\sqrt{2}$, $\sqrt{3}$, π, e, and so on; these remaining points [much more numerous than the few examples just given would suggest] are filled in by the real numbers.

It is possible to define the rational numbers in terms of the integers, and the real numbers in terms of the rationals. But it is not easy, and we shall not do it here. We discuss the characteristic properties of each system [natural numbers, rational numbers, real numbers] based on appropriate sets of axioms; the axioms form a minimum list of essential properties, and all others are derived from these. Sections 2, 4, and 5 stress the axioms, while 1 and 3 are devoted to various computations required in the main text.

I.1 SUMMATION

Sums like

$$1 + 2 + 3 + \cdots + n \tag{1}$$

or

$$1 + 2^2 + 3^2 + \cdots + n^2 \tag{2}$$

or

$$1 + r + r^2 + \cdots + r^n \tag{3}$$

arise often enough to deserve a shorthand notation. To start with a very simple example, the sum

$$1^2 + 2^2 + 3^2 + 4^2$$

is written

$$\sum_{j=1}^{4} j^2, \quad \text{or simply} \quad \sum_{1}^{4} j^2,$$

and called "the sum of j^2 from $j = 1$ to 4." The symbol $\sum$ is a Greek capital S, standing for *sum;* the letter j under the $\sum$ is called the "index of summation," the numbers 1 and 4 are the "limits of summation," and the expression j^2 is the "general term" of the sum. To evaluate the sum, substitute in the general term each value of the index of summation between the given limits, and add the resulting terms. In the present example, where the limits are $j = 1$ and $j = 4$ and the general term is j^2, we find that

$$j = 1 \text{ gives } 1^2,\ j = 2 \text{ gives } 2^2,\ j = 3 \text{ gives } 3^2,\ j = 4 \text{ gives } 4^2,$$

so

$$\sum_{j=1}^{4} j^2 = 1^2 + 2^2 + 3^2 + 4^2.$$

Some more examples:

EXAMPLES OF SUMMATION

$$\sum_{j=1}^{5} j^j = 1^1 + 2^2 + 3^3 + 4^4 + 5^5 \qquad \sum_{k=2}^{4} k^3 = 2^3 + 3^3 + 4^3$$

$$\sum_{j=0}^{4} r^j = r^0 + r^1 + r^2 + r^3 + r^4.$$

In this notation, the three sums (1)–(3) given above are written respectively, $\sum_{j=1}^{n} j$, $\sum_{j=1}^{n} j^2$, and $\sum_{j=0}^{n} r^j$.

These particular examples arise in computing areas, moments of inertia, and the like, and simple formulas for the sums would be very useful. A formula for (1) is easy to find, by adding the following two lines:

$$\sum_{j=1}^{n} j = 1 + 2 + \cdots + (n-1) + n$$

$$\sum_{j=1}^{n} j = n + (n-1) + \cdots + 2 + 1$$

$$2\sum_{j=1}^{n} j = \underbrace{(n+1) + (n+1) + \cdots + (n+1) + (n+1)}_{n \text{ terms}} = n(n+1).$$

Thus, dividing by 2,

$$\sum_{j=1}^{n} j = \frac{1}{2}n(n+1) = \frac{1}{2}n^2 + \frac{1}{2}n. \tag{4}$$

The sum (3) also has a convenient formula, which is discovered automatically in the attempt to factor $1 - r^{n+1}$. This polynomial in r vanishes when $r = 1$, so it must be divisible by $1 - r$. Carrying out the "long division," we find that the quotient is exactly the sum (3);

$$\frac{1 - r^{n+1}}{1 - r} = r^n + r^{n-1} + \cdots + 1 = \sum_{j=0}^{n} r^j.$$

Similarly, you can prove a formula used frequently in Chapter I,

$$a^n - b^n = (a - b)(a^{n-1} + a^{n-2}b + \cdots + ab^{n-2} + b^{n-1}). \tag{5}$$

The sum of squares $\sum_{j=1}^{n} j^2$ arises in computing the area bounded by a parabola [§5.1]. The trick leading to (4) does not appear to have any analog in the case of $\sum j^2$, so we are driven to an experimental approach.

Since we are dealing with sums of j^2, it is worth reviewing the simpler cases of $\sum j^1$ and $\sum j^0$:

$$\sum_{j=1}^{n} j^0 = \underbrace{1 + 1 + \cdots + 1}_{n \text{ terms}} = n$$

$$\sum_{j=1}^{n} j^1 = 1 + 2 + \cdots + n = \frac{1}{2}n^2 + \frac{1}{2}n.$$

Notice that $\sum_{j=1}^{n} j^0$ is given by a first degree polynomial in n, and $\sum_{j=1}^{n} j^1$ by a second degree polynomial in n. Two cases give scant evidence, but it is at least reasonable to guess that $\sum_{j=1}^{n} j^2$ is a third degree polynomial in n,

$$\sum_{j=1}^{n} j^2 = c_3n^3 + c_2n^2 + c_1n + c_0 .$$

If this were so, then we would have [with $n = 1$]

$$c_3 \cdot 1^3 + c_2 \cdot 1^2 + c_1 \cdot 1 + c_0 = \sum_{j=1}^{1} j^2 = 1^2$$

or

$$c_3 + c_2 + c_1 + c_0 = 1.$$

Similarly, setting $n = 2$, then $n = 3$, then $n = 4$, we get

$$c_3 \cdot 8 + c_2 \cdot 4 + c_1 \cdot 2 + c_0 = \sum_{j=1}^{2} j^2 = 5$$

$$c_3 \cdot 27 + c_2 \cdot 9 + c_1 \cdot 3 + c_0 = \sum_{j=1}^{3} j^2 = 14$$

$$c_3 \cdot 64 + c_2 \cdot 16 + c_1 \cdot 4 + c_0 = \sum_{j=1}^{4} j^2 = 30.$$

Solving these four simultaneous equations for c_0, c_1, c_2, c_3 is tedious, but not really difficult; you can check the solution

$$c_3 = 1/3, \qquad c_2 = 1/2, \qquad c_1 = 1/6, \qquad c_0 = 0.$$

Thus assuming that $\sum_{j=1}^{n} j^2$ is given by a cubic in n, it must be

$$\sum_{j=1}^{n} j^2 = \frac{1}{3}n^3 + \frac{1}{2}n^2 + \frac{1}{6}n. \tag{6}$$

From the way the coefficients were deduced, this formula is surely correct for $n = 1, 2, 3$, and 4. You will probably be convinced that it is true in general, if it also works for "randomly chosen" values of n, for example, $n = 6$:

$$\text{left-hand side} = \sum_{j=1}^{6} j^2 = 1^2 + 2^2 + \cdots + 6^2 = 91$$

$$\text{right-hand side} = \frac{1}{3}6^3 + \frac{1}{2}6^2 + \frac{1}{6}6 = 91 \qquad \text{[so it works]}$$

or $n = 10$:

$$\text{left-hand side} = \sum_{j=1}^{10} j^2 = 1^2 + 2^2 + \cdots + 10^2 = 385$$

$$\text{right-hand side} = \frac{1}{3}10^3 + \frac{1}{2}10^2 + \frac{1}{6}10 = 385 \qquad \text{[so it works].}$$

Of course, to be sure that we haven't chosen just those special cases that work, you should pick your own and try them. You would then be checking the formula by "experimental induction."

Even that would not constitute a mathematical proof. No matter *how* many special cases of a general formula are found to be correct, there is no assurance that we haven't by sheer luck avoided cases where it fails.

For example: Suppose you are asked whether, for every integer n, the number

$$p_n = n^2 + n + 41$$

is a prime number [that is, it cannot be written as a product of two positive integers $p_n = a \cdot b$, except in the trivial ways $1 \cdot p_n$ and $p_n \cdot 1$]. You would find this true for $n = 1, 2, 3, \ldots, 10$, and even higher values, up through $n = 39$. But it is actually false for $n = 40$, since

$$p_{40} = 40^2 + 40 + 41 = 40(40 + 1) + 41 = 41 \cdot 41.$$

Thus a claim that is valid for the first 39 cases fails at the fortieth. Experimental induction, though psychologically convincing, is logically inadequate.

The standard effective way to prove a formula like (6) is *mathematical* induction, outlined in the next section.

I.2 MATHEMATICAL INDUCTION, AND THE NATURAL NUMBERS

Formulas such as

$$6 \sum_{j=1}^{n} j^2 = 2n^3 + 3n^2 + n, \quad \text{for } n = 1, 2, 3, 4, \ldots \tag{1}$$

[used in §5.1 to compute an area] can generally be proved by *mathematical induction.* The proof consists in establishing two facts:

Fact (i): The formula is true for $n = 1$. This is easy to check; setting $n = 1$ in each side of the formula gives

$$\text{left-hand side} = 6 \sum_{j=1}^{1} j^2 = 6 \cdot 1^2 = 6$$

$$\text{right-hand side} = 2 \cdot 1^3 + 3 \cdot 1^2 + 1 = 6,$$

so (1) is true for $n = 1$.

Fact (ii): Supposing the formula is true for a particular integer N, then it is also true for the next integer $N + 1$.

This is checked as follows. When $n = N + 1$, the left-hand side of (1) becomes

$$6 \sum_{j=1}^{N+1} j^2 = 6 \left(\sum_{j=1}^{N} j^2 + (N + 1)^2 \right) \qquad \text{[obviously]}$$

$$= \left(6 \sum_{j=1}^{N} j^2\right) + 6N^2 + 12N + 6 \qquad \text{[algebra]}$$

$$= (2N^3 + 3N^2 + N) + 6N^2 + 12N + 6$$

[supposing that (1) is true for $n = N$]

$$= 2N^3 + 9N^2 + 13N + 6.$$

Next, setting $n = N + 1$ in the right-hand side of (1), we get

$$2(N + 1)^3 + 3(N + 1)^2 + (N + 1) = 2N^3 + 9N^2 + 13N + 6,$$

the same as the left-hand side. Thus supposing the formula is true for $n = N$, it is also true for $n = N + 1$.

It now follows that formula (1) is true for every integer n. It is true for $n = 1$, by Fact (i); then using Fact (ii) with $N = 1$ shows that the formula is true for $n = N + 1 = 2$; next, using (ii) with $N = 2$ shows that the formula is true for $n = N + 1 = 3$; and so on.

Fact (i) gives a starting place for the proof, at the "beginning" of the positive integers, and Fact (ii) carries the proof on from each integer to the next. This two-part method is called "proof by induction."

The following formulas (2) and (3) given in §I.1 can also be proved by induction.

Example 1. Prove by induction that

$$(1 - r) \sum_{j=0}^{n} r^j = 1 - r^{n+1}. \tag{2}$$

Solution. We show that

(i) The formula is true for $n = 1$:

$$\text{left-hand side} = (1 - r) \sum_{j=0}^{1} r^j = (1 - r)(r^0 + r^1) = 1 - r^2$$

$$\text{right-hand side} = 1 - r^{1+1} = 1 - r^2 \qquad [\text{it checks}].$$

(ii) *If* the formula is true for $n = N$, *then* it is true for $n = N + 1$: Beginning with the left-hand side, we get

$$(1 - r) \sum_{j=0}^{N+1} r^j = (1 - r)\left(\sum_{j=0}^{N} r^j + r^{N+1}\right)$$

$$= (1 - r) \sum_{j=0}^{N} r^j + (1 - r) r^{N+1}$$

$$= (1 - r^{N+1}) + (1 - r) r^{N+1}$$

[supposing (2) is true for $n = N$]

$$= 1 - r^{N+2}.$$

This agrees with the right-hand side of (2) for $n = N + 1$, so (ii) is proved. Facts (i) and (ii) together complete the proof by induction.

Example 2. To prove by induction that

$$2 \sum_{j=1}^{n} j = n(n + 1), \tag{3}$$

we prove:

(i) The formula is true for $n = 1$:

$$\text{left-hand side} = 2 \sum_{j=1}^{1} j = 2 \cdot 1 = 2$$

$$\text{right-hand side} = 1 \cdot (1 + 1) = 2.$$

(ii) *If* the formula is true for $n = N$, *then* it is true for $n = N + 1$: The left-hand side becomes, for $n = N + 1$,

$$\begin{aligned} 2\sum_{j=1}^{N+1} j &= 2\left(\sum_{j=1}^{N} j\right) + 2(N+1) \\ &= N(N+1) + 2(N+1) \qquad \text{[if (3) is true for } n = N\text{]} \\ &= N^2 + 3N + 2 \\ &= (N+1)(N+2), \end{aligned}$$

which agrees with the right-hand side of (3) for $n = N + 1$. Thus (i) and (ii) are established, and the proof by induction is complete.

Remark. The validity of proof by induction is clear, intuitively, but what is the formal justification? This should be based on some set of axioms. In fact, it is based on the following set of axioms for the "natural numbers" 1, 2, 3, 4,. . . .*

Axioms for the natural numbers

The natural numbers are a system with two operations, addition (+) and multiplication (×). If m and n are any positive integers then the sum $m + n$ is a well-defined integer, and the product $m \times n$ is a well-defined positive integer. These operations obey the following laws:

A +	(Associative law for +)	$l + (m + n) = (l + m) + n$
A ×	(Associative law for ×)	$l \times (m \times n) = (l \times m) \times n$
C +	(Commutative law for +)	$m + n = n + m$
C ×	(Commutative law for ×)	$m \times n = n \times m$
D	(Distributive law)	$l \times (m + n) = l \times m + l \times n$
U ×	(Multiplicative unit)	There is an integer (generally denoted 1) such that $1 \times m = m$ for every m.

Furthermore, the system satisfies the "infinity axiom":

INF for every n and m, $n + m$ is different from n.

Finally, most important of all, the *induction axiom:*

IND Let S be any set of natural numbers, and suppose that (i) 1 is in S and (ii) if N is in S, then $N + 1$ is also in S. It then follows that S must contain *all* natural numbers.

*The point of view implicit in the phrase "natural numbers" was put most emphatically by Leopold Kronecker's remark: "God created the integers, everything else is the work of man."

The properties expressed in the first six axioms (A + through U ×), which are learned in elementary school, form the basis for all algebraic manipulations with the natural numbers; this point is developed in §I.4 below. The "infinity axiom" is included to rule out smaller systems that are not the natural numbers, but nevertheless obey all the other axioms [see Problem 8 below]. The subtle induction axiom rules out larger systems [such as the rational numbers] but, more important, it provides the formal justification for proof by induction. Indeed, let S be the set of natural numbers n for which the formula in question is true [for example, the set of integers n such that formula (1) is true]. Then step (i) in the induction proof shows that 1 is in S [the formula is true for $n = 1$]; and step (ii) shows that if N is in S, then $N + 1$ is in S [if the formula is true for $n = N$, then also for $n = N + 1$]. Thus the set S contains all natural numbers [the formula is true for all natural numbers].

PROBLEMS

Most of Problems 2 and 3 are from I. S. Sominskii, *The Method of Mathematical Induction*, which is a good leisurely introduction to induction.

1. Some, *but not all*, of the following formulas are true for all positive integers. For those that are true, give a proof by induction. For those that are not, find which of the two inductive steps cannot be carried out, and find a positive integer n for which the formula is *not true*.

(a) $n = n + 1$

(b) $n = 2n - 1$

(c) $\sum_{j=1}^{n} j^2 = \frac{1}{3}(n^3 + n^2 + 1)$

(d) $\sum_{j=1}^{n} j^3 = \left[\sum_{j=1}^{n} j\right]^2$

(e) $\sum_{j=1}^{n} j = 2n - 1$

(f) $\sum_{j=1}^{n} j = \frac{1}{2}(n^2 + n + 1)$

2. Prove the following formulas by induction. Write out each sum "with dots," to see what it says.

(a) $\sum_{j=1}^{n} r^j = \dfrac{r - r^{n+1}}{1 - r}$ if $r \neq 1$

$$\left[\text{Written out ``with dots,'' } r + r^2 + \cdots + r^n = \frac{r - r^{n+1}}{1 - r},\ r \neq 1\right]$$

(b) $\sum_{j=1}^{n} jr^j = \dfrac{nr^{n+2} - (n + 1)r^{n+1} + r}{(1 - r)^2}$

(c) $\sum_{j=1}^{n} (-1)^{j-1} j^2 = (-1)^{n-1} \frac{n(n+1)}{2}$

(d) $\sum_{j=1}^{n} (2j-1)^2 = \frac{n(2n-1)(2n+1)}{3}$

(e) $\sum_{j=1}^{n} j(j+1) = \frac{n(n+1)(n+2)}{3}$

(f) $\sum_{j=1}^{n} j(j+1)(j+2) = \frac{n(n+1)(n+2)(n+3)}{4}$

(g) $\sum_{j=1}^{n} \frac{1}{(2j-1)(2j+1)} = \frac{n}{2n+1}$

(h) $\sum_{j=1}^{n} \frac{1}{(3j-2)(3j+1)} = \frac{n}{3n+1}$

(i) $\sum_{j=1}^{n} \frac{1}{(4j-3)(4j+1)} = \frac{n}{4n+1}$

(j) $\sum_{j=1}^{n} j \cdot (j!) = (n+1)! - 1$

(k) $\sum_{j=1}^{n} \sin(jx) = \frac{\sin \frac{n+1}{2} x}{\sin (x/2)} \sin \frac{nx}{2}$

[Hint: $2\left(\cos \frac{k+1}{2} x\right) \sin \frac{x}{2} = \left(\sin \frac{k+2}{2} x\right) - \sin \frac{kx}{2}$. Formula (k) is used in the study of *Fourier series.*]

(l) $\frac{1}{2} + \sum_{j=1}^{n} \cos(jx) = \frac{\sin \frac{2n+1}{2} x}{2 \sin (x/2)}$

3. Prove the following inequalities by induction.

(a) for $x > 0$, $(1+x)^m > (1+mx)$

(b) $\sum_{j=1}^{n} \frac{1}{n+j} \geq \frac{1}{2}$

(c) $\sum_{j=1}^{n} \frac{1}{j} \leq n$

(d) $\frac{4^n}{n+1} \leq \frac{(2n)!}{(n!)^2}$

4. (a) Guess at a formula for $\sum_{j=1}^{n} \frac{1}{j(j+1)}$, by writing out and simplifying the sum for $n = 1, 2, 3, 4$.
(b) Prove your formula by induction.

5. Repeat Problem 4 for $\sum_{j=1}^{n} (2j - 1)$.

6. For any numbers a, b_1, and b_2, the *distributive law* $a \cdot (b_1 + b_2) = a \cdot b_1 + a \cdot b_2$ is valid. Use this to prove by induction that for any numbers $a, b_1, \ldots, b_n$ the formula $a \cdot \sum_{j=1}^{n} b_j = \sum_{j=1}^{n} a \cdot b_j$ is valid. [Using dots, the formula is $a \cdot (b_1 + \cdots + b_n) = a \cdot b_1 + \cdots + a \cdot b_n$. Hint: $\sum_{j=1}^{k+1} b_j = (\sum_{j=1}^{k} b_j) + b_{k+1}$.]

7. In the following argument, we attempt to prove by induction that every group of students is either all male, or all female. Surely you will object to this conclusion, but what is wrong with the argument?

We argue by induction on the number of students in the group, call it n. When $n = 1$, obviously the group is either all male or all female. For step (ii) of the proof, suppose that no group with N students has mixed sexes, and suppose $S_1, S_2, \ldots, S_{N+1}$ is any group of $N + 1$ students. Out of this, form two groups of N students, namely,

$$S_1, S_2, \ldots, S_N,$$

and

$$S_2, \ldots, S_N, S_{N+1}.$$

Then, on the supposition that no group of N students is mixed, we find that $S_1, \ldots, S_N$ are either all males or all females, and $S_2, \ldots, S_{N+1}$ are likewise either all males or all females. Since S_N is in both groups, both groups have the same sex; hence all the students $S_1, \ldots, S_{N+1}$ have the same sex.

8. There is a system much simpler than the integers, the "odd-even" system. It has just two members, 1 ("odd") and 2 ("even"). The operations $+$ and $\times$ are defined as follows:

$$1 + 1 = 2 + 2 = 2, \qquad 1 + 2 = 2 + 1 = 1$$

$$1 \times 1 = 1, \qquad 1 \times 2 = 2 \times 1 = 2 \times 2 = 2.$$

[These are the familiar facts "odd + odd = even + even = even," and so on.] Prove that this system satisfies all the axioms for the integers, except for the "infinity axiom."

LAWS OF INEQUALITIES

I Reversible Operations

Addition

(1) $a \le b \iff a + c \le b + c$

Multiplication

(2) If $c > 0$, then $a \le b \iff ca \le cb$

(2′) If $c < 0$, then $a \le b \iff ca \ge cb$

Reciprocals

(3) If $a > 0$ and $b > 0$, then $a \le b \iff \dfrac{1}{a} \ge \dfrac{1}{b}$

II Non-Reversible Operations

Transitive Law

(4) $a \le b$ and $b \le c \Rightarrow a \le c$

Addition

(5) $a \le b$ and $c \le d \Rightarrow a + c \le b + d$

Multiplication

(6) $0 \le a \le b$ and $0 \le c \le d \Rightarrow ac \le bd$

Division

(7) $0 \le a \le b$ and $0 < c \le d \Rightarrow \dfrac{a}{d} \le \dfrac{b}{c}$

III Trichotomy

Given any a and b, exactly one of the following possibilities is true, and the other two are false: $a < b$, or $a = b$, or $a > b$.

IV Laws For the Absolute Value

(8) $|a| \le b \iff -b \le a \le b$

Triangle Inequality For every a and b,

(9) $|a + b| \le |a| + |b|$

Note: (1)–(8) all remain valid with $\le$ replaced by $<$ everywhere.

I.3 INEQUALITIES, AND THE RATIONAL NUMBERS

Calculus is based on the idea of limit; limits are based on the idea of approximations; and approximations require the use of inequalities.

The laws of inequalities [summarized in the table across page] are similar to the familiar algebraic manipulations with equalities. According to law (1) you can add the same number to both sides of a given inequality, or cancel the same number from both sides, and obtain an equivalent inequality. Similarly, by (2), you can multiply both sides by the same nonzero number, or cancel the same nonzero number from both sides. By (3), you can take reciprocals of both sides of an inequality between *positive* numbers. Each of these operations transforms the given inequality into a new, equivalent one. The operations are reversible in the sense that the given inequality implies the new one, and vice versa.

The nonreversible operations yield a new inequality from two given inequalities: You can add two inequalities, or multiply or divide two inequalities between positive numbers. These operations are nonreversible, since the new inequality does *not* imply the given ones.

In short, the familiar operations for equalities carry over to inequalities, *but with certain precautions:* The two laws (3) and (6) involving division hold only for *positive* numbers. Taking reciprocals *reverses the inequality* [Fig. 2]. Similarly, multiplication by a *negative* number *reverses* the inequality, as in (2′) and Fig. 3. Finally, from (7), a ratio of positive numbers is increased by increasing the numerator or *decreasing* the denominator, or both.

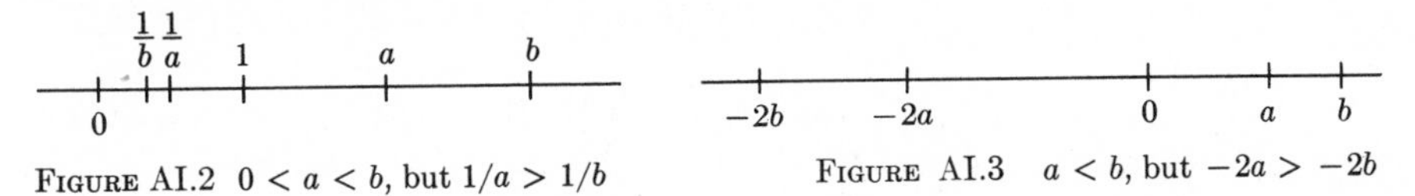

FIGURE AI.2 $0 < a < b$, but $1/a > 1/b$

FIGURE AI.3 $a < b$, but $-2a > -2b$

Example 1. Describe the set $S = \{x: -6x + 5 < 0\}$. *Solution.* Figure 4 suggests that $-6x + 5$ is negative iff $x > 5/6$; in other words, that S is the interval $(5/6, \infty)$. We prove this formally by solving the given inequality:

$$-6x + 5 < 0 \iff -6x < -5 \quad [\text{by (1) with } c = -5]$$

$$\iff x > \tfrac{5}{6} \quad [\text{by (2′) with } c = -6].$$

Hence x satisfies the given condition $-6x + 5 < 0$ iff x is in the interval $(5/6, \infty)$.

(0,5)

$(\frac{5}{6}, 0)$

$y = -6x + 5$

FIGURE AI.4

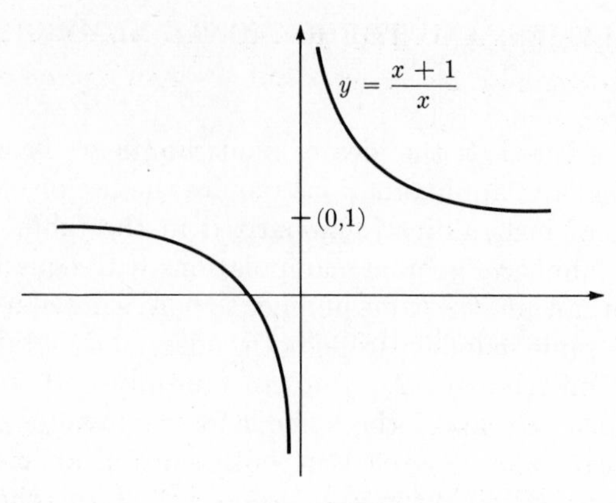

FIGURE AI.5

Example 2. Describe the set $S = \left\{x : \frac{x+1}{x} > 1\right\}$. *Solution.* The graph of $(x+1)/x$ [Fig. 5] suggests that S is simply the interval $(0,+\infty)$. To prove this formally, we might proceed as follows:

$$\frac{x+1}{x} > 1 \iff x+1 > x \qquad [\text{multiply by } x]$$

$$\iff 1 > 0. \qquad [\text{cancel } x]$$

Since $1 > 0$ is true, we conclude that $(x+1)/x > 1$ for every x, which is obviously false! The trouble lies in the first line; the claim that

$$\frac{x+1}{x} > 1 \iff x+1 > x$$

is valid *only* for $x > 0$; when $x < 0$, on the other hand, we are multiplying the inequality by a *negative* number, so the direction is reversed [see (2′)], and we get

$$\frac{x+1}{x} > 1 \iff x+1 < x \qquad [\text{if } x < 0].$$

Since $x + 1 < x$ is never true, it must be that $\frac{x+1}{x} > 1$ is false when $x < 0$.

Example 3. Suppose that $a > 0$ and $b > 0$. Prove that $a < b \iff a^2 < b^2$. *Solution.* Note that $b^2 - a^2 = (b+a)(b-a)$, and that $b + a > 0$. Thus [law (2)] multiplication by $b + a$ produces an equivalent inequality. Fitting this observation into a chain of equivalent inequalities yields

$$\begin{aligned} b > a &\iff b - a > 0 && [\text{adding } -a \text{ to both sides}] \\ &\iff (b+a)(b-a) > 0 && [\text{multiplying by } b + a] \\ &\iff b^2 - a^2 > 0 \\ &\iff b^2 > a^2. && Q.E.D. \end{aligned}$$

Many inequalities are deduced from the fact that squares are non-negative. This fact may be obvious, but we give a proof anyway, to illustrate the use of inequalities.

Theorem 1. *If $a \neq 0$, then $a^2 > 0$.*

Proof. By trichotomy, either $a > 0$ or $a < 0$ [since $a = 0$ is ruled out by the hypothesis of the theorem].

Suppose that

$$a > 0. \tag{10}$$

Then by (2) [for strict inequality, $<$], multiplying (10) by the positive number a yields

$$a^2 > a \cdot 0 = 0,$$

as desired.

Suppose on the other hand that

$$a < 0. \tag{11}$$

Then by (2′), multiplying (11) by the negative number a reverses the inequality, yielding

$$a^2 > 0.$$

Thus the theorem is true in every case.

Example 4. Prove that for $0 \leq x \leq 1$, the fraction

$$\frac{1 + x}{1 + x^2} \tag{12}$$

is ≤ 2. *Solution.* According to rule (7), a fraction involving positive numbers is increased by increasing the numerator or decreasing the denominator. Thus to see how large (12) can be, we may take the largest value of the numerator and the smallest value of the denominator. By Theorem 1, $x^2 \geq 0$ for all x, so the denominator $1 + x_2$ satisfies

$$1 + x^2 \geq 1.$$

Moreover, for the given values of x, the numerator $1 + x$ is clearly ≤ 2. Therefore [by (7)],

$$\frac{1 + x}{1 + x^2} \leq \frac{2}{1} = 2, \quad \text{for } 0 \leq x \leq 1.$$

As a final example, we prove formally that the approximations $(x^2 + c)/2x$ in Newton's method for extracting the square root of c [§1.6] are generally a little large.

Example 5. Prove that for any x and c,

$$\left(\frac{x^2 + c}{2x}\right)^2 \geq c. \tag{13}$$

Solution. We must suppose that $x \neq 0$, to avoid division by 0. Thus $(2x)^2 > 0$, [by Theorem 1], so (13) is equivalent to

$$\begin{aligned} & (x^2 + c)^2 \geq 4x^2c && \text{[by (2)]} \\ \Leftrightarrow\ & x^4 + 2x^2c + c^2 \geq 4x^2c && \text{[algebra]} \\ \Leftrightarrow\ & x^4 - 2x^2c + c^2 \geq 0 && \text{[by (1)]} \\ \Leftrightarrow\ & (x^2 - c)^2 \geq 0. && \text{[algebra]} \end{aligned}$$

But the last line is true, by Theorem 1; thus the inequality (13) is true.

Remark. One may legitimately ask, what is the status of the rules for inequalities outlined above? Are they axioms [basic assumptions about numbers] or theorems [to be proved from axioms] or what?

They are theorems, and they are proved on the basis of the axioms for an *ordered field*, a system larger and richer than the natural numbers. The next section lists these axioms, and derives the laws of inequalities from them.

PROBLEMS

1. For the given numbers a, b, and c, plot a, b, ac, and bc. Check that the result agrees with law (2) or (2′) for products.
(a) $a = 1, b = 5, c = \frac{1}{2}$
(b) $a = 1, b = 3, c = -\frac{1}{2}$
(c) $a = 1, b = -2, c = -1$
(d) $a = -1, b = -2, c = -2$

2. Repeat Problem 1 for the following choices of a, b, and c.
(a) $a = \frac{1}{2}, b = -\frac{1}{2}, c = -2$
(b) $a = -\frac{1}{3}, b = -\frac{2}{3}, c = 3$
(c) $a = -1, b = 2, c = -1$
(d) $a = 1, b = -2, c = 2$

3. For the following numbers a and b, plot a, b, $1/a$, and $1/b$. Notice that the relative positions of $1/a$ and $1/b$ are opposite to those of a and b, as required by (3).
(a) $a = 1, b = 2$
(b) $a = \frac{1}{2}, b = \frac{2}{3}$
(c) $a = 3, b = 2$
(d) $a = -1, b = -4$

4. (a) What should the law (3) for reciprocals say about $1/a$ and $1/b$ if $a < 0$ and $b < 0$?
(b) What should (3) say about $1/a$ and $1/b$ if $a < 0$ and $b > 0$?
(c) Prove the statement you gave in (a).
(d) Prove the statement you gave in (b).

5. Is it true, in general, that $a < b$ and $c < d \Rightarrow ac < bd$? If true, give a proof; if not, give examples showing that the statement is not generally true.

6. Identify each of the following sets as an interval, in two ways: (i) by sketching the appropriate graph, and (ii) by solving the given inequalities. [See Examples 1 and 2.]

(a) $\{x: 2x - 1 \leq 3\}$
(b) $\{t: -5t + 4 > 6\}$
(c) $\{s: -3s + 5 < 0\}$
(d) $\{y: 8y + 9 > 10\}$
(e) $\{x: 21x - 11 < 3\}$
(f) $\{x: 2x - 2 < 3\}$
(g) $\{x: 2 - 2x < 3\}$
(h) $\{x: 2x - 2 < 3 \text{ and } 2 - 2x < 3\}$
(i) $\{x: 2x - 5 > 3x + 1\}$
(j) $\{t: t < 2 - t < 5t\}$

7. Example 4 showed that $\dfrac{1+x}{1+x^2} \leq 2$ for $0 \leq x \leq 1$. Show similarly that $\dfrac{1+x}{1+x^2} \geq \dfrac{1}{2}$ for $0 \leq x \leq 1$. [The number 2 is called an *upper bound* for the function $f(x) = \dfrac{1+x}{1+x^2}$ on the interval [0,1], and 1/2 is called a *lower bound*].

UPPER, LOWER BOUND FOR FUNCTION

8. Find upper and lower bounds for $-\dfrac{1+x}{1+x^2}$ on the interval $0 \leq x \leq 1$. [Hint: From Problem 7, $\dfrac{1}{2} \leq \dfrac{1+x}{1+x^2} \leq 2$ for the given values of x. What happens if these inequalities are multiplied by -1?]

9. Suppose that $m \leq f(x) \leq M$ for $a \leq x \leq b$.

(a) Find an upper bound for $-f$ on $[a,b]$.
(b) Find a lower bound for $-f$ on $[a,b]$.

[Hint: See law (2′). Compare Problem 8.]

10. Suppose that $m \leq f(x) \leq M$ for $a \leq x \leq b$, where $m > 0$.

(a) Define g by $g(x) = 1/f(x)$, $a \leq x \leq b$. Find upper and lower bounds for g on $[a,b]$. [Compare Problem 9; see law (3).]
(b) Find upper and lower bounds for $h(x) = [f(x)]^2$ on $[a,b]$.
(c) Repeat part (a) if $m < 0$ and $M < 0$.
(d) Repeat part (b) if $m < 0$ and $M < 0$.

11. Suppose that f is strictly increasing on an interval I; thus, for a and b in I, $a < b \Rightarrow f(a) < f(b)$. Prove that actually $a < b \Leftrightarrow f(a) < f(b)$. [Hint: You have to prove that $f(a) < f(b) \Rightarrow a < b$. Suppose that $f(a) < f(b)$. By the law of trichotomy, if $a < b$ is false, then either $a = b$ or $a > b$. Show that each of these contradicts the supposition $f(a) < f(b)$.]

12. Suppose that $x_n > 0$ and $x_n{}^2 > c > 0$, and define x_{n+1} as in Newton's method for square roots [§1.6], $x_{n+1} = \dfrac{x_n{}^2 + c}{2x_n}$. Prove that $x_{n+1} < x_n$.

I.4 ORDERED FIELDS

The laws of inequalities are valid in any *ordered field.* The simplest ordered field is the *rational numbers,* which are obtained from the natural numbers by adjoining fractions, negatives, and zero.

Another important ordered field is the *real numbers* [§AI.5 below], the starting point for calculus.

The system in which the laws of inequalities work are described by the following axioms.

Axioms for an ordered field

Let F be a system with two operations, addition $(+)$ and multiplication $(\times)$, such that if a and b are any members of the system, then the sum $a + b$ and the product $a \times b$ are well-defined members. F is an *ordered field* if it satisfies the first six axioms for the natural numbers [the associative, commutative, distributive and multiplicative unit laws, §AI.2], as well as the following:

U + (Additive unit) There is a member (generally denoted 0) such that $0 + a = a$ for every a.

I + (Additive inverse) For every member a, there is a member $-a$ such that $a + (-a) = 0$.

I × (Multiplicative inverse) For every member $a \neq 0$, there is a member a^{-1} such that $a \times a^{-1} = 1$.

POS (Ordering, or Positivity, axiom) The system contains a set P, called the *positive* members, such that

(i) for every a, either $a = 0$, or a is in P, or $-a$ is in P, and only one of these three possibilities holds;

(ii) if a and b are in P, then $a + b$ and ab are in P.

We now show that the familiar rules of *equalities* follow from the first axioms down through I + and I ×; and that the laws of *inequalities* follow from these together with the Positivity axiom. A lot of this will appear to be "proving the obvious," and in fact that is why this particular material appears in an appendix. We are not proving things you didn't know, but rather showing how they fit into a general mathematical framework.

We will prove the following ten facts about ordered fields; the problems suggest still others which can be similarly proved. Taken all together, these should convince you that we have not left anything essential out of the axioms.

(1) $0 + a = a; \qquad 1 \times a = a$

(2) $(-a) + a = 0; \qquad a^{-1} \times a = 1 \quad \text{if } a \neq 0$

(3) $(b + c) \times a = b \times a + c \times a$

(4) $a + b = a + c \iff b = c$

(5) If $a \neq 0$, then $a \times b = a \times c \iff b = c$

[cancellation laws: (4) and (5)]

(6) $-(-a) = a; \qquad (a^{-1})^{-1} = a \quad \text{if } a \neq 0$

(7) $0 \times a = 0 = a \times 0$, for every a

(8) $(-a) \times b = -(a \times b)$

(9) $(-a) \times (-b) = a \times b$ [rule of signs]

(10) $a \times b = 0 \iff a = 0$ or $b = 0$

Formulas (1) and (2), asserting that the identities and inverses work on the left as well as the right, are obtained by combining the laws listed above, and those carried over from §I.2. We have

$$\begin{aligned} 0 + a &= a + 0 && \text{[by C +]} \\ &= a, && \text{[by U +]} \end{aligned}$$

proving the first half of (1). As for (2), we have

$$\begin{aligned} (-a) + a &= a + (-a) && \text{[C +]} \\ &= 0. && \text{[I +]} \end{aligned}$$

Similarly, for (3) [which asserts that multiplication can be distributed from the right as well as the left]:

$$\begin{aligned} (b + c) \times a &= a \times (b + c) && [\text{C} \times] \\ &= (a \times b) + (a \times c) && \text{[D]} \\ &= (b \times a) + (c \times a) && [\text{C} \times] \end{aligned}$$

proving (3).

In (4), the implication $b = c \Rightarrow a + b = a + c$ needs no justification; $b = c$ means that b and c are merely different symbols for the same numbers, so $a + b$ and $a + c$ are likewise different symbols for the same number. The other half of (4) follows from (2):

$$\begin{aligned} a + b = a + c &\Rightarrow (-a) + (a + b) = (-a) + (a + c) \\ &\Rightarrow (-a + a) + b = (-a + a) + c && \text{[A +]} \\ &\Rightarrow 0 + b = 0 + c && \text{[I +]} \\ &\Rightarrow b = c. && \text{[U +]} \end{aligned}$$

The proof of (5) is the same. The laws (6) for "double inverses" follow by cancellation: We already know that

$$(-a) + (-(-a)) = 0 \qquad [\text{I} +]$$

and

$$(-a) + a = 0. \qquad [\text{by (2)}]$$

Hence, $(-a) + (-(-a)) = (-a) + a$; cancelling $(-a)$, we obtain the first half of (6). The last half has an identical proof.

The proof of (7) is simple, but perhaps not obvious:

$$\begin{aligned} (0 \times a) + 0 &= 0 \times a && [\text{I} \times] \\ &= (0 \times 0) \times a && [\text{U} +] \\ &= (0 \times a) + (0 \times a). && [\text{by (3)}] \end{aligned}$$

Cancelling $(0 \times a)$ from the first and last lines leaves $0 = 0 \times a$, proving (7).

Cancellation is also involved in the proof of (8):

$$\begin{aligned} a \times b + (-a) \times b &= (a + (-a)) \times b && [\text{by (3)}] \\ &= 0 \times b && [\text{I} +] \\ &= 0 && [\text{by (7)}] \\ &= (a \times b) + (-(a \times b)). && [\text{I} +] \end{aligned}$$

Cancelling $a \times b$, we obtain (8).

Now the mysterious rule of signs (9) turns out to be a logical consequence of the other, eminently reasonable, rules:

$$\begin{aligned} -(a \times b) + (-a) \times (-b) &= (-a) \times b + (-a) \times (-b) && [\text{by (8)}] \\ &= (-a) \times (b + (-b)) && [\text{D}] \\ &= -a \times 0 && [\text{I} +] \\ &= 0 && [\text{by (7)}] \\ &= -(a \times b) + (a \times b). && [\text{by (2)}] \end{aligned}$$

Cancelling $-(a \times b)$ yields (9).

Rule (10) is used to solve equations by factoring; for example

$$\begin{aligned} (x + 1)(x + 2) = 0 &\Rightarrow x + 1 = 0 \text{ or } x + 2 = 0 \\ &\Rightarrow x = -1 \text{ or } x = -2, \end{aligned}$$

where the last line was obtained by adding -1 or -2 to the appropriate equation. The proof of (10) depends on the existence of inverses. If $a \times b = 0$, and $b \neq 0$, then

$$\begin{aligned} a \times b = 0 &\Rightarrow (a \times b) \times b^{-1} = 0 && [\text{by (7)}] \\ &\Rightarrow a \times (b \times b^{-1}) = 0 && [\text{A} +] \\ &\Rightarrow a \times 1 = 0 && [\text{I} \times] \\ &\Rightarrow a = 0. && [\text{U} \times] \end{aligned}$$

We have not yet mentioned subtraction and division; these operations are easily defined,

$$a - b = a + (-b); \qquad a/b = a \times (b^{-1}),$$

and their properties deduced from those of addition and multiplication. For example, the combination of fractions on a common denominator,

$$\frac{a}{b} + \frac{c}{d} = \frac{ad + bc}{bd},$$

is justified by "clearing the denominator" as follows:

$$\begin{aligned}
\left(\frac{a}{b} + \frac{c}{d}\right) \times (b \times d) &= (a \times b^{-1} + c \times d^{-1}) \times (b \times d) && \text{[definition]} \\
&= (a \times b^{-1}) \times (b \times d) + (c \times d^{-1}) \times (d \times b) && \text{[by (3) and C}\times\text{]} \\
&= a \times ((b^{-1} \times b) \times d) + c \times ((d^{-1} \times d) \times b) && \text{[A}\times\text{]} \\
&= a \times (1 \times d) + c \times (1 \times b) && \text{[I}\times\text{]} \\
&= a \times d + c \times b. && \text{[U}\times\text{]}
\end{aligned}$$

Hence, multiplying both sides by $(b \times d)^{-1}$,

$$\frac{a}{b} + \frac{c}{d} = (a \times d + c \times b) \times (b \times d)^{-1} = \frac{ad + cb}{bd}.$$

[In the first application of the associative law A $\times$ above, we left out a few steps which Problem 7 below asks you to supply.]

Any lingering doubts whether the axioms imply *all* of elementary arithmetic and algebra (up to, but not including the extraction of square roots) should be laid to rest by the problems below.

We turn now to the ordering axiom and its consequences, which are based on:

$a < b$ DEFINED

Definition 1. $a < b \Leftrightarrow b - a$ is positive.
$a < b$ can also be written $b > a$.

From POS (i) follows the *Law of Trichotomy: Exactly one of the three possibilities* $a < b$, $a = b$, $a > b$, *is valid;* these correspond respectively to the three possibilities $b - a$ is positive, $b - a = 0$, or $-(b - a) = -b + a = a - b$ is positive.

The remaining laws follow from POS (ii) and the rules of algebra, aided by an occasional trichotomy. Thus,

$$a < b \iff a + c < b + c; \tag{11}$$

for,

$$\begin{aligned} a < b &\iff b - a \quad \text{is positive} && \text{[Definition 1]} \\ &\iff (b + c) - (a + c) \quad \text{is positive} && \text{[algebra]} \\ &\iff a + c < b + c. && \text{[Definition 1]} \end{aligned}$$

Further, since $a = b \iff a + c = b + c$ [law of cancellation], we obtain

$$a \le b \iff a + c \le b + c. \tag{11a}$$

The cancellation law for multiplication has a similar proof: if c is positive, then

$$\begin{aligned} a < b &\iff b - a > 0 \\ &\Rightarrow c(b - a) > 0 \qquad \text{[POS (ii)]} \\ &\iff cb - ca > 0 \\ &\iff ca < cb. \end{aligned}$$

Similarly, $a > b \Rightarrow ca > cb$, and $a = b \Rightarrow ca = cb$; hence, using Trichotomy and a proof by contradiction, we find $ca < cb \Rightarrow a < b$, so

$$\text{if} \quad c > 0, \quad \text{then} \quad a < b \iff ca < cb \tag{12}$$

$$\text{if} \quad c > 0, \quad \text{then} \quad a \le b \iff ca \le cb. \tag{12a}$$

On the other hand, if $c < 0$, then $0 = c - c < 0 - c = -c$, so $-c > 0$ and we obtain

$$\begin{aligned} a < b &\iff b - a > 0 \\ &\iff -c(b - a) > 0 \qquad \text{[by (12)]} \\ &\iff ca - cb > 0 \\ &\iff ca > cb. \end{aligned}$$

Thus (12) implies the following law (12′), and similarly (12a) implies (12′a):

$$\text{if} \quad c < 0, \quad \text{then} \quad a > b \iff cb > ca \tag{12′}$$

$$\text{if} \quad c < 0, \quad \text{then} \quad a \ge b \iff cb \le ca. \tag{12′a}$$

The Law for Reciprocals is next. If $a > 0$ and $b > 0$, then so is $ab > 0$ [by POS (ii)]; hence

$$\begin{aligned} \frac{1}{a} > \frac{1}{b} \iff \frac{1}{a} - \frac{1}{b} > 0 &\iff \frac{b - a}{ab} > 0 \\ &\iff b - a > 0 \qquad \text{[by (12), with } c = ab\text{]} \\ &\iff a < b. \end{aligned}$$

Thus

$$\text{(13)} \qquad \text{if} \quad a > 0 \text{ and } b > 0, \quad \text{then} \quad a < b \iff \frac{1}{a} > \frac{1}{b}.$$

Further, since $\frac{1}{a} = \frac{1}{b} \iff ab\left(\frac{1}{a}\right) = ab\left(\frac{1}{b}\right) \iff b = a,$ we have

$$\text{(13a)} \qquad \text{if} \quad a > 0 \text{ and } b > 0, \quad \text{then} \quad a \leq b \iff \frac{1}{a} \geq \frac{1}{b}.$$

The *Transitive Law* goes as follows:

$$\begin{aligned} a < b \text{ and } b < c &\iff b - a > 0 \text{ and } c - b > 0 \\ &\Rightarrow (b - a) + (c - b) > 0 \qquad [\text{POS (ii)}] \\ &\iff c - a > 0 \qquad [\text{algebra}] \\ &\iff a < c. \end{aligned}$$

Thus

$$\text{(14)} \qquad a < b \text{ and } b < c \Rightarrow a < c.$$

The corresponding law

$$\text{(14a)} \qquad a \leq b \text{ and } b \leq c \Rightarrow a \leq c$$

follows, on consideration of the various cases where equality occurs. [See Problem 1.]

The remaining laws of addition, multiplication, and division are left as Problem 12 below.

Finally, we recall the definition of *absolute value,* and give its relation to inequalities. The absolute value $|a|$ of a real number a is defined as follows:

$$\text{(15)} \qquad |a| = \begin{cases} a & \text{if } a \text{ is positive} \\ 0 & \text{if } a = 0 \\ -a & \text{if } -a \text{ is positive.} \end{cases}$$

By Trichotomy, this defines $|a|$ uniquely, for every real a. From the definition, $|a| \geq 0$, and in fact $|a| > 0$ unless $a = 0$.

The fundamental fact relating inequalities and absolute value is:

$$\text{(16)} \qquad |a| < b \iff -b < a < b.$$

We prove this in each of the three cases suggested by (15):

Case (i): $a > 0$. Then $|a| = a$, so

$$\begin{aligned} |a| < b \;&\Leftrightarrow\; a < b \;\Rightarrow\; b > 0 && \text{[transitive law, since } a > 0\text{]} \\ &\Leftrightarrow\; -b < 0 \\ &\Rightarrow\; -b < a && \text{[transitive law]} \end{aligned}$$

so

$$|a| < b \;\Rightarrow\; -b < a < b.$$

The converse implication is obvious,

$$-b < a < b \;\Rightarrow\; a < b \;\Leftrightarrow\; |a| < b, \qquad \text{[since } a > 0\text{]}$$

so (16) is true in Case (i).

Case (ii): $a = 0$. Then $|a| = 0$, so

$$|a| < b \;\Leftrightarrow\; 0 < b \;\Rightarrow\; -b < 0,$$

so $|0| < b$ implies both $0 < b$ and $-b < 0$, that is

$$|0| < b \;\Rightarrow\; -b < 0 < b;$$

and the converse is obvious:

$$-b < 0 < b \;\Rightarrow\; 0 < b \;\Leftrightarrow\; |0| < b.$$

Case (iii): $a < 0$. Then $-a$ is positive, so $|a| = -a$, and $|-a| = -a$, so $|a| = |-a|$ and

$$\begin{aligned} |a| < b \;&\Leftrightarrow\; |-a| < b \\ &\Leftrightarrow\; -b < -a < b && \text{[Case (i)]} \\ &\Leftrightarrow\; -a - (-b) > 0 \text{ and } b - (-a) > 0 \\ &\Leftrightarrow\; b - a > 0 \text{ and } a - (-b) > 0 \\ &\Leftrightarrow\; b > a > -b. \end{aligned}$$

This completes the proof of (16). Similarly,

$$|a| \leq b \;\Leftrightarrow\; -b \leq a \leq b. \tag{17}$$

Finally, we obtain the *Triangle Inequality:*

$$|a + b| \leq |a| + |b|. \tag{18}$$

For, applying (17) with $b = |a|$ yields

$$-|a| \leq a \leq |a|.$$

Similarly,

$$-|b| \leq b \leq |b|.$$

Hence, adding these inequalities [Problem 12],

$$-(|a| + |b|) \leq a + b \leq |a| + |b|,$$

which is equivalent to the Triangle Inequality (8).

PROBLEMS

In each problem, use any preceding problems in this section, and any results proved in the text.

1. Prove that $1 \times a = a$.

2. Prove that $a^{-1} \times a = 1$, if $a \neq 0$.

3. Prove the cancellation law for $a \neq 0$: $a \times b = a \times c \Leftrightarrow b = c$.

4. Prove that $(a^{-1})^{-1} = a$ if $a \neq 0$.

5. Prove that $(-1) \times a = -a$.

6. Prove that $(a - b) + c = a - (b - c)$, using the definition of subtraction, $a - b = a + (-b)$.

7. (a) Prove that $(a \times b) \times (c \times d) = a \times ((b \times c) \times d)$. [Begin by letting $c \times d$ play the role of c in Axiom A $\times$.]
(b) Prove that $(a \times b) \times (c \times d) = ((a \times b) \times c) \times d$. [Formulas such as these are required since numbers can be multiplied only two at a time. Thus, *abc* is computed either as $(ab)c$ or $a(bc)$; by the associative law, both ways give the same answer. Similarly, *abcd* must be computed as $(a \times b) \times (c \times d)$ or $a \times ((b \times c) \times d)$, or $((a \times b) \times c) \times d$, or $(a \times (b \times c)) \times d$, or .... Again, by the associative law, all these different groupings yield the same result.]

8. Prove that $a/b = c/d \Leftrightarrow a \times d = b \times c$ ["cross-multiplying"; assume, of course, that $b \neq 0$ and $d \neq 0$.]

9. Solve the following equations carefully, justifying each step by an axiom or a formula derived above.
(a) $3x + 4 = 5$
(b) $1/x + 2 = 3$
(c) $2x - 5 = 6$
(d) $1/x + 1/(x - 1) = 3$

10. (a) Prove that $(x + a) \times (x + b) = x^2 + (a + b)x + ab$, where x^2 denotes $x \times x$. [Note: The sum of three numbers $A + B + C$ denotes either $(A + B) + C$, or $A + (B + C)$, which by the associative law are the same.]
(b) Solve $x^2 + 3x + 2 = 0$. [Factor, and use (10).]
(c) Solve $x^2 - 3x + 2 = 0$.

11. Which of the axioms A + through I $\times$ are satisfied by:
(a) The positive integers
(b) The integers (positive, negative, and 0)
(c) The positive real numbers
(d) The set of all finite (terminating) decimals $d.d_1d_2\ldots d_n$
(e) The rational numbers

12. Prove the following.
(a) $a < b$ and $c < d \Rightarrow a + c < b + d$
(b) $a \leq b$ and $c \leq d \Rightarrow a + c \leq b + d$
(c) $0 < a < b$ and $0 < c < d \Rightarrow a \times c < b \times d$ [Use (12) and the Transitive Law.]
(d) $0 \leq a \leq b$ and $0 \leq c \leq d \Rightarrow ac \leq bd$
(e) $0 < a < b$ and $0 < c < d \Rightarrow a/d < b/c$
(f) $0 \leq a \leq b$ and $0 < c \leq d \Rightarrow a/d \leq b/c$

I.5 THE LEAST UPPER BOUND AXIOM, AND THE REAL NUMBERS

Many quantities occurring naturally in geometry, such as the diagonal of a unit square, the circumference of a unit circle, or the base e of the natural logarithms, are not given by rational numbers. Therefore the rational number system has been enlarged to form the so-called *real numbers* [see the introduction to this appendix]. The basic property of the real numbers which guarantees the existence of the desired "irrational" numbers is formulated as the *Least Upper Bound Axiom*. To state the axiom, we need two definitions:

UPPER BOUND OF A SET S

Definition 1. Let S be a set of numbers. A number M is called an *upper bound* for S if every member of S is $\leq M$.

Example 1. $S = [0,1]$. Obviously, 1 is an upper bound; so are 2, π, 10^{10}, etc.

Example 2. $S = \{x : x^2 \leq 2\}$. For this set, 2, $1\frac{1}{2}$, 1.42, 1.415, and so on, are all upper bounds. We give a careful proof that $1\frac{1}{2}$ is an upper bound.
Let x be any member of S. Then

$$x^2 \leq 2. \qquad \text{[by definition of } S\text{]}$$

Also, $(1\frac{1}{2})^2 = \frac{9}{4} > 2$. Hence

$$x^2 < (1\tfrac{1}{2})^2. \qquad \text{[by the Transitive Law]}$$

If $x > 0$, then

$$x < 1\tfrac{1}{2}. \qquad \text{[Example 3, §AI.3]}$$

And if $x \leq 0$, then clearly $x < 1\frac{1}{2}$. Thus every member of S is less than $1\frac{1}{2}$, in other words, $1\frac{1}{2}$ is an upper bound for S.

Example 3. $S =$ all the positive numbers. This set S has *no* upper bound. Suppose there were an upper bound M. Then since 1 is in S, we must have

$M \geq 1$, and hence M is positive. But then $M + 1$ is a positive number, so $M + 1$ is in S. Since M is supposed to be an upper bound for S, we must have $M \geq M + 1$. This is obviously impossible, so there cannot be an upper bound M.

Definition 2. Let S be a set of numbers. A number L is called a *least upper bound* of S if it satisfies these two conditions:

LEAST UPPER BOUND OF A SET S

(i) L is an upper bound of S.

(ii) No number smaller than L is an upper bound of S.

The least upper bound of a set S, if it exists, is denoted l.u.b.S. Another notation for the least upper bound is sup S, read "soup of S" or "soup S"; sup is short for *supremum.*

Example 4. If S is a finite interval, then l.u.b.S is the right endpoint of S. Thus l.u.b.$[a,b] = b$, l.u.b.$(a,b) = b$. Notice that l.u.b.S may or may not belong to S. For example, l.u.b.$[a,b]$ belongs to $[a,b]$, but l.u.b.(a,b) does not belong to (a,b).

☆☆☆ LEAST UPPER BOUND AXIOM

Now we can state the ***Least Upper Bound Axiom*** (LUB): *If S is any set of real numbers, and S is not empty, and S has an upper bound, then S has a least upper bound.*

There is a similar definition of greatest lower bound (g.l.b.), and a corresponding Greatest Lower Bound property. We leave the formulation to you in Problems 3–5 below. (The GLB property is not called an axiom, because it follows from LUB.)

Two interpretations

In LUB, we are given a set S, and we know that there is some number M greater than or equal to every member of S. Thus if we imagine the points of S displayed along a line as in Fig. 6, none of them is to the right of M. Now, suppose a point Q starts at M and moves steadily to the left toward S, so long as it does not pass any points of S. The Least Upper Bound Axiom says that Q *stops at a certain number* L. As long as Q is to the right of L, it can keep moving without passing any points of S; but if Q were to the left of L, then it would have passed some point of S.

FIGURE AI.6 The set S is marked by heavy lines

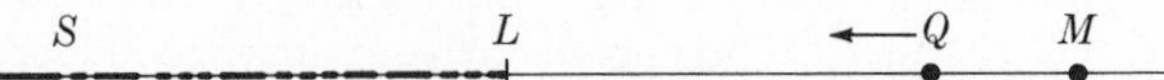

FIGURE AI.6 The set S is marked by heavy lines

Figure 6 shows why the two conditions imposed on S in the statement of LUB are necessary. If S has no upper bound, then there is no place to start Q; and if S is empty, then Q keeps moving indefinitely to the left and does not come to rest anywhere.

Figure 7 illustrates a second interpretation. Take the set U consisting of all upper bounds for S. Then U looks very much like an infinite interval; for if u is an upper bound for S, then all points to the right of u are also upper bounds for S. The least upper bound property says that the set U of upper bounds for S *is* indeed an interval. It is the closed interval $[L, \infty)$, and its left endpoint L is the *least* upper bound.

Typical applications of LUB are the definitions of $\sqrt{2}$, π, the natural logarithm, and e:

IMPORTANT NUMBERS DEFINED AS LEAST UPPER BOUNDS

(1) $\sqrt{2} = \text{l.u.b. } \{x : x^2 < 2\}$ [Fig. 8]

(2) $\pi = \text{l.u.b. } \{L : L$ is the length of a polygon inscribed in a circle of diameter $1\}$ [Fig. 9]

(3) $\log b = \text{l.u.b.} \{s : s = \sum_1^n \frac{1}{x_j} (x_j - x_{j-1})$ for some partition $1 = x_0 < x_1 < \cdots < x_n = b$ of the interval $[1,b]\}$ [§5.1 and Fig. 10]

(4) $e = \text{l.u.b. } \{x : \log x < 1\}$ [Fig. 11]

The numbers defined in (2) and (3) can be taken as definitions requiring no further justification; but in the case of (1), we must ask whether the number $\sqrt{2}$ defined there satisfies $(\sqrt{2})^2 = 2$, and in (4) whether $\log e = 1$. These questions are answered by the Intermediate Value Theorem [§AII.5].

Remark. We have stressed just one of the axioms for the real numbers, taking the others for granted. The complete list of essential properties is easily given:

The real numbers are an ordered field [§AI.4] *in which the Least Upper Bound Axiom holds.*

S L U

FIGURE AI.7 The set S is marked by heavy lines

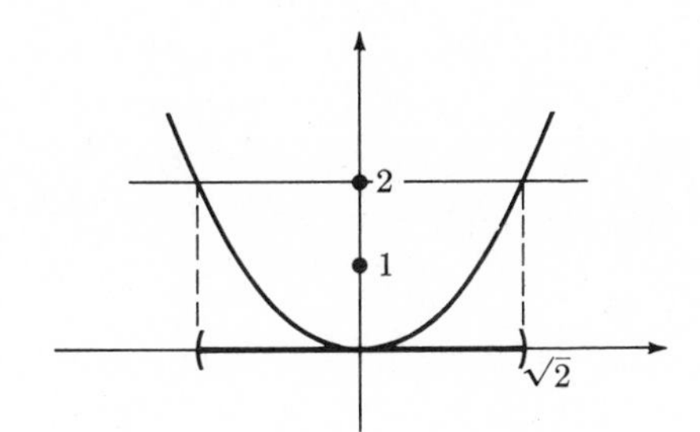

FIGURE AI.8 $S = \{x : x^2 < 2\}$

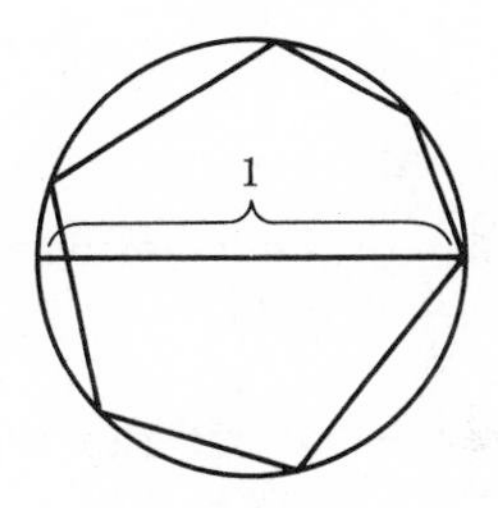

FIGURE AI.9 $S = \{$all lengths of polygons inscribed in a circle of diameter 1$\}$

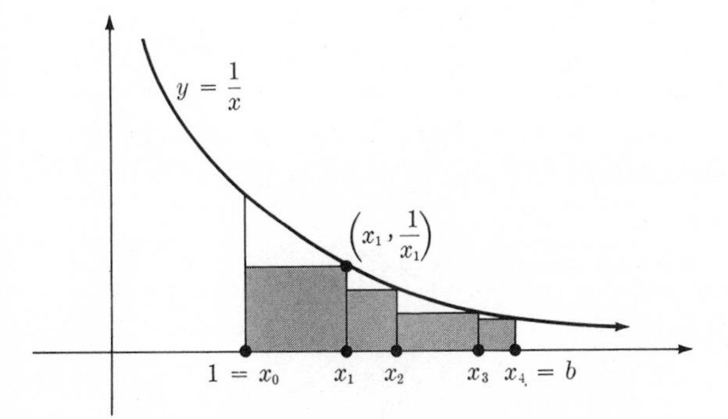

FIGURE AI.10 Sum of areas of inscribed rectangles is $\sum_1^4 \frac{1}{x_j}(x_j - x_{j-1})$. Our set S is the set of all such sums, for all possible choices of the partition $1 = x_0 < x_1 < \cdots < x_n = b$.

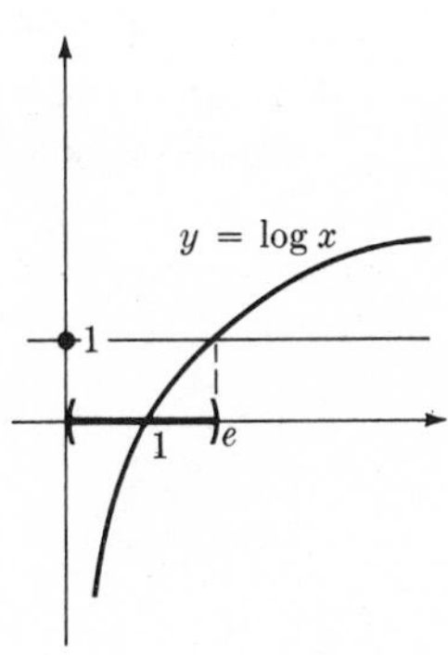

AI.11 $S = \{x : \log x < 1\}$

Section AI.4 shows how to derive the laws of algebra and inequalities from the axioms for an ordered field. This section suggests how to use the Least Upper Bound Axiom; the crucial applications for calculus are carried out in §AII.5 [Intermediate Value Theorem, Maximum Value Theorem], §AII.8 [definition of integral], §10.3 [convergence of monotone sequences], and §11.4 [radius of convergence theorem for power series].

To complete the picture, we need one further step: To show *from the axioms* how the natural numbers fit into the real numbers. This is done in the next section.

PROBLEMS

1. Which of the following sets S has (have) a l.u.b.?
   (a) $S = (0,1)$
   (b) $S = \{n : n = k^2 \text{ for some integer } k\}$
   (c) $S = \{x : x \text{ is an integer and } 0 < x^2 < 1\}$
   (d) $S = \{s : s = 1 - 1/k^2 \text{ for some integer } k\}$

2. Prove that l.u.b.$(0,1) = 1$. [Show that if $L < 1$, then L is not an upper bound for $(0,1)$; and if $L > 1$, then L is not a *least* upper bound.]

3. Define "lower bound."

4. Define "greatest lower bound" (g.l.b.).

5. (a) State the Greatest Lower Bound property (GLB).
   (b) Prove the GLB property, using the LUB axiom.

6. The number π is defined to be l.u.b. S, where S is the set of all lengths of polygons inscribed in a circle of diameter 1. To justify this definition, we have to prove that S has an upper bound. Figure 12 shows that 4 is an upper bound. Explain why. [This is more a problem in intuitive geometry than in numbers.]

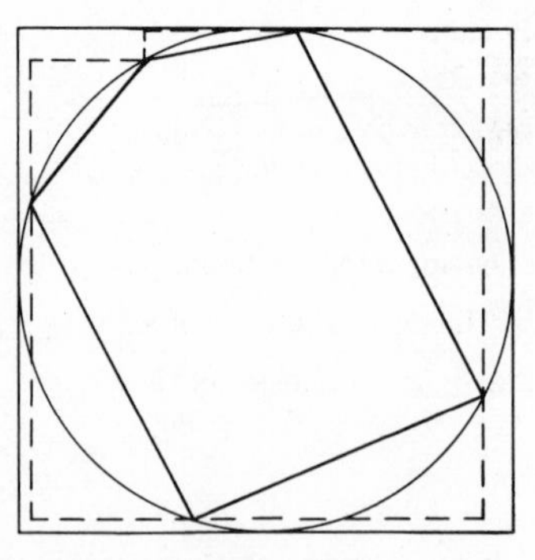

FIGURE AI.12

I.6 THE INTEGERS AS REAL NUMBERS: ARCHIMEDEAN PROPERTY

The particular real numbers $1, 1 + 1 = 2, 2 + 1 = 3$, etc., are called *positive integers.* It seems clear that these numbers form a well-defined set, call it N, and that N should satisfy all the axioms for the natural numbers [§AI.2]. Accepting this for the moment, we will prove:

The Archimedean Property of the Real Numbers: *Given any real number r, there is a positive integer n such that $n > r$.*

This property enters subtly in certain proofs in Chapter X [Infinite Sequences], where we say e.g. "Given $\epsilon > 0$, choose a positive integer $n > 1/\epsilon$"; the Archimedean Property guarantees that there really is such an n. It also enters in §5.2, where we use the fact that between every two real numbers there is an irrational number [see Problem 5 below for the proof that there is].

The Archimedean Property says simply that *the set of positive integers N has no upper bound in the real numbers;* it is therefore no surprise that the proof requires the Least Upper Bound Axiom. We proceed "by contradiction": Suppose that N *has* an upper bound. Then it has a least upper bound L:

$$L > n \quad \text{for every positive integer } n. \tag{1}$$

Since L is a *least* upper bound, $L - 1$ is *not* an upper bound, i.e. there is an integer n_0 such that

$$n_0 > L - 1.$$

It follows that

$$n_0 + 1 > L. \tag{2}$$

But $n_0 + 1$, like n_0, is a positive integer, so (2) contradicts (1). This contradiction shows that the supposition "N has an upper bound" is false, so the proof of the Archimedean Property is complete.

One fussy detail will complete our picture of the real numbers; we need a proper definition of the set N of positive integers, based only on the axioms for the real numbers. N should have the properties

(i) 1 is in N

(ii) if n is in N, so is $n + 1$

(iii) N contains no more than properties (i) and (ii) require.

INDUCTIVE SET

To define such a set N, we introduce the notion of an *inductive set*. We call a set of real numbers *inductive* if it satisfies (i) and (ii). Thus, for example, the set P of positive numbers is inductive: 1 is positive [since it's a square] and if r is positive, so is $r + 1$. The set of negative numbers, on the other hand, is *not* inductive since it does not contain 1, violating condition (i). The interval [0,1] is not inductive, since it violates condition (ii).

To achieve (i) and (ii), we want an inductive set. To achieve (iii), we want the *smallest* inductive set. The following definition picks this out.

POSITIVE INTEGERS DEFINED

Definition. A real number n is a *positive integer* if n belongs to every inductive set. The collection of all these numbers is denoted N.

In particular, 1 is in every inductive set, so 1 is a positive integer.

Since the positive numbers P form an inductive set, every positive integer is in P, i.e. every positive integer really is a positive number.

The inductive axiom [§AI.2] holds for the system N, since it is built right into the definition: Let S be any set of real numbers satisfying (i) and (ii); then S is inductive, so by definition it contains all the positive integers. Using this fact, it is easy to prove [by induction] that the sum and product of any two positive integers is again a positive integer, and then the remaining axioms for the natural numbers [§AI.2] follow directly from the axioms for the reals.

This completes our discussion of the real numbers.

PROBLEMS

1. Let n be a positive integer. Prove that $n + 1$ is a positive integer. [Hint: Any inductive set containing n must also contain $n + 1$.]

2. Let n and m be positive integers.
(a) Prove that $m + n$ is a positive integer. [For fixed m, use induction on n, and Problem 1.]
(b) Prove that $m \times n$ is a positive integer. [Use induction on n, and part (a).]

3. Let a and b be real numbers, with $b - a > 1$, $a \geq 0$, $b \geq 0$. Show that the interval (a,b) contains an integer. [Hint: Suppose, on the contrary, that (a,b) does *not* contain an integer, and let S = all positive integers n which are $<b$. Show that S is inductive; note that if $n < b$, then $n \leq a$, since (a,b) is supposed to contain no integer.]

4. This problem shows that every open interval (a,b) contains at least one rational number. [Note that this follows from Problem 3, if $0 \leq a$ and $b - a > 1$.]

(a) Prove it in case $0 \leq a < b$. [Expand the interval until its length is >1; why is there an integer m such that $mb - ma > 1$?]

(b) Prove it in the general case. [Shift the interval to the right of 0, if necessary. There is an integer m such that $a + m \geq 0$ (why?); apply part (a) to $(a + m, b + m)$.]

5. Every open interval (a,b), $a < b$, contains an irrational number. [Hint: Apply Problem 4 to the interval $(a/\sqrt{2},\ b/\sqrt{2})$.]

6. This problem is useful in proving the so-called *well-ordering* of the positive integers [Problem 7 below]; it proves that if m and n are distinct positive integers, and $m < n$, then $n - m \geq 1$. Let $S =$ {all positive integers n such that there is no positive integer m between $n - 1$ and n}.

(a) Prove that S contains 1. [Hint: $\{x: x \geq 1\}$ is an inductive set, so it contains every integer.]

(b) Prove that if n is in S, so is $n + 1$. [Hint: Suppose, on the contrary, that there *is* a positive integer p between n and $n + 1$. Show that if p is deleted from the set of positive integers, the remaining positive integers form an inductive set. Why is this a contradiction?]

(c) Conclude that if $m < n$, then $m \leq n - 1$, so $n - m \geq 1$. [Here m and n are any positive integers.]

7. [*Well-ordering of the positive integers*] Prove that *every non-empty set S of positive integers contains a smallest member.* [Hint: Use Problem 6. Let $L =$ g.l.b. S, and suppose that L is *not* in S. Since L is a lower bound for S, and $L + 1$ is not (why?) S contains an integer n between L and $L + 1$, $L < n < L + 1$; and S contains another integer m, $L < m < n < L + 1$ (why?). This contradicts Problem 6.]

8. [*Decimal expansions*] Let a be any positive real number.

(a) Prove that there is a unique integer $n \geq 0$ such that $a - 1 < n \leq a$. [Apply Problem 7 to the set of integers which are $>(a - 1)$.]

(b) Prove that there is a unique integer a_1 such that $a - 1/10 < n + a_1/10 \leq a$, and $a \leq a_1 \leq 9$.

(c) Prove that for every m, there is a unique integer a_m such that $0 \leq a_m \leq 9$ and

$$a - 10^{-m} < n + \sum_{j=1}^{m} a_j 10^{-j} \leq a.$$

How to Prove the Basic Propositions of Calculus

AII.1 LIMITS

The idea of limit is introduced, informally, in §1.1, and is used in crucial places throughout the text. Thus the first step in securing calculus on a firm logical foundation is the analytic definition of limit.

The basic idea is that of approximation: $\lim_{x \to x_0} f(x) = A$ means, roughly, that $f(x)$ is a good approximation to A whenever x is close enough to x_0. The actual definition is little more than a careful quantitative version of this statement.

By ancient tradition, two Greek letters are used in this context:

ϵ (epsilon) stands for the maximum allowable error in the approximation, and

δ (delta) stands for the allowable distance between x and x_0 corresponding to the error ϵ;

when the distance from x to x_0 is less than δ, then the error in the approximation of A by $f(x)$ is less than the maximum allowable, ϵ.

Definition 1. Let f be defined in an open interval containing x_0, except perhaps at x_0 itself. The number A is the *limit* of $f(x)$ as x approaches x_0 (in symbols, $A = \lim_{x \to x_0} f(x)$) if and only if

LIMIT DEFINED

for every $\epsilon > 0$, there is a $\delta > 0$ such that

$$0 < |x - x_0| < \delta \Rightarrow |f(x) - A| < \epsilon.$$

The definition takes the form of a challenge: Given ϵ, go find a δ. To meet the challenge requires a certain facility with inequalities; you may find it necessary to study §AI.3 for this background.

Example 1. Prove that $\lim_{x \to 1} 2x = 2$. *Solution.* In Definition 1 we have $x_0 = 1$, $f(x) = 2x$. Thus, given $\epsilon > 0$, we must fill in the question mark in the implication

$$0 < |x - 1| < (?) \Rightarrow |2x - 2| < \epsilon.$$

In this particular case, we simply solve the inequality $|2x - 2| < \epsilon$ for x:

$$|2x - 2| < \epsilon \Leftrightarrow 2|x - 1| < \epsilon \Leftrightarrow |x - 1| < \epsilon/2.$$

It follows that

$$0 < |x - 1| < \epsilon/2 \Rightarrow |2x - 2| < \epsilon,$$

and Definition 1 is satisfied with $\delta = \epsilon/2$.

It is helpful to visualize Definition 1 on the graph of f [Fig. 1]. The condition

$$|f(x) - A| < \epsilon \tag{1}$$

means that the point $(x,f(x))$ on the graph lies between the two horizontal lines $y = A + \epsilon$ and $y = A - \epsilon$. This is a restriction on $f(x)$, to be satisfied by making an appropriate restriction on x. The condition

$$0 < |x - x_0| < \delta \tag{2}$$

means that x lies in one of the two intervals of length δ to either side of x_0, but $x \neq x_0$; we have to make (1) hold by imposing on x a restriction of the form (2). In Example 1, the relation between ϵ and $\delta = \epsilon/2$ is clear from Fig. 2; over the "punctured interval" $0 < |x - 1| < \epsilon/2$, the graph of f lies between the lines $y = 2 - \epsilon$ and $y = 2 + \epsilon$.

Example 2. If $x_0 > 0$, prove that $\lim_{x \to x_0} (1/x) = 1/x_0$. [The case $x_0 < 0$ is Problem 1(d) below.] *Solution.* Draw a picture [Fig. 3] showing the range in which $1/x$ must lie, namely between $(1/x_0) - \epsilon$ and $(1/x_0) + \epsilon$. From the figure you can see that x should lie between $\dfrac{1}{(1/x_0) + \epsilon}$ and $\dfrac{1}{(1/x_0) - \epsilon}$, and that we should pick

$$\delta = x_0 - \frac{1}{1/x_0 + \epsilon} = \frac{x_0^2\epsilon}{1 + x_0\epsilon}.$$

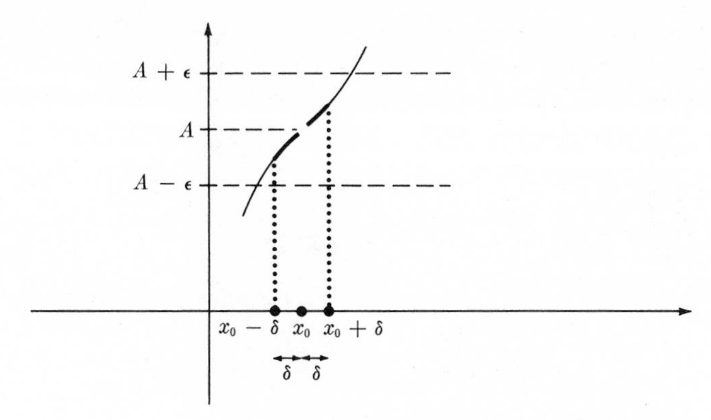

FIGURE AII.1 $\epsilon > 0$ is given; $\delta > 0$ must then be chosen so that the part of the graph for $0 < |x - x_0| < \delta$ (marked heavy in the figure) lies between the lines $y = A - \epsilon$ and $y = A + \epsilon$

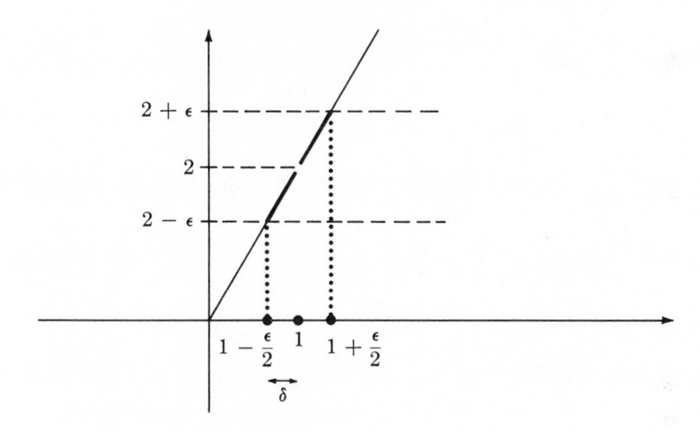

FIGURE AII.2 $f(x) = 2x;$ $f(x) = 2 \pm \epsilon \Rightarrow x = 1 \pm \epsilon/2$

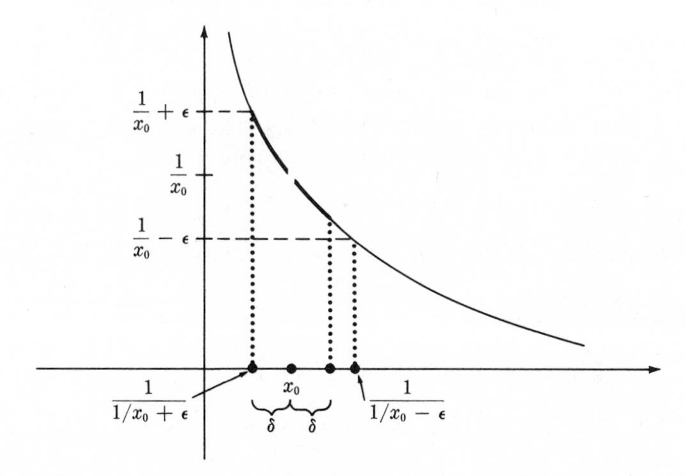

FIGURE AII.3

All that remains is to prove, *without* referring to the figure, that this choice of δ works. We begin by "solving" the inequality $|x - x_0| < \delta$ for $1/x - 1/x_0$, as follows:

$$|x - x_0| < \delta \iff x_0 - \delta < x < x_0 + \delta \qquad \text{[clear]}$$

$$\iff \frac{1}{x_0 - \delta} > \frac{1}{x} > \frac{1}{x_0 + \delta}$$

[since all numbers involved are positive]

$$\iff \frac{1}{x_0} + \epsilon > \frac{1}{x} > \frac{1}{2x_0 - \dfrac{1}{1/x_0 + \epsilon}} = \frac{1/x_0 + \epsilon}{1 + 2x_0\epsilon}$$

$$\left[\text{simple algebra, since } \delta = x_0 - \frac{1}{1/x_0 + \epsilon}\right]$$

$$\iff \epsilon > \frac{1}{x} - \frac{1}{x_0} > \frac{-\epsilon}{1 + 2x_0\epsilon}. \qquad \text{[subtract } 1/x_0 \text{ from each term]}$$

But $\dfrac{-\epsilon}{1 + 2x_0\epsilon} > -\epsilon$, so the above chain of inequalities proves that

$$0 < |x - x_0| < \delta \implies \epsilon > \frac{1}{x} - \frac{1}{x_0} > -\epsilon \iff \left|\frac{1}{x} - \frac{1}{x_0}\right| < \epsilon,$$

when

$$\delta = x_0 - \frac{1}{1/x_0 + \epsilon} = \frac{x_0^2\epsilon}{1 + x_0\epsilon}.$$

Since this number δ is positive, Definition 1 is satisfied.

More examples of this type can be found in the problems below. However, the following general theorems make it unnecessary, in most cases, to use the "ϵ, δ" definition in evaluating limits.

SIMPLE BUT IMPORTANT SPECIAL CASE OF LIMIT

Theorem 1. *Let m and b be any constants. Then* $\lim_{x \to x_0} (mx + b) = mx_0 + b$ *for every* x_0.

Proof. For every $\epsilon > 0$, we must fill in

$$0 < |x - x_0| < \textbf{(?)} \implies |(mx + b) - (mx_0 + b)| < \epsilon.$$

When $m \neq 0$, we simply "solve" the last inequality for $|x - x_0|$:

$$|(mx + b) - (mx_0 + b)| = |m| \cdot |x - x_0|,$$

so

$$|(mx + b) - (mx_0 + b)| < \epsilon \iff |m| \cdot |x - x_0| < \epsilon$$
$$\iff |x - x_0| < \epsilon/|m|.$$

Thus

$$0 < |x - x_0| < \epsilon/|m| \Rightarrow |(mx + b) - (mx_0 + b)| < \epsilon,$$

and the proof is complete for the case $m \neq 0$; take $\delta = \epsilon/|m|$ in Definition 1.

When $m = 0$, the proof is even simpler; in this case, we have $|(mx + b) - (mx_0 + b)| = 0$ for *every* x, so we can use any positive number for the δ in Definition 1. For instance, take $\delta = 1$; you can see that

$$0 < |x - x_0| < 1 \Rightarrow |(0x + b) - (0x_0 + b)| = 0 < \epsilon. \quad Q.E.D.$$

Theorem 2. *If* $\lim_{x \to x_0} f(x) = A$ *and* $\lim_{x \to x_0} g(x) = B$, *then*

$$\lim_{x \to x_0} (f(x) + g(x)) = A + B \tag{3}$$

$$\lim_{x \to x_0} (f(x)g(x)) = A \cdot B \tag{4}$$

$$\lim_{x \to x_0} (cf(x)) = cA, \quad \textit{for any constant } c. \tag{5}$$

Sketchy proof of (3): Given $\epsilon > 0$, we must make $f(x) + g(x)$ approximate $A + B$ with an error $< \epsilon$. This is easy; let $f(x)$ approximate A and $g(x)$ approximate B, each with an error $< (\epsilon/2)$.

Formal proof of (3): Let $\epsilon > 0$ be given. Since $\lim_{x \to x_0} f(x) = A$, and the number $\epsilon_1 = \epsilon/2$ is positive, Definition 1 assures us a number $\delta_1 > 0$ such that

$$0 < |x - x_0| < \delta_1 \Rightarrow |f(x) - A| < \epsilon/2. \tag{6}$$

Similarly, there is a number δ_2 (not necessarily the same as δ_1) such that

$$0 < |x - x_0| < \delta_2 \Rightarrow |g(x) - B| < \epsilon/2. \tag{7}$$

Let δ be the smaller of δ_1 and δ_2.

Suppose that $0 < |x - x_0| < \delta$. Then also $0 < |x - x_0| < \delta_1$, so $|f(x) - A| < \epsilon/2$, by (6). Similarly, $0 < |x - x_0| < \delta_2$, so $|g(x) - B| < \epsilon/2$, by (7). Hence

$$\begin{aligned} |(f(x) + g(x)) - (A + B)| &= |(f(x) - A) + (g(x) - B)| \\ &\leq |f(x) - A| + |g(x) - B| \\ &< \epsilon/2 + \epsilon/2 = \epsilon. \end{aligned}$$

Thus

$$0 < |x - x_0| < \delta \;\Rightarrow\; |(f(x) + g(x)) - (A + B)| < \epsilon,$$

which proves (3): $\lim_{x \to x_0} (f(x) + g(x)) = A + B.$

Sketchy proof of (4): $\lim_{x \to x_0} (f(x)g(x)) = AB.$ For each $\epsilon > 0$, we must make $|f(x)g(x) - AB| < \epsilon$. Since

$$f(x)g(x) - AB = (f(x) - A)g(x) + A(g(x) - B),$$

the problem is solved if we can make $|(f(x) - A)\cdot g(x)| < \epsilon/2$ and $|A(g(x) - B)| < \epsilon/2$. For this, we must make $|f(x) - A|$ small [done in (10) below], $|g(x)|$ not too large [see (9) below] and $|g(x) - B|$ small [see (8) below].

Formal proof of (4): Given $\epsilon > 0$, consider the number $\epsilon_1 = \dfrac{\epsilon}{2|A| + 1}$.

The significance of this number will appear at the end of the proof, but at least one thing is clear from the outset, namely, $\epsilon_1 > 0$. Therefore, since $\lim_{x \to x_0} g(x) = B$, there is by definition a number $\delta_1 > 0$ such that

$$0 < |x - x_0| < \delta_1 \;\Rightarrow\; |g(x) - B| < \frac{\epsilon}{2|A| + 1}. \tag{8}$$

This "controls" $|g(x) - B|$, and $|g(x)|$ as well: $|g(x)| = |B + (g(x) - B)| \leq |B| + |g(x) - B|$, so it follows from (8) that

$$0 < |x - x_0| < \delta_1 \;\Rightarrow\; |g(x)| \leq |B| + \frac{\epsilon}{2|A| + 1} = M, \tag{9}$$

where for brevity we have set $|B| + \dfrac{\epsilon}{2|A| + 1} = M$. Notice that $M > 0$, so the number $\epsilon_2 = \epsilon/2M$ is positive. Since $\lim_{x \to x_0} f(x) = A$, there is a number $\delta_2 > 0$ such that

$$0 < |x - x_0| < \delta_2 \;\Rightarrow\; |f(x) - A| < \epsilon/2M. \tag{10}$$

Let δ be the smaller of δ_1 and δ_2.

Suppose that $0 < |x - x_0| < \delta$. Then $0 < |x - x_0| < \delta_1$, and $0 < |x - x_0| < \delta_2$, so

$$\begin{aligned}
|f(x)g(x) - AB| &= |(f(x) - A)g(x)| + |A(g(x) - B)| \\
&\leq |f(x) - A|\cdot|g(x)| + |A|\cdot|g(x) - B| \\
&< \frac{\epsilon}{2M}\cdot M + |A|\frac{\epsilon}{2|A| + 1} = \left(\frac{1}{2} + \frac{|A|}{2|A| + 1}\right)\epsilon \\
&\qquad\qquad \text{[by (8), (9), and (10)]} \\
&< \epsilon. \qquad\qquad \text{[since } |A|/(2|A| + 1) < 1/2\text{]}
\end{aligned}$$

This proves (4): $\lim_{x \to x_0} f(x)g(x) = AB$.

Remark. If $A \neq 0$, the right-hand side in (8) can be simplified to $\epsilon/2|A|$; we made it $\epsilon/(2|A| + 1)$ to avoid a possible division by zero.

Proof of (5): $\lim_{x \to x_0} (cf(x)) = cA$. Apply (4), with $g(x) = c$. By Theorem 1 (with $m = 0$), $\lim_{x \to x_0} g(x) = c$. Hence by (4), $\lim_{x \to x_0} (cf(x)) = cA$.

This completes the proof of Theorem 2.

The limit of any polynomial function can be obtained from Theorems 1 and 2. For example,

$$\lim_{x \to x_0} x^2 = \lim_{x \to x_0} (x \cdot x) = x_0 \cdot x_0 = x_0^2 \qquad \text{[by (4)]}$$

$$\lim_{x \to x_0} cx^2 = c \lim_{x \to x_0} x^2 = cx_0^2 \qquad \text{[by (5)]}$$

$$\begin{aligned}
\lim_{x \to x_0} (c_2x^2 + c_1x + c_0) &= \lim_{x \to x_0} (c_2x^2) + \lim_{x \to x_0} (c_1x + c_0) \qquad \text{[by (3)]} \\
&= c_2x_0^2 + c_1x_0 + c_0
\end{aligned}$$

$$\begin{aligned}
\lim_{x \to x_0} (cx^3) = \lim_{x \to x_0} (x \cdot cx^2) &= (\lim_{x \to x_0} x)(\lim_{x \to x_0} cx^2) \qquad \text{[by (4)]} \\
&= x_0 \cdot cx_0^2 = cx_0^3
\end{aligned}$$

$$\begin{aligned}
\lim_{x \to x_0} (c_3x^3 + c_2x^2 + c_1x + c_0) &= \lim_{x \to x_0} (c_3x^3) + \lim_{x \to x_0} (c_2x^2 + c_1x + c_0) \\
&\qquad\qquad \text{[by (3)]} \\
&= c_3x_0^3 + c_2x_0^2 + c_1x_0 + c_0.
\end{aligned}$$

Generally, then, *the limit as $x \to x_0$ of any polynomial is obtained simply by setting $x = x_0$*:

$$\lim_{x \to x_0} (c_n x^n + c_{n-1}x^{n-1} + \cdots + c_1 x + c_0)$$
$$= c_n x_0^n + c_{n-1}x_0^{n-1} + \cdots + c_1 x_0 + c_0 . \quad (11)$$

This is obvious from the above examples; a complete proof can be given by mathematical induction.

The next step might naturally be a theorem about $\lim f(x)/g(x)$. However, it is more efficient to prove first the following result on limits of composite functions [Proposition C, page 149]:

lim(f(g)) = f(lim g)

Theorem 3. *Suppose that*

(i) $\lim_{x \to x_0} g(x) = B$ *and* (ii) $\lim_{y \to B} f(y) = f(B)$.

Then $\lim_{x \to x_0} f(g(x)) = f(B)$.

The theorem says that under the given conditions, the symbol $\lim_{x \to x_0}$ can be "taken inside f"; $\lim_{x \to x_0} f(g(x)) = f(\lim_{x \to x_0} g(x))$.

Sketchy proof. By (ii), $f(g(x))$ is close to $f(B)$ when $g(x)$ is close to B; and by (i), $g(x)$ is close to B when x is close to x_0.

Formal proof. Let $\epsilon > 0$ be given. Since $\lim_{y \to B} f(y) = f(B)$, there is a $\delta_1 > 0$ such that

$$0 < |y - B| < \delta_1 \Rightarrow |f(y) - f(B)| < \epsilon. \quad (12)$$

In other words, $|f(y) - f(B)| < \epsilon$ whenever $0 < |y - B| < \delta_1$. Moreover, when $0 = |y - B|$ (i.e. when $y = B$), we have $|f(y) - f(B)| = 0 < \epsilon$, so the following implication is actually true:

$$|y - B| < \delta_1 \Rightarrow |f(y) - f(B)| < \epsilon. \quad (13)$$

Next, since $\lim_{x \to x_0} g(x) = B$, there is a $\delta > 0$ such that

$$0 < |x - x_0| < \delta \Rightarrow |g(x) - B| < \delta_1 . \quad (14)$$

Setting $y = g(x)$ in (13), and combining this with (14), gives

$$0 < |x - x_0| < \delta \Rightarrow |f(g(x)) - f(B)| < \epsilon.$$

Hence, $\lim_{x \to x_0} f(g(x)) = f(B)$. *Q.E.D.*

Example 3. Take $f(y) = 1/y$, and $B \neq 0$. By Example 2 (if $B > 0$) or Problem 1(d) (if $B < 0$),

$$\lim_{y \to B} f(y) = \lim_{y \to B} \frac{1}{y} = \frac{1}{B} = f(B).$$

Hence if $\lim_{x \to x_0} g(x) = B$, then by Theorem 3,

$$\lim_{x \to x_0} \frac{1}{g(x)} = \lim_{x \to x_0} f(g(x)) = f(B) = \frac{1}{B}.$$

Example 3, together with the formula for the limit of a product, gives the expected result for the limit of a quotient:

Theorem 4. *If* $\lim_{x \to x_0} f(x) = A$, *and* $\lim_{x \to x_0} g(x) = B$, *and* $B \neq 0$, *then*

$$\lim_{x \to x_0} \frac{f(x)}{g(x)} = \frac{A}{B}.$$

$$\lim \left(\frac{f}{g}\right) = \frac{\lim f}{\lim g}$$

Proof. By Example 3, $\lim_{x \to x_0} \frac{1}{g(x)} = \frac{1}{B}$. Hence by (4),

$$\lim_{x \to x_0} \frac{f(x)}{g(x)} = \lim_{x \to x_0} \left(f(x) \cdot \frac{1}{g(x)} = A \cdot \frac{1}{B} \right) = \frac{A}{B}. \quad Q.E.D.$$

Combining Theorem 4 with (11), we get the limit of the ratio of any two polynomials P and Q:

$$\lim_{x \to x_0} \frac{P(x)}{Q(x)} = \frac{P(x_0)}{Q(x_0)}, \qquad \text{if } Q(x_0) \neq 0.$$

Theorems 1, 2, and 4 together comprise Proposition A, page 109.

PROBLEMS

1. Use Definition 1 directly to prove the following:

(a) $\lim_{x \to x_0} 2x = 2x_0$

(b) $\lim_{x \to x_0} cx = cx_0$ if $c \neq 0$

(c) $\lim_{x \to 0} |x| = 0$

(d) $\lim_{x \to x_0} \frac{1}{x} = \frac{1}{x_0}$, where $x_0 < 0$
[Compare Example 2.]

(e) $\lim_{x \to 3} x^2 = 9$

2. (a) Find $\lim_{x \to -1} |x|$, and prove your result.

 (b) Find $\lim_{x \to x_0} |x|$ for $x_0 \neq 0$, and prove your result.

3. Prove that $\lim_{x \to x_0} g(x) = A \Leftrightarrow \lim_{x \to x_0} [g(x) - A] = 0$. [Hint: Write out each statement according to the definition of limit.]

4. Define a function f as follows: $f(0) = 0$, $f(x) = 0$ if x is irrational, and $f(x) = 1/q$ if $x = p/q$ with p an integer, q a positive integer, and p/q in lowest terms.

 (a) Try to sketch the graph of f.

 (b) For what numbers x_0 does $\lim_{x \to x_0} f(x)$ exist? Prove your answer.

AII.2 MORE LIMITS

We give the behavior of limits with respect to inequalities, establish a criterion used to prove that certain limits exist (the "Poe Principle"), and define several variations of the idea of limits.

$f \leq g \Rightarrow \lim f \leq \lim g$

Theorem 5. *Suppose that $f(x) \leq g(x)$ wherever both f and g are defined, and that $\lim_{x \to x_0} f(x) = A$, $\lim_{x \to x_0} g(x) = B$. Then $A \leq B$.*

Sketchy proof. Suppose, on the contrary, that $A > B$. Then, when x is very near to x_0, we would have $f(x)$ approximating A very closely, and $g(x)$ approximating B very closely [Fig. 4]; this requires $g(x) > f(x)$, contradicting the hypothesis $f(x) \leq g(x)$.

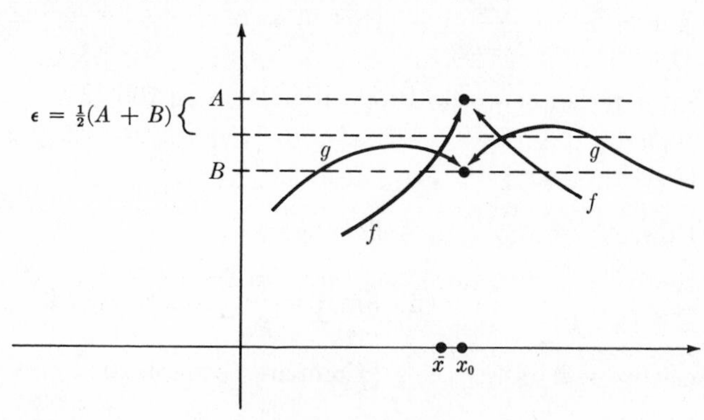

FIGURE AII.4 $\lim f = A > B = \lim g$ forces f to be larger than g at points near x_0

Formal proof by contradiction. If $A > B$, then $\frac{1}{2}(A - B) > 0$. Since $\lim_{x \to x_0} f(x) = A$ and $\lim_{x \to x_0} g(x) = B$, there are numbers $\delta_1 > 0$ and $\delta_2 > 0$ such that

$$0 < |x - x_0| < \delta_1 \Rightarrow |f(x) - A| < \tfrac{1}{2}(A - B)$$

$$0 < |x - x_0| < \delta_2 \Rightarrow |g(x) - B| < \tfrac{1}{2}(A - B).$$

Take any number $\bar{x}$ satisfying both conditions $0 < |\bar{x} - x_0| < \delta_1$ and $0 < |\bar{x} - x_0| < \delta_2$; for example, take $\bar{x} = x_0 + \frac{1}{2}[\min(\delta_1, \delta_2)]$. Then

$$|f(\bar{x}) - A| < \tfrac{1}{2}(A - B), \quad \text{so} \quad -\tfrac{1}{2}(A - B) < f(\bar{x}) - A,$$

or

$$f(\bar{x}) > \tfrac{1}{2}(A + B).$$

At the same time,

$$|g(\bar{x}) - B| < \tfrac{1}{2}(A - B), \quad \text{so} \quad g(\bar{x}) - B < \tfrac{1}{2}(A - B),$$

or

$$g(\bar{x}) < \tfrac{1}{2}(A + B).$$

It follows that $g(\bar{x}) < f(\bar{x})$, contradicting the hypothesis that $f(x) \leq g(x)$ wherever both f and g are defined. *Q.E.D.*

There is a subtle point in the proof of Theorem 5. When we take $\bar{x}$ as above and compare $f(\bar{x})$ to $g(\bar{x})$, it is essential that both $f(\bar{x})$ and $g(\bar{x})$ are defined. But $f(x)$ *is* defined when $0 < |x - x_0| < \delta_1$, in view of the relation

$$0 < |x - x_0| < \delta_1 \Rightarrow |f(x) - A| < \epsilon.$$

For, saying that $|f(x) - A| < \epsilon$ implies clearly that $f(x)$ is defined. Similarly, $g(x)$ is defined when $0 < |x - x_0| < \delta_2$, so there is indeed a point $\bar{x}$ where both $f(\bar{x})$ and $g(\bar{x})$ are defined, and the necessary inequalities are fulfilled.

LIMITS ARE UNIQUELY DEFINED

Corollary. *A given function f cannot have two different limits at* x_0.

The proof is by contradiction. If f has two different limits, let A denote the larger, and B denote the smaller. Applying Theorem 5 with $f = g$, we obviously have $f \leq g$ (since actually $f = g$); hence $A = \lim_{x \to x_0} f(x)$ is $\leq B = \lim_{x \to x_0} g(x)$. But this contradicts our choice of A as the *larger* of the two (hypothetical) limits; therefore, it is impossible to have two different limits. *Q.E.D.*

Thus when f has a limit at x_0, the symbol $\lim_{x \to x_0} f(x)$ denotes a single number (a fact that you had probably taken for granted anyway).

Another "obvious" theorem is the Poe Principle [Fig. 5]:

POE PRINCIPLE

Theorem 6. *Suppose that* $f(x) \leq g(x) \leq h(x)$, *and* $\lim_{x\to x_0} f(x) = L$, $\lim_{x\to x_0} h(x) = L$. *Then also* $\lim_{x\to x_0} g(x) = L$.

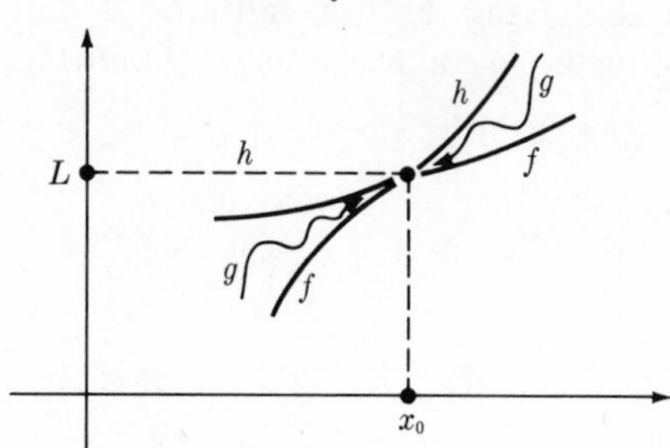

FIGURE AII.5 Poe Principle: $\lim_{x\to x_0} h(x) = L$

Proof. Let $\epsilon > 0$ be given. By assumption, there is a $\delta_1 > 0$ such that

$$0 < |x - x_0| < \delta_1 \Rightarrow |f(x) - L| < \epsilon \tag{1}$$

and a $\delta_2 > 0$ such that

$$0 < |x - x_0| < \delta_2 \Rightarrow |h(x) - L| < \epsilon. \tag{2}$$

Let $\delta = \min(\delta_1, \delta_2)$. Then if $0 < |x - x_0| < \delta$, we have both

$$L - \epsilon < f(x) < L + \epsilon \qquad [\text{from } (1)]$$

and

$$L - \epsilon < h(x) < L + \epsilon. \qquad [\text{from } (2)]$$

Hence

$$L - \epsilon < f(x) \leq g(x) \leq h(x) < L + \epsilon,$$

whence $L - \epsilon < g(x) < L + \epsilon$. Thus

$$0 < |x - a| < \delta \Rightarrow |g(x) - L| < \epsilon. \quad Q.E.D.$$

Other forms of limit

Limits "from the right" or "from the left," or limits "at $\pm\infty$", which are introduced informally in §1.1, have been left out of the discussion above, for two reasons. They are far less important than $\lim_{x\to x_0}$; and the theorems and proofs for these other forms are virtually the same as for $\lim_{x\to x_0}$, so they make good exercises for reviewing the proofs above. As a sample, we define the *limit from the right at the point a:*

$\lim_{x \to a+} f(x) = L$ iff, for every $\epsilon > 0$, there is a $\delta > 0$ such that

$\lim_{x \to a+} f(x)$

$$a < x < a + \delta \Rightarrow |f(x) - L| < \epsilon.$$

As another sample, we define the *limit at* $+\infty$:

$\lim_{x \to +\infty} f(x) = L$ iff, for every $\epsilon > 0$, there is a number N such that

$\lim_{x \to +\infty} f(x)$

$$x > N \Rightarrow |f(x) - L| < \epsilon.$$

Further definitions and proofs are left as problems.

PROBLEMS

1. Suppose that $f(x) < g(x)$ wherever both f and g are defined. Does it follow that $\lim_{x \to x_0} f(x) < \lim_{x \to x_0} g(x)$? [Hint: Consider $f(x) = x^2$, $g(x) = 2x^2$, both defined only for $x \neq 0$.]

2. Prove that $\lim_{x \to 0} x^2 \sin(1/x) = 0$. [Apply the Poe Principle with $f(x) = -x^2$ and $h(x) = x^2$.]

3. (a) Prove that $\lim_{x \to 0+} \dfrac{x}{|x|} = 1$.

(b) Define $\lim_{x \to a-} f(x)$.

(c) Prove that $\lim_{x \to 0-} \dfrac{x}{|x|} = -1$.

4. For $\lim_{x \to a+}$, state and prove:

(a) Theorem 1
(b) Theorem 2
(c) Theorem 3
(d) Theorem 4
(e) Theorem 5
(f) Theorem 6

[*Caution:* The statement of Theorem 3 involves $\lim_{x \to a+} g(x)$ and $\lim_{y \to B} f(y)$, *not* $\lim_{y \to B+} f(y)$.]

5. Define $\lim_{x \to a-}$, and repeat Problem 4 for this form of limit.

6. Prove that $\lim_{x \to +\infty} (1/x) = 0$.

7. Repeat Problem 4, using $\lim_{x \to +\infty}$ instead of $\lim_{x \to a+}$.

8. Define $\lim_{x \to -\infty}$, and prove that $\lim_{x \to -\infty} (1/x^2) = 0$.

9. Repeat Problem 4 for $\lim_{x\to-\infty}$.

10. Evaluate $\lim_{x\to 0+} f(x)$ and $\lim_{x\to 0-} f(x)$ for the following functions, *if* these limits exist.

(a) $f(x) = \sin(1/x)$ (c) $f(x) = x\sin(1/x)$

(b) $f(x) = e^{-1/x}$ (d) $f(x) = \dfrac{1}{x}\sin\dfrac{1}{x}$

[In (b), use the fact that for $t > 0$, $e^t > t$ and $e^{-t} < 1/t$.]

11. Prove the THEOREM: $\lim_{x\to x_0} f(x) = A$ iff $\lim_{x\to x_0+} f(x) = A$ and $\lim_{x\to x_0-} f(x) = A$.

12. From Problem 11, prove that $\lim_{x\to x_0} f(x)$ does not exist if $\lim_{x\to x_0+} f(x) \neq \lim_{x\to x_0-} f(x)$. [A careful argument requires the corollary of Theorem 5.]

13. (a) Using Problems 6–9, prove that

$$\lim_{x\to\pm\infty} (1 + c_1/x + \cdots + c_n/x^n) = 1.$$

(b) Prove that $\lim_{x\to\pm\infty} \dfrac{a_n x^n + \cdots + a_0}{b_m x^m + \cdots + b_0} \begin{cases} = a_n/b_m & \text{if } n = m \\ = 0 & \text{if } n < m \\ \text{does not exist} & \text{if } n > m. \end{cases}$

[Assume that $a_n \neq 0$ and $b_m \neq 0$.]

14. The statement "$\lim_{x\to+\infty} f(x) = +\infty$" means that for every $M > 0$, no matter how *large*, there is an N such that $x > N \Rightarrow f(x) > M$.

(a) Prove that $\lim_{x\to+\infty} ax^n = +\infty$, if $a > 0$ and $n = 1, 2, \ldots$.

(b) Define "$\lim_{x\to+\infty} f(x) = -\infty$".

(c) Prove that $\lim_{x\to+\infty} ax^n = -\infty$ if $a < 0$ and $n = 1, 2, \ldots$.

(d) Define "$\lim_{x\to-\infty} f(x) = +\infty$" and "$\lim_{x\to-\infty} f(x) = -\infty$".

(e) Discuss $\lim_{x\to-\infty} ax^n$, for a real and $n = 1, 2, 3, \ldots$.

(f) Discuss $\lim_{x\to+\infty} \dfrac{a_n x^n + \cdots + a_0}{b_m x^m + \cdots + b_0}$ for $n > m$.

(g) Discuss $\lim_{x\to-\infty} \dfrac{a_n x^n + \cdots + a_0}{b_m x^m + \cdots + b_0}$ for $n > m$.

[In parts (f) and (g) assume that $a_n \neq 0$ and $b_m \neq 0$.]

AII.3 DERIVATIVES AND TANGENTS

If f is defined in an open interval containing a, and if

$$\lim_{x \to x_0} \frac{f(x) - f(x_0)}{x - x_0}$$

exists, then this limit is called the *derivative of f at x_0*, denoted $f'(x_0)$. The elementary rules for computing derivatives follow directly from the theorems on limits in the previous section; they are derived in Chapter II, and need not be repeated here. However, the following simple fact about derivatives was used without proof in our discussion of maxima and minima [Proposition G, page 234]:

Theorem 7. *If $f'(c) > 0$, then there is a number $\delta > 0$ such that*

$$c - \delta < x < c \Rightarrow f(x) < f(c) \tag{1}$$

and

$$c < x < c + \delta \Rightarrow f(x) > f(c). \tag{2}$$

Proof. The conclusion follows almost immediately from the definition of limit. By assumption,

$$\lim_{x \to c} \frac{f(x) - f(c)}{x - c} = f'(c) > 0.$$

Take $\epsilon = f'(c)$ in the definition of limit; it follows that there is a $\delta > 0$ such that

$$0 < |x - c| < \delta \Rightarrow -f'(c) < \frac{f(x) - f(c)}{x - c} - f'(c) < f'(c).$$

Hence, for $0 < |x - c| < \delta$, adding $f'(c)$ to the inequalities above yields

$$0 < \frac{f(x) - f(c)}{x - c}. \tag{3}$$

Now let $0 < x - c < \delta$; multiplying (3) by the positive number $x - c$ yields

$$0 < f(x) - f(c),$$

which proves (2). On the other hand, when $-\delta < x - c < 0$, multiplying (3) by the *negative* number $x - c$ *reverses* the inequality, yielding

$$0 > f(x) - f(c),$$

which proves (1). *Q.E.D.*

Characterization of the tangent line

The tangent line has been our main interpretation of the derivative (Chapter I), but so far we have only an intuitive idea of tangents. Now we give a definition of tangent line, and establish its connection with the derivative.

The definition is based on the idea of approximation: The tangent at $(x_0, f(x_0))$ should be that line which approximates the graph of f best at the point $(x_0, f(x_0))$. In particular, the line should pass through $(x_0, f(x_0))$, so (if it is not vertical) it has an equation $y = f(x_0) + m(x - x_0)$, where m is the slope. Figure 6 shows three such lines.

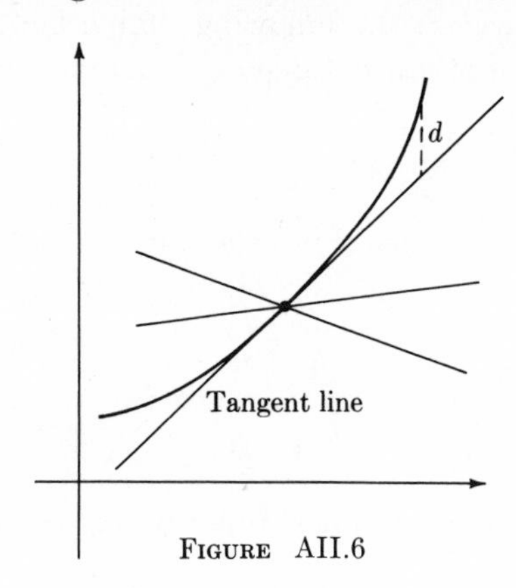

FIGURE AII.6

The fact that the tangent line approximates the graph best at $(x_0, f(x_0))$ can be expressed in terms of an appropriate limit relation, involving the vertical distance d between the tangent line and the graph, given by

$$d(x) = f(x) - [f(x_0) + m(x - x_0)].$$

All the lines in Fig. 6 satisfy the relation $\lim_{x \to x_0} d(x) = 0$. In order to select the correct slope m, we need the stronger condition

$$\lim_{x \to x_0} \frac{d(x)}{x - x_0} = 0.$$

TANGENT LINE **Definition 2.** A line with equation $y = f(x_0) + m(x - x_0)$ is *tangent* to the graph of f at the point $(x_0, f(x_0))$ if and only if

$$\lim_{x \to x_0} \frac{f(x) - [f(x_0) + m(x - x_0)]}{x - x_0} = 0. \tag{4}$$

With this definition, we can state and prove the precise connection between derivatives and tangent lines.

Theorem 8. *The graph of* $y = f(x_0) + m(x - x_0)$ *is tangent to the graph of* f *at* $(x_0, f(x_0))$ *if and only if* f *is differentiable at* x_0, *and* $f'(x_0) = m$.

The proof is simple. Observe that

$$\frac{f(x) - [f(x_0) + m(x - x_0)]}{x - x_0} = \frac{f(x) - f(x_0)}{x - x_0} - m, \quad \text{if } x \neq x_0,$$

and

$$\lim_{x \to x_0} \frac{f(x) - f(x_0)}{x - x_0} = m \quad \Leftrightarrow \quad \lim_{x \to x_0} \left(\frac{f(x) - f(x_0)}{x - x_0} - m \right) = 0$$

[Problem 3, §AII.1]. Hence the relation (4) is equivalent to $f'(x_0) = m$. *Q.E.D.*

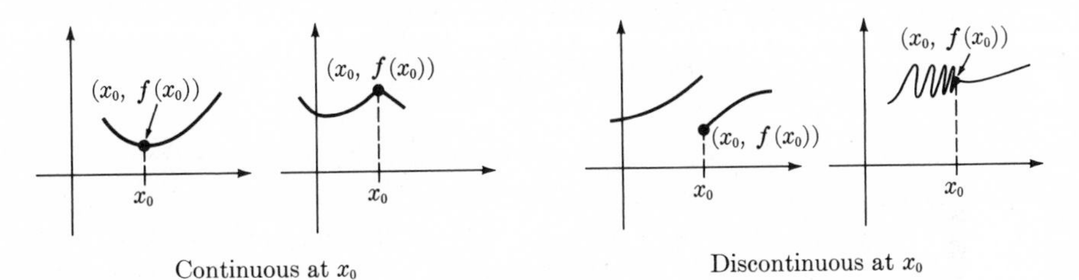

FIGURE AII.7

AII.4 CONTINUOUS FUNCTIONS

A continuous function is, roughly speaking, one whose graph does not jump around too erratically [Fig. 7].

Speaking a little less roughly, a function f is continuous at a point x_0 if $f(x)$ is close to $f(x_0)$ whenever x is close to x_0. The precise analytic version of this idea is:

Definition 3. Let f be a function with domain D, and let x_0 be in D. Then f is *continuous at* x_0 if and only if for every $\epsilon > 0$, there is a $\delta > 0$ such that

$$x \text{ in } D \text{ and } |x - x_0| < \delta \quad \Rightarrow \quad |f(x) - f(x_0)| < \epsilon.$$

Continuity is essential in the Maximum Value Theorem (Chapter IV), the Inverse Function Theorem (Chapter I), and the Fundamental Theorem of Calculus (Chapter V).

Our study of continuity begins with some elementary properties following directly from the definition, and goes deeper in the next section, where we prove two of the theorems mentioned above.

Definition 3 is strikingly similar to Definition 1 for limits. In fact, simply comparing the two definitions leads immediately to the following result:

Theorem 9. *If the domain of f includes an open interval about x_0, then*

$$f \text{ is continuous at } x_0 \iff \lim_{x \to x_0} f(x) = f(x_0).$$

If the domain of f is a closed interval $[a,b]$ or a half-closed interval $[a,b)$, then

$$f \text{ is continuous at } a \iff \lim_{x \to a+} f(x) = f(a).$$

If the domain of f is a closed interval $[a,b]$ or a half-closed interval $(a,b]$, then

$$f \text{ is continuous at } b \iff \lim_{x \to b-} f(x) = f(b).$$

Thus every rational function P/Q (P and Q polynomials) is continuous at x_0 if $Q(x_0) \neq 0$. [See §AII.1.]

A function can be continuous at x_0 without having a derivative at x_0; the absolute value function $|x|$ is continuous at 0, but has no derivative there. But a function can *not* have a derivative at x_0 without being continuous there:

Theorem 10. *If $f'(x_0)$ exists, then f is continuous at x_0.*

Proof.

$$\lim_{x \to x_0} (f(x) - f(x_0)) = \lim_{x \to x_0} \left(\frac{f(x) - f(x_0)}{x - x_0} \cdot (x - x_0) \right)$$

$$= \lim_{x \to x_0} \frac{f(x) - f(x_0)}{x - x_0} \lim_{x \to x_0} (x - x_0)$$

$$= f'(x_0) \cdot 0 = 0. \qquad Q.E.D.$$

Most of the familiar theorems for limits have analogs for continuous functions.

Theorem 11. *Suppose f and g are continuous at x_0. Then $f+g$ and fg are continuous at x_0, and if $g(x_0) \neq 0$, then f/g is continuous at x_0. If c is any constant, then cf is continuous at x_0.*

SUM, PRODUCT, QUOTIENT, COMPOSITION OF CONTINUOUS FUNCTIONS

Theorem 12. *If g is continuous at x_0, and f is continuous at $g(x_0)$, then $f \circ g$ is continuous at x_0.*

For the functions generally considered in calculus (i.e. functions defined on intervals), Theorems 11 and 12 follow immediately from Theorems 2–4 on limits, together with Theorem 9 relating limits and continuity. For example, if f and g are defined on an open interval containing x_0 and continuous at x_0, you can prove the continuity of $f+g$ at x_0 as follows:

$$\lim_{x \to x_0} (f(x) + g(x)) = \lim_{x \to x_0} f(x) + \lim_{x \to x_0} g(x) \qquad \text{[Theorem 2]}$$

$$= f(x_0) + g(x_0) \qquad [f \text{ and } g \text{ are continuous at } x_0]$$

$$= (f+g)(x_0);$$

hence $f+g$ is continuous at x_0, by Theorem 9.

In the general case, where the domains of f and g are arbitrary, Theorems 1 and 2 can be proved simply by copying the corresponding proofs for limits, paying attention to the domains. As an example, we prove Theorem 12 by copying the proof of Theorem 3:

Let $\epsilon > 0$ be given. Since f is continuous at $g(x_0)$, there is a $\delta_1 > 0$ such that

$$f(y) \text{ defined and } |y - x_0| < \delta_1 \quad \Rightarrow \quad |f(y) - f(g(x_0))| < \epsilon. \tag{1}$$

Since $\delta_1 > 0$, and g is continuous at x_0, there is a $\delta > 0$ such that

$$g(x) \text{ defined and } |x - x_0| < \delta \quad \Rightarrow \quad |g(x) - g(x_0)| < \delta_1. \tag{2}$$

Now, "x is in the domain of $f \circ g$" means precisely that $g(x)$ is defined and $f(g(x))$ is defined; hence setting $y = g(x)$ in (1), and combining this with (2), yields

$$x \text{ is in the domain of } f \circ g \text{ and } |x - x_0| < \delta \quad \Rightarrow \quad |f(g(x)) - f(g(x_0))| < \epsilon.$$

This proves that $f \circ g$ is continuous at x_0.

PROBLEMS

1. Use Definition 3 directly to prove the following.
 (a) If f is continuous at x_0, and c is any constant, then cf is continuous at x_0. [You might consider separately the cases $c = 0$ and $c \neq 0$.]
 (b) If f and g are continuous at x_0, so is fg.

2. Prove the following.
(a) If g is continuous at x_0 and $g(x_0) \neq 0$, then $1/g$ is continuous at x_0. [Use Theorem 12.]
(b) If f and g are continuous at x_0, and $g(x_0) \neq 0$, then f/g is continuous at x_0. [Use Problem 1(b).]

3. Prove that if $f(x) = x^{1/3}$, then f is continuous at 0.

4. Prove that if $f(x) = x$ for $x \geq 0$, and is undefined for $x < 0$, then f is continuous at 0.

5. Suppose that f is a function whose domain consists of the integers. Prove that f is continuous at each point of its domain. [Hint: In this case, what happens in Definition 3 when you chose $\delta \leq 1$?]

USEFUL FACT **6.** Prove the following THEOREM, which is useful in connection with integrals, and with arc length: *Let f be defined on (a,b) and continuous at a point x_0 in that interval.* As in Fig. 8, set

$$M(t) = \text{l.u.b.}\ \{f(x) : |x - x_0| < |t|\}$$
$$m(t) = \text{g.l.b.}\ \{f(x) : |x - x_0| < |t|\}.$$

Then $\lim_{t \to 0} M(t) = f(x_0)$ and $\lim_{t \to 0} m(t) = f(x_0)$. [Hint: Given $\epsilon > 0$, let $\epsilon' = \epsilon/2$. For this ϵ', choose the corresponding δ as in Definition 3, and show that when $0 < |t| < \delta$, then

$$f(x_0) - \epsilon < f(x_0) - \epsilon/2 \leq m(t) \leq M(t) \leq f(x_0) + \epsilon/2 < f(x_0) + \epsilon.$$

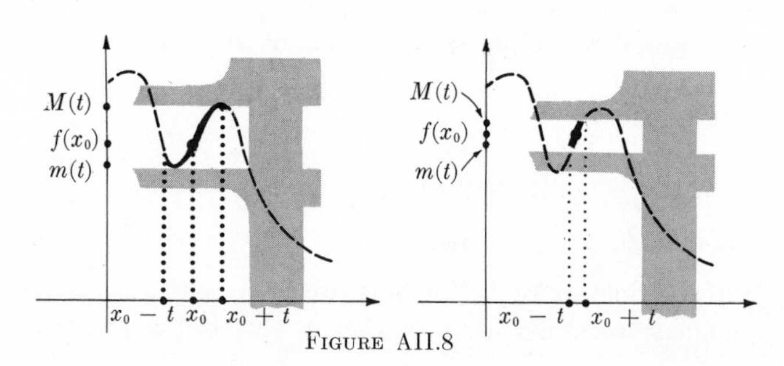

FIGURE AII.8

AII.5 FUNCTIONS CONTINUOUS ON A CLOSED FINITE INTERVAL

The theorems in §AII.4 concerned functions continuous at a single point x_0. Now we consider functions f that are continuous throughout a closed finite interval $[a,b]$. That is,

$$\lim_{x \to x_0} f(x) = f(x_0) \quad \text{for } a < x_0 < b;$$
$$\lim_{x \to a+} f(x) = f(a); \qquad \lim_{x \to b-} f(x) = f(b). \tag{1}$$

For such functions f we prove two basic "existence theorems":

> *There exist a maximum and a minimum value of f.* [Maximum Value Theorem; Fig. 9.]

> *For every number v between $f(a)$ and $f(b)$, there exists a point c where* $f(c) = v$. [Intermediate Value Theorem; Fig. 10.]

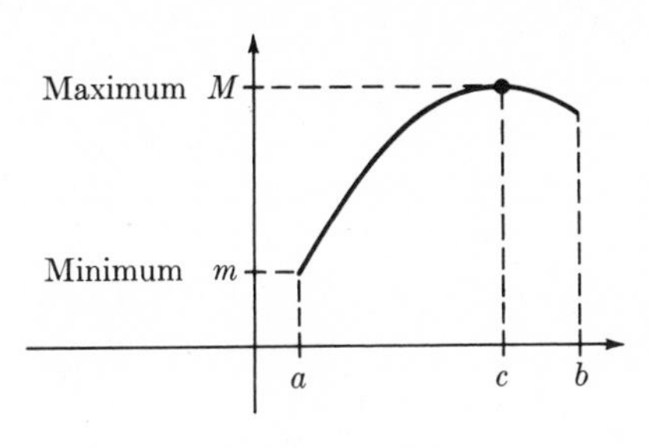

FIGURE AII.9

FIGURE AII.10 $f(c) = v$

These two theorems have a lot in common:

They sound a little vague. They don't say exactly *where* are the maximum and minimum, or *where* is the point c such that $f(c) = v$. However:

They are very important. Chapter IV was based on the Maximum Value Theorem, and Chapter III was based on a corollary of the Mean Value Theorem, which in turn is derived from the Maximum Value Theorem. The *Intermediate Value Theorem* is essential in the study of inverse functions [§AII.6].

Although there are various ways to prove these theorems, *all proofs rest,* directly or indirectly, *on the least upper bound property of the real numbers.*

The proofs given here use infinite sequences (Chapter X). We therefore require the following relation between sequence limits and function limits:

Lemma 1. *Let f be continuous on $[a,b]$, and let $\{x_n\}$ be a convergent sequence with $a \leq x_n \leq b$ for all n. Then*

$$\lim_{n\to\infty} f(x_n) = f(\lim_{n\to\infty} x_n).$$

Proof. Since $a \leq x_n \leq b$, it follows [Theorem 3, §10.2] that $a \leq \lim_{n\to\infty} x_n \leq b$. In view of the limit relations in (1), it follows from Theorem 8, §10.4 (and its analog for $\lim_{x\to a+}$ and $\lim_{x\to b-}$) that $\lim_{n\to\infty} f(x_n) = f(\lim_{n\to\infty} x_n)$. *Q.E.D.*

The chief idea in our proof of both main theorems is illustrated in Fig. 11. Let v be a given number, let S_v be the set of points where $f \geq v$, and let c be the least upper bound of S_v. Then $f(c) = v$, unless $f(b) > v$ [as in Fig. 11(b)] or worse, S_v is empty [Fig. 11(c)].

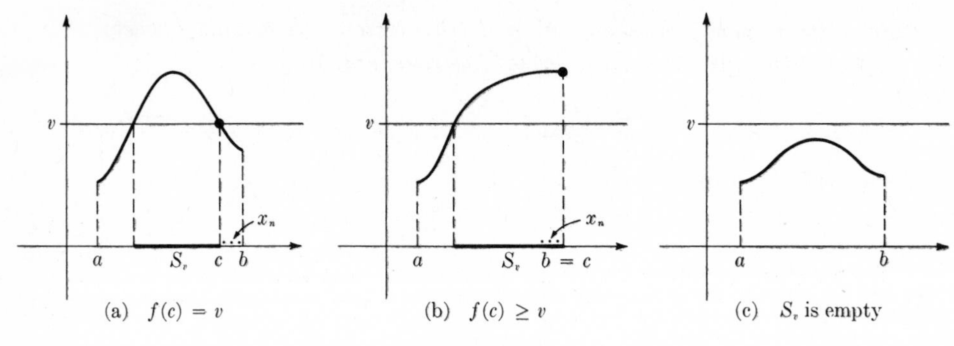

(a) $f(c) = v$ (b) $f(c) \geq v$ (c) S_v is empty

FIGURE AII.11

Lemma 2. *Let f be continuous on $[a,b]$, and let v be a given number. Let* $S_v = \{x : a \leq x \leq b \text{ and } f(x) \geq v\}$. *Then:*

(i) *If S_v is not empty, and* $c = \text{l.u.b. } S_v$, *then* $f(c) \geq v$.

(ii) *If, further,* $v \geq f(b)$, *then* $f(c) = v$.

Proof of (i): S_v is not empty (by hypothesis) and has the upper bound b; therefore it has a least upper bound c. For every positive integer n, $c - 1/n$ is *not* an upper bound for S_v (because c is the *least* upper bound). Hence for every n there is an x_n in S_v [Fig. 11(b)] such that

$$c - 1/n < x_n . \tag{1}$$

Moreover,

$$f(x_n) \geq v \qquad [\text{because } x_n \text{ is in } S_v] \tag{2}$$

and

$$x_n \leq c. \qquad [\text{since } c \text{ is an upper bound for } S_v] \tag{3}$$

From (1), (3), and the Poe Principle [§10.2, Problem 7], $\lim_{n\to\infty} x_n = c$.

Hence

$$f(c) = \lim_{n\to\infty} f(x_n). \qquad [\text{Lemma 1}]$$

This limit, combined with the inequality (2), gives the desired conclusion:

$$f(c) \geq v. \qquad [\text{Theorem 3, §10.2}] \tag{4}$$

Proof of (ii): Now we suppose further that $v \geq f(b)$ [Fig. 11(a)]. If $v = f(b)$, then clearly $b = \text{l.u.b. } S_v = c$, so $f(c) = f(b) = v$ as required.

Suppose then that $v > f(b)$. We proved in (i) that $f(c) \geq v$; it remains to be shown that $f(c) \leq v$. First of all, note that $c < b$; for, $c \leq b$ by definition, and $c \neq b$, since $f(b) < v \leq f(c)$ [by (i)]. Thus there is a sequence $\{x_n\}$, with $c < x_n < b$ and $\lim_{n\to\infty} x_n = c$; for instance, $x_n = c + (b - c)/n$. Since $x_n > c = \text{l.u.b. } S_v$, the points x_n are *not* in S_v; thus, by the definition of S_v, $f(x_n) < v$. Therefore,

$$f(c) = \lim_{n\to\infty} f(x_n) \leq v,$$

and this with (4) proves that $f(c) = v$. *Q.E.D.*

☆☆☆ INTERMEDIATE VALUE THEOREM

We have practically proved one of our main theorems.

Theorem 13. *Let f be continuous on the finite closed interval $[a,b]$. Then for each number v between $f(a)$ and $f(b)$, there is a point c in $[a,b]$ where $f(c) = v$.*

Proof: Case 1. $f(a) \geq v \geq f(b)$. Define S_v as in Lemma 2; then S_v is not empty [since $f(a) \geq v$], and $v \geq f(b)$, so by Lemma 2, $f(c) = v$, where $c = \text{l.u.b. } S_v$.

Case 2. $f(a) \leq v \leq f(b)$. Consider the function $g = -f$. Then $g(a) \geq -v \geq g(b)$, and g is continuous on $[a,b]$. Therefore, by Case 1, $g(c) = -v$ for some point c in $[a,b]$. But $-v = g(c) = -f(c)$ is equivalent to $v = f(c)$. *Q.E.D.*

The Intermediate Value Theorem proves the existence of roots, with ease.

CUBE ROOTS

Example 1. Prove that every positive real number v has a cube root $v^{1/3}$.
Solution. Consider the function $f(x) = x^3$; we must prove that there is a number c such that $f(c) = v$. Let $a = 0$ and $b = v + 1$. Then $f(a) = 0 < v$, and $f(b) = (v+1)^3 = v^3 + 3v^2 + 3v + 1 > v$. By the Intermediate Value Theorem, there is a number c such that $f(c) = v$, i.e. $c^3 = v$, or $c = v^{1/3}$.

We should add that there is *only one* number c such that $c^3 = v$. For if $c_1^3 = v$ and $c_2^3 = v$, then $c_1 < c_2$ is impossible (for $c_1 < c_2 \Rightarrow c_1^3 < c_2^3$), and $c_1 > c_2$ is equally impossible, so $c_1 = c_2$.

We come at last to the *Maximum Value Theorem: If f is continuous on $[a,b]$, then f has a maximum value $f(c)$.* The proof takes two steps: first we prove that f is *bounded* (Lemma 3), then that it has a maximum value.

Lemma 3. *If f is continuous on $[a,b]$, then f is bounded above on $[a,b]$; there is a number M such that $f(x) \leq M$ for all x in $[a,b]$.*

Proof by contradiction. Suppose that f is not bounded. Then none of the sets

$$S_n = \{x: a \leq x \leq b \text{ and } f(x) \geq n\}$$

is empty. Let $c_n = \text{l.u.b.}\{S_n\}$ [Fig. 12]. Then $c_n \geq c_{n+1}$; for S_n contains S_{n+1}, so c_n is an upper bound for S_{n+1}; hence $c_n \geq c_{n+1} = \text{l.u.b. } S_{n+1}$. Thus c_n is a monotone decreasing sequence, bounded below by a. Hence $c = \lim\limits_{n\to\infty} c_n$ exists, and

$$f(c) = \lim_{n\to\infty} f(c_n). \qquad \text{[Lemma 1]} \tag{5}$$

On the other hand, $f(c_n) \geq n$ [by Lemma 2(i)], so

$$\text{the sequence } \{f(c_n)\} \text{ is unbounded.} \tag{6}$$

But (5) and (6) are contradictory, for an unbounded sequence cannot converge [Theorem 5, §10.3]. This contradiction completes the proof of Lemma 3, and we are ready at last to prove:

☆☆☆ MAXIMUM VALUE THEOREM

Theorem 14. *A continuous function f on a closed finite interval $[a,b]$ assumes a maximum (and also a minimum) on $[a,b]$.*

Proof. The range of f is certainly not empty, and by Lemma 2 it has an upper bound. Let

$$M = \text{l.u.b. (range of } f).$$

We must prove that $f(c) = M$ for some c in $[a,b]$. Suppose, on the contrary, that $f(x) = M$ for *no* x in $[a,b]$. Then

$$M - f(x) \text{ is never zero, for } a \leq x \leq b. \tag{7}$$

But M is the *least* upper bound of the range of f, so $M - 1/n$ is *not* an upper bound, and

$$\text{for every integer } n > 0, \text{ there is an } x_n \text{ such that } f(x_n) > M - \frac{1}{n}. \tag{8}$$

In view of (7) and Theorem 11, the function

$$g(x) = \frac{1}{M - f(x)}, \qquad a \leq x \leq b,$$

is continuous on $[a,b]$. In view of (8), $\dfrac{1}{M - f(x_n)} > n$, so g is unbounded.

But this contradicts the continuity of g, by Lemma 3.

This completes the proof that f has a maximum value M. One could prove analogously that f has a minimum; however, it is easy to deduce this by the following trick. Observe that $-f$ is continuous on $[a,b]$, so $-f$ assumes a maximum $-f(\bar{c})$ at some point $\bar{c}$. Then $f(\bar{c})$ is a minimum for f, since

$$-f(\bar{c}) \geq -f(x) \quad \Leftrightarrow \quad f(\bar{c}) \leq f(x). \qquad \textit{Q.E.D.}$$

We have now filled in all the missing links in the proofs of the Mean Value Theorem [§4.2] and its many corollaries.

We can sum up the Intermediate Value Theorem and the Maximum Value Theorem in a single statement:

Theorem 15. *If f is a continuous function whose domain is a closed finite interval $[a,b]$, then the range of f is the closed finite interval $[m,M]$, where $m = \min_{[a,b]} f$ and $M = \max_{[a,b]} f$.*

The proof is left to you in Problem 2. Figure 13 illustrates the result. The range is an *interval* because of the Intermediate Value Theorem; this interval includes its endpoints because of the Maximum Value Theorem.

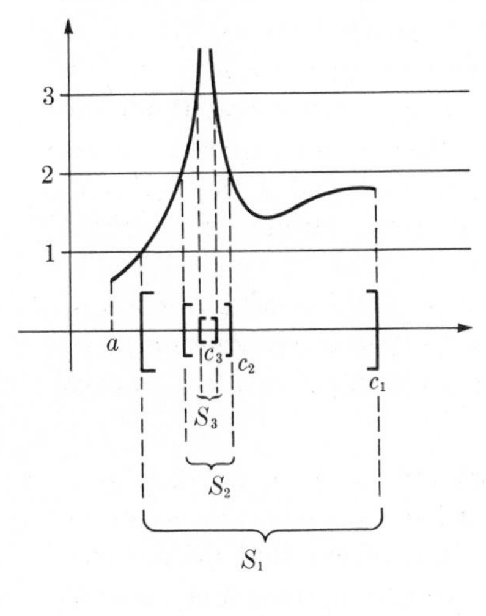

FIGURE AII.12

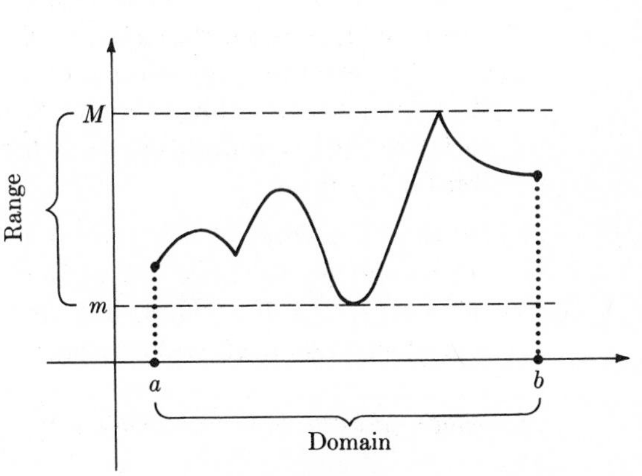

FIGURE AII.13

PROBLEMS

1. Which of the following "theorems" are true, and which are false?
   (a) If f is a continuous function whose domain is an open finite interval (a,b), then the range of f is an open interval.
   (b) If f is a continuous function whose domain is an open finite interval (a,b), then the range of f is a finite interval.
   (c) If f is a continuous function whose domain is an interval, then the range of f is an interval.

2. Prove Theorem 15. [The range of f obviously excludes all values either larger than M or less than m. Prove that it includes every value in the interval $[m,M]$.]

nth ROOTS

3. Prove that if v is any real number, positive or not, then there is a real number c such that $c^3 = v$.

4. Given a positive number v and an integer n, prove that there is a positive number c such that $c^n = v$.

INTERMEDIATE VALUE THEOREM

5. Suppose f is continuous on a closed finite interval $[a,b]$, and differentiable on the open interval (a,b). Where can the maximum values of f occur? Prove your answer.

6. In the sixteenth century Simon Stevin gave the following procedure for finding the roots r of a polynomial P. [The roots of P are the numbers r such that $P(r) = 0$.] Suppose that $P(n_1) < 0$ for at least one integer, and $P(n_2) > 0$ for another, $n_2 > n_1$. Then there is an integer n such that $P(n) \leq 0$ and $P(n+1) > 0$. Divide the interval between n and $n+1$ by nine equally spaced points, denoted in decimal notation by $n.1, n.2, \ldots, n.9$. Then there is an integer a_1, $0 \leq a_1 < 9$, such that $P(n.a_1) \leq 0$ and $P(n.a_1 + \frac{1}{10}) > 0$. Once more, divide the interval between $n.a_1$ and $n.a_1 + \frac{1}{10}$ by nine equally spaced points, and find an integer a_2 such that $0 \leq a_2 < 9$ and $P(n.a_1a_2) \leq 0$, $P(n.a_1a_2 + \frac{1}{100}) > 0$. Continuing thus, you obtain a decimal $r = n.a_1a_2a_3\ldots$ such that $P(r) = 0$.

   Convert this procedure into a proof of the following version of the Intermediate Value Theorem: *Given a continuous function on* $[0,1]$ *such that* $f(0) < 0$ *and* $f(1) > 0$, *there is a number* r, $0 < r < 1$, *such that* $f(r) = 0$.

7. Prove the THEOREM: *If* $f(x) = x^n + a_{n-1}x^{n-1} + \cdots + a_1x + a_0$ *is a polynomial of* odd *degree* n, *then* f *has at least one real root* r, i.e. *a point* r *such that* $f(r) = 0$. [Hint: The first step is to prove that f is positive somewhere, and negative somewhere else. Write $f(x) = x^n g(x)$, $x \neq 0$,

   where $g(x) = 1 + \dfrac{a_{n-1}}{x} + \cdots + \dfrac{a_0}{x^n}$, and note that $\lim\limits_{x \to \pm\infty} g(x) = 1$;

   hence $g(x)$ is positive for $|x|$ sufficiently large. What does this imply about the sign of f?]

8. Prove the THEOREM: *If* $f(x) = a_nx^n + \cdots + a_0$ *is a polynomial of* even *degree* n, *then* f *has* (i) *a minimum point over* R, *if* $a_n > 0$, *or* (ii) *a maximum point over* R, *if* $a_n < 0$. [Hint: Write $f(x) = a_nx^ng(x)$, as in Problem 7. If $a_n > 0$, find a large interval $[-A,A]$ outside of which $f(x) \geq f(0)$; then $\min_{[-A,A]} f = \min_{(-\infty,\infty)} f$.]

9. Prove the following THEOREM: *Let* f' *exist throughout* $[a,b]$, *and let* $f'(a) < v < f'(b)$. *Then there is a point* c *in* (a,b) *where* $f'(c) = v$. [Hints: (i) Note that f' is not assumed continuous, so you cannot apply the Intermediate Value Theorem to f'. (ii) One possibility is to apply both the Intermediate Value and Mean Value Theorems to f. (iii) Another possibility is to apply the Minimum Value Theorem to the function $g(x) = f(x) - xv$. Prove that the minimum cannot occur at an endpoint.]

MEAN VALUE THEOREM FOR DERIVATIVES

10. Prove the THEOREM: *For every* x *in* (a,b), *let* $f'(x) \neq 0$. *Then* f *is either increasing on* (a,b) *or decreasing on* (a,b). [Use the previous problem and Proposition F, Chapter III.]

AII.6 INVERSE FUNCTIONS

Inverse functions were introduced in §0.4, where we showed the following:

(I) A function f has an inverse g if and only if each horizontal line intersects the graph of f at most once.

(II) If f has an inverse g, then

$$y = f(x) \quad \Leftrightarrow \quad x = g(y).$$

As a consequence, the domain of g equals the range of f; and f and g satisfy $g(f(x)) = x$ and $f(g(y)) = y$ for x in the domain of f and y in the domain of g.

The next encounter with inverse functions was in §2.5, where we stated:

Proposition D. *Suppose that the domain of* f *is an open interval* I, *and either*

(i) $f'(x) > 0$ *for all points* x *in* I, *or*

(ii) $f'(x) < 0$ *for all points* x *in* I.

Then f *is one-to-one* (*so it has an inverse*), *and the inverse function is* ***differentiable***.

This section proves two versions of the Inverse Function Theorem. The second version is essentially Proposition D; the first version, somewhat simpler to prove, is the following:

Theorem 16. *Let f be a continuous and strictly increasing function whose domain is the closed interval $[a,b]$. Then f assumes every value in the interval $[f(a), f(b)]$ exactly once, and there is a unique inverse function g defined on $[f(a), f(b)]$. This function g is strictly increasing and continuous. Finally, if f is differentiable at any point $\bar{x}$ where $a < \bar{x} < b$, and if $f'(\bar{x}) \neq 0$, then g is differentiable at $f(\bar{x})$ and $g'(f(\bar{x})) = 1/f'(\bar{x})$.*

In Theorem 16, the fact that the domain of the inverse function is an interval is required in the proof of differentiability, for we can take $\lim\limits_{y \to y_0} \dfrac{g(y) - g(y_0)}{y - y_0}$ only when g is defined in an interval about y_0. The continuity of the inverse function g also enters into the proof of differentiability, as you will see.

We now establish one by one the many claims in the Inverse Function Theorem.

Proof of Theorem 16

1. *f assumes each value in $[f(a), f(b)]$*, by the Intermediate Value Theorem.

2. *The range of f is precisely the interval $[f(a), f(b)]$.* From **1,** the range of f includes this interval. And, conversely, this interval includes the range, because f is increasing; $a \leq x \leq b \Rightarrow f(a) \leq f(x) \leq f(b)$.

3. *f assumes each value in its range exactly once*, again because f is increasing. It is impossible to have $f(x_1) = f(x_2)$ for two different values x_1 and x_2. For if $x_1 < x_2$, then $f(x_1) < f(x_2)$, and if $x_1 > x_2$, then $f(x_1) > f(x_2)$, contradicting $f(x_1) = f(x_2)$.

4. *f has an inverse function g, defined on $[f(a), f(b)]$, such that $g(f(x)) = x$ and $f(g(y)) = y$*, for all x and y in the appropriate intervals. For by **3,** each horizontal line intersects the graph of f at most once, so f has an inverse g; the domain of g is the range of f, which is given in **2**.

5. *g is strictly increasing.* Let $y_1 < y_2$; then $g(y_1) < g(y_2)$. For, assuming the contrary, we would have

$$g(y_1) \geq g(y_2) \quad \Rightarrow \quad f(g(y_1)) \geq f(g(y_2)) \qquad [f \text{ is increasing}]$$
$$\Rightarrow \quad y_1 \geq y_2 , \qquad [\text{since } f(g(y)) = y]$$

contradicting the hypothesis $y_1 < y_2$.

6. *g is continuous.* Let y_0 be given, and $\epsilon > 0$ be given. Suppose at first that y_0 is not an endpoint. We must find a $\delta > 0$ such that

$$y_0 - \delta < y < y_0 + \delta \quad \Rightarrow \quad g(y_0) - \epsilon < g(y) < g(y_0) + \epsilon.$$

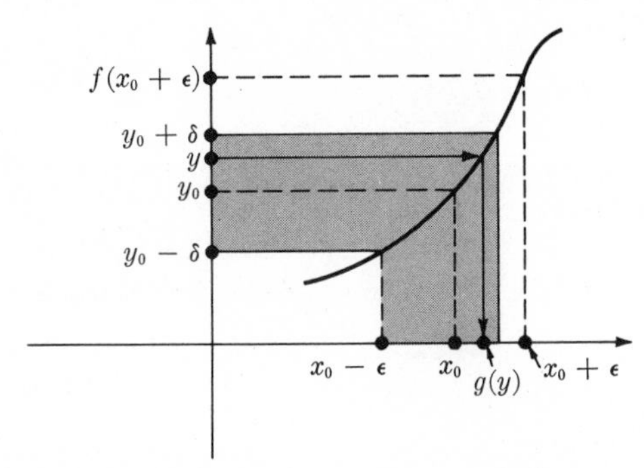

FIGURE AII.14 If $y_0 - \delta < y < y_0 + \delta$, then $x_0 - \epsilon < g(y) < x_0 + \epsilon$

As Fig. 14 suggests, we can take δ to be the smaller of

$$f(x_0 + \epsilon) - y_0 \qquad \text{and} \qquad y_0 - f(x_0 - \epsilon),$$

where $y_0 = f(x_0)$. Then $f(x_0 - \epsilon) \leq y_0 - \delta$ and $y_0 + \delta \leq f(x_0 + \epsilon)$, so

$$y_0 - \delta < y < y_0 + \delta \quad \Rightarrow \quad f(x_0 - \epsilon) < y < f(x_0 + \epsilon)$$
$$\Rightarrow \quad g(f(x_0 - \epsilon)) < g(y) < g(f(x_0 + \epsilon))$$
$$[g \text{ is increasing}]$$
$$\Rightarrow \quad x_0 - \epsilon < g(y) < x_0 + \epsilon \qquad [g(f(x)) = x]$$
$$\Rightarrow \quad g(y_0) - \epsilon < g(y) < g(y_0) + \epsilon$$
$$[y_0 = f(x_0), \text{ so } g(y_0) = x_0]$$

as desired. [We have tacitly assumed that $x_0 - \epsilon$ and $x_0 + \epsilon$ lie in $[a,b]$, so that $f(x_0 - \epsilon)$ and $f(x_0 + \epsilon)$ are defined. If this were not so, we could replace ϵ by the smaller number $\epsilon_1 = \min(x_0 - a, b - x_0)$. Then the above proof works for ϵ_1. Let δ be the number corresponding to ϵ_1. Then

$$y_0 - \delta < y < y_0 + \delta \quad \Rightarrow \quad |g(y) - g(y_0)| < \epsilon_1 < \epsilon, \quad \text{as required.}$$

We have assumed that $f(a) < y_0 < f(b)$. The proof when y_0 is an endpoint is similar but simpler, and can be omitted.]

7. *If* $f'(\bar{x}) \neq 0$, *then* $g'(f(\bar{x}))$ *exists and equals* $1/f'(\bar{x})$. Let $\bar{y} = f(\bar{x})$. We want to show that

$$\lim_{y \to \bar{y}} \frac{g(y) - g(\bar{y})}{y - \bar{y}} = \frac{1}{f'(\bar{x})}.$$

Consider the function D defined by

$$D(x) = \frac{f(x) - f(\bar{x})}{x - \bar{x}} \qquad \text{if } x \neq \bar{x},$$

$$D(\bar{x}) = f'(\bar{x}).$$

Then, by the definition of derivative, the definition of $D(\bar{x})$, and the choice of $\bar{y}$, we have

$$\lim_{x \to \bar{x}} D(x) = f'(\bar{x}) = D(\bar{x}) = D(g(\bar{y})). \tag{1}$$

Hence

$$\begin{aligned}
f'(\bar{x}) &= D(\bar{x}) && \text{[definition of } D\text{]}\\
&= D(g(\bar{y}))\\
&= D(\lim_{y \to \bar{y}} g(y)) && \text{[by 6]}\\
&= \lim_{y \to \bar{y}} D(g(y)) && \text{[by (1) and Theorem 3]}\\
&= \lim_{y \to \bar{y}} \frac{f(g(y)) - f(g(\bar{y}))}{g(y) - g(\bar{y})} && \text{[definition of } D\text{]}\\
&= \lim_{y \to \bar{y}} \frac{y - \bar{y}}{g(y) - g(\bar{y})}. && [f(g(y)) = y]
\end{aligned}$$

Since $f'(\bar{x}) \neq 0$, we may take reciprocals, obtaining

$$\frac{1}{f'(\bar{x})} = \lim_{y \to \bar{y}} \frac{g(y) - g(\bar{y})}{y - \bar{y}} = g'(\bar{y}). \quad \text{Q.E.D.}$$

We often require the inverse to a function defined on an open interval, for instance, the inverse of e^x on $(-\infty, \infty)$. Then Theorem 16 does not apply, and we need a slightly different version of the inverse function theorem.

Theorem 17. *Let f be strictly increasing on an open interval, finite or infinite. Then the range of f is an open interval (finite or infinite), and on this range there is defined a unique inverse function g. This function is strictly increasing and continuous, and $f(g(y)) = y$. If f is differentiable at $\bar{x}$ and $f'(\bar{x}) \neq 0$, then g is differentiable at $f(\bar{x})$ and $g'(f(\bar{x})) = 1/f'(\bar{x})$.*

The proof is exactly like the proof of Theorem 16, except that in step **1** we prove that the range of f is an open interval I. Then in the remaining steps we replace the closed bounded interval $[f(a), f(b)]$ by the open interval I.

Depending on f, the range can take any of four different forms: it can be bounded above and below, or above but not below, or below and not above, or neither below nor above. We consider the case where the range is bounded above but not below. In that case, there is a least upper bound to the range, $c = \text{l.u.b. } \{f(x)\}$, and we can show in two steps that the range of f is the open interval $(-\infty, c)$.

(a) The range of f lies in $(-\infty, c)$. By definition of c, the range does not contain any numbers greater than c. It does not contain c either; for if $f(x) = c$, then since f is defined on an open interval, it is defined for an $x_1 > x$, and $f(x_1) > f(x) = c$, contradicting the definition of c. Thus the range of f lies in the open interval $(-\infty, c)$.

(b) The range of f includes $(-\infty, c)$. For if y lies in this interval, then $y < c$, so y is not an upper bound for the range, and there is a $y_1 > y$ with $f(x_1) = y_1$; and since the range is not bounded below, there is a $y_2 < y$ with $f(x_2) = y_2$. Then from the Intermediate Value Theorem, there is an x with $f(x) = y$. *Q.E.D.*

Example 1. If $f(x) = e^x$, then the range of f is unbounded above, and its greatest lower bound is 0. Hence there is a differentiable inverse function g defined on the open interval $(0, \infty)$. This is the log function.

Proposition D is an easy consequence of Theorem 17. Suppose that $f' > 0$ on I; then f is continuous and strictly increasing, so there is a differentiable inverse g. In the case where $f' < 0$ on I, we need only consider $-f$; $(-f)' = -f' > 0$, so $-f$ has a differentiable inverse h. If we define $g(y) = h(-y)$, then g is the required inverse of f, since

$$y = f(x) \quad \Leftrightarrow \quad -y = -f(x) \quad \Leftrightarrow \quad x = h(-y) \quad \Leftrightarrow \quad x = g(y).$$

PROBLEMS

In Problems 1–3, *assume* the existence of a function exp with domain $(-\infty, \infty)$, range $(0, \infty)$, and satisfying the equations

$$\exp' = \exp \qquad \text{and} \qquad \exp(x + y) = \exp(x) \cdot \exp(y).$$

1. Let log be defined as the inverse function of exp. Prove that $\log'(x) = 1/x$. [Hint: You can use the chain rule, together with Theorem 17 guaranteeing that log′ exists.]

2. Let log be defined as the inverse function of exp. Prove that $\log(ab) = \log a + \log b$.

3. Define a^x by $a^x = e^{x \log a}$ for $a > 0$. Prove that $a^{x+y} = a^x a^y$.

4. Find the derivative of the function inverse to f on $(0, \infty)$, if $f(x) = x^n$, n a positive integer.

5. Suppose f is strictly increasing, but not continuous. What parts of the inverse function theorem fail? [Hint: Sketch a strictly increasing function with a discontinuity at one point.]

6. Let $f(x) = x^3$ on $(-\infty, \infty)$; then f is differentiable everywhere. Prove that in this case the inverse function is *not* differentiable everywhere.

AII.7 UNIFORM CONTINUITY

The concept of uniform continuity has no simple geometric interpretation, and no immediate practical application. Like the Mean Value Theorem, its value and its real meaning appear in the proofs of other more geometric or more directly applicable theorems. In this book, the main application is in the proof that every continuous function is integrable [§AII.8].

UNIFORM CONTINUITY

Definition 4. A function f defined on a set S is *uniformly continuous* on S iff, for each $\epsilon > 0$, there is a $\delta > 0$ such that

$$|x - \bar{x}| < \delta \quad \Rightarrow \quad |f(x) - f(\bar{x})| < \epsilon$$

whenever x and $\bar{x}$ are in S.

You can prove that if f is *uniformly* continuous on its domain D, then it is continuous at each point of D. The converse is not true; the function

$f(x) = 1/x$ is *not* uniformly continuous on its domain $D = \{x: x \neq 0\}$, but *is* continuous at each point of D.

The adverb "uniformly" is applied because a single restriction on $|x - \bar{x}|$ [namely $|x - \bar{x}| < \delta$] makes $|f(x) - f(\bar{x})|$ uniformly small throughout S [$|f(x) - f(\bar{x})| < \epsilon$].

One very simple condition guarantees uniform continuity:

Theorem 18. *If $|f'(x)| \leq M$ for every x in an interval I, then f is uniformly continuous on I.*

Proof. By the Mean Value Theorem,

$$|f(x) - f(\bar{x})| = |f'(c)(x - \bar{x})| = |f'(c)|\,|x - \bar{x}| \leq M|x - \bar{x}|;$$

hence

$$|x - \bar{x}| < \frac{\epsilon}{M} \quad \Rightarrow \quad |f(x) - f(\bar{x})| < \epsilon,$$

and Definition 4 is satisfied with $\delta = \epsilon/M$. *Q.E.D.*

Thus assuming a uniform bound on $|f'|$ (i.e. a single bound valid throughout the interval), we easily deduce the uniform continuity of f.

The theorem below is much more powerful, for we assume only that f is continuous at each point in I. But we have to assume a little more about I, namely, that it is *closed* and *bounded*, just as in the Maximum Value Theorem and the Intermediate Value Theorem.

UNIFORM CONTINUITY: MAIN THEOREM

Theorem 19. *If f is continuous at every point of the closed bounded interval $[a,b]$, then f is uniformly continuous on that interval.*

Proof. We suppose that the theorem is false, and use the Bolzano-Weierstrass theorem [§10.5] to locate the source of the difficulty. This leads to a contradiction, so the theorem is true after all. Here are the details.

If the theorem is false, then for *some* $\epsilon > 0$ there is *no* $\delta > 0$ such that

$$|x - \bar{x}| < \delta \quad \Rightarrow \quad |f(x) - f(\bar{x})| < \epsilon. \tag{1}$$

In particular, the implication (1) is false when $\delta = 1/n$, so there are numbers x_n and $\bar{x}_n$ such that

$$|x_n - \bar{x}_n| < 1/n, \tag{2}$$

but

$$|f(x_n) - f(\bar{x}_n)| \geq \epsilon. \tag{3}$$

By the Bolzano-Weierstrass theorem, there is a subsequence $\{x_{n_k}\}$ of the x_n such that $n_k \geq k$ and $\lim_{k\to\infty} x_{n_k} = c$ for some point c in $[a,b]$. It follows from (2) that

$$x_{n_k} - \frac{1}{n_k} < \bar{x}_{n_k} < x_{n_k} + \frac{1}{n_k};$$

hence $\lim_{k\to\infty} \bar{x}_{n_k} = \lim_{k\to\infty} x_{n_k} = c$. Since f is continuous at c, we have

$$\lim_{k\to\infty} [f(x_{n_k}) - f(\bar{x}_{n_k})] = f(c) - f(c) = 0,$$

and this contradicts (3). *Q.E.D.*

Example 1. Prove that every function continuous on $[a,b]$ is bounded on $[a,b]$. *Solution.* We proved this in §AII.5, but now we will do it again, using Theorem 19. Since f is uniformly continuous, there is a $\delta > 0$ such that

$$|x - \bar{x}| < \delta \Rightarrow |f(x) - f(\bar{x})| < 1. \tag{4}$$

Now divide $[a,b]$ into n intervals of length $< 2\delta$ by the points $x_j = a + j\dfrac{b-a}{n}$ where $\dfrac{b-a}{n} < 2\delta$. Let M be the maximum of $|f(x_0)|,\ldots, |f(x_n)|$. Then for every point x in $[a,b]$, one of the points x_j satisfies $|x_j - x| < \delta$. Hence by (4),

$$\begin{aligned}|f(x)| = |f(x_j) + [f(x) - f(x_j)]| &\leq |f(x_j)| + |f(x) - f(x_j)| \\ &\leq M + 1. \qquad \textit{Q.E.D.}\end{aligned}$$

PROBLEMS

1. Each of the following functions is uniformly continuous on the given interval. In each case, state whether Theorem 18 applies, Theorem 19 applies, both apply, or neither applies. If neither applies, prove the uniform continuity.
   (a) $f(x) = x^3$ on $[-1,1]$
   (b) $f(x) = x$ on $(-\infty, \infty)$
   (c) $f(x) = x$ on $[0,1)$
   (d) $f(x) = |x|$ on $[-1,1]$
   (e) $f(x) = |x|$ on $(-1,1)$

2. Let $f(x) = 1/x$ on $(0, \infty)$.
   (a) Sketch the graph of f; pick a general point $(x, 1/x)$ on the graph; sketch a horizontal line passing one unit above this point; label the intersection of this horizontal line with the graph as $(x_0, 1/x_0)$. [Note that as x gets close to zero, the distance δ from x to x_0 also goes to zero. This suggests that f is not uniformly continuous. The remainder of this problem proves this fact in detail.]

(b) Show that $\frac{1}{x_0} - \frac{1}{x} = \epsilon$ when $x_0 = x - \frac{\epsilon x^2}{1 + \epsilon x}$.

(c) Show that $x_0 > 0$ [with x_0 as in part (a)].

(d) Show that if $\delta > \frac{\epsilon x^2}{1 + \epsilon x}$, then there is a number $\bar{x} > 0$ such that

$|x - \bar{x}| < \delta$ but $\left|\frac{1}{x} - \frac{1}{\bar{x}}\right| > \epsilon$. [Note that you can take $\bar{x} < x_0 =$

$x - \frac{\epsilon x^2}{1 + \epsilon x}$; compare part (b).]

(e) Show that f is not uniformly continuous on $(0, \infty)$. [Note from part (d) that the number δ in Definition 4 would have to be

$\leq \left(\frac{\epsilon x^2}{1 + \epsilon x}\right)$ for *every* $x > 0$.]

3. Does Theorem 18 imply Theorem 19? Does Theorem 19 imply Theorem 18? Explain.

4. Let $f(x) = 1/x$. On which of the following intervals is f uniformly continuous?
(a) $(0,1)$ (b) $(0, \infty)$ (c) $[1,2]$ (d) $[1, \infty)$

5. Let $f(x) = 1/(1 + x^2)$. Is f uniformly continuous on the whole real line?

6. Let $f(x) = x^{1/3}$.
(a) Does Theorem 18 show that f is uniformly continuous on $[-1,1]$?
(b) Does Theorem 19 show that f is uniformly continuous on $[-1,1]$?
(c) Does either theorem show that f is uniformly continuous on $(-\infty, \infty)$?
(d) Prove that f is uniformly continuous on $(-\infty, \infty)$.

7. Suppose f is any function whose domain is the integers. Prove that f is uniformly continuous. [Hint: Try $\delta = 1/2$.]

8. Suppose that f is uniformly continuous on its domain D. Prove that f is continuous at each point of D.

AII.8 INTEGRALS OF CONTINUOUS FUNCTIONS

The theory of integration in Chapter V rests on two basic propositions about functions f that are continuous on a closed finite interval $[a,b]$:

Proposition I. *Partition $[a,b]$ into n equal subintervals by points $a = x_0 < x_1 < \cdots < x_n = b$, chose ξ_j anywhere in $[x_{j-1}, x_j]$, and form the Riemann sum*

$$S_n = \sum_{j=1}^{n} f(\xi_j)\,\Delta x_j \qquad (\text{where } \Delta x_j = x_j - x_{j-1}).$$

Then, as $n \to \infty$, the sequence S_n converges to a definite number, denoted $\int_a^b f$, and called the integral of f over $[a,b]$.

Proposition J. *If we define $A(t) = \int_a^t f$, then $A'(t) = f(t)$.*

In neither proposition can we drop the assumption that f be continuous. The curious example

$$f(x) = \begin{cases} 1 & \text{if } x \text{ is rational} \\ 0 & \text{if } x \text{ is irrational} \end{cases} \tag{1}$$

gives a function for which the Riemann sums do *not* converge to a definite number, so $\int_a^b f$ is not defined.

The trick in proving Proposition I is to find the number to which the Riemann sums converge. The two main ideas used here are the abstract *least upper bound axiom* [§AI.5], and the geometric interpretation of the integral as the area under the graph. [The definition of arc length, §AI.9, uses the least upper bound axiom in a very similar way.]

sup f, inf f

Definition 5. Suppose that f is bounded on an interval $[a,b]$, i.e. there are numbers m and M such that

$$m \le f(x) \le M \quad \text{for all points in } [a,b].$$

Let R denote the range of f,

$$R = \{y : y = f(x) \text{ for some } x \text{ in } [a,b]\}.$$

Then the *sup* and *inf* of f over the interval $[a,b]$ are defined by

$$\sup_{[a,b]} f = \text{l.u.b. } R \qquad \text{and} \qquad \inf_{[a,b]} f = \text{g.l.b. } R.$$

Thus $\sup_{[a,b]} f$ is the smallest of all upper bounds for f on $[a,b]$, and $\inf_{[a,b]} f$ is the largest of all lower bounds [Fig. 15].

Definition 6. Let f be bounded on $[a,b]$, and let P be a partition of $[a,b]$ by the numbers

$$a = x_0 < x_1 < \cdots < x_n = b.$$

We then define the *lower Riemann sum:*

$$L(f, P) = \sum_{j=1}^{n} \left(\inf_{[x_{j-1}, x_j]} f \right) \Delta x_j .$$

LOWER RIEMANN SUM

When $f \geq 0$, the lower Riemann sum represents the area of a region contained *below* the graph of f; if the partition is very fine (i.e. each Δx_j is small), this lower sum should, intuitively, be very close to the area of the region under the graph.

Example 1. If $f = c$ (a constant), then it is easy to see that $L(f,P) = c \cdot (b - a)$ for every partition P. For in this case, $\inf_{[x_{j-1}, x_j]} f = c$, so

$$L(f,P) = \sum_{j=1}^{n} c\Delta x_j = c \sum_{j=1}^{n} \Delta x_j = c \cdot (b - a).$$

Example 2. If f is the unusual function given by (1) above, then $L(f,P) = 0$ for every partition P; for in this case, every interval $[x_{j-1}, x_j]$ contains an irrational number ξ, where $f(\xi) = 0$, so $\inf_{[x_{j-1}, x_j]} f = 0$, and hence

$$L(f,P) = \sum_{j=1}^{n} 0 \cdot \Delta x_j = 0.$$

Example 3. Suppose that $f(x) = c$ except at one point, say at $x = a$, where we have $f(a) = c' \neq c$ [Fig. 16]. Consider first the case where $c' > c$. Then $\inf_{[x_{j-1}, x_j]} f = c$ for every j, so

$$L(f,P) = \sum_{j=1}^{n} c\Delta x_j = c \cdot (b - a).$$

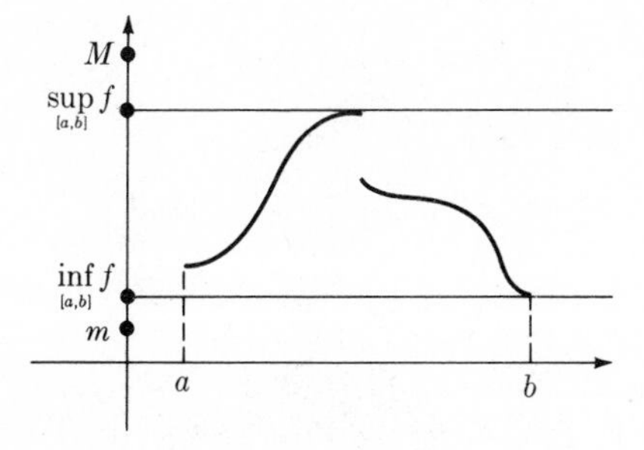

FIGURE AII.15

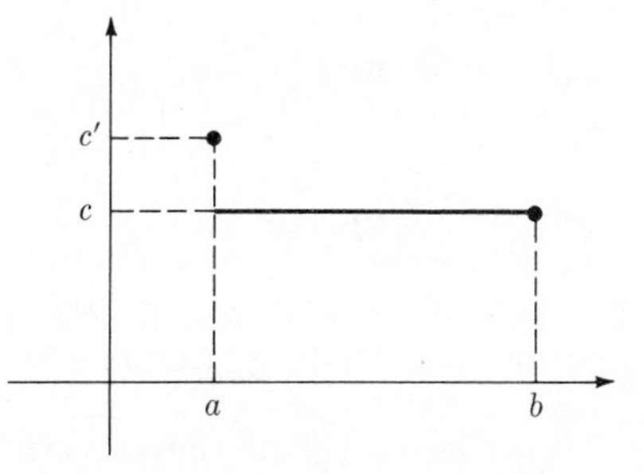

FIGURE AII.16 $f(a) = c'$, $f(x) = c$ for $a < x \leq b$

If, on the other hand, $f(a) = c' < c$, we get $\inf_{[x_0,x_1]} f = c'$, so

$$L(f,P) = c'\Delta x_1 + \sum_{j=2}^{n} c\Delta x_j = c'\cdot(x_1 - a) + c\cdot(b - x_1).$$

In this case, all lower sums are $< c\cdot(b - a)$; but when x_1 is nearly equal to a, then $L(f,P)$ is nearly equal to $c\cdot(b - a)$.

Example 4. If $f(x) = x$ for all x in $[0,1]$, then a lower sum is $\sum_{j=1}^{b} x_{j-1}\Delta x_j$. It is clear from Fig. 17 that (i) all these sums are less than the actual area 1/2, but that (ii) a partition of $[0,1]$ into n equal subintervals yields the lower sum $1/2 - 1/2n$. Thus the actual area, 1/2, is the *least upper bound* of the lower sums; it is an upper bound by (i), and the least upper bound by (ii).

The last two examples suggest that for a reasonable function f, the area under the graph may not be given by any lower sum, but it should be given by the *least upper bound* of the lower sums. This suggests the following concept of *lower integral:*

Definition 7. Let f be bounded on $[a,b]$. Then the *lower integral of* f, denoted $\underline{\int_a^b} f$, is the least upper bound of all lower Riemann sums formed with all possible partitions,

LOWER INTEGRAL

$$\underline{\int_a^b} f = \operatorname*{l.u.b.}_{P} L(f,P).$$

Example 5. If $f = c$ (a constant), then $L(f,P) = c\cdot(b - a)$ for every P [Example 1], so

$$\underline{\int_a^b} f = \text{l.u.b. } L(f,P) = c\cdot(b - a).$$

In other words, $\underline{\int_a^b} c = c\cdot(b - a)$.

Example 6. If f is defined by (1), then $L(f,P) = 0$ for every P, so $\underline{\int_a^b} f = 0$.

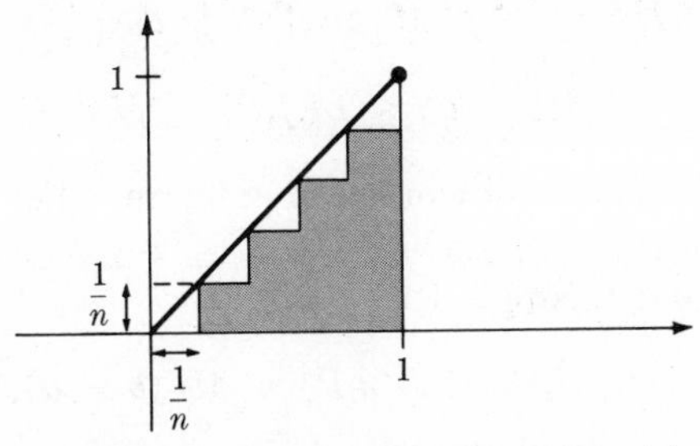

FIGURE AII.17 There are n little triangles of base $1/n$ and height $1/n$; the sum of their areas is $n \cdot (\frac{1}{2}n^2) = n/2$

Example 7. Changing the value of f at a single point does not affect the lower integral. To illustrate this, we compare the function $f = c$ in Example 5 to the function defined by

$$f(x) = c \quad \text{if } a < x \leq b; \qquad f(a) = c'.$$

When $c' > c$, we found in Example 3 that $L(f,P) = c \cdot (b - a)$ for every b, so $\underline{\int_a^b} f = \sup L(f,P) = c \cdot (b - a)$. When $c' < c$, then $L(f,P) < c \cdot (b - a)$ for every P; but $L(f,P)$ is arbitrarily close to $c \cdot (b - a)$ for partitions where x_1 is very close to a, so the *least upper bound* of all the $L(f,P)$ is again $c \cdot (b - a)$. Thus $\underline{\int_a^b} f = c \cdot (b - a)$, exactly as for the constant function in Example 5.

Our course now is the following: We prove certain lemmas about lower sums, and corresponding theorems about lower integrals; we then prove Propositions I and J above. The following two lemmas about lower sums are practically obvious:

Lemma 1. *If $f \leq g$, then $L(f,P) \leq L(g,P)$.*

Proof. This is clear from a figure. [Draw your own.] For a formal proof, notice that when $f \leq g$, it follows that

$$\inf_{[x_{j-1},x_j]} f \leq \inf_{[x_{j-1},x_j]} g$$

and since $\Delta x_j > 0$, we find

$$\left(\inf_{[x_{j-1},x_j]} f\right)\Delta x_j \leq \left(\inf_{[x_{j-1},x_j]} g\right)\Delta x_j .$$

Adding these inequalities from $j = 1$ to $j = n$ gives

$$L(f,P) = \sum_{j=1}^{n} \left(\inf_{[x_{j-1},x_j]} f\right)\Delta x_j \leq \sum_{j=1}^{n} \left(\inf_{[x_{j-1},x_j]} g\right)\Delta x_j = L(g,P). \quad Q.E.D.$$

Lemma 2. *If* $m \leq f \leq M$ *on* $[a,b]$, *then*

$$m(b - a) \leq L(f,P) \leq M(b - a).$$

Proof. This is fairly obvious from a figure. For a formal proof, consider the constant function $g(x) = M$, $a \leq x \leq b$. Then $f \leq g$ on $[a,b]$, so by Lemma 1 and Example 1,

$$L(f,P) \leq L(g,P) = M \cdot (b - a).$$

Similarly, using $h(x) = m$, we get

$$m \cdot (b - a) = L(h,P) \leq L(f,P). \quad Q.E.D.$$

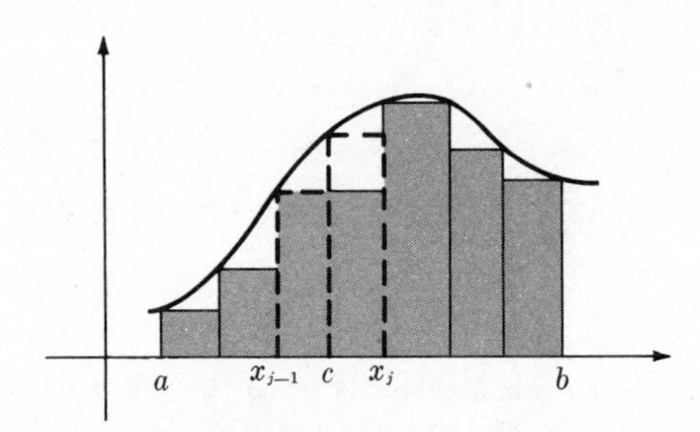

FIGURE AII.18 $L(f,P)$ is shaded; $L(f,P^*)$ replaces the single shaded rectangle over $[x_{j-1}, x_j]$ by two dashed rectangles

Lemma 3. *Let* P *be a partition of* $[a,b]$, *and form a new partition* P^* *by adjoining a new point* c [Fig. 18]. *Then* $L(f,P) \leq L(f,P^*)$.

Proof. Suppose that c lies in $[x_{j-1}, x_j]$. Then $L(f, P^*)$ differs from $L(f,P)$ in that the term

$$\left(\inf_{[x_{j-1},x_j]} f\right)(x_j - x_{j-1}) \tag{2}$$

is replaced by

$$\left(\inf_{[x_{j-1},c]} f\right)(c - x_{j-1}) + \left(\inf_{[c,x_j]} f\right)(x_j - c). \tag{3}$$

But the intervals $[x_{j-1}, c]$ and $[c,x_j]$ are both contained in $[x_{j-1}, x_j]$, so

$$\inf_{[x_{j-1},c]} f \geq \inf_{[x_{j-1},x_j]} f \quad \text{and} \quad \inf_{[c,x_j]} f \geq \inf_{[x_{j-1},x_j]} f.$$

Therefore, the sum (3) is greater than or equal to

$$\left(\inf_{[x_{j-1},x_j]} f\right)[(c - x_{j-1}) + (x_j - c)] = \left(\inf_{[x_{j-1},x_j]} f\right)(x_j - x_{j-1}),$$

which is the expression in (2). Thus $L(f,P^*)$ is obtained from $L(f,P)$ by replacing (2) with the larger sum (3), so $L(f,P^*) \geq L(f,P)$. *Q.E.D.*

Next we prove some theorems about lower integrals. We need to know exactly what Definition 7 means:

(i) $\underline{\int_a^b} f \geq L(f,P)$ for every partition P

$$\left[\underline{\int_a^b} f \text{ is an upper bound for the lower sums}\right].$$

(ii) If M is any number such that $L(f,P) \leq M$ for all P, then also $\underline{\int_a^b} f \leq M$ $\left[\underline{\int_a^b} f \text{ is the } \textit{least} \text{ upper bound}\right]$.

Our first theorem is simple; when f and g are positive, it corresponds to the fact that larger regions have larger areas [Fig. 19].

Theorem 20. *If f and g are bounded and $f \leq g$, then*

$$\underline{\int_a^b} f \leq \underline{\int_a^b} g.$$

Proof. By Lemma 1 and Definition 2, the following inequalities are valid for every partition P:

$$L(f,P) \leq L(g,P) \leq \underline{\int_a^b} g.$$

Applying (ii) above with $M = \underline{\int_a^b} g$, we see that $\underline{\int_a^b} f \leq \underline{\int_a^b} g$. *Q.E.D.*

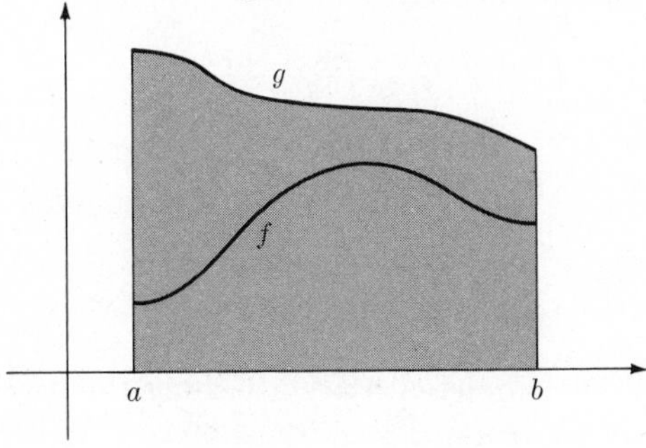

FIGURE AII.19 When $f \leq g$, then $\underline{\int_a^b} f \leq \underline{\int_a^b} g$

Example 8. Suppose that $m \le f \le M$ on $[a,b]$, where m and M are constants. Then applying Theorem 20, and using Example 5 to evaluate the integrals of m and M, we find

$$m(b-a) = \underline{\int_a^b} m \le \underline{\int_a^b} f \le \underline{\int_a^b} M = M(b-a).$$

The next theorem corresponds to the geometric idea that the area of the sum of two disjoint regions is the sum of the areas of the regions [Fig. 20].

LOWER INTEGRALS ARE "ADDITIVE OVER INTERVALS"

Theorem 21. *Let f be bounded on $[a,b]$, and let $a < c < b$. Then*

$$\underline{\int_a^c} f + \underline{\int_c^b} f = \underline{\int_a^b} f.$$

Proof. Let P_1 be any partition of $[a,c]$, and P_2 any partition of $[c,b]$. Then P_1 and P_2 together constitute a partition P^* of $[a,b]$, as in Fig. 21. Clearly, $L(f,P_1) + L(f,P_2) = L(f,P^*)$, so by (i) above,

$$L(f,P_1) + L(f,P_2) = L(f,P^*) \le \underline{\int_a^b} f,$$

and therefore

$$L(f,P_1) \le \underline{\int_a^b} f - L(f,P_2). \tag{4}$$

Now keep P_2 fixed. The inequality (4) holds for every partition P_1 of $[a,c]$, so by (ii) above,

$$\underline{\int_a^c} f \le \underline{\int_a^b} f - L(f,P_2), \qquad \text{or} \qquad L(f,P_2) \le \underline{\int_a^b} f - \underline{\int_a^c} f.$$

This inequality holds for all partitions P_2 of $[c,b]$, so by (ii) again,

$$\underline{\int_c^b} f \le \underline{\int_a^b} f - \underline{\int_a^c} f,$$

or

$$\underline{\int_a^c} f + \underline{\int_c^b} f \le \underline{\int_a^b} f. \tag{5}$$

The proof will be completed by establishing the reverse inequality to (5). Let P be any partition $a = x_0 < x_1 < \cdots < x_n = b$ of $[a,b]$. The point c is probably *not* one of the partition points in P; in this case, form a new partition P^* by adjoining c to P. Then by Lemma 3,

$$L(f,P) \le L(f,P^*).$$

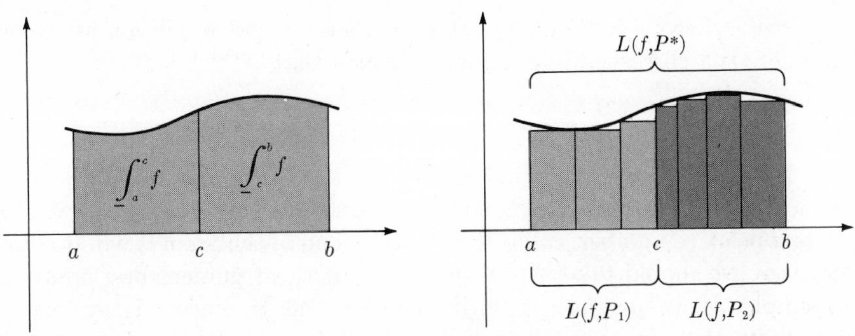

FIGURE AII.20 $\underline{\int_a^b} f = \underline{\int_a^c} + \underline{\int_c^b} f$

FIGURE AII.21 P_1 and P_2 combine to form P^*; $L(f,P^*) = L(f,P_1) + L(f,P_2)$

Now, P^* comes from a partition P_1 of $[a,c]$ and a partition P_2 of $[c,b]$, as in Fig. 21, so we have

$$L(f,P) \leq L(f,P^*) = L(f,P_1) + L(f,P_2) \leq \underline{\int_a^c} f + \underline{\int_c^b} f.$$

Therefore, by (ii),

$$\underline{\int_a^b} f \leq \underline{\int_a^c} f + \underline{\int_c^b} f.$$

This, together with (5), proves that

$$\underline{\int_a^b} f = \underline{\int_a^c} f + \underline{\int_c^b} f. \qquad Q.E.D.$$

Next, we prove that the Riemann sums for a continuous function f converge; in fact, they converge to $\underline{\int_a^b} f$. The crucial fact in the proof is that a continuous function on a closed finite interval $[a,b]$ is *uniformly continuous;* for every $\epsilon > 0$, there is a $\delta > 0$ such that for all x and y in $[a,b]$,

$$|x - y| < \delta \Rightarrow |f(x) - f(y)| < \epsilon.$$

Theorem 22. *Let f be continuous on $[a,b]$. For each integer n, partition $[a,b]$ into n equal subintervals. In each subinterval $[x_{j-1}, x_j]$ choose a point ξ_j and form the Riemann sum*

$$S_n = \sum_{j=1}^{n} f(\xi_j)\,\Delta x_j. \tag{6}$$

RIEMANN SUMS FOR CONTINUOUS FUNCTION CONVERGE

Then $\lim_{n\to\infty} S_n = \underline{\int_a^b} f.$

Proof. Let an arbitrary positive number ϵ be given. We must show that there is a corresponding number N_ϵ such that

$$n > N_\epsilon \quad \Rightarrow \quad \left| S_n - \underline{\int}_a^b f \right| < \epsilon.$$

We do this by applying the condition of uniform continuity, but with a *different* positive number, call it ϵ_1. [At the end of the proof it will become clear that we should take $\epsilon_1 = \epsilon/(b-a)$, but the argument and notation are simpler if we just use ϵ_1 until the very end.] Since f is uniformly continuous, there is a positive number δ_1 such that for all x and y in $[a,b]$,

$$|x - y| < \delta_1 \quad \Rightarrow \quad |f(x) - f(y)| < \epsilon_1. \tag{7}$$

We now require the number n of subintervals to be so large that each has length less than δ_1; thus we choose $N_\epsilon = (b-a)/\delta_1$. Then if $n > N_\epsilon$, and x is any point in the jth subinterval $[x_{j-1}, x_j]$, we have [Fig. 22]

$$|x - \xi_j| \leq \Delta x_j = \frac{b-a}{n} < \frac{b-a}{N_\epsilon} = \delta_1,$$

and therefore, by (7), $|f(x) - f(\xi_j)| < \epsilon_1$. Thus

$$f(\xi_j) - \epsilon_1 < f < f(\xi_j) + \epsilon_1 \text{ on the interval } [x_{j-1}, x_j].$$

It follows [Example 8] that

$$(f(\xi_j) - \epsilon_1)\Delta x_j \leq \underline{\int}_{x_{j-1}}^{x_j} f \leq (f(\xi_j) + \epsilon_1)\Delta x_j.$$

Adding these inequalities for $j = 1, \ldots, n$ yields

$$\left(\sum_{j=1}^n f(\xi_j)\,\Delta x_j\right) - \epsilon_1 \sum_{j=1}^n \Delta x_j \leq \sum \underline{\int}_{x_{j-1}}^{x_j} f \leq \left(\sum_{j=1}^n f(\xi_j)\,\Delta x_j\right) + \epsilon_1 \sum_{j=1}^n \Delta x_j.$$

Now recall the definition (6) of the Riemann sum S_n, note that $\sum_{j=1}^n \Delta x_j = b - a$ [sum of the lengths of the subintervals], and apply Theorem 21 to obtain

$$S_n - \epsilon_1(b-a) \leq \underline{\int}_a^b f \leq S_n + \epsilon_1(b-a),$$

for $n > N_\epsilon$. Thus, taking $\epsilon_1 = \dfrac{\epsilon}{b-a}$, we get the desired conclusion:

$$n > N_\epsilon \quad \Rightarrow \quad \left| S_n - \underline{\int}_a^b f \right| < \epsilon. \qquad Q.E.D.$$

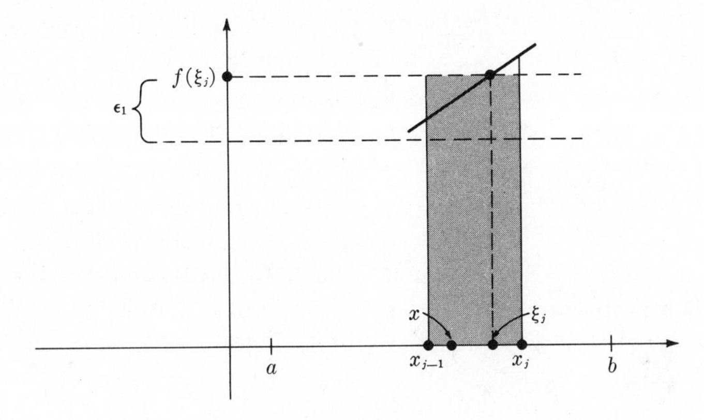

FIGURE AII.22 $|x - \xi_j| \leq x_j - x_{j-1} = \Delta x_j < \delta_1$, so $|f(x) - f(\xi_j)| < \epsilon_1$

Recall that the lower integral $\underline{\int_a^b} f$ is defined in terms of lower sums; intuitively speaking, these are approximations from below [Figs. 17, 18, 21]. The sums S_n in Theorem 22, by contrast, are not required to approximate either from above or from below. If $f(\xi_j)$ is the minimum of f over $[x_{j-1}, x_j]$, then S_n approximates from below; but $f(\xi_j)$ may just as well be the maximum, and then S_n approximates from above. Thus, when $\lim S_n$ exists, independent of the choice of the ξ_j, it is not appropriate to keep the "lower bar" on the integral, and we write simply

$$\int_a^b f = \lim S_n .$$

In particular, by Theorem 22, this limit exists when f is continuous on $[a,b]$, and the limit agrees with the lower integral, so

$$\underline{\int_a^b} f = \int_a^b f \quad \text{when } f \text{ is continuous.}$$

It thus follows from Theorem 21 that

$$\int_a^c f + \int_c^b f = \int_a^b f,$$

when $a < c < b$ and f is continuous on $[a,b]$. Similarly, if g is continuous, and $f \leq g$, then Theorem 20 shows that

$$\int_a^b f \leq \int_a^b g.$$

We come at last to the Fundamental Theorem of Calculus.

☆☆☆
FUNDAMENTAL THEOREM OF CALCULUS

Theorem 23. *Let f be continuous on $[a,b]$. Set*

$$F(t) = \int_a^t f, \qquad a \leq t \leq b.$$

Then F is continuous on $[a,b]$, and

$$F'(t) = f(t), \qquad a < t < b.$$

Proof. We prove the formula for $F'(t)$, and leave the proof of continuity as a problem.

Suppose that $a < t < b$. Let $h > 0$. Then

$$\frac{F(t+h) - F(t)}{h} = \frac{1}{h}\left[\int_a^{t+h} f - \int_a^t f\right] = \frac{1}{h}\int_t^{t+h} f \tag{8}$$

by Theorem 21. Let

$$m(h) = \inf_{[t-h,t+h]} f \qquad \text{and} \qquad M(h) = \sup_{[t-h,t+h]} f.$$

Then $m(h) \leq f \leq M(h)$ on the interval $[t, t+h]$, so by Example 8,

$$hm(h) \leq \int_t^{t+h} f \leq hM(h).$$

Dividing by the positive number h, and recalling (8), we thus obtain

$$m(h) \leq \frac{F(t+h) - F(t)}{h} \leq M(h). \tag{9}$$

We have assumed $h > 0$; a similar proof establishes (9) when $h < 0$ as well, if we take the sup, $M(h)$, and the inf, $m(h)$, over the interval $[t - |h|, t + |h|]$. Since f is assumed continuous at the point t, we have [Problem 6, §A2.4]

$$\lim_{h\to 0} m(h) = f(t) \qquad \text{and} \qquad \lim_{h\to 0} M(h) = f(t).$$

Therefore, applying the Poe Principle to (9), we get the desired formula for $F'(t)$:

$$F'(t) = \lim_{h\to 0} \frac{F(t+h) - F(t)}{h} = f(t). \qquad Q.E.D.$$

PROBLEMS

Problems 1–3 fill in the small gaps left in the discussion above. The remaining problems delve more deeply into Riemann integration.

LOWER INTEGRATION IS NOT LINEAR

1. (a) Let f be the unusual function defined by (1) at the beginning of this section, and let $g = -f$. Prove that $\underline{\int}_a^b g = -1$.

(b) Prove that $\underline{\int}_a^b (f+g) \neq \underline{\int}_a^b f + \underline{\int}_a^b g$. [Thus lower integration is not a linear operation.]

2. Let f be bounded on $[a,b]$, that is $m \le f(x) \le M$ for $a \le x \le b$. Set $F(t) = \underline{\int}_a^t f, \quad a \le t \le b$.

(a) Prove that $\lim_{t\to a+} F(t) = 0$. $\left[\text{Hint: } m(t-a) \le \underline{\int}_a^t f \le M(t-a).\right]$

(b) Prove that $\lim_{t\to b-} F(t) = F(b)$. [Hint: Show that

$$m(b-t) \le \underline{\int}_a^b f - \underline{\int}_a^t f \le M(b-t).]$$

(c) Prove that if $a < t_0 < b$, then $\lim_{t\to t_0} F(t) = F(t_0)$. [This proves that the function F in Theorem 23 is continuous on $[a,b]$.]

3. Prove formula (9) in the text for $h < 0$.

UPPER SUMS

4. If f is bounded on $[a,b]$, and P is a partition, the corresponding *upper sum* is defined by

$$U(f,P) = \sum_{j=1}^{n} \left(\sup_{[x_{j-1},x_j]} f \right) x_j .$$

(a) Prove that for any interval $[a,b]$, $\sup_{[a,b]} f = -\inf_{[a,b]} (-f)$.

(b) Prove that upper and lower sums are related by $U(f,P) = -L(-f,P)$.

UPPER INTEGRALS

5. Upper sums are defined in Problem 4; upper integrals are defined by

$$\overline{\int}_a^b f = \inf_P U(f,P).$$

(a) Explain why we take $\inf U(f,P)$, not $\sup U(f,P)$.

(b) Prove that $\overline{\int}_a^b f = -\underline{\int}_a^b (-f)$.

(c) Prove that if $f \le g$, then $\overline{\int}_a^b f \le \overline{\int}_a^b g$.

(d) Prove that if $a < c < b$, then $\overline{\int_a^c} f + \overline{\int_c^b} f = \overline{\int_a^b} f$. [Use part (b) and Theorem 21; or rework the proof of Theorem 21 to cover upper integrals.]

6. Find a function f such that $\overline{\int_a^b} f \neq \underline{\int_a^b} f$.

7. Find bounded functions f and g such that

$$\overline{\int_a^b} (f+g) \neq \overline{\int_a^b} f + \overline{\int_a^b} g. \qquad \text{[See Problem 1.]}$$

8. (a) Prove that for any interval $[a,b]$, $\inf\limits_{[a,b]} (f+g) \geq \inf\limits_{[a,b]} f + \inf\limits_{[a,b]} g$.

(b) Prove that if f and g are bounded on $[a,b]$, then $L(f+g, P) \geq L(f,P) + L(g,P)$.

(c) Prove that if f and g are bounded on $[a,b]$, then

$$\underline{\int_a^b} (f+g) \geq \underline{\int_a^b} f + \underline{\int_a^b} g.$$

(d) Prove that if f and g are bounded on $[a,b]$, then

$$\overline{\int_a^b} (f+g) \leq \overline{\int_a^b} f + \overline{\int_a^b} g. \qquad \text{[Use Problem 5(b).]}$$

(e) If $c > 0$, then $\underline{\int_a^b} cf = c \underline{\int_a^b} f$ and $\overline{\int_a^b} cf = c \overline{\int_a^b} f$.

9. This problem proves that if f is *continuous*, then $\underline{\int_a^b} f = \overline{\int_a^b} f$. Let f be continuous, and define

FOR A CONTINUOUS FUNCTION, UPPER, LOWER INTEGRALS AGREE

$$\underline{F}(t) = \underline{\int_a^t} f, \qquad \bar{F}(t) = \overline{\int_a^t} f, \qquad a \leq t \leq b.$$

(a) Explain why Theorem 23 proved, actually, that $\underline{F}$ is continuous on $[a,b]$ and $\underline{F}'(t) = f(t)$, $a < t < b$.

(b) Prove that $\bar{F}$ is continuous on $[a,b]$ and $\bar{F}'(t) = f(t)$, $a < t < b$. [Hint: Apply part (a), replacing f by $-f$.]

(c) Prove that $\bar{F}(t) = \underline{F}(t) + C$, $a < t < b$.

(d) Prove that the constant C in part (c) is 0.

(e) Prove that $\underline{\int_a^b} f = \overline{\int_a^b} f$.

10. A function f is called *Riemann integrable* on $[a,b]$ if $\overline{\int_a^b} f = \underline{\int_a^b} f$. **RIEMANN INTEGRABLE FUNCTIONS**

The common value of these upper and lower integrals is denoted $\int_a^b f$.

[Problem 9 proved that every continuous function is Riemann integrable.] Let f and g be Riemann integrable, and prove the following:

(a) If $f \leq g$, then $\int_a^b f \leq \int_a^b g$. $\left[\text{Hint: } \int_a^b f = \underline{\int_a^b} f.\right]$

(b) $\int_a^b (f+g) = \left(\int_a^b f\right) + \left(\int_a^b g\right)$. [See Problems 8(c) and 8(d).]

(c) $\int_a^b (-f) = -\int_a^b f$. [See Problem 5(b).]

(d) If $c > 0$, then $\int_a^b cf = c\int_a^b f$. [See Problem 8(e).]

(e) If $c < 0$, then $\int_a^b cf = c\int_a^b f$. [See parts (c) and (d).]

(f) Assuming f is integrable on $[a,c]$ and on $[c,b]$ where $a < c < b$, prove that

$$\int_a^b f = \int_a^c f + \int_c^b f.$$

[For more on Riemann integration, see M. Spivak, *Calculus.*]

AII.9 ARC LENGTH

The definition of arc length, like that of area under a curve (the integral) is based directly on the least upper bound axiom. Before defining the length, we have to state exactly what we mean by a curve.

Definition 8. A *curve in the plane* is an ordered pair of continuous functions on some interval I. If the functions are denoted by f and g, then the curve is denoted by (f,g). **CURVE: FORMAL DEFINITION**

This abstract definition has a concrete interpretation. We think of the interval I on which f and g are defined as an interval of time, and think of the curve as traced out by a moving particle: At time t the particle is at the point $(f(t), g(t))$. [See Chapter VII.]

Example 1. Let $f(t) = t$, $g(t) = 1 - t^2$. Then at time t, the particle is at the point $(t, 1 - t^2)$. Figure 23 shows its position at times $t = -1, -1/2, 0, 1/2, 1, 1\frac{1}{2}$, and suggests the general course of the particle. The arrow indicates the direction in which the particle moves.

Now that we know what a curve is, we approach the definition of length by way of the following considerations.

(a) It is reasonable to say that the length of a segment equals the distance between its endpoints; in other words, the length of the segment from (x_1, y_1) to (x_2, y_2) is $\sqrt{(x_1 - x_2)^2 + (y_1 - y_2)^2}$.

(b) It is reasonable to approximate the length of a curve by the sum of the lengths of inscribed segments, as in Fig. 24 below. (The segments form an inscribed polygon, and the sum of their lengths is the length of the polygon.)

(c) It is reasonable to expect that the length of each inscribed segment is $\leq$ the length of the corresponding piece of the curve ("a straight line is the shortest distance between two points"), and hence that the length of each inscribed polygon is $\leq$ the length of the curve. In other words, it is reasonable to expect that the length of the curve is an upper bound of the lengths of the inscribed polygons.

(d) It is reasonable to expect that the length of an inscribed polygon will be arbitrarily close to the length of the curve if only the segments are short enough. Thus the length of the curve must be the *least* upper bound of the lengths of the inscribed polygons.

Of course, there is something to check before we take the least upper bound: Is there any upper bound at all? It turns out that there are curves so badly wrinkled that there is *no* upper bound for the lengths of the inscribed polygons [Problem 3 below]. So we have to restrict our attention to curves for which there is indeed such an upper bound, called *rectifiable* curves. For these curves, we can (and do) define arc length exactly as in (d) above. [It may be tempting to think of (a)–(d) as "proving" the definition of arc length. You can resist this temptation, however, when you realize that in (a)–(d) we talked about lengths of segments and curves, whereas length is just what we are trying to define!]

To make the definition, we need the following analytic formulation of the "length of an inscribed polygon."

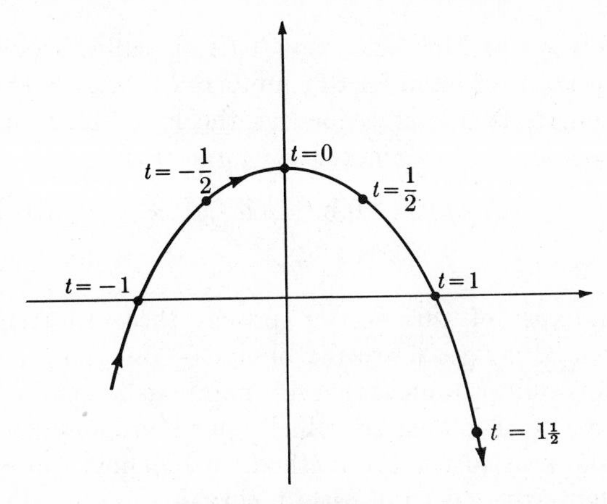

FIGURE AII.23

Definition 9. Let (f,g) be any plane curve, with f and g defined on $[a,b]$. Let P be any partition of $[a,b]$ given by points $t_0, \ldots, t_n$, where $a = t_0 < t_1 < \cdots < t_n = b$. Then we define

$$L(P,f,g) = \sum_{1}^{n} \sqrt{[f(t_j) - f(t_{j-1})]^2 + [g(t_j) - g(t_{j-1})]^2}\,.$$

In this definition, we think of $(f(t_j), g(t_j))$ as the endpoints of the segments of an inscribed polygon, and of $L(P,f,g)$ as the sum of the lengths of the segments. [See Fig. 24.]

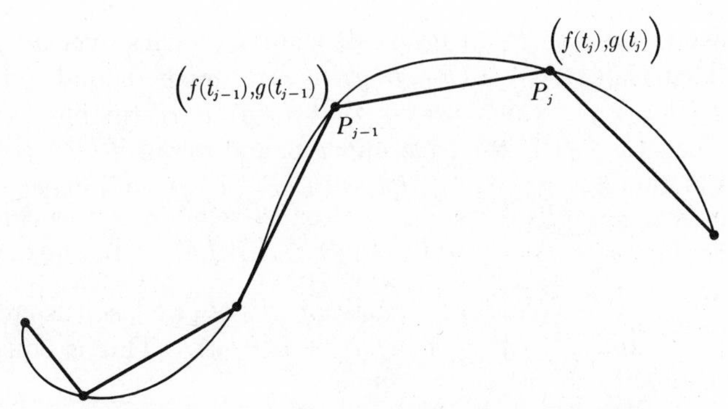

FIGURE AII.24 The length of the segment $P_{j-1}P_j$ is

$$\sqrt{\left(f(t_j)-f(t_{j-1})\right)^2+\left(g(t_j)-g(t_{j-1})\right)^2}$$

LENGTH OF RECTIFIABLE CURVE

Definition 10. The curve (f,g) is called *rectifiable* on $[a,b]$ if there is an upper bound for the numbers $L(P,f,g)$. If the curve is rectifiable then its length is defined as the least upper bound of the numbers $L(P,f,g)$, and is denoted $\mathrm{L}_a^b(f,g)$. Briefly,

$$\mathrm{L}_a^b(f,g) = \text{l.u.b. } \{L(P,f,g) : P \text{ is a partition of } [a,b]\}.$$

The rest of this section proves three theorems about arc length. Theorem 24 is "geometrically obvious," but the proof is not easy. Theorem 25 is quite simple; it serves mainly as a tool in the proof of our most useful result, Theorem 26, which relates arc length to differentiation, and thus allows us to use the methods of indefinite integration [Chapter VI] to compute the length of certain curves.

LENGTH OF WHOLE IS SUM OF LENGTHS OF PARTS

Theorem 24. *Suppose (f,g) is a rectifiable curve on $[a,b]$ and suppose $a < c < b$. Then (f,g) is rectifiable on $[a,c]$ and on $[c,b]$, and*

$$\mathrm{L}_a^b(f,g) = \mathrm{L}_a^c(f,g) + \mathrm{L}_c^b(f,g).$$

Proof. We show first that $\mathrm{L}_a^c(f,g) + \mathrm{L}_c^b(f,g) \le \mathrm{L}_a^b(f,g)$, and then the reverse inequality, $\mathrm{L}_a^b(f,g) \le \mathrm{L}_a^c(f,g) + \mathrm{L}_c^b(f,g)$.

Let P_1 and P_2 be any partitions of $[a,c]$ and $[c,b]$. Then P_1 and P_2 together form a partition P of $[a,b]$. [See Fig. 25.] Since $\mathrm{L}_a^b(f,g)$ is an upper bound for the $L(P,f,g)$, we have

$$L(P_1,f,g) + L(P_2,f,g) = L(P,f,g) \le \mathrm{L}_a^b(f,g),$$

which can be rewritten as

$$L(P_1,f,g) \le \mathrm{L}_a^b(f,g) - L(P_2,f,g). \tag{1}$$

Now, imagine P_2 to be fixed, while P_1 varies over all partitions of $[a,c]$. Then $\mathrm{L}_a^b(f,g) - L(P_2,f,g)$ is an upper bound for all the numbers $L(P_1,f,g)$, which proves that (f,g) is rectifiable on $[a,c]$. Moreover, since $\mathrm{L}_a^c(f,g)$ is the *least* upper bound of the $L(P_1,f,g)$, it follows from (1) that $\mathrm{L}_a^c(f,g) \le \mathrm{L}_a^b(f,g) - L(P_2,f,g)$, and hence

$$L(P_2,f,g) \le \mathrm{L}_a^b(f,g) - \mathrm{L}_a^c(f,g). \tag{2}$$

Now, (2) is valid for all partitions P_2 of $[c,b]$, so the curve is rectifiable on $[c,b]$, and $\mathrm{L}_c^b(f,g) \le \mathrm{L}_a^b(f,g) - \mathrm{L}_a^c(f,g)$. This is equivalent to the first inequality we set out to prove,

$$\mathrm{L}_a^c(f,g) + \mathrm{L}_c^b(f,g) \le \mathrm{L}_a^b(f,g). \tag{3}$$

[This proof resembles a powerful method in the game of bridge, called a "cross ruff."]

We now obtain the reverse inequality to (3). Let P be any partition of $[a,b]$. If c happens to be a point of this partition, then P is equivalent to partitions P_1 of $[a,c]$ and P_2 of $[c,b]$ [see Fig. 25], and we have

$$L(P,f,g) = L(P_1, f, g) + L(P_2, f, g) \leq \mathrm{L}_a^c(f,g) + \mathrm{L}_c^b(f,g). \quad \text{(4a)}$$

If c is *not* a point of the partition P, we form a new partition P^* by adjoining c to the points of P. Then, as Fig. 26 suggests, in changing from $L(P,f,g)$ to $L(P^*,f,g)$, we replace the length of the side AB by the sum of the lengths of the other two sides of the triangle ABC. Since any side of a triangle is shorter than the sum of the lengths of the other two sides [see §7.2], we see that $L(P,f,g) \leq L(P^*,f,g)$. Now the partition P^* corresponds to partitions P_1 and P_2 of $[a,c]$ and $[c,b]$ such that $L(P^*,f,g) = L(P_1, f, g) + L(P_2, f, g)$; hence

$$\begin{aligned} L(P,f,g) \leq L(P^*,f,g) &= L(P_1, f, g) + L(P_2, f, g) \\ &\leq \mathrm{L}_a^c(f,g) + \mathrm{L}_c^b(f,g). \end{aligned} \quad \text{(4b)}$$

Combining (4a) and (4b) shows that $L(P,f,g) \leq \mathrm{L}_a^c(f,g) + \mathrm{L}_c^b(f,g)$ whether the partition P contains c or not, and hence

$$\mathrm{L}_a^b(f,g) = \text{l.u.b. } L(P,f,g) \leq \mathrm{L}_a^c(f,g) + \mathrm{L}_c^b(f,g). \quad \text{(5)}$$

Theorem 24 follows from (3) and (5) together. *Q.E.D.*

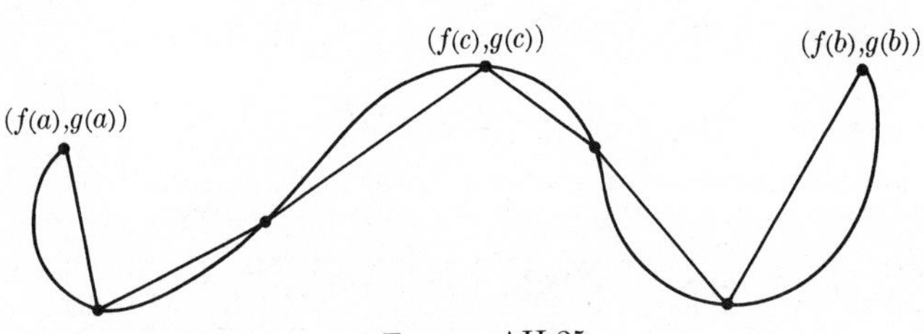

FIGURE AII.25

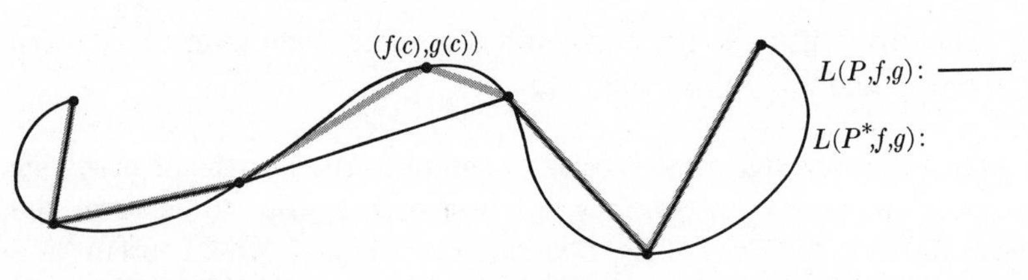

FIGURE AII.26

Now we define $\mathrm{L}_b^a(f,g) = -\mathrm{L}_a^b(f,g)$. [Recall a similar definition in Chapter V, $\int_b^a f = -\int_a^b f$.] Then we may drop the restrictions $a < b$ and $a < c < b$, and Theorem 24 remains true as long as (f,g) is rectifiable on an interval containing a, b, and c.

CURVES WITH BOUNDED DERIVATIVE ARE RECTIFIABLE

Theorem 25. *Let f and g be differentiable on the open interval (a,b), and suppose that the derivatives f' and g' are bounded. Then (f,g) is rectifiable, and*

$$\sqrt{\inf_{(a,b)}(f')^2 + \inf_{(a,b)}(g')^2} \leq \frac{1}{b-a}\mathrm{L}_a^b(f,g)$$

$$\leq \sqrt{\sup_{(a,b)}(f')^2 + \sup_{(a,b)}(g')^2}\,. \qquad (6)$$

[See §AII.8 for the definition of inf and sup.]

Proof. If P is any partition, then by the Mean Value Theorem,

$$L(P,f,g) = \sum_1^n \sqrt{[f(t_j) - f(t_{j-1})]^2 + [g(t_j) - g(t_{j-1})]^2}$$

$$= \sum_1^n (t_j - t_{j-1})\sqrt{f'(c_j)^2 + g'(d_j)^2}$$

$$\leq \sum_1^n (t_j - t_{j-1})\sqrt{\sup(f')^2 + \sup(g')^2}$$

$$= (b-a)\sqrt{\sup(f')^2 + \sup(g')^2}\,.$$

This inequality is valid for every partition P; taking the l.u.b. over all P, we obtain

$$\mathrm{L}_a^b(f,g) \leq (b-a)\sqrt{\sup(f')^2 + \sup(g')^2}\,.$$

In the same way, we find $L(P,f,g) \geq (b-a)\sqrt{\inf(f')^2 + \inf(g')^2}$, and Theorem 25 is proved.

The inequality (6) remains valid if a and b are interchanged, for then both $b - a$ and $\mathrm{L}_a^b(f,g)$ change sign.

Our final result makes it easy to compute the lengths of many simple curves. The idea is to consider not just $\mathrm{L}_a^b(f,g)$, but to consider the arc length *function* $s(t) = \mathrm{L}_a^t(f,g)$, defined for $a \leq t \leq b$. Theorem 26 evaluates the derivative $s'(t)$. If this function s' is amenable to the methods of indefinite integration given in Chapter VI, then we can find a function

F such that $s = F + c$ for some constant c, and so evaluate the length: $\mathrm{L}_a^b(f,g) = s(b) = s(b) - s(a) = F(b) - F(a)$.

[The letter s is a traditional notation for the arc length function, perhaps because *sinus* is Latin for "arc."]

Theorem 26. *If (f,g) is rectifiable on $[a,b]$, and f' and g' are continuous at c, then the function $s(t) = \mathrm{L}_a^t(f,g)$ has for $t = c$ the derivative*

ARC LENGTH DERIVATIVE

$$s'(c) = \sqrt{f'(c)^2 + g'(c)^2}\,.$$

Proof. For $h > 0$, let

$$m(f,h) = \inf_{[c-|h|,c+|h|]} (f')^2 \quad \text{and} \quad M(f,h) = \sup_{[c-|h|,c+|h|]} (f')^2.$$

From Theorem 24,

$$s(c+h) - s(c) = \mathrm{L}_a^{c+h}(f,g) - \mathrm{L}_a^c(f,g) = \mathrm{L}_a^{c+h}(f,g),$$

and hence, from Theorem 25, when $h > 0$, we obtain

$$\sqrt{m(f,h) + m(g,h)} \le \frac{s(c+h) - s(c)}{h} \le \sqrt{M(f,h) + M(g,h)}. \quad (7)$$

Paying attention to the signs, we get the same inequality for $h < 0$. Since f' and g' are continuous at c, so are $(f')^2$ and $(g')^2$, and hence [§AII.4, Problem 6]

$$\lim_{h\to 0} [m(f,h) + m(g,h)] = f'(c)^2 + g'(c)^2$$

and

$$\lim_{h\to 0} [M(f,h) + M(g,h)] = f'(c)^2 + g'(c)^2.$$

Since the square root function is continuous, it follows that both extremes in the inequalities (7) tend to $\sqrt{f'(c)^2 + g'(c)^2}$ as $h \to 0$, and hence $s'(c) = \sqrt{f'(c)^2 + g'(c)^2}$.

PROBLEMS

1. Suppose that $a < c < b$, that f and g are continuous on $[a,b]$, that f' and g' are continuous on (a,b), and that $f'(c)^2 + g'(c)^2 \neq 0$. Show that

$$\lim_{\delta\to 0+} \frac{\sqrt{[f(c+\delta) - f(c-\delta)]^2 + [g(c+\delta) - g(c-\delta)]^2}}{\mathrm{L}_{c-\delta}^{c+\delta}(f,g)} = 1.$$

$\dfrac{\text{CHORD LENGTH}}{\text{ARC LENGTH}} \to 1$

In other words, the ratio of chord length to arc length tends to 1. [You can use the Mean Value Theorem in numerator and denominator.]

2. On the interval $[-1,1]$, set $f(t) = t^2$ and $g(t) = t^3$.

(a) Sketch the curve. [Note that at $t = 0$ the curve is parallel to the x axis.]

(b) From the sketch, do you think the result in Problem 1 holds when $c = 0$?

(c) Evaluate $\lim_{\delta \to 0+} \dfrac{\sqrt{[f(\delta) - f(-\delta)]^2 + [g(\delta) - g(-\delta)]^2}}{\mathrm{L}_{-\delta}^{\delta}(f,g)}$.

3. On the interval $[-1,0]$, set $f(t) = t$ and $g(t) = t \cos(1/t)$, $g(0) = 0$. Prove that (f,g) is *not* rectifiable. [Hint: Make partitions P using the points $t_j = -1/j\pi$. Sketch the curve.]

AII.10 L'HÔPITAL'S RULE FOR THE ∞/∞ CASE

The general discussion of L'Hôpital's rule is in §7.6. Assuming you are familiar with that, we now fill in the missing proof of the "∞/∞" case.

L'HÔPITAL'S RULE: ∞/∞ CASE

Theorem 27. *If* $\lim_{t \to a+} f(t) = +\infty$ *and* $\lim_{t \to a+} g(t) = +\infty$, *and if f and g are differentiable in an interval to the right of a, and if*

$$\lim_{t \to a+} \frac{g'(t)}{f'(t)} = L, \quad \textit{then} \quad \lim_{t \to a+} \frac{g(t)}{f(t)} = L.$$

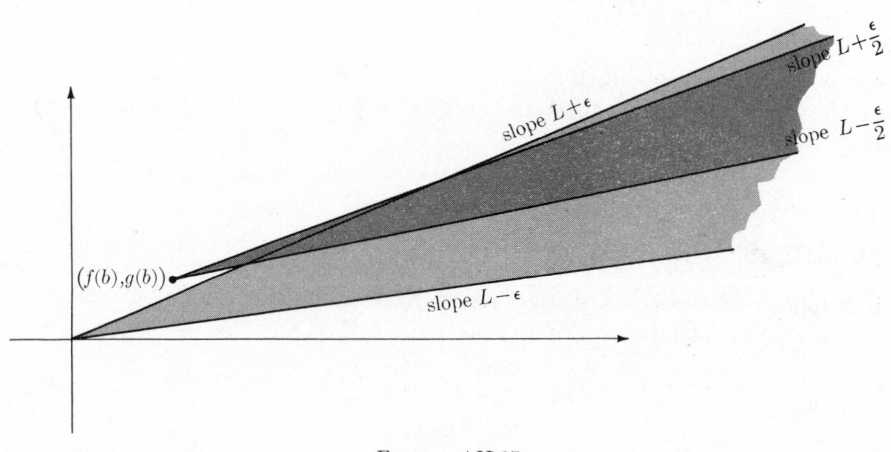

FIGURE AII.27

Figure 27 illustrates the proof. Let $\epsilon > 0$ be given, and choose b so that

$$L - \frac{\epsilon}{2} < \frac{g'(t)}{f'(t)} < L + \frac{\epsilon}{2} \quad \text{for } a < t < b. \tag{1}$$

Then for these values of t, the tangent line to the curve traced by $(f(t), g(t))$ has slope between $L - \epsilon/2$ and $L + \epsilon/2$. Thus the curve lies in the darkly shaded angle in Fig. 27 with vertex at $(f(b), g(b))$. Since $f(t) \to +\infty$, the curve moves to the right and eventually remains in the lightly shaded angle, where $L - \epsilon < y/x < L + \epsilon$, i.e. where $L - \epsilon < g(t)/f(t) < L + \epsilon$.

The formal proof is based on Lemma 1, §7.6. Choose $b > a$ as in (1); by the Lemma, we have

$$\frac{g(t) - g(b)}{f(t) - f(b)} = \frac{g'(c)}{f'(c)},$$

and hence from (1),

$$L - \frac{\epsilon}{2} < \frac{g(t) - g(b)}{f(t) - f(b)} < L + \frac{\epsilon}{2}, \quad \text{for } t > b. \tag{2}$$

We want to obtain information about $g(t)/f(t)$, so we rewrite the middle term in (2) as

$$\frac{g(t) - g(b)}{f(t) - f(b)} = \left[\frac{g(t)}{f(t)} - \frac{g(b)}{f(t)}\right]\left[1 - \frac{f(b)}{f(t)}\right]^{-1},$$

in which $g(t)/f(t)$ appears explicitly. Since $f(t) \to +\infty$, we find, when t is sufficiently close to a, that $f(b)/f(t) < 1$, so $1 - f(b)/f(t) > 0$. Then we can solve the inequalities (2) for $g(t)/f(t)$, obtaining

$$\left(1 - \frac{f(b)}{f(t)}\right)\left(L - \frac{\epsilon}{2}\right) + \frac{g(b)}{f(t)} < \frac{g(t)}{f(t)} < \left(1 - \frac{f(b)}{f(t)}\right)\left(L + \frac{\epsilon}{2}\right) + \frac{g(b)}{f(t)}. \tag{3}$$

Now, as $t \to a+$, the left-hand member of (3) has the limit $L - \epsilon/2$, and the right-hand member has the limit $L + \epsilon/2$. Thus there is a $\delta > 0$ such that, for $a < t < \delta$, the left-hand member is $> (L - \epsilon)$, and the right-hand member is $< (L + \epsilon)$. It follows that $L - \epsilon < \dfrac{g(t)}{f(t)} < L + \epsilon$ when $a < t < \delta$, and Theorem 27 is proved.

Part One: Calculus of One Variable

CHAPTER 0

§0.1

1. (a) $|-2| = 2$ (b) $|-3/-5| = 3/5$ (c) $|8/3| = 8/3$
(d) $|-(-2/3)| = 2/3$ (e) $|-\sqrt{2}| = \sqrt{2}$ (f) $|-(-1/8)| = 1/8$

2. (a) $|2-3| = 1$ (b) $|-2-3| = 5$ (c) $|2-(-3)| = 5$
(d) $|-2-(-3)| = 1$

4. (a) $(1,2)$ (b) Two points, 0 and 1 (d) $(-\infty,+\infty)$ (j) $[1,2]$

5. (a) $(-\infty,-1)$ together with $(0,+\infty)$ (c) Two points, -1 and 0

6. (e) $(a-b, a+b)$ (g) $(1,2)$ together with $(4,5)$
(h) $(a-b, a)$ together with $(a, a+b)$ (i) $(\frac{1}{2},\infty)$
(k) $[2,3]$ (1) Two points, $1\frac{1}{2}$ and $3\frac{1}{2}$

8. (a) $-1\frac{1}{2}$ (b) $-1\frac{1}{2}$ (e) 0

§0.3

3. (b), (d): $m = -1$ (f) $m = -2/3$

4. (a) Parallel (b) Perpendicular (c) Neither (d) Parallel
(e) Perpendicular (f) Parallel

5. (a) Vertical (b) $m = 0$ (c) $m = 2/15$ (d) $m = -1/8$

6. (a) $y = 2x$ (b) $y = 4x - 2$

7. (a) $y = -\frac{1}{3}x + 10$ (b) $y = 10x - 100$

8. $\text{slope}(L_1) = \frac{1}{2}$, $\text{slope}(L_2) = -2$

9. (a) $y = -x$ (b) $y = \frac{5}{4}x - \frac{17}{8}$ (c) $y = 0$

10. $3y + 4x = 0$

§0.4

1. (b) Domain $(-\infty,\infty)$, range $(-\infty,\infty)$
(d) Domain $\{t: t \neq 0\}$, range contains only -1 and $+1$
(f) Domain $\{t: t \neq 1\}$, range $\{y: y > 0\}$
(h) Domain $(-\infty,\infty)$, range $(-\infty,\infty)$

2. (b) $b(a(3)) = b(2) = 7$ (f) $\dfrac{1}{|\sqrt{x+1}-1|}$
(d) $f(c(1)) = f(\frac{1}{3}) = \frac{3}{2}$

4. (a) Functions a, b, c, and h have inverses; the others do not.
(c) $a^{\leftarrow}(y) = y^2 - 1$; $b^{\leftarrow}(y) = (y+1)^{1/3}$; $c^{\leftarrow}(t) = 1/t - 2$;
$h^{\leftarrow}(x) = \frac{1}{5}(x+2)$

6. (a) $\text{erf}^{\leftarrow}(0.2) = 0.2$; $\text{erf}^{\leftarrow}(0.8) = 0.9$

9. Functions a, e, g, and i are even; c, d, and j are odd; the others are neither.

10. (a) "Even" means symmetric in the y axis.
(b) "Odd" means symmetric in the origin.

Chapter 0: Answers to Review Problems

1. (a) and (b) 0 1 2 3 4 (c) −3 −1 0

(e) 0 $\frac{1}{2}$ 1 $1\frac{1}{2}$ (f) 0 1

2. (a) -2 (b) -4 (c) $+2$ (d) -20

3. (a), (b), (d), and (e) are true for all x and y;
(c) is false, e.g. when $x = -1$.

4.

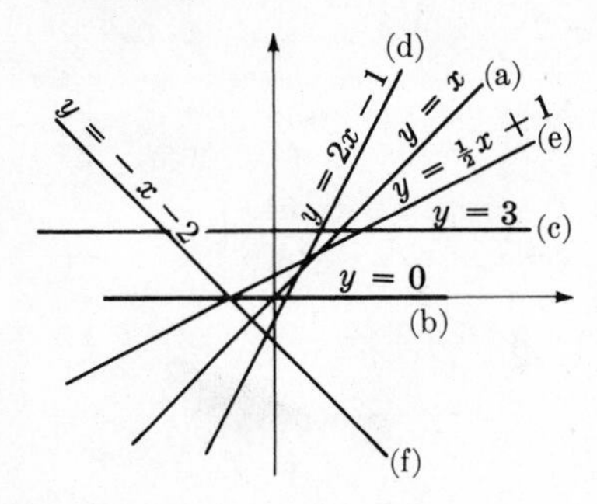

5. (a) $y - 2 = \frac{1}{4}(x - 1)$; slope $\frac{1}{4}$
(b) $y - 1 = \frac{1}{2}(x - 1)$; slope $\frac{1}{2}$ (c) $y - 1 = -(x - 1)$; slope -1

6. (a) $(-\frac{1}{2},0)$ (b) $(0,1)$ (c) $(-1,-1)$
(d) No points of intersection

7. (a) $P_1' = (0,1)$, $P_2' = (-1,5)$
(b) $P_1' = (-1,0)$, $P_2' = (-5,-1)$ (c) $P_1' = (1,0)$, $P_2' = (5,1)$

8. (a) $D = (-\infty,+\infty)$, $R = (-\infty,1]$ (b) $D = (-\infty,+\infty) = R$
(c) $D = [-1,+\infty)$, $R = [0,\infty)$ (d) $D = (-\infty,+\infty)$, $R = [0,\infty)$

9. (a) $f^{\leftarrow}(y) = \sqrt{y}$, $y \geq 0$ (d) $f^{\leftarrow}(y) = 2/y$, $y \neq 0$
(b) No inverse (e) $f^{\leftarrow}(y) = y^2$, $y \geq 0$
(c) $f^{\leftarrow}(y) = y^{1/3}$, $-\infty < y < \infty$ (f) $f^{\leftarrow}(y) = (y + 1)/3$

CHAPTER I

§1.1

1. (a) Type $-\infty, +\infty$ at $x = 1$

(b) $\lim\limits_{x\to 1} \dfrac{(x-1)^2}{x^2-1} = 0$, -1 is type $+\infty, -\infty$

(c) $\lim\limits_{x\to 3} \dfrac{x^3-27}{x^2-9} = \dfrac{9}{2}$; type $-\infty, +\infty$ at $x = -3$

(d) Type $+\infty, +\infty$ at $x = 0$ (e) Type $+\infty, -\infty$ at $x = -\frac{1}{2}$

(i) Type $-\infty, +\infty$ at $x = 0$

(j) $\lim_{x\to 1-} 2\dfrac{x-1}{|x-1|} = -2, \quad \lim_{x\to 1+} 2\dfrac{x-1}{|x-1|} = 2$

(m) $\lim_{x\to 6-} f(x) = 56, \quad \lim_{x\to 6+} f(x) = 1$

2. (a) $\lim_{x\to\pm\infty} \dfrac{x^2-1}{(x-1)^2} = +1$ (i) $\lim_{x\to\pm\infty} \dfrac{x+1/x}{1+x^2} = 0$

(j) $\lim_{x\to+\infty} 2\dfrac{x-1}{|x-1|} = 2, \quad \lim_{x\to-\infty} 2\dfrac{x-1}{|x-1|} = -2$

(m) $\lim_{x\to+\infty} f(x) = +\infty, \quad \lim_{x\to-\infty} f(x) = +\infty$

4. (a) 12 (b) 18 (c) 3 (d) m (e) 15 (f) $-1/9$
5. (a) $4x_0$ (b) $6x_0$ (c) x_0 (d) m (e) $4x_0 + 3$ (f) $-1/x_0^2$

§1.2

1. (a) $f'(x_0) = 2x_0, \; f'(2) = 4$ (b) $f'(x_0) = -1/(x_0+1)^2$
(c) $f'(x_0) = -\frac{1}{2}x_0^{-3/2}$ (e) $f'(x_0) = -2x_0/(x_0^2+1)^2$
4. (b) As near as can be guessed from the given data, $\sin'(0) = 1$.
5. As near as can be guessed from the given data, $\cos'(0) = 0$.
7. (b) $f'(x_0) = 4a_4x_0^3 + 3a_3x_0^2 + 2a_2x_0 + a_1$
8. (b) $f'(x_0) = -3x_0^{-4}$ **12.** $y = x - 1/4$ and $y = -x - 1/4$

§1.3

2. (a) $F = (0,1)$; $y = -1$ (e) $F = (0,2)$; $y = 0$
3. The vertical line $\{(x,y) : x = -1/4a^2x_0\}$

§1.4

1. (a) $f'(x) = -2x + 3$, $V = (3/2, 17/4)$; opens downward
(b) $f'(x) = 10x - 2$, $V = (1/5, 4/5)$; opens upward
2. (a) $f'(x) = -3x^2 + 1$ (b) $f'(x) = 3x^2 - 3$. See figures below.
3. (a) $f'(x) = 2ax + b$; $2ax + b > 0 \Leftrightarrow 2ax > -b \Leftrightarrow x > -b/2a$ (since $2a > 0$). [For a review of inequalities, see §AI.3.]
4. (a) $2ax + b > 0 \Leftrightarrow 2ax > -b \Leftrightarrow x < -b/2a$ (the inequality is reversed because we are dividing by the negative number $2a$).
5. (b) $F = (-b/2a, (4ac - b^2 + 1)/4a)$ (c) $y = (4ac - b^2 - 1)/4a$
(d) $D = (x, (4ac - b^2 - 1)/4a)$
7. (b) (0,0) (c) (0,0) (d) (0,1) (e) $(-1/3, 2/27)$
(f) $(-b/3a, f(-b/3a))$

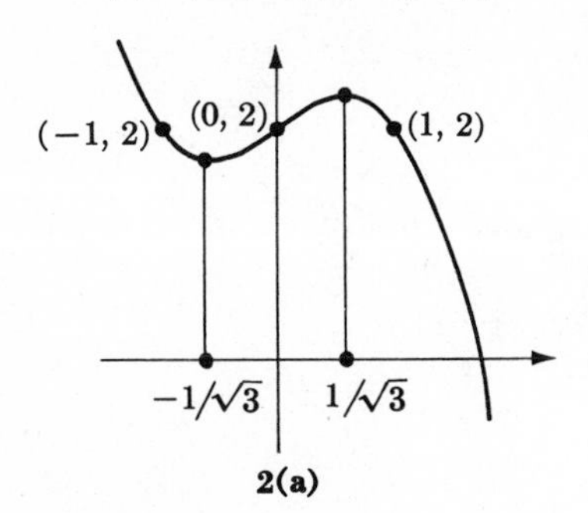

2(a)

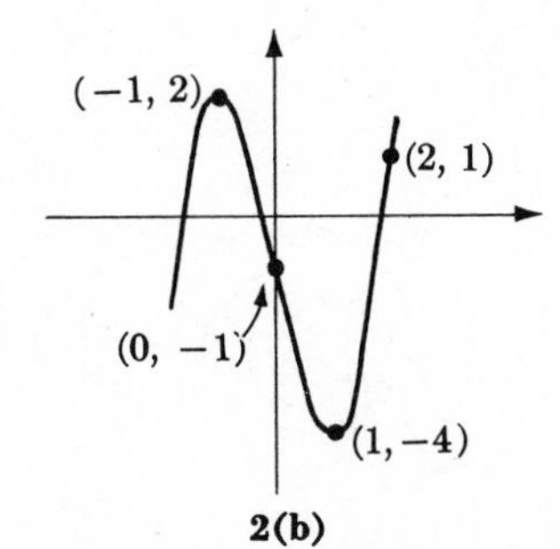

2(b)

§1.5

1. (a) $20/3 \times 20/3 \times 5/3$ **2.** (a) $\frac{1}{3}$ feet (=4 inches) **3.** (a) $P_0 = (1,1)$
4. Circle of diameter $400/\pi$
5. Straightaways are 100 meters long; the curves have radius $100/\pi$
6. (a) $10 \times 10 \times 20$ (b) radius $20/\pi$; length 20

§1.6

2. Taking $x_0 = 3$, you get $x_1 = 19/6 = 3.166\ldots.$
Checking, $(3.16)^2 < 10 < (3.17)^2$.
3. (b) If $x_0 = 2$, then $x_1 = 9/4$, $x_2 = 161/72 = 2.236\ldots.$
(c) If $x_0 = 3$, then $x_1 = 17.5/6 = 2.91\ldots.$
(d) Take $x_0 = 1.8$ (slightly larger than $\sqrt{3} = 1.73\ldots$); then
$x_1 = 6.38/3.6 = 1.77\ldots.$
5. $x_{n+1} = (c + 2x_n)/3x_n^2$. Taking $c = 3$, $x_0 = 1$, you get $x_1 = 5/3$,
$x_2 = 331/225 = 1.47\ldots.$ It is difficult to work with x_2, so we start again with a more convenient nearby number $x_0 = 1.5 = 3/2$, yielding $x_1 = 39/27 = 1.444\ldots.$
7. $x_1 = 3/4$, $x_2 = 3/4 - 17/252 = 0.682\ldots.$

Appendix

2. (c) The answer to Problem 3(c), §1.6, suggests $x_0 = 2.9$. Then
$x_1 = 16.91/5.8$. From (3),

$$x_1 - \sqrt{8.5} = \frac{(8.41 - 8.5)^2}{5.8(2.9 + \sqrt{8.5})^2} < \frac{(.1)^2}{5^3} < .0001,$$

so $x_1 = 2.9155\ldots$ is accurate to four places.

§1.7

1. (a) $v(t) = 32t + 5$ (b) $a(t) = 32$ **2.** (c) $s'(1/5)$ is approximately 405.
3. $s'(3)$ is approximately -1.1. **4.** s in (iii) matches v in (ii).
8. (a) $M'(x) = 2$, a constant (b) $M'(x) = D$
9. (b) The left end ($x = 0$) is denser.
10. (c) It should appear to be proportional. **11.** (c) Q' is not proportional to Q.

§1.8

1. (a) $dy/dx = 6x$ (c) $ds/dt = 2$
(b) $dz/dy = 10y + 2$ (d) $dt/ds = 1/2$

CHAPTER II

§2.1

1. (a) $15x^2 - 15x^4$ (b) $8x - 8$ (d) $32t^3 - 8t$ (g) $\frac{1}{5}(3x^2 - 18x) - \frac{2}{3}$
3. $f'(x) = x^2 - x - 2 = (x - 2)(x + 1)$ (a) 2 and -1 (d) $(-\infty, -1)$ and $(2, +\infty)$
4. (a) $6x(x^2 + 1)^2$ (b) $2(5x^4 + 3x^2 + 1)(20x^3 + 6x)$
(c) $n(x + 1)^{n-1}$ (d) $3[(x^2 + 1)^2 + 2]^2 2(x^2 + 1) \cdot 2x$

§2.2

1. (a) $-(x+1)^{-2}$ (c) $\dfrac{2t}{(t^2+1)^2}$ (d) $\dfrac{2z^3-3z^2+1}{(z-1)^2} = 2z+1$
(b) $(x+1)^{-2}$

(e) $\dfrac{-2x^5-6x^4-12x^3-2x^2-4x+2}{(x^4+x^2+1)^2}$ (f) $\dfrac{ad-bc}{(ct+d)^2}$

(g) $-y^{-2}-2y^{-3}-3y^{-4}$ (h) 0 (i) $\dfrac{t(a+1)-2(at+b)}{(a+1)^3}$

2. (a) $6x(x^2+1)^2$ (b) $-12x(x^2+1)^{-4}$
(c) $-12(10x^5+5x^4-2)^{-5}(50x^4+20x^3)$

3. (a) $2x(x^2+1)^{-3}-6x(x^2-1)(x^2+1)^{-4} = -4x(x^2-2)(x^2+1)^{-4}$
(b) $3(3x^2+2x)^2(6x+2)(2x^2+2x+2)^{-5}$
$+(3x^2+2x)^3(-5)(2x^2+2x+2)^{-6}(4x+2)$
(c) $4u(u^2-3)(u^2+3)^{-1}(u+1)^{-2}+(u^2-3)^2(-2u)(u^2+3)^{-2}(u+1)^{-2}$
$-2(u^2-3)^2(u^2+3)^{-1}(u+1)^{-3}$

4. (a) $8x_0(x_0^2+1)^3$ (b) $2(x_0+1)(x_0^2+1)^{-4}-8x_0(x_0+1)^2(x_0^2+1)^{-5}$
(c) $4(x_0-1)/(x_0+1)^3$ (d) $-3(x_0^3-x_0)^{-4}(3x_0^2-1)$

5. (a) $f'(1)=0, f'(-1)=0; f'(x)>0$ for $|x|<1, f'(x)<0$ for $|x|>1$
(b) $f'(0)=0, f'(-2)=0$. Neither f nor f' is defined at $x=-1$, where f has an asymptote of type $-\infty, +\infty$. $f'(x)<0$ for $-2<x<0, x\neq 1$.

§2.3

1. (a) $f'(x)=\cos^2 x-\sin^2 x$ (b) $f'(x)=\sin x+x\cos x$
(c) $f'(t)=(t\cos t-\sin t)/t^2$

5. $f'(x)=\sqrt{2}\cos x, g'(x)=-\sqrt{2}\sin x$. Since $\cos^2 x+\sin^2 x=1$, we have $\cos x=\sin x$ only if $2\cos^2 x=1$, so $\cos x=\pm 1/\sqrt{2}$, $\sin x=\pm 1/\sqrt{2}$. Therefore, at a point of intersection, either $f'=+1$ and $g'=-1$, or $f'=-1$ and $g'=+1$.

Appendix

1. $\sin\pi = 0 = \sin(-\pi)$, $\cos\pi=-1=\cos(-\pi)$, $\sin(-\pi/2)=-1$, $\cos(-\pi/2)=0$

6. (a) zeros at $x=\pm\pi/6, \pm\pi/2, \pm 5\pi/6, \pm 7\pi/6, \ldots$
(b) zeros at $x=\pm 3\pi/2, \pm 9\pi/2, \pm 15\pi/2, \ldots$
(c) zeros at $x=0, \pm\pi/2, \pm\pi, \pm 3\pi/2, \ldots$;
asymptotes at $x=\pm\pi/4, \pm 3\pi/4, \ldots$
(d) zeros at $x=0, \pm 1, \pm 2, \ldots$; asymptotes at $x=\pm 1/2, \pm 3/2, \ldots$
(g) zeros at $x=-\pi/4, -\pi/4\pm\pi, -\pi/4\pm 2\pi, \ldots$

§2.4

1. (a) $\sin(\sqrt{x})$ (b) $\sin(\sqrt{t})$ (c) $\sin 1$ (d) $\sqrt{\sin y}$
(i) $\sin(t+1)$ (k) $\sqrt{s^2+1}$

2. (a) $g(x)=2x, f(y)=\sin y, (f\circ g)'(x)=\cos(2x)\cdot 2$
(e) $g(x)=\sin x, f(y)=y^2+2, (f\circ g)'(x)=2\sin x\cos x$
(f) $g(x)=\cos x, f(y)=\sec y, (f\circ g)'(x)=\sec(\cos x)\tan(\cos x)\cdot[-\sin x]$

3. (a) $-2\sin(2x)$ (b) $-2x\sin(x^2)$ (d) $10(4x-2)^{-2}\sin\left(\dfrac{3x+1}{4x-2}\right)$

(e) $2\cos 2x$ (f) $(2x+5)\cos(x^2+5x)$ (h) $3(x^2+1)\sec^2(x^3+3x)$

4. (a) $-2(\cos 2x)^{-3}(-\sin 2x)\cdot 2$

(b) $-\sin(\sin x)\cos x$

(c) $-\sin(\sin 5x)\cdot(\cos 5x)\cdot 5$

(d) $\dfrac{\sin 2x-(x+2)2\cos 2x}{\sin^2 2x}$

§2.5

4. (b) $6x^2(4x^6-4x^3+2)^{-1}$ (c) $b/[b^2+(x-a)^2]$

6. (a) $(-\sin\theta)\arcsin\theta+\dfrac{\cos\theta}{\sqrt{1-\theta^2}}+1$

(b) $\dfrac{2}{1+(2x+1)^2}$ (d) $x[\arcsin(1-x^2)]^{-3/2}/|x|\sqrt{2-x^2}$

(c) $x/\sqrt{x^2+1}$ (e) $[-6\sin(2\arcsin(3t))]/\sqrt{1-9t^2}$

§2.6

1. (a) $1/x$ (b) $1/x$ (c) $2/x$ (d) $(2\log x)/x$ (e) $-\tan x$

(f) $-2\tan(2x+1)$ (g) $1/(x\log x)$

7. (b) Area of a trapezoid is base times the midpoint height, $(b-a)\times\dfrac{2}{a+b}$

9. $A'(t)=t^3$

§2.7

1. (a) $3e^{3x}$ (c) $(2x+5)e^{x^2+5x}$

(e) $e^x\cos(e^x)$ (g) $a^x\log a$ (i) $-\exp(t)\tan(\exp(t))$

(k) $\arcsin x+x/\sqrt{1-x^2}+\pi(a^2-x^2)^{\pi-1}(-2x)$

(m) $ae^{ax}\cos bx-be^{ax}\sin bx$ (p) $\exp(e^{\exp x})e^{\exp x}\exp x$.

2. Since exp is the inverse of log, we have $\log(\exp x)=x$. Since $\log'(x)=1/x$ is always positive, exp has a derivative [by the inverse function theorem]. Differentiating the formula $\log(\exp x)=x$ therefore gives, by the chain rule,

$$\frac{1}{\exp(x)}\exp'(x)=1,\qquad \text{so } \exp'(x)=\exp(x).$$

7. $x=\log_a y\iff a^x=y$, by Definition 3.

§2.9

1. $-\sin x e^{\cos x}$

3. $1-2\cos x\sin x$

5. $\dfrac{2}{1-x^2}$

7. $\dfrac{1}{\sqrt{x^2-1}}$

9. $\dfrac{-1}{x\sqrt{1+x^2}}$

11. $\dfrac{(ax^2+bx+c)-2x(2ax+b)}{(ax^2+bx+c)^3}$

13. ke^{kx}

15. $2ke^{2kx}$

18. $e^x \exp(e^x)$

20. $\frac{1}{x}\exp(1 + \log x)$

21. $\exp(\log^2 x)\, 2\,\frac{\log x}{x}$

22. $\left[\frac{\cos x}{x} - (\log x)(\sin x)\right] \exp[(\log x)(\cos x)]$

23. $1/x + 1/(x - 1)$

25. $\frac{-\sin x + \cos x + \sec^2 x}{\cos x + \sin x + \tan x}$

26. $\frac{1}{\cos\theta \sin\theta}$

27. $-\frac{3}{x}\sin(3\log x)$

29. $\frac{-x}{\sqrt{1 - x^2}}$

30. $\frac{-x}{\sqrt{1 - x^2}}$

35. $\frac{-\csc\theta\cot\theta}{1 + \csc^2\theta}$

36. (a) $\sec\theta$ (c) $(a^2 + b^2)e^{bx}\cos ax$
(e) x^2e^x

§2.10

1. $\frac{dy}{dt}(t_0) = 3$

2. $e^{xy}\left[x\frac{dy}{dt} + y\frac{dx}{dt}\right] = \cos x\frac{dx}{dt}$

3. $a^{-2}x\, dx/dt + b^{-2}y\, dy/dt = 0$

6. $dy/dt = -(2x + 3y)/(3x + 2y)$

7. (b) $\frac{dx}{dt}(t_0) = -3^{21/2}a/b$

8. $\frac{dy}{dt} = \frac{dr}{dt}\sin\theta + r\cos\theta\frac{d\theta}{dt} = 3\frac{\sqrt{3}}{2} + 6\cdot\frac{1}{2}(-3) = \frac{3\sqrt{3} - 18}{2}$

10. (c) $dy/dx = -\cot\theta$ (d) $dy/dx = -x/\sqrt{l^2 - x^2}$. (This can be deduced from part (c) or from the relation $x^2 + y^2 = l^2$.)

11. (a) $x^2 + y^2 = 13^2$, $2x\frac{dx}{dt} + 2y\frac{dy}{dt} = 0$; at time t_0, $x = 5$, $dx/dt = 2$,

$y = \sqrt{13^2 - 5^2} = 12$, so $dy/dt = -5/6$.

(b) Let z be the height of the man above ground. By similar triangles, $z = \frac{2}{3}y$ for all times t; therefore, at the given time t_0,

$$\frac{dz}{dt} = \frac{2}{3}\frac{dy}{dt} = -\frac{5}{9}.$$

13. $V = \frac{\pi}{3}h^3(\tan\alpha)^2$, $2 = \frac{dV}{dt} = \pi h^2\tan^2\alpha\frac{dh}{dt}$, $\frac{dh}{dt} = \frac{2}{\pi h^2\tan^2\alpha}$

15. $V = 12\pi r_0^2$, $\frac{dV}{dt} = 24\pi r_0\frac{dr}{dt} = 3\pi r_0$

17. Let x be the position of train A, and y the position of train B, so $dx/dt = 40$ and $dy/dt = 60$. The distance 0 between A and B satisfies

$$D^2 = \frac{1}{16} + (x-y)^2, \qquad \text{so} \qquad 2D\frac{dD}{dt} = 2(x-y)\left(\frac{dx}{dt} - \frac{dy}{dt}\right).$$

(a) $dD/dt = 0$ (b) $dD/dt = -4\sqrt{5}$

§2.11

1. (d) $2\frac{b}{a}\sqrt{a^2 - x^2}$ **3.** (a) e^2 (d) $(ex/x_0)^2$

CHAPTER III

§3.1

1. (a)

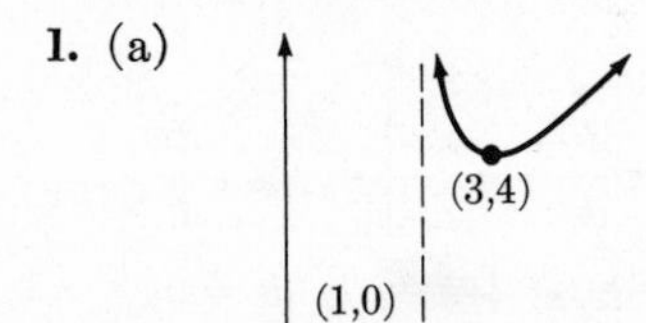

(b)

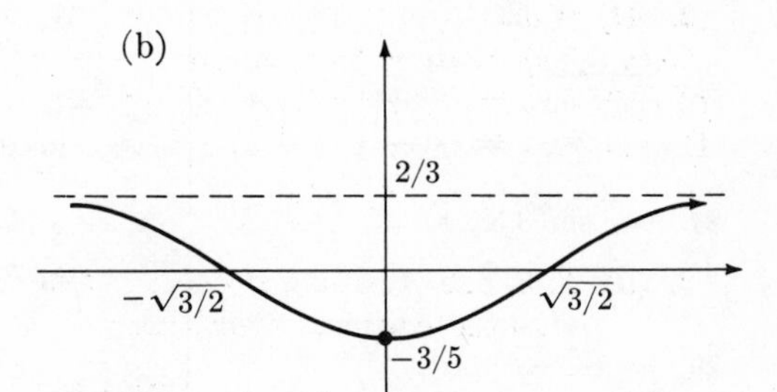

3.

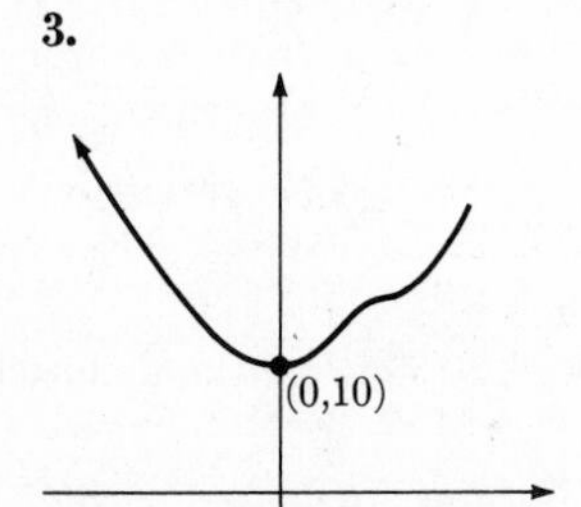

5.

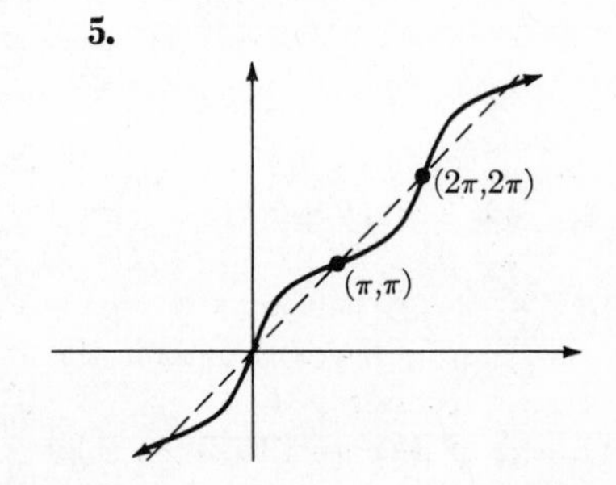

7.

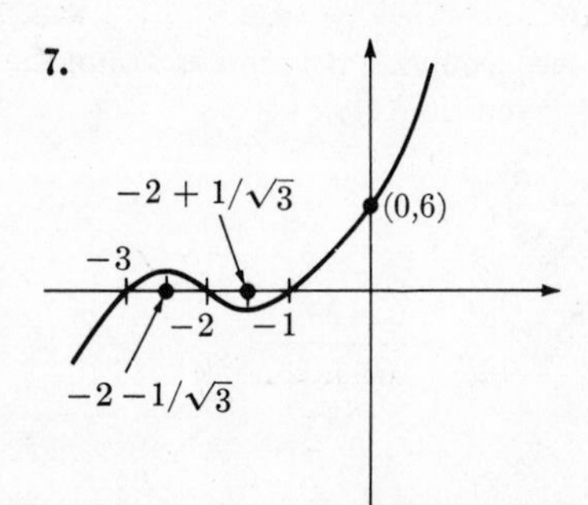

9.

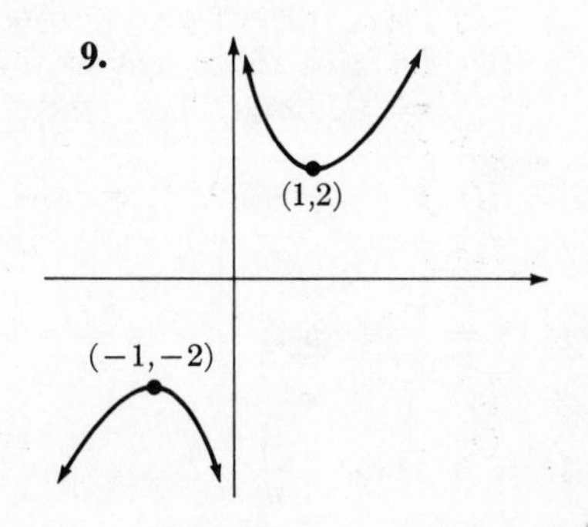

11.

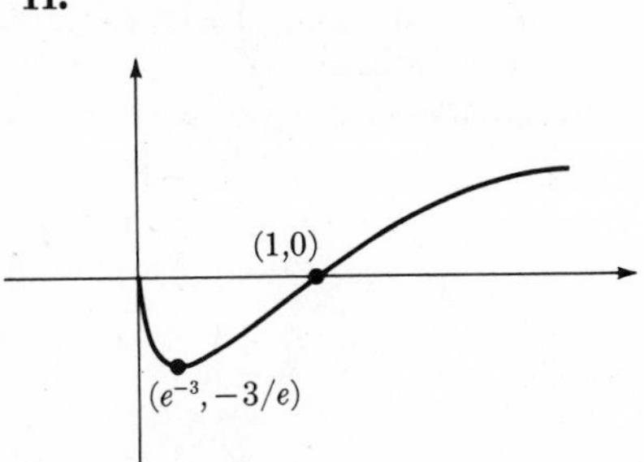

13.

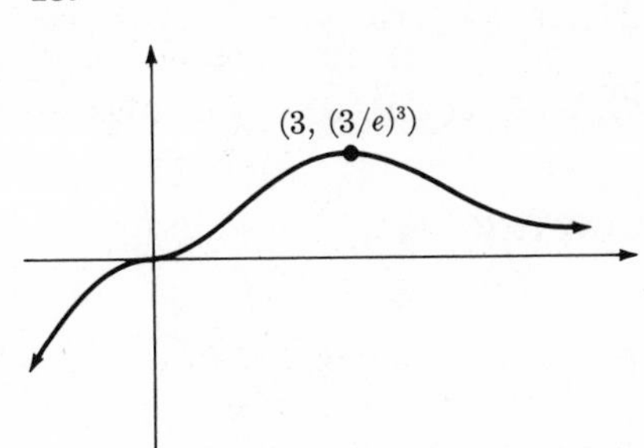

§3.2

2. $v(t) = 32t$
5. (a) $f(x) = \frac{1}{2}x^2 + 1$ (c) $f(x) = x^4 - x^3 + x^2 - x + 10$
10. (a) $f(x) = x^{-1} + C$ (c) $f(x) = \arctan x + C$
11. (a) $f(x) = x^{-1} - 1$ (c) $f(x) = \arctan x - \pi/2$

§3.3

3. (a) After one year the percentage remaining is $e^{-10^{-9}\log 2/4.4}$, practically 100% since the exponent is practically 0.

4. (a) $k = \frac{1}{10}\log(1.03)$ (b) $\dfrac{10\log 2}{\log(1.03)}$
6. (b) $A(t) = (t + \sqrt{3})^2$
8. $a = -k$ and $b = 0$

§3.4

2. (e) Concave for $x < -1$; convex for $x > -1$ (f) Convex everywhere
(g) Convex for $x < 0$; concave for $x > 0$
(h) $f''(x) = 2x^{-5/3}(\frac{14}{9}x^2 + 1)$; concave for $x < 0$, convex for $x > 0$
(i) $f''(x) = 1/x$, $x \neq 0$; concave for $x < 0$, convex for $x > 0$
(j) $f''(x) = x(5 + 6\log|x|)$; convex for $x > e^{-5/6}$ and for $-e^{-5/6} < x < 0$, concave for $x < -e^{-5/6}$ or $0 < x < e^{-5/6}$
(l) $f''(x) = -16(3x^2 + 1)(x^2 - 1)^{-3}$; convex for $|x| < 1$, concave for $|x| > 1$
(n) $f''(x) = \frac{1}{2}\sec^2(x/2)\tan(x/2)$ has the same sign as $\tan(x/2)$. Convex for $0 < x < \pi/4$, $\pi/2 < x < 3\pi/4, \ldots$ Concave for $-\pi/4 < x < 0$, $\pi/4 < x < \pi/2$, $3\pi/4 < x < \pi, \ldots$
3. (e) $f''(x) = -xe^{-x^2/2}$, convex for $x < 0$, concave for $x > 0$
(f) $f''(x) = -3x^2e^{-x^3}$; concave everywhere
4. (a) $f'''(t) = \omega^3 e^{\omega t}$ (b) $\omega = \sqrt{c}$
9. (c) $f(x) = 2 + 3x + 2x^2$

§3.5

1. (a) $f(t) = \cos\omega t$ (c) $f(t) = -2\cos\omega t$
(e) $f(t) = B\sin\omega t$, B arbitrary (g) $f(t) = \cos\omega t + B\sin\omega t$, B arbitrary

2. $f(t) = \left(A_0 \cos \omega t_0 - \dfrac{B_0}{\omega} \sin \omega t_0\right) \cos \omega t + \left(A_0 \sin \omega t_0 + \dfrac{B_0}{\omega} \cos \omega t_0\right) \sin \omega t_0$

7. (d) $k = mA^2/M^2$ (e) Maximum compression M^2/A

CHAPTER IV

§4.1

1. (a) $\max\limits_{[-2,2]} f = 1/2$, $\min\limits_{[-2,2]} f = -1/2$ (e) $\max\limits_{[-2,2]} f = \dfrac{e^4 - 1}{e^4 + 1} = -\min\limits_{[-2,2]} f$

(g) $\max\limits_{[-1,3]} f = 3$, $\min\limits_{[-1,3]} f = 0$

2. (a) Maximum, by Theorem 4 (b) No result (d) No result

3. (a) $\max \sin x = 1$, $\min \sin x = -1$ (c) Neither max nor min

7. Let x = number moved from A to B. Let m be the smallest of the three numbers $20a + 50b$, $20c + 30b$, $50c + 30a$. Then the minimum cost occurs when

$$\begin{aligned} x &= 0 && \text{if } m = 20a + 50b \\ x &= 20 && \text{if } m = 20c + 30b \\ x &= 50 && \text{if } m = 50c + 30a. \end{aligned}$$

9. $r = \frac{1}{2}V^{1/3}$, $h = 4V^{1/3}/\pi$. (Notice that twice as much material goes into the sides as into the top and bottom; the same is true in Example 5.)

11. (a) $(1,1)$ and $(-1,-1)$

(b) This reduces to the solution of the equation $P(x) = x^4 + x - 1 = 0$. Since $P(0) = -1$ and $P(1) = +1$, there is a solution x_0 such that $0 < x_0 < 1$ and $P(x_0) = 0$. Since this is the only zero of $P(x)$ for $x > 0$, the corresponding point $P_0 = (x_0, 1/x_0)$ on the graph is the closest point on the upper right half of the hyperbola. Since every point on the lower left half of the hyperbola is at least 1 unit from $(1,0)$, and P_0 is less than 1 unit from $(1,0)$, it follows that P_0 is the closest point on the whole parabola. (If a decimal expansion of x_0 were required, it could be obtained by Newton's method, §1.6.)

13. $r/h = 1/\sqrt{2}$

15. $4b = 3a$, i.e. the same amount of fence is used in each direction. The minimum amount of fence is $3a + 4b = 4\sqrt{3A}$.

17. Area r^2, dimensions $r/\sqrt{2} \times \sqrt{2}r$

19. At the closest point $P_1 = (x_1, y_1)$, the equation $(x_1 - x_0) + (y_1 - y_0)f'(x_1) = 0$ is satisfied. Since the tangent line of the graph has slope $m_1 = f'(x_1)$, and the line through P_0 and P_1 has slope $m_2 = (y_1 - y_0)/(x_1 - x_0)$, it follows that $m_1 m_2 = -1$.

21. $R = e^{1/2}r$ **23.** $A/B = 1/8$

§4.2

1. (a) $\sqrt{2}$ (c) $1/\sqrt{3}$ (e) $\arccos(2/\pi)$ (g) $-1/\sqrt{2}$

(i) Does not apply

3. (a) s.l.max at $x = 0$ (c) s.l.max at $x = 1/\sqrt{3}$, s.l.min at $x = -1/\sqrt{3}$ (*not* max or min over the whole interval)

(e) s.l.max at $x = \pi/2, \pi/2 + 2\pi, \pi/2 - 2\pi, \pi/2 - 4\pi$;
s.l.min at $x = -\pi/2, -\pi/2 - 2\pi, -\pi/2 + 2\pi, -\pi/2 + 4\pi$

(f) s.l.max at $t = 1/2\pi, 1/4\pi, 1/6\pi, \ldots$; s.l.min at $t = 1/\pi, 1/3\pi, 1/5\pi, \ldots$

CHAPTER V

§5.1

3. (c) $\frac{3}{n}\sum_{j=1}^{n}\frac{1}{(1+3(j-1)/n)^2} - \frac{3}{n}\sum_{j=1}^{n}\frac{1}{(1+3j/n)^2} = \frac{3}{n}\left(1-\frac{1}{16}\right)$

12. $\frac{1}{5}\sum_{j=1}^{5}\sqrt{1-j^2/25} = \frac{1}{25}(\sqrt{24}+\sqrt{21}+4+3+0) = .66\ldots$ is a lower sum,

and $\frac{1}{5}\sum_{j=0}^{4}\sqrt{1-j^2/25} = \frac{1}{25}(5+\sqrt{24}+\sqrt{21}+4+3) = .86\ldots$ is an upper sum. They differ by 1/5, and the actual integral lies between them, so their average, .76 . . . , is accurate within $\pm(1/10)$.

§5.2

2. $\int_1^b \frac{1}{x}\,dx$ **4.** $\int_0^b e^{-t}\,dt$ **6.** $\int_0^1 u^4\,du$ **8.** $\int_1^2 \frac{1}{x^2}\,dx$

§5.3

1. (d) $(5^{100}-3^{100})/100$ (h) $1-1/e$
2. (a) $1+1/2+1/3$ (c) $4r^3/3$ (e) π
5. All are true except for (e).

§5.4

1. (b) $\frac{1}{3}(e^{30}-1)$ (d) 0

5. (b) $\int_a^b \left(y_1 + \frac{y_3-y_1}{b-a}(x-a) - y_1 - \frac{y_2-y_1}{b-a}(x-a)\right)dx$

$$= \frac{1}{2}(y_3-y_2)(b-a)$$

7. $\int_0^a \pi\frac{b^2}{a^2}(a^2-x^2)\,dx = \frac{2}{3}\pi ab^2$ is half the total volume.

22. (a) $\int_1^{100}\frac{c}{x^2}\,dx = -\left.\frac{c}{x}\right|_1^{100} = \frac{99c}{100}$

23. (a) $\int_{s_0}^{s_1} -ks\,ds = -k\left.\frac{s^2}{2}\right|_{s_0}^{s_1} = -\frac{1}{2}k(s_1^2-s_0^2)$. Taking s_0 fixed, we get $\frac{1}{2}ks_0^2$ is constant, and for any s_1, the work is $-\frac{1}{2}ks_1^2+$ constant.

CHAPTER VI

§6.1

2. (a) $\int u^a\,du = \frac{u^{a+1}}{a+1}+c$ (b) $\int\frac{du}{u} = \log|u|+c$ (c) $\int e^u\,du = e^u+c$

(d) $\int \sin u\,du = -\cos u + c$ (e) $\int \cos u\,du = \sin u + c$

(f) $\int \sec^2 u\, du = \tan u + c$ (g) $\int \csc^2 u\, du = -\cot u + c$
(h) $\int \sec u \tan u\, du = \sec u + c$ (i) $\int \csc u \cot u\, du = -\csc u + c$

(j) $\displaystyle\int \frac{du}{1+u^2} = \arctan u + c$

3. (a) $\frac{2}{3}x^3 + 3x + c$; 15 (c) $\dfrac{(2x-1)^3}{6} + c$; $-\dfrac{185}{3}$

(d) $x + x^2/2 + x^3/3 + x^4/4 + x^5/5 + c$; 137/60

4. (a) $\pi^2/4$ (c) $\tan\theta - \cot\theta + c$ (e) $u^2/2 + u^4/4 + c$

7. (d) $\displaystyle\int \sum_0^n a_j \cos(j\theta)\, d\theta = a_0\theta + \sum_1^n \frac{a_j}{j} \sin(j\theta) + c$

8. (a) See §5.4. (d) $\displaystyle\int_a^b \pi x^{-2}\, dx = -\pi x^{-1}\Big|_a^b = \pi\left(\frac{1}{a} - \frac{1}{b}\right)$

10. (b) $\displaystyle\int_0^2 x^3\, dx = \frac{x^4}{4}\Big|_0^2 = 4$ (d) $(3/2, 6/5)$ **13.** 25γ

§6.2

1. (a) $\frac{1}{2}\sin 2x + c$ (c) $\frac{1}{5}e^{5x} + c$ (e) $\frac{1}{10}\tan 10\theta + c$

2. (a) $\sin(3+\theta) + c$ (c) $\frac{1}{7}\log|3+7x| + c$

(e) $\dfrac{2}{b}\sqrt{a+bt} + c$ (g) $\dfrac{2}{9}(1+3x)^{3/2}$

(j) $\dfrac{-1}{a}\cos(a\theta + b) + c$

3. (a) $\frac{1}{2}\sin(\theta^2) + c$ (c) $\frac{1}{12}e^{3x^4} + c$ (e) $-\frac{1}{2}e^{-x^2} + c$

4. (a) $-\frac{1}{2}\cos(\theta^2+1) + c$ (c) $\frac{1}{3}e^{x^3+3x} + c$

(e) $\frac{1}{2}\log|x^2+2x| + c$ (g) $-\dfrac{1}{2n+2}(a^2-x^2)^{n+1}$

5. (a) $\frac{1}{3}\sin^3\theta + c$ (c) $\dfrac{1}{n+1}\sin^{n+1}\theta + c$

(e) $-\dfrac{1}{an+a}\cos^{n+1} ax + c$ (g) $\sin\theta - \frac{1}{3}\sin^3\theta + c$

6. (a) $\frac{1}{2}\log(a^2+t^2) + c$ (d) $\log|\sec\theta| + c$
(g) $\log|\sec\theta + \tan\theta| + c$

7. (b) $\dfrac{1}{a}\arctan\left(\dfrac{t-b}{a}\right) + c$ (e) $\dfrac{2}{\sqrt{4AC-B^2}}\arctan\left(\dfrac{2Ax+B}{\sqrt{4AC-B^2}}\right) + c$

8. (b) $\frac{1}{2}\theta + \frac{1}{12}\sin 6\theta + c$ (d) $\frac{5}{2}\theta - \frac{1}{2}\sin 2\theta + \frac{3}{16}\sin 8\theta + c$
(e) $\frac{3}{8}\theta + \frac{1}{4}\sin 2\theta - \frac{1}{32}\sin 4\theta + c$

9. (a) $\frac{2}{45}(3x+5)^{5/2} - \frac{10}{27}(3x+5)^{3/2} + c$ (c) $\frac{5}{11}(t-1)^{11/5} + \frac{5}{6}(t-1)^{6/5} + c$

(e) $\dfrac{(2x+3)^{\pi}}{8}\left[\dfrac{(2x+3)^3}{3+\pi} - \dfrac{6(2x+3)^2}{2+\pi} + \dfrac{9(2x+3)}{1+\pi}\right] + c$

10. (a) $n = 1$ (c) $n = -1$ (e) $n = 3/2$

12. (a) $\dfrac{1}{a}\arctan\dfrac{b}{a}$ (c) $1/3$ (e) $\dfrac{2}{\sqrt{3}}\left(\arctan\left(\dfrac{2b+1}{\sqrt{3}}\right) - \dfrac{\pi}{6}\right)$

13. (b) π/a (c) 2 (e) 0 **18.** $\dfrac{\pi}{6c^2}[(1 + 4c^2b^2)^{3/2} - 1]$

16. $2(10^{3/2} - 1)/27$

Appendix

1. $0 = x^2 + y^2 - 2x - 4y = (x-1)^2 + (y-2)^2 - 5$, or $\sqrt{(x-1)^2 + (y-2)^2} = \sqrt{5}$. Center (1,2) and radius $\sqrt{5}$.

2. Empty set **3.** The single point $(-1,-2)$

§6.3

1. (a) $-t^2e^{-t} - 2te^{-t} - 2e^{-t} + c$ (c) $-\frac{1}{2}\sin x \cos x + \frac{1}{2}x + c$

(e) $-\dfrac{1}{3a}\sin^2 ax \cos ax - \dfrac{2}{3a}\sin ax \cos ax + c$

2. (a) $\dfrac{x}{a}\sin ax + \dfrac{1}{a^2}\cos ax + c$ (c) $\theta\tan\theta + \log|\cos\theta| + c$

(d) $\dfrac{x^2}{a}\sin ax + \dfrac{2x}{a^2}\cos ax - \dfrac{2}{a^3}\sin ax + c$

3. (a) $x\arctan x - \frac{1}{2}\log(1 + x^2) + c$ (b) $x\arcsin x + \sqrt{1 - x^2} + c$

(d) $x \operatorname{arcsec} x - \log|x + \sqrt{x^2 - 1}| + c$ (e) $\dfrac{x^2+1}{2}\arctan x - \dfrac{x}{2} + c$

(h) See Problem 9.

4. (a) $-\frac{1}{2}(x^2 + 1)e^{-x^2} + c$ (b) $\frac{1}{4}(x^4 + 1)\arctan x^2 - \frac{1}{4}x^2 + c$

6. (c) $\displaystyle\int x^n \sin ax\, dx = -\frac{x^n}{a}\cos ax + \frac{n}{a}\int x^{n-1}\cos ax\, dx$

9. (b) $\displaystyle\int x^m (\log x)^n\, dx = \frac{x^{m+1}}{m+1}(\log x)^n - \frac{n}{m+1}\int x^m (\log x)^{n-1}\, dx$

10. (a) $\displaystyle\int (\cos ax)^m\, dx = \frac{1}{am}(\cos ax)^{m-1}\sin ax + \frac{m-1}{am}\int (\cos ax)^{m-2}\, dx$

12. (a) $\displaystyle\int x^m e^{-ax^2} dx = -\frac{x^{m-1}}{2a}e^{-ax^2} + \frac{m-1}{2a}\int x^{m-2}e^{-ax^2}\, dx$

14. (a) 1 (b) 2 (c) 6 **16.** (a) $\sqrt{\pi}/4$

§6.4

1. (a) $x - 1 + \dfrac{2}{x+2}$ (b) $1 + \dfrac{3x-2}{x^2-3x+2}$ (f) $\dfrac{x}{(x^2+9)^2}$

2. (a) $\frac{1}{2}x^2 - x + 2\log|x+2| + c$ (c) $\sqrt{3}\arctan\left(\dfrac{x+1}{\sqrt{3}}\right) + c$

(e) $\frac{1}{3}x^3 - x^2 + 3x - 4\log|x+1| - (x+1)^{-1} + c$ (g) $-1/(x+2) + c$

3. (a) $\dfrac{1/(1-n)}{a(ax+b)^{n-1}} + c$ (c) $\dfrac{1}{2q^3}\dfrac{u}{u^2+1} + \dfrac{1}{2q^3}\arctan u + c;\quad u = \dfrac{x-p}{q}$

(e) $\dfrac{1}{2}\log[(x-p)^2 + q^2]$

4. (a) $\dfrac{A}{x-1} + \dfrac{B}{(x-1)^2} + \dfrac{C}{x+1} + \dfrac{D}{(x+1)^2}$ (c) $\dfrac{Ax+B}{x^2+1} + \dfrac{Cx+D}{(x^2+1)^2}$

(e) $\dfrac{Ax+B}{x^2+2x+2} + \dfrac{Cx+D}{(x^2+2x+2)^2}$

5. (a) $\dfrac{1}{(x^2-1)^2} = \dfrac{1}{4}\left[\dfrac{-1}{x-1} + \dfrac{1}{(x-1)^2} + \dfrac{1}{x+1} + \dfrac{1}{(x+1)^2}\right]$

(c) $\dfrac{1}{(x^2+16)^3}$ cannot be further decomposed.

(e) $\dfrac{x^3}{(x^2+2x+2)^2} = \dfrac{x-2}{x^2+2x+2} + \dfrac{2x+4}{(x^2+2x+2)^2}$

6. (a) $A\log|x-1| + B/(x-1) + C\log|x+1| + D/(x+1)$
(c) $A\arctan x + B\log(x^2+1) + (Cx+D)/(x^2+1)$
(e) $A\arctan(x+1) + B\log(x^2+2x+2) + (Cx+D)/(x^2+2x+2)$

7. (a) $\dfrac{1}{4}\left(\log\left|\dfrac{x+1}{x-1}\right| - \dfrac{1}{x-1} - \dfrac{1}{x+1}\right) + c$

(c) $\dfrac{x}{64(x^2+16)^2} + \dfrac{3}{8\cdot 16^2}\dfrac{x}{x^2+16} + \dfrac{3}{2\cdot 16^3}\arctan(x/4) + c$

(e) $\dfrac{1}{2}\log(x^2+2x+2) + \dfrac{x}{x^2+2x+2} - 2\arctan(x+1) + c$

8. (a) $x + \frac{5}{4}\log|x-1| - 1/2(x-1) - \frac{1}{4}\log|x+1| + c$

(c) $\log\left|\dfrac{\cos\theta+2}{\cos\theta+1}\right| + c$ (e) $\frac{5}{2}\log|u+3| - \log|u+2| + \frac{1}{2}\log|u-1| + c$

§6.5

1. $-\frac{1}{6}\cos 3\theta + \frac{1}{2}\cos\theta + c$ **3.** $\dfrac{1}{2a}\cos a\theta\sin a\theta + \dfrac{1}{2}\theta + c$

5. $\frac{1}{3}\tan^3\theta - \tan\theta + \theta + c$ **7.** $\tan\theta + c$

11. $\dfrac{1}{\sqrt{2}} \log \left|\dfrac{\tan(x/2) + 1 - \sqrt{2}}{\tan(x/2) + 1 + \sqrt{2}}\right|$

§6.6

17. $\dfrac{1}{2}\pi a^2 b + b(h-a)\sqrt{2ah - h^2} + a^2 b \arcsin\left(\dfrac{h-a}{a}\right)$

18. $\sqrt{17} + \frac{1}{4}\log|4 + \sqrt{17}|$

20. $2\displaystyle\int_{-\pi/2}^{\pi/2} 2\pi(a + r\sin\theta)r^2\cos^2\theta\, d\theta = 4\pi ar^2 \int_{-\pi/2}^{\pi/2} \cos^2\theta\, d\theta = 2\pi^2 ar^2$

[The integral of the *odd* function $\sin\theta\cos^2\theta$ from $-b$ to b is zero.]

§6.7

1. (a) $y = \log \dfrac{1}{c - e^x}$ (c) $y = (\frac{2}{3}x + c)^{3/2}$

(e) $-y\cos y + \sin y = \log x + x^2/2 + c$

2. (a) $y = \pm Ce^{-\cos x}$, $C > 0$ (c) $y = \pm C\sec x$, $C > 0$

5. (a) $f(x) = -\log(2 - e^x)$ (b) $\lim\limits_{x\to-\infty} f(x) = -\log 2$

(c) Vertical asymptote at $x = \log 2$. Notice that the solution is not defined for $x \geq \log 2$.

6. (a) $v(t) = Ce^{-(k/m)t}$ (b) No, since $v(t) > 0$ for all t.

(c) No; $v(t) = v_0 e^{-(k/m)t}$, so the car rolls $\displaystyle\int_0^\infty v_0 e^{-(k/m)t}\, dt = v_0 m/k$. Notice that the distance it rolls increases with the initial velocity v_0 and the mass m, and decreases with the friction constant k.

10. (a) $\dfrac{20 - Q}{A - Q/120} = \dfrac{20}{A} e^{kt(1/6-A)}$, if $A \neq 1/6$. (b) Solve part (a) for Q.

§6.8

1. (a) $f(x) = Ce^{-x^2/2}$ (c) $f(x) = -\dfrac{1}{x}\cos x + \dfrac{C}{x}$.

2. (a) $F(x) = 0$ for all x.

(c) $f(x) = \dfrac{1 - \cos x}{x}$ makes $\lim\limits_{x\to 0} f(x) = -\cos'(0) = -\sin(0) = 0$.

It is true, but not easy to prove without further theory, that the function

$$f(x) = \begin{cases} \dfrac{1 - \cos x}{x} & x \neq 0 \\ 0 & x = 0 \end{cases}$$

is differentiable at $x = 0$, and solves $xf'(x) + f(x) = \sin x$ for *all* x.

(f) There *are no* solutions with $f(0) = 1$. This is obvious from $xf'(x) + f(x) = \sin x$; substituting $x = 0$ yields $0 + f(0) = 0$.

3. (a) $f(x) = \frac{3}{2}x^4 + x^2\log|x| + cx^2$; $c = -7/2$ makes $f(1) = -2$.

(c) $y = x^2(\log|x| + c)$; $c = -\log 2$ makes $y = 0$ when $x = 2$.

(e) $y = \frac{1}{2}(x+1)^4 + c(x+1)^2$; $c = -1$ makes $y = 4$ when $x = 1$.

(g) $F'(x) = \dfrac{(x-1)^2}{3} + \dfrac{C_1}{x-1}$, $F(x) = \dfrac{(x-1)^3}{9} + C_1 \log|x-1| + C_2$.

$C_2 = 8/9$, $C_1 = -1/3$ makes $F(2) = 1$ and $F'(2) = 0$.

6. (a) $s' = (-rs/g) + rC$; the term $-rs/g$ is the rate at which the substance flows out, and the term rC is the rate at which it flows in with the incoming water.

(b) $s(t) = gC + e^{-rt/g}(s_0 - gC)$.

9. $I(t) = \dfrac{E}{L^2\omega^2 + R^2}(R\sin\omega t - L\omega\cos\omega t) + ce^{-Rt/L}$.

§6.9

4. (a) $y = Cx^{-2}$ (c) $y^2 - x^2 = c$ (d) $y = -\frac{1}{2}\log x + c$.

(e) The "family of slopes" of the equipotential lines is $\dfrac{dy}{dx} = 2cx = 2\dfrac{y}{x}$. The lines of force are $y^2 + \frac{1}{2}x^2 = c$.

CHAPTER VII

§7.1

1.

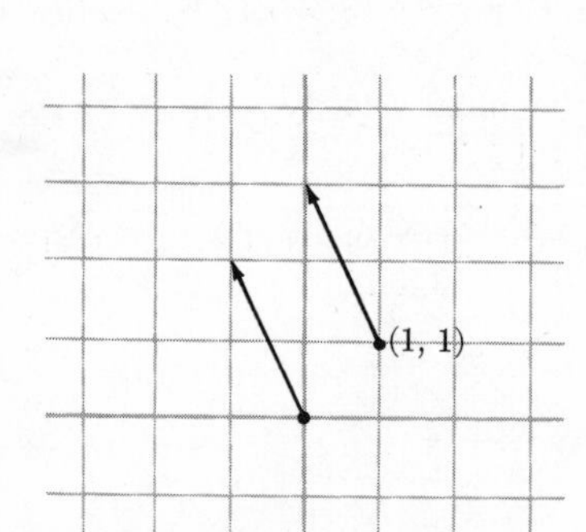

2.

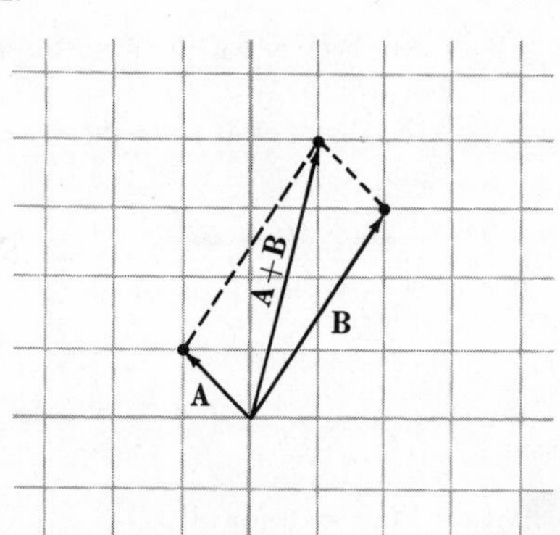

3. The arrows for $\mathbf{A} + \mathbf{B}$ and $\mathbf{B} - \mathbf{A}$ intersect at the midpoints.

4. (a) Let $\mathbf{A} = [a_1, a_2]$, $\mathbf{B} = [b_1, b_2]$, $\mathbf{C} = [c_1, c_2]$. Then

$$\begin{aligned}(\mathbf{A}+\mathbf{B})+\mathbf{C} &= [a_1+b_1, a_2+b_2] + [c_1, c_2]\\ &= [a_1+b_1+c_1, a_2+b_2+c_2]\\ &= [a_1, a_2] + [b_1+c_1, b_2+c_2]\\ &= \mathbf{A} + (\mathbf{B}+\mathbf{C}).\end{aligned}$$

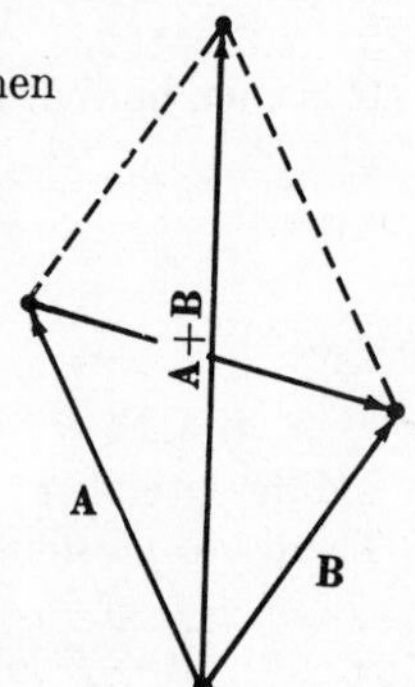

§7.2

1. (a) $|\mathbf{V}| = \sqrt{2} = |\mathbf{W}|$, $|\mathbf{V}| = 1$ (d) $\pi/4$; π

3. (a) $\mathbf{A}\cdot\mathbf{A} = a_1^2 + a_2^2 = |\mathbf{A}|^2$,
$\mathbf{A}\cdot\mathbf{B} = a_1b_1 + a_2b_2 = b_1a_1 + b_2a_2 = \mathbf{B}\cdot\mathbf{A}$.

12. (b) $\mathbf{W} = [1/\sqrt{2}, -1/\sqrt{2}]$ (c) $\mathbf{W} = [-1,0]$ (e) $\mathbf{W} = [-\sqrt{2}, 2\sqrt{2}]$.

§7.3

1. (a) $x = 1 + t$, $y = 3 - 2t$; $y = 5 - 2x$, slope -2.
(c) $x = 0$, $y = t$; vertical line (e) $x = 1 + t$, $y = 3$; the standard equation is $y = 3$, and the slope is 0.

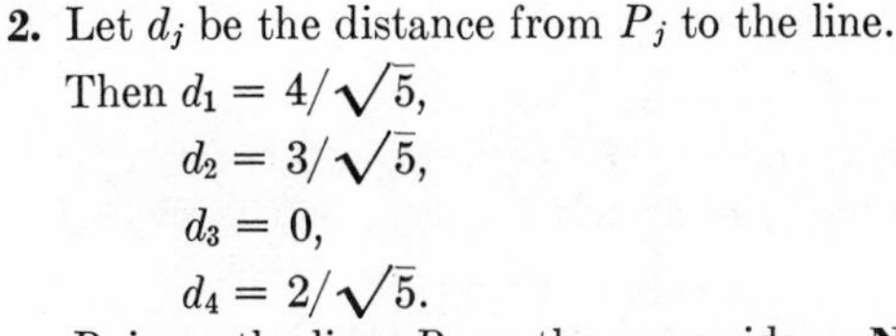

2. Let d_j be the distance from P_j to the line.
Then $d_1 = 4/\sqrt{5}$,
$d_2 = 3/\sqrt{5}$,
$d_3 = 0$,
$d_4 = 2/\sqrt{5}$.
P_3 is on the line, P_4 on the same side as $\mathbf{N}$, P_1 and P_2 on the opposite side.

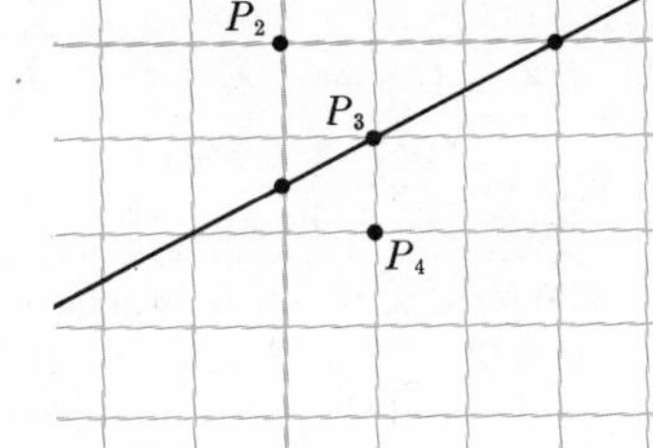

3. (a) $x = 1 + t$, $y = 3 - (3\frac{1}{2})t$

4. The normals are $[1,1]$ and $[1,-2]$, so

$$\cos\theta = \frac{[1,1]\cdot[1,-2]}{\sqrt{2}\sqrt{5}} = \frac{-1}{\sqrt{10}}$$

5. $x + y = 0$.

6. (b) parallel when $(a_1b_1 + a_2b_2)^2 = (a_1^2 + b_1^2)(a_2^2 + b_2^2)$, or $a_1b_2 - a_2b_1 = 0$; perpendicular when $a_1b_1 + a_2b_2 = 0$.

7. (a) $\cos\theta = \dfrac{[10,-1]\cdot[11,-1]}{\sqrt{101}\sqrt{122}}$

14. $\mathbf{A}\cdot\mathbf{U} = c_1\mathbf{U}\cdot\mathbf{V} + c_2\mathbf{V}\cdot\mathbf{U} = c_1$, $\mathbf{A}\cdot\mathbf{V} = c_1\mathbf{U}\cdot\mathbf{V} + c_2\mathbf{V}\cdot\mathbf{V} = c_2$

§7.4

2. (a) $x = t\sin\dfrac{\pi t}{1800}$, $y = t\cos\dfrac{\pi t}{1800}$

3. (a) $\left(\dfrac{1-t^2}{1+t^2}\right)^2 + \left(\dfrac{2t}{1+t^2}\right)^2 = 1$ (b) Counterclockwise
(c) $(-1,0)$ is omitted, but is approached in the limit as $t \to \pm\infty$.
(d) Once (e) Take any rational value of t.

4. On the path, $x = 1 - t^2 = 1 - y^2$. Axis of the parabola is horizontal.

§7.5

1. (c) $\mathbf{V}(t) = \dfrac{2}{(1+t^2)^2}[-2t, 1 - t^2]$, $\mathbf{A}(t) = \dfrac{4}{(1+t^2)^3}[3t^2 - 1, t^3 - 3t]$

8. $\frac{1}{2}\sqrt{7{,}850}$ mph, about 44 mph.

§7.6

1. (b) $+\infty$

(d) L'Hôpital's rule doesn't apply. [The limit is zero, by the Poe Principle, since $-1/x < (\sin x)/x < 1/x$ for $x > 0$.]

(f) 0 (h) 1 (j) 1/2

§7.7

1. (a) $\mathbf{V} = kr[-\sin kt, \cos kt]$, $\mathbf{A} = -k^2r[\cos kt, \sin kt]$,
$|\mathbf{V}| = \text{speed} = kr = ds/dt$, $\kappa = 1/r$, from formula (8)

(d) $\mathbf{V} = [-1 - \sin t, \cos t]$, $\mathbf{A} = [\cos t, -\sin t]$,

$$|\mathbf{V}| = \sqrt{2 + 2\sin t}\,, \quad \kappa = \frac{1}{2\sqrt{2}(1 + \sin t)}$$

3. [Warning: $\sqrt{1 + \sin t} = \sqrt{(\cos t/2 + \sin t/2)^2} = |\cos t/2 + \sin t/2| = \cos t/2 + \sin t/2$ for $0 \le t \le 3\pi/2$, and $= -(\cos t/2 + \sin t/2)$ for $3\pi/2 \le t \le 2\pi$. Evaluate the integral separately for these two intervals.]

4. $$\kappa(t) = \frac{3(2\sin t + 3)}{(13 + 12\sin t)^{3/2}}, \qquad \kappa'(t) = \left[\frac{6(-14 - 6\sin t)}{(13 + 12\sin t)^{5/2}}\right]\cos t$$

The term in brackets is negative for all t, so κ' is negative for $|t| < \pi/2$, positive for $\pi/2 < |t| < 3\pi/2$. Therefore, the minimum of κ is at $t = \pi/2$ (highest point), and the maximum is at $\kappa = 3\pi/2$ (lowest point).

§7.8

1.

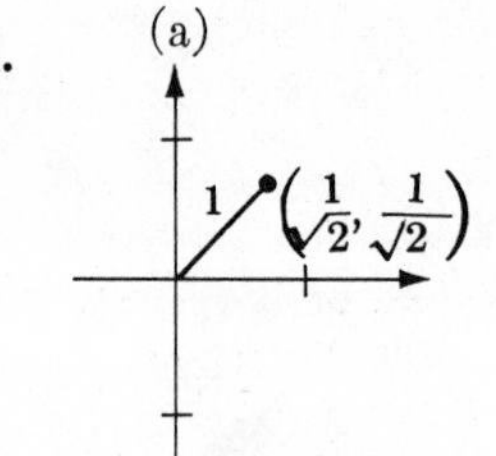

(b) $\left(-\frac{1}{\sqrt{2}}, \frac{1}{\sqrt{2}}\right)$ 1 $\frac{\pi}{4}$

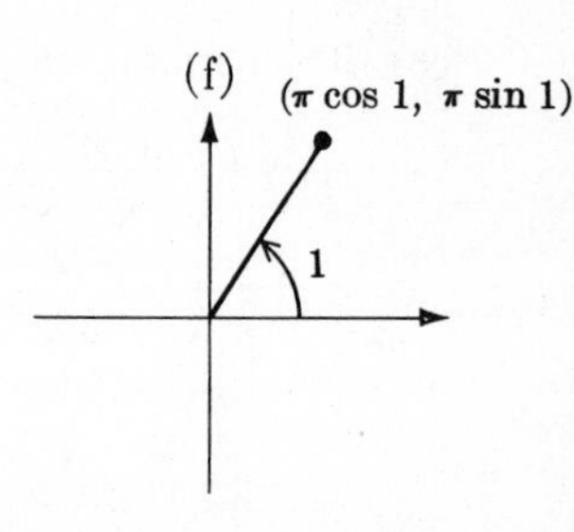

2. (a) $r = 0$, $0 < \theta < \infty$.

(c) Let $\theta_0 = \arcsin(-2/\sqrt{5})$. Then coordinates are $r = \sqrt{5}/2$, $\theta = \theta_0$, $\theta \pm 2\pi$, $\theta_0 \pm 4\pi, \ldots$ or $r = -\sqrt{5}$, $\theta - \theta_0 \pm \pi$, $\theta_0 \pm 3\pi$, $\theta_0 \pm 5\pi, \ldots$.

3. (a) Circle of radius 1/2, center at (1/2,0). (c) Spiral

(e) Heart-shaped figure

5. (a) $e = 1/2$, $d = 1$, $\varphi = \pi/2$; ellipse (b) $(0, -1)$ and $(0, 1/3)$

(c) Directrix is $y = 1$. (d) $1\frac{1}{3}$

9. (a) $$r = \frac{1}{1 + \cos\theta + 2\sin\theta} = \frac{1}{1 + \sqrt{5}\cos(\theta - \varphi)}, \text{ where } \cos\varphi = 1/\sqrt{5},$$

$\sin\varphi = 2/\sqrt{5}$, $e = \sqrt{5}$, $d = 1/\sqrt{5}$. [See §3.5.]

10. (e) $r\cos(\theta - \varphi) = d$; $x\cos\varphi + y\sin\varphi = d$.

§7.9

1. (a) πc^2 (c) $3\pi/2$

2. (a) $2\pi c$ (c) 8 [Note that $\sqrt{x^2} = |x|$, not $= x$.]

§7.10

1. $a = 1/2 = b$, $r(t) = \frac{1}{2}e^{kt} + \frac{1}{2}e^{-kt} = \cosh(kt)$. [§2.8.] A sort of spiralling motion, staying away from the origin.

2. $r(t) = \dfrac{k-1}{2k}e^{kt} + \dfrac{k+1}{2k}e^{-kt}$. Another sort of spiral; it passes through the origin iff $0 < k < 1$.

CHAPTER VIII

§8.1

1. (a) $\bar{1} = 1, \quad |1| = 1, \quad 1^{-1} = 1$ (c) $\overline{-i} = i, \quad |-i| = 1, \quad (-i)^{-1} = i$
(e) $\overline{1+i} = 1 - i, \quad |1+i| = \sqrt{2}, \quad (1+i)^{-1} = \frac{1}{2} - \frac{1}{2}i$

2. (a) $i + (-i) = 0$, $i \cdot (-i) = 1$
(b) $(1+i) + (1+i) = 2 + 2i$, $(1+i) \cdot (1+i) = 2i$.

4. (a) $(a+ib)(x+iy) = ax - by + i(bx + ay)$
$= xa - yb + i(xb + ya) = (x+iy)(a+ib)$

§8.2

2. (c) Circle of radius 2 about the point $(-1,0)$
(d) Perpendicular bisector of the segment from $(0,1)$ to $(2,0)$
(e) Part of the parabola $y^2 = 2x - 1$ where $x \geq 0$ (f) Ellipse

§8.3

1. (a) $\arg(1+i) = \dfrac{\pi}{4} \pm 2k\pi$ (f) $\arg\left(-\dfrac{3}{2} + i\dfrac{\sqrt{3}}{2}\right) = \dfrac{5\pi}{6}$

(c) $\arg(1+3i) = \arctan 3 \pm 2k\pi$

3. $e^{z+2\pi i} = e^z e^{2\pi i} = e^z(\cos 2\pi + i \sin 2\pi) = e^z$ **5.** $\pm 1, \pm i$

6. $\pm(\sqrt{3}\, 2^{-5/6} + i2^{-5/6}), \pm 2^{1/6}i, \pm(-\sqrt{3}\, 2^{-5/6} + i2^{-5/6})$

7. (b) $z^2 = \sqrt{3}e^{i5\pi/6}, \ \sqrt{3}e^{i7\pi/6}; \quad z = \pm 3^{1/4}e^{i5\pi/12}, \ \pm 3^{1/4}e^{i7\pi/12}$

§8.4

1. (a) e^{-2it} and e^{2it} (c) e^{-t} (e) $e^{-t}e^{it}$ and $e^{-t}e^{-it}$

7. (b) $r^{-2}e^{rt}(rt - 1)$

11. (a) $\displaystyle\int_0^\infty e^{ixt}e^{-t}\,dt = \frac{1}{1-ix} = \frac{1+ix}{1+x^2}$ (c) $\displaystyle\int_{-\infty}^\infty e^{ixt}e^{-|t|}\,dt = \frac{2}{1+x^2}$

§8.5

1. (a) $(D+1)e^{-t} = 0$ (c) $(D+1)^2te^{-t} = 0$
(e) $(D^2 + i)e^{at} = (a^2 + i)e^{at}$

2. (a) $f(t) = c_1e^{2it} + c_1e^{-2it} = C_1 \cos 2t + C_2 \sin 2t$
(c) $f(t) = c_1e^{(2+i)t} + c_2e^{(2-i)t} = e^{2t}(C_1 \cos t + C_2 \sin t)$
(e) $f(t) = c_1 + c_2e^{4t}$

3. (a) $f(t) = \frac{1}{2}\sin 2t$ (c) $f(t) = e^{2t}(\cos t - \sin t)$ (e) $f(t) = 1$ for all t

6. (a) $q(t) = e^{-Rt/2L}(c_1e^{i\omega_0 t} + c_2e^{-i\omega_0 t})$, where $\omega_0 = \sqrt{(1/LC) - (R/2L)^2}$ may be real or imaginary.

§8.6

1. (a) $f(t) = c_1 e^{3t} + e^{-3^{1/3}/2}[c_2 \cos(3^{5/6}\, t/2) + c_3 \sin(3^{5/6}\, t/2)]$
(c) $f(t) = e^{it}(c_1 + c_2 t + c_3 t^2)$ (e) $f(t) = c_1 + c_2 e^t + c_3 e^{2t}$

2 (a) $f(t) = -1/27 + f_h$; for f_h see Problem 1(a)

(c) $f(t) = \dfrac{i-1}{4} e^t + f_h$ (e) $f(t) = (-e^{2t}/19) - (1/27) + f_h$
(g) $f(t) = -\frac{1}{5} \cos 2t + c_1 e^t + c_2 e^{-t}$

3. (b) $f(t) = -\frac{1}{5} \cos 2t + (3/5)(e^t + e^{-t}) = -\frac{1}{5} \cos 2t + \frac{6}{5} \cosh t$

4. (a) $f(t) = \dfrac{i-1}{4} e^t + e^{it}\left[\dfrac{5-i}{4} - \dfrac{3it}{2} - \dfrac{i+1}{4} t^2\right]$

5. (a) $q(t) = CE + q_h$; for q_h see Problem 6(a), §8.5

(b) $q(t) = \dfrac{R\omega \sin \omega t}{(1/C - L\omega^2)^2 + R^2\omega^2} + \dfrac{(1/C - L\omega^2) \cos \omega t}{(1/C - L\omega^2)^2 + R^2\omega^2} + q_h$

CHAPTER IX

1. (a) 1.16 $[a = 1, b = 1.02]$ (c) $1/2 + \sqrt{3}\pi/48 = .61 \ldots$

2. (b) $2 + 110(-.01) = 0.9$ $[a = 1, b = 0.99]$

3. (b) $\gamma(T) = 4\pi^2 L/T^2$, $\gamma(T + h) - \gamma(T) = (\text{approx}) h\gamma'(T) = -(8\pi^2 L/T^3)h$

6. (a) $|R| \le |b - a|^2 \max_{[a,b]} |f''| = b^2 \max_{[0,b]} |\sin x| \le b^3$

§9.2

1. (a) $\cos h = 1 - \dfrac{1}{2} h^2 + \dfrac{1}{4!} h^4 - \cdots + \begin{Bmatrix} 0 \text{ if } n \text{ is odd} \\ (-1)^k h^{2k}/(2k)! \text{ if } n = 2k \end{Bmatrix} + R_{n+1}$,

$|R_{n+1}| \le \dfrac{|h|^{n+1}}{(n+1)!}$

(c) $e^h = 1 + h + \cdots + \dfrac{1}{n!} h^n + R_{n+1}$, $|R_{n+1}| \le \dfrac{|h|^{n+1}}{(n+1)!} e^h < 3 \dfrac{|h|^{n+1}}{(n+1)!}$

if $0 \le h < 1$, and $|R_{n+1}| \le \dfrac{|h|^{n+1}}{(n+1)!}$ if $-1 < h < 0$

(e) $\log(1 + h) = h - \dfrac{1}{2} h^2 + \dfrac{1}{3} h^3 - \cdots + (-1)^{n+1} \dfrac{h^n}{n} + R_{n+1}$,

$|R_{n+1}| \le \dfrac{|h|^{n+1}}{n+1}$ if $0 \le h < 1$, $|R_{n+1}| \le \dfrac{|h|^{n+1}}{(n+1)(1+h)^{n+1}}$ if $-1 < h < 0$

(g) $(1 + h)^s = 1 + sh + \dfrac{s(s-1)}{2} h^2 + \cdots + \dfrac{s(s-1)\cdots(s-n+1)}{n!} h^n + R_{n+1}$;

$$|R_{n+1}| \leq \left|\frac{s(s-1)\cdots(s-n)}{(n+1)!}\right| \text{ if } 0 \leq h < 1 \text{ and } n \geq s$$

$$|R_{n+1}| \leq \frac{|s(s-1)\cdots(s-n)|}{(n+1)!(1+h)^{n-s+1}} \text{ if } -1 < h < 0 \text{ and } n \geq s$$

Analogous answers for $n < s$.

5. From Problem 1(a) with $n = 3$, $\cos(0.1) = 1 - .005 + R_4$, where $|R_4| \leq (.1)^4/4! < 1/2 \times 10^{-5}$.

8. (a) $\sin^2 h = h^2 - \frac{1}{3}h^4 + R_5$

9. $\sin^2 h = h^2 - \frac{1}{3}h^4 + \left[\frac{h^6}{36} + 2hR_5 - \frac{1}{3}h^3R_5 + R_5{}^2\right]$. Notice that R_5 here is a *different* remainder from the R_5 in Problem 8(a). Notice also that the term in square brackets here is $\leq$ (constant times h^6), so this expansion agrees with the one in 8(a) up through the term in h^4.

14. $f(h) = h - \frac{h^3}{3} + R_4$, $|R_4| \leq \frac{|h|^4}{4!} \max_{[-1,1]} |12t - 8t^3|\, e^{-t^2} \leq \frac{1}{2}h^4$, for $|h| < 1$

§9.3

1. The function f is convex and has positive slope on [0,1], and $f(1/2) < 0$, so $1 \geq x_n > x_{n+1} > \bar{x} > 1/2$.

5. Only in (d) do the approximations work.

§9.4

1. (b) $\log 2 =$ (approx) $\frac{1}{12}(1 + \frac{16}{5} + \frac{4}{3} + \frac{16}{7} + \frac{1}{2}) = \frac{1747}{2520} = 0.6932\ldots$, with an error known to be $< 1/1820 < 5 \times 10^{-4}$. [Actually, a table will show that the error is $< (2 \times 10^{-4})$; the only way to show this is to compare our number 0.6932 to an approximation with proven accuracy to four decimals.]

2. (b) Length $=$ (approx) $\frac{\pi}{12}\left[b + 4\sqrt{\frac{1}{2}a^2 + \frac{1}{2}b^2} + a\right]$. Notice that when $a = b$, this gives $\pi a/2$, which is *exactly* the arc length of a quarter circle of radius a.

3. (b) $f(x) = \frac{1}{1+x^2}$, $f^{(4)}(x) = 24\frac{1 - 10x^2 + 5x^4}{(1+x^2)^5}$. On [0,1] the numerator $|1 - 10x^2 + 5x^4| \leq 4$, so $|f^{(4)}(x)| \leq 4 \times 24$, and the error in Simpson's rule is $< \left(\frac{1}{2}\right)^5 \frac{4 \times 24}{90n^4} = \frac{1}{30n^4} < 10^{-4}$ if $n \geq 5$. Thus 10 subdivisions would suffice. [A more precise estimate of $|f^{(4)}|$ would show that 8 subdivisions suffice.]

4. If $f(x) = x^2$, then $\int_a^b x^2\,dx = \frac{b-a}{2}[b^2 + a^2] = \frac{(b-a)^3}{12}\max|f''|$.

CHAPTER X

§10.1

1. (b) $|n^{-1/3} - 0| < \epsilon$ when $n > 1/\epsilon^3$

(d) $\left|\sqrt{\dfrac{n+1}{n}} - 1\right| < \epsilon$ when $n > \dfrac{1}{\epsilon^2 + 2\epsilon}$

2. (a) $\dfrac{1}{n(n+1)} \leq \dfrac{1}{n}$ (c) $\left|\dfrac{r^n}{n}\right| \leq \dfrac{1}{n}$

(e) $10^{-2^n} \leq 10^{-n} = r^n$ where $r = 1/10 < 1$

§10.2

1. (a) 3 (c) 0 (e) $1/(1-r)$ **3.** (d) $h\dfrac{1+r}{1-r}$

6. Informal proof, for $B > 0$: If $|b - B| < \frac{1}{2}|B|$, then b lies in the open interval from $\frac{1}{2}B$ to $\frac{3}{2}B$.

§10.3

1. (b) Bounded, not monotone (d) Monotone, not bounded
(f) Bounded, not monotone
(h) Monotone (easy to see) and bounded (but don't feel bad if you couldn't tell; wait until the next chapter). (j) Bounded and monotone
(l) Bounded, not necessarily monotone; for example, if $f(x) = \sin x$ on $0 \leq x \leq 2\pi$, then $s_1 < s_2$ but $s_2 > s_3$.

§10.4

1. (b) If $x_n = 1/n$, then $f(x_n) = n$ and $\lim_{x\to\infty} f(x_n)$ does not exist [Theorem 5].
(d) If $x_n = (1/e)^n$, then $\lim x_n = 0$, but $\lim \log x_n = \lim(-n)$ does not exist.
3. (a) 1 (c) 1 (e) 0 (g) 0 (i) 1

§10.5

1. (b) $\lim\sup x_n = 1$, $\lim\inf x_n = 0$ (d) $\lim\sup a_n = 8$, $\lim\inf a_n = 1$
2. (b) and (c)

CHAPTER XI

§11.2

1. (a) $1\frac{1}{2}$ (c) 1 (e) $\log 2$ (g) $\log 3$
2. (a) 1/2 (c) 123/999 (e) $2\frac{71}{100} + \frac{8281}{9999} \times 10^{-2}$

§11.3

1. (a) Divergent (c) Convergent **2.** (b) Convergent (d) Divergent
3. (a) Convergent (d) Convergent **4.** Convergent **6.** Convergent
8. Sum is 1 **10.** Divergent **11.** (a) $x > 1$ (c) all x (e) no x
(f) $x < e$ (Note from Fig. 10.2(e) that $(1 + 1/n)^n < e$; see also Problem 17.)
(g) $x < 4$ (h) all x (i) $x \leq 1$
17. (b) What about $\sum 1/n$? **19.** degree$(P) <$ degree$(Q) - 1$

§11.4

1. all x **3.** $x = 0, \pm\pi, \pm 2\pi, \ldots$ **5.** $|x| < 4$ **7.** $|x + 1| < 1$

12. (a) Alternating, not convergent (c) Not alternating
(e) Alternating, not convergent

13. (a) $|x| < 1$ (c) $-1 \leq x < 1$ (e) $-1 \leq x < 1$
(f) $x \neq \pi/2, \pi/2 \pm 2\pi, \pi/2 \pm 4\pi, \ldots$

16. $J_0(1) = 0.7656 \pm 0.0005$

§11.5

3. (c) $R = \infty$ **4.** (a) See Problem 11

6. (a) $(1+b)^{1/3}\left[1 + \dfrac{(x-b)}{3(1+b)} - \sum_{1}^{\infty} 3^{-n} \dfrac{1\cdot 2\cdot 5\cdots(3n-4)}{n!}\left(\dfrac{b-x}{1+b}\right)^n\right]$

12. $1/p! - 1/(p+1)! = p/(p+1)!$

13. (b) $Y^{(0)}(x) = J_0(x)\log x - \sum_{j=1}^{\infty} \dfrac{(-1)^j}{(j!)^2}\left(1 + \dfrac{1}{2} + \cdots + \dfrac{1}{j}\right)\left(\dfrac{x}{2}\right)^{2j}$

15. (a) $x^{-1}(e^x - 1)$

§11.7

2. $(1-x)^{-3} = \dfrac{1}{2}\sum_{j=0}^{\infty} (j+2)(j+1)x^j$

5. $f(x) = 1 + x + x^2 + \frac{4}{3}x^3 + \frac{7}{6}x^4 + \cdots$

Answers to Selected Problems

Part Two: Calculus of Several Variables

CHAPTER I

§1.1

3. (c) $\cos\theta = \frac{1}{2}\sqrt{5/13}$

5. (a), (e), and (f) are parallel to (1,1,−1).

6. (a) All vectors of the form $(0,y,z)$ [the yz plane]

7. (b) (3,−1,1) is orthogonal to (1,2,−1) and (−1,0,3)

§1.2

3. (a) Area $= |\mathbf{P}_2 \times \mathbf{P}_4| = 2\sqrt{2}$ (b) Area $= \frac{1}{2}|\mathbf{P}_2 \times \mathbf{P}_4| = \sqrt{2}$

§1.3

1. (a) $\mathbf{P} = (-5,-6,8) + t(6,9,-1)$; $x = -5 + 6t$, $y = -6 + 9t$, $z = 8 - t$;
$\mathbf{M} = (-2, -3/2, 15/2)$

2. (a) $\mathbf{N} = (1,1,1)$; $\mathbf{P}_0 = (1,0,0)$ (b) $\mathbf{N} = (2,3,-1)$; $\mathbf{P}_0 = (0,0,-2)$

3. (a) $x + y + z = 1$ (b) $z = 0$ **6.** (7,−13,2)

10. (a) (−1,3,−6)
(b) The whole line lies in the plane; thus the points of intersection are precisely the points on the given line. (c) There are no points of intersection.

12. (a) $\mathbf{P} = (-5,2,1) + t(1,-1,0)$ (b) $\mathbf{P}_2 = (-\frac{3}{2},-\frac{3}{2},1)$

13. (a) $\mathbf{P} = \mathbf{P}_1 + t\mathbf{N} = (x_1 + ta, y_1 + tb, z_1 + tc)$

(b) $\mathbf{P}_0 = \mathbf{P}_1 + \dfrac{d - \mathbf{P}_1\cdot\mathbf{N}}{\mathbf{N}\cdot\mathbf{N}}\mathbf{N}$

§1.4

1. (b) $\cos\theta = 1/2, \theta = \pi/3$ (d) $\theta = 0$ (f) $\cos\theta = -\sqrt{330}/110$
2. (a) A single point, $(-\frac{1}{2}, 2\frac{1}{4})$ (c) The whole line $\mathbf{X} = t(3,5)$
   (d) No points of intersection
3. (a) $\mathbf{X} = (1,3) \pm \dfrac{2}{\sqrt{5}}(-2,-1)$

   (c) If $2r^2 < 85$, there is no intersection; if $2r^2 = 85$, the intersection is a single point $\mathbf{P}_0 = (-3, \frac{9}{2}, \frac{7}{2}, -1)$; and if $2r^2 > 85$, it is the two points

   $\mathbf{P}_0 \pm \dfrac{1}{2}\sqrt{\dfrac{2r^2 - 85}{7}}\,(2,-1,3,0)$. 4. (b) $r = \sqrt{85/2}$

8. (a) $\mathbf{B} = (5,2)$ (c) $\mathbf{B} = (-2,1,0)$ (d) $\mathbf{B} = (-1,2,-1,0)$
9. (a) $\mathbf{B} = (1,-3,1)$

§1.5

1. (a) Dependent (b) Basis (c) Dependent (Theorem 9)
   (d) Independent, no basis; $(1,0) \neq t(1,-1)$ (e) Dependent: $t(0,0) = \mathbf{0}$
   (f) Dependent (g) Independent, but not a basis (h) Basis (Theorem 12)
   (k) Dependent (l) Dependent: $\mathbf{A} + (-1)\mathbf{B} + (-1)\mathbf{C} + (-1)\mathbf{D} = \mathbf{0}$
4. (b) and (c) $\mathbf{C} = (0,0,1)$ 5. (b) $\mathbf{C} = (0,0,1)$
12. (a) $\mathbf{A} = (1,0,0)$, $\mathbf{B} = (0,1,0)$, $\mathbf{C} = (0,0,1)$
    (c) $\mathbf{A} = \mathbf{B} = \mathbf{C} = (1,0,0)$ span the x axis (d) $\mathbf{A} = \mathbf{B} = \mathbf{C} = \mathbf{0}$ span the origin

CHAPTER II

§2.1

1. (a) $\boldsymbol{\gamma}'(t) = \left(-\dfrac{1}{2}\sin\dfrac{t}{2}, \dfrac{1}{2}\cos\dfrac{t}{2}, 1\right)$

4. (a) $\mathbf{P}_1 = (0,4,2)$ and $\mathbf{P}_2 = (-4,12,18)$
   (b) If α is the angle between the curve and the plane at $\mathbf{P}_1$, then $\sin\alpha = \cos(\pi/2 - \alpha) = 28/\sqrt{4326}$.
5. (a) The curves intersect at the points $(0,1,0)$ and $(0,-1,0)$.
   (b) At $\mathbf{P}_0 = (0,1,0) = \boldsymbol{\gamma}(\pi/2) = \bar{\boldsymbol{\gamma}}(0)$, we have the perpendicular line $\boldsymbol{\delta}(t) = \mathbf{P}_0 + t\mathbf{A}_0$, where $\mathbf{A}_0 = \boldsymbol{\gamma}'(\pi/2) \times \bar{\boldsymbol{\gamma}}'(0)$, and at $\mathbf{P}_1 = (0,-1,0) = \boldsymbol{\gamma}(-\pi/2) = \bar{\boldsymbol{\gamma}}(\pi)$, set $\boldsymbol{\delta}(t) = \mathbf{P}_1 + t\mathbf{A}_1$, with $\mathbf{A}_1 = \boldsymbol{\gamma}'(-\pi/2) \times \bar{\boldsymbol{\gamma}}'(\pi)$.
7. $(\mathbf{P} - (a,0,0)) \cdot (0,a,b) = 0$, or $ay + bz = 0$. The plane contains the x axis.
9. (c) No (d) No 10. (c) No (d) Yes

§2.2

1. (a) $\boldsymbol{\gamma}''(t) = (-2\sin t, -3\cos t, 0)$, $\mathbf{F} = 5\boldsymbol{\gamma}''$
   (b) $\boldsymbol{\gamma}''(t) = (2e^t, 2e^{-t}, -5\cos t)$, $\mathbf{F} = (m\boldsymbol{\gamma}')' = (0, 4e^{-2t}, 5(\cos t + \sin t)e^{-t})$
   (d) $\boldsymbol{\gamma}'' = \mathbf{F} = \mathbf{0}$ (e) $\boldsymbol{\gamma}'' = \mathbf{0}$, $\mathbf{F} = \mathbf{B}$
2. It spirals around an elliptical cylinder.
3. $\mathbf{F}\cdot\bar{\mathbf{F}} = 0$, so the forces are at right angles.
4. $\boldsymbol{\gamma}(t) = (1 - t, 2 - 3t, 3 + et^2/2m)$ 5. $\boldsymbol{\gamma}(t) = ((1 - \cos t)/m, 0, t^2/2m)$
10. (a) The electron hits the face when $t = d/z_1$, at the point

$$\left(\frac{mx_1}{qh}\sin\frac{qhd}{mz_1}, \frac{mx_1}{qh}\left(1 - \cos\frac{qhd}{mz_1}\right), d\right). \qquad (*)$$

(b) The point (*) lies on the line through $(0,0,d)$ in the direction

$$\left(\sin\frac{qhd}{mz_1}, 1 - \cos\frac{qhd}{mz_1}, 0\right).$$

11. (c) $x = Ae^{r_1t} + Be^{r_2t}$, $y = (1 + 3r_1)Ae^{r_1t} + (1 + 3r_2)Be^{r_2t}$, where $r_1 = \dfrac{-5 + \sqrt{7}}{9}$, $r_2 = \dfrac{-5 - \sqrt{7}}{9}$.

§2.3

1. (b) Since $\boldsymbol{\gamma}(1) = (2,1,1)$, the tangent line is given by $\mathbf{P} = (2,1,1) + t\boldsymbol{\gamma}'(1)$.

2. (a) $\mathbf{T} = \boldsymbol{\gamma}'(\pi/8)/|\boldsymbol{\gamma}'(\pi/8)| = (1/\sqrt{17})(1,0,-4)$.

3. (a) $\pi\sqrt{17}$ (b) $\int_0^1 \sqrt{5 + 4t^2}\, dt$

4. (a) $\ell(t) = t\sqrt{17}$ (b) $g(s) = s/\sqrt{17}$

(c) $\mathbf{G}(s) = \left(\dfrac{s}{\sqrt{17}}, \sin\dfrac{4s}{\sqrt{17}}, \cos\dfrac{4s}{\sqrt{17}}\right)$

5. (a) $\ell(t) = e^{3t} - e^{-3t} = 2\sinh 3t$

(c) $\mathbf{G}(s) = (e^{\operatorname{arcsinh}(s/2)}, e^{-\operatorname{arcsinh}(s/2)}, \sqrt{2}\operatorname{arcsinh}(s/2))$

7. (a) $\mathbf{T} = \left(-\dfrac{a}{c}\sin\dfrac{s}{c}, \dfrac{a}{c}\cos\dfrac{s}{c}, \dfrac{b}{c}\right)$ (b) $\mathbf{T}' = \left(-\dfrac{a}{c^2}\cos\dfrac{s}{c}, -\dfrac{a}{c^2}\sin\dfrac{s}{c}, 0\right)$;

$$\kappa = |\mathbf{T}'| = \frac{|a|}{c^2} = \frac{|a|}{a^2 + b^2}.$$ In the rest of the problem, assume $a > 0$.

$$\mathbf{N} = \frac{1}{\kappa}\mathbf{T}' = \left(-\cos\frac{s}{c}, -\sin\frac{s}{c}, 0\right).$$

(c) $\mathbf{B} = \mathbf{T} \times \mathbf{N} = \left(\dfrac{b}{c}\sin\dfrac{s}{c}, -\dfrac{b}{c}\cos\dfrac{s}{c}, \dfrac{a}{c}\right)$

(d) $\mathbf{B}' = \left(\dfrac{b}{c^2}\cos\dfrac{s}{c}, \dfrac{b}{c^2}\sin\dfrac{s}{c}, 0\right) = \tau\mathbf{N}$. Comparing with part (b), $\tau = -b/c^2$.

CHAPTER III

§3.1

2. The functions in (a), (c), (e), and (f) are continuous at 0; those in (b) and (d) are not.

9. The given set is open. If $\mathbf{P}_0 = (x_0, y_0)$ is in S, then the disk $D = \{\mathbf{P}: |\mathbf{P} - \mathbf{P}_0| < |y_0|\}$ is in S, as shown in the figure. To prove it,

$$|\mathbf{P} - \mathbf{P}_0| < |y_0| \Leftrightarrow (x - x_0)^2 + (y - y_0)^2$$
$$\Rightarrow (y - y_0)^2 < y_0^2$$
$$\Rightarrow y \neq 0.$$

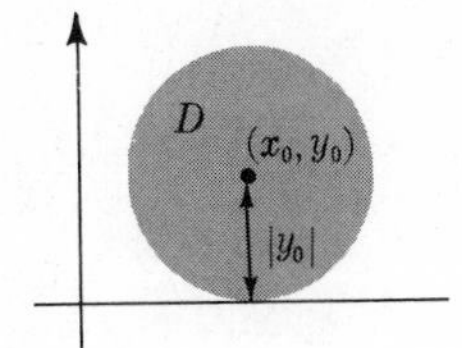

12. Apply Problem 11 to the function $F = (b - f)(f - a)$.

13. (a) Let $f(\mathbf{P}) = |\mathbf{P}|^2 = x^2 + y^2$, and apply Problem 12 with $a = -1$ and $b = r^2$.

(b) Take f as in part (a), and apply Problem 12 with $a = 1$, $b = 4$.

(c) Take f as in the hint, $a = 0$, $b = 1$. Since $f(x,y)$ is not defined for $y \geq 1$, it never assumes a value ≥ 1; hence $0 < f < 1 \Leftrightarrow 0 < f$.

(e) $f(x,y) = \sin^2 x$, $a = -1$, $b = 1/2$; or, $f(x,y) = \sin x$, $a = -1/\sqrt{2}$, $b = 1/\sqrt{2}$.

§3.2

2. (a) $(1,-1)$ (b) $(0,0)$ (c) $(1,-1)$

(d) $z = -\frac{5}{4}x + \frac{1}{4}$, so the slope is $(-\frac{5}{4},0)$ (e) $z = -\frac{1}{3}x - \frac{2}{3}y$, $\mathbf{M} = (-\frac{1}{3},-\frac{2}{3})$

(f) No slope

5 and **6.** (a) Slope $\mathbf{M} = (2,3)$, maximum at $(2/\sqrt{13}, 3/\sqrt{13})$, minimum at $(-2/\sqrt{13}, -3/\sqrt{13})$

(b) Slope $\mathbf{M} = (10,-5)$, max at $(2/\sqrt{5}, -1/\sqrt{5})$, min at $(-2/\sqrt{5}, 1/\sqrt{5})$

(c) Slope $\mathbf{M} = (-5,4)$, max at $(-5/\sqrt{41}, 4/\sqrt{41})$, min at $(5/\sqrt{41}, -4/\sqrt{41})$

(d) Slope $\mathbf{M} = (-1,-1)$, max at $(-1/\sqrt{2}, -1/\sqrt{2})$, min at $(1/\sqrt{2}, 1/\sqrt{2})$

7. (a) The max occurs at the lower right-hand corner $(1,-2)$. Its value is $10\cdot 1 - 3\cdot(-2) = 16$.

(b) A max of 20 occurs at the upper left-hand corner $(-1,2)$.

8. (a) $10x - 3y$ attains a max of 51 at the point $A = (6,3)$.

(b) $-4x + 5y$ attains a max of -1 at the point $B = (\frac{2}{3},\frac{1}{3})$.

Appendix

1. The profit function now is $\mathbf{P}(p,a) = \frac{3}{4}p + a$, with slope $\mathbf{M} = (\frac{3}{4},1)$. Plotting this on Fig. 3.20 shows that the corresponding line of support passes through the intersection of $p + a = 100$ and $a = 75$, i.e. $(25,75)$. He buys 25 bushels of peaches and 75 bushels of apples, spending \$125.

3. (a) $w \geq r$ (more washers than refrigerators)

$r \geq s$

$w \geq 10$, $r \geq 5$, $s \geq 3$ (orders to be filled)

$w + r + s = 50$; $\mathbf{P} = 30w + 60r + 50s$

(b) $w = 50 - r - s$

$w \geq r \Leftrightarrow 2r + s \leq 50$

$w \geq 10 \Leftrightarrow r + s \leq 40$

$r \geq s$, $r \geq 5$, $s \geq 3$, as before; $\mathbf{P}(r,s) = 1500 + 30r + 20s$

(d) The max occurs at $r = 50/3$, $s = 50/3$, $w = 50/3$, $\mathbf{P} = 2333\frac{1}{3}$

7. $f(x,y) = x + y$, $f_1(x,y) = -x$, $f_2(x,y) = -y$

§3.3

1. (a) $\nabla f = (ye^{xy}, xe^{xy})$, continuous everywhere

(b) $\nabla f = (x/\sqrt{x^2+y^2}, y/\sqrt{x^2+y^2}) = \mathbf{P}/|\mathbf{P}|$, continuous except at $\mathbf{P} = \mathbf{0}$

(c) $\nabla f = ((-x \sin\sqrt{x^2+y^2})/\sqrt{x^2+y^2}, (-y\sin\sqrt{x^2+y^2})/\sqrt{x^2+y^2})$ $= (-\mathbf{P}\sin|\mathbf{P}|)/|\mathbf{P}|$; continuous except at $\mathbf{P} = \mathbf{0}$

(d) $\nabla f = (-y/x^2, 1/x)$, continuous except on the y axis

(e) $\nabla f = \left(\int_0^y e^{t^2}\,dt, xe^{y^2}\right)$, continuous everywhere

(f) $\nabla f = (y/(x^2+y^2), -x/(x^2+y^2))$

2. (b) The contour line is $xy = 2$, $y = 2/x$, $\nabla f = (y,x) = (2,1)$

(c) The contour line is $4x^2 - y^2 = 0$, or $|y| = 2|x|$, and the gradient is $(8x,-2y) = (8,-4)$.

(d) Contour line $x^2 + y = 3$, gradient $(2x,1) = (2,1)$

3. (a) $\nabla f = (2x,2y) = (2,2), f(1,1) = 2$, so the equation is
$z = 2 + ((x,y) - (1,1))\cdot(2,2) = 2x + 2y - 2.$
(b) $\nabla f = (y,x) = (1,1), f(1,1) = 1$, so the equation is
$z = 1 + ((x,y) - (1,1))\cdot(1,1) = x + y - 1.$

4. (a) $\nabla f = \left(\arctan\frac{x}{y} + \frac{xy}{x^2+y^2}, \frac{-x^2}{x^2+y^2}\right) = \left(\frac{\pi}{4} + \frac{1}{2}, -\frac{1}{2}\right)$ when $x = 0, y = 1$;
$\nabla f\cdot \mathbf{U} = 1/2.$
(b) $\nabla f\cdot \mathbf{U} = \pi/4 + 1/2$
(c) $\nabla f = 2e^{-x^2-y^2}(-x,-y) = (0,0)$ when $x = 0,\ y = 0$; $\nabla f\cdot \mathbf{U} = 0$
(d) $\nabla f = 2e^{-2}(-1,1)$; $\nabla f\cdot \mathbf{U} = 2e^{-2}/5$

6. (a) The xy plane has equation $z = 0$; $z = 1 - x^2 - y^2 = 0 \Leftrightarrow x^2 + y^2 = 1$, so the intersection is the unit circle.

7. (a) $\cos^2 t + \sin^2 t = 1$
(b) $\boldsymbol{\gamma}'(0) = (0,1)$; $\nabla f(1,0) = (2,0) \perp \boldsymbol{\gamma}'(0)$
(c) $\boldsymbol{\gamma}'(t) = (-\sin t, \cos t)$; $\nabla f(x,y) = (2x,2y)$; $\nabla f(\boldsymbol{\gamma}(t)) = (2\cos t, 2\sin t)$;
$\nabla f(\boldsymbol{\gamma}(t))\cdot\boldsymbol{\gamma}'(t) = 0$

9. (a) $\nabla f(x,y) = (2xy^3, 3x^2y^2)$; $\nabla f(1,1) = (2,3)$; $z = 1 + 2(x-1) + 3(y-1)$
(b) The equation of the plane can be written $2x + 3y - z = 4$, so $\mathbf{N} = (2,3,-1)$ is a normal.

§3.4

1. (a) (i) $$\frac{\partial w}{\partial r} = \frac{\partial w}{\partial x}\frac{\partial x}{\partial r} + \frac{\partial w}{\partial y}\frac{\partial y}{\partial r} = 2xe^r - 2e^s \sin r = 2e^{2r} - 2e^s \sin r$$

$$\frac{\partial w}{\partial s} = \frac{\partial w}{\partial x}\frac{\partial x}{\partial s} + \frac{\partial w}{\partial y}\frac{\partial y}{\partial s} = 2x\cdot 0 + 2e^s \cos r = 2e^s \cos r$$

(ii) $$w = e^{2r} + 2e^s\cos r,\qquad \frac{\partial w}{\partial r} = 2e^{2r} - 2e^s \sin r,\qquad \frac{\partial w}{\partial s} = 2e^s\cos r$$

(b) (i) $$\frac{\partial z}{\partial r} = \frac{\partial z}{\partial x}\frac{\partial x}{\partial r} + \frac{\partial z}{\partial y}\frac{\partial y}{\partial r} = \frac{2x}{x^2+y^2}\cos\theta + \frac{2y}{x^2+y^2}\sin\theta = \frac{2}{r}$$

$$\frac{\partial z}{\partial \theta} = \frac{\partial z}{\partial x}\frac{\partial x}{\partial \theta} + \frac{\partial z}{\partial y}\frac{\partial y}{\partial \theta} = \frac{2x}{x^2+y^2}(-r\sin\theta) + \frac{2y}{x^2+y^2}(r\cos\theta) = 0$$

(ii) $$z = \log r^2,\qquad \frac{\partial z}{\partial r} = \frac{2r}{r^2} = \frac{2}{r},\qquad \frac{\partial z}{\partial \theta} = 0$$

(e) (i) $$\frac{\partial u}{\partial x} = \frac{du}{dz}\frac{\partial z}{\partial x} = 2z\cdot a,\qquad \frac{\partial u}{\partial y} = \frac{du}{dz}\frac{\partial z}{\partial y} = 2z(-1)$$

(ii) $u = (ax - y)^2 + e^2$, $\partial u/\partial x = 2(ax - y)a$, $\partial u/\partial y = -2(ax - y)$

2. Scheme: $(x,t) \longrightarrow x + ct = u \longrightarrow f(u) = w.$

$$\frac{\partial F}{\partial x} = \frac{\partial w}{\partial x} = \frac{dw}{du}\frac{\partial u}{\partial x} = f'(u)\cdot 1;\qquad \frac{\partial F}{\partial t} = \frac{\partial w}{\partial t} = \frac{dw}{du}\frac{\partial u}{\partial t} = f'(u)\cdot c$$

5. Let $u = x + y$ and $v = x - y$.

$$\frac{\partial z}{\partial x} = \frac{\partial z}{\partial u}\frac{\partial u}{\partial x} + \frac{\partial z}{\partial v}\frac{\partial z}{\partial x} = \frac{\partial z}{\partial u}\cdot 1 + \frac{\partial z}{\partial v}\cdot 1,$$

$$\frac{\partial z}{\partial y} = \frac{\partial z}{\partial u}\frac{\partial u}{\partial y} + \frac{\partial z}{\partial v}\frac{\partial v}{\partial y} = \frac{\partial z}{\partial u}\cdot 1 + \frac{\partial z}{\partial v}\cdot(-1),$$

$$\frac{\partial z}{\partial x}\frac{\partial z}{\partial y} = \left(\frac{\partial z}{\partial u} + \frac{\partial z}{\partial v}\right)\left(\frac{\partial z}{\partial u} - \frac{\partial z}{\partial v}\right) = \left(\frac{\partial z}{\partial u}\right)^2 - \left(\frac{\partial z}{\partial v}\right)^2$$

10. (a) From (*), noting $(\partial T/\partial V)_T = 0$, we get

$$T\frac{\partial^2 S}{\partial V\,\partial T} = \frac{\partial^2 E}{\partial V\,\partial T};$$

and from (**),

$$1\left(\frac{\partial S}{\partial V}\right)_T + T\frac{\partial^2 S}{\partial T\,\partial V} = \frac{\partial^2 E}{\partial T\,\partial V} + \left(\frac{\partial P}{\partial T}\right)_V.$$

Subtracting, $(\partial S/\partial V)_T = (\partial P/\partial T)_V$.

(b) This is just the chain rule.

(c) $\partial P/\partial P = 1$; and when P and T are independent variables, then $\partial T/\partial P = 0$.

(d) $0 = \left(\frac{\partial P}{\partial T}\right)_P = \left(\frac{\partial P}{\partial T}\right)_V\left(\frac{\partial T}{\partial T}\right)_P + \left(\frac{\partial P}{\partial V}\right)_T\left(\frac{\partial V}{\partial T}\right)_P = \left(\frac{\partial P}{\partial T}\right)_V + \left(\frac{\partial P}{\partial V}\right)_T\left(\frac{\partial V}{\partial T}\right)_P$

§3.5

1. (a) $y = \varphi(x) = \dfrac{1}{x}$ (b) $y = +\sqrt{\dfrac{3 - x^2}{2}}$

(c) $y = -\sqrt{x}$ (not $\sqrt{x}$) (d) $y = \dfrac{1}{x}\log\left(\dfrac{2}{x}\right)$

2. (a) $R = \{(x,y) : |x - 1| < \frac{1}{2}|y - 1| < 1\}$ will do.

(b) $R = \{(x,y) : |x + 1| < \sqrt{3} - 1,\ |y - 1| < 1\}$ is best possible.

(c) $R = \{(x,y) : |x - 1| < 1,\ |y + 1| < 1\}$ is best possible.

(d) $R = \{(x,y) : |x - 1| < \frac{1}{2},\ |y - \log 2| < 3\}$ will do.

3. (a) $\left(1 + \dfrac{dy}{dx}\right)e^{xy} + (x + y)e^{xy}\left(y + x\dfrac{dy}{dx}\right) = 0,$

$$\frac{dy}{dx}(1 + xy + x^2) + 1 + xy + y^2 = 0, \quad \frac{dy}{dx} = -\frac{1 + xy + y^2}{1 + xy + x^2}.$$

The point (x_0, y_0) lies on the given contour line $f = 2e$; hence the value of the implicit function at x_0 is y_0, and the derivative at that point is

$$\frac{dy}{dx} = -\frac{1 + 1 + 1}{1 + 1 + 1} = -1.$$

§3.6

1. (a) $f_x = \cos(x + y) - x\sin(x + y),\quad f_{xy} = -\sin(x + y) - x\cos(x + y)$

(b) $f_y = -x\sin(x + y),\quad f_{yx} = -\sin(x + y) - x\cos(x + y)$

(c) $D_1{}^2D_2 f = f_{yxx} = -2\cos(x+y) + x\sin(x+y)$
(d) $f_{xx} = -2\sin(x+y) - x\cos(x+y)$,
$f_{xxy} = -2\cos(x+y) + x\sin(x+y)$
(e) $\partial^3 f/\partial x\, \partial y\, \partial x = f_{xyx} = -2\cos(x+y) + x\sin(x+y)$

§3.7

1. (a) $f_x = y\cos xy,\quad f_y = x\cos xy,\quad f_{xx} = -y^2\sin xy,$
$f_{xy} = \cos xy - xy\sin xy,\quad f_{yy} = -x^2\sin xy,\quad f_2(h,k) = hk$
(b) $f_2(h,k) = 0 - h - \pi k - hk$ (c) $f_2(h,k) = 1 - h^2 - k^2$

2. (c) $f_3(h,k) = 1 - h^2 - k^2$

3. (a) $$f_x(x,y) = \begin{cases} \dfrac{xy\cos xy - \sin xy}{x^2}, & x \neq 0 \\ 0, & x = 0 \end{cases} \qquad f_y(x,y) = \begin{cases} \cos xy, & x \neq 0 \\ 1, & x = 0 \end{cases}$$

$$f_{xx}(0,0) = \lim_{h\to 0} \frac{f_x(h,0) - f_x(0,0)}{h} = \lim \frac{0}{h^3} = 0$$

$$f_{xy}(0,0) = \lim_{h\to 0} \frac{f_x(0,h) - f_x(0,0)}{h} = 0$$

$$f_{yy}(0,0) = \lim_{h\to 0} \frac{f_y(0,h) - f_y(0,0)}{h} = 0.$$

So $f_2(h,k) = k$.

§3.8

1. (a) $\nabla f = (2x+2,\ 6y+4) = 0$ when $x = -1$, $y = -2/3$.
$f_{xx} = 2$, $f_{xy} = 0$, $f_{yy} = 6$, so there is a *minimum* at $(-1,-2/3)$.
(b) $\nabla f = (3x^2 - 3y,\ 3y^2 - 3x) = 0$ when $y = x^2$ and $x = y^2$.
This implies $y = y^4$, $y = 0$ or 1; thus there are two critical points, $(0,0)$ and $(1,1)$. Since $f_{xx} = 6x$, $f_{xy} = -3$, $f_{yy} = 6y$, the origin is a *saddle point*, and $(1,1)$ is a *minimum point*.
(c) Let $g(x,y) = 3x^2 + y^2 + 2x + 1$, and $f = \log g$.

Then $\nabla f = \dfrac{1}{g}\nabla g = \dfrac{1}{g}(6x+2,\ 2y) = 0$ when $x = -\dfrac{1}{3}$, $y = 0$.

$$f_{xx} = \frac{6g - (6x+2)g_x}{g^2},\quad f_{xy} = \frac{-(6x+2)g_y}{g^2},\quad f_{yy} = \frac{2g - 2yg_y}{g^2},$$

and evaluating at the given point $(-\frac{1}{3},0)$ gives $g = \frac{2}{3}$, $g_x = g_y = 0$, so $f_{xx} = 9$, $f_{xy} = 0$, $f_{yy} = 3$; hence $(-\frac{1}{3},0)$ is a local *minimum point*.
(d) $f_{xx} = 2$, $f_{xy} = 1$, $f_{yy} = 0$; saddle point
(e) $f_{xx} = 2$, $f_{xy} = f_{yy} = 0$; no case of Theorem 14 applies.

2. (a) From Problem 1(d), the critical point $(0,0)$ is a saddle point for $f(x,y) = x^2 + xy$, so the contour lines are hyperbolas with asymptotes $\{(x,y) : f(x,y) = f(0,0) = 0\}$. $x^2 + xy = 0 \iff x = 0$ or $x + y = 0$, so these are the asymptotes.
(b) $\nabla f = (2x + 2y + 1,\ 2x + 2y - 1) = 0$ for *no* x and y. Hence there are no critical points, and $f = c$ is a parabola. To find the axis, we look for the minimum of $|\nabla f|^2 = 2[(2x+2y)^2 + 1]$, which occurs on the line $x = -y$.

(c) $\nabla f = (-2y + 1, 2y - 2x - 1) = 0$ when $y = 1/2$, $x = 0$. Since $f_{xx} = 0$, $f_{xy} = -2$, $f_{yy} = 2$, we have a saddle point. The asymptotes of the contour lines are

$$\{(x,y): f(x,y) = f(0,\tfrac{1}{2}) = -\tfrac{1}{4}\}.$$

Since this should yield two lines passing through $(0,\frac{1}{2})$, we try to factor $f(x,y) + \frac{1}{4} = y^2 - 2xy + x - y + \frac{1}{4}$ into $(y - ax - \frac{1}{2})(y - cx - \frac{1}{2})$. This leads to asymptotes with equations

$$y = \tfrac{1}{2}, \qquad y = 2x + \tfrac{1}{2}.$$

(d) $\nabla f = (4x + 2y + 1, 2x + 10y - 2) = 0$ when $y = 1/6$, $x = -1/3$. $f_{xx} = 4$, $f_{xy} = 2$, $f_{yy} = 10$, so there is a minimum point, and the contour lines are ellipses with center $(-\frac{1}{3},\frac{1}{6})$.

3. (a) Because the critical point is a saddle, any maxima or minima must occur on the boundary. They can be found as in Example 2, or geometrically by looking at the contour lines; the max and min on the unit circle should occur along the lines bisecting the asymptotes, hence at $(\cos\theta, \sin\theta)$ where $\theta = \pi/8$, $5\pi/8$, $9\pi/8$, or $-3\pi/8$. The first and third points give a maximum, and the second and fourth give a minimum.

(b) Since $f(x,y) = x^2 + y$ has no critical points, we look at the boundary. Since the contour lines are parabolas opening downward, it is easy to see that the minimum occurs at $(0,-1)$, and the maximum occurs where the parabola $x^2 + y = c$ is tangent to the circle $x^2 + y^2 = 1$. On $x^2 + y = c$ we have $2x + dy/dx = 0$; and on the circle, $2x + 2y(dy/dx) = 0$. At a point of tangency, thus, we have $dy/dx = -2x = -x/y$. Hence $x = 0$, or $y = \frac{1}{2}$. This leads to the points $(0,\pm 1)$ and $(\pm\sqrt{3}/2, 1/2)$. Since $f(0,1) = 1$, $f(0,-1) = -1$, and $f(\pm\sqrt{3}/2, 1/2) = 5/4$, we find a minimum at $(0,-1)$ and maxima at $(\pm\sqrt{3}/2, 1/2)$.

(c) The point $(-\frac{1}{3},0)$ found in Problem 1(c) lies inside the disk D, so this is a candidate for a minimum. The problem is simplified by noting that $\log z$ is a monotone function for $z > 0$, and $3x^2 + y^2 + 2x + 10 > 0$ on the disk D; hence it is enough to locate the maxima and minima of $f(x,y) = 3x^2 + y^2 + 2x + 10$. There is an interior minimum at $(-\frac{1}{3},0)$, and a maximum on the boundary which can be found as in part (b), or as in Exercise 2.

CHAPTER IV

§4.1

1. (a)
$$\int_0^\pi \int_0^y y \sin x \, dx \, dy = \int_0^\pi \Big[-y \cos x\Big]_0^y \, dy$$

$$= \int_0^\pi [-y\cos y + y]\, dy = -y \sin y\Big|_0^\pi + \int_0^\pi \sin y \, dy + \frac{1}{2} y^2\Big|_0^\pi$$

$$= -\cos y\Big|_0^\pi + \frac{1}{2}\pi^2 = 2 + \frac{1}{2}\pi^2$$

(b) $5/6$ (c) $e^3 + e$ (d) $1/3$ (e) $e^2 - 3$ (f) $1 - \pi/4$

3. (a) $\int_0^{\pi}\int_x^{\pi} y \sin x \, dy \, dx = \int_0^{\pi} \frac{1}{2} y^2 \sin x \Big|_x^{\pi} dx = \int_0^{\pi} \frac{1}{2}(\pi^2 - x^2) \sin x \, dx$

$$= -\frac{1}{2}(\pi^2 - x^2)\cos x\Big|_0^{\pi} - \int_0^{\pi} x \cos x \, dx$$

$$= \frac{1}{2}\pi^2 - x \sin x\Big|_0^{\pi} + \int_0^{\pi} \sin x \, dx = 2 + \frac{1}{2}\pi^2$$

4. (a) $f_x = e^{xy}, \quad f_y = \int_1^x s e^{sy} \, ds$

(b) $g_x = 2xe^{x^2 y}, \quad g_y = \int_1^{x^2} s e^{sy} \, ds$

(c) $h_x = \sin xy, \quad h_y = \int_0^x s \cos sy \, ds$

(d) $i_x = 3x^2 \sin(x^3 y), \quad i_y = \int_0^{x^3} s \cos sy \, ds$

(e) $j_x = \sin((x + y)y), \quad j_y = \sin((x + y)y) + \int_0^{x+y} s \cos sy \, ds$

(f) $k_x = \varphi'(x) \sin(\varphi(x)y), \quad k_y = \int_0^{\varphi(x)} s \cos sy \, ds$

(g) $\ell_x = \sin(\varphi y)\varphi_x, \quad \ell_y = \sin(\varphi y)\varphi_y + \int_0^{\varphi} s \cos sy \, ds$

(h) $m_x = \sin(\varphi_2 y) \frac{\partial \varphi_2}{\partial x} - \sin(\varphi_1 y) \frac{\partial \varphi_1}{\partial x}$

$$m_y = \sin(\varphi_2 y) \frac{\partial \varphi_2}{\partial y} - \sin(\varphi_1 y) \frac{\partial \varphi_1}{\partial y} + \int_{\varphi_1}^{\varphi_2} s \sin(sy) \, ds$$

§4.2

1. (a) $M_y = 2x$, $N_x = 2x$, so it's closed. To find f, apply (2) with $x_0 = 0$, $y_0 = 0$:

$$f(x,y) = f(0,0) + \int_0^x s \cdot 0 \, ds + \int_0^y x^2 \, dt = f(0,0) + 0 + x^2 y.$$

In general, thus, $f(x,y) = c + x^2 y$. As a check, $f_x = 2xy$, $f_y = x^2$.

(b) $M_y = 2x$, $M_x = -2x$; not closed

(c) $M_y = e^{x+y}$, $M_x = e^{x+y}$, so it's closed; it's the gradient of $f(x,y) = e^{x+y}$

(d) $M_y = x \cos(xy)$, $N_x = -y \sin(xy)$

(e) $M_y = e^y$, $N_x = e^y$; closed. Potential is $f(x,y) = xe^y + c$

(f) $M_y = -\sin(x + y)$, $N_x = -\sin(x + y)$; closed. Potential is $\sin(x + y)$

(g) $M_y = -2yx/(x^2+y^2)^2$, $N_x = -2xy/(x^2+y^2)^2$; closed.

$$f(x,y) = \int_1^x \frac{s}{s^2+0}\,ds + \int_0^y \frac{t}{x^2+t^2}\,dt$$
$$= \log x + \tfrac{1}{2}\log(x^2+y^2) - \tfrac{1}{2}\log(x^2+0) = \tfrac{1}{2}\log(x^2+y^2)$$

2. (a) $ye^{xy}\,dx + xe^{xy}\,dy = 0$, $f(x,y) = e^{xy} = c$, $xy = \log c$, $y = (\log c)/x$
(b) $xe^{xy}\,dx + ye^{xy}\,dy = 0$ is not exact
(c) $x\,dx + y\,dy = 0$ is exact, $f(x,y) = \frac{1}{2}(x^2+y^2) = c \quad (c > 0)$; $y = \pm\sqrt{2c - x^2}$
(d) $x^2\,dx + y^2\,dy = 0$ is exact, $f(x,y) = \frac{1}{3}(x^3+y^3) = c$, $y = (3c - x^3)^{1/3}$
(e) $y^2\,dx + x^2\,dy = 0$; not exact
(f) $(x-y)dx + (y^2-x)dy = 0$ is exact, $f(x,y) = \frac{1}{2}x^2 - xy + y^3/3 = c$; not easy to solve for y

§4.3

1. (a) $x = t$, $y = t^2$, so $\displaystyle\int x^2y\,dx + \int xy^2\,dy = \int_0^1 (t^4\,dt + t^5\cdot 2t\,dt) = \frac{1}{5} + \frac{2}{7}$

(b) $\frac{1}{4}(c_1^3c_2 + c_1c_2^3)$ (c) $\displaystyle -\frac{c_1^2}{c_2}\cos c_2 + \left(\left(\frac{c_1}{c_2}\right)^2 + 1\right)\sin c_2$

(d) 0 (e) $\frac{1}{2}(x_1 - x_2)(y_1 + y_2)$

2. (a) $x = t$, $y = t^3$, $0 \le t \le 1$
(b) $x = x_1 + t(x_2 - x_1)$, $y = y_1 + t(y_2 - y_1)$, $0 \le t \le 1$
(c) $x = \cos(\pi - t)$, $y = \sin(\pi - t)$, $0 \le t \le \pi$. ($x = t$, $y = \sqrt{1-t^2}$, $-1 \le t \le 1$, is not good since it is not differentiable at $t = \pm 1$.)
(d) On γ^1: $x = t$, $y = 0$, $0 \le t \le 2$
On γ^2: $x = 2 - 2t$, $y = t$, $0 \le t \le 1$, On γ^3: $x = 0$, $y = -t$, $0 \le t \le 1$

3. (a) 0 (b) $x_2 - x_1 + y_1 - y_2$ (c) -2 (d) 0

4. (a) -1 (b) 1 (c) 0

(d) $\displaystyle\frac{1}{|b|} - \frac{1}{|a|}$ (assume that 0 does not lie in (a,b); otherwise the integral is not defined). (e) $1/|b| - 1/|a|$.

6. (a) Intuitively, the rate of flow is 1. To calculate it, parameterize the curve: $x = 0$, $y = t$, $0 \le t \le 1$. Then since $M = 1$ and $N = 0$ for the given flow,

$$\int_\gamma M\,dy - N\,dx = \int_\gamma 1\,dy - 0\,dx = \int_0^1 dt = 1.$$

(b) 0 (c) 1 (d) -1 (e) 1

7. (a) 0 (b) 0 (c) $\pi/8$ (d) $-2/3$ (e) $\pi/4$

17. (a) 1 (b) -1 (c) 2 (d) 0 (e) 2 (f) -1

§4.4

1. (a) $\gamma(t) = (r\cos t, r\sin t)$, $-\pi \le t \le \pi$;

$$\int -N\,dx + M\,dy$$
$$= -\int x\,dx + y\,dy = -r\int_{-\pi}^{\pi} (-\cos t\sin t + \cos t\sin t)dt = 0;$$
$$\iint_S (M_x + N_y)\,dx\,dy = 0$$

(b) $\iint_S (M_x + N_y)\,dx\,dy = 2\cdot(\text{area of } S) = 2(b-a)(c-d)$;

$$\int_\gamma -N\,dx + M\,dy = \int_\gamma -y\,dx + x\,dy = \int_a^b (d-c)\,dx + \int_c^d (b-a)\,dy$$

$$= 2(d-c)(b-a)$$

(c) $M_x + N_y \equiv 0$, so $(-N,M)$ is closed; in fact, $(-N,M) = (0,1) = \nabla f$, where $f(x,y) = y$. Thus $\int_\gamma -N\,dx + M\,dy = \int_\gamma \nabla f = 0$ since $\boldsymbol{\gamma}$ is closed.

(d) $M_x + N_y \equiv 1$, so $\iint_S (M_x + N_y)\,dx\,dy = 4$;

$$\int_\gamma x\,dy = \int_{-1}^1 [1 - (-1)]\,dy = 4$$

(e) $\iint_S M_x + N_y = 1, \quad \int_\gamma x\,dy = \int_0^1 (1-0)\,dy = 1$

(f) $M_x + N_y \equiv 0$, so $\iint_S (M_x + N_y)\,dx\,dy = 0$. Further, $(-N,M) = (-x,0) = \nabla f(x,y)$, where $f(x,y) = -\frac{1}{2}x^2$, so $\int_\gamma -N\,dx + M\,dy = 0$ for any closed chain $\boldsymbol{\gamma}$.

2. $\int_\gamma x\,dy = \iint_S 1\,dx\,dy \qquad (M = 0,\ N = x,\ N_x - M_y = 1)$

$-\int_\gamma y\,dx = \iint_S 1\,dx\,dy \qquad (M = -y,\ N = 0,\ N_x - M_y = 1)$

$\frac{1}{2}\int_\gamma (x\,dy - y\,dx) = \iint_S 1\,dx\,dy \qquad$ (Add the first two integrals.)

6. (a) For every disk D, $\iint_D (M_x + N_y)\,dx\,dy = \int_\gamma (-N\,dx + M\,dy) = 0$; hence $M_x + N_y \equiv 0$.

(b) If $\mathbf{F} = \nabla f = (f_x, f_y)$ is incompressible, then its divergence is zero, by (a), so $0 = (f_x)_x + (f_y)_y = f_{xx} + f_{yy}$.

§4.5

1. (a) $\int_0^1 \int_0^{\sqrt{1-y^2}} x\,dx\,dy = \int_0^{\pi/2}\int_0^1 r\cos\theta\, r\,dr\,d\theta = \int_0^{\pi/2} \frac{1}{3}\cos\theta\,d\theta = \frac{1}{3}$

(b) $\int_{-\pi}^{\pi}\int_{r_1}^{r_2} \frac{r^2\cos\theta\sin\theta}{r^2}\, r\,dr\,d\theta = \frac{r_2^2 - r_1^2}{2}\int_{-\pi}^{\pi} \frac{\cos 2\theta}{2}\,d\theta = 0$

(c) $\int_0^{\pi/2}\int_0^R e^{r^2} r\,dr\,d\theta = \frac{\pi}{2}\left[\frac{1}{2}e^{r^2}\right]_0^R = \frac{\pi}{4}(e^{R^2}-1)$ **2.** (a) See Problem 1(c).

4. (a) $x = u + v = \varphi(u,v)$, $y = u - v = \psi(u,v)$,
$S = \{(x,y): -1 \le x \le 1, -1 \le y \le x\}$,
$\bar{S} = \{(u,v): 0 \le v \le 1, v - 1 \le u \le 1 - v\}$

(b) $\int_{-1}^{1}\int_{-1}^{x} \frac{dy\,dx}{x-y-2} = \int_0^1 \int_{v-1}^{1-v} \frac{2}{2v-2}\,du\,dv = \int_0^1 (-2)\,dv = -2$

CHAPTER V

§5.1

2. (a) $f_{xx} = 2, f_{yy} = -4, f_{zz} = 2$ (c) $f_{xx} = 12x, f_{yy} = -6x, f_{zz} = -6x$

3. (a) $\nabla f = (2x,2y,4z) = (2,0,0)$ at the given point.
Equation is $(\mathbf{P} - \mathbf{P}_0)\cdot(2,0,0) = 0$, or $x = 1$ (a vertical plane).
(b) Equation is $(\mathbf{P} - \mathbf{P}_0)\cdot(0,1,0) = 0$, or $y = 0$ (the xz plane).
(c) Equation is $2(x-1) + 3(y+1) - 32(z-2) = 0$.

4. (a) $\nabla f = (2x_1, 2x_2) = (2,2)$ at the given point. Equation is $2(x_1 - 1) + 2(x_2 - 1) = 0$, or $x_1 + x_2 = 2$. (This tangent hyperplane is a line.)
(b) Equation: $x_1 + x_2 + x_3 + x_4 = 4$.

5. (a) $\nabla f = (yz,xz,xy) = (6,3,2)$ at the given point. This gives the direction of maximum temperature increase.

8. (a) $F_r = f_x \cos\theta + f_y \sin\theta$, $F_\theta = -f_x r \sin\theta + f_y r \cos\theta$, $F_z = f_z$.
(b) See Problem 11, §3.6.

§5.2

1. (a) $f_z = 1 + e^{x^2+y^2+z^2}(2z\cos(xyz) - xy\sin(xyz))$,
$f_z(0,0,0) = 1 \ne 0$; hence the implicit function theorem applies.
(b) $z = \varphi(0,0) \Leftrightarrow f(0,0,z) = 1$; but $f(0,0,0) = 1$, so $\varphi(0,0) = 0$.

2. (a) $f_x = \sin(x+y+z) + x\cos(x+y+z)$, so $f_x(2,-1,-1) = 2 \ne 0$.

3. (a) $(z-\varphi)^2 = 0 \Leftrightarrow z - \varphi = 0 \Leftrightarrow z = \varphi$.
(c) The implicit function theorem is a "one-way" theorem; if $f_z \ne 0$, then $f = c$ defines z as a differentiable function of x and y. It does *not* claim the converse: if $f = c$ defines z as a differentiable function of x and y, then $f_z \ne 0$.

§5.3

1. Let $x_1 = x$, $x_2 = y$, $x_3 = z$, to simplify notation. $f(x,y,z) = xe^{y+z}$, $f_x = e^{y+z}$, $f_y = f$, $f_z = f$, $f_{xx} \equiv 0$, $f_{yy} = f$, $f_{zz} = f$, $f_{xy} = f_{xz} = e^{y+z}$, $f_{yz} = f$.
(a) $f(0) = 0$ and $\nabla f(0) = (1,0,0)$, so $f_1(0,\mathbf{H}) = h_1$
(b) $f(1,-1,1) = 1$, $\nabla f(1,-1,1) = (1,1,1)$, etc. Thus $f_2(\mathbf{A},\mathbf{H}) = 1 + h_1 + h_2 + h_3 + \frac{1}{2}(h_2^2 + h_3^2) + (h_1h_2 + h_1h_3 + h_2h_3)$
(c) At $\mathbf{A} = (0,0,1)$ we have $f = 0$, $\nabla f = (e,0,0)$, $f_{xx} = f_{yy} = f_{zz} = f_{yz} = 0$, $f_{xy} = f_{xz} = e$, $0 = f_{xxx} = f_{yyy} = f_{zzz} = f_{xxy} = f_{xxz}$, $f_{xyy} = f_{xyz} = f_{xzz} = e$, and all other third order partial derivatives of f vanish at $\mathbf{A}$. Hence

$$f_3(\mathbf{A},\mathbf{H}) = eh_1 + e(h_1h_2 + h_1h_3) + \frac{1}{3!}(3eh_1h_2^2 + 6eh_1h_2h_3 + 3eh_1h_3^2).$$

§5.4

1. (a) $\int_0^1 (ae^{ta} + be^{tb} + ce^{tc})\,dt = e^a + e^b + e^c$

(b) $\int_0^1 [(t^3 + t^4)\cdot 2t + (t^4 + t^2)3t^2 + (t^2 + t^3)4t^3]\,dt$

2. (d) $(100^2 + 50^2 + 80^2)^{-1/2} - (35)^{-1/2}$ (e) $-(35)^{-1/2}$

6. (a) $\mathbf{F}$ closed $\Leftrightarrow D_2f_3 = D_3f_2,\quad D_3f_1 = D_1f_3,\quad D_1f_2 = D_2f_1 \Leftrightarrow \nabla\times\mathbf{F} = \mathbf{0}$

(b) $\nabla\times(\nabla f) = (D_2D_3f - D_3D_2f,\, D_3D_1f - D_1D_3f,\, D_1D_2f - D_2D_1f) = \mathbf{0}$

§5.5

1. (c) $\iint_S |\mathbf{N}_\sigma| = \int_0^\pi \int_0^{2\pi} \sin v\,du\,dv = 4\pi$

(d) $\iint_S \mathbf{F}\cdot\mathbf{N}_\sigma = \int_0^\pi \int_0^{2\pi} -\sin^2 v\cos u\,du\,dv = 0$ since $\int_0^{2\pi} \cos u\,du = 0$. Intuitively, $\iint_\sigma \mathbf{F}$ gives the rate of flow from the negative to the positive side of the sphere; in the parametrization here, this is from the outside to the inside. The flow into the sphere on the side where $x < 0$ is exactly balanced by the flow out through the sphere on the side where $x > 0$, so the net rate of flow is zero.

2. (b) 8π (double covering) (c) 2π (covers half the sphere) (d) 16π

3. (b) Triangle (half the parallelogram in part (a).) Area is $\int_0^1 \int_0^{1-u} |\mathbf{A}\times\mathbf{B}|\,dv\,du = \frac{1}{2}|\mathbf{A}\times\mathbf{B}|$. **5.** (b) $\iint_S \mathbf{A}\cdot\mathbf{N}_\sigma = \pm\pi r^2|\mathbf{A}|$; $\iint_S \mathbf{B}\cdot\mathbf{N}_\sigma = 0$

6. (a) $\iint_\sigma \nabla\times\mathbf{F} = 0$ (b) $-1/2$ (c) π **7.** $\int_0^\pi \int_0^{2\pi} |\mathbf{N}_\sigma|\,du\,dv$

8. Area $= \iint_S |\mathbf{N}_\sigma| = \iint_S \sqrt{1 + |\nabla f|^2}$ **9.** (a) $\sqrt{10}$ (b) $\sqrt{26}$ (c) $\pi\sqrt{10}$

§5.6

1. (a) $\int_2^3 \int_1^2 \int_0^1 x\,dx\,dy\,dz = \int_2^3 \int_1^2 \tfrac{1}{2}\,dy\,dz = \tfrac{1}{2}$ (region V is a cube)

(b) 13/24 (c) $\frac{4}{3}\cdot 124 + 48$ (d) $\frac{1}{3} + \cos(\pi + 1) - \cos(\pi/2 + 1)$

2. (a) 0 (b) $\frac{16}{3}a^3$ (volume of the intersection of two cylinders of radius a)

(c) 1/6 (volume of a pyramid of height 1, with area of base = 1/2)

(d) $\frac{4}{3}\pi(R^2 - r^2)^{3/2}$ (e) $\frac{2}{3}\pi R^3 + \frac{8}{9}R^3$ [Use polar coordinates.]

3. (a) $\rho\pi a^2 b$ (b) $\pi\rho b a^3$ (c) 0 (d) $\frac{1}{2}\rho b^2\pi a^2$

§5.7

1. (a) $\nabla\cdot\mathbf{F} = 0$. Rate of flow into any region from the back equals the rate of flow out of the front (assuming axes drawn as in Fig. 5.31).

(b) $\nabla\cdot\mathbf{F} = 1$. Rate of flow in at the back is less than rate of flow out at the front. (c) 0 (d) 3 (e) $1/|\mathbf{P}|^2$

2. (a) both 0 (b) both $\frac{4}{3}\pi r^3$ (c) both 0 (d) both $4\pi r^3$

Table of Natural Logarithms

N	0	1	2	3	4	5	6	7	8	9
1.0	0000	0100	0198	0296	0392	0488	0583	0677	0770	0862
1.1	0953	1044	1133	1222	1310	1398	1484	1570	1655	1740
1.2	1823	1906	1989	2070	2151	2231	2311	2390	2469	2546
1.3	2624	2700	2776	2852	2927	3001	3075	3148	3221	3293
1.4	3365	3436	3507	3577	3646	3716	3784	3853	3920	3988
1.5	4055	4121	4187	4253	4318	4383	4447	4511	4574	4637
1.6	4700	4762	4824	4886	4947	5008	5068	5128	5188	5247
1.7	5306	5365	5423	5481	5539	5596	5653	5710	5766	5822
1.8	5878	5933	5988	6043	6098	6152	6206	6259	6313	6366
1.9	6419	6471	6523	6575	6627	6678	6729	6780	6831	6881
2.0	6931	6981	7031	7080	7129	7178	7227	7275	7324	7372
2.1	7419	7467	7514	7561	7608	7655	7701	7747	7793	7839
2.2	7885	7930	7975	8020	8065	8109	8154	8198	8242	8286
2.3	8329	8372	8416	8459	8502	8544	8587	8629	8671	8713
2.4	8755	8796	8838	8879	8920	8961	9002	9042	9083	9123
2.5	9163	9203	9243	9282	9322	9361	9400	9439	9478	9517
2.6	9555	9594	9632	9670	9708	9746	9783	9821	9858	9895
2.7	9933	9969	*0006	*0043	*0080	*0116	*0152	*0188	*0225	*0260
2.8	1.0296	0332	0367	0403	0438	0473	0508	0543	0578	0613
2.9	0647	0682	0716	0750	0784	0818	0852	0886	0919	0953
3.0	1.0986	1019	1053	1086	1119	1151	1184	1217	1249	1282
3.1	1314	1346	1378	1410	1442	1474	1506	1537	1569	1600
3.2	1632	1663	1694	1725	1756	1787	1817	1848	1878	1909
3.3	1939	1969	2000	2030	2060	2090	2119	2149	2179	2208
3.4	2238	2267	2296	2326	2355	2384	2413	2442	2470	2499
3.5	1.2528	2556	2585	2613	2641	2669	2698	2726	2754	2782
3.6	2809	2837	2865	2892	2920	2947	2975	3002	3029	3056
3.7	3083	3110	3137	3164	3191	3218	3244	3271	3297	3324
3.8	3350	3376	3403	3429	3455	3481	3507	3533	3558	3584
3.9	3610	3635	3661	3686	3712	3737	3762	3788	3813	3838
4.0	1.3863	3888	3913	3938	3962	3987	4012	4036	4061	4085
4.1	4110	4134	4159	4183	4207	4231	4255	4279	4303	4327
4.2	4351	4375	4398	4422	4446	4469	4493	4516	4540	4563
4.3	4586	4609	4633	4656	4679	4702	4725	4748	4770	4793
4.4	4816	4839	4861	4884	4907	4929	4951	4974	4996	5019
4.5	1.5041	5063	5085	5107	5129	5151	5173	5195	5217	5239
4.6	5261	5282	5304	5326	5347	5369	5390	5412	5433	5454
4.7	5476	5497	5518	5539	5560	5581	5602	5623	5644	5665
4.8	5686	5707	5728	5748	5769	5790	5810	5831	5851	5872
4.9	5892	5913	5933	5953	5974	5994	6014	6034	6054	6074
5.0	1.6094	6114	6134	6154	6174	6194	6214	6233	6253	6273
5.1	6292	6312	6332	6351	6371	6390	6409	6429	6448	6467
5.2	6487	6506	6525	6544	6563	6582	6601	6620	6639	6658
5.3	6677	6696	6715	6734	6752	6771	6790	6808	6827	6845
5.4	6864	6882	6901	6919	6938	6956	6974	6993	7011	7029

N	0	1	2	3	4	5	6	7	8	9
5.5	1.7047	7066	7084	7102	7120	7138	7156	7174	7192	7210
5.6	7228	7246	7263	7281	7299	7317	7334	7352	7370	7387
5.7	7405	7422	7440	7457	7475	7492	7509	7527	7544	7561
5.8	7579	7596	7613	7630	7647	7664	7681	7699	7716	7733
5.9	7750	7766	7783	7800	7817	7834	7851	7867	7884	7901
6.0	1.7918	7934	7951	7967	7984	8001	8017	8034	8050	8066
6.1	8083	8099	8116	8132	8148	8165	8181	8197	8213	8229
6.2	8245	8262	8278	8294	8310	8326	8342	8358	8374	8390
6.3	8405	8421	8437	8453	8469	8485	8500	8516	8532	8547
6.4	8563	8579	8594	8610	8625	8641	8656	8672	8687	8703
6.5	1.8718	8733	8749	8764	8779	8795	8810	8825	8840	8856
6.6	8871	8886	8901	8916	8931	8946	8961	8976	8991	9006
6.7	9021	9036	9051	9066	9081	9095	9110	9125	9140	9155
6.8	9169	9184	9199	9213	9228	9242	9257	9272	9286	9301
6.9	9315	9330	9344	9359	9373	9387	9402	9416	9430	9445
7.0	1.9459	9473	9488	9502	9516	9530	9544	9559	9573	9587
7.1	9601	9615	9629	9643	9657	9671	9685	9699	9713	9727
7.2	9741	9755	9769	9782	9796	9810	9824	9838	9851	9865
7.3	9879	9892	9906	9920	9933	9947	9961	9974	9988	*0001
7.4	2.0015	0028	0042	0055	0069	0082	0096	0109	0122	0136
7.5	2.0149	0162	0176	0189	0202	0215	0229	0242	0255	0268
7.6	0281	0295	0308	0321	0334	0347	0360	0373	0386	0399
7.7	0412	0425	0438	0451	0464	0477	0490	0503	0516	0528
7.8	0541	0554	0567	0580	0592	0605	0618	0630	0643	0656
7.9	0669	0681	0694	0707	0719	0732	0744	0757	0769	0782
8.0	2.0794	0807	0819	0832	0844	0857	0869	0882	0894	0906
8.1	0919	0931	0943	0956	0968	0980	0992	1005	1017	1029
8.2	1041	1054	1066	1078	1090	1102	1114	1126	1138	1150
8.3	1163	1175	1187	1199	1211	1223	1235	1247	1258	1270
8.4	1282	1294	1306	1318	1330	1342	1353	1365	1377	1389
8.5	2.1401	1412	1424	1436	1448	1459	1471	1483	1494	1506
8.6	1518	1529	1541	1552	1564	1576	1587	1599	1610	1622
8.7	1633	1645	1656	1668	1679	1691	1702	1713	1725	1736
8.8	1748	1759	1770	1782	1793	1804	1815	1827	1838	1849
8.9	1861	1872	1883	1894	1905	1917	1928	1939	1950	1961
9.0	2.1972	1983	1994	2006	2017	2028	2039	2050	2061	2072
9.1	2083	2094	2105	2116	2127	2138	2148	2159	2170	2181
9.2	2192	2203	2214	2225	2235	2246	2257	2268	2279	2289
9.3	2300	2311	2322	2332	2343	2354	2364	2375	2386	2396
9.4	2407	2418	2428	2439	2450	2460	2471	2481	2492	2502
9.5	2.2513	2523	2534	2544	2555	2565	2576	2586	2597	2607
9.6	2618	2628	2638	2649	2659	2670	2680	2690	2701	2711
9.7	2721	2732	2742	2752	2762	2773	2783	2793	2803	2814
9.8	2824	2834	2844	2854	2865	2875	2885	2895	2905	2915
9.9	2925	2935	2946	2956	2966	2976	2986	2996	3006	3016

Use $\ln 10 = 2.30259$ to find logarithms of numbers greater than 10 or less than 1. *Example:* $\ln 220 = \ln 2.2 + 2 \ln 10 = 0.7885 + 2(2.30259) = 5.3937$.

Index

1 2 3 4 5 6 7 8 9 10 11 12 13 14 15 16 17 18 19 20 21 22 23 24 25 80 79 78 77 76 75 74 73 72